Lecture Notebook

to accompany

Life
The Science of Biology
SEVENTH EDITION

PURVES • SADAVA • ORIANS • HELLER

Sinauer Associates, Inc.

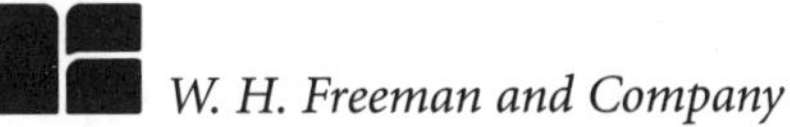
W. H. Freeman and Company

Cover photo © Steve Bloom/stevebloom.com.

Lecture Notebook to accompany *Life: The Science of Biology,* **Seventh Edition**

Address editorial correspondence to:
Sinauer Associates, Inc.
23 Plumtree Road
Sunderland, MA, 01375 U.S.A.
Fax: 413-549-1118
Internet: www.sinauer.com; publish@sinauer.com

Address orders to:
VHPS/W. H. Freeman & Co. Order Department
16365 James Madison Highway,
U.S. Route 15, Gordonsville, VA, 22942 U.S.A.
Internet: www.whfreeman.com

ISBN 0-7167-5812-1
Printed in the U.S.A.

4 3 2 1

Contents

Part 5 The Evolution of Diversity

Part 6 The Biology of Flowering Plants

Part 7 The Biology of Animals

Part 8 Ecology and Biogeography

1 An Evolutionary Framework for Biology

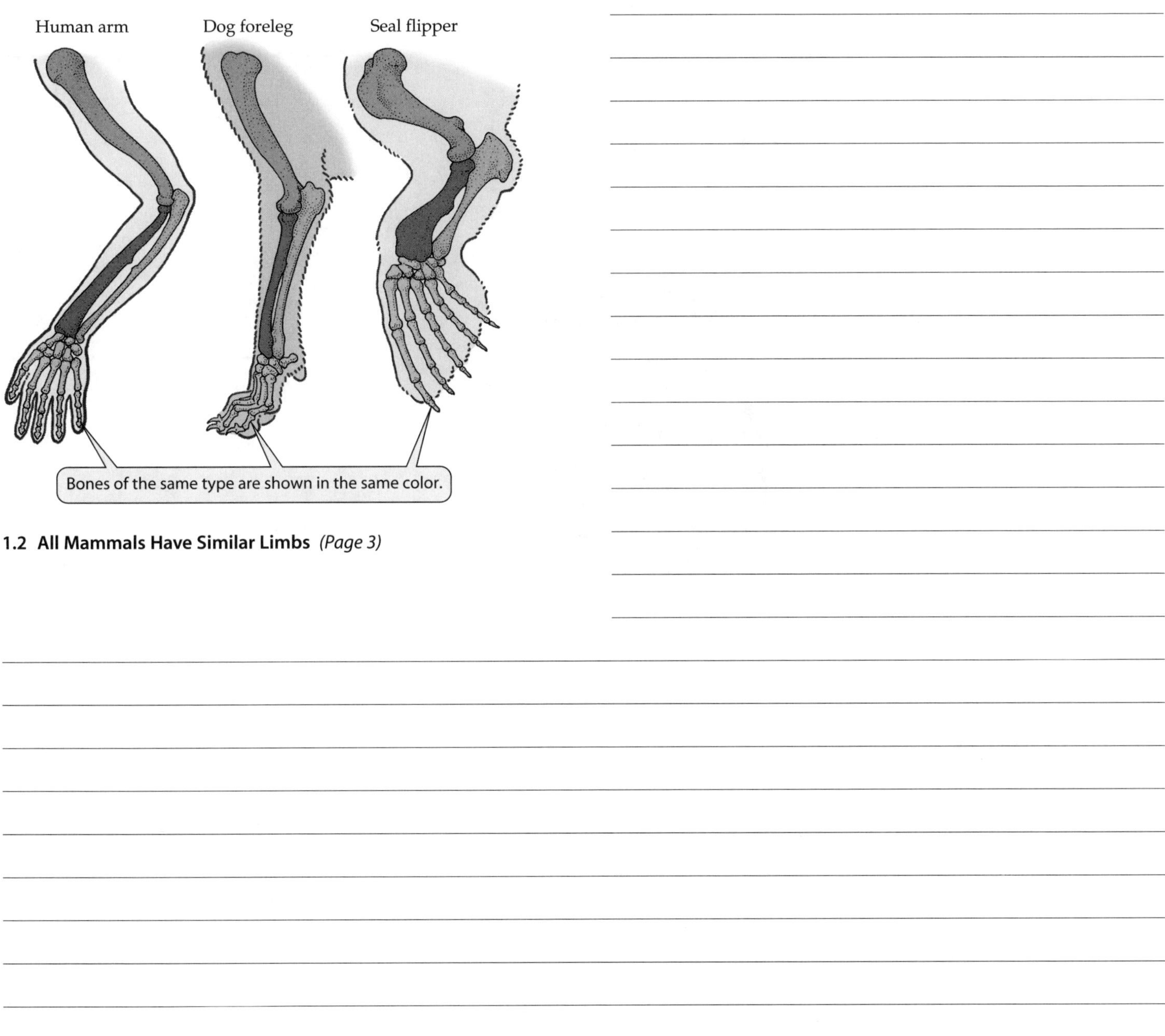
Human arm
Dog foreleg
Seal flipper
Bones of the same type are shown in the same color.

1.2 All Mammals Have Similar Limbs *(Page 3)*

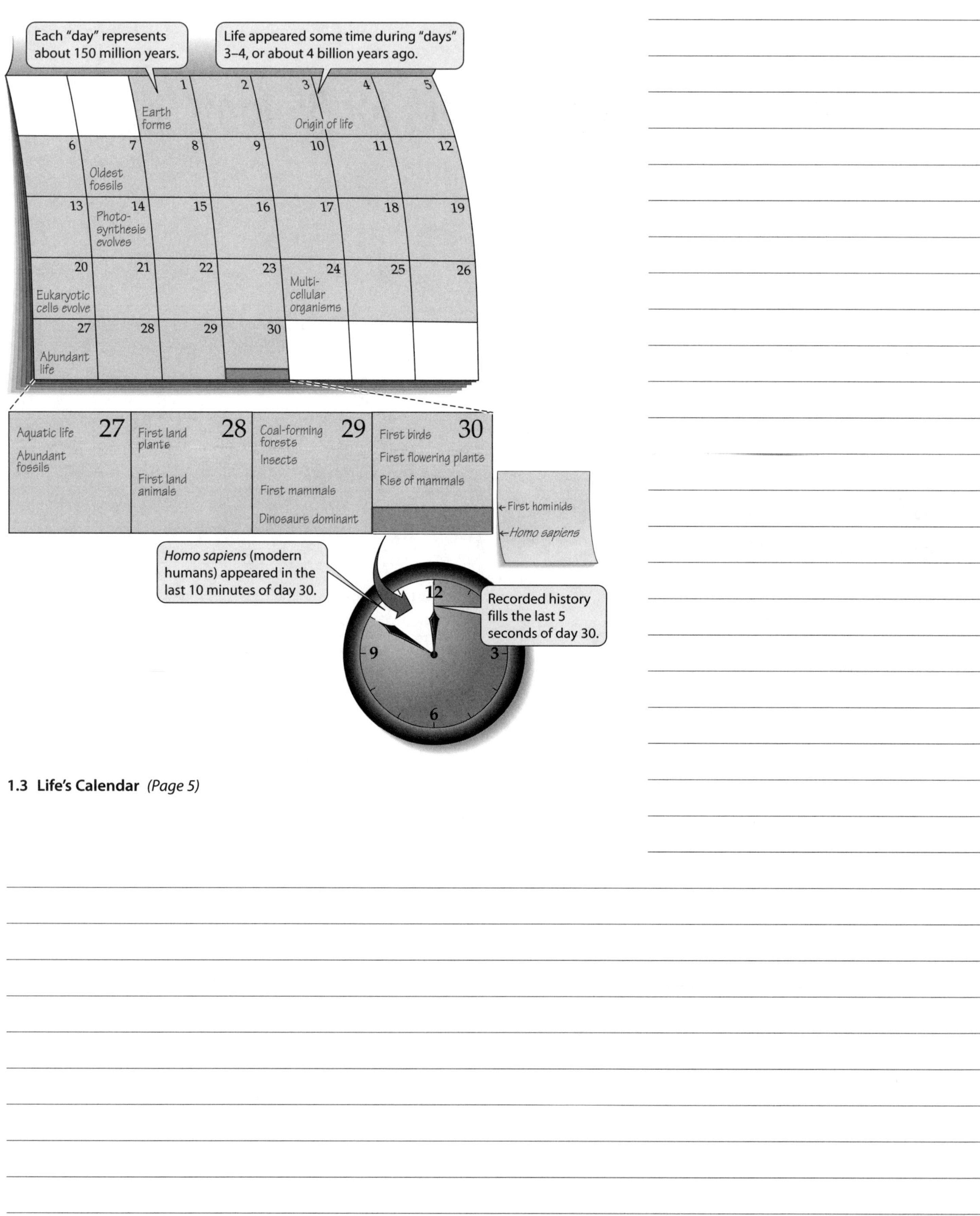
Each "day" represents about 150 million years.
Life appeared some time during "days" 3–4, or about 4 billion years ago.
1
Earth forms
2
3
4
5
Origin of life
6
7
Oldest fossils
8
9
10
11
12
13
14
Photo-synthesis evolves
15
16
17
18
19
20
Eukaryotic cells evolve
21
22
23
24
Multi-cellular organisms
25
26
27
Abundant life
28
29
30
Aquatic life
Abundant fossils
27
First land plants
First land animals
28
Coal-forming forests
Insects
First mammals
Dinosaurs dominant
29
First birds
First flowering plants
Rise of mammals
30
First hominids
Homo sapiens
Homo sapiens (modern humans) appeared in the last 10 minutes of day 30.
Recorded history fills the last 5 seconds of day 30.
12
3
6
9

1.3 Life's Calendar *(Page 5)*

Atoms
Molecule
Molecules are made up of **atoms**, and in turn can be organized into cells.
Cell (neuron)
Cells of many types are the working components of living organisms.
Tissue (ganglion)
A **tissue** is a group of many cells with similar and coordinated functions.
Organ (brain)
Organs combine several tissues that function together. Organs form systems, such as the nervous system.

1.6 From Molecules to the Biosphere: The Hierarchy of Life *(Page 7)*

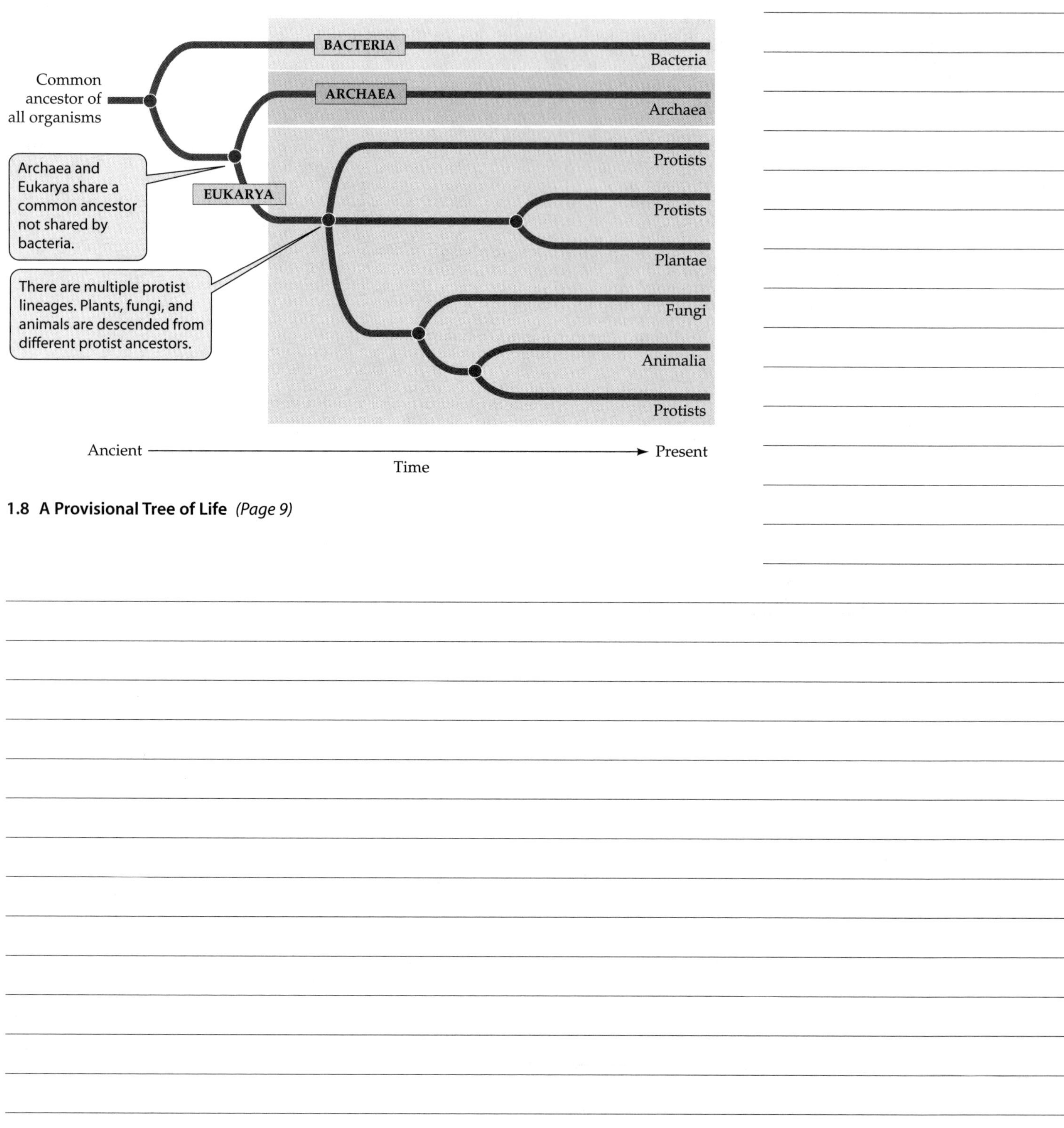
BACTERIA
Bacteria
Common ancestor of all organisms
ARCHAEA
Archaea
Archaea and Eukarya share a common ancestor not shared by bacteria.
EUKARYA
Protists
Protists
Plantae
There are multiple protist lineages. Plants, fungi, and animals are descended from different protist ancestors.
Fungi
Animalia
Protists
Ancient
Present
Time

1.8 A Provisional Tree of Life *(Page 9)*

EXPERIMENT

Hypothesis: Susceptibility to UV-B radiation has contributed to the disappearance of some frogs from high-elevation ponds.

METHOD Establish 3 identical artificial tanks at each of 2 elevations (1,365 meters and 1,600 meters). Set up 6 trays in each tank. Place equal numbers of embryos of one of the two frog species in each tray. In each tank, 2 trays receive unfiltered sunlight; 2 receive sunlight filtered to remove UV-B; and 2 receive filtered sunlight that allowed UV-B transmission. Count the number of surviving individuals 3 times a week for 4 weeks.

RESULTS The probability of dying was much greater for individuals of the species that had disappeared from high elevations (*Litoria verreauxii*) than for individuals of the species surviving there (*Crinia signifera*).

Unfiltered sunlight
Filtered, UV-B blocked
Filtered, UV-B allowed
Elevation 1,365 meters 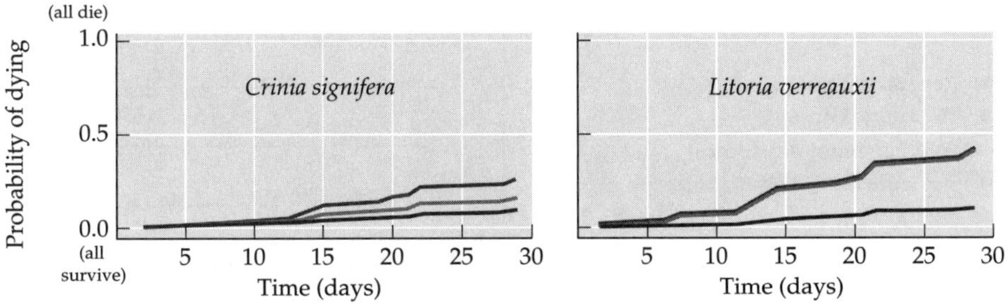
(all die)
1.0
0.5
0.0
(all survive)
Probability of dying
Crinia signifera
Litoria verreauxii
5 10 15 20 25 30
Time (days)
Elevation 1,600 meters
Probability of dying
1.0
0.5
0.0
Crinia signifera
Litoria verreauxii
5 10 15 20 25 30
Time (days)

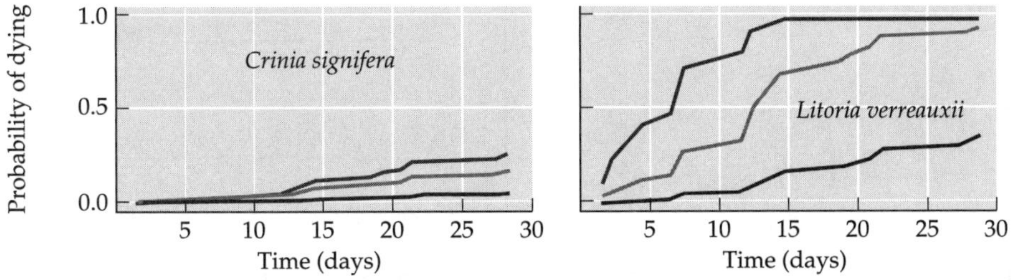

Conclusion: The results support the hypothesis that suceptibility to UV-B radiation has contributed to the disappearance of *Litoria verreauxii* from high elevations.

1.9 A Controlled Experiment Tests the Effects of UV-B *(Page 11)*

COMPARATIVE METHOD

Hypothesis: Airborne pesticides from agricultural fields and urban air pollutants are contributing to the decline of amphibian populations.

PREDICTION If pesticides and urban air pollutants are factors in amphibian population declines, populations close to and downwind from agricultural and urban areas should have decreased more strikingly than populations upwind and farther away from those sources of air pollutants.

METHOD Census (count) and then compare persistence of populations of species of amphibians at suitable habitat sites that lie upwind and downwind of major agricultural and urban areas.

RESULTS Populations of some species, as illustrated here by *Rana aurora*, persist in areas upwind of or remote from sources of urban and agricultural pollutants, but this amphibian is largely absent from areas close to or downwind of air pollution sources. (Distributions of three other species of *Rana* were similar to that of *R. aurora*.)

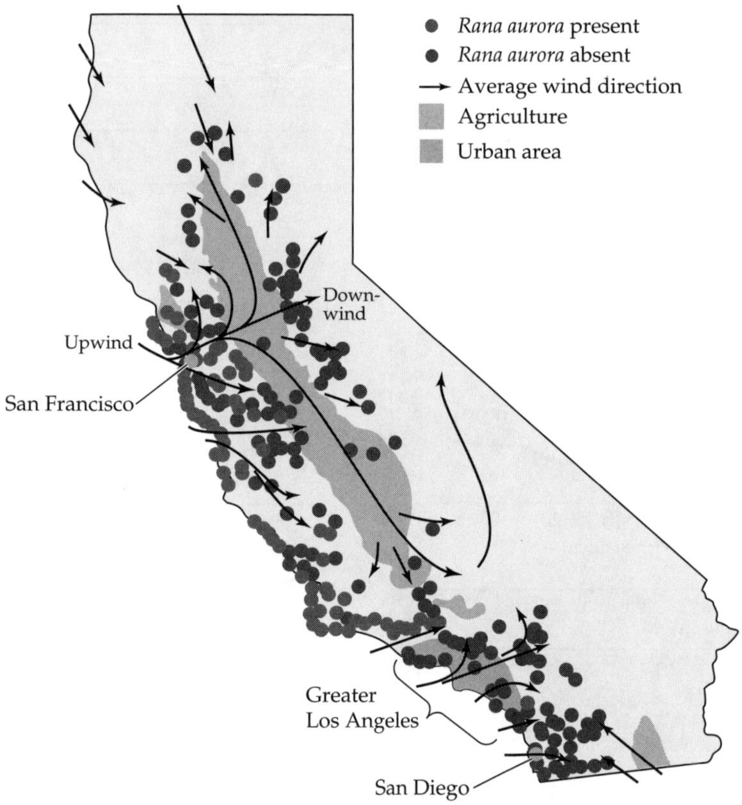

Rana aurora present
Rana aurora absent
Average wind direction
Agriculture
Urban area
Down-wind
Upwind
San Francisco
Greater Los Angeles
San Diego

Conclusion: Airborne agricultural pesticides and urban air pollutants are contributing to declines in populations of some amphibian species.

1.10 Using the Comparative Method to Test a Hypothesis *(Page 12)*

2 Life and Chemistry: Small Molecules

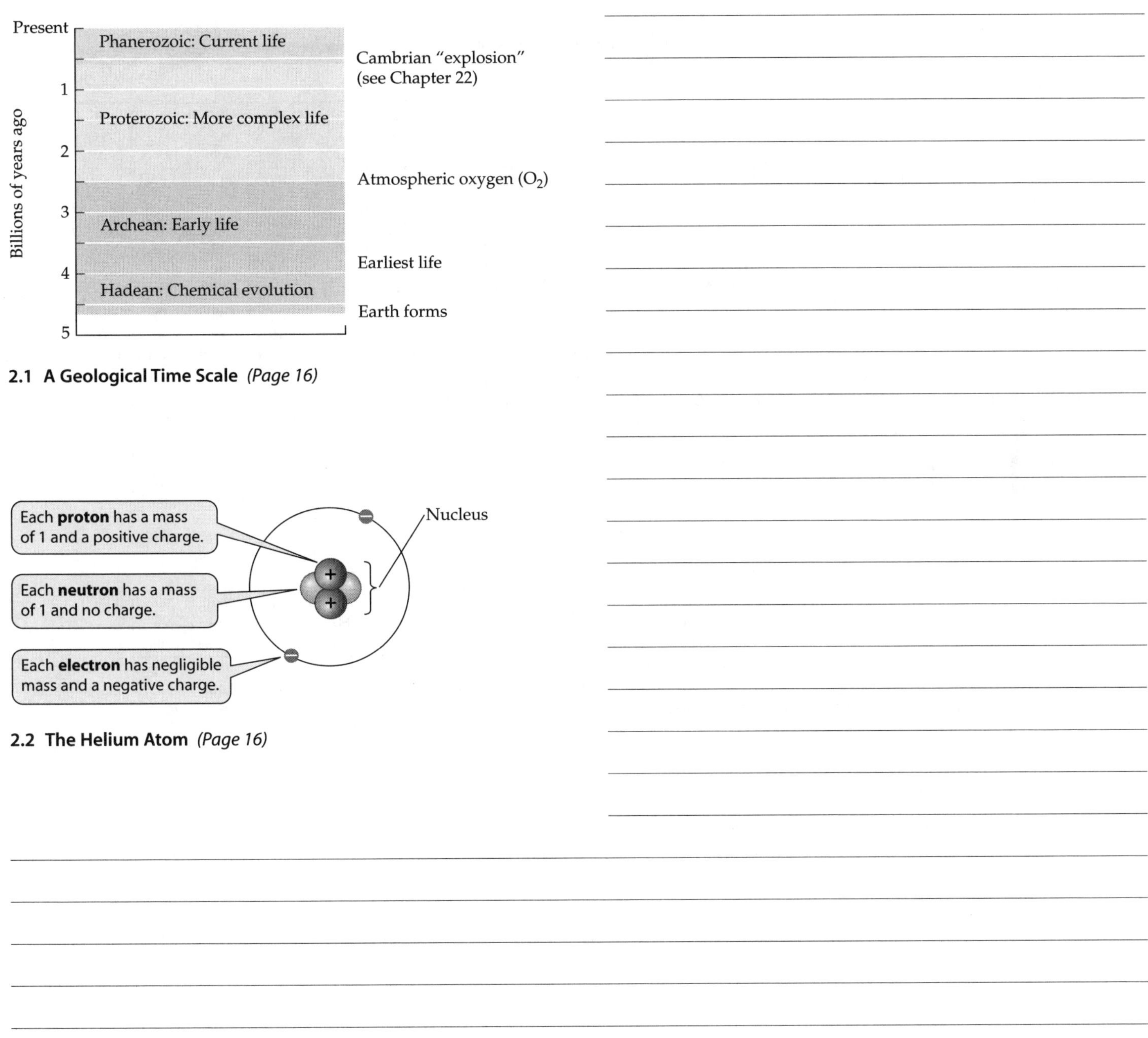

Present
Phanerozoic: Current life
Cambrian "explosion"
(see Chapter 22)
1
Proterozoic: More complex life
2
Atmospheric oxygen (O_2)
3
Archean: Early life
Earliest life
4
Hadean: Chemical evolution
Earth forms
5
Billions of years ago

2.1 A Geological Time Scale *(Page 16)*

Each **proton** has a mass of 1 and a positive charge.
Each **neutron** has a mass of 1 and no charge.
Each **electron** has negligible mass and a negative charge.
Nucleus

2.2 The Helium Atom *(Page 16)*

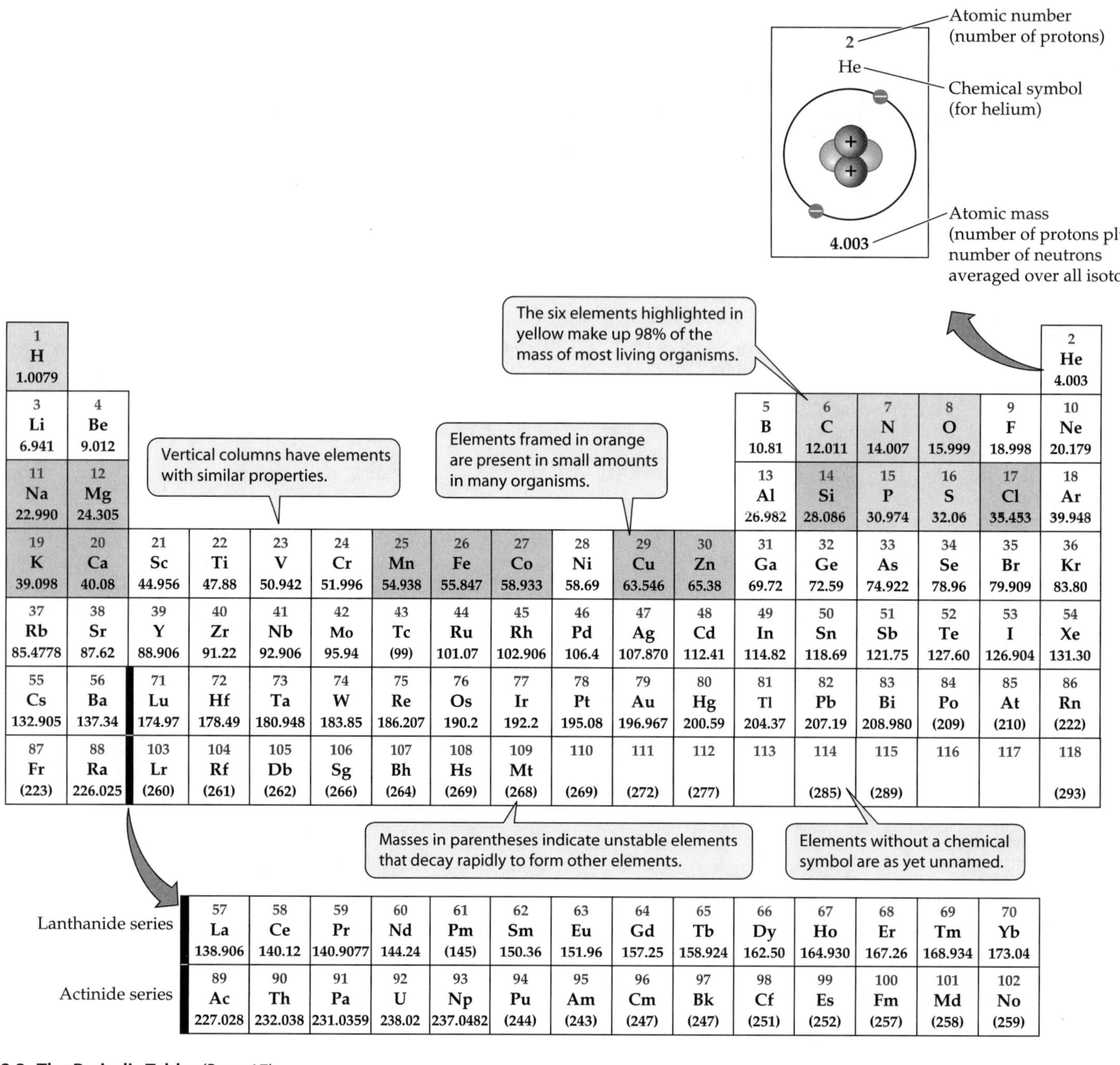
2
He
4.003
Atomic number (number of protons)
Chemical symbol (for helium)
Atomic mass (number of protons plus number of neutrons averaged over all isotopes)
The six elements highlighted in yellow make up 98% of the mass of most living organisms.
Vertical columns have elements with similar properties.
Elements framed in orange are present in small amounts in many organisms.
Masses in parentheses indicate unstable elements that decay rapidly to form other elements.
Elements without a chemical symbol are as yet unnamed.
Lanthanide series
Actinide series

1 H 1.0079																	2 He 4.003
3 Li 6.941	4 Be 9.012											5 B 10.81	6 C 12.011	7 N 14.007	8 O 15.999	9 F 18.998	10 Ne 20.179
11 Na 22.990	12 Mg 24.305											13 Al 26.982	14 Si 28.086	15 P 30.974	16 S 32.06	17 Cl 35.453	18 Ar 39.948
19 K 39.098	20 Ca 40.08	21 Sc 44.956	22 Ti 47.88	23 V 50.942	24 Cr 51.996	25 Mn 54.938	26 Fe 55.847	27 Co 58.933	28 Ni 58.69	29 Cu 63.546	30 Zn 65.38	31 Ga 69.72	32 Ge 72.59	33 As 74.922	34 Se 78.96	35 Br 79.909	36 Kr 83.80
37 Rb 85.4778	38 Sr 87.62	39 Y 88.906	40 Zr 91.22	41 Nb 92.906	42 Mo 95.94	43 Tc (99)	44 Ru 101.07	45 Rh 102.906	46 Pd 106.4	47 Ag 107.870	48 Cd 112.41	49 In 114.82	50 Sn 118.69	51 Sb 121.75	52 Te 127.60	53 I 126.904	54 Xe 131.30
55 Cs 132.905	56 Ba 137.34	71 Lu 174.97	72 Hf 178.49	73 Ta 180.948	74 W 183.85	75 Re 186.207	76 Os 190.2	77 Ir 192.2	78 Pt 195.08	79 Au 196.967	80 Hg 200.59	81 Tl 204.37	82 Pb 207.19	83 Bi 208.980	84 Po (209)	85 At (210)	86 Rn (222)
87 Fr (223)	88 Ra 226.025	103 Lr (260)	104 Rf (261)	105 Db (262)	106 Sg (266)	107 Bh (264)	108 Hs (269)	109 Mt (268)	110 (269)	111 (272)	112 (277)	113	114 (285)	115 (289)	116	117	118 (293)

Lanthanide series	57 La 138.906	58 Ce 140.12	59 Pr 140.9077	60 Nd 144.24	61 Pm (145)	62 Sm 150.36	63 Eu 151.96	64 Gd 157.25	65 Tb 158.924	66 Dy 162.50	67 Ho 164.930	68 Er 167.26	69 Tm 168.934	70 Yb 173.04
Actinide series	89 Ac 227.028	90 Th 232.038	91 Pa 231.0359	92 U 238.02	93 Np 237.0482	94 Pu (244)	95 Am (243)	96 Cm (247)	97 Bk (247)	98 Cf (251)	99 Es (252)	100 Fm (257)	101 Md (258)	102 No (259)

2.3 The Periodic Table *(Page 17)*

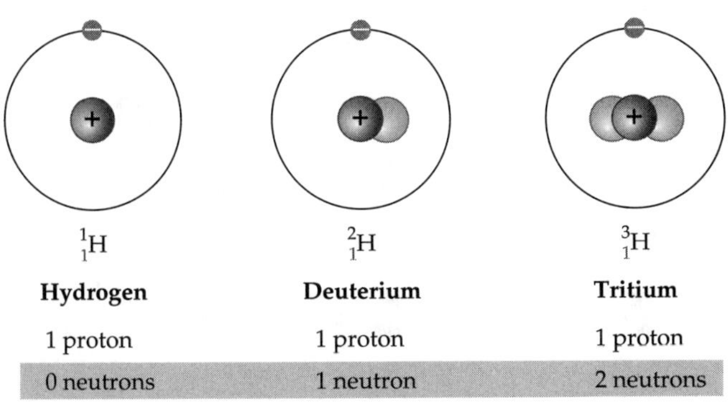
^1_1H
Hydrogen
1 proton
0 neutrons
^2_1H
Deuterium
1 proton
1 neutron
^3_1H
Tritium
1 proton
2 neutrons

2.4 Isotopes Have Different Numbers of Neutrons *(Page 18)*

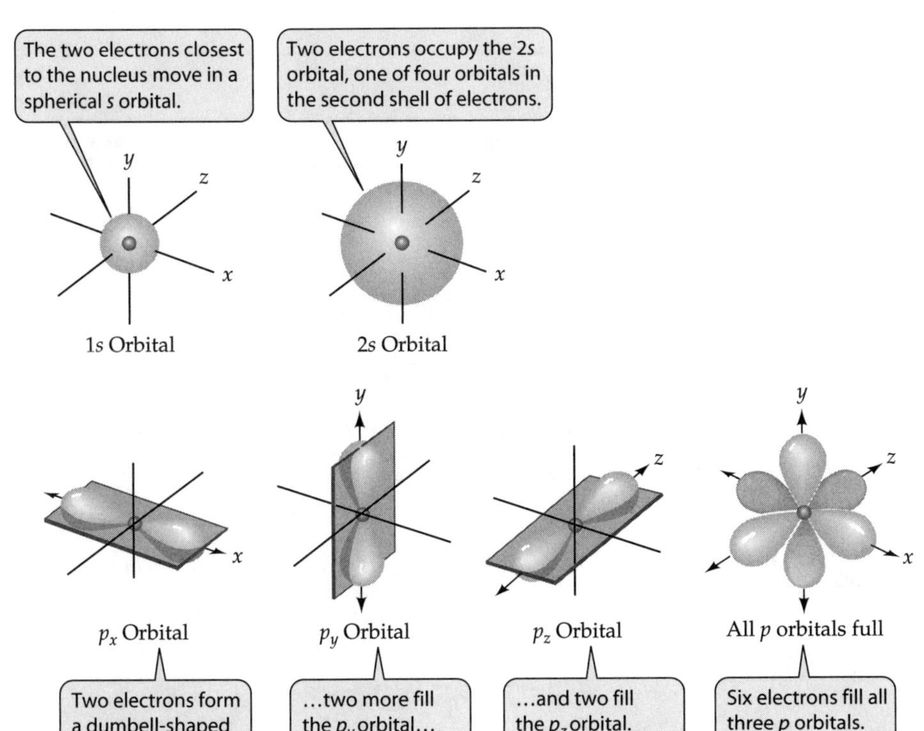
The two electrons closest to the nucleus move in a spherical s orbital.
Two electrons occupy the 2s orbital, one of four orbitals in the second shell of electrons.
y
z
x
1s Orbital
2s Orbital
p_x Orbital
p_y Orbital
p_z Orbital
All p orbitals full
Two electrons form a dumbell-shaped x-axis (p_x) orbital…
…two more fill the p_y orbital…
…and two fill the p_z orbital.
Six electrons fill all three p orbitals.

2.6 Electron Orbitals *(Page 19)*

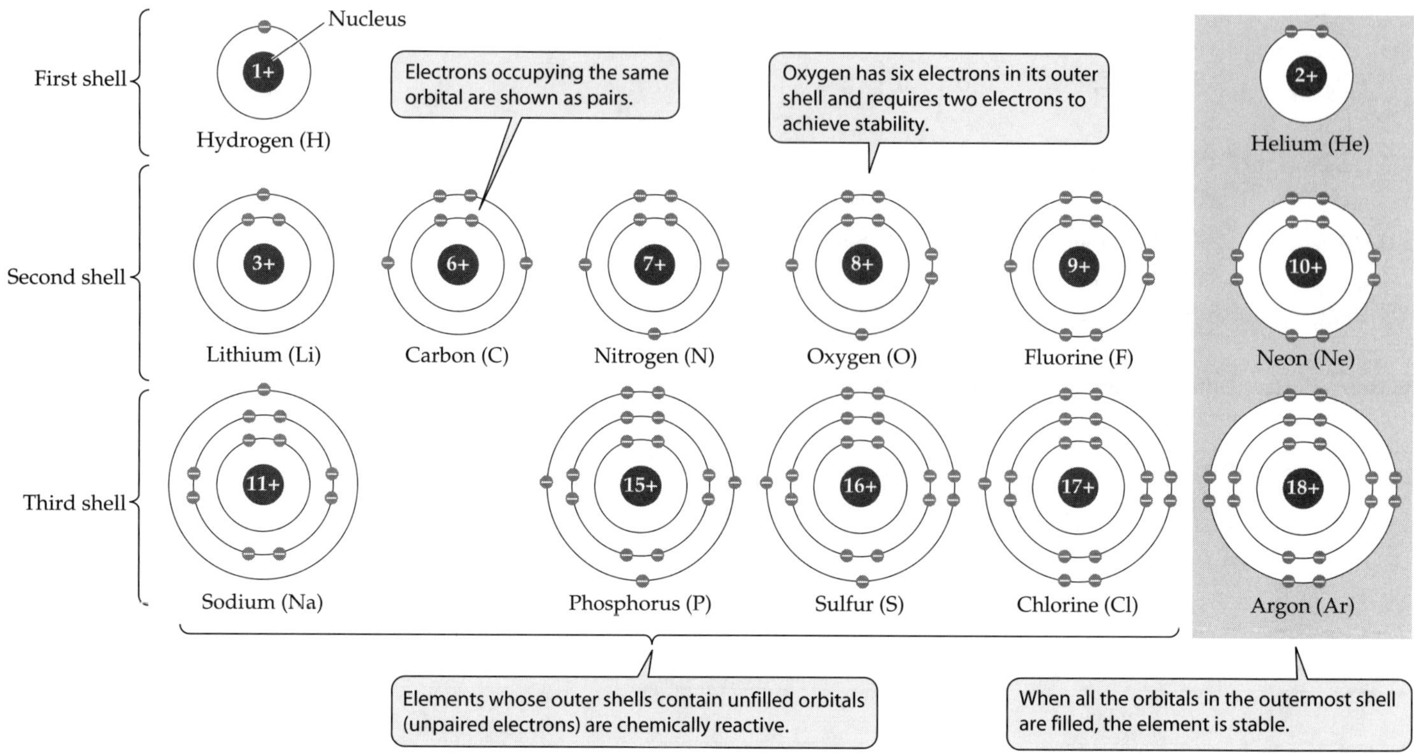
Nucleus
First shell
Second shell
Third shell
Hydrogen (H)
Electrons occupying the same orbital are shown as pairs.
Oxygen has six electrons in its outer shell and requires two electrons to achieve stability.
Helium (He)
Lithium (Li)
Carbon (C)
Nitrogen (N)
Oxygen (O)
Fluorine (F)
Neon (Ne)
Sodium (Na)
Phosphorus (P)
Sulfur (S)
Chlorine (Cl)
Argon (Ar)
1+
2+
3+
6+
7+
8+
9+
10+
11+
15+
16+
17+
18+
Elements whose outer shells contain unfilled orbitals (unpaired electrons) are chemically reactive.
When all the orbitals in the outermost shell are filled, the element is stable.

2.7 Electron Shells Determine the Reactivity of Atoms *(Page 20)*

2.1 *Chemical Bonds and Interactions*

NAME	BASIS OF INTERACTION	STRUCTURE	BOND ENERGY[a] (KCAL/MOL)
Covalent bond	Sharing of electron pairs	H O; | ‖; —N—C—	50–110
Hydrogen bond	Sharing of H atom	—N(H)—H$^{\delta+}$ ••••• $^{\delta-}$O=C—	3–7
Ionic bond	Attraction of opposite charges	—N(H)(H)—H ⊕ ⊖ O—C(=O)—	3–7
Hydrophobic interaction	Interaction of nonpolar substances in the presence of polar substances	—C(H)(H)—C(H)(H)—H∘∘∘∘∘H—C(H)(H)—C(H)(H)—	1–2
van der Waals interaction	Interaction of electrons of nonpolar substances	H—H ··· C(H)(H)(H)(H)	1

[a]*Bond energy* is the amount of energy needed to separate two bonded or interacting atoms under physiological conditions.

(Page 20)

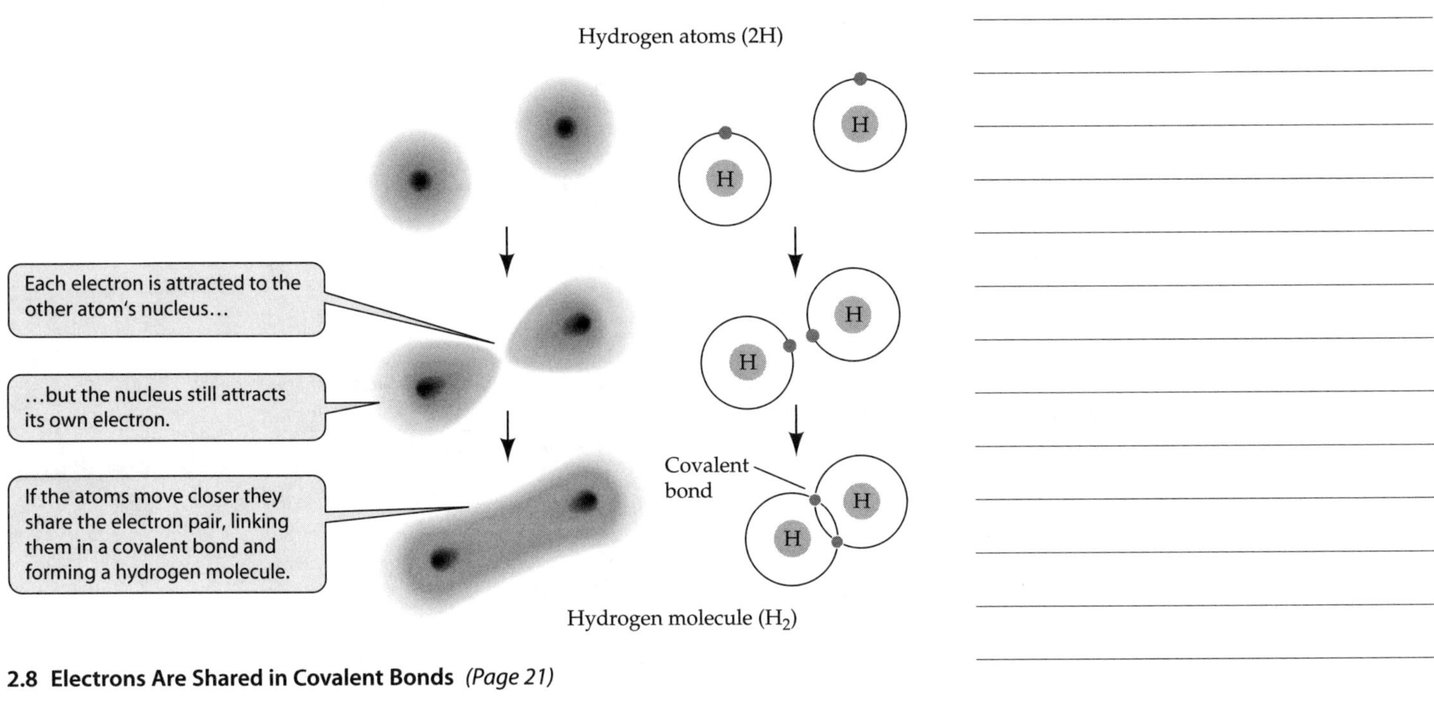
Hydrogen atoms (2H)
H
H
Each electron is attracted to the other atom's nucleus...
...but the nucleus still attracts its own electron.
If the atoms move closer they share the electron pair, linking them in a covalent bond and forming a hydrogen molecule.
Covalent bond
Hydrogen molecule (H_2)

2.8 Electrons Are Shared in Covalent Bonds *(Page 21)*

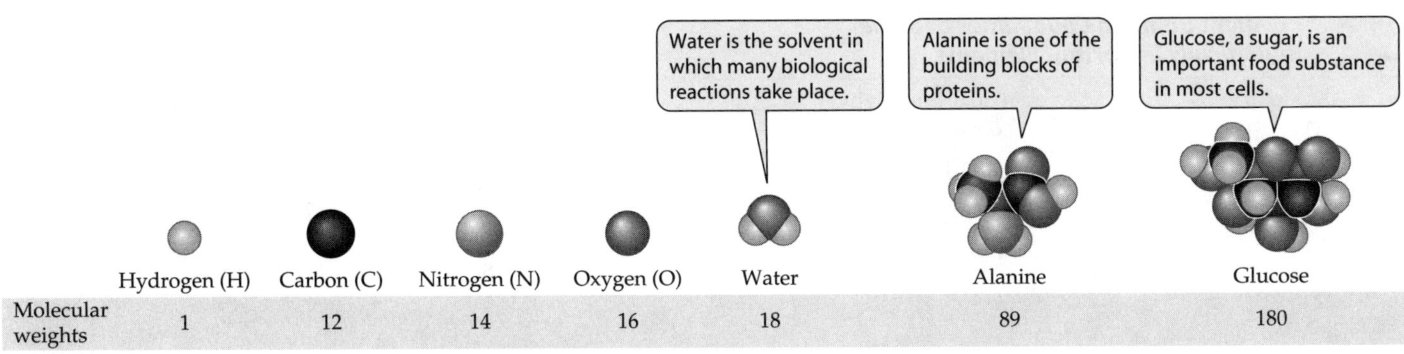
Water is the solvent in which many biological reactions take place.
Alanine is one of the building blocks of proteins.
Glucose, a sugar, is an important food substance in most cells.
Hydrogen (H)
Carbon (C)
Nitrogen (N)
Oxygen (O)
Water
Alanine
Glucose
Molecular weights
1
12
14
16
18
89
180

2.9 Weights and Sizes of Atoms and Molecules *(Page 21)*

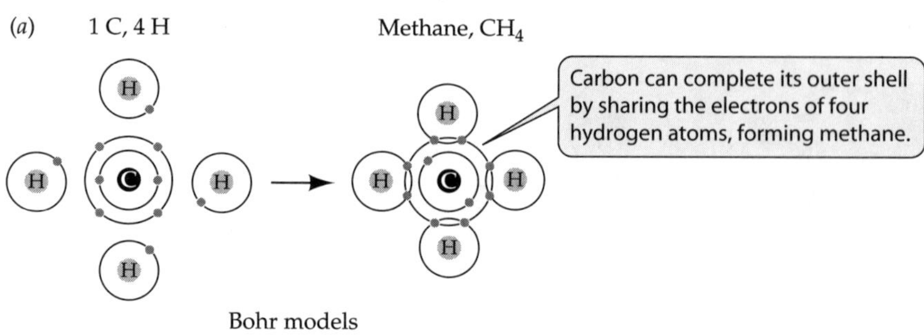
(a)
1 C, 4 H
Methane, CH_4
Carbon can complete its outer shell by sharing the electrons of four hydrogen atoms, forming methane.
Bohr models

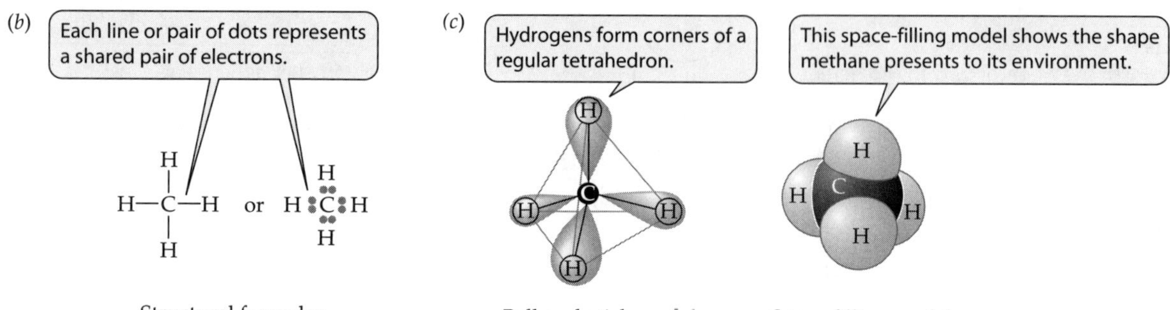
(b)
Each line or pair of dots represents a shared pair of electrons.
Structural formulas
(c)
Hydrogens form corners of a regular tetrahedron.
This space-filling model shows the shape methane presents to its environment.
Ball-and-stick model
Space-filling model

2.10 Covalent Bonding with Carbon *(Page 22)*

2.2 *Covalent Bonding Capabilities of Some Biologically Important Elements*

ELEMENT	USUAL NUMBER OF COVALENT BONDS
Hydrogen (H)	1
Oxygen (O)	2
Sulfur (S)	2
Nitrogen (N)	3
Carbon (C)	4
Phosphorus (P)	5

(Page 22)

2.3 *Some Electronegativities*

ELEMENT	ELECTRONEGATIVITY
Oxygen (O)	3.5
Chlorine (Cl)	3.1
Nitrogen (N)	3.0
Carbon (C)	2.5
Phosphorus (P)	2.1
Hydrogen (H)	2.1
Sodium (Na)	0.9
Potassium (K)	0.8

(Page 23)

(a)

δ^+ H—O δ^- — H δ^+

Water has polar covalent bonds.

(b)

Unshared pairs of electrons are not part of water's covalent bond.

Water's bonding electrons are shared unequally; electron density is greatest around the oxygen atom.

2.11 The Polar Covalent Bond in the Water Molecule *(Page 23)*

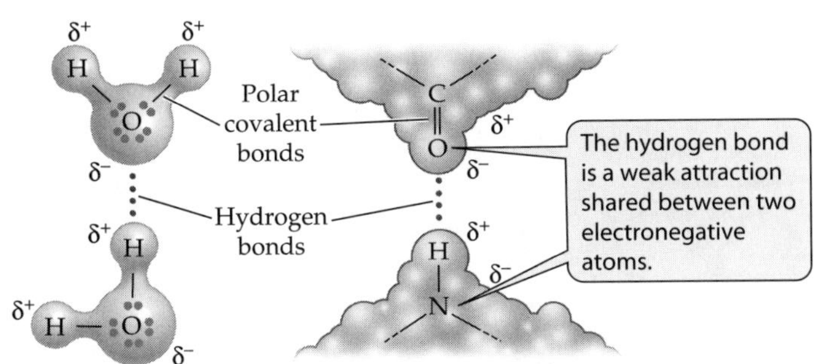
δ+
δ+
H
H
O
Polar
covalent
bonds
C
O
δ+
δ−
The hydrogen bond is a weak attraction shared between two electronegative atoms.
δ−
Hydrogen
bonds
δ+
H
δ+
H
N
δ−
δ+
H
O
δ−
Two water molecules
Two parts of one large molecule (or two large molecules)

2.12 Hydrogen Bonds Can Form between or within Molecules *(Page 23)*

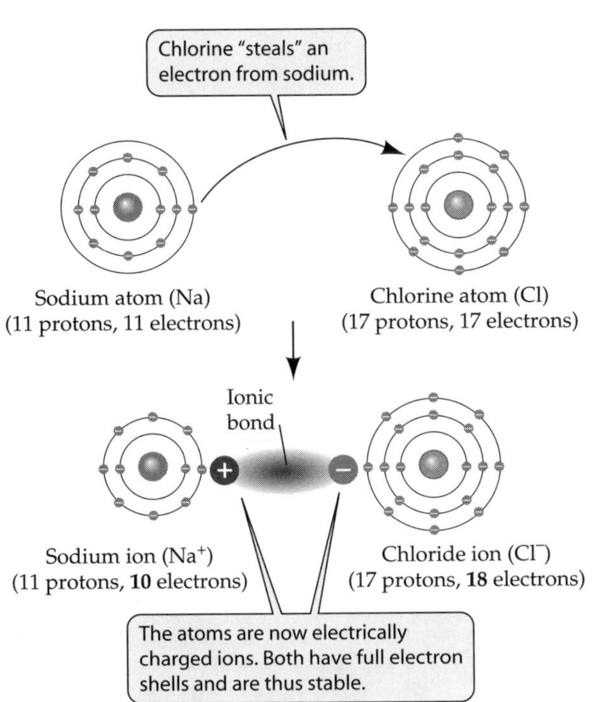
Chlorine "steals" an electron from sodium.
Sodium atom (Na)
(11 protons, 11 electrons)
Chlorine atom (Cl)
(17 protons, 17 electrons)
Ionic
bond
Sodium ion (Na^+)
(11 protons, **10** electrons)
Chloride ion (Cl^-)
(17 protons, **18** electrons)
The atoms are now electrically charged ions. Both have full electron shells and are thus stable.

2.13 Formation of Sodium and Chloride Ions *(Page 24)*

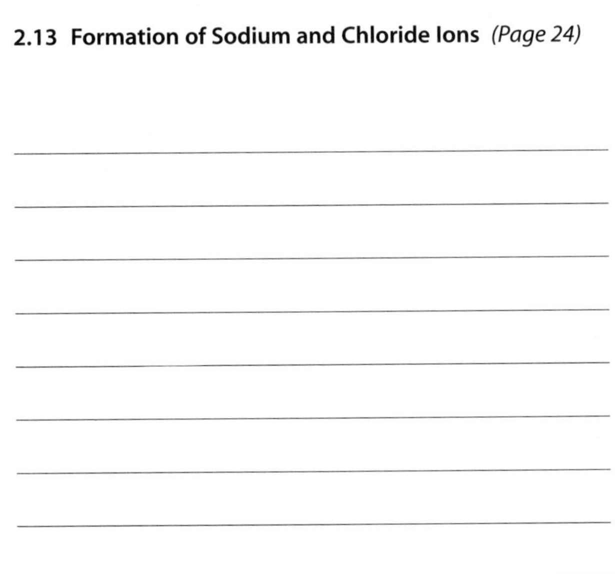

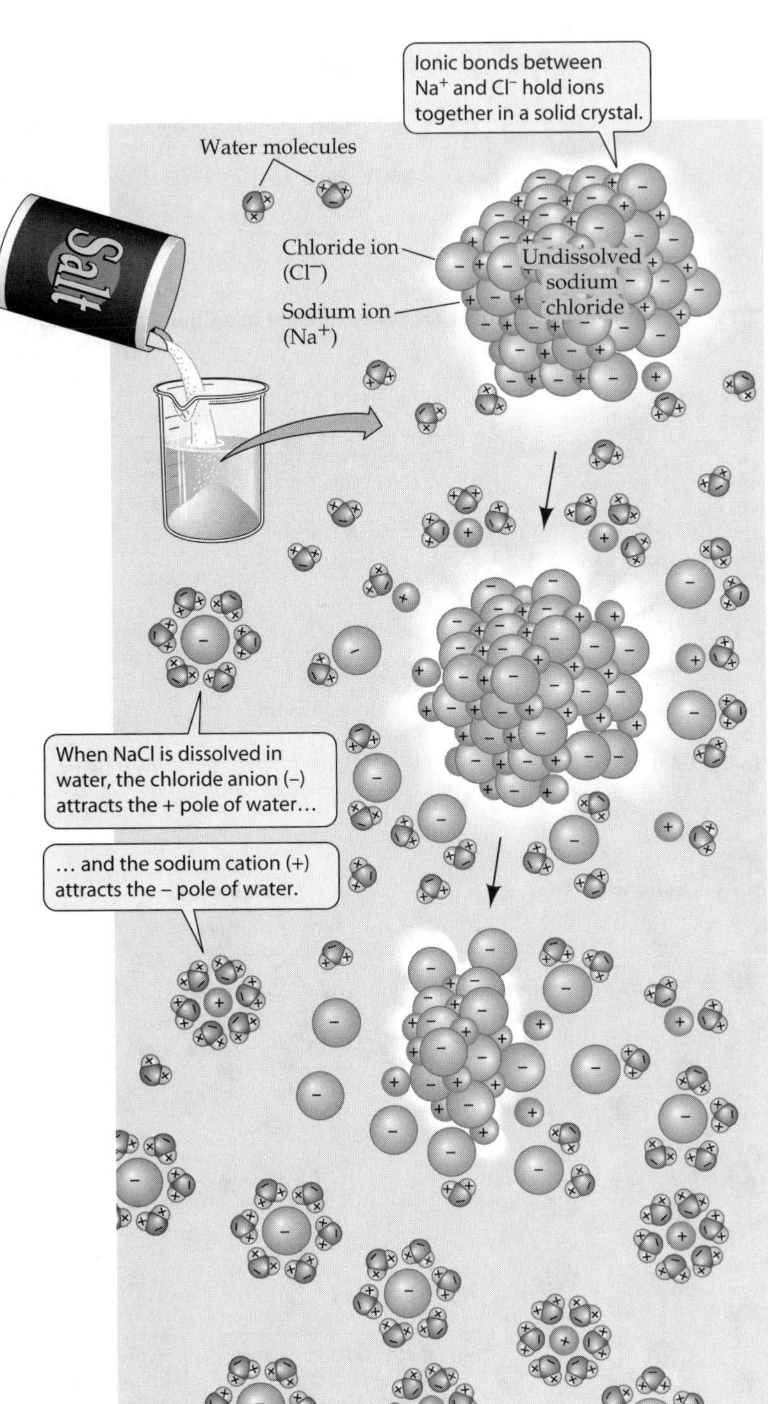
Ionic bonds between Na^+ and Cl^- hold ions together in a solid crystal.
Water molecules
Salt
Chloride ion
(Cl^-)
Sodium ion
(Na^+)
Undissolved
sodium
chloride
When NaCl is dissolved in water, the chloride anion (–) attracts the + pole of water…
… and the sodium cation (+) attracts the – pole of water.

2.14 Water Molecules Surround Ions *(Page 25)*

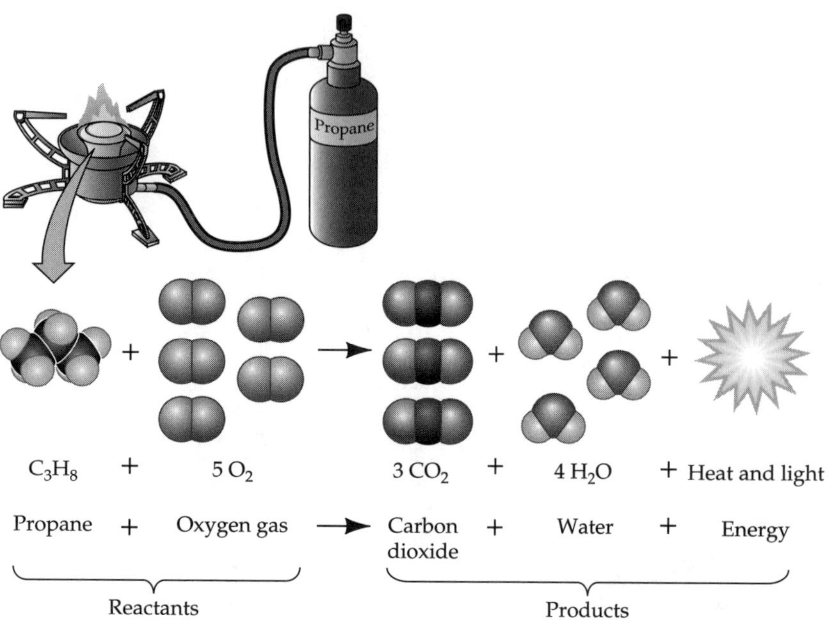

Propane
C_3H_8 + $5\ O_2$ $3\ CO_2$ + $4\ H_2O$ + Heat and light
Propane + Oxygen gas → Carbon dioxide + Water + Energy
Reactants
Products

2.15 Bonding Partners and Energy May Change in a Chemical Reaction *(Page 25)*

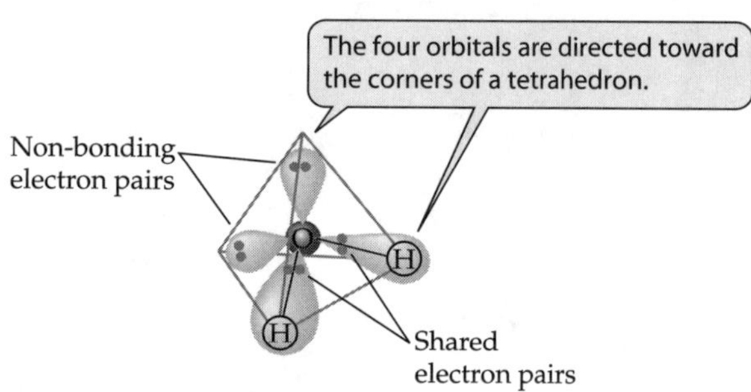

The four orbitals are directed toward the corners of a tetrahedron.
Non-bonding electron pairs
H
H
Shared electron pairs

In-Text Art *(Page 26)*

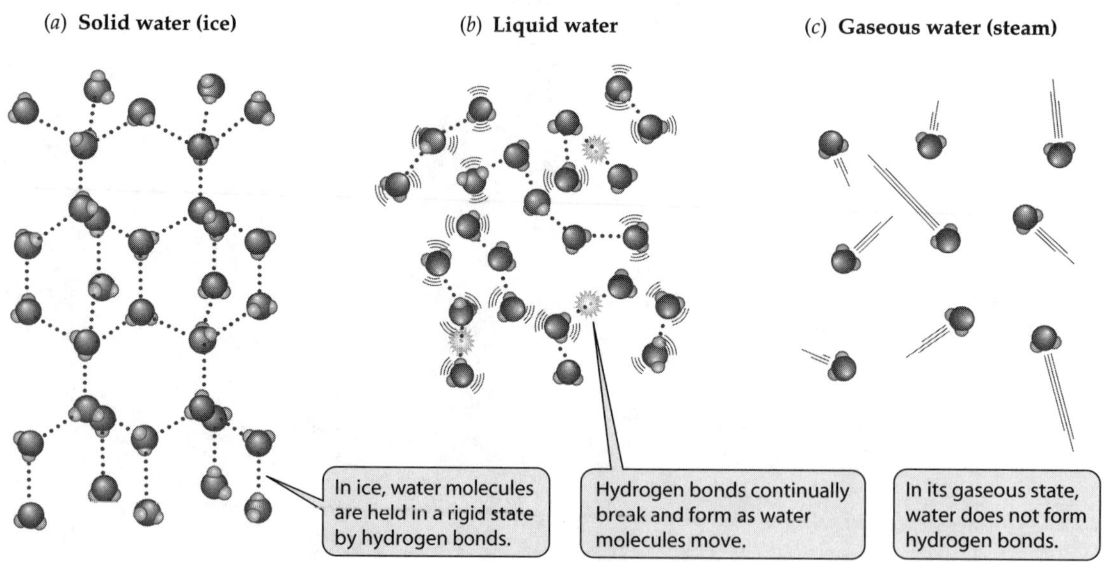

(a) Solid water (ice)
(b) Liquid water
(c) Gaseous water (steam)
In ice, water molecules are held in a rigid state by hydrogen bonds.
Hydrogen bonds continually break and form as water molecules move.
In its gaseous state, water does not form hydrogen bonds.

2.16 Hydrogen Bonds Hold Water Molecules Together *(Page 27)*

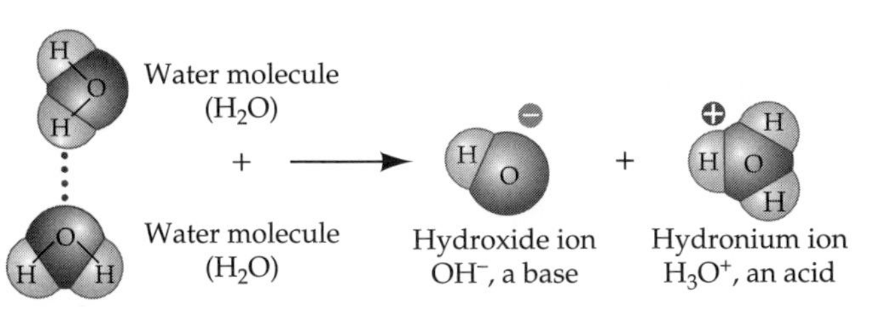

Water molecule
(H_2O)
+
Water molecule
(H_2O)
Hydroxide ion
OH^-, a base
+
Hydronium ion
H_3O^+, an acid

In-Text Art *(Page 29)*

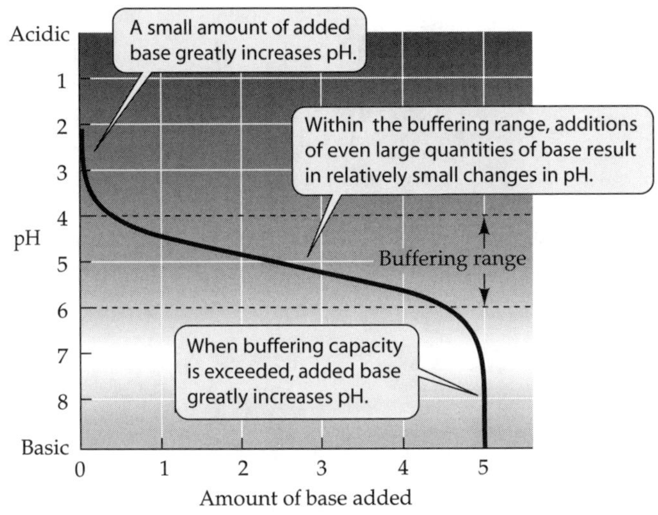

Acidic
A small amount of added base greatly increases pH.
Within the buffering range, additions of even large quantities of base result in relatively small changes in pH.
pH
Buffering range
When buffering capacity is exceeded, added base greatly increases pH.
Basic
Amount of base added

2.19 Buffers Minimize Changes in pH *(Page 31)*

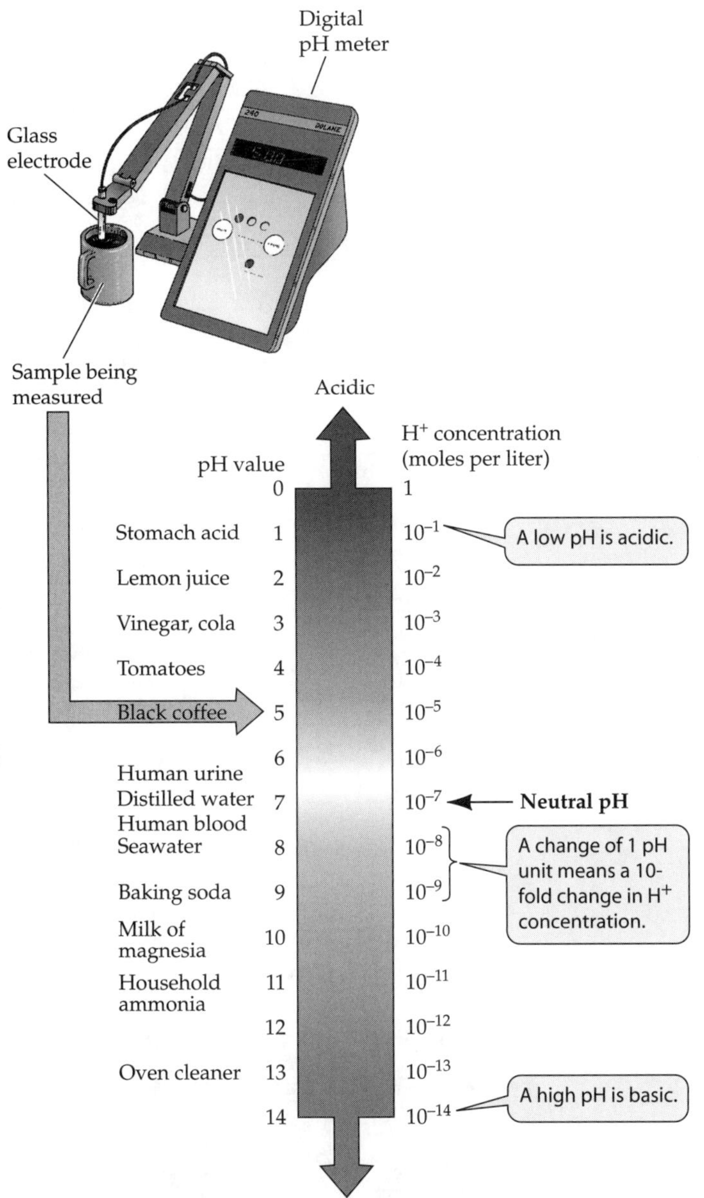

Digital pH meter
Glass electrode
Sample being measured
Acidic
pH value
H^+ concentration (moles per liter)
Stomach acid
Lemon juice
Vinegar, cola
Tomatoes
Black coffee
Human urine
Distilled water
Human blood
Seawater
Baking soda
Milk of magnesia
Household ammonia
Oven cleaner
A low pH is acidic.
Neutral pH
A change of 1 pH unit means a 10-fold change in H^+ concentration.
A high pH is basic.
Basic

2.18 pH Values of Some Familiar Substances *(Page 30)*

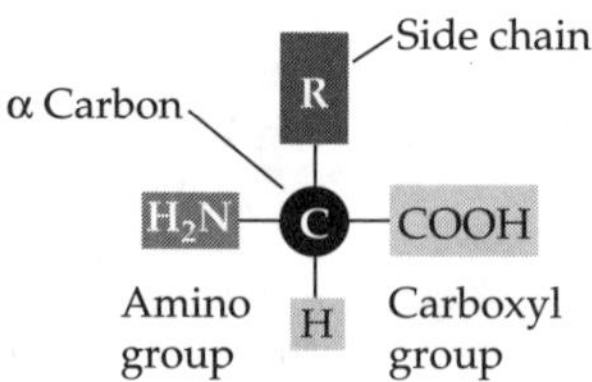
Side chain
α Carbon
R
H_2N
C
COOH
H
Amino group
Carboxyl group

In-Text Art *(Page 31)*

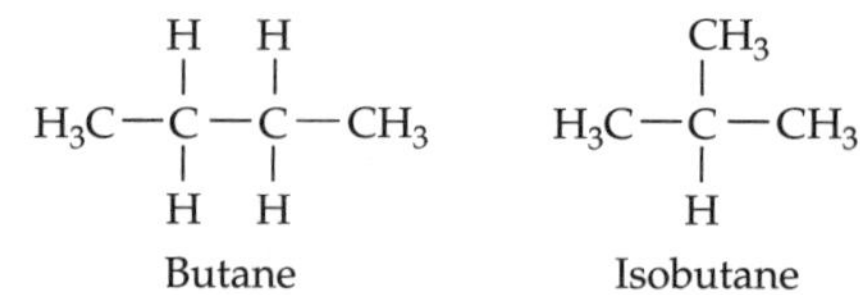
Butane
Isobutane

In-Text Art *(Page 32)*

This part = R

Functional group	Class of compounds	Structural formula	Example
Hydroxyl —OH	Alcohols	R—OH	CH_3CH_2—OH Ethanol
Aldehyde —CHO	Aldehydes	R—C(=O)H	CH_3—C(=O)H Acetaldehyde
Keto >CO	Ketones	R—C(=O)—R	CH_3—C(=O)—CH_3 Acetone
Carboxyl —COOH	Carboxylic acids	R—C(=O)OH	CH_3—C(=O)OH Acetic acid
Amino $—NH_2$	Amines	R—N(H)H	CH_3—N(H)H Methylamine
Phosphate $—OPO_3^{2-}$	Organic phosphates	R—O—P(=O)(O^-)—O^-	HO—C(=O)—CH(OH)—CH_2—O—P(=O)(O^-)—O^- 3-Phosphoglyceric acid
Sulfhydryl —SH	Thiols	R—SH	HO—CH_2—CH_2—SH Mercaptoethanol

2.20 Some Functional Groups Important to Living Systems *(Page 31)*

(*a*)

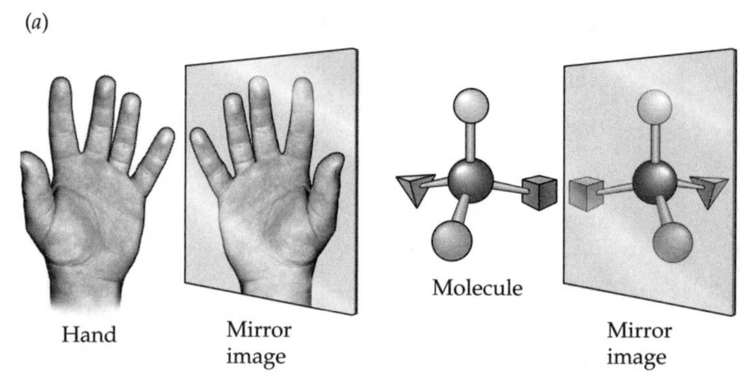
Hand
Mirror image
Molecule
Mirror image

(*b*)

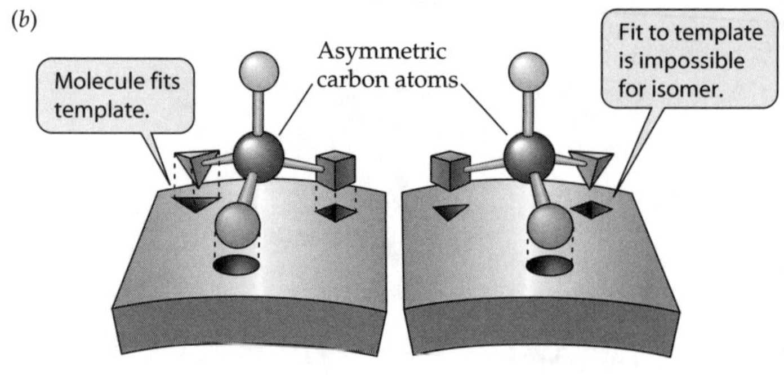
Molecule fits template.
Asymmetric carbon atoms
Fit to template is impossible for isomer.

2.21 Optical Isomers *(Page 32)*

3 Life and Chemistry: Large Molecules

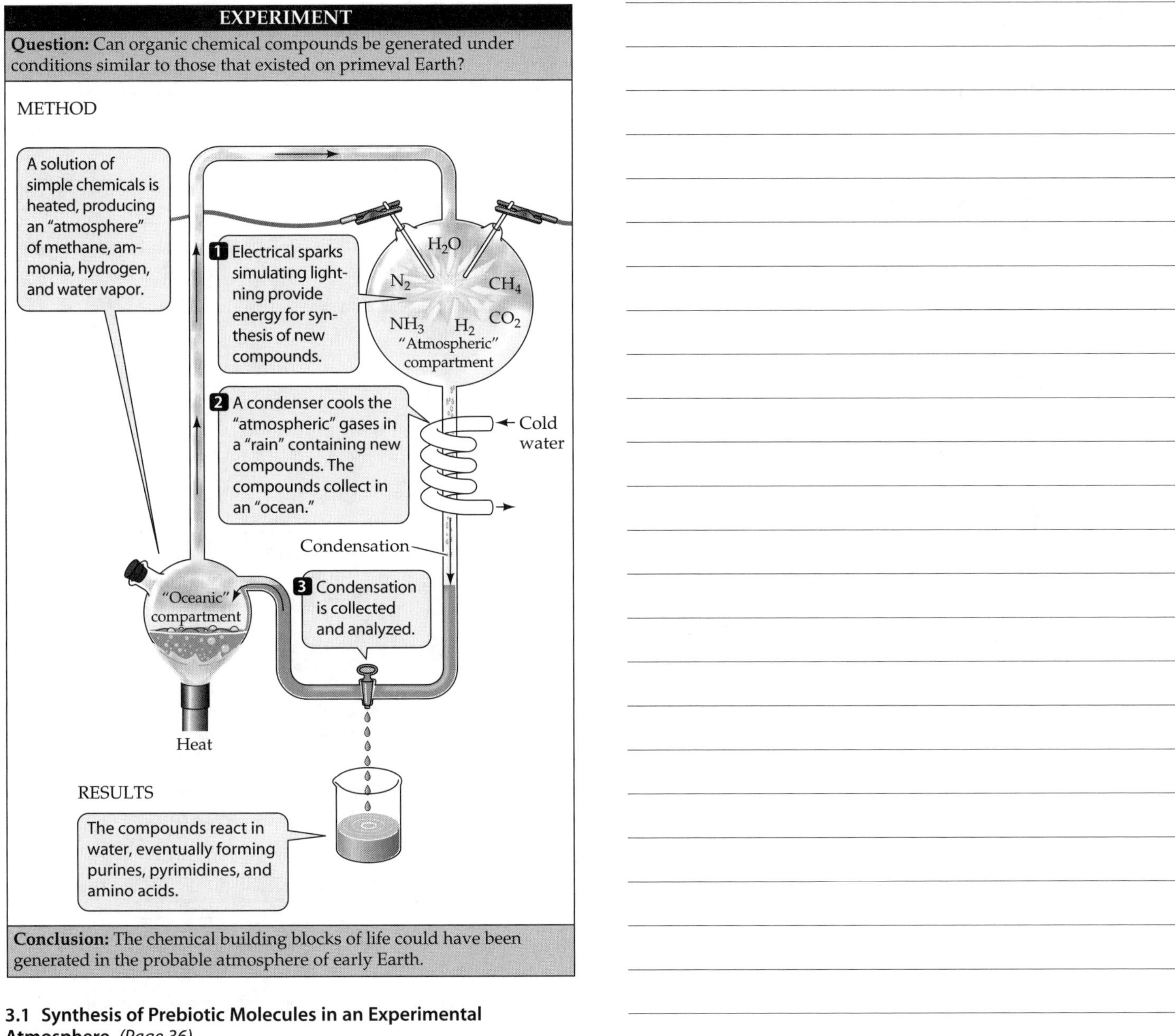
EXPERIMENT
Question: Can organic chemical compounds be generated under conditions similar to those that existed on primeval Earth?
METHOD
A solution of simple chemicals is heated, producing an "atmosphere" of methane, ammonia, hydrogen, and water vapor.
1 Electrical sparks simulating lightning provide energy for synthesis of new compounds.
H_2O
N_2
CH_4
NH_3
H_2
CO_2
"Atmospheric" compartment
2 A condenser cools the "atmospheric" gases in a "rain" containing new compounds. The compounds collect in an "ocean."
Cold water
Condensation
"Oceanic" compartment
3 Condensation is collected and analyzed.
Heat
RESULTS
The compounds react in water, eventually forming purines, pyrimidines, and amino acids.
Conclusion: The chemical building blocks of life could have been generated in the probable atmosphere of early Earth.

3.1 Synthesis of Prebiotic Molecules in an Experimental Atmosphere *(Page 36)*

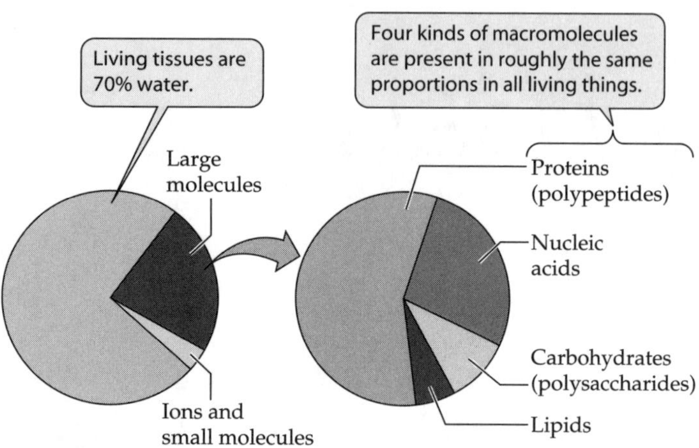
Living tissues are 70% water.
Four kinds of macromolecules are present in roughly the same proportions in all living things.
Large molecules
Proteins (polypeptides)
Nucleic acids
Carbohydrates (polysaccharides)
Lipids
Ions and small molecules

3.2 Substances Found in Living Tissues *(Page 37)*

3.1 *The Building Blocks of Organisms*

MONOMER	SIMPLE POLYMER	COMPLEX POLYMER (MACROMOLECULE)
Amino acid	Peptide or oligopeptide	Polypeptide (protein)
Nucleotide	Oligonucleotide	Nucleic acid
Monosaccharide (sugar)	Oligosaccharide	Polysaccharide (carbohydrate)

(Page 37)

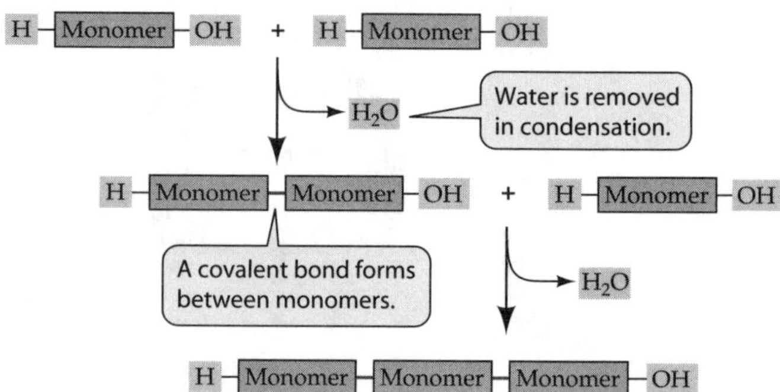
(a) Condensation
H Monomer OH + H Monomer OH
H2O
Water is removed in condensation.
H Monomer Monomer OH + H Monomer OH
A covalent bond forms between monomers.
H2O
H Monomer Monomer Monomer OH

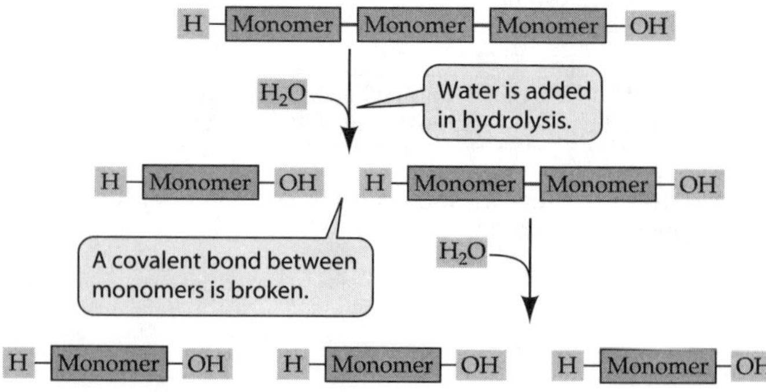
(b) Hydrolysis
H Monomer Monomer Monomer OH
H2O
Water is added in hydrolysis.
H Monomer OH H Monomer Monomer OH
A covalent bond between monomers is broken.
H2O
H Monomer OH H Monomer OH H Monomer OH

3.3 Condensation and Hydrolysis of Polymers *(Page 38)*

3.2 The Twenty Amino Acids Found in Proteins

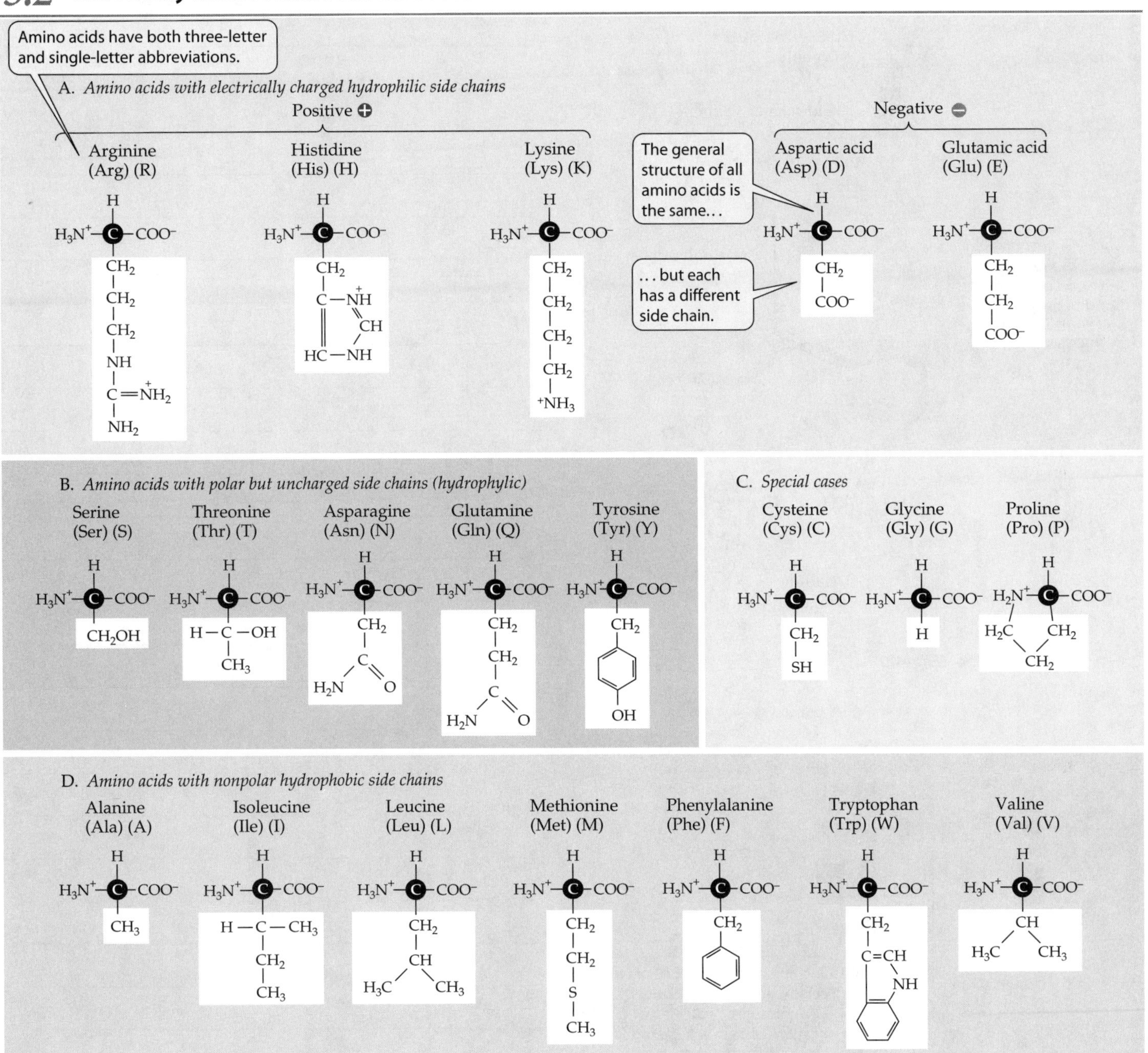
Amino acids have both three-letter and single-letter abbreviations.
A. Amino acids with electrically charged hydrophilic side chains
Positive
Negative
Arginine (Arg) (R)
Histidine (His) (H)
Lysine (Lys) (K)
The general structure of all amino acids is the same...
... but each has a different side chain.
Aspartic acid (Asp) (D)
Glutamic acid (Glu) (E)
B. Amino acids with polar but uncharged side chains (hydrophylic)
Serine (Ser) (S)
Threonine (Thr) (T)
Asparagine (Asn) (N)
Glutamine (Gln) (Q)
Tyrosine (Tyr) (Y)
C. Special cases
Cysteine (Cys) (C)
Glycine (Gly) (G)
Proline (Pro) (P)
D. Amino acids with nonpolar hydrophobic side chains
Alanine (Ala) (A)
Isoleucine (Ile) (I)
Leucine (Leu) (L)
Methionine (Met) (M)
Phenylalanine (Phe) (F)
Tryptophan (Trp) (W)
Valine (Val) (V)

(Page 39)

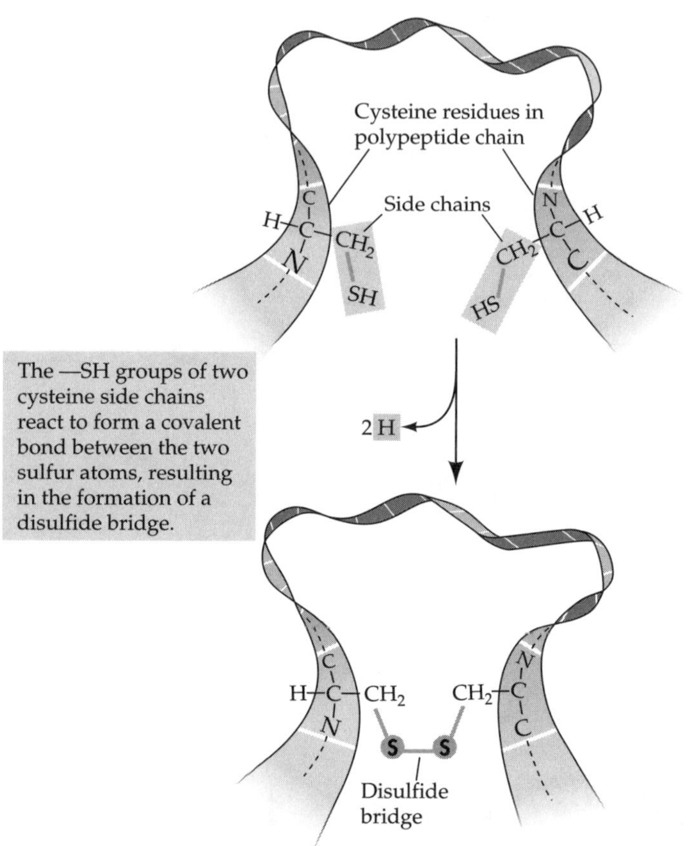
Cysteine residues in polypeptide chain
Side chains
CH_2
SH
CH_2
HS
The —SH groups of two cysteine side chains react to form a covalent bond between the two sulfur atoms, resulting in the formation of a disulfide bridge.
2 H
CH_2
CH_2
S
S
Disulfide bridge

3.4 A Disulfide Bridge *(Page 40)*

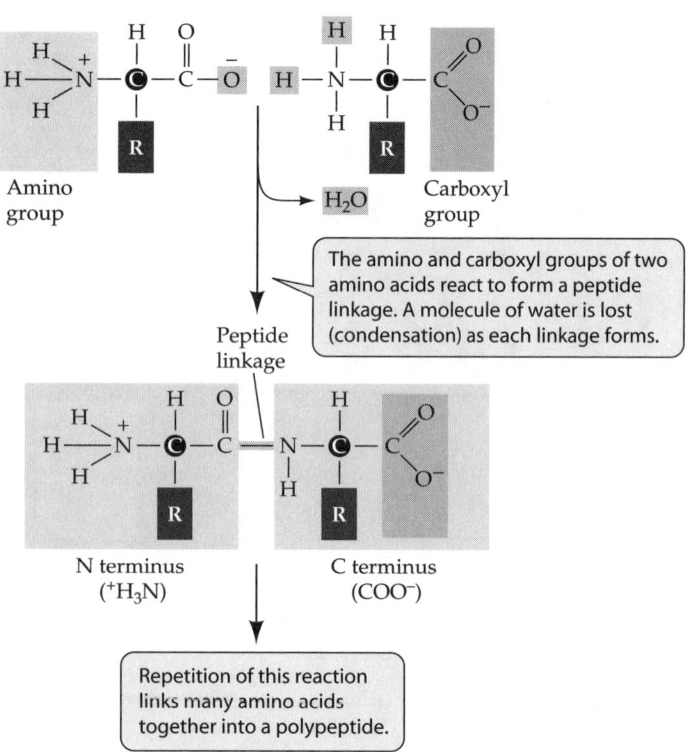
Amino group
H_2O
Carboxyl group
The amino and carboxyl groups of two amino acids react to form a peptide linkage. A molecule of water is lost (condensation) as each linkage forms.
Peptide linkage
N terminus (^+H_3N)
C terminus (COO^-)
Repetition of this reaction links many amino acids together into a polypeptide.

3.5 Formation of Peptide Linkages *(Page 40)*

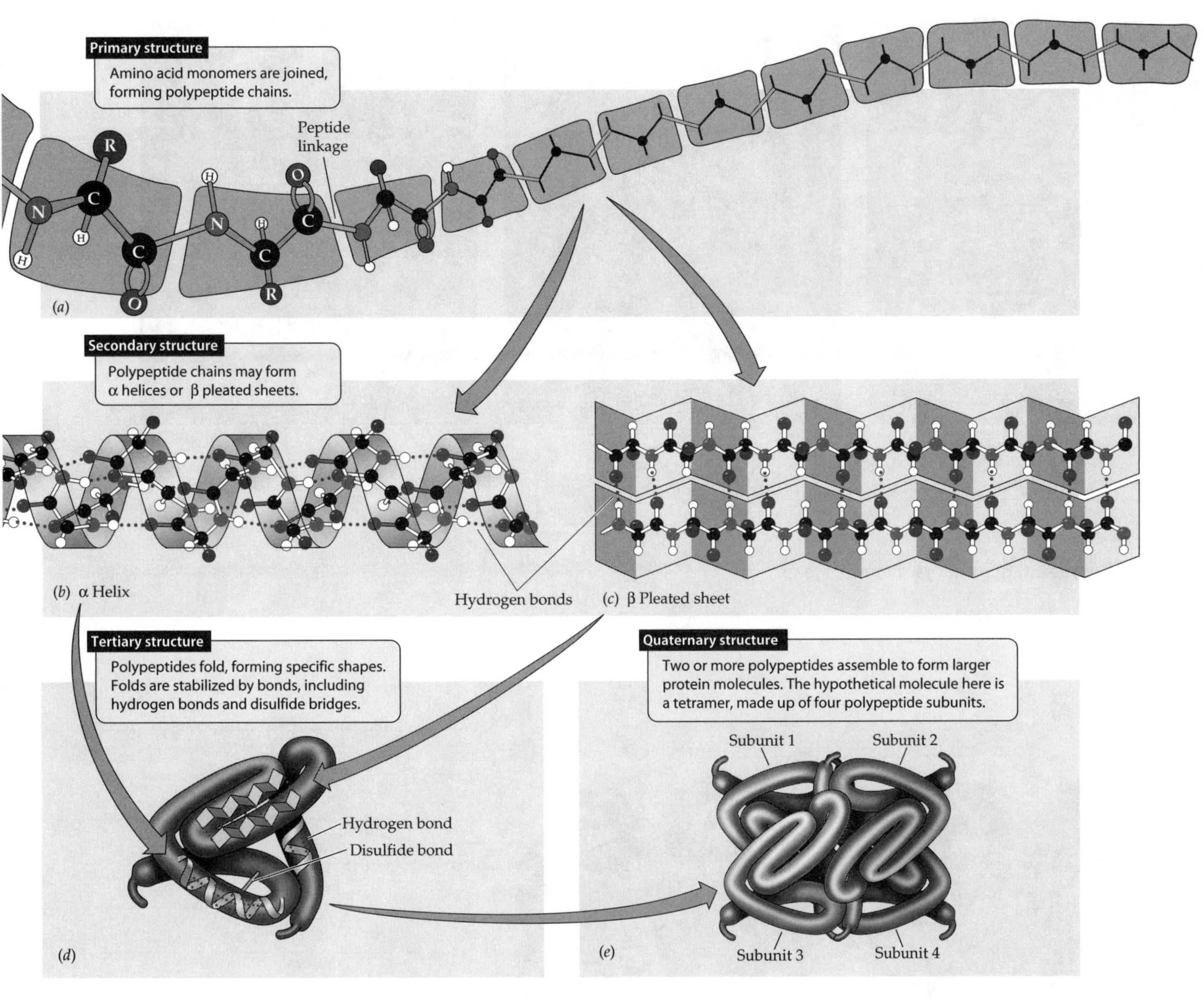
Primary structure
Amino acid monomers are joined, forming polypeptide chains.
Peptide linkage
R
N
C
H
C
O
N
H
C
O
C
R
(a)
Secondary structure
Polypeptide chains may form α helices or β pleated sheets.
(b) α Helix
Hydrogen bonds
(c) β Pleated sheet
Tertiary structure
Polypeptides fold, forming specific shapes. Folds are stabilized by bonds, including hydrogen bonds and disulfide bridges.
Hydrogen bond
Disulfide bond
(d)
Quaternary structure
Two or more polypeptides assemble to form larger protein molecules. The hypothetical molecule here is a tetramer, made up of four polypeptide subunits.
Subunit 1
Subunit 2
Subunit 3
Subunit 4
(e)

3.6 The Four Levels of Protein Structure *(Page 42)*

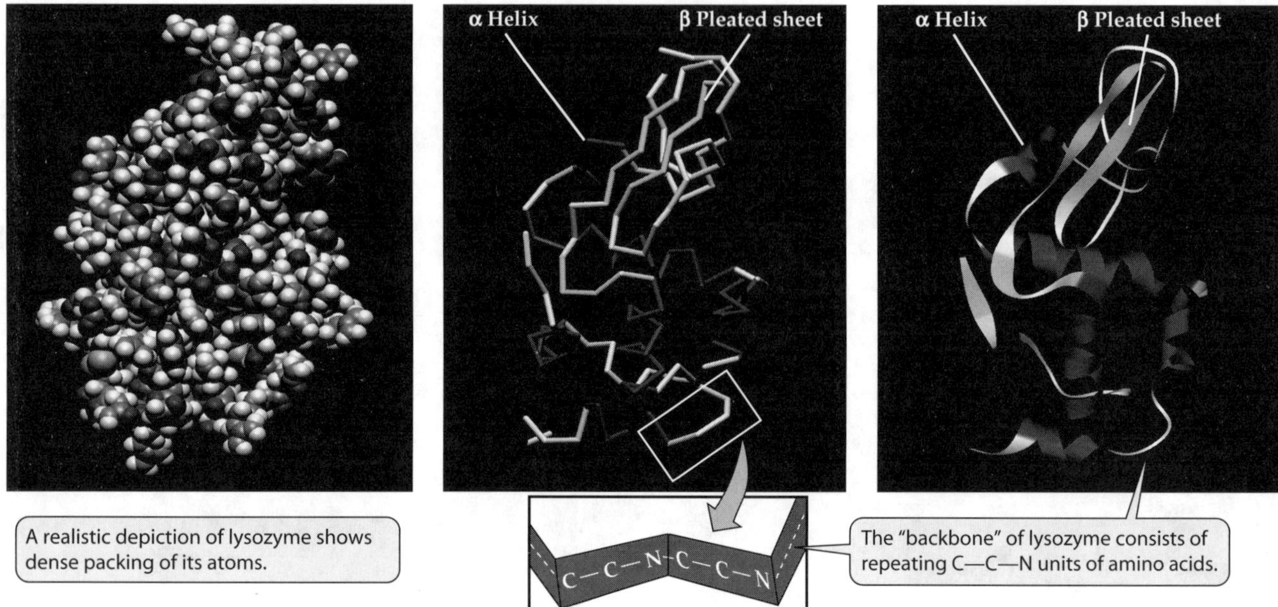
α Helix
β Pleated sheet
α Helix
β Pleated sheet
A realistic depiction of lysozyme shows dense packing of its atoms.
C—C—N—C—C—N
The "backbone" of lysozyme consists of repeating C—C—N units of amino acids.

3.7 Three Representations of Lysozyme *(Page 43)*

(*a*)

(*b*)

α Subunits

β Subunits

Heme

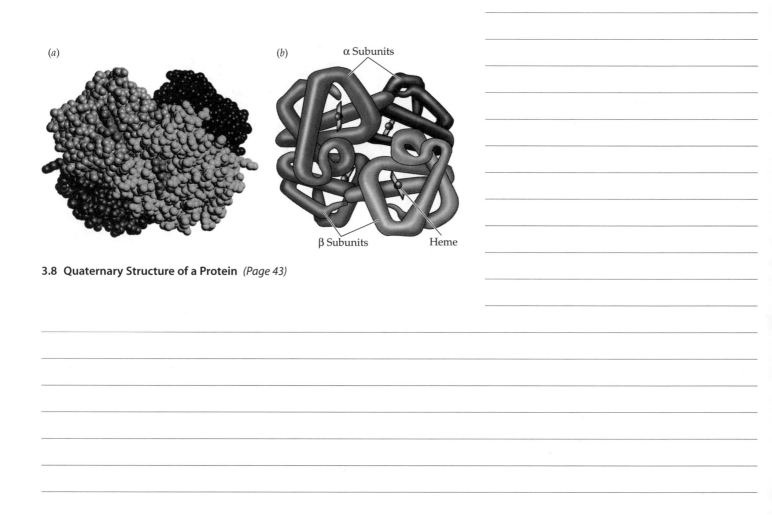

3.8 Quaternary Structure of a Protein *(Page 43)*

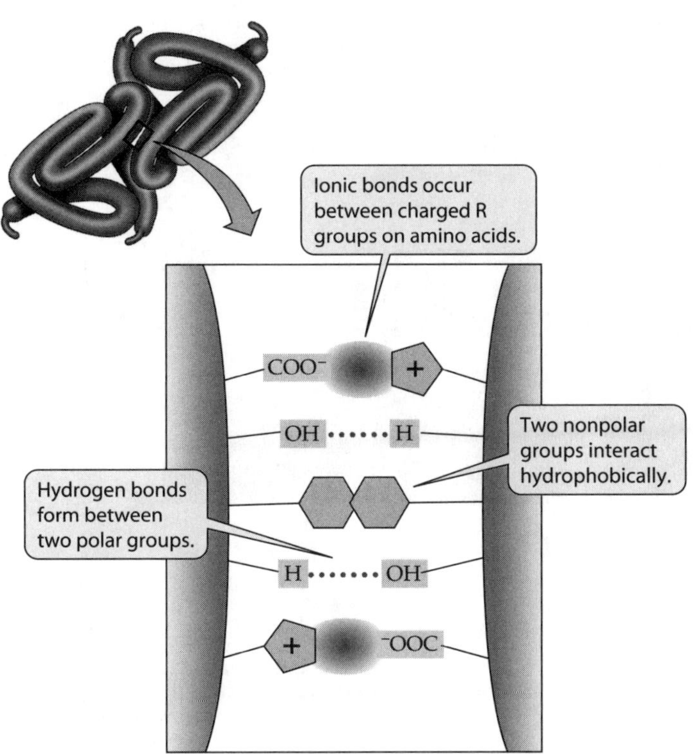
Ionic bonds occur between charged R groups on amino acids.
COO⁻
+
OH ······ H
Two nonpolar groups interact hydrophobically.
Hydrogen bonds form between two polar groups.
H ······· OH
+
⁻OOC

3.9 Noncovalent Interactions between Proteins and Other Molecules *(Page 44)*

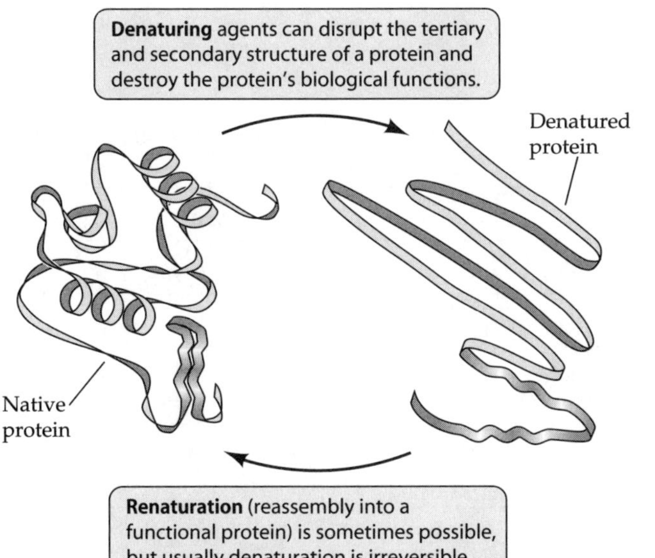
Denaturing agents can disrupt the tertiary and secondary structure of a protein and destroy the protein's biological functions.
Denatured protein
Native protein
Renaturation (reassembly into a functional protein) is sometimes possible, but usually denaturation is irreversible.

3.11 Denaturation Is the Loss of Tertiary Protein Structure and Function *(Page 45)*

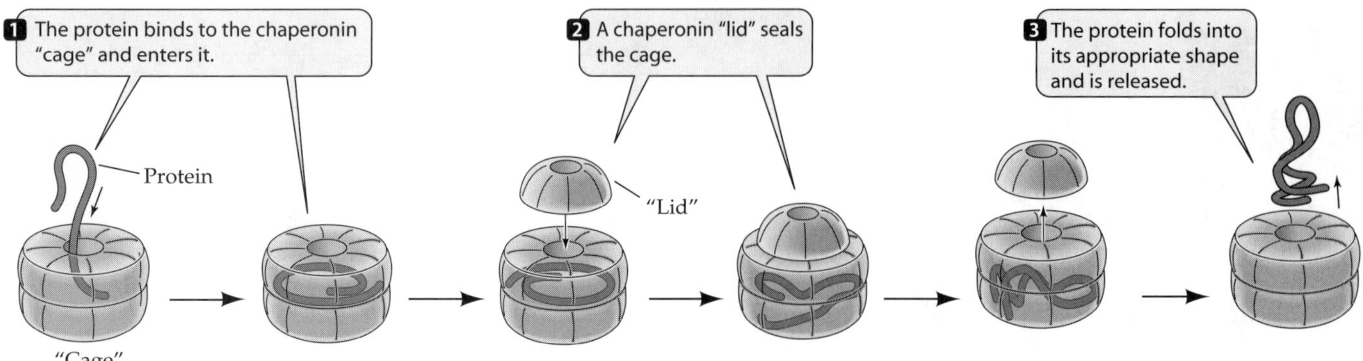
1 The protein binds to the chaperonin "cage" and enters it.
Protein
"Cage"
2 A chaperonin "lid" seals the cage.
"Lid"
3 The protein folds into its appropriate shape and is released.

3.12 Chaperonins Protect Proteins from Inappropriate Binding *(Page 46)*

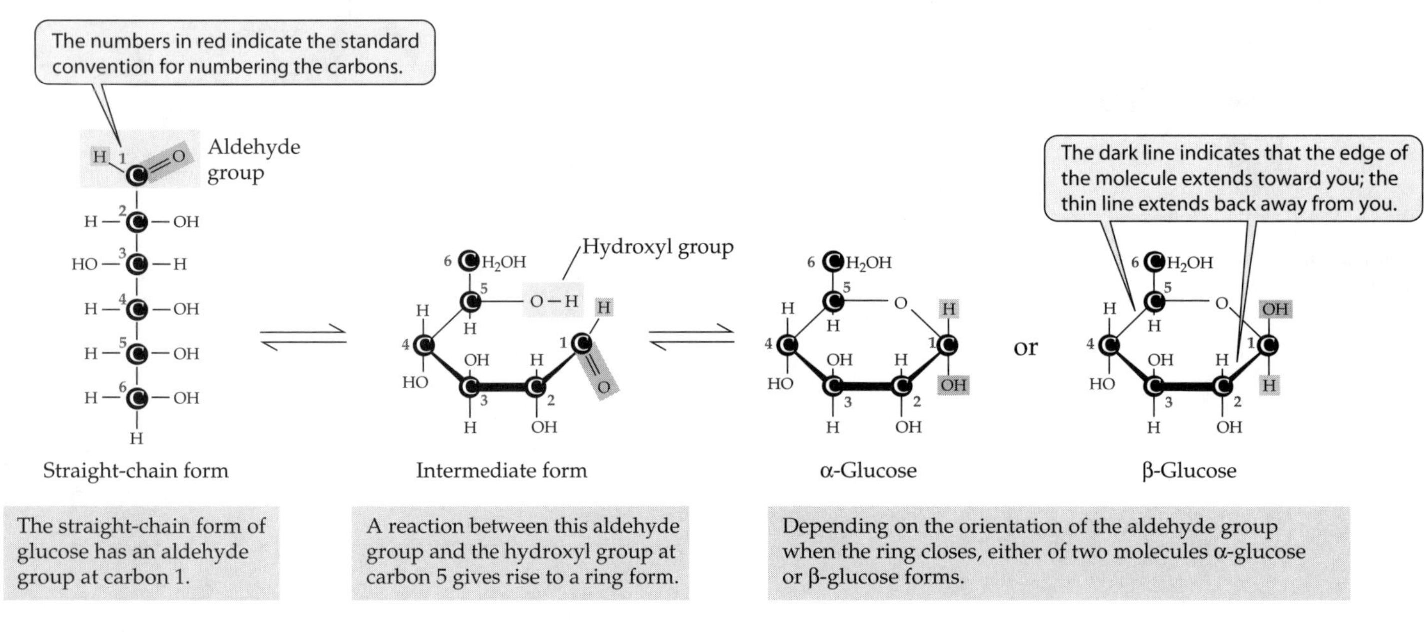
The numbers in red indicate the standard convention for numbering the carbons.
Aldehyde group
Hydroxyl group
The dark line indicates that the edge of the molecule extends toward you; the thin line extends back away from you.
or
Straight-chain form
Intermediate form
α-Glucose
β-Glucose
The straight-chain form of glucose has an aldehyde group at carbon 1.
A reaction between this aldehyde group and the hydroxyl group at carbon 5 gives rise to a ring form.
Depending on the orientation of the aldehyde group when the ring closes, either of two molecules α-glucose or β-glucose forms.

3.13 Glucose: From One Form to the Other *(Page 47)*

Three-carbon sugar

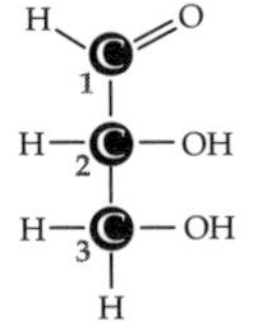
H
O
1
H
2
OH
H
3
OH
H

Glyceraldehyde

Glyceraldehyde is the smallest monosaccharide and exists only as the straight-chain form.

Five-carbon sugars (pentoses)

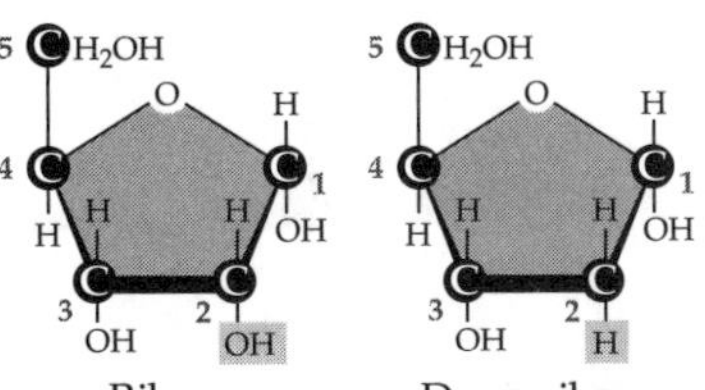
5 CH₂OH
O
H
4
1
H
H
H
OH
3
2
OH
OH
Ribose
Deoxyribose

Ribose and deoxyribose each have five carbons, but very different chemical properties and biological roles.

Six-carbon sugars (hexoses)

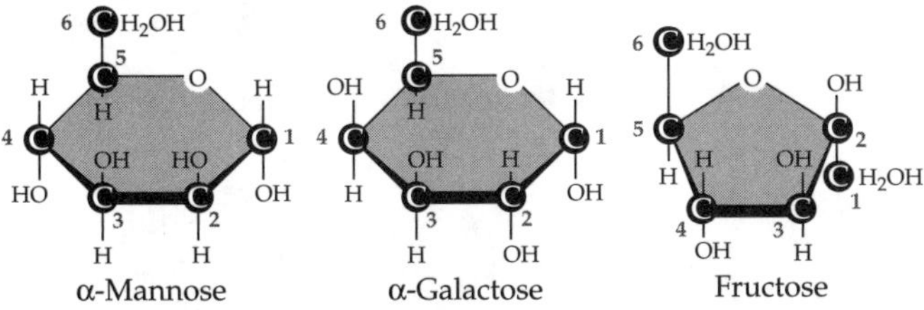
6 CH₂OH
5
O
H
OH
HO
α-Mannose
α-Galactose
Fructose
1
2
3
4

These hexoses are isomers. All have the formula $C_6H_{12}O_6$, but each has distinct chemical properties and biological roles.

3.14 Monosaccharides Are Simple Sugars *(Page 47)*

The presence of a carbon atom (C) at a junction such as this is implied.

Maltose is produced when an α-1,4 glycosidic linkage forms between two glucose molecules. The hydroxyl group on carbon 1 of one glucose in the α (down) position reacts with the hydroxyl group on carbon 4 of the other glucose.

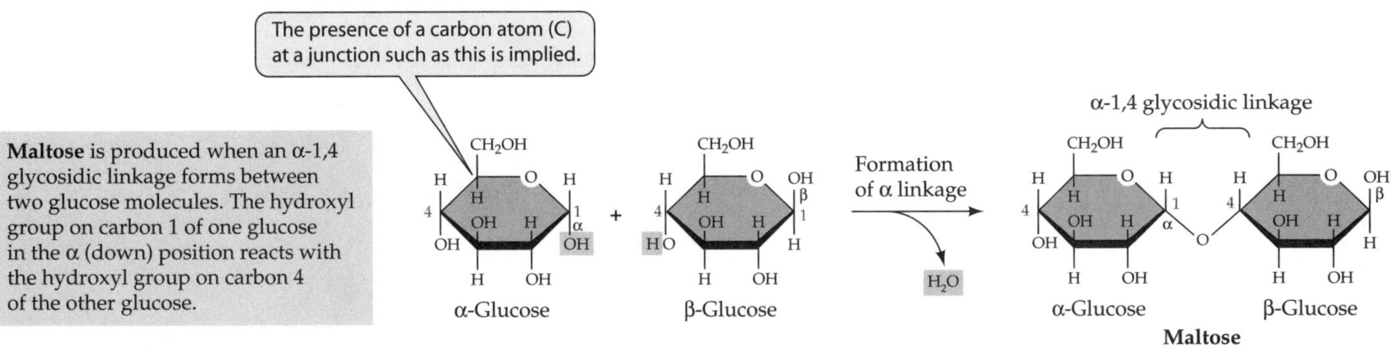
α-1,4 glycosidic linkage
CH₂OH
O
H
OH
α
β
+
Formation of α linkage
H₂O
α-Glucose
β-Glucose
α-Glucose
β-Glucose
Maltose

In **cellobiose**, two glucoses are linked by a β-1,4 glycosidic linkage.

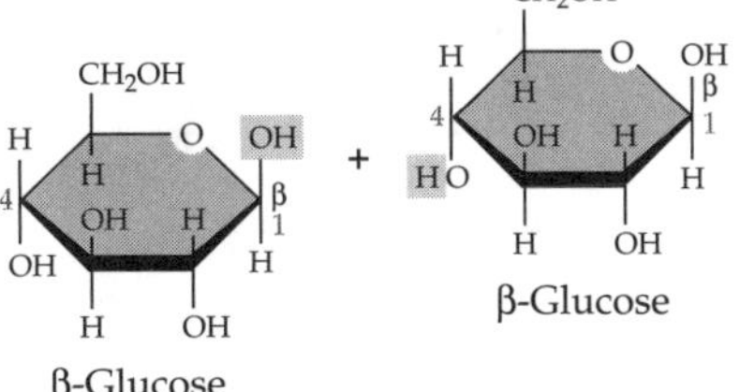
CH₂OH
O
OH
H
β
1
4
+
β-Glucose

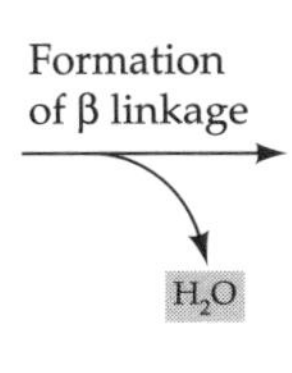
CH₂OH
O
OH
HO
β
β-Glucose

Formation of β linkage

H_2O

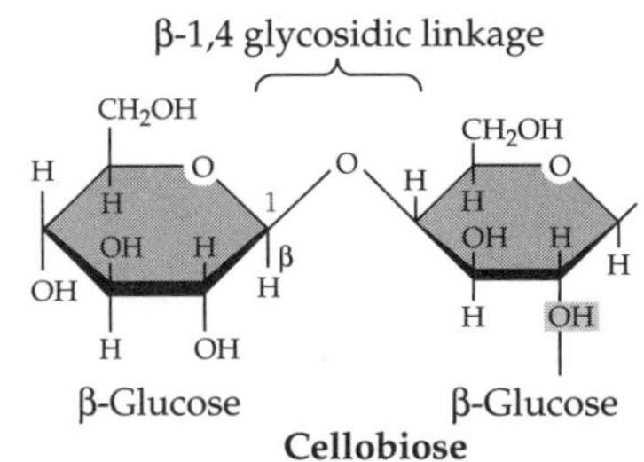
β-1,4 glycosidic linkage
CH₂OH
O
H
OH
β
β-Glucose
β-Glucose
Cellobiose

3.15 Disaccharides Are Formed by Glycosidic Linkages *(Page 48)*

(*a*) **Molecular structure**

Cellulose

Hydrogen bonding to other cellulose molecules can occur at these points.

Cellulose is an unbranched polymer of glucose with β-1,4 glycosidic linkages that are chemically very stable.

Starch and glycogen

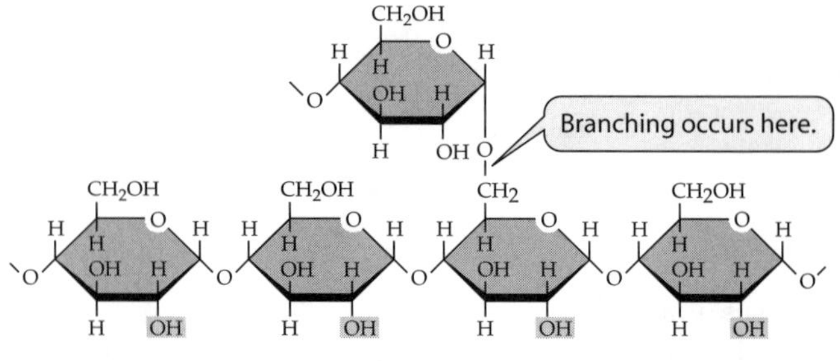
Branching occurs here.

Glycogen and starch are polymers of glucose with α-1,4 glycosidic linkages. α-1,6 glycosidic linkages produce branching at carbon 6.

(*b*) **Macromolecular structure**

Linear (cellulose)

Parallel cellulose molecules from hydrogen-bonds, resulting in thin fibrils.

Branched (starch)

Branching limits the number of hydrogen bonds that can form in starch molecules, making starch less compact than cellulose.

Highly branched (glycogen)

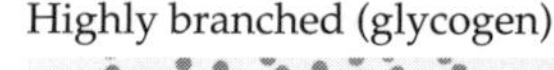

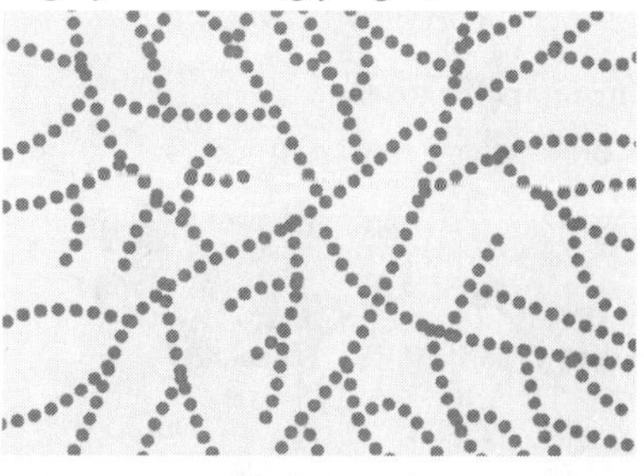

The high amount of branching in glycogen makes its solid deposits more compact than starch.

3.16 Representative Polysaccharides *(Page 49)*

(*a*) **Sugar phosphate**

Fructose 1,6 bisphosphate is involved in the reactions that liberate energy from glucose. (The numbers in its name refer to the carbon sites of phosphate bonding; *bis-* indicates that two phosphates are present.)

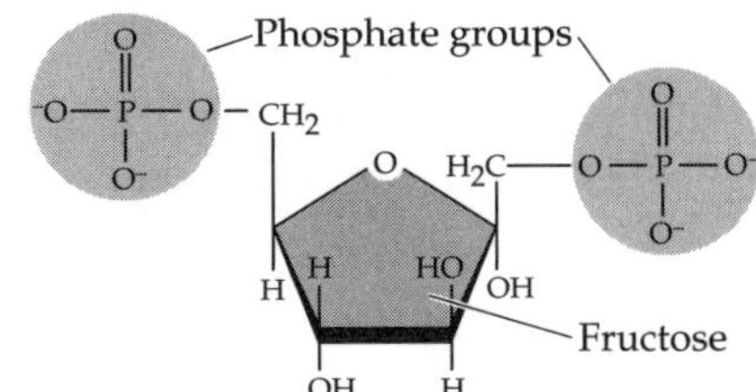
Phosphate groups
Fructose
Fructose 1,6 bisphosphate

(*b*) **Amino sugars**

The monosaccharides glucosamine and galactosamine are amino sugars with an amino group in place of a hydroxyl group.

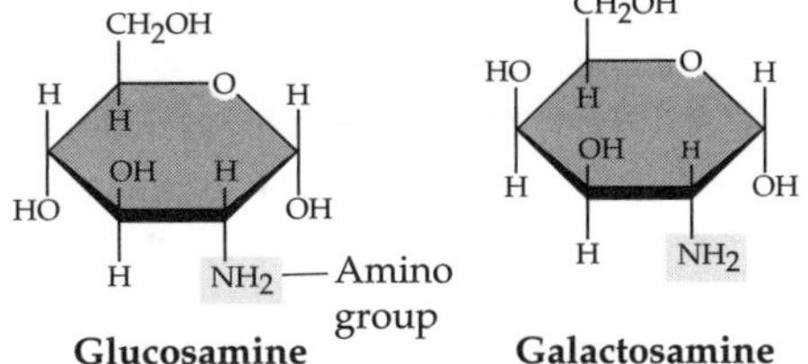
Amino group
Glucosamine
Galactosamine

(*c*) **Chitin**

Chitin is a polymer of *N*-acetylglucosamine; *N*-acetyl groups provide additional sites for hydrogen bonding between the polymers.

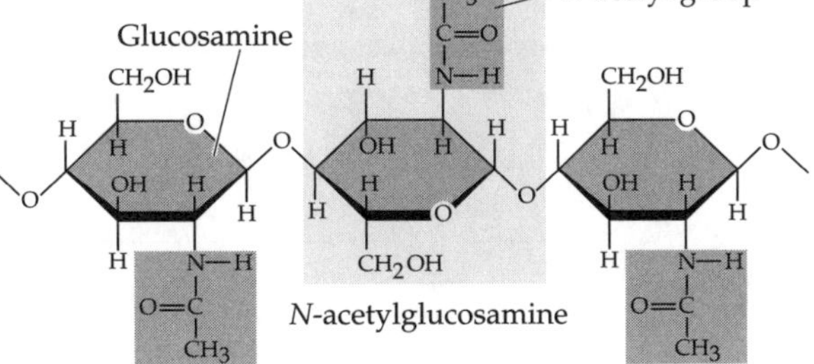
N-acetyl group
Glucosamine
N-acetylglucosamine
Chitin

3.17 Chemically Modified Carbohydrates *(Page 50)*

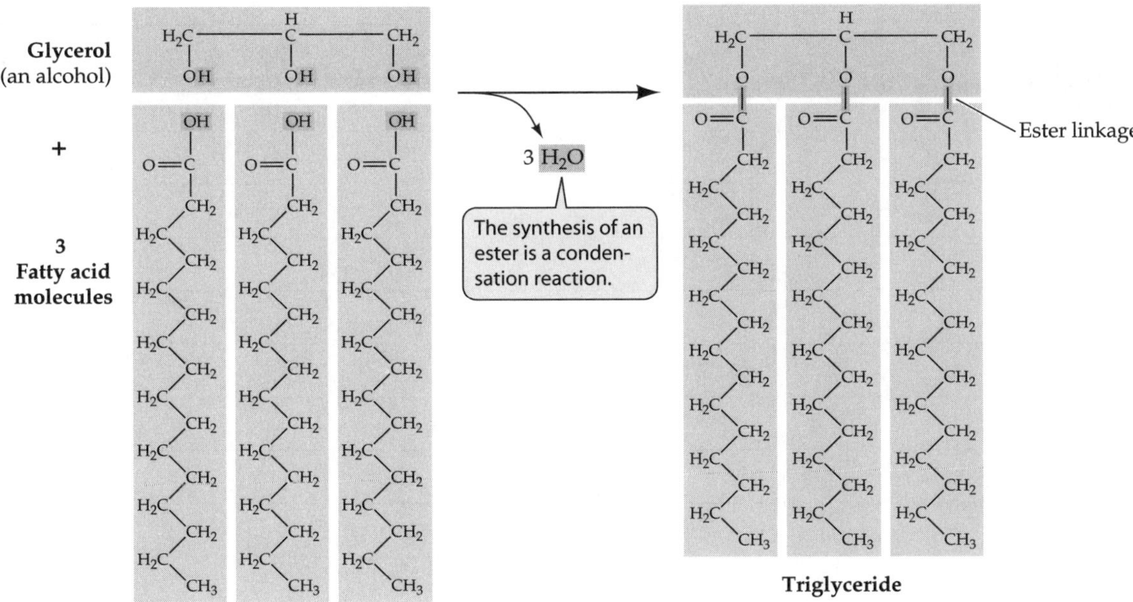
Glycerol
(an alcohol)
+
3
Fatty acid
molecules
3 H_2O
The synthesis of an ester is a condensation reaction.
Ester linkage
Triglyceride

3.18 Synthesis of a Triglyceride *(Page 51)*

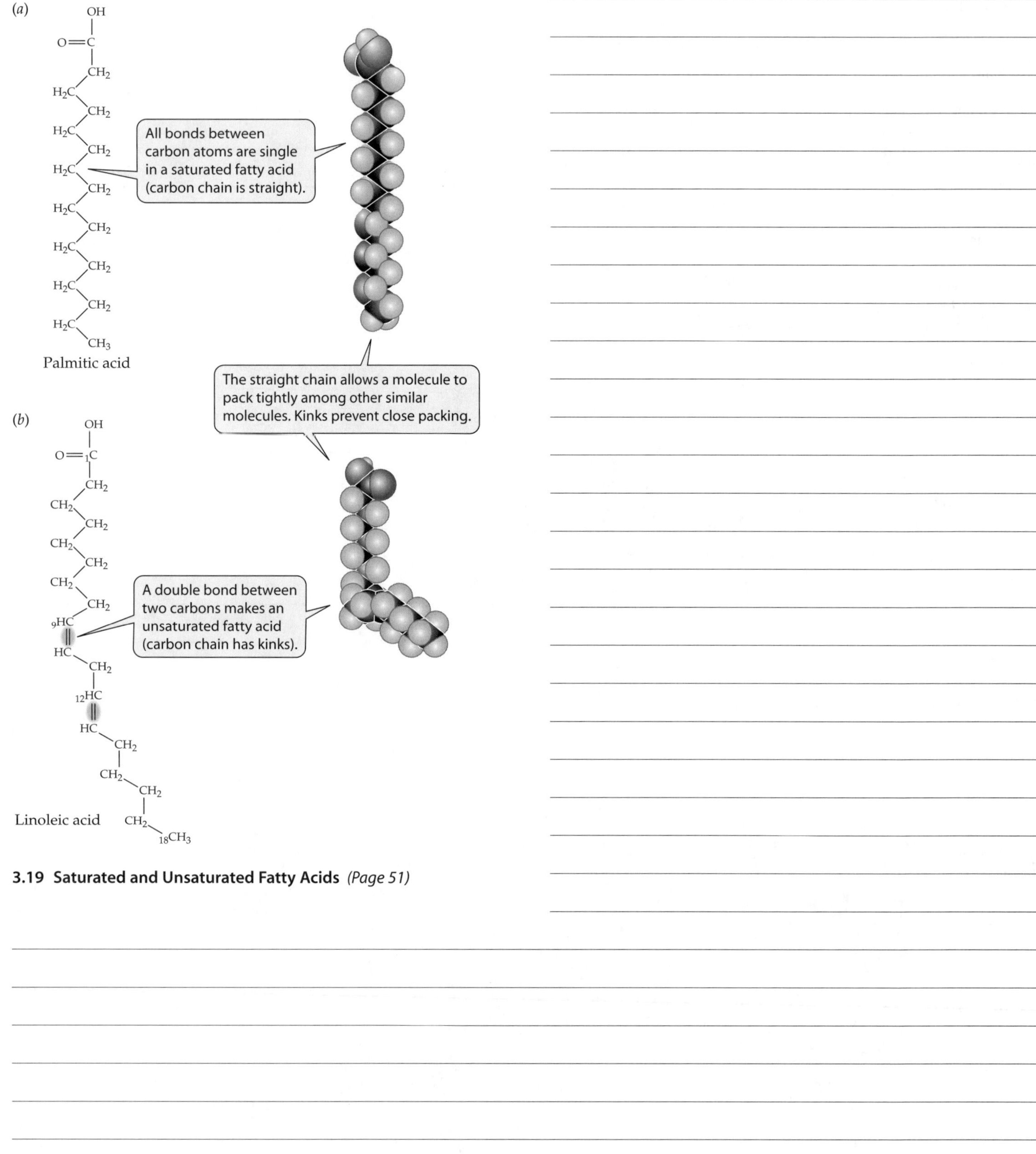
(a)
OH
O=C
CH2
H2C
CH3
Palmitic acid
All bonds between carbon atoms are single in a saturated fatty acid (carbon chain is straight).
The straight chain allows a molecule to pack tightly among other similar molecules. Kinks prevent close packing.
(b)
OH
O=1C
CH2
9HC
HC
12HC
18CH3
Linoleic acid
A double bond between two carbons makes an unsaturated fatty acid (carbon chain has kinks).

3.19 Saturated and Unsaturated Fatty Acids *(Page 51)*

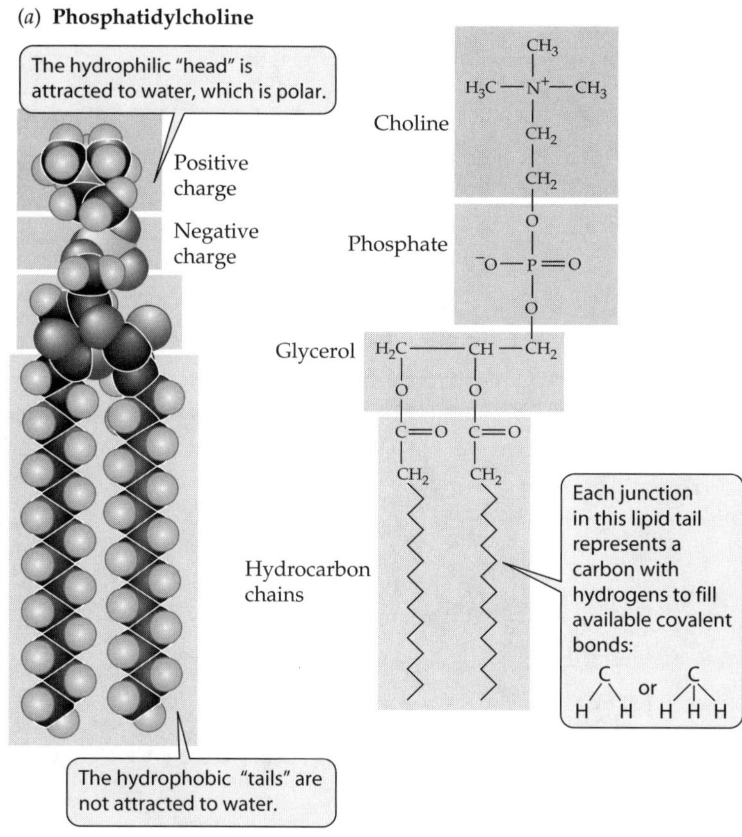
(a) Phosphatidylcholine
The hydrophilic "head" is attracted to water, which is polar.
Positive charge
Negative charge
Choline
Phosphate
Glycerol
Hydrocarbon chains
Each junction in this lipid tail represents a carbon with hydrogens to fill available covalent bonds:
or
The hydrophobic "tails" are not attracted to water.

(*b*) **Membrane phospholipid,** generalized symbol

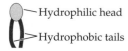

Hydrophilic head

Hydrophobic tails

3.20 Phospholipid Structure *(Page 52)*

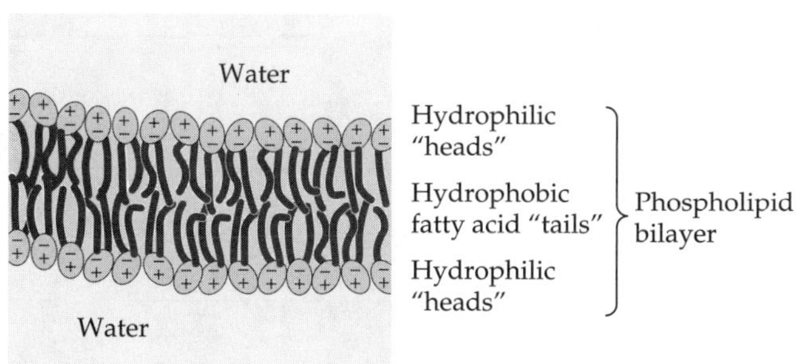
Water
Hydrophilic "heads"
Hydrophobic fatty acid "tails"
Hydrophilic "heads"
Phospholipid bilayer
Water

3.21 Phospholipids Form a Bilayer *(Page 52)*

$$\mathrm{H_2C{=}\overset{\displaystyle CH_3}{\overset{|}{C}}{-}\underset{\displaystyle H}{C}{=}CH_2}$$

In-Text Art *(Page 52)*

β-Carotene

Vitamin A

Vitamin A

3.22 β-Carotene is the Source of Vitamin A *(Page 53)*

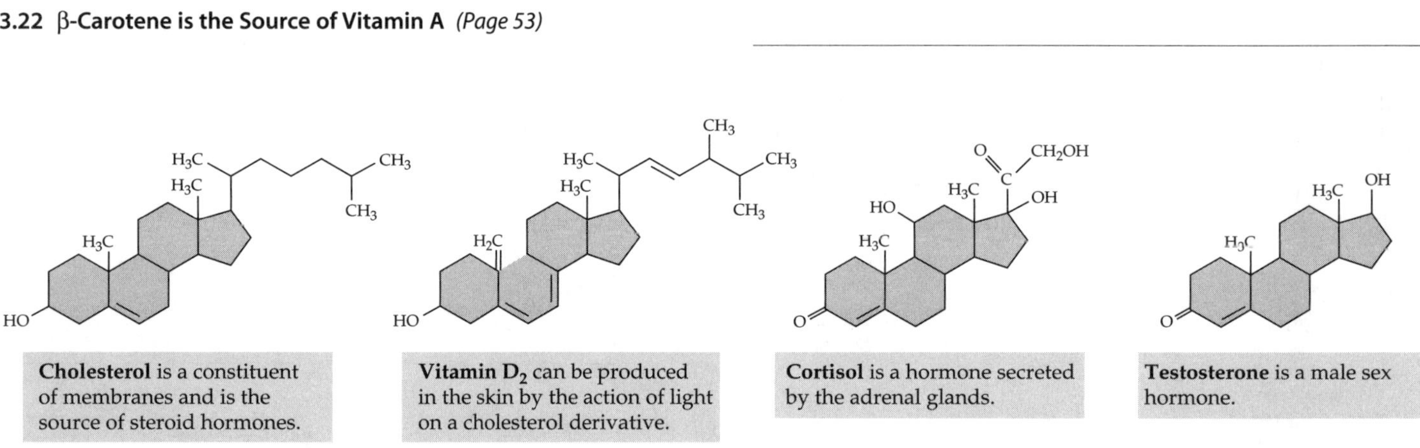

Cholesterol is a constituent of membranes and is the source of steroid hormones.
Vitamin D₂ can be produced in the skin by the action of light on a cholesterol derivative.
Cortisol is a hormone secreted by the adrenal glands.
Testosterone is a male sex hormone.

3.23 All Steroids Have the Same Ring Structure *(Page 53)*

$$\underbrace{H_3C-(CH_2)_{14}-\overset{\overset{\displaystyle O}{\|}}{C}}_{\text{Fatty acid}}-\underbrace{O-CH_2-(CH_2)_{28}-CH_3}_{\text{Alcohol}}$$

Ester linkage

In-Text Art *(Page 54)*

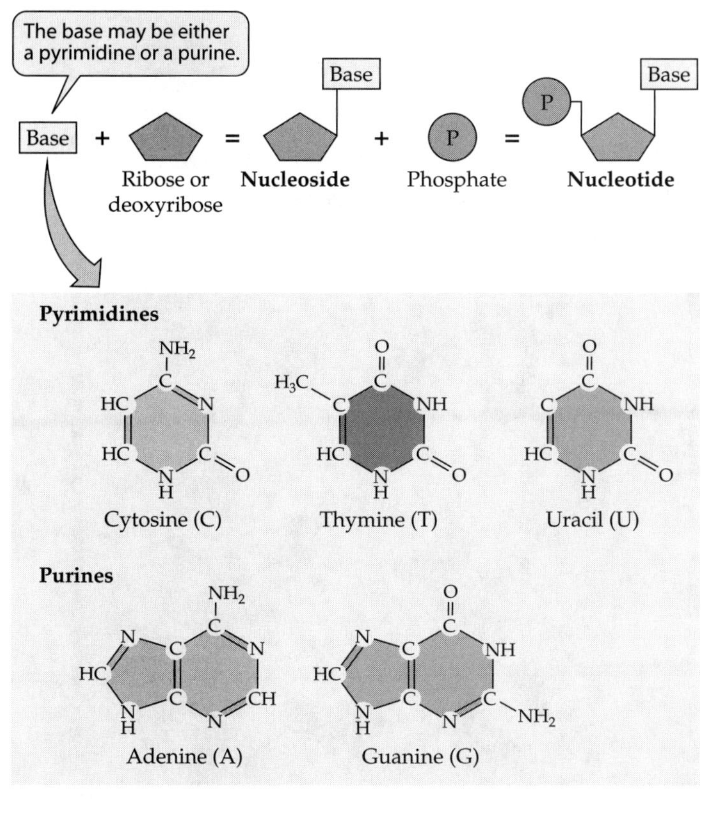
The base may be either a pyrimidine or a purine.
Base
Ribose or deoxyribose
Nucleoside
Phosphate
Nucleotide
P
Pyrimidines
Cytosine (C)
Thymine (T)
Uracil (U)
Purines
Adenine (A)
Guanine (G)

3.24 Nucleotides Have Three Components *(Page 54)*

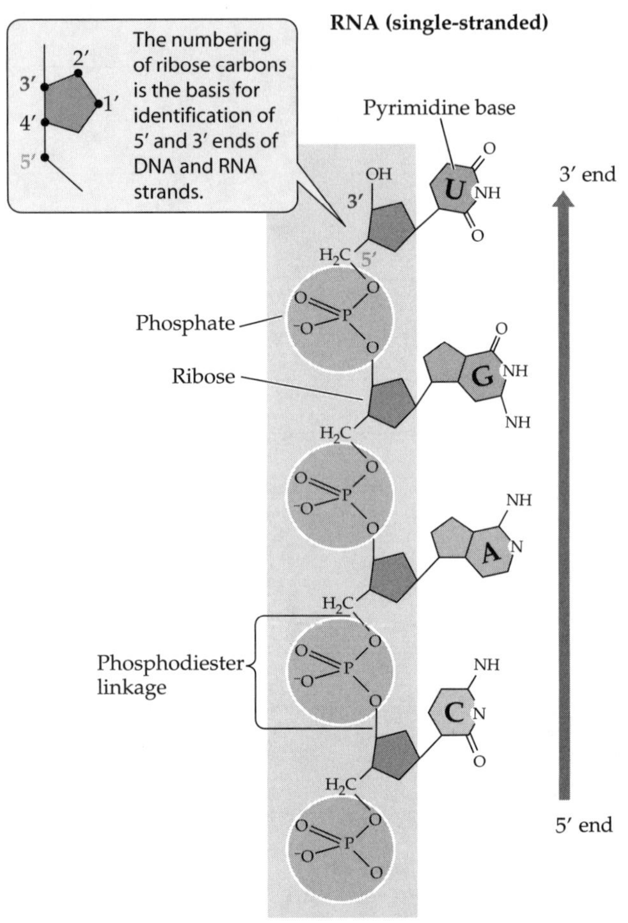

RNA (single-stranded)
The numbering of ribose carbons is the basis for identification of 5′ and 3′ ends of DNA and RNA strands.
2′
3′
1′
4′
5′
Pyrimidine base
3′ end
5′ end
OH
3′
U
NH
O
H2C
5′
Phosphate
Ribose
G
NH
A
N
Phosphodiester linkage
C
N
P

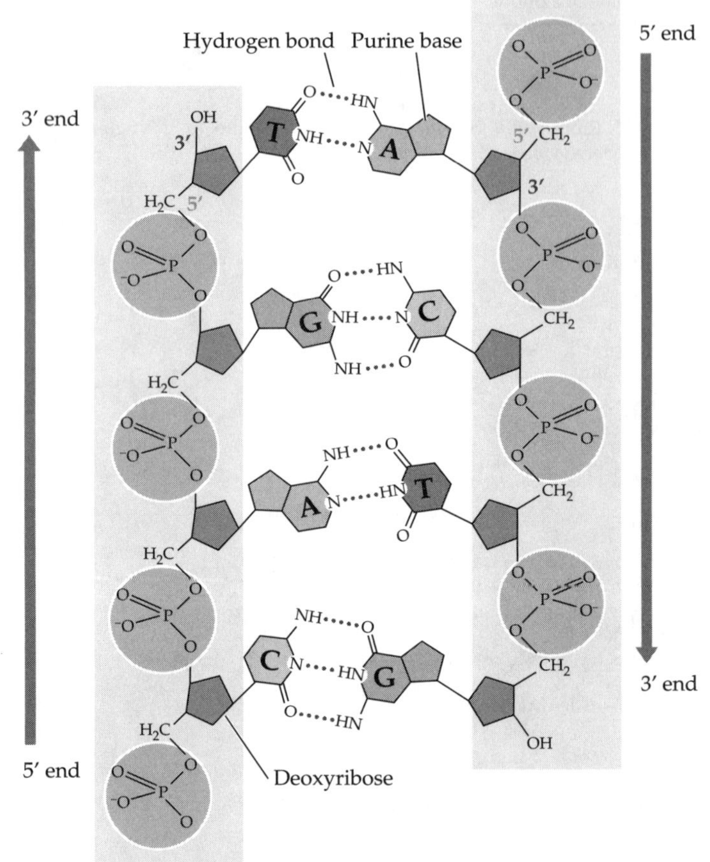

DNA (double-stranded)
Hydrogen bond
Purine base
5′ end
3′ end
3′ end
5′ end
T
A
G
C
A
T
C
G
CH2
H2C
OH
Deoxyribose

In RNA, the bases are attached to ribose. The bases in RNA are the purines adenine (A) and guanine (G) and the pyrimidines cytosine (C) and uracil (U).

In DNA, the bases are attached to deoxyribose, and the base thymine (T) is found instead of uracil. Hydrogen bonds between purines and pyrimidines hold the two strands of DNA together.

3.25 Distinguishing Characteristics of DNA and RNA *(Page 55)*

3.3 *Distinguishing RNA from DNA*

NUCLEIC ACID	SUGAR	BASES
RNA	Ribose	Adenine Cytosine Guanine Uracil
DNA	Deoxyribose	Adenine Cytosine Guanine Thymine

(Page 55)

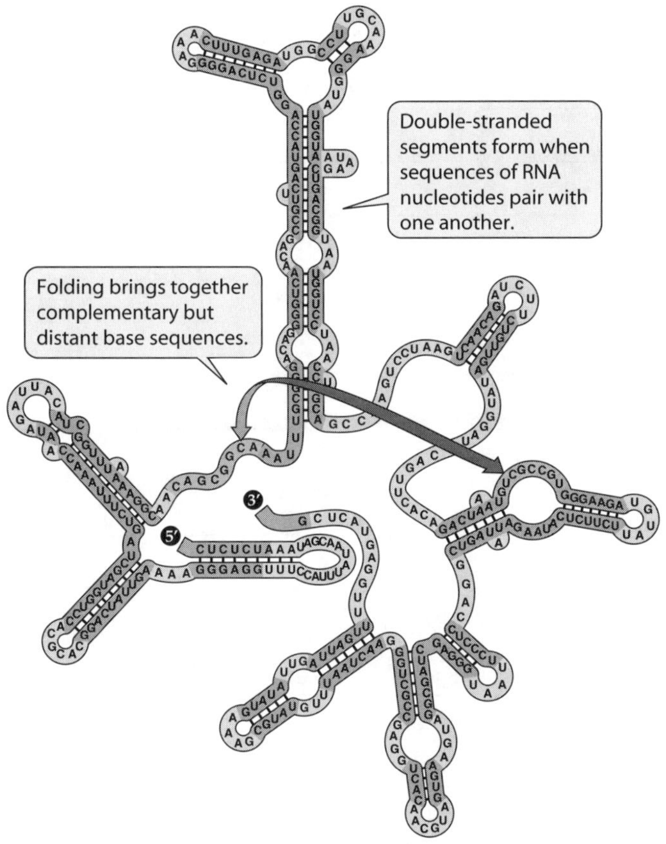

Double-stranded segments form when sequences of RNA nucleotides pair with one another.

Folding brings together complementary but distant base sequences.

3.26 Hydrogen Bonding in RNA *(Page 56)*

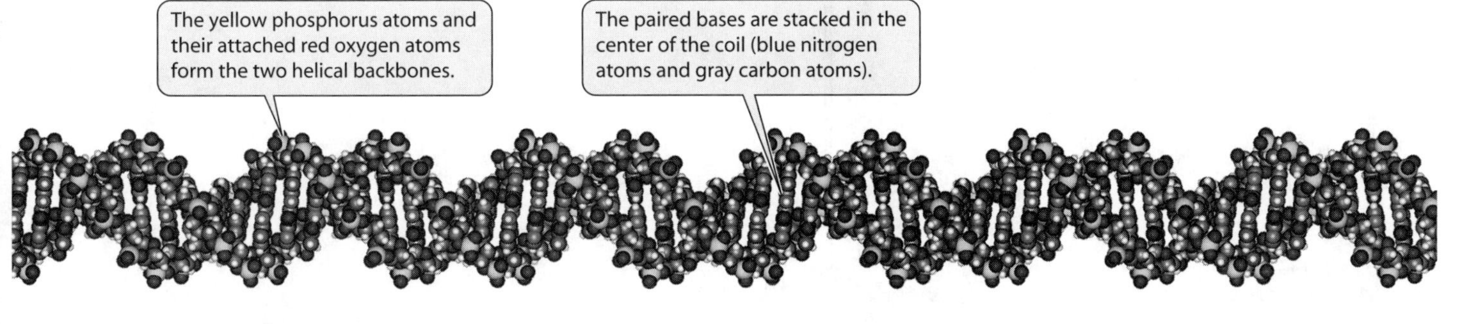
The yellow phosphorus atoms and their attached red oxygen atoms form the two helical backbones.
The paired bases are stacked in the center of the coil (blue nitrogen atoms and gray carbon atoms).

3.27 The Double Helix of DNA *(Page 56)*

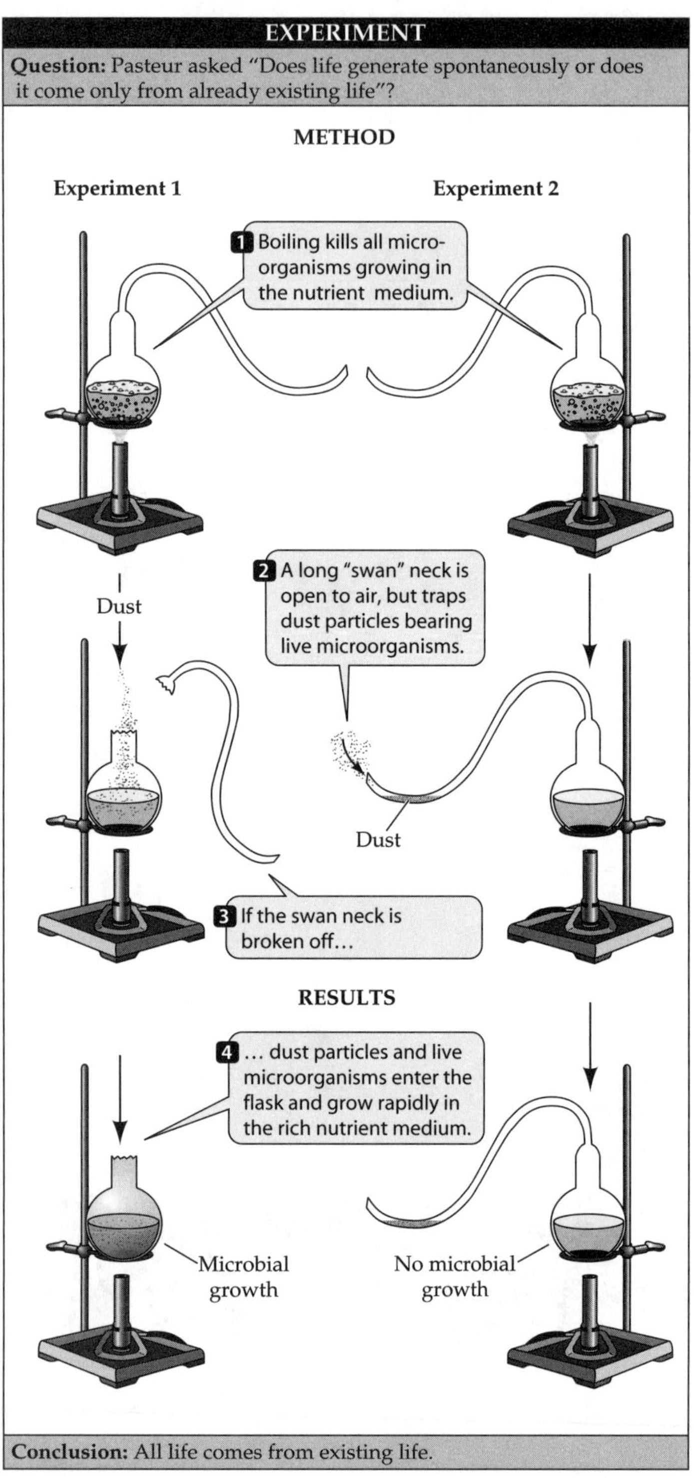
EXPERIMENT
Question: Pasteur asked "Does life generate spontaneously or does it come only from already existing life"?
METHOD
Experiment 1
Experiment 2
1 Boiling kills all micro-organisms growing in the nutrient medium.
Dust
2 A long "swan" neck is open to air, but traps dust particles bearing live microorganisms.
Dust
3 If the swan neck is broken off…
RESULTS
4 … dust particles and live microorganisms enter the flask and grow rapidly in the rich nutrient medium.
Microbial growth
No microbial growth
Conclusion: All life comes from existing life.

3.28 Disproving the Spontaneous Generation of Life *(Page 58)*

4 Cells: The Basic Units of Life

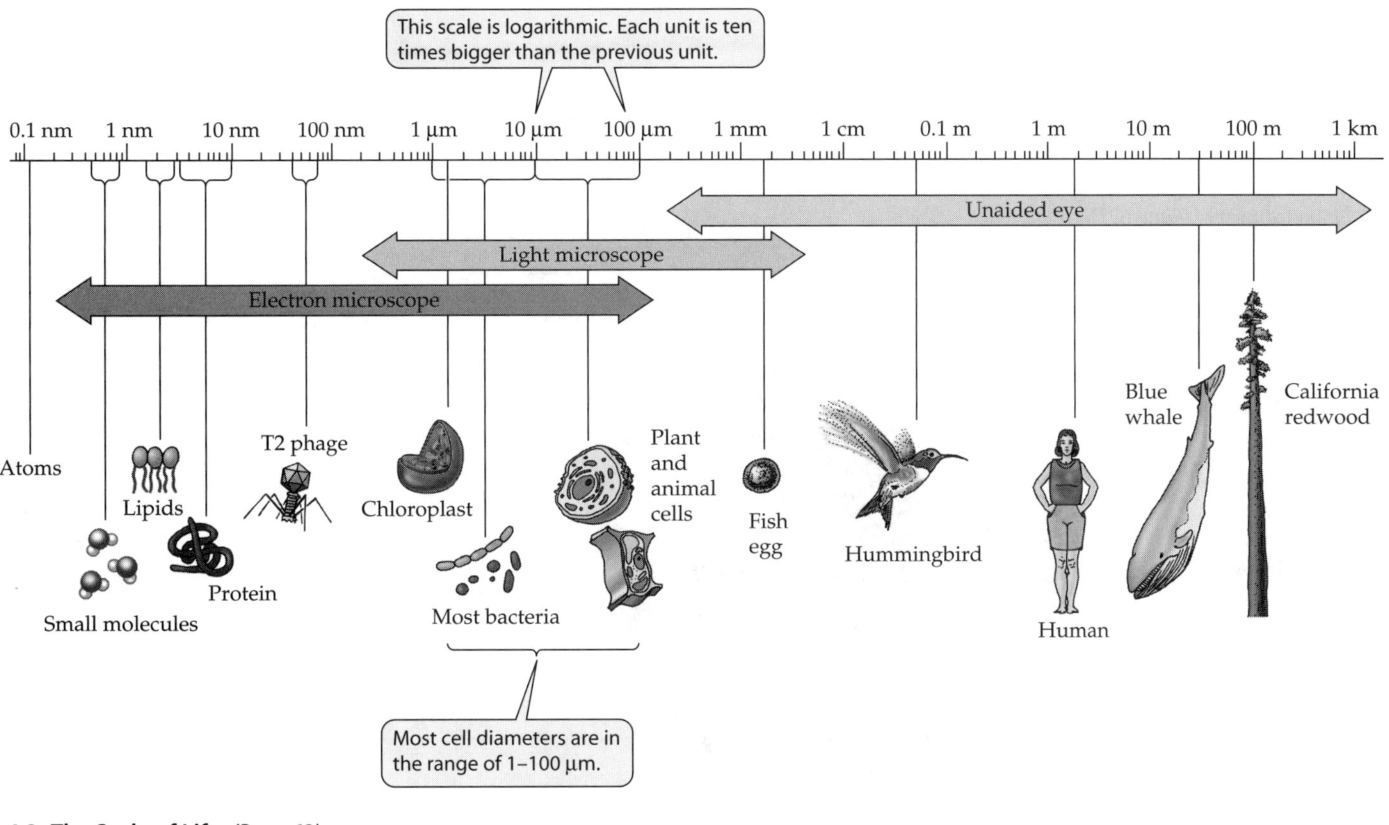

This scale is logarithmic. Each unit is ten times bigger than the previous unit.
0.1 nm
1 nm
10 nm
100 nm
1 µm
10 µm
100 µm
1 mm
1 cm
0.1 m
1 m
10 m
100 m
1 km
Unaided eye
Light microscope
Electron microscope
Atoms
Lipids
Small molecules
Protein
T2 phage
Chloroplast
Most bacteria
Plant and animal cells
Fish egg
Hummingbird
Human
Blue whale
California redwood
Most cell diameters are in the range of 1–100 µm.

4.2 The Scale of Life *(Page 63)*

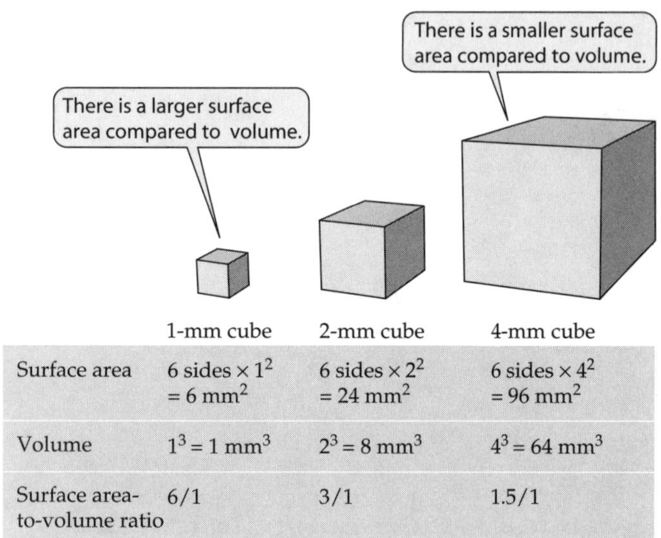
There is a smaller surface area compared to volume.
There is a larger surface area compared to volume.

	1-mm cube	2-mm cube	4-mm cube
Surface area	6 sides × 1^2 = 6 mm^2	6 sides × 2^2 = 24 mm^2	6 sides × 4^2 = 96 mm^2
Volume	1^3 = 1 mm^3	2^3 = 8 mm^3	4^3 = 64 mm^3
Surface area-to-volume ratio	6/1	3/1	1.5/1

4.3 Why Cells Are Small *(Page 63)*

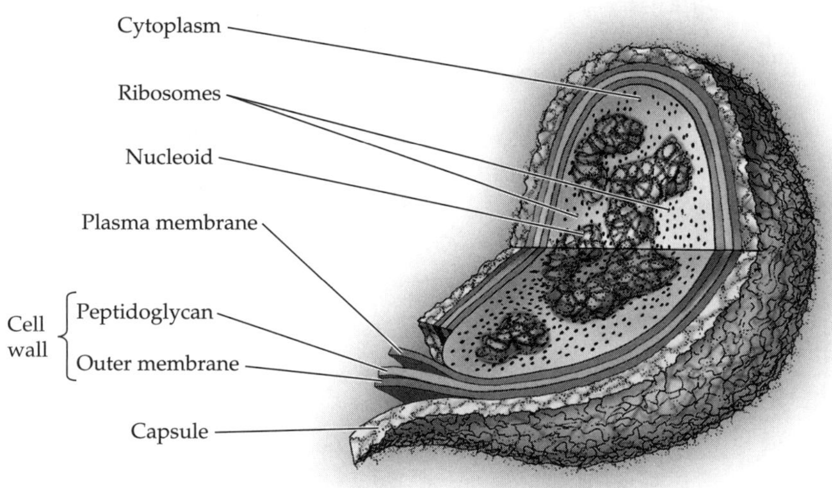
Cytoplasm
Ribosomes
Nucleoid
Plasma membrane
Cell wall
Peptidoglycan
Outer membrane
Capsule

4.5 A Prokaryotic Cell *(Page 66)*

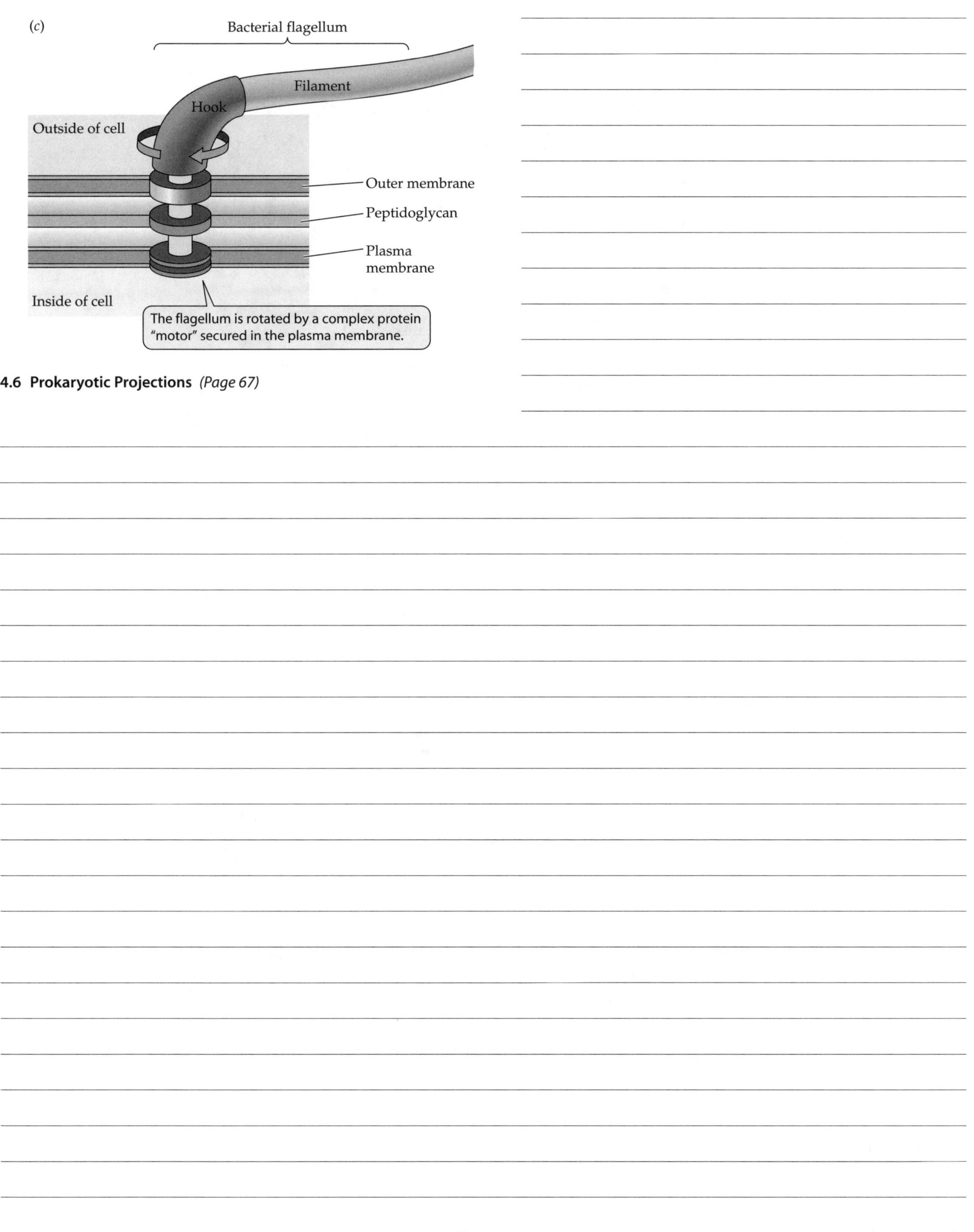
(c)
Bacterial flagellum
Filament
Hook
Outside of cell
Outer membrane
Peptidoglycan
Plasma membrane
Inside of cell
The flagellum is rotated by a complex protein "motor" secured in the plasma membrane.

4.6 Prokaryotic Projections *(Page 67)*

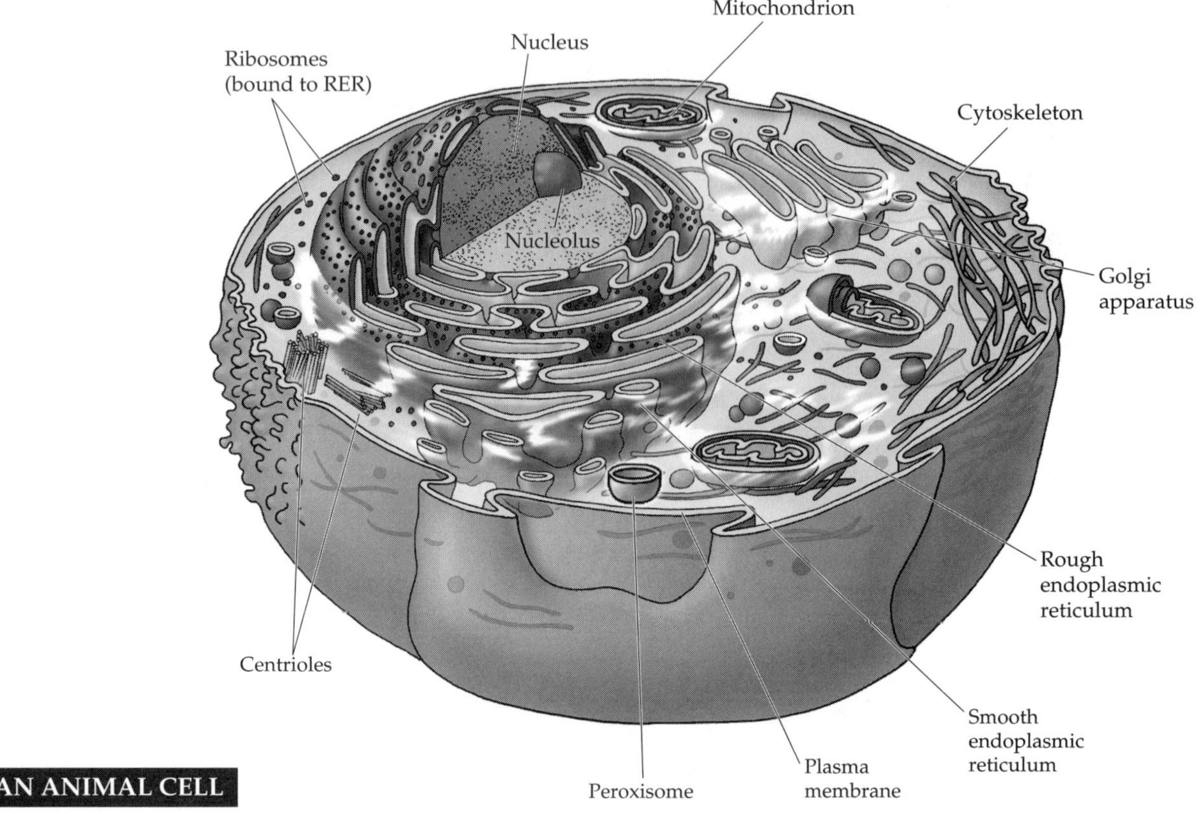
Mitochondrion
Nucleus
Ribosomes
(bound to RER)
Cytoskeleton
Nucleolus
Golgi
apparatus
Rough
endoplasmic
reticulum
Centrioles
Smooth
endoplasmic
reticulum
Plasma
membrane
Peroxisome

AN ANIMAL CELL

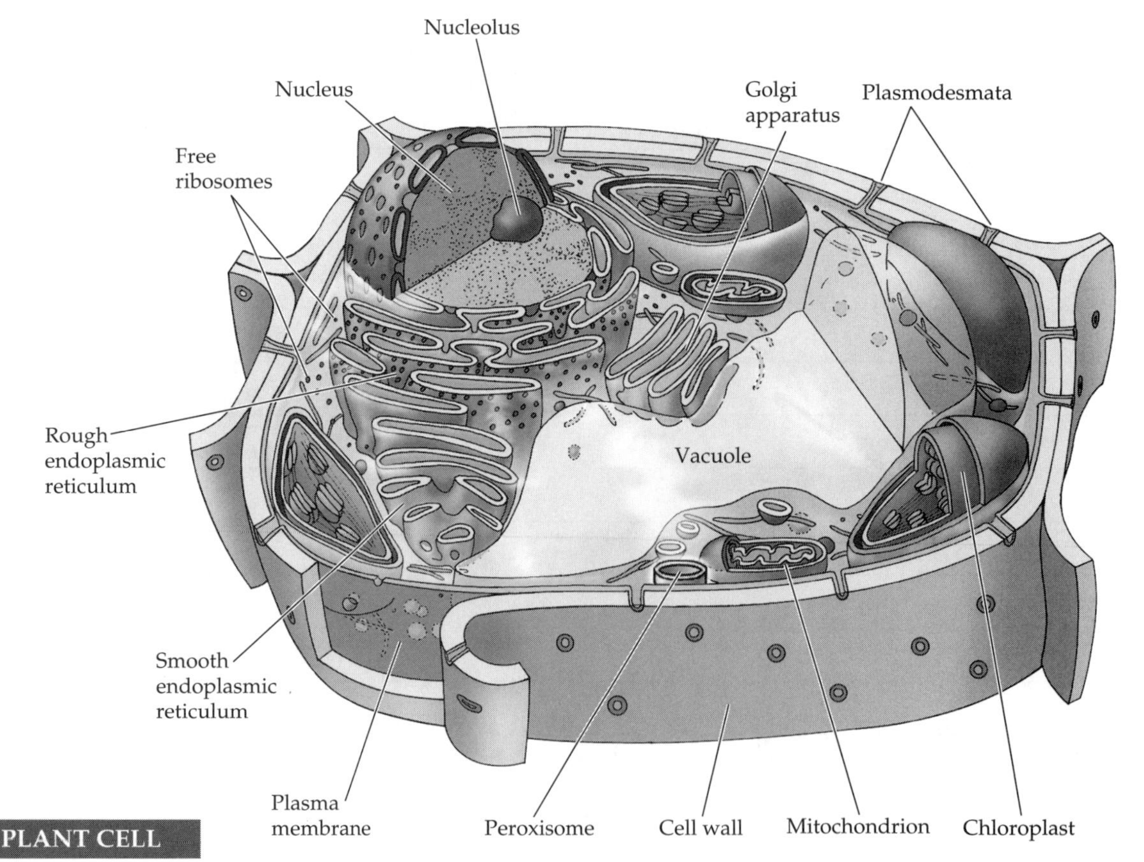
Nucleolus
Nucleus
Golgi
apparatus
Plasmodesmata
Free
ribosomes
Rough
endoplasmic
reticulum
Vacuole
Smooth
endoplasmic
reticulum
Plasma
membrane
Peroxisome
Cell wall
Mitochondrion
Chloroplast

A PLANT CELL

4.7 Eukaryotic Cells *(Pages 68–69)*

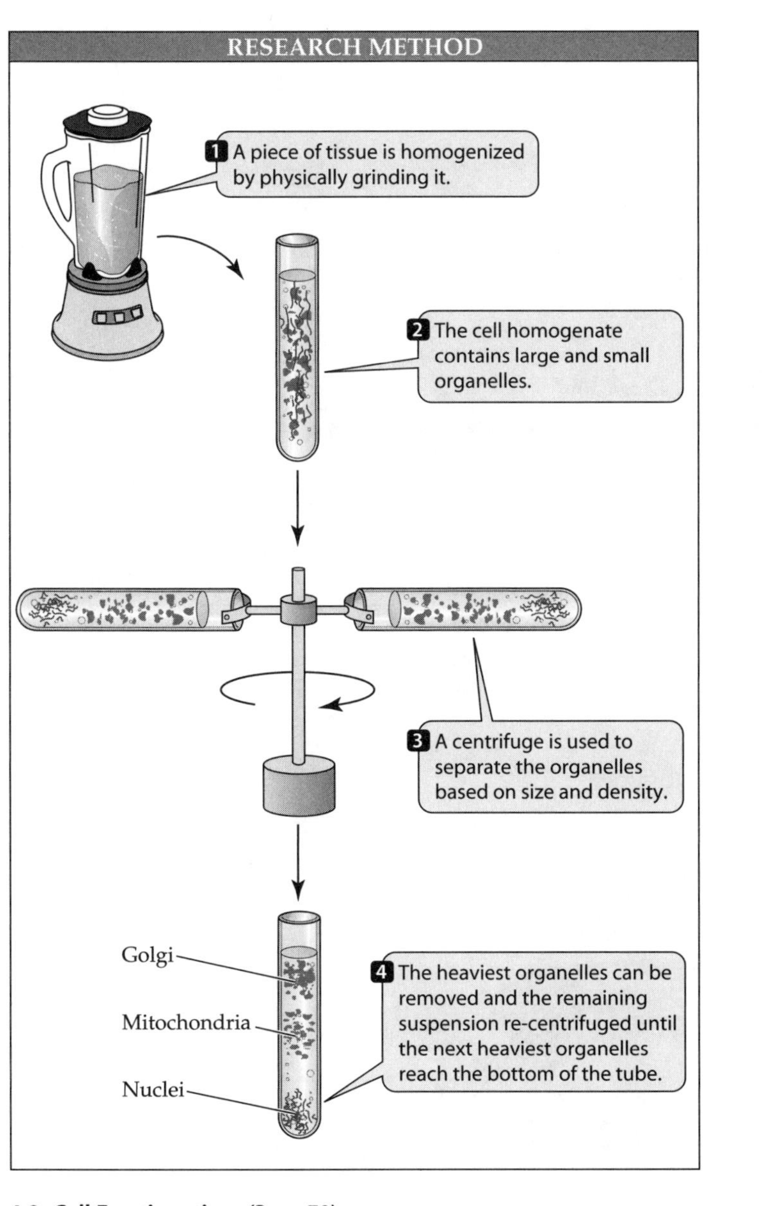
RESEARCH METHOD
1 A piece of tissue is homogenized by physically grinding it.
2 The cell homogenate contains large and small organelles.
3 A centrifuge is used to separate the organelles based on size and density.
Golgi
Mitochondria
Nuclei
4 The heaviest organelles can be removed and the remaining suspension re-centrifuged until the next heaviest organelles reach the bottom of the tube.

4.8 Cell Fractionation *(Page 70)*

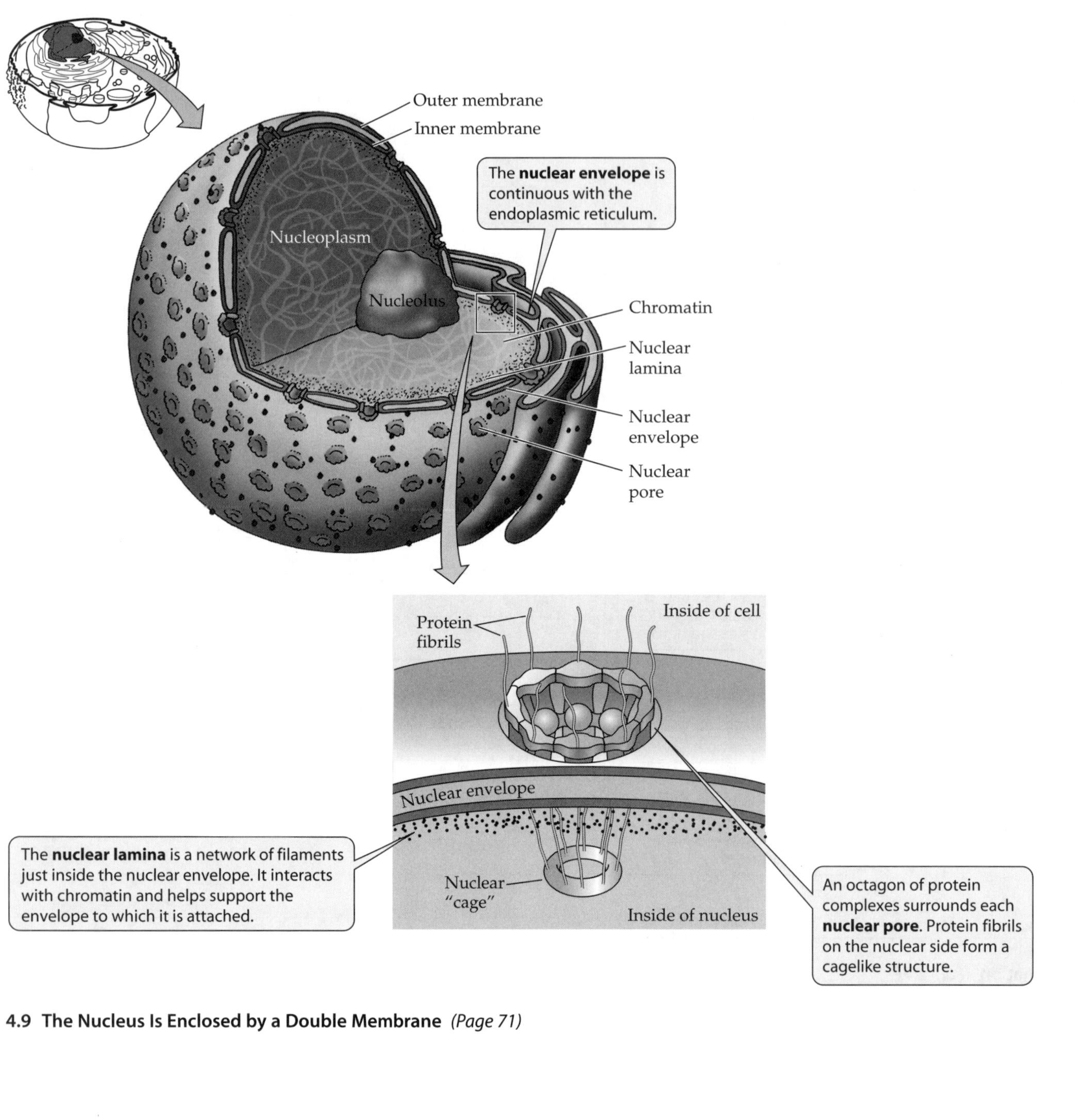
Outer membrane
Inner membrane
The **nuclear envelope** is continuous with the endoplasmic reticulum.
Nucleoplasm
Nucleolus
Chromatin
Nuclear lamina
Nuclear envelope
Nuclear pore
Protein fibrils
Inside of cell
Nuclear envelope
Nuclear "cage"
Inside of nucleus
The **nuclear lamina** is a network of filaments just inside the nuclear envelope. It interacts with chromatin and helps support the envelope to which it is attached.
An octagon of protein complexes surrounds each **nuclear pore**. Protein fibrils on the nuclear side form a cagelike structure.

4.9 The Nucleus Is Enclosed by a Double Membrane *(Page 71)*

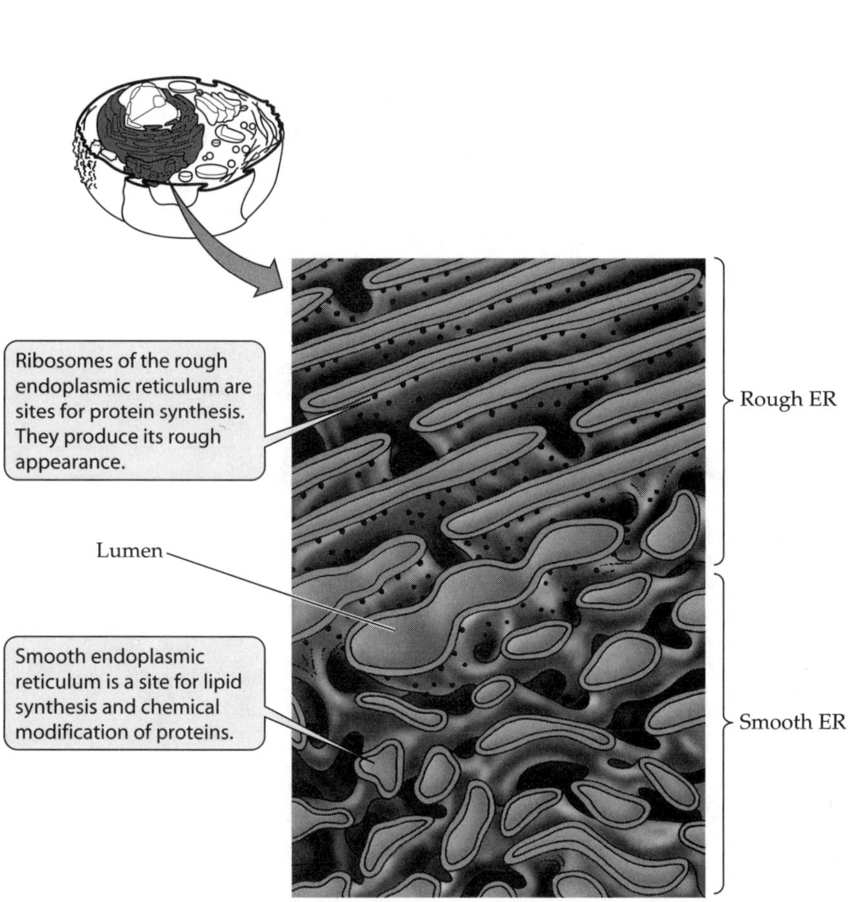

Ribosomes of the rough endoplasmic reticulum are sites for protein synthesis. They produce its rough appearance.

Rough ER

Lumen

Smooth endoplasmic reticulum is a site for lipid synthesis and chemical modification of proteins.

Smooth ER

4.11 The Endoplasmic Reticulum *(Page 72)*

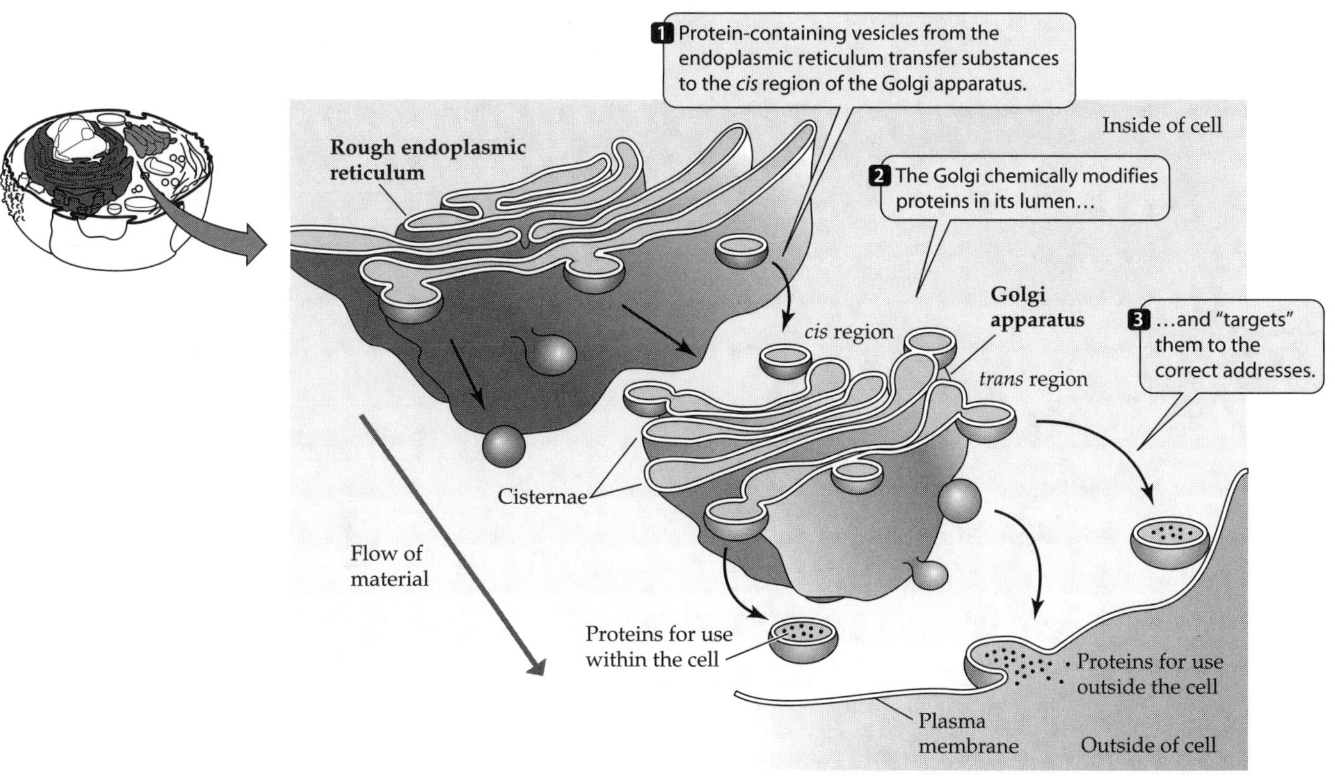
1 Protein-containing vesicles from the endoplasmic reticulum transfer substances to the *cis* region of the Golgi apparatus.
Inside of cell
Rough endoplasmic reticulum
2 The Golgi chemically modifies proteins in its lumen…
Golgi apparatus
cis region
trans region
3 …and "targets" them to the correct addresses.
Cisternae
Flow of material
Proteins for use within the cell
Proteins for use outside the cell
Plasma membrane
Outside of cell

4.12 The Golgi Apparatus *(Page 73)*

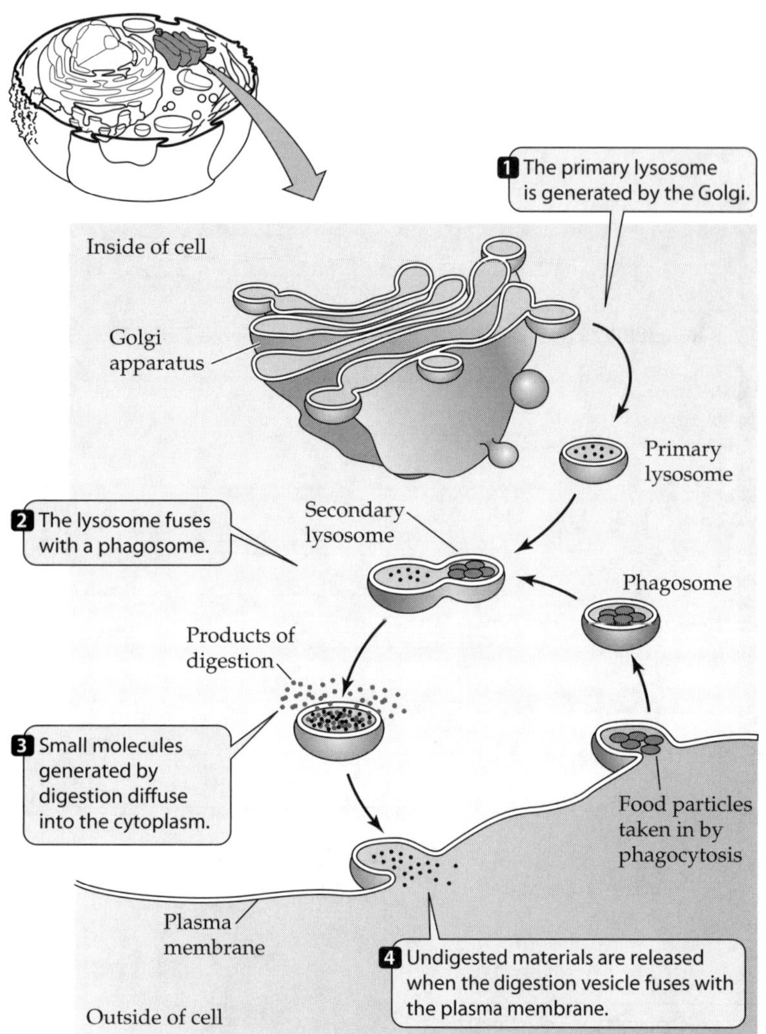

1 The primary lysosome is generated by the Golgi.
Inside of cell
Golgi apparatus
Primary lysosome
2 The lysosome fuses with a phagosome.
Secondary lysosome
Phagosome
Products of digestion
3 Small molecules generated by digestion diffuse into the cytoplasm.
Food particles taken in by phagocytosis
Plasma membrane
4 Undigested materials are released when the digestion vesicle fuses with the plasma membrane.
Outside of cell

4.13 Lysosomes Isolate Digestive Enzymes from the Cytoplasm *(Page 74)*

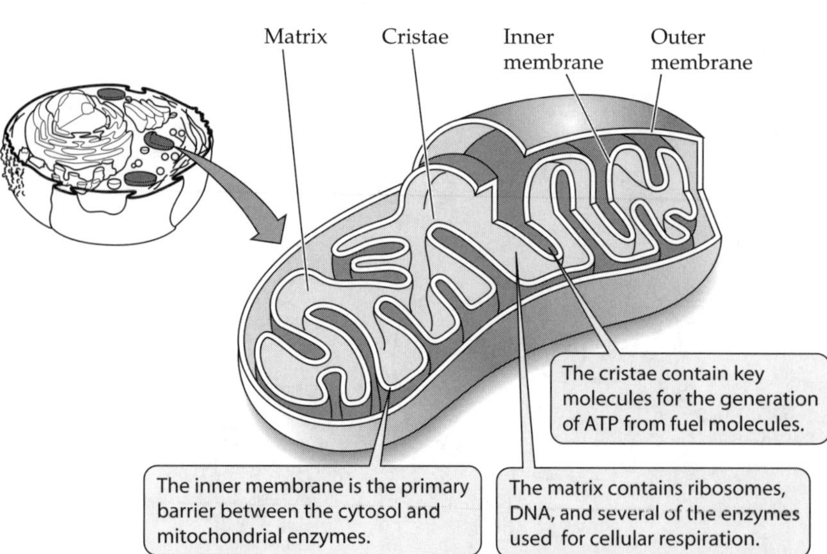

Matrix
Cristae
Inner membrane
Outer membrane
The cristae contain key molecules for the generation of ATP from fuel molecules.
The inner membrane is the primary barrier between the cytosol and mitochondrial enzymes.
The matrix contains ribosomes, DNA, and several of the enzymes used for cellular respiration.

4.14 A Mitochondrion Converts Energy from Fuel Molecules into ATP *(Page 75)*

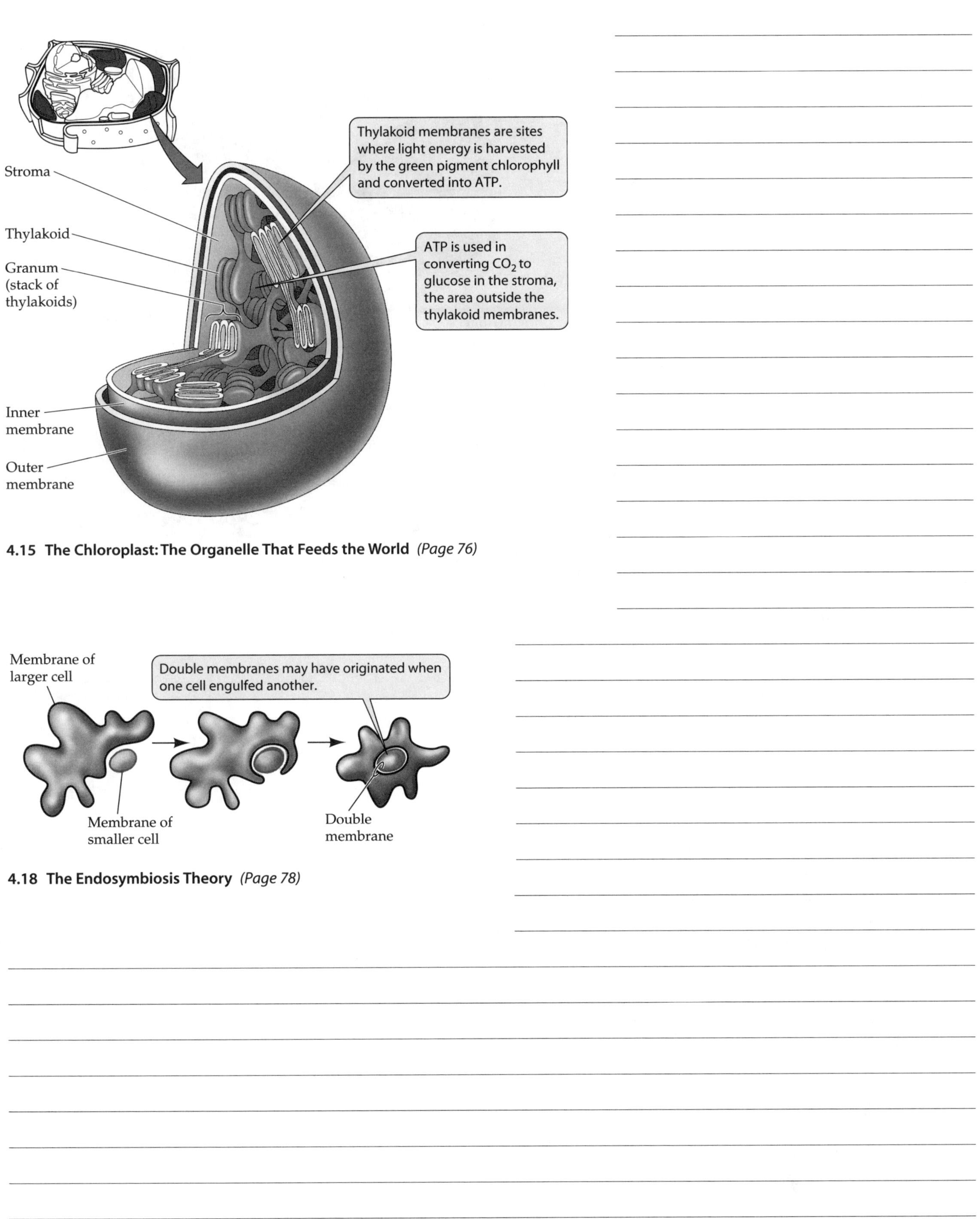
Stroma
Thylakoid
Granum (stack of thylakoids)
Inner membrane
Outer membrane
Thylakoid membranes are sites where light energy is harvested by the green pigment chlorophyll and converted into ATP.
ATP is used in converting CO_2 to glucose in the stroma, the area outside the thylakoid membranes.
Membrane of larger cell
Double membranes may have originated when one cell engulfed another.
Membrane of smaller cell
Double membrane

4.15 The Chloroplast: The Organelle That Feeds the World *(Page 76)*

4.18 The Endosymbiosis Theory *(Page 78)*

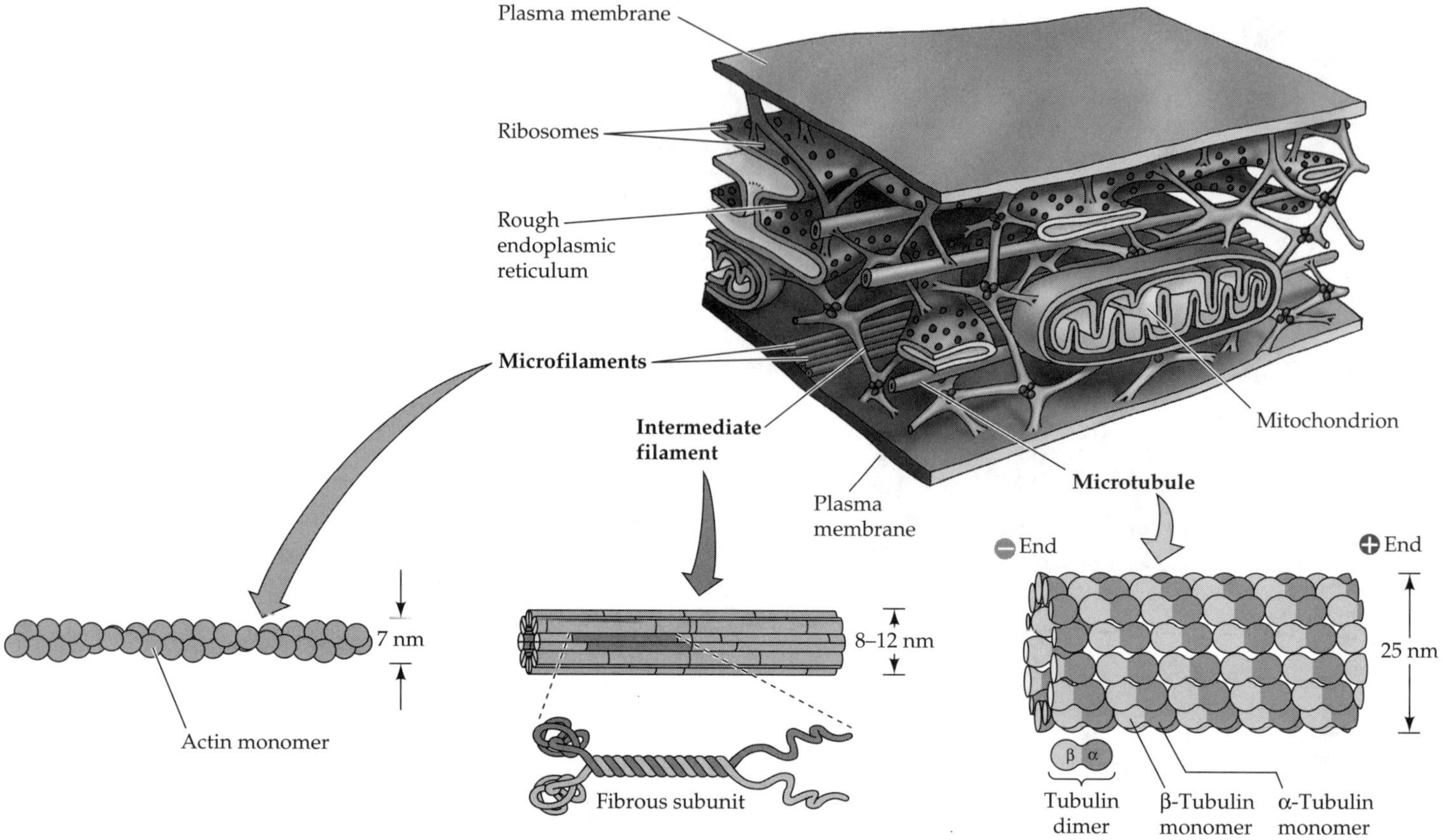

Plasma membrane
Ribosomes
Rough endoplasmic reticulum
Microfilaments
Intermediate filament
Plasma membrane
Mitochondrion
Microtubule
7 nm
Actin monomer
8–12 nm
Fibrous subunit
− End
+ End
25 nm
β α
Tubulin dimer
β-Tubulin monomer
α-Tubulin monomer

4.21 The Cytoskeleton *(Page 80)*

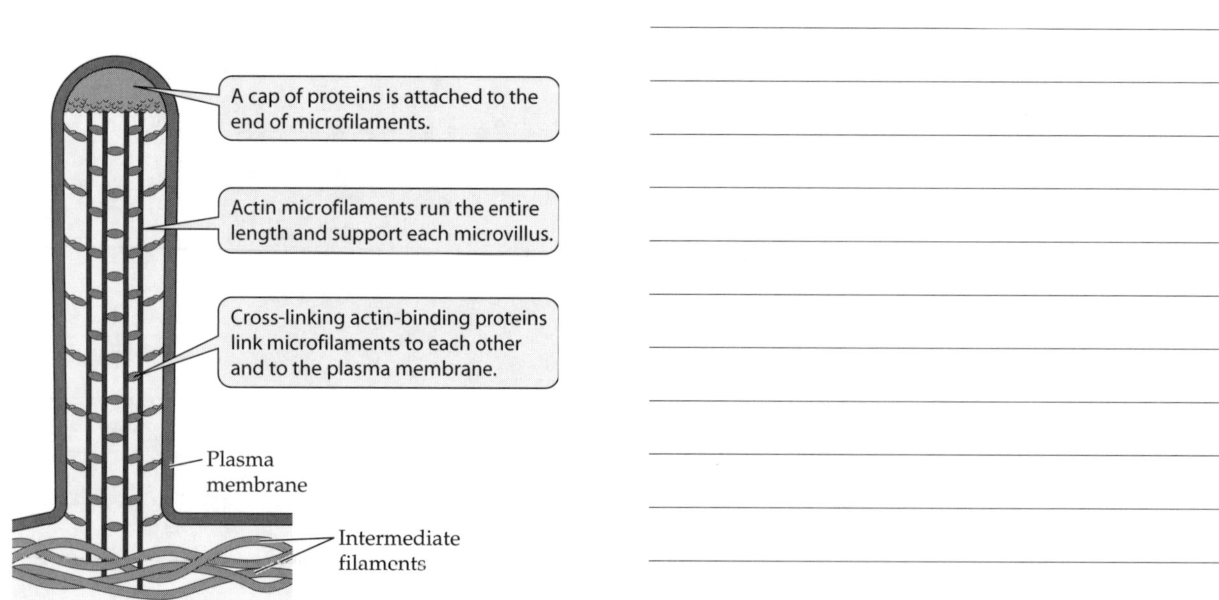
A cap of proteins is attached to the end of microfilaments.
Actin microfilaments run the entire length and support each microvillus.
Cross-linking actin-binding proteins link microfilaments to each other and to the plasma membrane.
Plasma membrane
Intermediate filaments

4.22 Microfilaments for Support *(Page 81)*

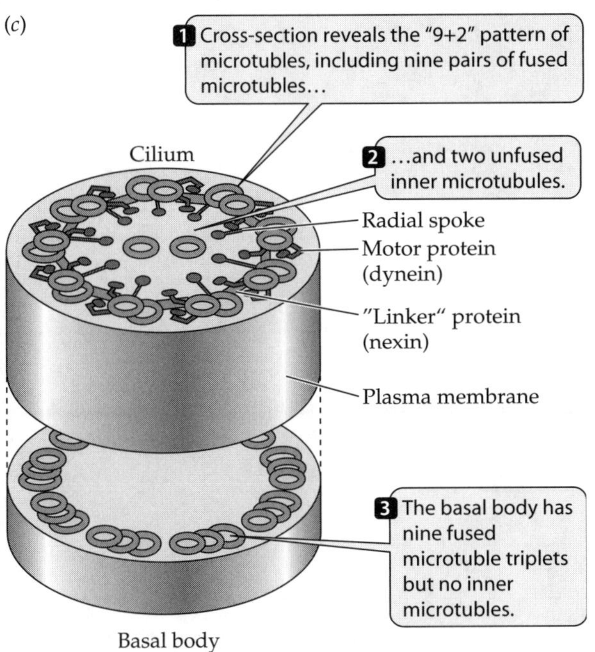
(c)
1 Cross-section reveals the "9+2" pattern of microtubules, including nine pairs of fused microtubules…
Cilium
2 …and two unfused inner microtubules.
Radial spoke
Motor protein (dynein)
"Linker" protein (nexin)
Plasma membrane
3 The basal body has nine fused microtubule triplets but no inner microtubules.
Basal body

4.23 Cilia are Made up of Microtubules *(Page 82)*

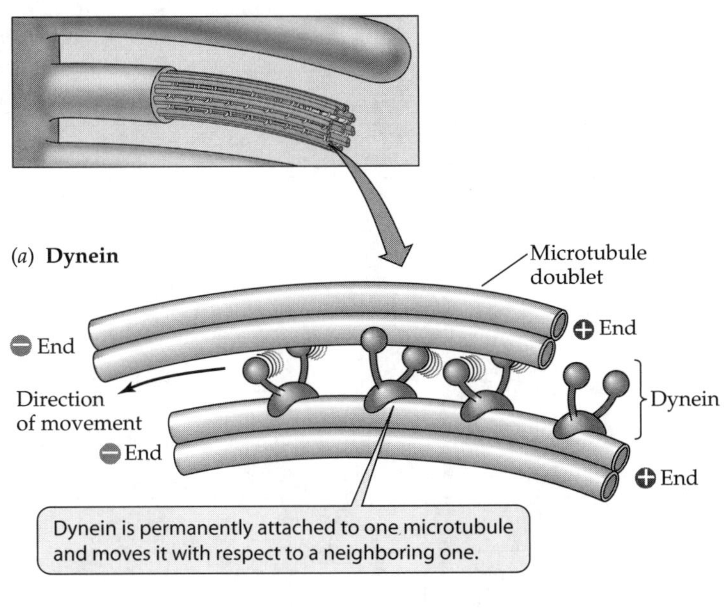
(a) Dynein
Microtubule doublet
End
End
Direction of movement
End
Dynein
End
Dynein is permanently attached to one microtubule and moves it with respect to a neighboring one.

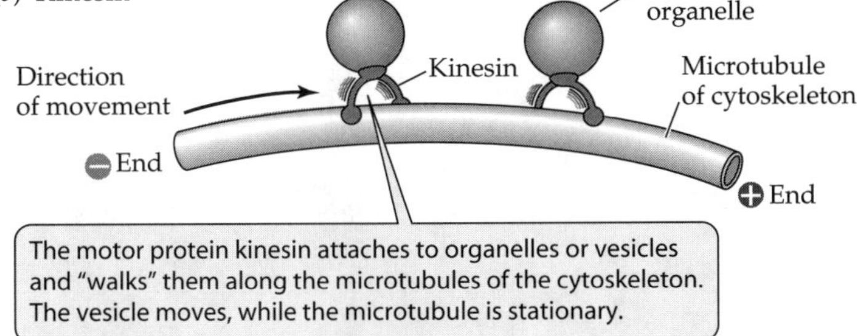
(b) Kinesin
Vesicle or organelle
Direction of movement
Kinesin
Microtubule of cytoskeleton
End
End
The motor protein kinesin attaches to organelles or vesicles and "walks" them along the microtubules of the cytoskeleton. The vesicle moves, while the microtubule is stationary.

4.24 Motor Proteins Use Energy from ATP to Move Things *(Page 83)*

The ECM is composed of a tangled complex of enormous molecules made of proteins and long polysaccharide chains.
Proteoglycan
Collagen

4.26 An Extracellular Matrix *(Page 84)*

5 Cellular Membranes

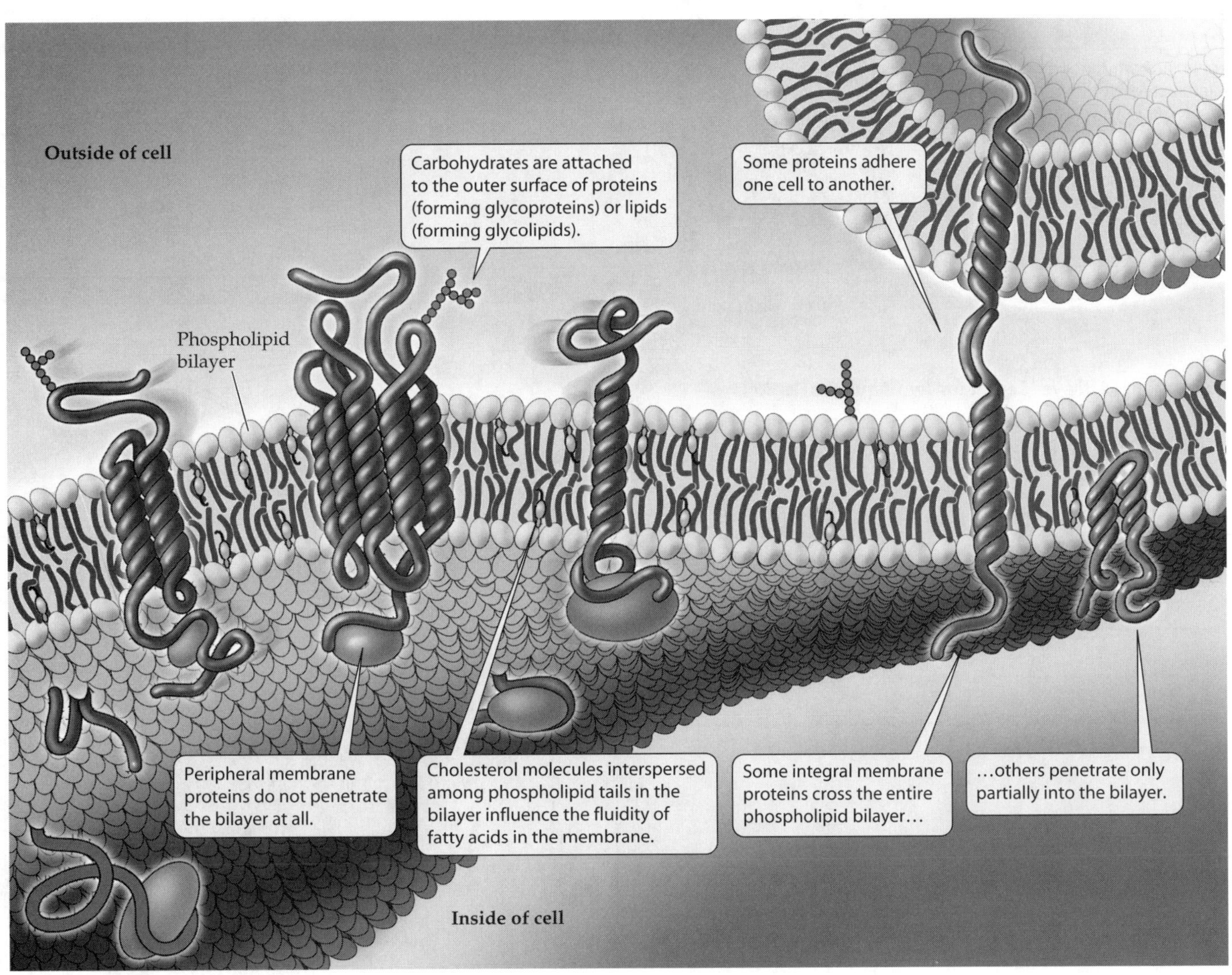

Outside of cell
Carbohydrates are attached to the outer surface of proteins (forming glycoproteins) or lipids (forming glycolipids).
Some proteins adhere one cell to another.
Phospholipid bilayer
Peripheral membrane proteins do not penetrate the bilayer at all.
Cholesterol molecules interspersed among phospholipid tails in the bilayer influence the fluidity of fatty acids in the membrane.
Some integral membrane proteins cross the entire phospholipid bilayer…
…others penetrate only partially into the bilayer.
Inside of cell

5.1 The Fluid Mosaic Model *(Page 88)*

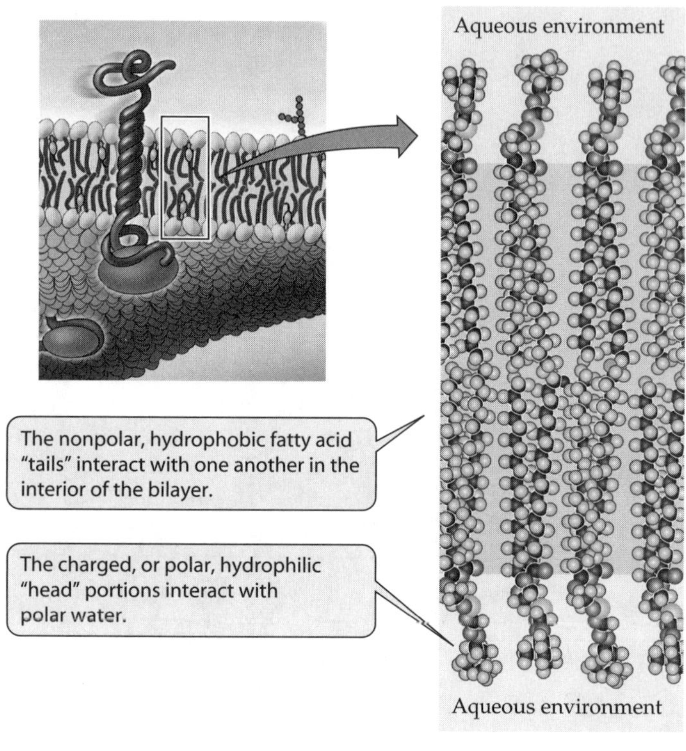
Aqueous environment
The nonpolar, hydrophobic fatty acid "tails" interact with one another in the interior of the bilayer.
The charged, or polar, hydrophilic "head" portions interact with polar water.
Aqueous environment

5.2 A Phospholipid Bilayer Separates Two Aqueous Regions
(Page 89)

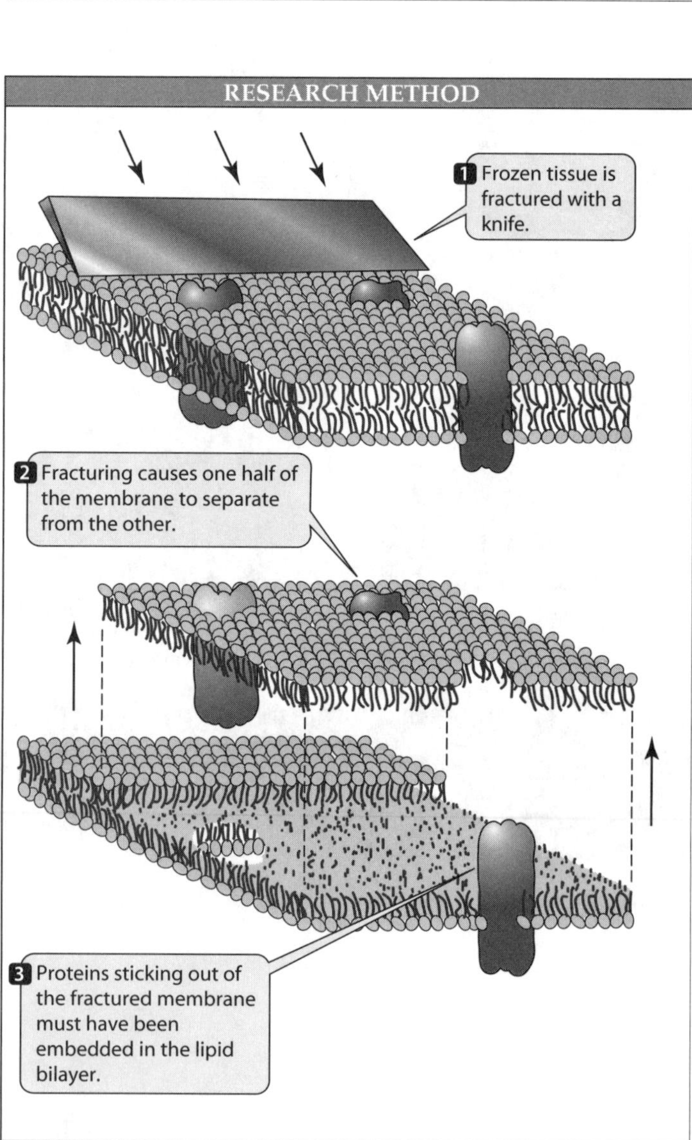
RESEARCH METHOD
1 Frozen tissue is fractured with a knife.
2 Fracturing causes one half of the membrane to separate from the other.
3 Proteins sticking out of the fractured membrane must have been embedded in the lipid bilayer.

5.3 Membrane Proteins Revealed by the Freeze-Fracture Technique
(Page 90)

Hydrophilic R groups in exposed parts of the protein interact with water.

Outside of cell

Aqueous environment (extracellular)

Hydrophobic interior of bilayer

Aqueous environment (cytoplasmic)

Inside of cell

Hydrophobic R groups in this part of the protein interact with the hydrophobic core of the membrane, away from water.

5.4 Interactions of Integral Membrane Proteins *(Page 90)*

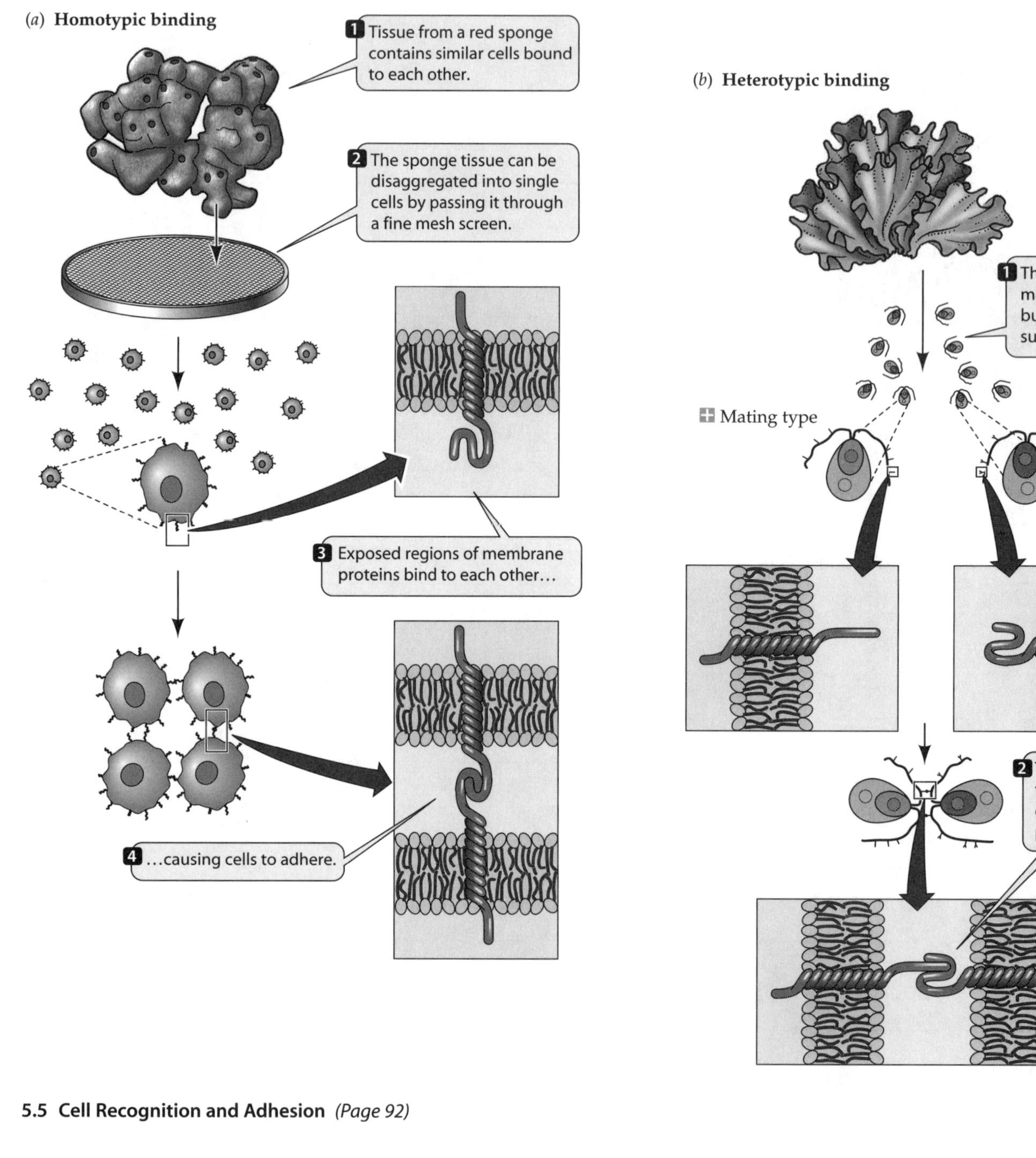
(a) Homotypic binding
1 Tissue from a red sponge contains similar cells bound to each other.
2 The sponge tissue can be disaggregated into single cells by passing it through a fine mesh screen.
3 Exposed regions of membrane proteins bind to each other…
4 …causing cells to adhere.
(b) Heterotypic binding
1 These gametes from a marine alga look identical but have different cell surface proteins.
+ Mating type
− Mating type
2 The gametes adhere to each other by complementary protein binding.

5.5 Cell Recognition and Adhesion *(Page 92)*

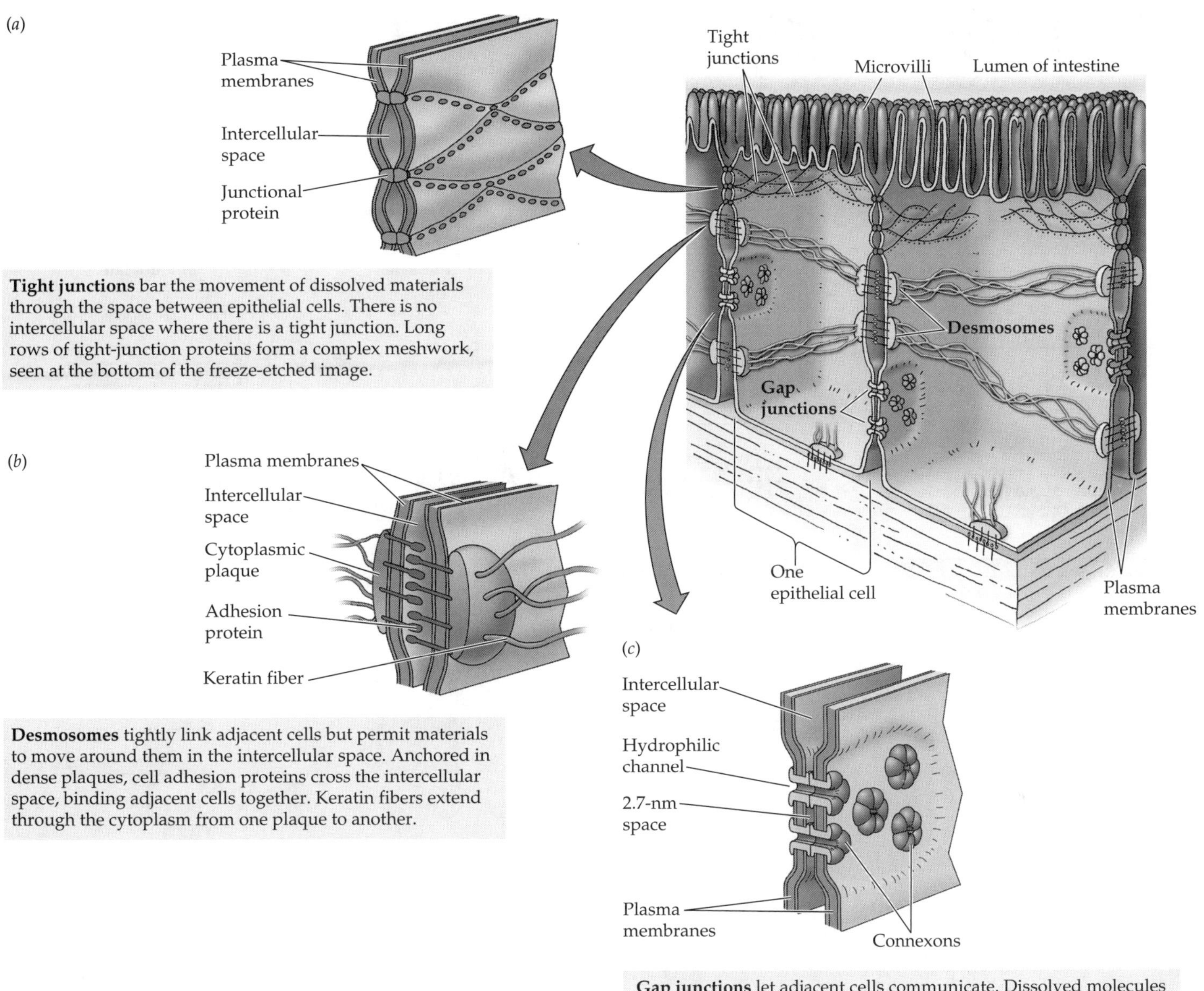
(a)
Plasma membranes
Intercellular space
Junctional protein
Tight junctions
Microvilli
Lumen of intestine
Desmosomes
Gap junctions
One epithelial cell
Plasma membranes
(b)
Plasma membranes
Intercellular space
Cytoplasmic plaque
Adhesion protein
Keratin fiber
(c)
Intercellular space
Hydrophilic channel
2.7-nm space
Plasma membranes
Connexons

Tight junctions bar the movement of dissolved materials through the space between epithelial cells. There is no intercellular space where there is a tight junction. Long rows of tight-junction proteins form a complex meshwork, seen at the bottom of the freeze-etched image.

Desmosomes tightly link adjacent cells but permit materials to move around them in the intercellular space. Anchored in dense plaques, cell adhesion proteins cross the intercellular space, binding adjacent cells together. Keratin fibers extend through the cytoplasm from one plaque to another.

Gap junctions let adjacent cells communicate. Dissolved molecules and electric signals may pass from one cell to the other through the channel formed by two connexons extending from adjacent cells.

5.6 Junctions Link Animal Cells Together *(Page 93)*

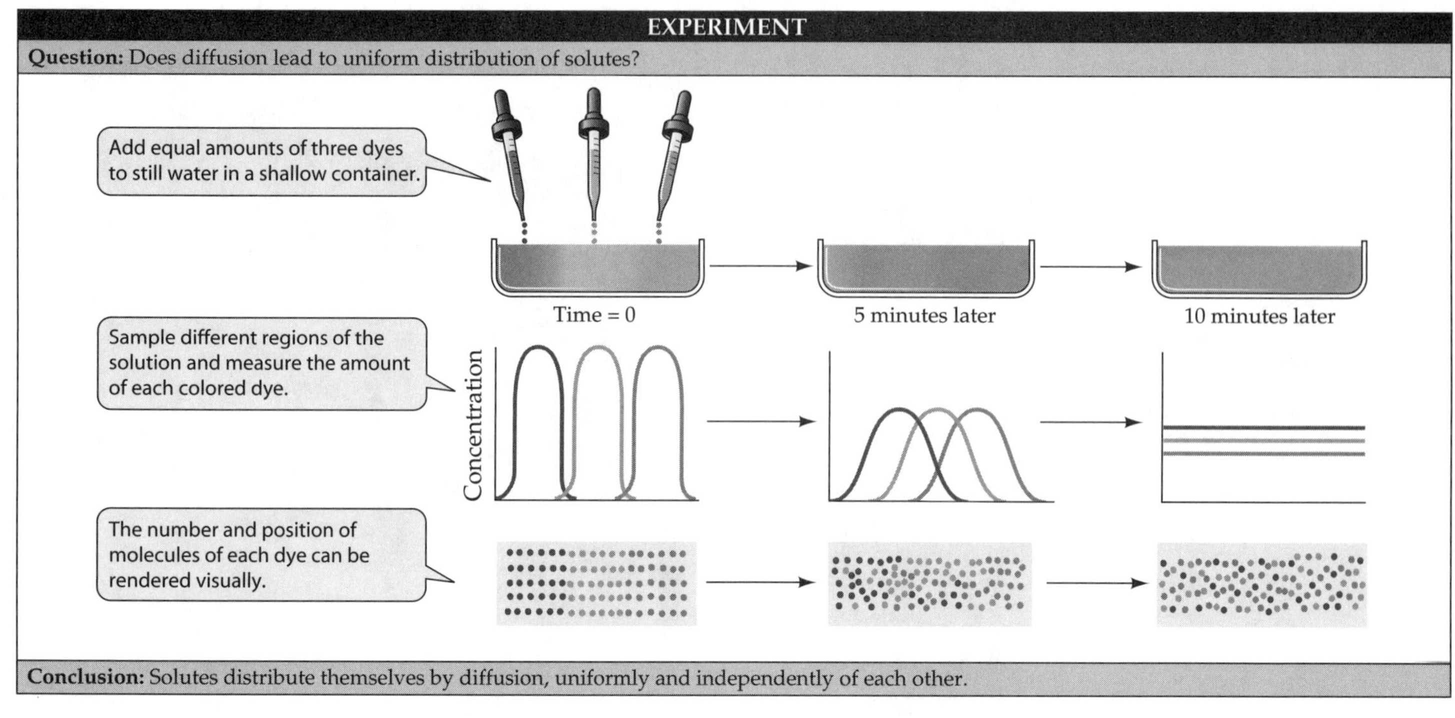
EXPERIMENT
Question: Does diffusion lead to uniform distribution of solutes?
Add equal amounts of three dyes to still water in a shallow container.
Time = 0
5 minutes later
10 minutes later
Sample different regions of the solution and measure the amount of each colored dye.
Concentration
The number and position of molecules of each dye can be rendered visually.
Conclusion: Solutes distribute themselves by diffusion, uniformly and independently of each other.

5.7 Diffusion Leads to Uniform Distribution of Solutes *(Page 95)*

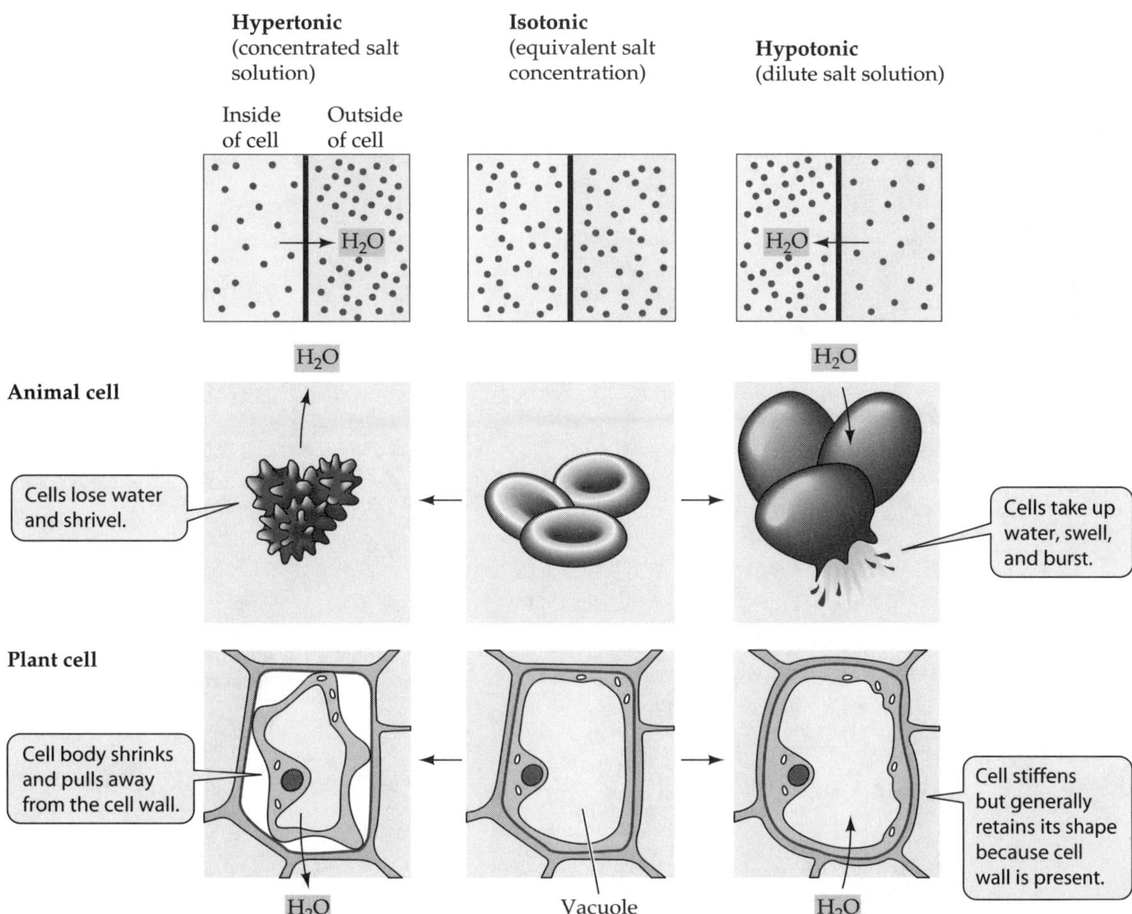

Hypertonic
(concentrated salt solution)
Isotonic
(equivalent salt concentration)
Hypotonic
(dilute salt solution)
Inside of cell
Outside of cell
H_2O
Animal cell
Cells lose water and shrivel.
Cells take up water, swell, and burst.
Plant cell
Cell body shrinks and pulls away from the cell wall.
Vacuole
Cell stiffens but generally retains its shape because cell wall is present.

5.8 Osmosis Modifies the Shapes of Cells *(Page 96)*

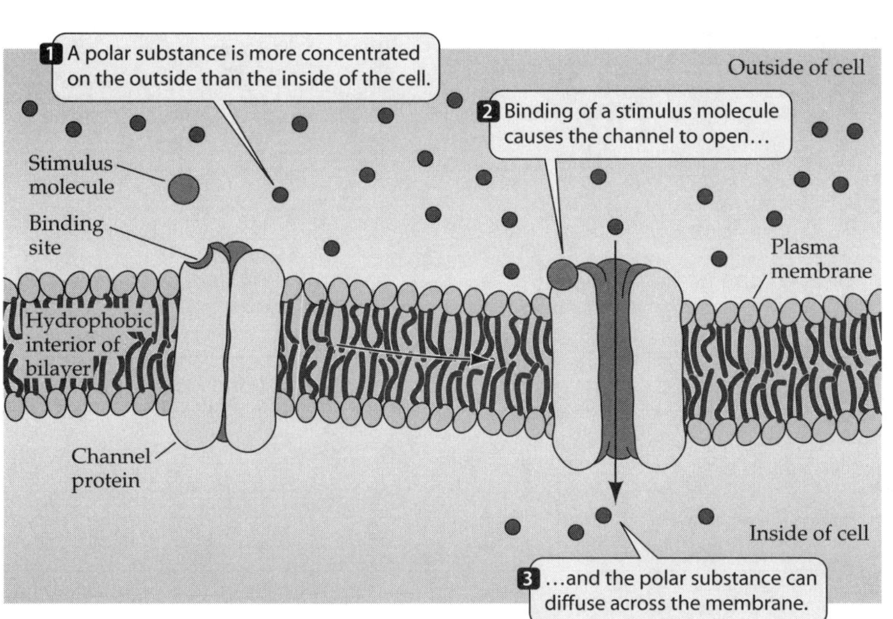

1 A polar substance is more concentrated on the outside than the inside of the cell.
Outside of cell
2 Binding of a stimulus molecule causes the channel to open…
Stimulus molecule
Binding site
Plasma membrane
Hydrophobic interior of bilayer
Channel protein
Inside of cell
3 …and the polar substance can diffuse across the membrane.

5.9 A Gated Channel Protein Opens in Response to a Stimulus *(Page 97)*

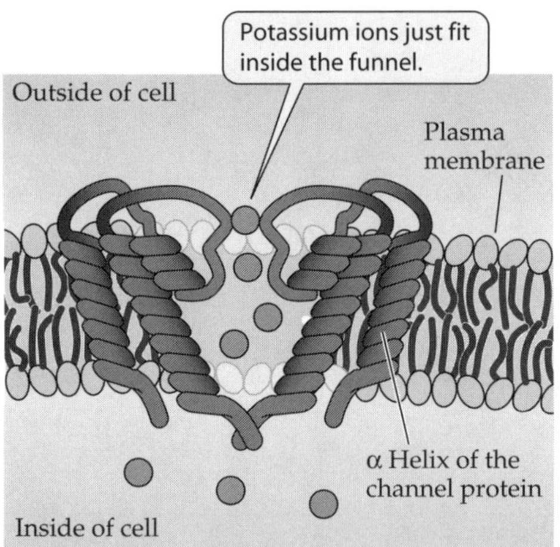
Potassium ions just fit inside the funnel.
Outside of cell
Plasma membrane
α Helix of the channel protein
Inside of cell

5.10 **The K$^+$ Channel** *(Page 98)*

(a)

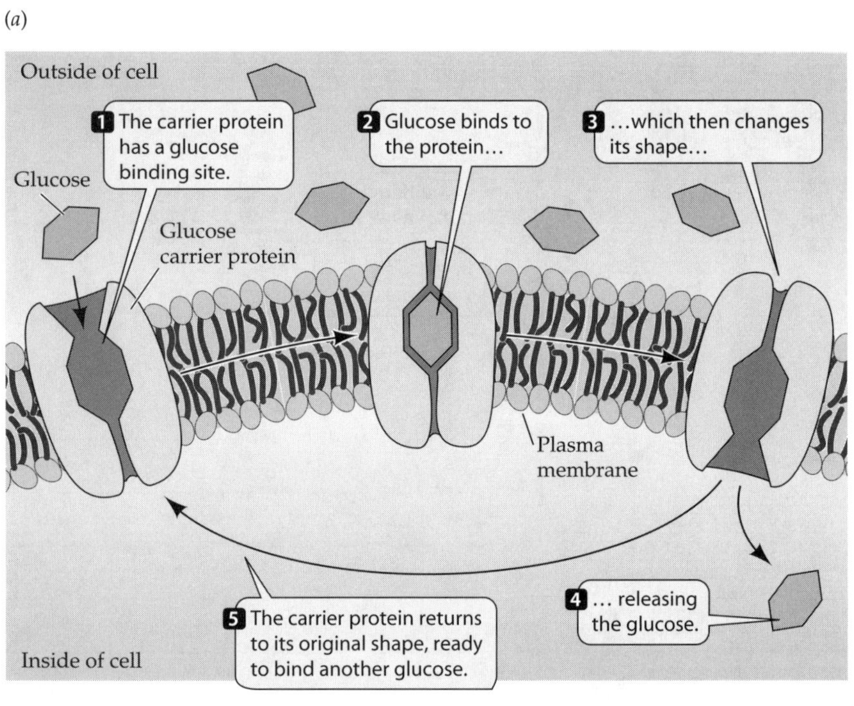
Outside of cell
1 The carrier protein has a glucose binding site.
2 Glucose binds to the protein…
3 …which then changes its shape…
Glucose
Glucose carrier protein
Plasma membrane
4 … releasing the glucose.
5 The carrier protein returns to its original shape, ready to bind another glucose.
Inside of cell

(b)

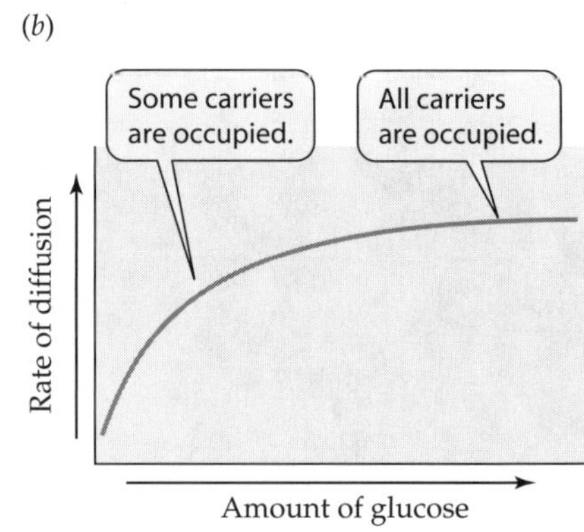
Some carriers are occupied.
All carriers are occupied.
Rate of diffusion
Amount of glucose

5.11 **A Carrier Protein Facilitates Diffusion** *(Page 99)*

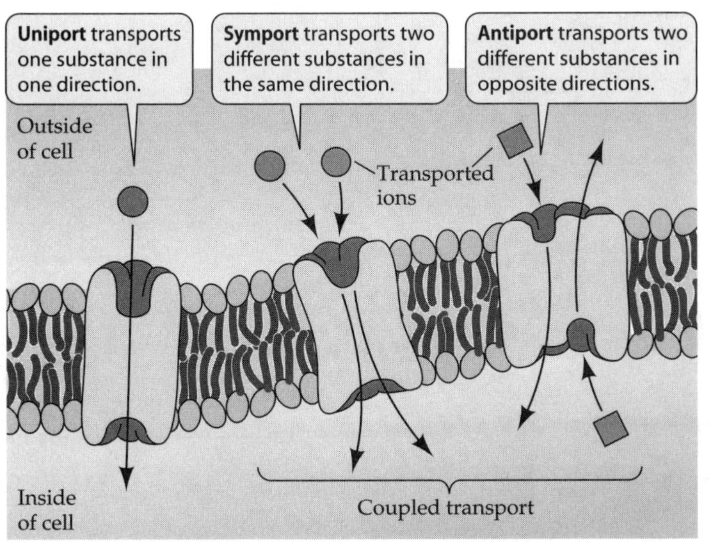
Uniport transports one substance in one direction.
Symport transports two different substances in the same direction.
Antiport transports two different substances in opposite directions.
Outside of cell
Transported ions
Inside of cell
Coupled transport

5.12 **Three Types of Proteins for Active Transport** *(Page 99)*

5.1 *Membrane Transport Mechanisms*

TRANSPORT MECHANISM	EXTERNAL ENERGY REQUIRED?	DRIVING FORCE	MEMBRANE PROTEIN REQUIRED?	SPECIFICITY
Simple diffusion	No	With concentration gradient	No	Not specific
Facilitated diffusion	No	With concentration gradient	Yes	Specific
Active transport	Yes	ATP hydrolysis (primary) (against concentration gradient)	Yes	Specific

(Page 100)

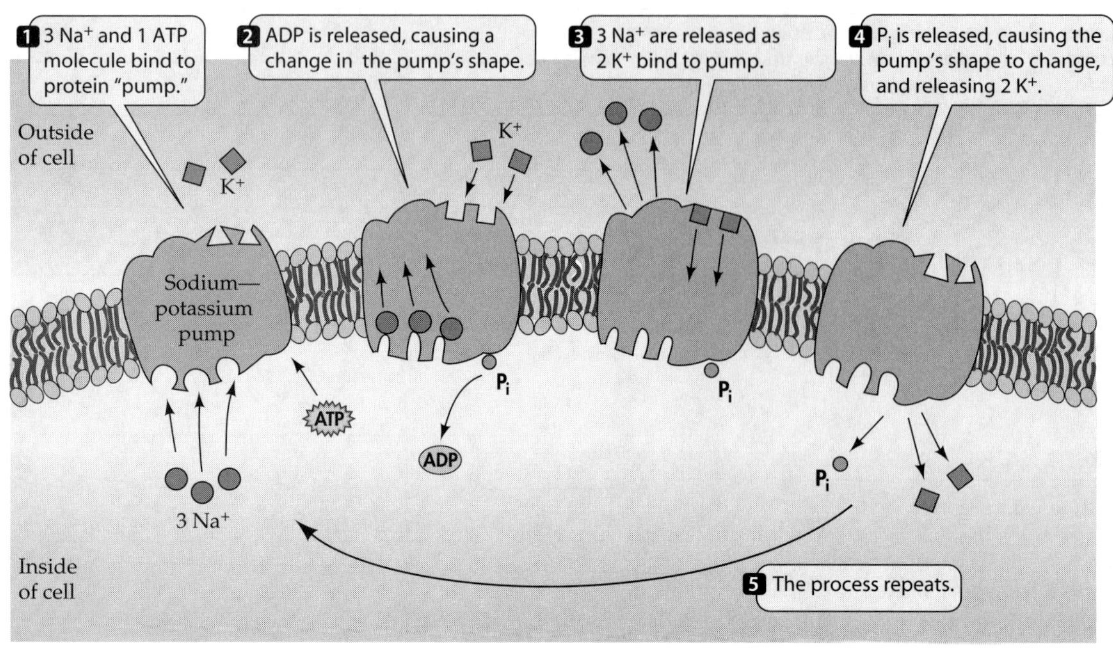
1 3 Na^+ and 1 ATP molecule bind to protein "pump."
2 ADP is released, causing a change in the pump's shape.
3 3 Na^+ are released as 2 K^+ bind to pump.
4 P_i is released, causing the pump's shape to change, and releasing 2 K^+.
Outside of cell
K^+
K^+
Sodium–potassium pump
ATP
P_i
ADP
P_i
P_i
3 Na^+
Inside of cell
5 The process repeats.

5.13 Primary Active Transport: The Sodium–Potassium Pump *(Page 101)*

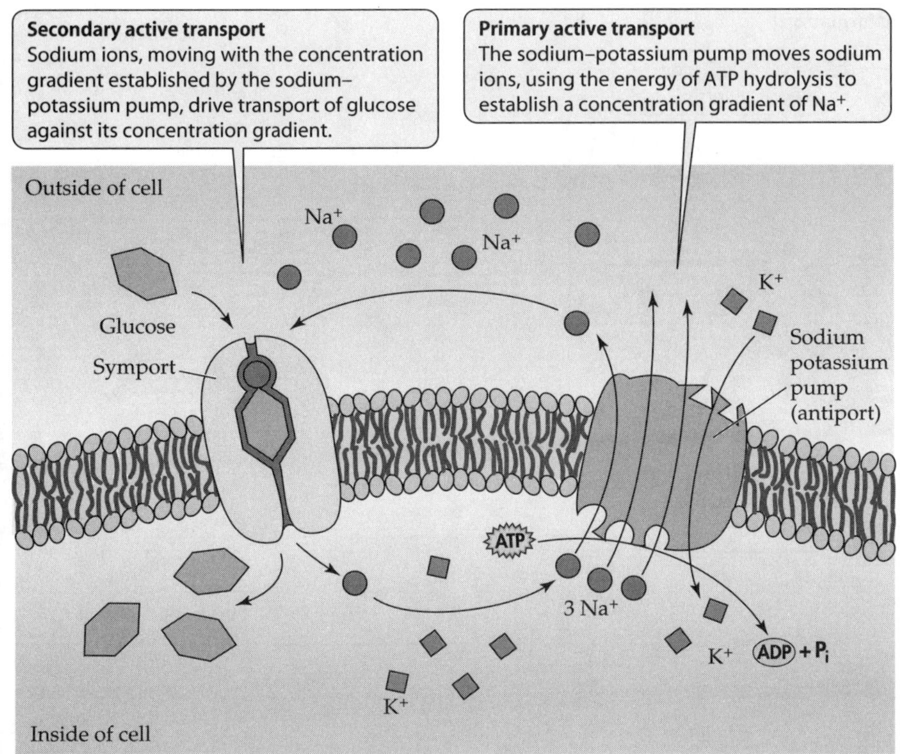

Secondary active transport
Sodium ions, moving with the concentration gradient established by the sodium–potassium pump, drive transport of glucose against its concentration gradient.
Primary active transport
The sodium–potassium pump moves sodium ions, using the energy of ATP hydrolysis to establish a concentration gradient of Na+.
Outside of cell
Na+
Na+
K+
Glucose
Symport
Sodium potassium pump (antiport)
ATP
3 Na+
K+
ADP + Pi
K+
Inside of cell

5.14 Secondary Active Transport *(Page 101)*

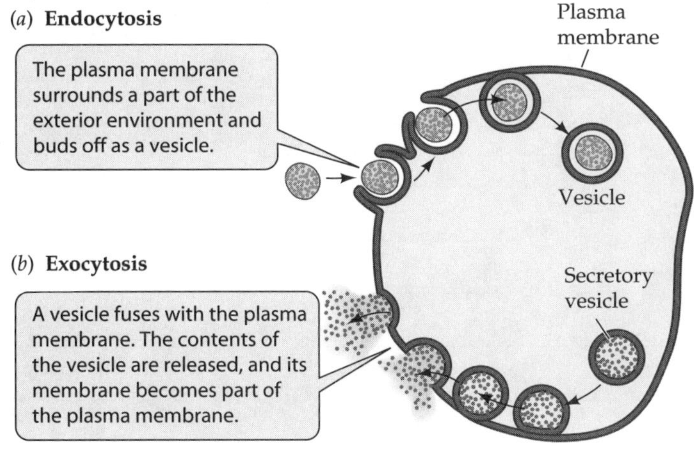

(a) Endocytosis
The plasma membrane surrounds a part of the exterior environment and buds off as a vesicle.
Plasma membrane
Vesicle
(b) Exocytosis
A vesicle fuses with the plasma membrane. The contents of the vesicle are released, and its membrane becomes part of the plasma membrane.
Secretory vesicle

5.15 Endocytosis and Exocytosis *(Page 101)*

(*a*) **Information processing**

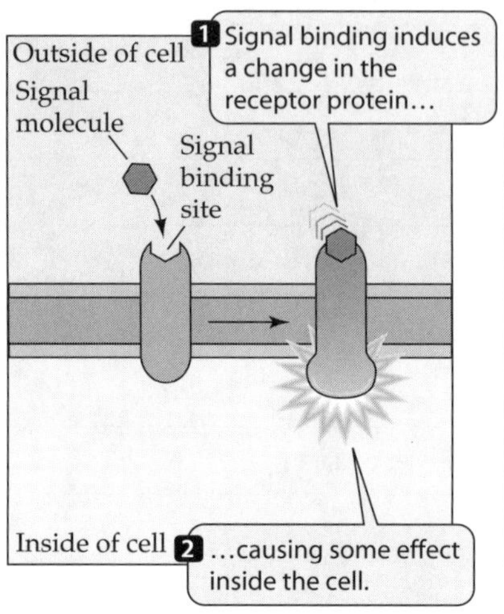
Outside of cell
Signal molecule
Signal binding site
1 Signal binding induces a change in the receptor protein…
Inside of cell
2 …causing some effect inside the cell.

(*b*) **Energy transformation**

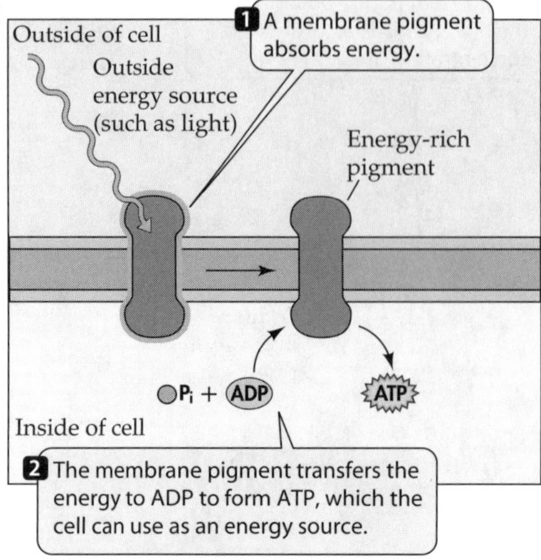
Outside of cell
Outside energy source (such as light)
1 A membrane pigment absorbs energy.
Energy-rich pigment
Pi + ADP
ATP
Inside of cell
2 The membrane pigment transfers the energy to ADP to form ATP, which the cell can use as an energy source.

(*c*) **Organizing chemical reactions**

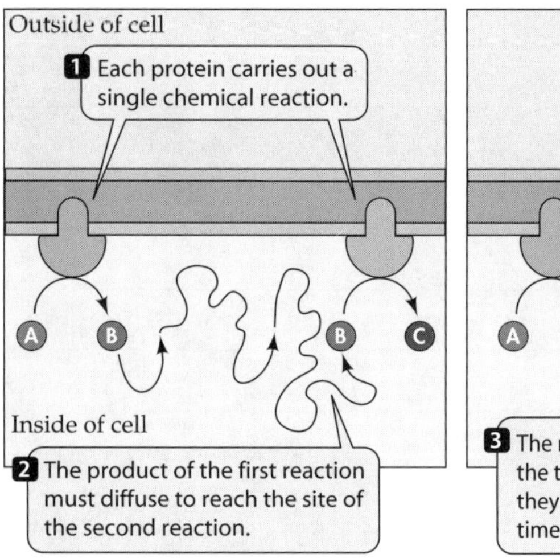
Outside of cell
1 Each protein carries out a single chemical reaction.
A
B
B
C
Inside of cell
2 The product of the first reaction must diffuse to reach the site of the second reaction.

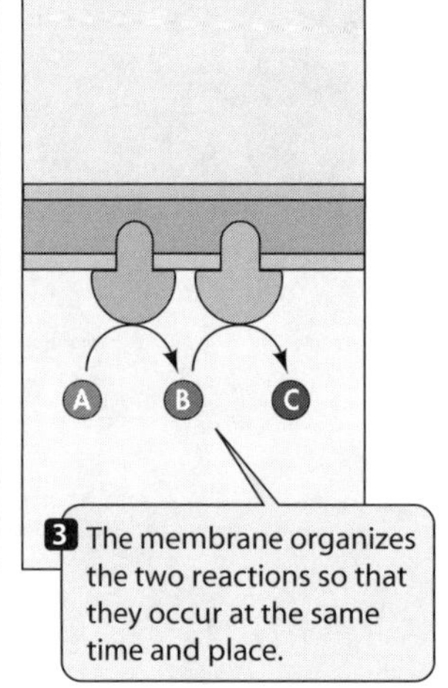
A
B
C
3 The membrane organizes the two reactions so that they occur at the same time and place.

5.17 More Membrane Functions *(Page 103)*

6 Energy, Enzymes, and Metabolism

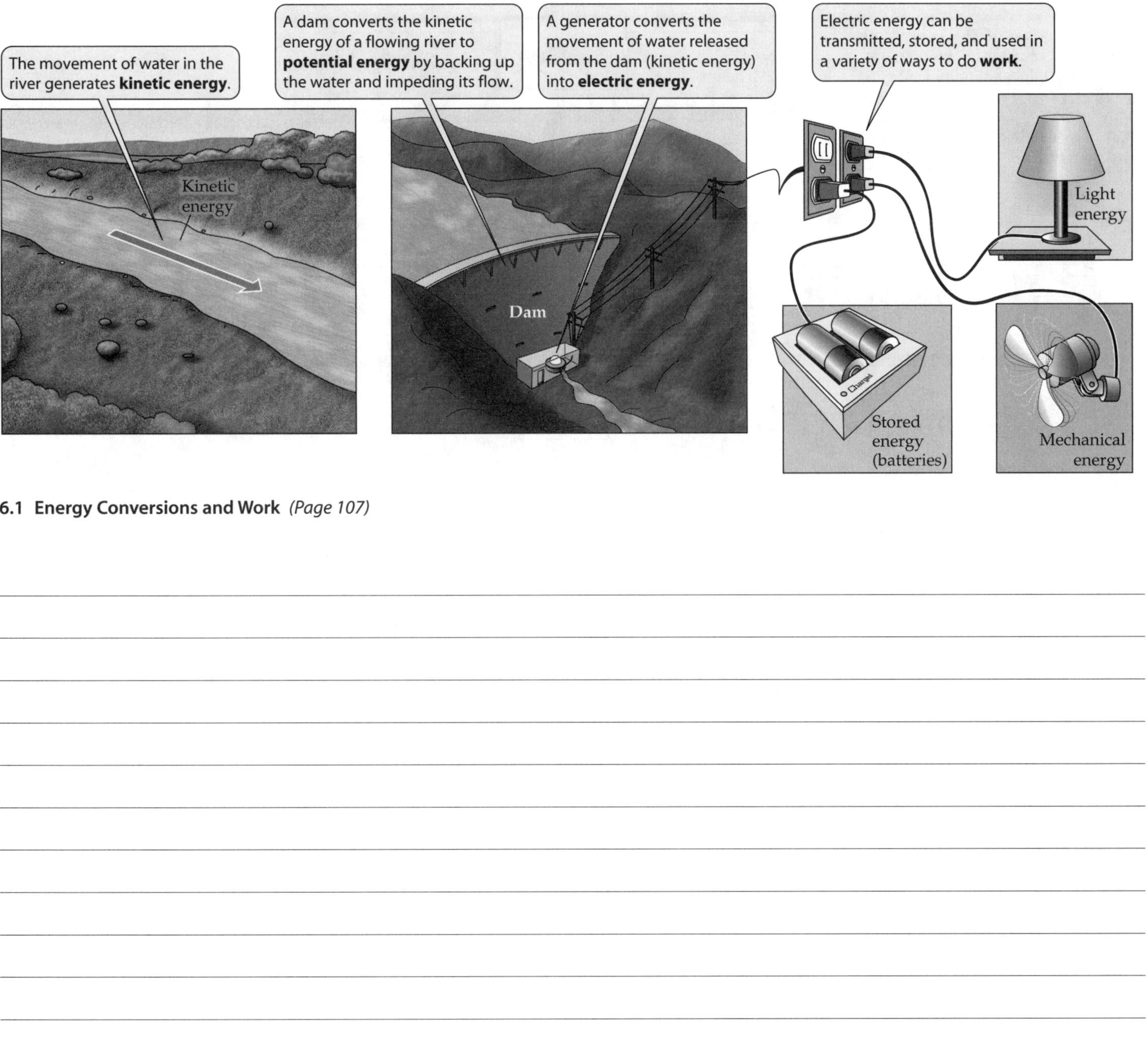
The movement of water in the river generates **kinetic energy**.
A dam converts the kinetic energy of a flowing river to **potential energy** by backing up the water and impeding its flow.
A generator converts the movement of water released from the dam (kinetic energy) into **electric energy**.
Electric energy can be transmitted, stored, and used in a variety of ways to do **work**.
Kinetic energy
Dam
Light energy
Stored energy (batteries)
Mechanical energy

6.1 Energy Conversions and Work *(Page 107)*

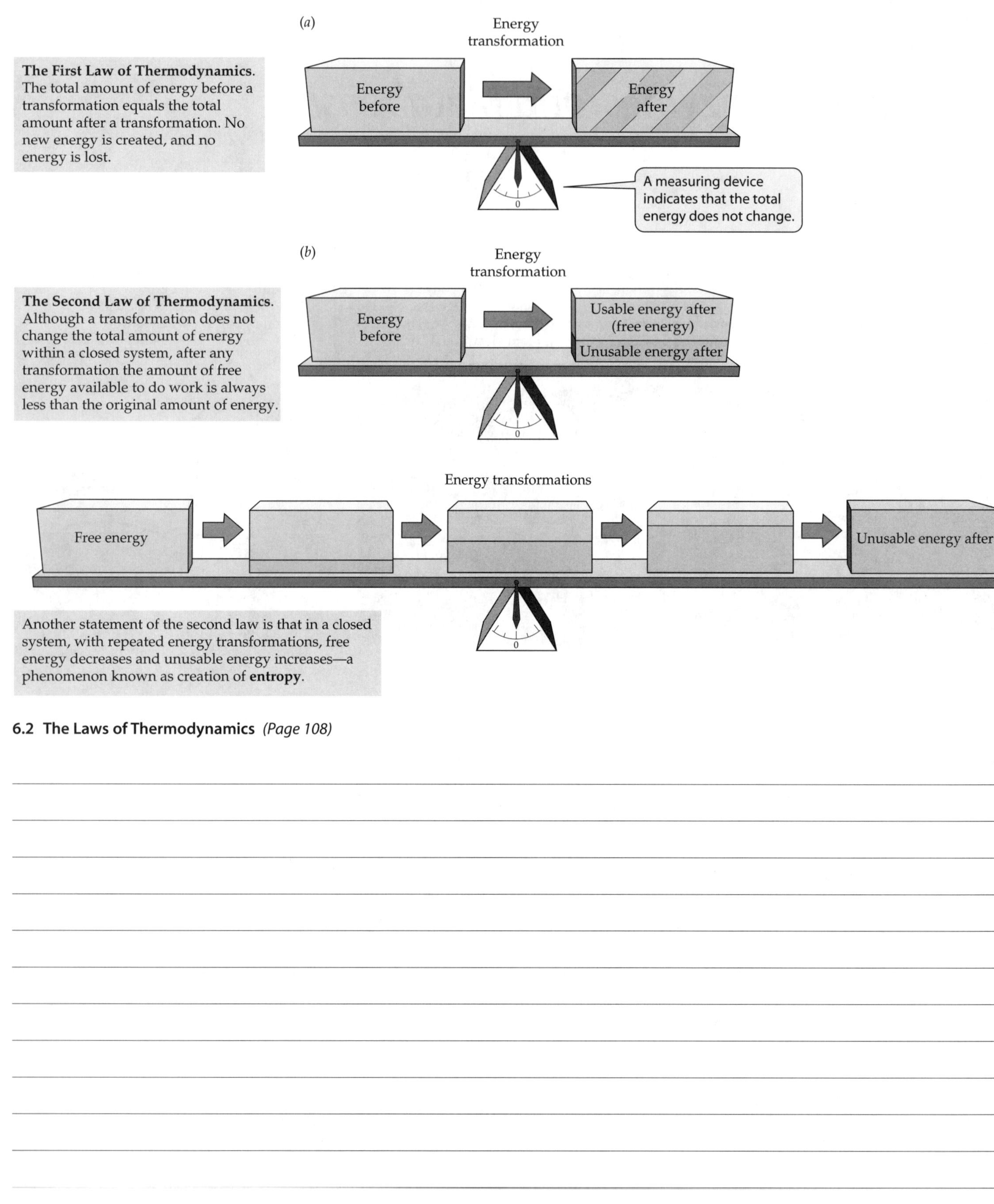
(a)
Energy transformation
The First Law of Thermodynamics. The total amount of energy before a transformation equals the total amount after a transformation. No new energy is created, and no energy is lost.
Energy before
Energy after
0
A measuring device indicates that the total energy does not change.
(b)
Energy transformation
The Second Law of Thermodynamics. Although a transformation does not change the total amount of energy within a closed system, after any transformation the amount of free energy available to do work is always less than the original amount of energy.
Energy before
Usable energy after (free energy)
Unusable energy after
0
Energy transformations
Free energy
Unusable energy after
0
Another statement of the second law is that in a closed system, with repeated energy transformations, free energy decreases and unusable energy increases—a phenomenon known as creation of entropy.

6.2 The Laws of Thermodynamics *(Page 108)*

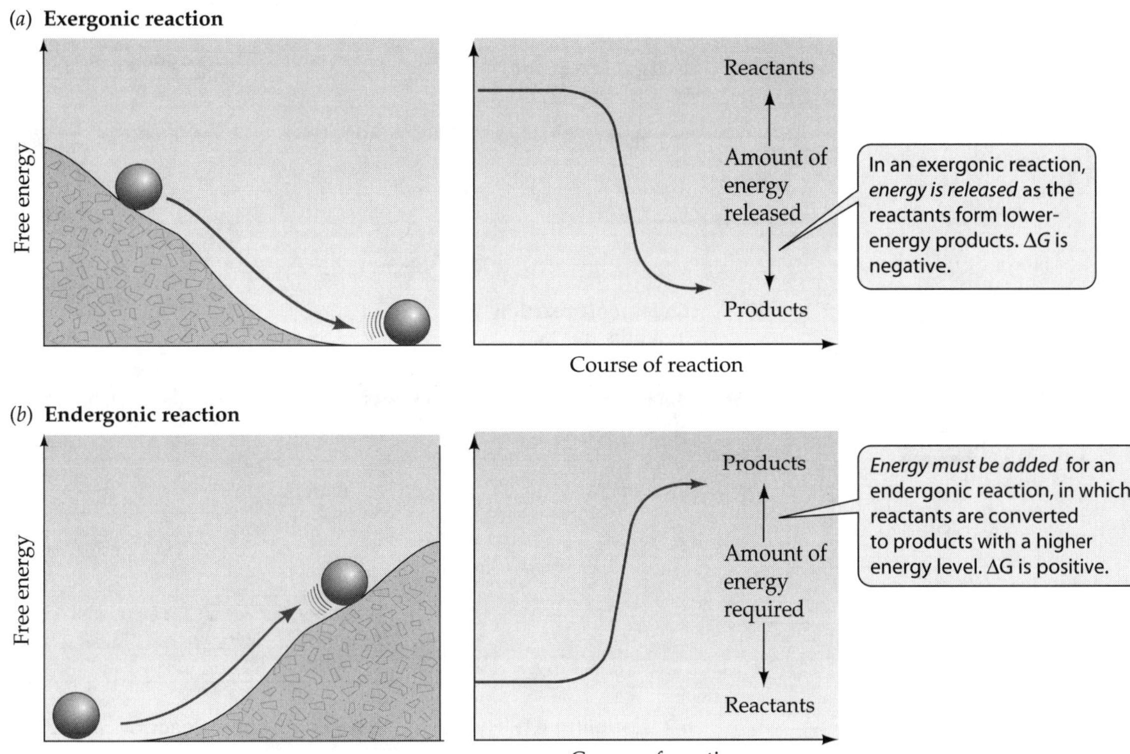
(a) Exergonic reaction
Free energy
Reactants
Amount of energy released
Products
Course of reaction
In an exergonic reaction, energy is released as the reactants form lower-energy products. ΔG is negative.
(b) Endergonic reaction
Free energy
Products
Amount of energy required
Reactants
Course of reaction
Energy must be added for an endergonic reaction, in which reactants are converted to products with a higher energy level. ΔG is positive.

6.3 Exergonic and Endergonic Reactions *(Page 110)*

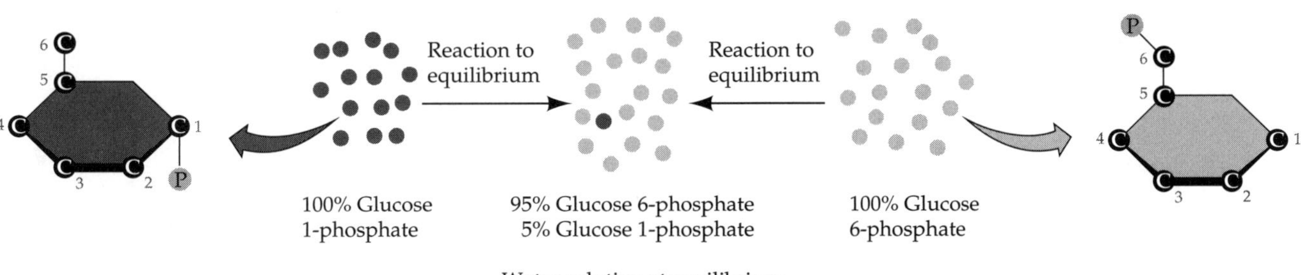
Reaction to equilibrium
Reaction to equilibrium
100% Glucose 1-phosphate
95% Glucose 6-phosphate
5% Glucose 1-phosphate
100% Glucose 6-phosphate
Water solution at equilibrium

6.4 Concentration at Equilibrium *(Page 111)*

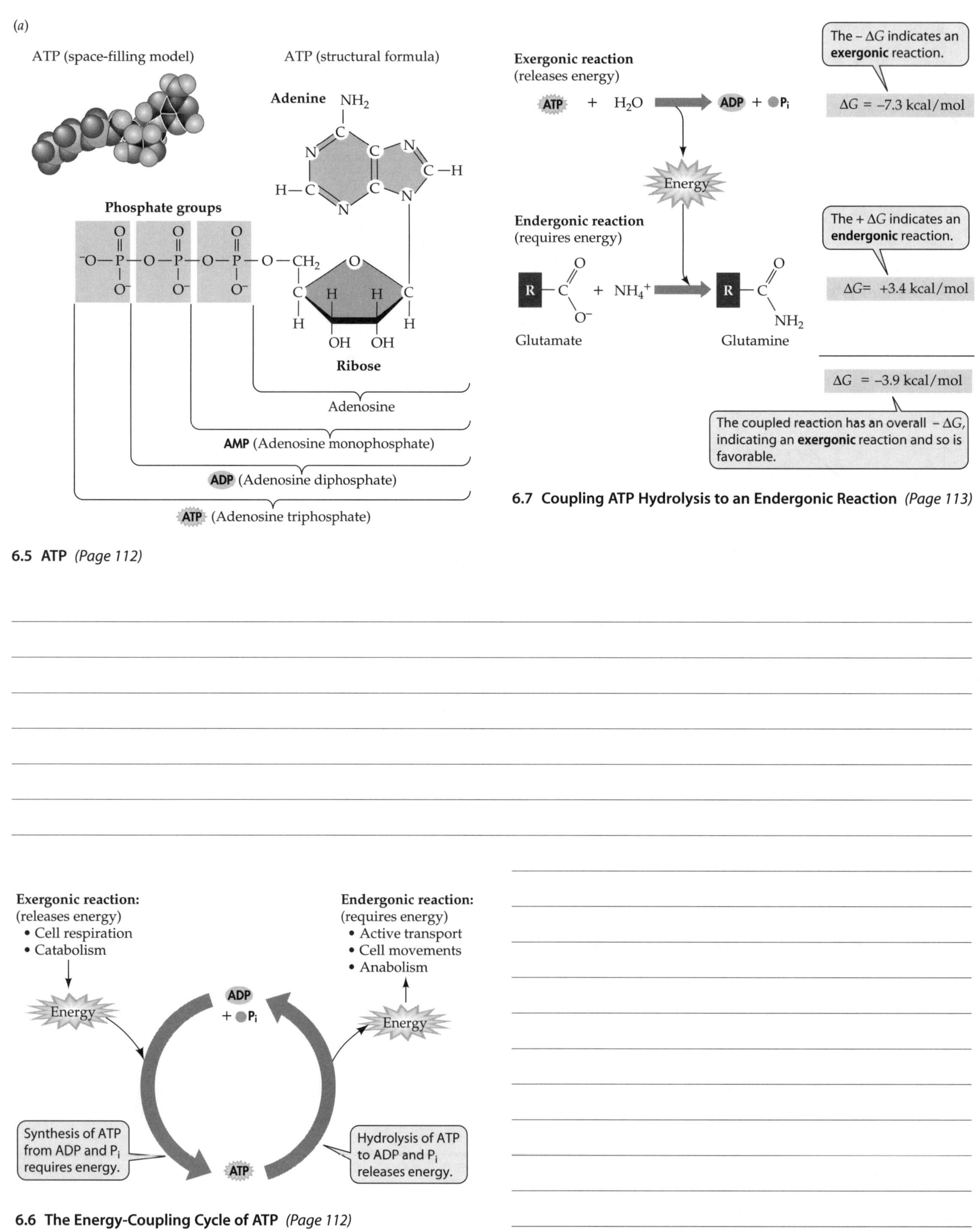
(a)
ATP (space-filling model)
ATP (structural formula)
Adenine
Phosphate groups
Ribose
Adenosine
AMP (Adenosine monophosphate)
ADP (Adenosine diphosphate)
ATP (Adenosine triphosphate)
Exergonic reaction
(releases energy)
ATP + H2O → ADP + Pi
Energy
Endergonic reaction
(requires energy)
Glutamate
Glutamine
The – ΔG indicates an exergonic reaction.
ΔG = –7.3 kcal/mol
The + ΔG indicates an endergonic reaction.
ΔG= +3.4 kcal/mol
ΔG = –3.9 kcal/mol
The coupled reaction has an overall – ΔG, indicating an exergonic reaction and so is favorable.
Exergonic reaction:
(releases energy)
• Cell respiration
• Catabolism
Endergonic reaction:
(requires energy)
• Active transport
• Cell movements
• Anabolism
Energy
ADP + Pi
ATP
Synthesis of ATP from ADP and Pi requires energy.
Hydrolysis of ATP to ADP and Pi releases energy.

6.5 ATP *(Page 112)*

6.7 Coupling ATP Hydrolysis to an Endergonic Reaction *(Page 113)*

6.6 The Energy-Coupling Cycle of ATP *(Page 112)*

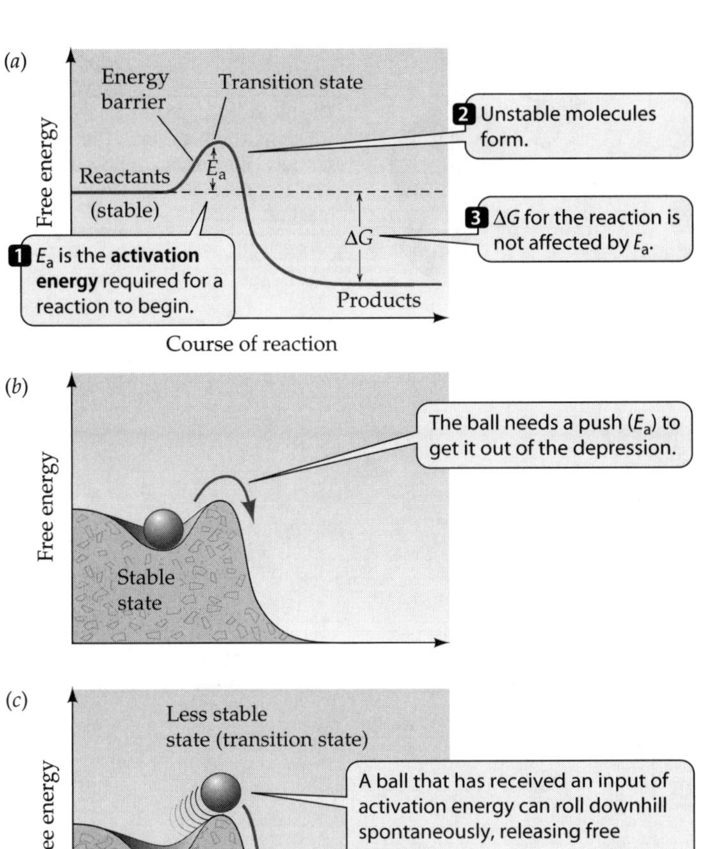
(a)
Energy barrier
Transition state
Free energy
Reactants
(stable)
E_a
ΔG
Products
Course of reaction
1 E_a is the **activation energy** required for a reaction to begin.
2 Unstable molecules form.
3 ΔG for the reaction is not affected by E_a.
(b)
Free energy
Stable state
The ball needs a push (E_a) to get it out of the depression.
(c)
Free energy
Less stable state (transition state)
A ball that has received an input of activation energy can roll downhill spontaneously, releasing free energy.

6.8 Activation Energy Initiates Reactions *(Page 114)*

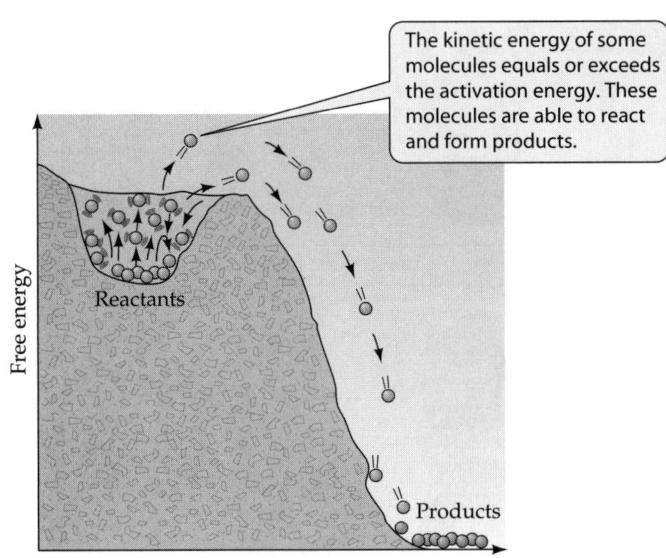
The kinetic energy of some molecules equals or exceeds the activation energy. These molecules are able to react and form products.
Free energy
Reactants
Products
Course of reaction

6.9 Over the Energy Barrier *(Page 114)*

Product

Active site

Substrates fit precisely into the active site…

…but nonsubstrate does not.

Enzyme

Enzyme–substrate complex

Enzyme

The breakdown of the enzyme–substrate complex yields the product. The enzyme is now available to catalyze another reaction.

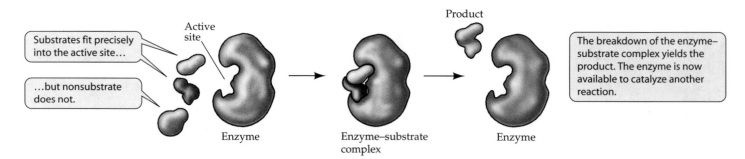

6.10 Enzyme and Substrate *(Page 115)*

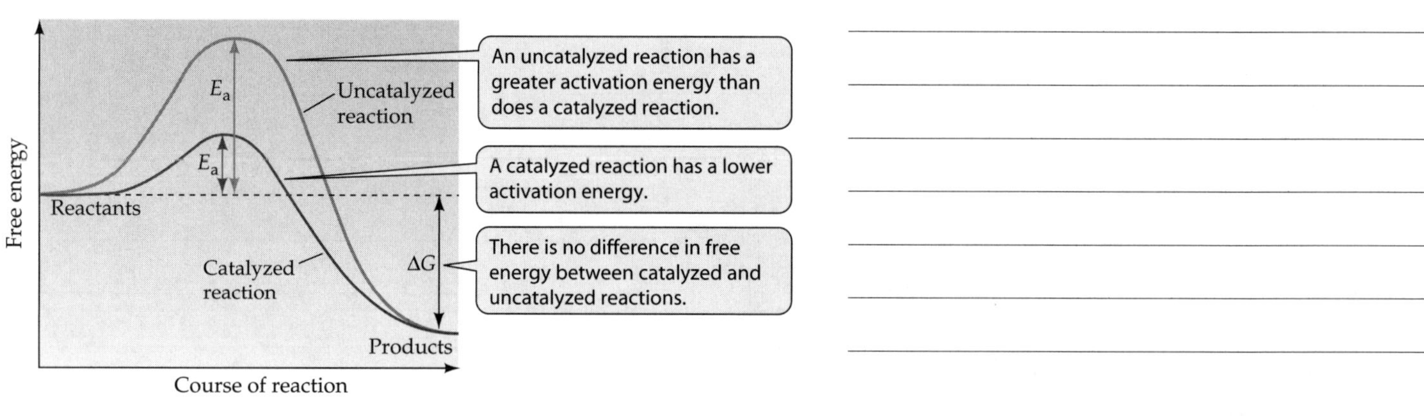
Free energy
E_a
E_a
Uncatalyzed reaction
Reactants
Catalyzed reaction
ΔG
Products
Course of reaction
An uncatalyzed reaction has a greater activation energy than does a catalyzed reaction.
A catalyzed reaction has a lower activation energy.
There is no difference in free energy between catalyzed and uncatalyzed reactions.

6.11 Enzymes Lower the Energy Barrier *(Page 115)*

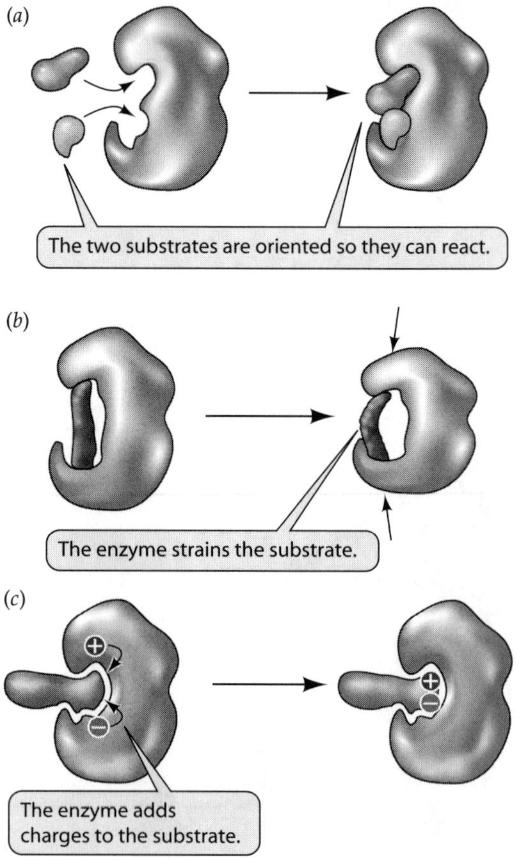

(*a*)

The two substrates are oriented so they can react.

(*b*)

The enzyme strains the substrate.

(*c*)

The enzyme adds charges to the substrate.

6.12 Life at the Active Site *(Page 116)*

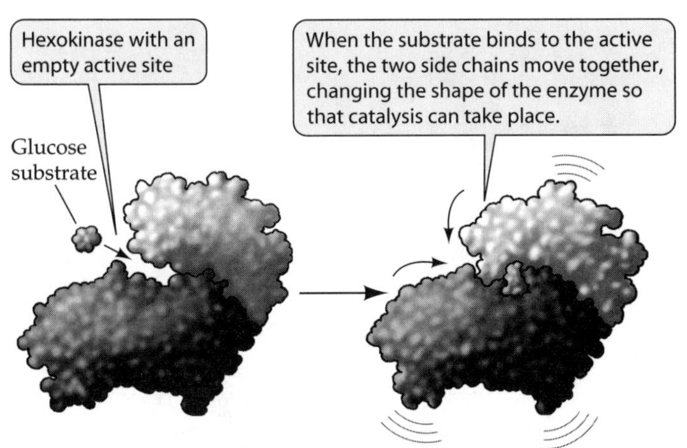
Hexokinase with an empty active site
When the substrate binds to the active site, the two side chains move together, changing the shape of the enzyme so that catalysis can take place.
Glucose substrate

6.14 Some Enzymes Change Shape When Substrate Binds to Them *(Page 117)*

6.1 *A Few Examples of Nonprotein "Partners" of Enzymes*

TYPE OF MOLECULE	ROLE IN CATALYZED REACTIONS
Cofactors	
Iron (Fe^{2+} or Fe^{3+})	Oxidation/reduction
Copper (Cu^{+} or Cu^{2+})	Oxidation/reduction
Zinc (Zn^{2+})	Helps bind NAD
Coenzymes	
Biotin	Carries —COO^-
Coenzyme A	Carries —CH_2—CH_3
NAD	Carries electrons
FAD	Carries electrons
ATP	Provides/extracts energy
Prosthetic groups	
Heme	Binds ions, O_2, and electrons; contains iron cofactor
Flavin	Binds electrons
Retinal	Converts light energy

(Page 118)

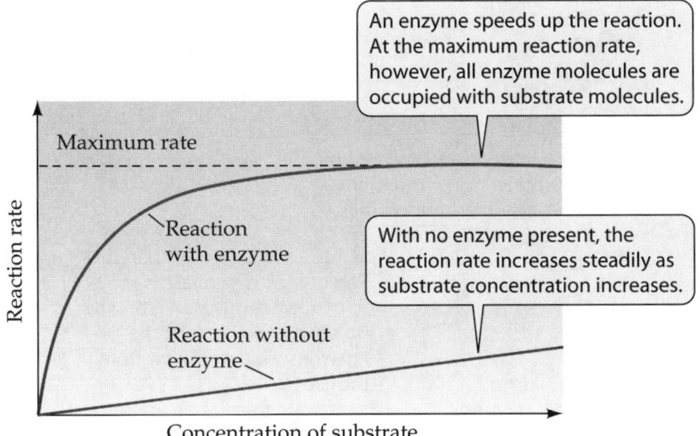
An enzyme speeds up the reaction. At the maximum reaction rate, however, all enzyme molecules are occupied with substrate molecules.
Maximum rate
Reaction with enzyme
With no enzyme present, the reaction rate increases steadily as substrate concentration increases.
Reaction without enzyme
Reaction rate
Concentration of substrate

6.16 Catalyzed Reactions Reach a Maximum Rate *(Page 118)*

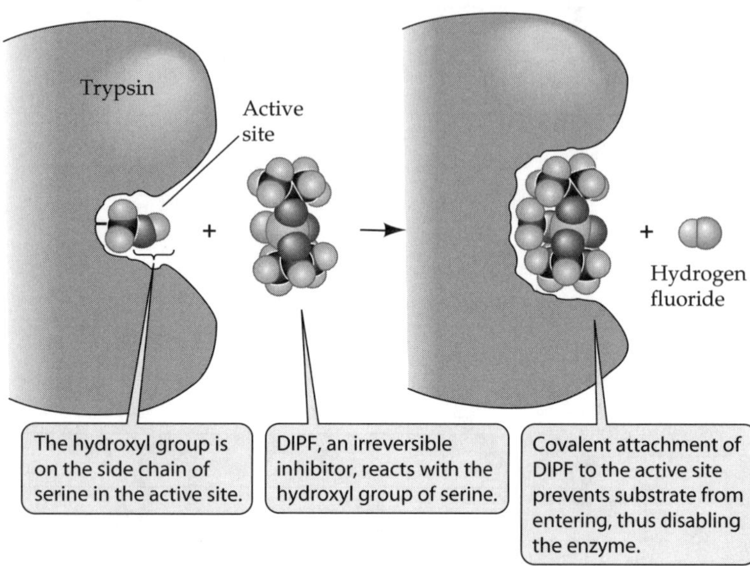
Trypsin
Active site
Hydrogen fluoride
The hydroxyl group is on the side chain of serine in the active site.
DIPF, an irreversible inhibitor, reacts with the hydroxyl group of serine.
Covalent attachment of DIPF to the active site prevents substrate from entering, thus disabling the enzyme.

6.17 Irreversible Inhibition *(Page 119)*

(a) **Competitive inhibition**

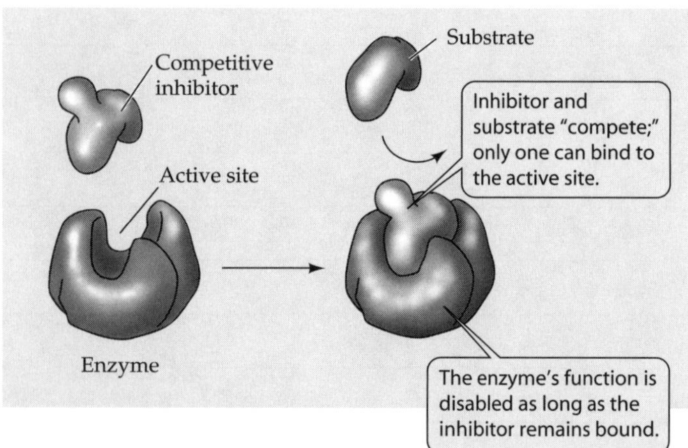
Substrate
Competitive inhibitor
Inhibitor and substrate "compete;" only one can bind to the active site.
Active site
Enzyme
The enzyme's function is disabled as long as the inhibitor remains bound.

(b) **Noncompetitive inhibition**

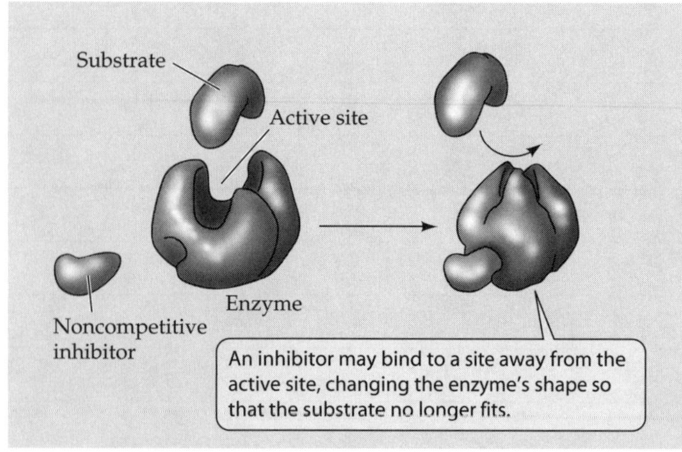
Substrate
Active site
Enzyme
Noncompetitive inhibitor
An inhibitor may bind to a site away from the active site, changing the enzyme's shape so that the substrate no longer fits.

Competitive inhibition of succinate dehydrogenase

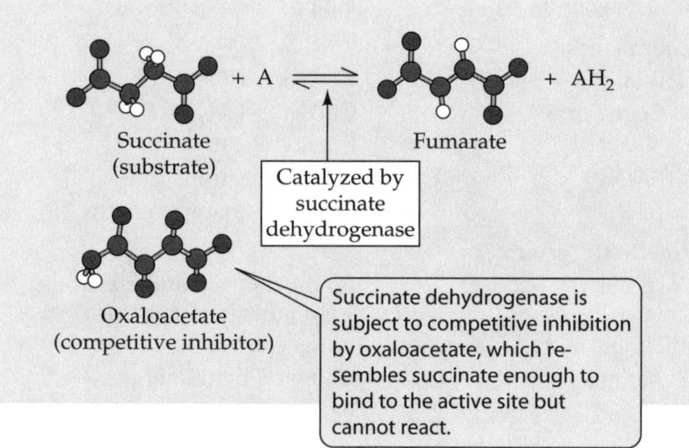
+ A
+ AH2
Succinate (substrate)
Fumarate
Catalyzed by succinate dehydrogenase
Oxaloacetate (competitive inhibitor)
Succinate dehydrogenase is subject to competitive inhibition by oxaloacetate, which resembles succinate enough to bind to the active site but cannot react.

Noncompetitive inhibition of threonine dehydratase

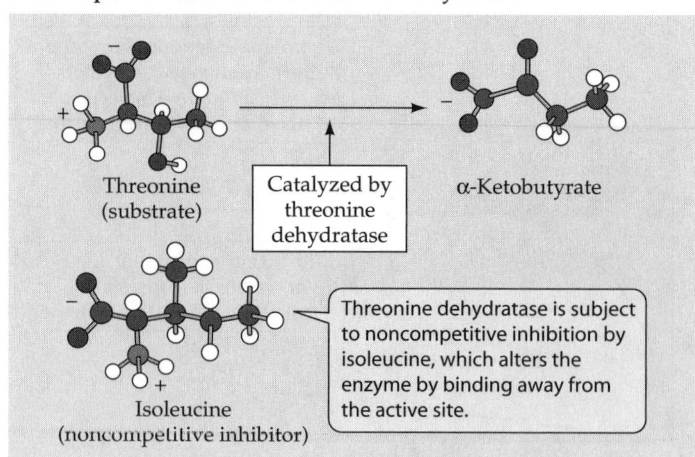
Threonine (substrate)
Catalyzed by threonine dehydratase
α-Ketobutyrate
Isoleucine (noncompetitive inhibitor)
Threonine dehydratase is subject to noncompetitive inhibition by isoleucine, which alters the enzyme by binding away from the active site.

6.18 Reversible Inhibition *(Page 120)*

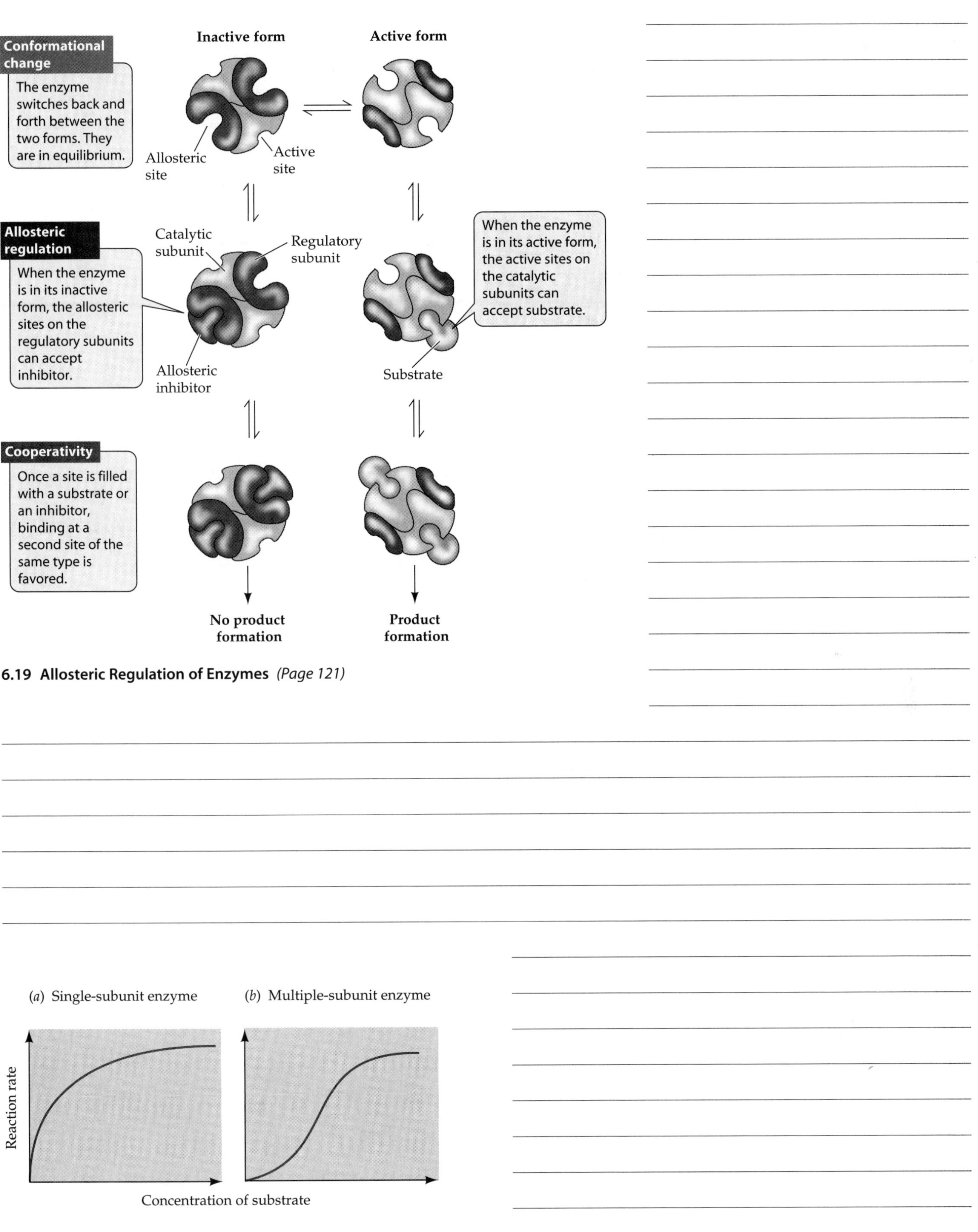

Inactive form
Active form
Conformational change
The enzyme switches back and forth between the two forms. They are in equilibrium.
Allosteric site
Active site
Allosteric regulation
When the enzyme is in its inactive form, the allosteric sites on the regulatory subunits can accept inhibitor.
Catalytic subunit
Regulatory subunit
Allosteric inhibitor
When the enzyme is in its active form, the active sites on the catalytic subunits can accept substrate.
Substrate
Cooperativity
Once a site is filled with a substrate or an inhibitor, binding at a second site of the same type is favored.
No product formation
Product formation

6.19 Allosteric Regulation of Enzymes *(Page 121)*

(a) Single-subunit enzyme
(b) Multiple-subunit enzyme
Reaction rate
Concentration of substrate

6.20 Allostery and Reaction Rate *(Page 121)*

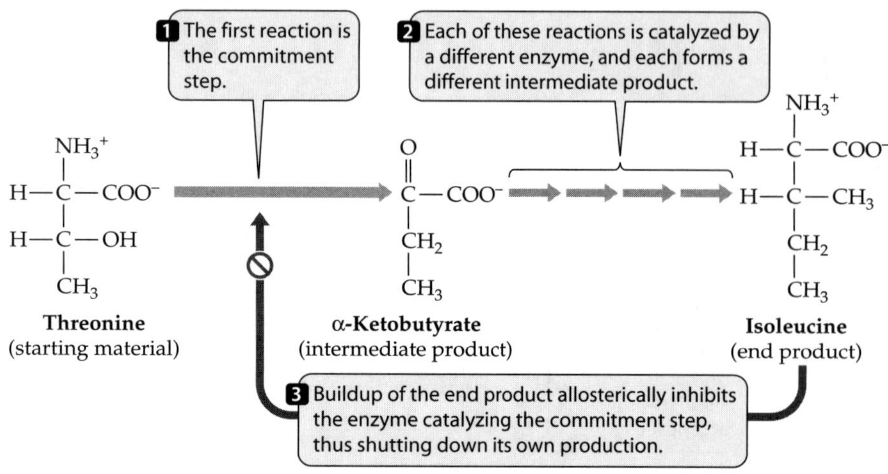
1 The first reaction is the commitment step.
2 Each of these reactions is catalyzed by a different enzyme, and each forms a different intermediate product.
Threonine
(starting material)
α-Ketobutyrate
(intermediate product)
Isoleucine
(end product)
3 Buildup of the end product allosterically inhibits the enzyme catalyzing the commitment step, thus shutting down its own production.

6.21 Inhibition of Metabolic Pathways *(Page 122)*

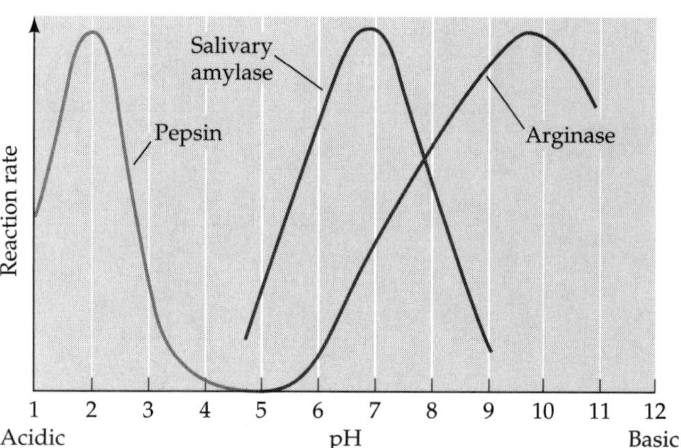
Salivary amylase
Pepsin
Arginase
Reaction rate
1 2 3 4 5 6 7 8 9 10 11 12
Acidic
pH
Basic

6.22 pH Affects Enzyme Activity *(Page 122)*

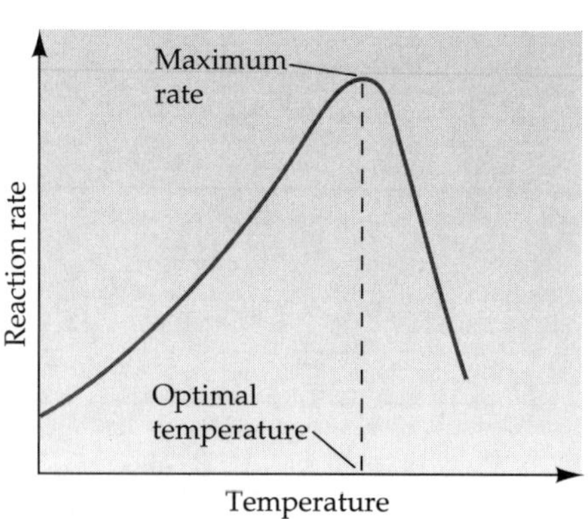
Maximum rate
Reaction rate
Optimal temperature
Temperature

6.23 Temperature Affects Enzyme Activity *(Page 123)*

7 Cellular Pathways that Harvest Chemical Energy

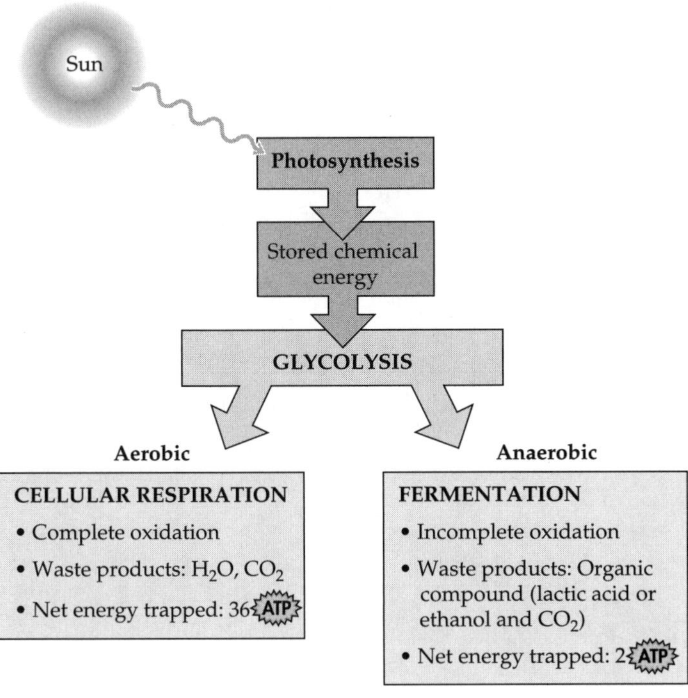

Sun
Photosynthesis
Stored chemical energy
GLYCOLYSIS
Aerobic
Anaerobic
CELLULAR RESPIRATION
• Complete oxidation
• Waste products: H_2O, CO_2
• Net energy trapped: 36 ATP
FERMENTATION
• Incomplete oxidation
• Waste products: Organic compound (lactic acid or ethanol and CO_2)
• Net energy trapped: 2 ATP

7.1 Energy for Life *(Page 126)*

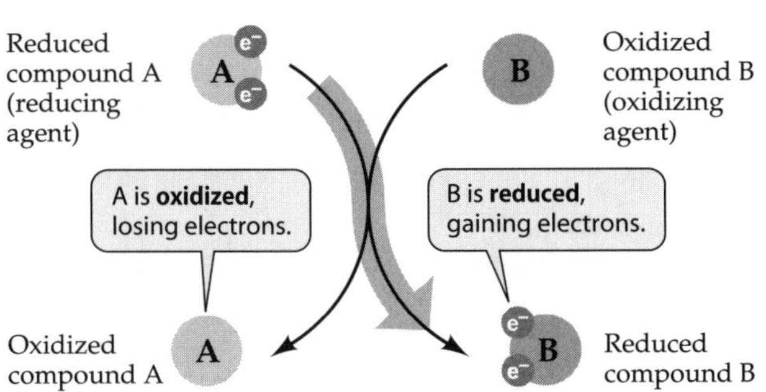

Reduced compound A (reducing agent)
Oxidized compound B (oxidizing agent)
A is **oxidized**, losing electrons.
B is **reduced**, gaining electrons.
Oxidized compound A
Reduced compound B

7.2 Oxidation and Reduction Are Coupled *(Page 127)*

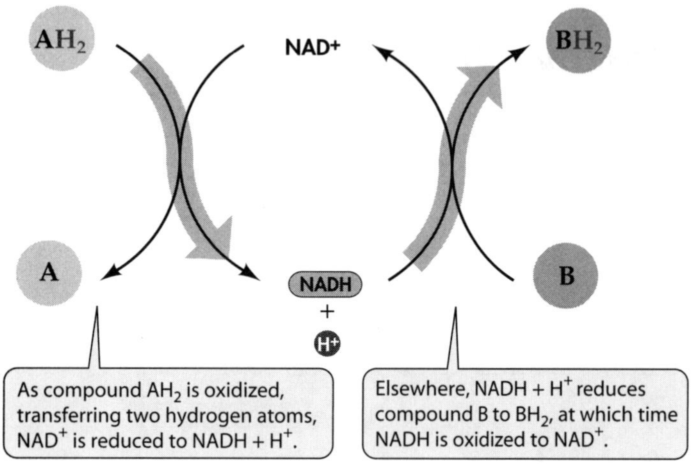
AH2
NAD+
BH2
A
NADH
+
H+
B
As compound AH2 is oxidized, transferring two hydrogen atoms, NAD+ is reduced to NADH + H+.
Elsewhere, NADH + H+ reduces compound B to BH2, at which time NADH is oxidized to NAD+.

7.3 NAD Is an Energy Carrier *(Page 127)*

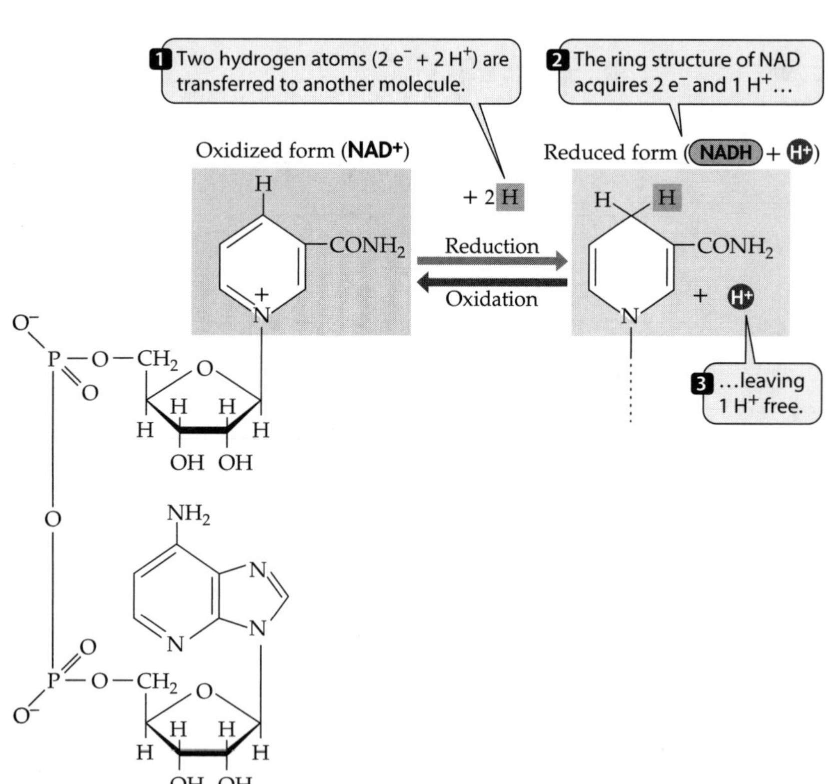
1 Two hydrogen atoms (2 e− + 2 H+) are transferred to another molecule.
2 The ring structure of NAD acquires 2 e− and 1 H+…
Oxidized form (NAD+)
Reduced form (NADH + H+)
+ 2 H
CONH2
Reduction
Oxidation
CONH2
+ H+
3 …leaving 1 H+ free.
NH2
OH OH
OH OH

7.4 Oxidized and Reduced Forms of NAD *(Page 127)*

7.1 *Cellular Locations for Energy Pathways in Eukaryotes and Prokaryotes*

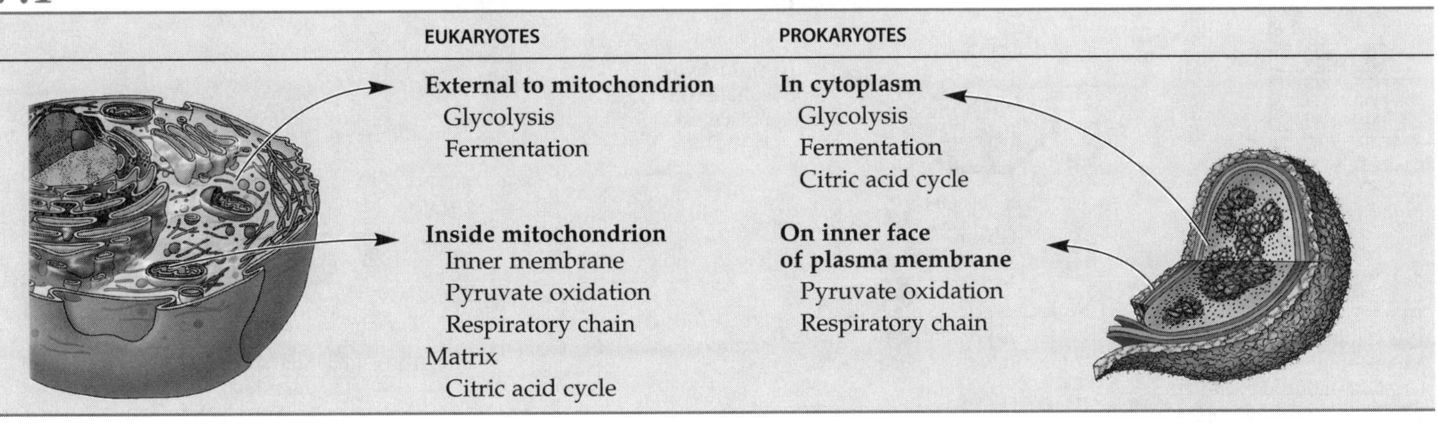

EUKARYOTES	PROKARYOTES
External to mitochondrion Glycolysis Fermentation	**In cytoplasm** Glycolysis Fermentation Citric acid cycle
Inside mitochondrion Inner membrane Pyruvate oxidation Respiratory chain Matrix Citric acid cycle	**On inner face of plasma membrane** Pyruvate oxidation Respiratory chain

(Page 128)

(*a*) **Glycolysis and cellular respiration**

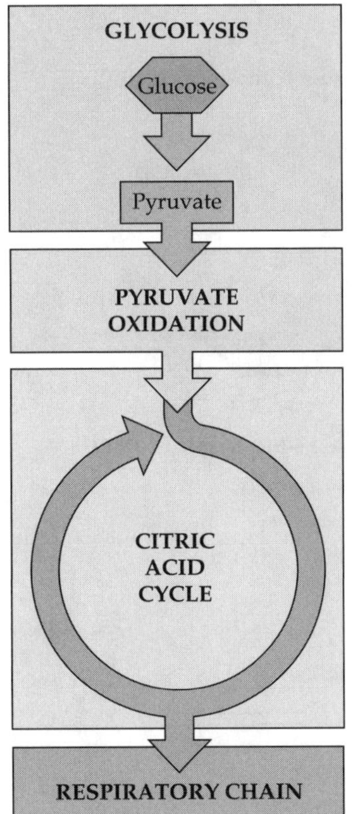
GLYCOLYSIS
Glucose
Pyruvate
PYRUVATE OXIDATION
CITRIC ACID CYCLE
RESPIRATORY CHAIN

(*b*) **Glycolysis and fermentation**

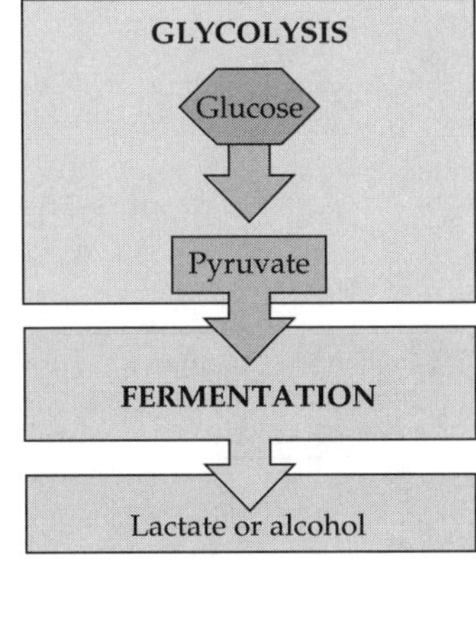
GLYCOLYSIS
Glucose
Pyruvate
FERMENTATION
Lactate or alcohol

7.5 Energy-Producing Metabolic Pathways *(Page 128)*

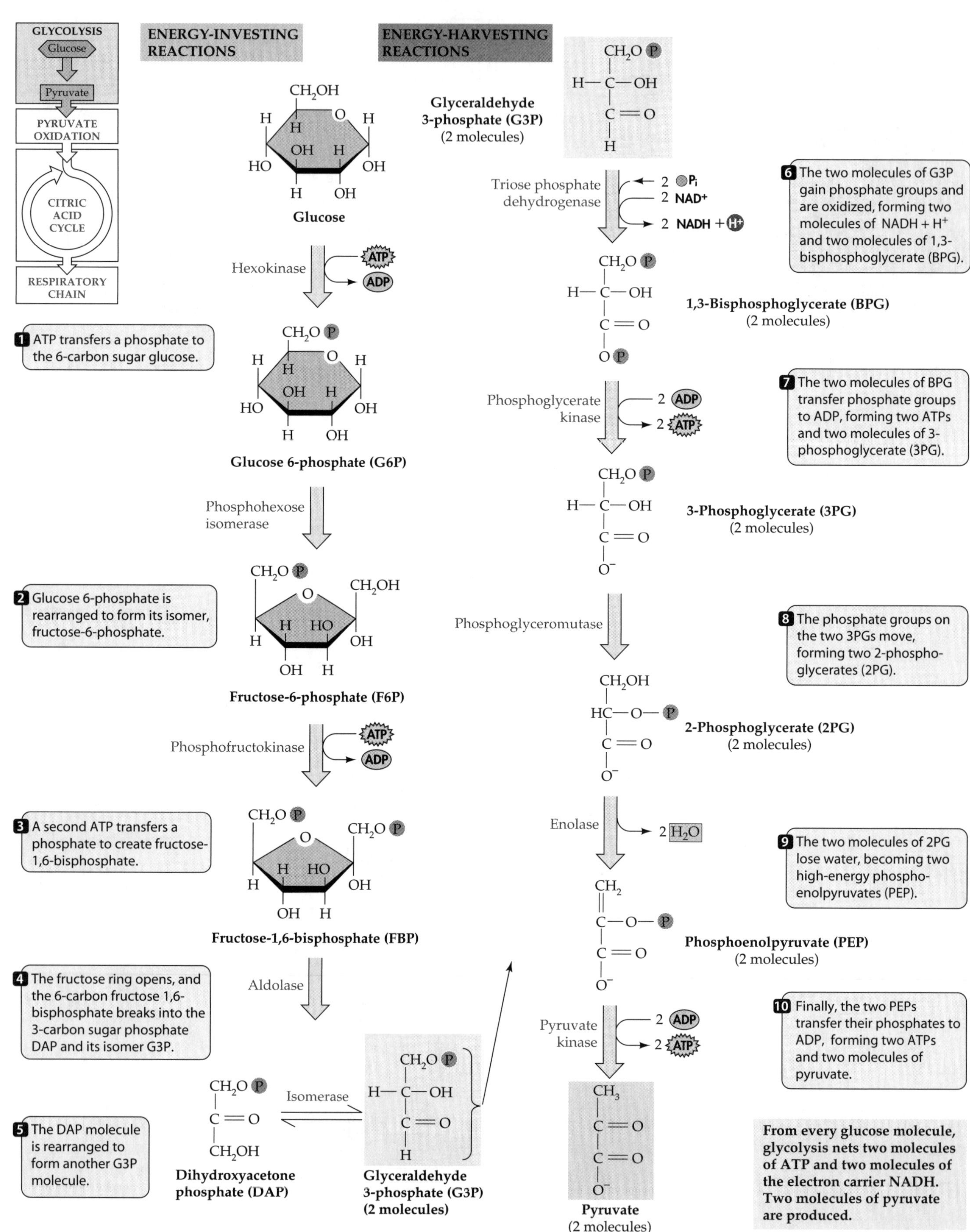

GLYCOLYSIS
Glucose
Pyruvate
PYRUVATE OXIDATION
CITRIC ACID CYCLE
RESPIRATORY CHAIN
ENERGY-INVESTING REACTIONS
ENERGY-HARVESTING REACTIONS
Glucose
Hexokinase
ATP
ADP
1 ATP transfers a phosphate to the 6-carbon sugar glucose.
Glucose 6-phosphate (G6P)
Phosphohexose isomerase
2 Glucose 6-phosphate is rearranged to form its isomer, fructose-6-phosphate.
Fructose-6-phosphate (F6P)
Phosphofructokinase
3 A second ATP transfers a phosphate to create fructose-1,6-bisphosphate.
Fructose-1,6-bisphosphate (FBP)
Aldolase
4 The fructose ring opens, and the 6-carbon fructose 1,6-bisphosphate breaks into the 3-carbon sugar phosphate DAP and its isomer G3P.
Isomerase
5 The DAP molecule is rearranged to form another G3P molecule.
Dihydroxyacetone phosphate (DAP)
Glyceraldehyde 3-phosphate (G3P) (2 molecules)
Glyceraldehyde 3-phosphate (G3P) (2 molecules)
Triose phosphate dehydrogenase
2 Pi
2 NAD+
2 NADH + H+
6 The two molecules of G3P gain phosphate groups and are oxidized, forming two molecules of NADH + H+ and two molecules of 1,3-bisphosphoglycerate (BPG).
1,3-Bisphosphoglycerate (BPG) (2 molecules)
Phosphoglycerate kinase
2 ADP
2 ATP
7 The two molecules of BPG transfer phosphate groups to ADP, forming two ATPs and two molecules of 3-phosphoglycerate (3PG).
3-Phosphoglycerate (3PG) (2 molecules)
Phosphoglyceromutase
8 The phosphate groups on the two 3PGs move, forming two 2-phosphoglycerates (2PG).
2-Phosphoglycerate (2PG) (2 molecules)
Enolase
2 H2O
9 The two molecules of 2PG lose water, becoming two high-energy phosphoenolpyruvates (PEP).
Phosphoenolpyruvate (PEP) (2 molecules)
Pyruvate kinase
2 ADP
2 ATP
10 Finally, the two PEPs transfer their phosphates to ADP, forming two ATPs and two molecules of pyruvate.
Pyruvate (2 molecules)
From every glucose molecule, glycolysis nets two molecules of ATP and two molecules of the electron carrier NADH. Two molecules of pyruvate are produced.

7.6 Glycolysis Converts Glucose to Pyruvate *(Page 129)*

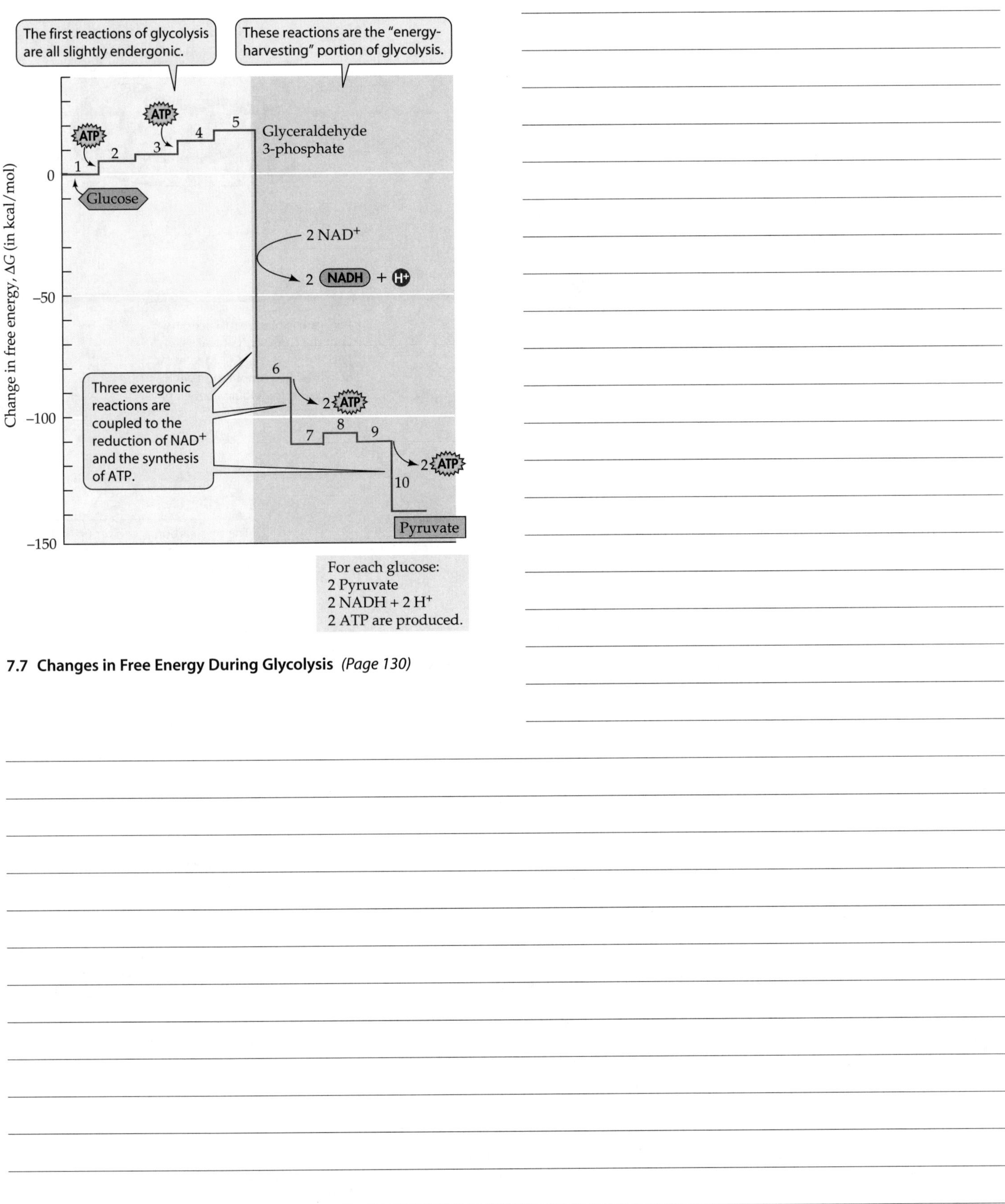
The first reactions of glycolysis are all slightly endergonic.
These reactions are the "energy-harvesting" portion of glycolysis.
ATP
ATP
1
2
3
4
5
Glyceraldehyde 3-phosphate
Glucose
2 NAD+
2 NADH + H+
Change in free energy, ΔG (in kcal/mol)
0
–50
–100
–150
6
2 ATP
7
8
9
2 ATP
10
Three exergonic reactions are coupled to the reduction of NAD+ and the synthesis of ATP.
Pyruvate
For each glucose:
2 Pyruvate
2 NADH + 2 H+
2 ATP are produced.

7.7 Changes in Free Energy During Glycolysis *(Page 130)*

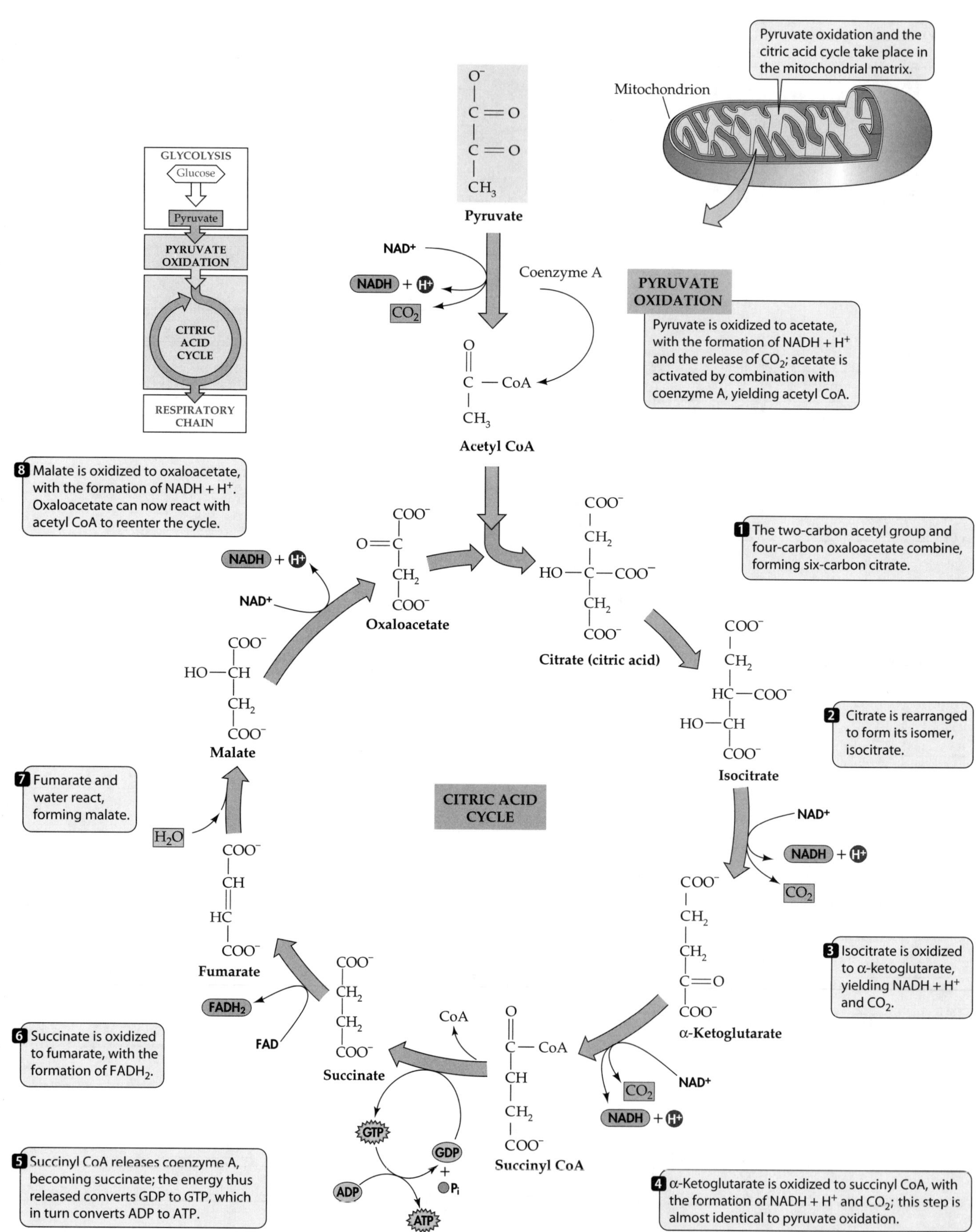

Pyruvate oxidation and the citric acid cycle take place in the mitochondrial matrix.
Mitochondrion
GLYCOLYSIS
Glucose
Pyruvate
PYRUVATE OXIDATION
CITRIC ACID CYCLE
RESPIRATORY CHAIN
Pyruvate
NAD+
NADH + H+
CO2
Coenzyme A
Acetyl CoA
PYRUVATE OXIDATION
Pyruvate is oxidized to acetate, with the formation of NADH + H+ and the release of CO2; acetate is activated by combination with coenzyme A, yielding acetyl CoA.
8 Malate is oxidized to oxaloacetate, with the formation of NADH + H+. Oxaloacetate can now react with acetyl CoA to reenter the cycle.
Oxaloacetate
Citrate (citric acid)
1 The two-carbon acetyl group and four-carbon oxaloacetate combine, forming six-carbon citrate.
Malate
Isocitrate
2 Citrate is rearranged to form its isomer, isocitrate.
7 Fumarate and water react, forming malate.
H2O
CITRIC ACID CYCLE
Fumarate
FADH2
FAD
α-Ketoglutarate
3 Isocitrate is oxidized to α-ketoglutarate, yielding NADH + H+ and CO2.
6 Succinate is oxidized to fumarate, with the formation of FADH2.
Succinate
CoA
Succinyl CoA
GTP
GDP
Pi
ADP
ATP
5 Succinyl CoA releases coenzyme A, becoming succinate; the energy thus released converts GDP to GTP, which in turn converts ADP to ATP.
4 α-Ketoglutarate is oxidized to succinyl CoA, with the formation of NADH + H+ and CO2; this step is almost identical to pyruvate oxidation.

7.8 Pyruvate Oxidation and the Citric Acid Cycle *(Page 132)*

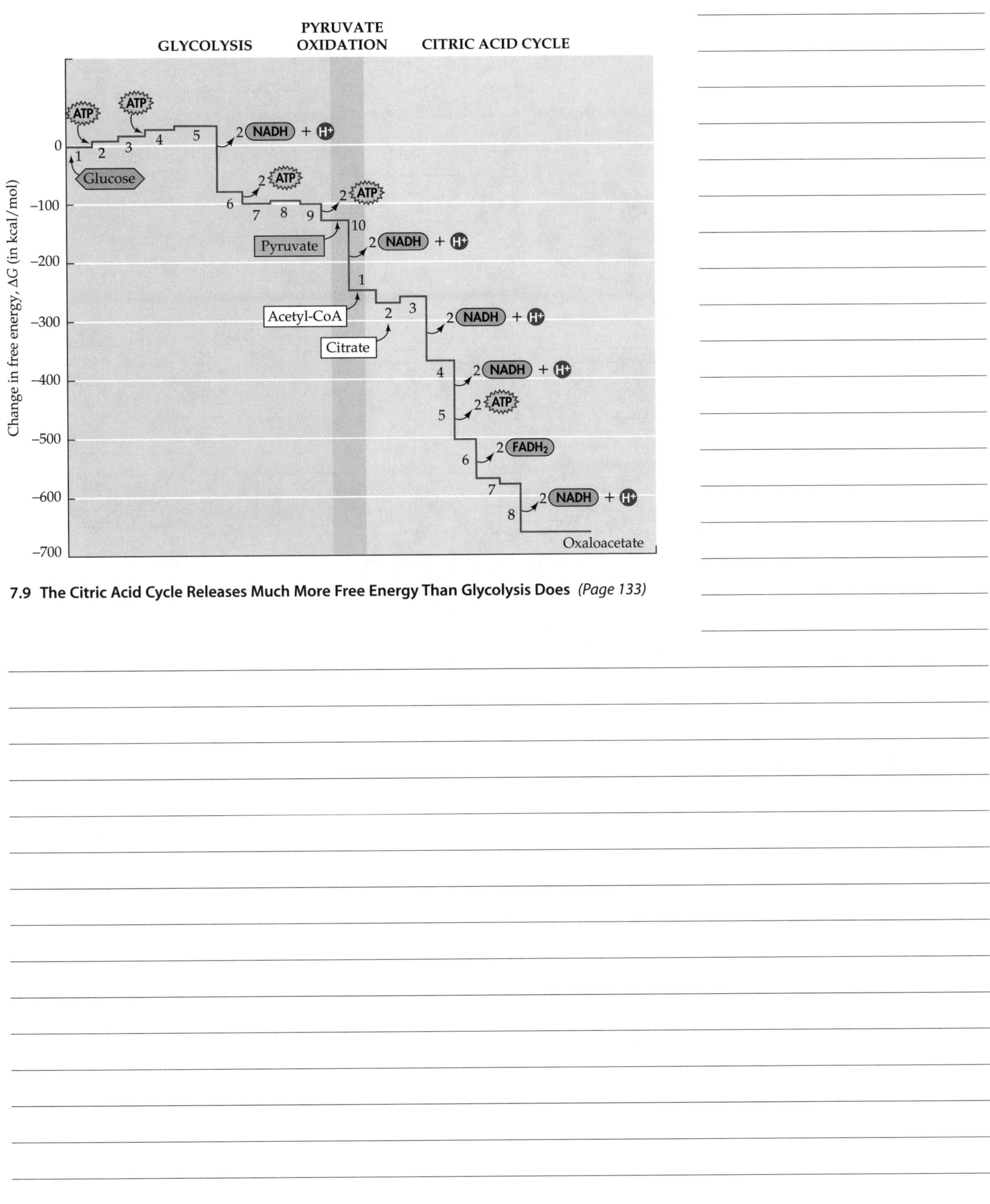
GLYCOLYSIS
PYRUVATE OXIDATION
CITRIC ACID CYCLE
Change in free energy, ΔG (in kcal/mol)
0
–100
–200
–300
–400
–500
–600
–700
ATP
ATP
1
2
3
4
5
Glucose
2 NADH + H+
6
2 ATP
7
8
9
2 ATP
10
Pyruvate
2 NADH + H+
1
Acetyl-CoA
2
3
Citrate
2 NADH + H+
4
2 NADH + H+
5
2 ATP
6
2 FADH2
7
8
2 NADH + H+
Oxaloacetate

7.9 The Citric Acid Cycle Releases Much More Free Energy Than Glycolysis Does *(Page 133)*

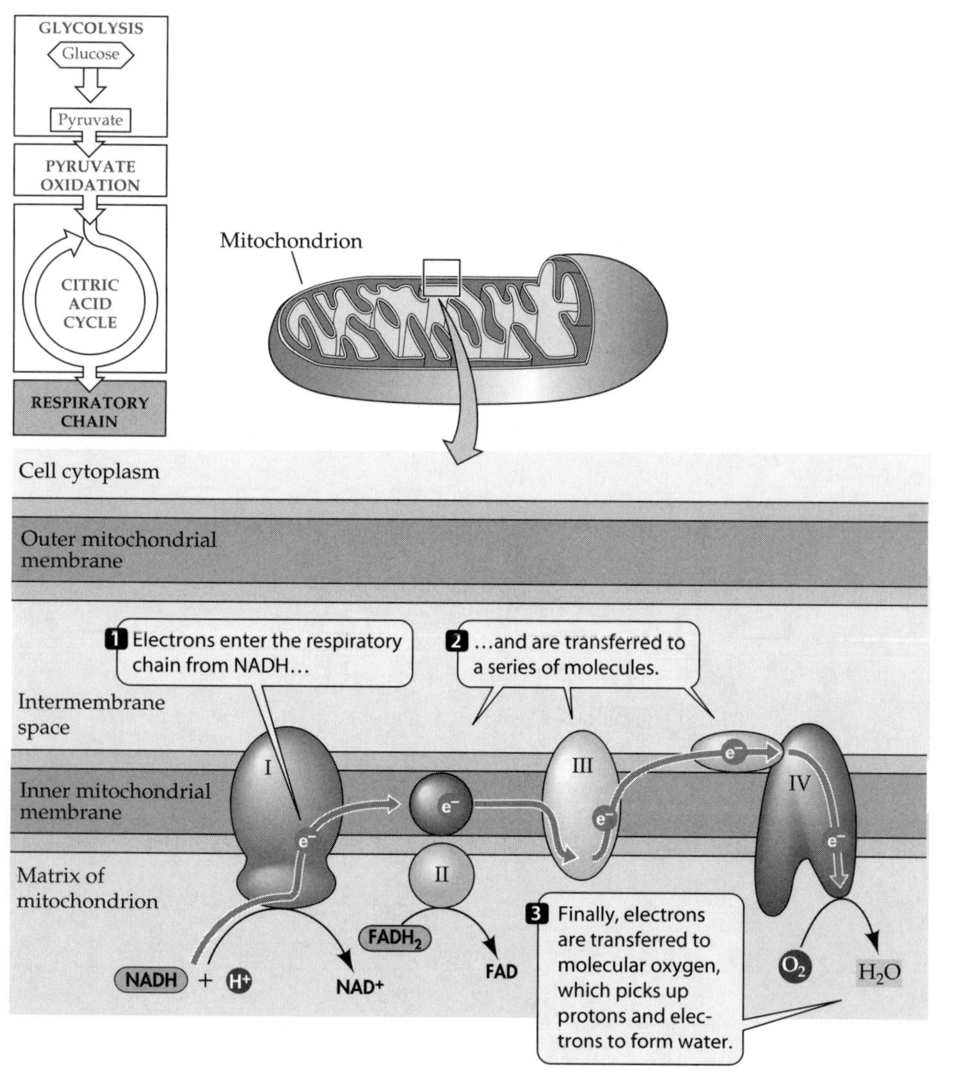
GLYCOLYSIS
Glucose
Pyruvate
PYRUVATE OXIDATION
CITRIC ACID CYCLE
RESPIRATORY CHAIN
Mitochondrion
Cell cytoplasm
Outer mitochondrial membrane
1 Electrons enter the respiratory chain from NADH...
2 ...and are transferred to a series of molecules.
Intermembrane space
Inner mitochondrial membrane
Matrix of mitochondrion
I
II
III
IV
e⁻
FADH2
FAD
NADH + H+
NAD+
3 Finally, electrons are transferred to molecular oxygen, which picks up protons and electrons to form water.
O2
H2O

7.10 The Oxidation of NADH + H$^+$ *(Page 135)*

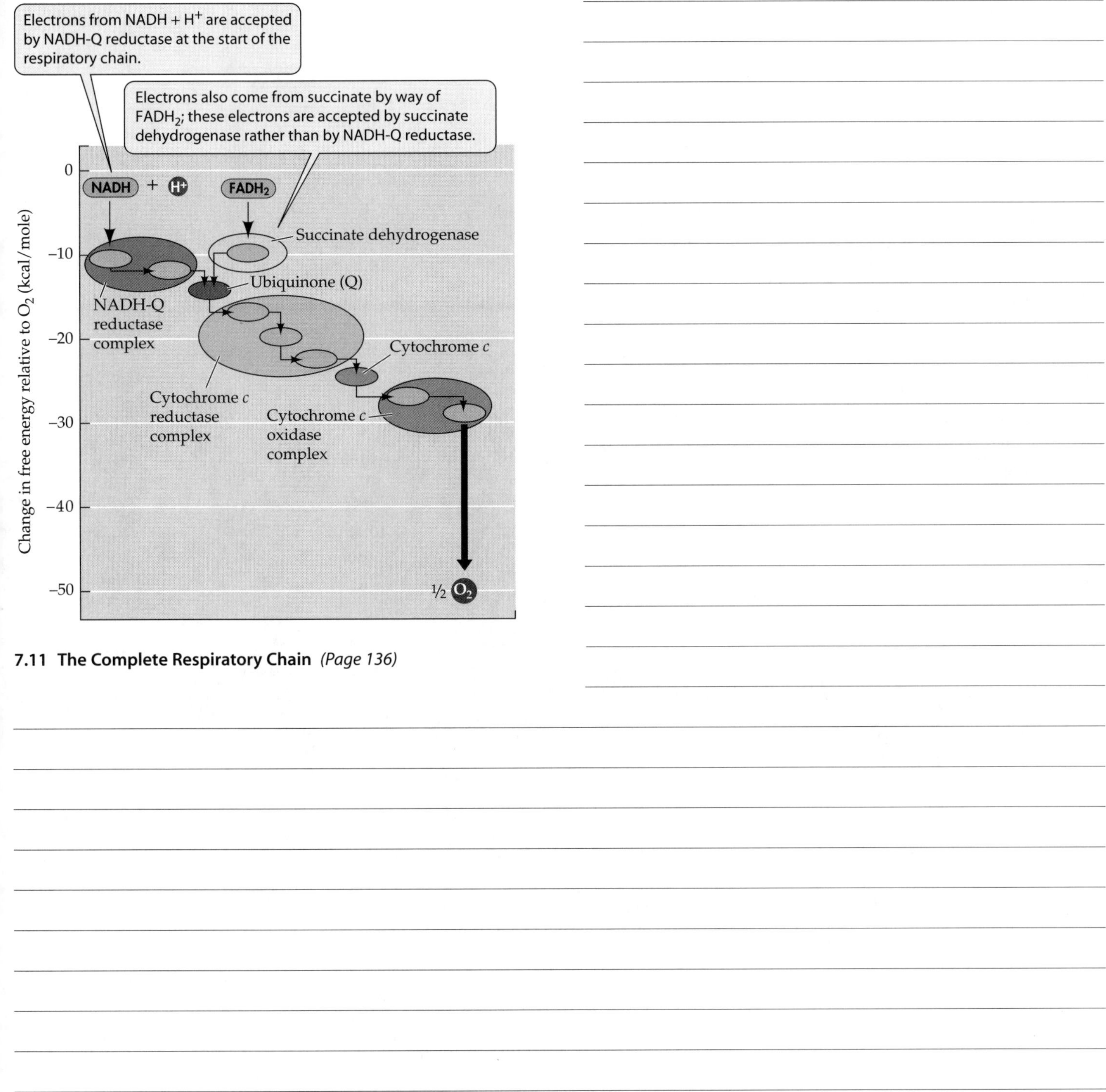
Electrons from NADH + H+ are accepted by NADH-Q reductase at the start of the respiratory chain.
Electrons also come from succinate by way of FADH2; these electrons are accepted by succinate dehydrogenase rather than by NADH-Q reductase.
NADH
+
H+
FADH2
Succinate dehydrogenase
Ubiquinone (Q)
NADH-Q reductase complex
Cytochrome c
Cytochrome c reductase complex
Cytochrome c oxidase complex
½ O2
Change in free energy relative to O2 (kcal/mole)
0
−10
−20
−30
−40
−50

7.11 The Complete Respiratory Chain *(Page 136)*

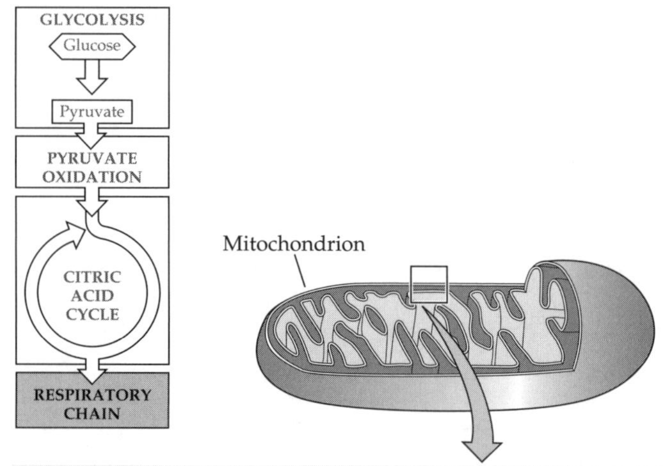
GLYCOLYSIS
Glucose
Pyruvate
PYRUVATE OXIDATION
CITRIC ACID CYCLE
RESPIRATORY CHAIN
Mitochondrion
Cell cytoplasm
Outer mitochondrial membrane
ELECTRON TRANSPORT
ATP SYNTHESIS
Intermembrane space
NADH-Q reductase
Cytochrome c reductase
Ubiquinone
Cytochrome c
Cytochrome c oxidase
ATP synthase
High concentration of H^+
Inner mitochondrial membrane
Mitochondrial matrix
Low concentration of H^+
NADH + H^+
NAD^+
$FADH_2$
FAD
H_2O
2H + ½ O_2
ADP + P_i
ATP

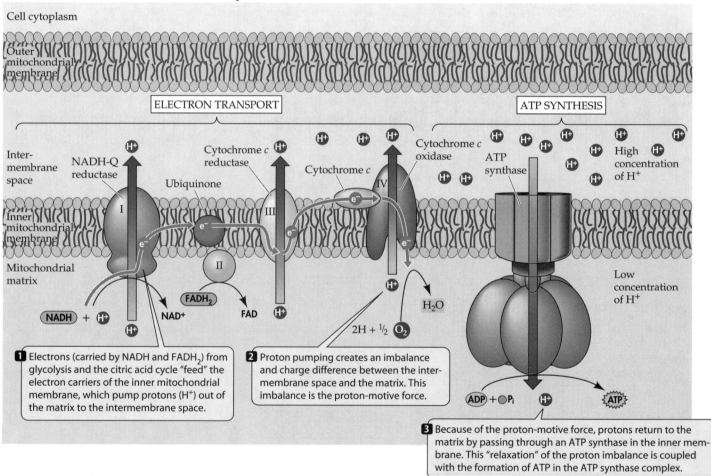

1 Electrons (carried by NADH and $FADH_2$) from glycolysis and the citric acid cycle "feed" the electron carriers of the inner mitochondrial membrane, which pump protons (H^+) out of the matrix to the intermembrane space.

2 Proton pumping creates an imbalance and charge difference between the intermembrane space and the matrix. This imbalance is the proton-motive force.

3 Because of the proton-motive force, protons return to the matrix by passing through an ATP synthase in the inner membrane. This "relaxation" of the proton imbalance is coupled with the formation of ATP in the ATP synthase complex.

7.12 A Chemiosmotic Mechanism Produces ATP *(Page 136)*

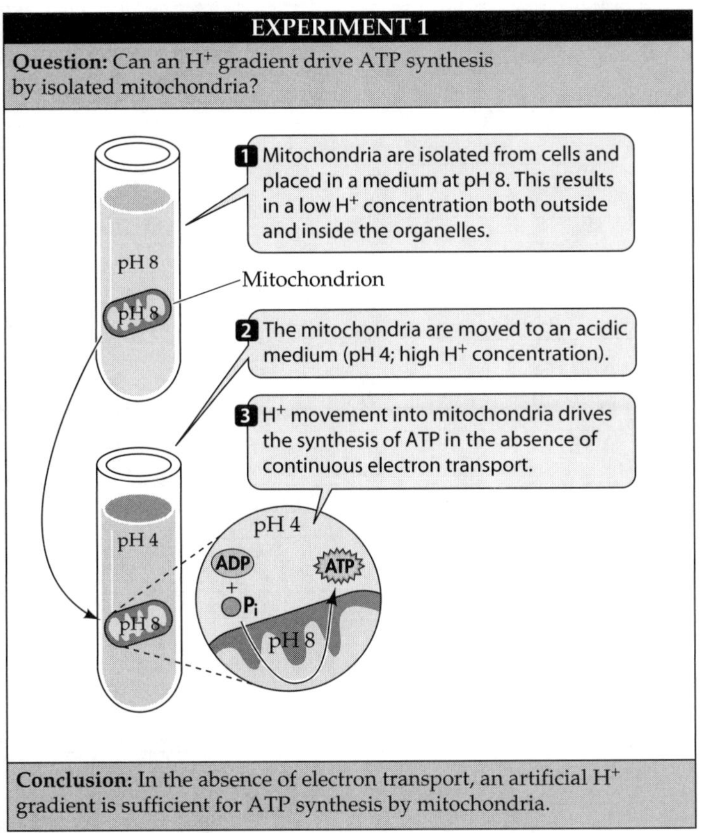
EXPERIMENT 1
Question: Can an H+ gradient drive ATP synthesis by isolated mitochondria?
1 Mitochondria are isolated from cells and placed in a medium at pH 8. This results in a low H+ concentration both outside and inside the organelles.
pH 8
Mitochondrion
pH 8
2 The mitochondria are moved to an acidic medium (pH 4; high H+ concentration).
3 H+ movement into mitochondria drives the synthesis of ATP in the absence of continuous electron transport.
pH 4
pH 4
ADP
+
Pi
ATP
pH 8
pH 8
Conclusion: In the absence of electron transport, an artificial H+ gradient is sufficient for ATP synthesis by mitochondria.

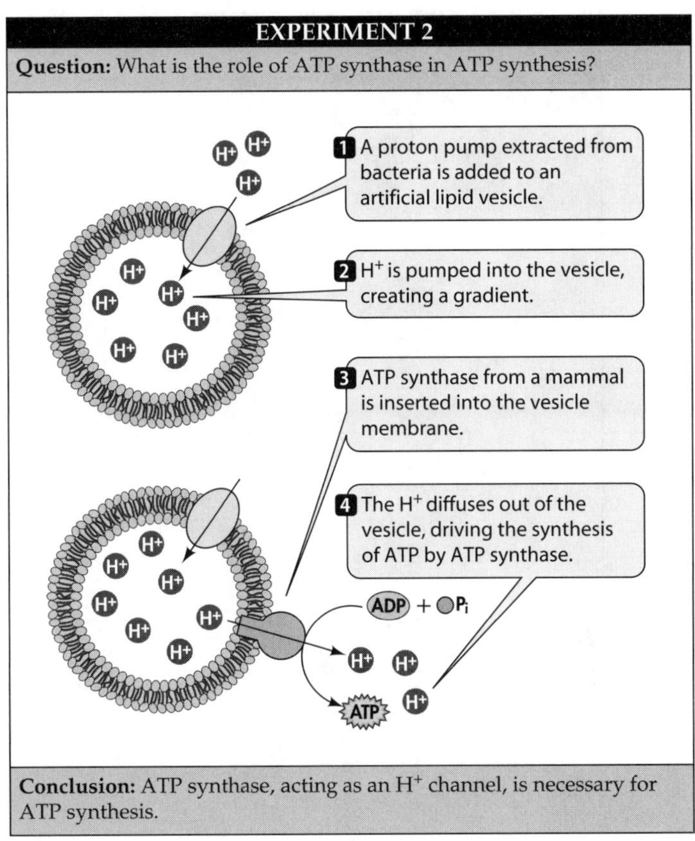
EXPERIMENT 2
Question: What is the role of ATP synthase in ATP synthesis?
1 A proton pump extracted from bacteria is added to an artificial lipid vesicle.
2 H+ is pumped into the vesicle, creating a gradient.
3 ATP synthase from a mammal is inserted into the vesicle membrane.
4 The H+ diffuses out of the vesicle, driving the synthesis of ATP by ATP synthase.
H+
ADP + Pi
ATP
Conclusion: ATP synthase, acting as an H+ channel, is necessary for ATP synthesis.

7.13 Two Experiments Demonstrate the Chemiosmotic Mechanism *(Page 138)*

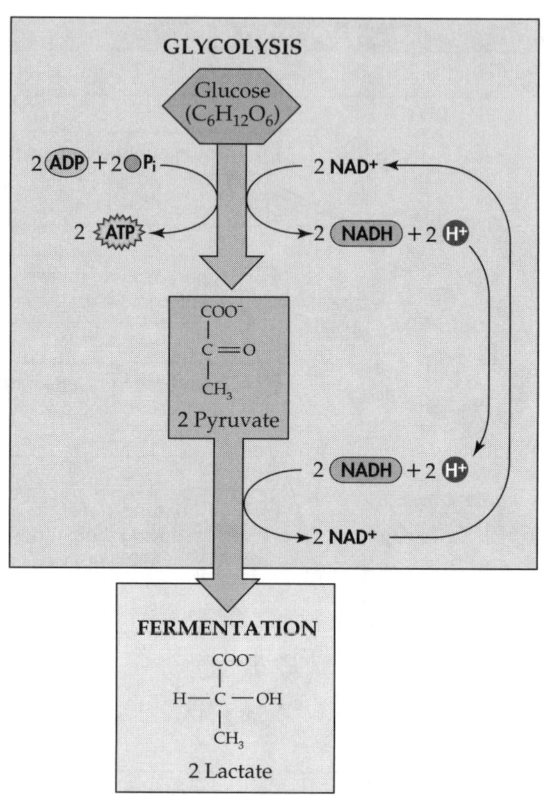
GLYCOLYSIS
Glucose
($C_6H_{12}O_6$)
2 ADP + 2 Pi
2 NAD+
2 ATP
2 NADH + 2 H+
COO^-
C=O
CH_3
2 Pyruvate
2 NADH + 2 H+
2 NAD+
FERMENTATION
COO^-
H—C—OH
CH_3
2 Lactate

7.14 Lactic Acid Fermentation *(Page 139)*

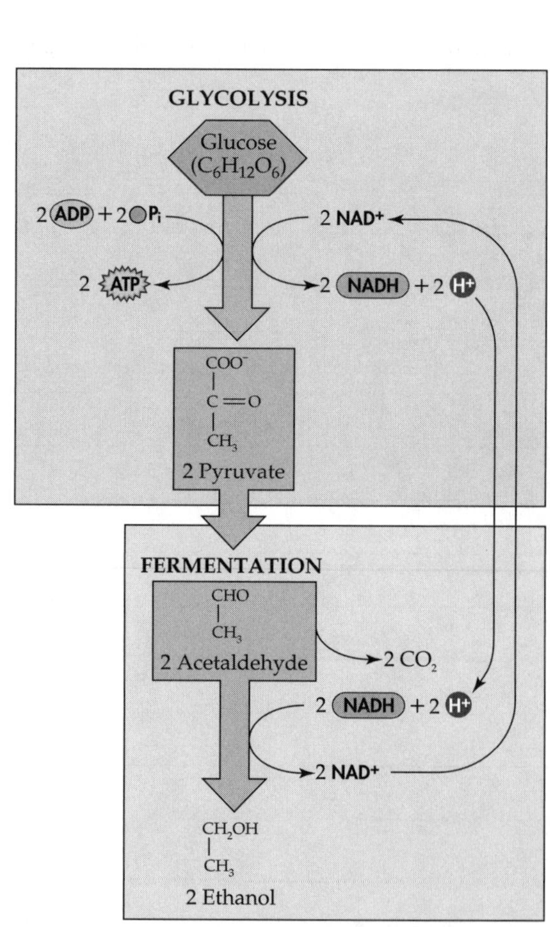
GLYCOLYSIS
Glucose
($C_6H_{12}O_6$)
2 ADP + 2 Pi
2 NAD+
2 ATP
2 NADH + 2 H+
COO^-
C=O
CH_3
2 Pyruvate
FERMENTATION
CHO
CH_3
2 Acetaldehyde
2 CO_2
2 NADH + 2 H+
2 NAD+
CH_2OH
CH_3
2 Ethanol

7.15 Alcoholic Fermentation *(Page 139)*

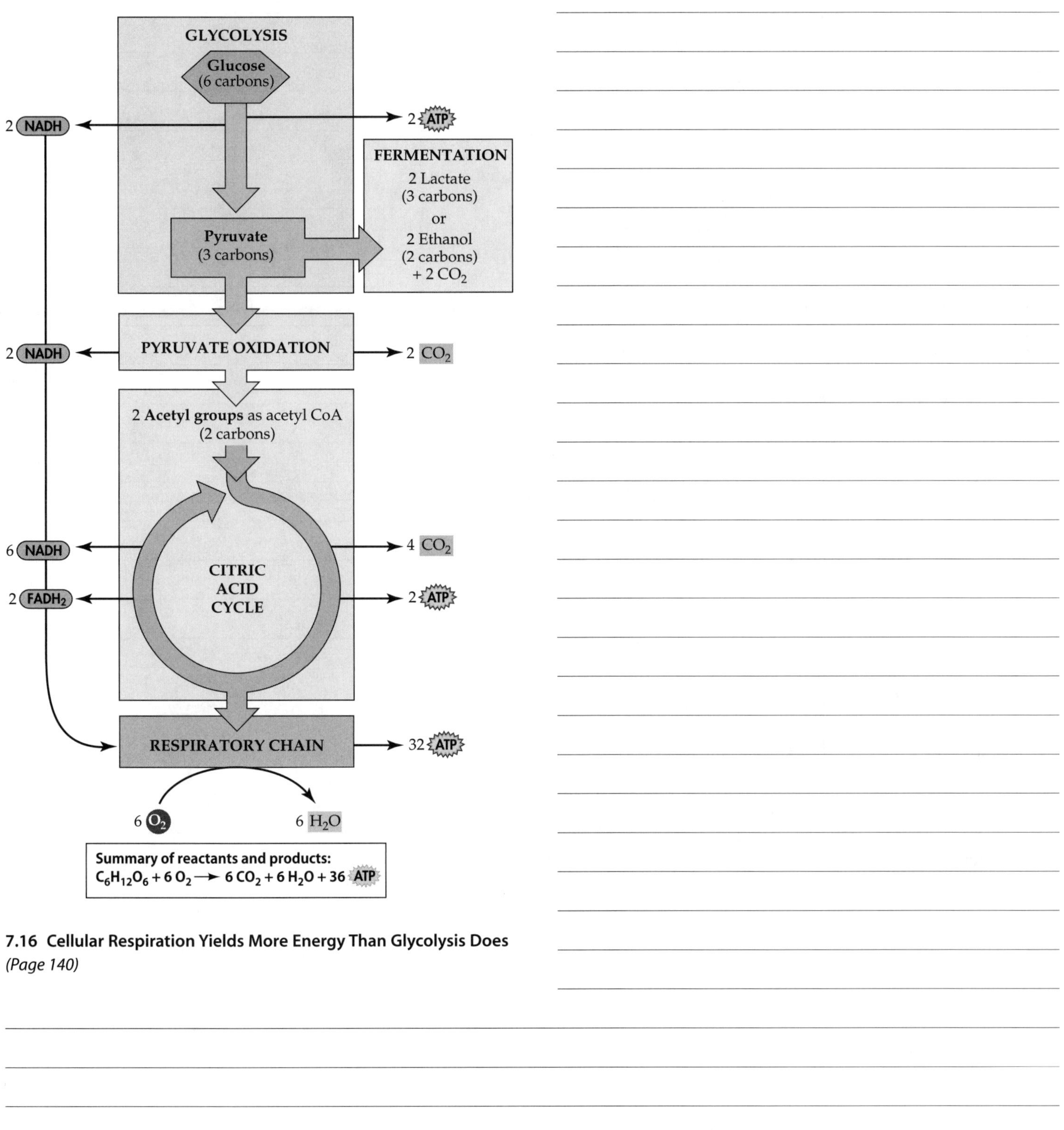

GLYCOLYSIS
Glucose
(6 carbons)
2 NADH
2 ATP
FERMENTATION
2 Lactate
(3 carbons)
or
2 Ethanol
(2 carbons)
+ 2 CO_2
Pyruvate
(3 carbons)
2 NADH
PYRUVATE OXIDATION
2 CO_2
2 Acetyl groups as acetyl CoA
(2 carbons)
6 NADH
4 CO_2
2 $FADH_2$
CITRIC
ACID
CYCLE
2 ATP
RESPIRATORY CHAIN
32 ATP
6 O_2
6 H_2O
Summary of reactants and products:
$C_6H_{12}O_6 + 6\ O_2 \rightarrow 6\ CO_2 + 6\ H_2O + 36$ ATP

7.16 Cellular Respiration Yields More Energy Than Glycolysis Does
(Page 140)

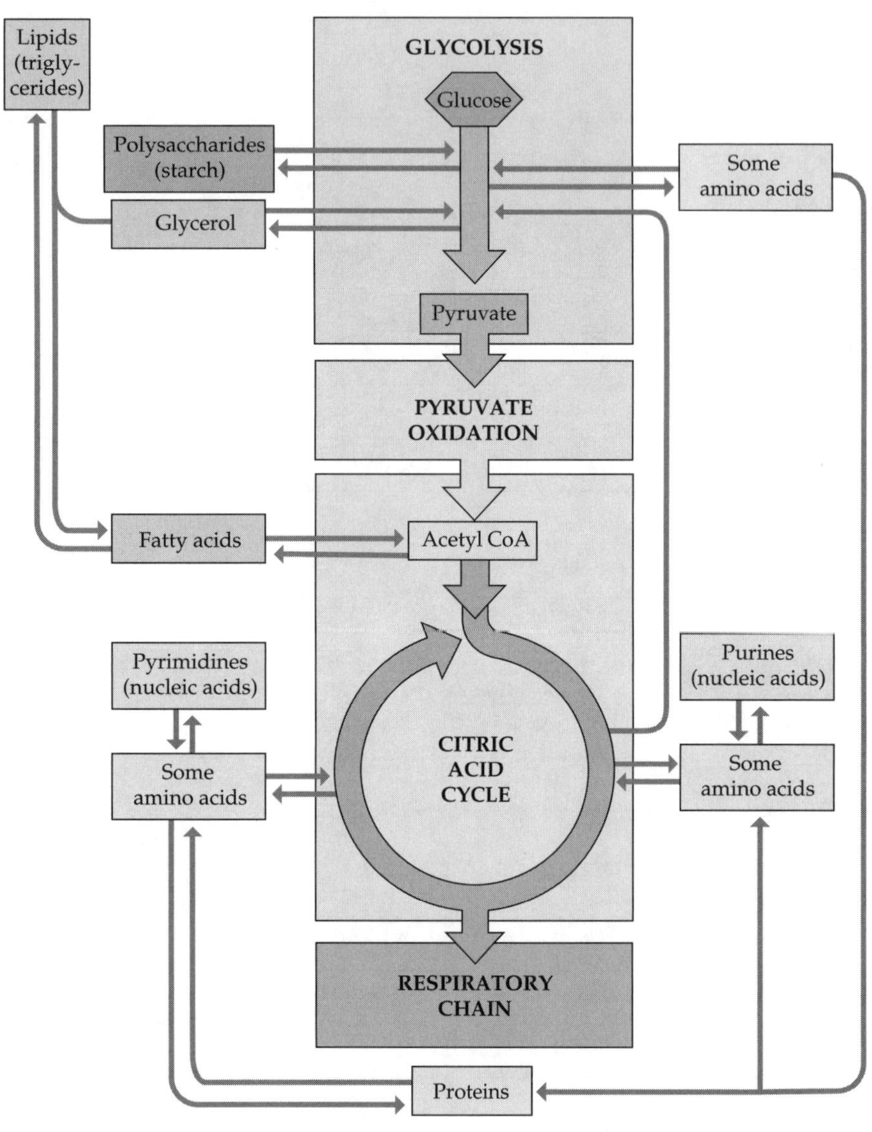
Lipids (trigly-cerides)
GLYCOLYSIS
Glucose
Polysaccharides (starch)
Glycerol
Some amino acids
Pyruvate
PYRUVATE OXIDATION
Fatty acids
Acetyl CoA
Pyrimidines (nucleic acids)
Purines (nucleic acids)
Some amino acids
CITRIC ACID CYCLE
Some amino acids
RESPIRATORY CHAIN
Proteins

7.17 Relationships Among the Major Metabolic Pathways of the Cell *(Page 141)*

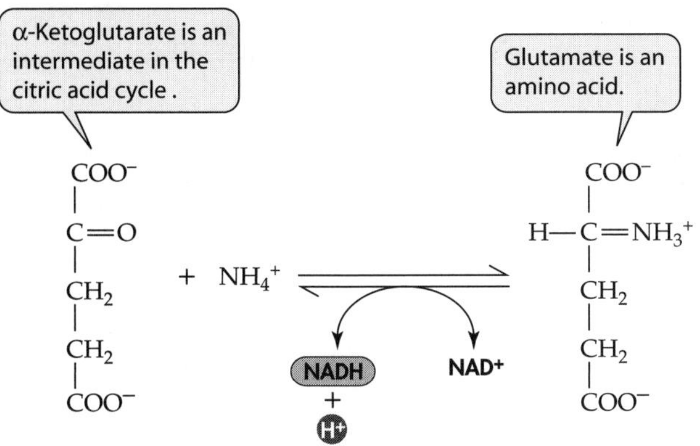
α-Ketoglutarate is an intermediate in the citric acid cycle .
Glutamate is an amino acid.
COO⁻
C=O
CH2
CH2
COO⁻
+ NH4+
NADH
+
H+
NAD+
COO⁻
H—C=NH3+
CH2
CH2
COO⁻

7.18 Coupling Metabolic Pathways *(Page 141)*

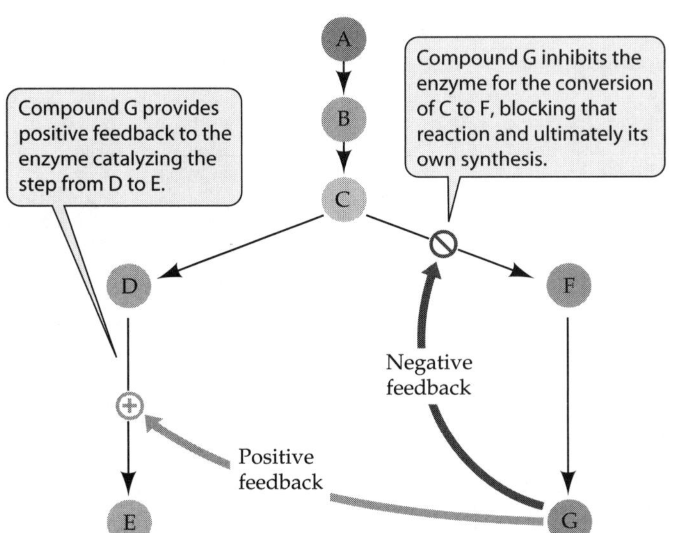
A
B
C
D
E
F
G
Compound G provides positive feedback to the enzyme catalyzing the step from D to E.
Compound G inhibits the enzyme for the conversion of C to F, blocking that reaction and ultimately its own synthesis.
Negative feedback
Positive feedback

7.19 Regulation by Negative and Positive Feedback *(Page 142)*

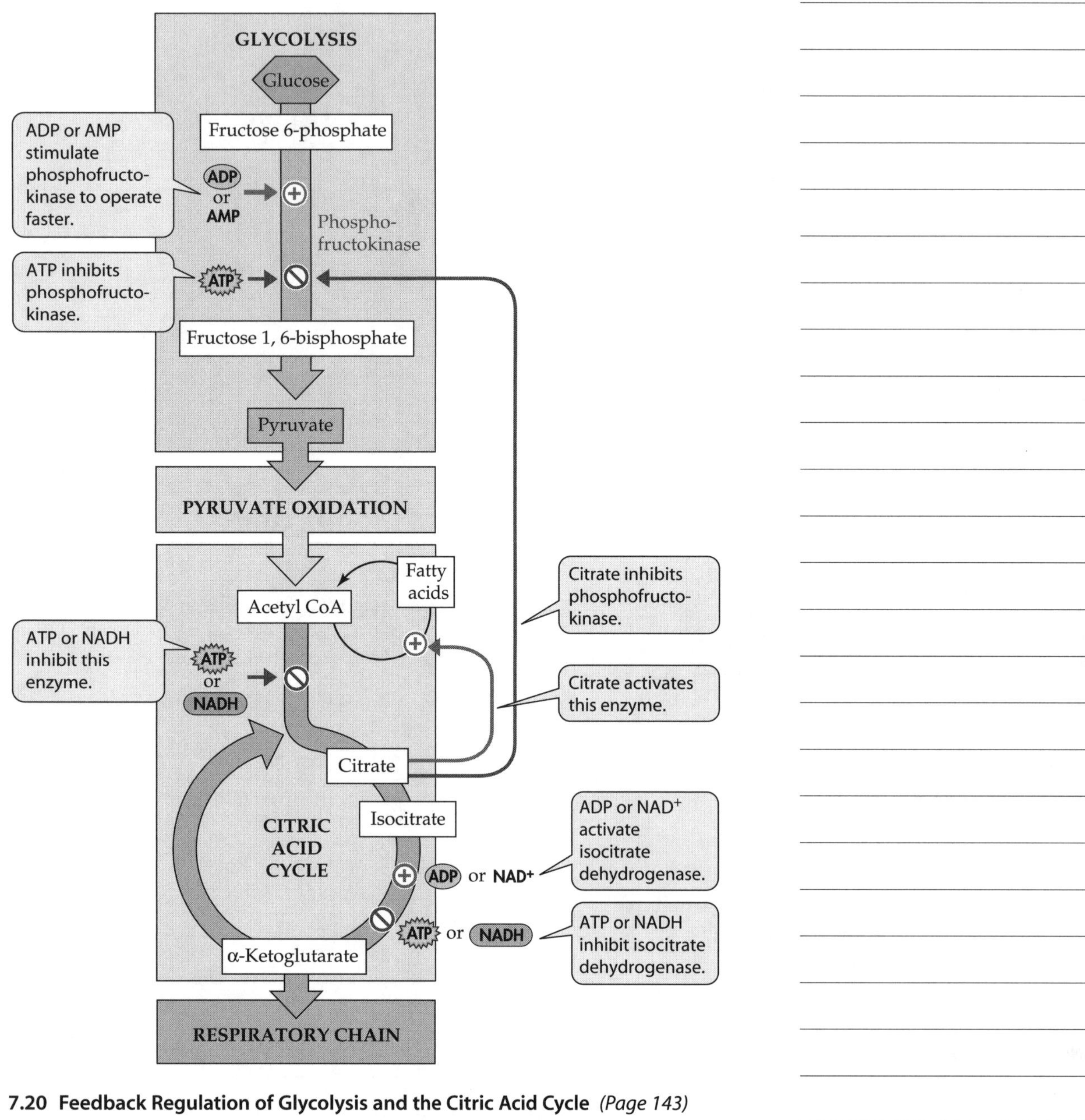
GLYCOLYSIS
Glucose
Fructose 6-phosphate
ADP or AMP stimulate phosphofructo-kinase to operate faster.
ADP or AMP
Phospho-fructokinase
ATP inhibits phosphofructo-kinase.
ATP
Fructose 1, 6-bisphosphate
Pyruvate
PYRUVATE OXIDATION
Fatty acids
Acetyl CoA
Citrate inhibits phosphofructo-kinase.
ATP or NADH inhibit this enzyme.
ATP or NADH
Citrate activates this enzyme.
Citrate
Isocitrate
CITRIC ACID CYCLE
ADP or NAD+ activate isocitrate dehydrogenase.
ADP or NAD+
ATP or NADH
ATP or NADH inhibit isocitrate dehydrogenase.
α-Ketoglutarate
RESPIRATORY CHAIN

7.20 Feedback Regulation of Glycolysis and the Citric Acid Cycle *(Page 143)*

8 Photosynthesis: Energy from the Sun

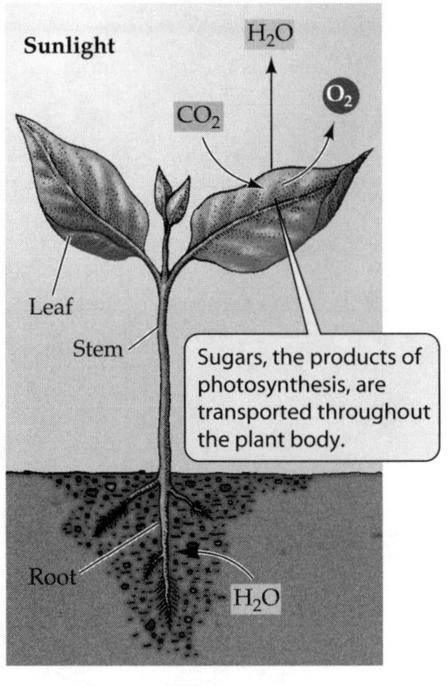

Sunlight
H_2O
CO_2
O_2
Leaf
Stem
Sugars, the products of photosynthesis, are transported throughout the plant body.
Root
H_2O

8.1 The Ingredients for Photosynthesis *(Page 146)*

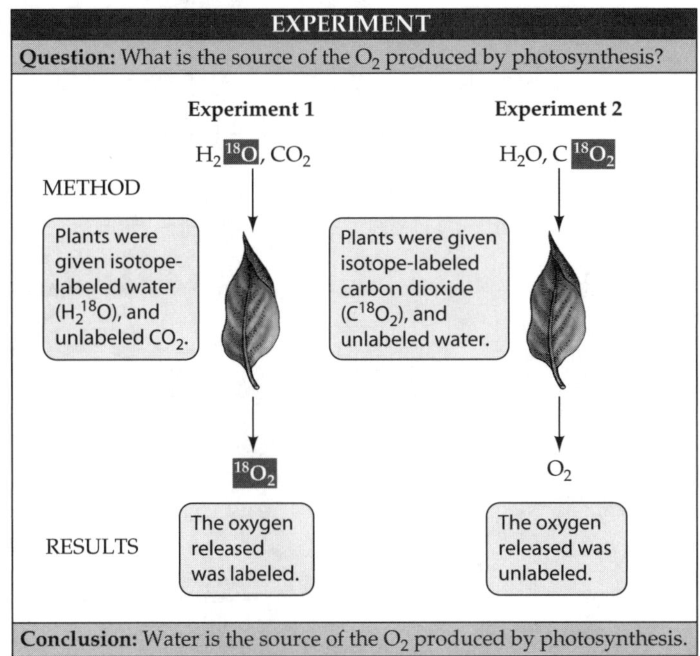

EXPERIMENT
Question: What is the source of the O_2 produced by photosynthesis?
Experiment 1
Experiment 2
$H_2{}^{18}O$, CO_2
H_2O, $C^{18}O_2$
METHOD
Plants were given isotope-labeled water ($H_2{}^{18}O$), and unlabeled CO_2.
Plants were given isotope-labeled carbon dioxide ($C^{18}O_2$), and unlabeled water.
$^{18}O_2$
O_2
RESULTS
The oxygen released was labeled.
The oxygen released was unlabeled.
Conclusion: Water is the source of the O_2 produced by photosynthesis.

8.2 Water Is the Source of the Oxygen Produced by Photosynthesis *(Page 146)*

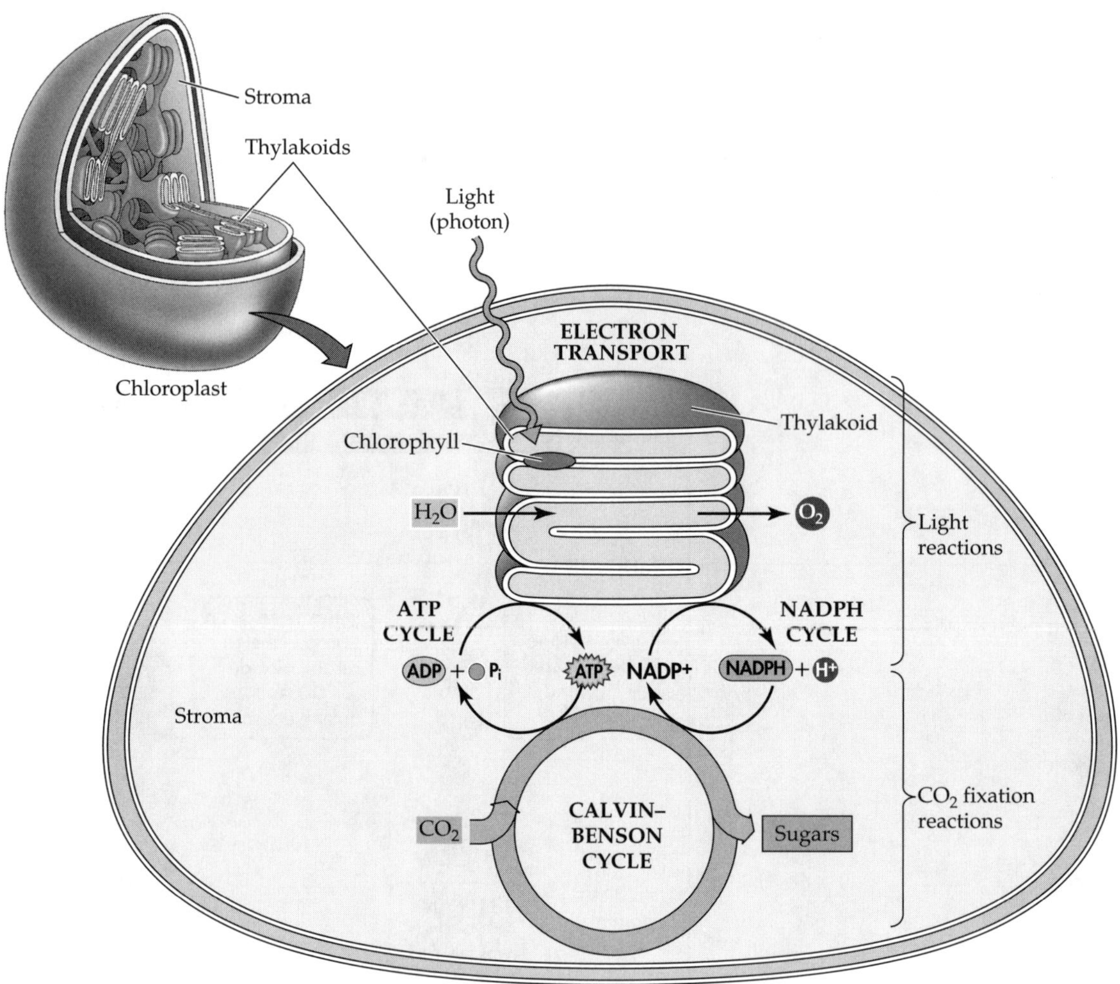
Stroma
Thylakoids
Chloroplast
Light
(photon)
ELECTRON
TRANSPORT
Thylakoid
Chlorophyll
H_2O
O_2
Light
reactions
ATP
CYCLE
NADPH
CYCLE
ADP + P_i
ATP
$NADP^+$
NADPH + H^+
Stroma
CO_2
CALVIN–
BENSON
CYCLE
Sugars
CO_2 fixation
reactions

8.3 An Overview of Photosynthesis *(Page 147)*

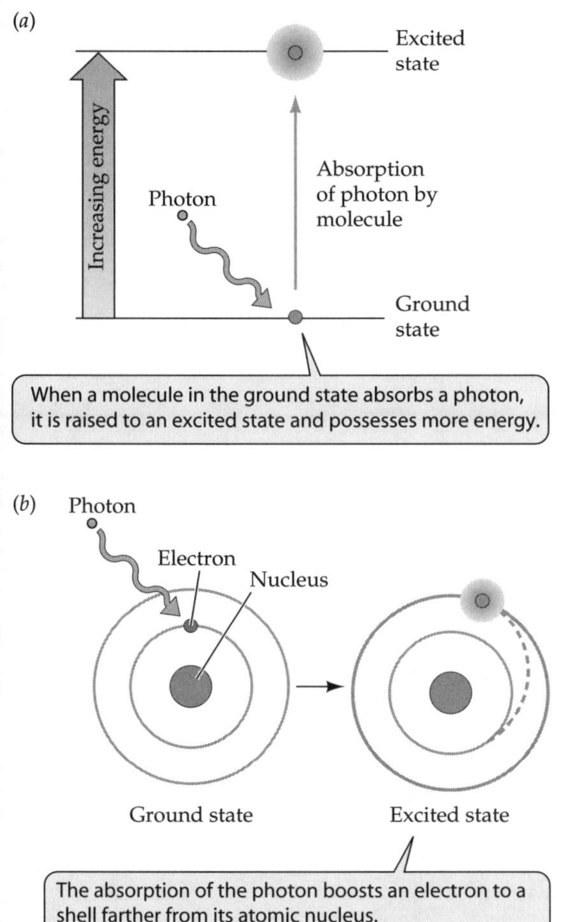
(a)
Excited state
Increasing energy
Photon
Absorption of photon by molecule
Ground state
When a molecule in the ground state absorbs a photon, it is raised to an excited state and possesses more energy.
(b) Photon
Electron
Nucleus
Ground state
Excited state
The absorption of the photon boosts an electron to a shell farther from its atomic nucleus.

8.4 Exciting a Molecule *(Page 148)*

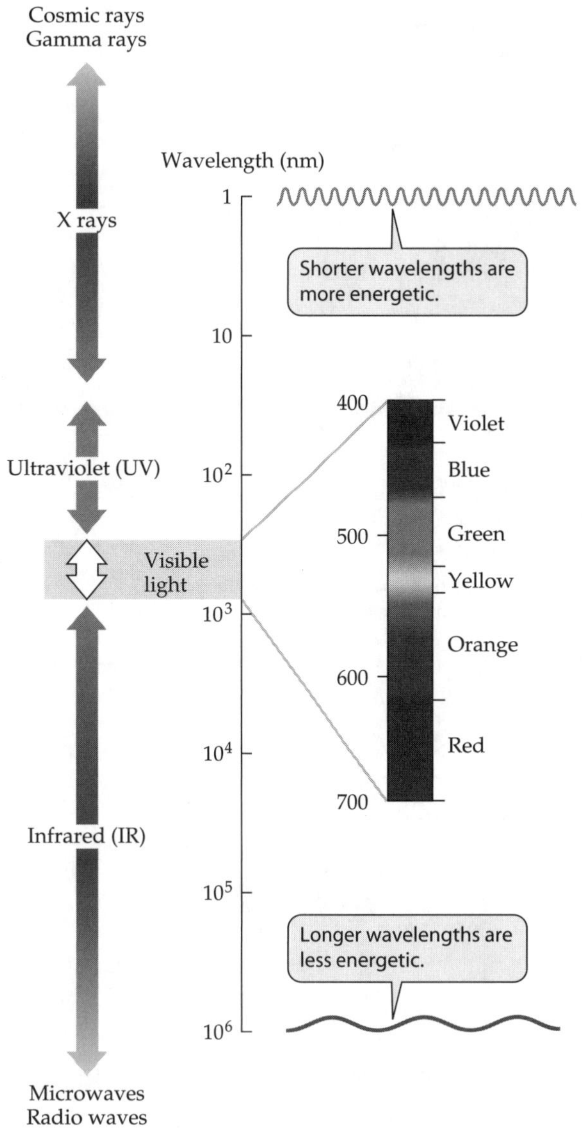

Cosmic rays
Gamma rays
X rays
Ultraviolet (UV)
Visible light
Infrared (IR)
Microwaves
Radio waves
Wavelength (nm)
1
10
10^2
10^3
10^4
10^5
10^6
Shorter wavelengths are more energetic.
Longer wavelengths are less energetic.
400
500
600
700
Violet
Blue
Green
Yellow
Orange
Red

8.5 The Electromagnetic Spectrum *(Page 148)*

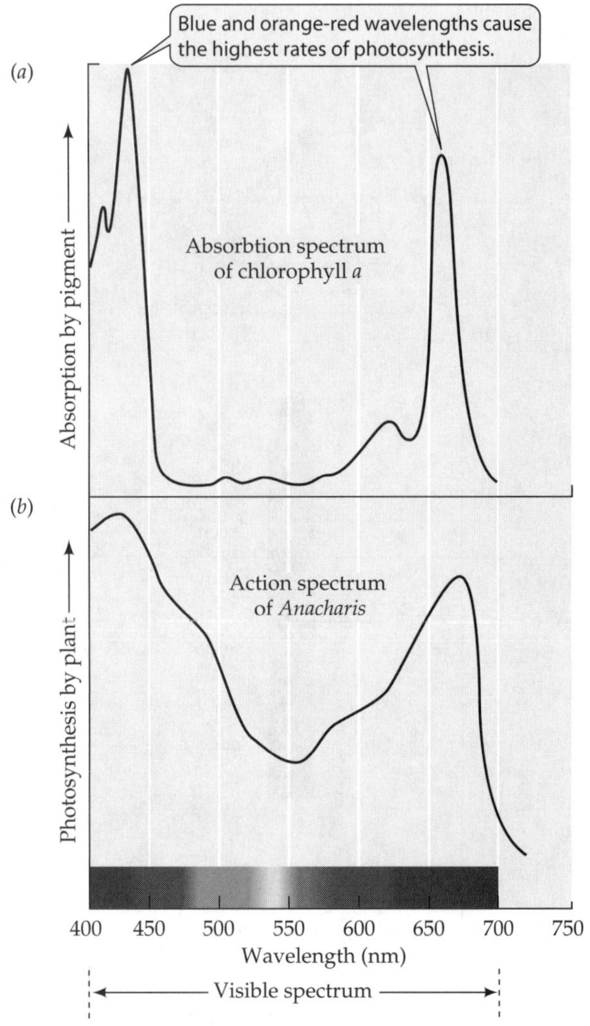
Blue and orange-red wavelengths cause the highest rates of photosynthesis.
(a)
Absorption by pigment
Absorbtion spectrum of chlorophyll a
(b)
Photosynthesis by plant
Action spectrum of Anacharis
400 450 500 550 600 650 700 750
Wavelength (nm)
Visible spectrum

8.6 Absorption and Action Spectra *(Page 149)*

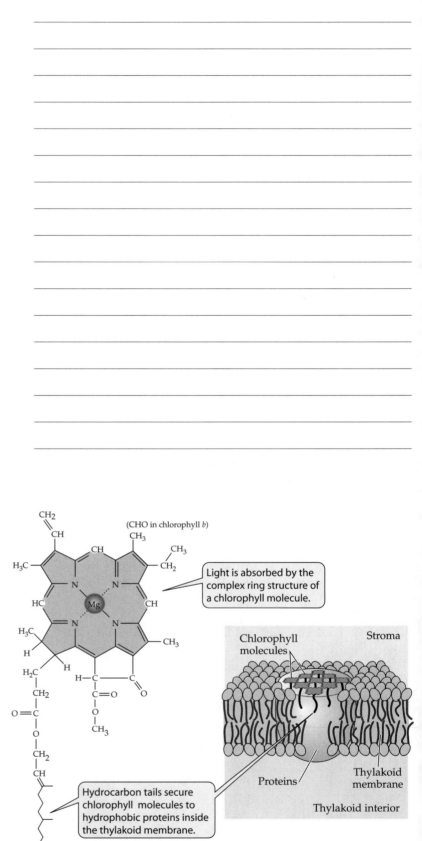

(CHO in chlorophyll *b*)

Light is absorbed by the complex ring structure of a chlorophyll molecule.

Chlorophyll molecules

Stroma

Proteins

Thylakoid membrane

Thylakoid interior

Hydrocarbon tails secure chlorophyll molecules to hydrophobic proteins inside the thylakoid membrane.

8.7 The Molecular Structure of Chlorophyll *(Page 150)*

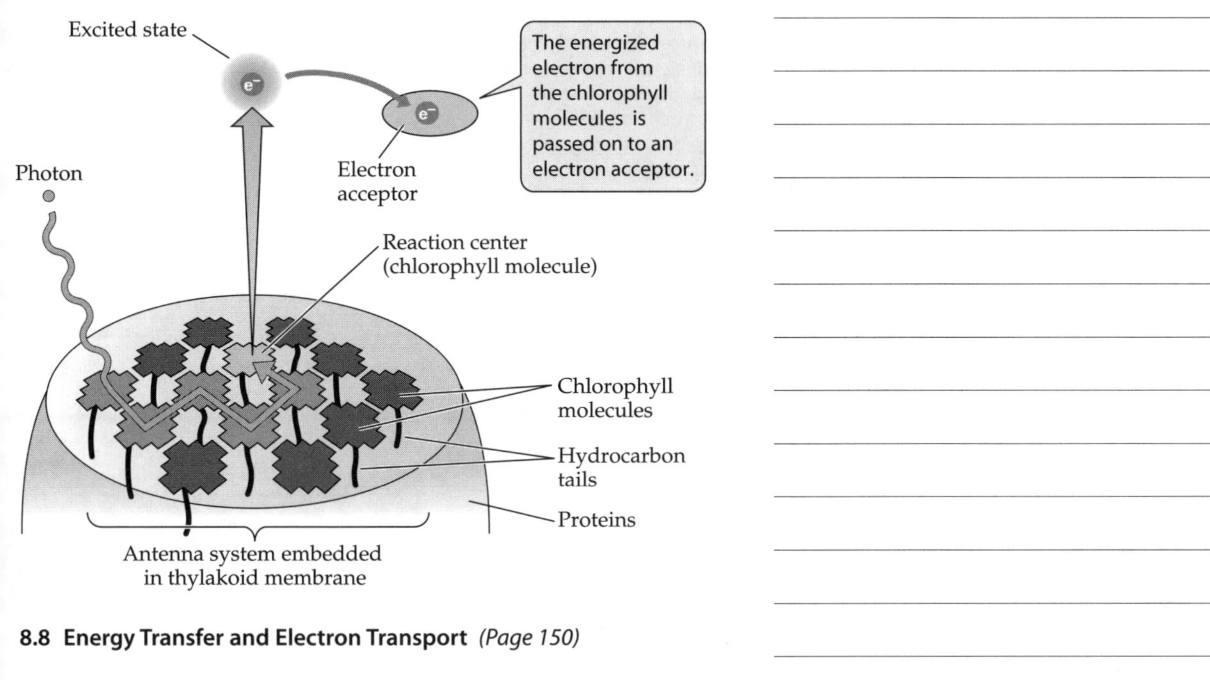
Excited state
The energized electron from the chlorophyll molecules is passed on to an electron acceptor.
Photon
Electron acceptor
Reaction center (chlorophyll molecule)
Chlorophyll molecules
Hydrocarbon tails
Proteins
Antenna system embedded in thylakoid membrane

8.8 Energy Transfer and Electron Transport *(Page 150)*

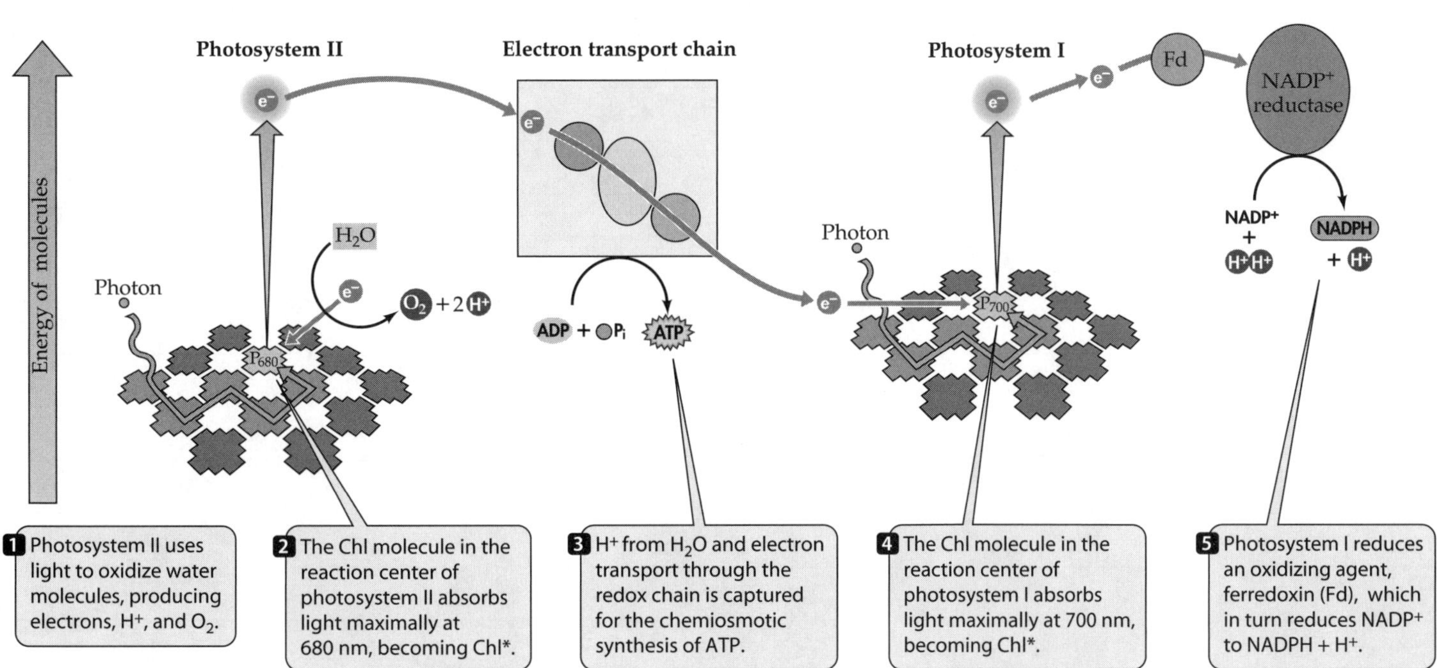
Energy of molecules
Photosystem II
Electron transport chain
Photosystem I
Fd
NADP+ reductase
Photon
H2O
O2 + 2H+
P680
ADP + Pi
ATP
P700
NADP+ + H+ H+
NADPH + H+
1 Photosystem II uses light to oxidize water molecules, producing electrons, H+, and O2.
2 The Chl molecule in the reaction center of photosystem II absorbs light maximally at 680 nm, becoming Chl*.
3 H+ from H2O and electron transport through the redox chain is captured for the chemiosmotic synthesis of ATP.
4 The Chl molecule in the reaction center of photosystem I absorbs light maximally at 700 nm, becoming Chl*.
5 Photosystem I reduces an oxidizing agent, ferredoxin (Fd), which in turn reduces NADP+ to NADPH + H+.

8.9 Noncyclic Electron Transport Uses Two Photosystems *(Page 151)*

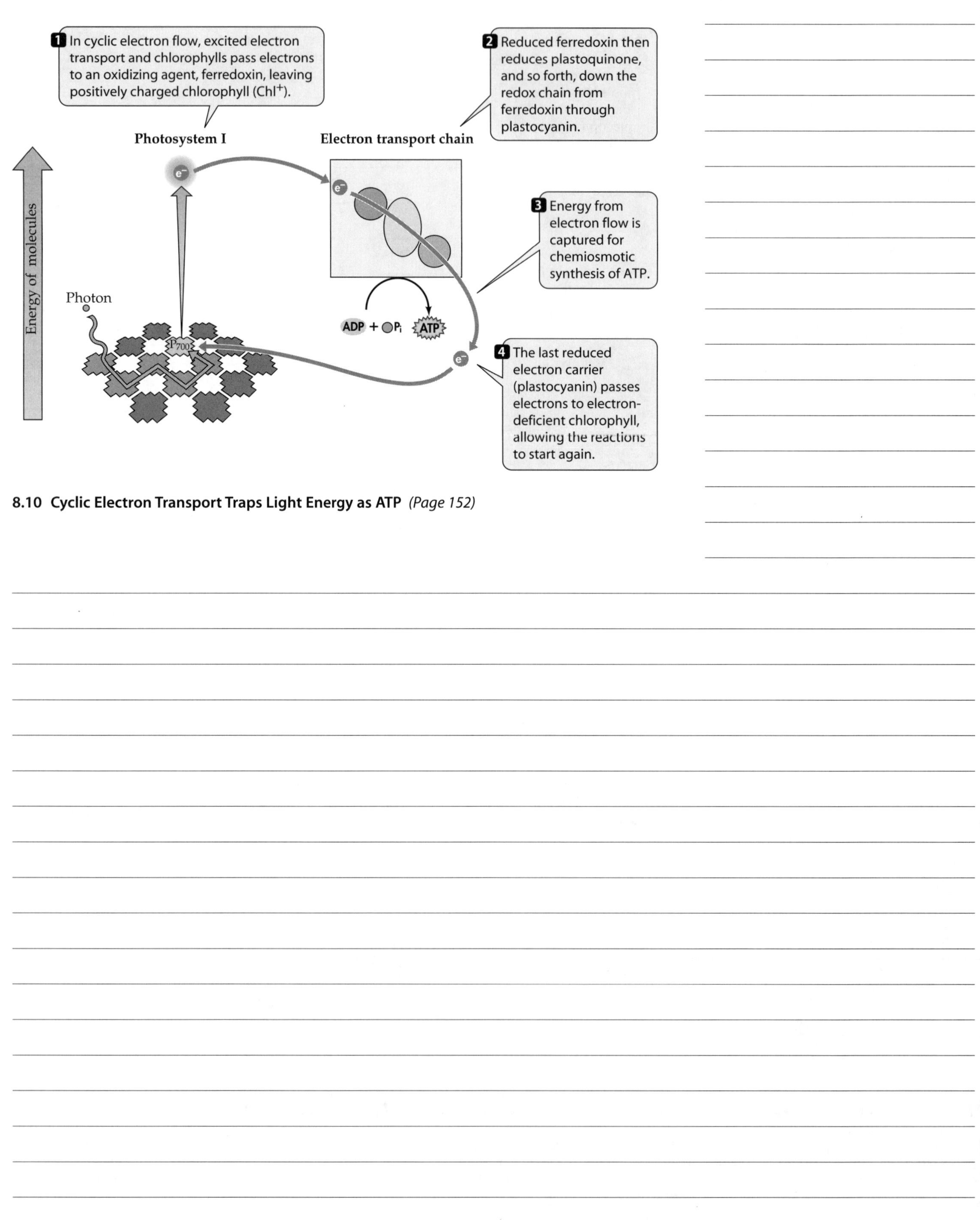
1 In cyclic electron flow, excited electron transport and chlorophylls pass electrons to an oxidizing agent, ferredoxin, leaving positively charged chlorophyll (Chl+).
2 Reduced ferredoxin then reduces plastoquinone, and so forth, down the redox chain from ferredoxin through plastocyanin.
Photosystem I
Electron transport chain
Energy of molecules
e−
e−
3 Energy from electron flow is captured for chemiosmotic synthesis of ATP.
Photon
ADP + Pi
ATP
P700
e−
4 The last reduced electron carrier (plastocyanin) passes electrons to electron-deficient chlorophyll, allowing the reactions to start again.

8.10 Cyclic Electron Transport Traps Light Energy as ATP *(Page 152)*

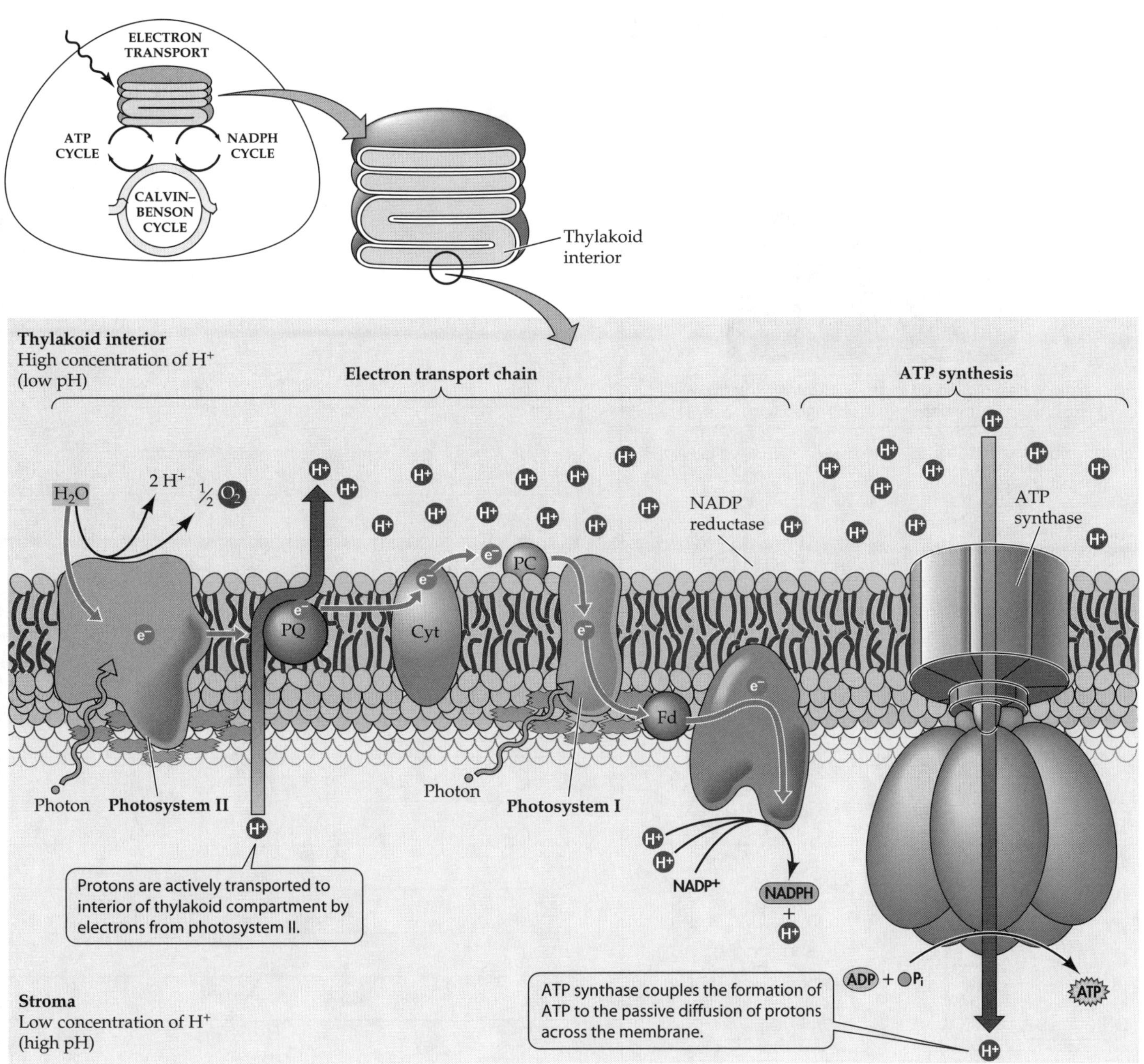
ELECTRON TRANSPORT
ATP CYCLE
NADPH CYCLE
CALVIN–BENSON CYCLE
Thylakoid interior
Thylakoid interior
High concentration of H+
(low pH)
Electron transport chain
ATP synthesis
H2O
2 H+
½ O2
NADP reductase
ATP synthase
PQ
Cyt
PC
Fd
Photon
Photosystem II
Photon
Photosystem I
NADP+
NADPH + H+
Protons are actively transported to interior of thylakoid compartment by electrons from photosystem II.
ATP synthase couples the formation of ATP to the passive diffusion of protons across the membrane.
ADP + Pi
ATP
Stroma
Low concentration of H+
(high pH)

8.11 Chloroplasts Form ATP Chemiosmotically *(Page 153)*

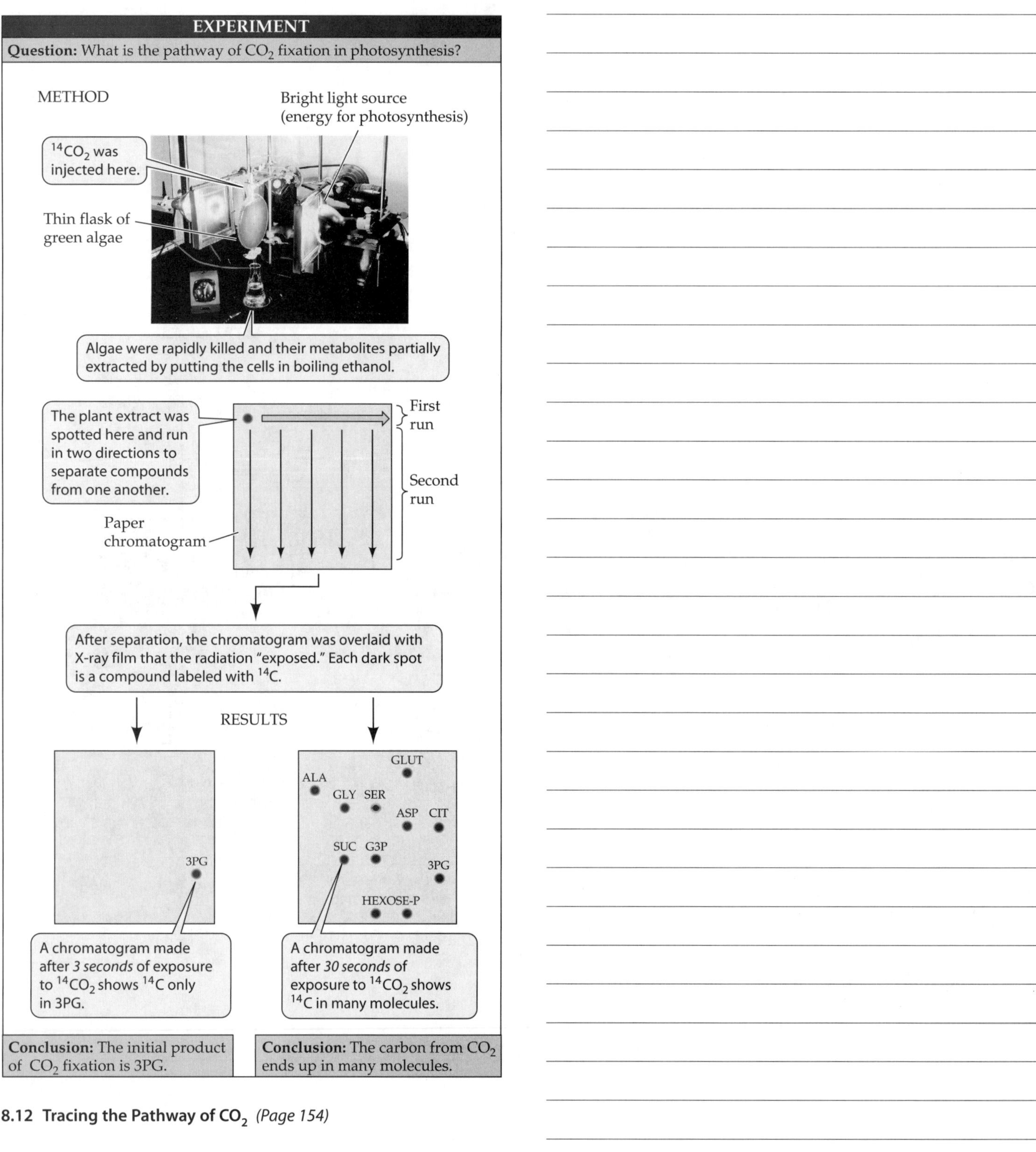
EXPERIMENT
Question: What is the pathway of CO_2 fixation in photosynthesis?
METHOD
Bright light source (energy for photosynthesis)
$^{14}CO_2$ was injected here.
Thin flask of green algae
Algae were rapidly killed and their metabolites partially extracted by putting the cells in boiling ethanol.
The plant extract was spotted here and run in two directions to separate compounds from one another.
First run
Second run
Paper chromatogram
After separation, the chromatogram was overlaid with X-ray film that the radiation "exposed." Each dark spot is a compound labeled with ^{14}C.
RESULTS
3PG
GLUT
ALA
GLY SER
ASP CIT
SUC G3P
3PG
HEXOSE-P
A chromatogram made after *3 seconds* of exposure to $^{14}CO_2$ shows ^{14}C only in 3PG.
A chromatogram made after *30 seconds* of exposure to $^{14}CO_2$ shows ^{14}C in many molecules.
Conclusion: The initial product of CO_2 fixation is 3PG.
Conclusion: The carbon from CO_2 ends up in many molecules.

8.12 Tracing the Pathway of CO_2 *(Page 154)*

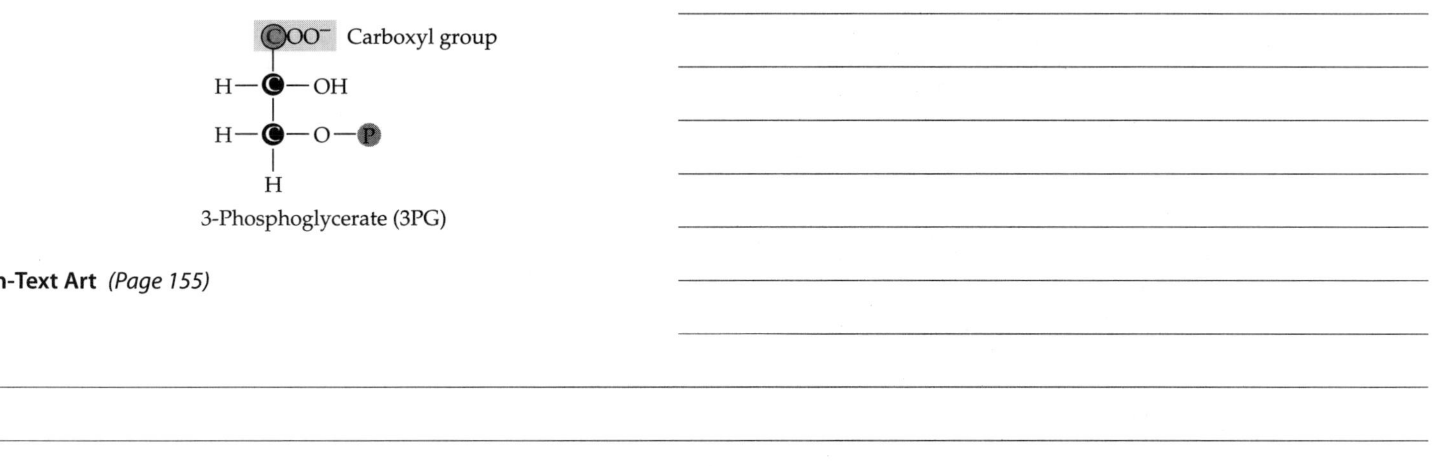
COO⁻ Carboxyl group
H—C—OH
H—C—O—P
H
3-Phosphoglycerate (3PG)

In-Text Art *(Page 155)*

ELECTRON TRANSPORT
ATP CYCLE
NADPH CYCLE
CALVIN–BENSON CYCLE

START

1 CO_2 combines with its acceptor, RuBP, forming 3PG.

6 CO_2

5 RuMP is converted to RuBP in a reaction requiring ATP. RuBP is ready to accept another CO_2.

6 RuBP

12 3PG

Carbon fixation

6 ADP

6 ATP

12 ATP

12 ADP

CALVIN–BENSON CYCLE

6 RuMP

Regeneration of RuBP

Reduction and sugar production

2 3PG is reduced to G3P in a two-step reaction requiring ATP and NADPH + H^+.

12 NADPH + 12 H^+

12 $NADP^+$

12 P_i

10 G3P

12 G3P

2 G3P

4 The remaining five-sixths of the G3P is processed in the complex reactions that produce RuMP.

3 About one-sixth of the G3P is used to make sugars—the output of the cycle.

Sugars

Other carbon compounds

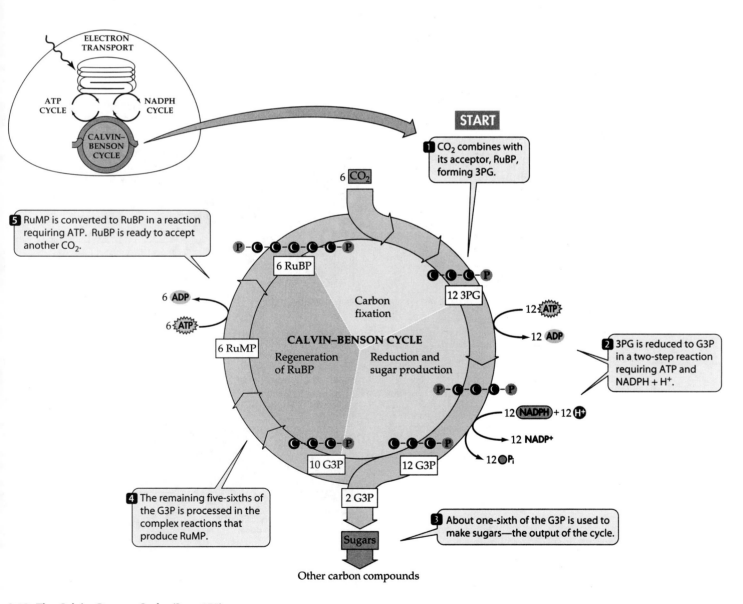

8.13 The Calvin–Benson Cycle *(Page 155)*

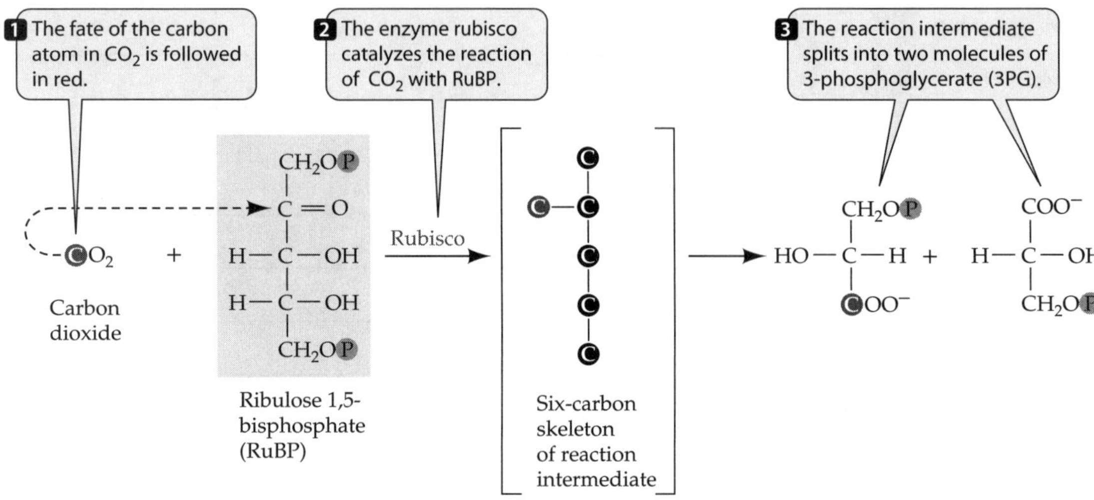
1 The fate of the carbon atom in CO_2 is followed in red.
2 The enzyme rubisco catalyzes the reaction of CO_2 with RuBP.
3 The reaction intermediate splits into two molecules of 3-phosphoglycerate (3PG).
CO_2
Carbon dioxide
+
CH_2OP
C=O
H—C—OH
H—C—OH
CH_2OP
Ribulose 1,5-bisphosphate (RuBP)
Rubisco
Six-carbon skeleton of reaction intermediate
CH_2OP
HO—C—H
COO^-
+
COO^-
H—C—OH
CH_2OP

8.14 RuBP Is the Carbon Dioxide Acceptor *(Page 156)*

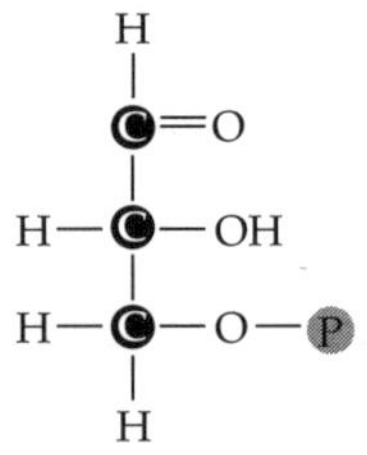

Glyceraldehyde 3-phosphate (G3P)

In-Text Art *(Page 156)*

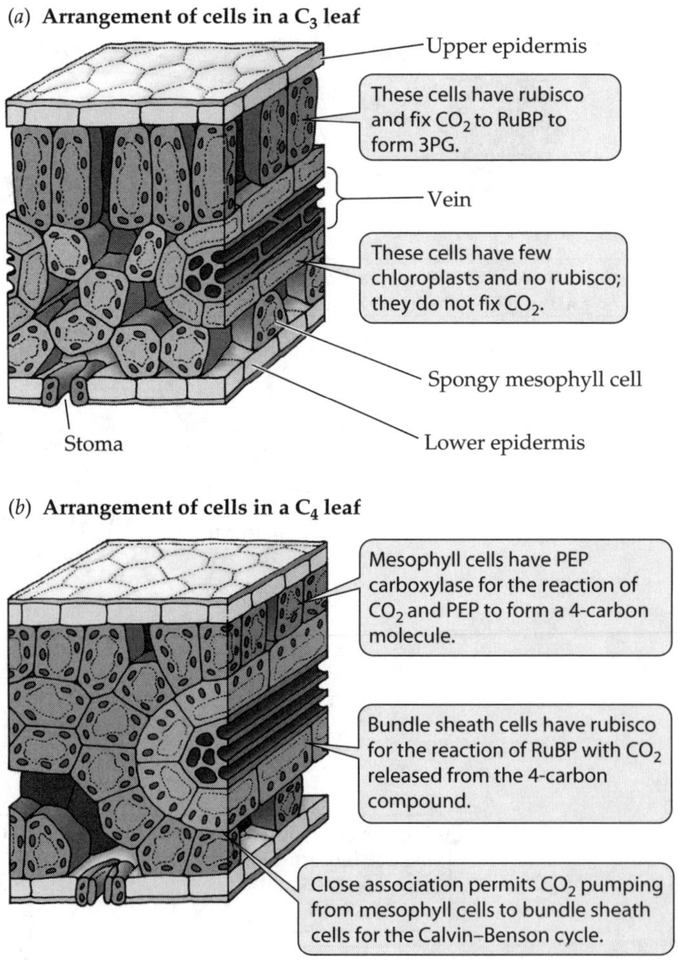

(*a*) **Arrangement of cells in a C_3 leaf**

Upper epidermis

These cells have rubisco and fix CO_2 to RuBP to form 3PG.

Vein

These cells have few chloroplasts and no rubisco; they do not fix CO_2.

Spongy mesophyll cell

Stoma

Lower epidermis

(*b*) **Arrangement of cells in a C_4 leaf**

Mesophyll cells have PEP carboxylase for the reaction of CO_2 and PEP to form a 4-carbon molecule.

Bundle sheath cells have rubisco for the reaction of RuBP with CO_2 released from the 4-carbon compound.

Close association permits CO_2 pumping from mesophyll cells to bundle sheath cells for the Calvin–Benson cycle.

8.16 Leaf Anatomy of C_3 and C_4 Plants *(Page 157)*

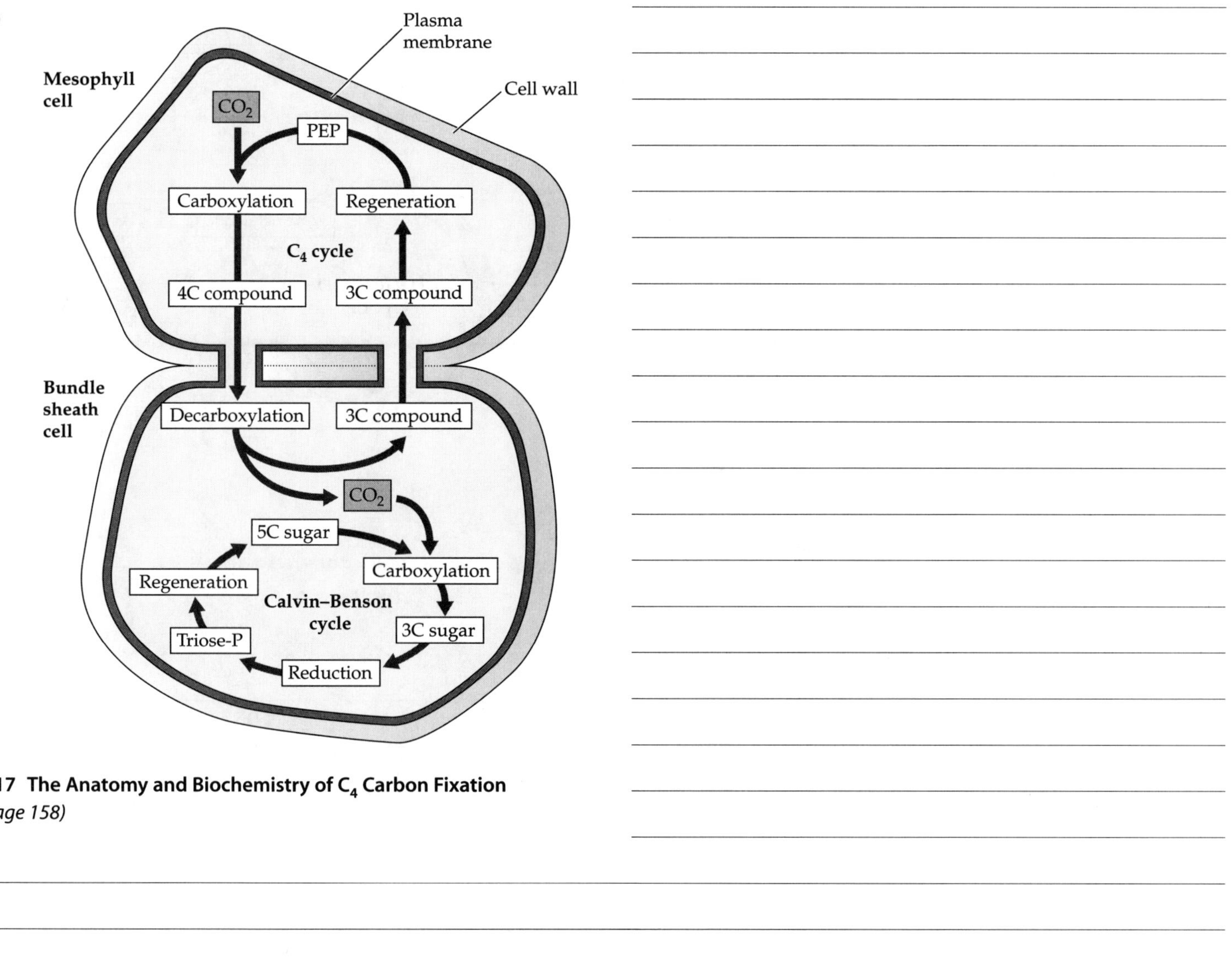
(b)
Mesophyll cell
Plasma membrane
Cell wall
CO_2
PEP
Carboxylation
Regeneration
C_4 cycle
4C compound
3C compound
Bundle sheath cell
Decarboxylation
3C compound
CO_2
5C sugar
Carboxylation
Regeneration
Calvin–Benson cycle
3C sugar
Triose-P
Reduction

8.17 The Anatomy and Biochemistry of C_4 Carbon Fixation
(Page 158)

8.1 *Comparison of Photosynthesis in C_3 and C_4 Plants*

VARIABLE	C_3 PLANTS	C_4 PLANTS
Photorespiration	Extensive	Minimal
Perform Calvin–Benson cycle?	Yes	Yes
Primary CO_2 acceptor	RuBP	PEP
CO_2-fixing enzyme	Rubisco (RuBP carboxylase/oxygenase)	PEP carboxylase and rubisco
First product of CO_2 fixation	3PG (3-carbon compound)	Oxaloacetate (4-carbon compound)
Affinity of carboxylase for CO_2	Moderate	High
Photosynthetic cells of leaf	Mesophyll	Mesophyll + bundle sheath
Classes of chloroplasts	One	Two

(Page 159)

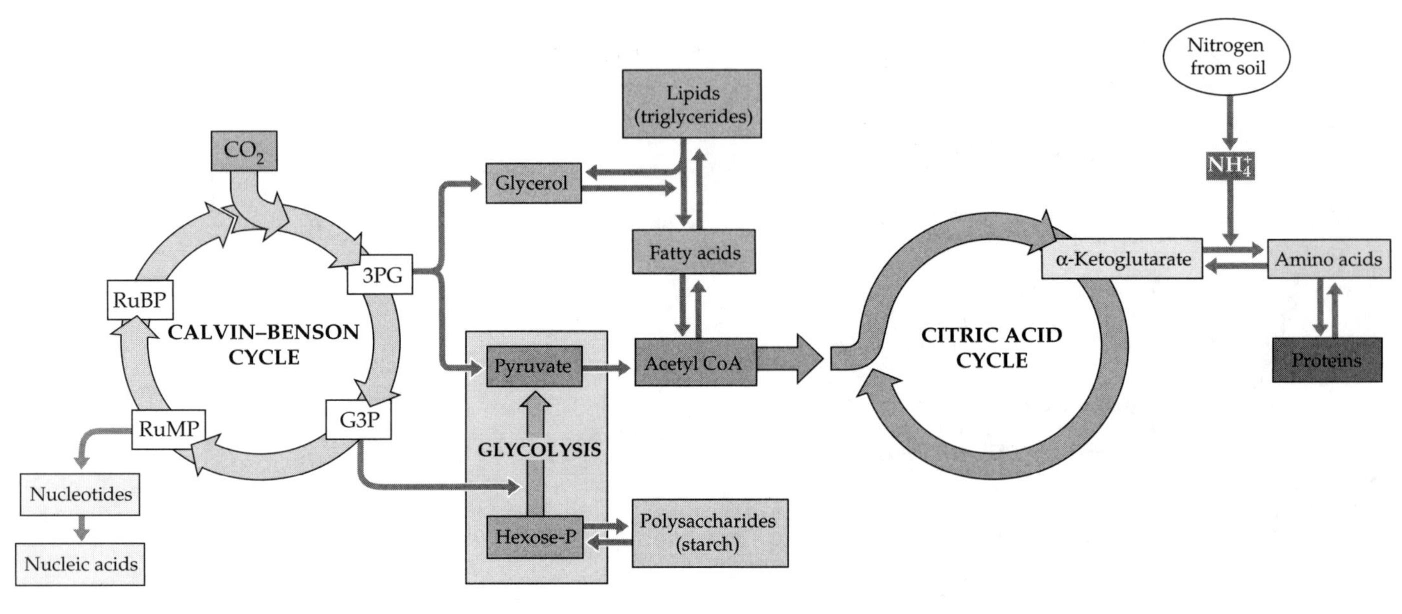
Nitrogen from soil
Lipids (triglycerides)
CO_2
Glycerol
NH_4^+
Fatty acids
α-Ketoglutarate
Amino acids
3PG
RuBP
CALVIN–BENSON CYCLE
CITRIC ACID CYCLE
Pyruvate
Acetyl CoA
Proteins
RuMP
G3P
GLYCOLYSIS
Nucleotides
Hexose-P
Polysaccharides (starch)
Nucleic acids

8.18 Metabolic Interactions in a Plant Cell *(Page 160)*

9 Chromosomes, the Cell Cycle, and Cell Division

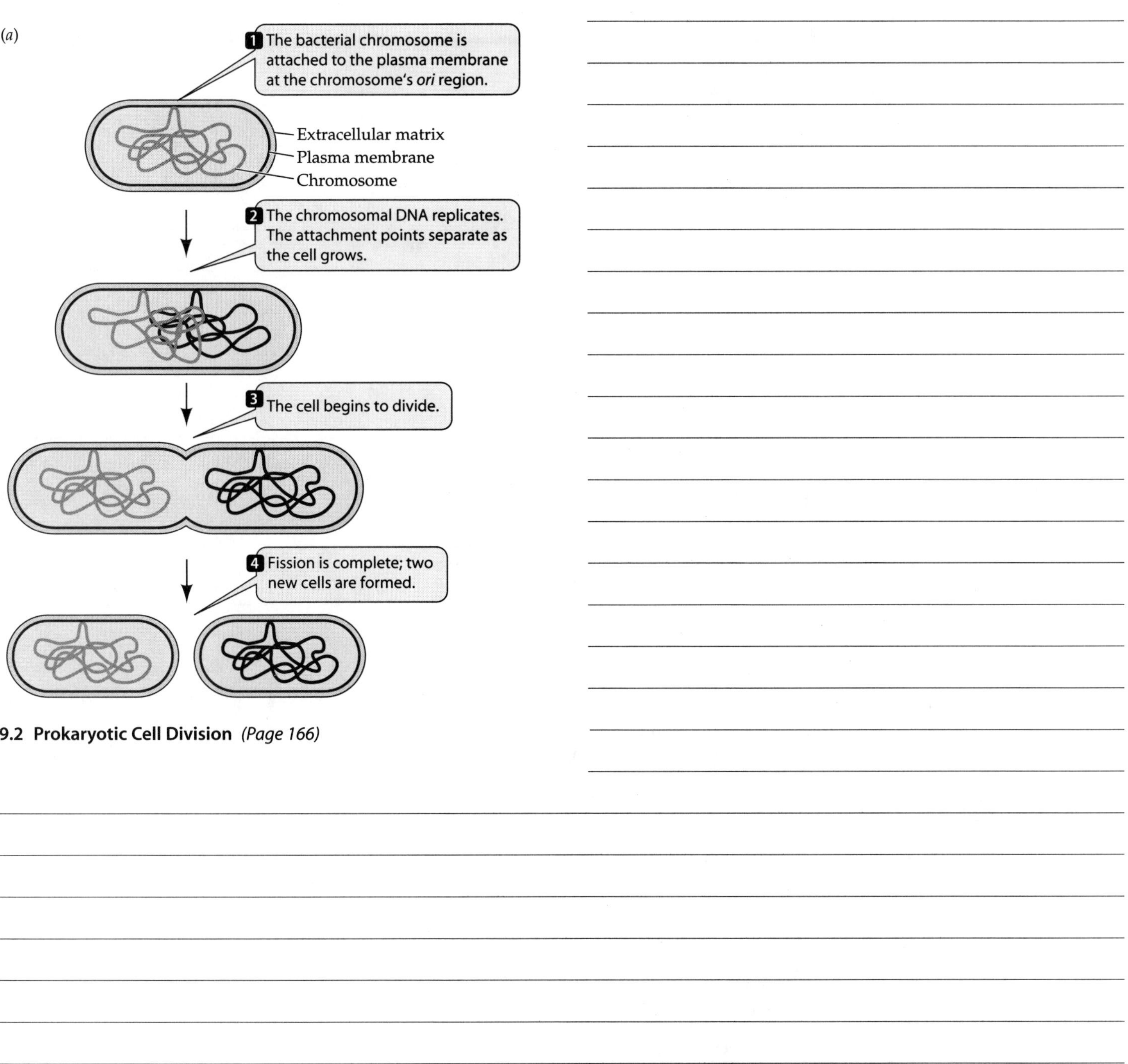

(a)
1 The bacterial chromosome is attached to the plasma membrane at the chromosome's *ori* region.
Extracellular matrix
Plasma membrane
Chromosome
2 The chromosomal DNA replicates. The attachment points separate as the cell grows.
3 The cell begins to divide.
4 Fission is complete; two new cells are formed.

9.2 Prokaryotic Cell Division *(Page 166)*

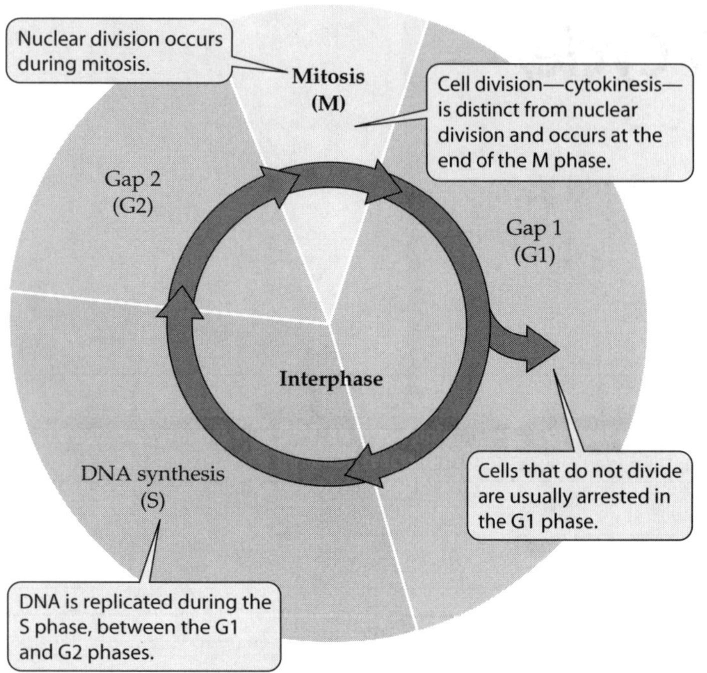
Nuclear division occurs during mitosis.
Mitosis (M)
Cell division—cytokinesis—is distinct from nuclear division and occurs at the end of the M phase.
Gap 2 (G2)
Gap 1 (G1)
Interphase
DNA synthesis (S)
Cells that do not divide are usually arrested in the G1 phase.
DNA is replicated during the S phase, between the G1 and G2 phases.

9.3 The Eukaryotic Cell Cycle *(Page 167)*

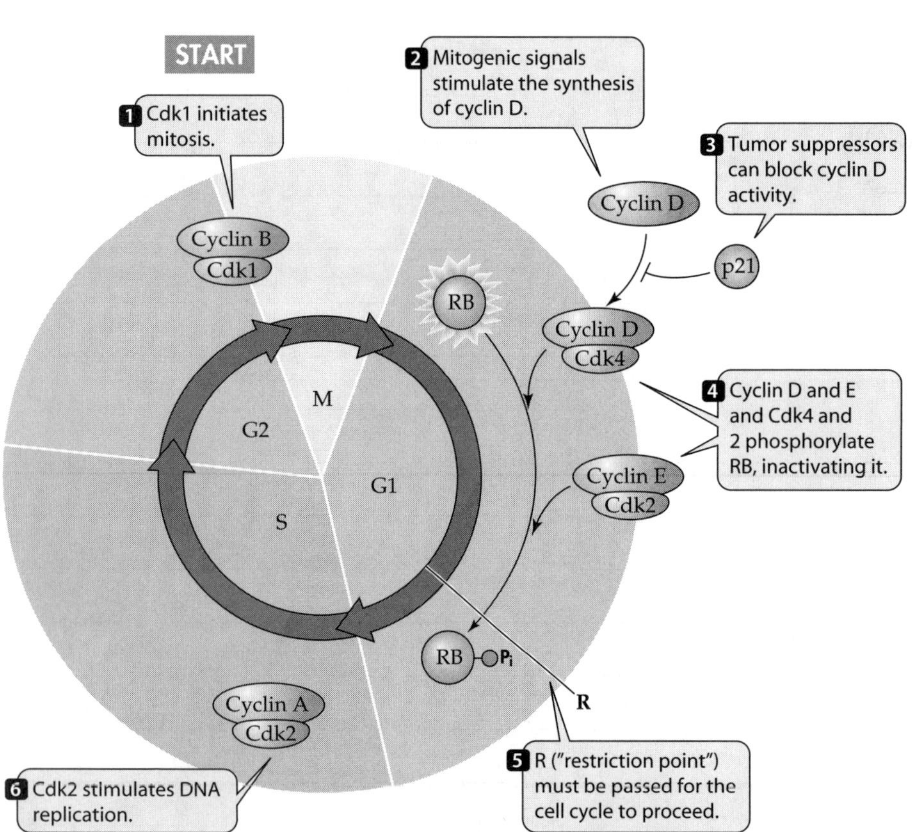
START
1 Cdk1 initiates mitosis.
2 Mitogenic signals stimulate the synthesis of cyclin D.
3 Tumor suppressors can block cyclin D activity.
Cyclin D
p21
Cyclin B
Cdk1
RB
Cyclin D
Cdk4
M
G2
G1
S
4 Cyclin D and E and Cdk4 and 2 phosphorylate RB, inactivating it.
Cyclin E
Cdk2
RB
Pi
R
Cyclin A
Cdk2
5 R ("restriction point") must be passed for the cell cycle to proceed.
6 Cdk2 stimulates DNA replication.

9.4 Cyclin-Dependent Kinases and Cyclins Trigger Transitions in the Cell Cycle *(Page 168)*

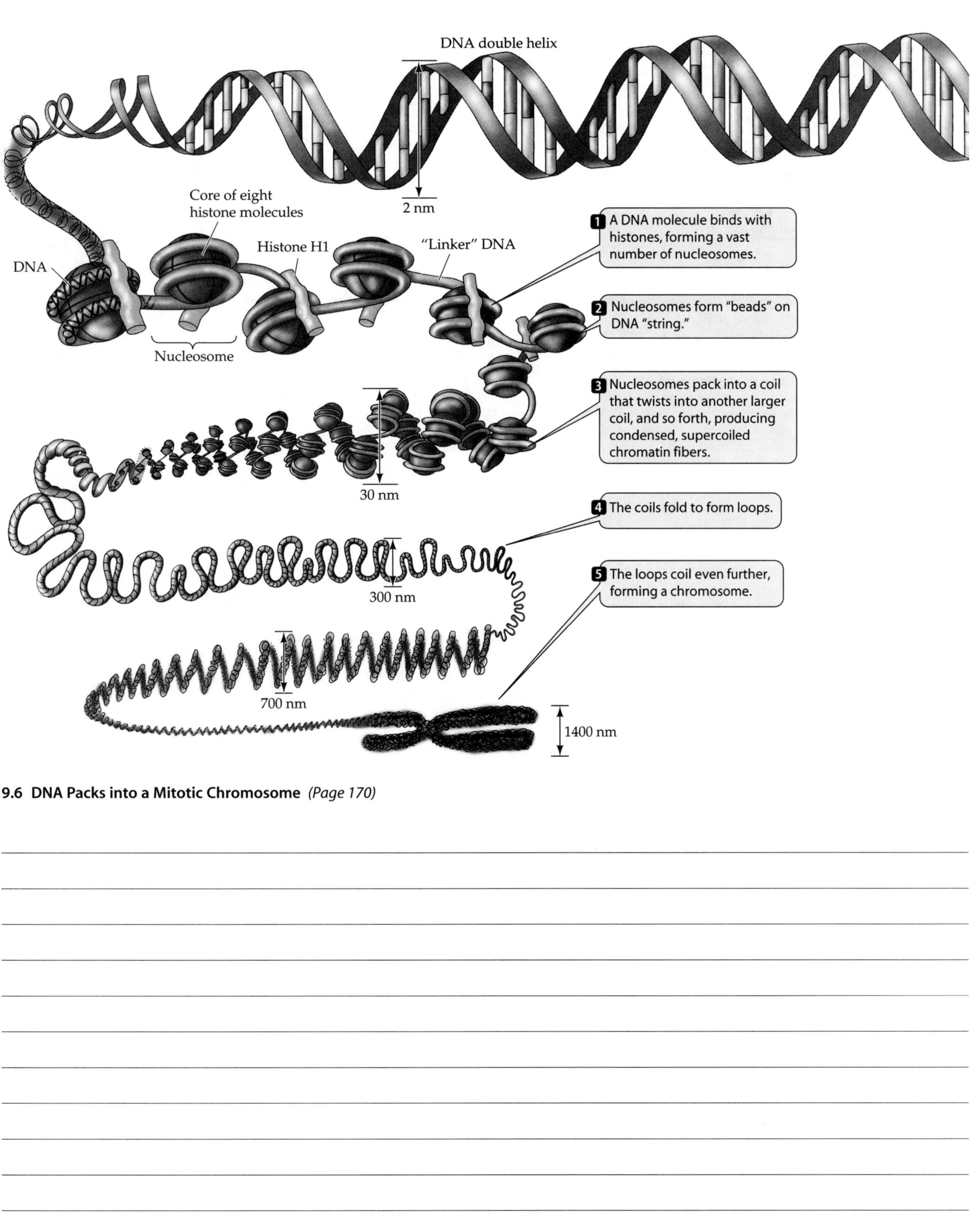
DNA double helix
2 nm
Core of eight histone molecules
Histone H1
"Linker" DNA
DNA
Nucleosome
1 A DNA molecule binds with histones, forming a vast number of nucleosomes.
2 Nucleosomes form "beads" on DNA "string."
3 Nucleosomes pack into a coil that twists into another larger coil, and so forth, producing condensed, supercoiled chromatin fibers.
30 nm
4 The coils fold to form loops.
300 nm
5 The loops coil even further, forming a chromosome.
700 nm
1400 nm

9.6 DNA Packs into a Mitotic Chromosome *(Page 170)*

(*a*)

Kinetochore microtubules attach to the kinetochores and to the spindle poles.

Kinetochore

Kinetochore microtubule

Mitotic center (centrosome)

Centriole

Polar microtubule

Polar microtubules extend from each pole of the spindle.

9.7 The Mitotic Spindle Consists of Microtubules *(Page 171)*

Interphase

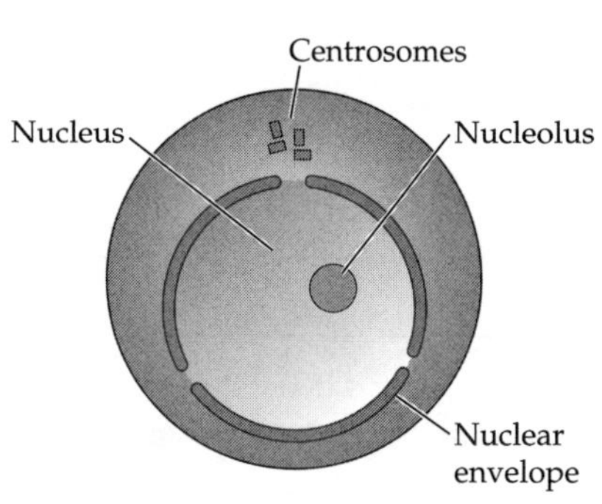
Centrosomes
Nucleus
Nucleolus
Nuclear envelope

1 During the S phase of interphase, the nucleus replicates its DNA and centrosomes.

Prophase

Developing spindle

Centrosome

Chromatids of chromosomes

2 The chromatin coils and supercoils, becoming more and more compact and eventually condensing into visible chromosomes. The chromosomes consist of identical, paired sister chromatids.

Prometaphase

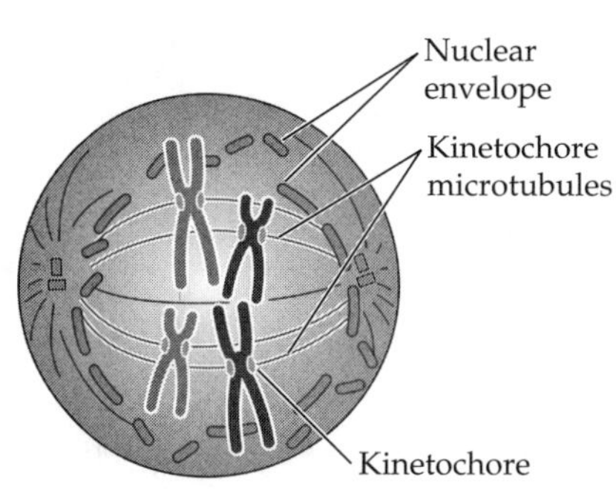
Nuclear envelope
Kinetochore microtubules
Kinetochore

3 The nuclear envelope breaks down. Kinetochore microtubules appear and connect the kinetochores to the microtubule organizing centers.

Metaphase

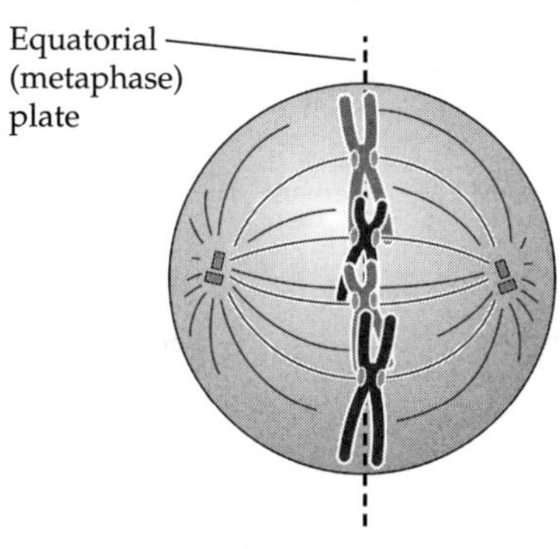
Equatorial (metaphase) plate

4 The centromeres (regions connecting paired chromatids) become aligned in a plane at the cell's equator.

Anaphase

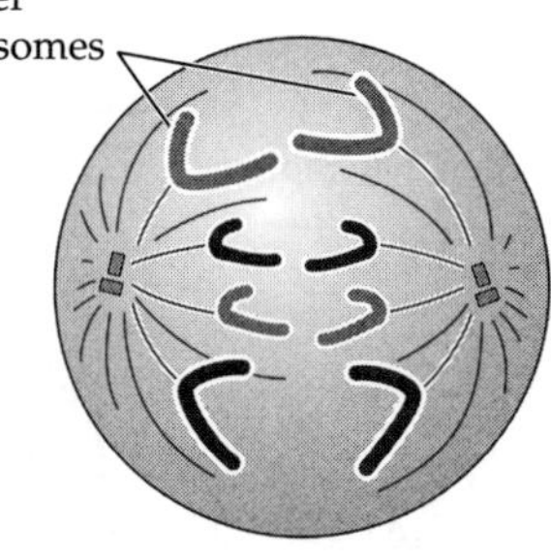
Daughter chromosomes

5 The paired sister chromatids separate, and the new daughter chromosomes begin to move toward the poles.

Telophase

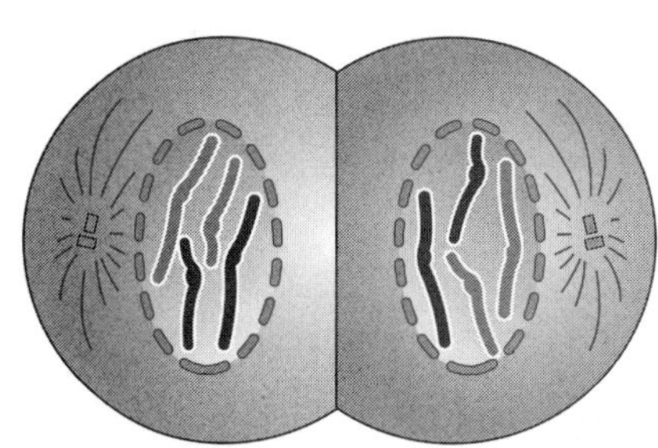

6 The daughter chromosomes reach the poles. Telophase passes into the next interphase as the nuclear envelopes and nucleoli re-form and the chromatin becomes diffuse.

9.8 Mitosis *(Pages 172–173)*

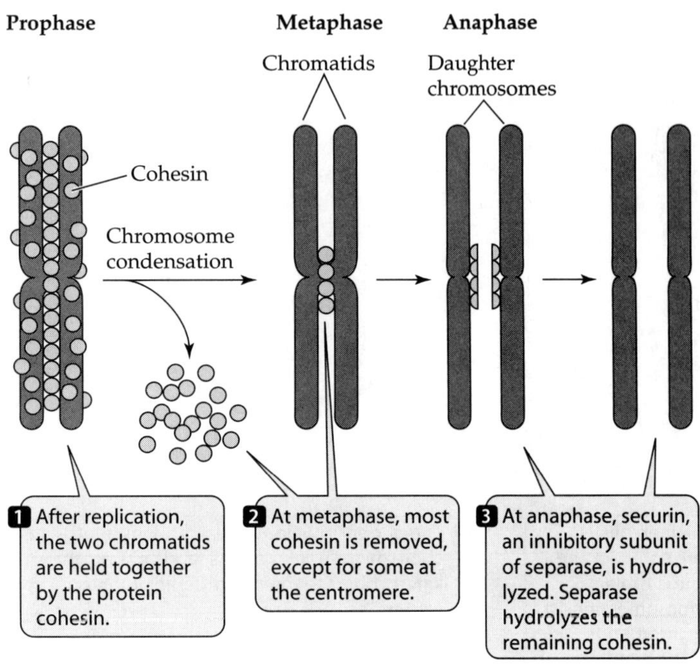
Prophase
Metaphase
Anaphase
Chromatids
Daughter chromosomes
Cohesin
Chromosome condensation
1 After replication, the two chromatids are held together by the protein cohesin.
2 At metaphase, most cohesin is removed, except for some at the centromere.
3 At anaphase, securin, an inhibitory subunit of separase, is hydrolyzed. Separase hydrolyzes the remaining cohesin.

9.9 Molecular Biology of Chromatid Attachment and Separation *(Page 173)*

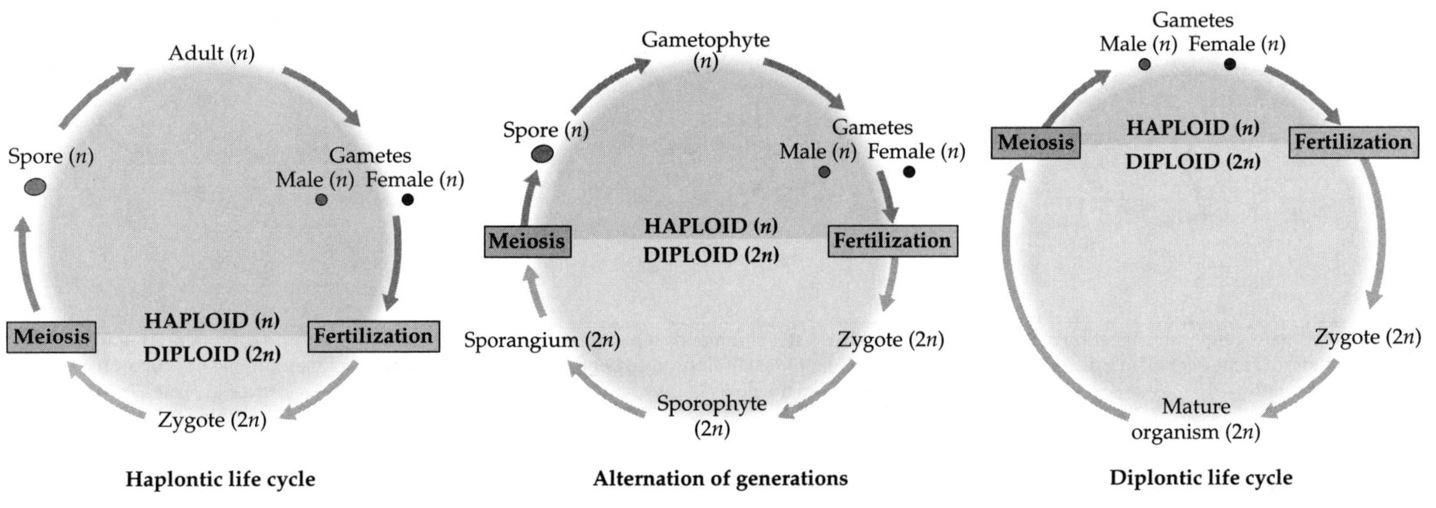
Adult (n)
Spore (n)
Gametes
Male (n) Female (n)
Meiosis
HAPLOID (n)
DIPLOID (2n)
Fertilization
Zygote (2n)
Haplontic life cycle
Gametophyte (n)
Spore (n)
Gametes
Male (n) Female (n)
Meiosis
HAPLOID (n)
DIPLOID (2n)
Fertilization
Sporangium (2n)
Zygote (2n)
Sporophyte (2n)
Alternation of generations
Gametes
Male (n) Female (n)
Meiosis
HAPLOID (n)
DIPLOID (2n)
Fertilization
Zygote (2n)
Mature organism (2n)
Diplontic life cycle

9.12 Fertilization and Meiosis Alternate in Sexual Reproduction *(Page 176)*

9.1 Numbers of Pairs of Chromosomes in Some Plant and Animal Species

COMMON NAME	SPECIES	NUMBER OF CHROMOSOME PAIRS
Mosquito	*Culex pipiens*	3
Housefly	*Musca domestica*	6
Toad	*Bufo americanus*	11
Rice	*Oryza sativa*	12
Frog	*Rana pipiens*	13
Alligator	*Alligator mississippiensis*	16
Rhesus monkey	*Macaca mulatta*	21
Wheat	*Triticum aestivum*	21
Human	*Homo sapiens*	23
Potato	*Solanum tuberosum*	24
Donkey	*Equus asinus*	31
Horse	*Equus caballus*	32
Dog	*Canis familiaris*	39
Carp	*Cyprinus carpio*	52

(Page 177)

MEIOSIS I

Early Prophase I

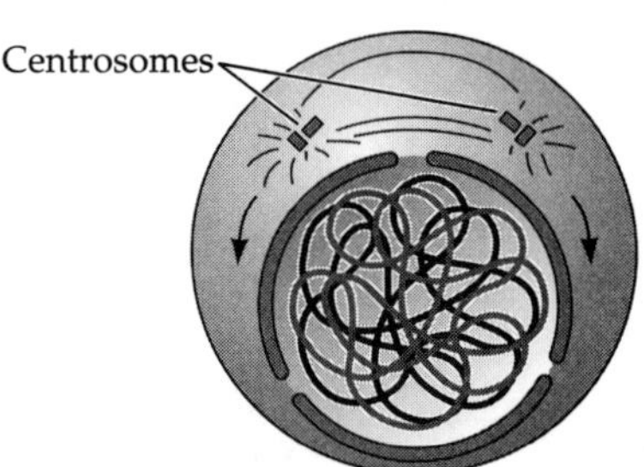
Centrosomes

1 The chromatin begins to condense following interphase.

Mid-Prophase I

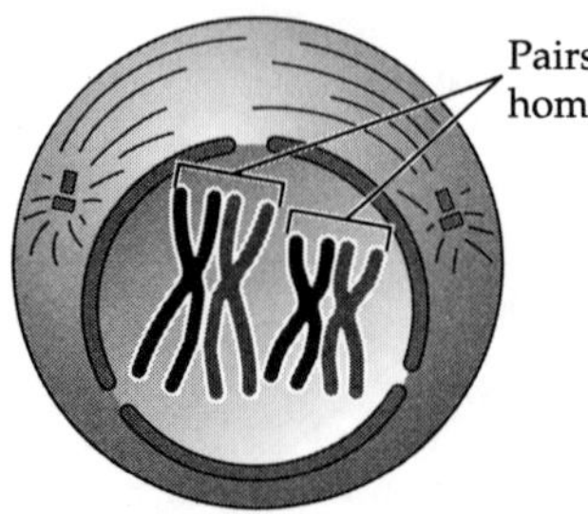
Pairs of homologs

2 Synapsis aligns homologs, and chromosomes condense. Homologs are shown in different colors indicating those coming from each parent. In reality, their differences are very small, usually comprising different alleles of some genes.

Late Prophase I–Prometaphase

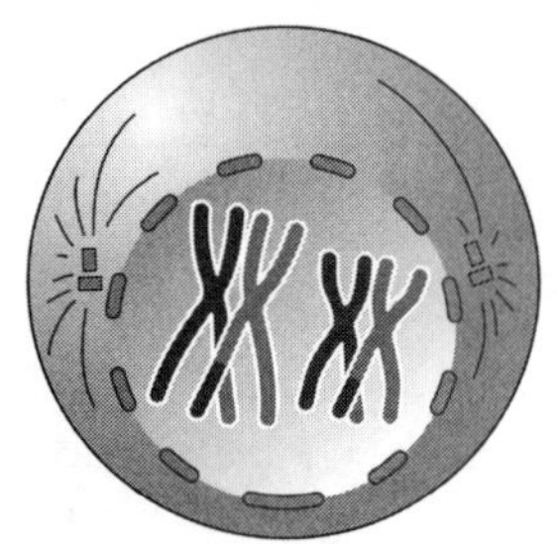

3 The chromosomes continue to coil and shorten. Crossing over results in an exchange of genetic material. In prometaphase the nuclear envelope breaks down.

MEIOSIS II

Prophase II

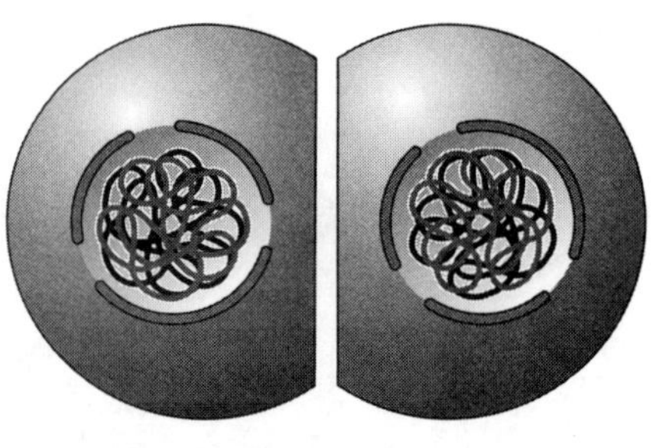

7 The chromosomes condense again, following a brief interphase (interkinesis) in which DNA does not replicate.

Metaphase II

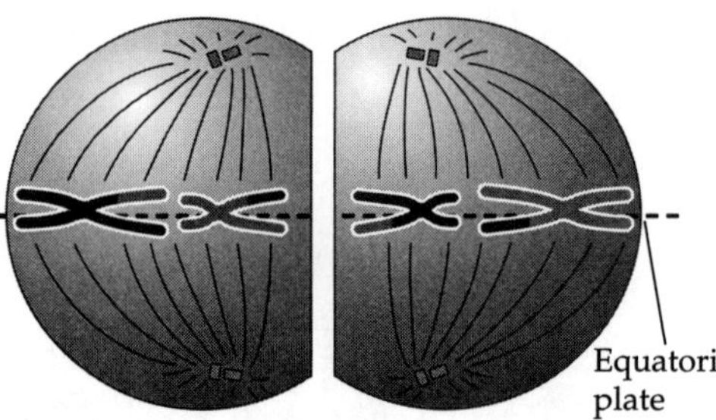
Equatorial plate

8 Kinetochores of the paired chromatids line up across the equatorial plates of each cell.

Anaphase II

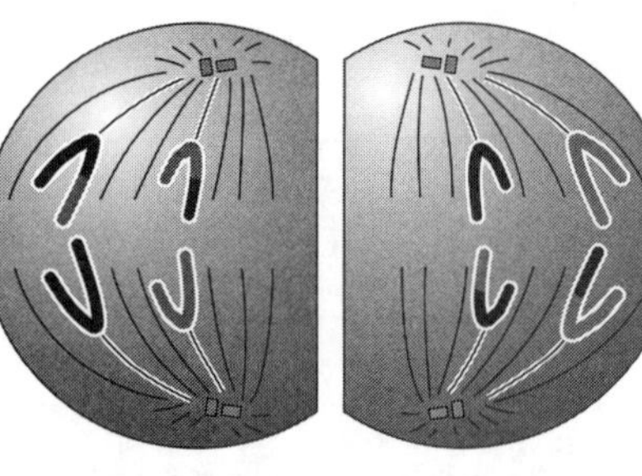

9 The chromatids finally separate, becoming chromosomes in their own right, and are pulled to opposite poles. Because of crossing over in prophase I, each new cell will have a different genetic makeup.

9.14 Meiosis *(Pages 178–179)*

Metaphase I

Equatorial plate

4 The homologous chromosomes line up on the equatorial (metaphase) plate.

Anaphase I

5 The homologous chromosomes (each with two chromatids) move to opposite poles of the cell.

Telophase I

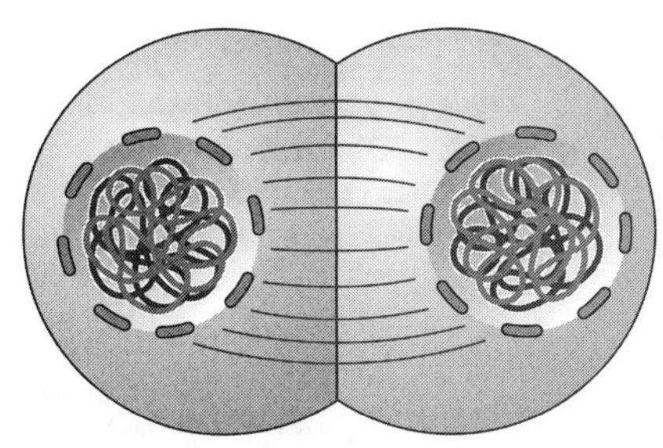

6 The chromosomes gather into nuclei, and the original cell divides.

Telophase II

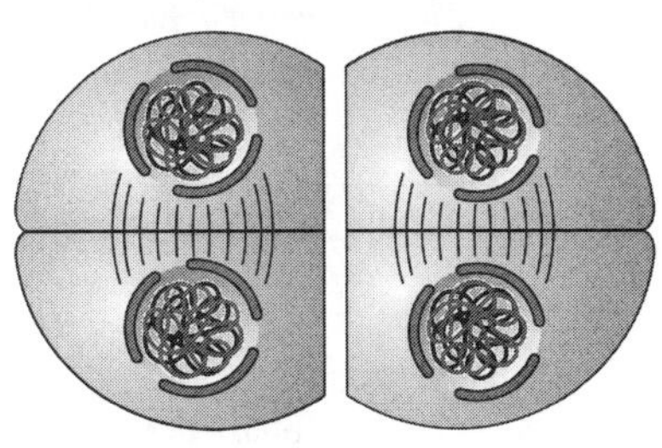

10 The chromosomes gather into nuclei, and the cells divide.

Products

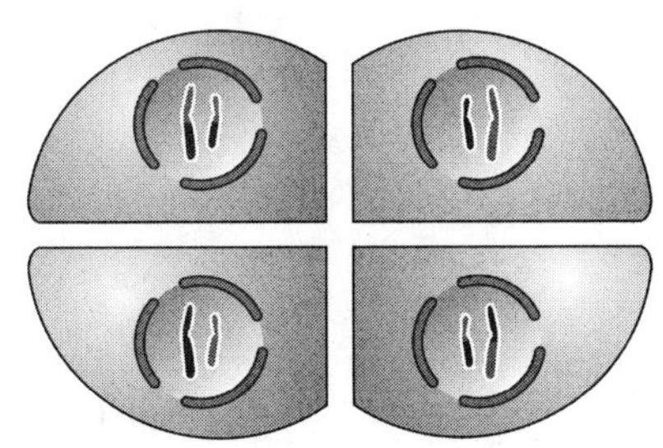

11 Each of the four cells has a nucleus with a haploid number of chromosomes.

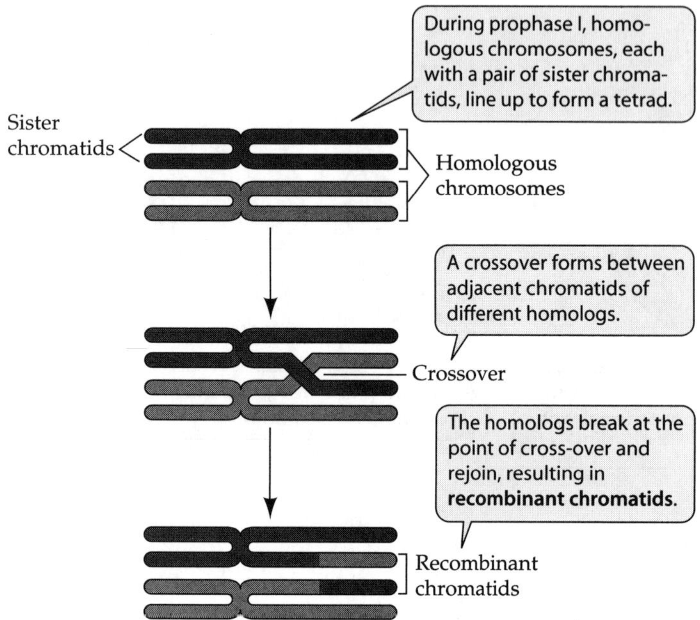
During prophase I, homologous chromosomes, each with a pair of sister chromatids, line up to form a tetrad.
Sister chromatids
Homologous chromosomes
A crossover forms between adjacent chromatids of different homologs.
Crossover
The homologs break at the point of cross-over and rejoin, resulting in **recombinant chromatids**.
Recombinant chromatids

9.16 Crossing Over Forms Genetically Diverse Chromosomes *(Page 180)*

MITOSIS

Parent cell (2*n*)

Prophase

No synapsis of homologous chromosomes

Metaphase

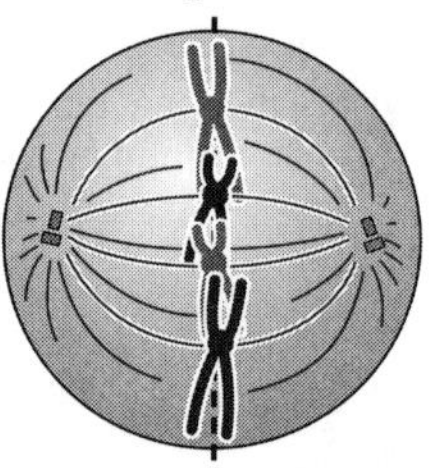

Individual chromosomes align at the equatorial plate.

Anaphase

Centromeres separate. Sister chromatids separate during anaphase, becoming daughter chromosomes.

MEIOSIS

Parent cell (2*n*)

Prophase I

Homologous chromosome pairs

Crossover

Synapsis and crossing over of homologs

Metaphase I

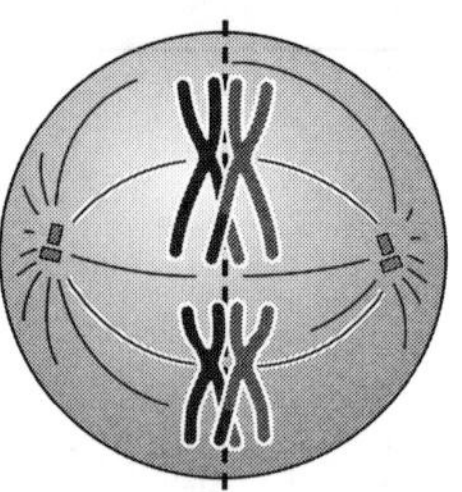

Homologous pairs align at the equatorial plate.

Anaphase I

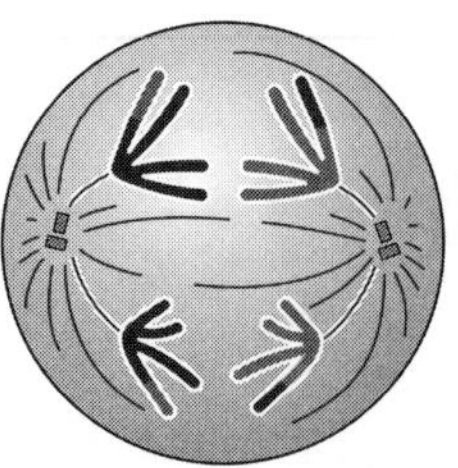

Centromeres do not separate; sister chromatids remain together during anaphase; homologs separate; DNA does not replicate before subsequent prophase.

9.17 Mitosis and Meiosis: A Comparison *(Pages 182-183)*

Two daughter cells (each 2*n*)

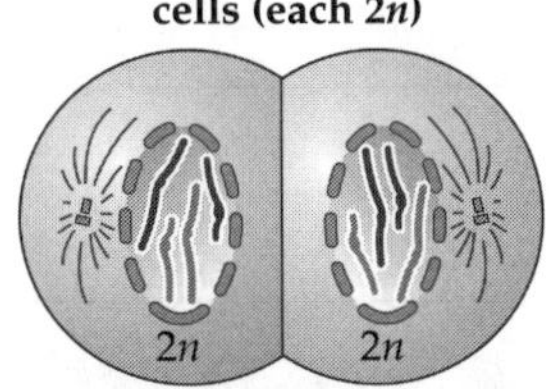
2n
2n

Mitosis is a mechanism for constancy: The parent nucleus produces two identical daughter nuclei.

Telophase I

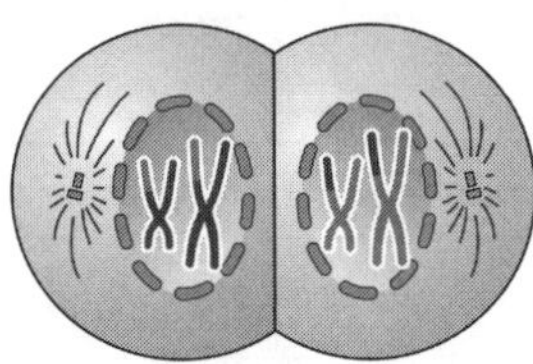

Metaphase II

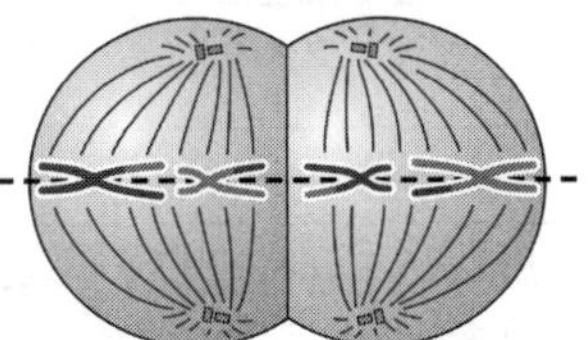

Four daughter cells (each *n*)

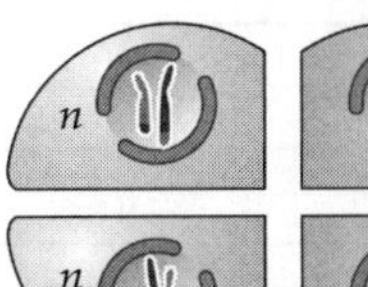
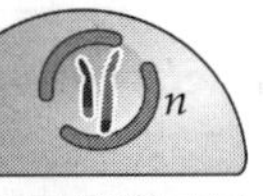
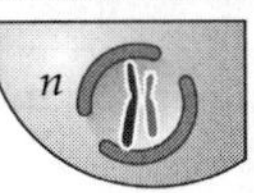

n
n
n
n

Chromatids separate.

Meiosis is a mechanism for diversity: The parent nucleus produces four different haploid daughter nuclei.

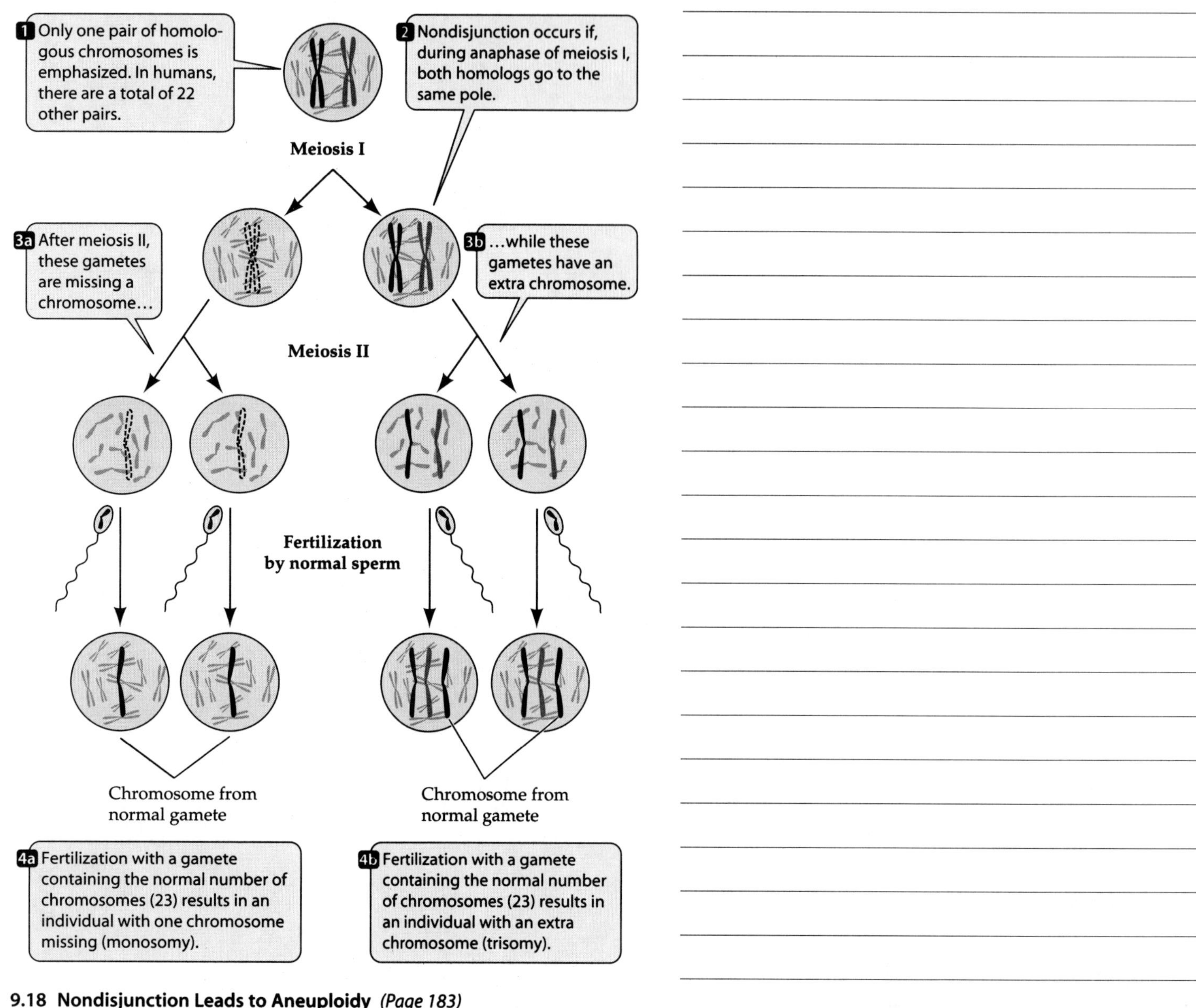

1 Only one pair of homologous chromosomes is emphasized. In humans, there are a total of 22 other pairs.
2 Nondisjunction occurs if, during anaphase of meiosis I, both homologs go to the same pole.
Meiosis I
3a After meiosis II, these gametes are missing a chromosome...
3b ...while these gametes have an extra chromosome.
Meiosis II
Fertilization by normal sperm
Chromosome from normal gamete
Chromosome from normal gamete
4a Fertilization with a gamete containing the normal number of chromosomes (23) results in an individual with one chromosome missing (monosomy).
4b Fertilization with a gamete containing the normal number of chromosomes (23) results in an individual with an extra chromosome (trisomy).

9.18 Nondisjunction Leads to Aneuploidy *(Page 183)*

9.2 *Two Different Ways for Cells to Die*

	NECROSIS	APOPTOSIS
Stimuli	Low O_2, toxins, ATP depletion, damage	Specific, genetically programmed physiological signals
ATP required	No	Yes
Cellular pattern	Swelling, organelle disruption, tissue death	Chromatin condensation, membrane blebbing, single-cell death
DNA breakdown	Random fragments	Nucleosome-sized fragments
Plasma membrane	Burst	Blebbed (see Figure 9.19)
Fate of dead cells	Ingested by phagocytes	Ingested by neighboring cells
Reaction in tissue	Inflammation	No inflammation

(Page 184)

10 Genetics: Mendel and Beyond

RESEARCH METHOD
Anatomy of a pea flower (shown in long section)
The **stigma**, where the pollen lands, is at the tip of the carpel.
Anthers at the tip of the stamen are the sites of pollen production.
Stamens are the male sex organs.
The **ovary** is the female sex organ.
Pea flower cross-pollination
Parent plant
Parent plant
Pollen
1 Using a brush, pollen is transferred from anthers of a purple flower to the stigma of a white flower whose anthers have been snipped off.
Pea pod
2 This cross-pollination produces seeds that are allowed to grow into new plants.
Seeds (peas)
3 Analysis of physical characteristics (see Table 10.1) of the offspring over 2 generations shows evidence of hereditary transmission from both parents.

10.1 A Controlled Cross between Two Plants *(Page 188)*

10.1 Mendel's Results from Monohybrid Crosses

PARENTAL GENERATION PHENOTYPES		F_2 GENERATION PHENOTYPES			
DOMINANT	RECESSIVE	DOMINANT	RECESSIVE	TOTAL	RATIO
Spherical seeds ×	Wrinkled seeds	5,474	1,850	7,324	2.96:1
Yellow seeds ×	Green seeds	6,022	2,001	8,023	3.01:1
Purple flowers ×	White flowers	705	224	929	3.15:1
Inflated pods ×	Constricted pods	882	299	1,181	2.95:1
Green pods ×	Yellow pods	428	152	580	2.82:1
Axial flowers ×	Terminal flowers	651	207	858	3.14:1
Tall stems (1 m) ×	Dwarf stems (0.3 m)	787	277	1,064	2.84:1

(Page 190)

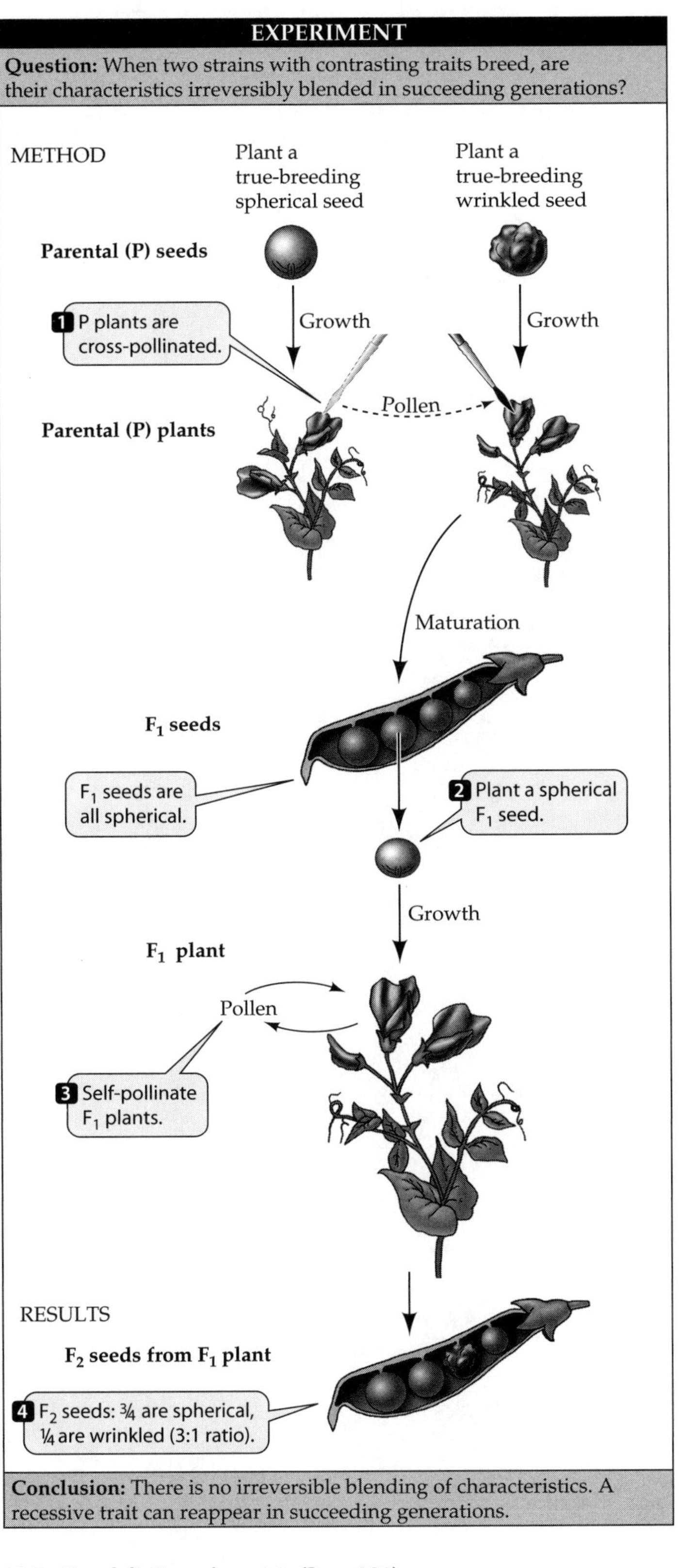
EXPERIMENT
Question: When two strains with contrasting traits breed, are their characteristics irreversibly blended in succeeding generations?
METHOD
Plant a true-breeding spherical seed
Plant a true-breeding wrinkled seed
Parental (P) seeds
Growth
Growth
1 P plants are cross-pollinated.
Pollen
Parental (P) plants
Maturation
F1 seeds
F1 seeds are all spherical.
2 Plant a spherical F1 seed.
Growth
F1 plant
Pollen
3 Self-pollinate F1 plants.
RESULTS
F2 seeds from F1 plant
4 F2 seeds: ¾ are spherical, ¼ are wrinkled (3:1 ratio).
Conclusion: There is no irreversible blending of characteristics. A recessive trait can reappear in succeeding generations.

10.3 Mendel's Experiment 1 *(Page 191)*

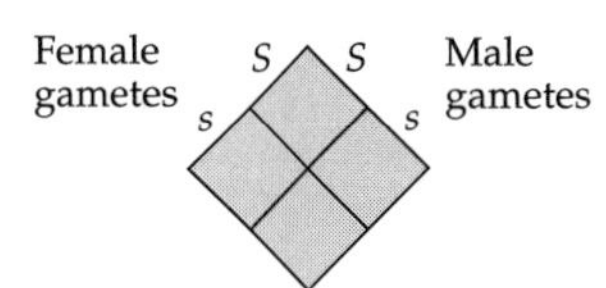
Female gametes
S
s
S
s
Male gametes

In-Text Art *(Page 192)*

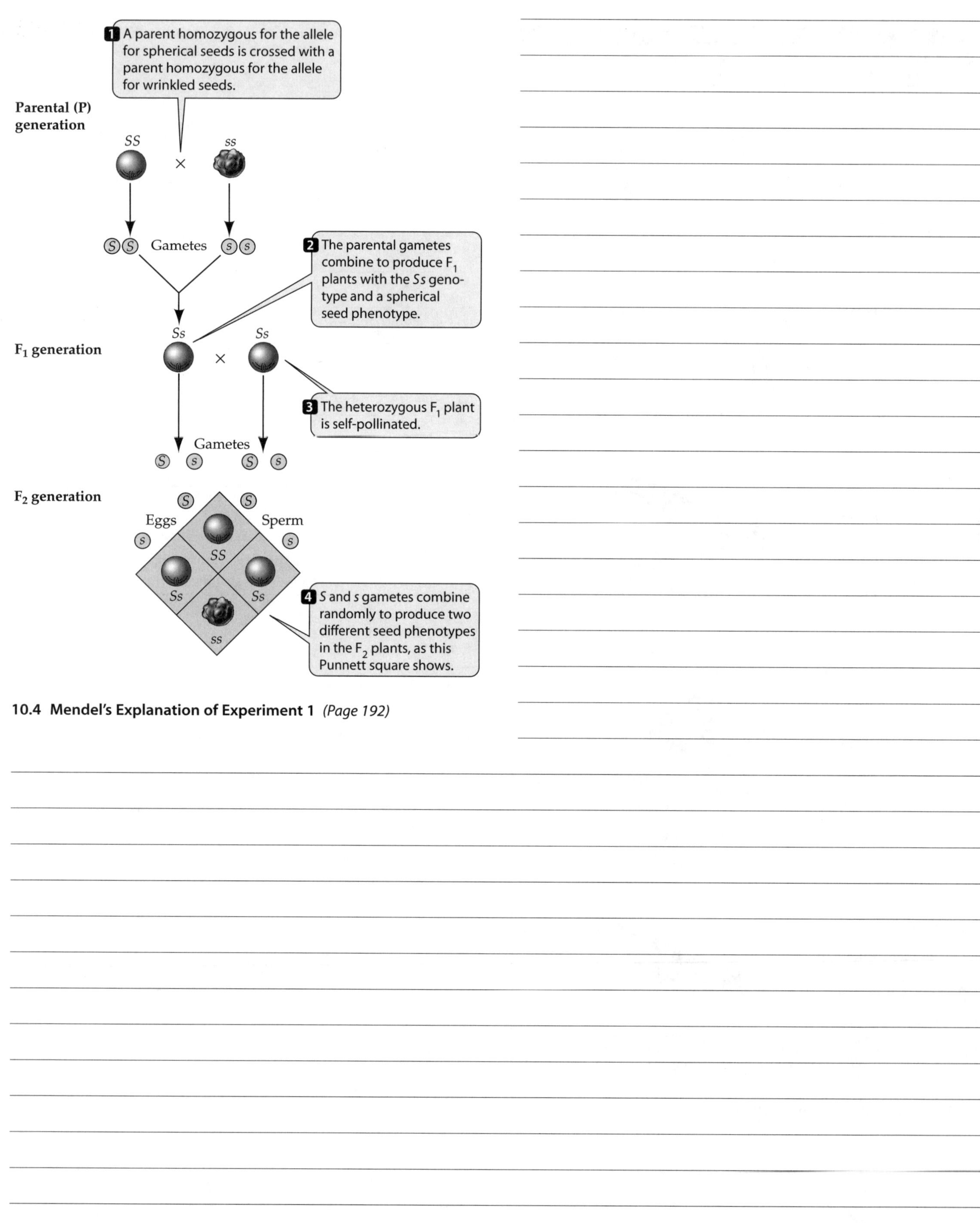
1 A parent homozygous for the allele for spherical seeds is crossed with a parent homozygous for the allele for wrinkled seeds.
Parental (P) generation
SS
ss
×
S S
Gametes
s s
2 The parental gametes combine to produce F_1 plants with the *Ss* genotype and a spherical seed phenotype.
F_1 generation
Ss
Ss
×
3 The heterozygous F_1 plant is self-pollinated.
Gametes
S s S s
F_2 generation
Eggs
Sperm
SS
Ss
Ss
ss
4 *S* and *s* gametes combine randomly to produce two different seed phenotypes in the F_2 plants, as this Punnett square shows.

10.4 Mendel's Explanation of Experiment 1 *(Page 192)*

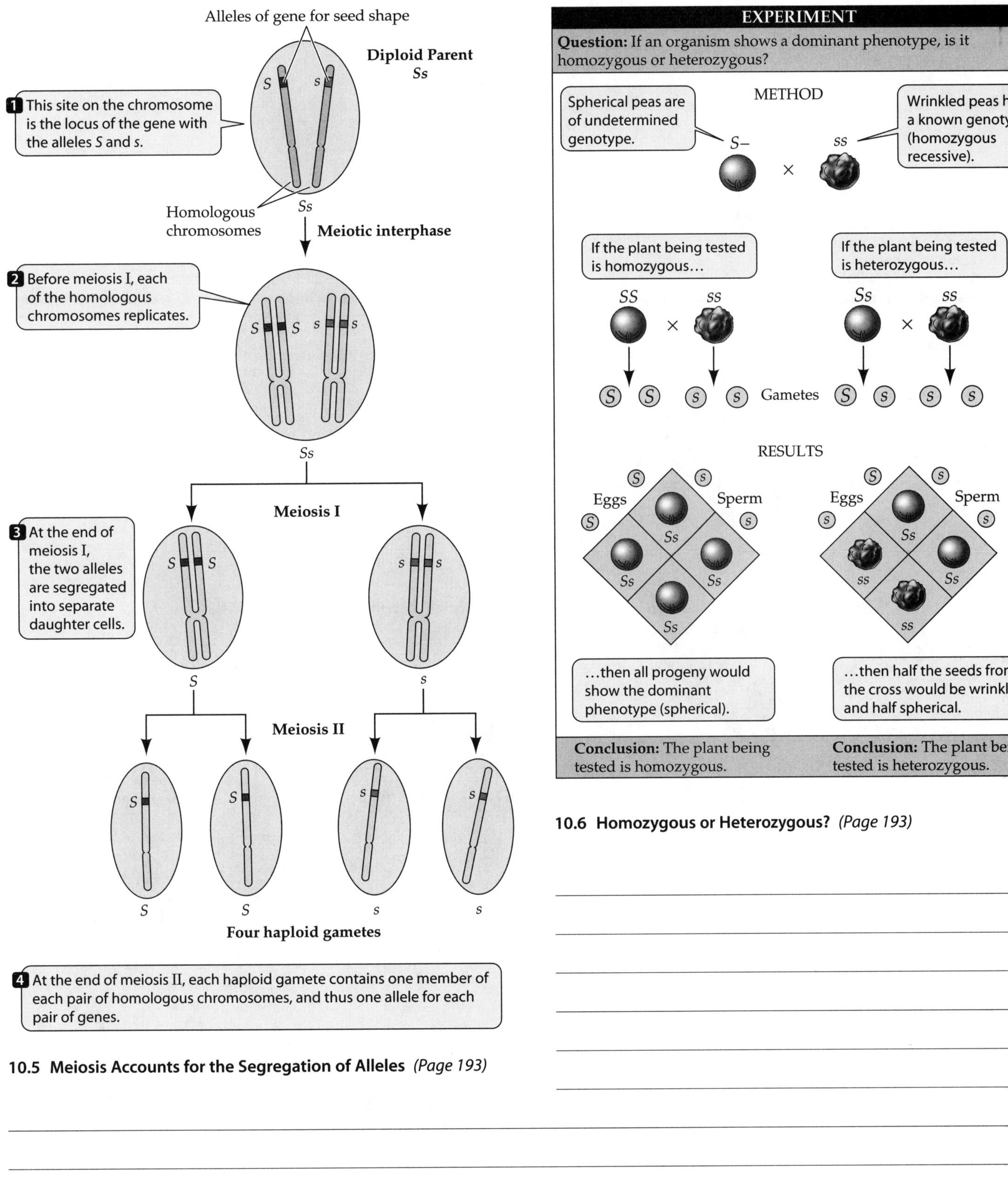
Alleles of gene for seed shape
Diploid Parent
Ss
S
s
1 This site on the chromosome is the locus of the gene with the alleles S and s.
Homologous chromosomes
Ss
Meiotic interphase
2 Before meiosis I, each of the homologous chromosomes replicates.
S S s s
Ss
Meiosis I
3 At the end of meiosis I, the two alleles are segregated into separate daughter cells.
S S
s s
S
s
Meiosis II
S S s s
S S s s
Four haploid gametes
4 At the end of meiosis II, each haploid gamete contains one member of each pair of homologous chromosomes, and thus one allele for each pair of genes.
EXPERIMENT
Question: If an organism shows a dominant phenotype, is it homozygous or heterozygous?
METHOD
Spherical peas are of undetermined genotype.
S–
ss
Wrinkled peas have a known genotype (homozygous recessive).
If the plant being tested is homozygous…
If the plant being tested is heterozygous…
SS
ss
Ss
ss
Gametes
RESULTS
Eggs
Sperm
Ss
ss
…then all progeny would show the dominant phenotype (spherical).
…then half the seeds from the cross would be wrinkled, and half spherical.
Conclusion: The plant being tested is homozygous.
Conclusion: The plant being tested is heterozygous.

10.6 Homozygous or Heterozygous? *(Page 193)*

10.5 Meiosis Accounts for the Segregation of Alleles *(Page 193)*

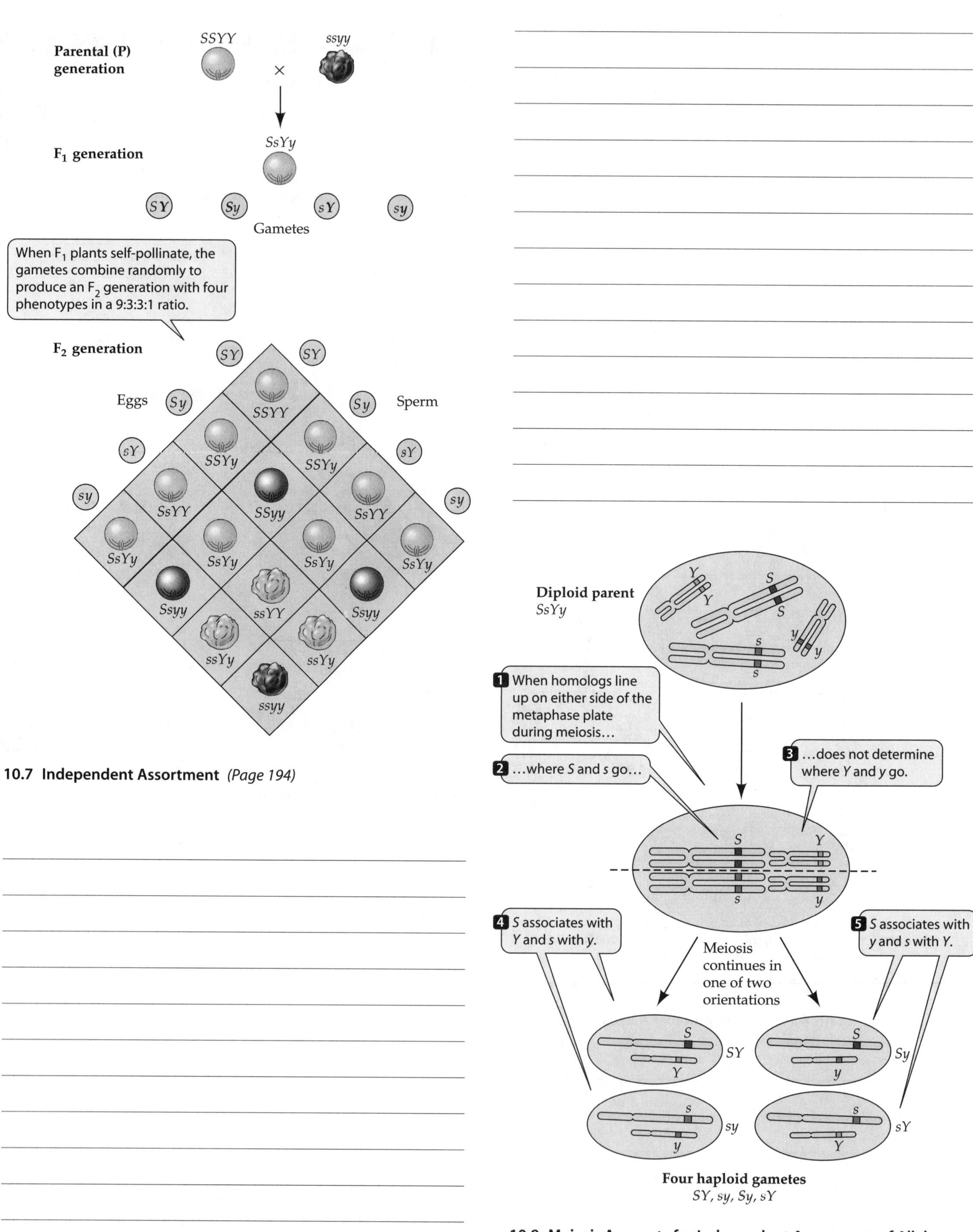

Parental (P) generation
SSYY
ssyy
×
F1 generation
SsYy
SY
Sy
sY
sy
Gametes
When F1 plants self-pollinate, the gametes combine randomly to produce an F2 generation with four phenotypes in a 9:3:3:1 ratio.
F2 generation
Eggs
Sperm
SSYY
SSYy
SsYY
SsYy
SSyy
Ssyy
ssYY
ssYy
ssyy
Diploid parent
SsYy
1 When homologs line up on either side of the metaphase plate during meiosis…
2 …where S and s go…
3 …does not determine where Y and y go.
4 S associates with Y and s with y.
5 S associates with y and s with Y.
Meiosis continues in one of two orientations
Four haploid gametes
SY, sy, Sy, sY

10.7 Independent Assortment (Page 194)

10.8 Meiosis Accounts for Independent Assortment of Alleles (Page 195)

1 Two coin tosses are independent events with an outcome probability (*P*) of ½ each.

2 This outcome is the result of these two independent events. The joint probability is ½ × ½ = ¼ (multiplication rule).

(*P*) = ½ *S*

S *P* = ½

P = ½ *s*

s *P* = ½

S *S*

S *s* ½ × ½ = ¼

s *S* ½ × ½ = ¼

S *s* ½ × ½ = ¼

s *s* ½ × ½ = ¼

There are two ways to arrive at a heterozygote, so we add the probabilities of the two individual outcomes: ¼ + ¼ = ½ (addition rule).

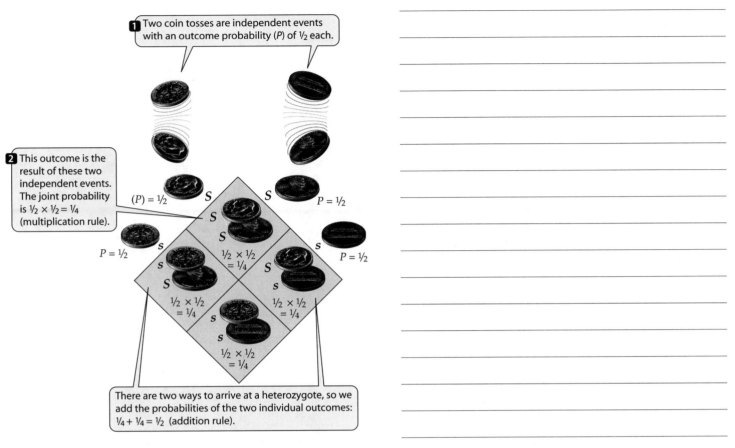

10.9 Using Probability Calculations in Genetics *(Page 196)*

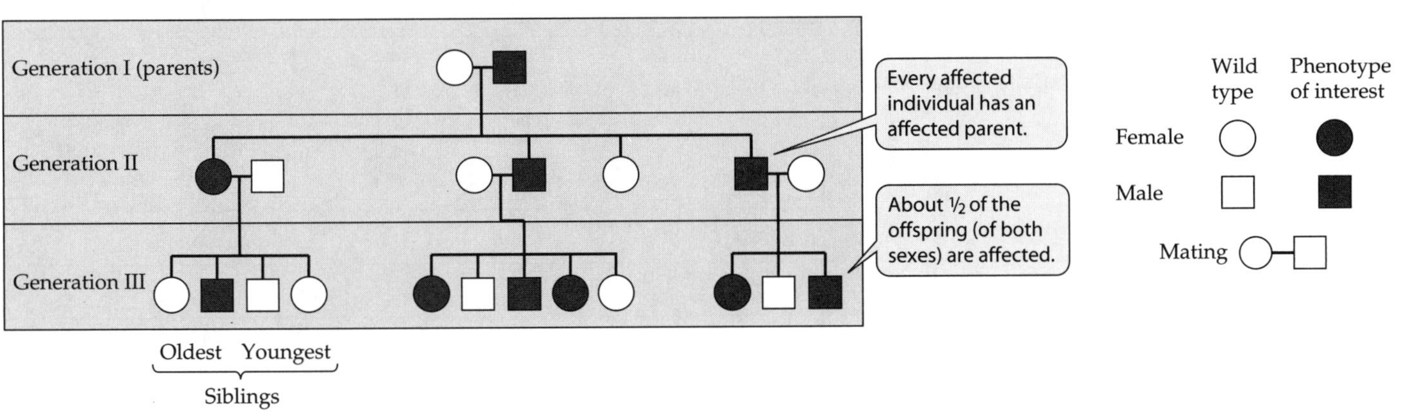
Generation I (parents)
Generation II
Generation III
Every affected individual has an affected parent.
About ½ of the offspring (of both sexes) are affected.
Oldest Youngest
Siblings
Wild type
Phenotype of interest
Female
Male
Mating

10.10 Pedigree Analysis and Dominant Inheritance *(Page 197)*

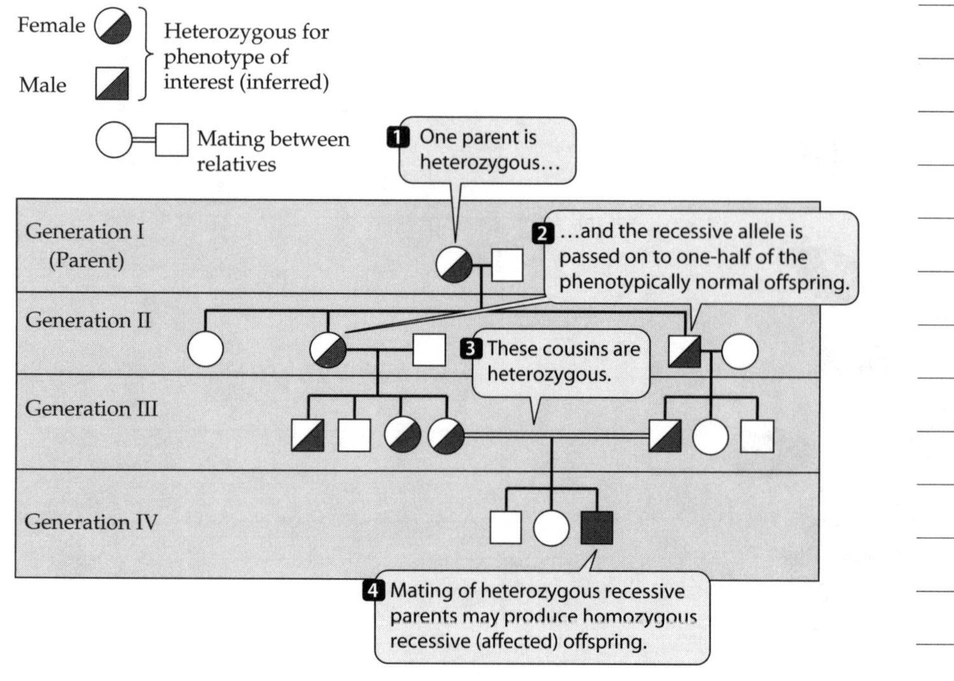
Female
Male
Heterozygous for phenotype of interest (inferred)
Mating between relatives
1 One parent is heterozygous…
2 …and the recessive allele is passed on to one-half of the phenotypically normal offspring.
3 These cousins are heterozygous.
4 Mating of heterozygous recessive parents may produce homozygous recessive (affected) offspring.
Generation I (Parent)
Generation II
Generation III
Generation IV

10.11 Recessive Inheritance *(Page 197)*

Possible genotypes	CC, Cc^{ch}, Cc^{h}, Cc	$c^{ch}c^{ch}$	$c^{ch}c^{h}$, $c^{ch}c$	$c^{h}c^{h}$, $c^{h}c$	cc
Phenotype	Dark gray	Chinchilla	Light gray	Himalayan	Albino

10.12 Inheritance of Coat Color in Rabbits *(Page 198)*

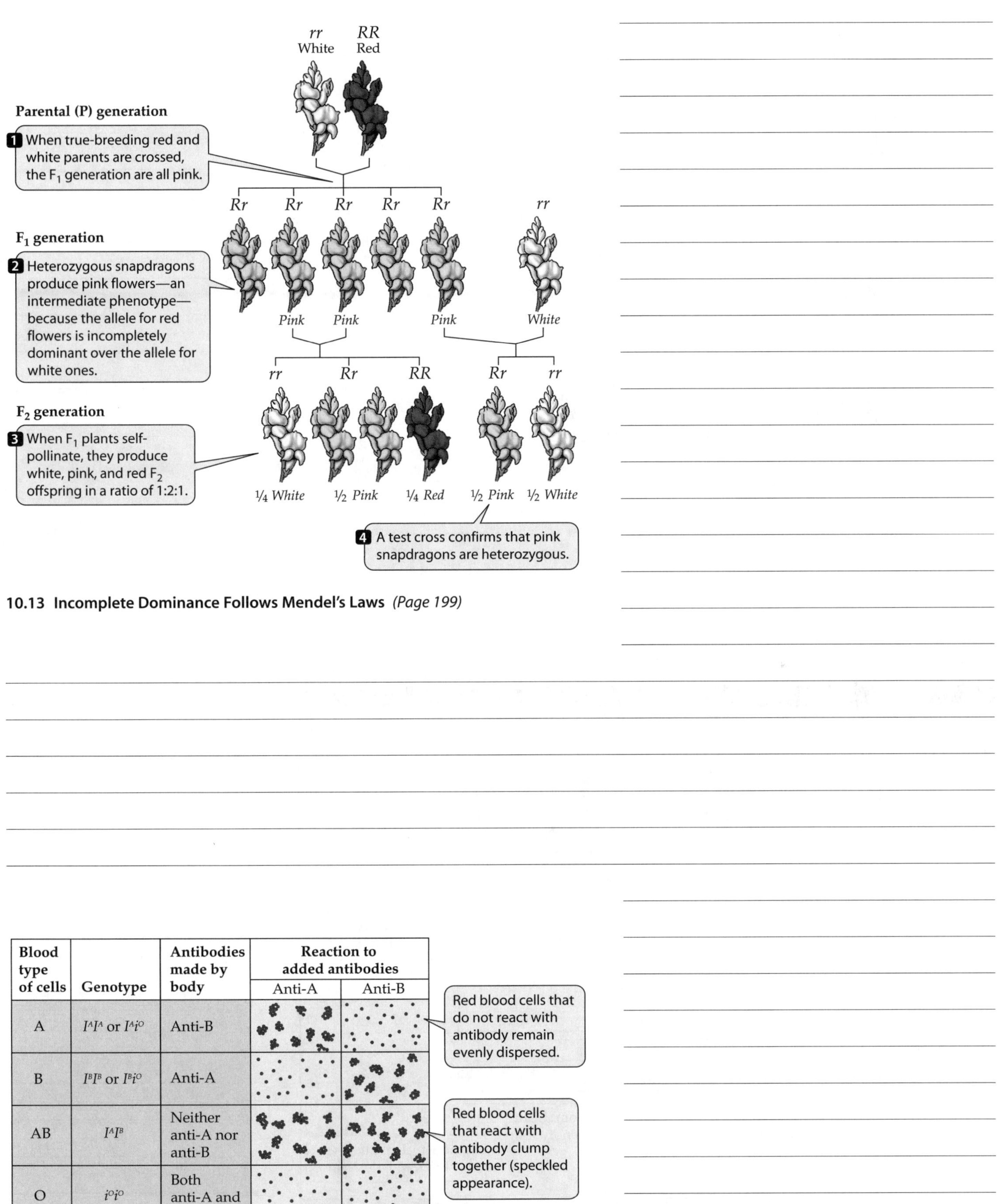
rr White
RR Red
Parental (P) generation
1 When true-breeding red and white parents are crossed, the F1 generation are all pink.
Rr Rr Rr Rr Rr rr
F1 generation
2 Heterozygous snapdragons produce pink flowers—an intermediate phenotype—because the allele for red flowers is incompletely dominant over the allele for white ones.
Pink Pink Pink White
rr Rr RR Rr rr
F2 generation
3 When F1 plants self-pollinate, they produce white, pink, and red F2 offspring in a ratio of 1:2:1.
¼ White ½ Pink ¼ Red ½ Pink ½ White
4 A test cross confirms that pink snapdragons are heterozygous.

10.13 Incomplete Dominance Follows Mendel's Laws *(Page 199)*

Blood type of cells	Genotype	Antibodies made by body	Reaction to added antibodies: Anti-A	Reaction to added antibodies: Anti-B
A	I^AI^A or I^Ai^O	Anti-B		
B	I^BI^B or I^Bi^O	Anti-A		
AB	I^AI^B	Neither anti-A nor anti-B		
O	i^Oi^O	Both anti-A and anti-B		

Red blood cells that do not react with antibody remain evenly dispersed.
Red blood cells that react with antibody clump together (speckled appearance).

10.14 ABO Blood Reactions Are Important in Transfusions *(Page 199)*

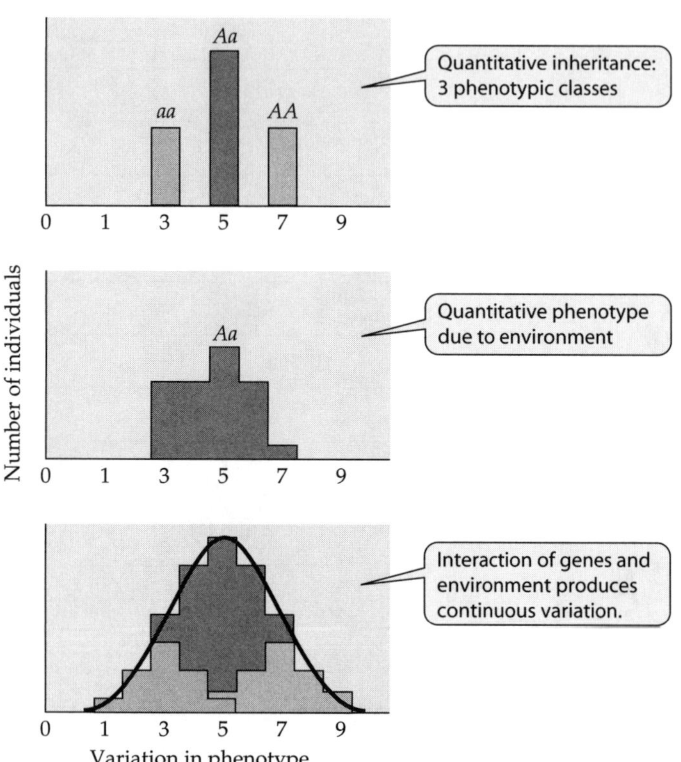
Aa
aa
AA
Quantitative inheritance: 3 phenotypic classes
0 1 3 5 7 9
Number of individuals
Aa
Quantitative phenotype due to environment
0 1 3 5 7 9
Interaction of genes and environment produces continuous variation.
0 1 3 5 7 9
Variation in phenotype

10.17 Quantitative Variation *(Page 201)*

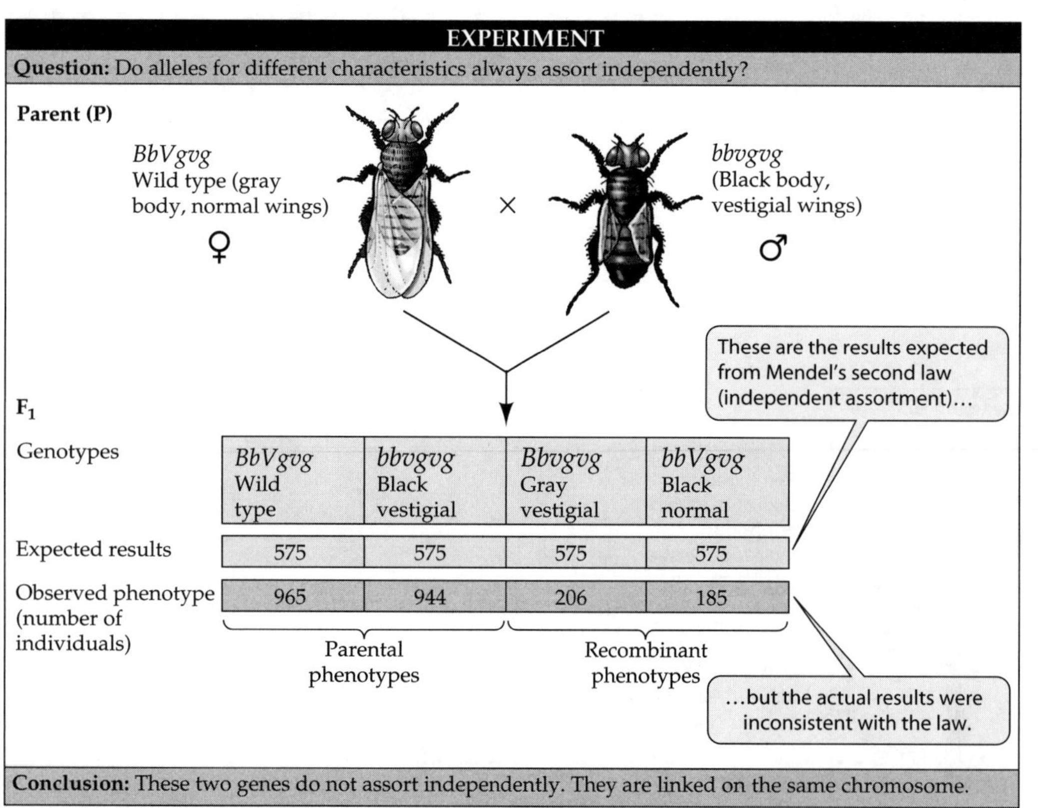
EXPERIMENT
Question: Do alleles for different characteristics always assort independently?
Parent (P)
BbVgvg
Wild type (gray body, normal wings)
♀
×
bbvgvg
(Black body, vestigial wings)
♂
These are the results expected from Mendel's second law (independent assortment)...
F1
Genotypes
BbVgvg Wild type
bbvgvg Black vestigial
Bbvgvg Gray vestigial
bbVgvg Black normal
Expected results
575 575 575 575
Observed phenotype (number of individuals)
965 944 206 185
Parental phenotypes
Recombinant phenotypes
...but the actual results were inconsistent with the law.
Conclusion: These two genes do not assort independently. They are linked on the same chromosome.

10.18 Some Alleles Do Not Assort Independently *(Page 202)*

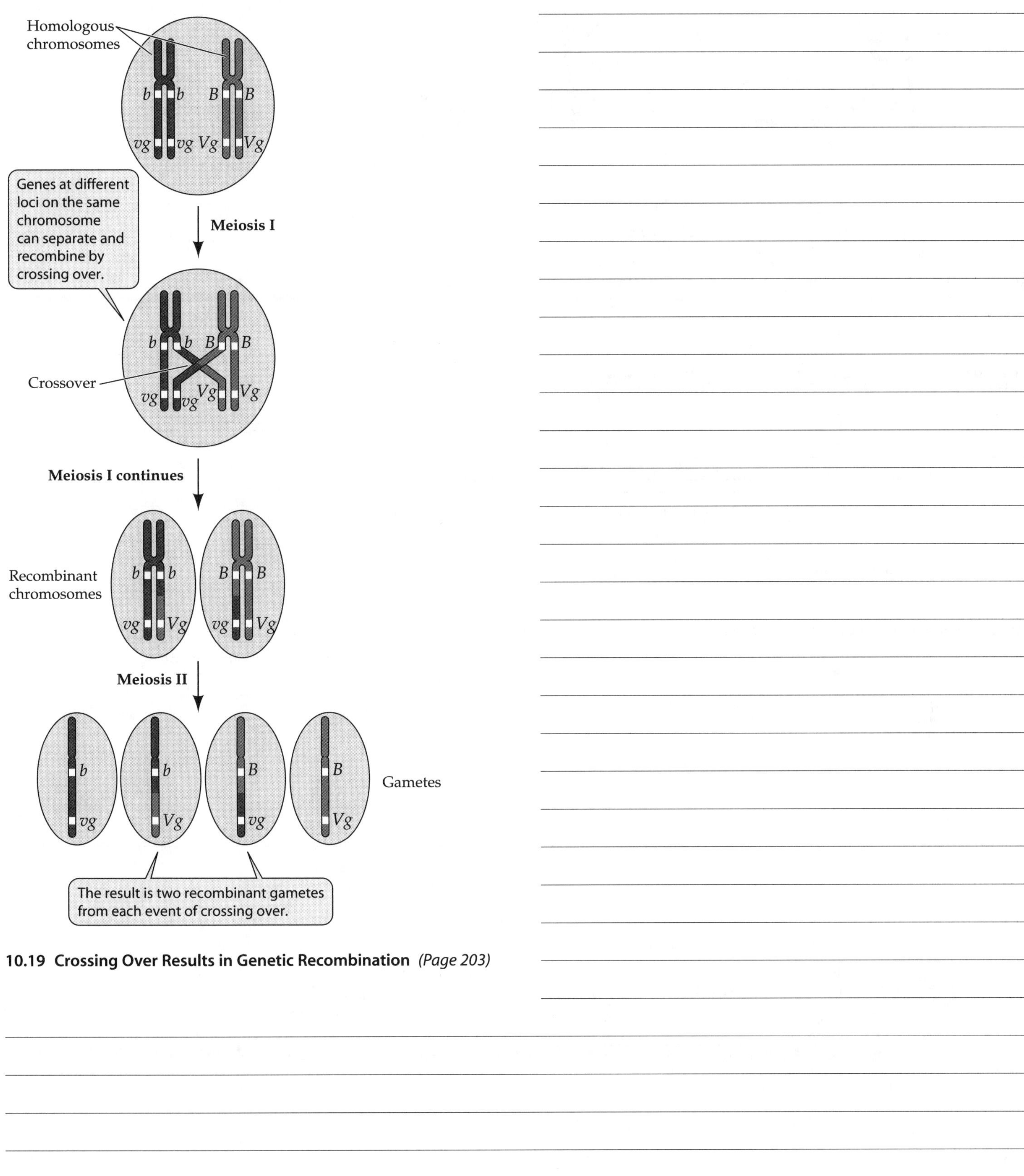
Homologous chromosomes
b b B B
vg vg Vg Vg
Genes at different loci on the same chromosome can separate and recombine by crossing over.
Meiosis I
b b B B
Crossover
vg vg Vg Vg
Meiosis I continues
Recombinant chromosomes
b b B B
vg Vg vg Vg
Meiosis II
b vg
b Vg
B vg
B Vg
Gametes
The result is two recombinant gametes from each event of crossing over.

10.19 Crossing Over Results in Genetic Recombination *(Page 203)*

Gray normal (wild type)

♀

B Vg

b vg

×

Black vestigial

♂

b vg

b vg

Recombination

Parental genotypes

Recombinant genotypes

b vg b vg	B Vg b vg	B vg b vg	b Vg b vg
Black vestigial	Wild type	Gray vestigial	Black normal
944	965	206	185

Number of individuals

Parental phenotypes

Recombinant (nonparental) phenotypes

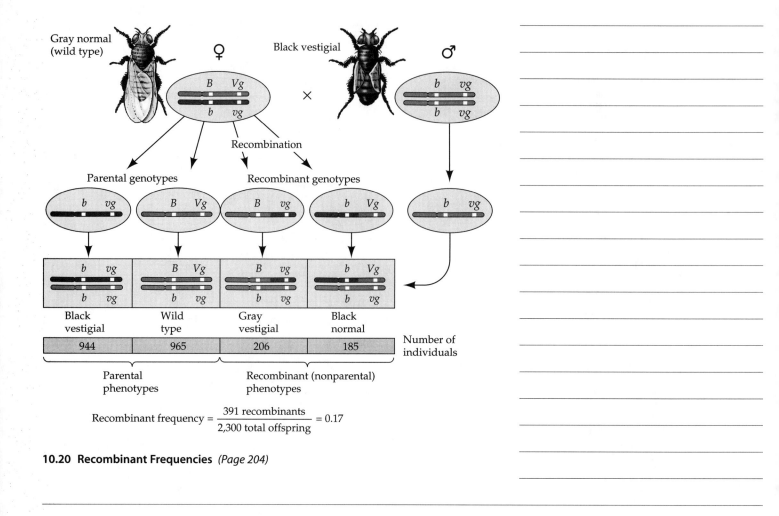

$$\text{Recombinant frequency} = \frac{\text{391 recombinants}}{\text{2,300 total offspring}} = 0.17$$

10.20 Recombinant Frequencies *(Page 204)*

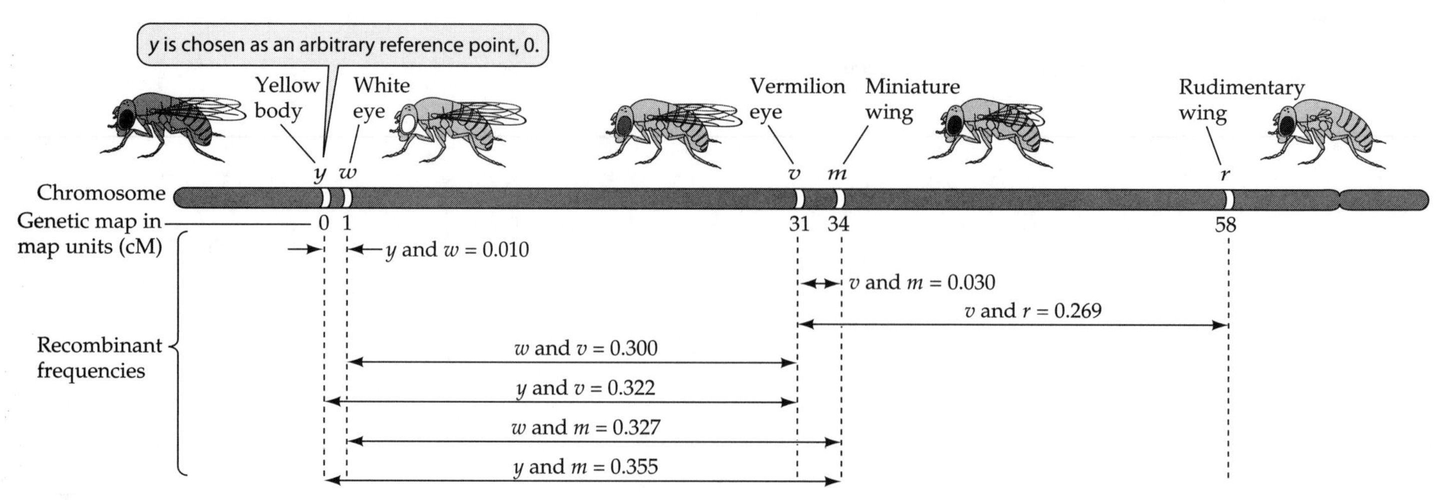

y is chosen as an arbitrary reference point, 0.
Yellow body
White eye
Vermilion eye
Miniature wing
Rudimentary wing
y w
v m
r
Chromosome
Genetic map in map units (cM)
0 1
31 34
58
y and w = 0.010
v and m = 0.030
v and r = 0.269
w and v = 0.300
y and v = 0.322
w and m = 0.327
y and m = 0.355
Recombinant frequencies

10.21 Steps toward a Genetic Map *(Page 204)*

1 At the outset, we have no idea of the individual distances, and there are several possible sequences (*a-b-c*, *a-c-b*, *b-a-c*).

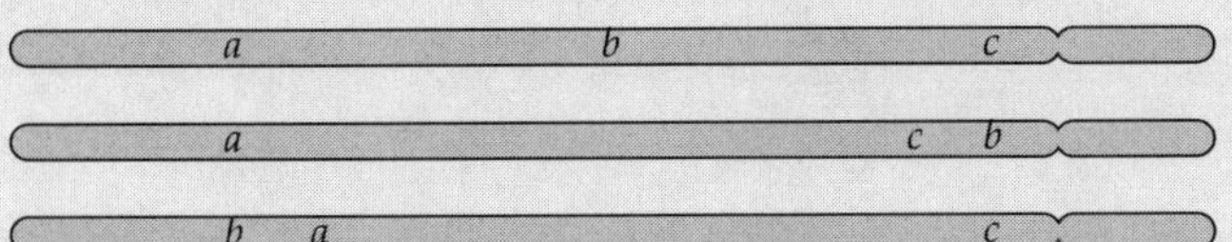
a
b
c
a
c
b
b
a
c

We make a cross *AABB* × *aabb*, and obtain an F_1 generation with a genotype *AaBb*. We test cross these *AaBb* individuals with *aabb*. Here are the genotypes of the first 1,000 progeny:

450 *AaBb*, 450 *aabb*, 50 *Aabb*, and 50 *aaBb*.

2 How far apart are the *a* and *b* genes? Well, what is the recombinant frequency? Which are the recombinant types, and which are the parental types?

Recombinant frequency (*a* to *b*) = (50 + 50)/1,000 = 0.1
So the map distance is
Map distance = 100 × recombinant frequency =
100 × 0.1 = 10 cM

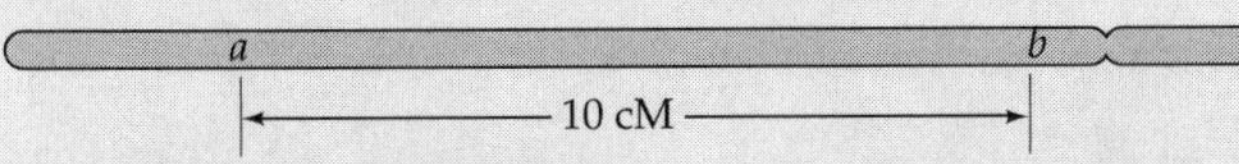
a
b
10 cM

3 Now we make a cross *AACC* × *aacc*, obtain an F_1 generation, and test cross it, obtaining:

460 *AaCc*, 460 *aacc*, 40 *Aacc*, and 40 *aaCc*.

How far apart are the *a* and *c* genes?

Recombinant frequency (*a* to *c*) = (40 + 40)/1,000 = 0.08
Map distance = 100 × recombinant frequency =
100 × 0.08 = 8 cM

a
c
8 cM

4 How far apart are the *b* and *c* genes?

We make a cross *BBCC* × *bbcc*, obtain an F_1 generation, and test cross it, obtaining:

490 *BbCc*, 490 *bbcc*, 10 *Bbcc*, and 10 *bbCc*.

Determine the map distance between *b* and *c*.

Recombinant frequency (*b* to *c*) = (10 + 10)/1,000 = 0.02
Map distance = 100 × recombinant frequency =
100 × 0.02 = 2 cM

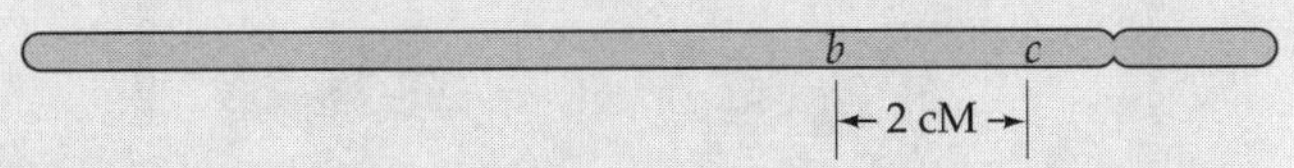
b
c
2 cM

5 Which of the three genes is between the other two?
Because *a* and *b* are the farthest apart, *c* must be between them.

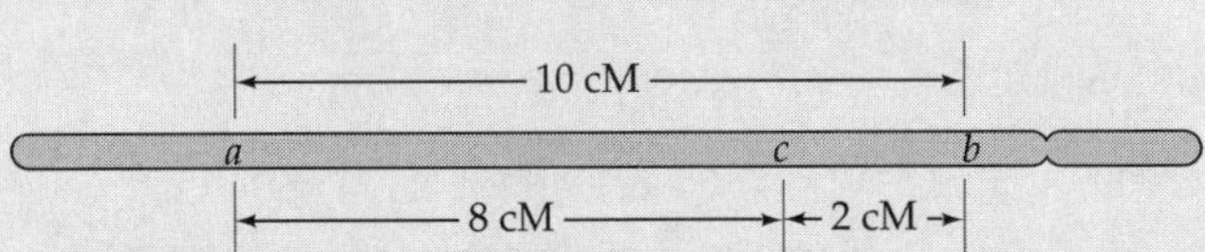
10 cM
a
c
b
8 cM
2 cM

These numbers add up perfectly, but in most real cases they don't add up perfectly because of multiple crossovers.

10.22 Map These Genes *(Page 205)*

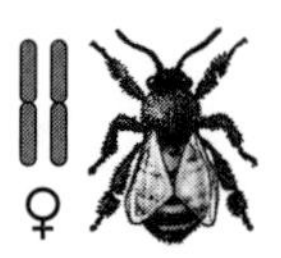
♀
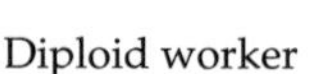
Diploid worker
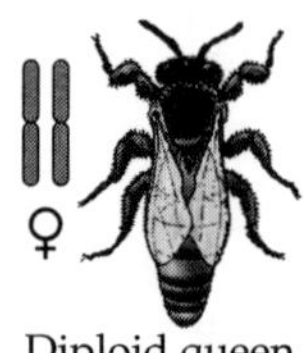
♀
Diploid queen
♂
Haploid drone

In-Text Art *(Page 205)*

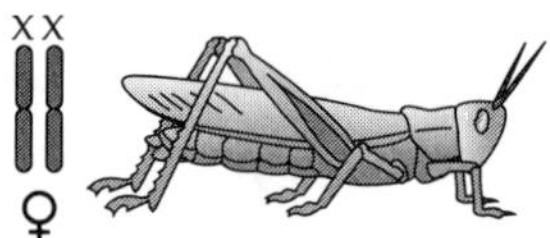
XX
♀

X
♂

In-Text Art *(Page 206)*

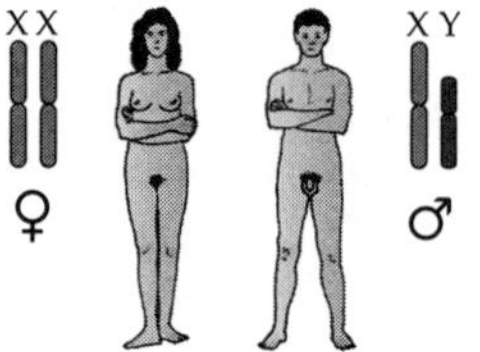
XX
♀
XY
♂

In-Text Art *(Page 206)*

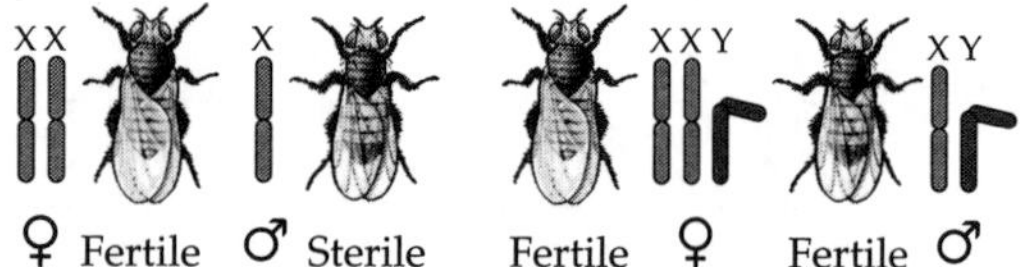
XX
♀ Fertile
X
♂ Sterile
XXY
Fertile ♀
XY
Fertile ♂

In-Text Art *(Page 206)*

ZW
♀

ZZ
♂

In-Text Art *(Page 207)*

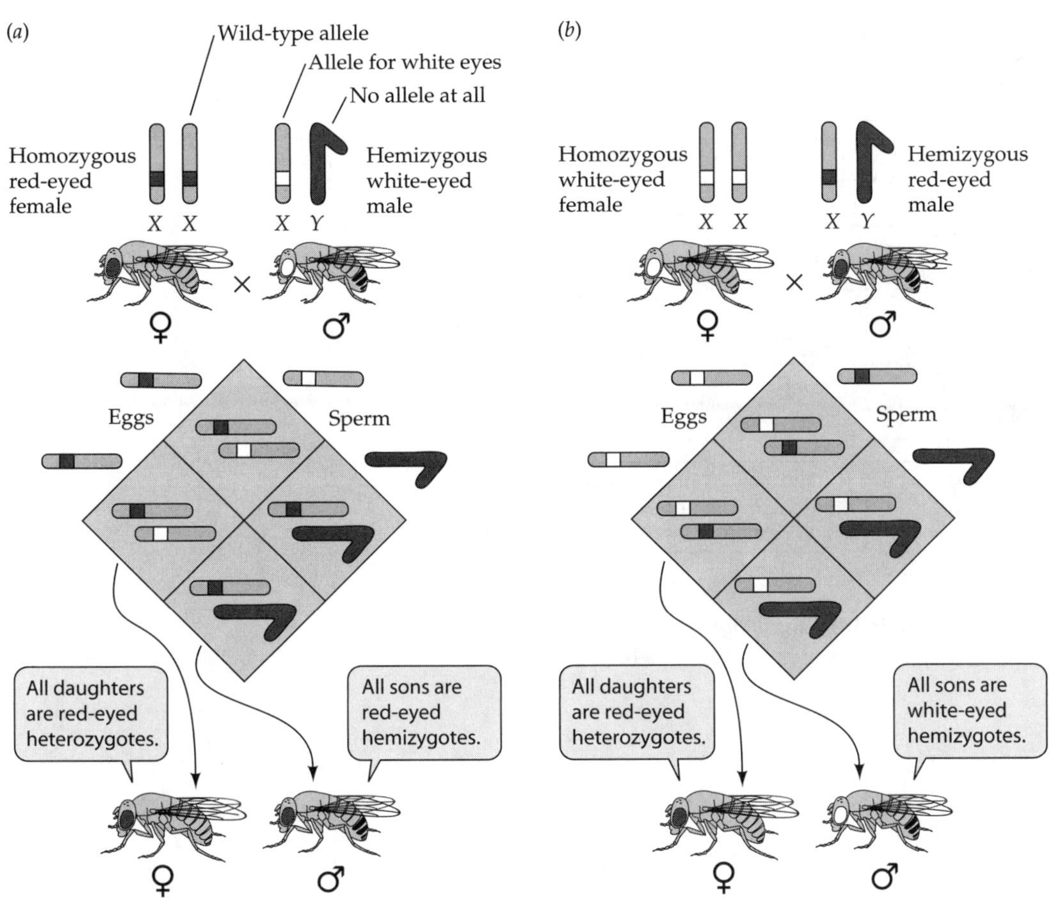
(a)
Wild-type allele
Allele for white eyes
No allele at all
Homozygous red-eyed female
X X
X Y
Hemizygous white-eyed male
♀
♂
Eggs
Sperm
All daughters are red-eyed heterozygotes.
All sons are red-eyed hemizygotes.
(b)
Homozygous white-eyed female
X X
X Y
Hemizygous red-eyed male
Eggs
Sperm
All daughters are red-eyed heterozygotes.
All sons are white-eyed hemizygotes.

10.23 Eye Color Is a Sex-Linked Trait in *Drosophila* *(Page 207)*

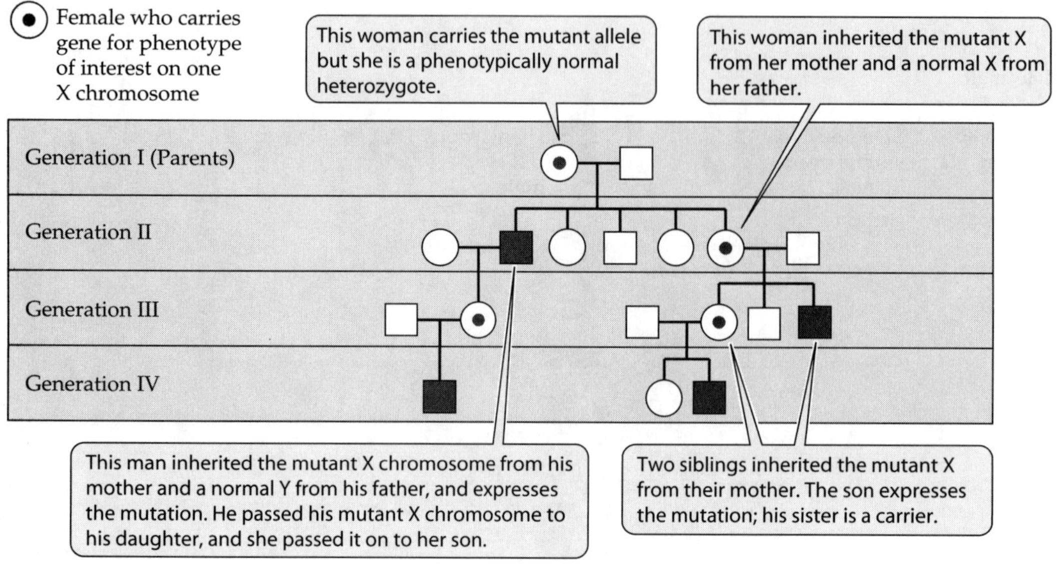
Female who carries gene for phenotype of interest on one X chromosome
This woman carries the mutant allele but she is a phenotypically normal heterozygote.
This woman inherited the mutant X from her mother and a normal X from her father.
Generation I (Parents)
Generation II
Generation III
Generation IV
This man inherited the mutant X chromosome from his mother and a normal Y from his father, and expresses the mutation. He passed his mutant X chromosome to his daughter, and she passed it on to her son.
Two siblings inherited the mutant X from their mother. The son expresses the mutation; his sister is a carrier.

10.24 Red-Green Color Blindness Is a Sex-Linked Trait In Humans *(Page 208)*

11 DNA and Its Role in Heredity

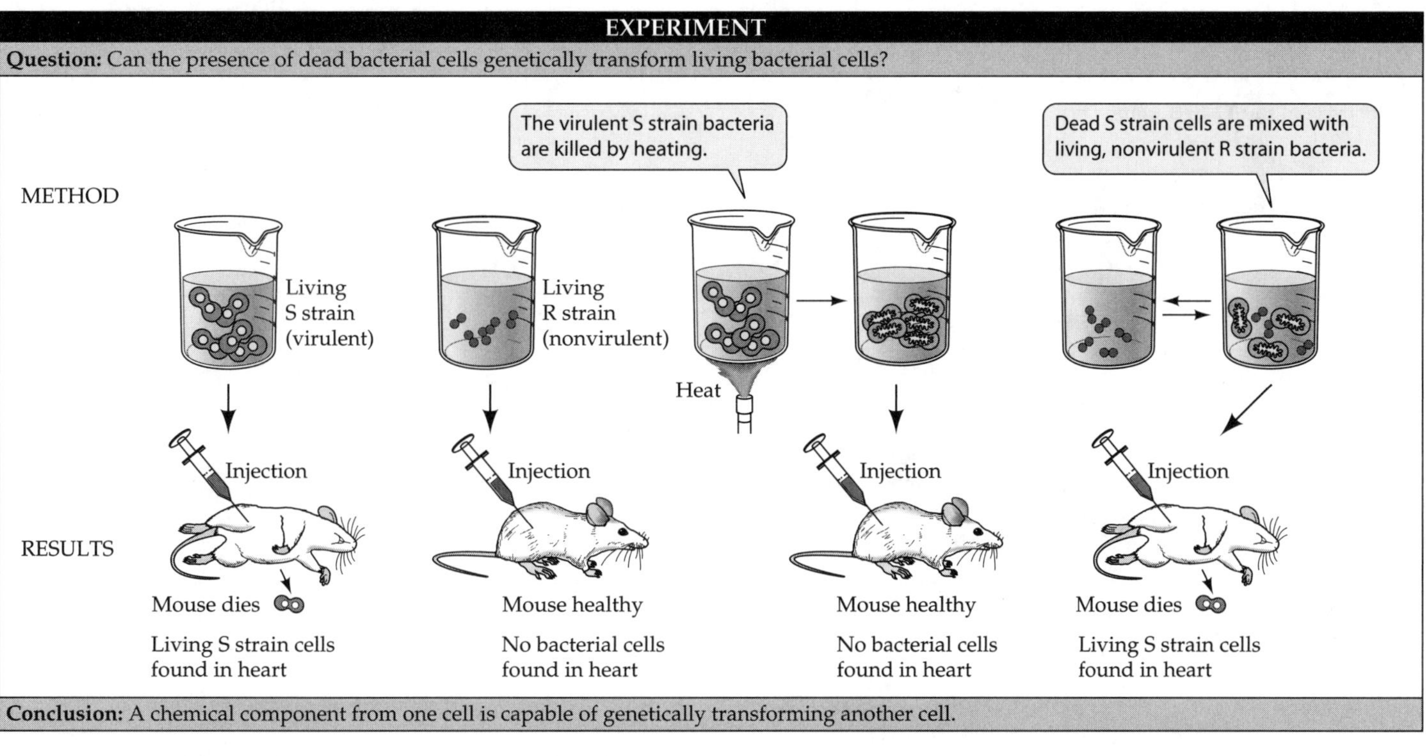

EXPERIMENT
Question: Can the presence of dead bacterial cells genetically transform living bacterial cells?
The virulent S strain bacteria are killed by heating.
Dead S strain cells are mixed with living, nonvirulent R strain bacteria.
METHOD
Living S strain (virulent)
Living R strain (nonvirulent)
Heat
Injection
Injection
Injection
Injection
RESULTS
Mouse dies
Living S strain cells found in heart
Mouse healthy
No bacterial cells found in heart
Mouse healthy
No bacterial cells found in heart
Mouse dies
Living S strain cells found in heart
Conclusion: A chemical component from one cell is capable of genetically transforming another cell.

11.1 Genetic Transformation of Nonvirulent Pneumococci *(Page 214)*

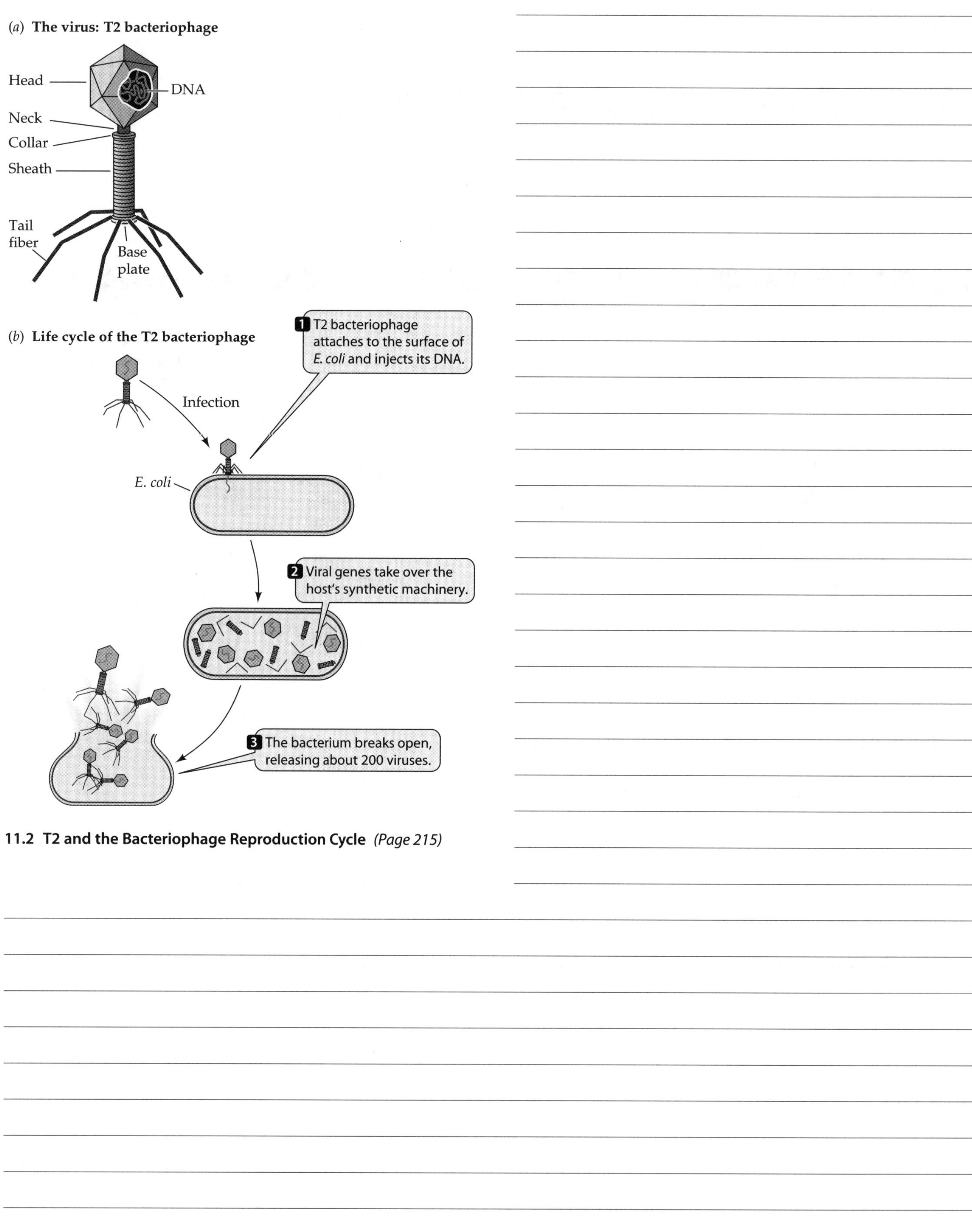
(a) The virus: T2 bacteriophage
Head
DNA
Neck
Collar
Sheath
Tail fiber
Base plate
(b) Life cycle of the T2 bacteriophage
Infection
1 T2 bacteriophage attaches to the surface of E. coli and injects its DNA.
E. coli
2 Viral genes take over the host's synthetic machinery.
3 The bacterium breaks open, releasing about 200 viruses.

11.2 T2 and the Bacteriophage Reproduction Cycle *(Page 215)*

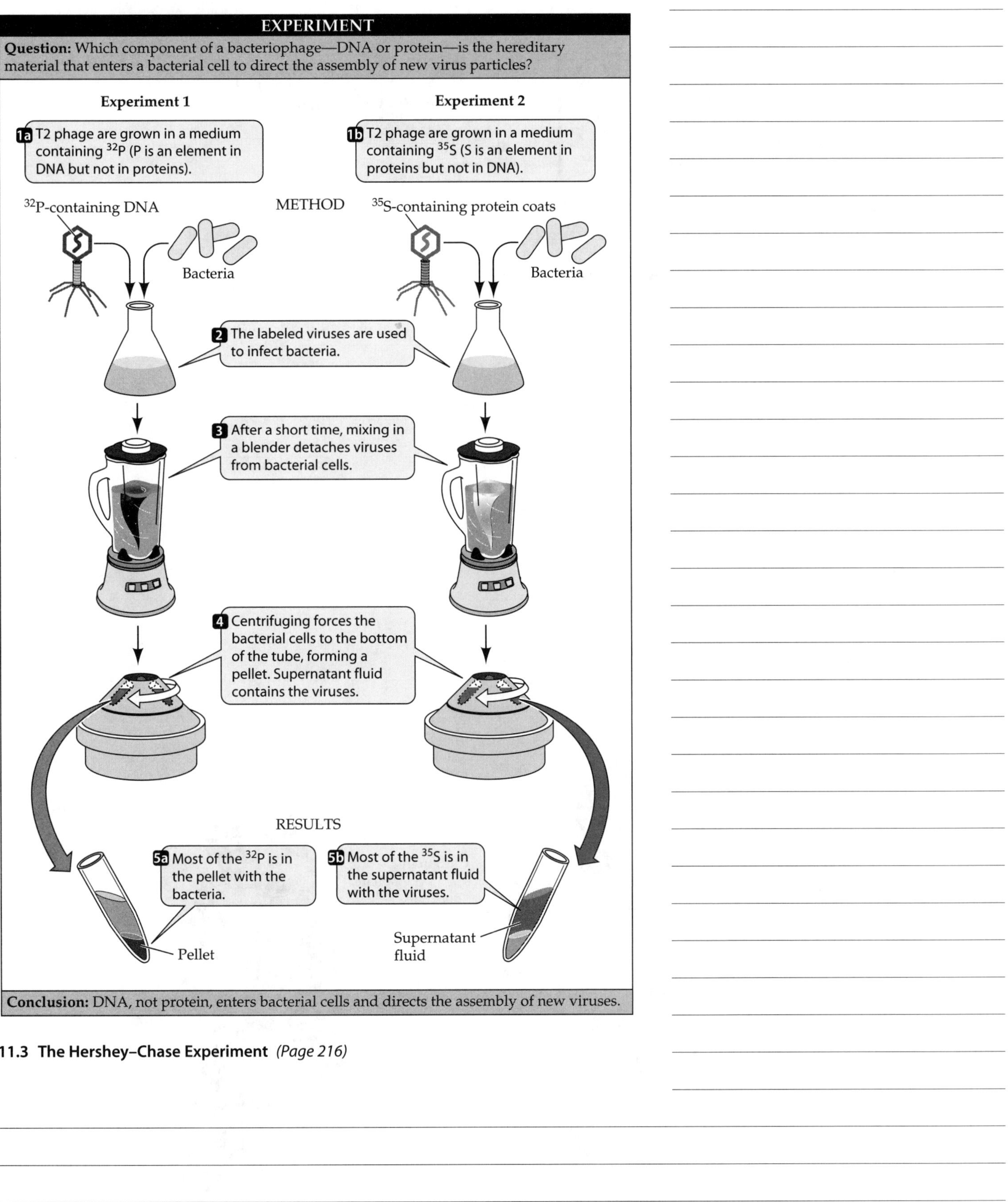
EXPERIMENT
Question: Which component of a bacteriophage—DNA or protein—is the hereditary material that enters a bacterial cell to direct the assembly of new virus particles?
Experiment 1
Experiment 2
1a T2 phage are grown in a medium containing 32P (P is an element in DNA but not in proteins).
1b T2 phage are grown in a medium containing 35S (S is an element in proteins but not in DNA).
32P-containing DNA
METHOD
35S-containing protein coats
Bacteria
Bacteria
2 The labeled viruses are used to infect bacteria.
3 After a short time, mixing in a blender detaches viruses from bacterial cells.
4 Centrifuging forces the bacterial cells to the bottom of the tube, forming a pellet. Supernatant fluid contains the viruses.
RESULTS
5a Most of the 32P is in the pellet with the bacteria.
5b Most of the 35S is in the supernatant fluid with the viruses.
Pellet
Supernatant fluid
Conclusion: DNA, not protein, enters bacterial cells and directs the assembly of new viruses.

11.3 The Hershey–Chase Experiment *(Page 216)*

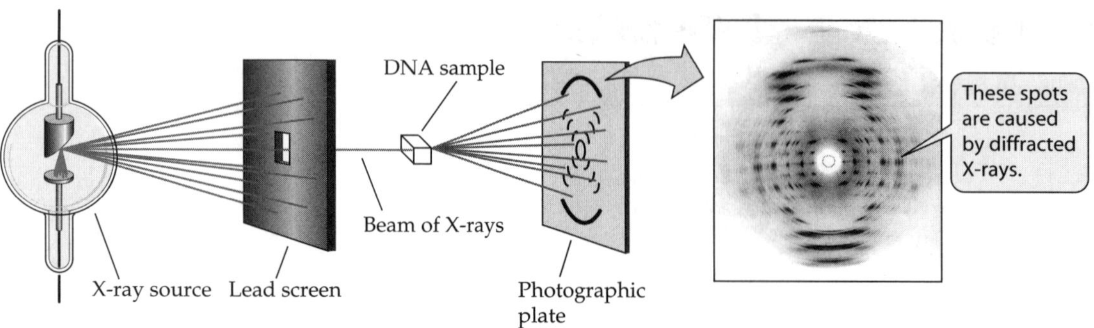
DNA sample
These spots are caused by diffracted X-rays.
Beam of X-rays
X-ray source
Lead screen
Photographic plate

11.4 X-Ray Crystallography Revealed the Basic Helical Structure of the DNA Molecule *(Page 217)*

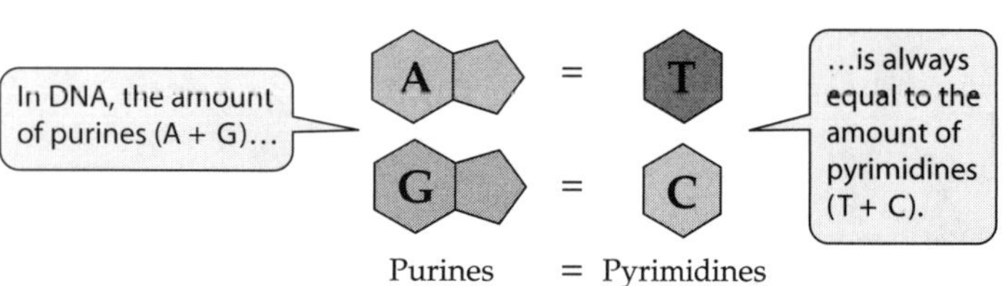
In DNA, the amount of purines (A + G)...
A
=
T
G
=
C
...is always equal to the amount of pyrimidines (T + C).
Purines
= Pyrimidines

11.5 Chargaff's Rule *(Page 217)*

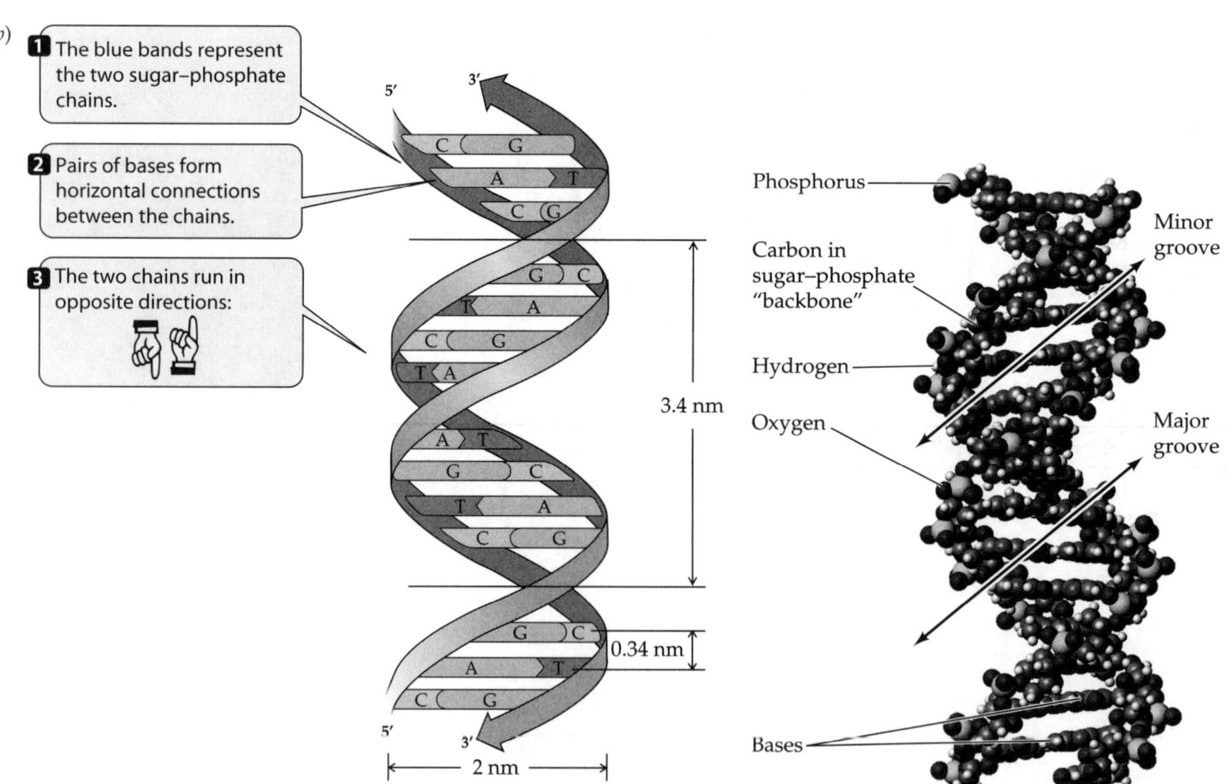
(b)
1 The blue bands represent the two sugar–phosphate chains.
2 Pairs of bases form horizontal connections between the chains.
3 The two chains run in opposite directions:
5′
3′
3.4 nm
0.34 nm
2 nm
Phosphorus
Carbon in sugar–phosphate "backbone"
Hydrogen
Oxygen
Bases
Minor groove
Major groove

11.6 DNA Is a Double Helix *(Page 218)*

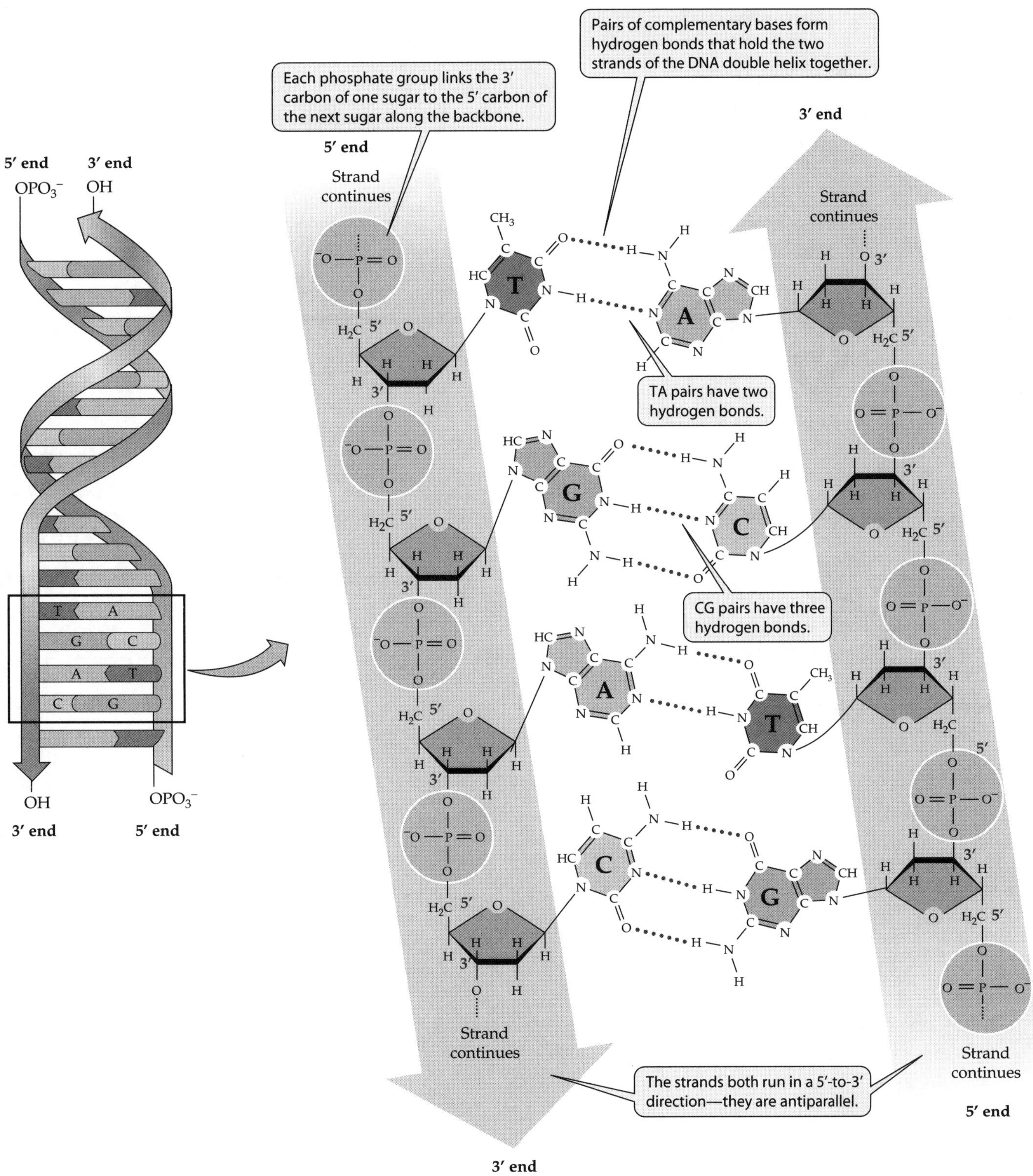

Pairs of complementary bases form hydrogen bonds that hold the two strands of the DNA double helix together.
Each phosphate group links the 3′ carbon of one sugar to the 5′ carbon of the next sugar along the backbone.
5′ end
3′ end
Strand continues
TA pairs have two hydrogen bonds.
CG pairs have three hydrogen bonds.
The strands both run in a 5′-to-3′ direction—they are antiparallel.

11.7 Base Pairing in DNA Is Complementary *(Page 219)*

Original DNA

After one round of replication

(a)

Semiconservative replication would produce molecules with both old and new DNA, but each molecule would contain one complete old strand and one new one.

(b)

Conservative replication would preserve the original molecule and generate an entirely new molecule.

(c)

Dispersive replication would produce two molecules with old and new DNA interspersed along each strand.

11.8 Three Models for DNA Replication *(Page 220)*

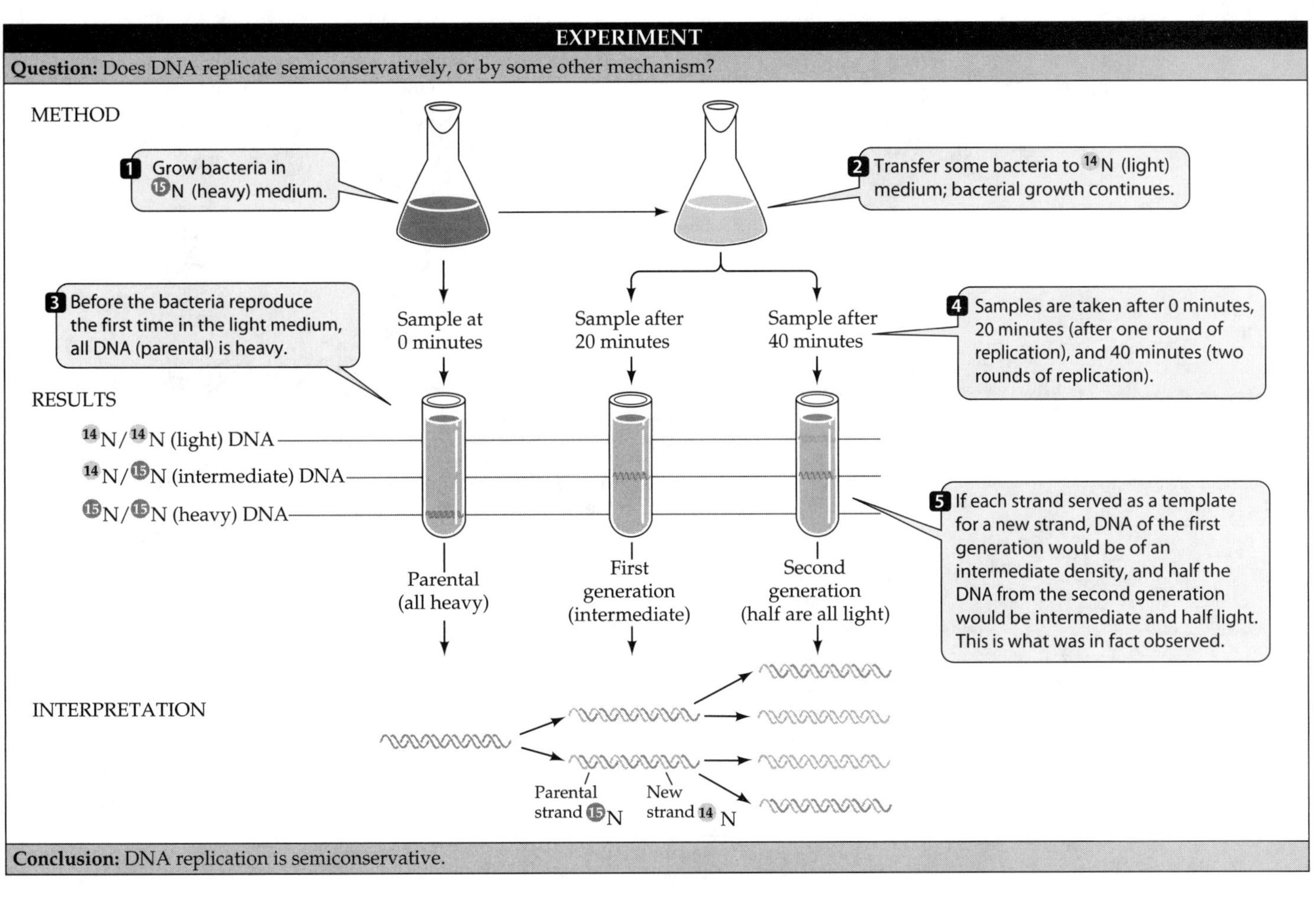
EXPERIMENT
Question: Does DNA replicate semiconservatively, or by some other mechanism?
METHOD
1 Grow bacteria in 15N (heavy) medium.
2 Transfer some bacteria to 14N (light) medium; bacterial growth continues.
3 Before the bacteria reproduce the first time in the light medium, all DNA (parental) is heavy.
Sample at 0 minutes
Sample after 20 minutes
Sample after 40 minutes
4 Samples are taken after 0 minutes, 20 minutes (after one round of replication), and 40 minutes (two rounds of replication).
RESULTS
14N/14N (light) DNA
14N/15N (intermediate) DNA
15N/15N (heavy) DNA
5 If each strand served as a template for a new strand, DNA of the first generation would be of an intermediate density, and half the DNA from the second generation would be intermediate and half light. This is what was in fact observed.
Parental (all heavy)
First generation (intermediate)
Second generation (half are all light)
INTERPRETATION
Parental strand 15N
New strand 14N
Conclusion: DNA replication is semiconservative.

11.9 The Meselson–Stahl Experiment *(Page 221)*

Growing strand
Template strand
5′ end
3′ end
OH
Phosphate
Sugar
Base
T A C G A T C G

1 Nucleotides are added to the "growing" end.

OH 3′ end
5′ end
OH 3′ end
5′ end

2 The enzyme DNA polymerase III adds the next deoxyribonucleotide, with the base C, to the —OH group at the 3′ end of the growing strand.

Growing strand
5′ end
3′ end
OH
OH 3′ end
Pyrophosphate ion
5′ end

3 Bonds linking the phosphate groups are broken, releasing energy to drive the reaction.

Phosphate ions

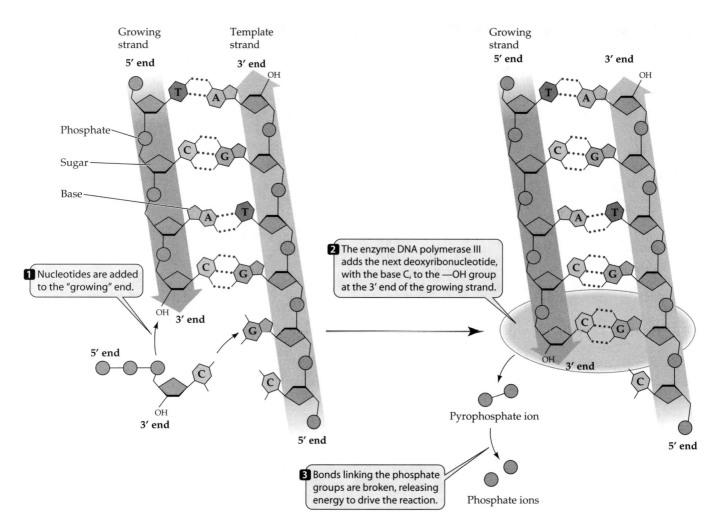

11.10 Each New DNA Strand Grows from its 5′ End to its 3′ End *(Page 223)*

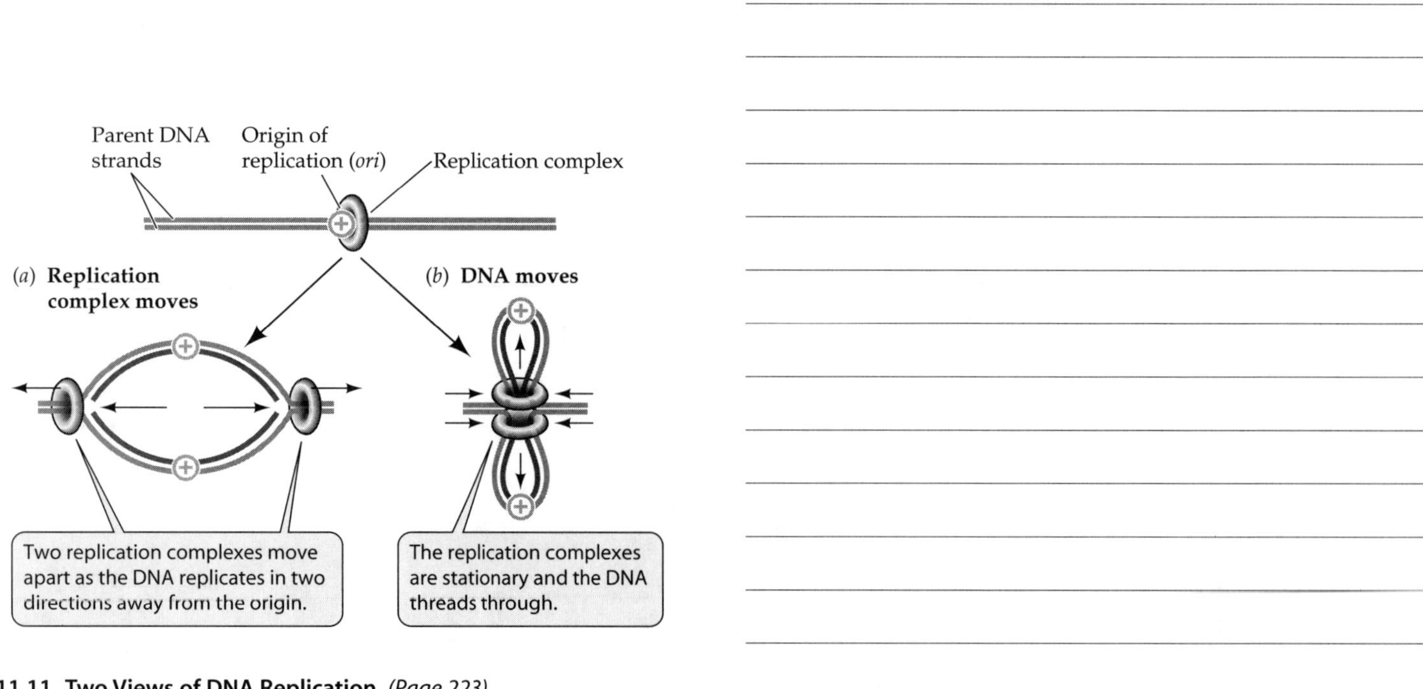
Parent DNA strands
Origin of replication (*ori*)
Replication complex
(*a*) Replication complex moves
(*b*) DNA moves
Two replication complexes move apart as the DNA replicates in two directions away from the origin.
The replication complexes are stationary and the DNA threads through.

11.11 Two Views of DNA Replication *(Page 223)*

(*a*) **Circular chromosome**

1 The origin of replication binds to the replication complex.

Replication complex

Ori

Ter

2 DNA is spooled through the complex, and comes out replicated.

Parental strand

New strand

3 Replication continues.

4 The two new DNA's are interlocked.

5 An enzyme, DNA topoisomerase, separates the two DNA's from each other.

(*b*) **Linear chromosome**

1 There are many origins of DNA replication.

Origin of replication

2 DNA is replicated from several origins simultaneously.

Replication forks

11.12 Replication in Small Circular and Large Linear Chromosomes *(Page 224)*

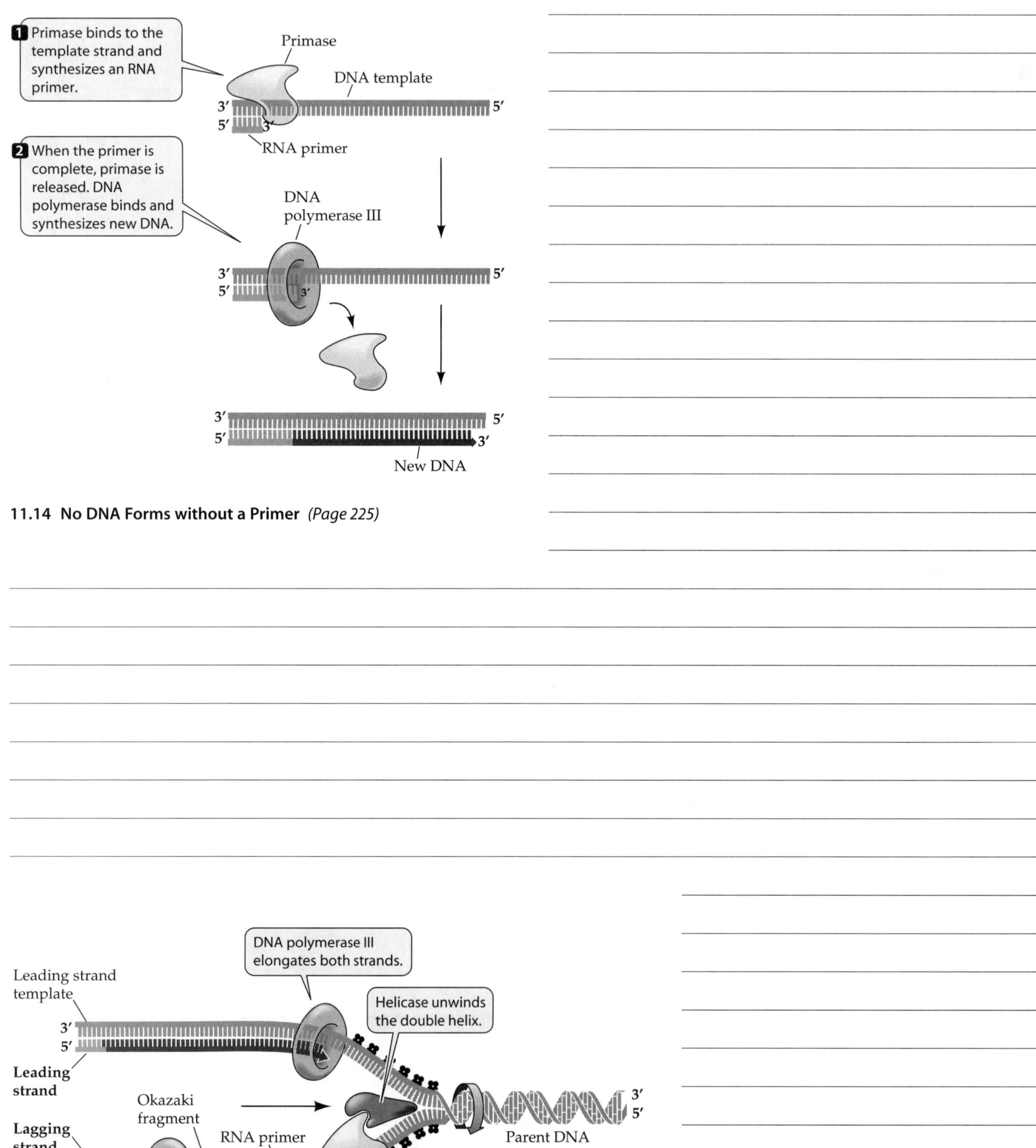

1 Primase binds to the template strand and synthesizes an RNA primer.
Primase
DNA template
3′
5′
3′
5′
RNA primer
2 When the primer is complete, primase is released. DNA polymerase binds and synthesizes new DNA.
DNA polymerase III
3′
5′
3′
5′
3′
5′
5′
3′
New DNA
DNA polymerase III elongates both strands.
Leading strand template
3′
5′
Leading strand
Helicase unwinds the double helix.
Okazaki fragment
3′
5′
Parent DNA
Lagging strand
RNA primer
3′
5′
Lagging strand template
Primase makes primer.
Single-strand DNA-binding proteins make the templates available to primase and DNA polymerase III.

11.14 No DNA Forms without a Primer *(Page 225)*

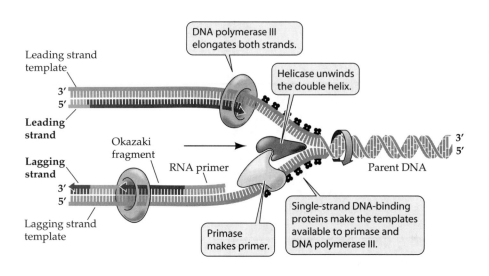

11.15 Many Proteins Collaborate at the Replication Fork *(Page 225)*

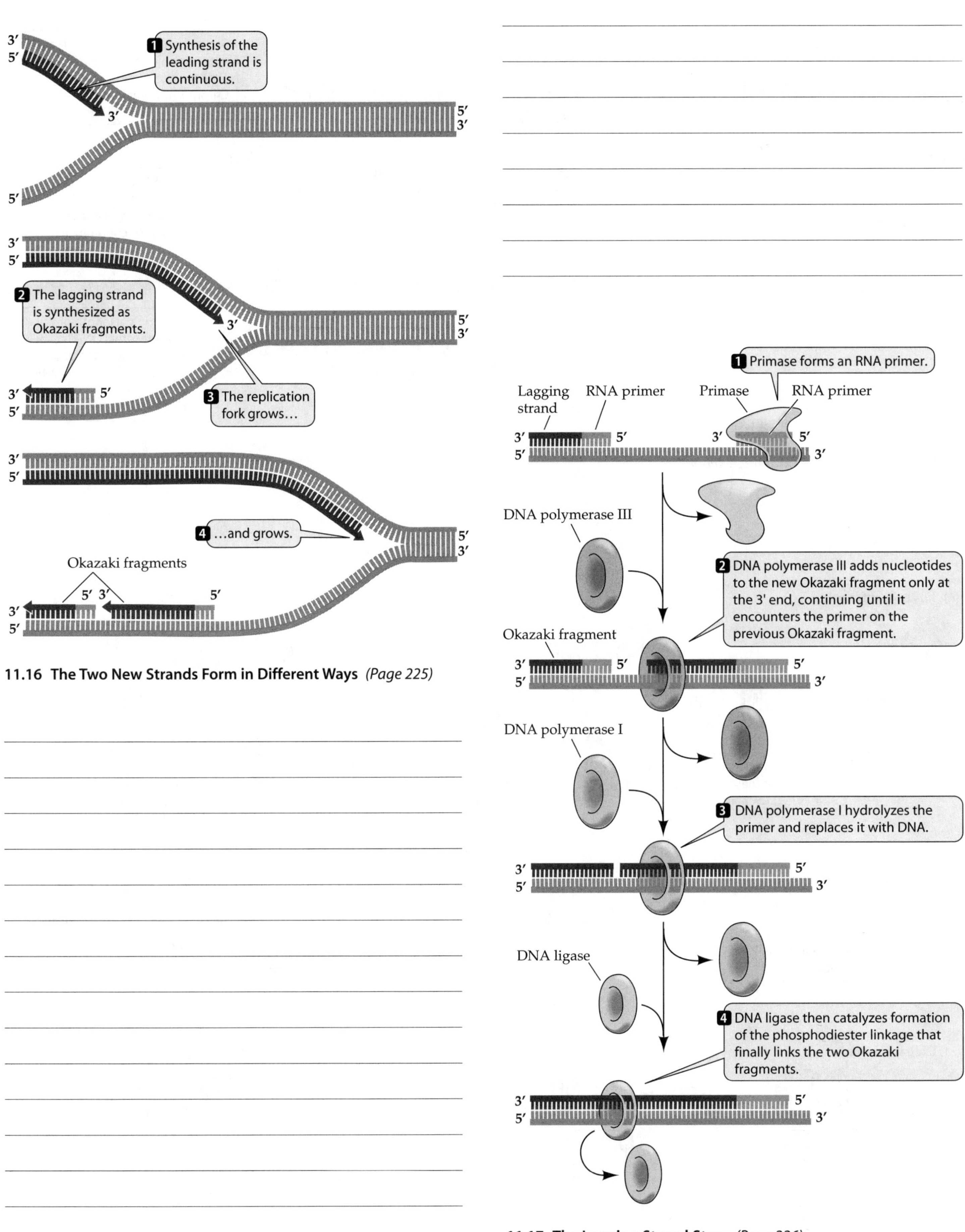
3′
5′
1 Synthesis of the leading strand is continuous.
3′
5′
3′
5′
3′
5′
2 The lagging strand is synthesized as Okazaki fragments.
3′
5′
3′
3′
5′
5′
3 The replication fork grows…
3′
5′
4 …and grows.
5′
3′
Okazaki fragments
5′ 3′
5′
3′
5′
1 Primase forms an RNA primer.
Lagging strand
RNA primer
Primase
RNA primer
3′
5′
5′
3′
5′
3′
DNA polymerase III
2 DNA polymerase III adds nucleotides to the new Okazaki fragment only at the 3' end, continuing until it encounters the primer on the previous Okazaki fragment.
Okazaki fragment
3′
5′
5′
5′
3′
DNA polymerase I
3 DNA polymerase I hydrolyzes the primer and replaces it with DNA.
3′
5′
5′
3′
DNA ligase
4 DNA ligase then catalyzes formation of the phosphodiester linkage that finally links the two Okazaki fragments.
3′
5′
5′
3′

11.16 The Two New Strands Form in Different Ways *(Page 225)*

11.17 The Lagging Strand Story *(Page 226)*

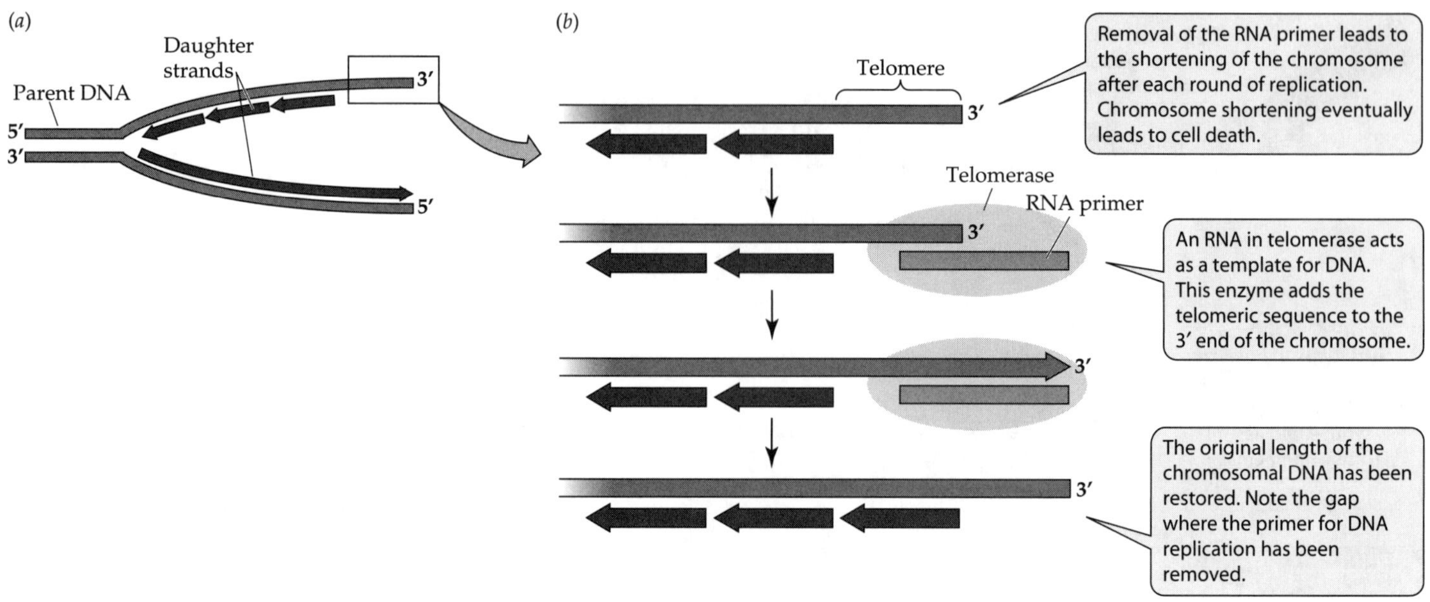
(a)
Daughter strands
Parent DNA
5′
3′
3′
5′
(b)
Telomere
3′
Removal of the RNA primer leads to the shortening of the chromosome after each round of replication. Chromosome shortening eventually leads to cell death.
Telomerase
RNA primer
3′
An RNA in telomerase acts as a template for DNA. This enzyme adds the telomeric sequence to the 3′ end of the chromosome.
3′
3′
The original length of the chromosomal DNA has been restored. Note the gap where the primer for DNA replication has been removed.

11.18 Telomeres and Telomerase *(Page 227)*

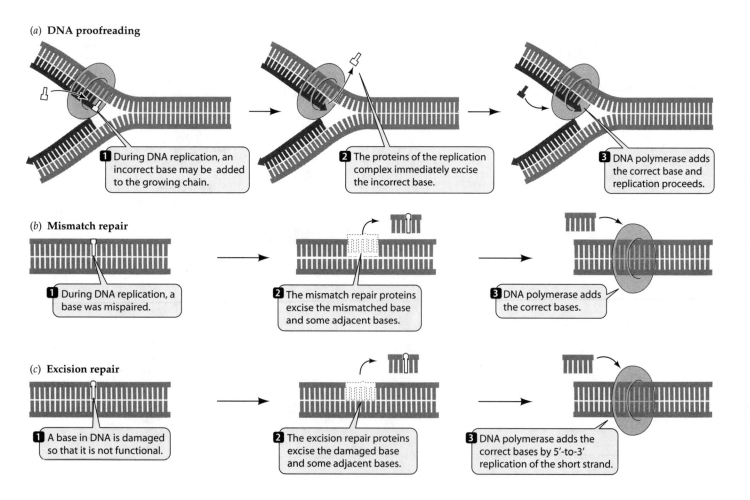

(*a*) **DNA proofreading**

1 During DNA replication, an incorrect base may be added to the growing chain.

2 The proteins of the replication complex immediately excise the incorrect base.

3 DNA polymerase adds the correct base and replication proceeds.

(*b*) **Mismatch repair**

1 During DNA replication, a base was mispaired.

2 The mismatch repair proteins excise the mismatched base and some adjacent bases.

3 DNA polymerase adds the correct bases.

(*c*) **Excision repair**

1 A base in DNA is damaged so that it is not functional.

2 The excision repair proteins excise the damaged base and some adjacent bases.

3 DNA polymerase adds the correct bases by 5′-to-3′ replication of the short strand.

11.19 DNA Repair Mechanisms *(Page 228)*

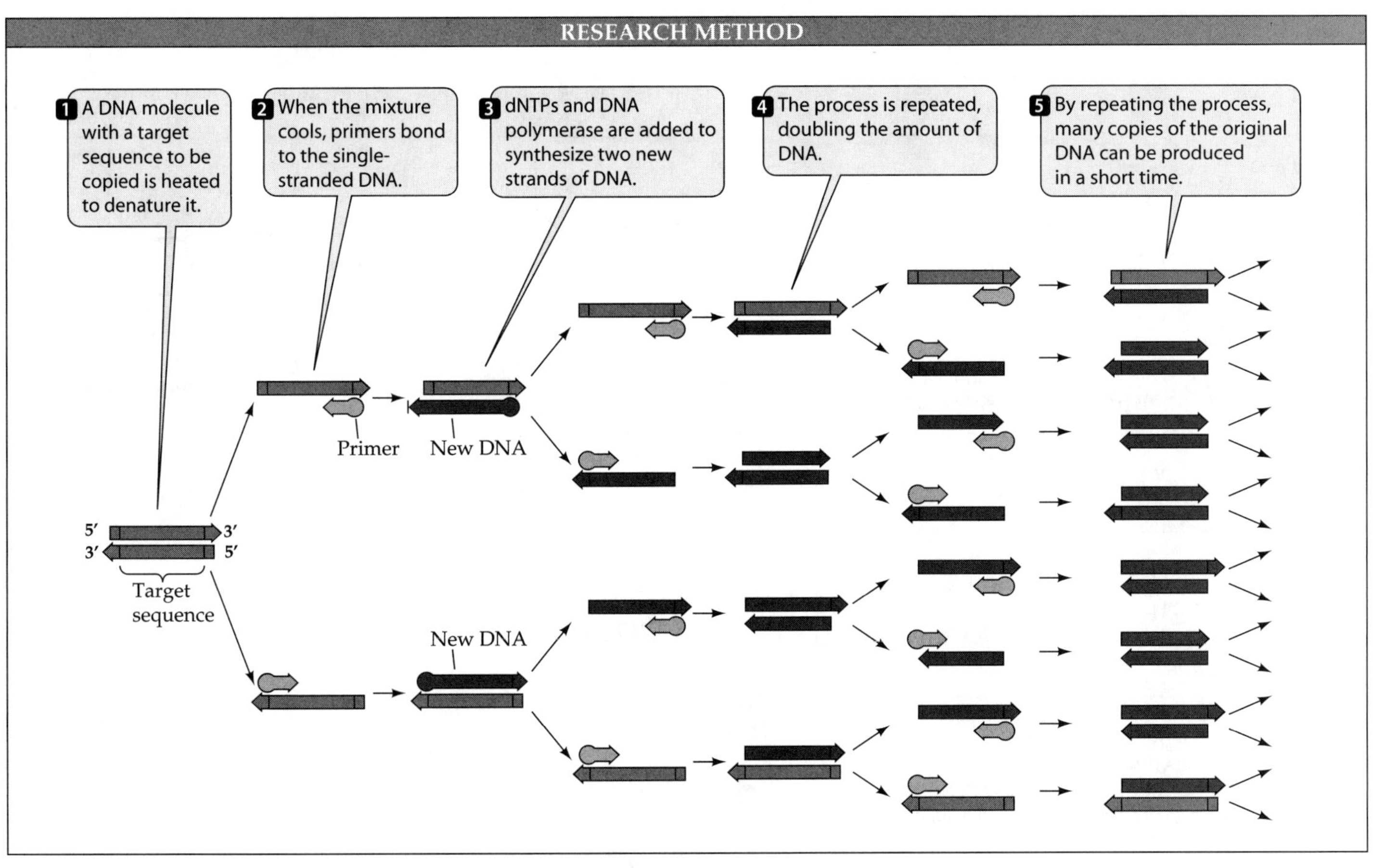
RESEARCH METHOD
1 A DNA molecule with a target sequence to be copied is heated to denature it.
2 When the mixture cools, primers bond to the single-stranded DNA.
3 dNTPs and DNA polymerase are added to synthesize two new strands of DNA.
4 The process is repeated, doubling the amount of DNA.
5 By repeating the process, many copies of the original DNA can be produced in a short time.
5′
3′
3′
5′
Target sequence
Primer
New DNA
New DNA

11.20 The Polymerase Chain Reaction *(Page 229)*

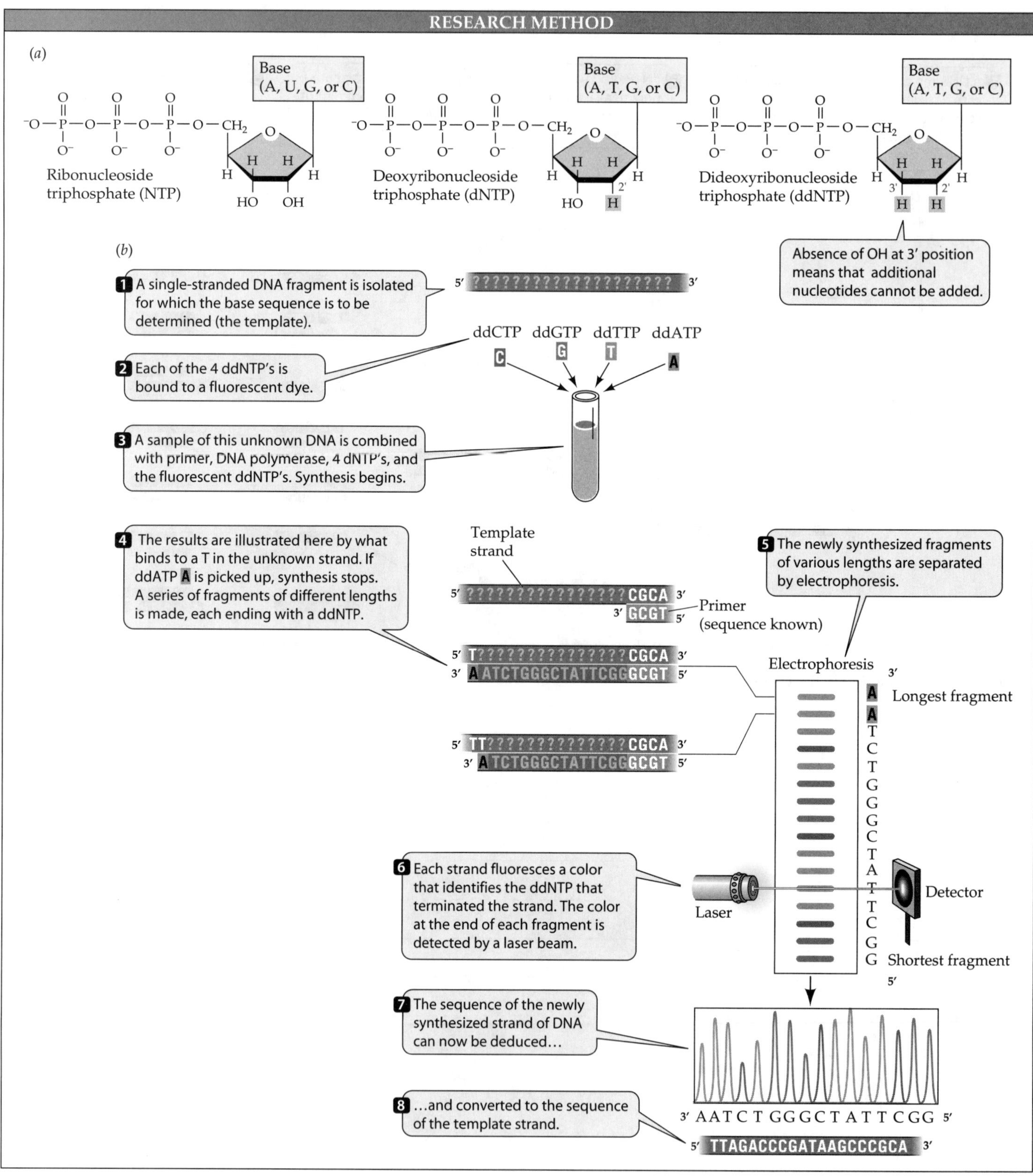
RESEARCH METHOD
(a)
Base (A, U, G, or C)
Ribonucleoside triphosphate (NTP)
Base (A, T, G, or C)
Deoxyribonucleoside triphosphate (dNTP)
Base (A, T, G, or C)
Dideoxyribonucleoside triphosphate (ddNTP)
Absence of OH at 3′ position means that additional nucleotides cannot be added.
(b)
1 A single-stranded DNA fragment is isolated for which the base sequence is to be determined (the template).
2 Each of the 4 ddNTP's is bound to a fluorescent dye.
ddCTP ddGTP ddTTP ddATP
3 A sample of this unknown DNA is combined with primer, DNA polymerase, 4 dNTP's, and the fluorescent ddNTP's. Synthesis begins.
4 The results are illustrated here by what binds to a T in the unknown strand. If ddATP A is picked up, synthesis stops. A series of fragments of different lengths is made, each ending with a ddNTP.
Template strand
5′ ????????????????CGCA 3′
3′ GCGT 5′
Primer (sequence known)
5′ T???????????????CGCA 3′
3′ AATCTGGGCTATTCGGGCGT 5′
5′ TT??????????????CGCA 3′
3′ ATCTGGGCTATTCGGGCGT 5′
5 The newly synthesized fragments of various lengths are separated by electrophoresis.
Electrophoresis
Longest fragment
Shortest fragment
Laser
Detector
6 Each strand fluoresces a color that identifies the ddNTP that terminated the strand. The color at the end of each fragment is detected by a laser beam.
7 The sequence of the newly synthesized strand of DNA can now be deduced…
8 …and converted to the sequence of the template strand.
3′ AATCTGGGCTATTCGG 5′
5′ TTAGACCCGATAAGCCCGCA 3′

11.21 **Sequencing DNA** *(Page 230)*

12 From DNA to Protein: Genotype to Phenotype

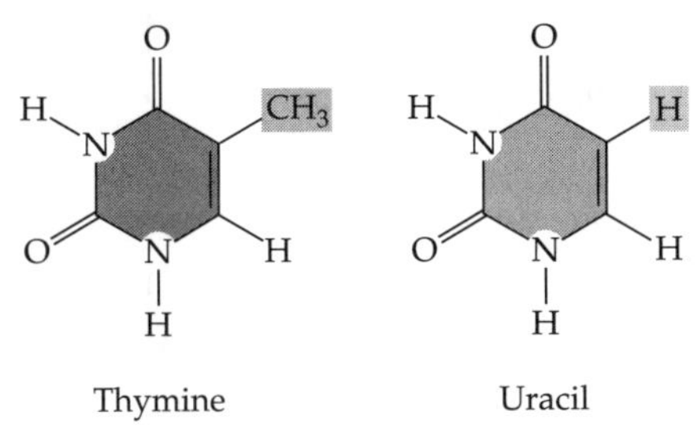

Thymine
Uracil

In-Text Art *(Page 236)*

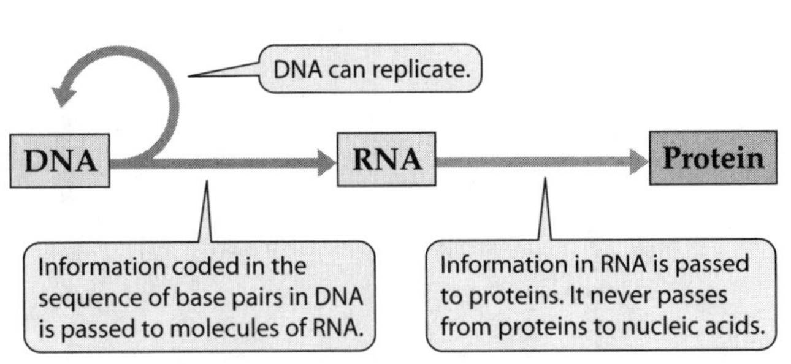

DNA can replicate.
DNA
RNA
Protein
Information coded in the sequence of base pairs in DNA is passed to molecules of RNA.
Information in RNA is passed to proteins. It never passes from proteins to nucleic acids.

12.2 The Central Dogma *(Page 236)*

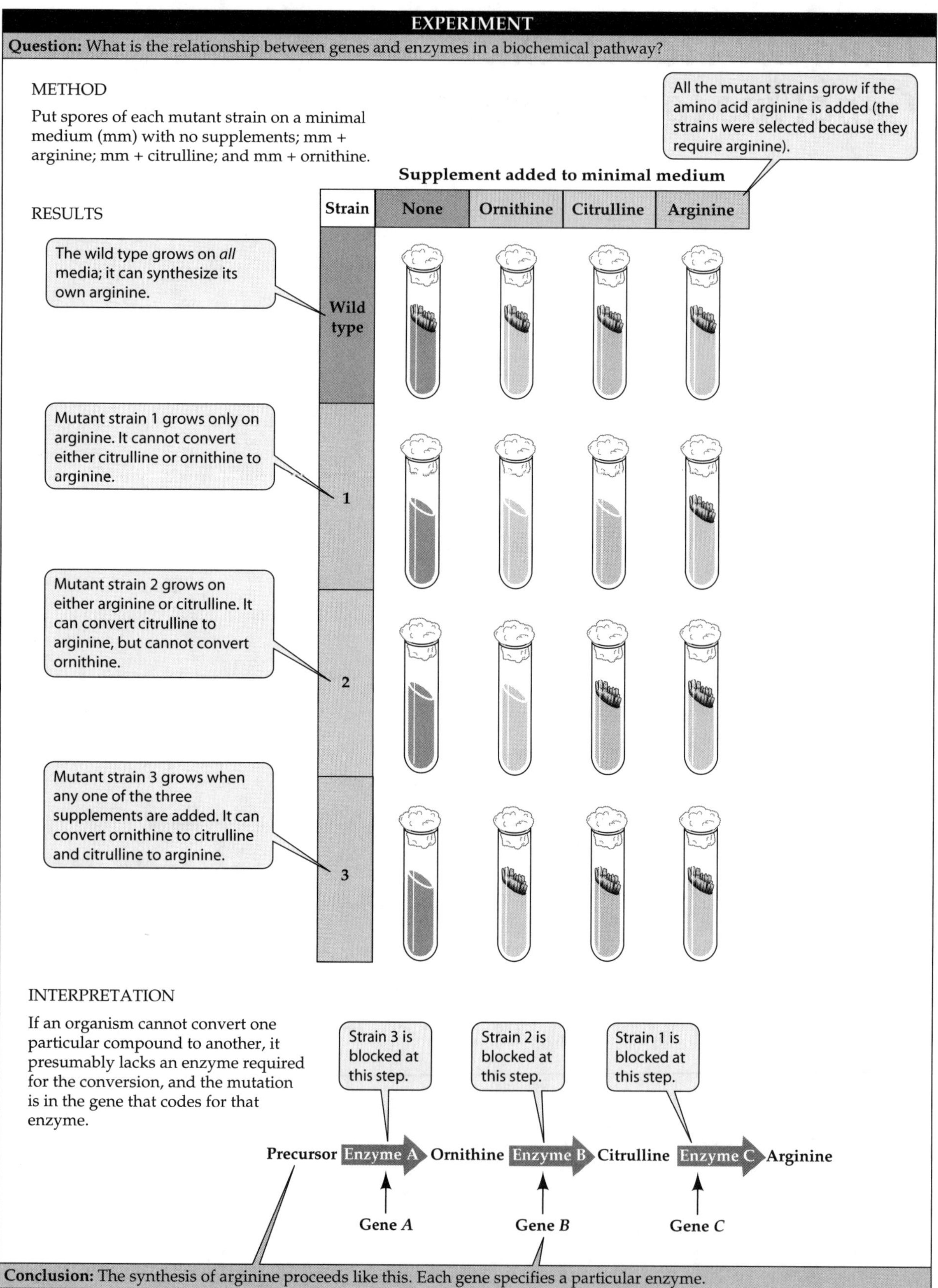
EXPERIMENT
Question: What is the relationship between genes and enzymes in a biochemical pathway?
METHOD
Put spores of each mutant strain on a minimal medium (mm) with no supplements; mm + arginine; mm + citrulline; and mm + ornithine.
All the mutant strains grow if the amino acid arginine is added (the strains were selected because they require arginine).
Supplement added to minimal medium
Strain
None
Ornithine
Citrulline
Arginine
RESULTS
The wild type grows on all media; it can synthesize its own arginine.
Wild type
Mutant strain 1 grows only on arginine. It cannot convert either citrulline or ornithine to arginine.
1
Mutant strain 2 grows on either arginine or citrulline. It can convert citrulline to arginine, but cannot convert ornithine.
2
Mutant strain 3 grows when any one of the three supplements are added. It can convert ornithine to citrulline and citrulline to arginine.
3
INTERPRETATION
If an organism cannot convert one particular compound to another, it presumably lacks an enzyme required for the conversion, and the mutation is in the gene that codes for that enzyme.
Strain 3 is blocked at this step.
Strain 2 is blocked at this step.
Strain 1 is blocked at this step.
Precursor
Enzyme A
Ornithine
Enzyme B
Citrulline
Enzyme C
Arginine
Gene A
Gene B
Gene C
Conclusion: The synthesis of arginine proceeds like this. Each gene specifies a particular enzyme.

12.1 One Gene, One Enzyme *(Page 235)*

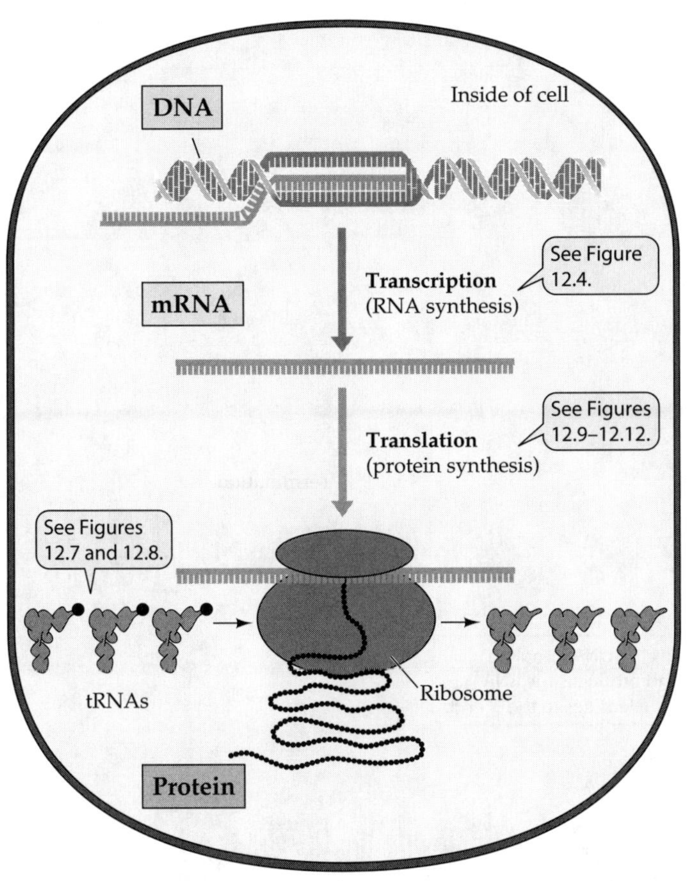
DNA
Inside of cell
See Figure 12.4.
mRNA
Transcription
(RNA synthesis)
See Figures 12.9–12.12.
Translation
(protein synthesis)
See Figures 12.7 and 12.8.
tRNAs
Ribosome
Protein

12.3 From Gene to Protein *(Page 237)*

RNA
RNA
Protein

In-Text Art *(Page 237)*

DNA
RNA
Protein

In-Text Art *(Page 237)*

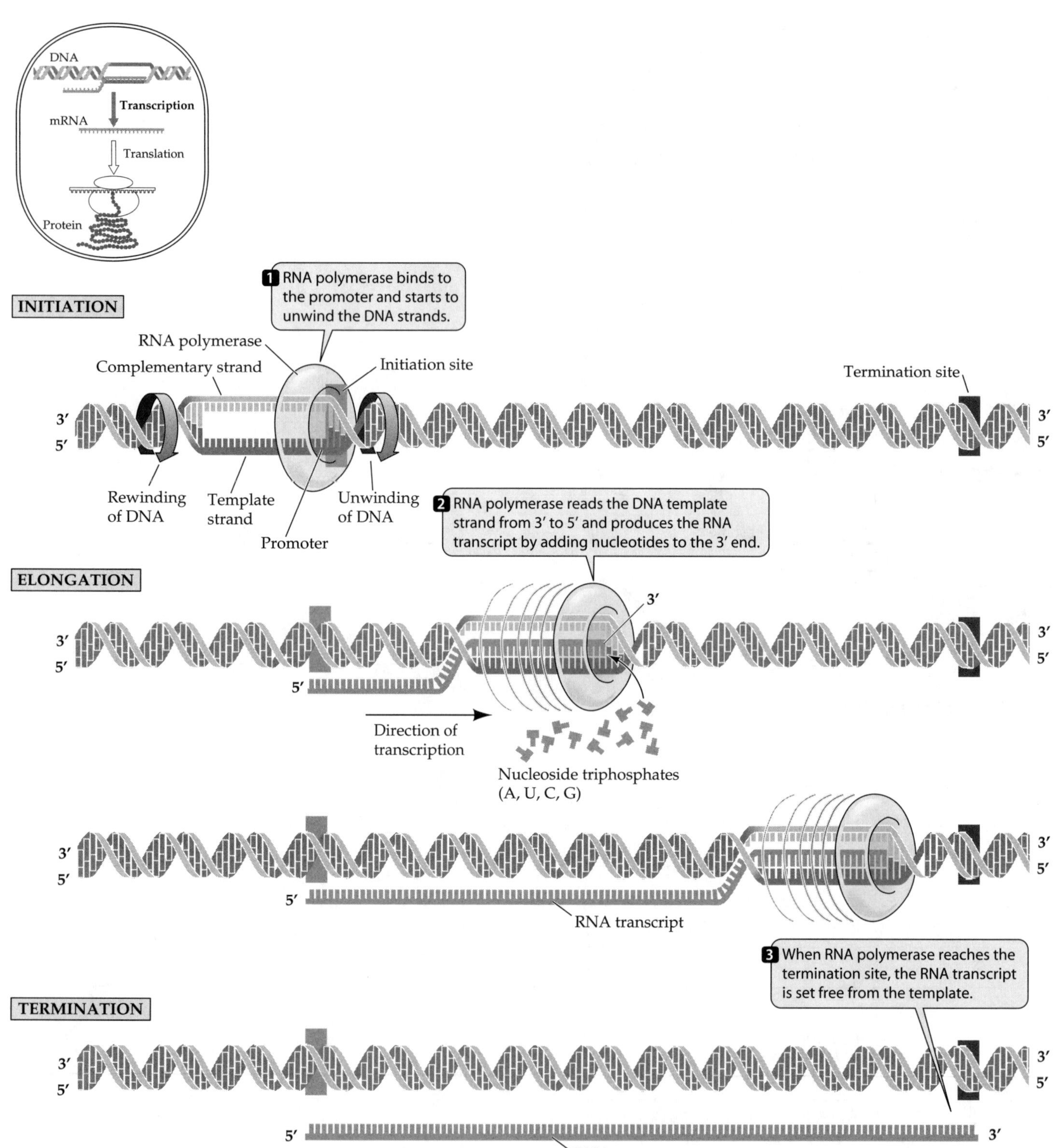
DNA
Transcription
mRNA
Translation
Protein
INITIATION
1 RNA polymerase binds to the promoter and starts to unwind the DNA strands.
RNA polymerase
Complementary strand
Initiation site
Termination site
3′
5′
Rewinding of DNA
Template strand
Promoter
Unwinding of DNA
2 RNA polymerase reads the DNA template strand from 3′ to 5′ and produces the RNA transcript by adding nucleotides to the 3′ end.
ELONGATION
Direction of transcription
Nucleoside triphosphates (A, U, C, G)
RNA transcript
3 When RNA polymerase reaches the termination site, the RNA transcript is set free from the template.
TERMINATION
RNA

12.4 DNA Is Transcribed into RNA *(Page 238)*

First letter	Second letter: U	Second letter: C	Second letter: A	Second letter: G	Third letter
U	UUU, UUC Phenylalanine UUA, UUG Leucine	UCU, UCC, UCA, UCG Serine	UAU, UAC Tyrosine UAA Stop codon UAG Stop codon	UGU, UGC Cysteine UGA Stop codon UGG Tryptophan	U C A G
C	CUU, CUC, CUA, CUG Leucine	CCU, CCC, CCA, CCG Proline	CAU, CAC Histidine CAA, CAG Glutamine	CGU, CGC, CGA, CGG Arginine	U C A G
A	AUU, AUC, AUA Isoleucine AUG Methionine; start codon	ACU, ACC, ACA, ACG Threonine	AAU, AAC Asparagine AAA, AAG Lysine	AGU, AGC Serine AGA, AGG Arginine	U C A G
G	GUU, GUC, GUA, GUG Valine	GCU, GCC, GCA, GCG Alanine	GAU, GAC Aspartic acid GAA, GAG Glutamic acid	GGU, GGC, GGA, GGG Glycine	U C A G

12.5 The Universal Genetic Code *(Page 239)*

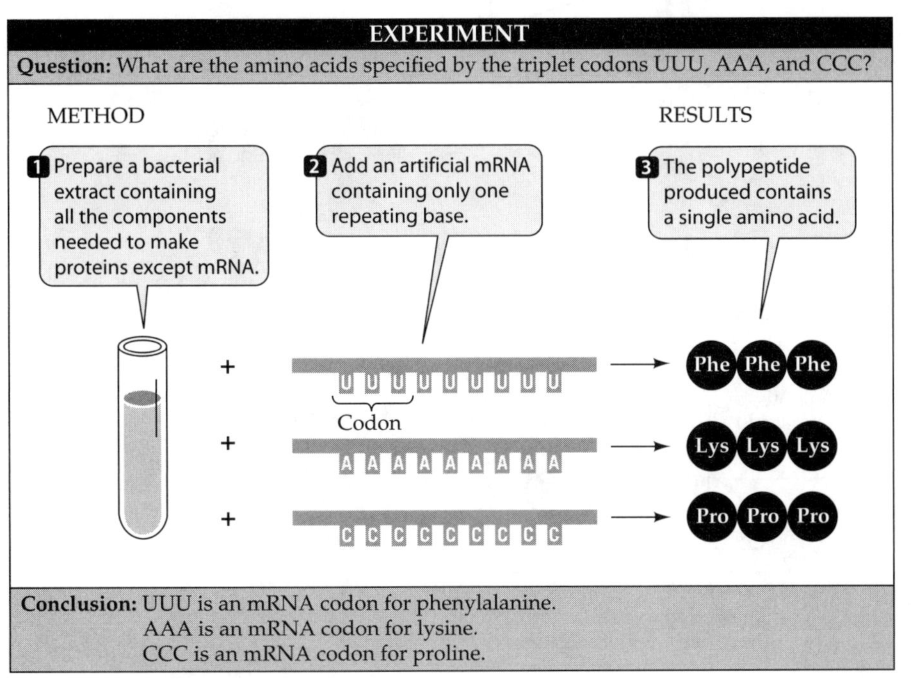
EXPERIMENT
Question: What are the amino acids specified by the triplet codons UUU, AAA, and CCC?
METHOD
RESULTS
1 Prepare a bacterial extract containing all the components needed to make proteins except mRNA.
2 Add an artificial mRNA containing only one repeating base.
3 The polypeptide produced contains a single amino acid.
+
U U U U U U U U U
Codon
Phe Phe Phe
A A A A A A A A A
Lys Lys Lys
C C C C C C C C C
Pro Pro Pro
Conclusion: UUU is an mRNA codon for phenylalanine.
AAA is an mRNA codon for lysine.
CCC is an mRNA codon for proline.

12.6 Deciphering the Genetic Code *(Page 240)*

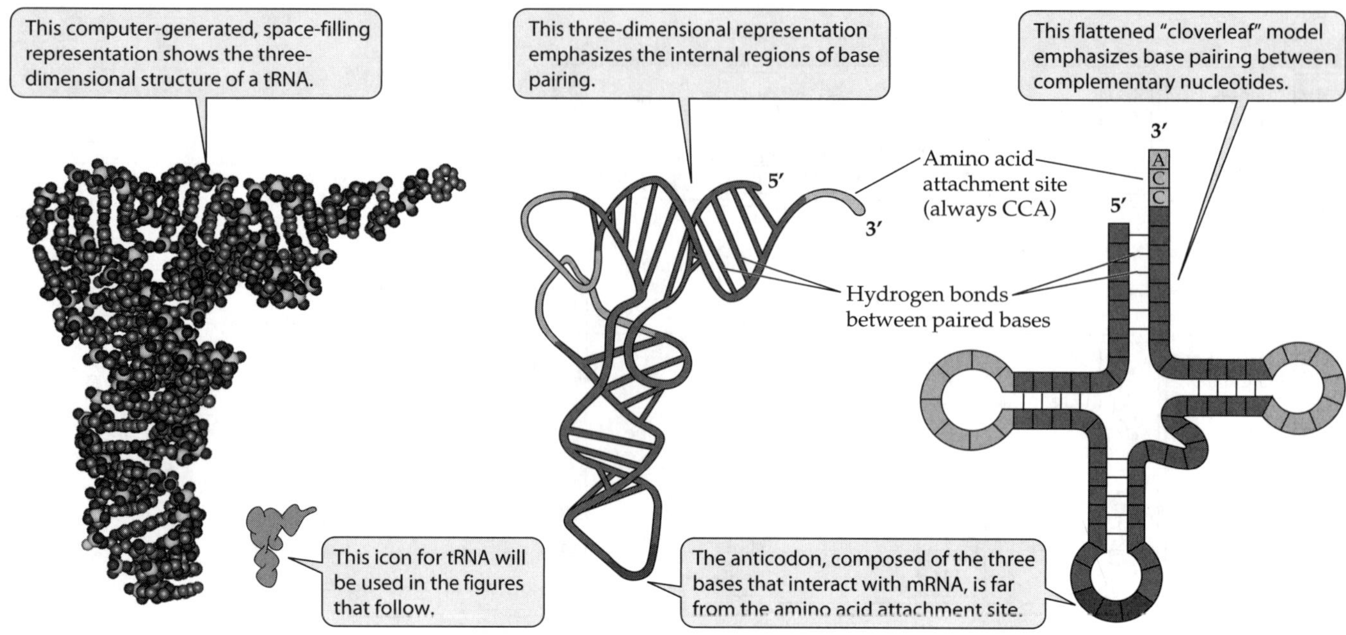
This computer-generated, space-filling representation shows the three-dimensional structure of a tRNA.
This three-dimensional representation emphasizes the internal regions of base pairing.
This flattened "cloverleaf" model emphasizes base pairing between complementary nucleotides.
5′
3′
Amino acid attachment site (always CCA)
3′
A
C
C
5′
Hydrogen bonds between paired bases
This icon for tRNA will be used in the figures that follow.
The anticodon, composed of the three bases that interact with mRNA, is far from the amino acid attachment site.

12.7 Transfer RNA *(Page 241)*

1 The enzyme activates the amino acid, catalyzing a reaction with ATP to form high energy AMP–amino acid and a pyrophosphate ion.

Specific amino acid (AA)

ATP

ATP

P_i

Pyrophosphate

Activating enzyme

Amino acid site

ATP site

tRNA site

AMP

Activated amino acid

Specific tRNA

2 The enzyme then catalyzes a reaction of the activated amino acid with the correct tRNA.

Charged tRNA

tRNA bonded to amino acid

AMP

AMP

4 The charged tRNA will deliver the appropriate amino acid to join the elongating polypeptide product of translation.

3 The specificity of the enzyme ensures that the correct amino acid and tRNA have been brought together.

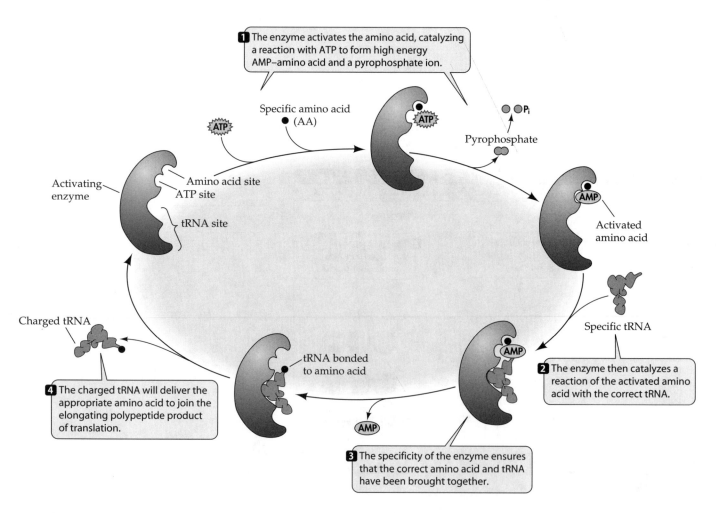

12.8 Charging a tRNA Molecule *(Page 243)*

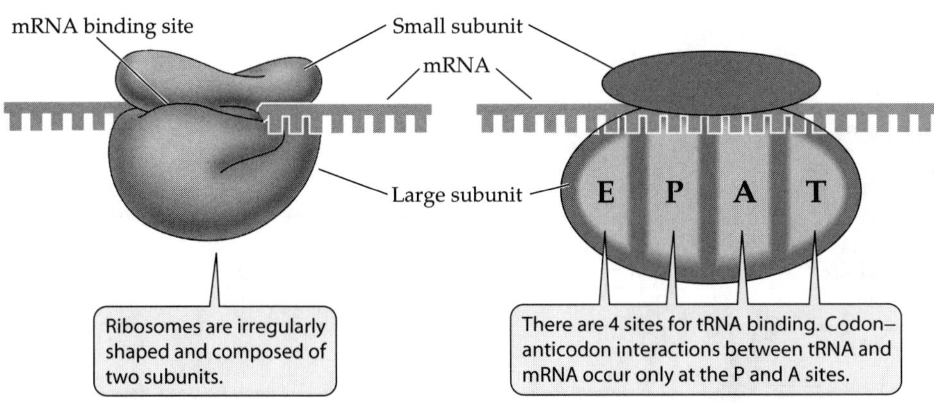
mRNA binding site
Small subunit
mRNA
Large subunit
E
P
A
T
Ribosomes are irregularly shaped and composed of two subunits.
There are 4 sites for tRNA binding. Codon–anticodon interactions between tRNA and mRNA occur only at the P and A sites.

12.9 Ribosome Structure *(Page 243)*

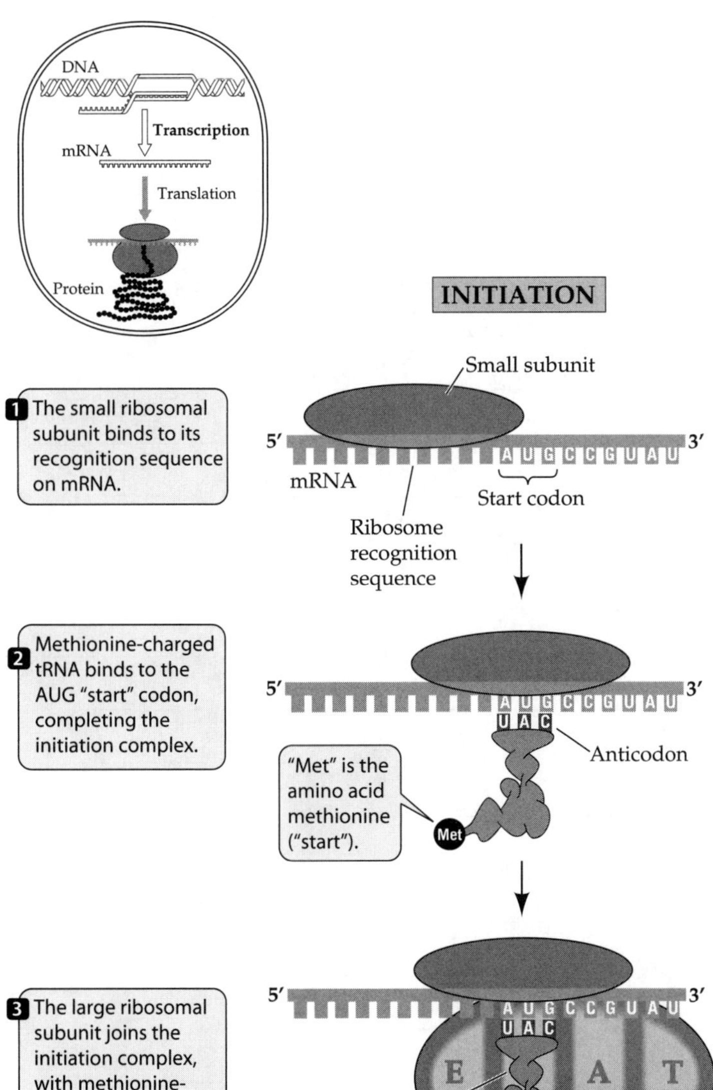
DNA
Transcription
mRNA
Translation
Protein
INITIATION
Small subunit
1 The small ribosomal subunit binds to its recognition sequence on mRNA.
5′
3′
A U G C C G U A U
mRNA
Start codon
Ribosome recognition sequence
2 Methionine-charged tRNA binds to the AUG "start" codon, completing the initiation complex.
U A C
Anticodon
"Met" is the amino acid methionine ("start").
Met
3 The large ribosomal subunit joins the initiation complex, with methionine-charged tRNA now occupying the P site.
E
A
T
P site

12.10 The Initiation of Translation *(Page 244)*

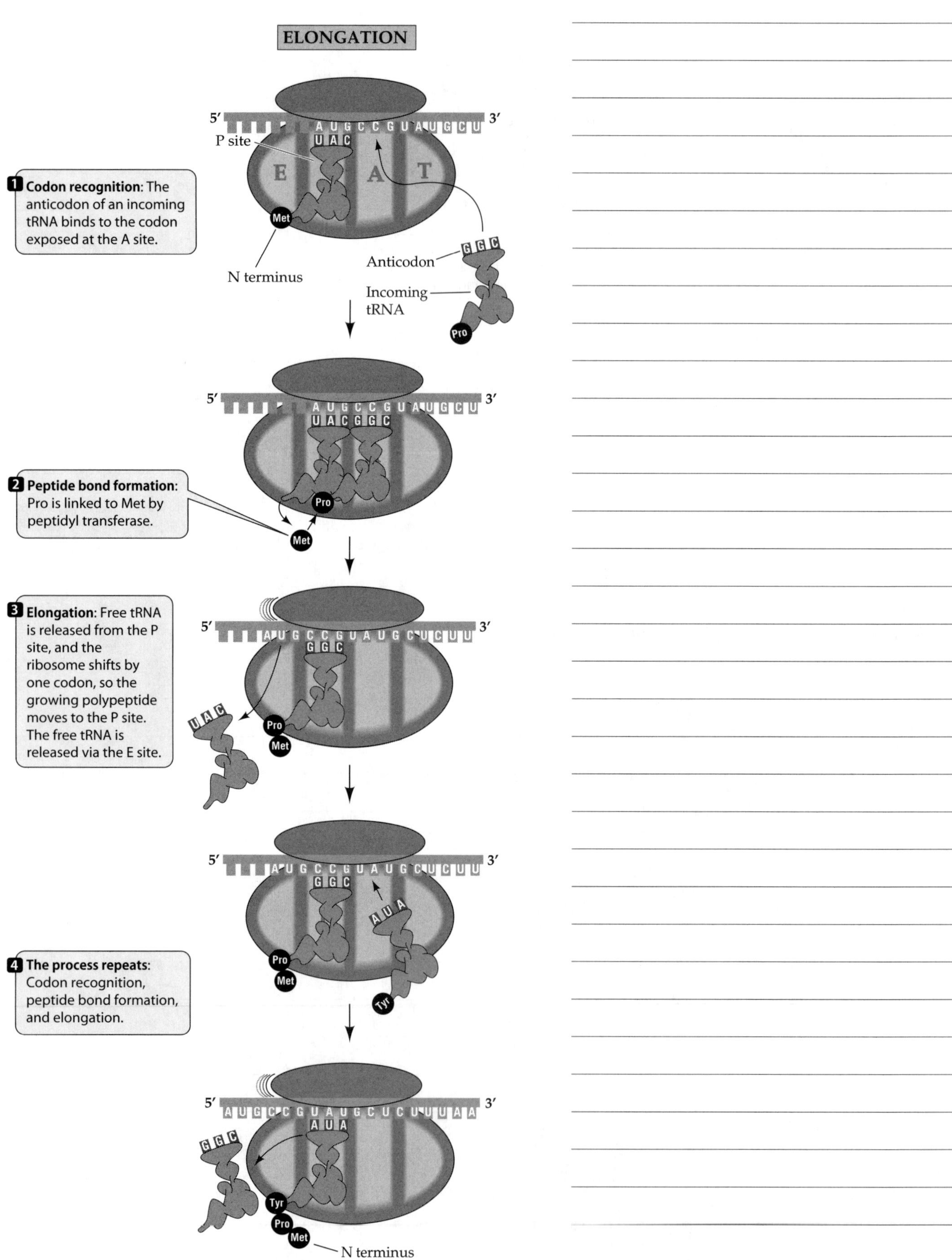

ELONGATION
5′
3′
A U G C C G U A U G C U
U A C
P site
E
A
T
Met
N terminus
Anticodon
G G C
Incoming tRNA
Pro
Codon recognition: The anticodon of an incoming tRNA binds to the codon exposed at the A site.
Peptide bond formation: Pro is linked to Met by peptidyl transferase.
Elongation: Free tRNA is released from the P site, and the ribosome shifts by one codon, so the growing polypeptide moves to the P site. The free tRNA is released via the E site.
The process repeats: Codon recognition, peptide bond formation, and elongation.
U A C
A U A
Tyr
N terminus

12.11 Translation: The Elongation Stage *(Page 245)*

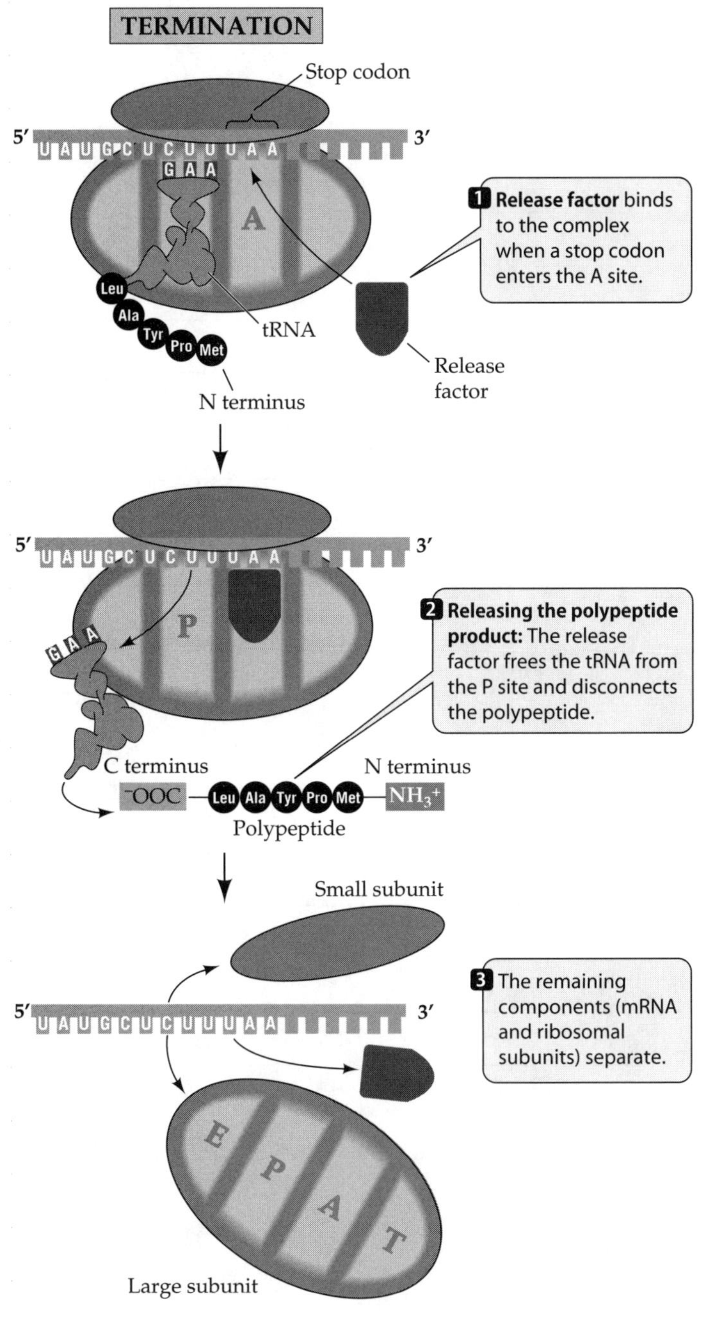
TERMINATION
Stop codon
5′
3′
U A U G C U C U U U A A
G A A
A
tRNA
Leu
Ala
Tyr
Pro
Met
N terminus
Release factor
1 Release factor binds to the complex when a stop codon enters the A site.
P
2 Releasing the polypeptide product: The release factor frees the tRNA from the P site and disconnects the polypeptide.
C terminus
N terminus
^{-}OOC
NH_3^+
Polypeptide
Small subunit
3 The remaining components (mRNA and ribosomal subunits) separate.
E
P
A
T
Large subunit

12.12 The Termination of Translation *(Page 246)*

12.1 *Signals that Start and Stop Transcription and Translation*

	TRANSCRIPTION	TRANSLATION
Initiation	Promoter sequence in DNA	AUG start codon in mRNA
Termination	Terminator sequence in DNA	UAA, UAG, or UGA stop codon in mRNA

(Page 246)

12.2 *Antibiotics that Inhibit Bacterial Protein Synthesis*

ANTIBIOTIC	STEP INHIBITED
Chloromycetin	Formation of peptide bonds
Erythromycin	Translocation of mRNA along ribosome
Neomycin	Interactions between tRNA and mRNA
Streptomycin	Initiation of translation
Tetracycline	Binding of tRNA to ribosome
Paromomycin	Validation of mRNA–tRNA match

(Page 247)

(a)

Initiation
Elongation
Termination
Large subunit
Ribosome
5′
Small subunit
mRNA
Polypeptide chain
Direction of translation
3′

12.13 A Polysome *(Page 247)*

Protein synthesis in cytosol
Inside of cell
Nucleus
To organelles
To rough endoplasmic reticulum
Golgi apparatus
Peroxisomes
Mitochondria
Plastids
Rough endoplasmic reticulum
Lysosome
Exocytosis
Plasma membrane
Outside of cell

12.14 Destinations for Newly Translated Polypeptides in a Eukaryotic Cell *(Page 248)*

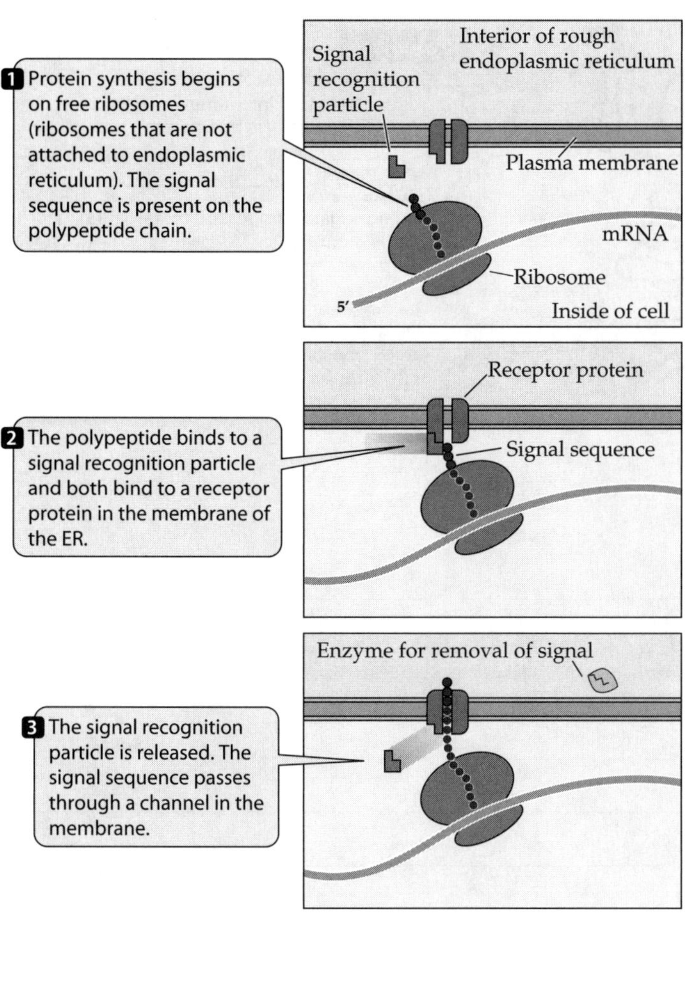
1 Protein synthesis begins on free ribosomes (ribosomes that are not attached to endoplasmic reticulum). The signal sequence is present on the polypeptide chain.
Signal recognition particle
Interior of rough endoplasmic reticulum
Plasma membrane
mRNA
Ribosome
5′
Inside of cell
2 The polypeptide binds to a signal recognition particle and both bind to a receptor protein in the membrane of the ER.
Receptor protein
Signal sequence
3 The signal recognition particle is released. The signal sequence passes through a channel in the membrane.
Enzyme for removal of signal

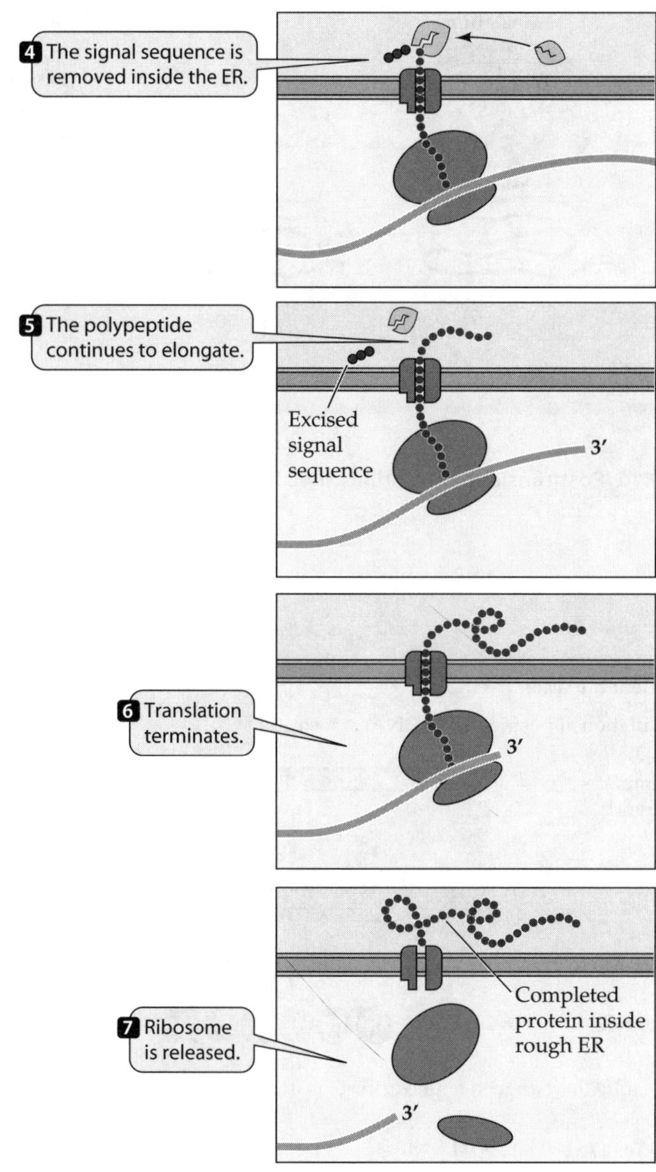
4 The signal sequence is removed inside the ER.
5 The polypeptide continues to elongate.
Excised signal sequence
3′
6 Translation terminates.
3′
7 Ribosome is released.
Completed protein inside rough ER
3′

12.15 A Signal Sequence Moves a Polypeptide into the ER *(Page 249)*

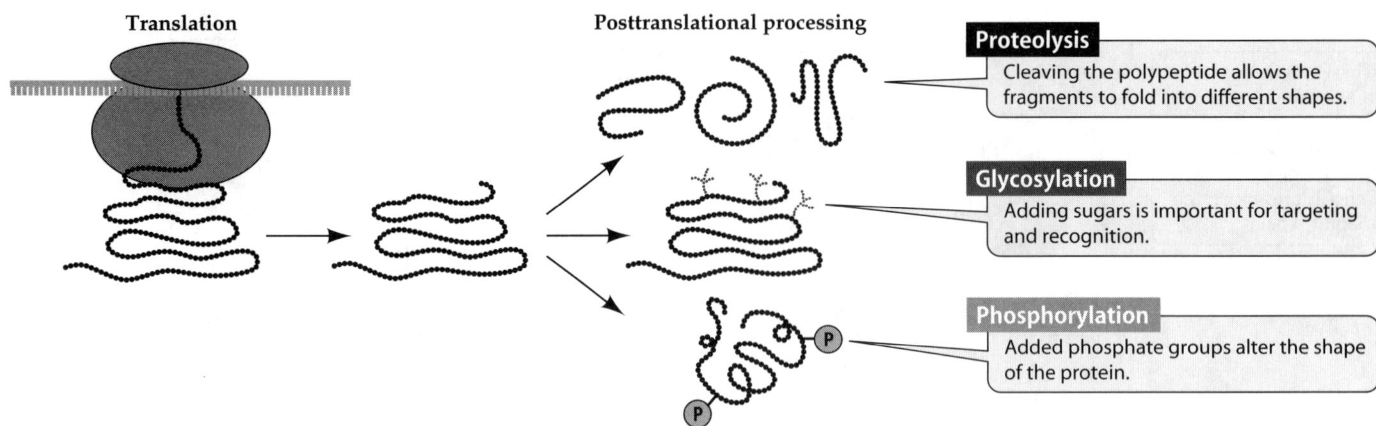

Translation
Posttranslational processing
Proteolysis
Cleaving the polypeptide allows the fragments to fold into different shapes.
Glycosylation
Adding sugars is important for targeting and recognition.
Phosphorylation
Added phosphate groups alter the shape of the protein.
P
P

12.16 Posttranslational Modifications to Proteins *(Page 250)*

Silent mutation

Mutation at position 12 in DNA: A instead of C

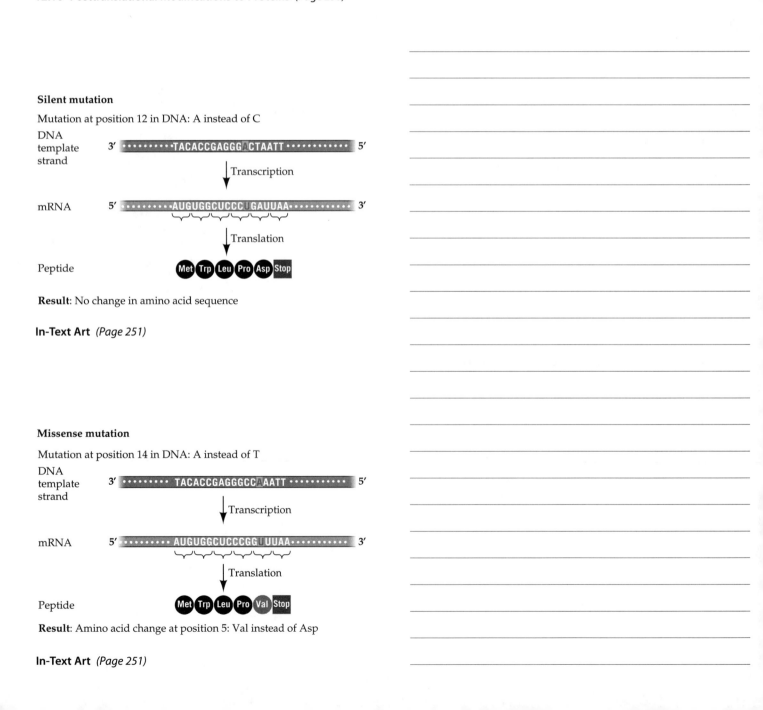

DNA template strand 3′ ••••••••• TACACCGAGGGACTAATT •••••••••••• 5′

Transcription

mRNA 5′ ••••••••• AUGUGGCUCCCUGAUUAA •••••••••••• 3′

Translation

Peptide Met Trp Leu Pro Asp Stop

Result: No change in amino acid sequence

In-Text Art *(Page 251)*

Missense mutation

Mutation at position 14 in DNA: A instead of T

DNA template strand 3′ ••••••••• TACACCGAGGGCCAAATT ••••••••••• 5′

Transcription

mRNA 5′ ••••••••• AUGUGGCUCCCGGUUUAA ••••••••••• 3′

Translation

Peptide Met Trp Leu Pro Val Stop

Result: Amino acid change at position 5: Val instead of Asp

In-Text Art *(Page 251)*

Nonsense mutation

Mutation at position 5 in DNA: T instead of C

DNA template strand
3′ TACATCGAGGGCCTAATT 5′
Transcription
mRNA
5′ AUGUAGCUCCCGGAUUAA 3′
Translation
Peptide
Met Stop

Result: Only one amino acid translated; no protein made

In-Text Art *(Page 252)*

Frame-shift mutation

Mutation by insertion of T between bases 6 and 7 in DNA

DNA template strand
3′ TACACCGAGGGCCTAATT 5′
DNA template strand
3′ TACACCTGAGGGCCTAATT 5′
Transcription
mRNA
5′ AUGUGGACUCCCGGAUUAA 3′
Translation
Peptide
Met Trp Thr Pro Gly Leu

Result: All amino acids changed beyond the insertion

In-Text Art *(Page 252)*

(a)
Deletion is the loss of a chromosome segment.
A B C D E F G
A B E F G
C D (lost)
(b)
Duplication and deletion result when homologous chromosomes break at different points…
…and swap segments.
A B C D E F G
A B C D E F G
A B E F G
A B C D C D E F G
(c)
Inversion results when a broken segment is inserted in reverse order.
A B C D E F G
A B E D C F G
(d)
Reciprocal translocation results when nonhomologous chromosomes exchange segments.
A B C D E F G
H I J K L M N O
A B L M N O
H I J K C D E F G

12.18 Chromosomal Mutations *(Page 253)*

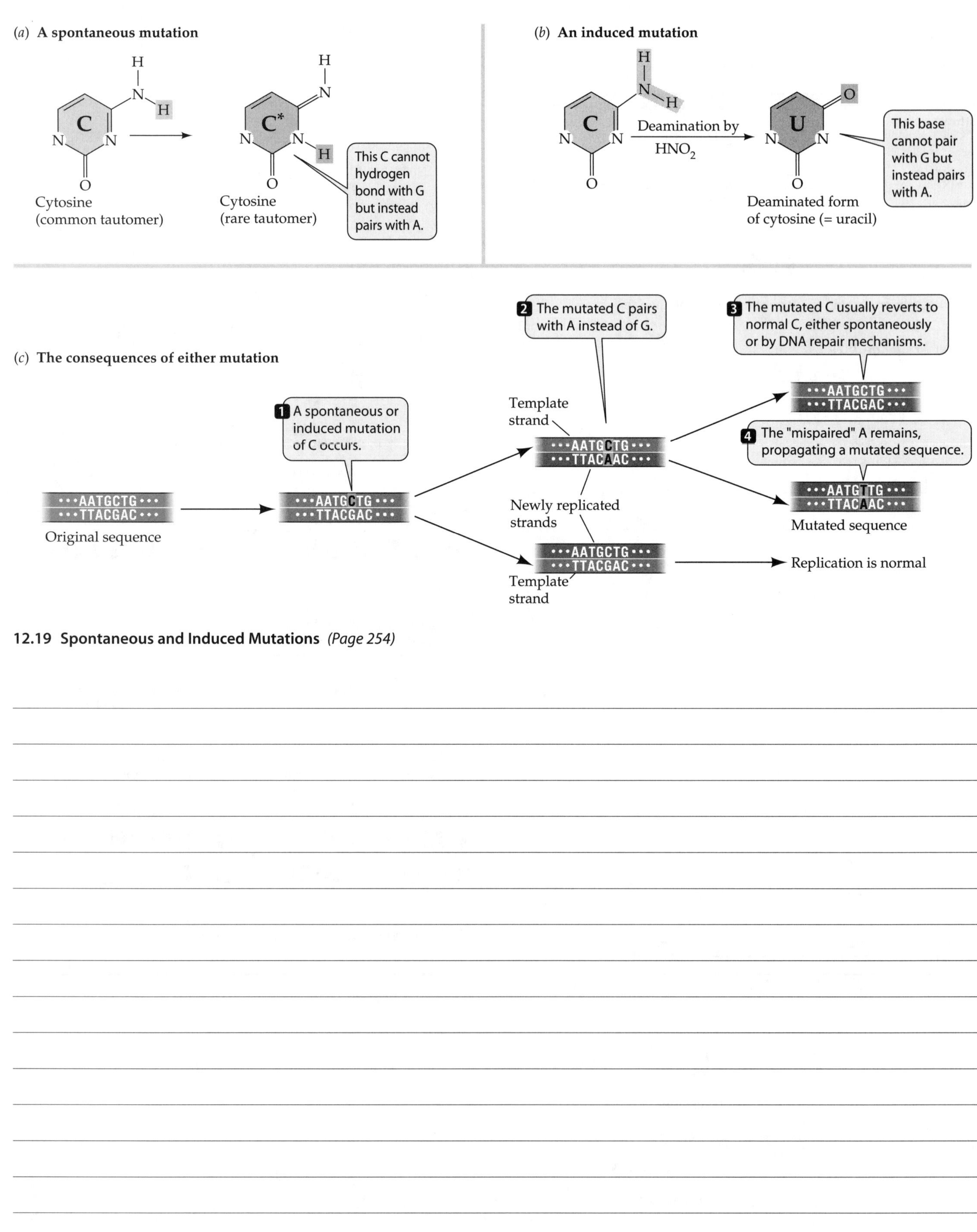
(a) A spontaneous mutation
C
N
N
H
O
Cytosine
(common tautomer)
C*
Cytosine
(rare tautomer)
This C cannot hydrogen bond with G but instead pairs with A.
(b) An induced mutation
Deamination by
HNO_2
U
This base cannot pair with G but instead pairs with A.
Deaminated form
of cytosine (= uracil)
(c) The consequences of either mutation
•••AATGCTG•••
•••TTACGAC•••
Original sequence
1 A spontaneous or induced mutation of C occurs.
2 The mutated C pairs with A instead of G.
Template strand
•••AATGCTG•••
•••TTACAAC•••
Newly replicated strands
Template strand
3 The mutated C usually reverts to normal C, either spontaneously or by DNA repair mechanisms.
4 The "mispaired" A remains, propagating a mutated sequence.
•••AATGTTG•••
•••TTACAAC•••
Mutated sequence
Replication is normal

12.19 **Spontaneous and Induced Mutations** *(Page 254)*

13 The Genetics of Viruses and Prokaryotes

13.1 Relative Sizes of Microorganisms

MICROORGANISM	TYPE	TYPICAL SIZE RANGE (μm^3)
Protists	Eukaryote	5,000–50,000
Photosynthetic bacteria	Prokaryote	5–50
Spirochetes	Prokaryote	0.1–2.0
Mycoplasmas	Prokaryote	0.01–0.1
Poxviruses	Virus	0.01
Influenza virus	Virus	0.0005
Poliovirus	Virus	0.00001

(Page 258)

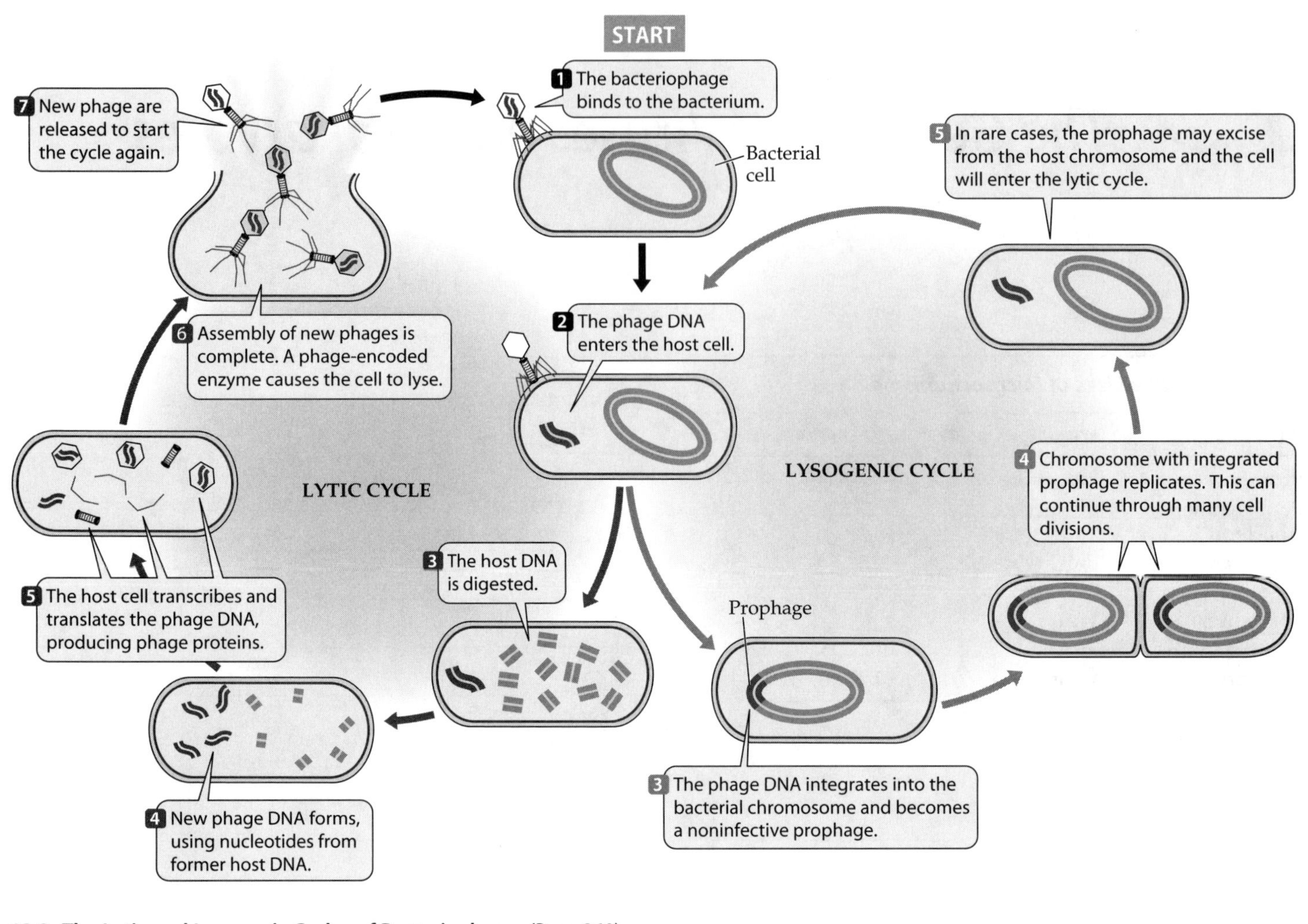
START
1 The bacteriophage binds to the bacterium.
Bacterial cell
2 The phage DNA enters the host cell.
LYTIC CYCLE
3 The host DNA is digested.
4 New phage DNA forms, using nucleotides from former host DNA.
5 The host cell transcribes and translates the phage DNA, producing phage proteins.
6 Assembly of new phages is complete. A phage-encoded enzyme causes the cell to lyse.
7 New phage are released to start the cycle again.
LYSOGENIC CYCLE
Prophage
3 The phage DNA integrates into the bacterial chromosome and becomes a noninfective prophage.
4 Chromosome with integrated prophage replicates. This can continue through many cell divisions.
5 In rare cases, the prophage may excise from the host chromosome and the cell will enter the lytic cycle.

13.2 The Lytic and Lysogenic Cycles of Bacteriophage *(Page 260)*

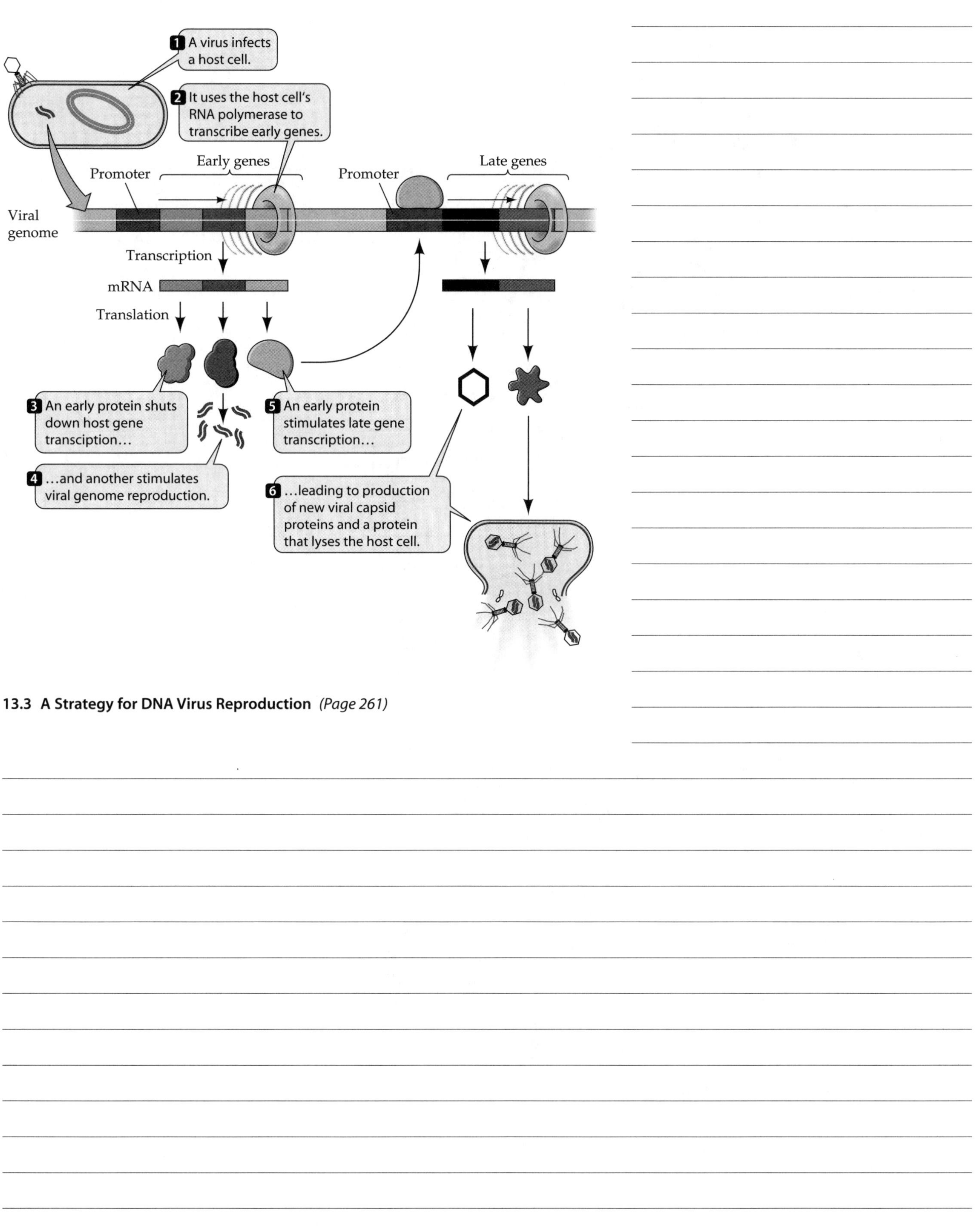
1 A virus infects a host cell.
2 It uses the host cell's RNA polymerase to transcribe early genes.
Promoter
Early genes
Promoter
Late genes
Viral genome
Transcription
mRNA
Translation
3 An early protein shuts down host gene transciption…
4 …and another stimulates viral genome reproduction.
5 An early protein stimulates late gene transciption…
6 …leading to production of new viral capsid proteins and a protein that lyses the host cell.

13.3 A Strategy for DNA Virus Reproduction *(Page 261)*

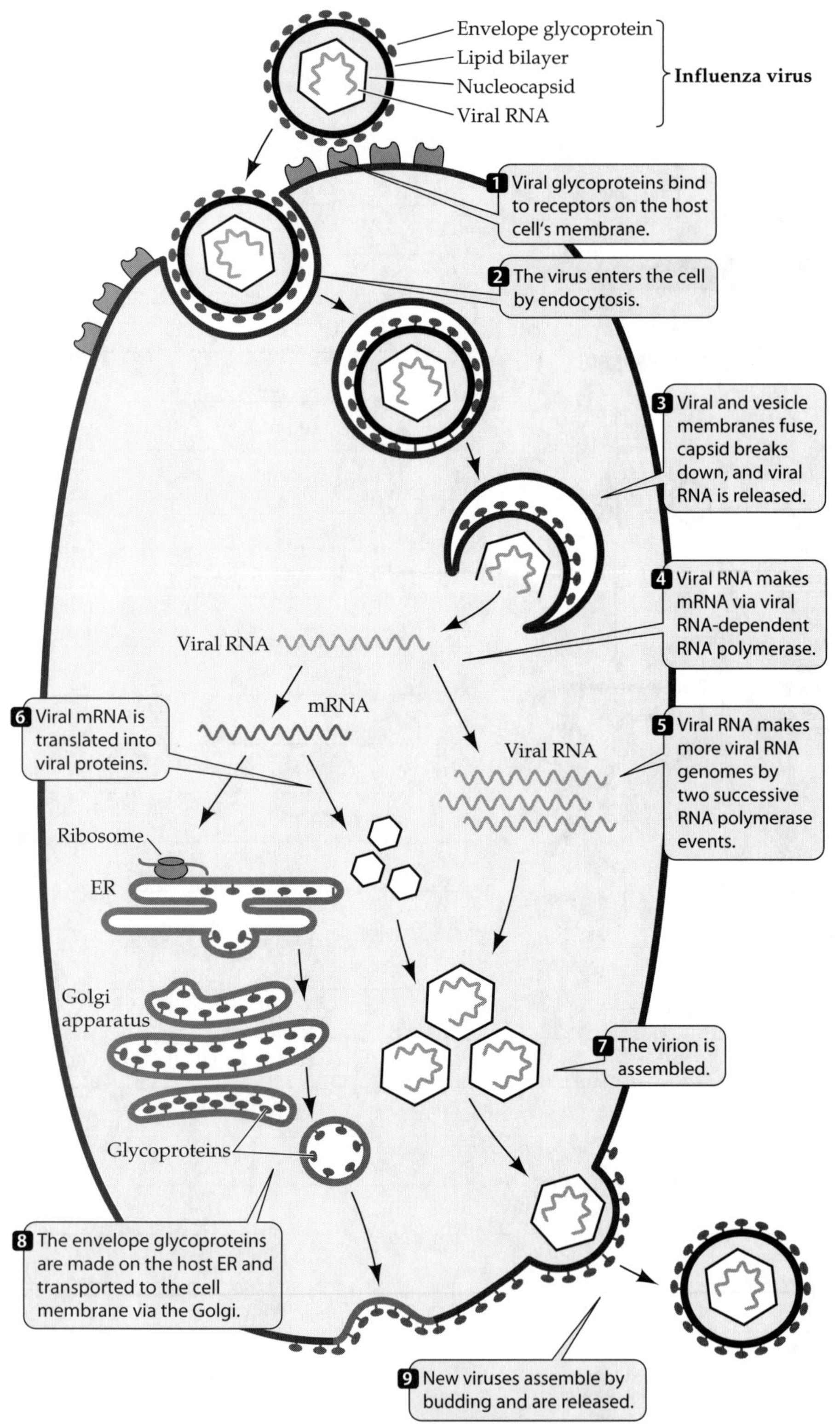
Envelope glycoprotein
Lipid bilayer
Nucleocapsid
Viral RNA
Influenza virus
1 Viral glycoproteins bind to receptors on the host cell's membrane.
2 The virus enters the cell by endocytosis.
3 Viral and vesicle membranes fuse, capsid breaks down, and viral RNA is released.
4 Viral RNA makes mRNA via viral RNA-dependent RNA polymerase.
Viral RNA
mRNA
6 Viral mRNA is translated into viral proteins.
Viral RNA
5 Viral RNA makes more viral RNA genomes by two successive RNA polymerase events.
Ribosome
ER
Golgi apparatus
7 The virion is assembled.
Glycoproteins
8 The envelope glycoproteins are made on the host ER and transported to the cell membrane via the Golgi.
9 New viruses assemble by budding and are released.

13.4 The Reproductive Cycle of the Influenza Virus *(Page 262)*

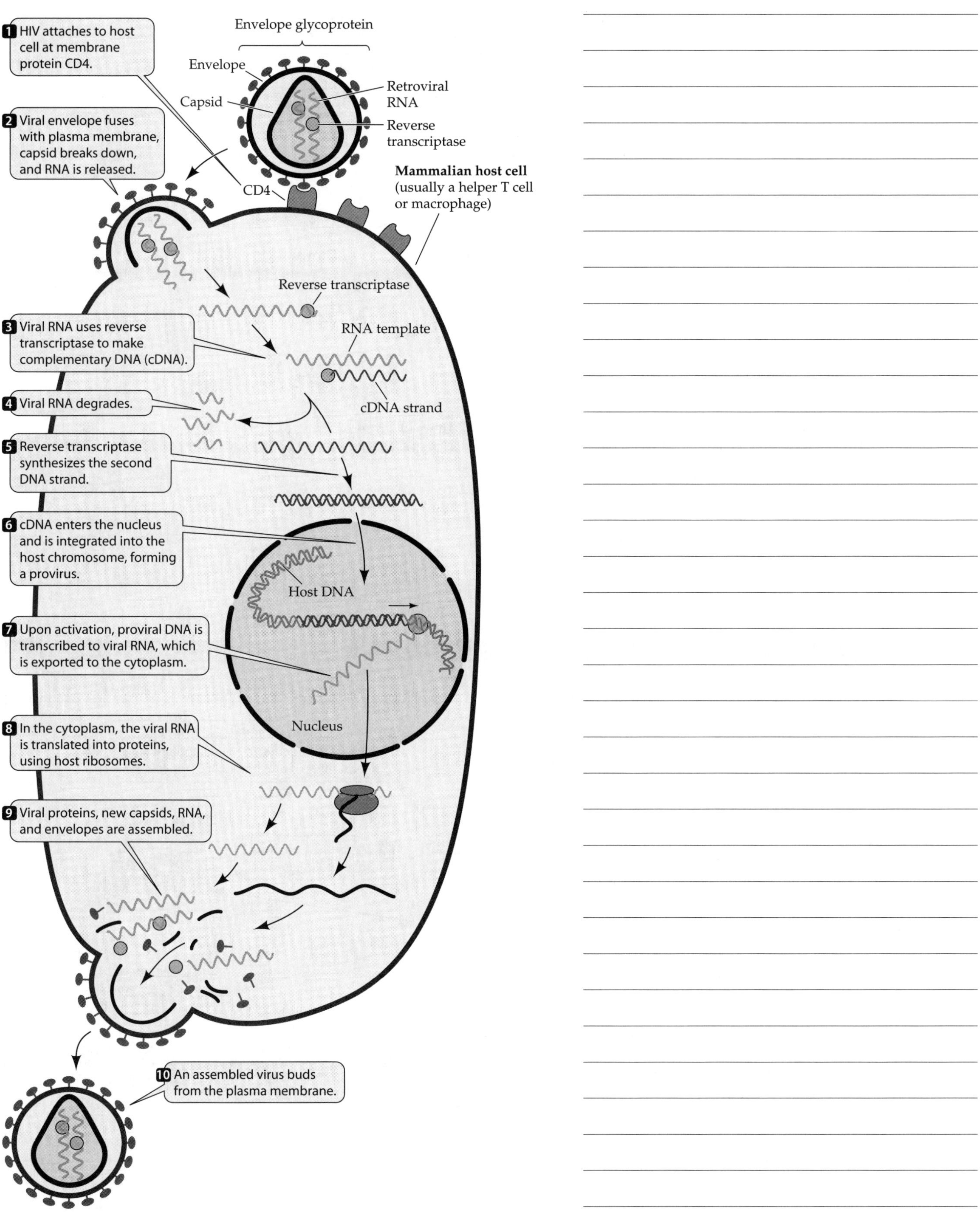

1 HIV attaches to host cell at membrane protein CD4.
Envelope glycoprotein
Envelope
Capsid
Retroviral RNA
Reverse transcriptase
2 Viral envelope fuses with plasma membrane, capsid breaks down, and RNA is released.
CD4
Mammalian host cell (usually a helper T cell or macrophage)
Reverse transcriptase
3 Viral RNA uses reverse transcriptase to make complementary DNA (cDNA).
RNA template
cDNA strand
4 Viral RNA degrades.
5 Reverse transcriptase synthesizes the second DNA strand.
6 cDNA enters the nucleus and is integrated into the host chromosome, forming a provirus.
Host DNA
7 Upon activation, proviral DNA is transcribed to viral RNA, which is exported to the cytoplasm.
Nucleus
8 In the cytoplasm, the viral RNA is translated into proteins, using host ribosomes.
9 Viral proteins, new capsids, RNA, and envelopes are assembled.
10 An assembled virus buds from the plasma membrane.

13.5 The Reproductive Cycle of HIV *(Page 263)*

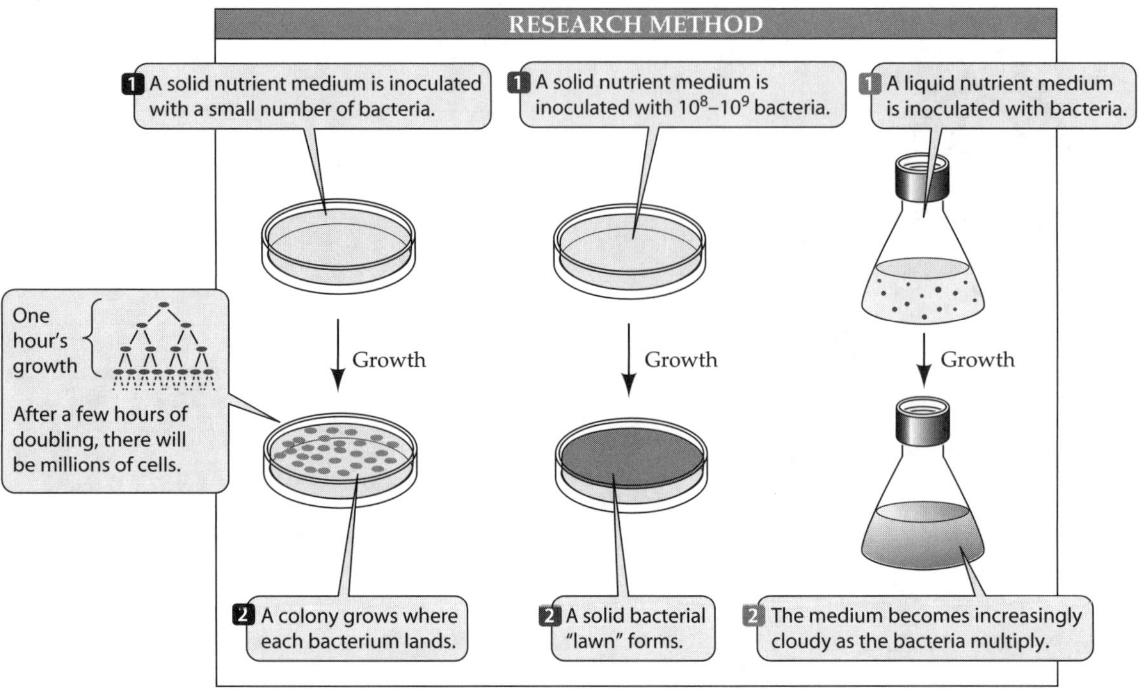

RESEARCH METHOD
1 A solid nutrient medium is inoculated with a small number of bacteria.
1 A solid nutrient medium is inoculated with 10^8–10^9 bacteria.
1 A liquid nutrient medium is inoculated with bacteria.
One hour's growth
After a few hours of doubling, there will be millions of cells.
Growth
Growth
Growth
2 A colony grows where each bacterium lands.
2 A solid bacterial "lawn" forms.
2 The medium becomes increasingly cloudy as the bacteria multiply.

13.6 Growing Bacteria in the Laboratory *(Page 264)*

EXPERIMENT

Question: Can bacteria exchange genetic material? When different auxotrophic (mutant) strains of bacteria are grown together, do new prototrophic (wild type) bacteria appear?

METHOD

Strain 1 of *E. coli* (*met*$^-$ *bio*$^-$ *thr*$^+$ *leu*$^+$) requires methionine and biotin for growth

1 Growth occurs on minimal medium + methionine and biotin.

2 There is no growth on minimal medium.

Strain 2 of *E. coli* (*met*$^+$ *bio*$^+$ *thr*$^-$ *leu*$^-$) requires threonine and leucine for growth

3 Growth occurs on minimal medium + threonine and leucine.

4 There is no growth on minimal medium.

5 Samples of strains 1 and 2 are combined and incubated together in liquid medium with methionine, biotin, threonine, and leucine.

RESULTS

6 Complete medium (many colonies grow).

7 On minimal medium with no supplements, a few colonies of prototrophic bacteria (*met*$^+$ *bio*$^+$ *thr*$^+$ *leu*$^+$) grow.

Conclusion: The prototrophic colonies growing on minimal medium could have arisen only by genetic recombination between the two different strains.

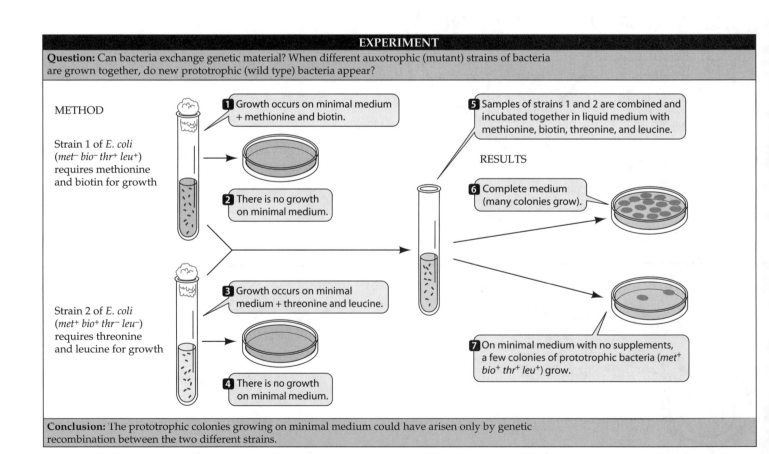

13.7 Lederberg and Tatum's Experiment *(Page 265)*

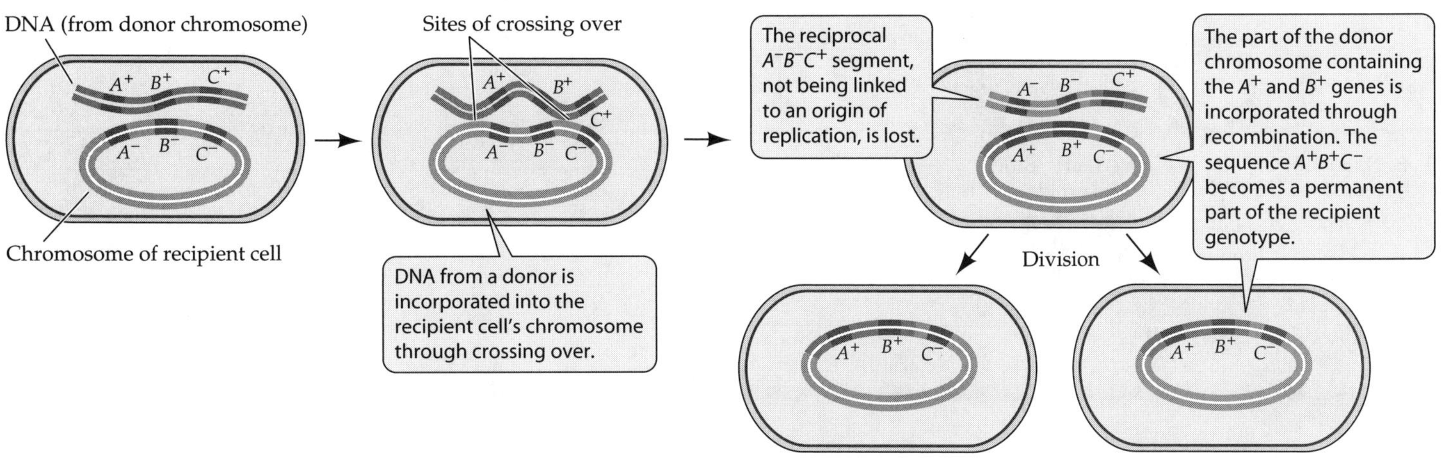
DNA (from donor chromosome)
A^+ B^+ C^+
A^- B^- C^-
Chromosome of recipient cell
Sites of crossing over
A^+ B^+ C^+
A^- B^- C^-
DNA from a donor is incorporated into the recipient cell's chromosome through crossing over.
The reciprocal $A^-B^-C^+$ segment, not being linked to an origin of replication, is lost.
A^- B^- C^+
A^+ B^+ C^-
The part of the donor chromosome containing the A^+ and B^+ genes is incorporated through recombination. The sequence $A^+B^+C^-$ becomes a permanent part of the recipient genotype.
Division
A^+ B^+ C^-
A^+ B^+ C^-

13.9 Recombination Following Conjugation *(Page 266)*

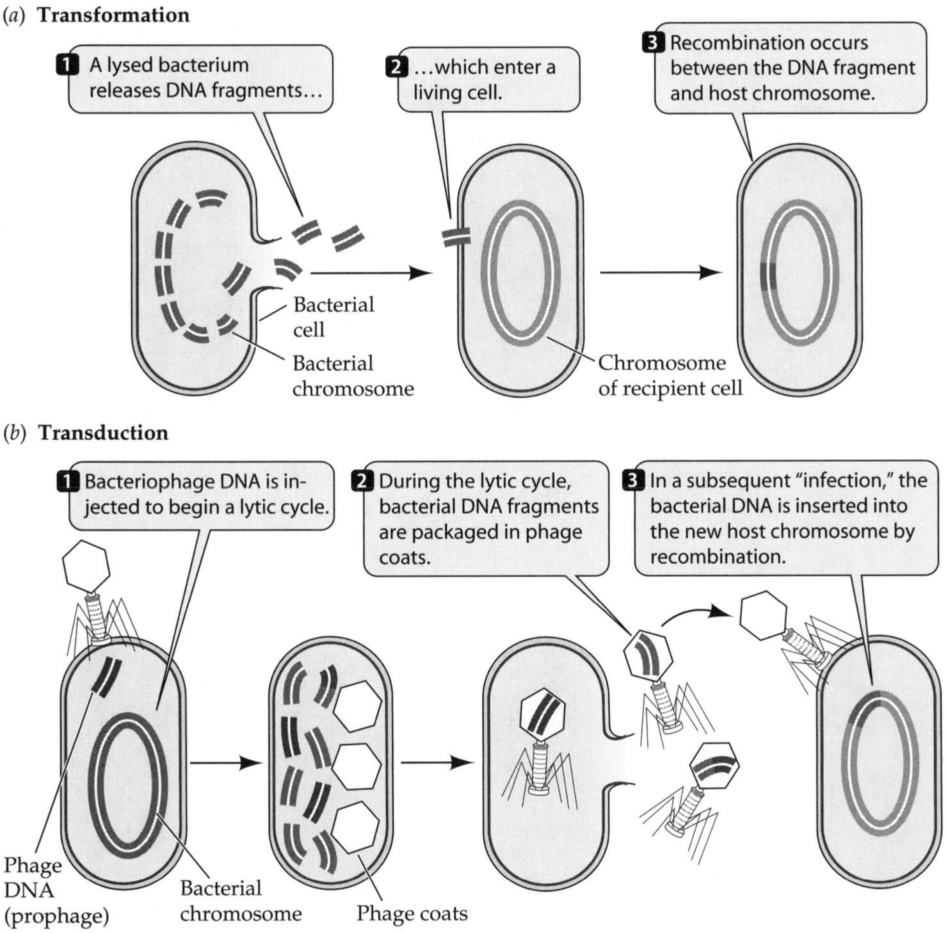

(*a*) **Transformation**

1 A lysed bacterium releases DNA fragments…

2 …which enter a living cell.

3 Recombination occurs between the DNA fragment and host chromosome.

Bacterial cell

Bacterial chromosome

Chromosome of recipient cell

(*b*) **Transduction**

1 Bacteriophage DNA is injected to begin a lytic cycle.

2 During the lytic cycle, bacterial DNA fragments are packaged in phage coats.

3 In a subsequent "infection," the bacterial DNA is inserted into the new host chromosome by recombination.

Phage DNA (prophage)

Bacterial chromosome

Phage coats

13.10 Transformation and Transduction *(Page 266)*

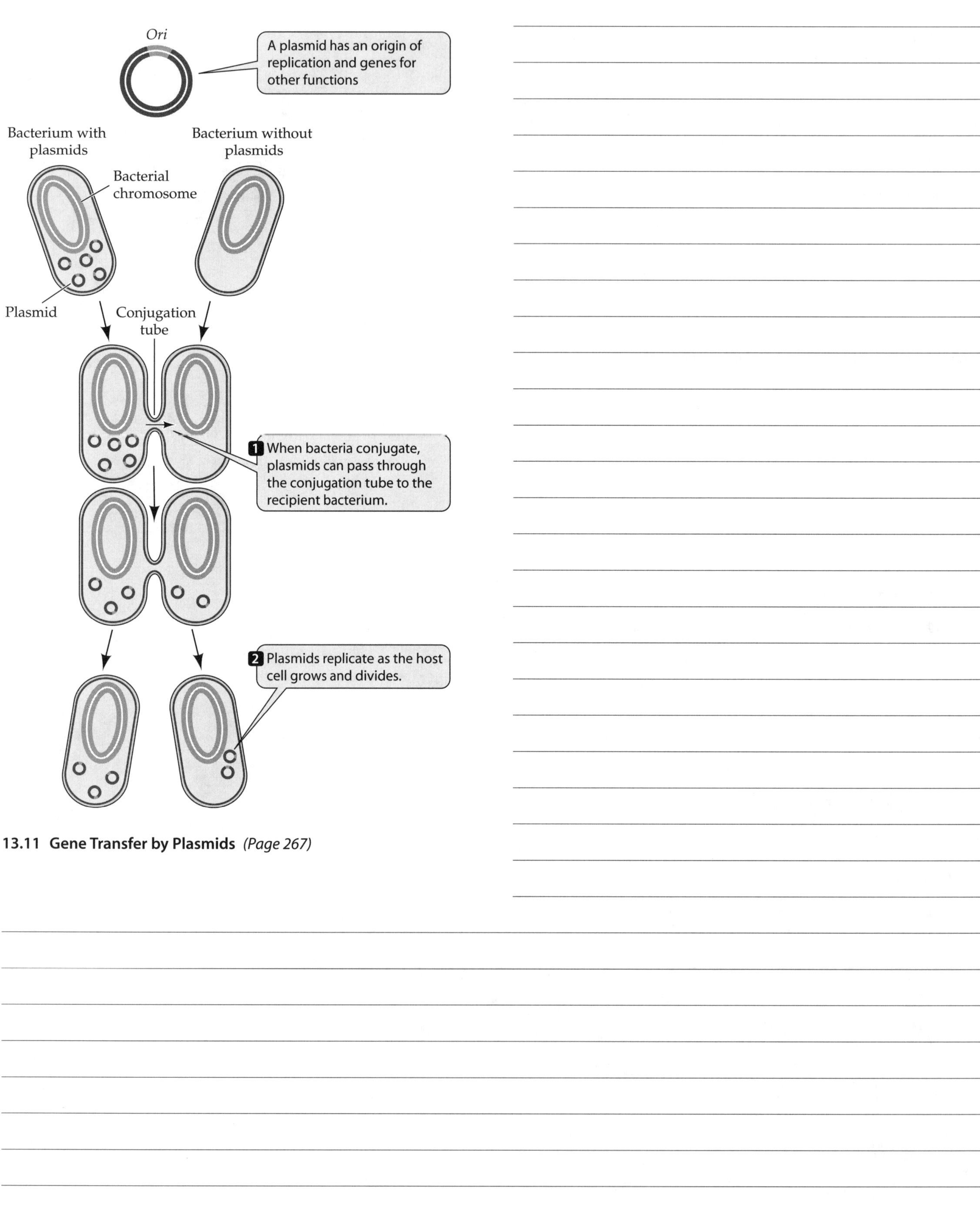
Ori
A plasmid has an origin of replication and genes for other functions
Bacterium with plasmids
Bacterium without plasmids
Bacterial chromosome
Plasmid
Conjugation tube
1 When bacteria conjugate, plasmids can pass through the conjugation tube to the recipient bacterium.
2 Plasmids replicate as the host cell grows and divides.

13.11 Gene Transfer by Plasmids *(Page 267)*

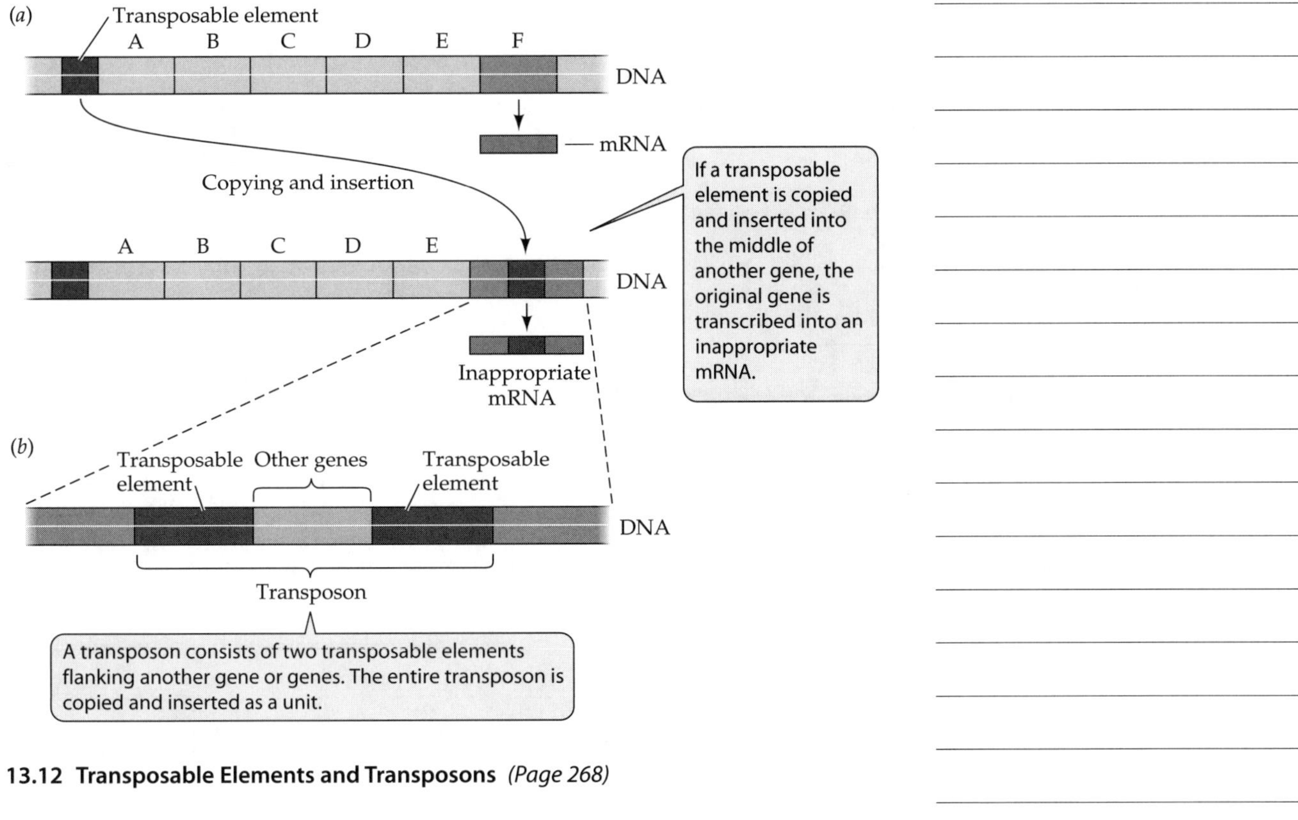
(a)
Transposable element
A B C D E F
DNA
mRNA
Copying and insertion
A B C D E
DNA
Inappropriate mRNA
If a transposable element is copied and inserted into the middle of another gene, the original gene is transcribed into an inappropriate mRNA.
(b)
Transposable element
Other genes
Transposable element
DNA
Transposon
A transposon consists of two transposable elements flanking another gene or genes. The entire transposon is copied and inserted as a unit.

13.12 Transposable Elements and Transposons *(Page 268)*

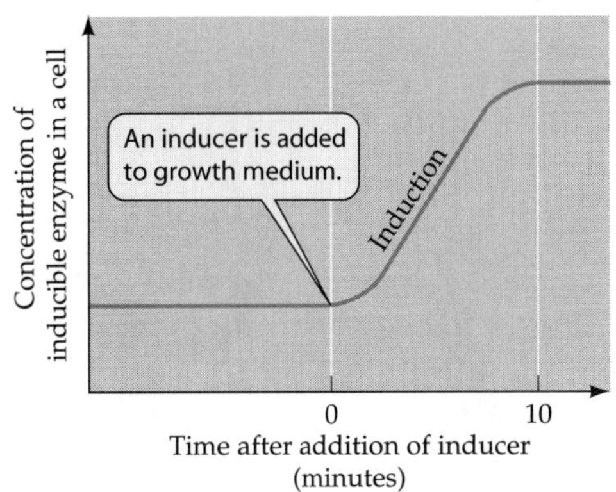
Concentration of inducible enzyme in a cell
An inducer is added to growth medium.
Induction
0
10
Time after addition of inducer (minutes)

13.13 An Inducer Stimulates the Synthesis of an Enzyme *(Page 269)*

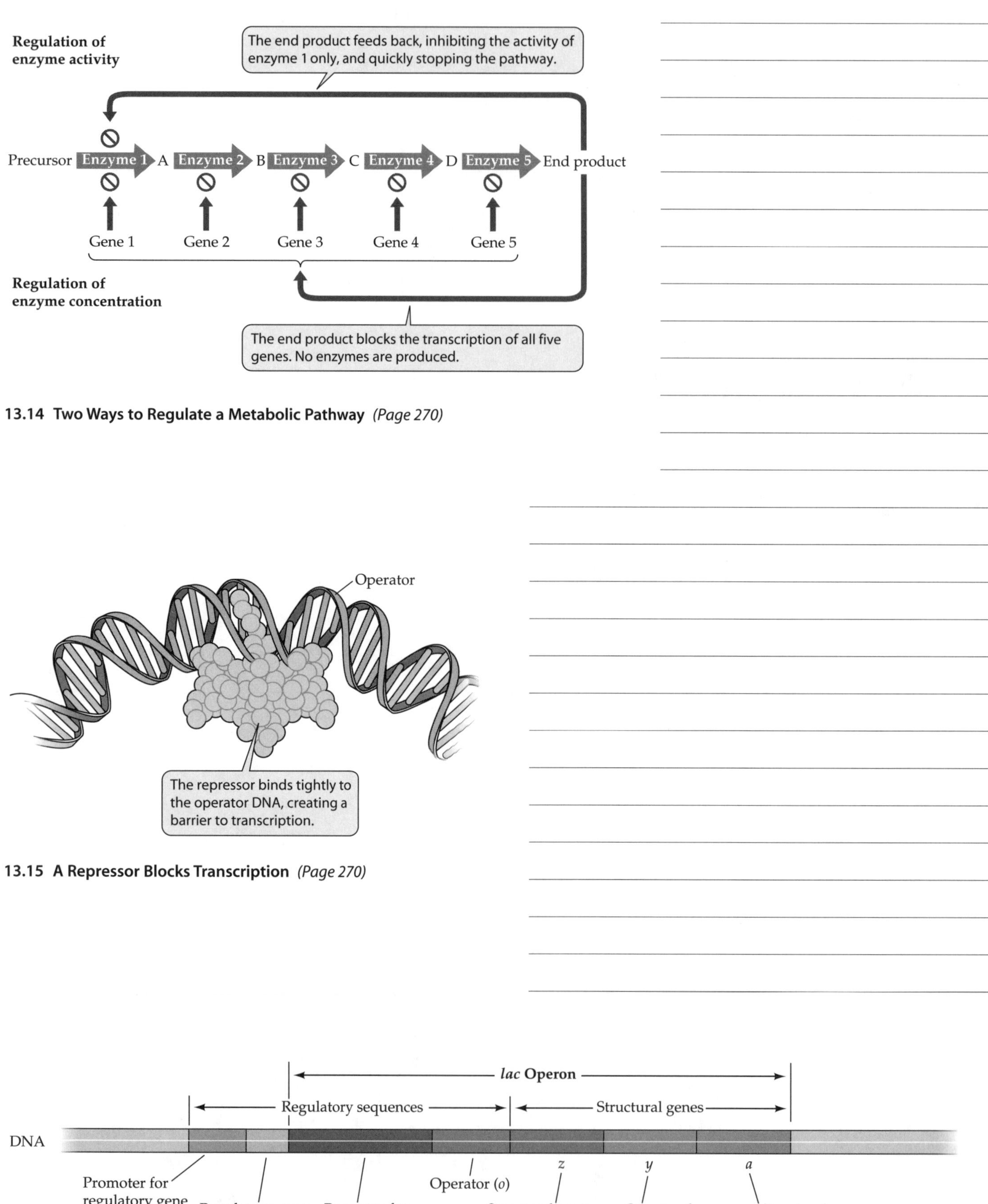
Regulation of enzyme activity
The end product feeds back, inhibiting the activity of enzyme 1 only, and quickly stopping the pathway.
Precursor
Enzyme 1
A
Enzyme 2
B
Enzyme 3
C
Enzyme 4
D
Enzyme 5
End product
Gene 1
Gene 2
Gene 3
Gene 4
Gene 5
Regulation of enzyme concentration
The end product blocks the transcription of all five genes. No enzymes are produced.

13.14 Two Ways to Regulate a Metabolic Pathway *(Page 270)*

Operator
The repressor binds tightly to the operator DNA, creating a barrier to transcription.

13.15 A Repressor Blocks Transcription *(Page 270)*

lac Operon
Regulatory sequences
Structural genes
DNA
Promoter for regulatory gene (p_i)
Regulatory gene (i) codes for repressor protein
Promoter for structural genes (p_{lac})
Operator (o)
z
y
a
Structural gene for β-galactosidase
Structural gene for β-galactoside permease
Structural gene for β-galactoside transacetylase

13.16 The *lac* Operon of *E. coli* *(Page 271)*

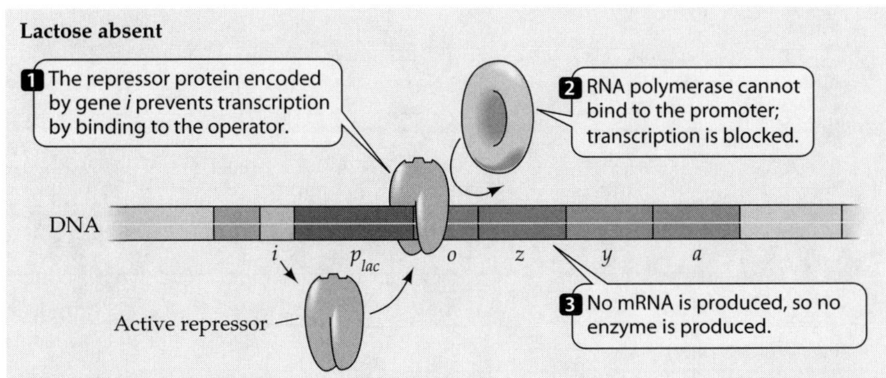
Lactose absent
1 The repressor protein encoded by gene *i* prevents transcription by binding to the operator.
2 RNA polymerase cannot bind to the promoter; transcription is blocked.
DNA
i
p_{lac}
o
z
y
a
Active repressor
3 No mRNA is produced, so no enzyme is produced.

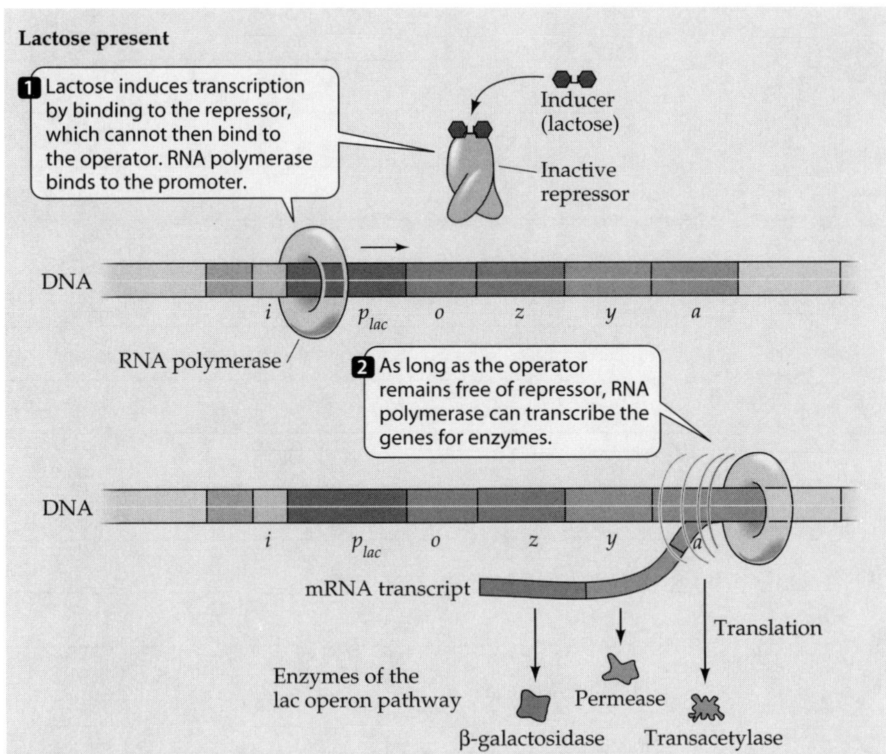
Lactose present
1 Lactose induces transcription by binding to the repressor, which cannot then bind to the operator. RNA polymerase binds to the promoter.
Inducer (lactose)
Inactive repressor
DNA
i
p_{lac}
o
z
y
a
RNA polymerase
2 As long as the operator remains free of repressor, RNA polymerase can transcribe the genes for enzymes.
DNA
mRNA transcript
Translation
Enzymes of the lac operon pathway
Permease
β-galactosidase
Transacetylase

13.17 The *lac* Operon: An Inducible System *(Page 271)*

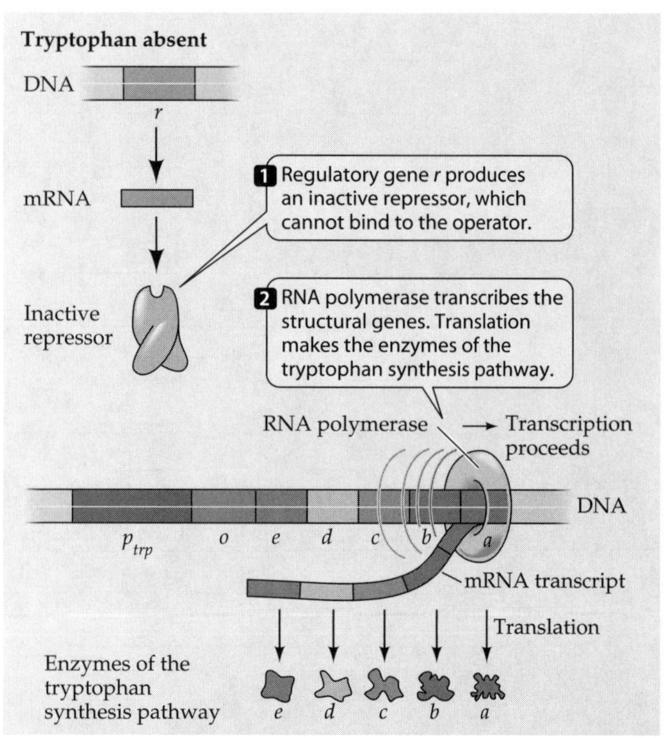
Tryptophan absent
DNA
r
mRNA
Inactive repressor
1 Regulatory gene r produces an inactive repressor, which cannot bind to the operator.
2 RNA polymerase transcribes the structural genes. Translation makes the enzymes of the tryptophan synthesis pathway.
RNA polymerase
Transcription proceeds
DNA
p_{trp}
o
e
d
c
b
a
mRNA transcript
Translation
Enzymes of the tryptophan synthesis pathway
e
d
c
b
a

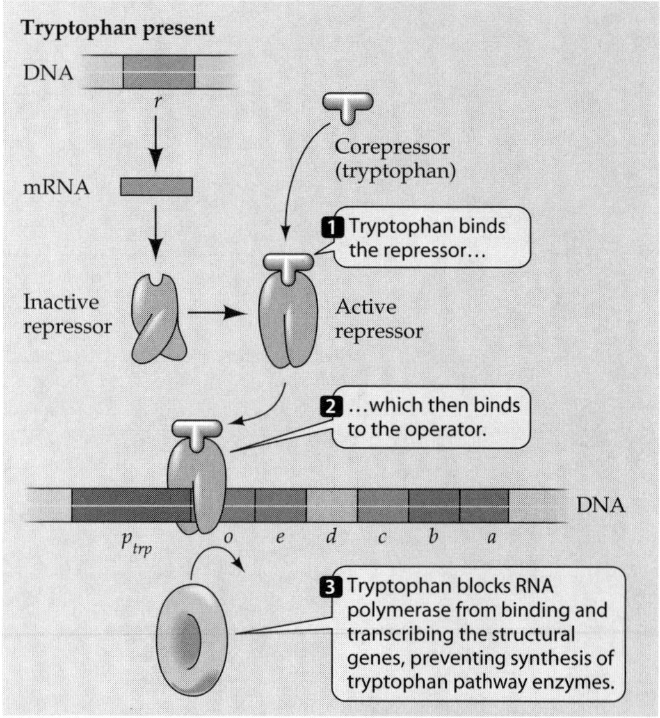
Tryptophan present
DNA
r
mRNA
Corepressor (tryptophan)
1 Tryptophan binds the repressor…
Inactive repressor
Active repressor
2 …which then binds to the operator.
DNA
p_{trp}
o
e
d
c
b
a
3 Tryptophan blocks RNA polymerase from binding and transcribing the structural genes, preventing synthesis of tryptophan pathway enzymes.

13.18 The *trp* Operon: A Repressible System *(Page 272)*

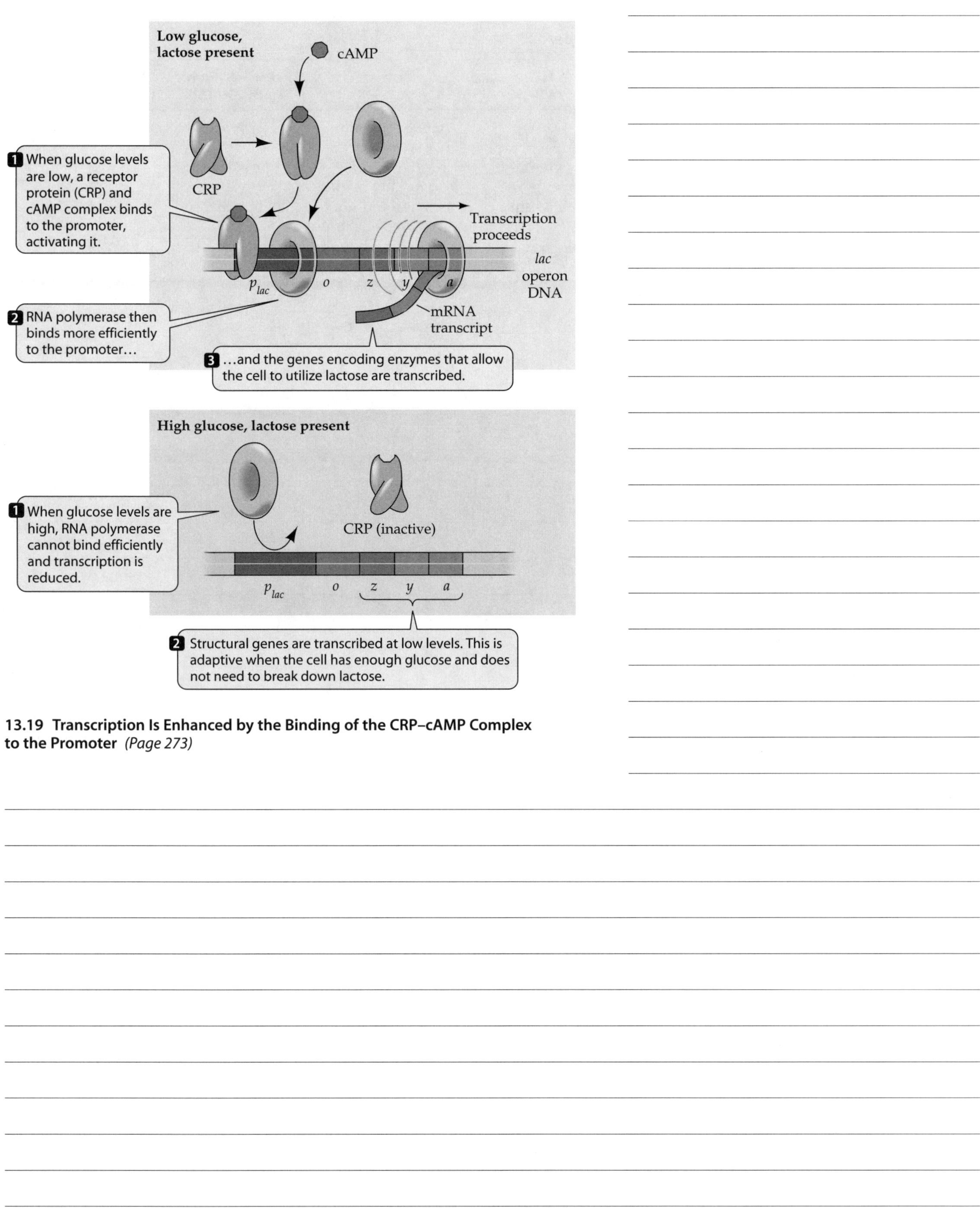
Low glucose, lactose present
cAMP
CRP
Transcription proceeds
lac operon DNA
p_{lac}
o
z
y
a
mRNA transcript
1 When glucose levels are low, a receptor protein (CRP) and cAMP complex binds to the promoter, activating it.
2 RNA polymerase then binds more efficiently to the promoter…
3 …and the genes encoding enzymes that allow the cell to utilize lactose are transcribed.
High glucose, lactose present
CRP (inactive)
p_{lac}
o
z
y
a
1 When glucose levels are high, RNA polymerase cannot bind efficiently and transcription is reduced.
2 Structural genes are transcribed at low levels. This is adaptive when the cell has enough glucose and does not need to break down lactose.

13.19 Transcription Is Enhanced by the Binding of the CRP–cAMP Complex to the Promoter *(Page 273)*

13.2 Positive and Negative Controls in the lac Operon[a]

GLUCOSE	cAMP LEVELS	RNA POLYMERASE BINDING TO PROMOTER	LACTOSE	LAC REPRESSOR	TRANSCRIPTION OF LAC GENES?	LACTOSE USED BY CELLS?
Present	Low	Absent	Absent	Active and bound to operator	No	No
Present	Low	Present, not efficient	Present	Inactive and not bound to operator	Low level	No
Absent	High	Present, very efficient	Present	Inactive and not bound to operator	High level	Yes
Absent	High	Absent	Absent	Active and bound to operator	No	No

[a]Negative controls are in red type.

(Page 274)

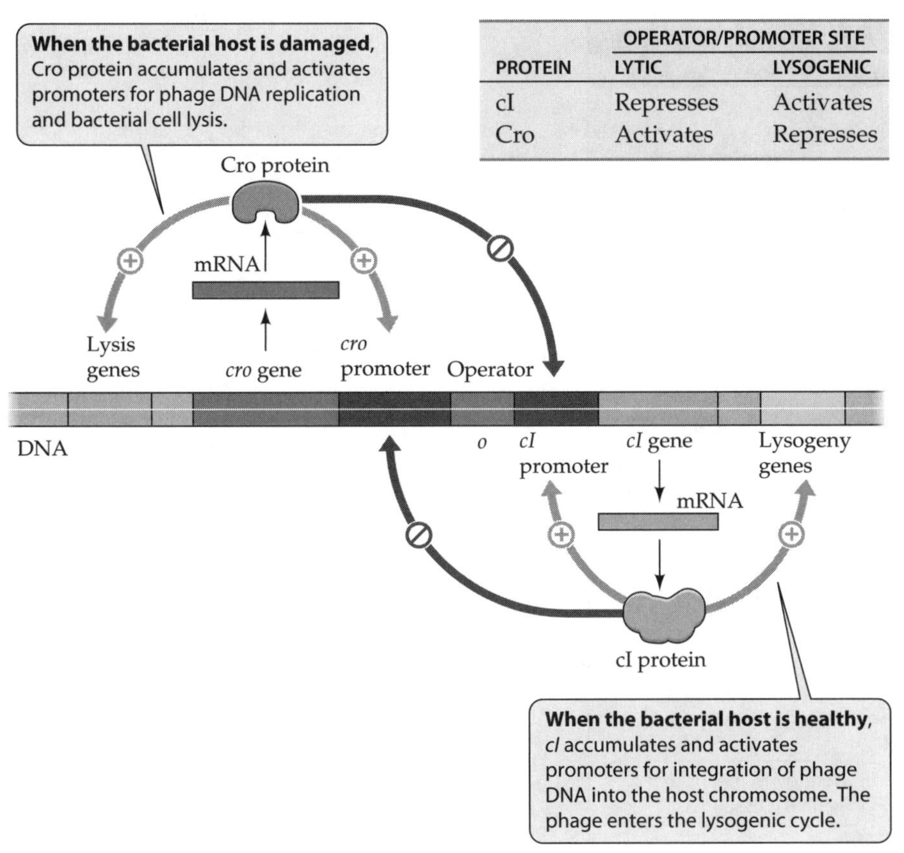
When the bacterial host is damaged, Cro protein accumulates and activates promoters for phage DNA replication and bacterial cell lysis.
Cro protein
mRNA
Lysis genes
cro gene
cro promoter
Operator
DNA
o
cI promoter
cI gene
mRNA
Lysogeny genes
cI protein
When the bacterial host is healthy, *cI* accumulates and activates promoters for integration of phage DNA into the host chromosome. The phage enters the lysogenic cycle.

	OPERATOR/PROMOTER SITE	
PROTEIN	LYTIC	LYSOGENIC
cI	Represses	Activates
Cro	Activates	Represses

13.20 Control of Phage λ Lysis and Lysogeny *(Page 274)*

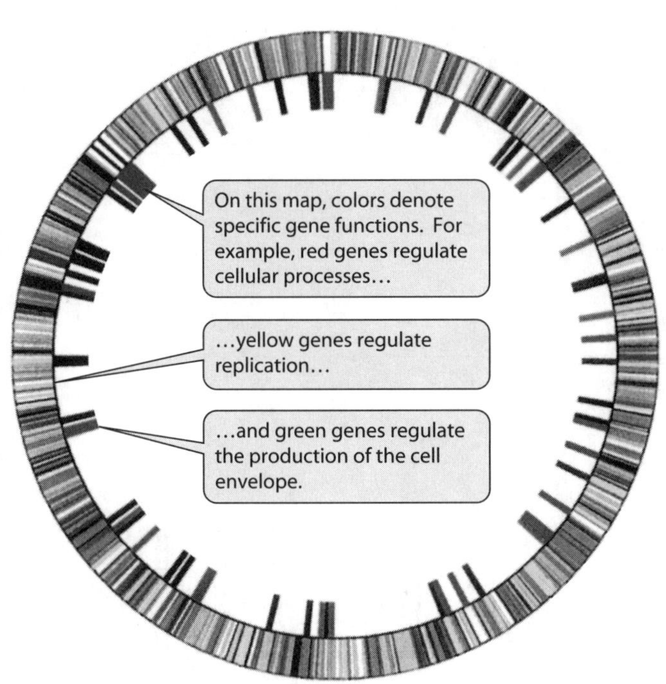
On this map, colors denote specific gene functions. For example, red genes regulate cellular processes…
…yellow genes regulate replication…
…and green genes regulate the production of the cell envelope.

13.21 Functional Organization of the Genome of *H. influenzae* *(Page 275)*

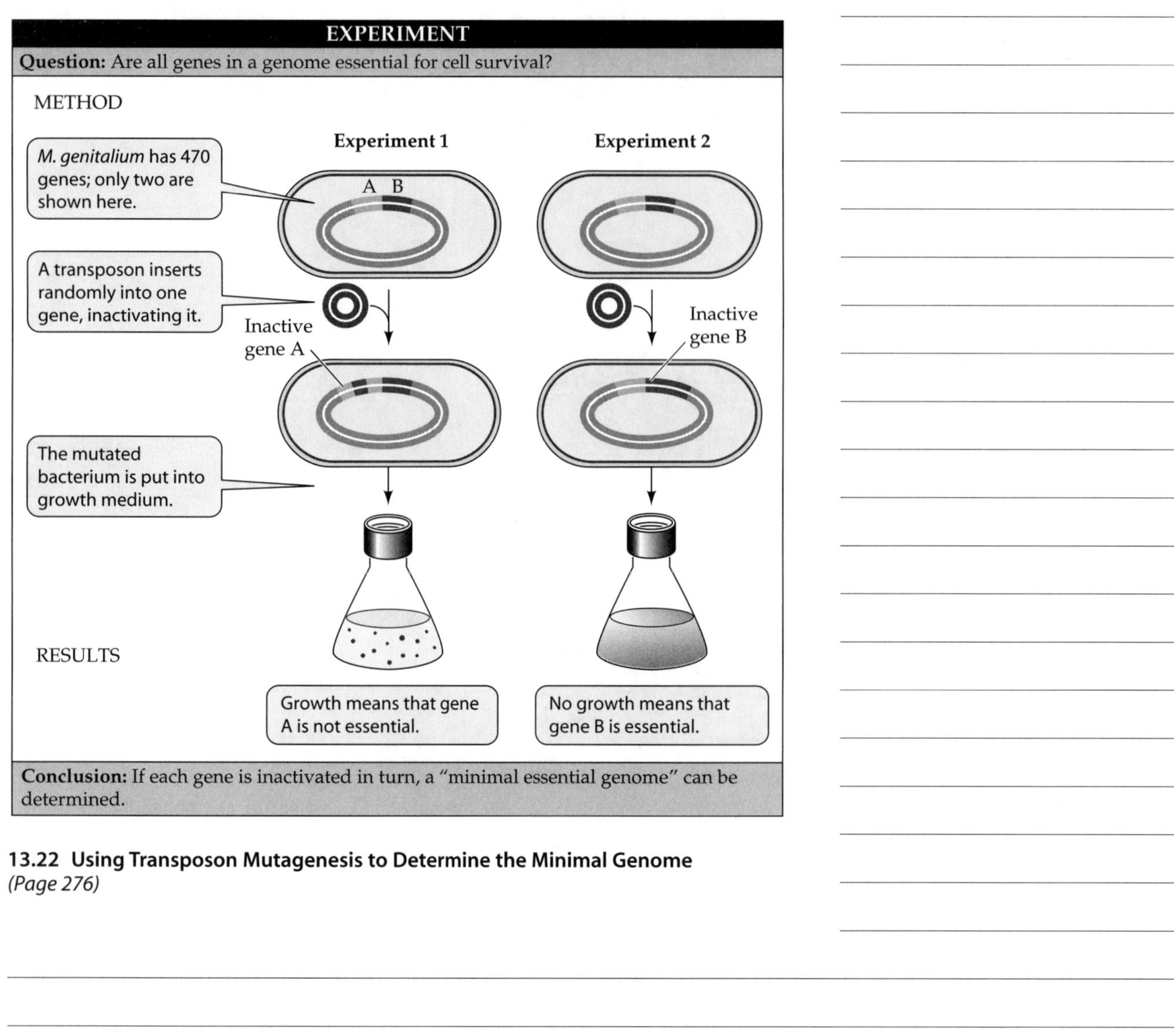
EXPERIMENT
Question: Are all genes in a genome essential for cell survival?
METHOD
Experiment 1
Experiment 2
M. genitalium has 470 genes; only two are shown here.
A B
A transposon inserts randomly into one gene, inactivating it.
Inactive gene A
Inactive gene B
The mutated bacterium is put into growth medium.
RESULTS
Growth means that gene A is not essential.
No growth means that gene B is essential.
Conclusion: If each gene is inactivated in turn, a "minimal essential genome" can be determined.

13.22 Using Transposon Mutagenesis to Determine the Minimal Genome
(Page 276)

14 The Eukaryotic Genome and Its Expression

14.1 A Comparison of Prokaryotic and Eukaryotic Genes and Genomes

CHARACTERISTIC	PROKARYOTES	EUKARYOTES
Genome size (base pairs)	10^4–10^7	10^8–10^{11}
Repeated sequences	Few	Many
Noncoding DNA within coding sequences	Rare	Common
Transcription and translation separated in cell	No	Yes
DNA segregated within a nucleus	No	Yes
DNA bound to proteins	Some	Extensive
Promoters	Yes	Yes
Enhancers/silencers	Rare	Common
Capping and tailing of mRNA	No	Yes
RNA splicing required (spliceosomes)	Rare	Common
Number of chromosomes in genome	One	Many

(Page 280)

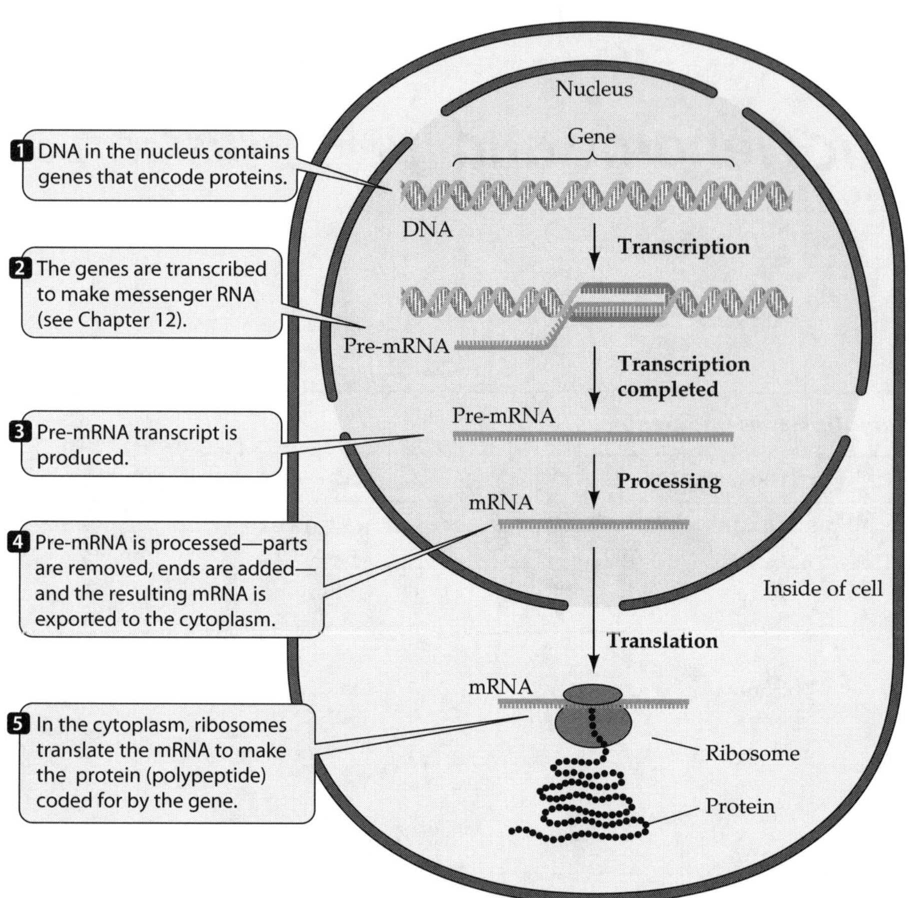

1 DNA in the nucleus contains genes that encode proteins.
2 The genes are transcribed to make messenger RNA (see Chapter 12).
3 Pre-mRNA transcript is produced.
4 Pre-mRNA is processed—parts are removed, ends are added—and the resulting mRNA is exported to the cytoplasm.
5 In the cytoplasm, ribosomes translate the mRNA to make the protein (polypeptide) coded for by the gene.
Nucleus
Gene
DNA
Transcription
Pre-mRNA
Transcription completed
Pre-mRNA
Processing
mRNA
Inside of cell
Translation
mRNA
Ribosome
Protein

14.1 Eukaryotic mRNA Is Transcribed in the Nucleus but Translated in the Cytoplasm
(Page 281)

14.2 *Comparison of the Genomes of* E. coli *and Yeast*

	E. COLI	YEAST
Genome length (base pairs)	4,640,000	12,068,000
Number of proteins	4,300	6,200
Proteins with roles in:		
Metabolism	650	650
Energy production/storage	240	175
Membrane transporters	280	250
DNA replication/repair/recombination	120	175
Transcription	230	400
Translation	180	350
Protein targeting/secretion	35	430
Cell structure	180	250

(Page 281)

14.3 *C. elegans* Genes Essential to Multicellularity

FUNCTION	PROTEIN/DOMAIN	NUMBER OF GENES
Transcription control	Zinc finger; homeobox	540
RNA processing	RNA binding domains	100
Nerve impulse transmission	Gated ion channels	80
Tissue formation	Collagens	170
Cell interactions	Extracellular domains; glycotransferases	330
Cell–cell signaling	G protein-linked receptors; protein kinases; protein phosphatases	1,290

(Page 282)

14.4 *Arabidopsis* Genes Unique to Plants

FUNCTION	NUMBER OF GENES
Cell wall and growth	420
Water channels	300
Photosynthesis	139
Defense and metabolism	94

(Page 283)

14.5 Comparison of the Rice and *Arabidopsis* Genomes

FUNCTION	PERCENTAGE OF GENOME	
	RICE	*ARABIDOPSIS*
Cell structure	9	10
Enzymes	21	20
Ligand binding	10	10
DNA binding	10	10
Signal transduction	3	3
Membrane transport	5	5
Cell growth and maintenance	24	22
Other functions	18	20

(Page 283)

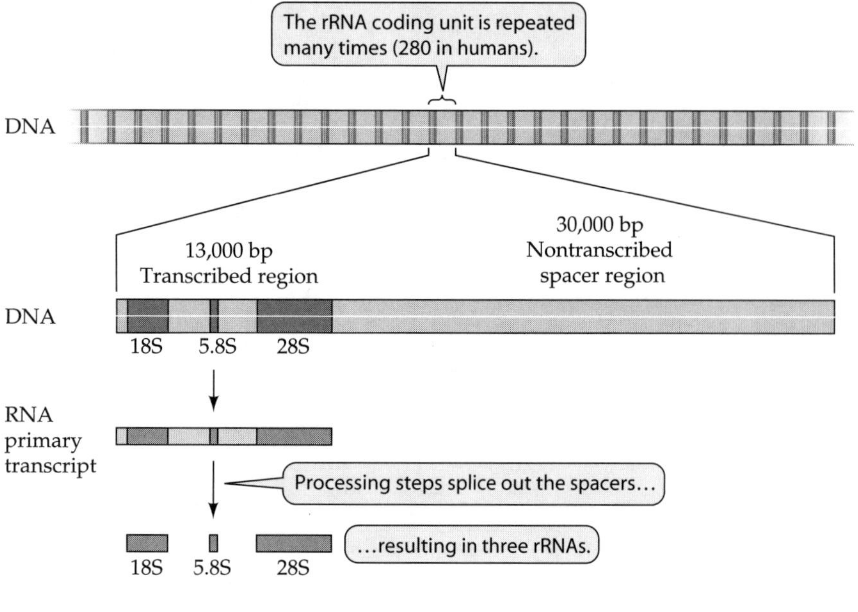
The rRNA coding unit is repeated many times (280 in humans).
DNA
13,000 bp
Transcribed region
30,000 bp
Nontranscribed
spacer region
DNA
18S
5.8S
28S
RNA primary transcript
Processing steps splice out the spacers…
…resulting in three rRNAs.
18S
5.8S
28S

14.2 A Moderately Repetitive Sequence Codes for rRNA *(Page 284)*

1 The transposon carries a gene for transposase, which catalyzes the movements of the DNA.

Transposon

DNA

AATG TTAC

Transposase gene

CATT GTAA

Inverted repeat

Inverted repeat

Protein-coding gene

DNA

Transposase

2 Transposase allows the DNA to loop out and move to a new location in the genome.

3 The transposon has inserted itself within a protein-coding gene, disrupting it…

Disrupted gene

Disrupted gene

DNA

4 …and resulting in a nonfunctional mRNA and/or protein.

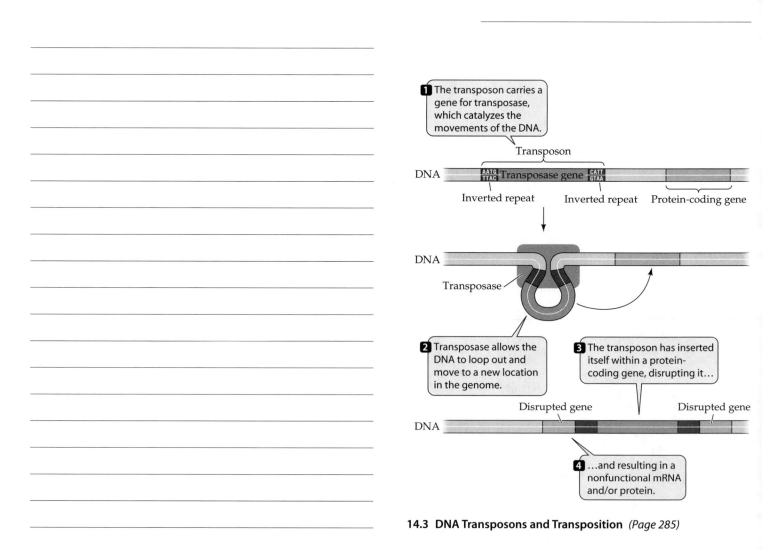

14.3 DNA Transposons and Transposition *(Page 285)*

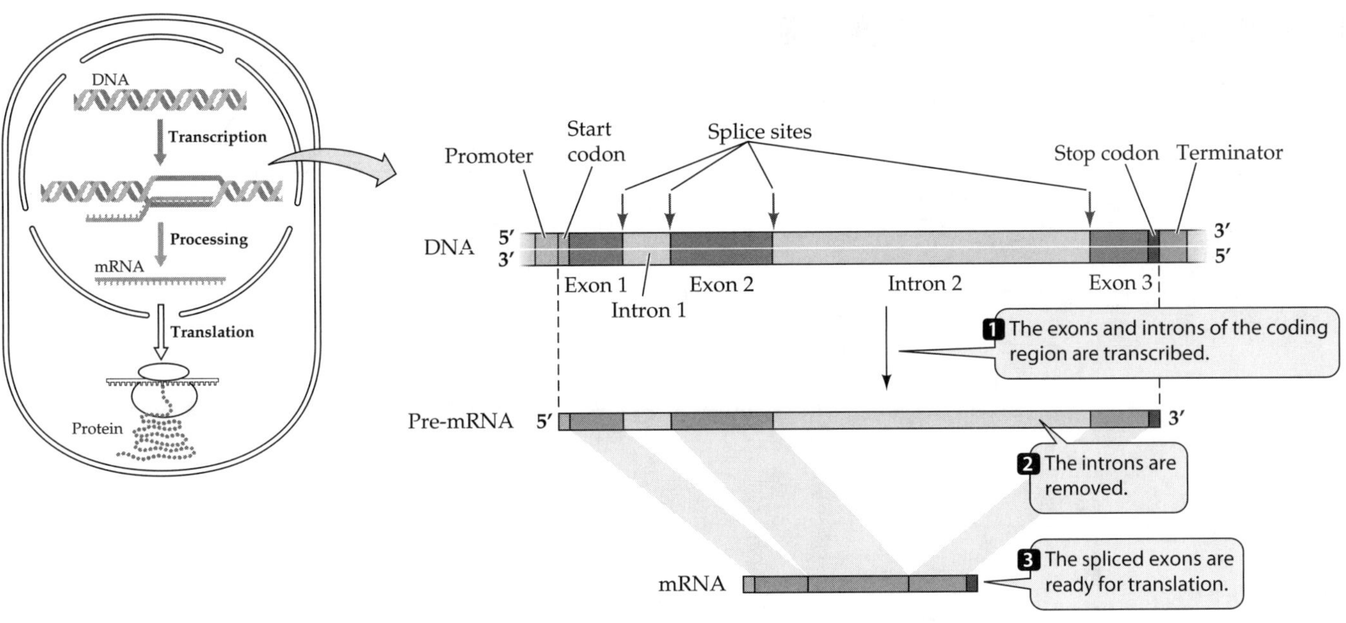

DNA
Transcription
Processing
mRNA
Translation
Protein
Promoter
Start codon
Splice sites
Stop codon
Terminator
DNA 5′ 3′
3′ 5′
Exon 1
Intron 1
Exon 2
Intron 2
Exon 3
1 The exons and introns of the coding region are transcribed.
Pre-mRNA 5′
3′
2 The introns are removed.
mRNA
3 The spliced exons are ready for translation.

14.4 The Structure and Transcription of a Eukaryotic Gene *(Page 286)*

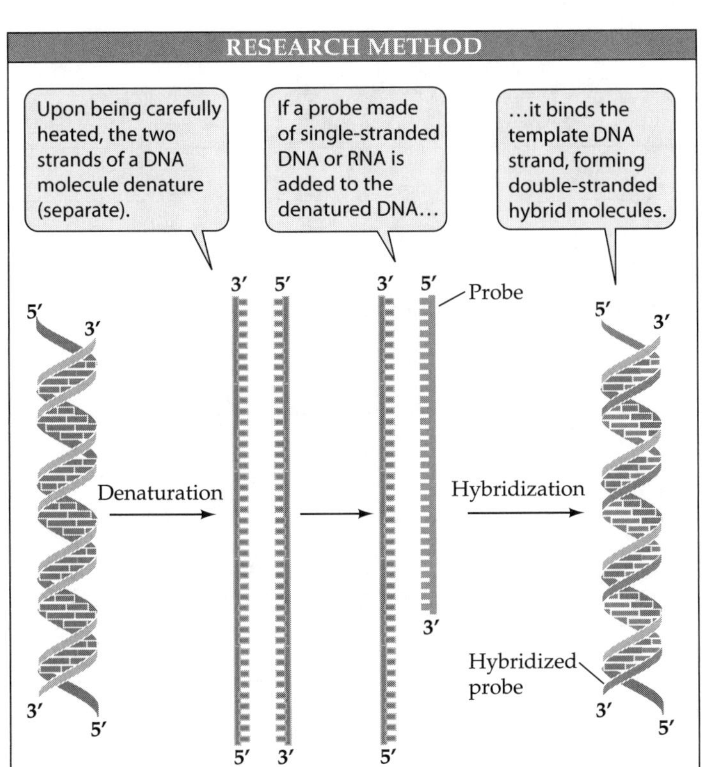

RESEARCH METHOD
Upon being carefully heated, the two strands of a DNA molecule denature (separate).
If a probe made of single-stranded DNA or RNA is added to the denatured DNA…
…it binds the template DNA strand, forming double-stranded hybrid molecules.
Denaturation
Probe
Hybridization
Hybridized probe

14.5 Nucleic Acid Hybridization *(Page 286)*

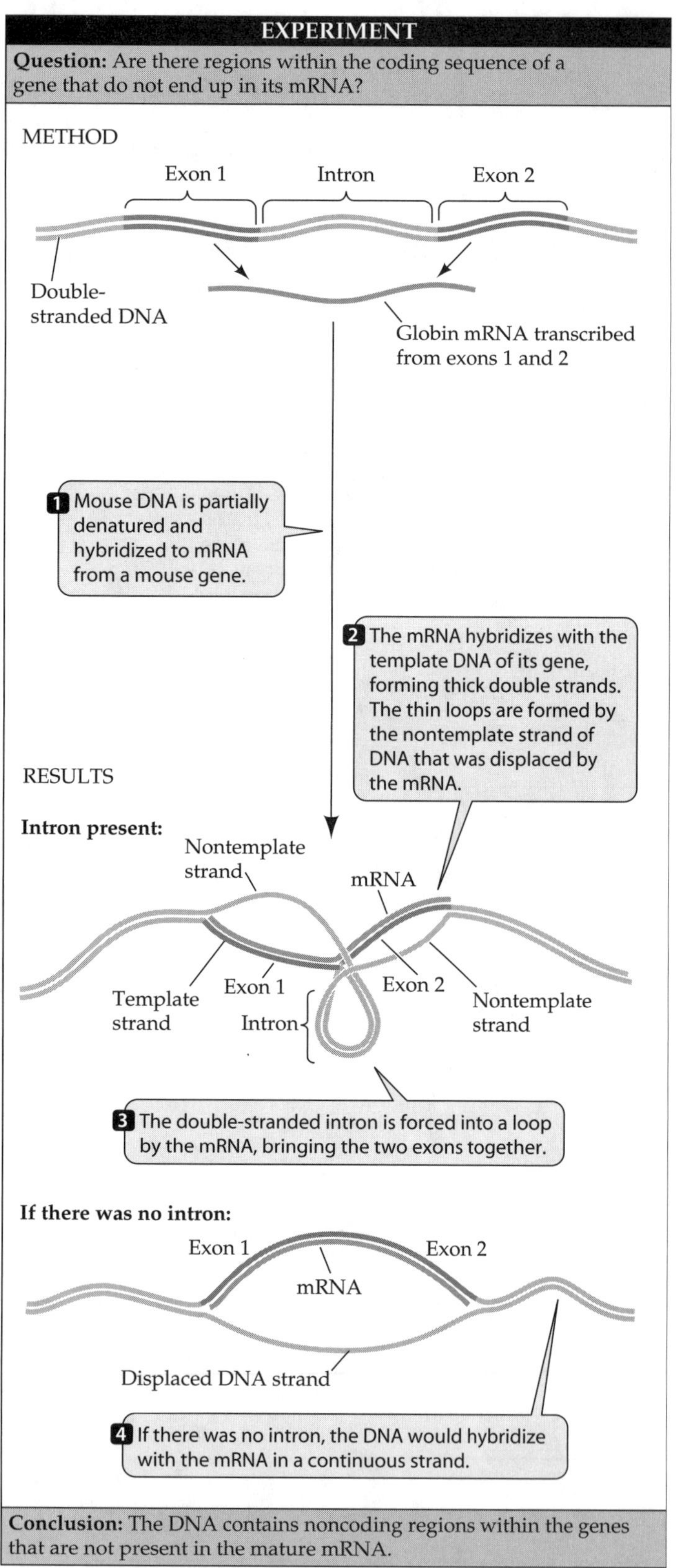
EXPERIMENT
Question: Are there regions within the coding sequence of a gene that do not end up in its mRNA?
METHOD
Exon 1
Intron
Exon 2
Double-stranded DNA
Globin mRNA transcribed from exons 1 and 2
1 Mouse DNA is partially denatured and hybridized to mRNA from a mouse gene.
2 The mRNA hybridizes with the template DNA of its gene, forming thick double strands. The thin loops are formed by the nontemplate strand of DNA that was displaced by the mRNA.
RESULTS
Intron present:
Nontemplate strand
mRNA
Template strand
Exon 1
Exon 2
Nontemplate strand
Intron
3 The double-stranded intron is forced into a loop by the mRNA, bringing the two exons together.
If there was no intron:
Exon 1
Exon 2
mRNA
Displaced DNA strand
4 If there was no intron, the DNA would hybridize with the mRNA in a continuous strand.
Conclusion: The DNA contains noncoding regions within the genes that are not present in the mature mRNA.

14.6 Nucleic Acid Hybridization Revealed Noncoding DNA
(Page 287)

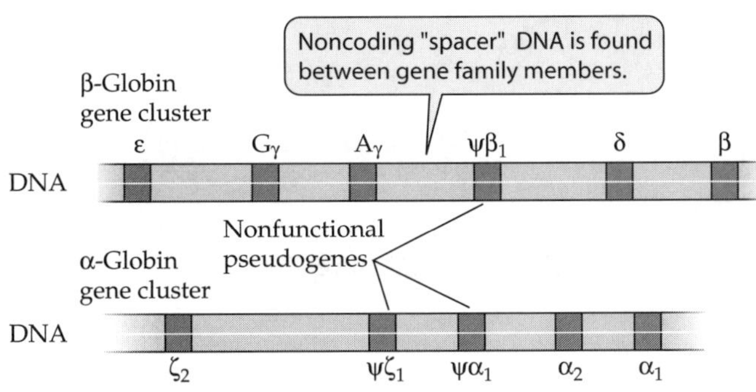
Noncoding "spacer" DNA is found between gene family members.
β-Globin gene cluster
ε
Gγ
Aγ
ψβ1
δ
β
DNA
Nonfunctional pseudogenes
α-Globin gene cluster
DNA
ζ2
ψζ1
ψα1
α2
α1

14.7 The Globin Gene Family *(Page 287)*

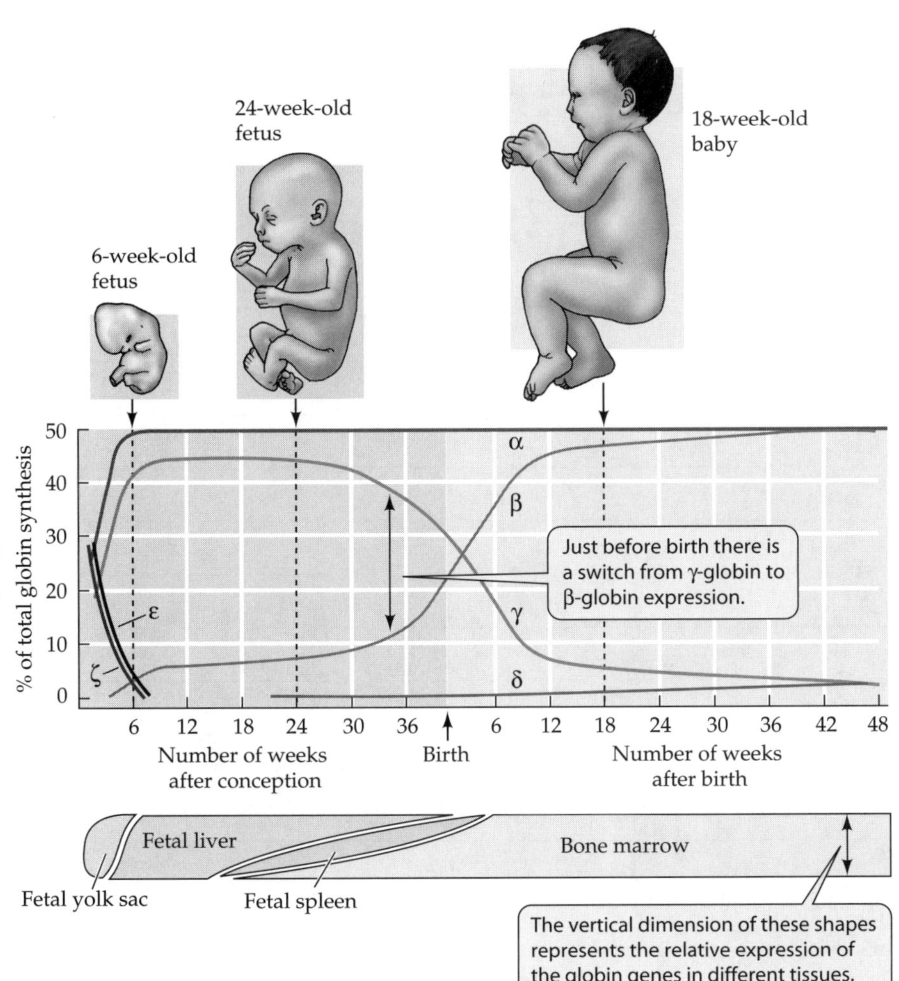
24-week-old fetus
18-week-old baby
6-week-old fetus
% of total globin synthesis
50
40
30
20
10
0
α
β
γ
δ
ε
ζ
Just before birth there is a switch from γ-globin to β-globin expression.
6 12 18 24 30 36 6 12 18 24 30 36 42 48
Number of weeks after conception
Birth
Number of weeks after birth
Fetal liver
Bone marrow
Fetal yolk sac
Fetal spleen
The vertical dimension of these shapes represents the relative expression of the globin genes in different tissues.

14.8 Differential Expression in the Globin Gene Family *(Page 288)*

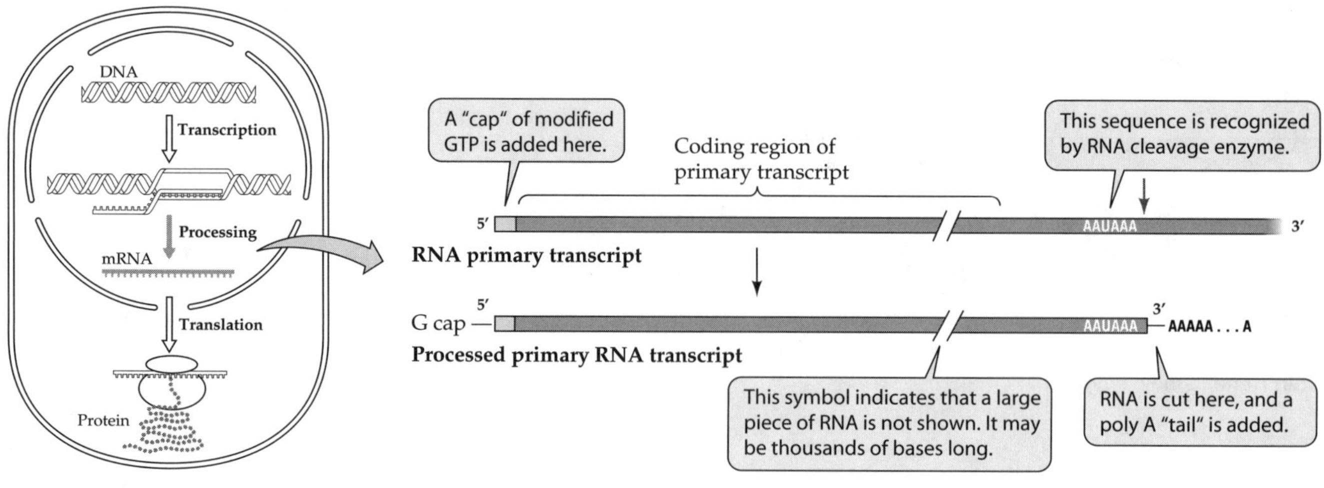
DNA
Transcription
Processing
mRNA
Translation
Protein
A "cap" of modified GTP is added here.
Coding region of primary transcript
This sequence is recognized by RNA cleavage enzyme.
5′
AAUAAA
3′
RNA primary transcript
G cap
5′
AAUAAA
3′
AAAAA . . . A
Processed primary RNA transcript
This symbol indicates that a large piece of RNA is not shown. It may be thousands of bases long.
RNA is cut here, and a poly A "tail" is added.

14.9 Processing the Ends of Eukaryotic Pre-mRNA *(Page 288)*

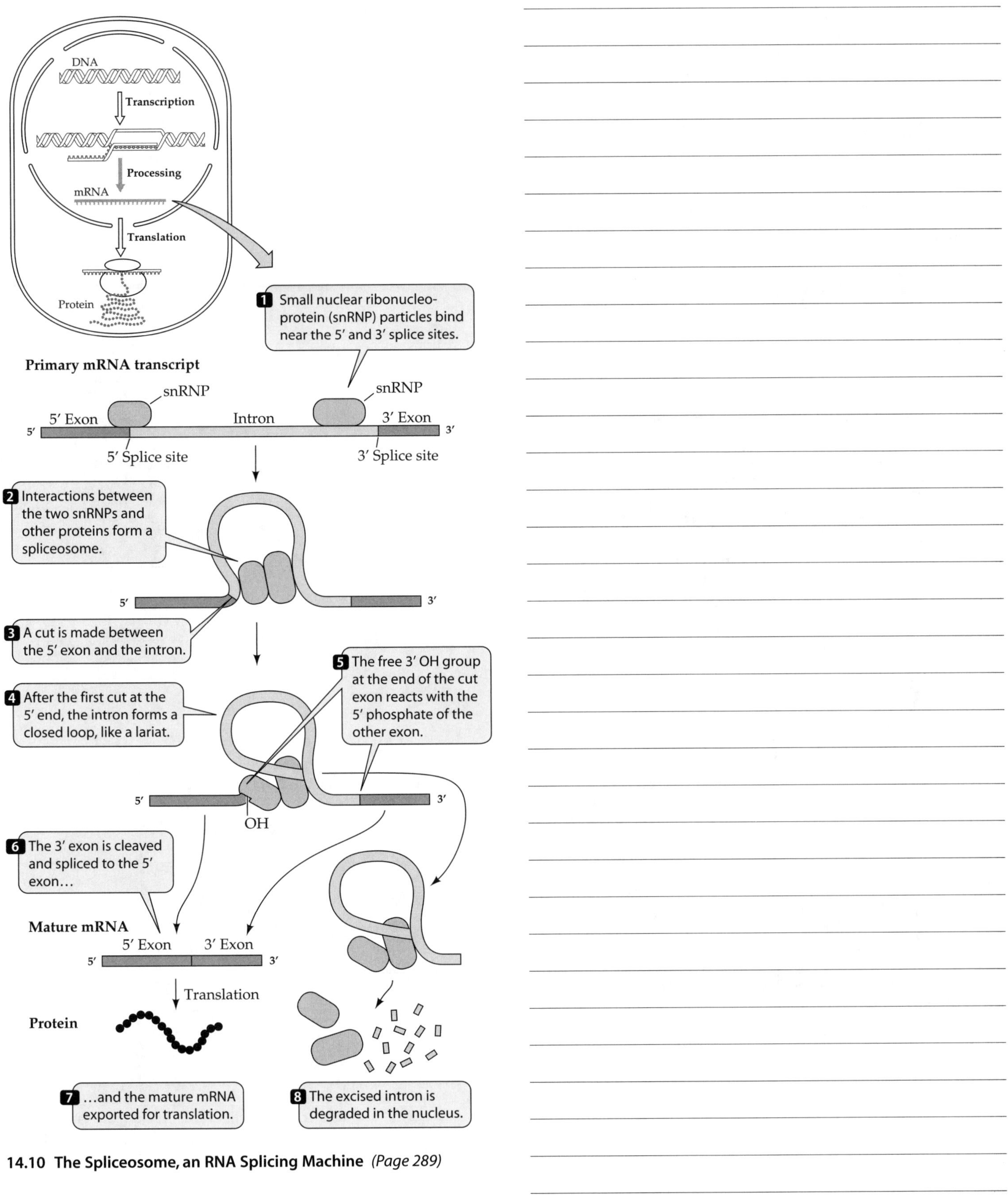
DNA
Transcription
Processing
mRNA
Translation
Protein
1 Small nuclear ribonucleo-protein (snRNP) particles bind near the 5′ and 3′ splice sites.
Primary mRNA transcript
snRNP
snRNP
5′ Exon
Intron
3′ Exon
5′
3′
5′ Splice site
3′ Splice site
2 Interactions between the two snRNPs and other proteins form a spliceosome.
5′
3′
3 A cut is made between the 5′ exon and the intron.
5 The free 3′ OH group at the end of the cut exon reacts with the 5′ phosphate of the other exon.
4 After the first cut at the 5′ end, the intron forms a closed loop, like a lariat.
5′
3′
OH
6 The 3′ exon is cleaved and spliced to the 5′ exon…
Mature mRNA
5′ Exon
3′ Exon
5′
3′
Translation
Protein
7 …and the mature mRNA exported for translation.
8 The excised intron is degraded in the nucleus.

14.10 The Spliceosome, an RNA Splicing Machine *(Page 289)*

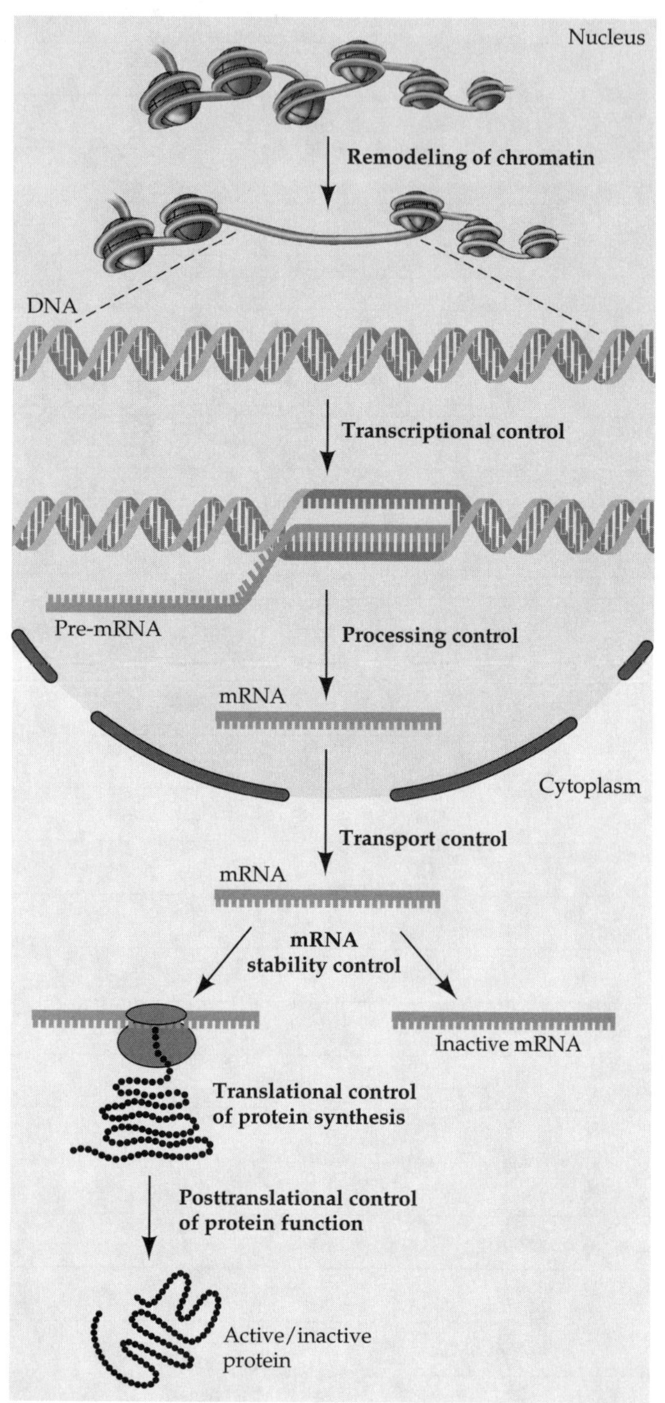
Nucleus
Remodeling of chromatin
DNA
Transcriptional control
Pre-mRNA
Processing control
mRNA
Cytoplasm
Transport control
mRNA
mRNA
stability control
Inactive mRNA
Translational control
of protein synthesis
Posttranslational control
of protein function
Active/inactive
protein

14.11 Potential Points for the Regulation of Gene Expression in Eukaryotes *(Page 290)*

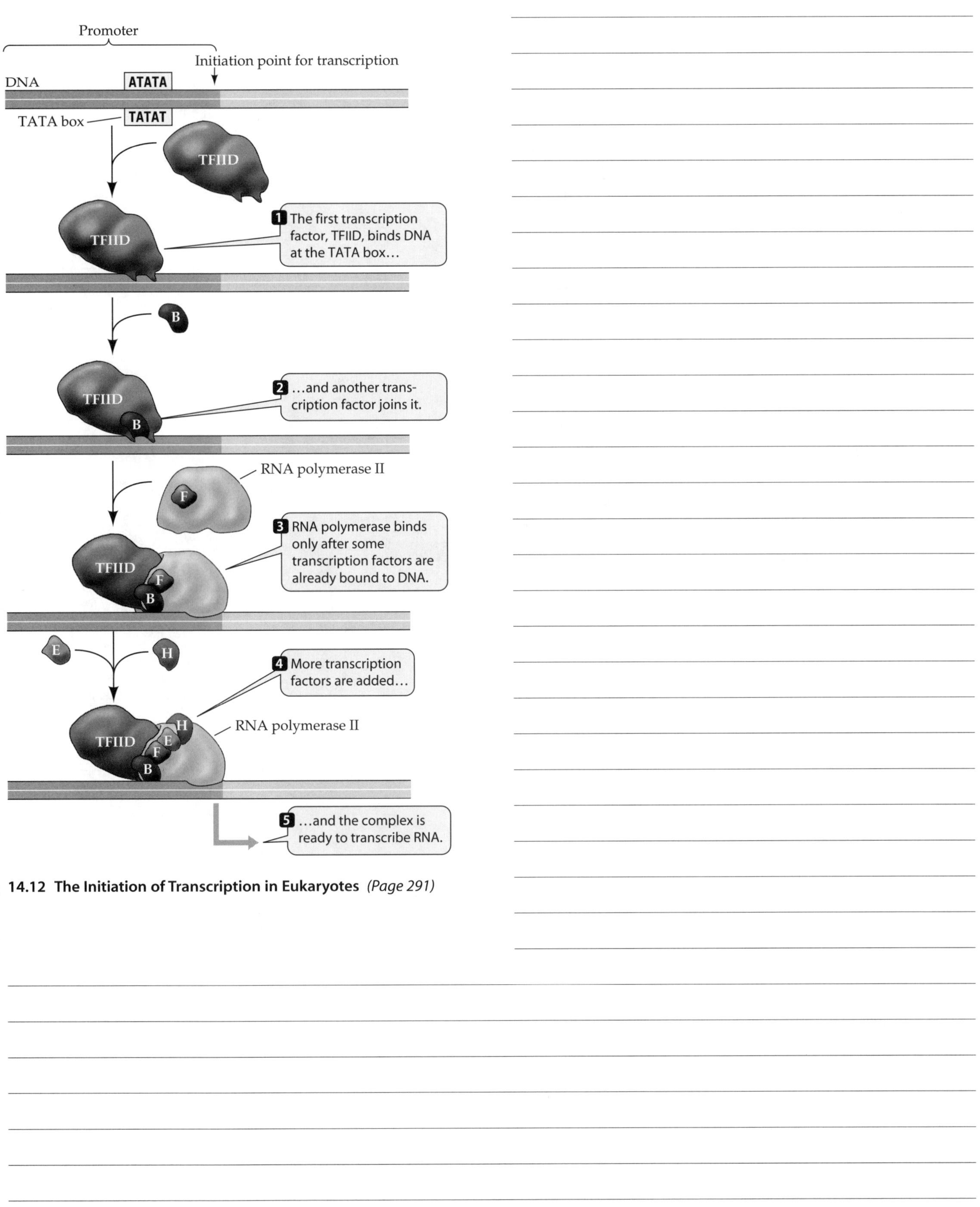
Promoter
Initiation point for transcription
DNA
ATATA
TATA box
TATAT
TFIID
1 The first transcription factor, TFIID, binds DNA at the TATA box…
B
2 …and another transcription factor joins it.
RNA polymerase II
F
3 RNA polymerase binds only after some transcription factors are already bound to DNA.
E
H
4 More transcription factors are added…
RNA polymerase II
5 …and the complex is ready to transcribe RNA.

14.12 The Initiation of Transcription in Eukaryotes *(Page 291)*

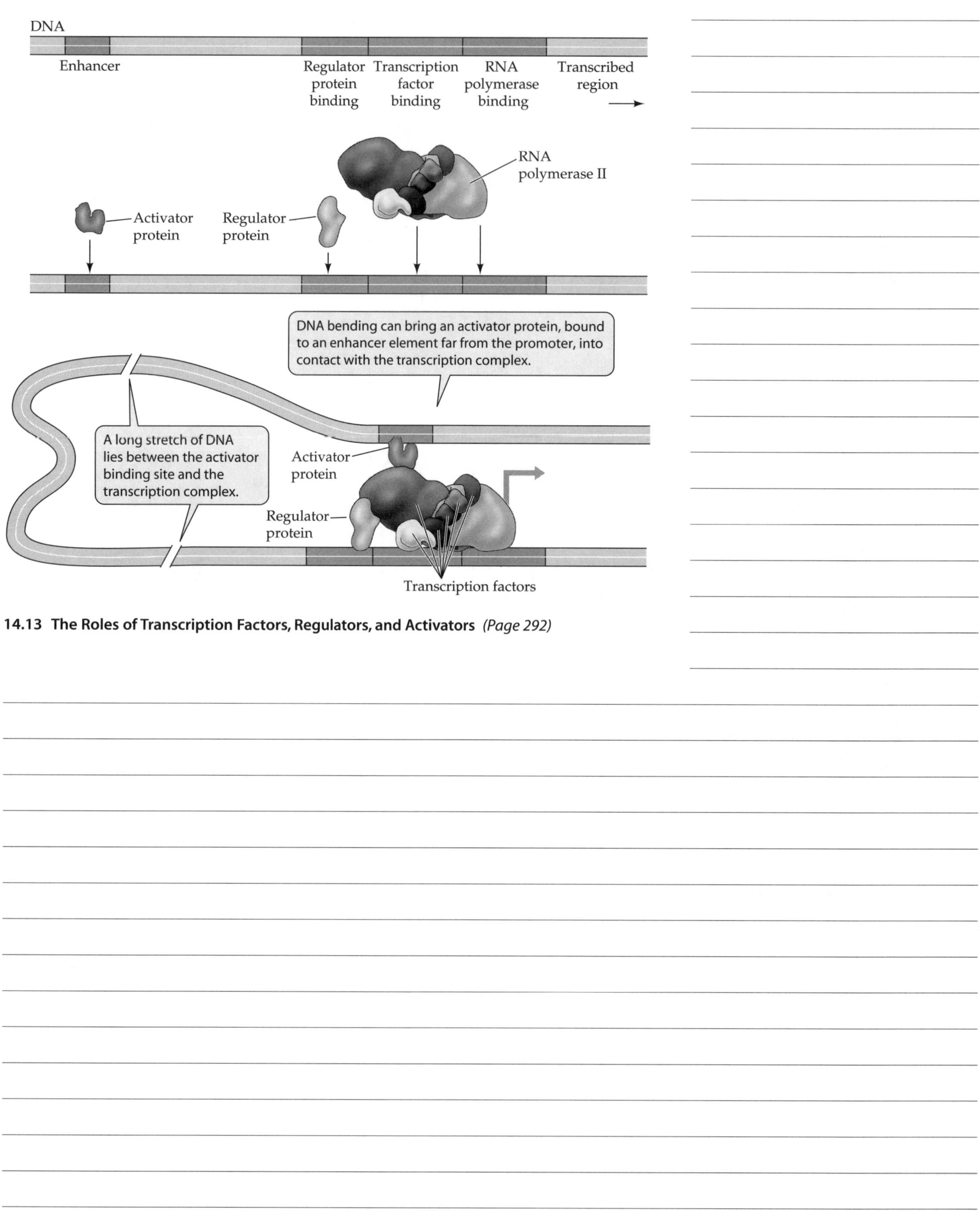
DNA
Enhancer
Regulator protein binding
Transcription factor binding
RNA polymerase binding
Transcribed region
RNA polymerase II
Activator protein
Regulator protein
DNA bending can bring an activator protein, bound to an enhancer element far from the promoter, into contact with the transcription complex.
A long stretch of DNA lies between the activator binding site and the transcription complex.
Activator protein
Regulator protein
Transcription factors

14.13 The Roles of Transcription Factors, Regulators, and Activators *(Page 292)*

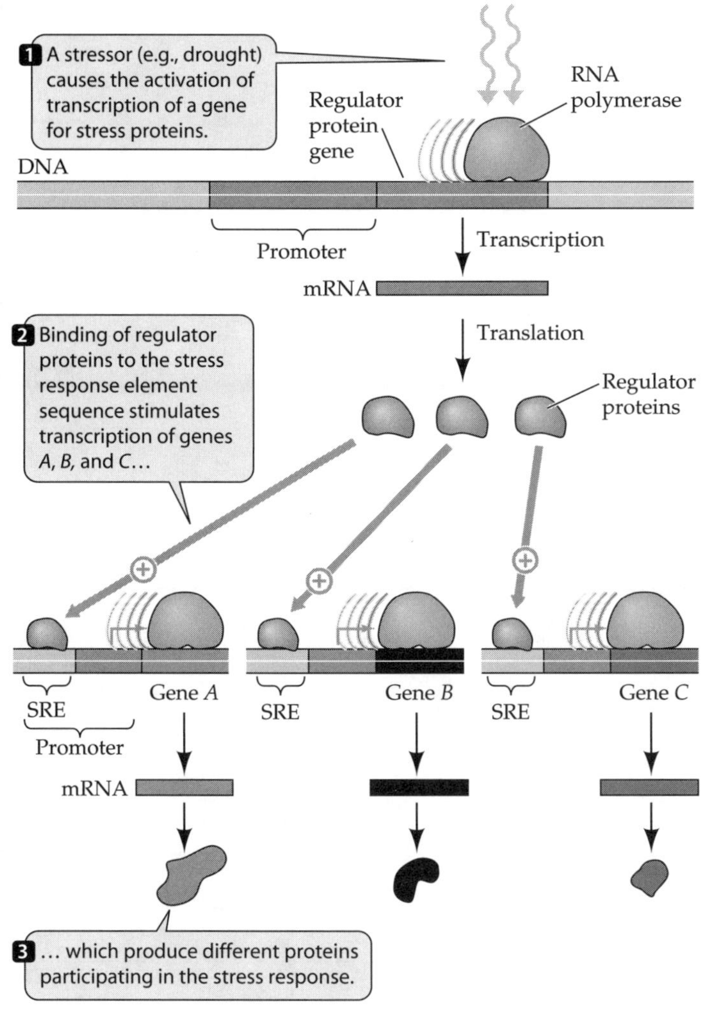
1 A stressor (e.g., drought) causes the activation of transcription of a gene for stress proteins.
RNA polymerase
Regulator protein gene
DNA
Promoter
Transcription
mRNA
2 Binding of regulator proteins to the stress response element sequence stimulates transcription of genes A, B, and C…
Translation
Regulator proteins
SRE
Gene A
SRE
Gene B
SRE
Gene C
Promoter
mRNA
3 … which produce different proteins participating in the stress response.

14.14 Coordinating Gene Expression *(Page 293)*

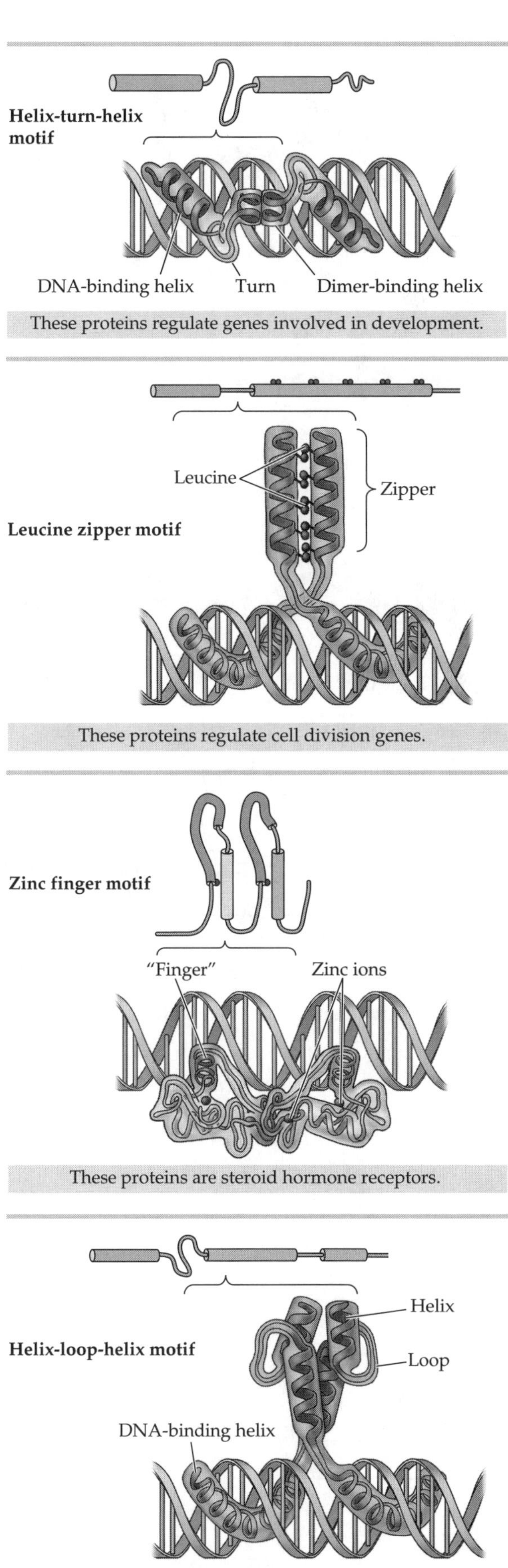
Helix-turn-helix motif
DNA-binding helix
Turn
Dimer-binding helix
These proteins regulate genes involved in development.
Leucine
Zipper
Leucine zipper motif
These proteins regulate cell division genes.
Zinc finger motif
"Finger"
Zinc ions
These proteins are steroid hormone receptors.
Helix
Helix-loop-helix motif
Loop
DNA-binding helix
These proteins regulate immune system genes.

14.15 Protein-DNA Interactions *(Page 293)*

INITIATION

Nucleosome

DNA

Histone proteins

Remodeling protein

1 Remodeling proteins bind, disaggregating the nucleosome.

2 Now the transcription complex can bind to begin transcription.

Transcription complex (see Figure 14.13)

mRNA

ELONGATION

1 A second remodeling protein can bind to the nucleosome…

Remodeling protein

mRNA

2 …allowing transcription without disaggregation.

14.16 Local Remodeling of Chromatin for Transcription *(Page 294)*

1 The *Xist* gene is on the X chromosome.

Xist gene

X chromosome

Transcription

RNAi

2 Transcription of the *Xist* gene makes an interference RNA.

3 The RNAi binds to the X chromosome that transcribed it.

4 Methylation and histone deacetylation attract chromosomal proteins.

14.18 A Model for X Chromosome Inactivation *(Page 295)*

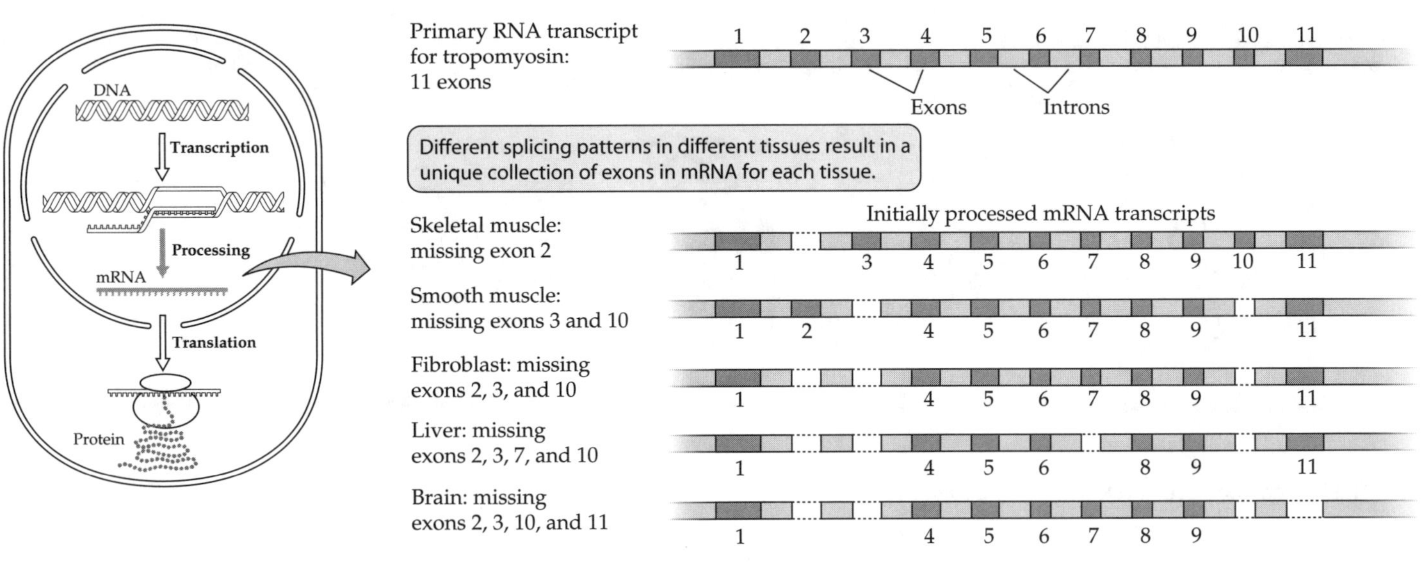
DNA
Transcription
Processing
mRNA
Translation
Protein
Primary RNA transcript for tropomyosin: 11 exons
1 2 3 4 5 6 7 8 9 10 11
Exons
Introns
Different splicing patterns in different tissues result in a unique collection of exons in mRNA for each tissue.
Initially processed mRNA transcripts
Skeletal muscle: missing exon 2
1 3 4 5 6 7 8 9 10 11
Smooth muscle: missing exons 3 and 10
1 2 4 5 6 7 8 9 11
Fibroblast: missing exons 2, 3, and 10
1 4 5 6 7 8 9 11
Liver: missing exons 2, 3, 7, and 10
1 4 5 6 8 9 11
Brain: missing exons 2, 3, 10, and 11
1 4 5 6 7 8 9

14.20 Alternative Splicing Results in Different mRNAs and Proteins *(Page 297)*

(a) DNA

ACCTCC

Transcription

mRNA

UGGAGG

1 The UGGAGG is translated to trp–arg.

Editing

Edited mRNA

UGGUUUAGG

2 The UGGUUUAGG is translated to trp–phe–arg.

(b) DNA

GTA

Transcription

mRNA

CAU

1 The CAU is translated to histidine.

Editing

Edited mRNA

UAU

2 The UAU is translated to tyrosine.

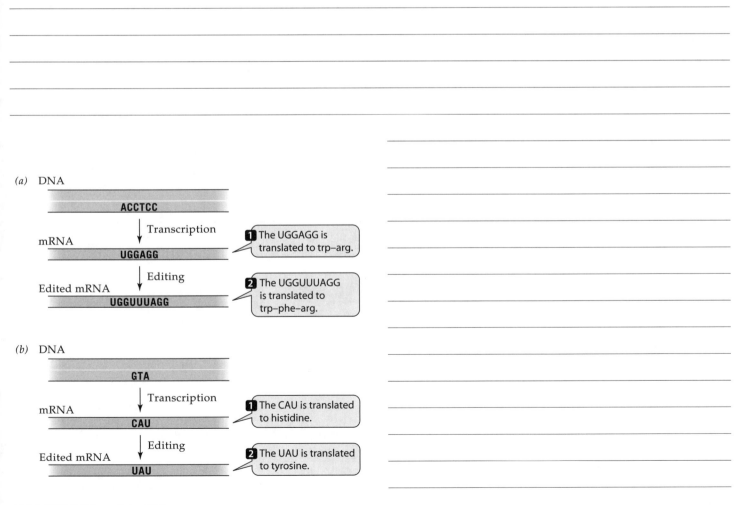

14.21 RNA Editing *(Page 297)*

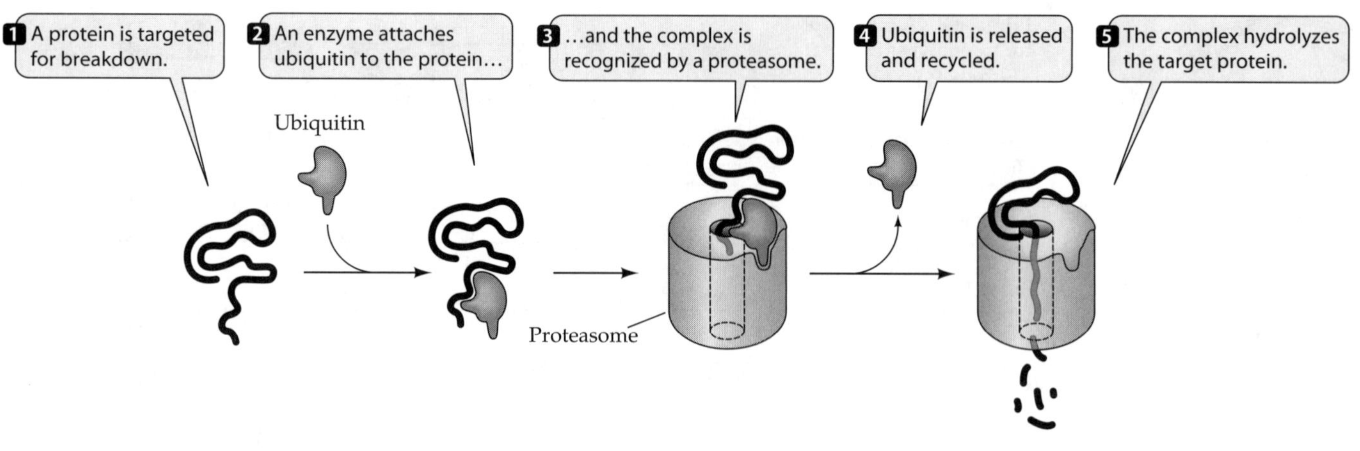
1 A protein is targeted for breakdown.
2 An enzyme attaches ubiquitin to the protein…
3 …and the complex is recognized by a proteasome.
4 Ubiquitin is released and recycled.
5 The complex hydrolyzes the target protein.
Ubiquitin
Proteasome

14.22 The Proteasome Breaks Down Proteins *(Page 298)*

15 Cell Signaling and Communication

Local signals

Autocrine signals bind to receptors on the cells that secrete them.
Paracrine signals bind to receptors on nearby cells.
Receptor
Secreting cell
Target cell
Not a target cell (no receptors)
Cells without receptors do not respond to a particular signal.

Distant signals

Circulating signals are transported by the circulatory system and bind to receptors on distant cells.
Target cell
Secreting cell
Circulatory vessel (e.g., a blood vessel)
Target cell

15.1 Chemical Signaling Systems *(Page 302)*

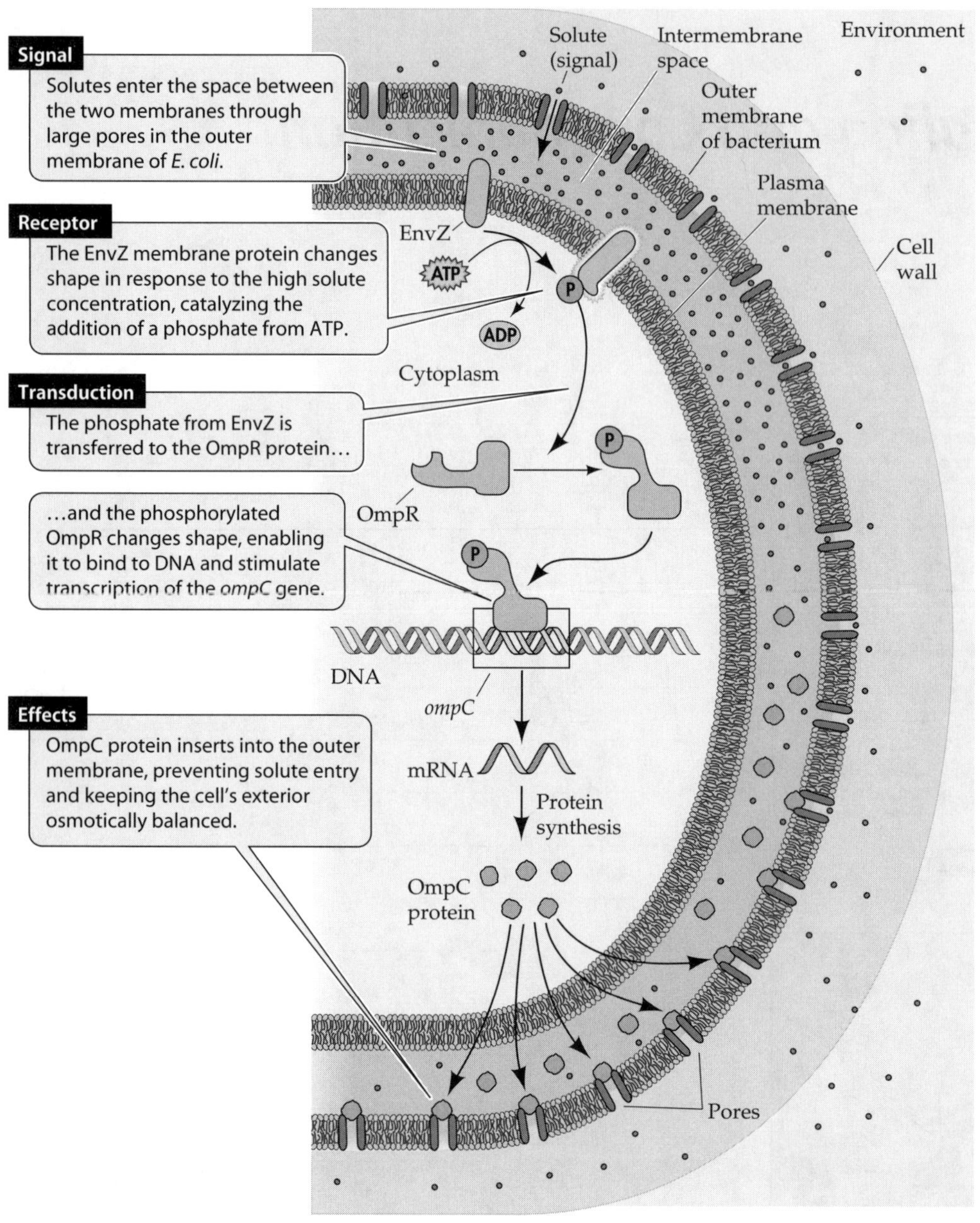
Signal
Solutes enter the space between the two membranes through large pores in the outer membrane of *E. coli*.
Receptor
The EnvZ membrane protein changes shape in response to the high solute concentration, catalyzing the addition of a phosphate from ATP.
Transduction
The phosphate from EnvZ is transferred to the OmpR protein...
...and the phosphorylated OmpR changes shape, enabling it to bind to DNA and stimulate transcription of the *ompC* gene.
Effects
OmpC protein inserts into the outer membrane, preventing solute entry and keeping the cell's exterior osmotically balanced.
Solute (signal)
Intermembrane space
Environment
Outer membrane of bacterium
Plasma membrane
Cell wall
EnvZ
ATP
P
ADP
Cytoplasm
OmpR
DNA
ompC
mRNA
Protein synthesis
OmpC protein
Pores

15.2 A Model Signal Transduction Pathway *(Page 303)*

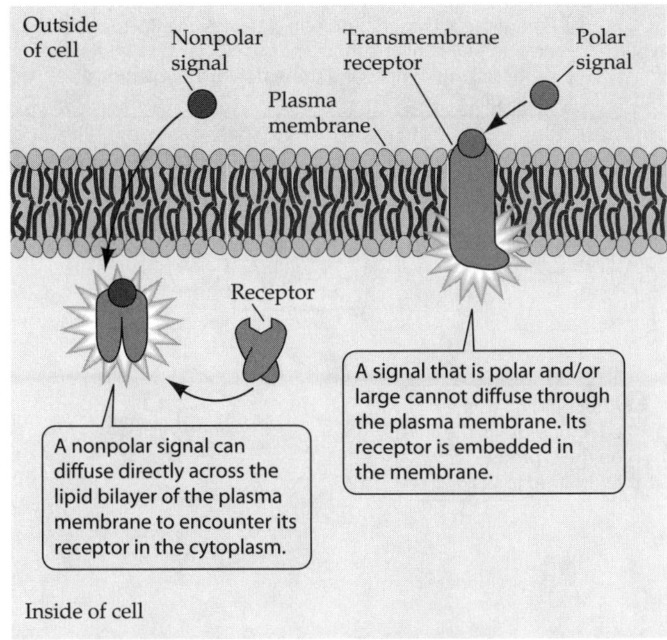
Outside of cell
Nonpolar signal
Transmembrane receptor
Polar signal
Plasma membrane
Receptor
A signal that is polar and/or large cannot diffuse through the plasma membrane. Its receptor is embedded in the membrane.
A nonpolar signal can diffuse directly across the lipid bilayer of the plasma membrane to encounter its receptor in the cytoplasm.
Inside of cell

15.4 Two Locations for Receptors *(Page 305)*

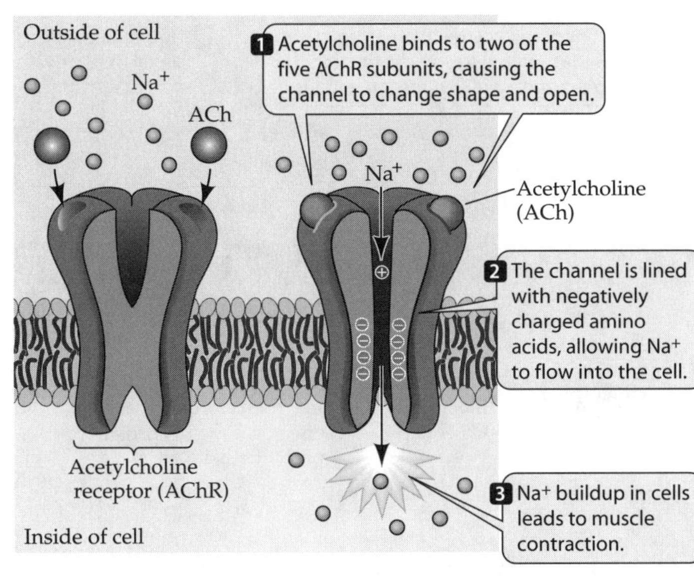
Outside of cell
Na+
ACh
1 Acetylcholine binds to two of the five AChR subunits, causing the channel to change shape and open.
Na+
Acetylcholine (ACh)
2 The channel is lined with negatively charged amino acids, allowing Na+ to flow into the cell.
Acetylcholine receptor (AChR)
3 Na+ buildup in cells leads to muscle contraction.
Inside of cell

15.5 A Gated Ion Channel *(Page 305)*

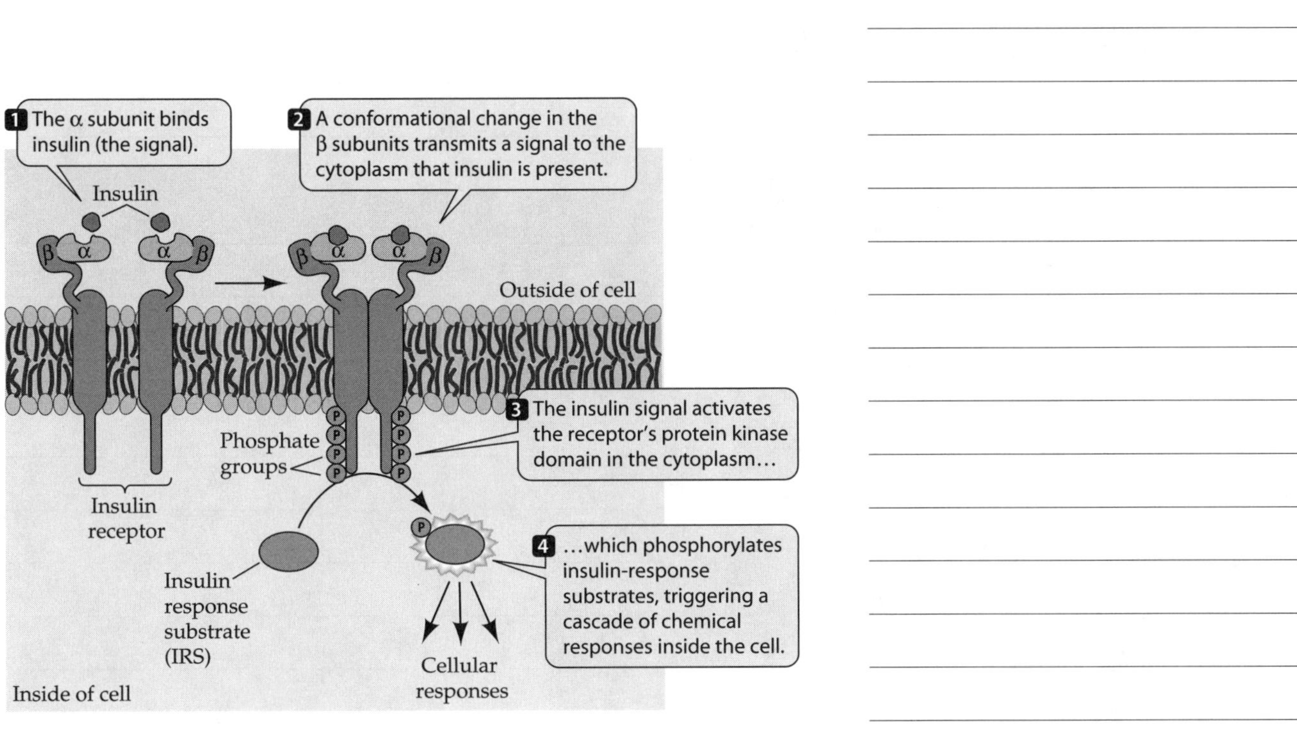
1 The α subunit binds insulin (the signal).
2 A conformational change in the β subunits transmits a signal to the cytoplasm that insulin is present.
Insulin
β α α β
Outside of cell
3 The insulin signal activates the receptor's protein kinase domain in the cytoplasm…
Phosphate groups
P
Insulin receptor
4 …which phosphorylates insulin-response substrates, triggering a cascade of chemical responses inside the cell.
Insulin response substrate (IRS)
Cellular responses
Inside of cell

15.6 A Protein Kinase Receptor *(Page 306)*

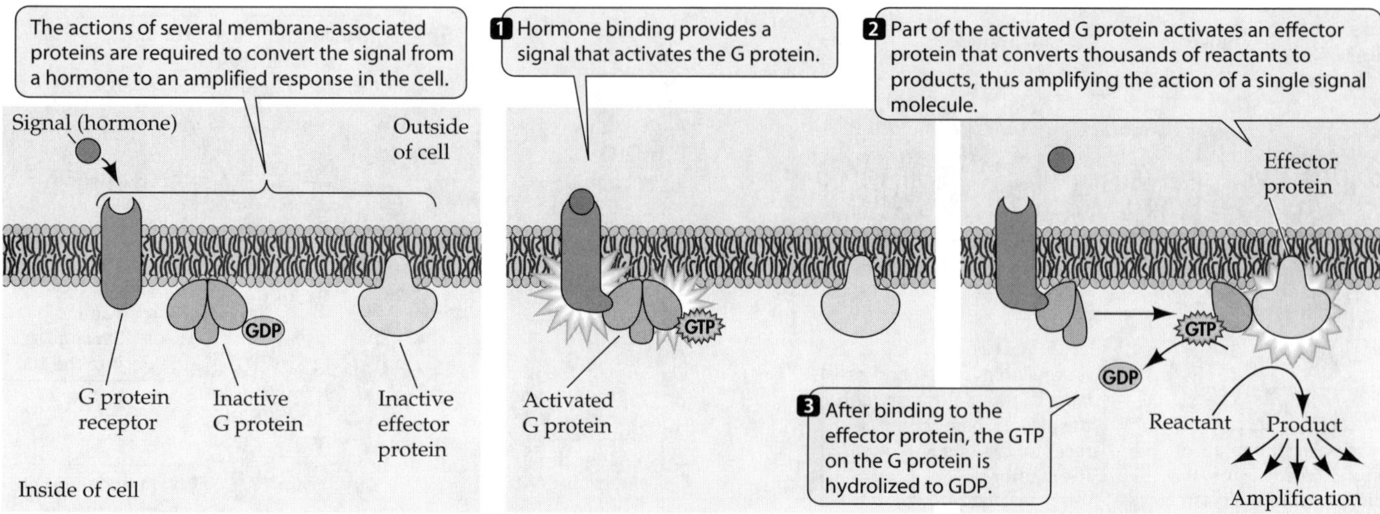
The actions of several membrane-associated proteins are required to convert the signal from a hormone to an amplified response in the cell.
Signal (hormone)
Outside of cell
G protein receptor
Inactive G protein
GDP
Inactive effector protein
Inside of cell
1 Hormone binding provides a signal that activates the G protein.
GTP
Activated G protein
2 Part of the activated G protein activates an effector protein that converts thousands of reactants to products, thus amplifying the action of a single signal molecule.
Effector protein
GTP
GDP
3 After binding to the effector protein, the GTP on the G protein is hydrolized to GDP.
Reactant
Product
Amplification

15.7 A G Protein-Linked Receptor *(Page 306)*

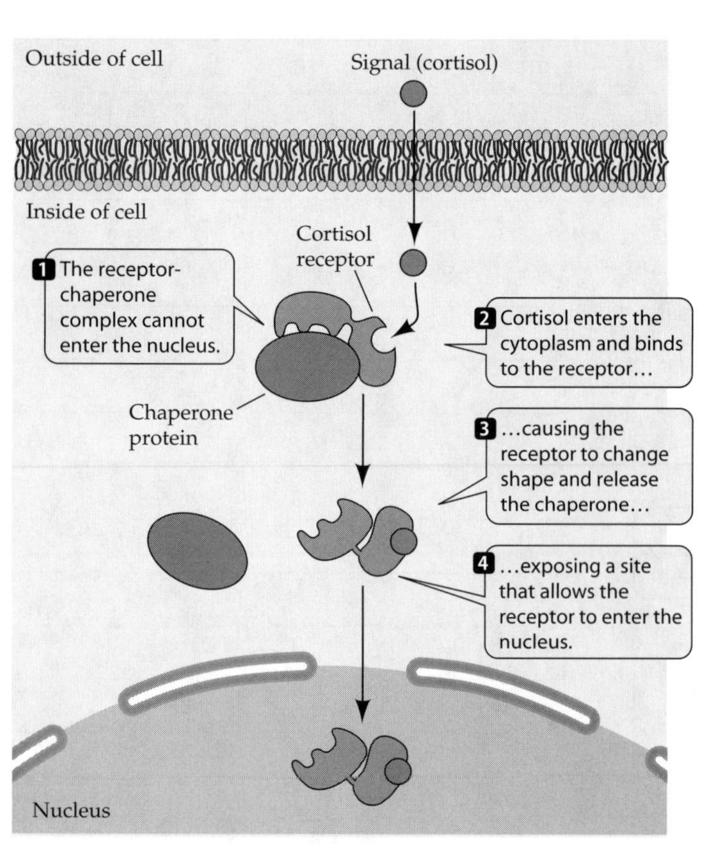
Outside of cell
Signal (cortisol)
Inside of cell
Cortisol receptor
1 The receptor-chaperone complex cannot enter the nucleus.
2 Cortisol enters the cytoplasm and binds to the receptor…
Chaperone protein
3 …causing the receptor to change shape and release the chaperone…
4 …exposing a site that allows the receptor to enter the nucleus.
Nucleus

15.8 A Cytoplasmic Receptor *(Page 307)*

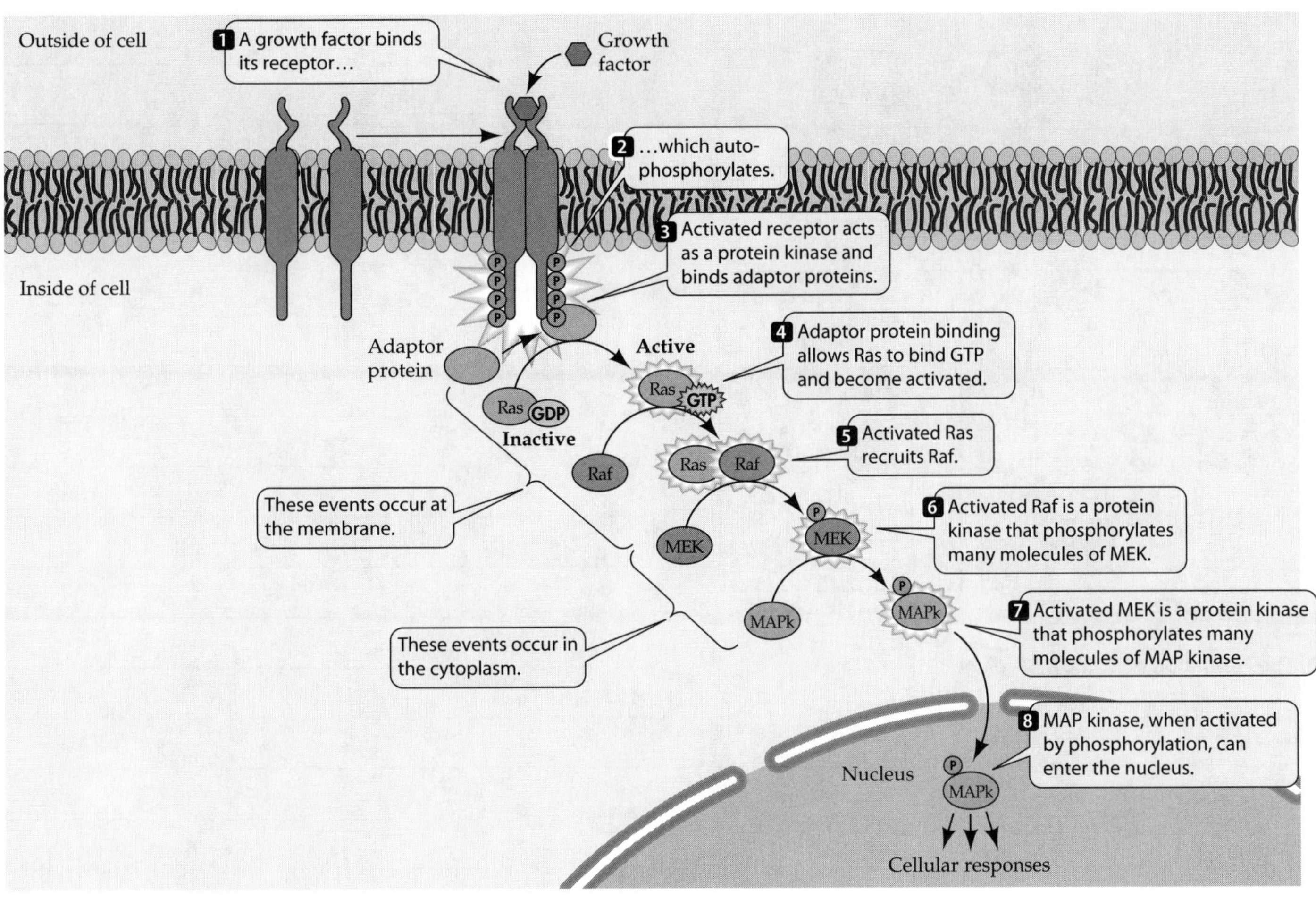
Outside of cell
1 A growth factor binds its receptor…
Growth factor
2 …which auto-phosphorylates.
3 Activated receptor acts as a protein kinase and binds adaptor proteins.
Inside of cell
Adaptor protein
Active
4 Adaptor protein binding allows Ras to bind GTP and become activated.
Ras
GTP
GDP
Inactive
Raf
5 Activated Ras recruits Raf.
These events occur at the membrane.
MEK
6 Activated Raf is a protein kinase that phosphorylates many molecules of MEK.
MAPk
7 Activated MEK is a protein kinase that phosphorylates many molecules of MAP kinase.
These events occur in the cytoplasm.
8 MAP kinase, when activated by phosphorylation, can enter the nucleus.
Nucleus
Cellular responses

15.9 A Protein Kinase Cascade *(Page 308)*

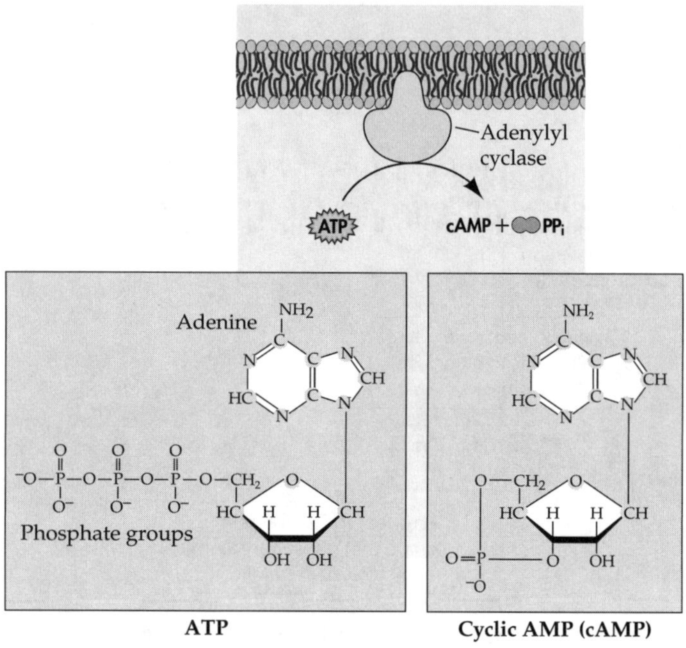
Adenylyl cyclase
ATP
cAMP + PP$_i$
Adenine
NH$_2$
Phosphate groups
OH OH
ATP
Cyclic AMP (cAMP)

15.10 The Formation of Cyclic AMP *(Page 309)*

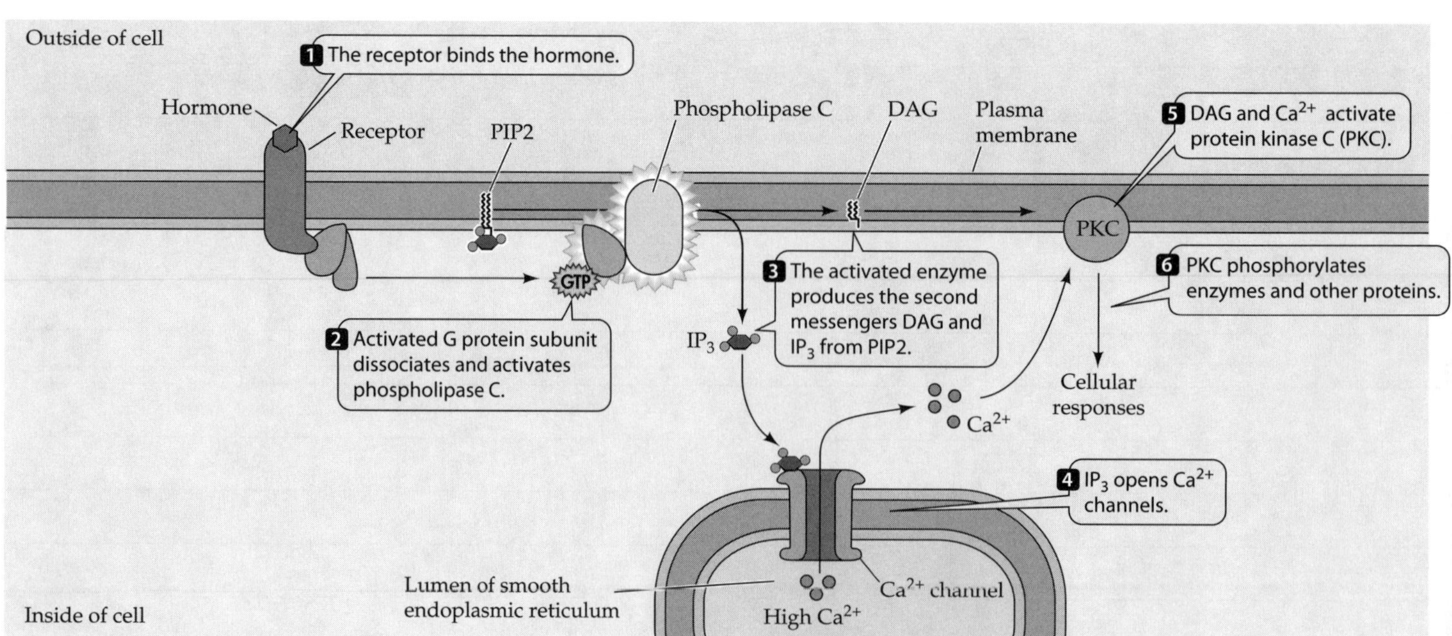
Outside of cell
1 The receptor binds the hormone.
Hormone
Receptor
PIP2
Phospholipase C
DAG
Plasma membrane
5 DAG and Ca^{2+} activate protein kinase C (PKC).
PKC
GTP
3 The activated enzyme produces the second messengers DAG and IP$_3$ from PIP2.
6 PKC phosphorylates enzymes and other proteins.
2 Activated G protein subunit dissociates and activates phospholipase C.
IP$_3$
Ca^{2+}
Cellular responses
4 IP$_3$ opens Ca^{2+} channels.
Lumen of smooth endoplasmic reticulum
Ca^{2+} channel
High Ca^{2+}
Inside of cell

15.11 The IP$_3$ and DAG Second Messenger System *(Page 310)*

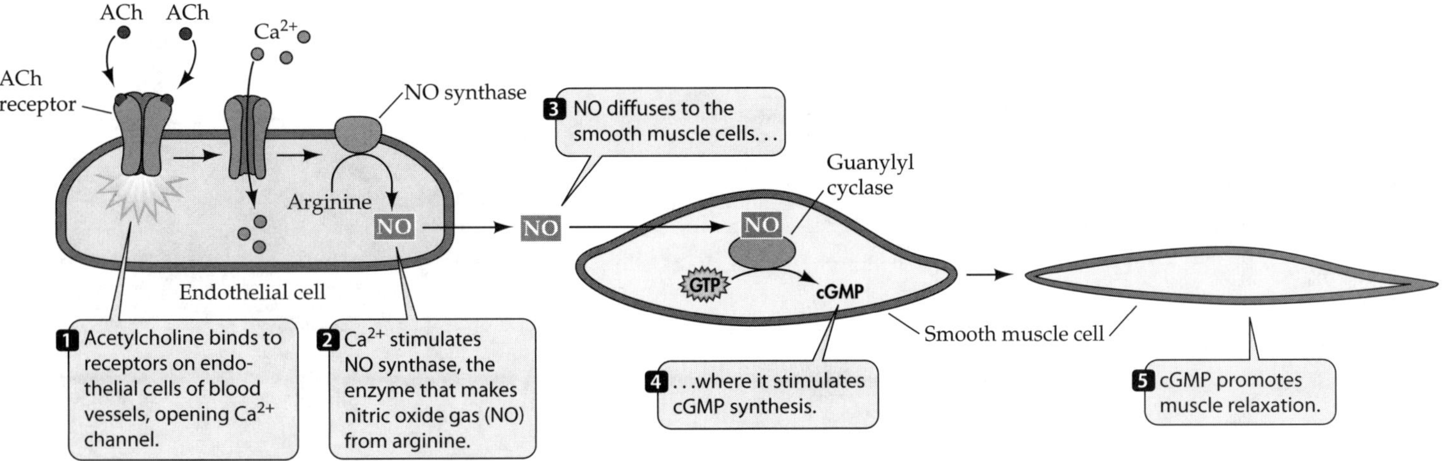
ACh
ACh
Ca^{2+}
ACh receptor
NO synthase
Arginine
NO
NO
NO
Endothelial cell
3 NO diffuses to the smooth muscle cells. . .
Guanylyl cyclase
GTP
cGMP
Smooth muscle cell
1 Acetylcholine binds to receptors on endothelial cells of blood vessels, opening Ca^{2+} channel.
2 Ca^{2+} stimulates NO synthase, the enzyme that makes nitric oxide gas (NO) from arginine.
4 . . .where it stimulates cGMP synthesis.
5 cGMP promotes muscle relaxation.

15.13 Nitric Oxide as a Second Messenger *(Page 312)*

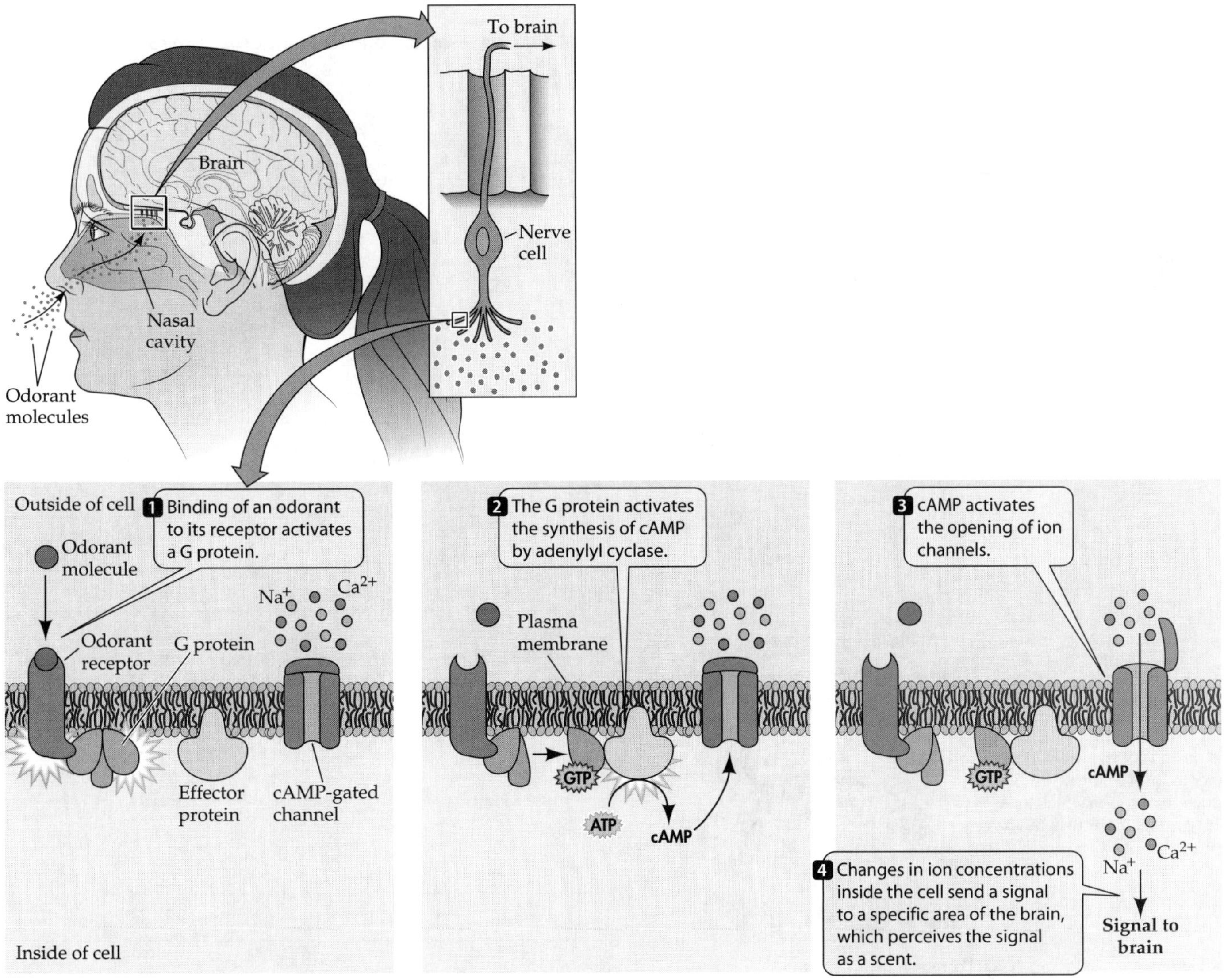

To brain
Brain
Nerve cell
Nasal cavity
Odorant molecules
Outside of cell
1 Binding of an odorant to its receptor activates a G protein.
Odorant molecule
Na^+
Ca^{2+}
Odorant receptor
G protein
Effector protein
cAMP-gated channel
Inside of cell
2 The G protein activates the synthesis of cAMP by adenylyl cyclase.
Plasma membrane
GTP
ATP
cAMP
3 cAMP activates the opening of ion channels.
GTP
cAMP
Na^+
Ca^{2+}
4 Changes in ion concentrations inside the cell send a signal to a specific area of the brain, which perceives the signal as a scent.
Signal to brain

15.14 A Signal Transduction Pathway Leads to the Opening of Ion Channels *(Page 313)*

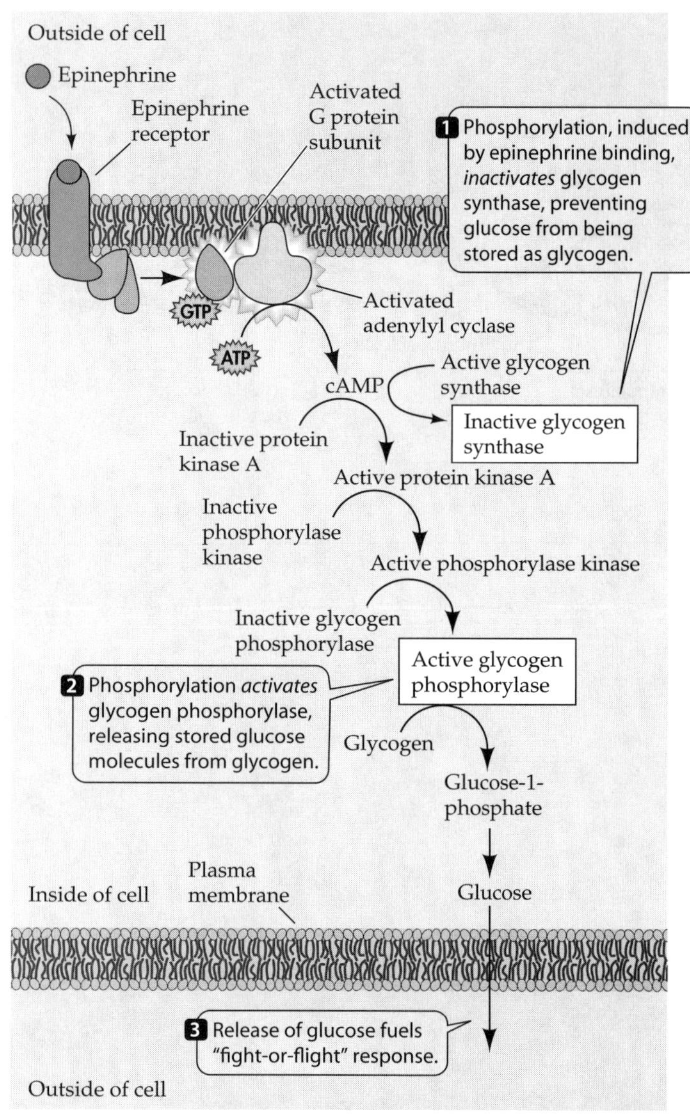

Outside of cell
Epinephrine
Epinephrine receptor
Activated G protein subunit
1 Phosphorylation, induced by epinephrine binding, *inactivates* glycogen synthase, preventing glucose from being stored as glycogen.
GTP
Activated adenylyl cyclase
ATP
cAMP
Active glycogen synthase
Inactive glycogen synthase
Inactive protein kinase A
Active protein kinase A
Inactive phosphorylase kinase
Active phosphorylase kinase
Inactive glycogen phosphorylase
Active glycogen phosphorylase
2 Phosphorylation *activates* glycogen phosphorylase, releasing stored glucose molecules from glycogen.
Glycogen
Glucose-1-phosphate
Inside of cell
Plasma membrane
Glucose
3 Release of glucose fuels "fight-or-flight" response.
Outside of cell

15.15 A Cascade of Reactions Leads to Altered Enzyme Activity *(Page 314)*

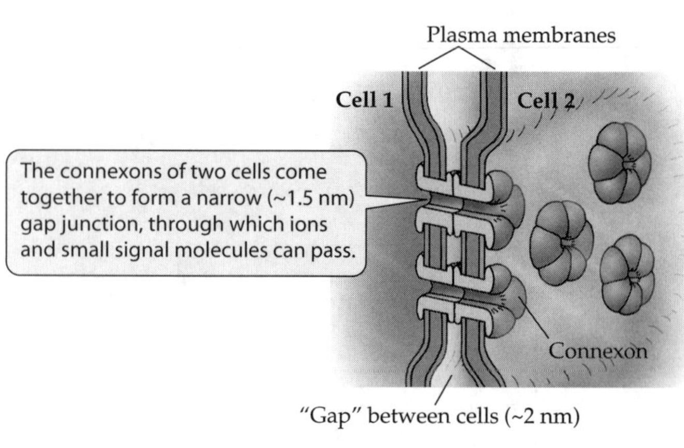

Plasma membranes
Cell 1
Cell 2
The connexons of two cells come together to form a narrow (~1.5 nm) gap junction, through which ions and small signal molecules can pass.
Connexon
"Gap" between cells (~2 nm)

15.16 Gap Junctions Connect Animal Cells *(Page 314)*

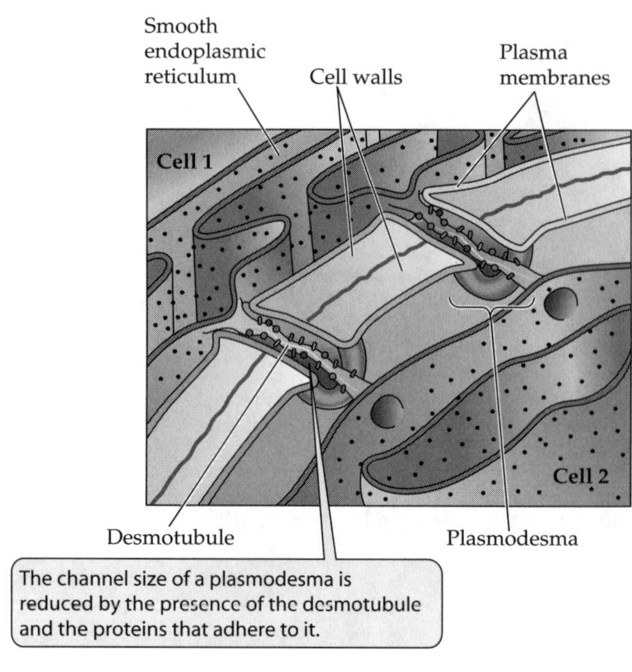

Smooth endoplasmic reticulum
Cell walls
Plasma membranes
Cell 1
Cell 2
Desmotubule
Plasmodesma
The channel size of a plasmodesma is reduced by the presence of the desmotubule and the proteins that adhere to it.

15.17 Plasmodesmata Connect Plant Cells *(Page 315)*

16 Recombinant DNA and Biotechnology

1 A restriction enzyme cleaves the incoming phage DNA at recognition sites.
2 Other enzymes degrade the phage DNA into smaller fragments.
Bacterial host cell
Host DNA
3 Methyl groups at the restriction sequence block the restriction enzyme and protect the bacterial DNA from being cleaved.

16.1 Bacteria Fight Invading Viruses with Restriction Enzymes
(Page 318)

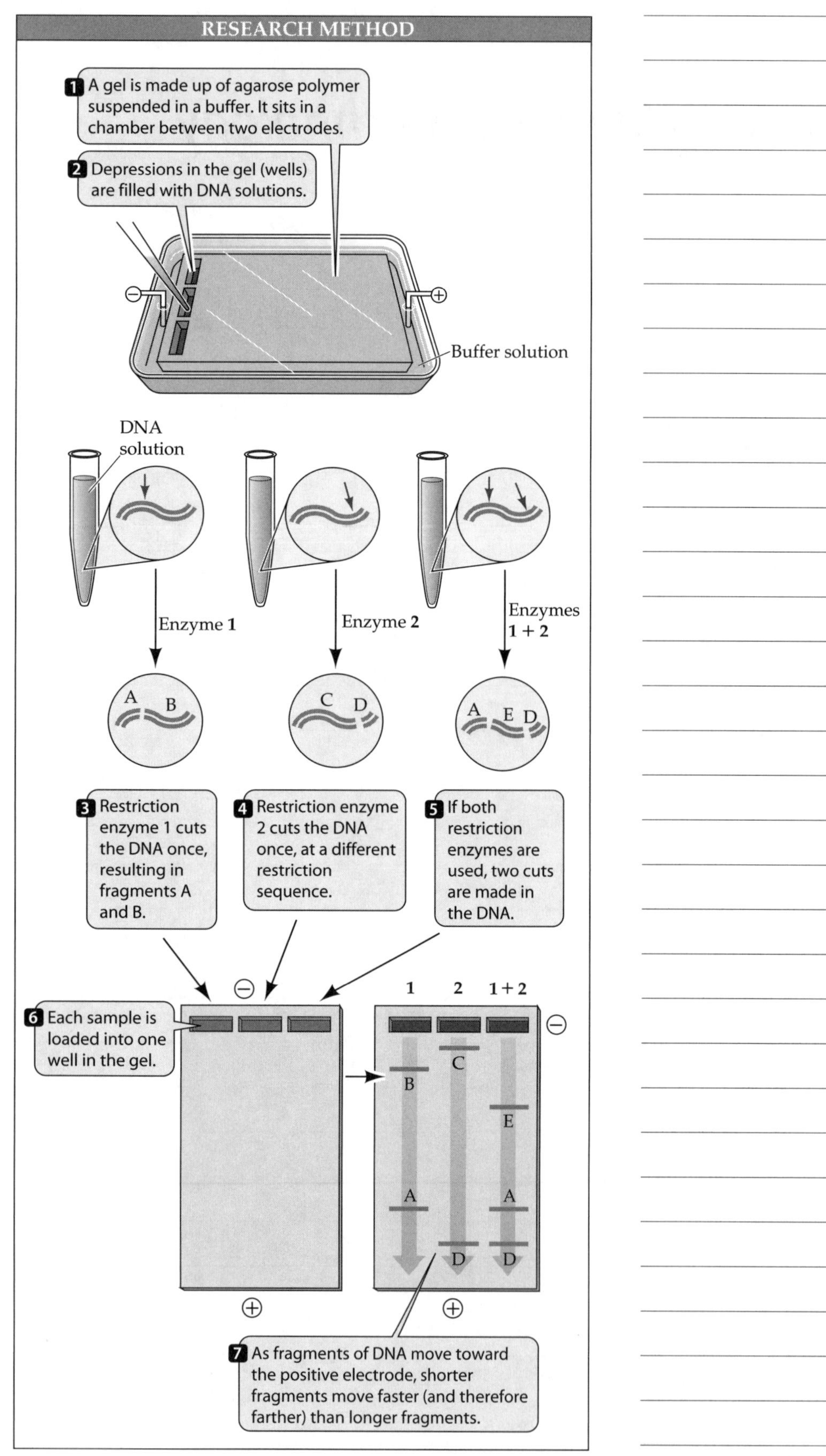

RESEARCH METHOD
1 A gel is made up of agarose polymer suspended in a buffer. It sits in a chamber between two electrodes.
2 Depressions in the gel (wells) are filled with DNA solutions.
Buffer solution
DNA solution
Enzyme 1
Enzyme 2
Enzymes 1 + 2
A B
C D
A E D
3 Restriction enzyme 1 cuts the DNA once, resulting in fragments A and B.
4 Restriction enzyme 2 cuts the DNA once, at a different restriction sequence.
5 If both restriction enzymes are used, two cuts are made in the DNA.
6 Each sample is loaded into one well in the gel.
1 2 1+2
B
C
E
A
A
D
D
7 As fragments of DNA move toward the positive electrode, shorter fragments move faster (and therefore farther) than longer fragments.

16.2 Separating Fragments of DNA by Gel Electrophoresis
(Page 319)

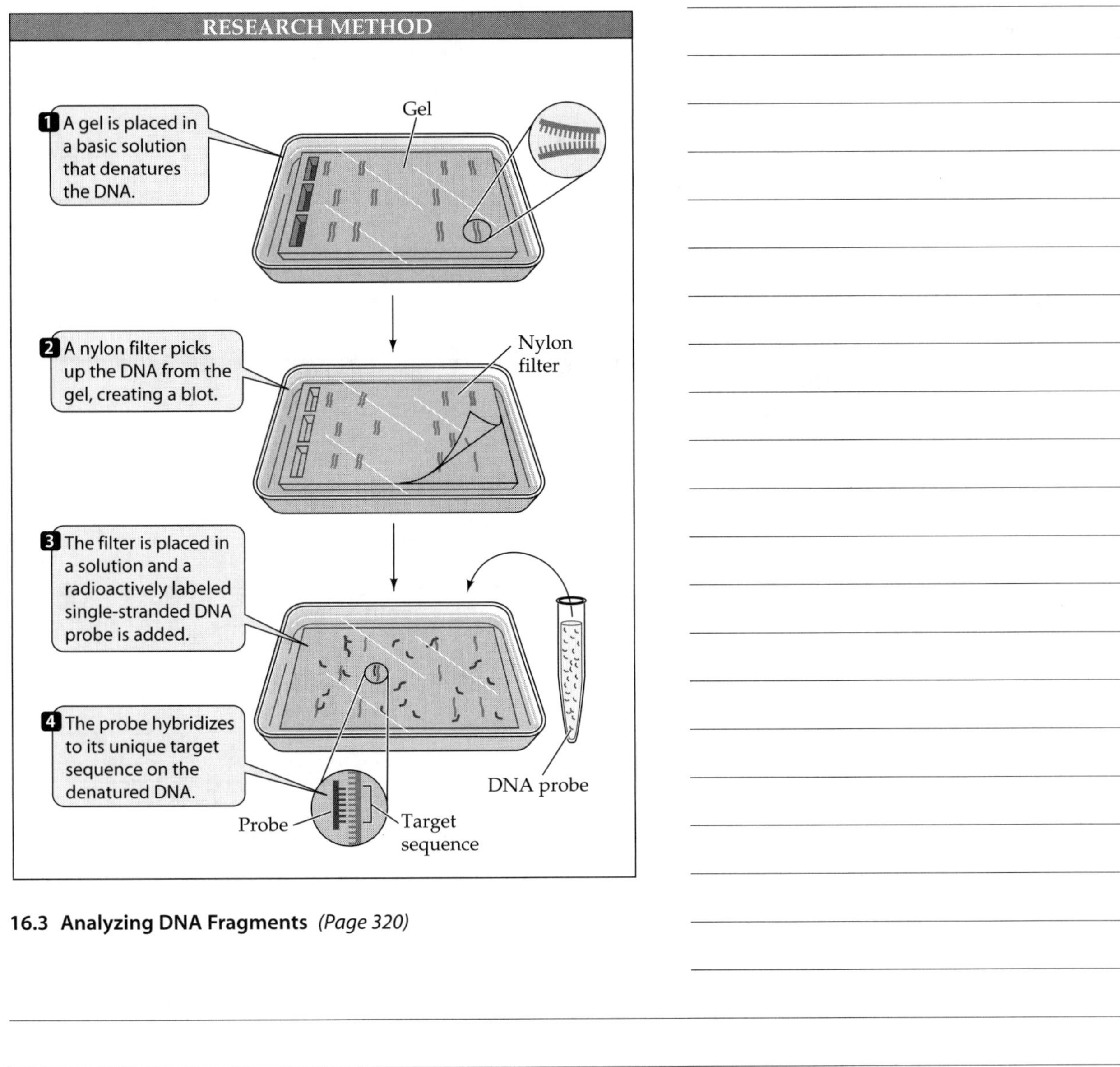
RESEARCH METHOD
1 A gel is placed in a basic solution that denatures the DNA.
Gel
2 A nylon filter picks up the DNA from the gel, creating a blot.
Nylon filter
3 The filter is placed in a solution and a radioactively labeled single-stranded DNA probe is added.
4 The probe hybridizes to its unique target sequence on the denatured DNA.
DNA probe
Probe
Target sequence

16.3 Analyzing DNA Fragments *(Page 320)*

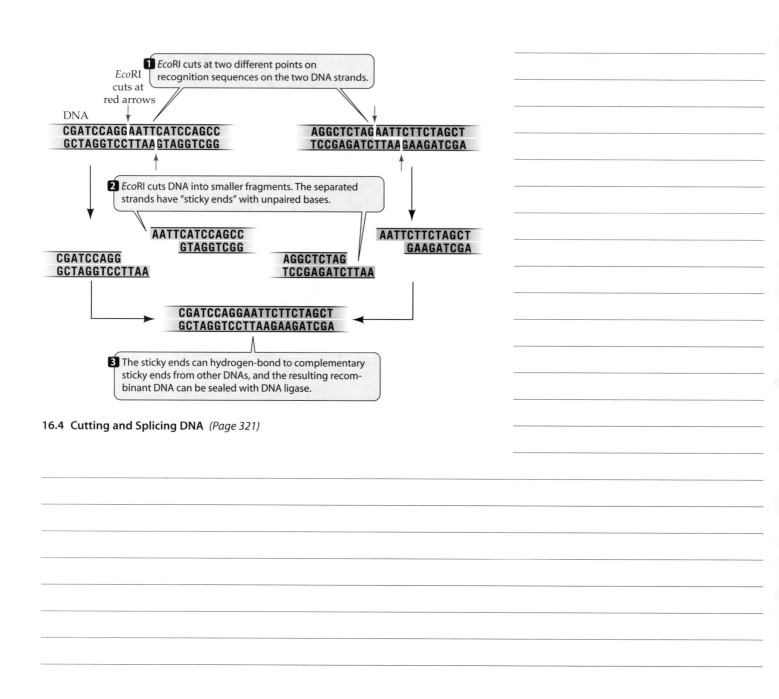

1 *Eco*RI cuts at two different points on recognition sequences on the two DNA strands.

*Eco*RI cuts at red arrows

DNA

CGATCCAGGAATTCATCCAGCC
GCTAGGTCCTTAAGTAGGTCGG

AGGCTCTAGAATTCTTCTAGCT
TCCGAGATCTTAAGAAGATCGA

2 *Eco*RI cuts DNA into smaller fragments. The separated strands have "sticky ends" with unpaired bases.

CGATCCAGG
GCTAGGTCCTTAA

AATTCATCCAGCC
GTAGGTCGG

AGGCTCTAG
TCCGAGATCTTAA

AATTCTTCTAGCT
GAAGATCGA

CGATCCAGGAATTCTTCTAGCT
GCTAGGTCCTTAAGAAGATCGA

3 The sticky ends can hydrogen-bond to complementary sticky ends from other DNAs, and the resulting recombinant DNA can be sealed with DNA ligase.

16.4 Cutting and Splicing DNA *(Page 321)*

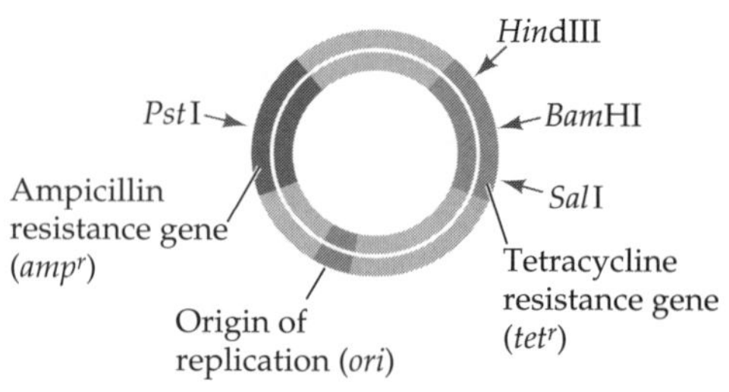

(*a*) Plasmid pBR322
Host: *E. coli*
*Hin*dIII
*Pst*I
*Bam*HI
*Sal*I
Ampicillin resistance gene (*amp*[r])
Tetracycline resistance gene (*tet*[r])
Origin of replication (*ori*)

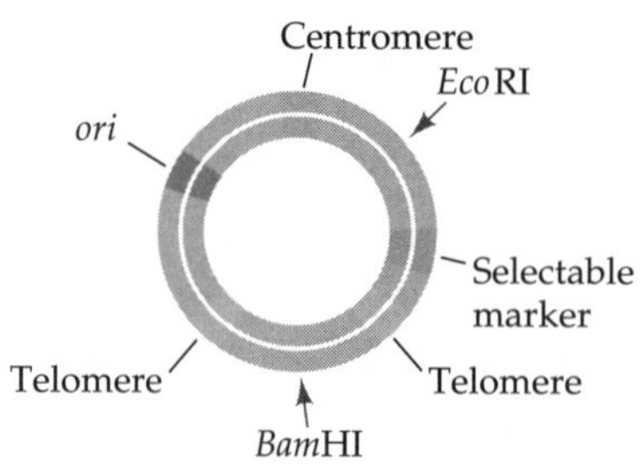

(*b*) Yeast artificial chromosome (YAC)
Host: yeast
Centromere
*Eco*RI
ori
Selectable marker
Telomere
Telomere
*Bam*HI

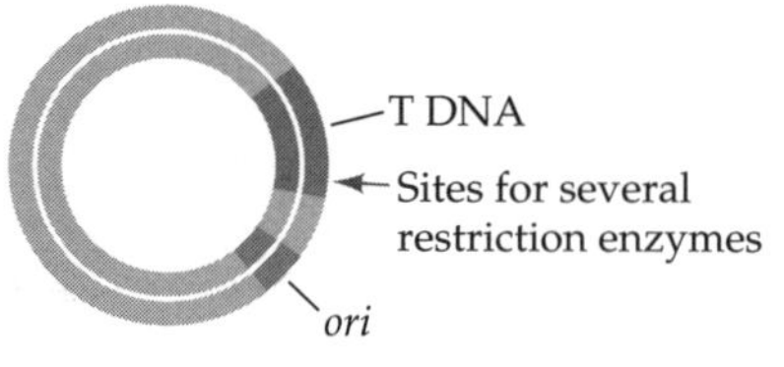

(*c*) Ti plasmid
Hosts: *Agrobacterium tumefaciens* (plasmid) and infected plants (T DNA)
T DNA
Sites for several restriction enzymes
ori

Recognition site for restriction enzymes

16.5 Vectors for Carrying DNA into Cells *(Page 322)*

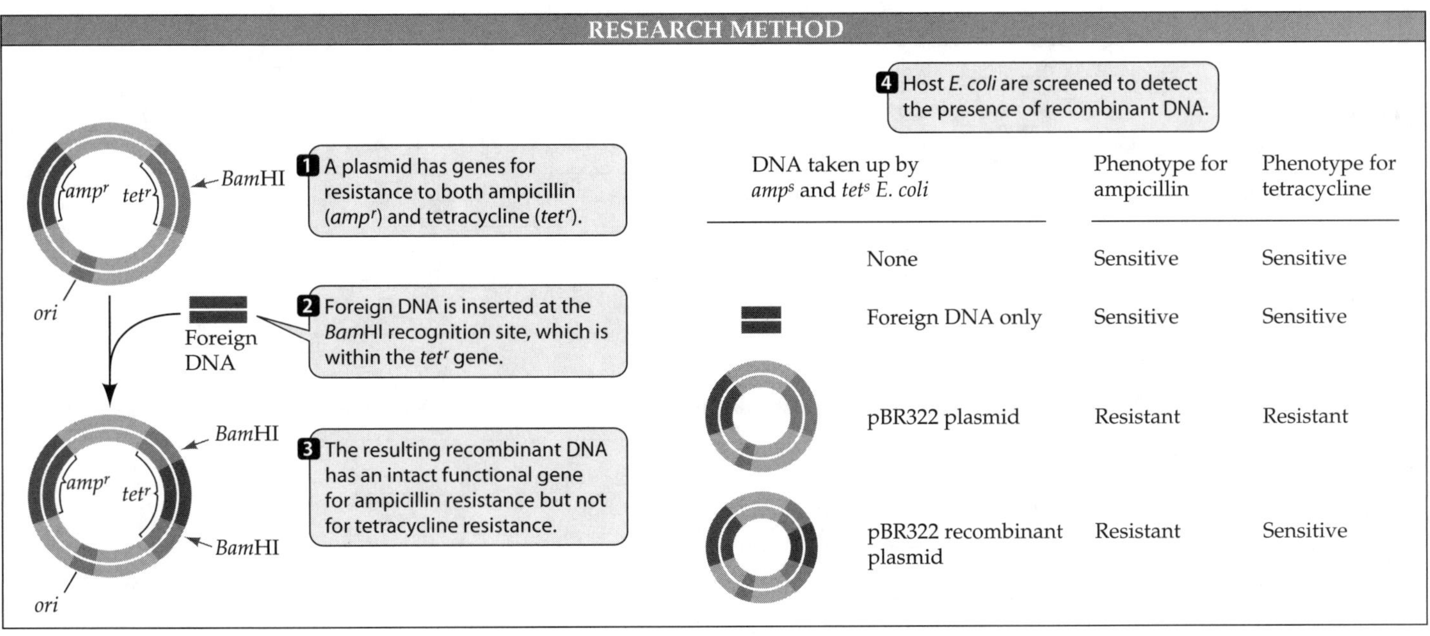
RESEARCH METHOD
ampr
tetr
BamHI
ori
1 A plasmid has genes for resistance to both ampicillin (ampr) and tetracycline (tetr).
Foreign DNA
2 Foreign DNA is inserted at the BamHI recognition site, which is within the tetr gene.
BamHI
ampr
tetr
BamHI
ori
3 The resulting recombinant DNA has an intact functional gene for ampicillin resistance but not for tetracycline resistance.
4 Host E. coli are screened to detect the presence of recombinant DNA.
DNA taken up by amps and tets E. coli
Phenotype for ampicillin
Phenotype for tetracycline
None
Sensitive
Sensitive
Foreign DNA only
Sensitive
Sensitive
pBR322 plasmid
Resistant
Resistant
pBR322 recombinant plasmid
Resistant
Sensitive

16.6 Marking Recombinant DNA by Inactivating a Gene *(Page 324)*

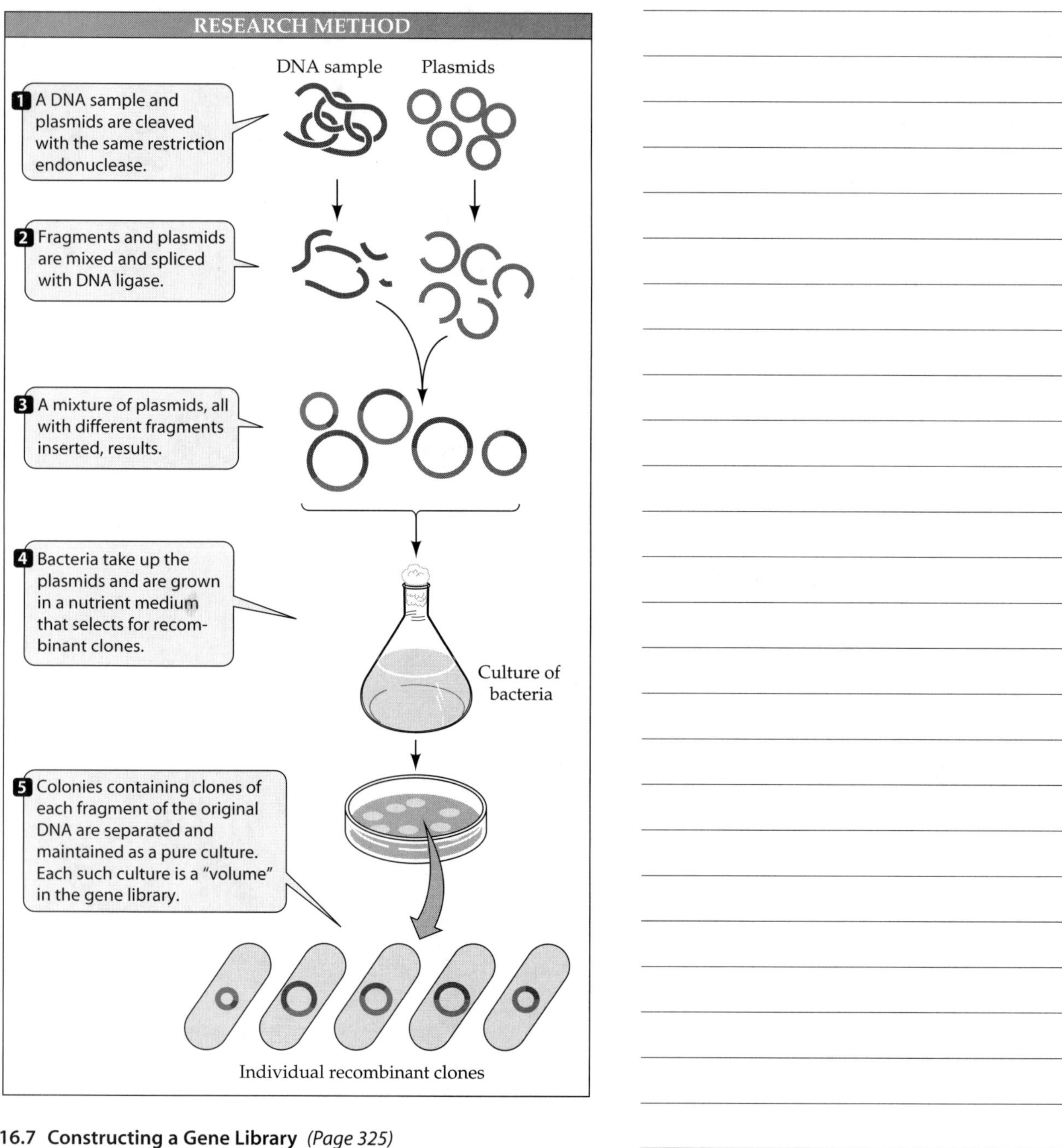
RESEARCH METHOD
DNA sample
Plasmids
1 A DNA sample and plasmids are cleaved with the same restriction endonuclease.
2 Fragments and plasmids are mixed and spliced with DNA ligase.
3 A mixture of plasmids, all with different fragments inserted, results.
4 Bacteria take up the plasmids and are grown in a nutrient medium that selects for recombinant clones.
Culture of bacteria
5 Colonies containing clones of each fragment of the original DNA are separated and maintained as a pure culture. Each such culture is a "volume" in the gene library.
Individual recombinant clones

16.7 Constructing a Gene Library *(Page 325)*

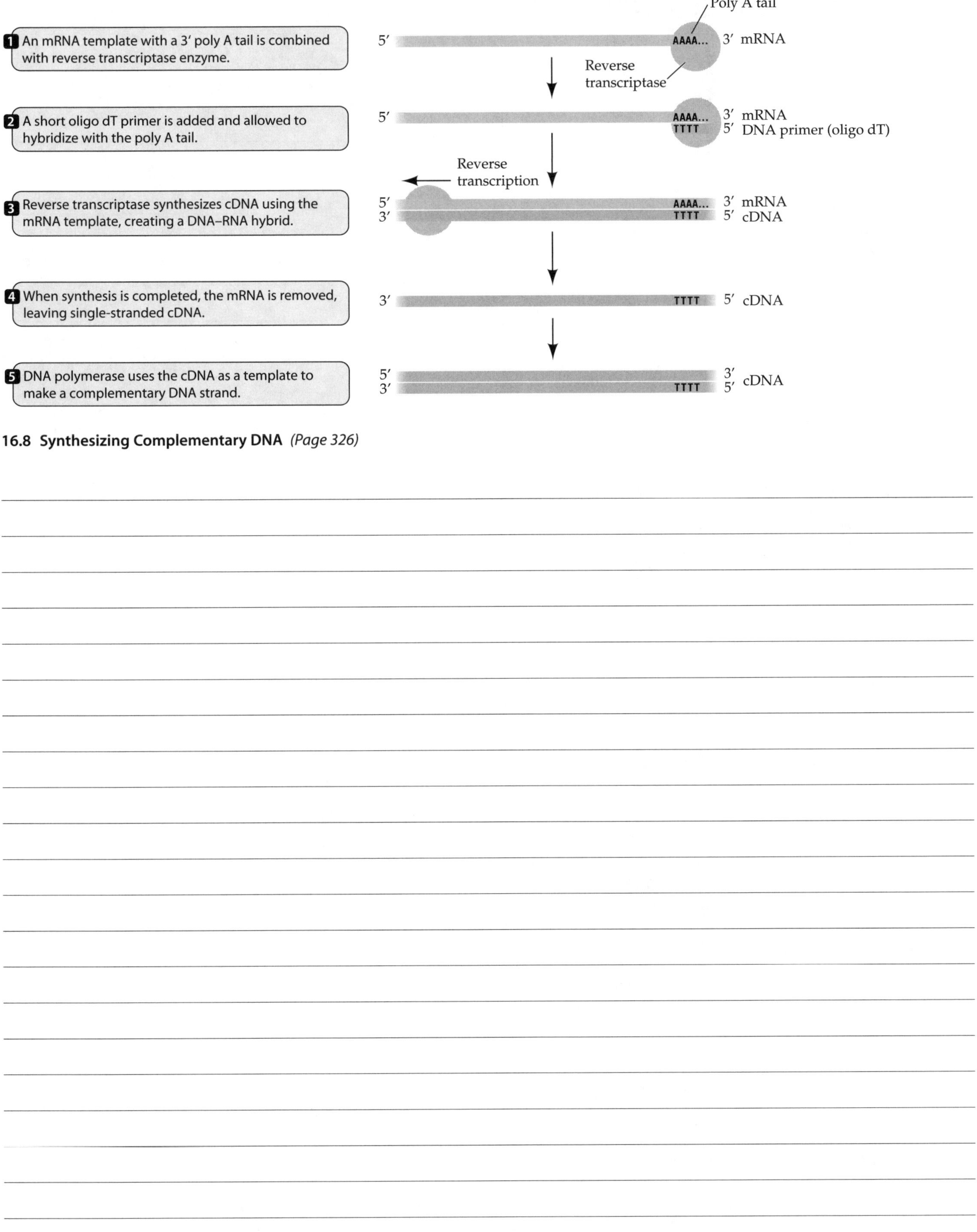
1 An mRNA template with a 3′ poly A tail is combined with reverse transcriptase enzyme.
2 A short oligo dT primer is added and allowed to hybridize with the poly A tail.
3 Reverse transcriptase synthesizes cDNA using the mRNA template, creating a DNA–RNA hybrid.
4 When synthesis is completed, the mRNA is removed, leaving single-stranded cDNA.
5 DNA polymerase uses the cDNA as a template to make a complementary DNA strand.
Poly A tail
5′
AAAA...
3′ mRNA
Reverse transcriptase
5′
AAAA...
TTTT
3′ mRNA
5′ DNA primer (oligo dT)
Reverse transcription
5′
3′
AAAA...
TTTT
3′ mRNA
5′ cDNA
3′
TTTT
5′ cDNA
5′
3′
TTTT
3′
5′
cDNA

16.8 Synthesizing Complementary DNA *(Page 326)*

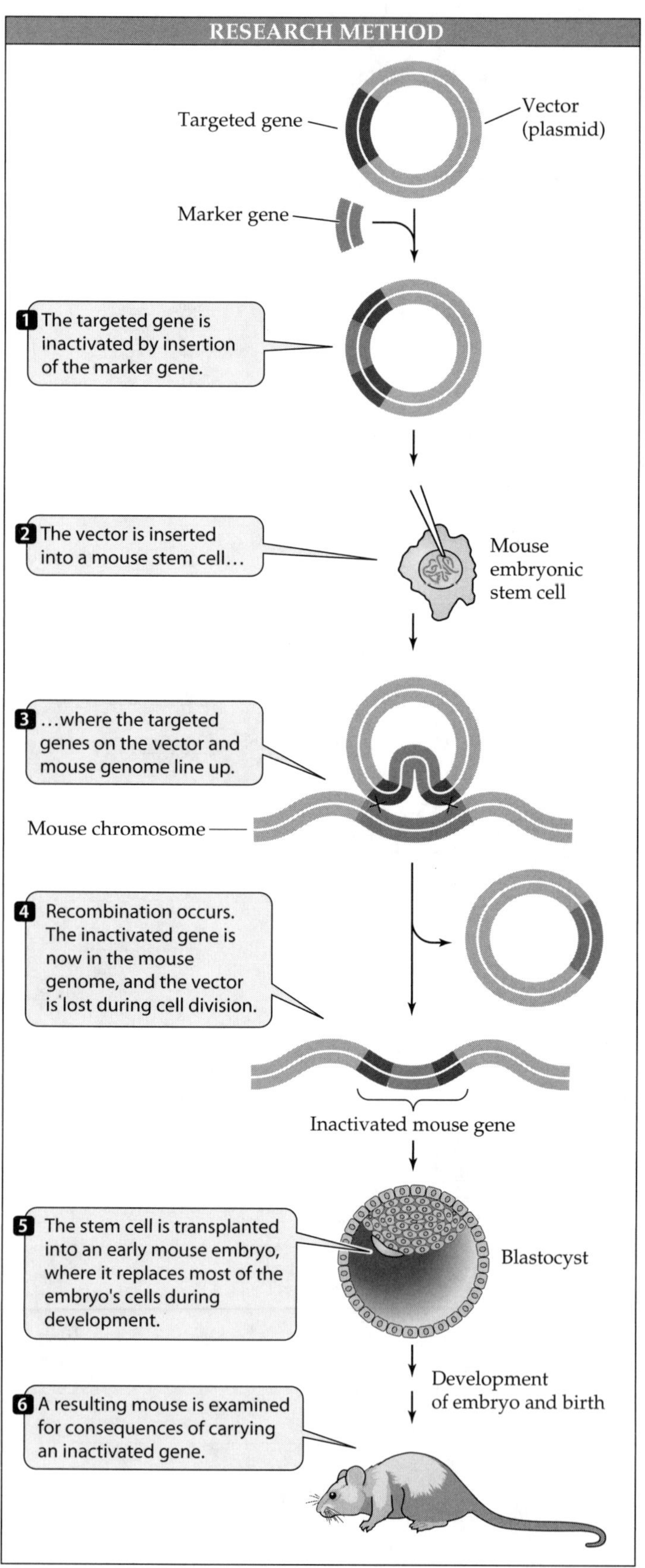
RESEARCH METHOD
Targeted gene
Vector (plasmid)
Marker gene
1 The targeted gene is inactivated by insertion of the marker gene.
2 The vector is inserted into a mouse stem cell…
Mouse embryonic stem cell
3 …where the targeted genes on the vector and mouse genome line up.
Mouse chromosome
4 Recombination occurs. The inactivated gene is now in the mouse genome, and the vector is lost during cell division.
Inactivated mouse gene
5 The stem cell is transplanted into an early mouse embryo, where it replaces most of the embryo's cells during development.
Blastocyst
Development of embryo and birth
6 A resulting mouse is examined for consequences of carrying an inactivated gene.

16.9 Making a Knockout Mouse *(Page 327)*

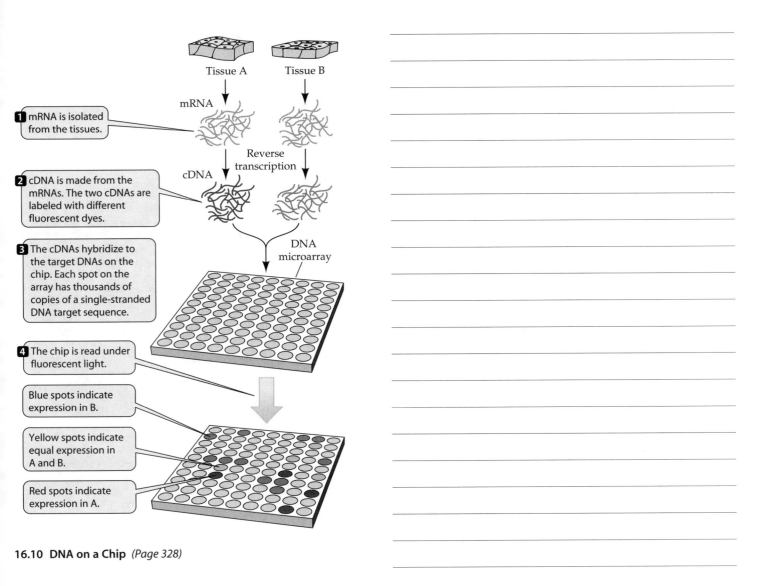

Tissue A
Tissue B
mRNA
1 mRNA is isolated from the tissues.
Reverse transcription
cDNA
2 cDNA is made from the mRNAs. The two cDNAs are labeled with different fluorescent dyes.
DNA microarray
3 The cDNAs hybridize to the target DNAs on the chip. Each spot on the array has thousands of copies of a single-stranded DNA target sequence.
4 The chip is read under fluorescent light.
Blue spots indicate expression in B.
Yellow spots indicate equal expression in A and B.
Red spots indicate expression in A.

16.10 DNA on a Chip *(Page 328)*

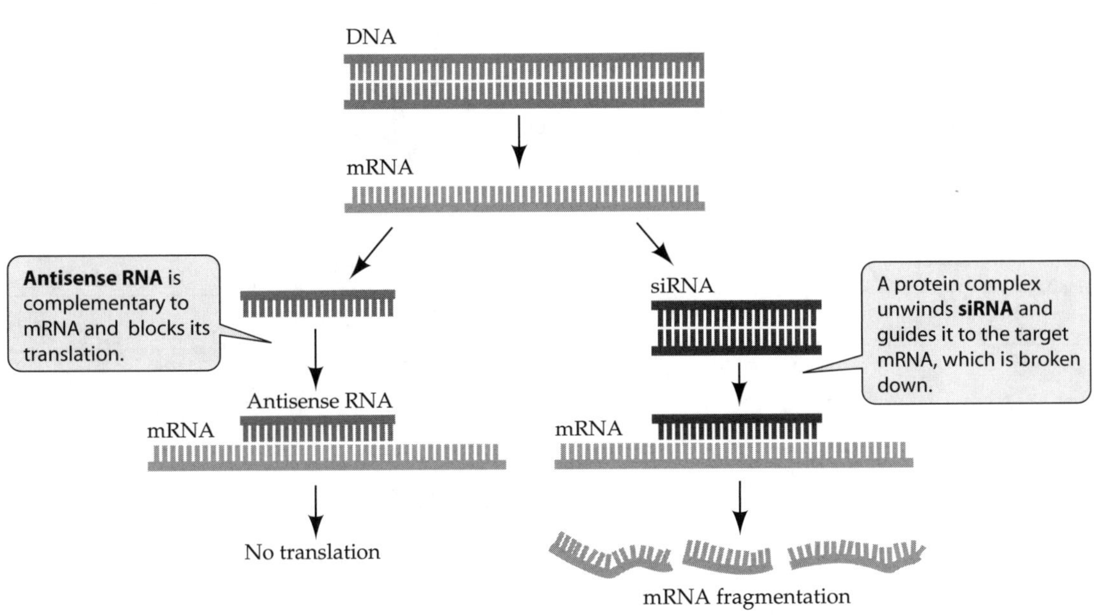
DNA
mRNA
Antisense RNA is complementary to mRNA and blocks its translation.
siRNA
A protein complex unwinds **siRNA** and guides it to the target mRNA, which is broken down.
Antisense RNA
mRNA
mRNA
No translation
mRNA fragmentation

16.11 Using Antisense RNA and RNAi to Block Translation of mRNA *(Page 329)*

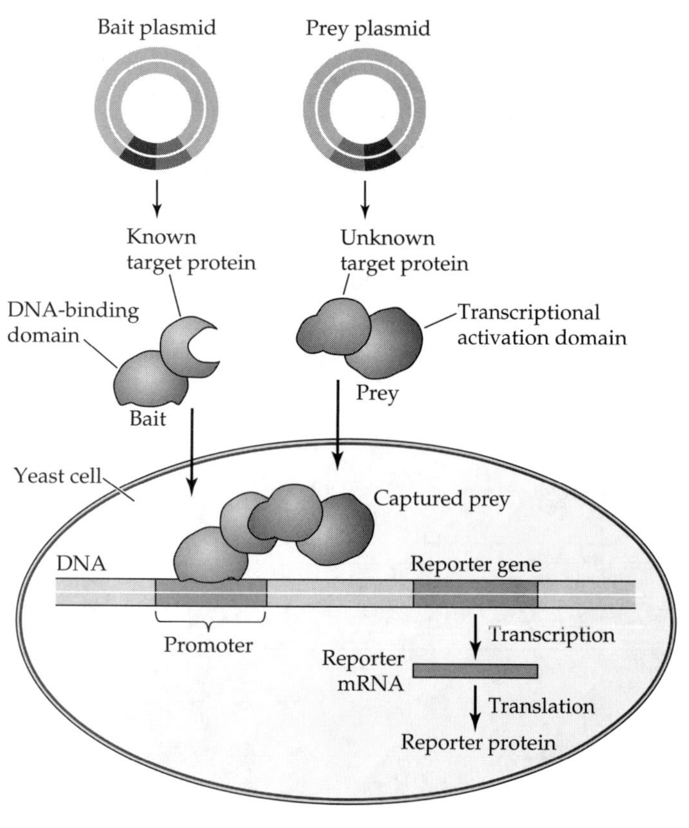
Bait plasmid
Prey plasmid
Known
target protein
Unknown
target protein
DNA-binding
domain
Transcriptional
activation domain
Bait
Prey
Yeast cell
Captured prey
DNA
Reporter gene
Promoter
Transcription
Reporter
mRNA
Translation
Reporter protein

16.12 The Two-Hybrid System *(Page 329)*

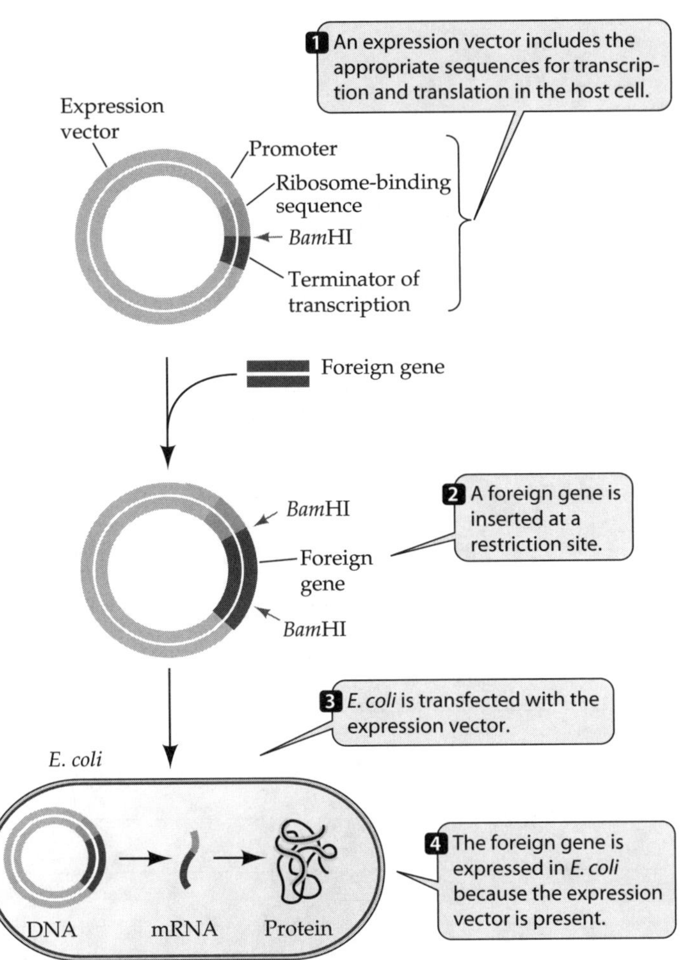
1 An expression vector includes the appropriate sequences for transcription and translation in the host cell.
Expression vector
Promoter
Ribosome-binding sequence
BamHI
Terminator of transcription
Foreign gene
BamHI
Foreign gene
BamHI
2 A foreign gene is inserted at a restriction site.
3 E. coli is transfected with the expression vector.
E. coli
DNA
mRNA
Protein
4 The foreign gene is expressed in E. coli because the expression vector is present.

16.13 An Expression Vector Allows a Foreign Gene to Be Expressed in a Host Cell *(Page 330)*

16.1 *Some Medically Useful Products of Biotechnology*

PRODUCT	USE
Colony-stimulating factor	Stimulates production of white blood cells in patients with cancer and AIDS
Erythropoietin	Prevents anemia in patients undergoing kidney dialysis and cancer therapy
Factor VIII	Replaces clotting factor missing in patients with hemophilia A
Growth hormone	Replaces missing hormone in people of short stature
Insulin	Stimulates glucose uptake from blood in people with insulin-dependent (Type I) diabetes
Platelet-derived growth factor	Stimulates wound healing
Tissue plasminogen activator	Dissolves blood clots after heart attacks and strokes
Vaccine proteins: Hepatitis B, herpes, influenza, Lyme disease, meningitis, pertussis, etc.	Prevent and treat infectious diseases

(Page 331)

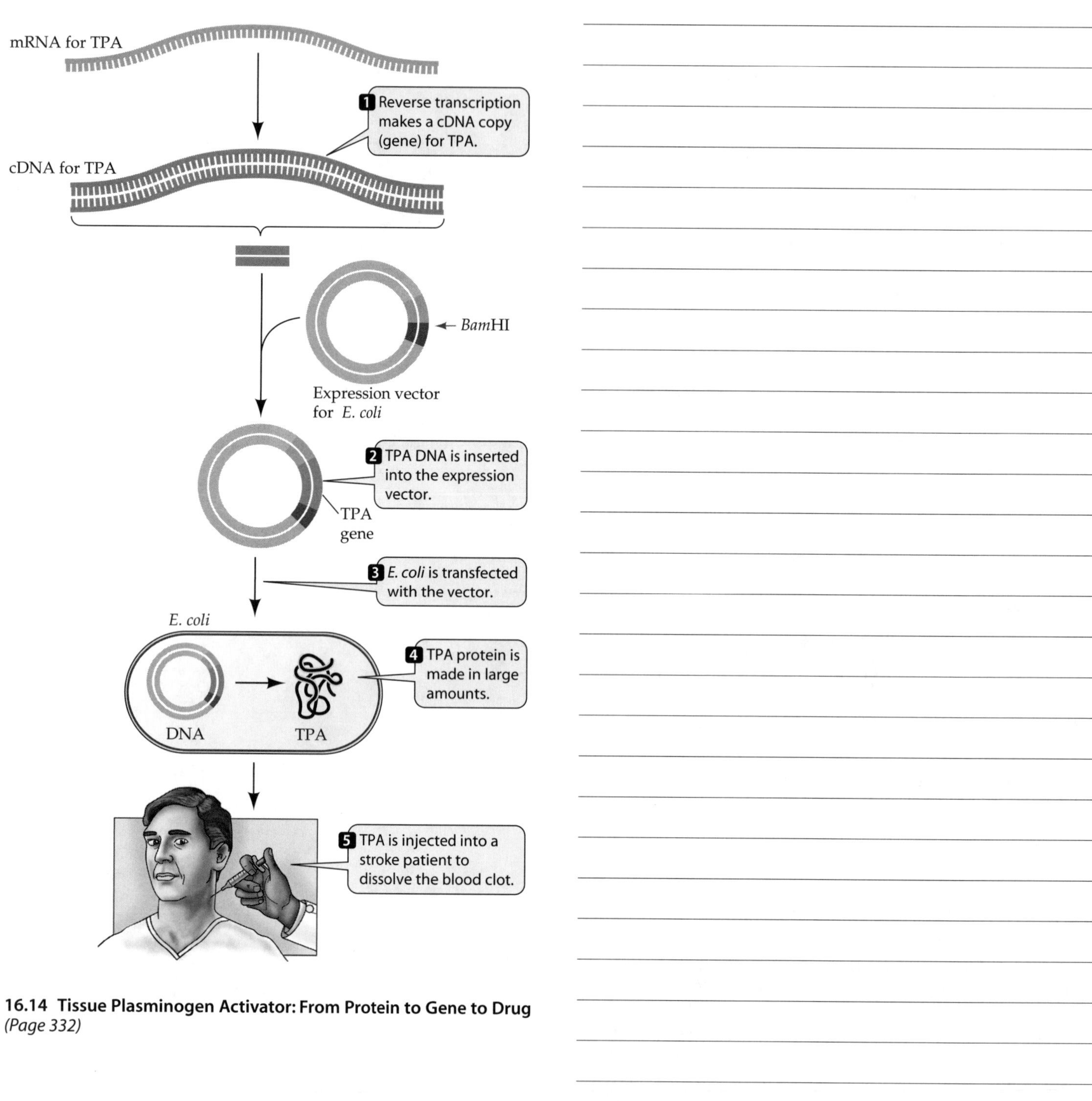

mRNA for TPA
1 Reverse transcription makes a cDNA copy (gene) for TPA.
cDNA for TPA
BamHI
Expression vector for E. coli
2 TPA DNA is inserted into the expression vector.
TPA gene
3 E. coli is transfected with the vector.
E. coli
4 TPA protein is made in large amounts.
DNA
TPA
5 TPA is injected into a stroke patient to dissolve the blood clot.

16.14 Tissue Plasminogen Activator: From Protein to Gene to Drug
(Page 332)

16.2 Agricultural Applications of Biotechnology under Development

PROBLEM	TECHNOLOGY/GENES
Improving the environmental adaptations of plants	Genes for drought tolerance, salt tolerance
Improving breeding	Male sterility for hybrid seeds
Improving nutritional traits	High-lysine seeds
Improving crops after harvest	Delay of fruit ripening; sweeter vegetables
Using plants as bioreactors	Plastics, oils, and drugs produced in plants
Controlling crop pests	Herbicide tolerance; resistance to viruses, bacteria, fungi, insects

(Page 333)

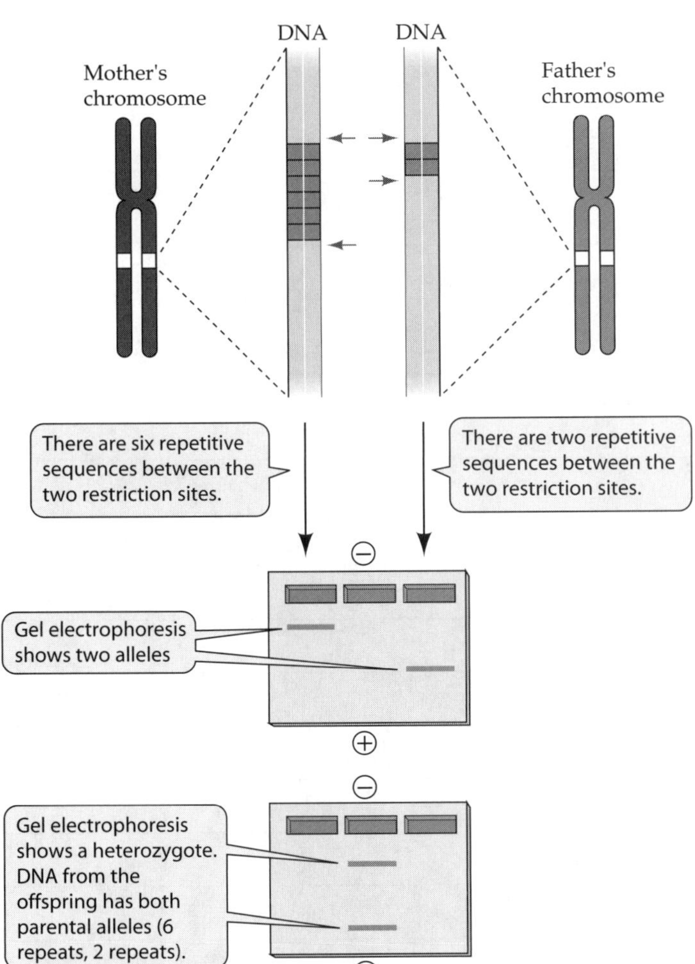
DNA
DNA
Mother's chromosome
Father's chromosome
There are six repetitive sequences between the two restriction sites.
There are two repetitive sequences between the two restriction sites.
Gel electrophoresis shows two alleles
Gel electrophoresis shows a heterozygote. DNA from the offspring has both parental alleles (6 repeats, 2 repeats).

16.17 DNA Fingerprinting *(Page 335)*

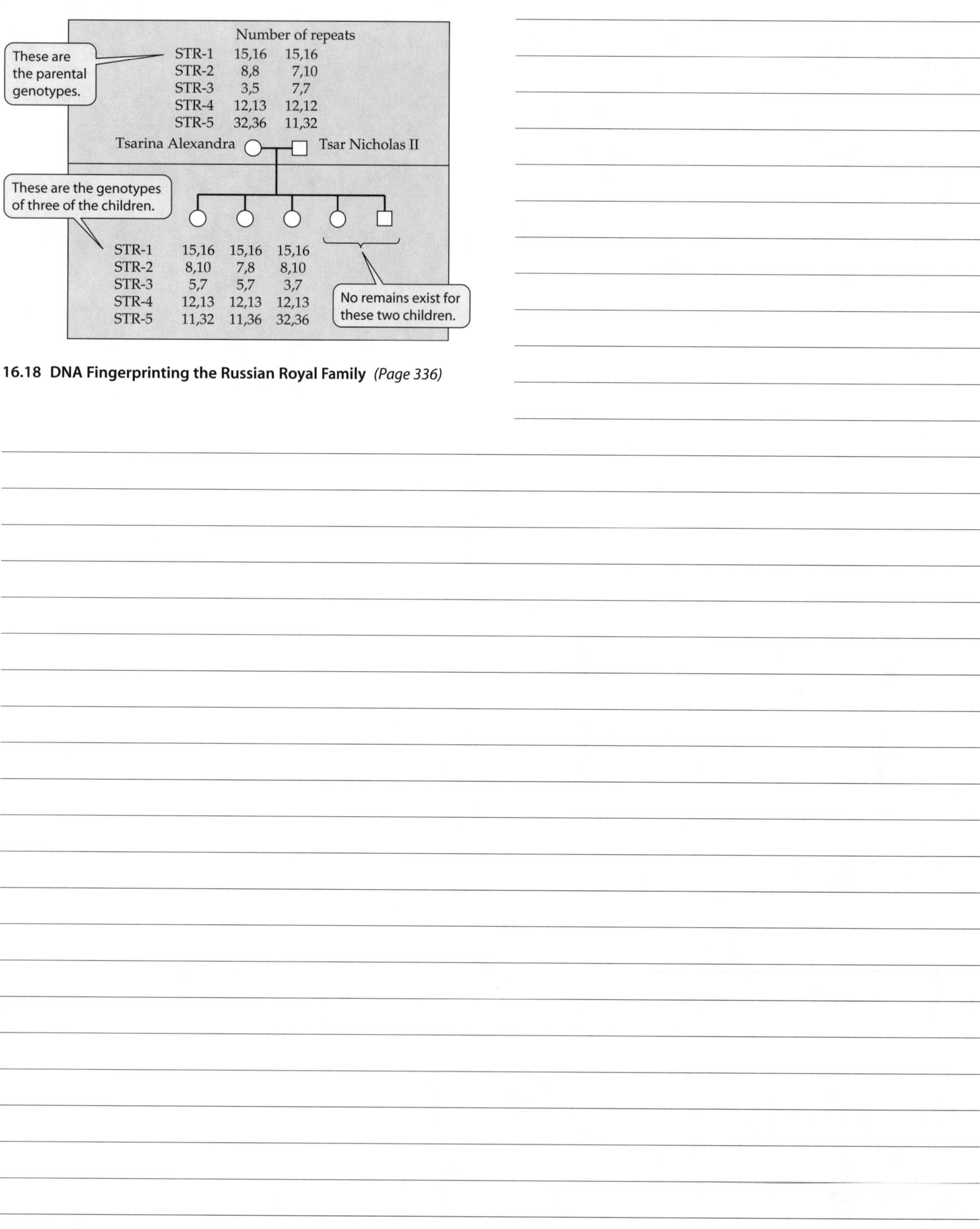
These are the parental genotypes.
Number of repeats
STR-1 15,16 15,16
STR-2 8,8 7,10
STR-3 3,5 7,7
STR-4 12,13 12,12
STR-5 32,36 11,32
Tsarina Alexandra
Tsar Nicholas II
These are the genotypes of three of the children.
STR-1 15,16 15,16 15,16
STR-2 8,10 7,8 8,10
STR-3 5,7 5,7 3,7
STR-4 12,13 12,13 12,13
STR-5 11,32 11,36 32,36
No remains exist for these two children.

16.18 DNA Fingerprinting the Russian Royal Family *(Page 336)*

17 Molecular Biology and Medicine

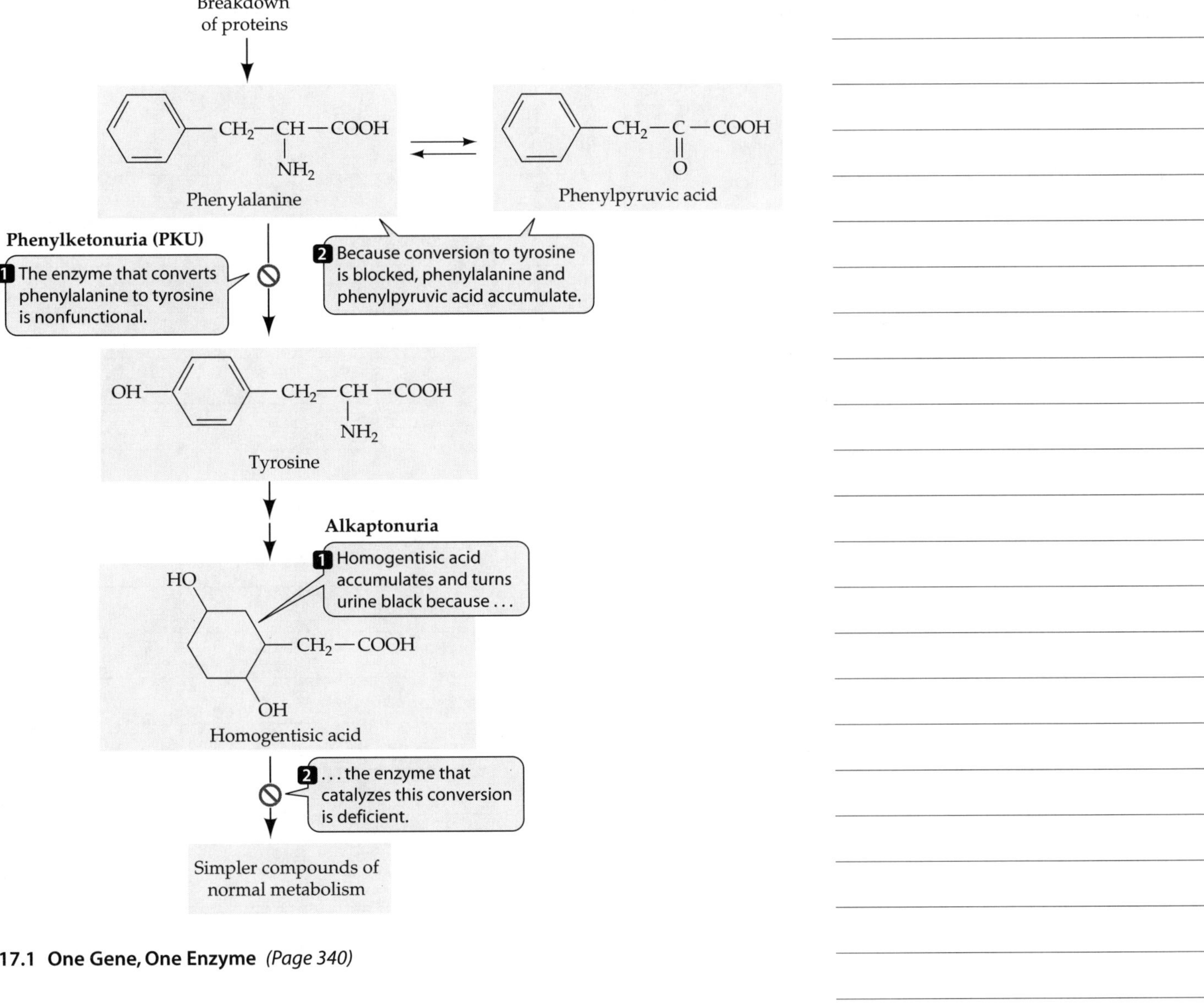

Breakdown of proteins
CH_2—CH—COOH
NH_2
Phenylalanine
CH_2—C—COOH
O
Phenylpyruvic acid
Phenylketonuria (PKU)
1 The enzyme that converts phenylalanine to tyrosine is nonfunctional.
2 Because conversion to tyrosine is blocked, phenylalanine and phenylpyruvic acid accumulate.
OH—
CH_2—CH—COOH
NH_2
Tyrosine
Alkaptonuria
1 Homogentisic acid accumulates and turns urine black because . . .
HO
CH_2—COOH
OH
Homogentisic acid
2 . . . the enzyme that catalyzes this conversion is deficient.
Simpler compounds of normal metabolism

17.1 One Gene, One Enzyme *(Page 340)*

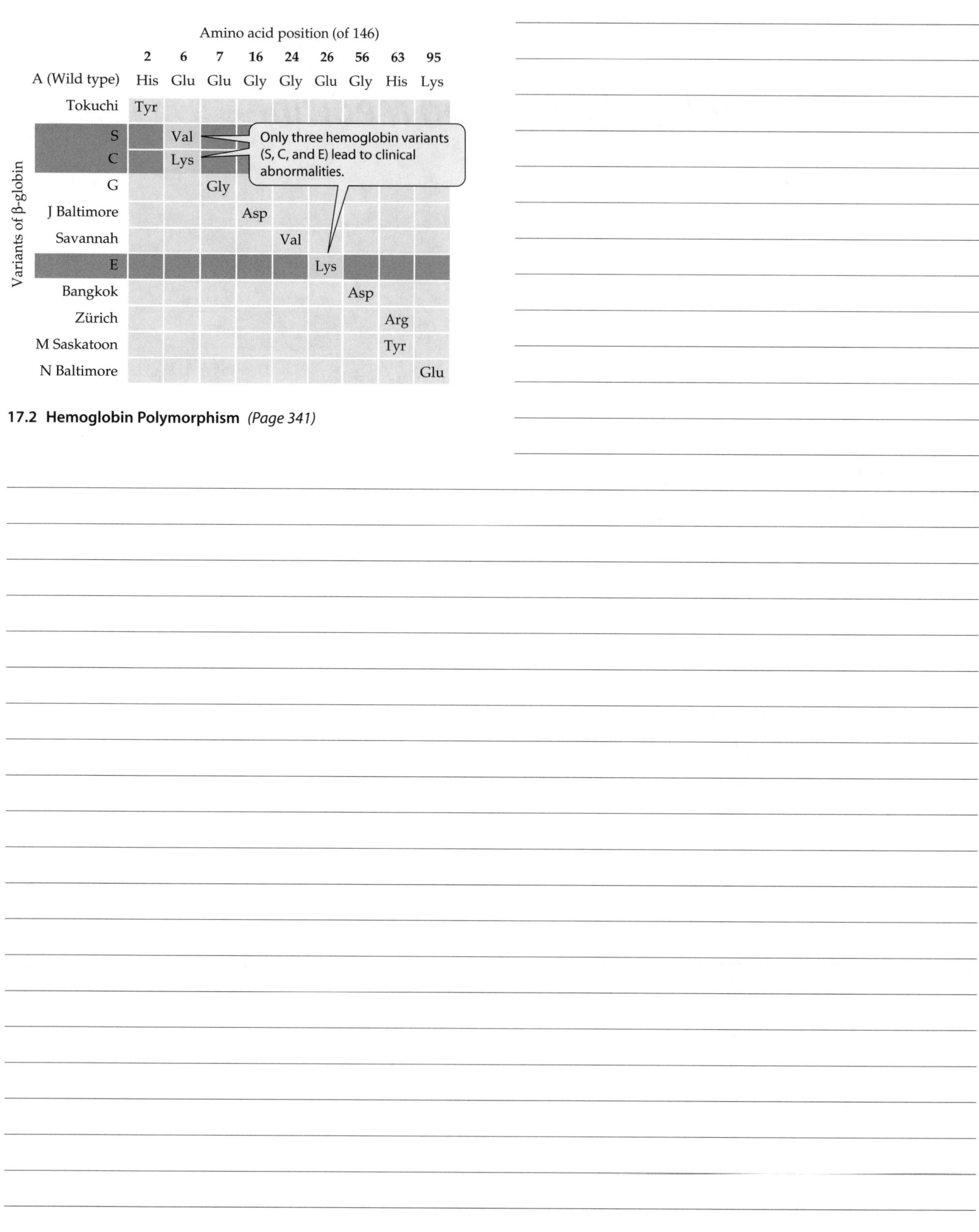
Amino acid position (of 146)
2 6 7 16 24 26 56 63 95
A (Wild type) His Glu Glu Gly Gly Glu Gly His Lys
Tokuchi Tyr
S Val
C Lys
G Gly
J Baltimore Asp
Savannah Val
E Lys
Bangkok Asp
Zürich Arg
M Saskatoon Tyr
N Baltimore Glu
Variants of β-globin
Only three hemoglobin variants (S, C, and E) lead to clinical abnormalities.

17.2 Hemoglobin Polymorphism *(Page 341)*

(*a*) **Hypercholesterolemia**

Normal liver cell: Cholesterol, as part of low-density lipoprotein (LDL), enters the cell after LDL binds to a receptor.

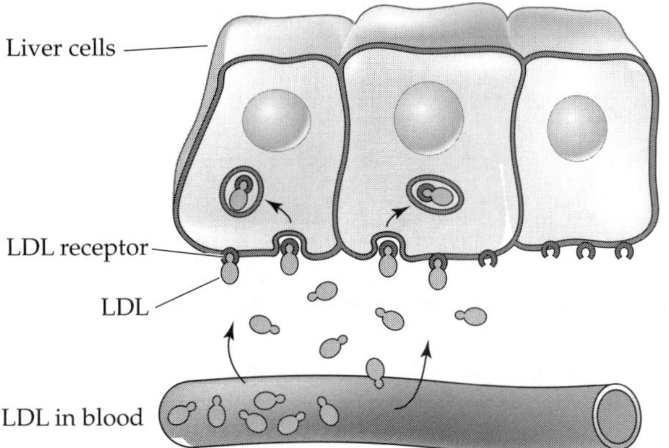
Liver cells
LDL receptor
LDL
LDL in blood

Familial hypercholesterolemia: Absence of a functional LDL receptor prevents cholesterol from entering the cells, and it accumulates in the blood.

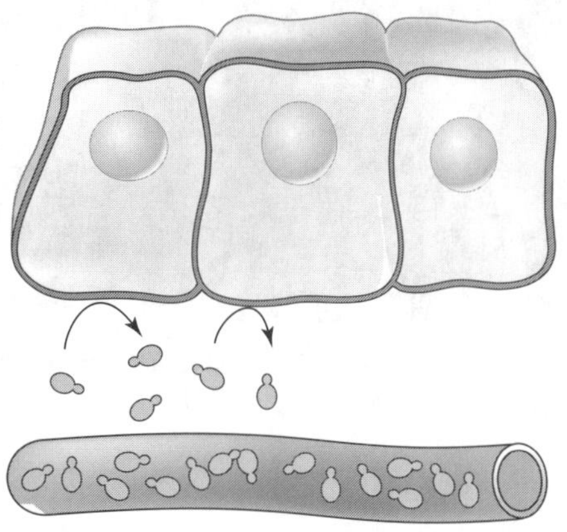

(*b*) **Cystic fibrosis**

Normal cell lining the airway: Cl^- leaves the cell through an ion channel. Water follows by osmosis, and moist thin mucus allows cilia to beat and sweep away foreign particles, including bacteria.

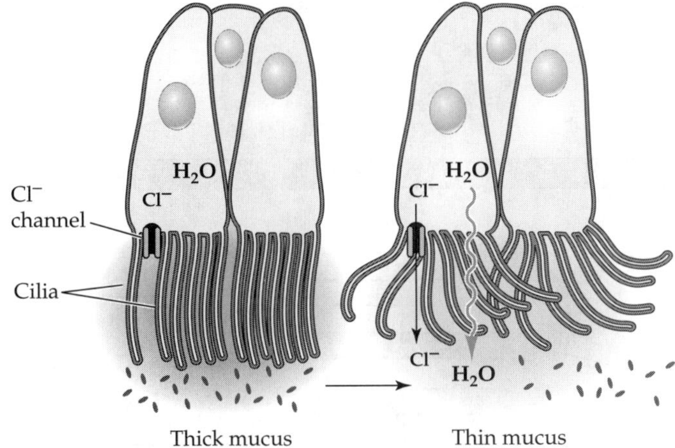
H_2O
Cl^-
Cl^- channel
Cilia
Thick mucus
H_2O
Cl^-
Cl^-
H_2O
Thin mucus

Cystic fibrosis: Lack of a Cl^- channel causes a thick, viscous mucus to form. Cilia cannot beat properly and remove bacteria; infections can easily take hold.

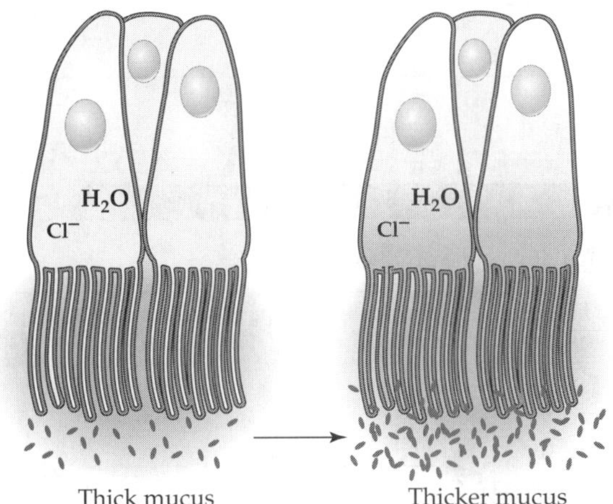
H_2O
Cl^-
Thick mucus
H_2O
Cl^-
Thicker mucus

17.3 Genetic Diseases of Membrane Proteins *(Page 342)*

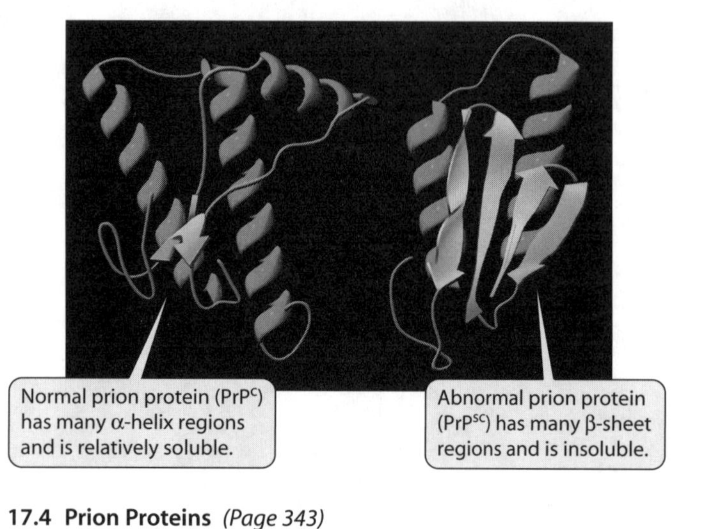
Normal prion protein (PrP^{C}) has many α-helix regions and is relatively soluble.
Abnormal prion protein (PrP^{SC}) has many β-sheet regions and is insoluble.

17.4 Prion Proteins *(Page 343)*

RESEARCH METHOD

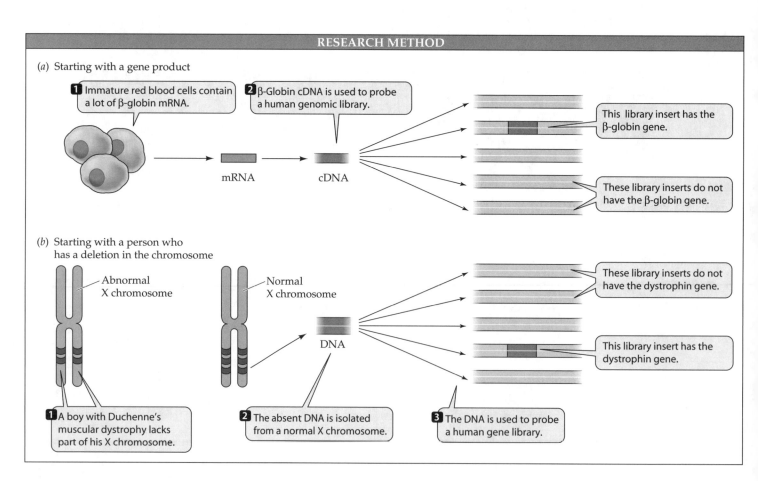

(*a*) Starting with a gene product

1 Immature red blood cells contain a lot of β-globin mRNA.

2 β-Globin cDNA is used to probe a human genomic library.

mRNA

cDNA

This library insert has the β-globin gene.

These library inserts do not have the β-globin gene.

(*b*) Starting with a person who has a deletion in the chromosome

Abnormal X chromosome

Normal X chromosome

DNA

These library inserts do not have the dystrophin gene.

This library insert has the dystrophin gene.

1 A boy with Duchenne's muscular dystrophy lacks part of his X chromosome.

2 The absent DNA is isolated from a normal X chromosome.

3 The DNA is used to probe a human gene library.

17.6 Strategies for Isolating Human Genes *(Page 345)*

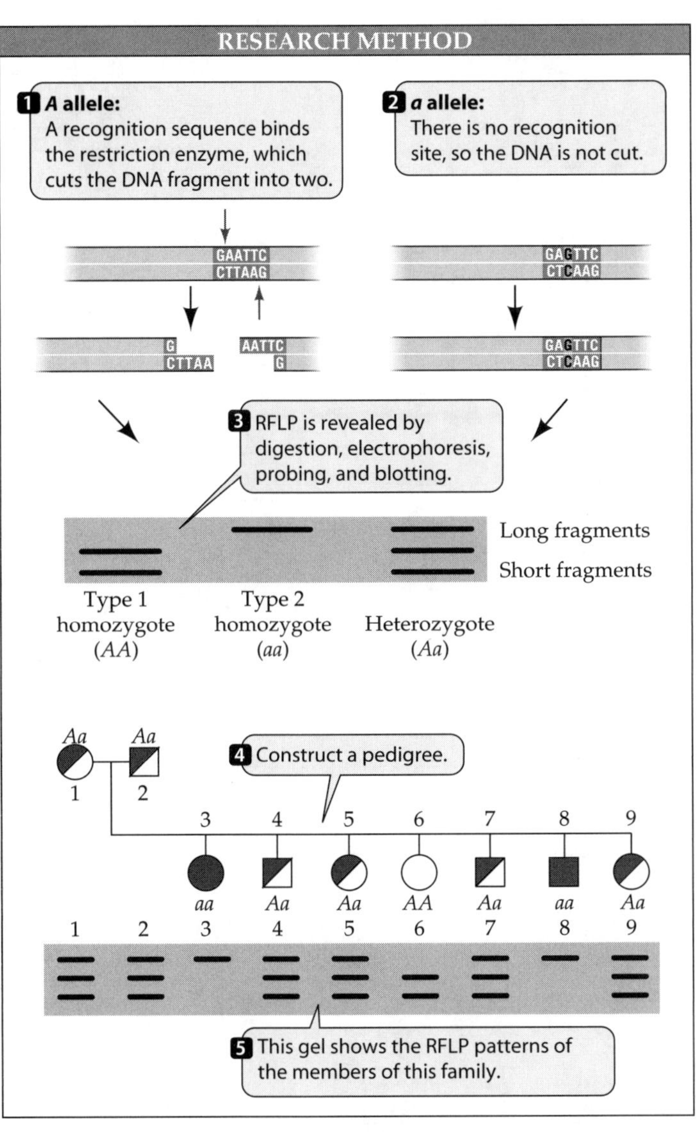
RESEARCH METHOD
1 A allele:
A recognition sequence binds the restriction enzyme, which cuts the DNA fragment into two.
2 a allele:
There is no recognition site, so the DNA is not cut.
GAATTC
CTTAAG
G
CTTAA
AATTC
G
GAGTTC
CTCAAG
3 RFLP is revealed by digestion, electrophoresis, probing, and blotting.
Long fragments
Short fragments
Type 1 homozygote (AA)
Type 2 homozygote (aa)
Heterozygote (Aa)
4 Construct a pedigree.
Aa
Aa
aa
AA
5 This gel shows the RFLP patterns of the members of this family.

17.7 RFLP Mapping *(Page 346)*

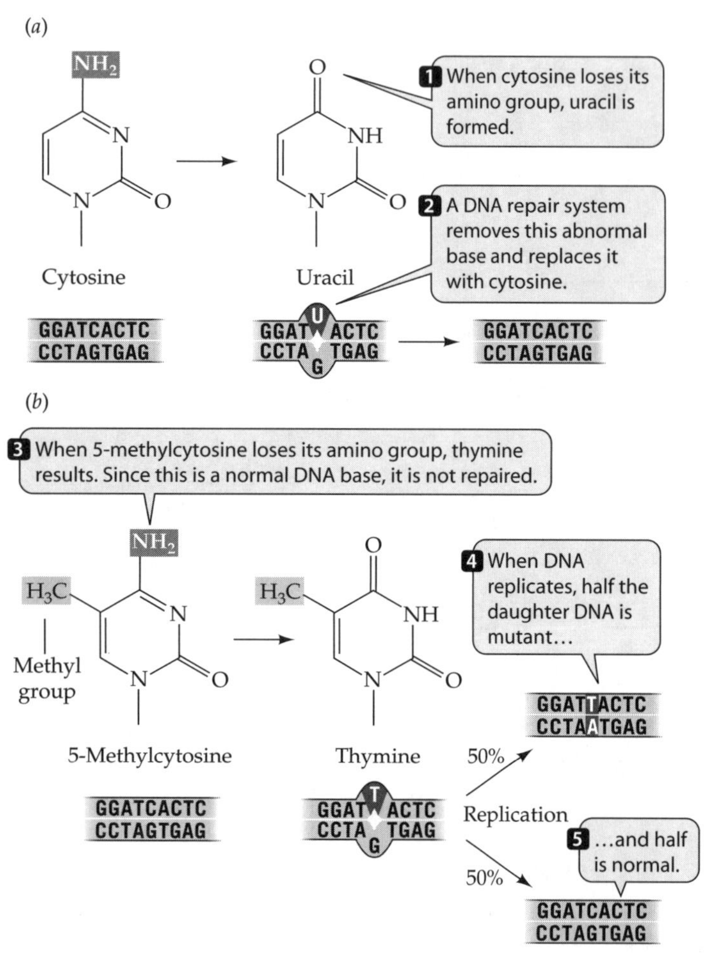

(a)
NH_2
N
N
O
Cytosine
O
NH
N
O
Uracil
1 When cytosine loses its amino group, uracil is formed.
2 A DNA repair system removes this abnormal base and replaces it with cytosine.
GGATCACTC
CCTAGTGAG
GGAT U ACTC
CCTA G TGAG
GGATCACTC
CCTAGTGAG
(b)
3 When 5-methylcytosine loses its amino group, thymine results. Since this is a normal DNA base, it is not repaired.
NH_2
H_3C
Methyl group
5-Methylcytosine
H_3C
Thymine
4 When DNA replicates, half the daughter DNA is mutant…
GGATTACTC
CCTAATGAG
50%
Replication
50%
5 …and half is normal.
GGATCACTC
CCTAGTGAG
GGAT T ACTC
CCTA G TGAG
GGATCACTC
CCTAGTGAG

17.8 5-Methylcytosine in DNA Is a "Hot Spot" for Mutagenesis *(Page 347)*

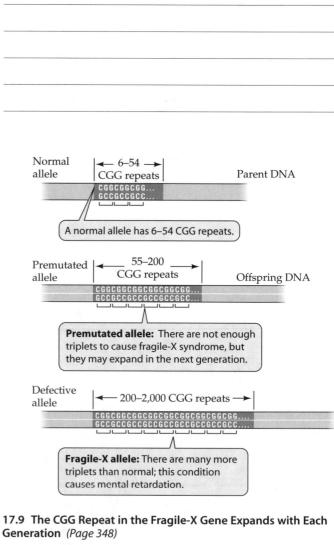

Normal allele
6–54 CGG repeats
Parent DNA
CGGCGGCGG...
GCCGCCGCC...
A normal allele has 6–54 CGG repeats.
Premutated allele
55–200 CGG repeats
Offspring DNA
CGGCGGCGGCGGCGGCGG...
GCCGCCGCCGCCGCCGCC...
Premutated allele: There are not enough triplets to cause fragile-X syndrome, but they may expand in the next generation.
Defective allele
200–2,000 CGG repeats
Fragile-X allele: There are many more triplets than normal; this condition causes mental retardation.

17.9 The CGG Repeat in the Fragile-X Gene Expands with Each Generation *(Page 348)*

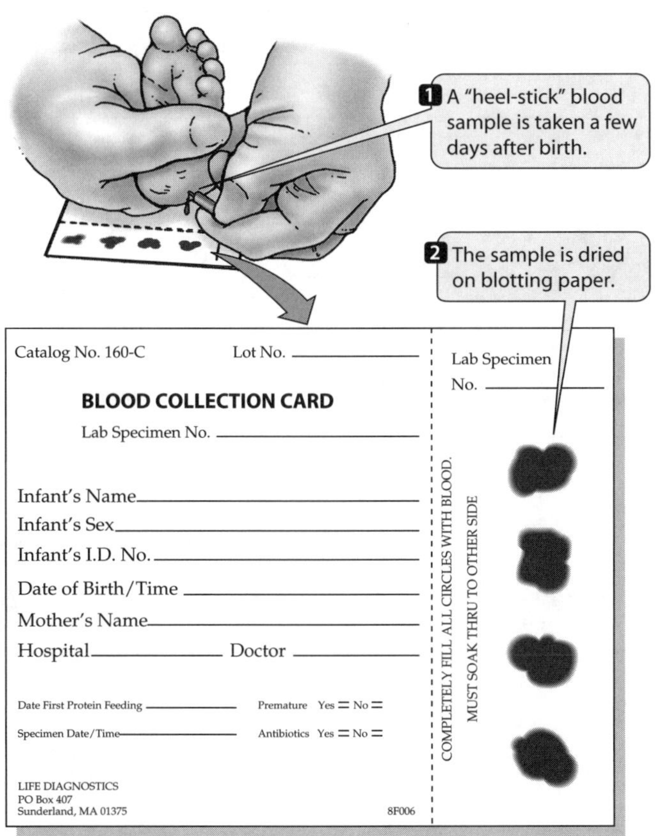
1 A "heel-stick" blood sample is taken a few days after birth.
2 The sample is dried on blotting paper.
Catalog No. 160-C Lot No.
BLOOD COLLECTION CARD
Lab Specimen No.
Infant's Name
Infant's Sex
Infant's I.D. No.
Date of Birth/Time
Mother's Name
Hospital Doctor
Date First Protein Feeding
Premature Yes No
Specimen Date/Time
Antibiotics Yes No
LIFE DIAGNOSTICS
PO Box 407
Sunderland, MA 01375
8F006
Lab Specimen No.
COMPLETELY FILL ALL CIRCLES WITH BLOOD.
MUST SOAK THRU TO OTHER SIDE

17.10 Genetic Screening of Newborns for Phenylketonuria
(Page 349)

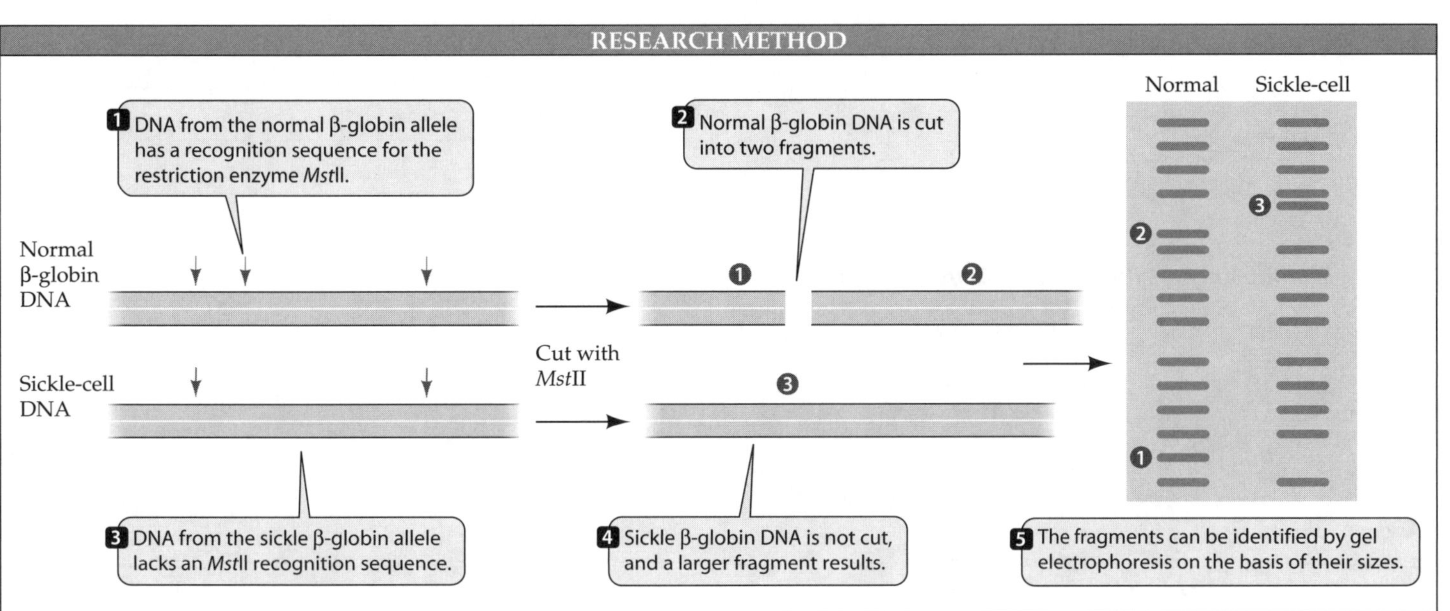
RESEARCH METHOD
1 DNA from the normal β-globin allele has a recognition sequence for the restriction enzyme *Mst*II.
Normal β-globin DNA
Sickle-cell DNA
Cut with *Mst*II
2 Normal β-globin DNA is cut into two fragments.
Normal
Sickle-cell
3 DNA from the sickle β-globin allele lacks an *Mst*II recognition sequence.
4 Sickle β-globin DNA is not cut, and a larger fragment results.
5 The fragments can be identified by gel electrophoresis on the basis of their sizes.

17.11 DNA Testing by Allele-Specific Cleavage *(Page 351)*

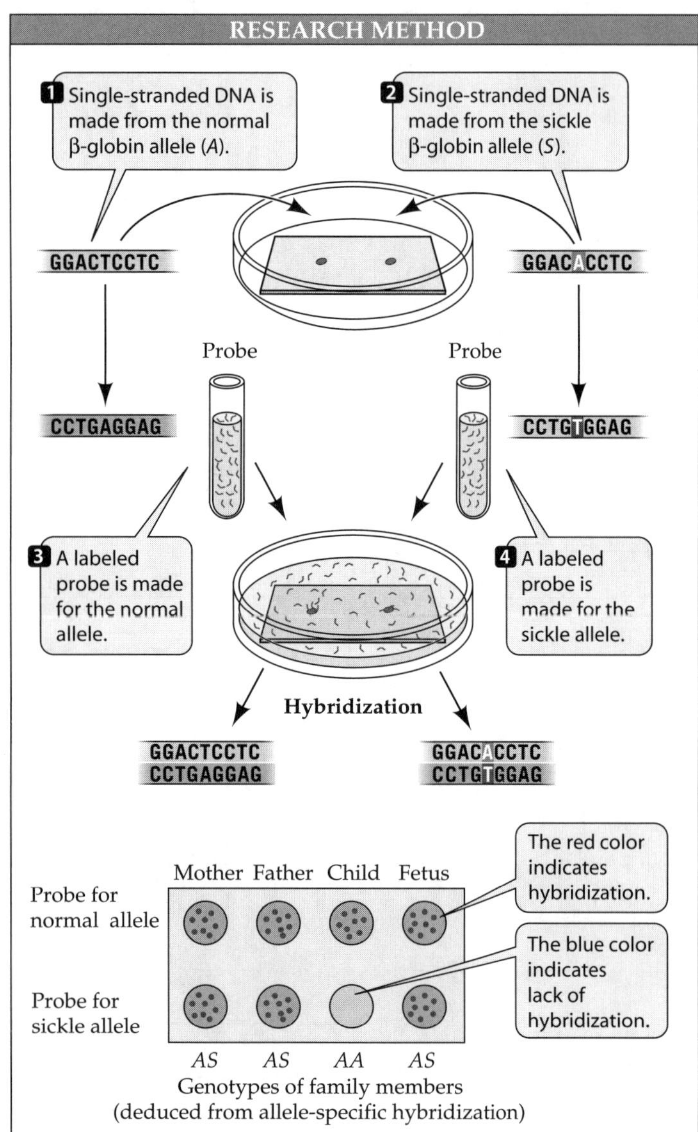
RESEARCH METHOD
1 Single-stranded DNA is made from the normal β-globin allele (A).
2 Single-stranded DNA is made from the sickle β-globin allele (S).
GGACTCCTC
GGACACCTC
Probe
Probe
CCTGAGGAG
CCTGTGGAG
3 A labeled probe is made for the normal allele.
4 A labeled probe is made for the sickle allele.
Hybridization
GGACTCCTC
CCTGAGGAG
GGACACCTC
CCTGTGGAG
Mother Father Child Fetus
Probe for normal allele
Probe for sickle allele
The red color indicates hybridization.
The blue color indicates lack of hybridization.
AS AS AA AS
Genotypes of family members
(deduced from allele-specific hybridization)

17.12 DNA Testing by Allele-Specific Oligonucleotide Hybridization
(Page 351)

17.1 *Human Cancers Known To Be Caused by Viruses*

CANCER	ASSOCIATED VIRUS
Liver cancer	Hepatitis B virus
Lymphoma, nasopharyngeal cancer	Epstein–Barr virus
T cell leukemia	Human T cell leukemia virus (HTLV-I)
Anogenital cancers	Papillomavirus
Kaposi's sarcoma	Kaposi's sarcoma herpesvirus

(Page 352)

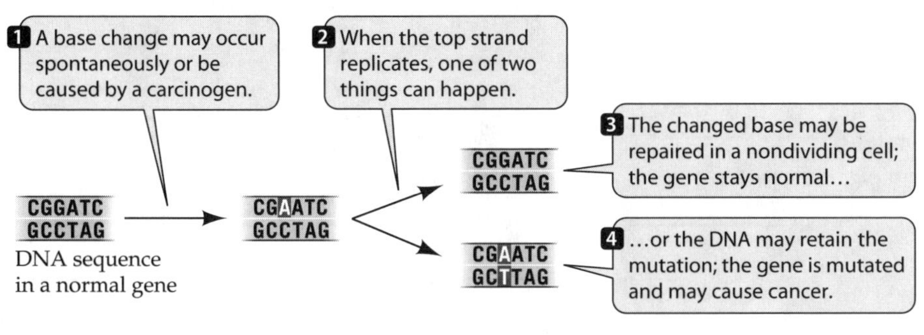
1 A base change may occur spontaneously or be caused by a carcinogen.
2 When the top strand replicates, one of two things can happen.
3 The changed base may be repaired in a nondividing cell; the gene stays normal…
4 …or the DNA may retain the mutation; the gene is mutated and may cause cancer.
CGGATC
GCCTAG
DNA sequence in a normal gene
CGAATC
GCCTAG
CGGATC
GCCTAG
CGAATC
GCTTAG

17.14 Dividing Cells Are Especially Susceptible to Genetic Damage *(Page 353)*

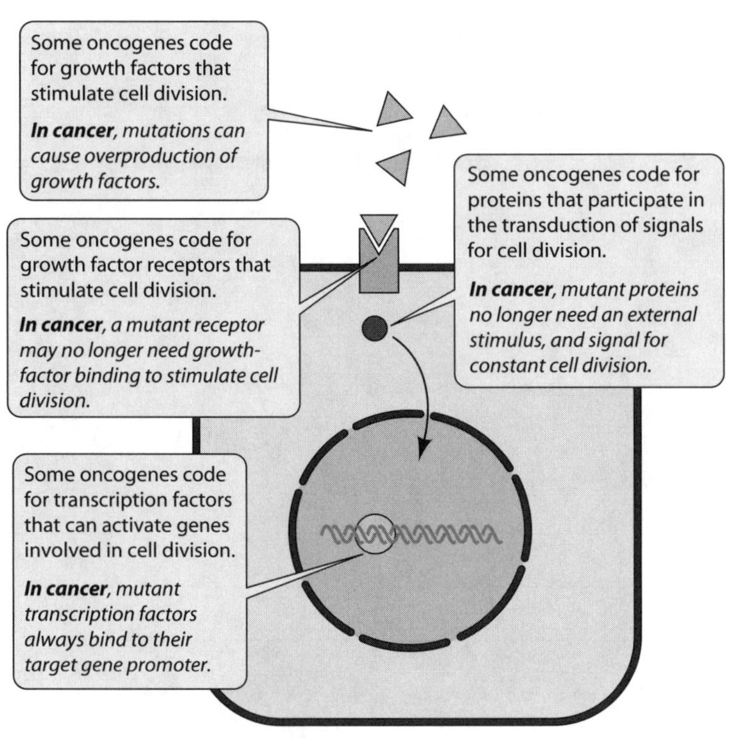
Some oncogenes code for growth factors that stimulate cell division.
In cancer, mutations can cause overproduction of growth factors.
Some oncogenes code for growth factor receptors that stimulate cell division.
In cancer, a mutant receptor may no longer need growth-factor binding to stimulate cell division.
Some oncogenes code for proteins that participate in the transduction of signals for cell division.
In cancer, mutant proteins no longer need an external stimulus, and signal for constant cell division.
Some oncogenes code for transcription factors that can activate genes involved in cell division.
In cancer, mutant transcription factors always bind to their target gene promoter.

17.15 Oncogene Products Stimulate Cell Division *(Page 353)*

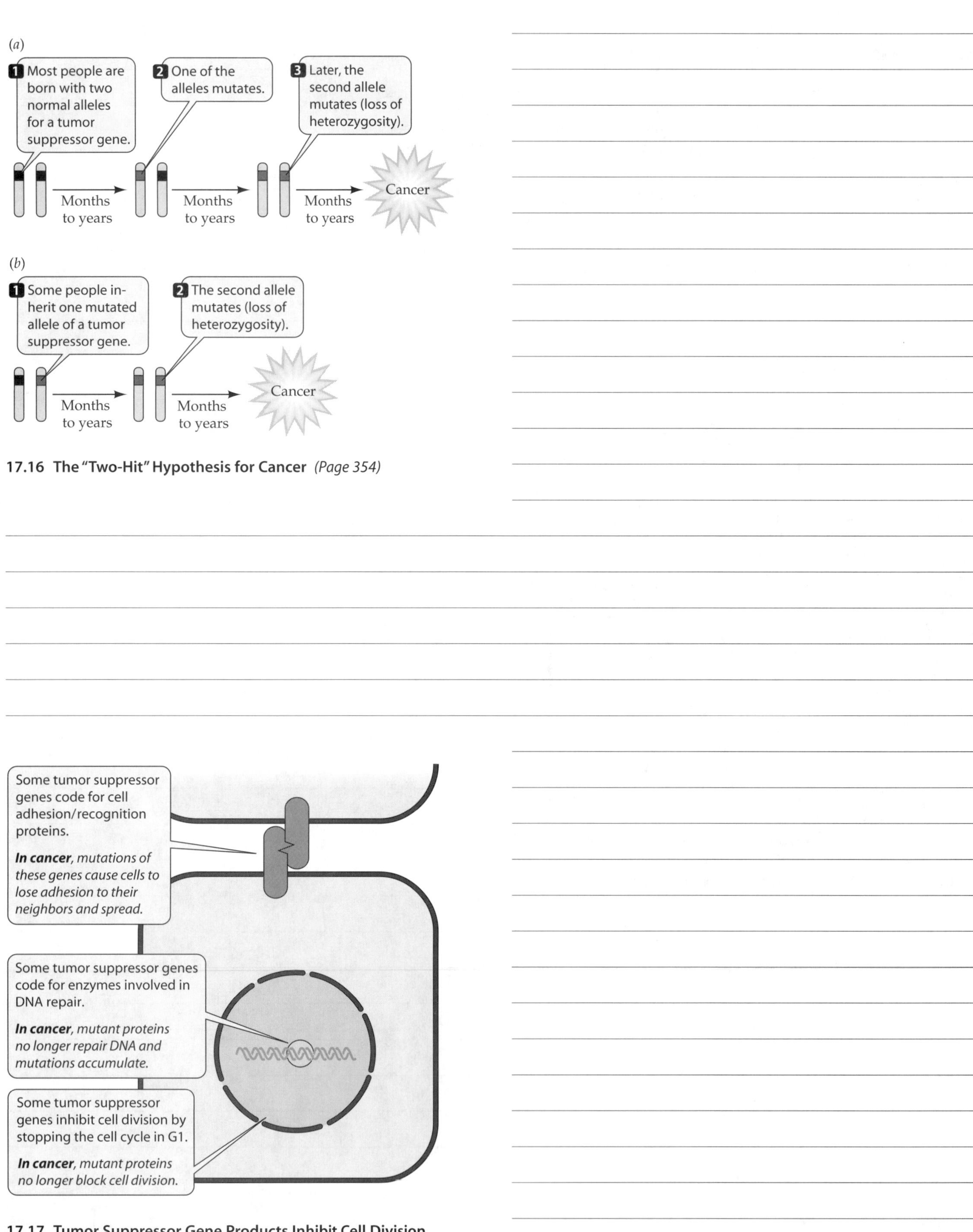
(a)
1 Most people are born with two normal alleles for a tumor suppressor gene.
2 One of the alleles mutates.
3 Later, the second allele mutates (loss of heterozygosity).
Months to years
Months to years
Months to years
Cancer
(b)
1 Some people inherit one mutated allele of a tumor suppressor gene.
2 The second allele mutates (loss of heterozygosity).
Months to years
Months to years
Cancer

17.16 The "Two-Hit" Hypothesis for Cancer *(Page 354)*

Some tumor suppressor genes code for cell adhesion/recognition proteins.
In cancer, mutations of these genes cause cells to lose adhesion to their neighbors and spread.
Some tumor suppressor genes code for enzymes involved in DNA repair.
In cancer, mutant proteins no longer repair DNA and mutations accumulate.
Some tumor suppressor genes inhibit cell division by stopping the cell cycle in G1.
In cancer, mutant proteins no longer block cell division.

17.17 Tumor Suppressor Gene Products Inhibit Cell Division and Cancer *(Page 354)*

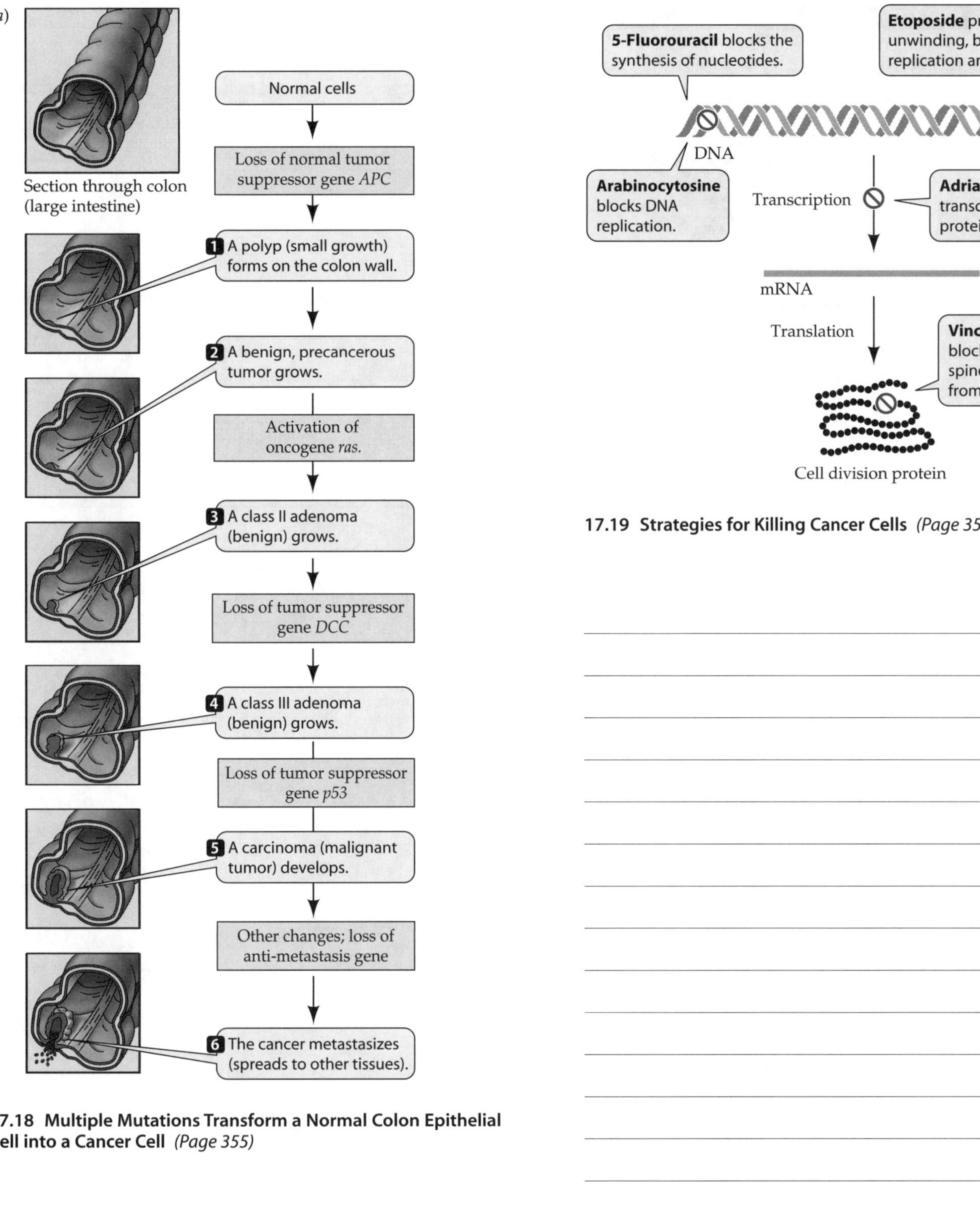
(a)
Section through colon (large intestine)
Normal cells
Loss of normal tumor suppressor gene APC
1 A polyp (small growth) forms on the colon wall.
2 A benign, precancerous tumor grows.
Activation of oncogene ras.
3 A class II adenoma (benign) grows.
Loss of tumor suppressor gene DCC
4 A class III adenoma (benign) grows.
Loss of tumor suppressor gene p53
5 A carcinoma (malignant tumor) develops.
Other changes; loss of anti-metastasis gene
6 The cancer metastasizes (spreads to other tissues).
5-Fluorouracil blocks the synthesis of nucleotides.
Etoposide prevents DNA from unwinding, blocking DNA replication and transcription.
DNA
Arabinocytosine blocks DNA replication.
Transcription
Adriamycin blocks transcription to inhibit protein synthesis.
mRNA
Translation
Vincristine and Taxol block the mitotic spindle microtubules from functioning.
Cell division protein

17.18 Multiple Mutations Transform a Normal Colon Epithelial Cell into a Cancer Cell *(Page 355)*

17.19 Strategies for Killing Cancer Cells *(Page 356)*

EXPERIMENT

Question: Can the introduction and expression of a normal allele help a patient who is homozygous for two defective alleles of an important gene?

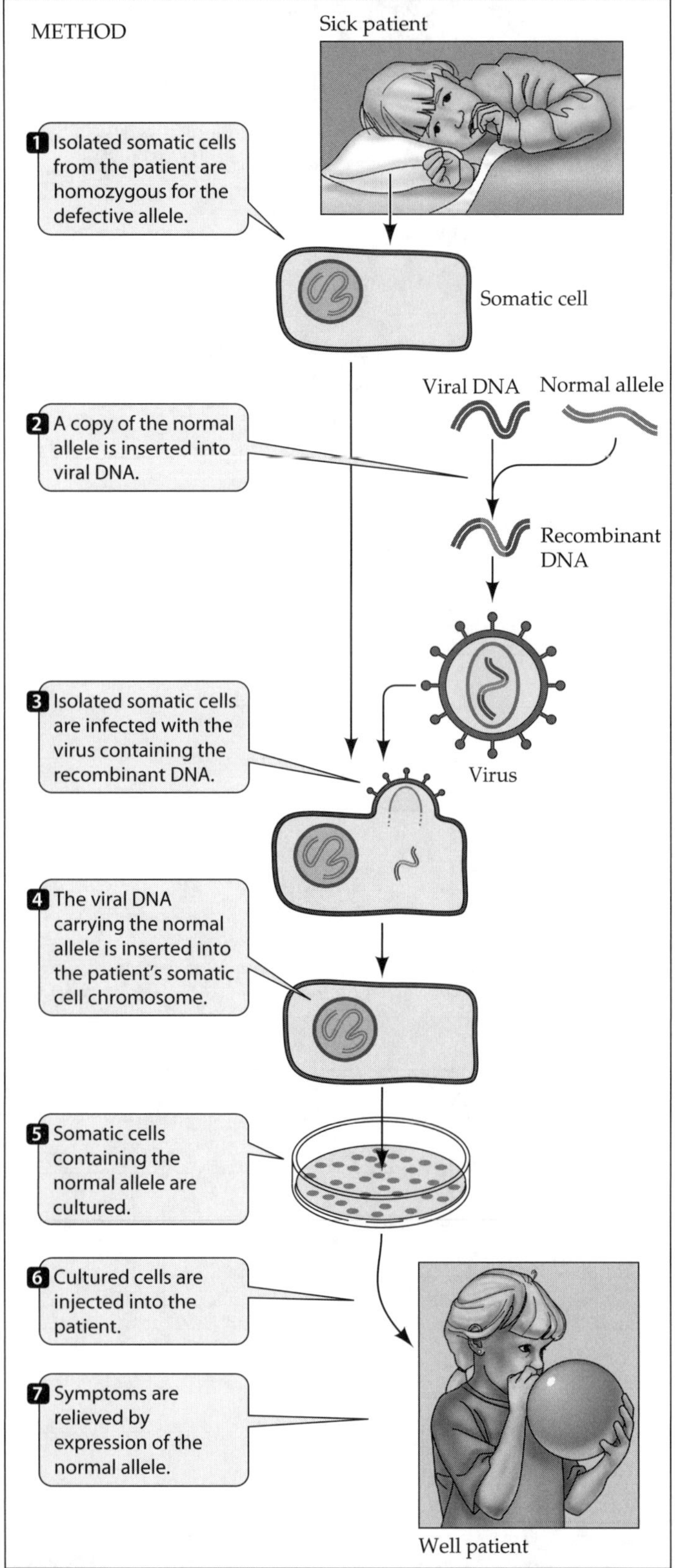

METHOD

Sick patient
1 Isolated somatic cells from the patient are homozygous for the defective allele.
Somatic cell
Viral DNA
Normal allele
2 A copy of the normal allele is inserted into viral DNA.
Recombinant DNA
3 Isolated somatic cells are infected with the virus containing the recombinant DNA.
Virus
4 The viral DNA carrying the normal allele is inserted into the patient's somatic cell chromosome.
5 Somatic cells containing the normal allele are cultured.
6 Cultured cells are injected into the patient.
7 Symptoms are relieved by expression of the normal allele.
Well patient

Conclusion: Gene therapy can be effective in relieving symptoms caused by a genetic disease.

17.20 Gene Therapy: The Ex Vivo Approach *(Page 357)*

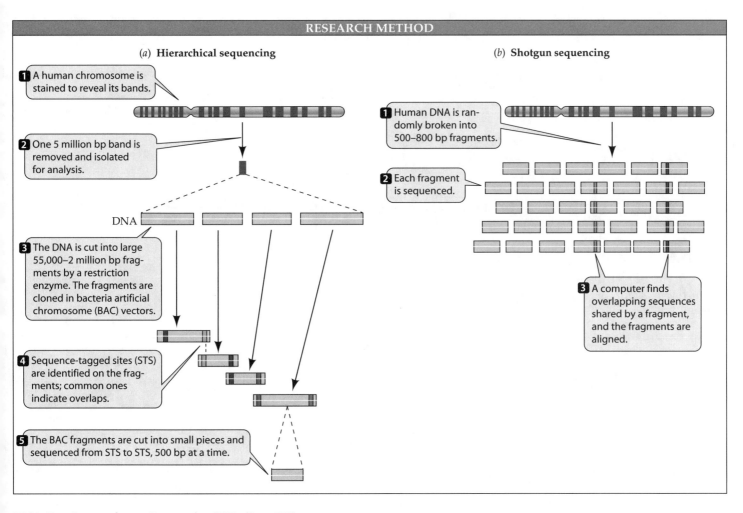

RESEARCH METHOD

(*a*) **Hierarchical sequencing**

1 A human chromosome is stained to reveal its bands.

2 One 5 million bp band is removed and isolated for analysis.

DNA

3 The DNA is cut into large 55,000–2 million bp fragments by a restriction enzyme. The fragments are cloned in bacteria artificial chromosome (BAC) vectors.

4 Sequence-tagged sites (STS) are identified on the fragments; common ones indicate overlaps.

5 The BAC fragments are cut into small pieces and sequenced from STS to STS, 500 bp at a time.

(*b*) **Shotgun sequencing**

1 Human DNA is randomly broken into 500–800 bp fragments.

2 Each fragment is sequenced.

3 A computer finds overlapping sequences shared by a fragment, and the fragments are aligned.

17.21 Two Approaches to Sequencing DNA *(Page 358)*

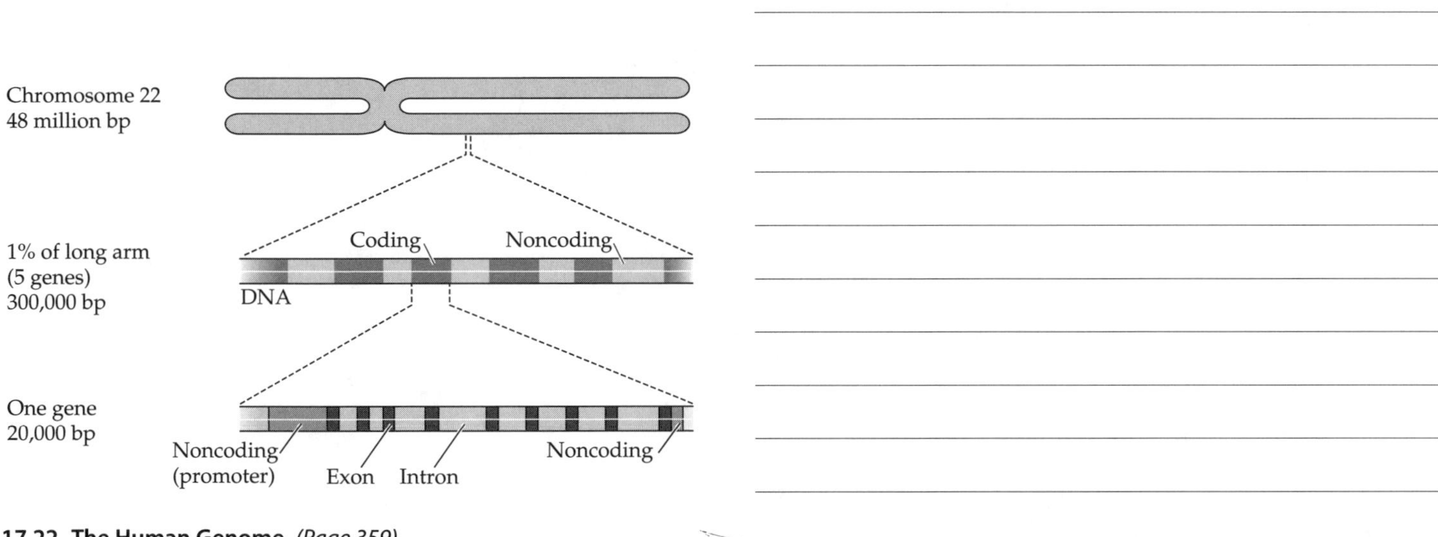
Chromosome 22
48 million bp
1% of long arm
(5 genes)
300,000 bp
Coding
Noncoding
DNA
One gene
20,000 bp
Noncoding
(promoter)
Exon
Intron
Noncoding

17.22 The Human Genome *(Page 359)*

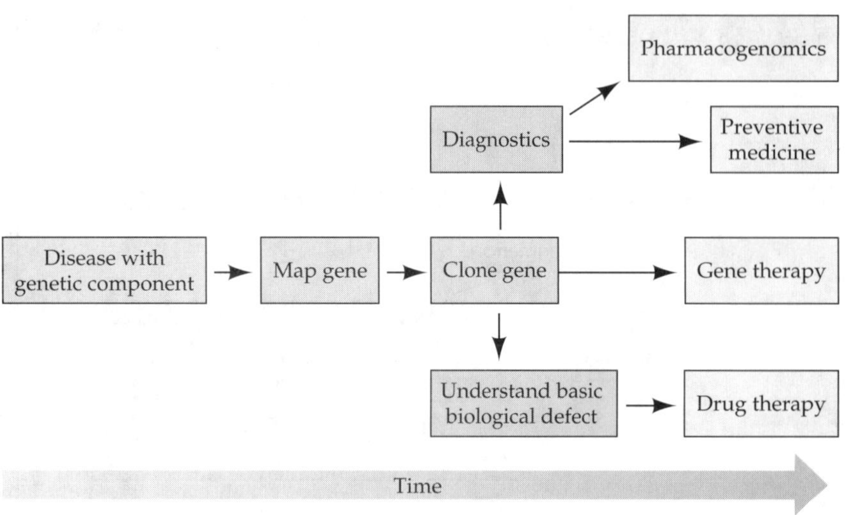
Pharmacogenomics
Diagnostics
Preventive medicine
Disease with genetic component
Map gene
Clone gene
Gene therapy
Understand basic biological defect
Drug therapy
Time

17.23 Is This the Future of Medicine? *(Page 360)*

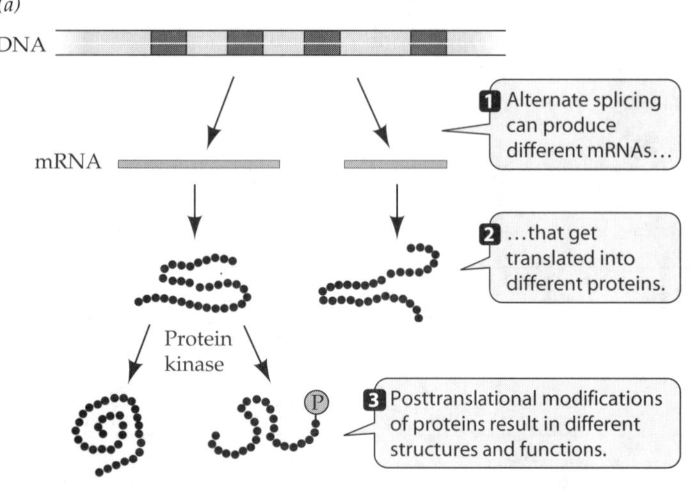
(a)
DNA
mRNA
1 Alternate splicing can produce different mRNAs…
2 …that get translated into different proteins.
Protein kinase
P
3 Posttranslational modifications of proteins result in different structures and functions.

17.24 Proteomics *(Page 361)*

18 Natural Defenses against Disease

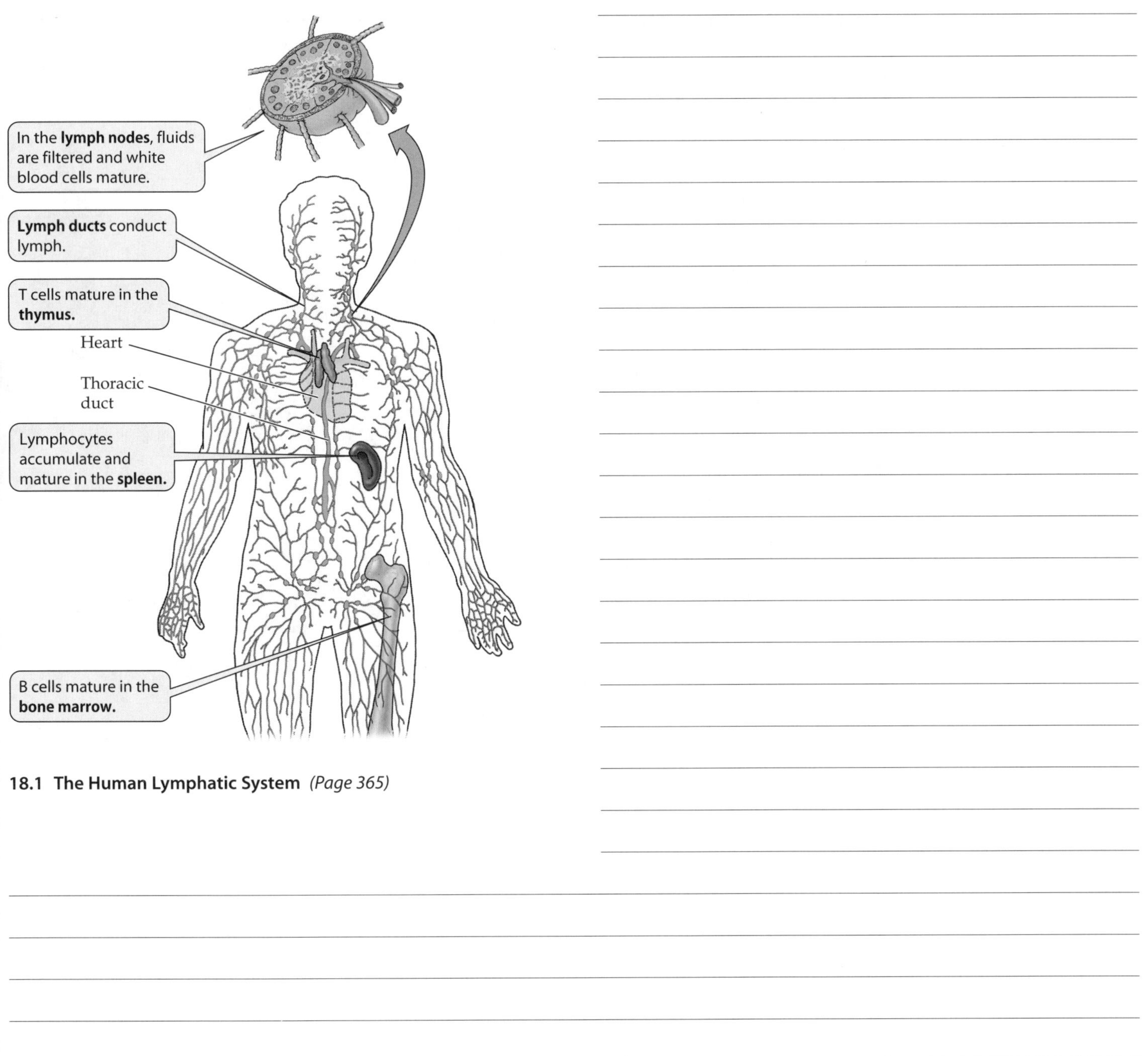

In the **lymph nodes**, fluids are filtered and white blood cells mature.
Lymph ducts conduct lymph.
T cells mature in the **thymus.**
Heart
Thoracic duct
Lymphocytes accumulate and mature in the **spleen.**
B cells mature in the **bone marrow.**

18.1 The Human Lymphatic System *(Page 365)*

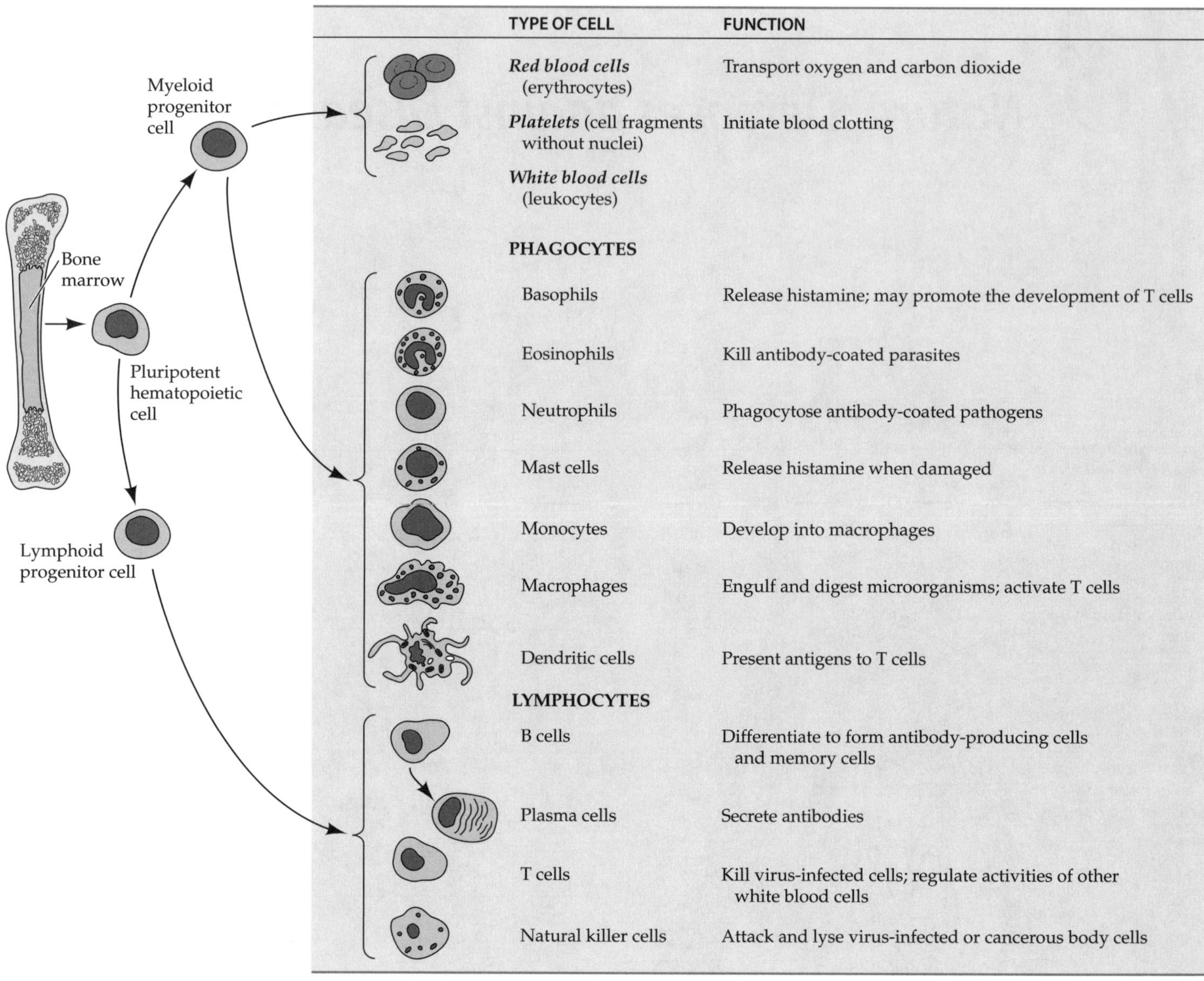
Myeloid progenitor cell
Bone marrow
Pluripotent hematopoietic cell
Lymphoid progenitor cell

TYPE OF CELL	FUNCTION
Red blood cells (erythrocytes)	Transport oxygen and carbon dioxide
Platelets (cell fragments without nuclei)	Initiate blood clotting
White blood cells (leukocytes)	
PHAGOCYTES	
Basophils	Release histamine; may promote the development of T cells
Eosinophils	Kill antibody-coated parasites
Neutrophils	Phagocytose antibody-coated pathogens
Mast cells	Release histamine when damaged
Monocytes	Develop into macrophages
Macrophages	Engulf and digest microorganisms; activate T cells
Dendritic cells	Present antigens to T cells
LYMPHOCYTES	
B cells	Differentiate to form antibody-producing cells and memory cells
Plasma cells	Secrete antibodies
T cells	Kill virus-infected cells; regulate activities of other white blood cells
Natural killer cells	Attack and lyse virus-infected or cancerous body cells

18.2 Blood Cells *(Page 366)*

18.1 *Human Nonspecific Defenses*

DEFENSIVE AGENT	FUNCTION
Surface barriers	
Skin	Prevents entry of pathogens and foreign substances
Acid secretions	Inhibit bacterial growth on skin
Mucous membranes	Prevent entry of pathogens
Mucous secretions	Trap bacteria and other pathogens in digestive and respiratory tracts
Nasal hairs	Filter bacteria in nasal passages
Cilia	Move mucus and trapped materials away from respiratory passages
Gastric juice	Concentrated HCl and proteases destroy pathogens in stomach
Acid in vagina	Limits growth of fungi and bacteria in female reproductive tract
Tears, saliva	Lubricate and cleanse; contain lysozyme, which destroys bacteria
Nonspecific cellular, chemical, and coordinated defenses	
Normal flora	Compete with pathogens; may produce substances toxic to pathogens
Fever	Body-wide response inhibits microbial multiplication and speeds body repair processes
Coughing, sneezing	Expels pathogens from upper respiratory passages
Inflammatory response (involves leakage of blood plasma and phagocytes from capillaries)	Limits spread of pathogens to neighboring tissues; concentrates defenses; digests pathogens and dead tissue cells; released chemical mediators attract phagocytes and specific defense lymphocytes to site
Phagocytes (macrophages and neutrophils)	Engulf and destroy pathogens that enter body
Natural killer cells	Attack and lyse virus-infected or cancerous body cells
Antimicrobial proteins	
Interferons	Released by virus-infected cells to protect healthy tissue from viral infection; mobilize specific defenses
Complement proteins	Lyse microorganisms, enhance phagocytosis, and assist in inflammatory response

(Page 367)

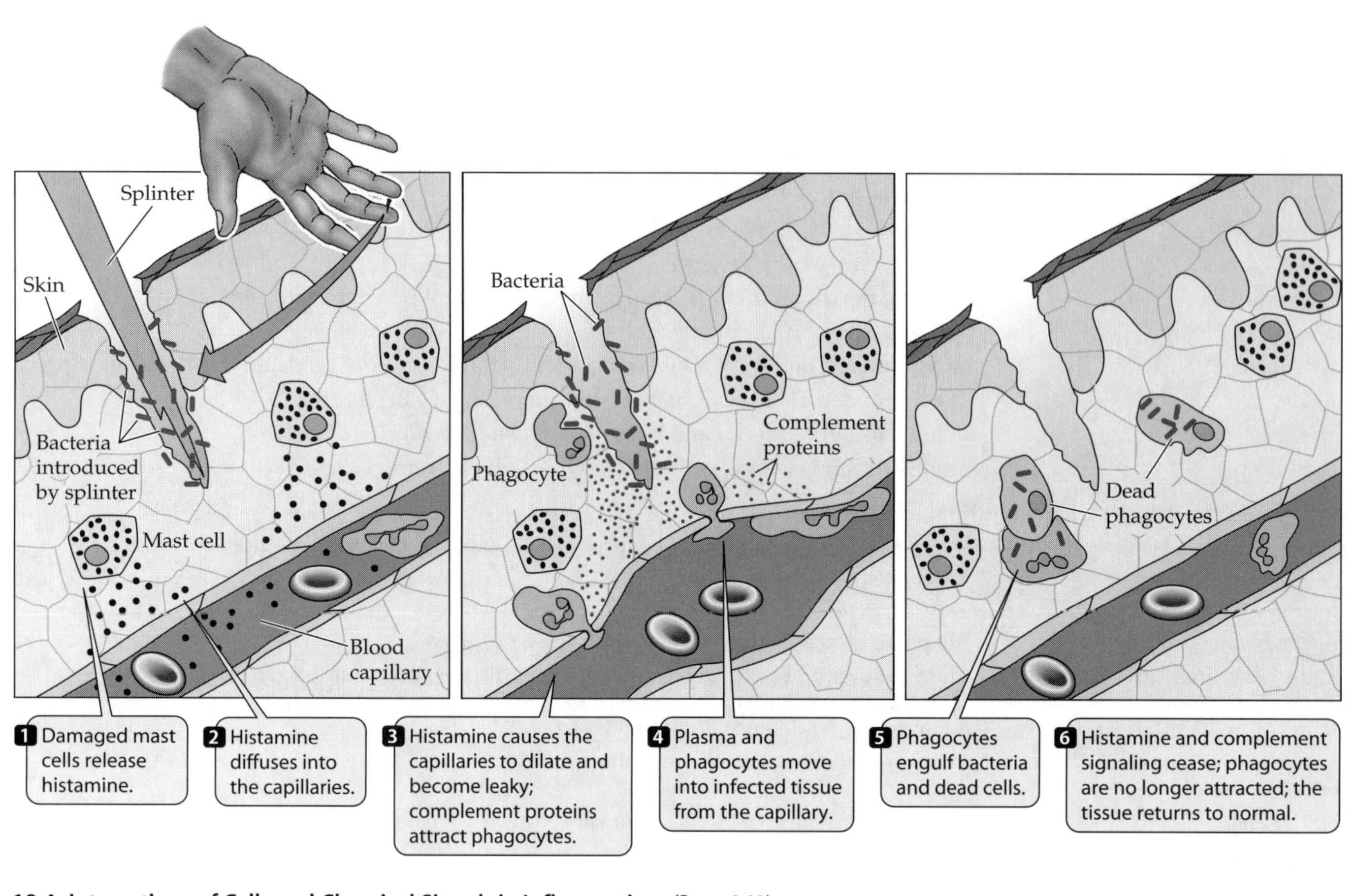
Splinter
Skin
Bacteria introduced by splinter
Mast cell
Blood capillary
Bacteria
Complement proteins
Phagocyte
Dead phagocytes
1 Damaged mast cells release histamine.
2 Histamine diffuses into the capillaries.
3 Histamine causes the capillaries to dilate and become leaky; complement proteins attract phagocytes.
4 Plasma and phagocytes move into infected tissue from the capillary.
5 Phagocytes engulf bacteria and dead cells.
6 Histamine and complement signaling cease; phagocytes are no longer attracted; the tissue returns to normal.

18.4 Interactions of Cells and Chemical Signals in Inflammation *(Page 369)*

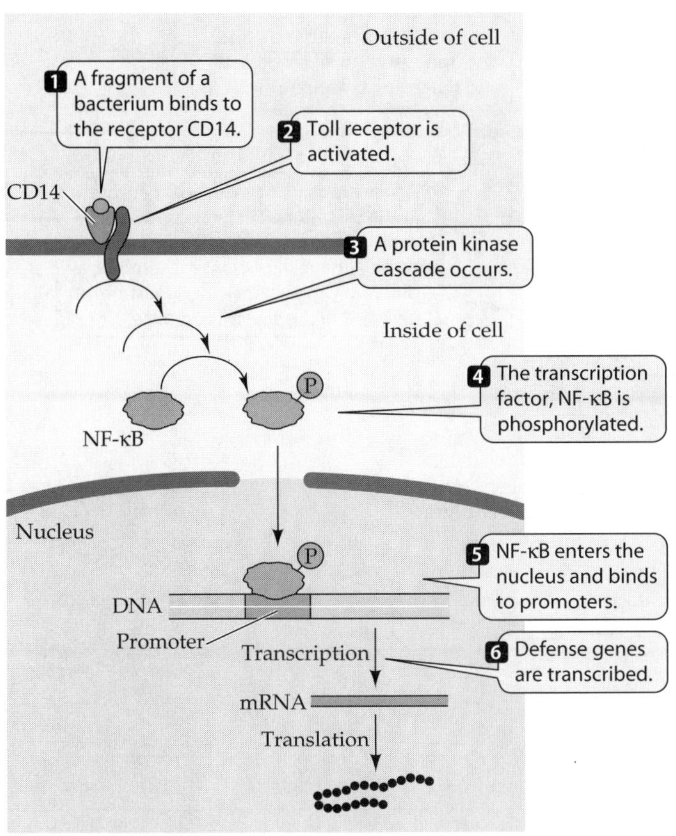
Outside of cell
1 A fragment of a bacterium binds to the receptor CD14.
2 Toll receptor is activated.
CD14
3 A protein kinase cascade occurs.
Inside of cell
4 The transcription factor, NF-κB is phosphorylated.
P
NF-κB
Nucleus
5 NF-κB enters the nucleus and binds to promoters.
DNA
Promoter
Transcription
6 Defense genes are transcribed.
mRNA
Translation

18.5 Cell Signaling and Defense *(Page 370)*

Each B cell makes a different, specific antibody and displays it on its cell surface.

1 This B cell makes an antibody that binds this specific determinant...

Antigenic determinant

Population of specific B cells

2 ...which stimulates the cell to divide...

3 ...resulting in a clone of cells.

Plasma cells

Memory cells

Antibodies

4 Some develop into plasma cells (effector cells) that secrete the same antibody as the parent cell.

5 A few develop into memory cells that divide at a low rate, perpetuating the clone.

18.7 Clonal Selection in B Cells *(Page 372)*

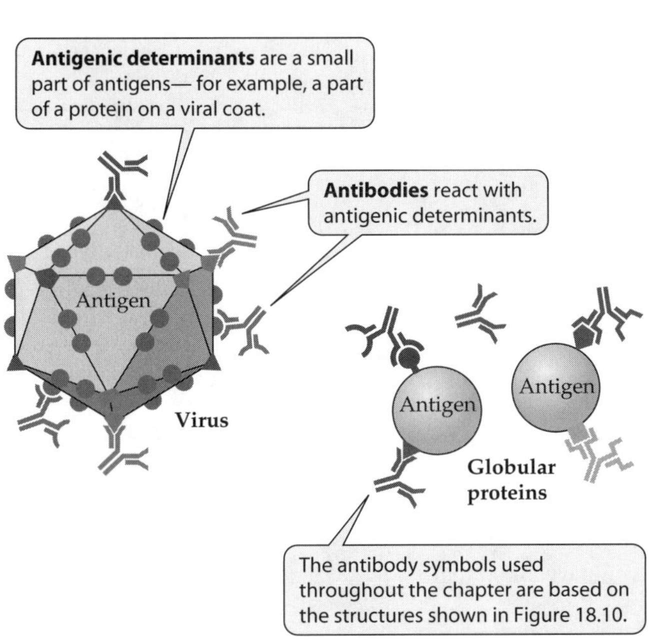
Antigenic determinants are a small part of antigens— for example, a part of a protein on a viral coat.
Antibodies react with antigenic determinants.
Antigen
Virus
Antigen
Antigen
Globular proteins
The antibody symbols used throughout the chapter are based on the structures shown in Figure 18.10.

18.6 Each Antibody Matches an Antigenic Determinant *(Page 371)*

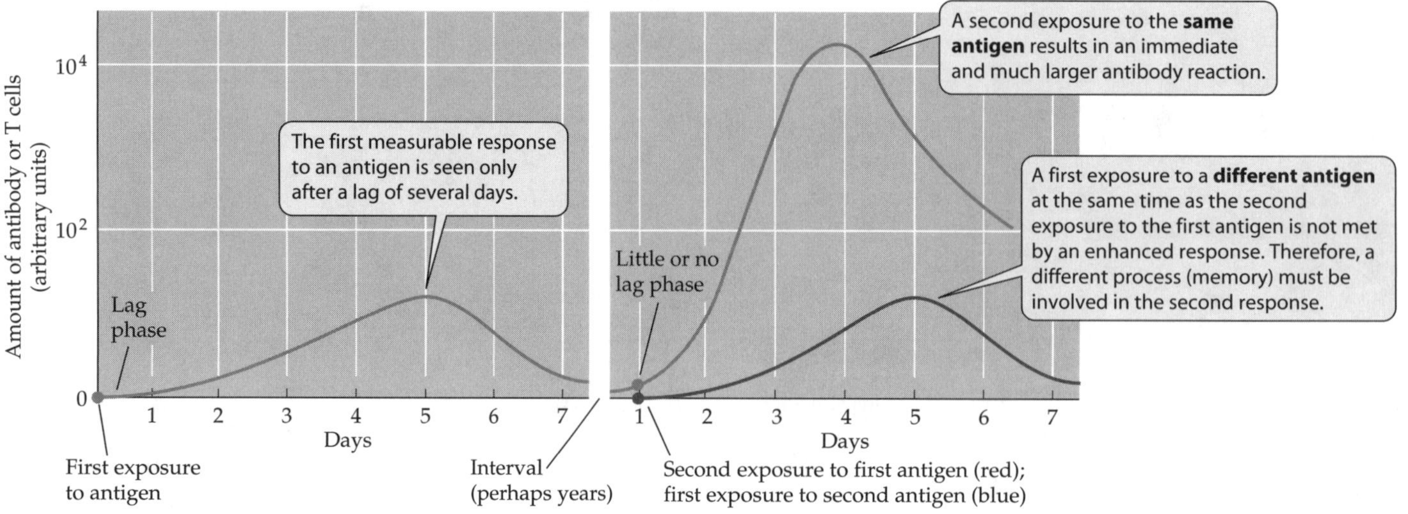
Amount of antibody or T cells (arbitrary units)
10^4
10^2
0
Lag phase
The first measurable response to an antigen is seen only after a lag of several days.
Little or no lag phase
A second exposure to the **same antigen** results in an immediate and much larger antibody reaction.
A first exposure to a **different antigen** at the same time as the second exposure to the first antigen is not met by an enhanced response. Therefore, a different process (memory) must be involved in the second response.
1 2 3 4 5 6 7
Days
First exposure to antigen
Interval (perhaps years)
Second exposure to first antigen (red); first exposure to second antigen (blue)

18.8 Immunological Memory *(Page 373)*

18.2 *Some Vaccines against Human Pathogens*

INFECTIOUS AGENT	DISEASE	VACCINATED POPULATION
Bacteria		
Bacillus anthracis	Anthrax	Exposed in biological warfare
Bordetella pertussis	Whooping cough	Children and adults
Clostridium tetani	Tetanus	Children
Corynebacterium diphtheriae	Diphtheria	Children
Haemophilus influenzae	Meningitis	Children
Mycobacterium tuberculosis	Tuberculosis	All people
Salmonella typhi	Typhoid fever	Areas exposed to agent
Streptococcus pneumoniae	Pneumonia	Elderly
Vibrio cholerae	Cholera	People in areas exposed to agent
Viruses		
Adenovirus	Respiratory disease	Military personnel
Hepatitis A	Liver disease	Areas exposed to agent
Hepatitis B	Liver disease, cancer	All people
Influenza virus	Flu	All people
Measles virus	Measles	Children and adolescents
Mumps virus	Mumps	Children and adolescents
Poliovirus	Polio	Children
Rabies virus	Rabies	Exposed to agent
Rubella virus	German measles	Children
Vaccinia virus	Smallpox	Laboratory workers, military personnel
Varicella-zoster virus	Chicken pox	Children

(Page 373)

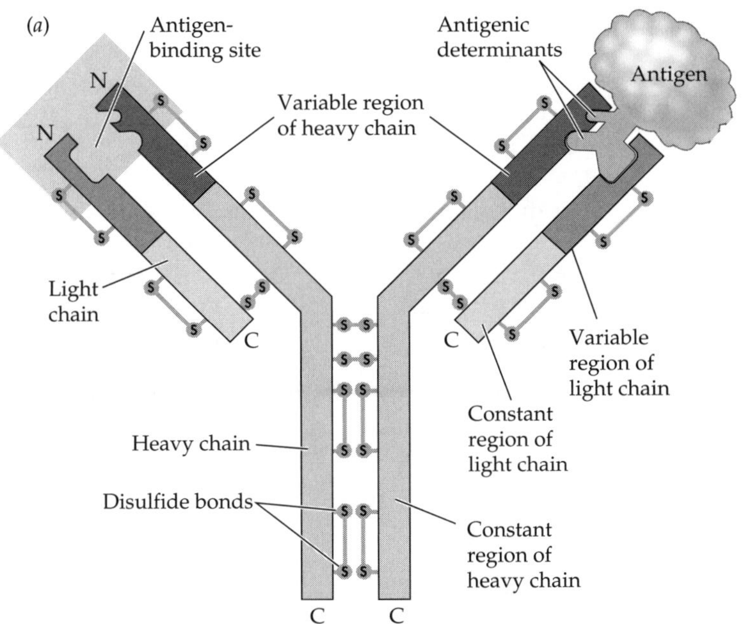
(a)
Antigen-binding site
N
N
Variable region of heavy chain
Antigenic determinants
Antigen
Light chain
C
C
Variable region of light chain
Constant region of light chain
Heavy chain
Disulfide bonds
Constant region of heavy chain
C C

18.10 **Structure of Immunoglobulins** *(Page 375)*

18.3 *Antibody Classes*

CLASS	GENERAL STRUCTURE	LOCATION	FUNCTION
IgG	Monomer	Free in plasma; about 80 percent of circulating antibodies	Most abundant antibody in primary and secondary responses; crosses placenta and provides passive immunization to fetus
IgM	Pentamer	Surface of B cell; free in plasma	Antigen receptor on B cell membrane; first class of antibodies released by B cells during primary response
IgD	Monomer	Surface of B cell	Cell surface receptor of mature B cell; important in B cell activation
IgA	Dimer	Monomer found in plasma; polymers in saliva, tears, milk, and other body secretions	Protects mucosal surfaces; prevents attachment of pathogens to epithelial cells
IgE	Monomer	Secreted by plasma cells in skin and tissues lining gastrointestinal and respiratory tracts	Found on mast cells and basophils; when bound to antigens, triggers release of histamine from mast cell or basophil that contributes to inflammation and some allergic responses

(Page 376)

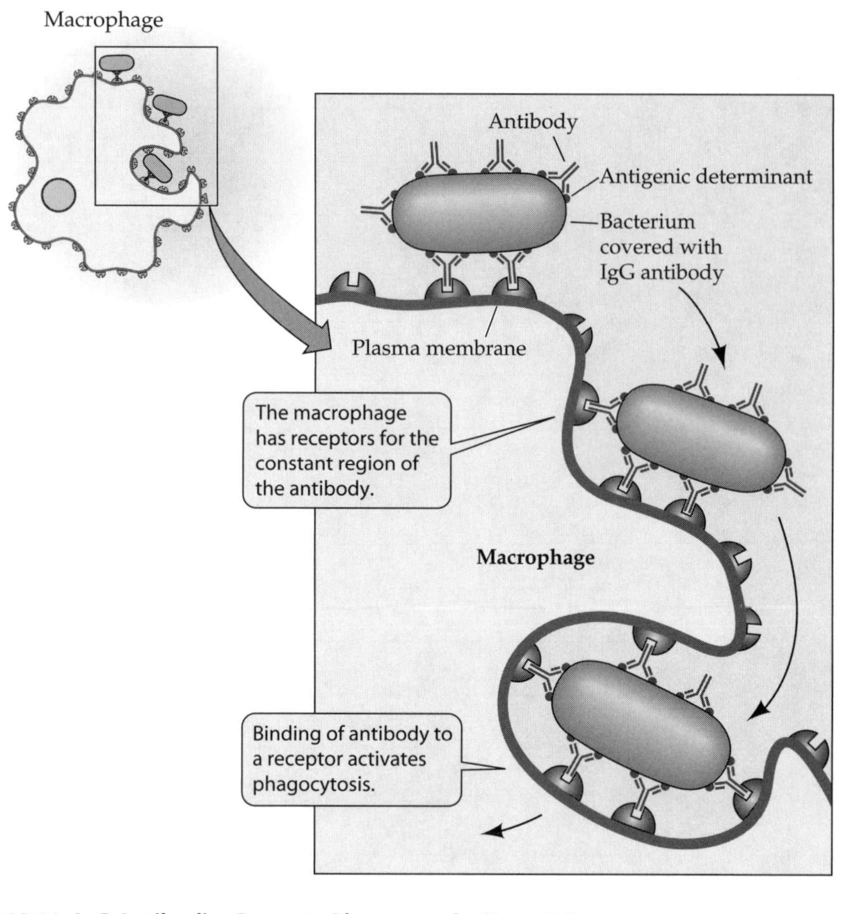
Macrophage
Antibody
Antigenic determinant
Bacterium covered with IgG antibody
Plasma membrane
The macrophage has receptors for the constant region of the antibody.
Macrophage
Binding of antibody to a receptor activates phagocytosis.

18.11 IgG Antibodies Promote Phagocytosis *(Page 376)*

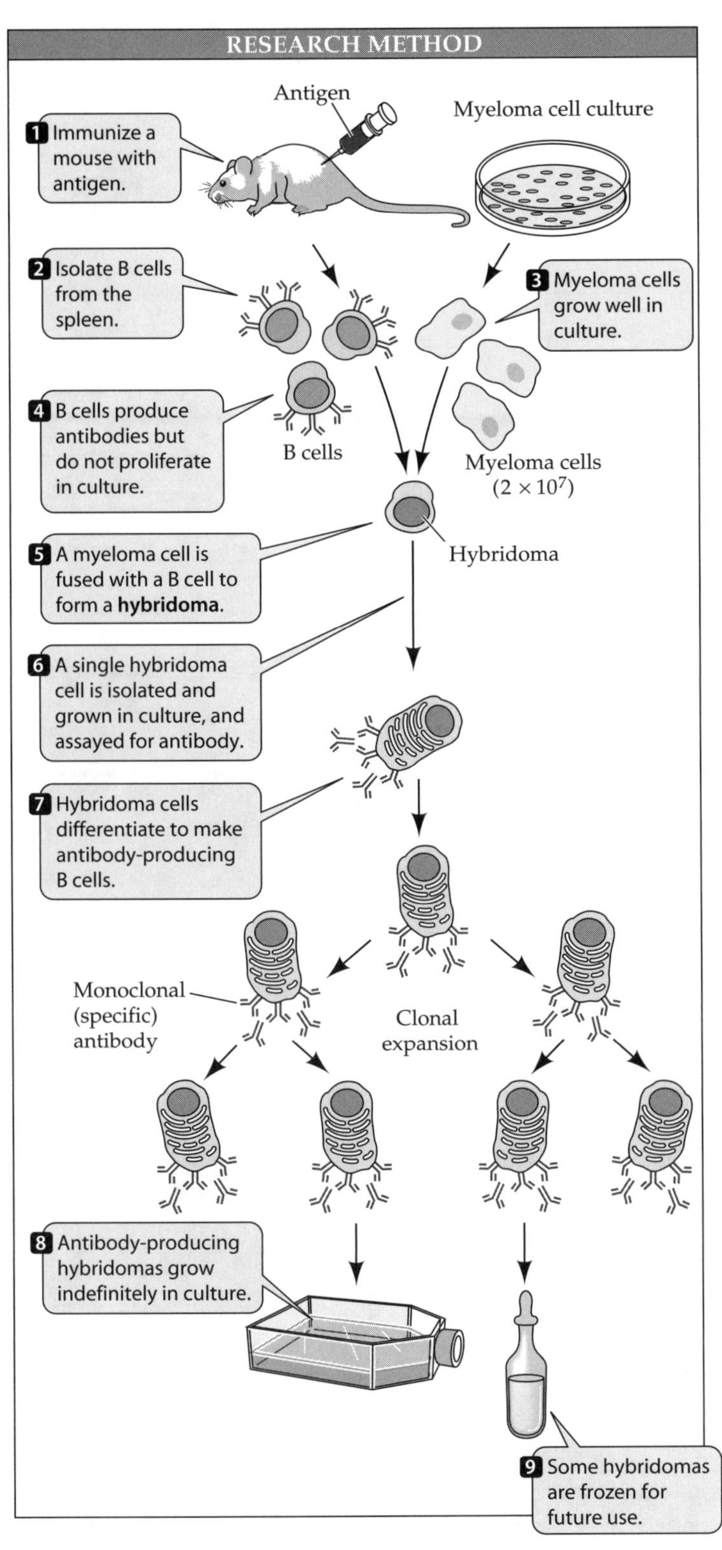
RESEARCH METHOD
Antigen
Myeloma cell culture
1 Immunize a mouse with antigen.
2 Isolate B cells from the spleen.
3 Myeloma cells grow well in culture.
4 B cells produce antibodies but do not proliferate in culture.
B cells
Myeloma cells (2×10^7)
Hybridoma
5 A myeloma cell is fused with a B cell to form a **hybridoma**.
6 A single hybridoma cell is isolated and grown in culture, and assayed for antibody.
7 Hybridoma cells differentiate to make antibody-producing B cells.
Monoclonal (specific) antibody
Clonal expansion
8 Antibody-producing hybridomas grow indefinitely in culture.
9 Some hybridomas are frozen for future use.

18.12 Creating Hybridomas for the Production of Monoclonal Antibodies *(Page 377)*

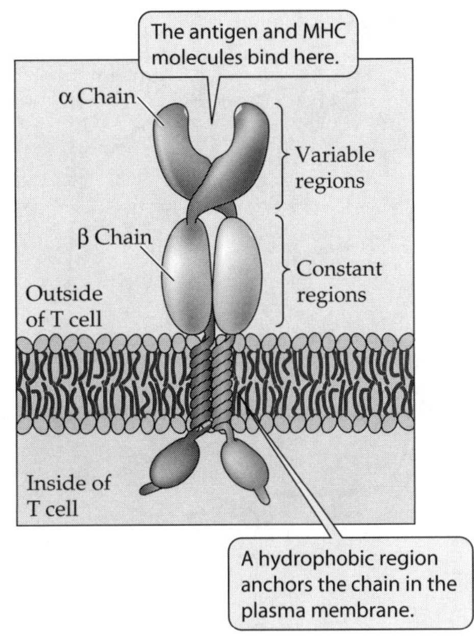
The antigen and MHC molecules bind here.
α Chain
Variable regions
β Chain
Constant regions
Outside of T cell
Inside of T cell
A hydrophobic region anchors the chain in the plasma membrane.

18.13 A T Cell Receptor *(Page 378)*

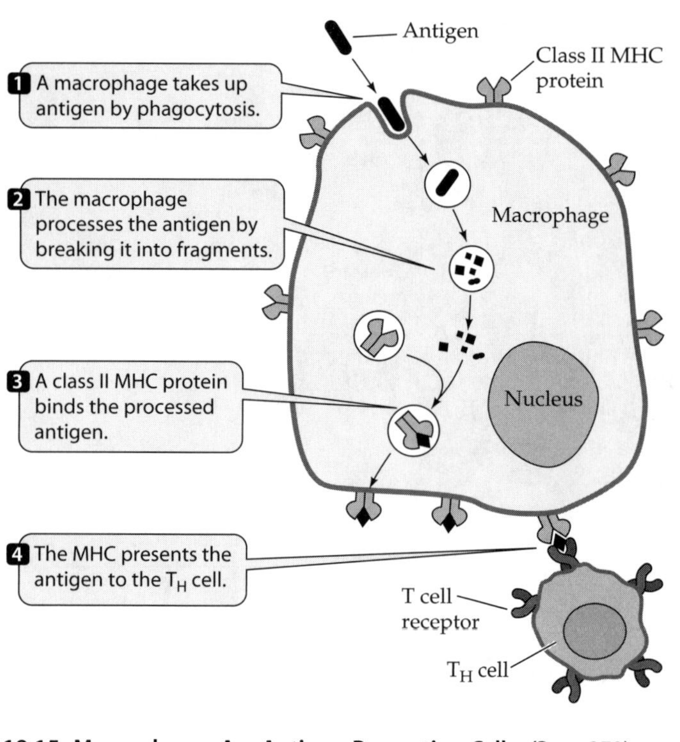
Antigen
Class II MHC protein
1 A macrophage takes up antigen by phagocytosis.
2 The macrophage processes the antigen by breaking it into fragments.
Macrophage
3 A class II MHC protein binds the processed antigen.
Nucleus
4 The MHC presents the antigen to the T_H cell.
T cell receptor
T_H cell

18.15 Macrophages Are Antigen-Presenting Cells *(Page 379)*

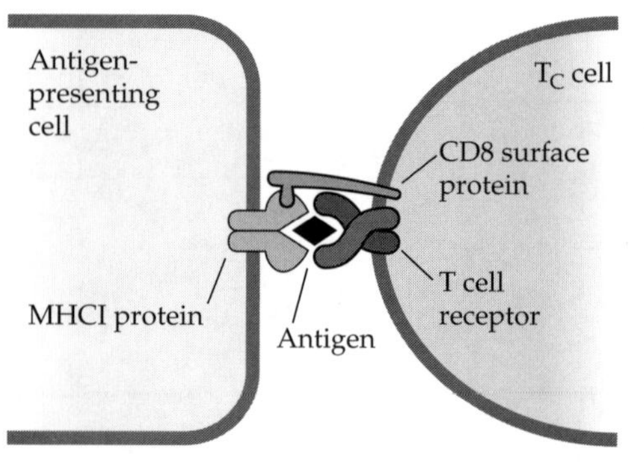
Antigen-presenting cell
T_C cell
CD8 surface protein
T cell receptor
MHCI protein
Antigen

Antigen-Presenting and T Cell Types

PRESENTING CELL TYPE	ANTIGEN PRESENTED	MHC CLASS	T CELL TYPE	T CELL SURFACE PROTEIN
Any cell	Intracellular protein fragment	Class I	Cytotoxic T cell (T_C)	CD8
Macrophages and B cells	Fragments from extracellular proteins	Class II	Helper T (T_H)	CD4

18.16 The Interaction between T Cells and Antigen-Presenting Cells *(Page 379)*

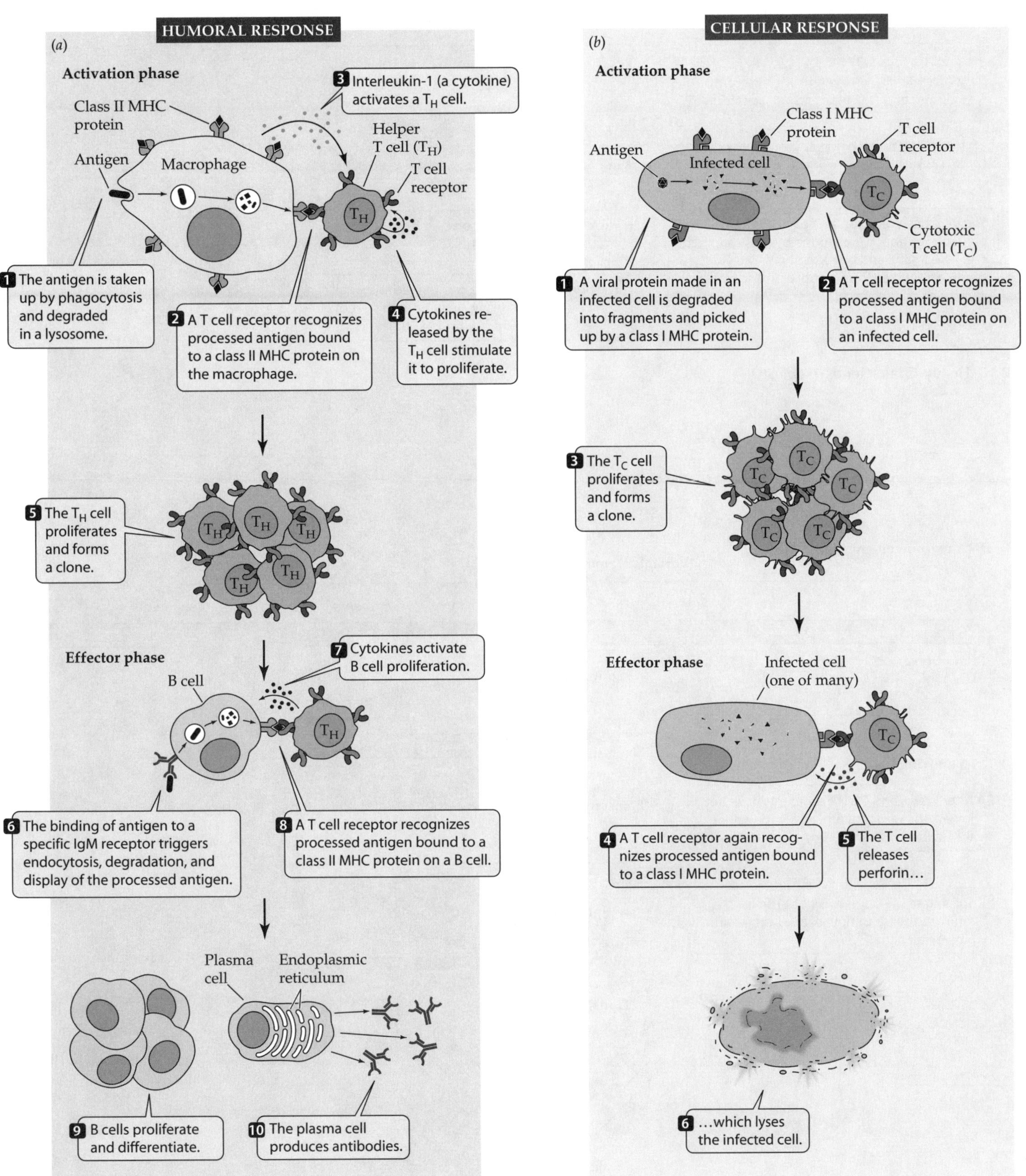
HUMORAL RESPONSE
(a)
Activation phase
Class II MHC protein
Antigen
Macrophage
3 Interleukin-1 (a cytokine) activates a T_H cell.
Helper T cell (T_H)
T cell receptor
T_H
1 The antigen is taken up by phagocytosis and degraded in a lysosome.
2 A T cell receptor recognizes processed antigen bound to a class II MHC protein on the macrophage.
4 Cytokines released by the T_H cell stimulate it to proliferate.
5 The T_H cell proliferates and forms a clone.
Effector phase
B cell
7 Cytokines activate B cell proliferation.
6 The binding of antigen to a specific IgM receptor triggers endocytosis, degradation, and display of the processed antigen.
8 A T cell receptor recognizes processed antigen bound to a class II MHC protein on a B cell.
Plasma cell
Endoplasmic reticulum
9 B cells proliferate and differentiate.
10 The plasma cell produces antibodies.
CELLULAR RESPONSE
(b)
Activation phase
Class I MHC protein
Antigen
Infected cell
T cell receptor
T_C
Cytotoxic T cell (T_C)
1 A viral protein made in an infected cell is degraded into fragments and picked up by a class I MHC protein.
2 A T cell receptor recognizes processed antigen bound to a class I MHC protein on an infected cell.
3 The T_C cell proliferates and forms a clone.
Effector phase
Infected cell (one of many)
4 A T cell receptor again recognizes processed antigen bound to a class I MHC protein.
5 The T cell releases perforin...
6 ...which lyses the infected cell.

18.17 Phases of the Humoral and Cellular Immune Responses *(Page 380)*

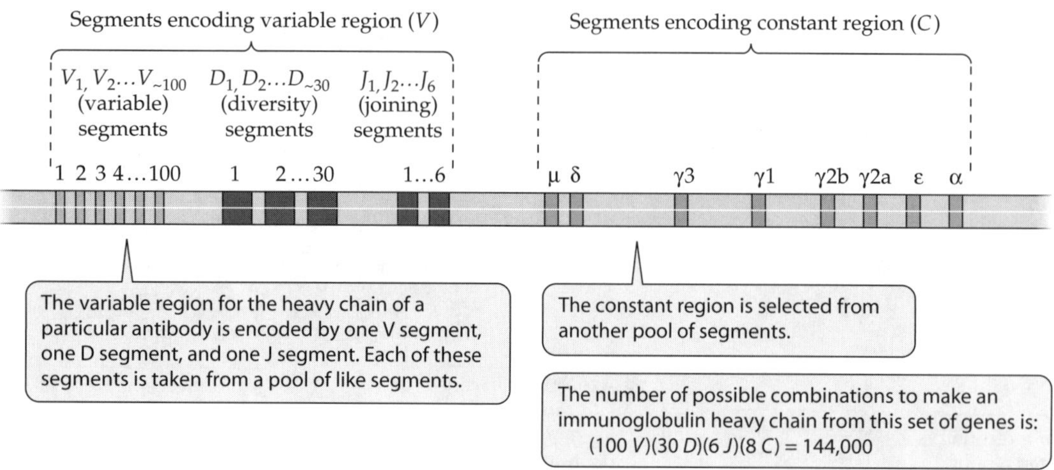
Segments encoding variable region (V)
Segments encoding constant region (C)
$V_1, V_2 \ldots V_{\sim 100}$ (variable) segments
$D_1, D_2 \ldots D_{\sim 30}$ (diversity) segments
$J_1, J_2 \ldots J_6$ (joining) segments
1 2 3 4…100
1 2…30
1…6
μ δ
γ3
γ1
γ2b γ2a
ε
α
The variable region for the heavy chain of a particular antibody is encoded by one V segment, one D segment, and one J segment. Each of these segments is taken from a pool of like segments.
The constant region is selected from another pool of segments.
The number of possible combinations to make an immunoglobulin heavy chain from this set of genes is:
(100 V)(30 D)(6 J)(8 C) = 144,000

18.18 Heavy-Chain Genes *(Page 382)*

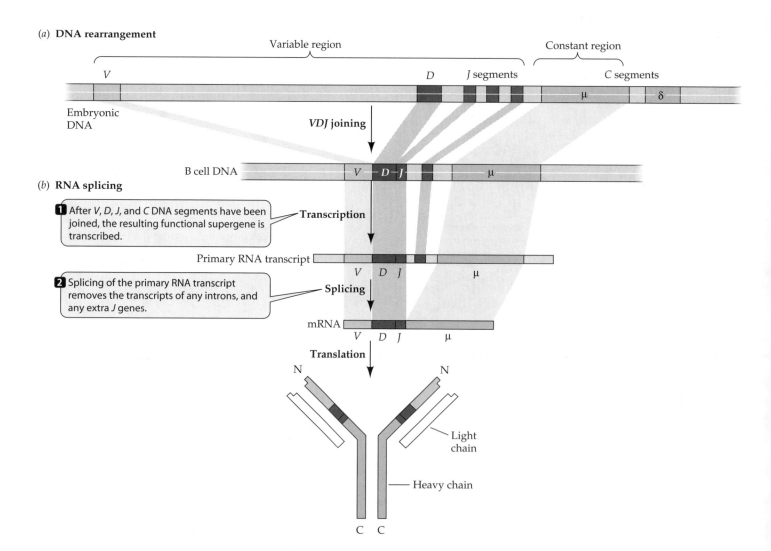

(*a*) **DNA rearrangement**

Variable region

Constant region

V

D

J segments

C segments

μ

δ

Embryonic DNA

***VDJ* joining**

B cell DNA

V *D* *J*

μ

(*b*) **RNA splicing**

1 After *V*, *D*, *J*, and *C* DNA segments have been joined, the resulting functional supergene is transcribed.

Transcription

Primary RNA transcript

V *D* *J* μ

2 Splicing of the primary RNA transcript removes the transcripts of any introns, and any extra *J* genes.

Splicing

mRNA

V *D* *J* μ

Translation

N N

Light chain

Heavy chain

C C

18.19 Heavy-Chain Gene Rearrangement and Splicing *(Page 383)*

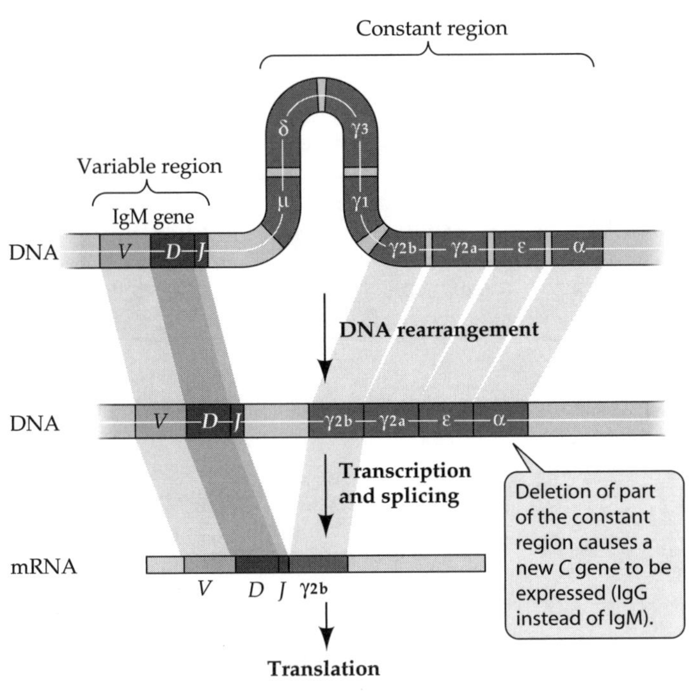
Constant region
Variable region
IgM gene
DNA
V
D
J
δ
γ3
μ
γ1
γ2b
γ2a
ε
α
DNA rearrangement
DNA
Transcription and splicing
mRNA
Deletion of part of the constant region causes a new *C* gene to be expressed (IgG instead of IgM).
Translation

18.20 Class Switching *(Page 384)*

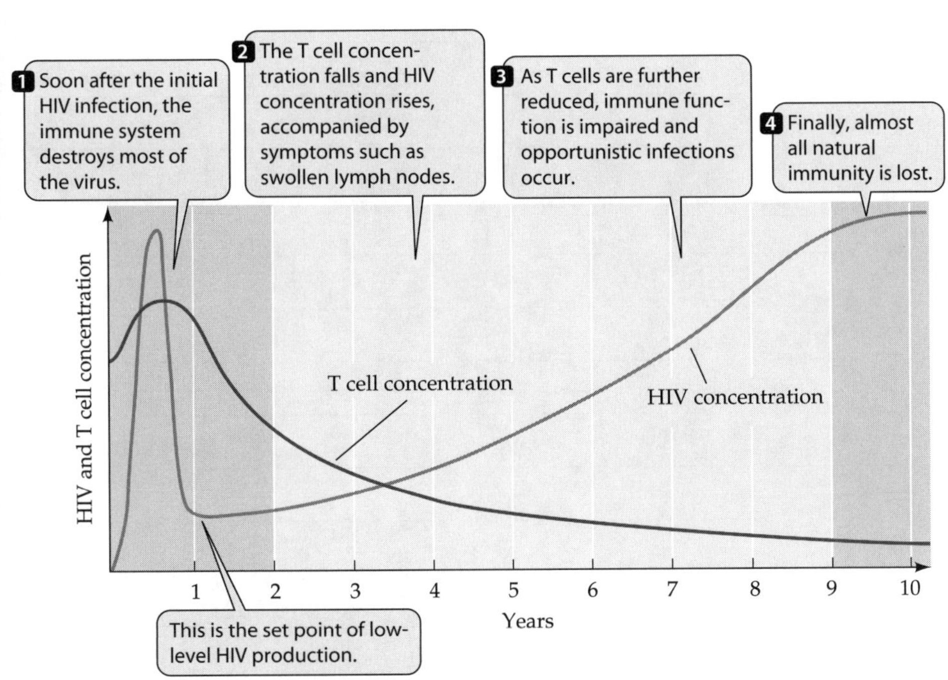
1 Soon after the initial HIV infection, the immune system destroys most of the virus.
2 The T cell concentration falls and HIV concentration rises, accompanied by symptoms such as swollen lymph nodes.
3 As T cells are further reduced, immune function is impaired and opportunistic infections occur.
4 Finally, almost all natural immunity is lost.
HIV and T cell concentration
T cell concentration
HIV concentration
1
2
3
4
5
6
7
8
9
10
Years
This is the set point of low-level HIV production.

18.21 The Course of an HIV Infection *(Page 385)*

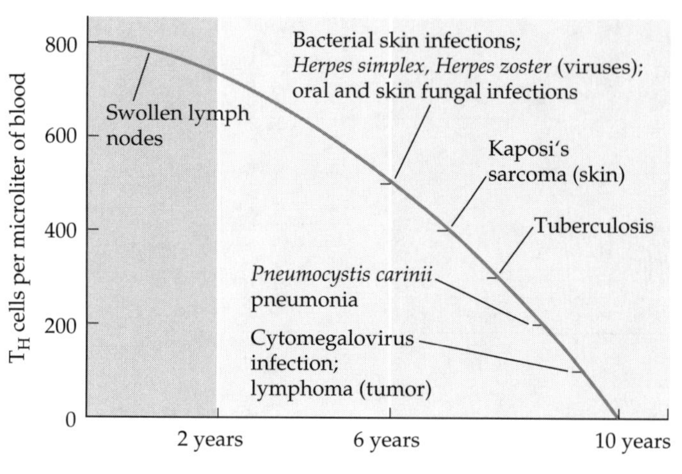

800
600
400
200
0
T_H cells per microliter of blood
2 years
6 years
10 years
Swollen lymph nodes
Bacterial skin infections; *Herpes simplex, Herpes zoster* (viruses); oral and skin fungal infections
Kaposi's sarcoma (skin)
Tuberculosis
Pneumocystis carinii pneumonia
Cytomegalovirus infection; lymphoma (tumor)

18.22 Relationship Between T_H Cell Count and Opportunistic Infections *(Page 386)*

19 Differential Gene Expression in Development

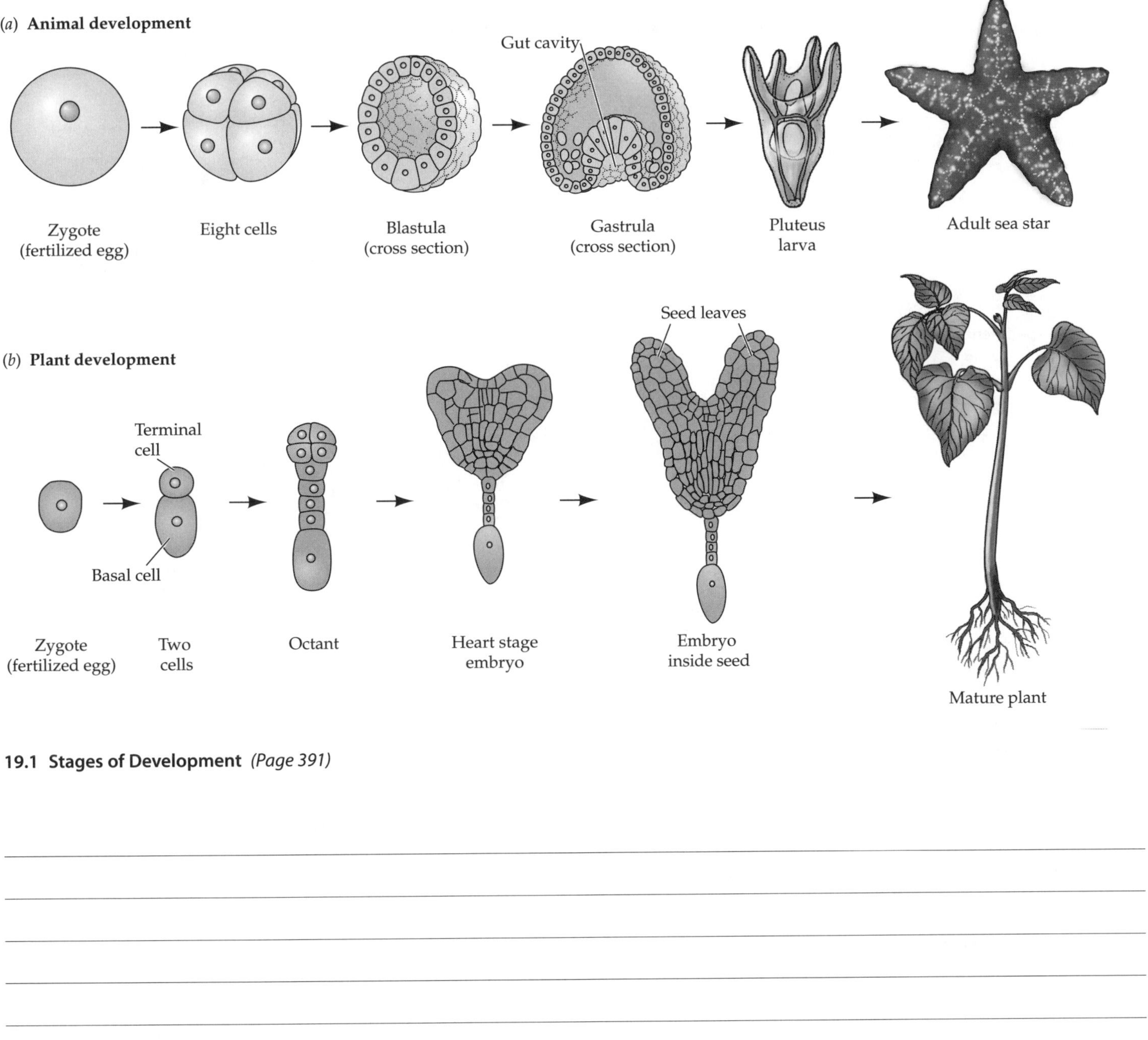

(a) Animal development
Gut cavity
Zygote (fertilized egg)
Eight cells
Blastula (cross section)
Gastrula (cross section)
Pluteus larva
Adult sea star
(b) Plant development
Terminal cell
Basal cell
Seed leaves
Zygote (fertilized egg)
Two cells
Octant
Heart stage embryo
Embryo inside seed
Mature plant

19.1 Stages of Development *(Page 391)*

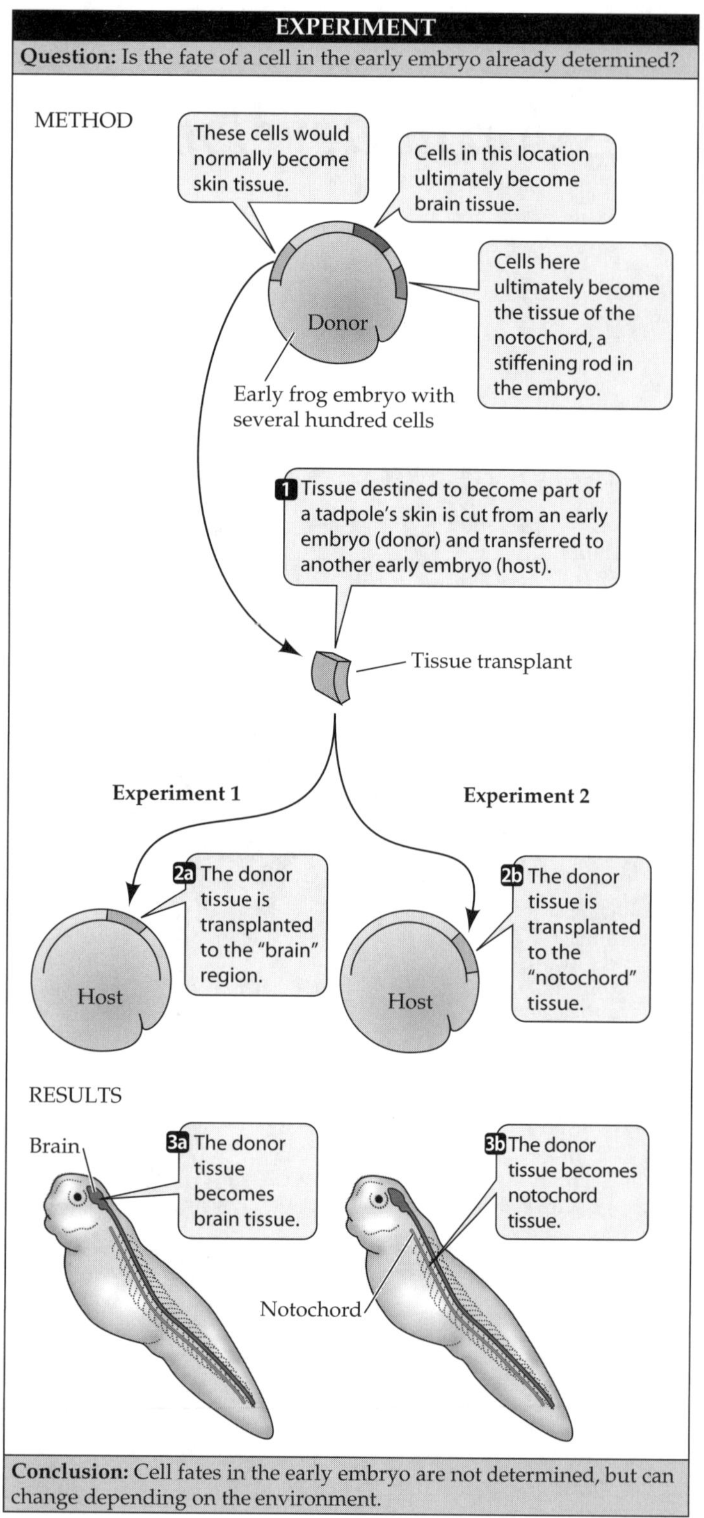
EXPERIMENT
Question: Is the fate of a cell in the early embryo already determined?
METHOD
These cells would normally become skin tissue.
Cells in this location ultimately become brain tissue.
Cells here ultimately become the tissue of the notochord, a stiffening rod in the embryo.
Donor
Early frog embryo with several hundred cells
1 Tissue destined to become part of a tadpole's skin is cut from an early embryo (donor) and transferred to another early embryo (host).
Tissue transplant
Experiment 1
Experiment 2
2a The donor tissue is transplanted to the "brain" region.
Host
2b The donor tissue is transplanted to the "notochord" tissue.
Host
RESULTS
Brain
3a The donor tissue becomes brain tissue.
3b The donor tissue becomes notochord tissue.
Notochord
Conclusion: Cell fates in the early embryo are not determined, but can change depending on the environment.

19.2 Developmental Potential in Early Frog Embryos *(Page 392)*

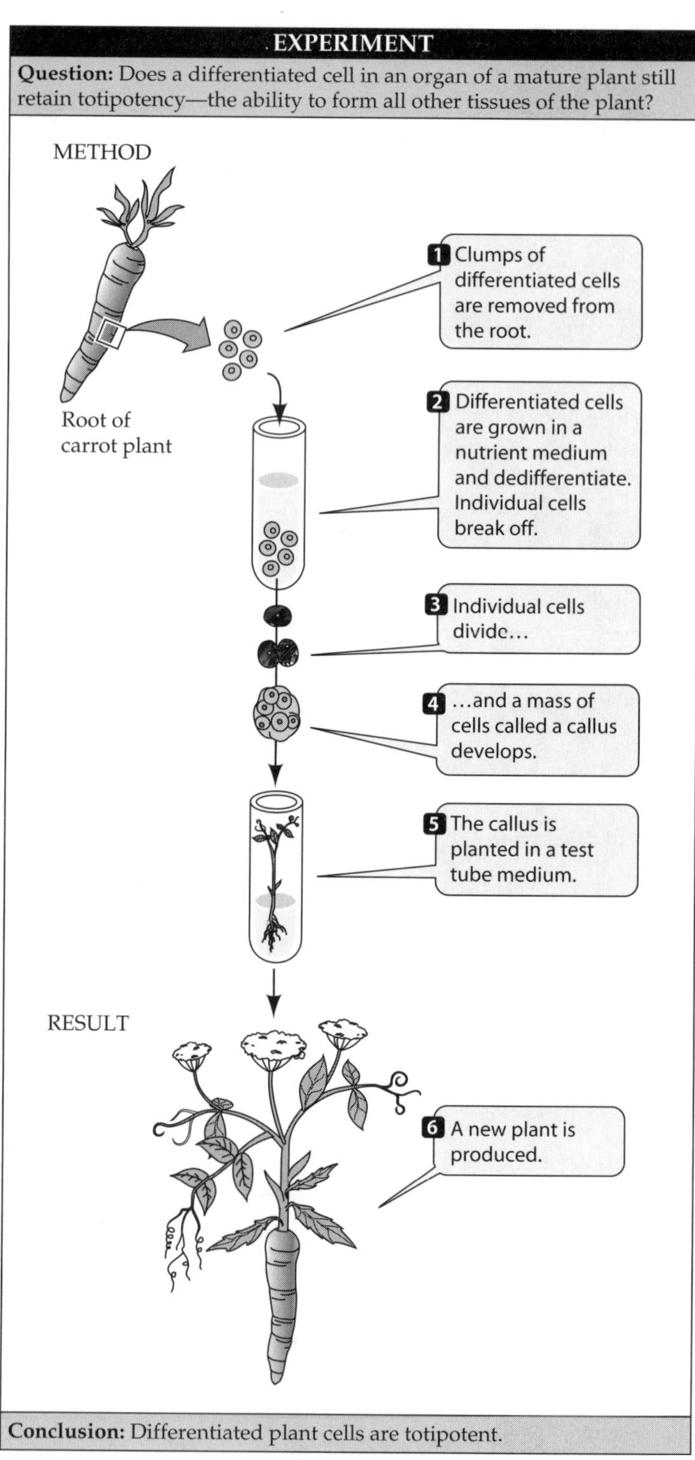
EXPERIMENT
Question: Does a differentiated cell in an organ of a mature plant still retain totipotency—the ability to form all other tissues of the plant?
METHOD
Root of carrot plant
1 Clumps of differentiated cells are removed from the root.
2 Differentiated cells are grown in a nutrient medium and dedifferentiate. Individual cells break off.
3 Individual cells divide…
4 …and a mass of cells called a callus develops.
5 The callus is planted in a test tube medium.
RESULT
6 A new plant is produced.
Conclusion: Differentiated plant cells are totipotent.

19.3 Cloning a Plant *(Page 394)*

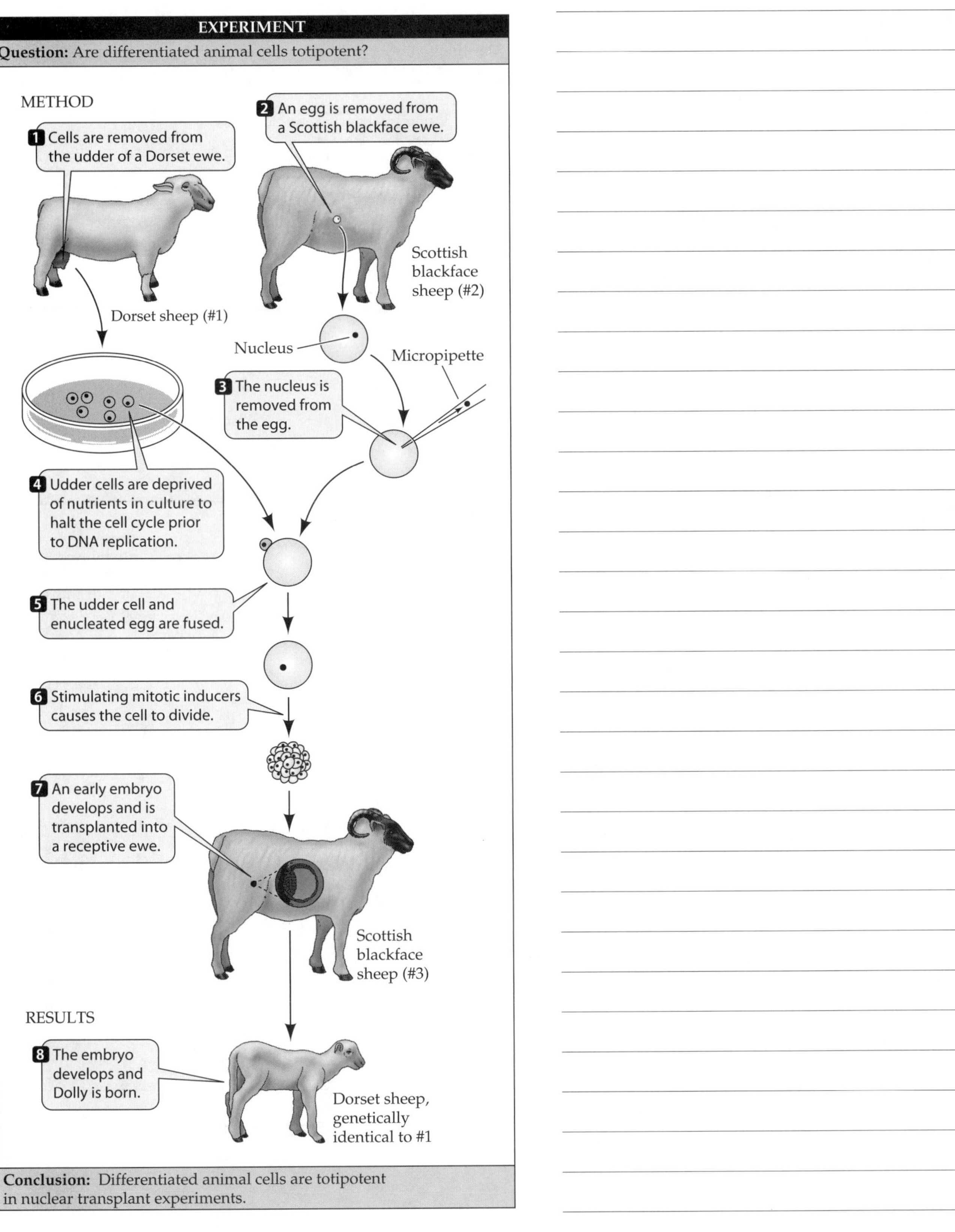
EXPERIMENT
Question: Are differentiated animal cells totipotent?
METHOD
1 Cells are removed from the udder of a Dorset ewe.
2 An egg is removed from a Scottish blackface ewe.
Scottish blackface sheep (#2)
Dorset sheep (#1)
Nucleus
Micropipette
3 The nucleus is removed from the egg.
4 Udder cells are deprived of nutrients in culture to halt the cell cycle prior to DNA replication.
5 The udder cell and enucleated egg are fused.
6 Stimulating mitotic inducers causes the cell to divide.
7 An early embryo develops and is transplanted into a receptive ewe.
Scottish blackface sheep (#3)
RESULTS
8 The embryo develops and Dolly is born.
Dorset sheep, genetically identical to #1
Conclusion: Differentiated animal cells are totipotent in nuclear transplant experiments.

19.4 A Clone and Her Offspring *(Page 395)*

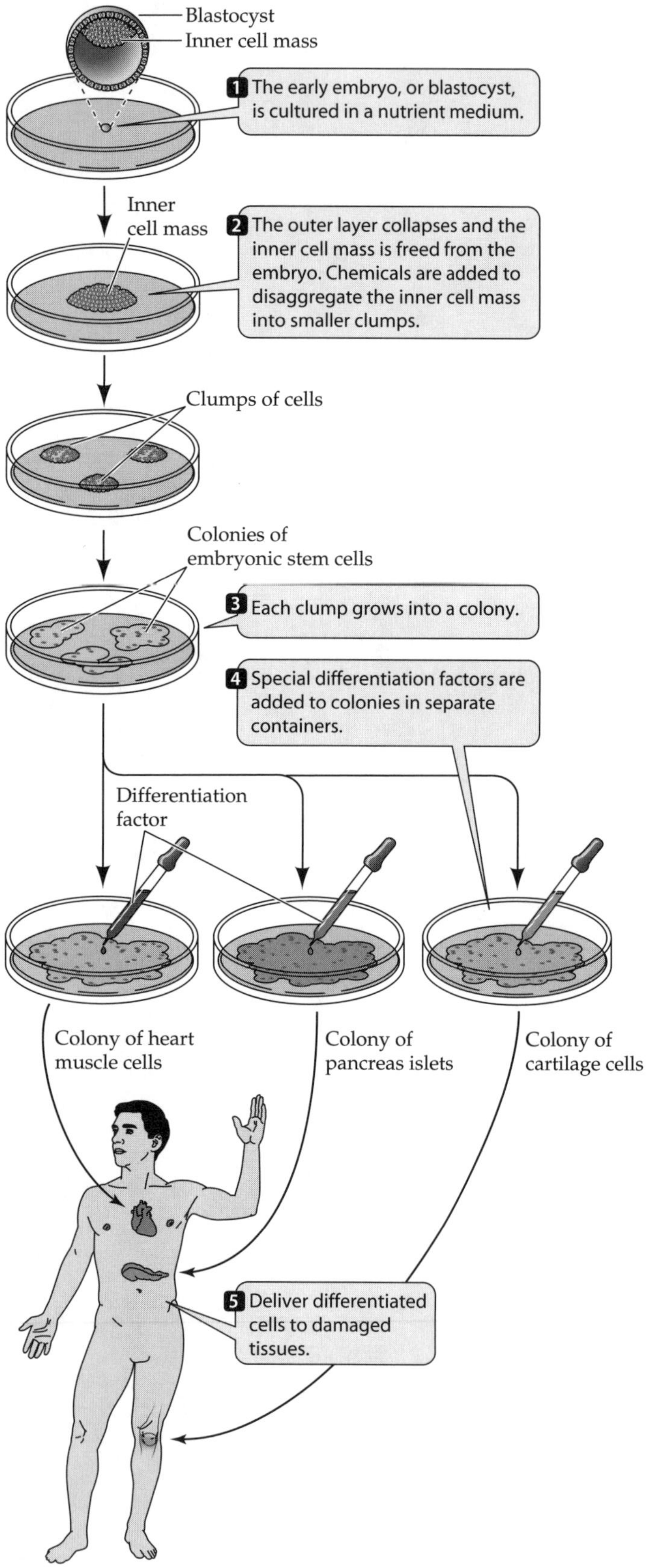

Blastocyst
Inner cell mass
1 The early embryo, or blastocyst, is cultured in a nutrient medium.
Inner cell mass
2 The outer layer collapses and the inner cell mass is freed from the embryo. Chemicals are added to disaggregate the inner cell mass into smaller clumps.
Clumps of cells
Colonies of embryonic stem cells
3 Each clump grows into a colony.
4 Special differentiation factors are added to colonies in separate containers.
Differentiation factor
Colony of heart muscle cells
Colony of pancreas islets
Colony of cartilage cells
5 Deliver differentiated cells to damaged tissues.

19.6 The Potential Use of Embryonic Stem Cells in Medicine *(Page 397)*

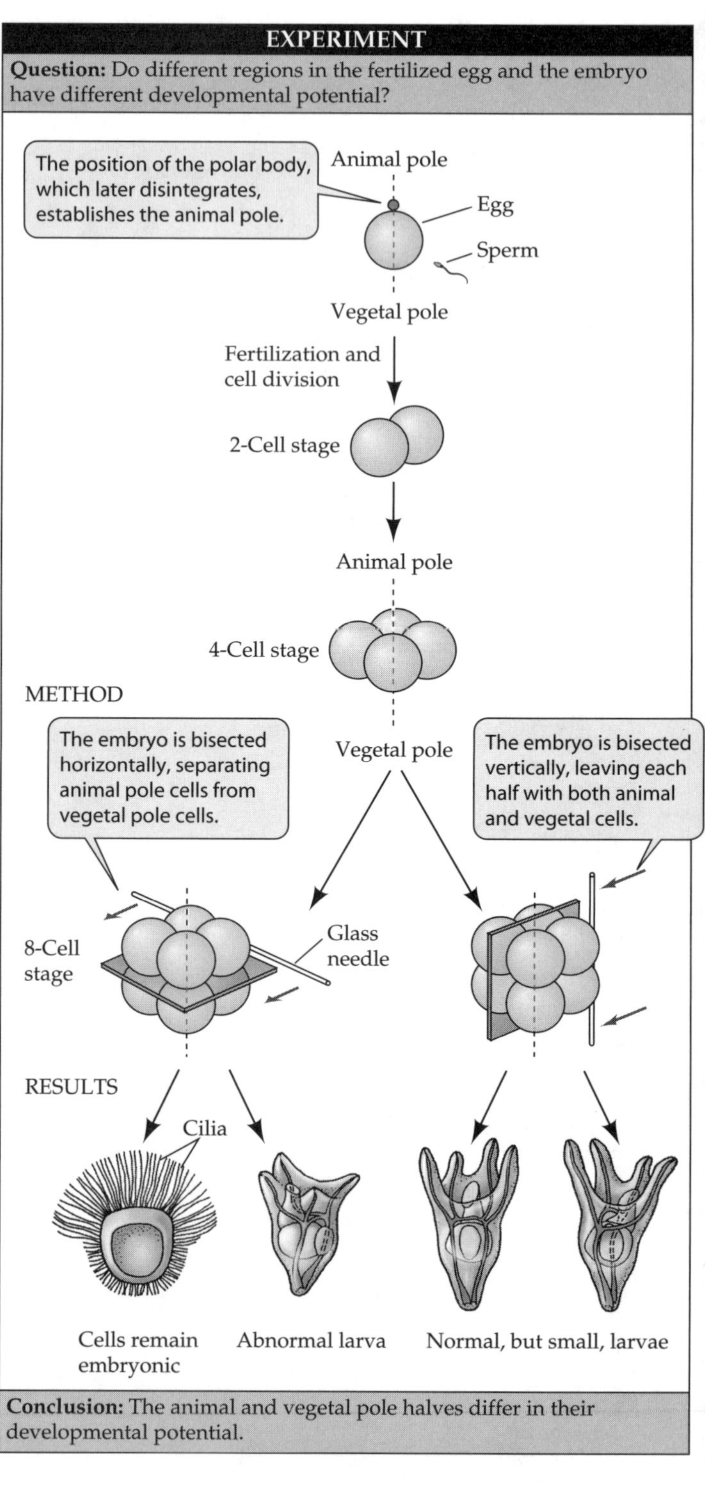
EXPERIMENT
Question: Do different regions in the fertilized egg and the embryo have different developmental potential?
The position of the polar body, which later disintegrates, establishes the animal pole.
Animal pole
Egg
Sperm
Vegetal pole
Fertilization and cell division
2-Cell stage
Animal pole
4-Cell stage
METHOD
Vegetal pole
The embryo is bisected horizontally, separating animal pole cells from vegetal pole cells.
The embryo is bisected vertically, leaving each half with both animal and vegetal cells.
8-Cell stage
Glass needle
RESULTS
Cilia
Cells remain embryonic
Abnormal larva
Normal, but small, larvae
Conclusion: The animal and vegetal pole halves differ in their developmental potential.

19.7 Asymmetry in the Early Embryo *(Page 398)*

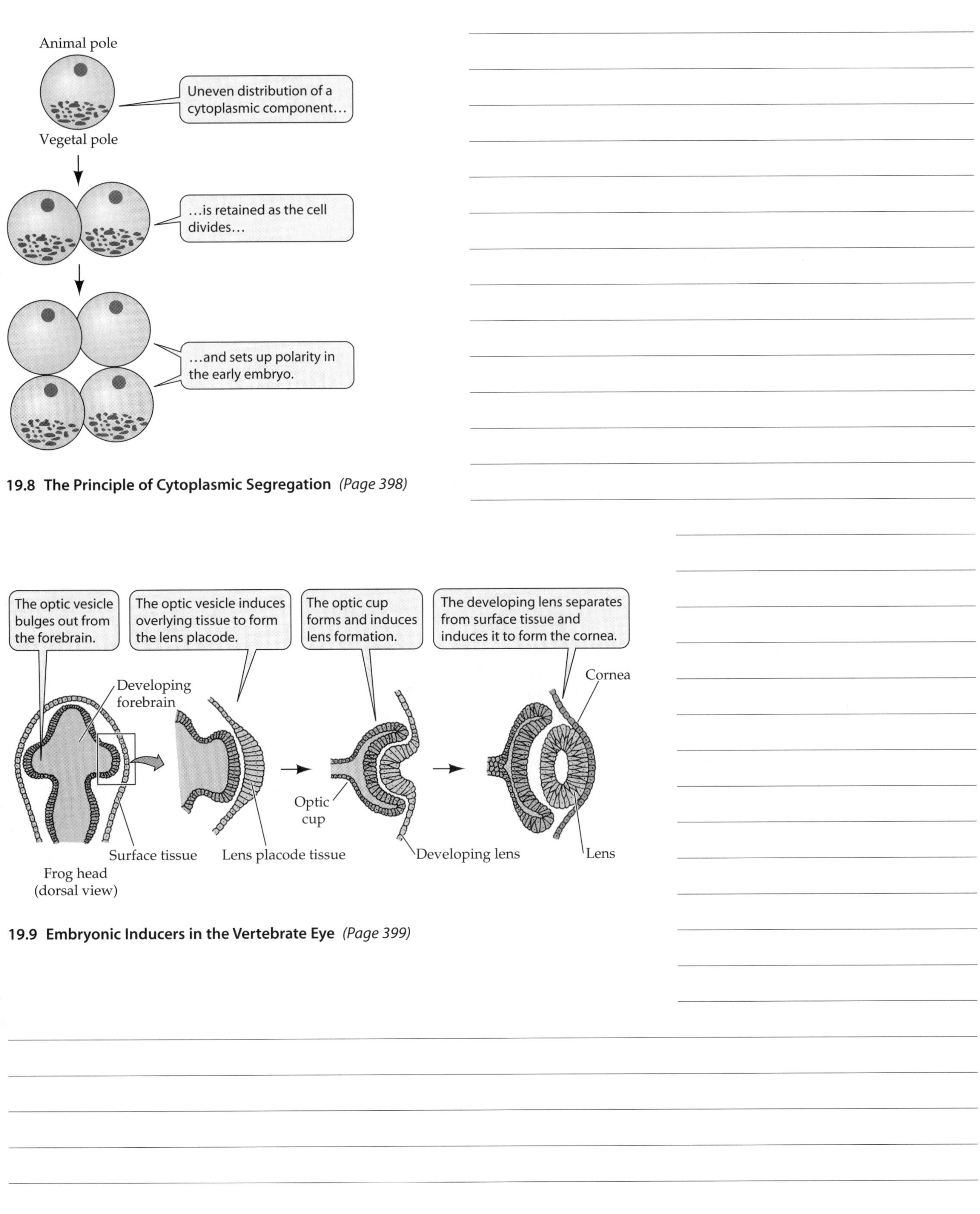
Animal pole
Vegetal pole
Uneven distribution of a cytoplasmic component…
…is retained as the cell divides…
…and sets up polarity in the early embryo.

19.8 **The Principle of Cytoplasmic Segregation** *(Page 398)*

The optic vesicle bulges out from the forebrain.
The optic vesicle induces overlying tissue to form the lens placode.
The optic cup forms and induces lens formation.
The developing lens separates from surface tissue and induces it to form the cornea.
Developing forebrain
Cornea
Optic cup
Surface tissue
Lens placode tissue
Developing lens
Lens
Frog head (dorsal view)

19.9 **Embryonic Inducers in the Vertebrate Eye** *(Page 399)*

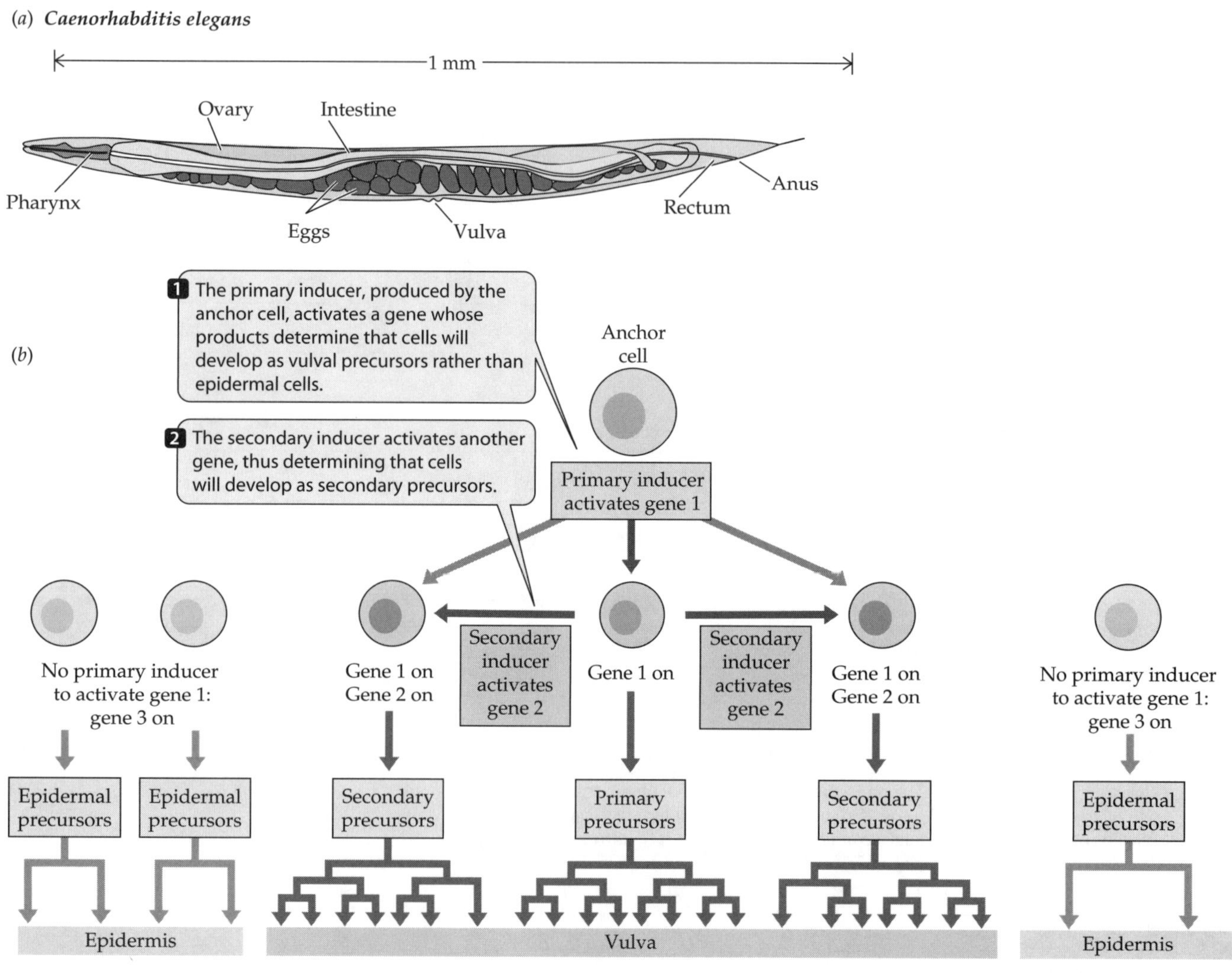
(a) Caenorhabditis elegans
1 mm
Ovary
Intestine
Pharynx
Eggs
Vulva
Rectum
Anus
(b)
1 The primary inducer, produced by the anchor cell, activates a gene whose products determine that cells will develop as vulval precursors rather than epidermal cells.
2 The secondary inducer activates another gene, thus determining that cells will develop as secondary precursors.
Anchor cell
Primary inducer activates gene 1
No primary inducer to activate gene 1: gene 3 on
Gene 1 on Gene 2 on
Secondary inducer activates gene 2
Gene 1 on
Secondary inducer activates gene 2
Gene 1 on Gene 2 on
No primary inducer to activate gene 1: gene 3 on
Epidermal precursors
Epidermal precursors
Secondary precursors
Primary precursors
Secondary precursors
Epidermal precursors
Epidermis
Vulva
Epidermis

19.10 Induction during Vulval Development in *C. elegans* *(Page 400)*

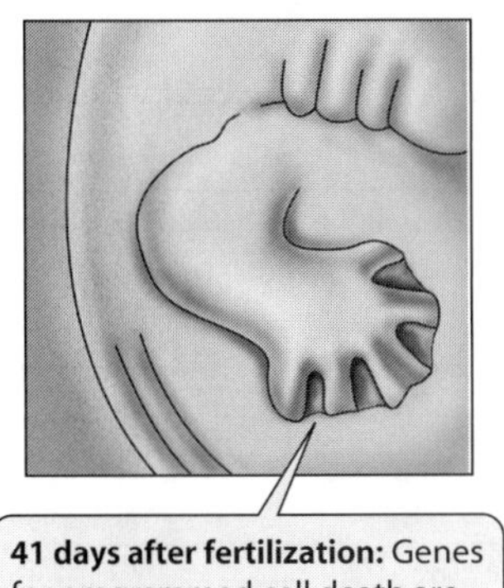
41 days after fertilization: Genes for programmed cell death are expressed in the tissue between the digits.

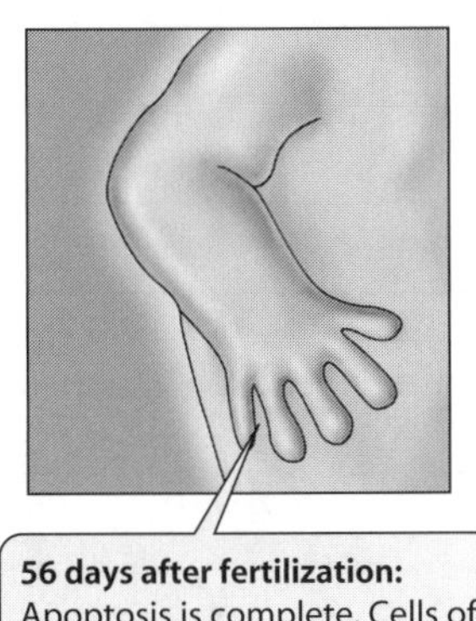
56 days after fertilization: Apoptosis is complete. Cells of the digits have absorbed the remains of the dead cells.

19.11 Apoptosis Removes the Tissue between Fingers *(Page 401)*

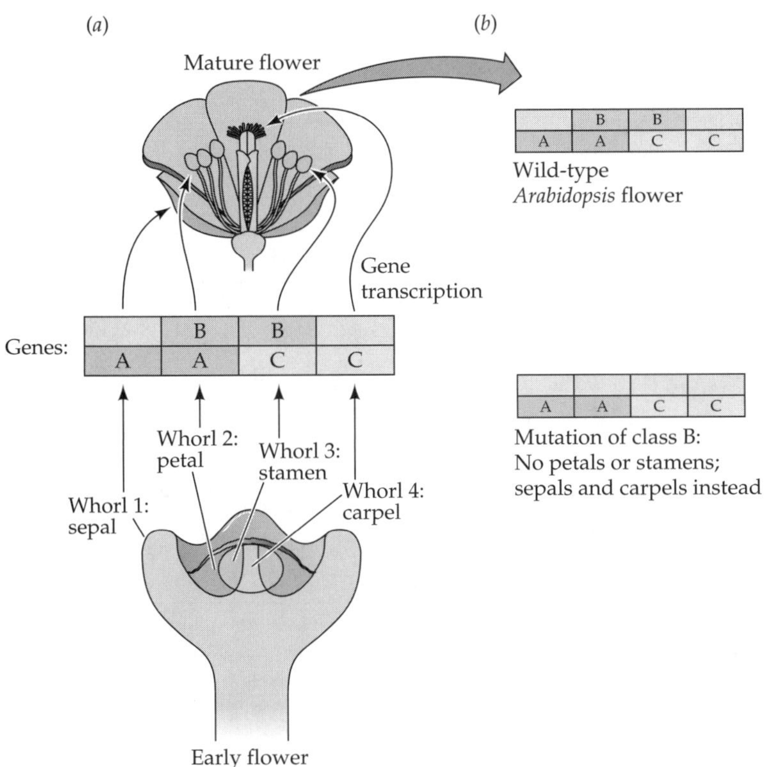
(a)
(b)
Mature flower
Gene transcription
Genes:
B B
A A C C
Whorl 1: sepal
Whorl 2: petal
Whorl 3: stamen
Whorl 4: carpel
Early flower differentiation (meristems)
B B
A A C C
Wild-type *Arabidopsis* flower
A A C C
Mutation of class B:
No petals or stamens;
sepals and carpels instead

B B
C C C C
Mutation of class A:
No petals or sepals;
stamens and carpels instead

B B
A A A A
Mutation of class C:
No stamens or carpels;
petals and sepals instead

19.12 Organ Identity Genes in *Arabidopsis* Flowers *(Page 402)*

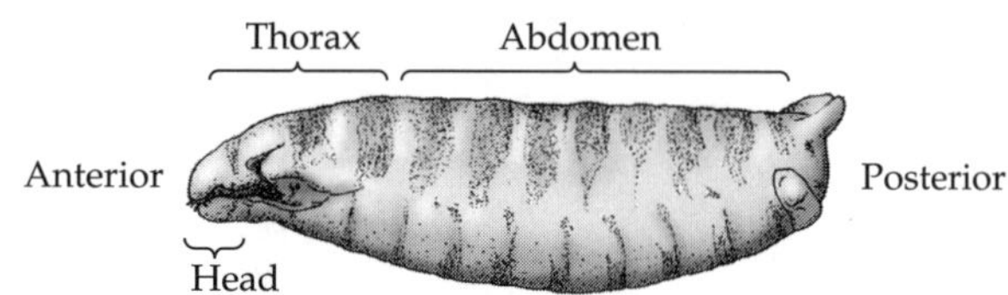
Thorax
Abdomen
Anterior
Posterior
Head

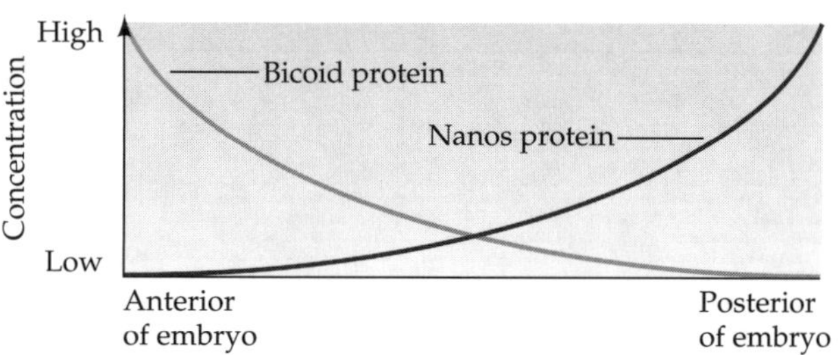
High
Concentration
Low
Bicoid protein
Nanos protein
Anterior of embryo
Posterior of embryo

19.14 Bicoid and Nanos Protein Gradients Provide Positional Information *(Page 404)*

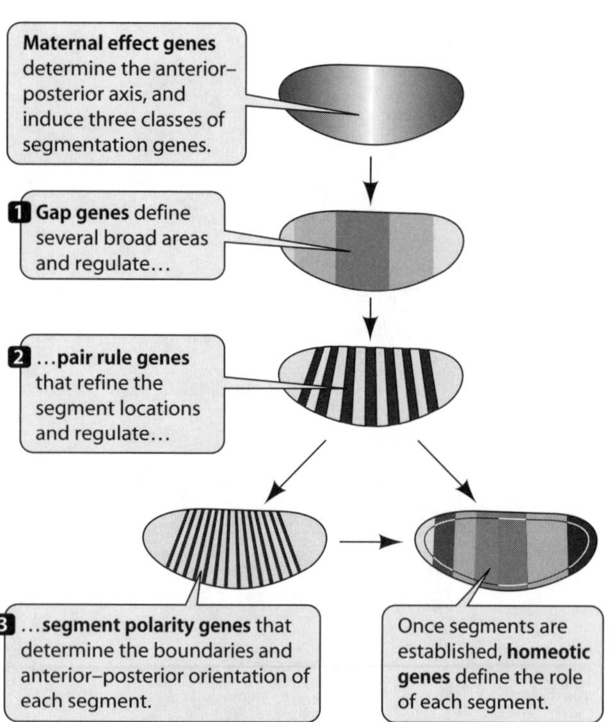
Maternal effect genes determine the anterior–posterior axis, and induce three classes of segmentation genes.
1 **Gap genes** define several broad areas and regulate…
2 …**pair rule genes** that refine the segment locations and regulate…
3 …**segment polarity genes** that determine the boundaries and anterior–posterior orientation of each segment.
Once segments are established, **homeotic genes** define the role of each segment.

19.15 A Gene Cascade Controls Pattern Formation in the *Drosophila* Embryo *(Page 404)*

20 Animal Development: From Genes to Organism

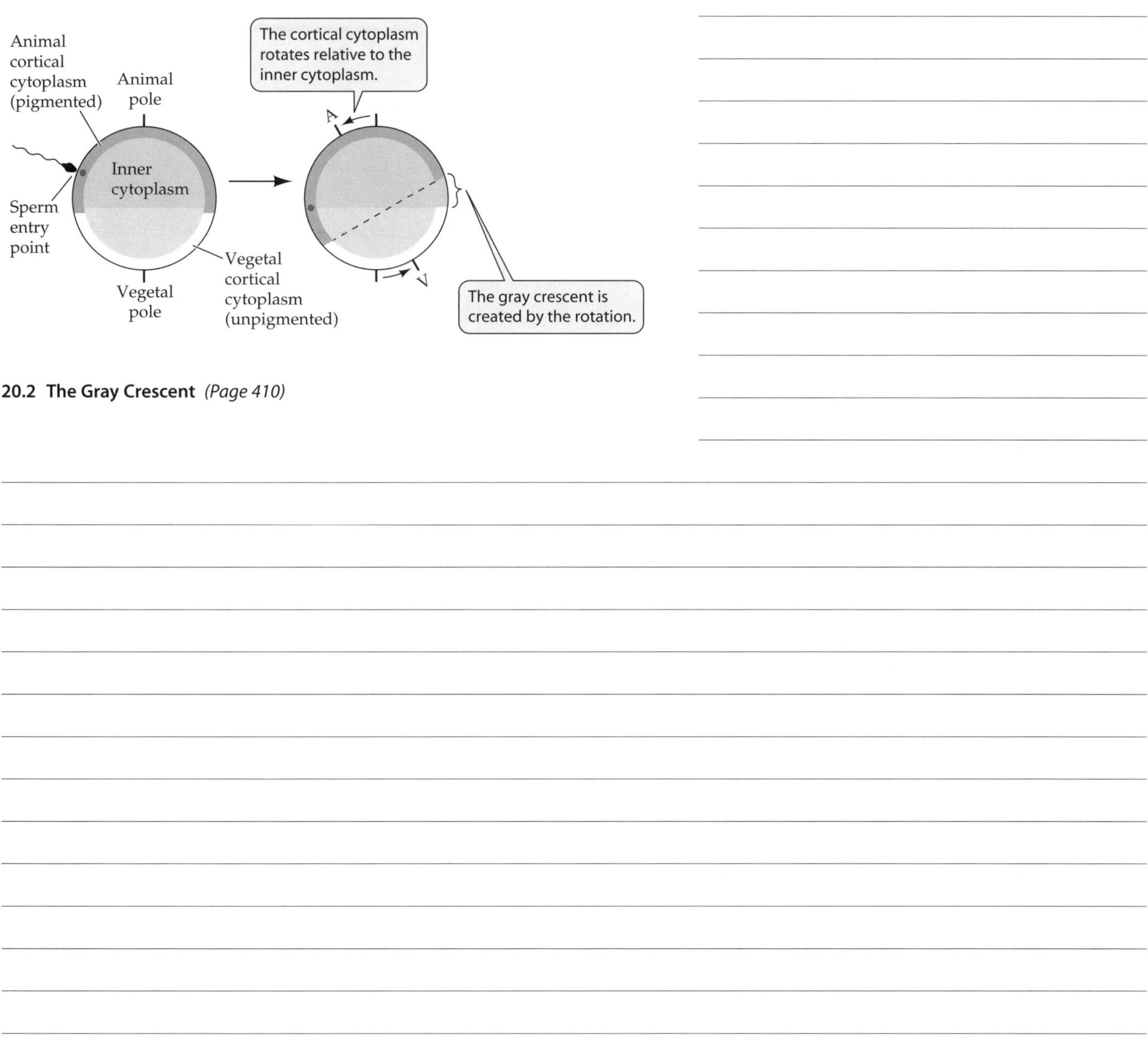

Animal cortical cytoplasm (pigmented)
Animal pole
Inner cytoplasm
Sperm entry point
Vegetal pole
Vegetal cortical cytoplasm (unpigmented)
The cortical cytoplasm rotates relative to the inner cytoplasm.
A
V
The gray crescent is created by the rotation.

20.2 The Gray Crescent *(Page 410)*

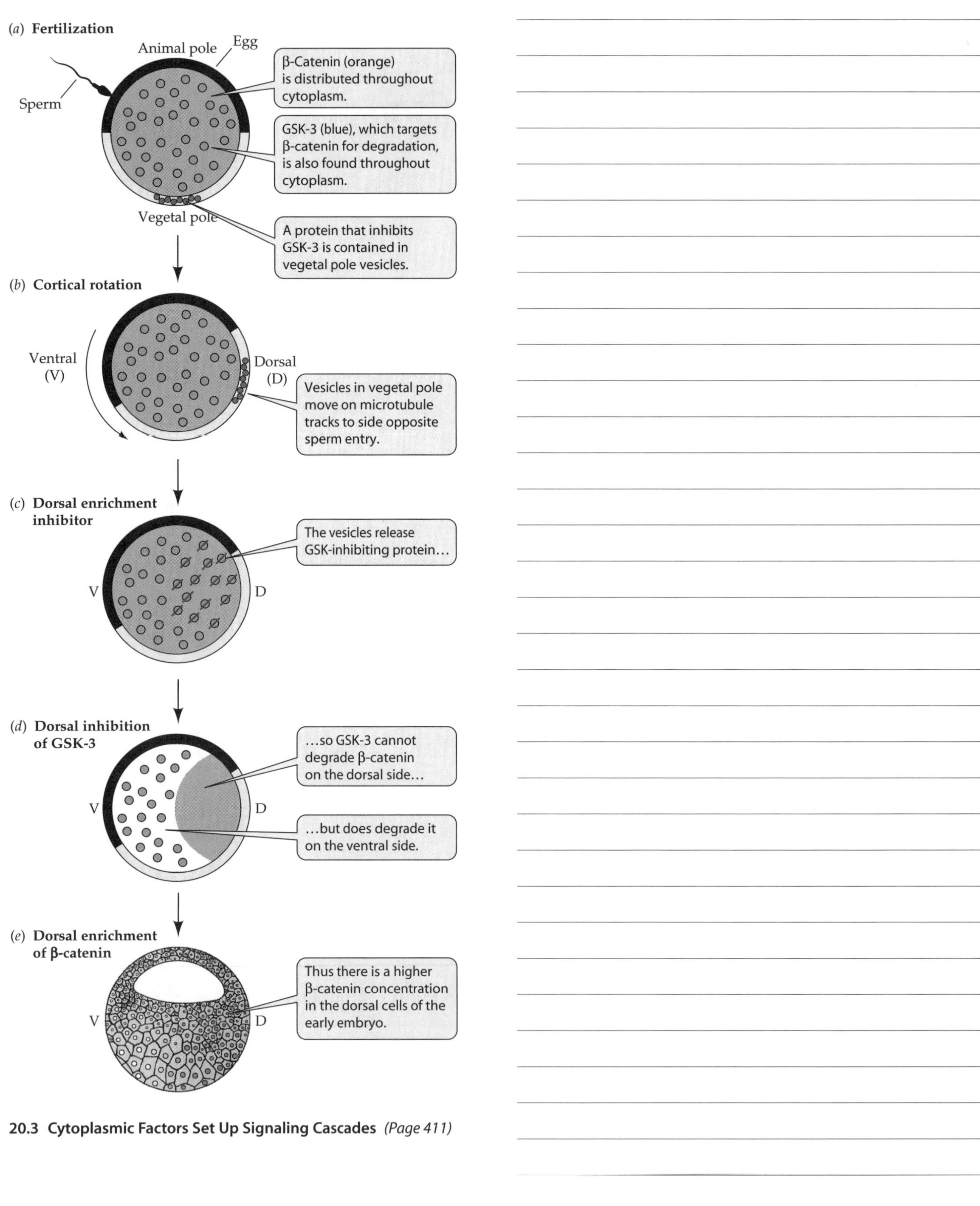
(a) Fertilization
Animal pole
Egg
Sperm
Vegetal pole
β-Catenin (orange) is distributed throughout cytoplasm.
GSK-3 (blue), which targets β-catenin for degradation, is also found throughout cytoplasm.
A protein that inhibits GSK-3 is contained in vegetal pole vesicles.
(b) Cortical rotation
Ventral (V)
Dorsal (D)
Vesicles in vegetal pole move on microtubule tracks to side opposite sperm entry.
(c) Dorsal enrichment inhibitor
V
D
The vesicles release GSK-inhibiting protein…
(d) Dorsal inhibition of GSK-3
V
D
…so GSK-3 cannot degrade β-catenin on the dorsal side…
…but does degrade it on the ventral side.
(e) Dorsal enrichment of β-catenin
V
D
Thus there is a higher β-catenin concentration in the dorsal cells of the early embryo.

20.3 Cytoplasmic Factors Set Up Signaling Cascades *(Page 411)*

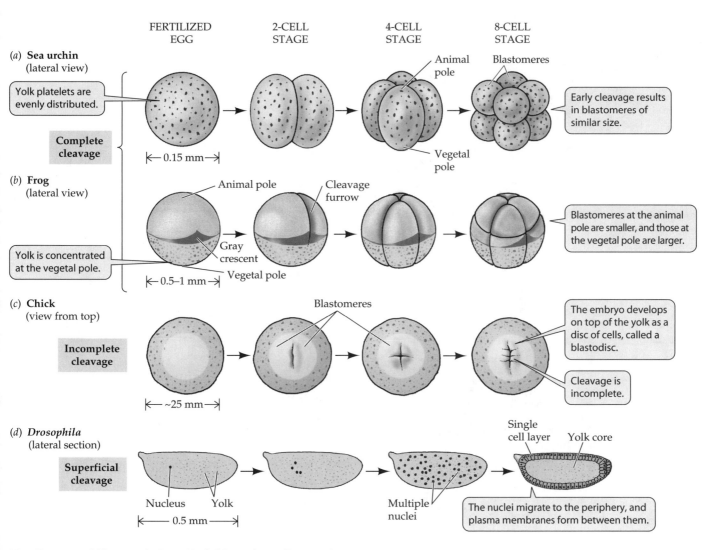

FERTILIZED EGG
2-CELL STAGE
4-CELL STAGE
8-CELL STAGE

(*a*) **Sea urchin**
(lateral view)
Yolk platelets are evenly distributed.
Animal pole
Blastomeres
Early cleavage results in blastomeres of similar size.
Complete cleavage
0.15 mm
Vegetal pole

(*b*) **Frog**
(lateral view)
Animal pole
Cleavage furrow
Blastomeres at the animal pole are smaller, and those at the vegetal pole are larger.
Gray crescent
Yolk is concentrated at the vegetal pole.
0.5–1 mm
Vegetal pole

(*c*) **Chick**
(view from top)
Blastomeres
The embryo develops on top of the yolk as a disc of cells, called a blastodisc.
Incomplete cleavage
Cleavage is incomplete.
~25 mm

(*d*) ***Drosophila***
(lateral section)
Single cell layer
Yolk core
Superficial cleavage
Nucleus
Yolk
0.5 mm
Multiple nuclei
The nuclei migrate to the periphery, and plasma membranes form between them.

20.4 Patterns of Cleavage in Four Model Organisms *(Page 412)*

(a)
Parallel plane
Plane of first cell division
A
Perpendicular plane
V

20.5 The Mammalian Zygote Becomes a Blastocyst *(Page 413)*

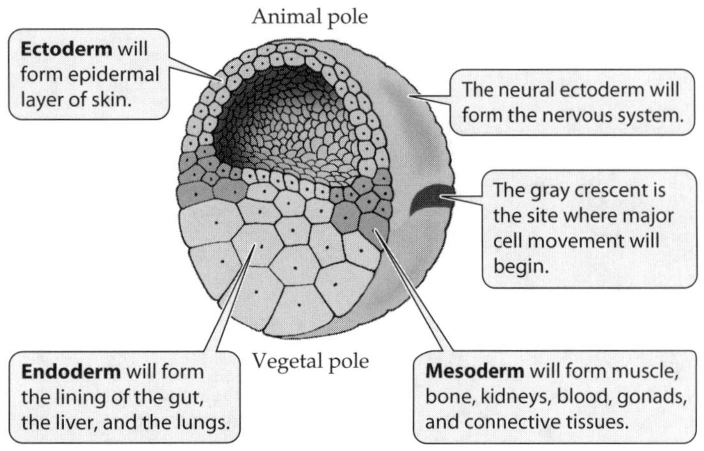
Animal pole
Ectoderm will form epidermal layer of skin.
The neural ectoderm will form the nervous system.
The gray crescent is the site where major cell movement will begin.
Endoderm will form the lining of the gut, the liver, and the lungs.
Vegetal pole
Mesoderm will form muscle, bone, kidneys, blood, gonads, and connective tissues.

20.6 Fate Map of a Frog Blastula *(Page 413)*

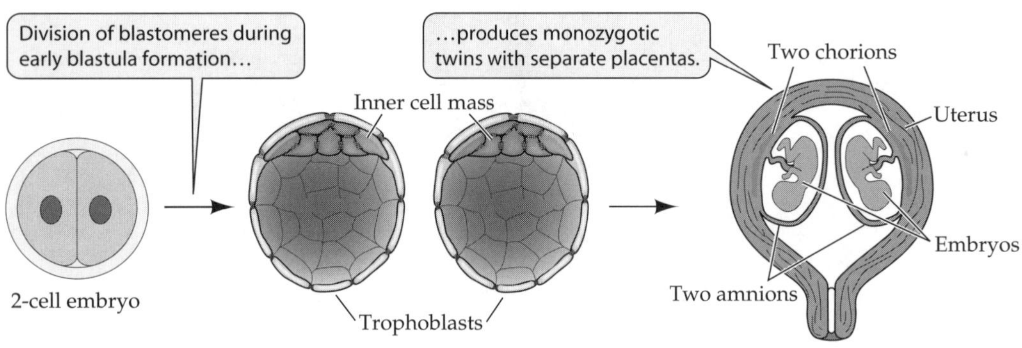
Division of blastomeres during early blastula formation…
…produces monozygotic twins with separate placentas.
Two chorions
Inner cell mass
Uterus
Embryos
Two amnions
2-cell embryo
Trophoblasts

20.7 Twinning in Humans *(Page 414)*

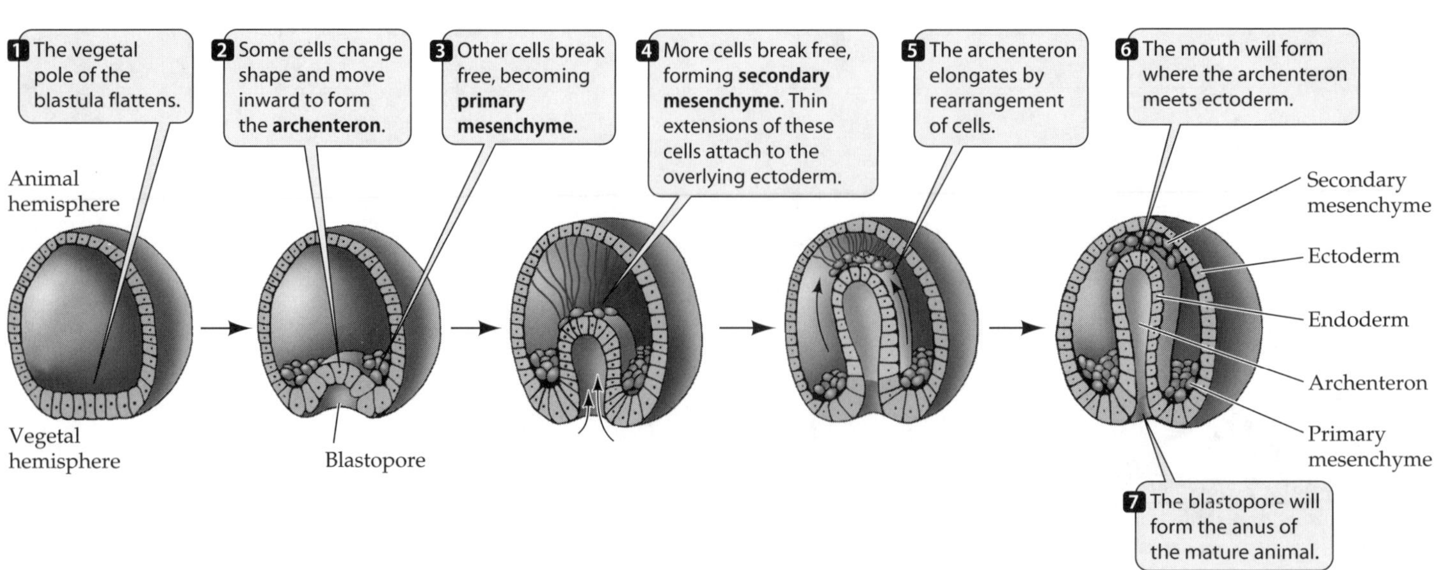
1 The vegetal pole of the blastula flattens.
2 Some cells change shape and move inward to form the **archenteron**.
3 Other cells break free, becoming **primary mesenchyme**.
4 More cells break free, forming **secondary mesenchyme**. Thin extensions of these cells attach to the overlying ectoderm.
5 The archenteron elongates by rearrangement of cells.
6 The mouth will form where the archenteron meets ectoderm.
Animal hemisphere
Vegetal hemisphere
Blastopore
Secondary mesenchyme
Ectoderm
Endoderm
Archenteron
Primary mesenchyme
7 The blastopore will form the anus of the mature animal.

20.8 Gastrulation in Sea Urchins *(Page 415)*

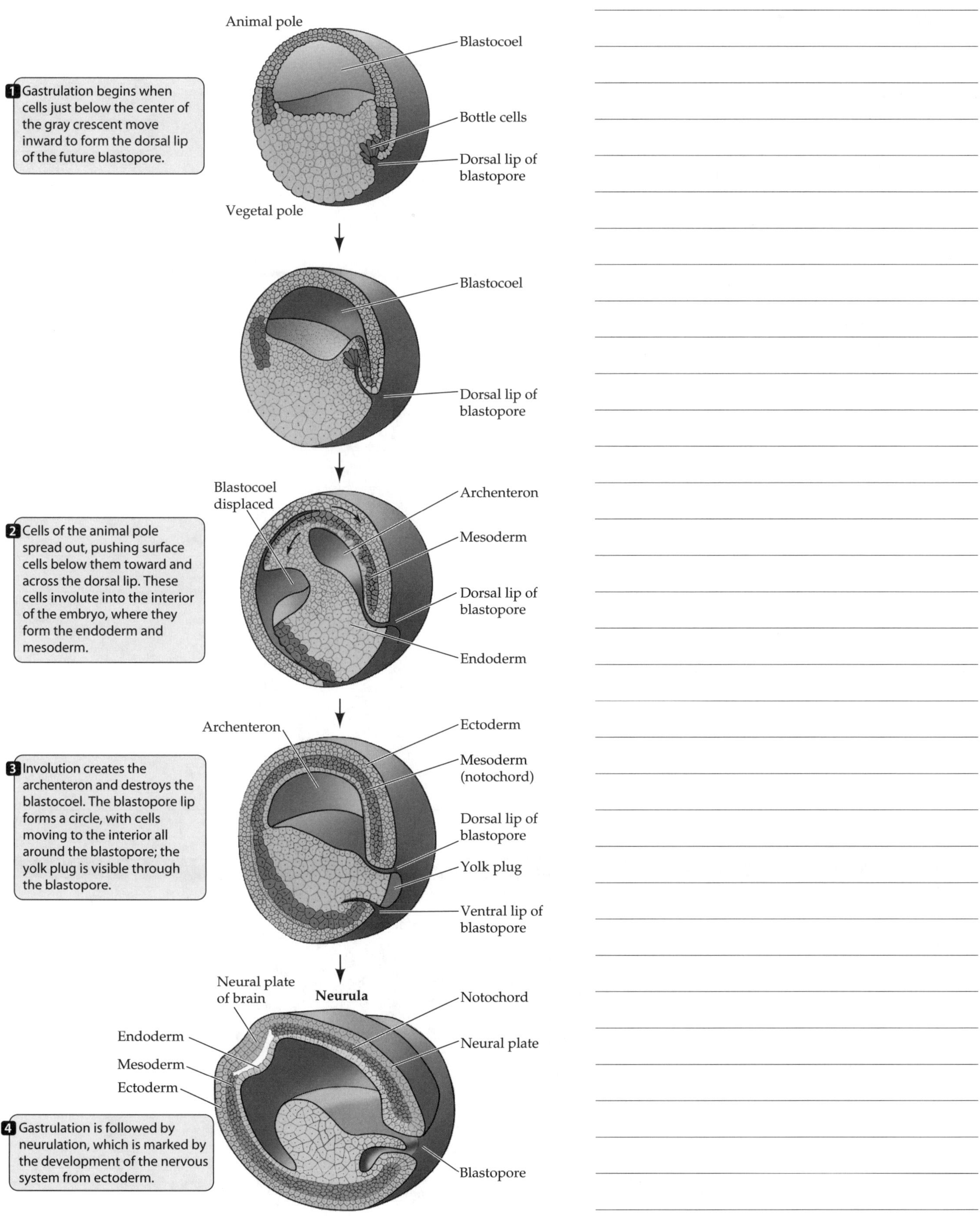
Animal pole
Blastocoel
1 Gastrulation begins when cells just below the center of the gray crescent move inward to form the dorsal lip of the future blastopore.
Bottle cells
Dorsal lip of blastopore
Vegetal pole
Blastocoel
Dorsal lip of blastopore
Blastocoel displaced
Archenteron
Mesoderm
2 Cells of the animal pole spread out, pushing surface cells below them toward and across the dorsal lip. These cells involute into the interior of the embryo, where they form the endoderm and mesoderm.
Dorsal lip of blastopore
Endoderm
Archenteron
Ectoderm
Mesoderm (notochord)
3 Involution creates the archenteron and destroys the blastocoel. The blastopore lip forms a circle, with cells moving to the interior all around the blastopore; the yolk plug is visible through the blastopore.
Dorsal lip of blastopore
Yolk plug
Ventral lip of blastopore
Neural plate of brain
Neurula
Notochord
Endoderm
Neural plate
Mesoderm
Ectoderm
4 Gastrulation is followed by neurulation, which is marked by the development of the nervous system from ectoderm.
Blastopore

20.9 Gastrulation in the Frog Embryo *(Page 416)*

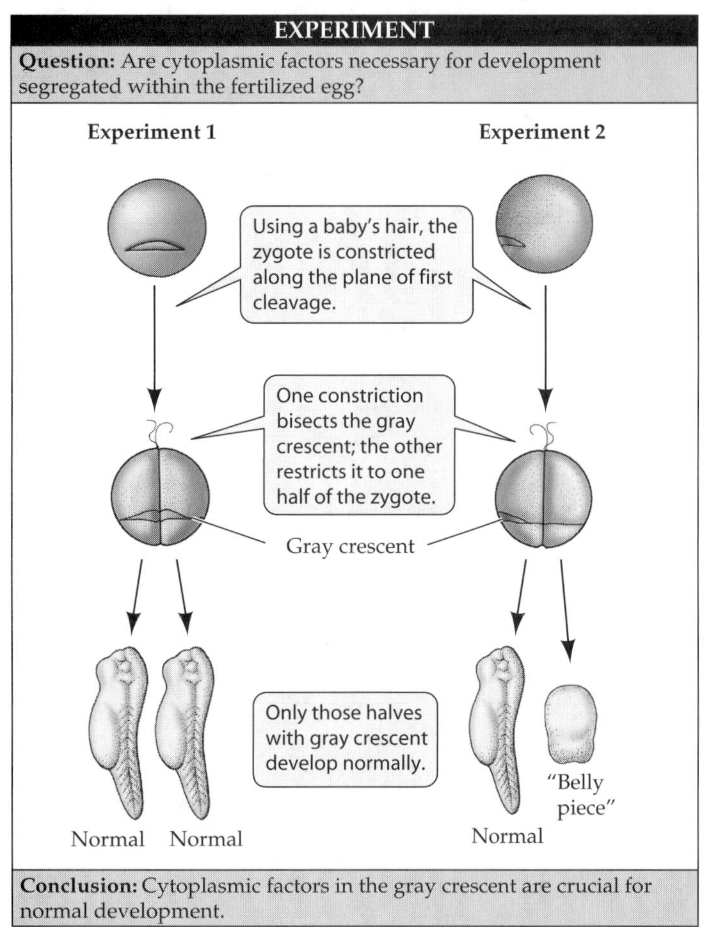
EXPERIMENT
Question: Are cytoplasmic factors necessary for development segregated within the fertilized egg?
Experiment 1
Experiment 2
Using a baby's hair, the zygote is constricted along the plane of first cleavage.
One constriction bisects the gray crescent; the other restricts it to one half of the zygote.
Gray crescent
Only those halves with gray crescent develop normally.
Normal
Normal
Normal
"Belly piece"
Conclusion: Cytoplasmic factors in the gray crescent are crucial for normal development.

20.10 Spemann's Experiment *(Page 417)*

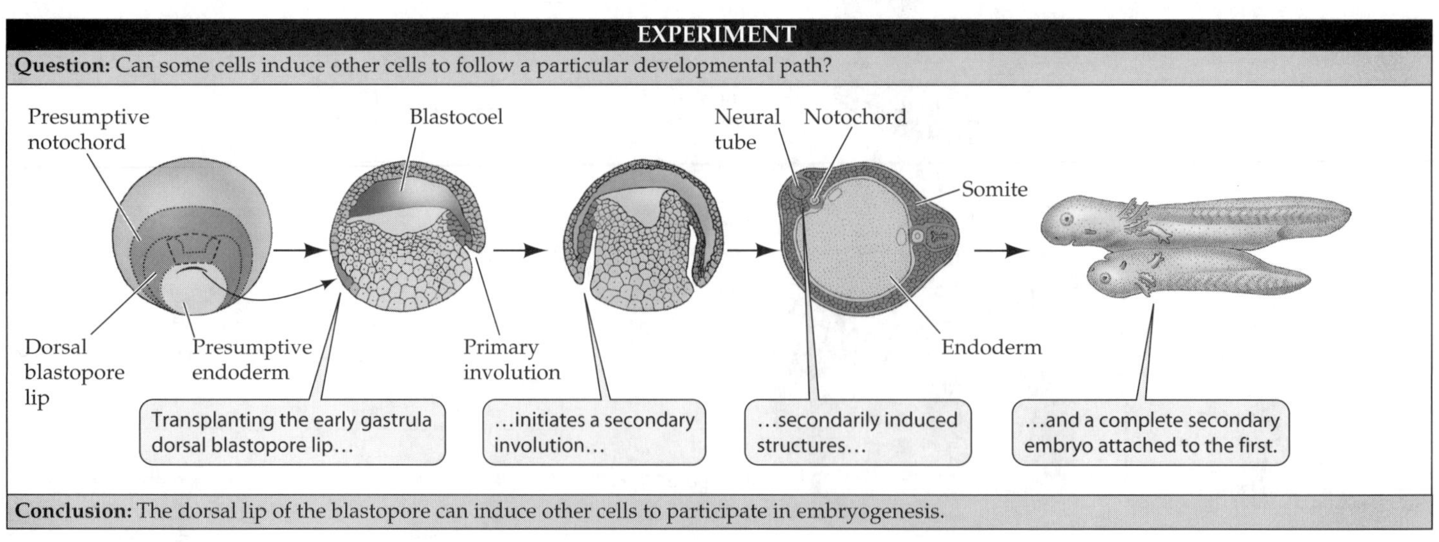
EXPERIMENT
Question: Can some cells induce other cells to follow a particular developmental path?
Presumptive notochord
Blastocoel
Neural tube
Notochord
Somite
Dorsal blastopore lip
Presumptive endoderm
Primary involution
Endoderm
Transplanting the early gastrula dorsal blastopore lip…
…initiates a secondary involution…
…secondarily induced structures…
…and a complete secondary embryo attached to the first.
Conclusion: The dorsal lip of the blastopore can induce other cells to participate in embryogenesis.

20.11 The Dorsal Lip Induces Embryonic Organization *(Page 417)*

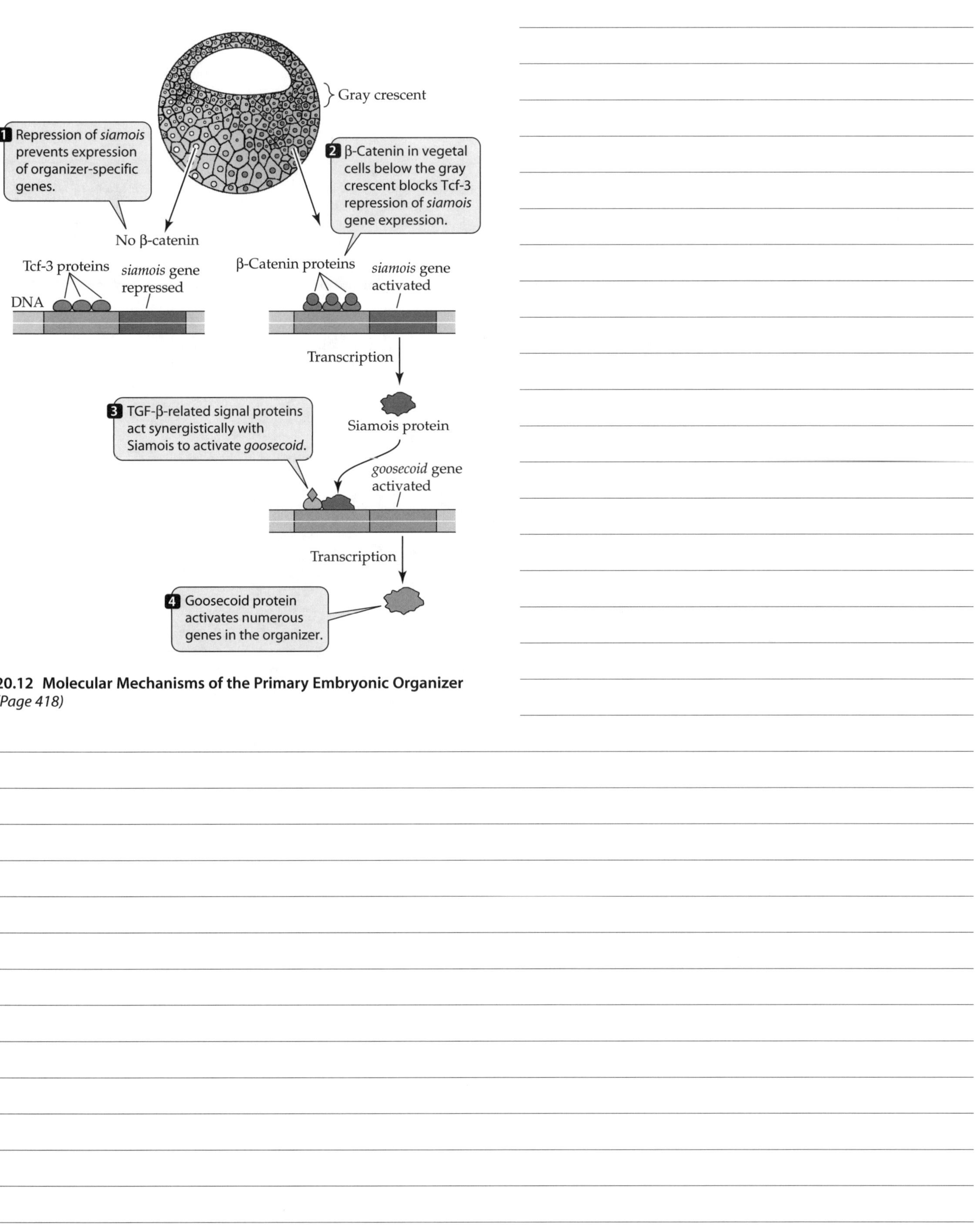
Gray crescent
1 Repression of siamois prevents expression of organizer-specific genes.
2 β-Catenin in vegetal cells below the gray crescent blocks Tcf-3 repression of siamois gene expression.
No β-catenin
Tcf-3 proteins
siamois gene repressed
DNA
β-Catenin proteins
siamois gene activated
Transcription
3 TGF-β-related signal proteins act synergistically with Siamois to activate goosecoid.
Siamois protein
goosecoid gene activated
Transcription
4 Goosecoid protein activates numerous genes in the organizer.

20.12 Molecular Mechanisms of the Primary Embryonic Organizer
(Page 418)

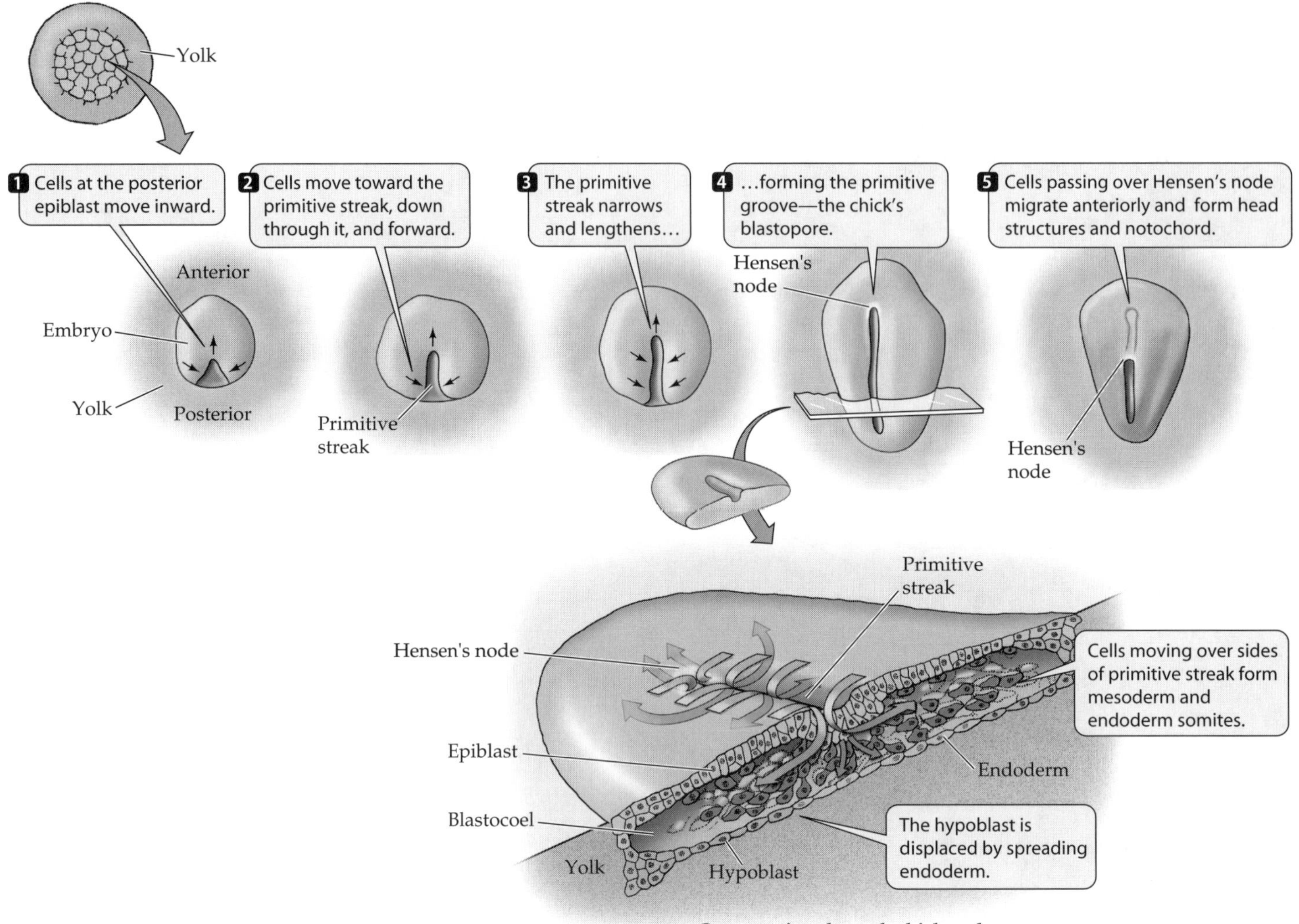
Chick embryo viewed from above
Yolk
1 Cells at the posterior epiblast move inward.
2 Cells move toward the primitive streak, down through it, and forward.
3 The primitive streak narrows and lengthens…
4 …forming the primitive groove—the chick's blastopore.
5 Cells passing over Hensen's node migrate anteriorly and form head structures and notochord.
Anterior
Embryo
Yolk
Posterior
Primitive streak
Hensen's node
Hensen's node
Primitive streak
Hensen's node
Cells moving over sides of primitive streak form mesoderm and endoderm somites.
Epiblast
Endoderm
Blastocoel
The hypoblast is displaced by spreading endoderm.
Yolk
Hypoblast
Cross section through chick embryo

20.13 Gastrulation in Birds *(Page 419)*

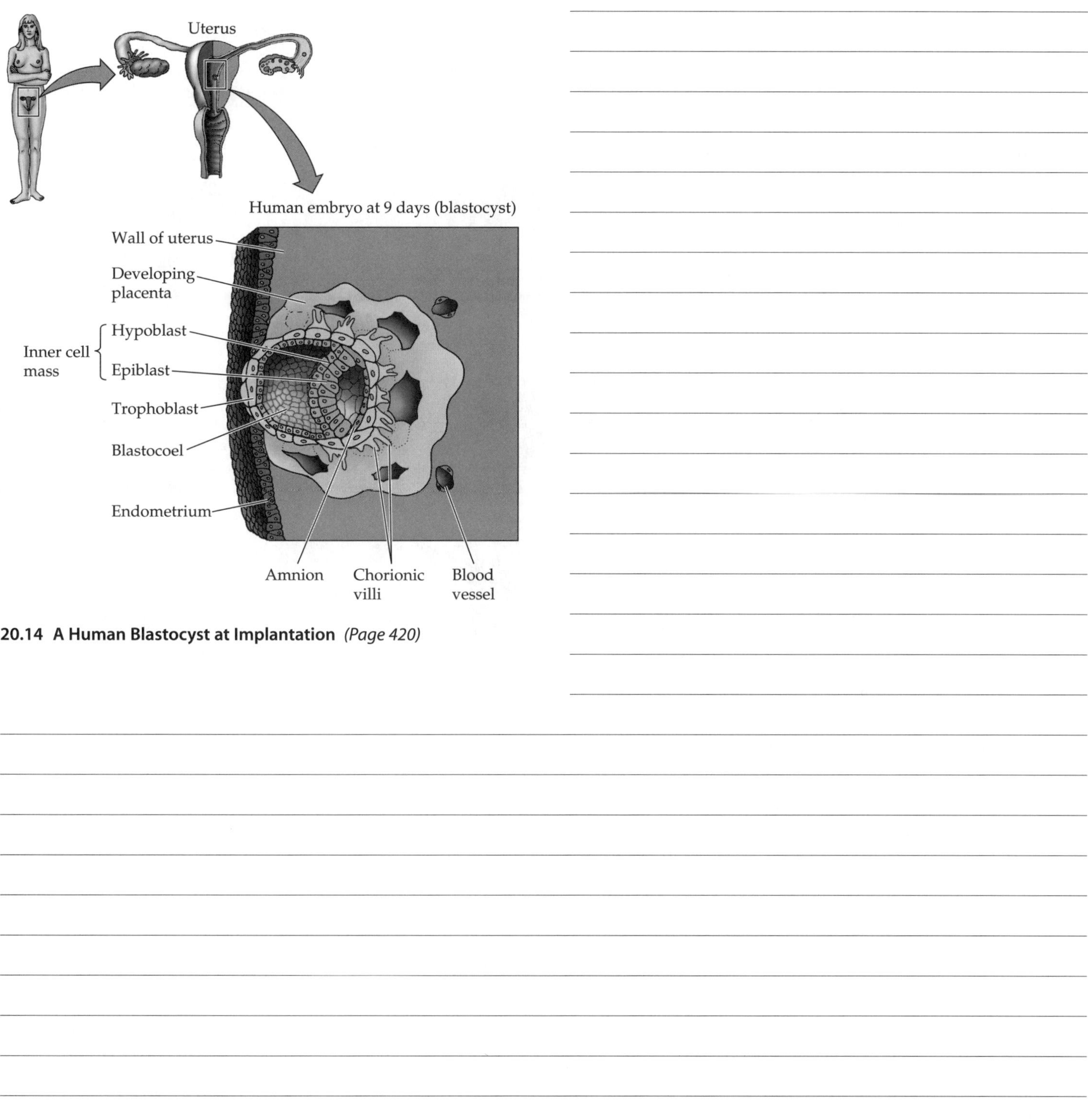
Uterus
Human embryo at 9 days (blastocyst)
Wall of uterus
Developing placenta
Inner cell mass
Hypoblast
Epiblast
Trophoblast
Blastocoel
Endometrium
Amnion
Chorionic villi
Blood vessel

20.14 A Human Blastocyst at Implantation *(Page 420)*

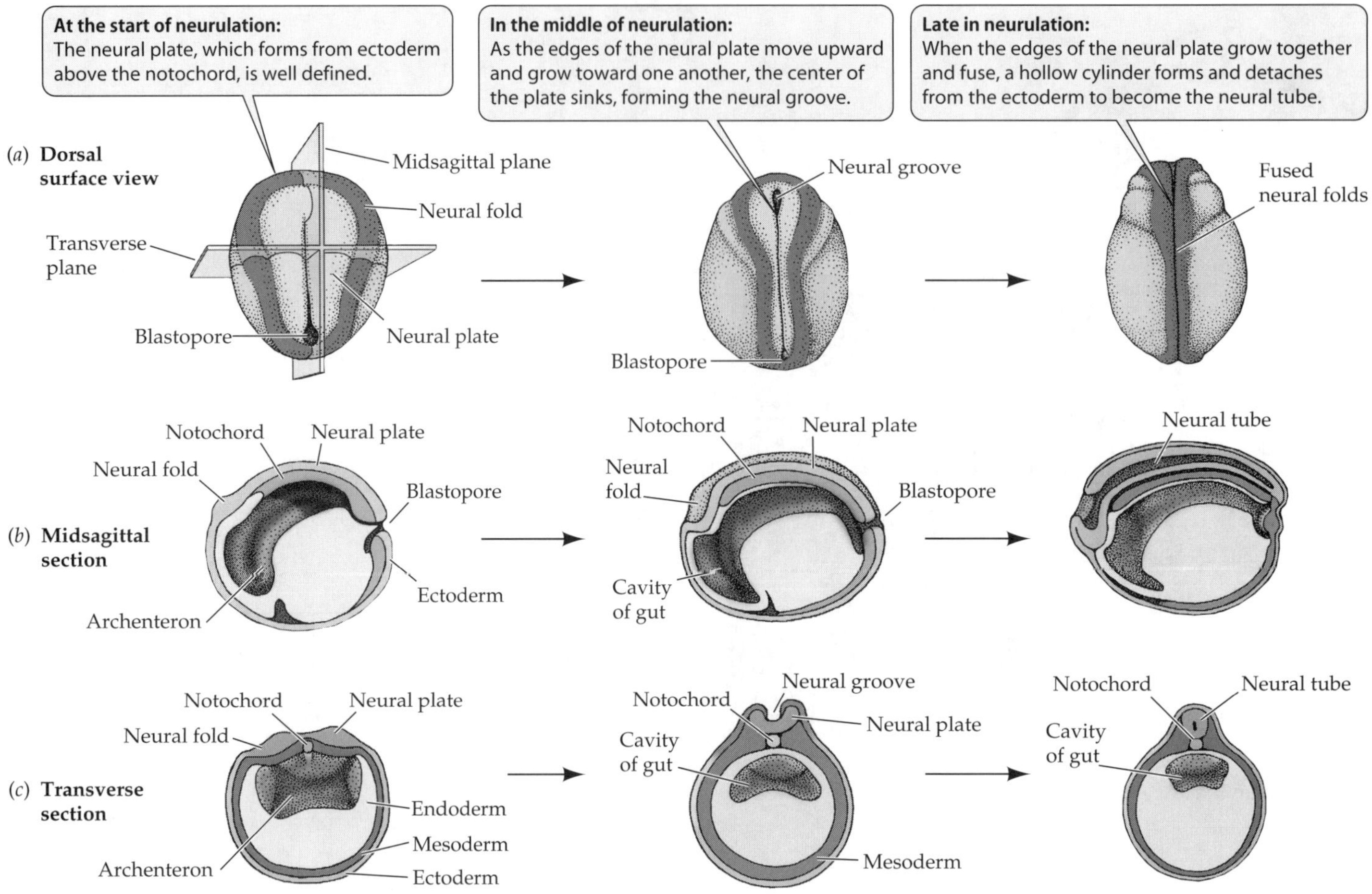

At the start of neurulation:
The neural plate, which forms from ectoderm above the notochord, is well defined.
In the middle of neurulation:
As the edges of the neural plate move upward and grow toward one another, the center of the plate sinks, forming the neural groove.
Late in neurulation:
When the edges of the neural plate grow together and fuse, a hollow cylinder forms and detaches from the ectoderm to become the neural tube.
(a) Dorsal surface view
Midsagittal plane
Neural fold
Transverse plane
Blastopore
Neural plate
Neural groove
Blastopore
Fused neural folds
(b) Midsagittal section
Notochord
Neural plate
Neural fold
Blastopore
Ectoderm
Archenteron
Notochord
Neural plate
Neural fold
Blastopore
Cavity of gut
Neural tube
(c) Transverse section
Notochord
Neural plate
Neural fold
Endoderm
Mesoderm
Ectoderm
Archenteron
Neural groove
Notochord
Neural plate
Cavity of gut
Mesoderm
Notochord
Neural tube
Cavity of gut

20.15 Neurulation in the Frog Embryo *(Page 421)*

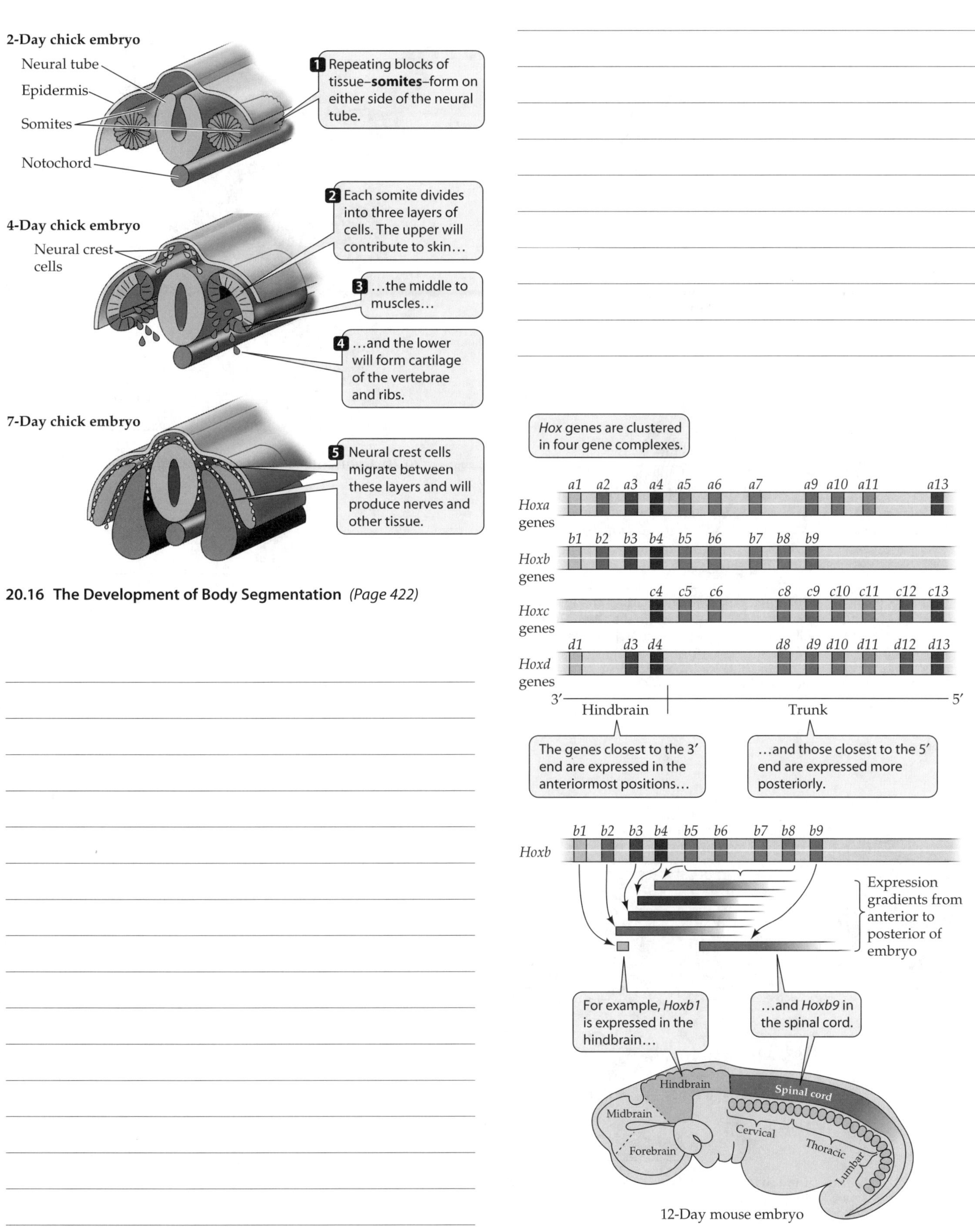
2-Day chick embryo
Neural tube
Epidermis
Somites
Notochord
1 Repeating blocks of tissue–**somites**–form on either side of the neural tube.
4-Day chick embryo
Neural crest cells
2 Each somite divides into three layers of cells. The upper will contribute to skin…
3 …the middle to muscles…
4 …and the lower will form cartilage of the vertebrae and ribs.
7-Day chick embryo
5 Neural crest cells migrate between these layers and will produce nerves and other tissue.
Hox genes are clustered in four gene complexes.
Hoxa genes
a1 a2 a3 a4 a5 a6 a7 a9 a10 a11 a13
Hoxb genes
b1 b2 b3 b4 b5 b6 b7 b8 b9
Hoxc genes
c4 c5 c6 c8 c9 c10 c11 c12 c13
Hoxd genes
d1 d3 d4 d8 d9 d10 d11 d12 d13
3′
5′
Hindbrain
Trunk
The genes closest to the 3′ end are expressed in the anteriormost positions…
…and those closest to the 5′ end are expressed more posteriorly.
Hoxb
b1 b2 b3 b4 b5 b6 b7 b8 b9
Expression gradients from anterior to posterior of embryo
For example, *Hoxb1* is expressed in the hindbrain…
…and *Hoxb9* in the spinal cord.
Hindbrain
Spinal cord
Midbrain
Forebrain
Cervical
Thoracic
Lumbar
12-Day mouse embryo

20.16 The Development of Body Segmentation *(Page 422)*

20.17 Hox Genes Control Body Segmentation *(Page 422)*

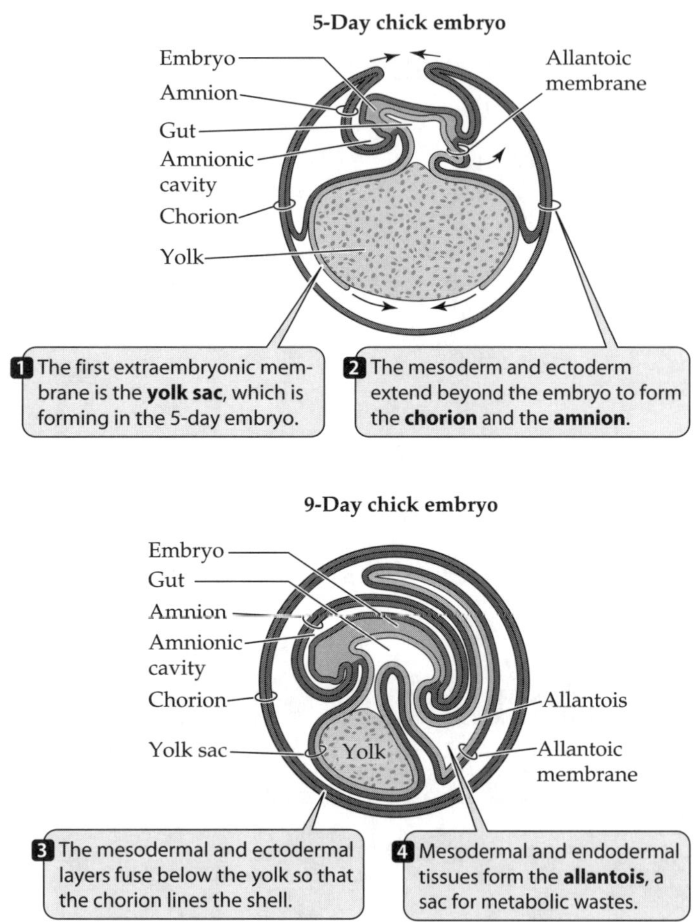
5-Day chick embryo
Embryo
Amnion
Gut
Amnionic cavity
Chorion
Yolk
Allantoic membrane
1 The first extraembryonic membrane is the **yolk sac**, which is forming in the 5-day embryo.
2 The mesoderm and ectoderm extend beyond the embryo to form the **chorion** and the **amnion**.
9-Day chick embryo
Embryo
Gut
Amnion
Amnionic cavity
Chorion
Yolk sac
Yolk
Allantois
Allantoic membrane
3 The mesodermal and ectodermal layers fuse below the yolk so that the chorion lines the shell.
4 Mesodermal and endodermal tissues form the **allantois**, a sac for metabolic wastes.

20.18 The Extraembryonic Membranes *(Page 423)*

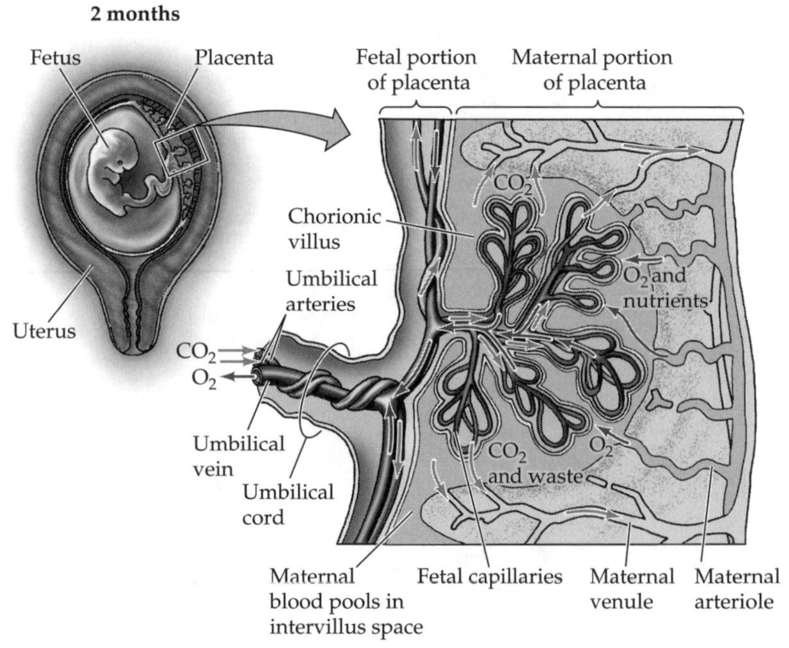
2 months
Fetus
Placenta
Uterus
Fetal portion of placenta
Maternal portion of placenta
Chorionic villus
Umbilical arteries
CO_2
O_2
Umbilical vein
Umbilical cord
CO_2
O_2 and nutrients
CO_2 and waste
O_2
Maternal blood pools in intervillus space
Fetal capillaries
Maternal venule
Maternal arteriole

20.19 The Mammalian Placenta *(Page 424)*

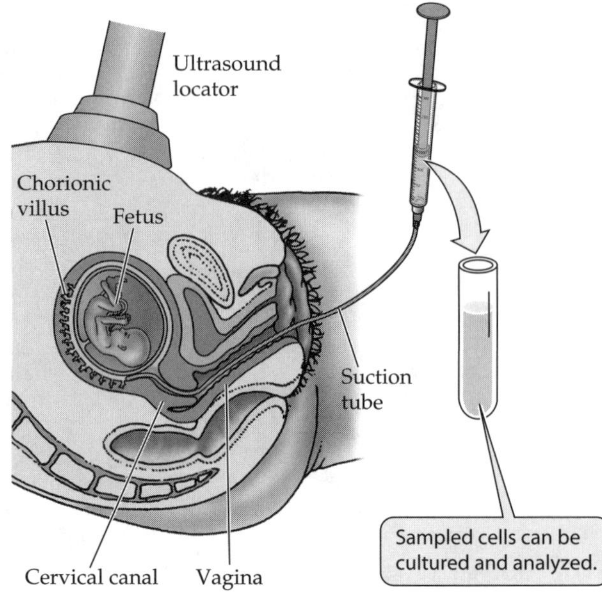
Ultrasound locator
Chorionic villus
Fetus
Suction tube
Cervical canal
Vagina
Sampled cells can be cultured and analyzed.

20.20 Chorionic Villus Sampling *(Page 425)*

21 Development and Evolutionary Change

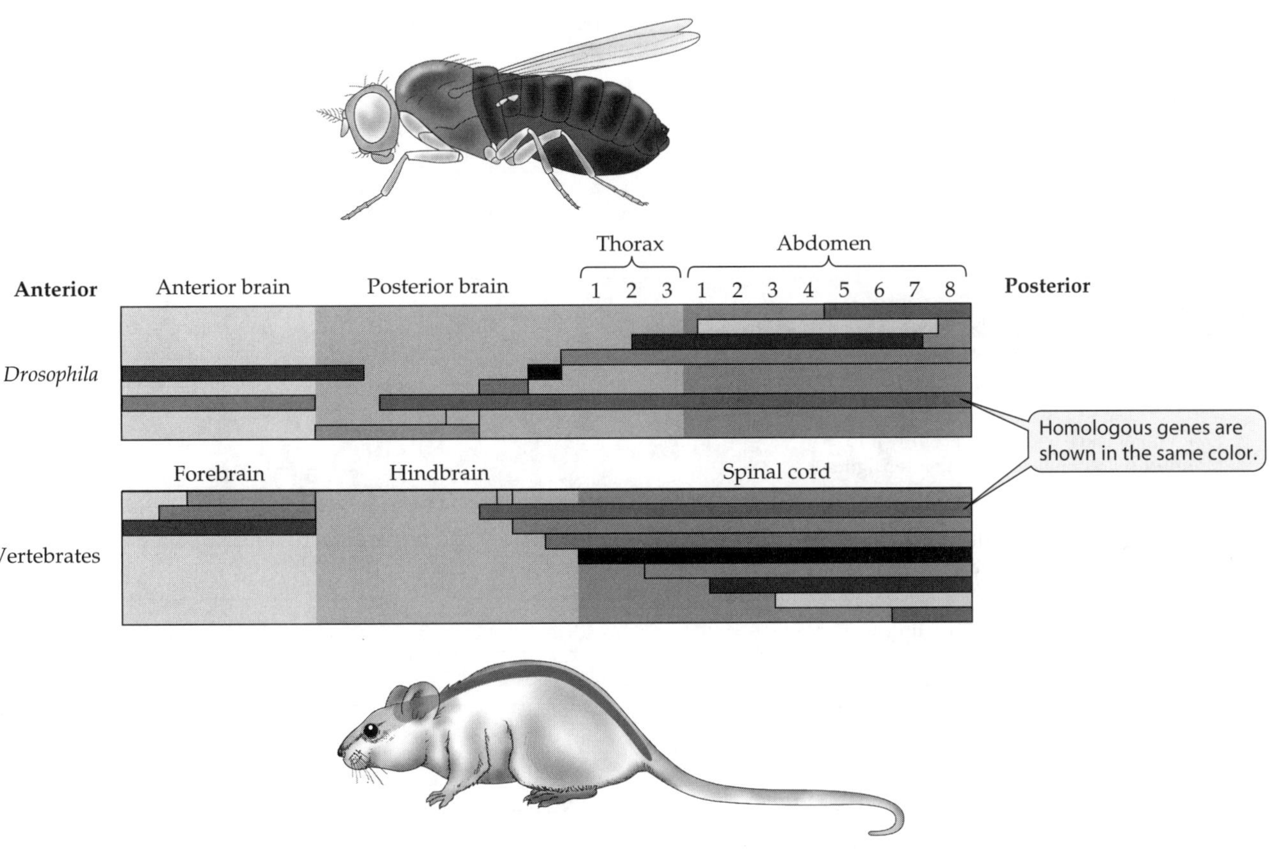

Thorax
Abdomen
Anterior
Anterior brain
Posterior brain
1 2 3 1 2 3 4 5 6 7 8
Posterior
Drosophila
Homologous genes are shown in the same color.
Forebrain
Hindbrain
Spinal cord
Vertebrates

Figure 21.3 Genes Show Similar Expression Patterns *(Page 431)*

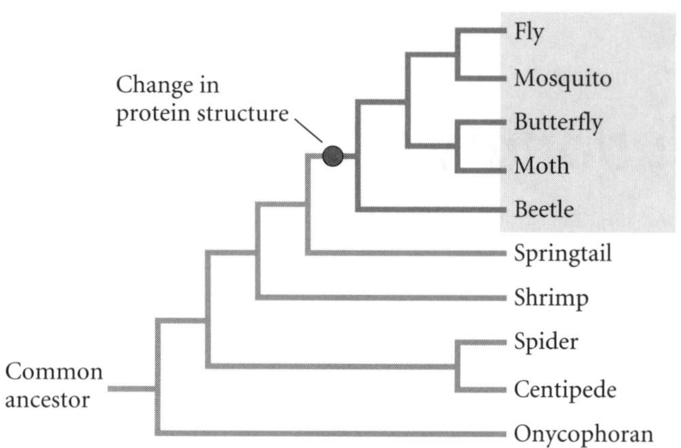
Change in
protein structure
Common
ancestor
Fly
Mosquito
Butterfly
Moth
Beetle
Springtail
Shrimp
Spider
Centipede
Onycophoran

Figure 21.5 A Mutation Changed the Number of Legs in Insects *(Page 433)*

EXPERIMENT	
Question: Will adding Gremlin protein (an inhibitor of BMP4) to a developing chick foot transform the chick foot into a ducklike form?	
METHOD	Open up chicken eggs and carefully add Gremlin-secreting beads to the interdigital (web) regions of one embryonic chick hindlimb. Add beads that do not contain Gremlin to the other hindlimbs (controls). Close the eggs and observe limb development.
RESULTS	In the hindlimbs in which Gremlin was secreted, the webbing does not undergo apoptosis, and the hindlimb resembles that of a duck. The control hindlimbs develop the normal chicken form.
Conclusion: Changes in *gremlin* gene expression could cause the changes in morphology, allowing duck hindlimbs to retain their webbing and chick limbs to lose it.	

Figure 21.7 Changing the Form of an Appendage *(Page 434)*

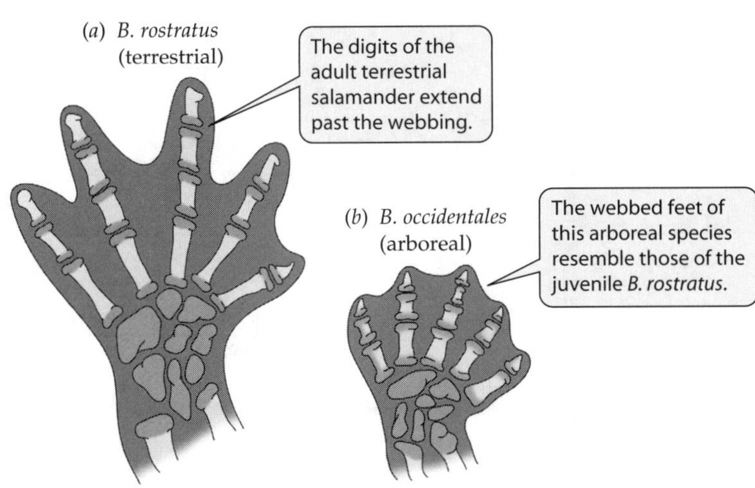
(a) B. rostratus (terrestrial)
The digits of the adult terrestrial salamander extend past the webbing.
(b) B. occidentales (arboreal)
The webbed feet of this arboreal species resemble those of the juvenile B. rostratus.

Figure 21.8 Heterochrony Created an Arboreal Salamander *(Page 434)*

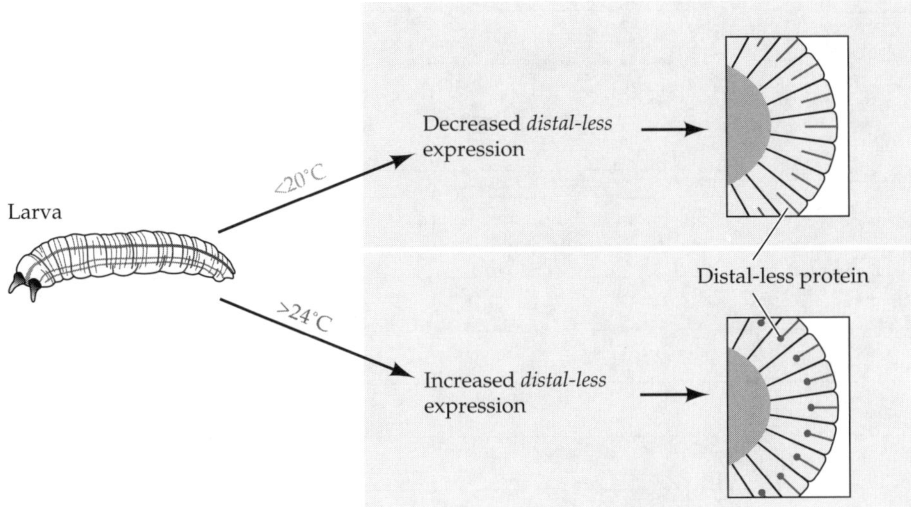
Larva
<20°C
>24°C
Decreased distal-less expression
Increased distal-less expression
Distal-less protein

Figure 21.9 Development of Eyespots in *Bicyclus anynana* Responds to Temperature *(Page 436)*

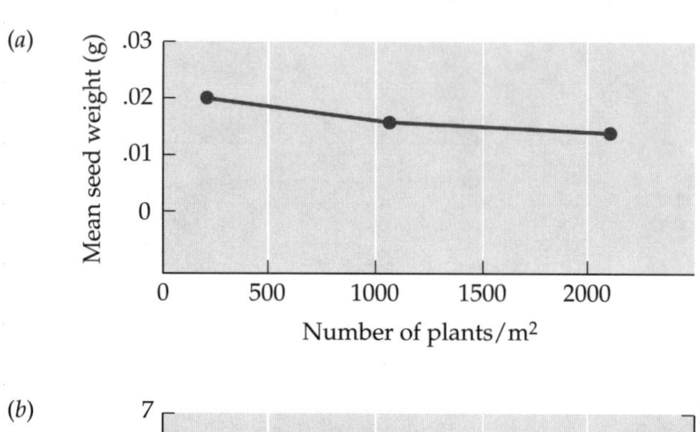
(a)
Mean seed weight (g)
.03
.02
.01
0
0
500
1000
1500
2000
Number of plants/m2

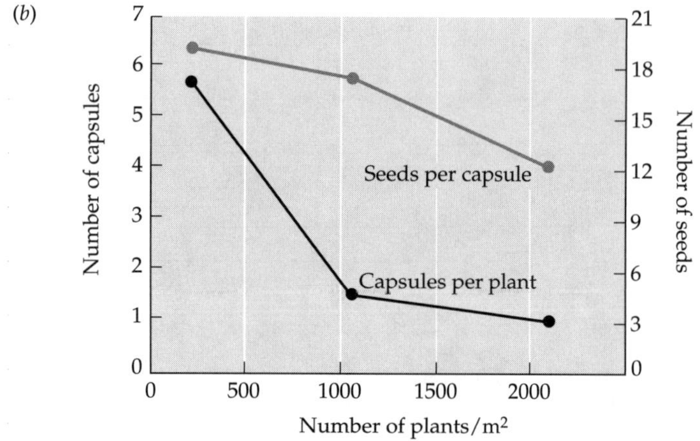
(b)
Number of capsules
7
6
5
4
3
2
1
0
21
18
15
12
9
6
3
0
Number of seeds
Seeds per capsule
Capsules per plant
0
500
1000
1500
2000
Number of plants/m2

Figure 21.13 Seed Production *(Page 438)*

22 The History of Life on Earth

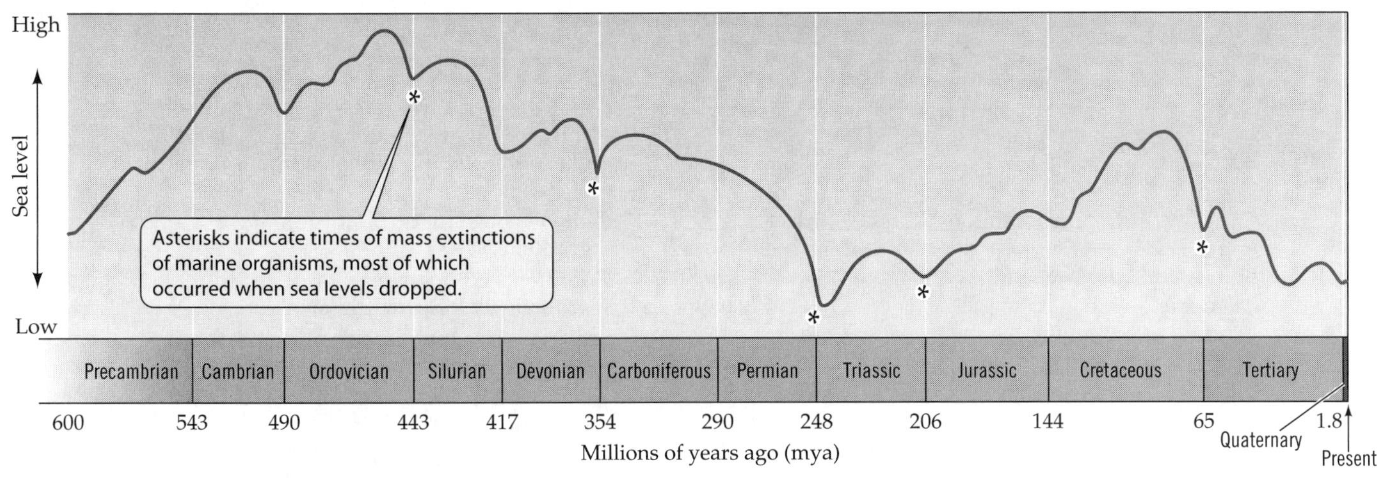
High
Sea level
Low
Asterisks indicate times of mass extinctions of marine organisms, most of which occurred when sea levels dropped.
Precambrian
Cambrian
Ordovician
Silurian
Devonian
Carboniferous
Permian
Triassic
Jurassic
Cretaceous
Tertiary
600 543 490 443 417 354 290 248 206 144 65 1.8
Quaternary
Present
Millions of years ago (mya)

22.2 Sea Levels Have Changed Repeatedly *(Page 446)*

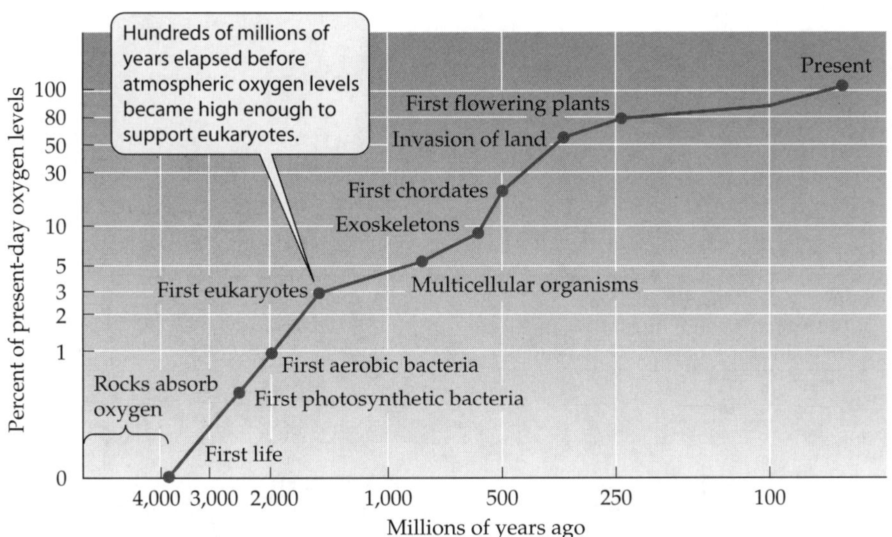
Hundreds of millions of years elapsed before atmospheric oxygen levels became high enough to support eukaryotes.
Present
First flowering plants
Invasion of land
First chordates
Exoskeletons
Multicellular organisms
First eukaryotes
First aerobic bacteria
First photosynthetic bacteria
Rocks absorb oxygen
First life
Percent of present-day oxygen levels
100 80 50 30 10 5 3 2 1 0
4,000 3,000 2,000 1,000 500 250 100
Millions of years ago

22.4 Larger Cells Need More Oxygen *(Page 447)*

22.1 Earth's Geological History

RELATIVE TIME SPAN	ERA	PERIOD	ONSET	MAJOR PHYSICAL CHANGES ON EARTH
	Cenozoic	Quaternary	1.8 mya[a]	Cold/dry climate; repeated glaciations
		Tertiary	65 mya	Continents near current positions; climate cools
	Mesozoic	Cretaceous	144 mya	Northern continents attached; Gondwana begins to drift apart; meteorite strikes Yucatán Peninsula
		Jurassic	206 mya	Two large continents form: Laurasia (north) and Gondwana (south); climate warm
		Triassic	248 mya	Pangaea slowly begins to drift apart; hot/humid climate
	Paleozoic	Permian	290 mya	Continents aggregate into Pangaea; large glaciers form; dry climates form in interior of Pangaea
		Carboniferous	354 mya	Climate cools; marked latitudinal climate gradients
		Devonian	417 mya	Continents collide at end of period; asteroid probably collides with Earth
		Silurian	443 mya	Sea levels rise; two large continents form; hot/humid climate
		Ordovician	490 mya	Gondwana moves over South Pole; massive glaciation, sea level drops 50 m
		Cambrian	543 mya	O_2 levels approach current levels
Precambrian	Precambrian		600 mya	O_2 level at >5% of current level
			1.5 bya[a]	O_2 level at >1% of current level
			3.8 bya	O_2 first appears in atmosphere
			4.5 bya	

[a]mya, million years ago; bya, billion years ago.

(Pages 444–445)

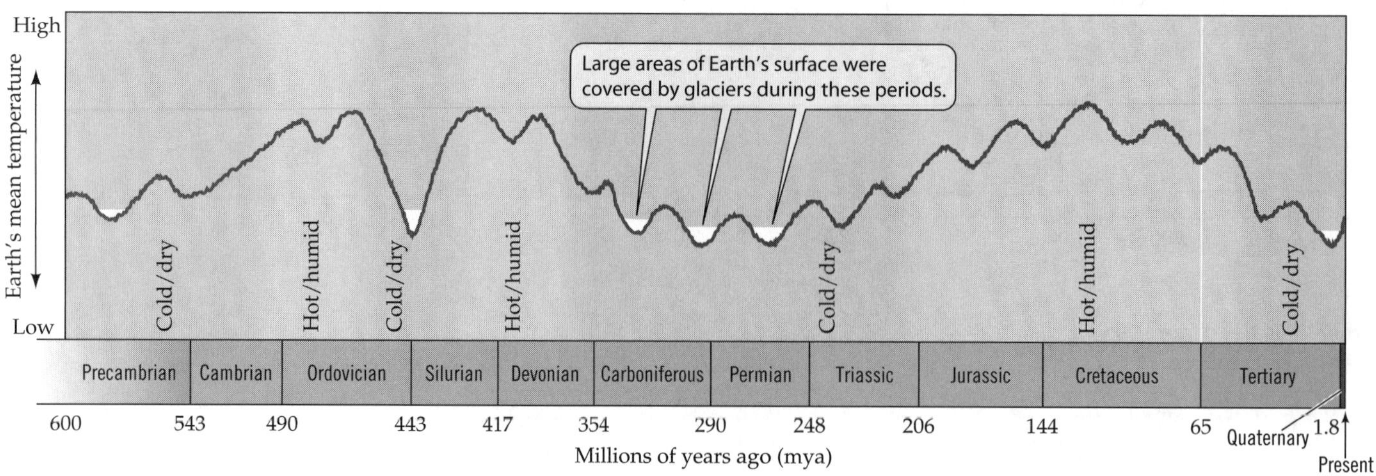

High
Low
Earth's mean temperature
Large areas of Earth's surface were covered by glaciers during these periods.
Cold/dry
Hot/humid
Cold/dry
Hot/humid
Cold/dry
Hot/humid
Cold/dry
Precambrian
Cambrian
Ordovician
Silurian
Devonian
Carboniferous
Permian
Triassic
Jurassic
Cretaceous
Tertiary
600
543
490
443
417
354
290
248
206
144
65
Quaternary
1.8
Present
Millions of years ago (mya)

22.5 Hot/Humid and Cold/Dry Conditions Have Alternated Over Earth's History (Page 447)

MAJOR EVENTS IN THE HISTORY OF LIFE
Humans evolve; many large mammals become extinct
Diversification of birds, mammals, flowering plants, and insects
Dinosaurs continue to diversify; flowering plants and mammals diversify. **Mass Extinction** at end of period (≈76% of species disappear)
Diverse dinosaurs; radiation of ray-finned fishes
Early dinosaurs; first mammals; marine invertebrates diversify; first flowering plants; **Mass Extinction** at end of period (≈65% of species disappear)
Reptiles diversify; amphibians decline; **Mass Extinction** at end of period (≈96% of species disappear)
Extensive "fern" forests; first reptiles; insects diversify
Fishes diversify; first insects and amphibians. **Mass Extinction** at end of period (≈75% of species disappear)
Jawless fishes diversify; first ray-finned fishes; plants and animals colonize land
Mass Extinction at end of period (≈75% of species disappear)
Most animal phyla present; diverse algae
Ediacaran fauna
Eukaryotes evolve; several animal phyla appear
Origin of life; prokaryotes flourish

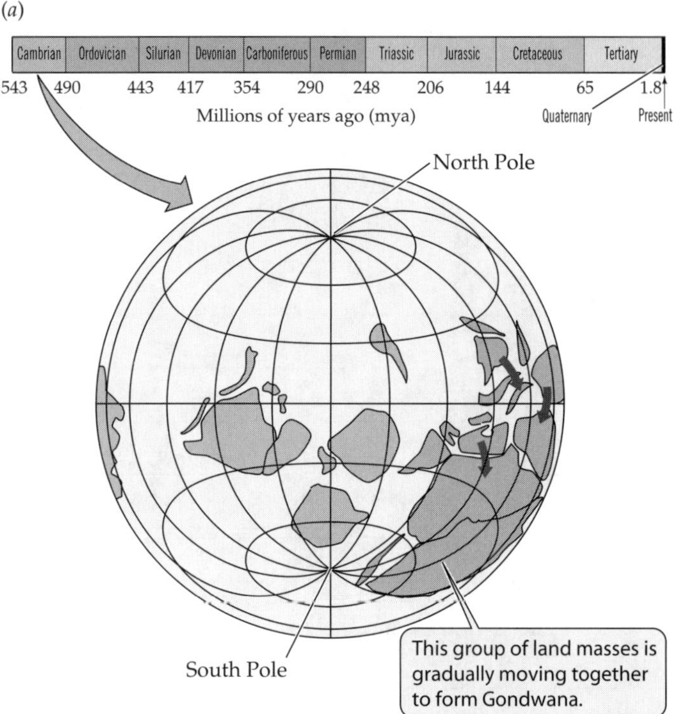
(a)
Cambrian
Ordovician
Silurian
Devonian
Carboniferous
Permian
Triassic
Jurassic
Cretaceous
Tertiary
543 490 443 417 354 290 248 206 144 65 1.8
Millions of years ago (mya)
Quaternary
Present
North Pole
South Pole
This group of land masses is gradually moving together to form Gondwana.

22.9 Cambrian Continents and Animals *(Page 450)*

(a)

Cambrian | Ordovician | Silurian | Devonian | Carboniferous | Permian | Triassic | Jurassic | Cretaceous | Tertiary

543 490 443 417 354 290 248 206 144 65 1.8

Millions of years ago (mya)

Quaternary

Present

Gondwana

During the Devonian period, the northern and southern continents were approaching one another.

22.11 Devonian Continents and Marine Communities *(Page 451)*

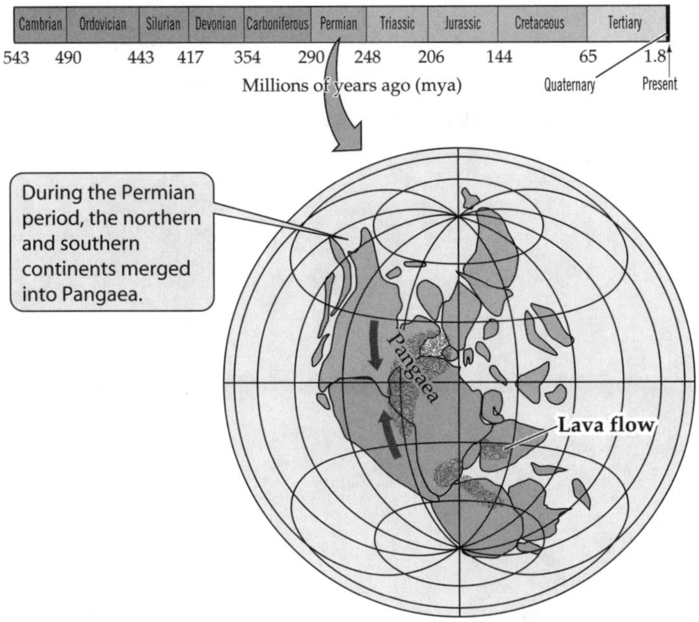

During the Permian period, the northern and southern continents merged into Pangaea.

22.13 Pangaea Formed in the Permian Period *(Page 453)*

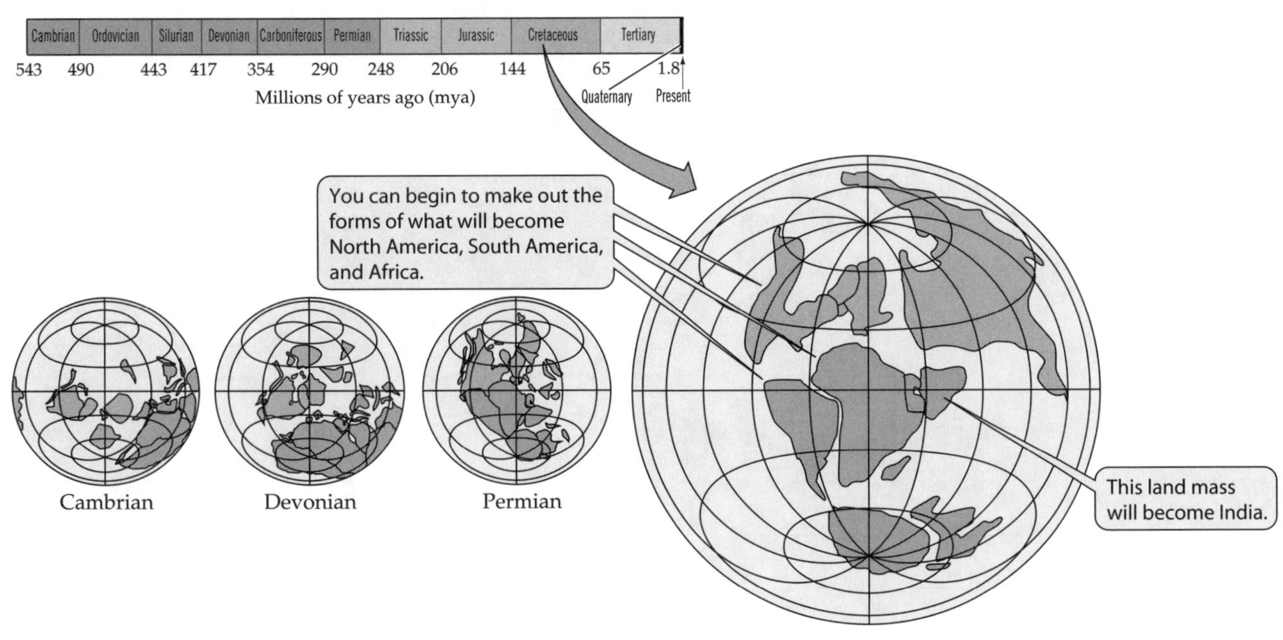
Cambrian
Ordovician
Silurian
Devonian
Carboniferous
Permian
Triassic
Jurassic
Cretaceous
Tertiary
543 490 443 417 354 290 248 206 144 65 1.8
Millions of years ago (mya)
Quaternary
Present
You can begin to make out the forms of what will become North America, South America, and Africa.
This land mass will become India.
Cambrian
Devonian
Permian

22.15 Positions of the Continents during the Cretaceous Period *(Page 453)*

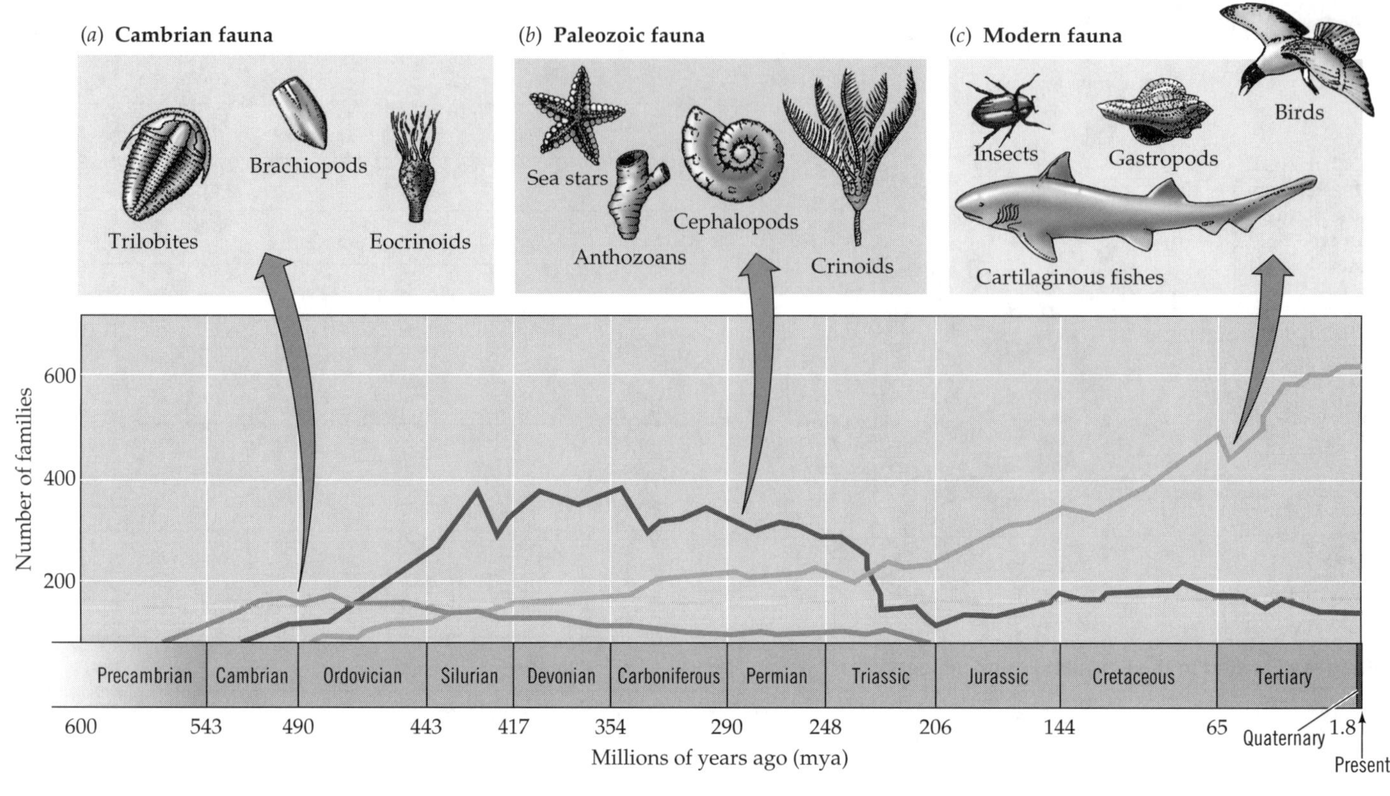
(a) Cambrian fauna
Trilobites
Brachiopods
Eocrinoids
(b) Paleozoic fauna
Sea stars
Anthozoans
Cephalopods
Crinoids
(c) Modern fauna
Insects
Gastropods
Birds
Cartilaginous fishes
Number of families
600
400
200
Precambrian
Cambrian
Ordovician
Silurian
Devonian
Carboniferous
Permian
Triassic
Jurassic
Cretaceous
Tertiary
600
543
490
443
417
354
290
248
206
144
65
Quaternary
1.8
Present
Millions of years ago (mya)

22.17 Evolutionary Faunas *(Page 455)*

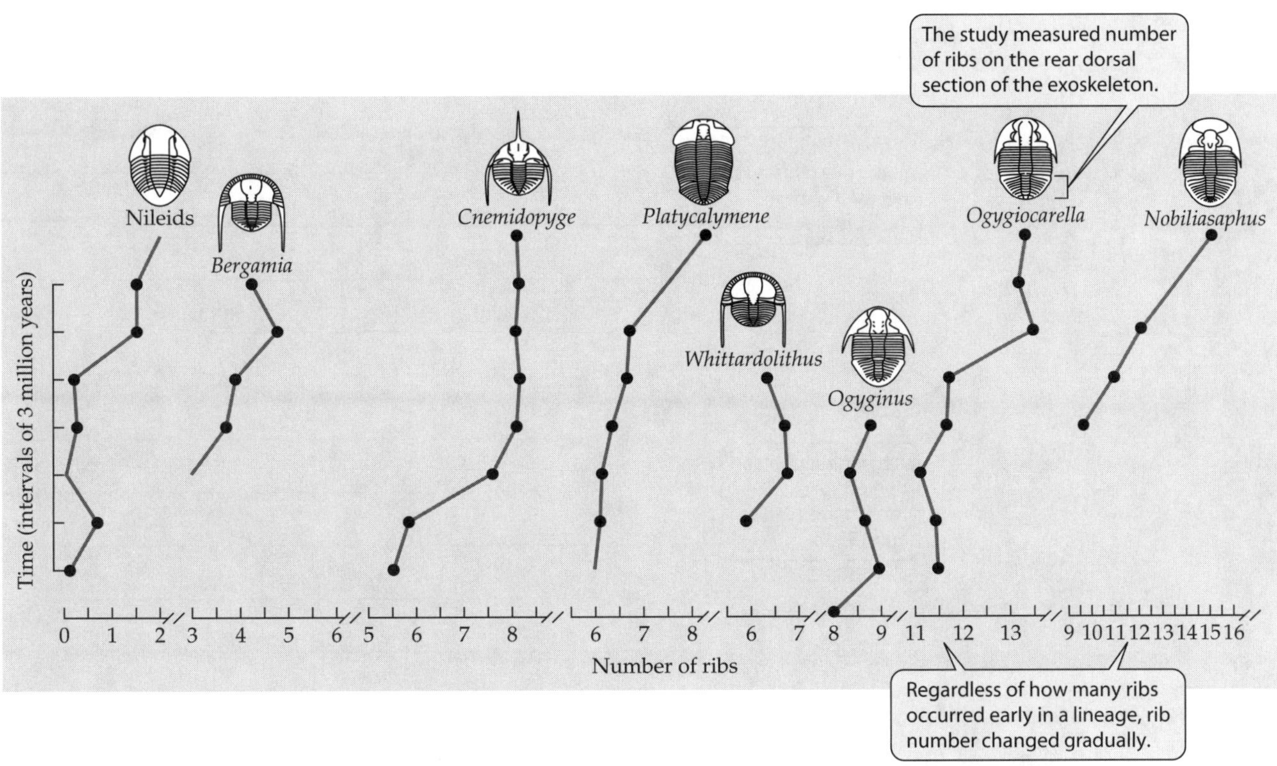

The study measured number of ribs on the rear dorsal section of the exoskeleton.
Nileids
Bergamia
Cnemidopyge
Platycalymene
Whittardolithus
Ogyginus
Ogygiocarella
Nobiliasaphus
Time (intervals of 3 million years)
Number of ribs
Regardless of how many ribs occurred early in a lineage, rib number changed gradually.

22.19 Rib Number Evolved Gradually in Trilobites *(Page 456)*

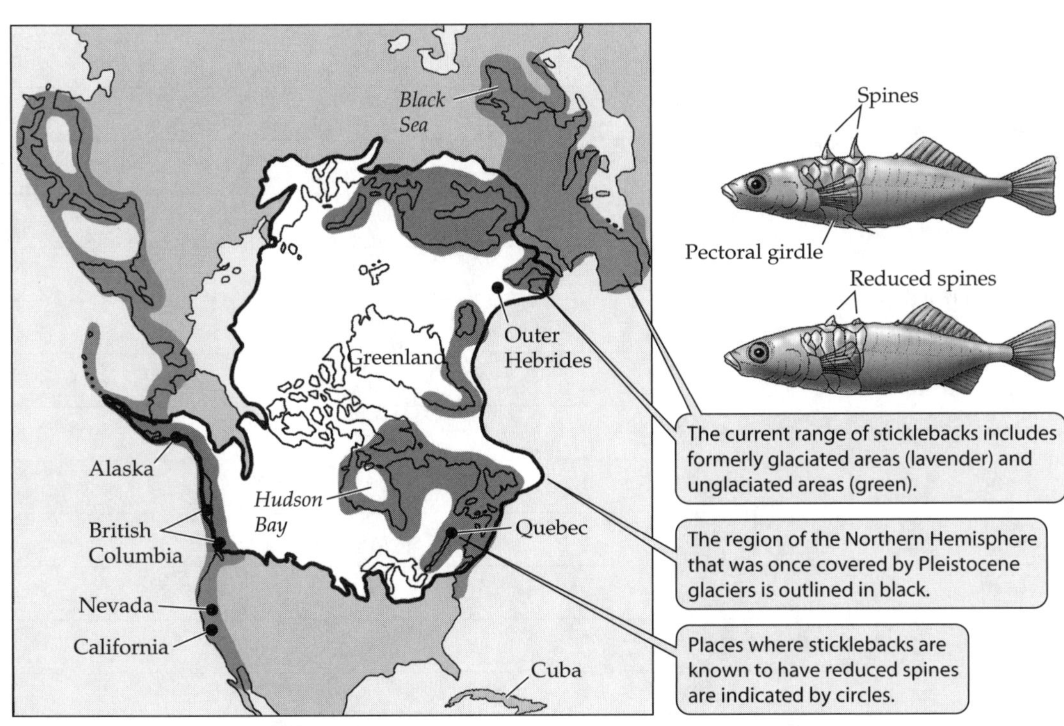

Black Sea
Greenland
Outer Hebrides
Alaska
Hudson Bay
British Columbia
Quebec
Nevada
California
Cuba
Spines
Pectoral girdle
Reduced spines
The current range of sticklebacks includes formerly glaciated areas (lavender) and unglaciated areas (green).
The region of the Northern Hemisphere that was once covered by Pleistocene glaciers is outlined in black.
Places where sticklebacks are known to have reduced spines are indicated by circles.

22.20 Natural Selection Acts on Stickleback Spines *(Page 456)*

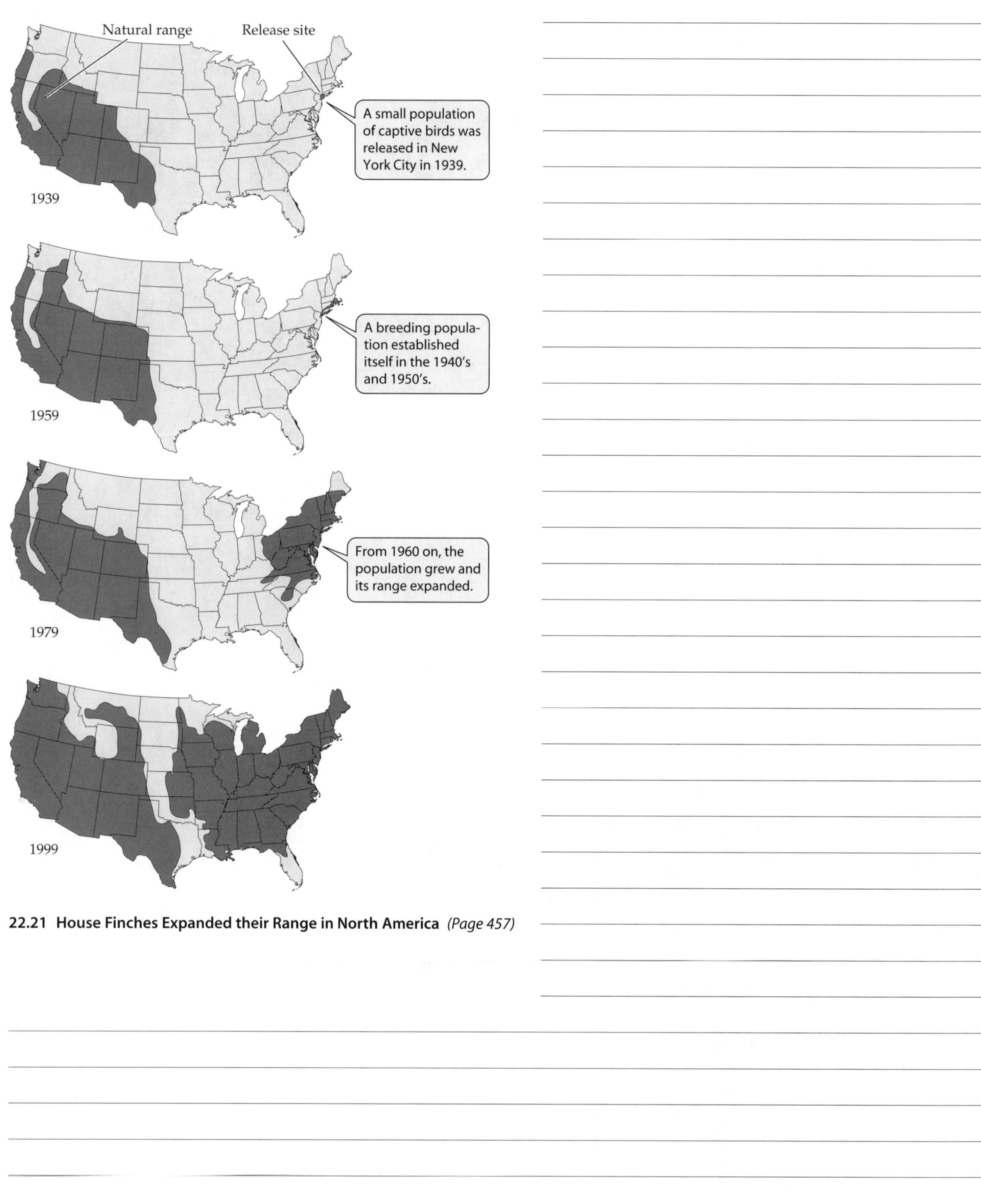
Natural range
Release site
A small population of captive birds was released in New York City in 1939.
1939
A breeding population established itself in the 1940's and 1950's.
1959
From 1960 on, the population grew and its range expanded.
1979
1999

22.21 House Finches Expanded their Range in North America *(Page 457)*

23 The Mechanisms of Evolution

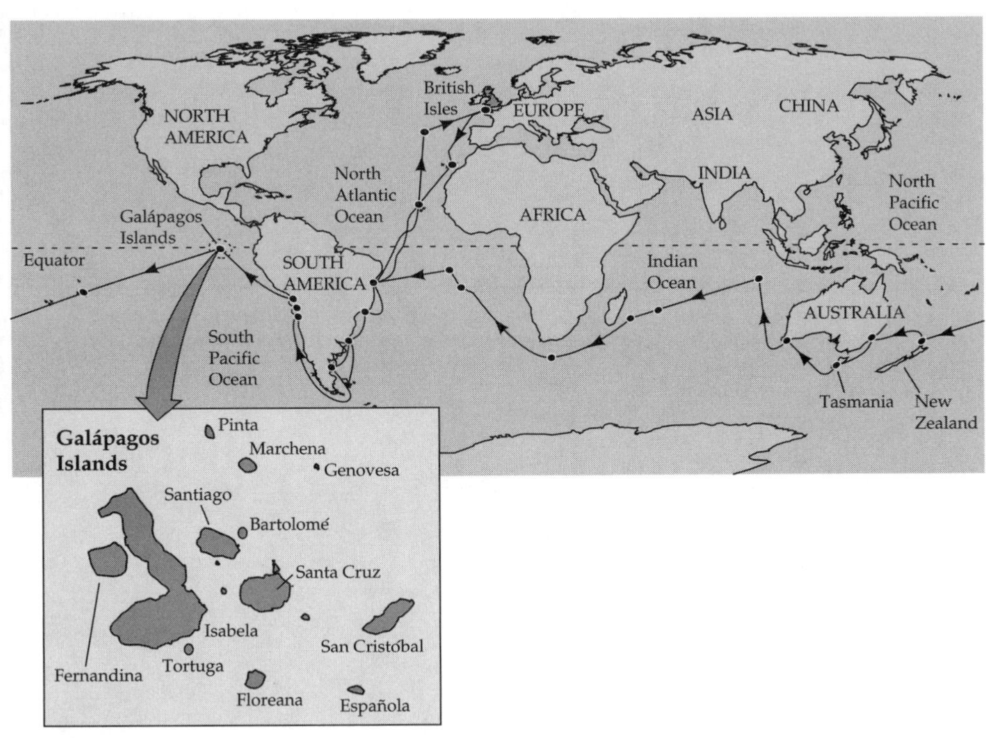

British Isles
EUROPE
ASIA
CHINA
NORTH AMERICA
INDIA
North Atlantic Ocean
AFRICA
North Pacific Ocean
Galápagos Islands
Equator
SOUTH AMERICA
Indian Ocean
AUSTRALIA
South Pacific Ocean
Tasmania
New Zealand
Galápagos Islands
Pinta
Marchena
Genovesa
Santiago
Bartolomé
Santa Cruz
Isabela
San Cristóbal
Fernandina
Tortuga
Floreana
Española

23.1 Darwin and the Voyage of the *Beagle* (Page 461)

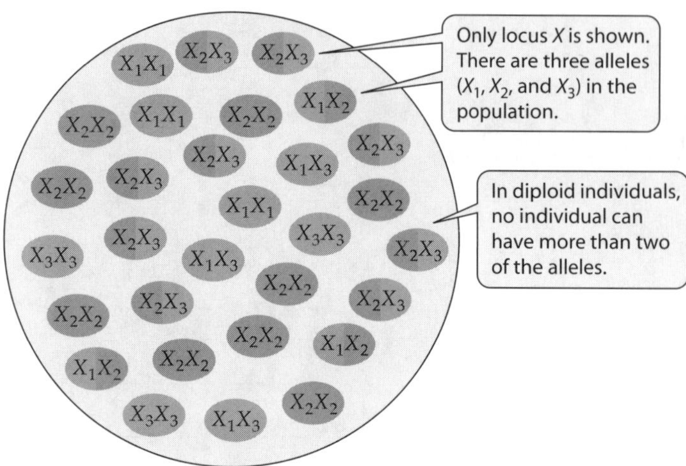
Only locus X is shown. There are three alleles (X_1, X_2, and X_3) in the population.
In diploid individuals, no individual can have more than two of the alleles.

23.3 A Gene Pool *(Page 463)*

European agriculturalists chose as parents for subsequent generations individual wild mustard plants that varied from the population's average by producing unusually large leaves, stems, buds, or flowers.

Brassica oleracea
(a common wild mustard)

Selection for terminal buds
Selection for lateral buds
Selection for stem
Selection for leaves
Selection for stems and flowers
Selection for flower clusters

Cabbage
Brussels sprouts
Kohlrabi
Kale
Broccoli
Cauliflower

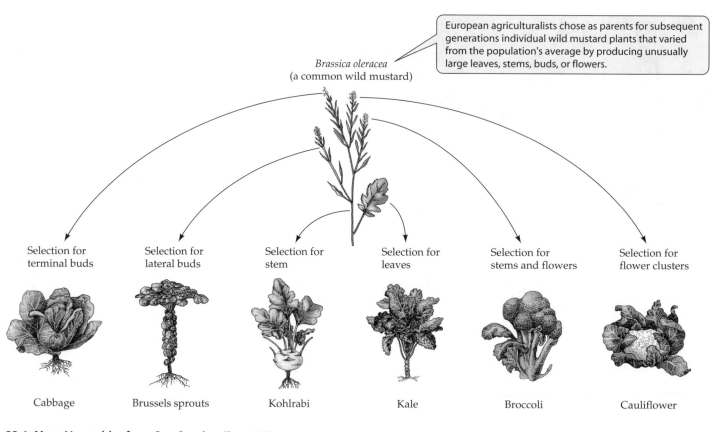

23.4 Many Vegetables from One Species *(Page 464)*

Leaves of a white oak (*Quercus alba*)

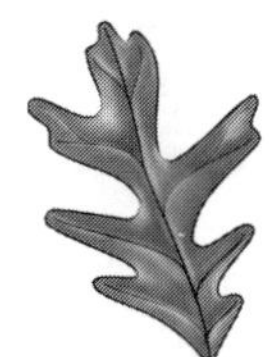

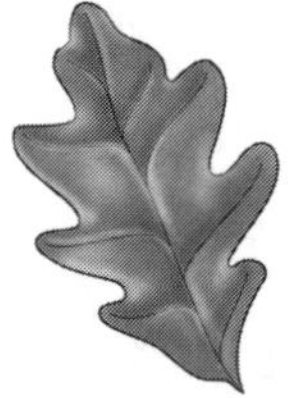

Grown in sun   Grown in shade

In-Text Art *(Page 464)*

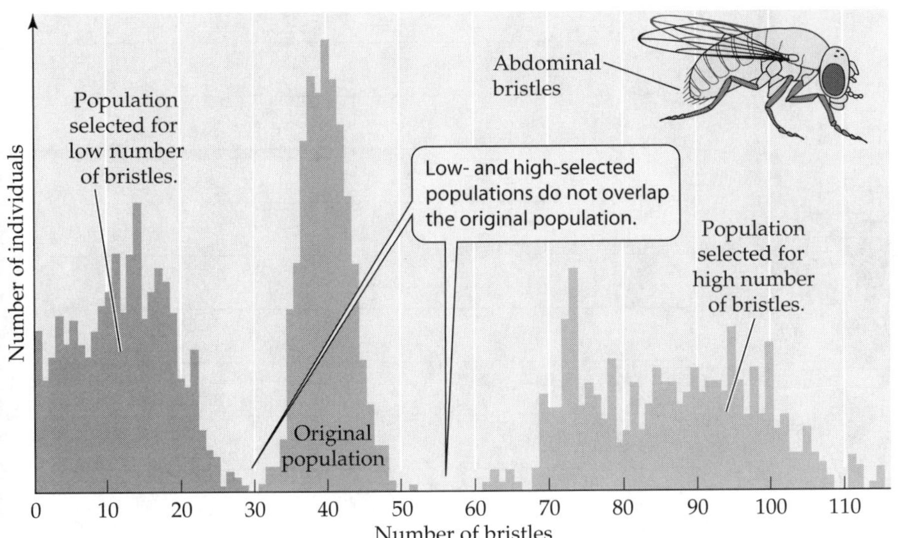

Number of individuals
Population selected for low number of bristles.
Abdominal bristles
Low- and high-selected populations do not overlap the original population.
Population selected for high number of bristles.
Original population
0
10
20
30
40
50
60
70
80
90
100
110
Number of bristles

23.5 Artificial Selection Reveals Genetic Variation *(Page 465)*

In any population:

$$\text{Frequency of allele } A = p = \frac{2N_{AA} + N_{Aa}}{2N} \qquad \text{Frequency of allele } a = q = \frac{2N_{aa} + N_{Aa}}{2N}$$

where N is the total number of individuals in the population.

For population 1 (mostly homozygotes):

$N_{AA} = 90$, $N_{Aa} = 40$, and $N_{aa} = 70$

so

$$p = \frac{180 + 40}{400} = 0.55$$

$$q = \frac{140 + 40}{400} = 0.45$$

For population 2 (mostly heterozygotes):

$N_{AA} = 45$, $N_{Aa} = 130$, and $N_{aa} = 25$

so

$$p = \frac{90 + 130}{400} = 0.55$$

$$q = \frac{50 + 130}{400} = 0.45$$

23.6 Calculating Allele Frequencies *(Page 466)*

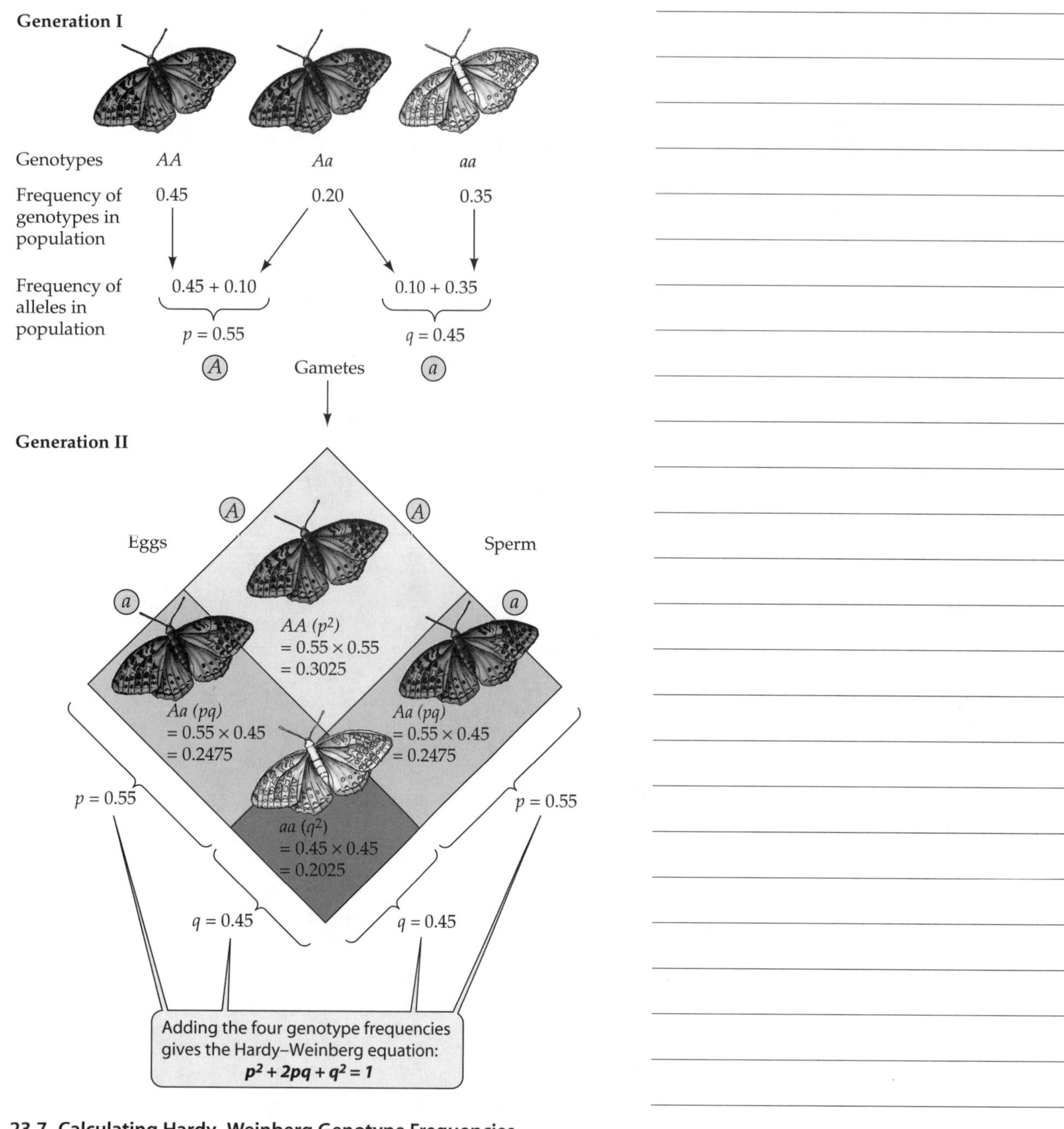
Generation I
Genotypes
AA
Aa
aa
Frequency of genotypes in population
0.45
0.20
0.35
Frequency of alleles in population
0.45 + 0.10
0.10 + 0.35
p = 0.55
q = 0.45
A
Gametes
a
Generation II
Eggs
Sperm
A
A
a
a
AA (p2)
= 0.55 × 0.55
= 0.3025
Aa (pq)
= 0.55 × 0.45
= 0.2475
Aa (pq)
= 0.55 × 0.45
= 0.2475
aa (q2)
= 0.45 × 0.45
= 0.2025
p = 0.55
p = 0.55
q = 0.45
q = 0.45
Adding the four genotype frequencies gives the Hardy–Weinberg equation:
p2 + 2pq + q2 = 1

23.7 Calculating Hardy–Weinberg Genotype Frequencies
(Page 467)

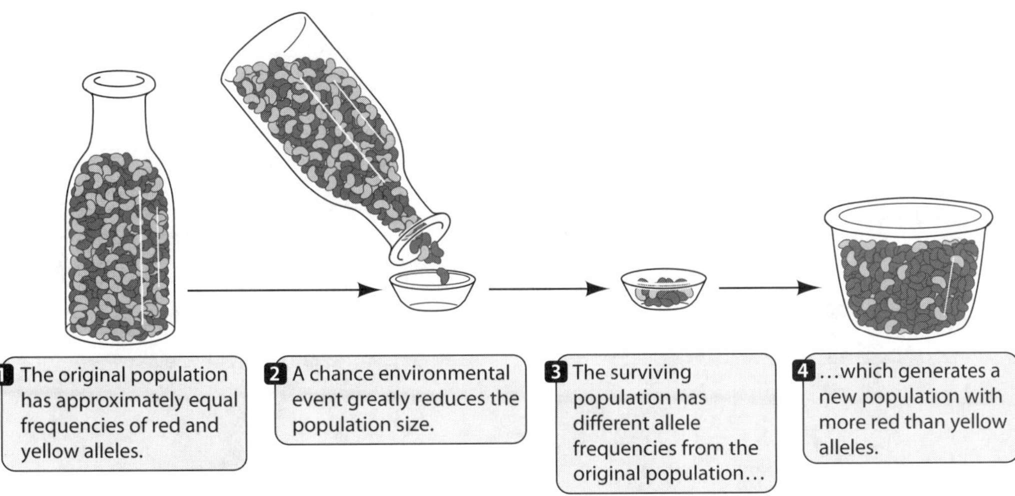
1 The original population has approximately equal frequencies of red and yellow alleles.
2 A chance environmental event greatly reduces the population size.
3 The surviving population has different allele frequencies from the original population…
4 …which generates a new population with more red than yellow alleles.

23.8 A Population Bottleneck *(Page 468)*

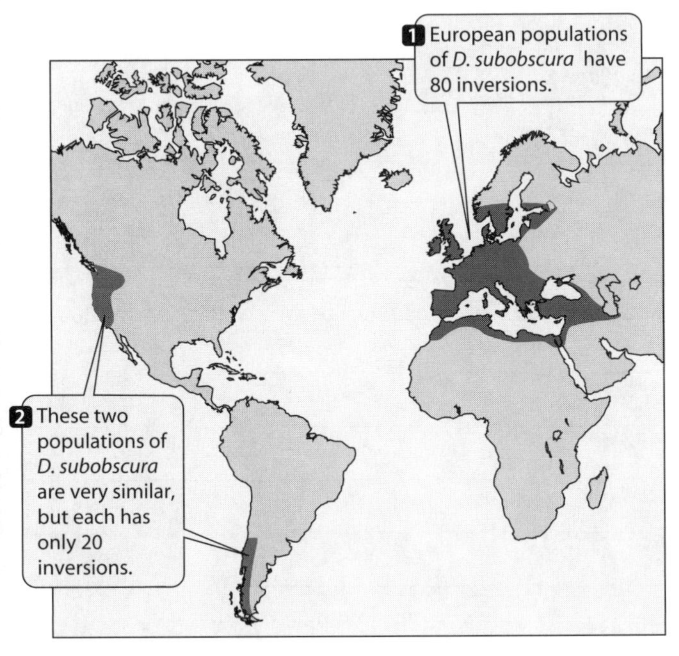
1 European populations of D. subobscura have 80 inversions.
2 These two populations of D. subobscura are very similar, but each has only 20 inversions.

23.10 A Founder Effect *(Page 469)*

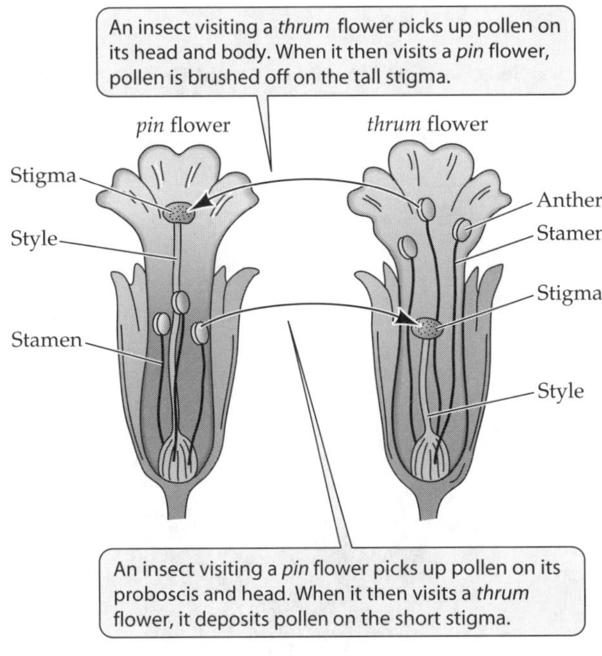
An insect visiting a thrum flower picks up pollen on its head and body. When it then visits a pin flower, pollen is brushed off on the tall stigma.
pin flower
thrum flower
Stigma
Style
Stamen
Anther
Stamen
Stigma
Style
An insect visiting a pin flower picks up pollen on its proboscis and head. When it then visits a thrum flower, it deposits pollen on the short stigma.

23.11 Flower Structure Fosters Nonrandom Mating *(Page 470)*

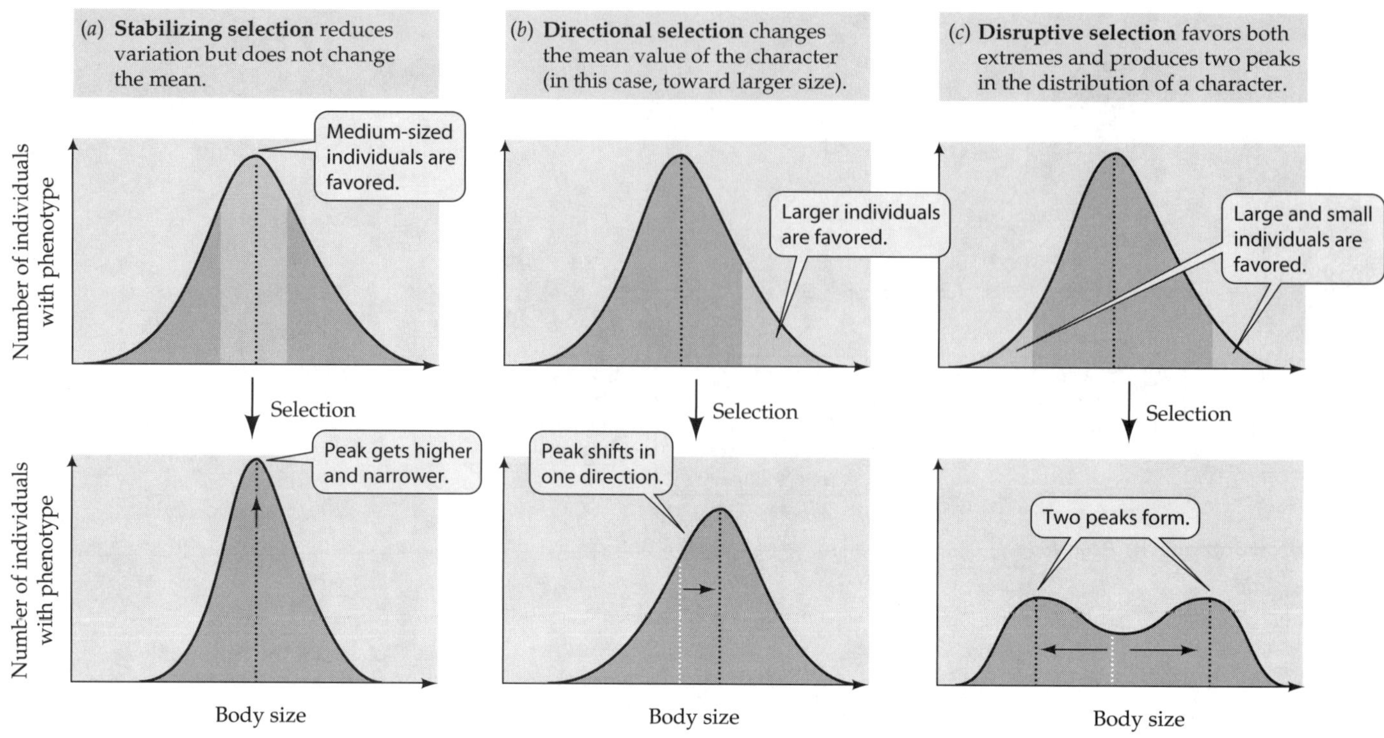
(a) Stabilizing selection reduces variation but does not change the mean.
(b) Directional selection changes the mean value of the character (in this case, toward larger size).
(c) Disruptive selection favors both extremes and produces two peaks in the distribution of a character.
Number of individuals with phenotype
Medium-sized individuals are favored.
Larger individuals are favored.
Large and small individuals are favored.
Selection
Peak gets higher and narrower.
Peak shifts in one direction.
Two peaks form.
Body size

23.12 Natural Selection Can Operate on Quantitative Variation in Several Ways *(Page 471)*

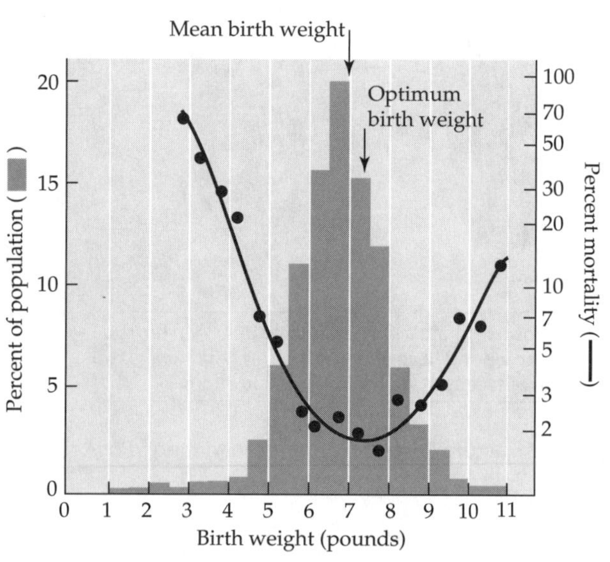
Mean birth weight
Optimum birth weight
Percent of population
Percent mortality
Birth weight (pounds)

23.13 Human Birth Weight Is Influenced by Stabilizing Selection *(Page 471)*

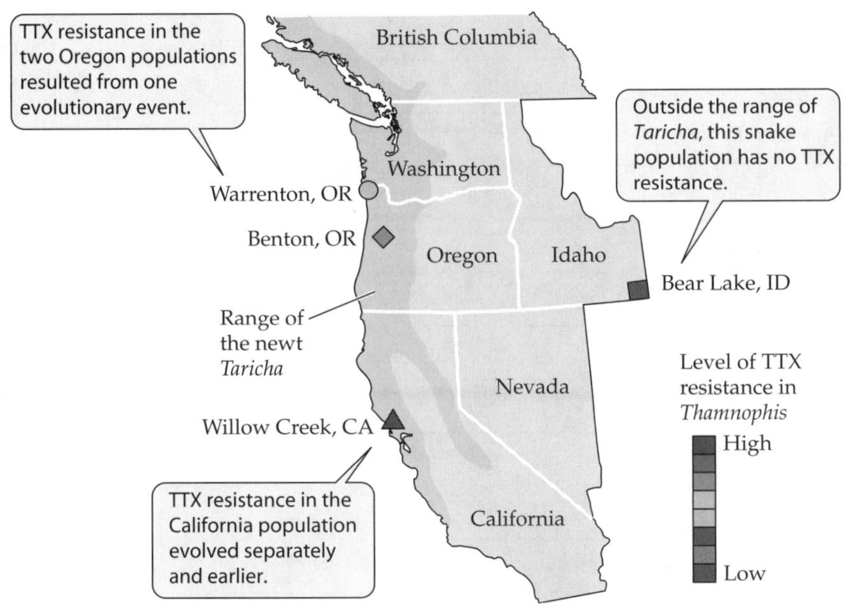
TTX resistance in the two Oregon populations resulted from one evolutionary event.
British Columbia
Washington
Oregon
Idaho
Nevada
California
Warrenton, OR
Benton, OR
Range of the newt *Taricha*
Willow Creek, CA
Outside the range of *Taricha*, this snake population has no TTX resistance.
Bear Lake, ID
Level of TTX resistance in *Thamnophis*
High
Low
TTX resistance in the California population evolved separately and earlier.

23.14 Resistance to TTX Is Associated with the Presence of Newts *(Page 472)*

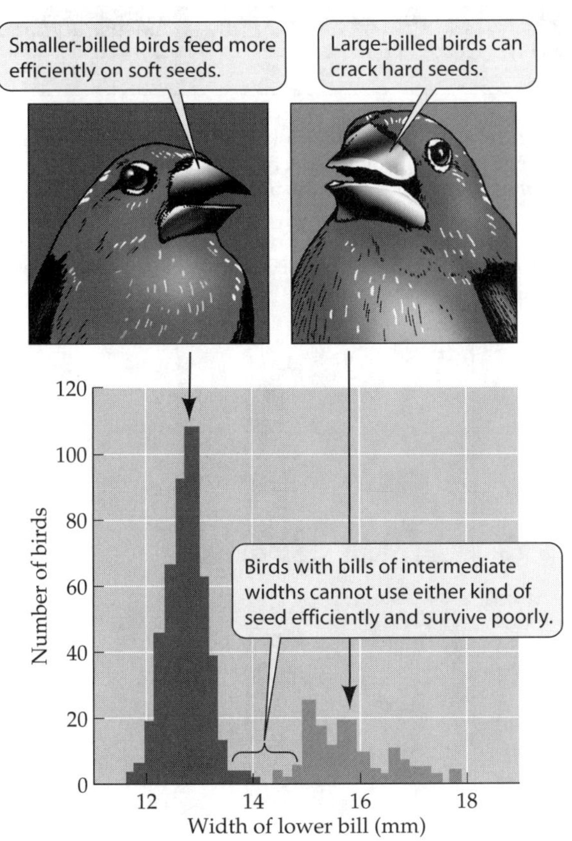
Smaller-billed birds feed more efficiently on soft seeds.
Large-billed birds can crack hard seeds.
Birds with bills of intermediate widths cannot use either kind of seed efficiently and survive poorly.
Number of birds
0
20
40
60
80
100
120
12
14
16
18
Width of lower bill (mm)

23.15 Disruptive Selection Results in a Bimodal Distribution *(Page 472)*

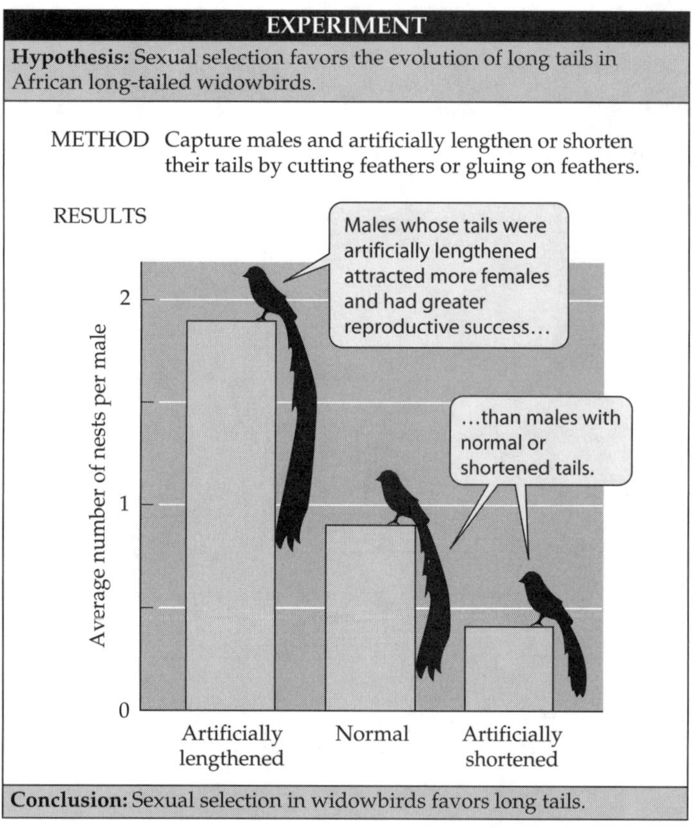
EXPERIMENT
Hypothesis: Sexual selection favors the evolution of long tails in African long-tailed widowbirds.
METHOD Capture males and artificially lengthen or shorten their tails by cutting feathers or gluing on feathers.
RESULTS
Males whose tails were artificially lengthened attracted more females and had greater reproductive success…
…than males with normal or shortened tails.
Average number of nests per male
0
1
2
Artificially lengthened
Normal
Artificially shortened
Conclusion: Sexual selection in widowbirds favors long tails.

23.16 The Longer the Tail, the Better the Male *(Page 473)*

(a)

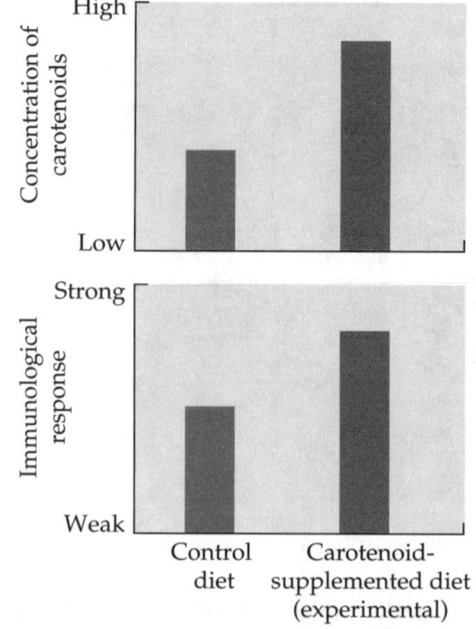
EXPERIMENT
Hypothesis: Having a bright red bill signals good health in a male zebra finch.
METHOD Provide carotenoids in drinking water for experimental, but not for control males. Challenge all males immunologically and measure response.
RESULTS Experimental males responded more strongly to the immunological challenge. They also developed brighter bills than control males.
High
Low
Concentration of carotenoids
Strong
Weak
Immunological response
Control diet
Carotenoid-supplemented diet (experimental)
Conclusion: Bill color is a clue to the health of a male zebra finch.

23.17 Bright Bills Signal Good Health *(Page 474)*

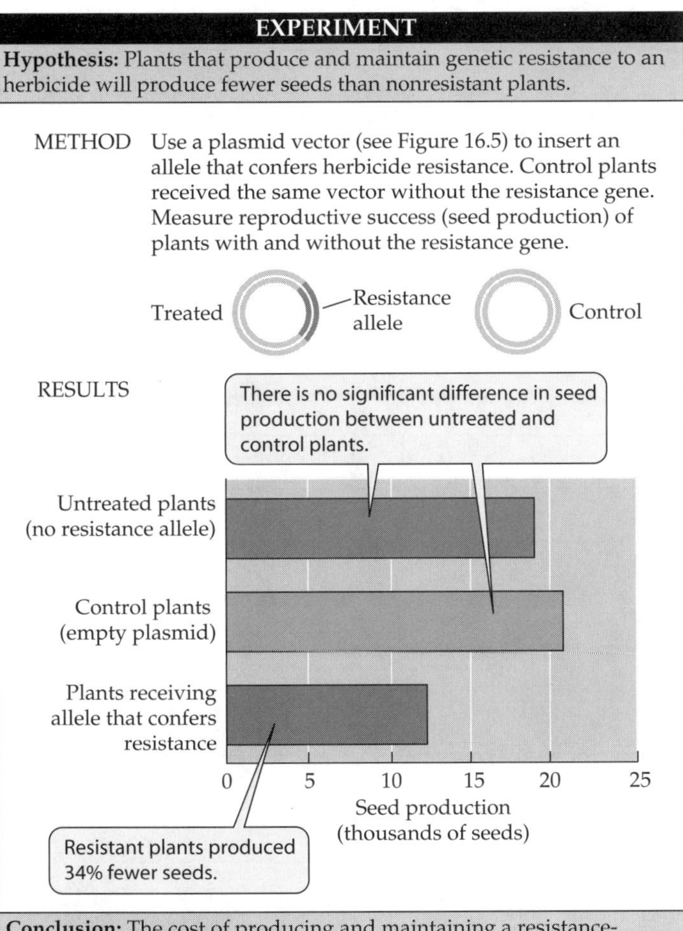
EXPERIMENT
Hypothesis: Plants that produce and maintain genetic resistance to an herbicide will produce fewer seeds than nonresistant plants.
METHOD Use a plasmid vector (see Figure 16.5) to insert an allele that confers herbicide resistance. Control plants received the same vector without the resistance gene. Measure reproductive success (seed production) of plants with and without the resistance gene.
Treated
Resistance allele
Control
RESULTS
There is no significant difference in seed production between untreated and control plants.
Untreated plants (no resistance allele)
Control plants (empty plasmid)
Plants receiving allele that confers resistance
0 5 10 15 20 25
Seed production (thousands of seeds)
Resistant plants produced 34% fewer seeds.
Conclusion: The cost of producing and maintaining a resistance-conferring compound is high.

23.18 Producing and Maintaining Resistance Is Costly *(Page 475)*

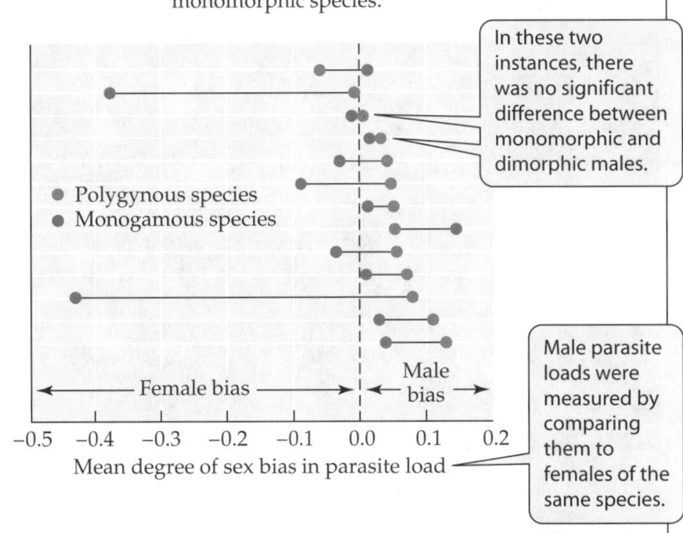
COMPARATIVE METHOD
Hypothesis: Males of sexually dimorphic species have greater parasite loads than males of monomorphic species.
METHOD Measure parasite loads in males. Compare parasite loads of males of polygynous, sexually dimorphic species with parasite loads in males of closely related monogamous species, which are essentially monomorphic.
RESULTS In all but two comparisons, males of dimorphic species had greater parasite loads than males of monomorphic species.
In these two instances, there was no significant difference between monomorphic and dimorphic males.
Polygynous species
Monogamous species
Female bias
Male bias
−0.5 −0.4 −0.3 −0.2 −0.1 0.0 0.1 0.2
Mean degree of sex bias in parasite load
Male parasite loads were measured by comparing them to females of the same species.
Conclusion: For male mammals, the cost of sexual dimorphism is an enhanced risk of parasites.

23.19 Sexually Selected Traits Impose Costs *(Page 475)*

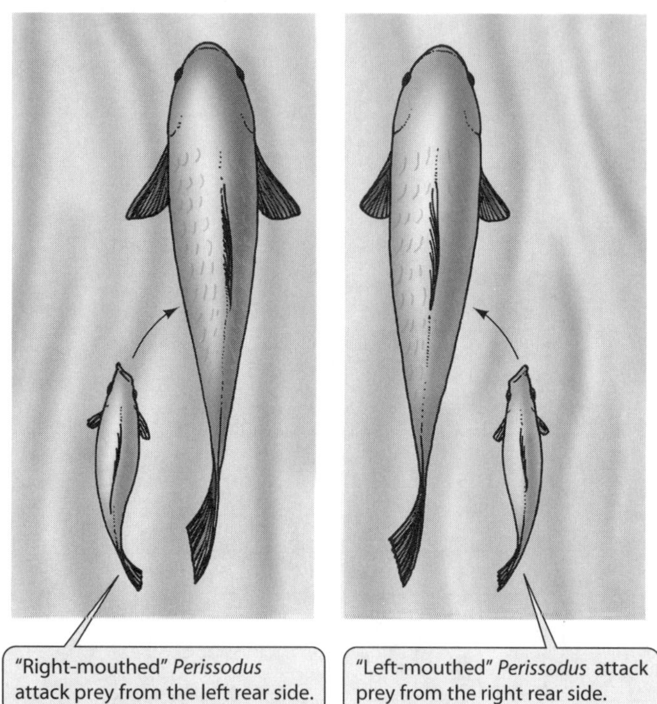
"Right-mouthed" *Perissodus* attack prey from the left rear side.
"Left-mouthed" *Perissodus* attack prey from the right rear side.

23.20 A Stable Polymorphism *(Page 476)*

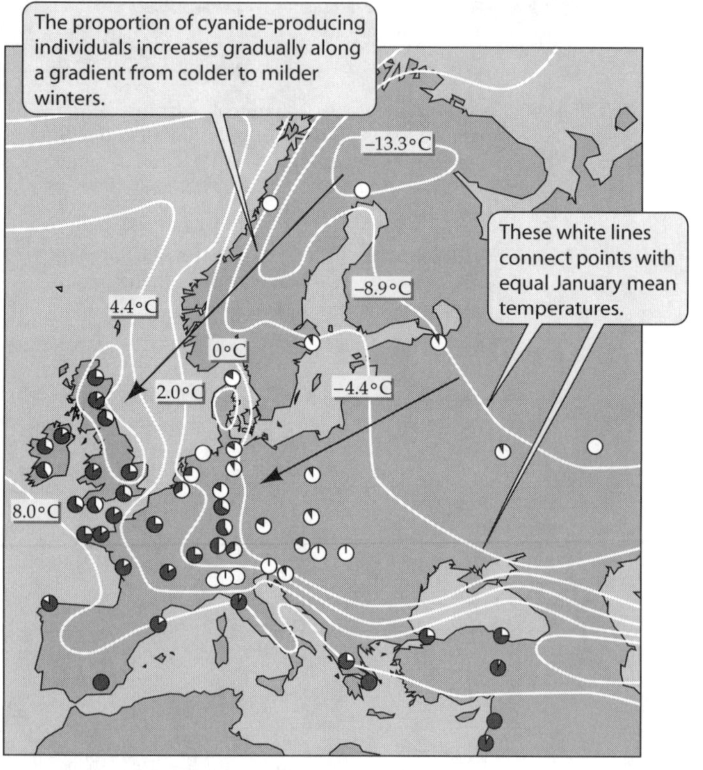
The proportion of cyanide-producing individuals increases gradually along a gradient from colder to milder winters.
–13.3°C
These white lines connect points with equal January mean temperatures.
–8.9°C
4.4°C
0°C
2.0°C
–4.4°C
8.0°C
White indicates proportion of plants not producing cyanide
Red indicates proportion of plants producing cyanide

23.21 Geographic Variation in Poisonous Clovers *(Page 477)*

24 Species and Their Formation

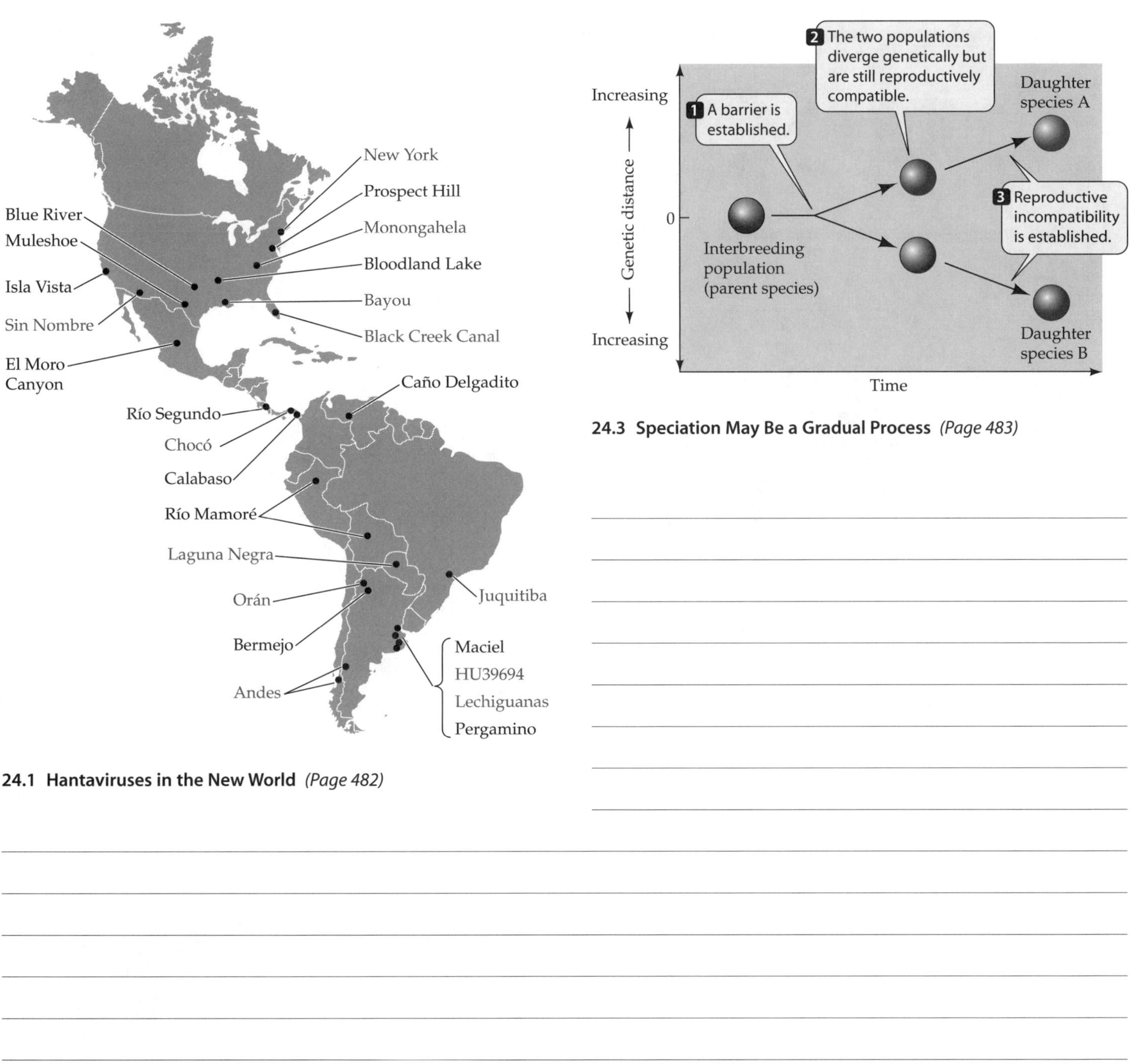
New York
Prospect Hill
Monongahela
Bloodland Lake
Bayou
Black Creek Canal
Blue River
Muleshoe
Isla Vista
Sin Nombre
El Moro Canyon
Caño Delgadito
Río Segundo
Chocó
Calabaso
Río Mamoré
Laguna Negra
Juquitiba
Orán
Bermejo
Andes
Maciel
HU39694
Lechiguanas
Pergamino
Increasing
Genetic distance
0
Increasing
Time
1 A barrier is established.
2 The two populations diverge genetically but are still reproductively compatible.
3 Reproductive incompatibility is established.
Interbreeding population (parent species)
Daughter species A
Daughter species B

24.1 Hantaviruses in the New World *(Page 482)*

24.3 Speciation May Be a Gradual Process *(Page 483)*

Time

A single species is distributed over a broad range.

Sea level rises and separates two populations. Populations adapt to differing environments on opposite sides of the barrier.

If the barrier is removed, the populations may recolonize the intervening area and mingle, but do not interbreed.

Range of overlap

24.4 Allopatric Speciation *(Page 484)*

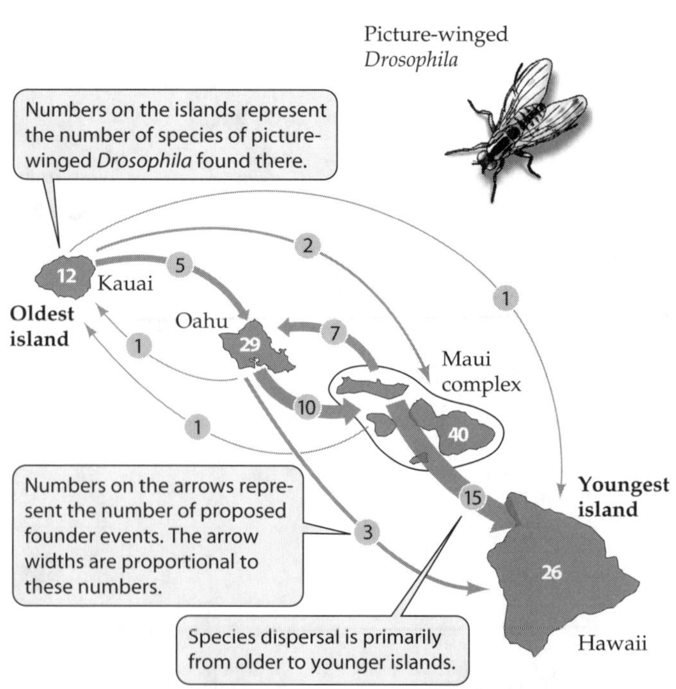

Picture-winged *Drosophila*
Numbers on the islands represent the number of species of picture-winged *Drosophila* found there.
12
Kauai
Oldest island
Oahu
29
Maui complex
40
Youngest island
26
Hawaii
5
2
1
7
1
10
1
15
3
Numbers on the arrows represent the number of proposed founder events. The arrow widths are proportional to these numbers.
Species dispersal is primarily from older to younger islands.

24.5 Founder Events Lead to Allopatric Speciation *(Page 484)*

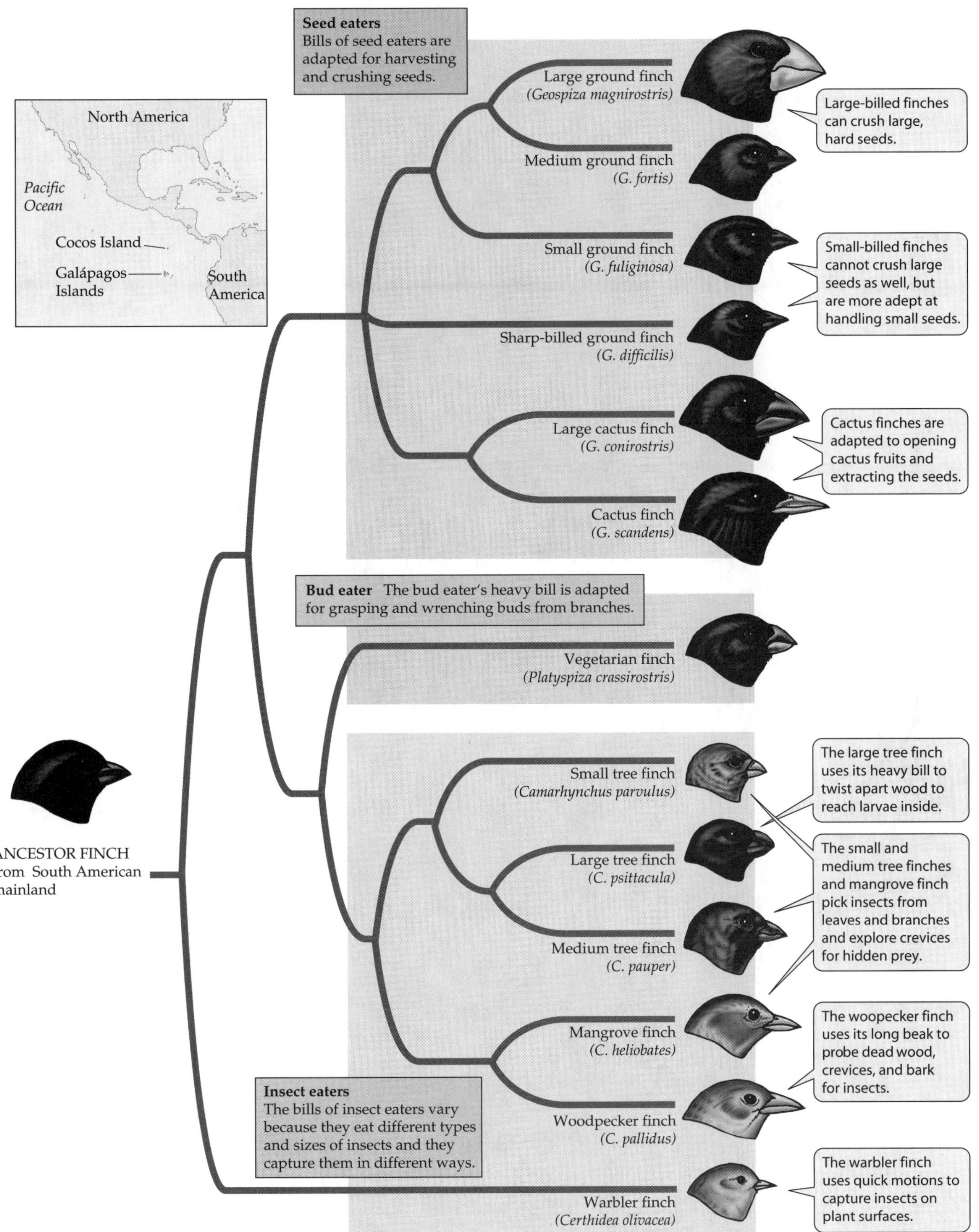
Seed eaters
Bills of seed eaters are adapted for harvesting and crushing seeds.
North America
Pacific Ocean
Cocos Island
Galápagos Islands
South America
Large ground finch
(Geospiza magnirostris)
Large-billed finches can crush large, hard seeds.
Medium ground finch
(G. fortis)
Small ground finch
(G. fuliginosa)
Small-billed finches cannot crush large seeds as well, but are more adept at handling small seeds.
Sharp-billed ground finch
(G. difficilis)
Large cactus finch
(G. conirostris)
Cactus finches are adapted to opening cactus fruits and extracting the seeds.
Cactus finch
(G. scandens)
Bud eater The bud eater's heavy bill is adapted for grasping and wrenching buds from branches.
Vegetarian finch
(Platyspiza crassirostris)
Small tree finch
(Camarhynchus parvulus)
The large tree finch uses its heavy bill to twist apart wood to reach larvae inside.
ANCESTOR FINCH from South American mainland
Large tree finch
(C. psittacula)
The small and medium tree finches and mangrove finch pick insects from leaves and branches and explore crevices for hidden prey.
Medium tree finch
(C. pauper)
Mangrove finch
(C. heliobates)
The woopecker finch uses its long beak to probe dead wood, crevices, and bark for insects.
Insect eaters
The bills of insect eaters vary because they eat different types and sizes of insects and they capture them in different ways.
Woodpecker finch
(C. pallidus)
Warbler finch
(Certhidea olivacea)
The warbler finch uses quick motions to capture insects on plant surfaces.

24.6 Allopatric Speciation among Darwin's Finches *(Page 485)*

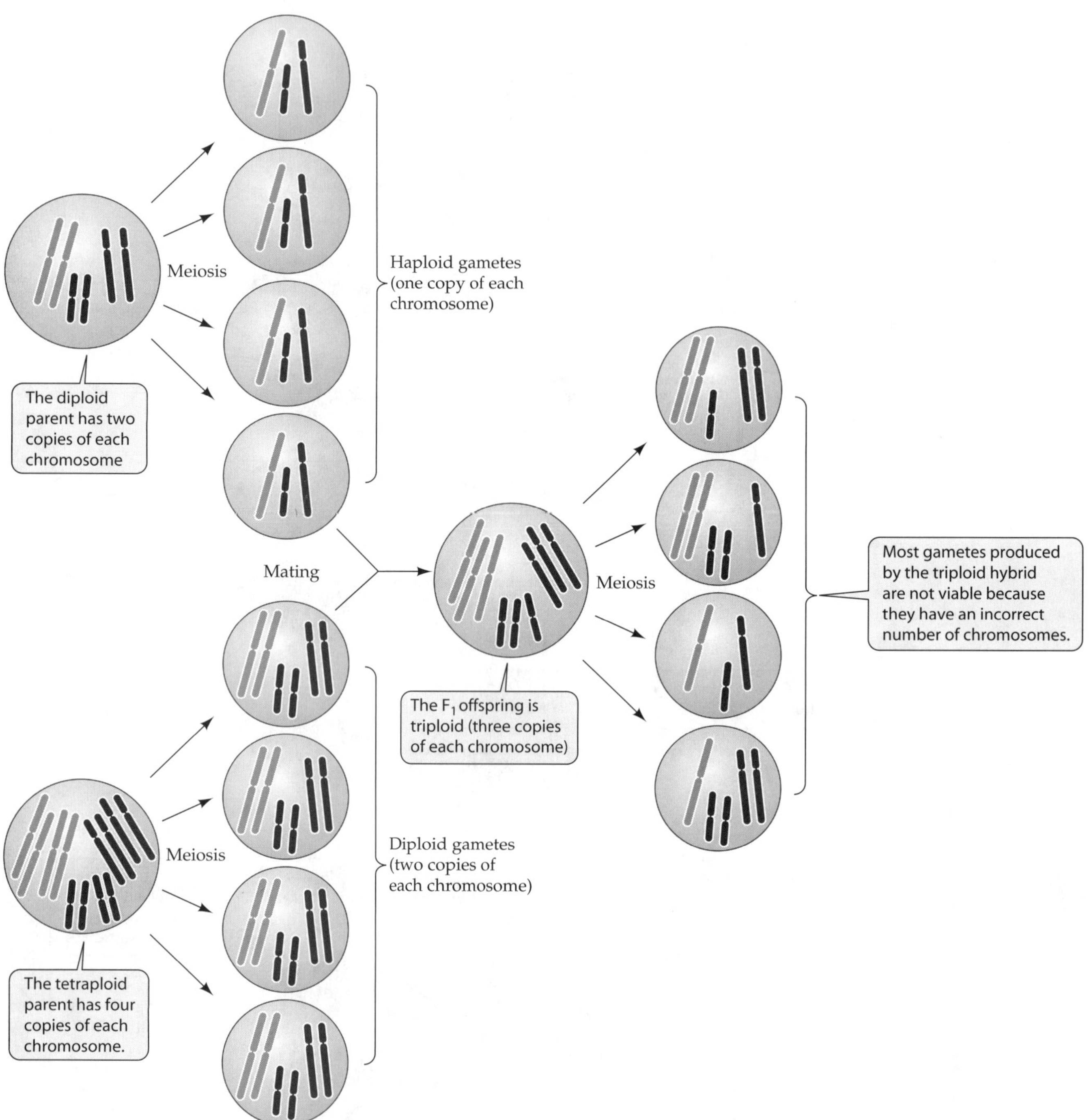
Meiosis
The diploid parent has two copies of each chromosome
Haploid gametes (one copy of each chromosome)
Mating
Meiosis
The tetraploid parent has four copies of each chromosome.
Diploid gametes (two copies of each chromosome)
The F1 offspring is triploid (three copies of each chromosome)
Meiosis
Most gametes produced by the triploid hybrid are not viable because they have an incorrect number of chromosomes.

24.7 Tetraploids Are Soon Reproductively Isolated from Diploids *(Page 486)*

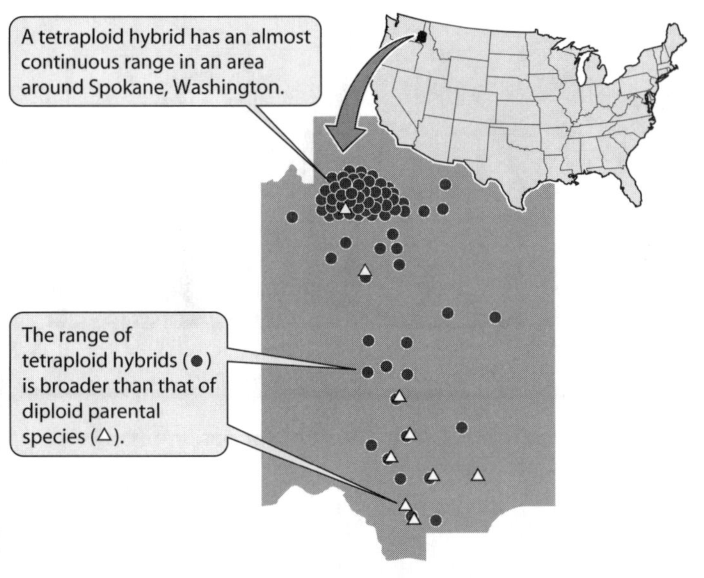
A tetraploid hybrid has an almost continuous range in an area around Spokane, Washington.
The range of tetraploid hybrids (●) is broader than that of diploid parental species (△).

24.8 Polyploids May Outperform Their Parent Species *(Page 487)*

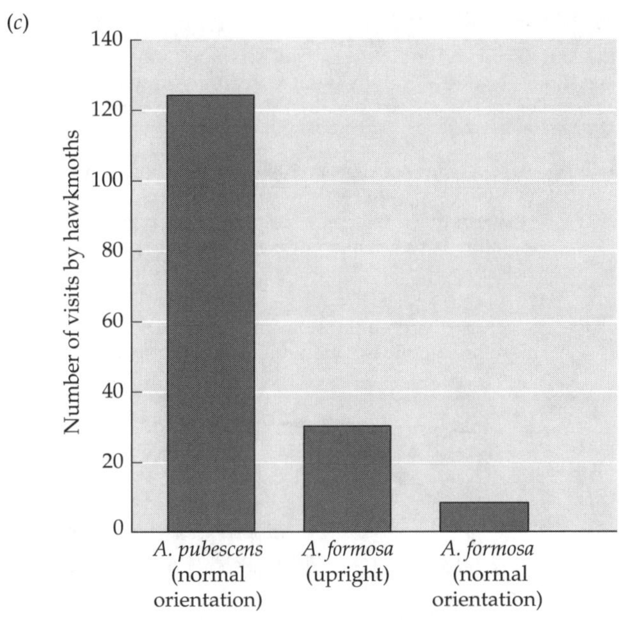
(c)
Number of visits by hawkmoths
140
120
100
80
60
40
20
0
A. pubescens (normal orientation)
A. formosa (upright)
A. formosa (normal orientation)

24.11 Hawkmoths Favor Flowers of One Columbine Species *(Page 489)*

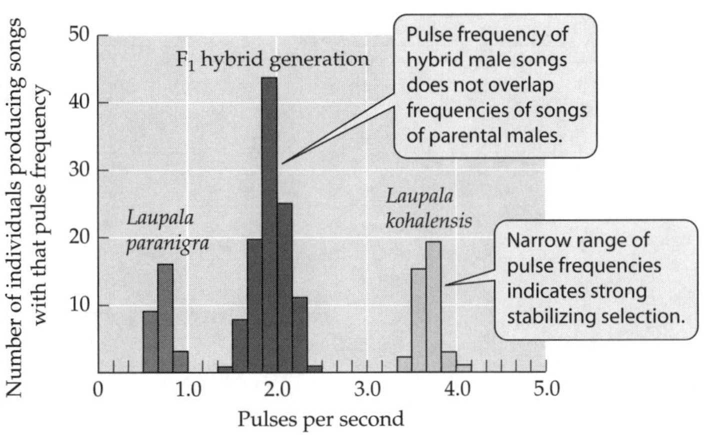
Number of individuals producing songs with that pulse frequency
50
40
30
20
10
F1 hybrid generation
Pulse frequency of hybrid male songs does not overlap frequencies of songs of parental males.
Laupala paranigra
Laupala kohalensis
Narrow range of pulse frequencies indicates strong stabilizing selection.
0
1.0
2.0
3.0
4.0
5.0
Pulses per second

24.10 Songs of Male Crickets are Genetically Determined *(Page 488)*

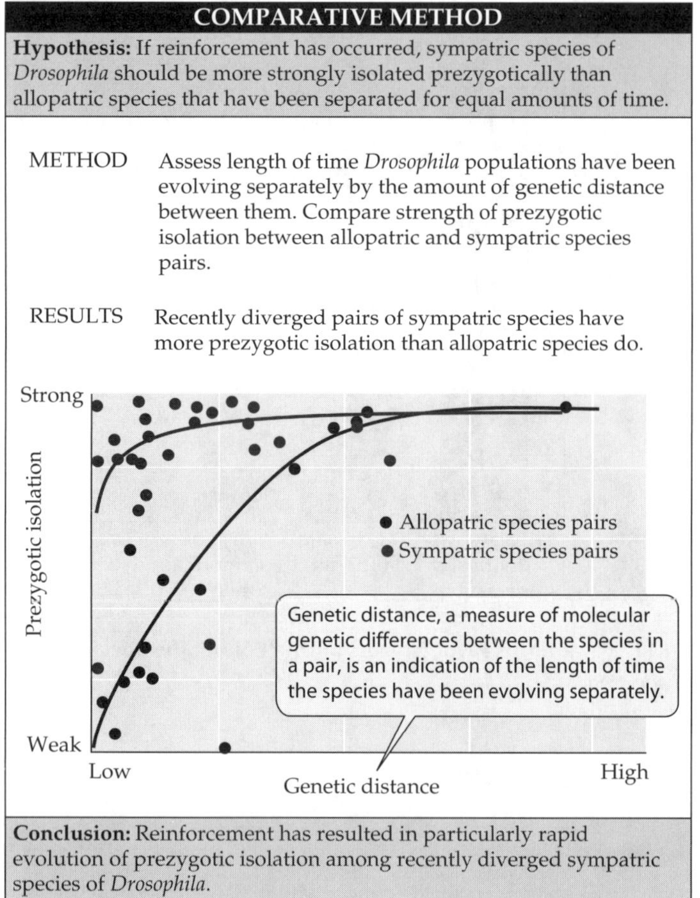
COMPARATIVE METHOD
Hypothesis: If reinforcement has occurred, sympatric species of Drosophila should be more strongly isolated prezygotically than allopatric species that have been separated for equal amounts of time.
METHOD Assess length of time Drosophila populations have been evolving separately by the amount of genetic distance between them. Compare strength of prezygotic isolation between allopatric and sympatric species pairs.
RESULTS Recently diverged pairs of sympatric species have more prezygotic isolation than allopatric species do.
Strong
Weak
Prezygotic isolation
Allopatric species pairs
Sympatric species pairs
Genetic distance, a measure of molecular genetic differences between the species in a pair, is an indication of the length of time the species have been evolving separately.
Low
High
Genetic distance
Conclusion: Reinforcement has resulted in particularly rapid evolution of prezygotic isolation among recently diverged sympatric species of Drosophila.

24.12 Prezygotic Barriers Can Evolve Rapidly *(Page 490)*

(c)

B. bombina
(fire-bellied
toad)
B. variegata
(yellow-bellied
toad)
Area of
overlap

24.13 Hybrid Zones May Be Long and Narrow *(Page 491)*

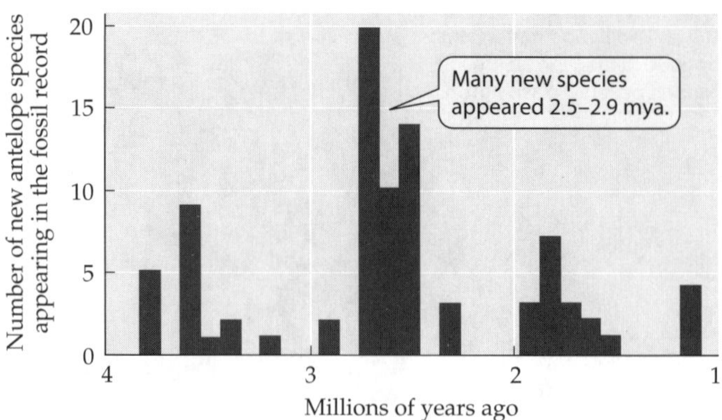
Number of new antelope species appearing in the fossil record
20
15
10
5
0
4
3
2
1
Millions of years ago
Many new species appeared 2.5–2.9 mya.

24.15 Climate Change Drove a Burst of Speciation among Antelopes *(Page 492)*

25 Reconstructing and Using Phylogenies

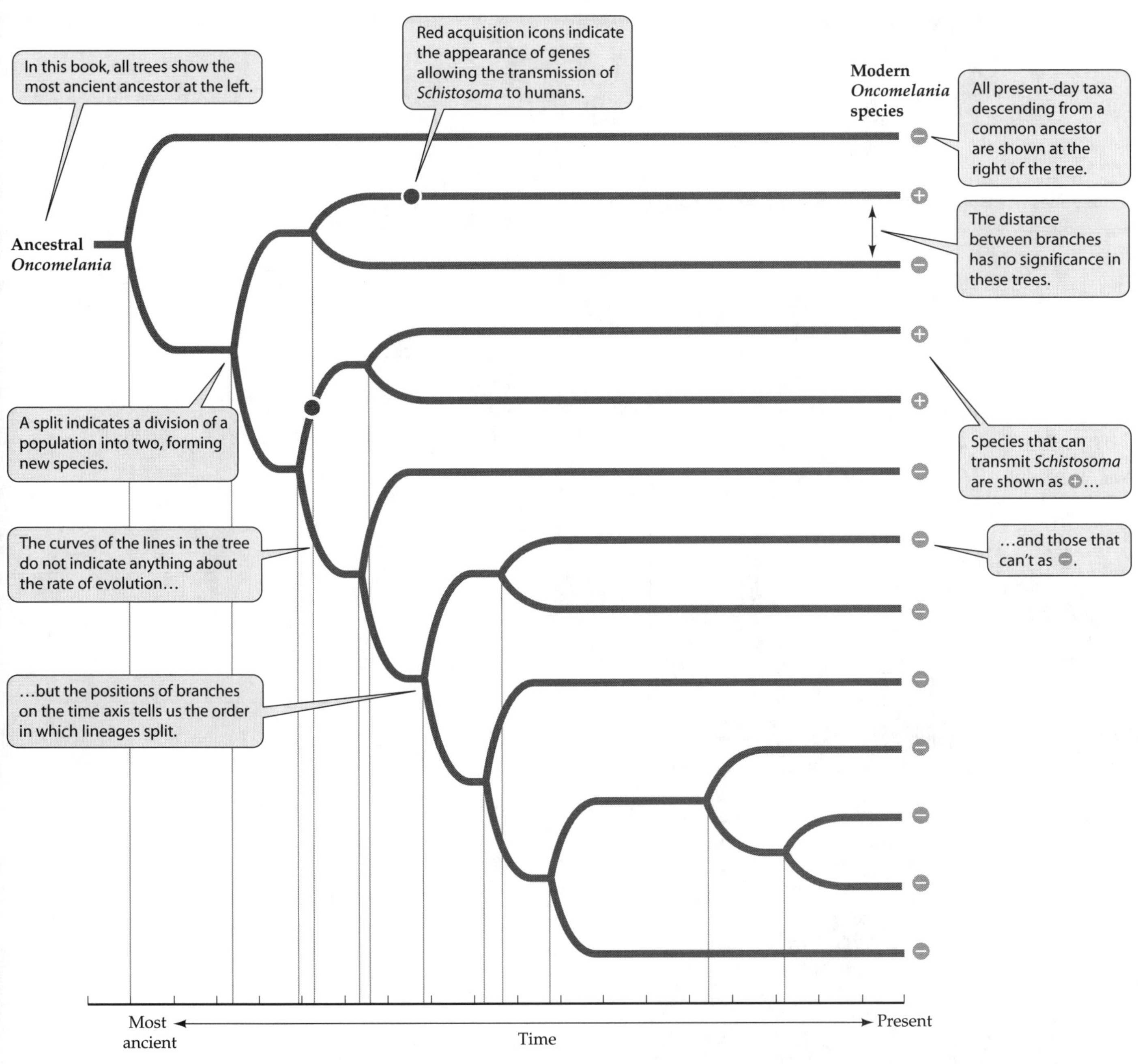

In this book, all trees show the most ancient ancestor at the left.
Red acquisition icons indicate the appearance of genes allowing the transmission of *Schistosoma* to humans.
Modern *Oncomelania* species
All present-day taxa descending from a common ancestor are shown at the right of the tree.
Ancestral *Oncomelania*
The distance between branches has no significance in these trees.
A split indicates a division of a population into two, forming new species.
Species that can transmit *Schistosoma* are shown as ⊕…
The curves of the lines in the tree do not indicate anything about the rate of evolution…
…and those that can't as ⊖.
…but the positions of branches on the time axis tells us the order in which lineages split.
Most ancient
Time
Present

25.1 How to Read a Phylogenetic Tree *(Page 497)*

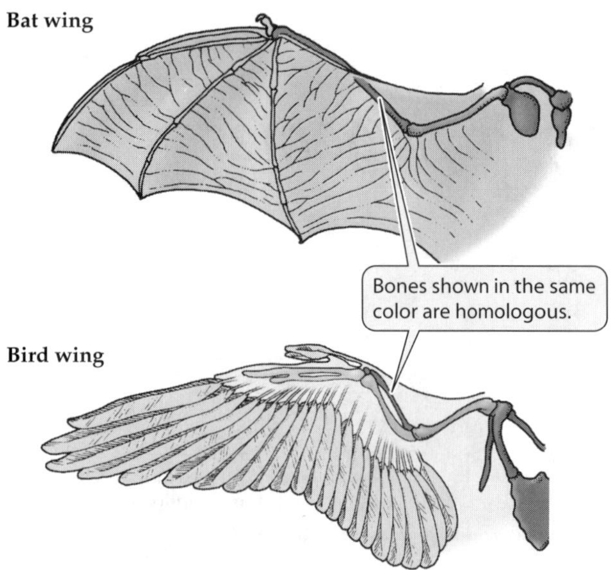
Bat wing
Bones shown in the same color are homologous.
Bird wing

25.2 The Bones Are Homologous, but the Wings Are Not *(Page 498)*

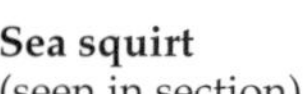
Sea squirt
(seen in section)

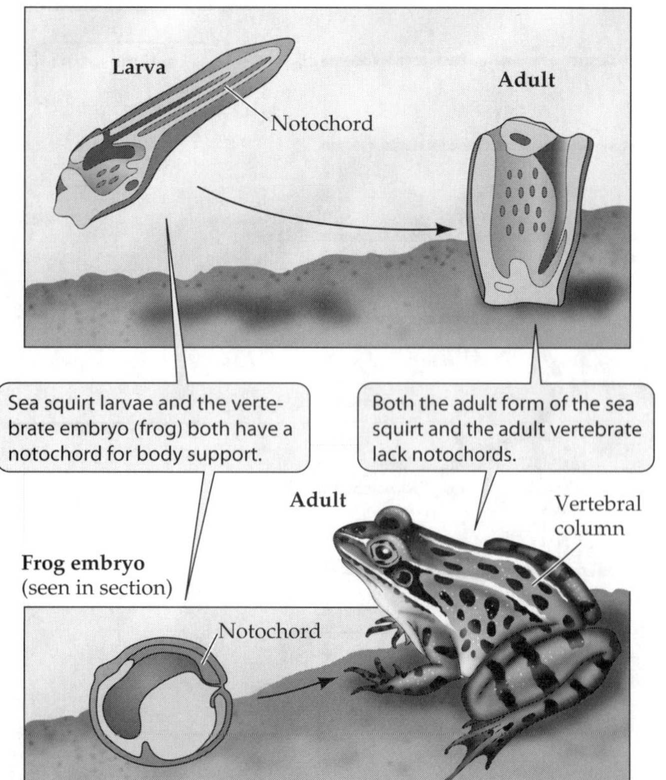
Larva
Notochord
Adult
Sea squirt larvae and the vertebrate embryo (frog) both have a notochord for body support.
Both the adult form of the sea squirt and the adult vertebrate lack notochords.
Adult
Vertebral column
Frog embryo
(seen in section)
Notochord

25.4 A Larva Reveals Evolutionary Relationships *(Page 500)*

25.1 *Eight Vertebrates Ordered According to Unique Shared Derived Traits*

	DERIVED TRAIT[a]							
TAXON	JAWS	LUNGS	CLAWS OR NAILS	GIZZARD	FEATHERS	FUR	MAMMARY GLANDS	KERATINOUS SCALES
Lamprey (outgroup)	–	–	–	–	–	–	–	–
Perch	+	–	–	–	–	–	–	–
Salamander	+	+	–	–	–	–	–	–
Lizard	+	+	+	–	–	–	–	+
Crocodile	+	+	+	+	–	–	–	+
Pigeon	+	+	+	+	+	–	–	+
Mouse	+	+	+	–	–	+	+	–
Chimpanzee	+	+	+	–	–	+	+	–

[a] A plus sign indicates the trait is present, a minus sign that it is absent.

(Page 501)

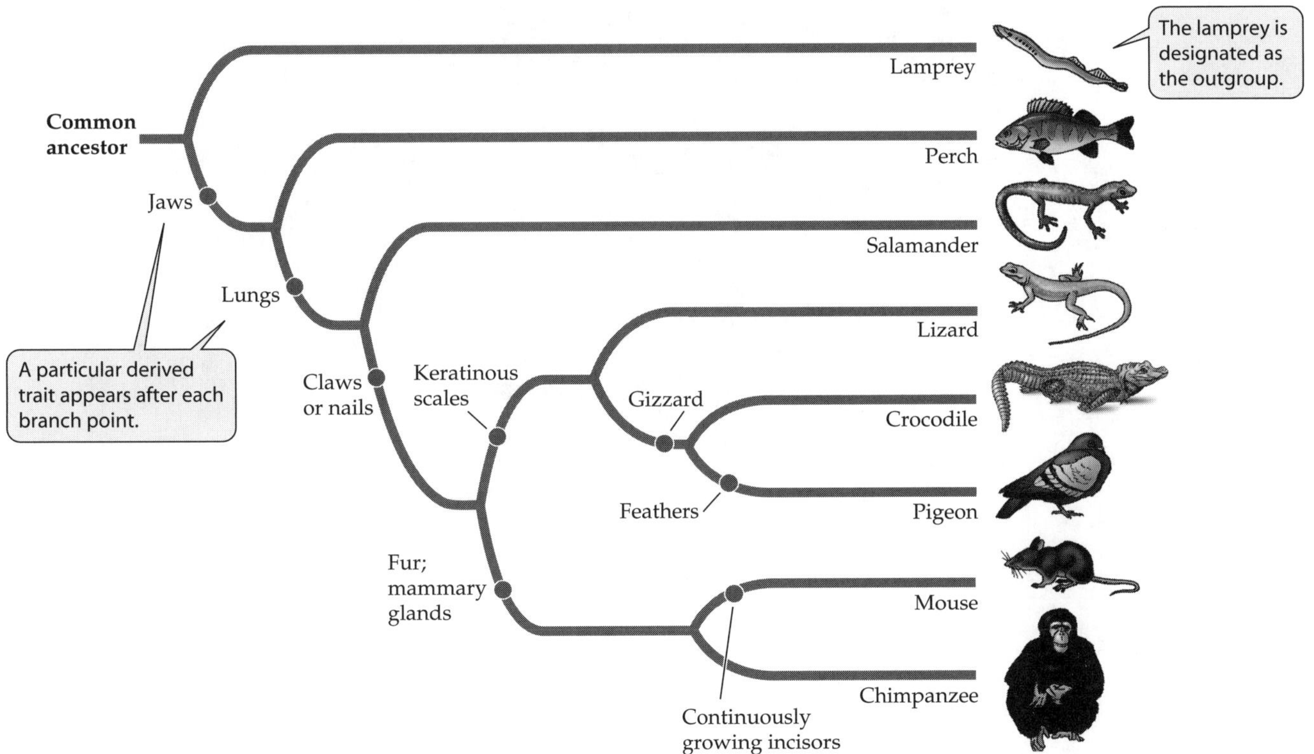
Common ancestor
Jaws
Lungs
A particular derived trait appears after each branch point.
Claws or nails
Keratinous scales
Gizzard
Feathers
Fur; mammary glands
Continuously growing incisors
Lamprey
Perch
Salamander
Lizard
Crocodile
Pigeon
Mouse
Chimpanzee
The lamprey is designated as the outgroup.

25.5 A Probable Phylogeny of Eight Vertebrates *(Page 501)*

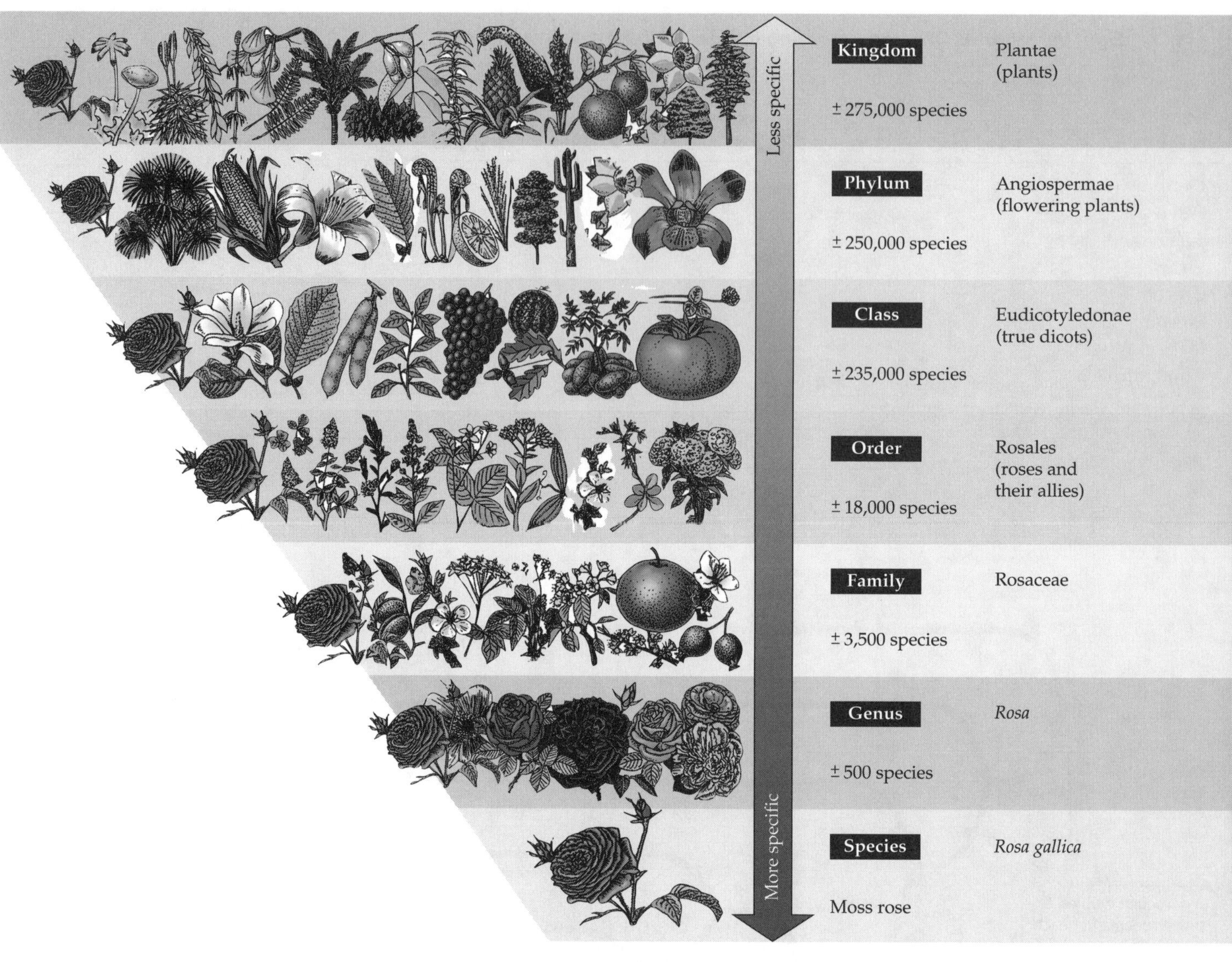
Less specific
More specific
Kingdom
Plantae
(plants)
± 275,000 species
Phylum
Angiospermae
(flowering plants)
± 250,000 species
Class
Eudicotyledonae
(true dicots)
± 235,000 species
Order
Rosales
(roses and
their allies)
± 18,000 species
Family
Rosaceae
± 3,500 species
Genus
Rosa
± 500 species
Species
Rosa gallica
Moss rose

25.6 Hierarchy in the Linnaean System *(Pages 502–503)*

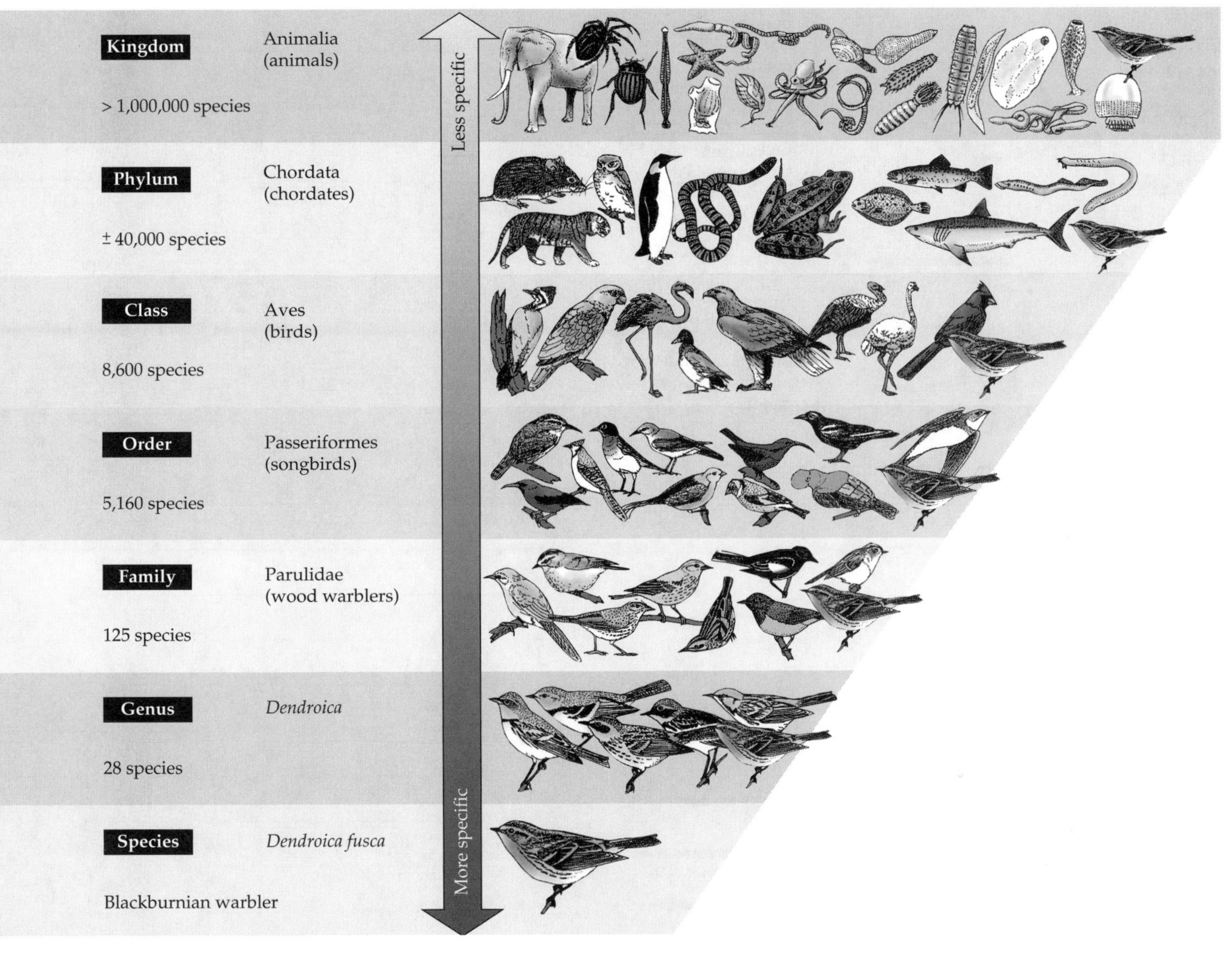
Kingdom
Animalia
(animals)
> 1,000,000 species
Phylum
Chordata
(chordates)
± 40,000 species
Class
Aves
(birds)
8,600 species
Order
Passeriformes
(songbirds)
5,160 species
Family
Parulidae
(wood warblers)
125 species
Genus
Dendroica
28 species
Species
Dendroica fusca
Blackburnian warbler
Less specific
More specific

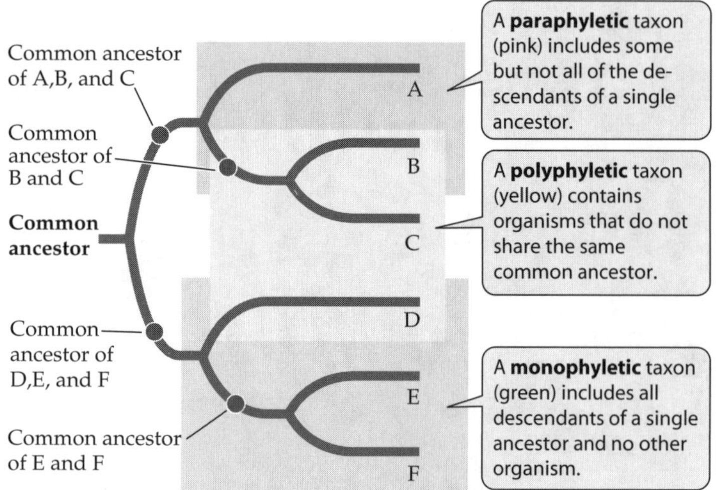
Common ancestor of A,B, and C
Common ancestor of B and C
Common ancestor
Common ancestor of D,E, and F
Common ancestor of E and F
A
B
C
D
E
F
A **paraphyletic** taxon (pink) includes some but not all of the descendants of a single ancestor.
A **polyphyletic** taxon (yellow) contains organisms that do not share the same common ancestor.
A **monophyletic** taxon (green) includes all descendants of a single ancestor and no other organism.

25.7 Monophyletic, Polyphyletic, and Paraphyletic Taxa *(Page 504)*

(*a*) **The evolutionary relationships**

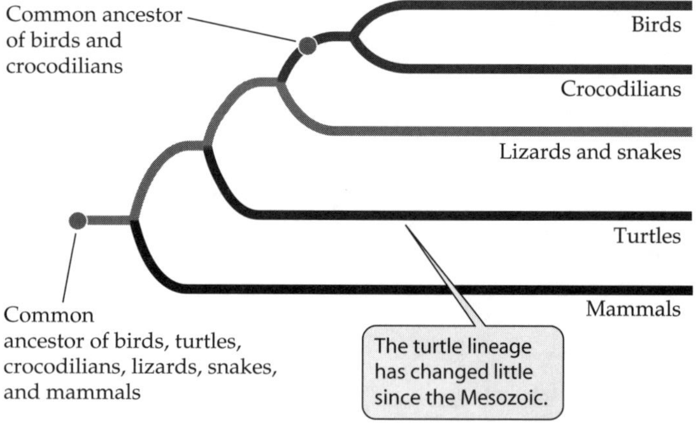
Common ancestor of birds and crocodilians
Birds
Crocodilians
Lizards and snakes
Turtles
Mammals
Common ancestor of birds, turtles, crocodilians, lizards, snakes, and mammals
The turtle lineage has changed little since the Mesozoic.

(*b*) **The traditional classification**

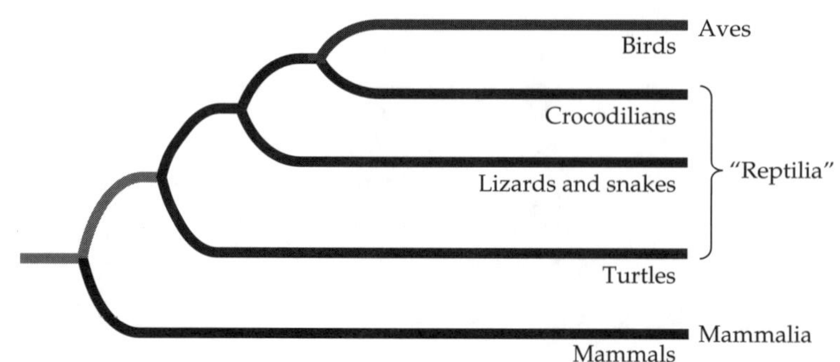
Aves
Birds
Crocodilians
Lizards and snakes
"Reptilia"
Turtles
Mammalia
Mammals

25.8 Phylogeny and Classification *(Page 505)*

Common ancestor of *Linanthus* species
L. nudatus
L. montanus
L. ciliatus
L. androsaceus
Self-compatibility evolves
L. "bicolor"
L. parviflorus
L. latisectus
L. liniflorus
L. acicularis
L. "bicolor"
L. jepsonii
L. "bicolor"
Self-compatibility arose in three separate lineages, fooling taxonomists into identifying all three species as *L. bicolor.*
Self-compatibility evolves

25.9 Phylogeny of a Section of the Phlox Genus *Linanthus* *(Page 506)*

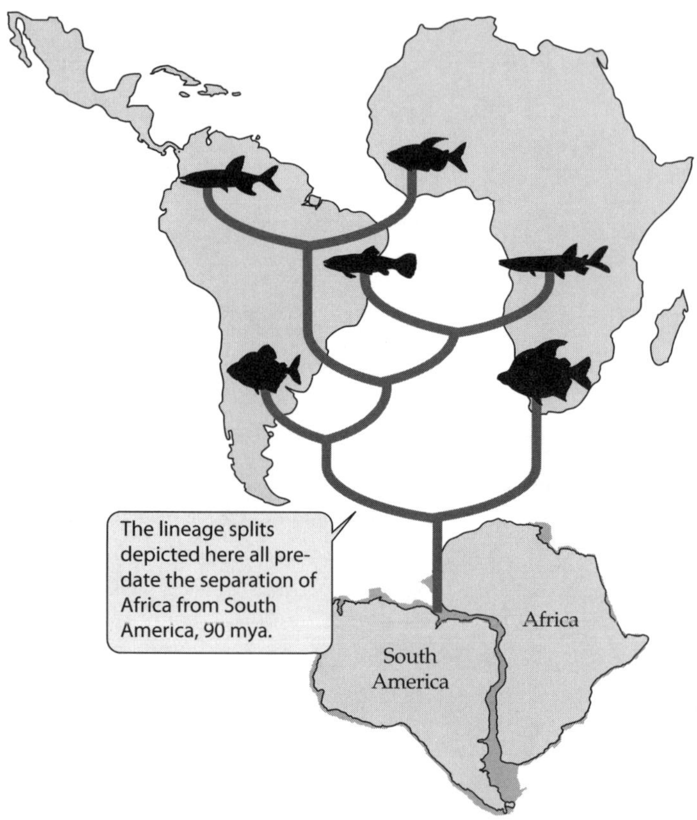
The lineage splits depicted here all predate the separation of Africa from South America, 90 mya.
Africa
South America

25.10 **Dating Lineage Splits** *(Page 507)*

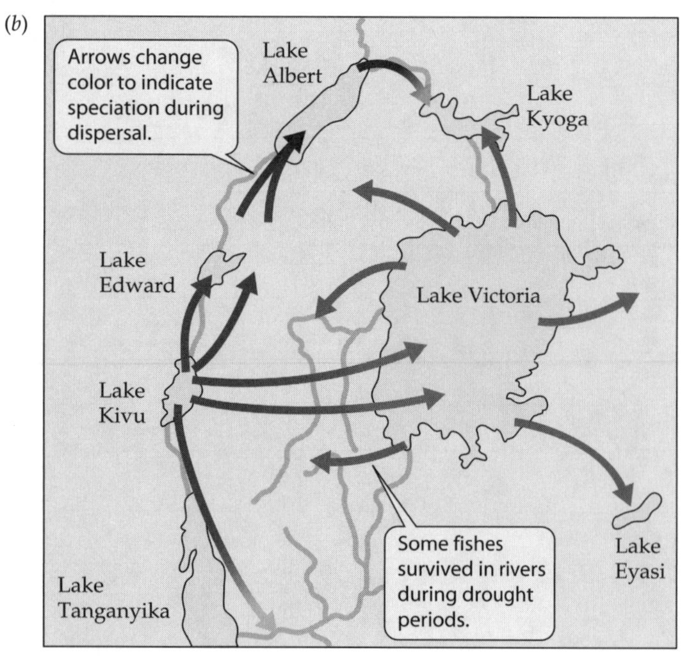
(b)
Arrows change color to indicate speciation during dispersal.
Lake Albert
Lake Kyoga
Lake Edward
Lake Victoria
Lake Kivu
Some fishes survived in rivers during drought periods.
Lake Eyasi
Lake Tanganyika
Lineages derived from Lake Kivu
Lake Victoria lineages

25.11 **Origins of the Cichlid Fishes of Lake Victoria** *(Page 507)*

26 Molecular and Genomic Evolution

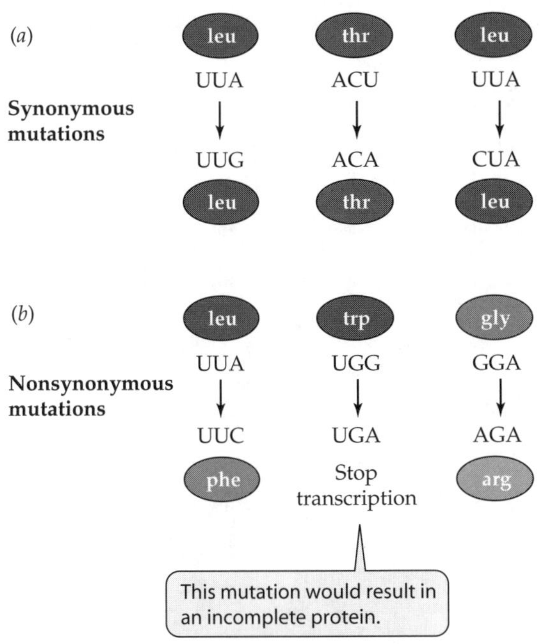

(a)
leu
thr
leu
UUA
ACU
UUA
Synonymous mutations
UUG
ACA
CUA
leu
thr
leu
(b)
leu
trp
gly
UUA
UGG
GGA
Nonsynonymous mutations
UUC
UGA
AGA
phe
Stop transcription
arg
This mutation would result in an incomplete protein.

26.1 When One Base Does or Doesn't Make a Difference *(Page 512)*

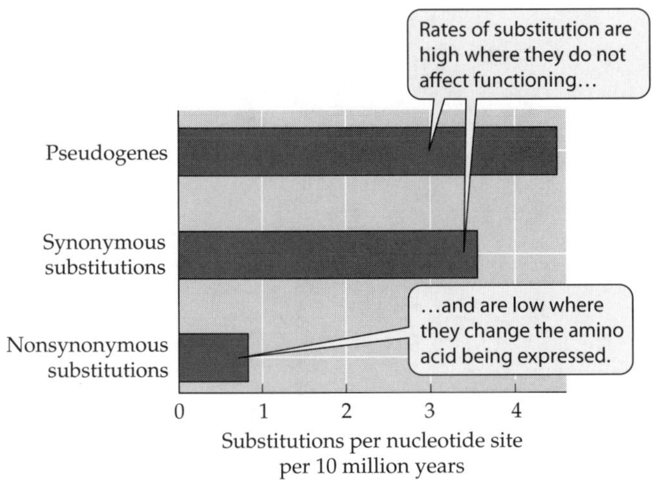

Rates of substitution are high where they do not affect functioning…
Pseudogenes
Synonymous substitutions
Nonsynonymous substitutions
…and are low where they change the amino acid being expressed.
0
1
2
3
4
Substitutions per nucleotide site per 10 million years

26.3 Rates of Base Substitution Differ *(Page 513)*

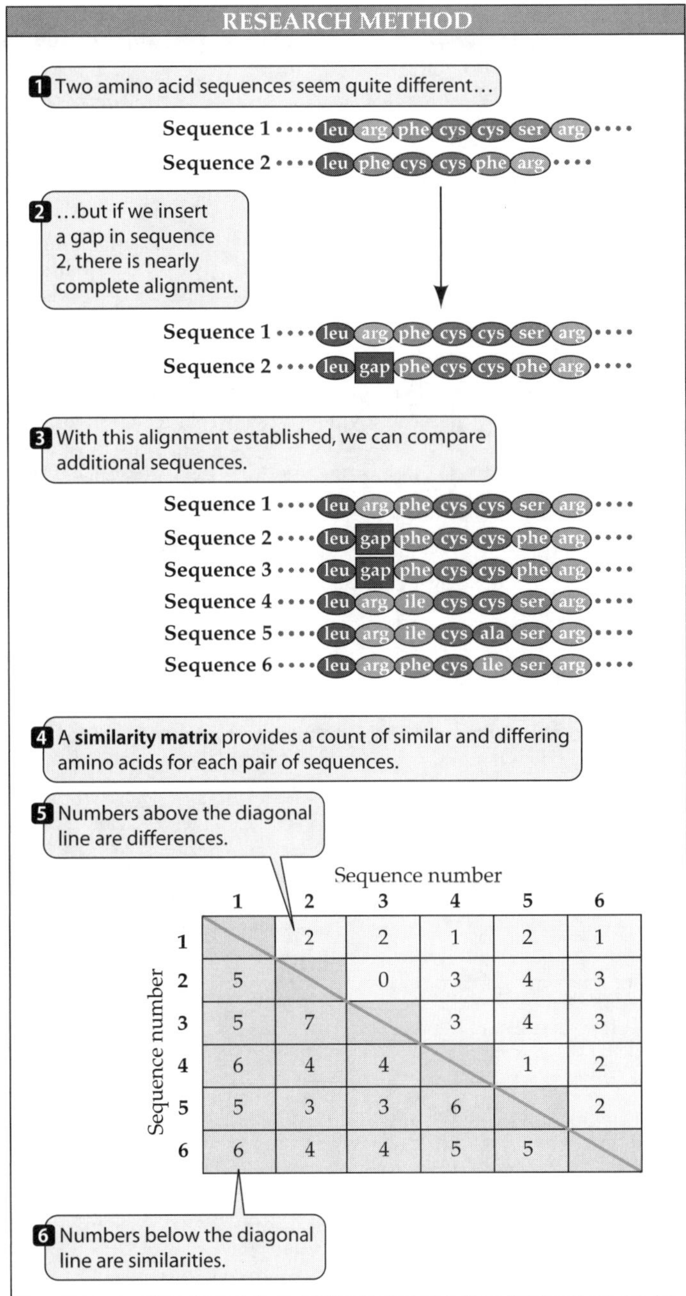

RESEARCH METHOD
1 Two amino acid sequences seem quite different…
Sequence 1 leu arg phe cys cys ser arg
Sequence 2 leu phe cys cys phe arg
2 …but if we insert a gap in sequence 2, there is nearly complete alignment.
Sequence 1 leu arg phe cys cys ser arg
Sequence 2 leu gap phe cys cys phe arg
3 With this alignment established, we can compare additional sequences.
Sequence 1 leu arg phe cys cys ser arg
Sequence 2 leu gap phe cys cys phe arg
Sequence 3 leu gap phe cys cys phe arg
Sequence 4 leu arg ile cys cys ser arg
Sequence 5 leu arg ile cys ala ser arg
Sequence 6 leu arg phe cys ile ser arg
4 A similarity matrix provides a count of similar and differing amino acids for each pair of sequences.
5 Numbers above the diagonal line are differences.
Sequence number
Sequence number
6 Numbers below the diagonal line are similarities.

	1	2	3	4	5	6
1		2	2	1	2	1
2	5		0	3	4	3
3	5	7		3	4	3
4	6	4	4		1	2
5	5	3	3	6		2
6	6	4	4	5	5	

26.2 Amino Acid Sequence Alignment *(Page 513)*

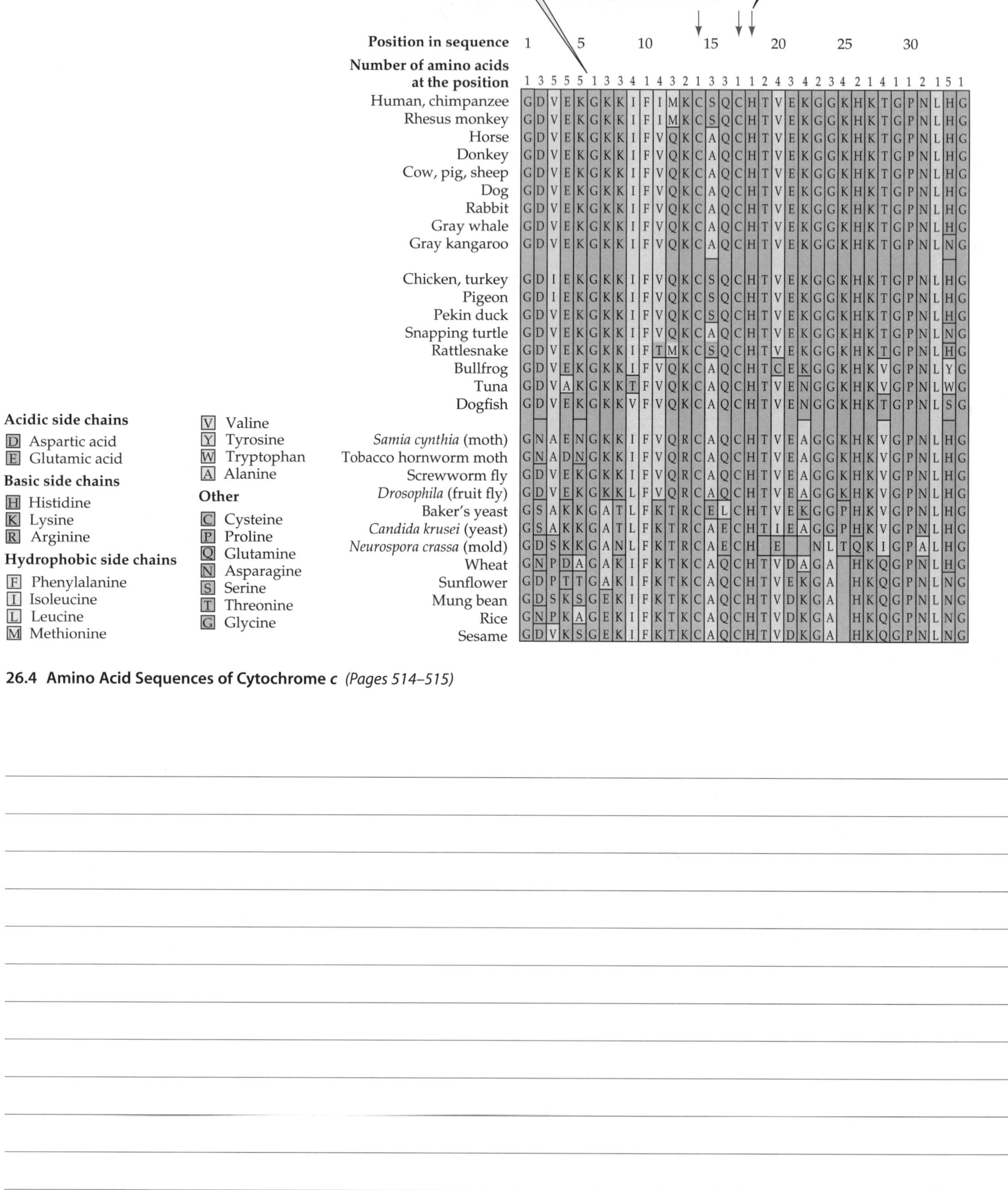
The number 1 indicates an invariant position in the cytochrome c molecule (i.e., all the organisms have the same amino acid in this position) and that the position is probably functionally very significant.
Side chains marked by red arrows interact with the heme group.
Position in sequence 1 5 10 15 20 25 30
Number of amino acids at the position 1 3 5 5 5 1 3 3 4 1 4 3 2 1 3 3 1 1 2 4 3 4 2 3 4 2 1 4 1 1 2 1 5 1
Human, chimpanzee GDVEKGKKIFIMKCSQCHTVEKGGKHKTGPNLHG
Rhesus monkey GDVEKGKKIFIMKCSQCHTVEKGGKHKTGPNLHG
Horse GDVEKGKKIFVQKCAQCHTVEKGGKHKTGPNLHG
Donkey GDVEKGKKIFVQKCAQCHTVEKGGKHKTGPNLHG
Cow, pig, sheep GDVEKGKKIFVQKCAQCHTVEKGGKHKTGPNLHG
Dog GDVEKGKKIFVQKCAQCHTVEKGGKHKTGPNLHG
Rabbit GDVEKGKKIFVQKCAQCHTVEKGGKHKTGPNLHG
Gray whale GDVEKGKKIFVQKCAQCHTVEKGGKHKTGPNLHG
Gray kangaroo GDVEKGKKIFVQKCAQCHTVEKGGKHKTGPNLNG
Chicken, turkey GDIEKGKKIFVQKCSQCHTVEKGGKHKTGPNLHG
Pigeon GDIEKGKKIFVQKCSQCHTVEKGGKHKTGPNLHG
Pekin duck GDVEKGKKIFVQKCSQCHTVEKGGKHKTGPNLHG
Snapping turtle GDVEKGKKIFVQKCAQCHTVEKGGKHKTGPNLNG
Rattlesnake GDVEKGKKIFTMKCSQCHTVEKGGKHKTGPNLHG
Bullfrog GDVEKGKKIFVQKCAQCHTCEKGGKHKVGPNLYG
Tuna GDVAKGKKTFVQKCAQCHTVENGGKHKVGPNLWG
Dogfish GDVEKGKKVFVQKCAQCHTVENGGKHKTGPNLSG
Samia cynthia (moth) GNAENGKKIFVQRCAQCHTVEAGGKHKVGPNLHG
Tobacco hornworm moth GNADNGKKIFVQRCAQCHTVEAGGKHKVGPNLHG
Screwworm fly GDVEKGKKIFVQRCAQCHTVEAGGKHKVGPNLHG
Drosophila (fruit fly) GDVEKGKKLFVQRCAQCHTVEAGGKHKVGPNLHG
Baker's yeast GSAKKGATLFKTRCELCHTVEKGGPHKVGPNLHG
Candida krusei (yeast) GSAKKGATLFKTRCAECHTIEAGGPHKVGPNLHG
Neurospora crassa (mold) GDSKKGANLFKTRCAECH E NLTQKIGPALHG
Wheat GNPDAGAKIFKTKCAQCHTVDAGA HKQGPNLHG
Sunflower GDPTTGAKIFKTKCAQCHTVEKGA HKQGPNLNG
Mung bean GDSKSGEKIFKTKCAQCHTVDKGA HKQGPNLNG
Rice GNPKAGEKIFKTKCAQCHTVDKGA HKQGPNLNG
Sesame GDVKSGEKIFKTKCAQCHTVDKGA HKQGPNLNG
Acidic side chains
D Aspartic acid
E Glutamic acid
Basic side chains
H Histidine
K Lysine
R Arginine
Hydrophobic side chains
F Phenylalanine
I Isoleucine
L Leucine
M Methionine
V Valine
Y Tyrosine
W Tryptophan
A Alanine
Other
C Cysteine
P Proline
Q Glutamine
N Asparagine
S Serine
T Threonine
G Glycine

26.4 Amino Acid Sequences of Cytochrome *c* *(Pages 514–515)*

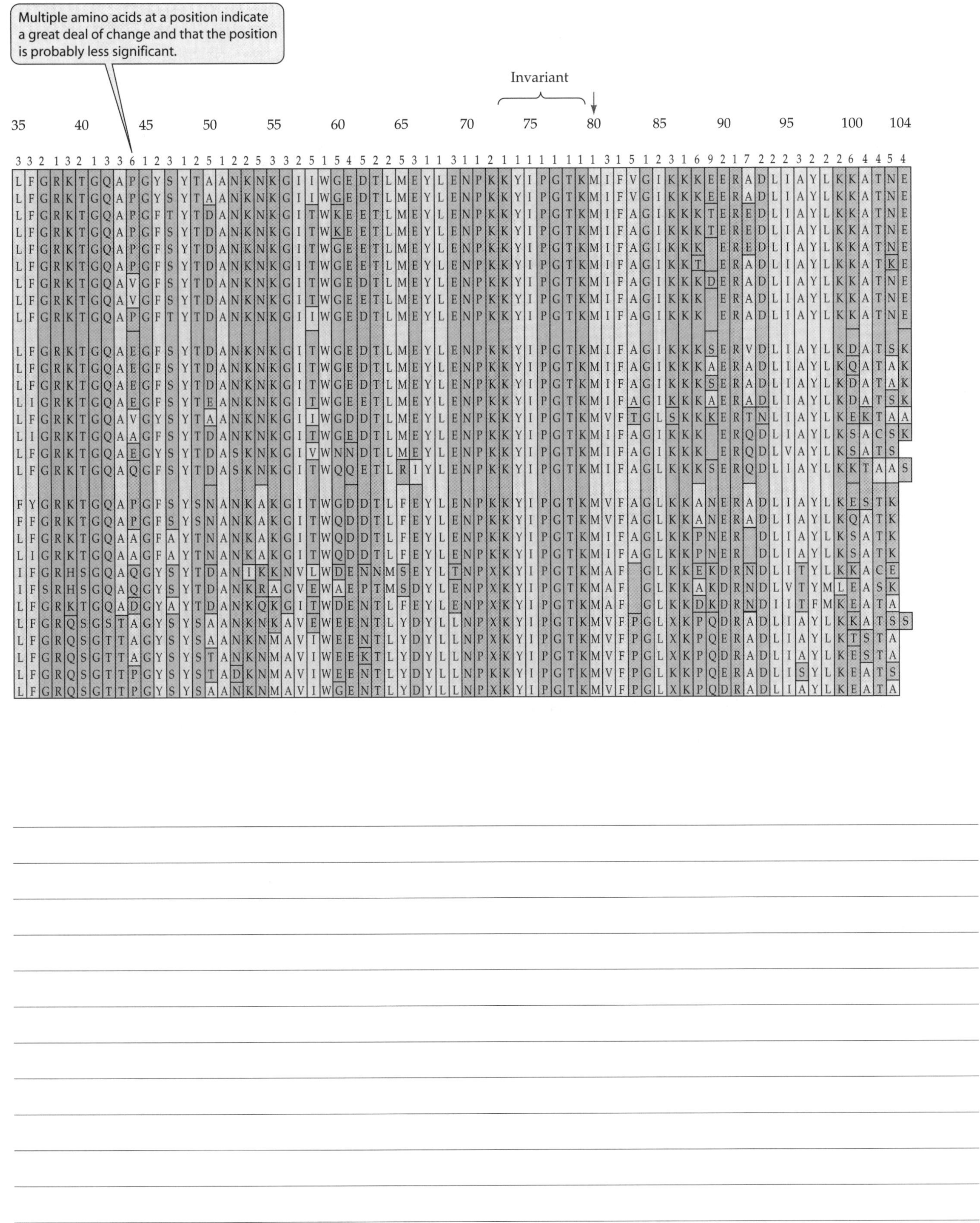
Multiple amino acids at a position indicate a great deal of change and that the position is probably less significant.
Invariant
35
40
45
50
55
60
65
70
75
80
85
90
95
100
104

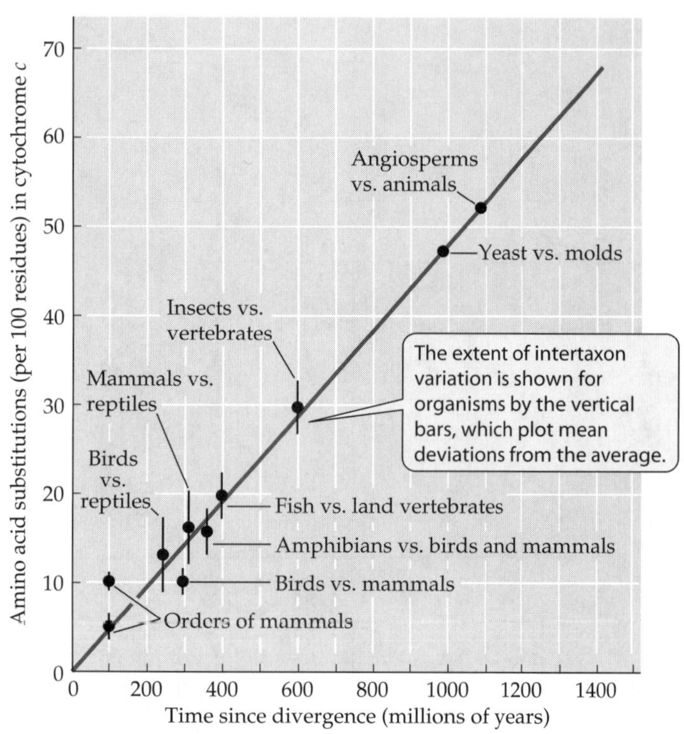
Amino acid substitutions (per 100 residues) in cytochrome c
Time since divergence (millions of years)
Angiosperms vs. animals
Yeast vs. molds
Insects vs. vertebrates
Mammals vs. reptiles
Birds vs. reptiles
The extent of intertaxon variation is shown for organisms by the vertical bars, which plot mean deviations from the average.
Fish vs. land vertebrates
Amphibians vs. birds and mammals
Birds vs. mammals
Orders of mammals

26.5 Cytochrome *c* Has Evolved at a Constant Rate *(Page 515)*

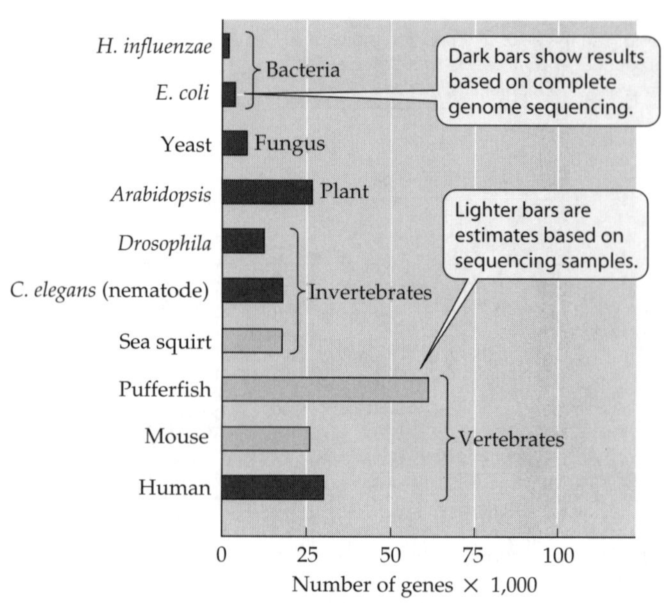
H. influenzae
E. coli
Bacteria
Dark bars show results based on complete genome sequencing.
Yeast
Fungus
Arabidopsis
Plant
Lighter bars are estimates based on sequencing samples.
Drosophila
C. elegans (nematode)
Invertebrates
Sea squirt
Pufferfish
Mouse
Vertebrates
Human
Number of genes × 1,000

26.7 Complex Organisms Have More Genes than Simpler Organisms *(Page 518)*

26.1 Similarity Matrix for Lysozyme in Mammals

SPECIES	LANGUR	BABOON	HUMAN	RAT	COW	HORSE
Langur*		14	18	38	32	65
Baboon	0		14	33	39	65
Human	0	1		37	41	64
Rat	0	1	0		55	64
Cow*	5	0	0	0		71
Horse	0	0	0	0	1	

Shown above the diagonal line is the number of amino acid sequence *differences* between the two species being compared; below the line are the number of sequences uniquely *shared* by the two species. Asterisks (*) indicate foregut-fermenting species.

(Page 516)

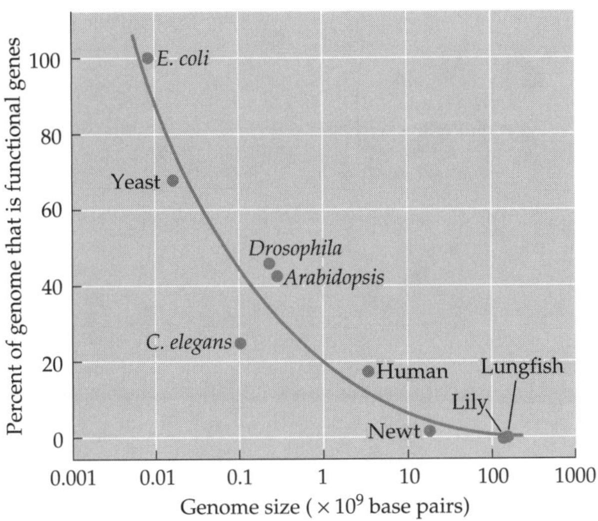

E. coli
Yeast
Drosophila
Arabidopsis
C. elegans
Human
Lungfish
Lily
Newt
Percent of genome that is functional genes
100
80
60
40
20
0
0.001
0.01
0.1
1
10
100
1000
Genome size (× 10⁹ base pairs)

26.8 A Large Proportion of DNA Is Noncoding *(Page 518)*

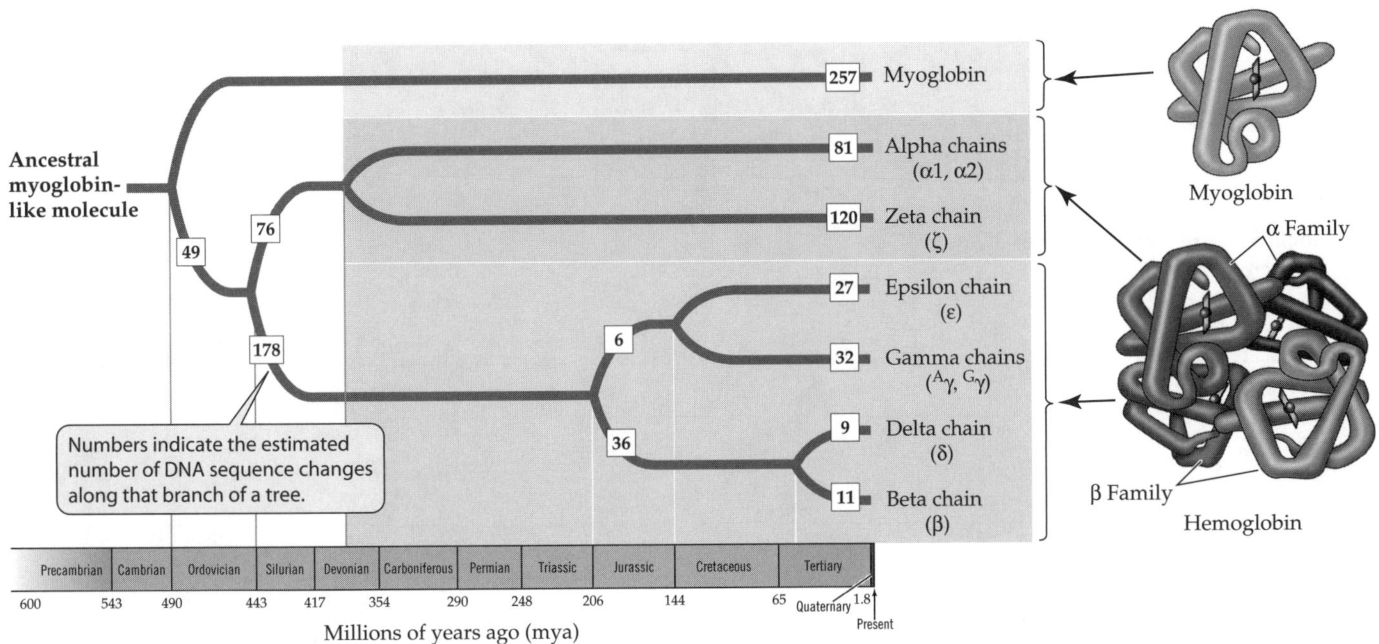

Ancestral myoglobin-like molecule
257 Myoglobin
81 Alpha chains (α1, α2)
120 Zeta chain (ζ)
27 Epsilon chain (ε)
32 Gamma chains (Aγ, Gγ)
9 Delta chain (δ)
11 Beta chain (β)
49
76
178
6
36
Numbers indicate the estimated number of DNA sequence changes along that branch of a tree.
Myoglobin
α Family
β Family
Hemoglobin
Precambrian
Cambrian
Ordovician
Silurian
Devonian
Carboniferous
Permian
Triassic
Jurassic
Cretaceous
Tertiary
Quaternary
600
543
490
443
417
354
290
248
206
144
65
1.8
Present
Millions of years ago (mya)

26.9 A Globin Family Gene Tree *(Page 519)*

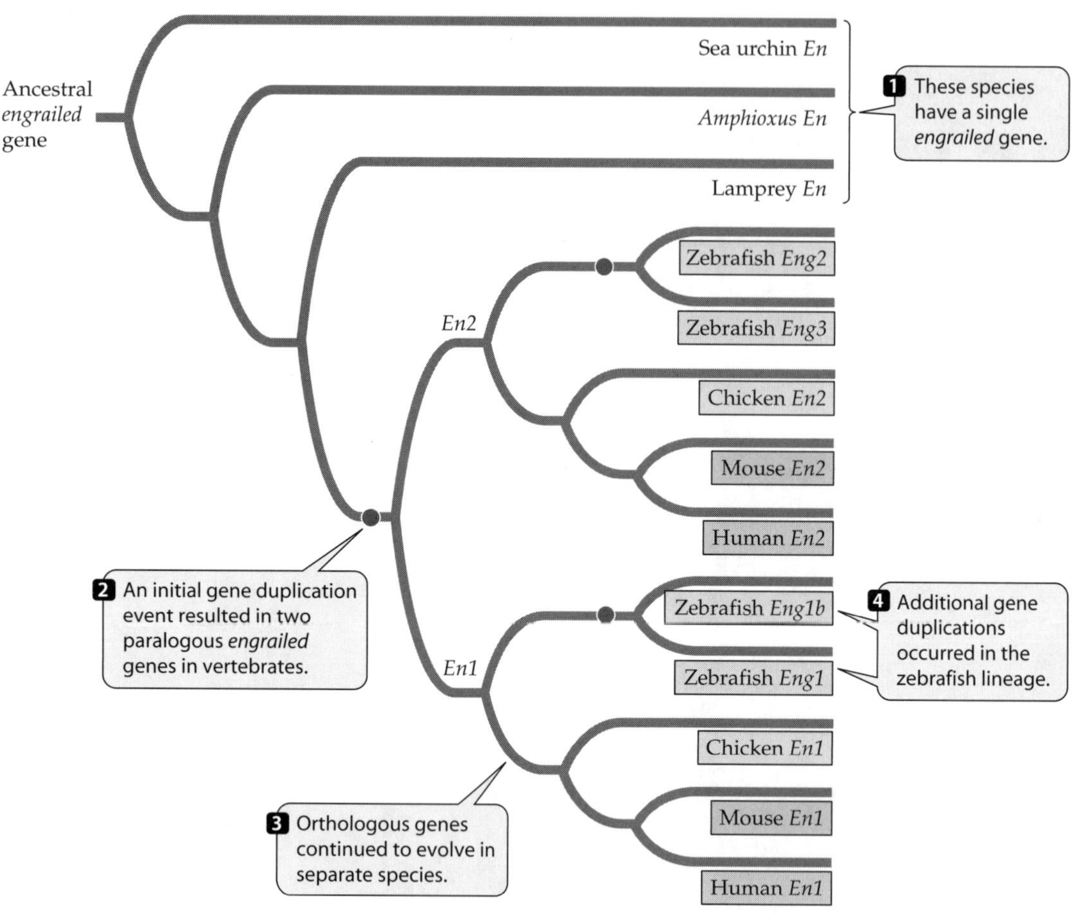
Ancestral
engrailed
gene
Sea urchin *En*
Amphioxus En
Lamprey *En*
1 These species have a single *engrailed* gene.
En2
Zebrafish *Eng2*
Zebrafish *Eng3*
Chicken *En2*
Mouse *En2*
Human *En2*
2 An initial gene duplication event resulted in two paralogous *engrailed* genes in vertebrates.
En1
Zebrafish *Eng1b*
Zebrafish *Eng1*
4 Additional gene duplications occurred in the zebrafish lineage.
Chicken *En1*
Mouse *En1*
Human *En1*
3 Orthologous genes continued to evolve in separate species.

26.10 Phylogeny of the *engrailed* Genes *(Page 520)*

27 Bacteria and Archaea: The Prokaryotic Domains

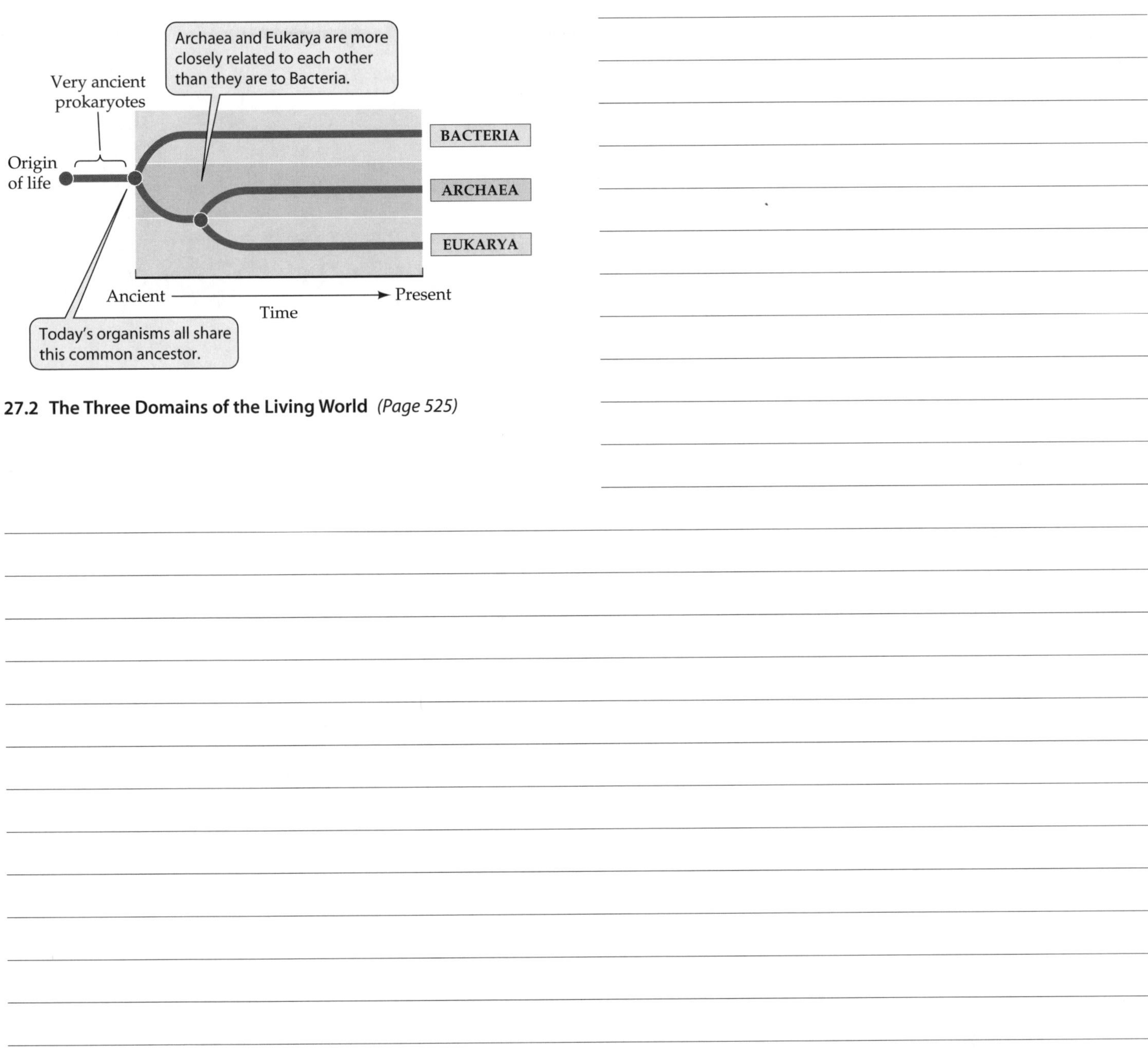

Archaea and Eukarya are more closely related to each other than they are to Bacteria.
Very ancient prokaryotes
Origin of life
BACTERIA
ARCHAEA
EUKARYA
Ancient
Present
Time
Today's organisms all share this common ancestor.

27.2 The Three Domains of the Living World *(Page 525)*

27.1 The Three Domains of Life on Earth

CHARACTERISTIC	DOMAIN		
	BACTERIA	ARCHAEA	EUKARYA
Membrane-enclosed nucleus	Absent	Absent	*Present*
Membrane-enclosed organelles	Absent	Absent	*Present*
Peptidoglycan in cell wall	*Present*	Absent	Absent
Membrane lipids	Ester-linked Unbranched	*Ether-linked* *Branched*	Ester-linked Unbranched
Ribosomes[a]	70S	70S	*80S*
Initiator tRNA	*Formylmethionine*	Methionine	Methionine
Operons	Yes	Yes	*No*
Plasmids	Yes	Yes	*Rare*
RNA polymerases	One	One[b]	Three
Ribosomes sensitive to chloramphenicol and streptomycin	*Yes*	No	No
Ribosomes sensitive to diphtheria toxin	*No*	Yes	Yes
Some are methanogens	No	*Yes*	No
Some fix nitrogen	Yes	Yes	*No*
Some conduct chlorophyll-based photosynthesis	Yes	No	Yes

[a] 70S ribosomes are smaller than 80S ribosomes.
[b] Archaeal RNA polymerase is similar to eukaryotic polymerases.

(Page 526)

(a)

Gram-positive bacteria have a uniformly dense cell wall consisting primarily of peptidoglycan.

Outside of cell
Cell wall (peptidoglycan)
Plasma membrane
Periplasmic space
Cytoplasm

(b)

Gram-negative bacteria have a very thin peptidoglycan layer and an outer membrane.

Outer membrane of cell wall
Peptidoglycan layer
Plasma membrane
Periplasmic space

27.6 The Gram Stain and the Bacterial Cell Wall *(Page 529)*

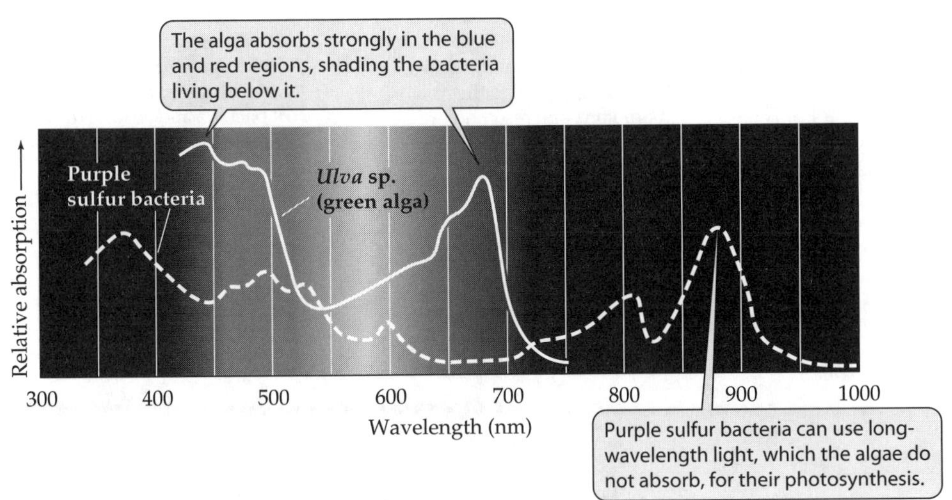
The alga absorbs strongly in the blue and red regions, shading the bacteria living below it.
Purple sulfur bacteria
Ulva sp. (green alga)
Relative absorption
300
400
500
600
700
800
900
1000
Wavelength (nm)
Purple sulfur bacteria can use long-wavelength light, which the algae do not absorb, for their photosynthesis.

27.7 Bacteriochlorophyll Absorbs Long- Wavelength Light *(Page 530)*

27.2 *How Organisms Obtain Their Energy and Carbon*

NUTRITIONAL CATEGORY	ENERGY SOURCE	CARBON SOURCE
Photoautotrophs (found in all three domains)	Light	Carbon dioxide
Photoheterotrophs (some bacteria)	Light	Organic compounds
Chemolithotrophs (some bacteria, many archaea)	Inorganic substances	Carbon dioxide
Chemoheterotrophs (found in all three domains)	Organic compounds	Organic compounds

(Page 530)

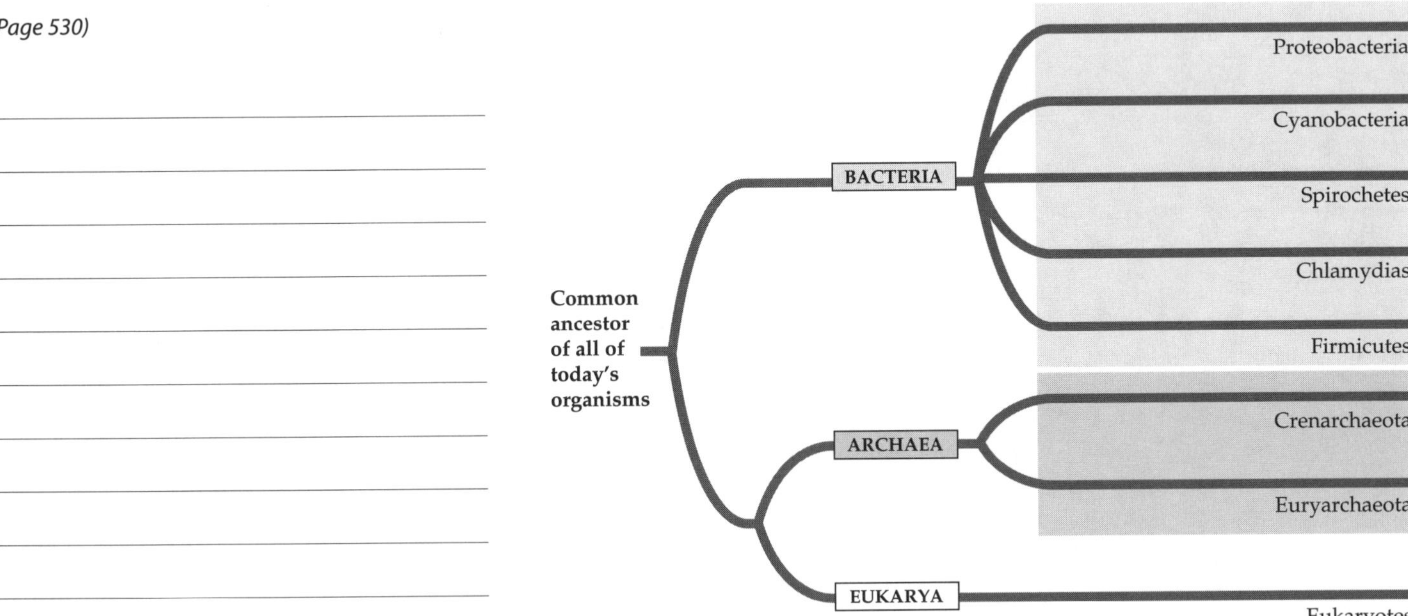
Common ancestor of all of today's organisms
BACTERIA
Proteobacteria
Cyanobacteria
Spirochetes
Chlamydias
Firmicutes
ARCHAEA
Crenarchaeota
Euryarchaeota
EUKARYA
Eukaryotes

27.8 Two Domains: A Brief Overview *(Page 534)*

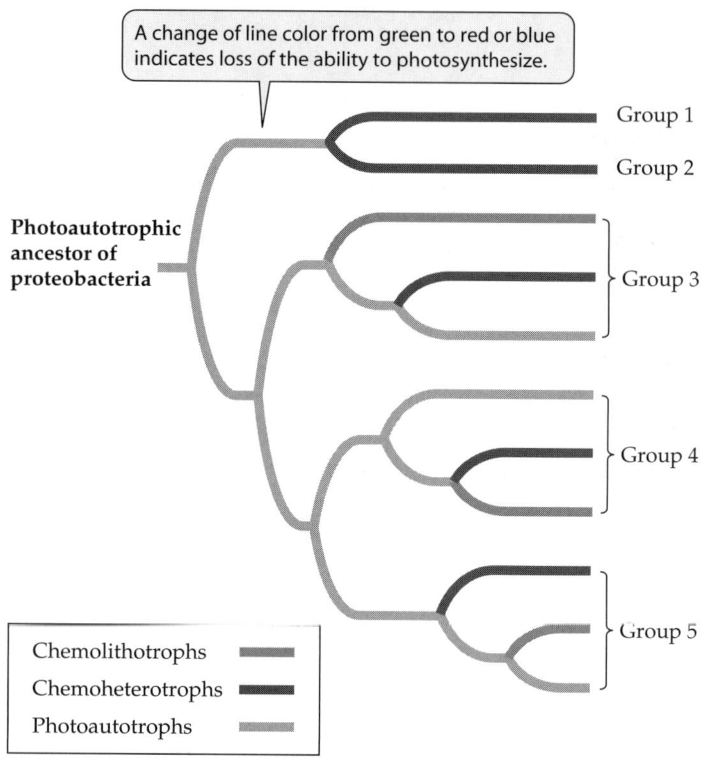
A change of line color from green to red or blue indicates loss of the ability to photosynthesize.
Group 1
Group 2
Photoautotrophic ancestor of proteobacteria
Group 3
Group 4
Group 5
Chemolithotrophs
Chemoheterotrophs
Photoautotrophs

27.9 The Evolution of Metabolism in the Proteobacteria *(Page 535)*

$$-\overset{\overset{\displaystyle O}{\|}}{C}-O-\underset{\underset{\displaystyle H}{|}}{\overset{\overset{\displaystyle H}{|}}{C}}-$$

In-Text Art *(Page 539)*

$$-\underset{\underset{\displaystyle H}{|}}{\overset{\overset{\displaystyle H}{|}}{C}}-O-\underset{\underset{\displaystyle H}{|}}{\overset{\overset{\displaystyle H}{|}}{C}}-$$

In-Text Art *(Page 539)*

Some archaea have long-chain hydrocarbons with glycerol at both ends, spanning the membrane and resulting in a lipid monolayer.

Other archaeal hydrocarbons fit the same membrane template as do the fatty acids of bacteria and eukaryotes, resulting in a lipid bilayer.

27.18 Membrane Architecture in Archaea *(Page 539)*

28 Protists and the Dawn of the Eukarya

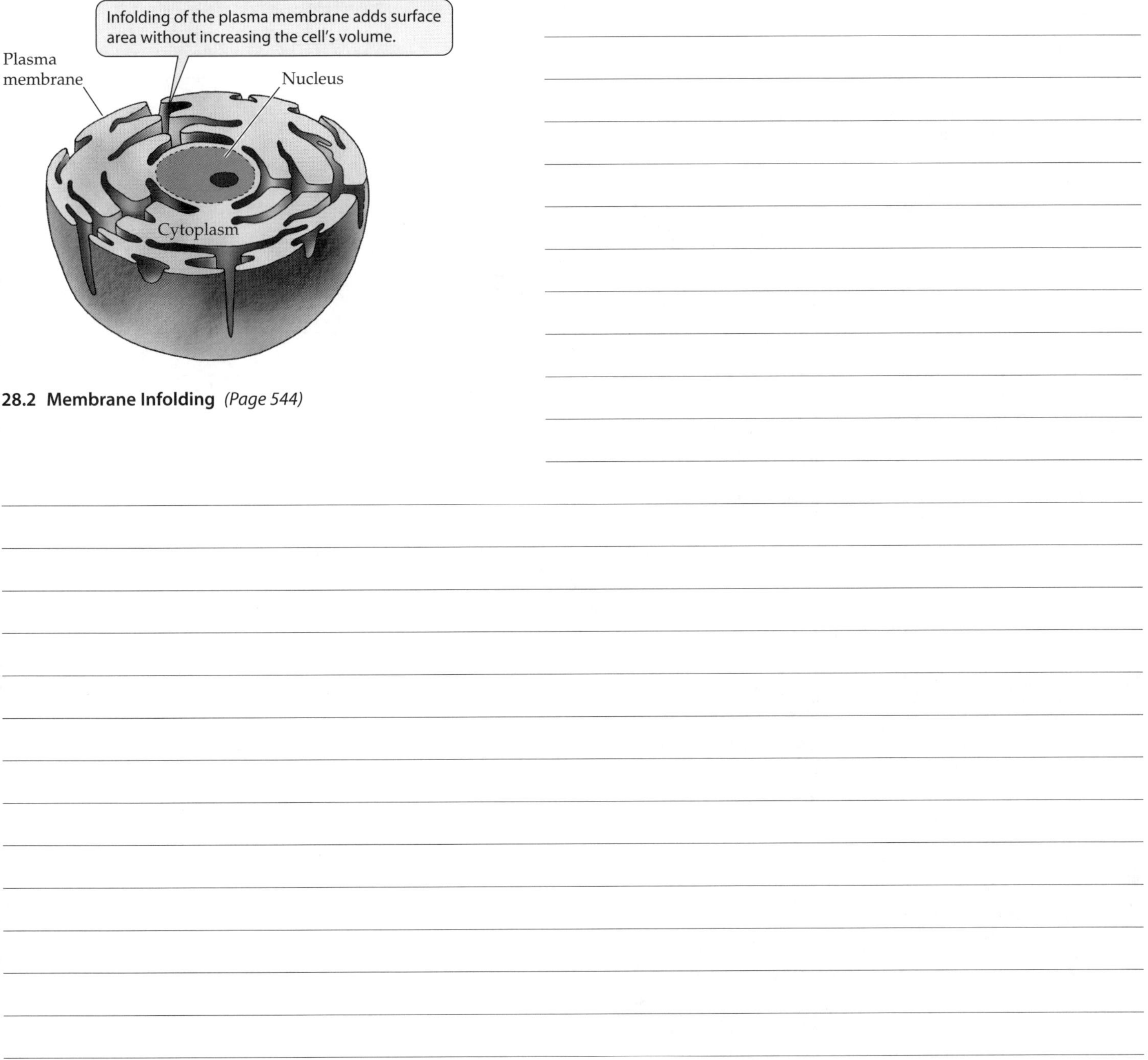
Infolding of the plasma membrane adds surface area without increasing the cell's volume.
Plasma membrane
Nucleus
Cytoplasm

28.2 Membrane Infolding *(Page 544)*

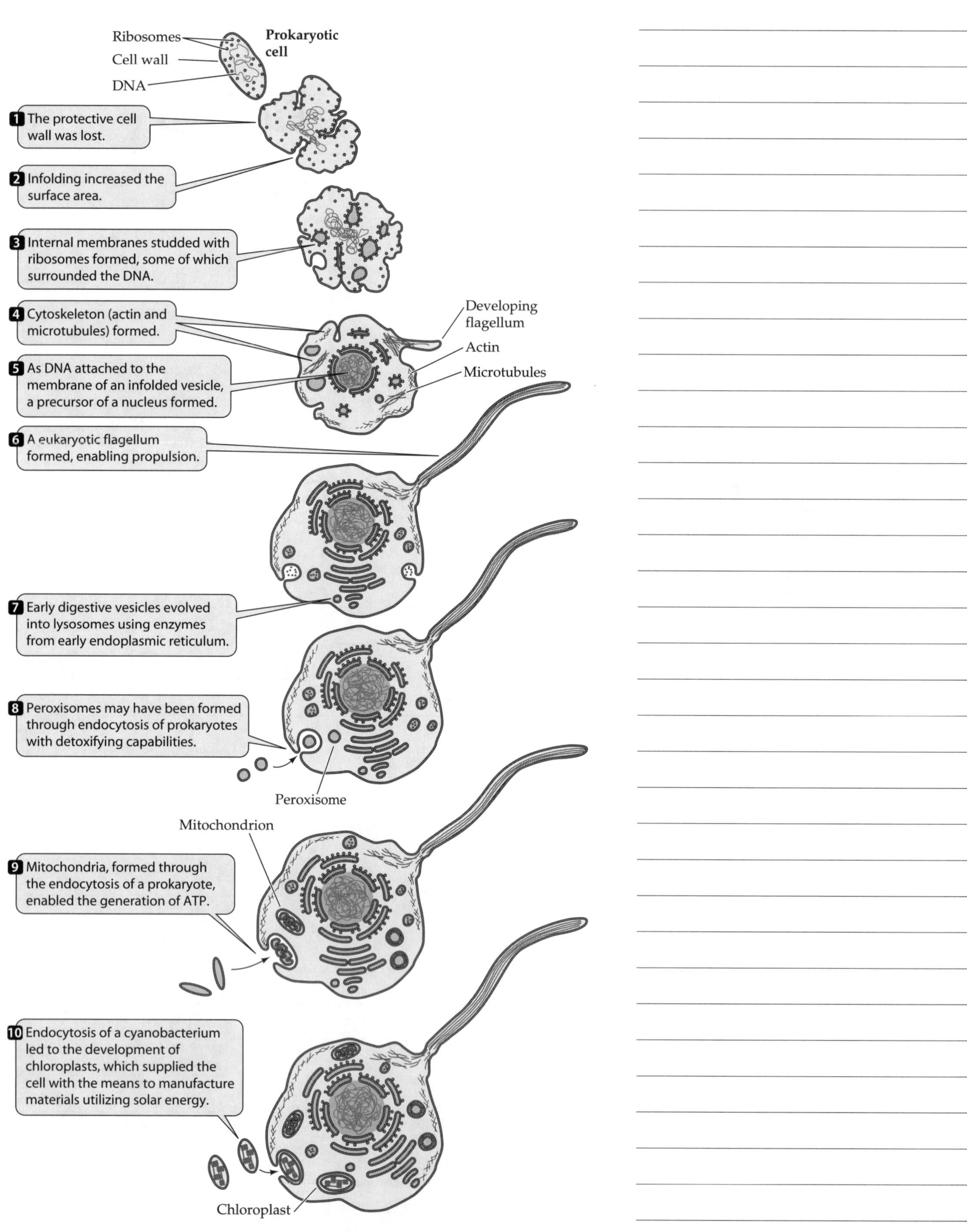
Ribosomes
Cell wall
DNA
Prokaryotic cell
1 The protective cell wall was lost.
2 Infolding increased the surface area.
3 Internal membranes studded with ribosomes formed, some of which surrounded the DNA.
4 Cytoskeleton (actin and microtubules) formed.
5 As DNA attached to the membrane of an infolded vesicle, a precursor of a nucleus formed.
Developing flagellum
Actin
Microtubules
6 A eukaryotic flagellum formed, enabling propulsion.
7 Early digestive vesicles evolved into lysosomes using enzymes from early endoplasmic reticulum.
8 Peroxisomes may have been formed through endocytosis of prokaryotes with detoxifying capabilities.
Peroxisome
Mitochondrion
9 Mitochondria, formed through the endocytosis of a prokaryote, enabled the generation of ATP.
10 Endocytosis of a cyanobacterium led to the development of chloroplasts, which supplied the cell with the means to manufacture materials utilizing solar energy.
Chloroplast

28.3 From Prokaryotic Cell to Eukaryotic Cell *(Page 545)*

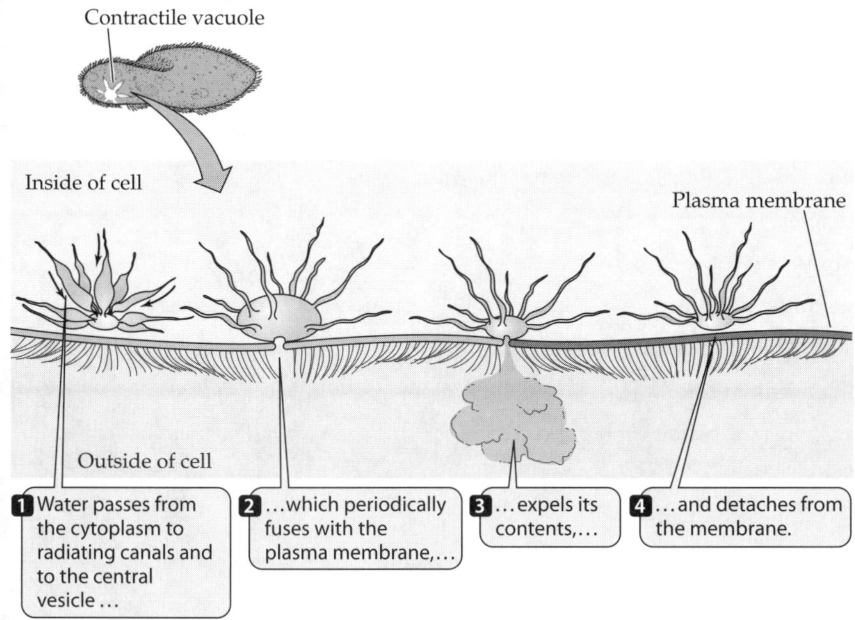
Contractile vacuole
Inside of cell
Plasma membrane
Outside of cell
1 Water passes from the cytoplasm to radiating canals and to the central vesicle . . .
2 . . .which periodically fuses with the plasma membrane,. . .
3 . . .expels its contents,. . .
4 . . .and detaches from the membrane.

28.5 Contractile Vacuoles Bail Out Excess Water *(Page 547)*

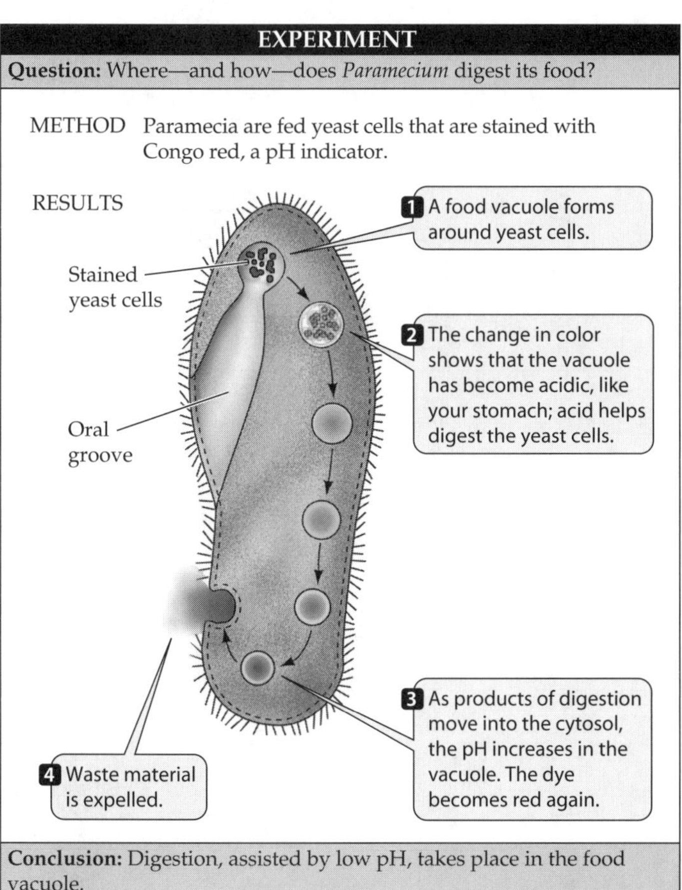
EXPERIMENT
Question: Where—and how—does Paramecium digest its food?
METHOD Paramecia are fed yeast cells that are stained with Congo red, a pH indicator.
RESULTS
Stained yeast cells
Oral groove
1 A food vacuole forms around yeast cells.
2 The change in color shows that the vacuole has become acidic, like your stomach; acid helps digest the yeast cells.
3 As products of digestion move into the cytosol, the pH increases in the vacuole. The dye becomes red again.
4 Waste material is expelled.
Conclusion: Digestion, assisted by low pH, takes place in the food vacuole.

28.6 Food Vacuoles Handle Digestion and Excretion *(Page 547)*

28.1 Major Protist Clades

GROUP	ATTRIBUTES	EXAMPLES
Diplomonads	Unicellular, no mitochondria, two nuclei, flagella	*Giardia*
Parabasalids	Unicellular, no mitochondria, flagella and undulating membrane	*Trichomonas*
Euglenozoans	Unicellular, with flagella	
Euglenoids	Mostly photoautotrophic	*Euglena*
Kinetoplastids	Have a single large mitochondrion	*Trypanosoma*
Alveolates	Unicellular; cavities (alveoli) below cell surface	
Dinoflagellates	Pigments give golden-brown color	*Gonyaulax*
Apicomplexans	Apical complex in spores for penetration of host	*Plasmodium*
Ciliates	Cilia; two types of nuclei	*Paramecium*
Stramenopiles	Two unequal flagella, one with hairs	
Diatoms	Unicellular; photoautotrophic; two-part cell walls; no flagellum	*Thalassiosira*
Brown algae	Multicellular; marine; photoautotrophic	*Fucus*, *Macrocystis*
Oomycetes (water molds, powdery mildews)	Mostly coenocytic; heterotrophic	*Saprolegnia*
Red algae	No flagella; photoautrophic; phycoerythrin and phycocyanin	*Chondrus*, *Polysiphonia*
Chlorophytes ("Green algae"[a])	Photoautotrophic	*Ulva*, *Volvox*
Choanoflagellates	Resemble sponge cells; heterotrophic; with flagella	*Codosiga*, *Choanoeca*

[a]The green algae do not constitute a clade. The chlorophytes are a clade of green algae; a different green algal lineage gave rise to the plant kingdom.

(Page 549)

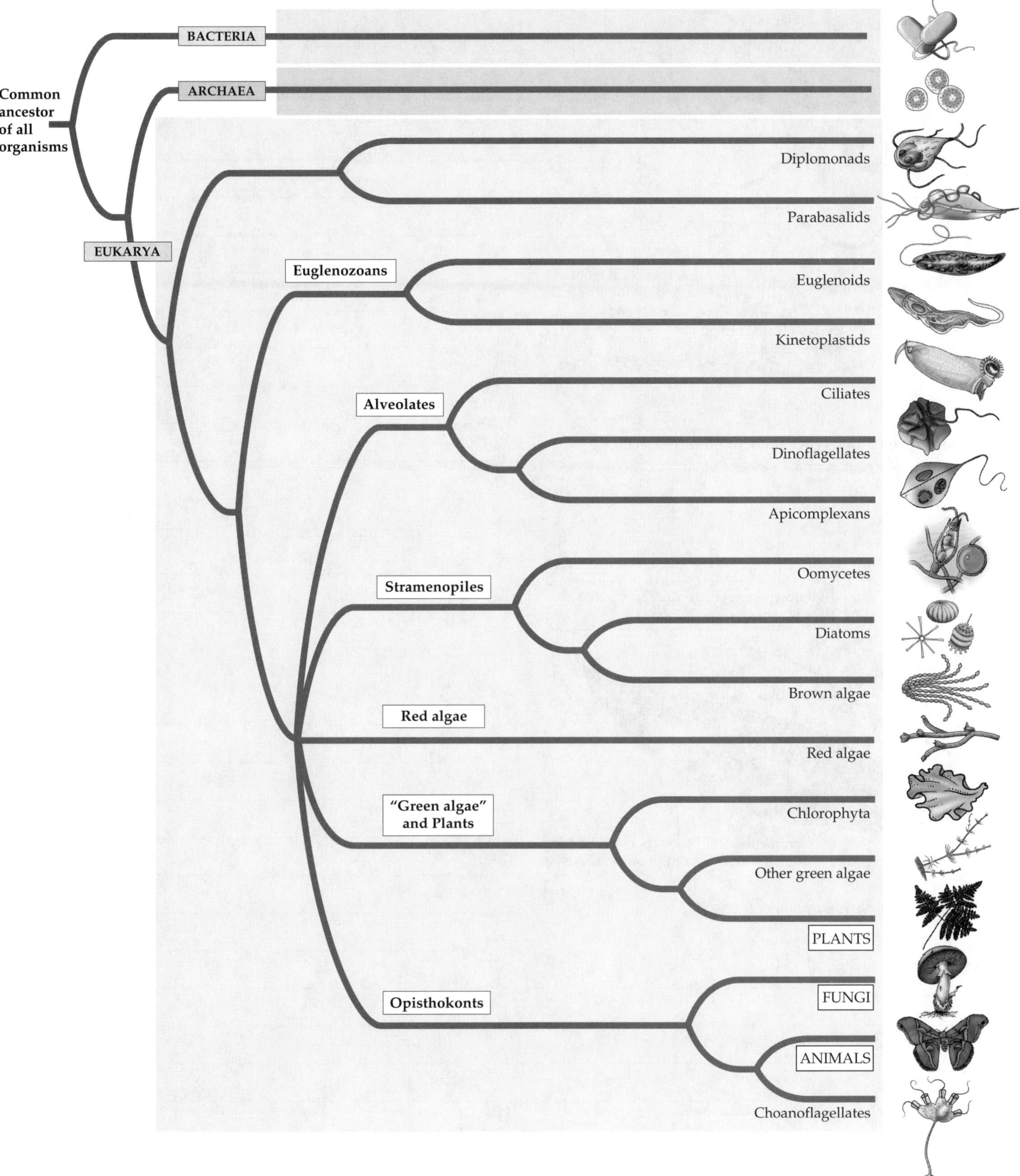
Common ancestor of all organisms
BACTERIA
ARCHAEA
EUKARYA
Diplomonads
Parabasalids
Euglenozoans
Euglenoids
Kinetoplastids
Alveolates
Ciliates
Dinoflagellates
Apicomplexans
Stramenopiles
Oomycetes
Diatoms
Brown algae
Red algae
Red algae
"Green algae" and Plants
Chlorophyta
Other green algae
PLANTS
Opisthokonts
FUNGI
ANIMALS
Choanoflagellates

28.9 Major Protist Groups in an Evolutionary Context *(Page 550)*

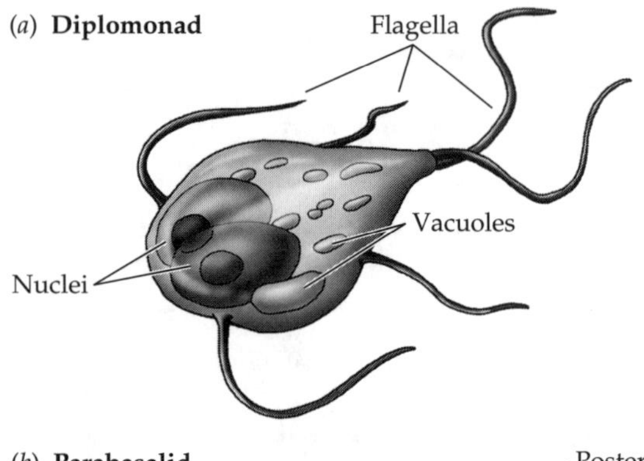
(a) Diplomonad
Flagella
Vacuoles
Nuclei

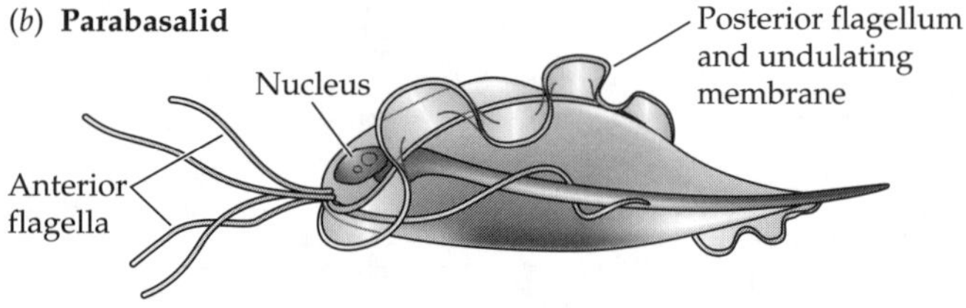
(b) Parabasalid
Nucleus
Posterior flagellum and undulating membrane
Anterior flagella

28.10 Two Protist Groups Lack Mitochondria *(Page 551)*

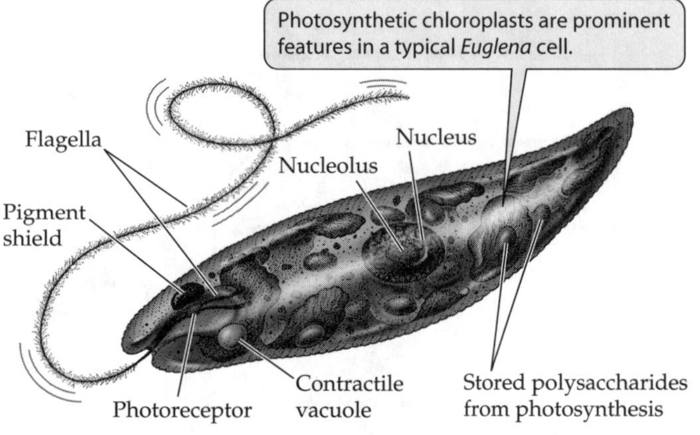
Photosynthetic chloroplasts are prominent features in a typical *Euglena* cell.
Flagella
Nucleolus
Nucleus
Pigment shield
Photoreceptor
Contractile vacuole
Stored polysaccharides from photosynthesis

28.11 A Photosynthetic Euglenoid *(Page 551)*

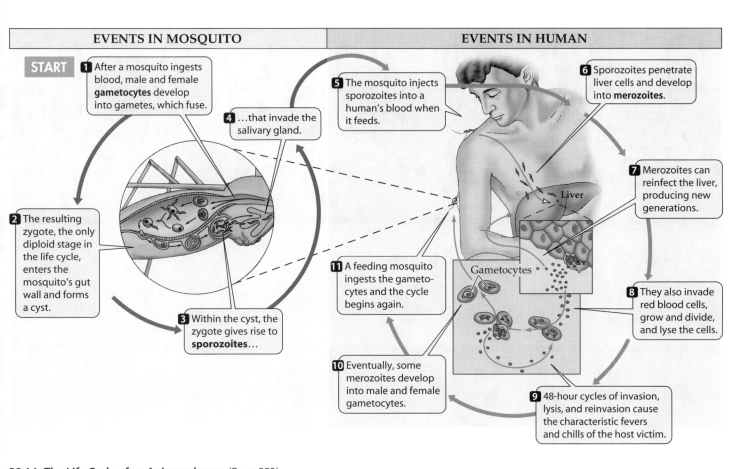

EVENTS IN MOSQUITO

EVENTS IN HUMAN

START

1 After a mosquito ingests blood, male and female **gametocytes** develop into gametes, which fuse.

2 The resulting zygote, the only diploid stage in the life cycle, enters the mosquito's gut wall and forms a cyst.

3 Within the cyst, the zygote gives rise to **sporozoites**...

4 ...that invade the salivary gland.

5 The mosquito injects sporozoites into a human's blood when it feeds.

6 Sporozoites penetrate liver cells and develop into **merozoites**.

7 Merozoites can reinfect the liver, producing new generations.

Liver

Gametocytes

8 They also invade red blood cells, grow and divide, and lyse the cells.

9 48-hour cycles of invasion, lysis, and reinvasion cause the characteristic fevers and chills of the host victim.

10 Eventually, some merozoites develop into male and female gametocytes.

11 A feeding mosquito ingests the gametocytes and the cycle begins again.

28.14 The Life Cycle of an Apicomplexan *(Page 553)*

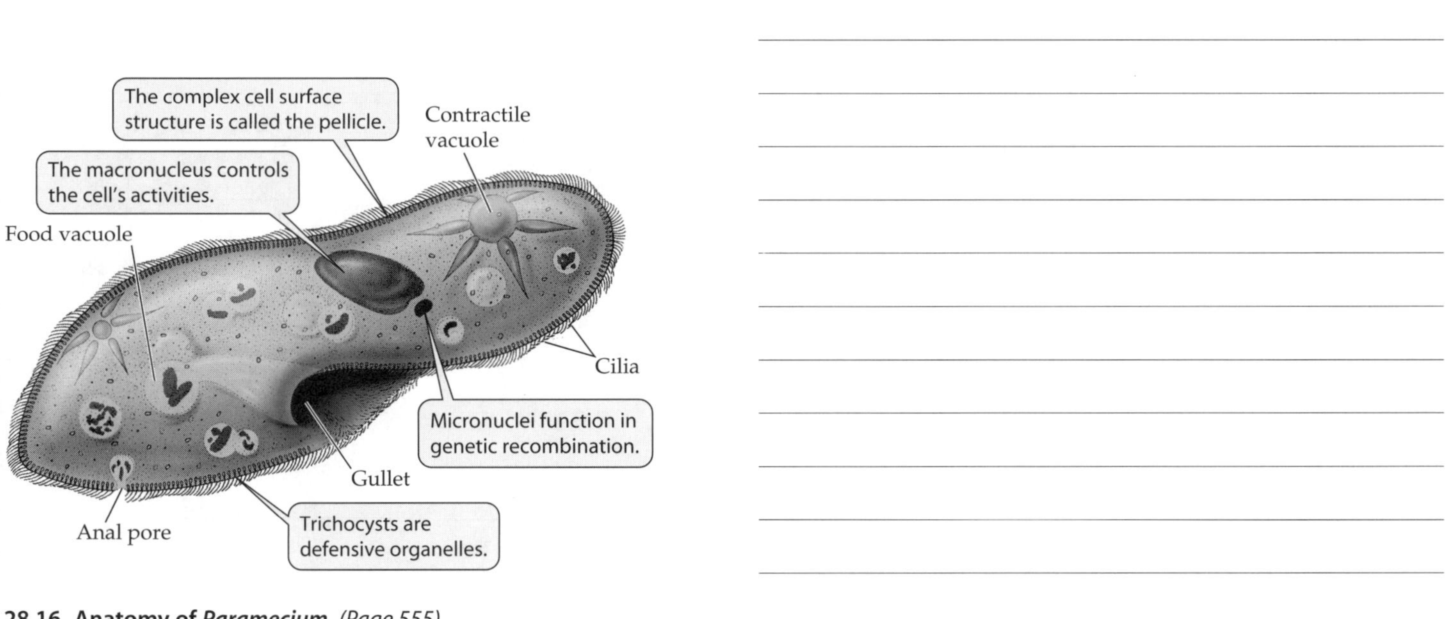

The complex cell surface structure is called the pellicle.
Contractile vacuole
The macronucleus controls the cell's activities.
Food vacuole
Cilia
Micronuclei function in genetic recombination.
Gullet
Anal pore
Trichocysts are defensive organelles.

28.16 Anatomy of *Paramecium* *(Page 555)*

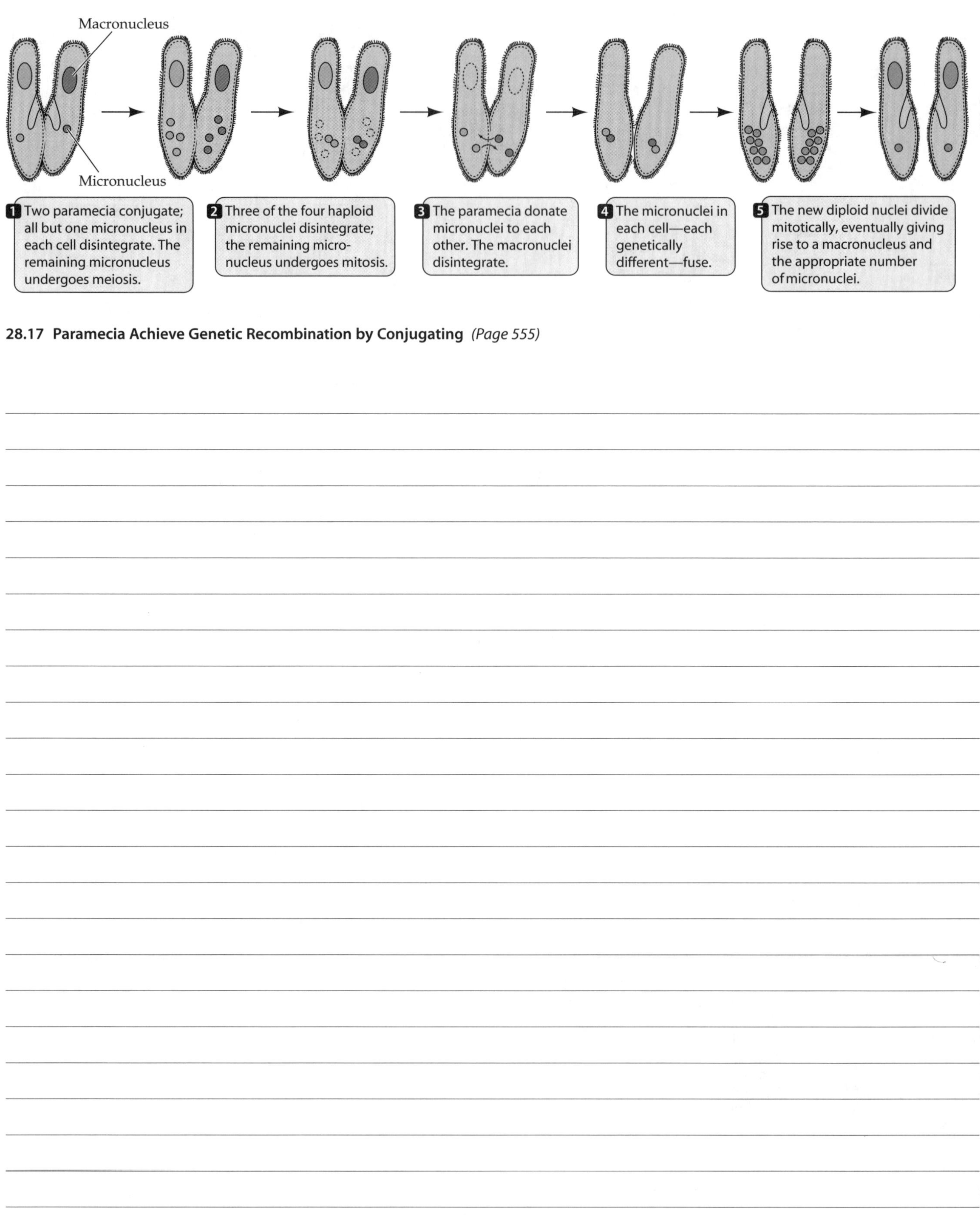
Macronucleus
Micronucleus
1 Two paramecia conjugate; all but one micronucleus in each cell disintegrate. The remaining micronucleus undergoes meiosis.
2 Three of the four haploid micronuclei disintegrate; the remaining micro-nucleus undergoes mitosis.
3 The paramecia donate micronuclei to each other. The macronuclei disintegrate.
4 The micronuclei in each cell—each genetically different—fuse.
5 The new diploid nuclei divide mitotically, eventually giving rise to a macronucleus and the appropriate number of micronuclei.

28.17 Paramecia Achieve Genetic Recombination by Conjugating *(Page 555)*

Sexual reproduction

1 The silicon-impregnated cell walls of diatoms are two-part "petri plates."

2 In sexual reproduction, the "petri plates" split, releasing gametes.

Gamete

Gamete from another individual

Gametes fuse

4 In asexual reproduction, the two parts of the cell wall separate, each becoming the top of a new "plate."

Meiosis

HAPLOID (*n*)
DIPLOID (2*n*)

Fertilization

Zygote

Asexual reproduction

3 Zygotes produced by sexual reproduction grow and lay down new full-size cell walls.

5 The offspring cells from the bottom "plate" become progressively smaller.

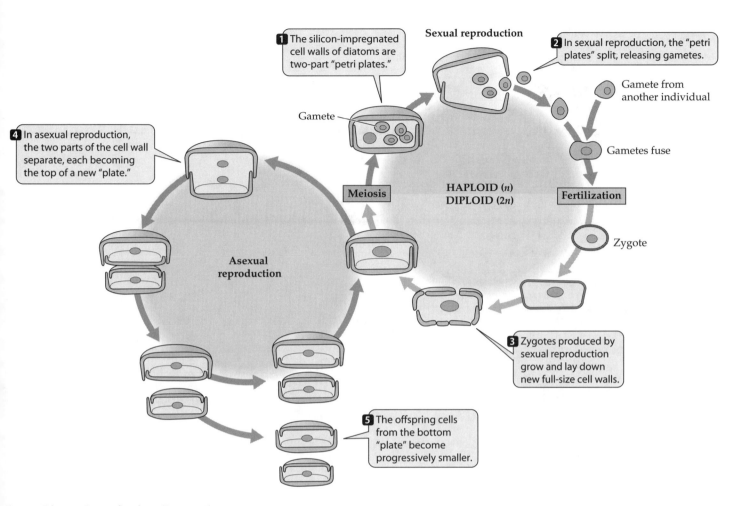

28.19 Diatom Reproduction *(Page 557)*

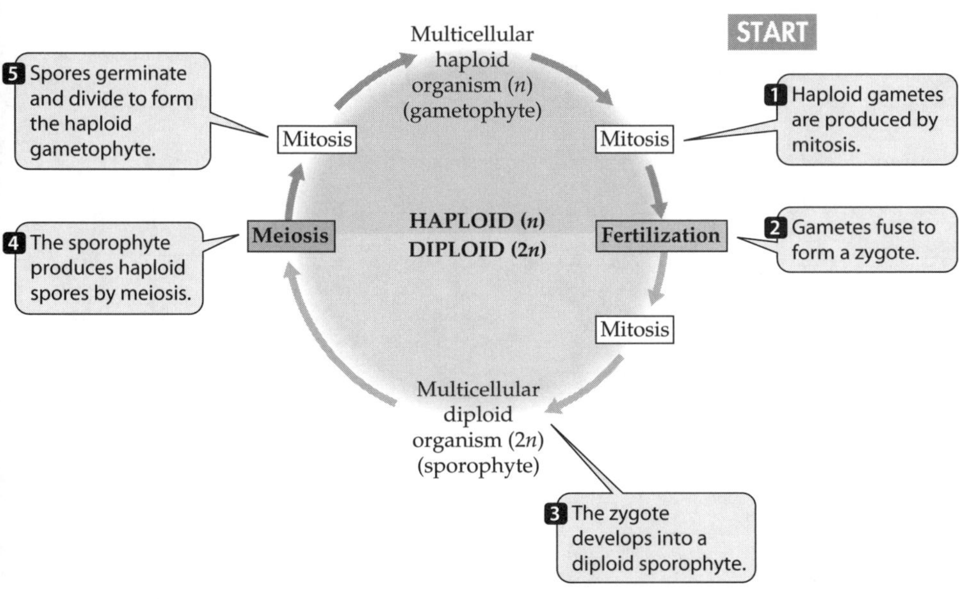
START
Multicellular haploid organism (*n*) (gametophyte)
5 Spores germinate and divide to form the haploid gametophyte.
Mitosis
Mitosis
1 Haploid gametes are produced by mitosis.
4 The sporophyte produces haploid spores by meiosis.
Meiosis
HAPLOID (*n*)
DIPLOID (2*n*)
Fertilization
2 Gametes fuse to form a zygote.
Mitosis
Multicellular diploid organism (2*n*) (sporophyte)
3 The zygote develops into a diploid sporophyte.

28.22 Alternation of Generations *(Page 559)*

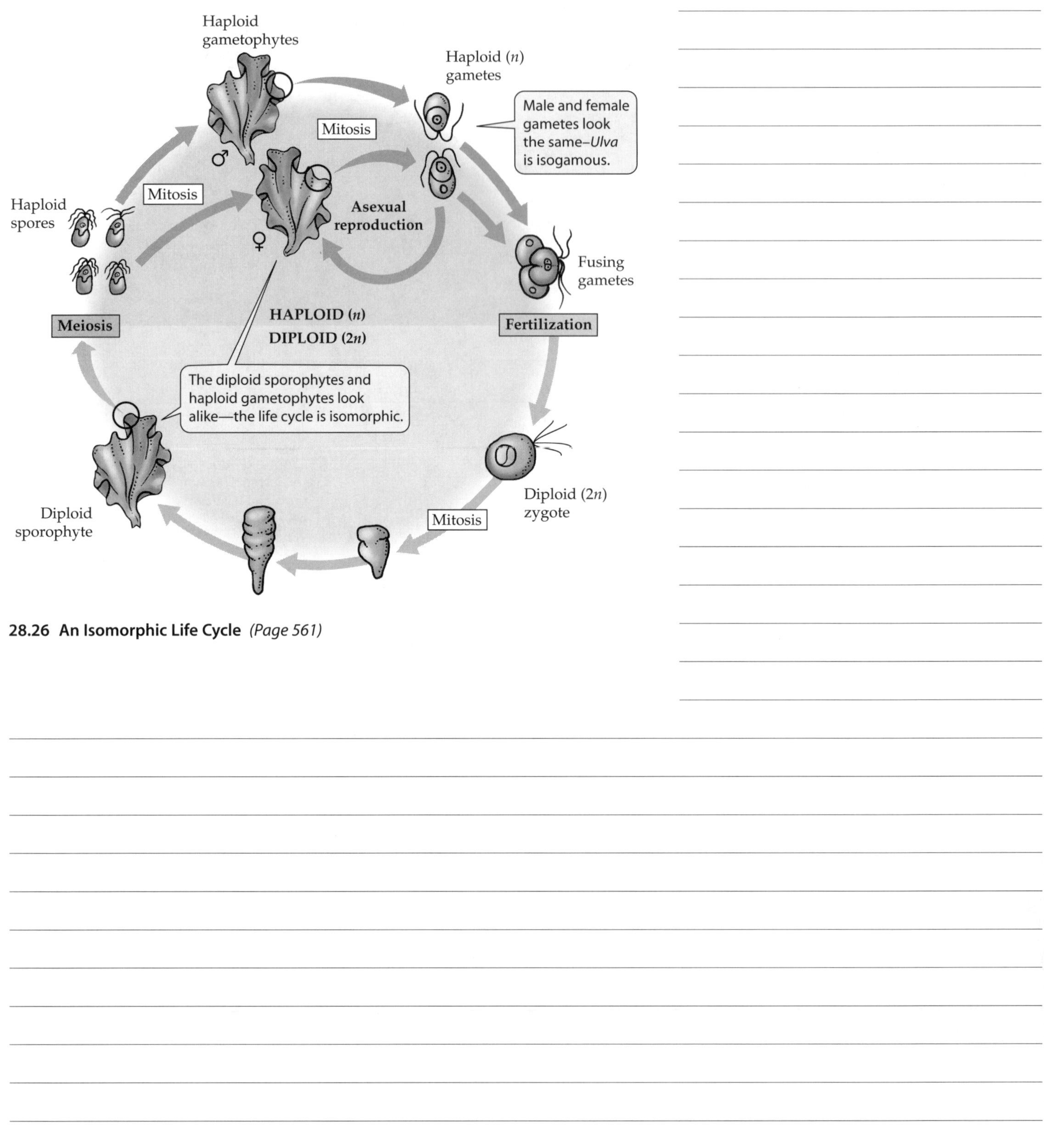
Haploid gametophytes
Haploid (n) gametes
Male and female gametes look the same–Ulva is isogamous.
Mitosis
♂
Mitosis
Haploid spores
Asexual reproduction
♀
Fusing gametes
Meiosis
HAPLOID (n)
DIPLOID (2n)
Fertilization
The diploid sporophytes and haploid gametophytes look alike—the life cycle is isomorphic.
Diploid (2n) zygote
Diploid sporophyte
Mitosis

28.26 An Isomorphic Life Cycle *(Page 561)*

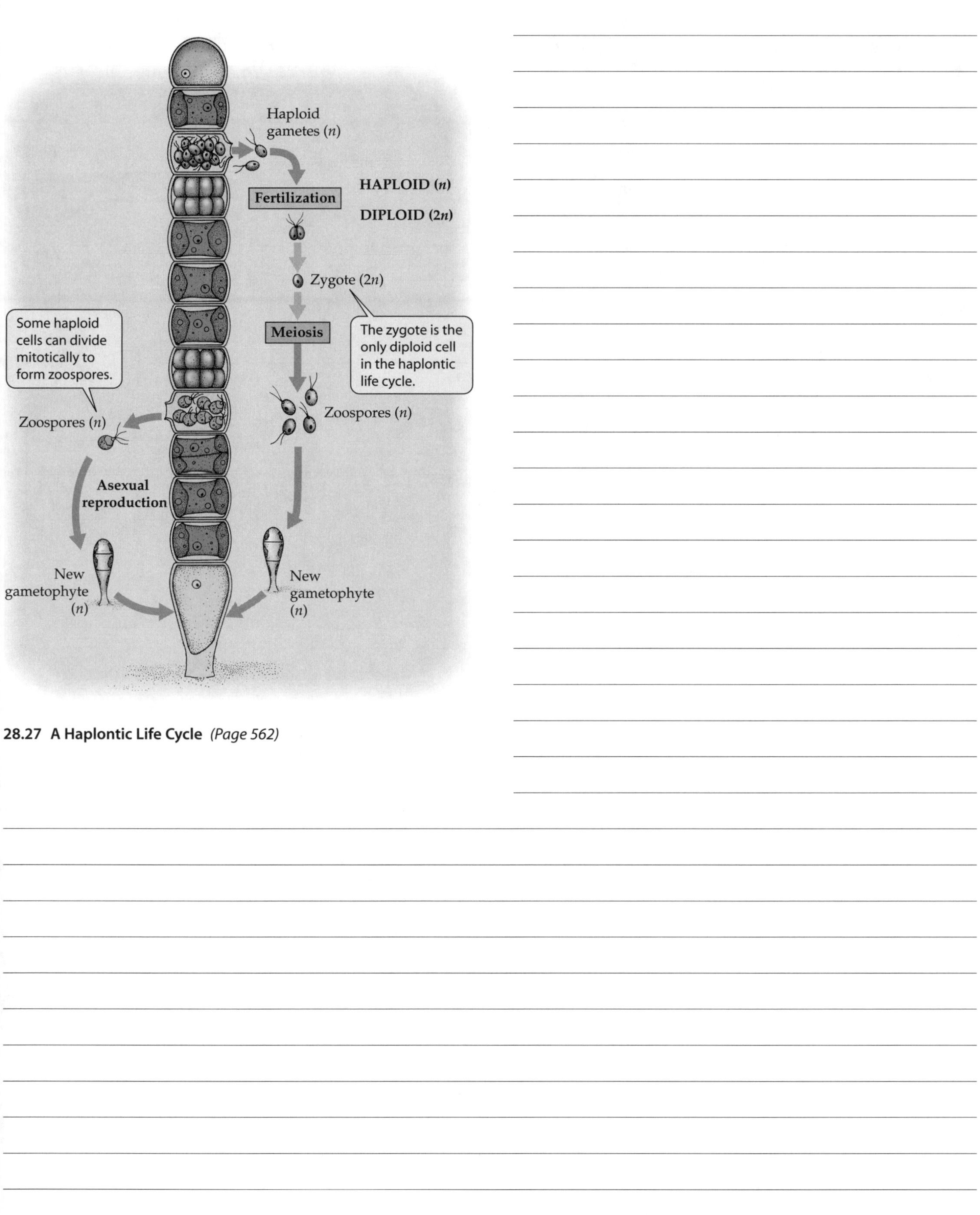
Haploid
gametes (n)
Fertilization
HAPLOID (n)
DIPLOID (2n)
Zygote (2n)
Meiosis
The zygote is the only diploid cell in the haplontic life cycle.
Some haploid cells can divide mitotically to form zoospores.
Zoospores (n)
Zoospores (n)
Asexual reproduction
New gametophyte (n)
New gametophyte (n)

28.27 A Haplontic Life Cycle *(Page 562)*

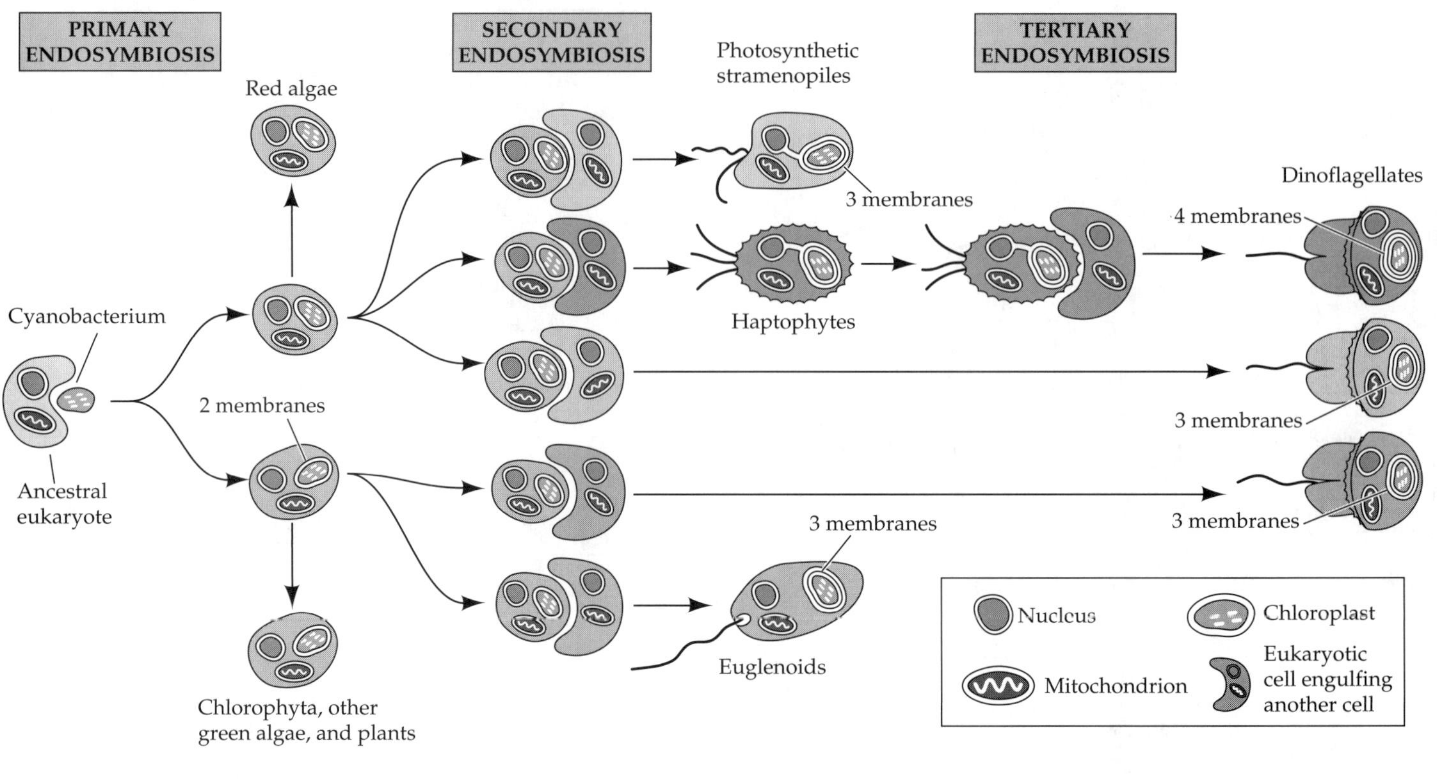
PRIMARY ENDOSYMBIOSIS
SECONDARY ENDOSYMBIOSIS
TERTIARY ENDOSYMBIOSIS
Red algae
Photosynthetic stramenopiles
3 membranes
Dinoflagellates
4 membranes
Haptophytes
Cyanobacterium
2 membranes
Ancestral eukaryote
3 membranes
3 membranes
3 membranes
Euglenoids
Chlorophyta, other green algae, and plants
Nucleus
Chloroplast
Mitochondrion
Eukaryotic cell engulfing another cell

28.29 A Chloroplast Family Tree *(Page 563)*

29 Plants without Seeds: From Sea to Land

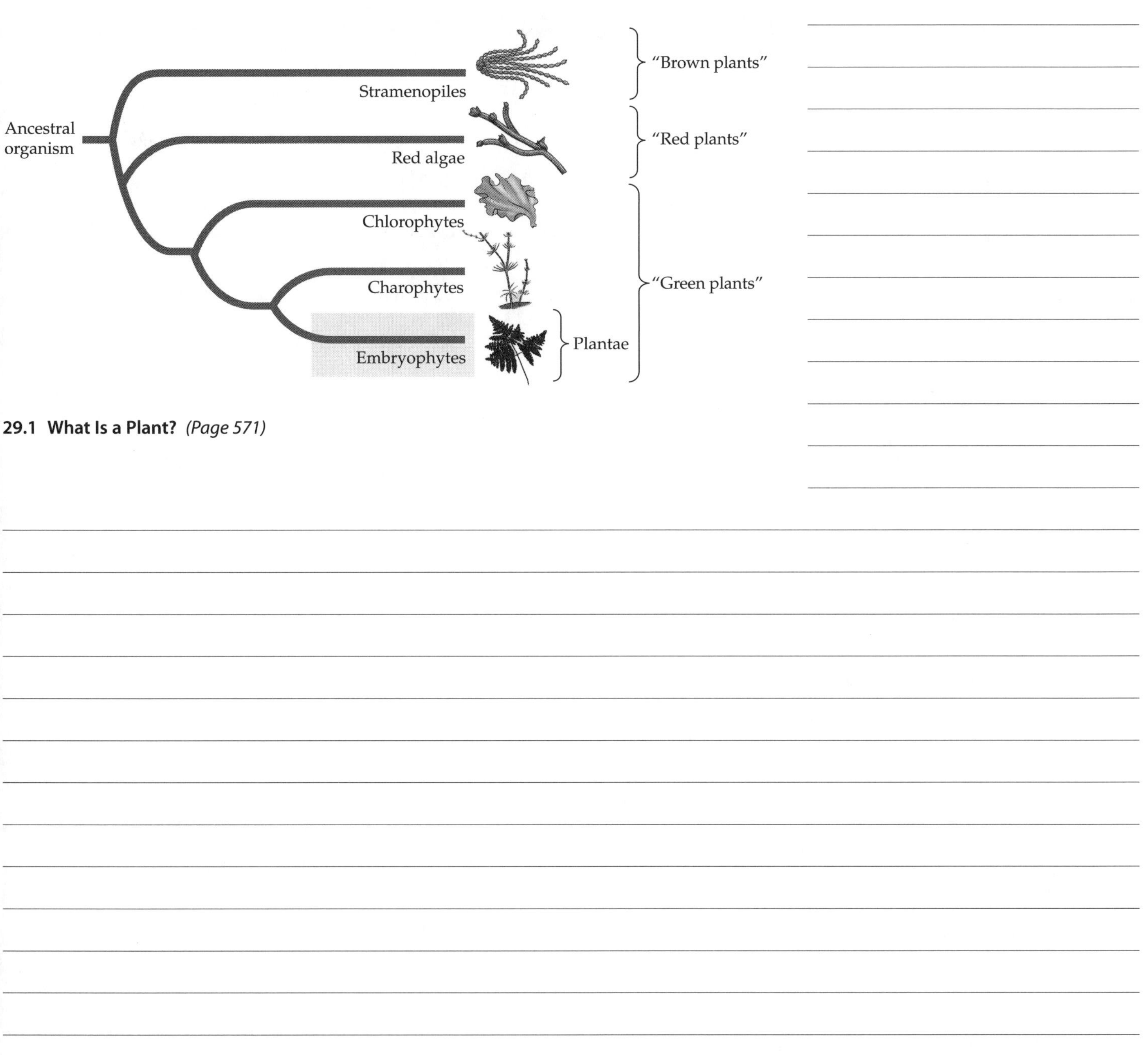
Ancestral organism
Stramenopiles
Red algae
Chlorophytes
Charophytes
Embryophytes
"Brown plants"
"Red plants"
"Green plants"
Plantae

29.1 What Is a Plant? *(Page 571)*

29.1 Classification of Plants[a]

PHYLUM	COMMON NAME	CHARACTERISTICS
Nontracheophytes		
Hepatophyta	Liverworts	No filamentous stage; gametophyte flat
Anthocerophyta	Hornworts	Embedded archegonia; sporophyte grows basally
Bryophyta	Mosses	Filamentous stage; sporophyte grows apically (from the tip)
Tracheophytes		
Nonseed tracheophytes		
Lycophyta	Club mosses	Microphylls in spirals; sporangia in leaf axils
Pteridophyta	Ferns and allies	Differentiation between main axis and side branches
Seed plants		
Gymnosperms		
Cycadophyta	Cycads	Compound leaves; swimming sperm; seeds on modified leaves
Ginkgophyta	Ginkgo	Deciduous; fan-shaped leaves; swimming sperm
Gnetophyta	Gnetophytes	Vessels in vascular tissue; opposite, simple leaves
Pinophyta	Conifers	Seeds in cones; needlelike or scalelike leaves
Angiosperms		
Angiospermae	Flowering plants	Endosperm; carpels; much reduced gametophytes; seeds in fruit

[a] No extinct groups are included in this classification.

(Page 572)

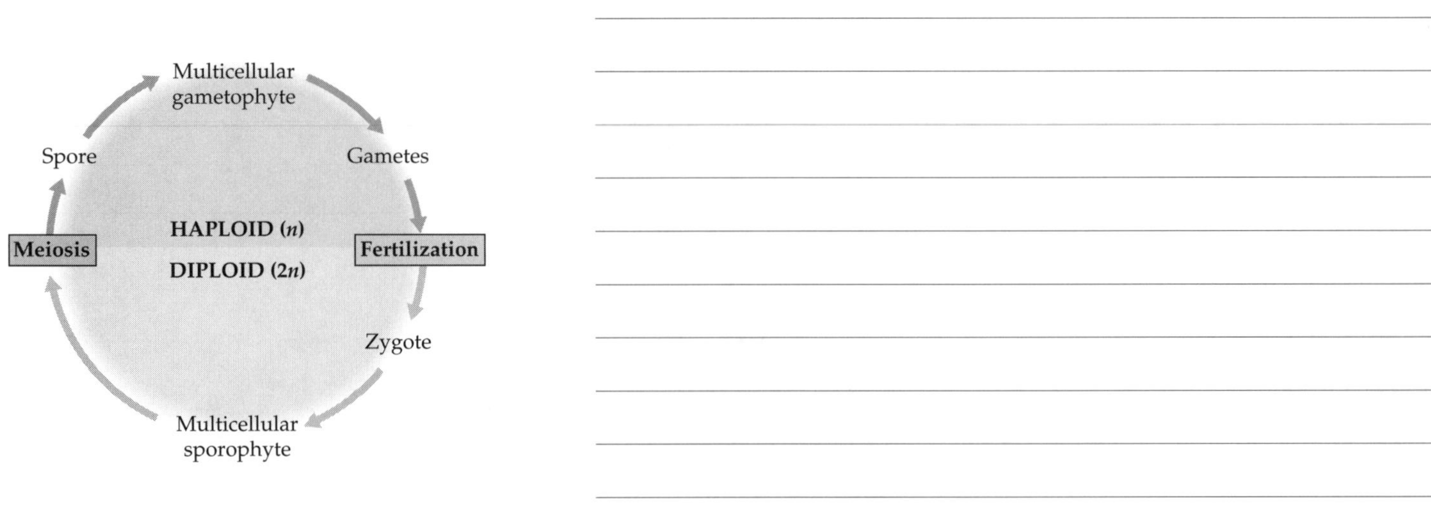
Multicellular gametophyte
Gametes
Spore
HAPLOID (n)
Meiosis
Fertilization
DIPLOID (2n)
Zygote
Multicellular sporophyte

29.2 Alternation of Generations *(Page 572)*

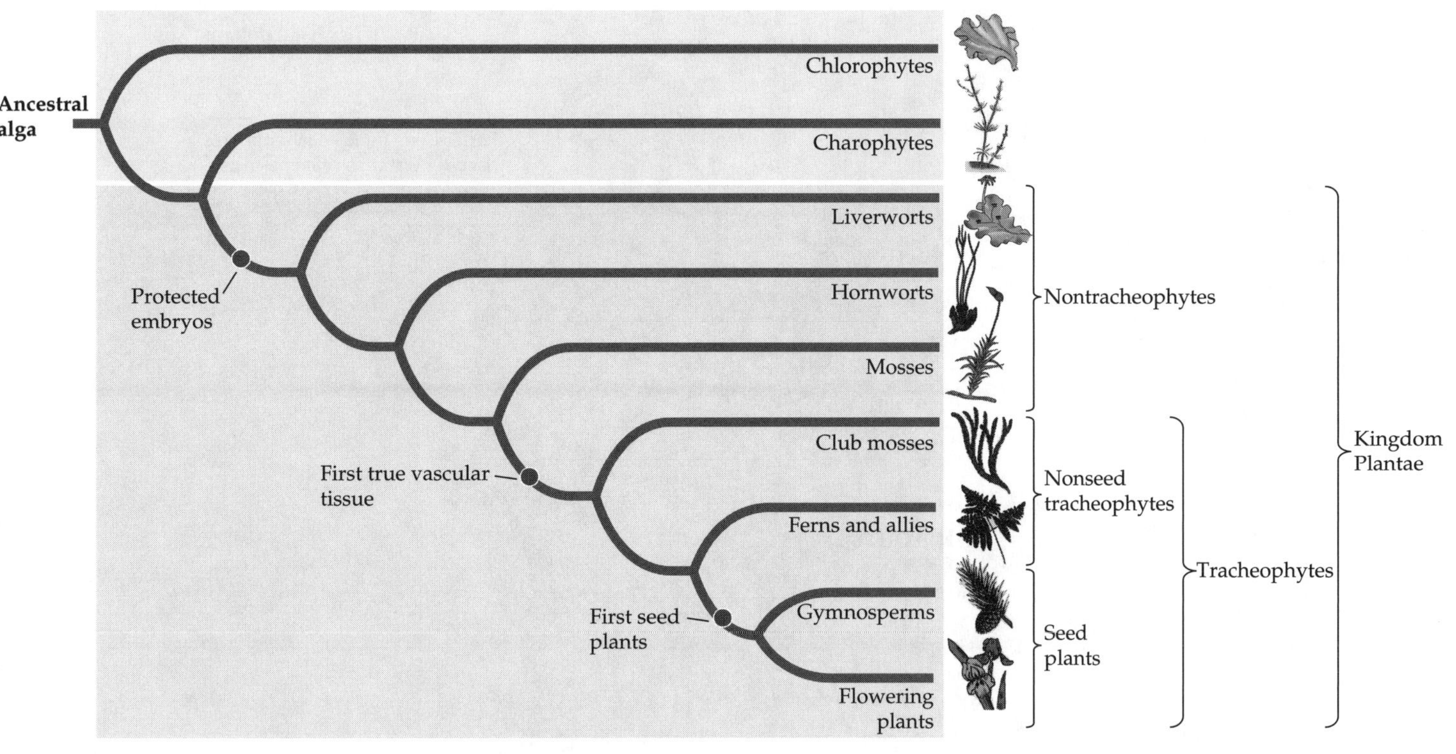
Ancestral alga
Chlorophytes
Charophytes
Liverworts
Hornworts
Mosses
Club mosses
Ferns and allies
Gymnosperms
Flowering plants
Protected embryos
First true vascular tissue
First seed plants
Nontracheophytes
Nonseed tracheophytes
Seed plants
Tracheophytes
Kingdom Plantae

29.4 From Green Algae to Plants *(Page 574)*

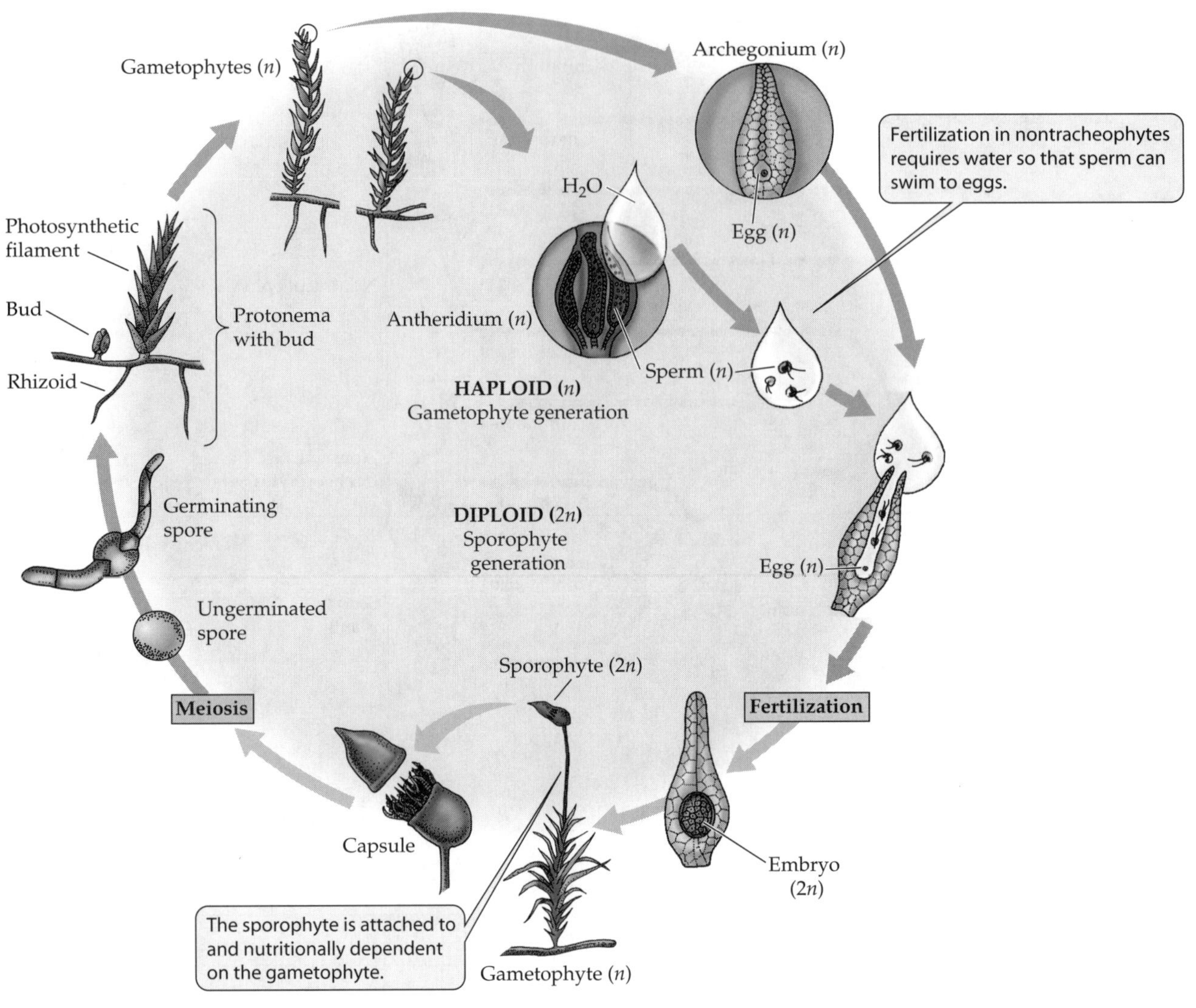
Gametophytes (n)
Archegonium (n)
Fertilization in nontracheophytes requires water so that sperm can swim to eggs.
H_2O
Egg (n)
Photosynthetic filament
Bud
Rhizoid
Protonema with bud
Antheridium (n)
Sperm (n)
HAPLOID (n)
Gametophyte generation
Germinating spore
DIPLOID ($2n$)
Sporophyte generation
Egg (n)
Ungerminated spore
Sporophyte ($2n$)
Meiosis
Fertilization
Capsule
Embryo ($2n$)
The sporophyte is attached to and nutritionally dependent on the gametophyte.
Gametophyte (n)

29.5 A Nontracheophyte Life Cycle *(Page 575)*

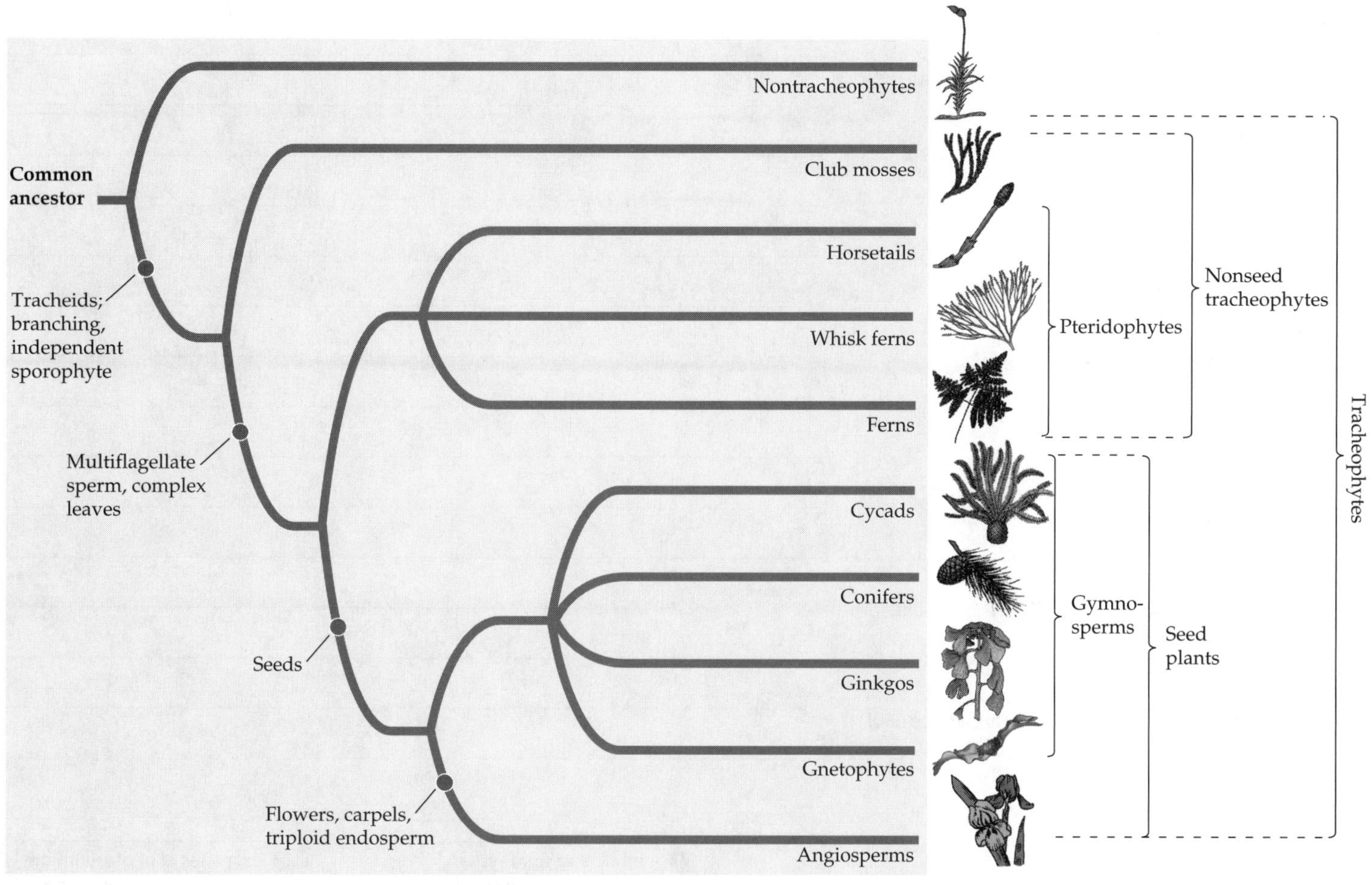
Common ancestor
Tracheids; branching, independent sporophyte
Multiflagellate sperm, complex leaves
Seeds
Flowers, carpels, triploid endosperm
Nontracheophytes
Club mosses
Horsetails
Whisk ferns
Ferns
Cycads
Conifers
Ginkgos
Gnetophytes
Angiosperms
Pteridophytes
Nonseed tracheophytes
Gymno- sperms
Seed plants
Tracheophytes

29.10 The Evolution of Today's Plants *(Page 579)*

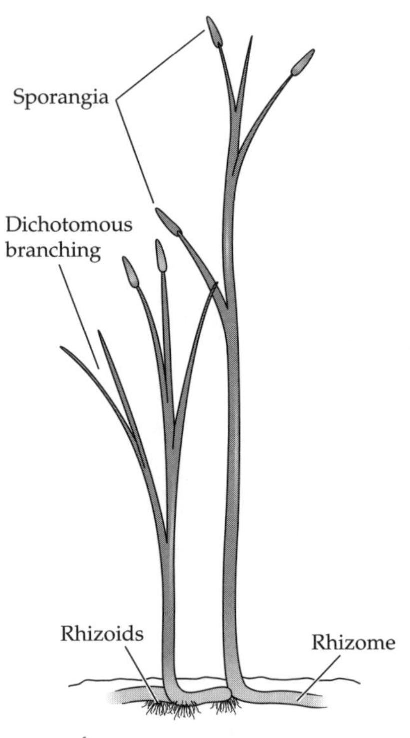
Sporangia
Dichotomous branching
Rhizoids
Rhizome

29.12 An Ancient Tracheophyte Relative *(Page 580)*

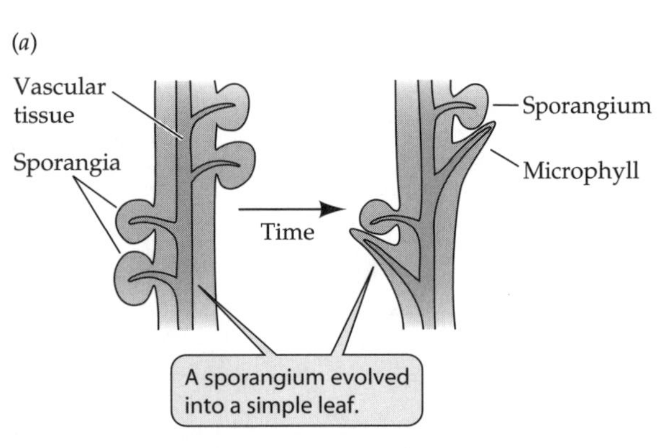
(a)
Vascular tissue
Sporangia
Sporangium
Microphyll
Time
A sporangium evolved into a simple leaf.

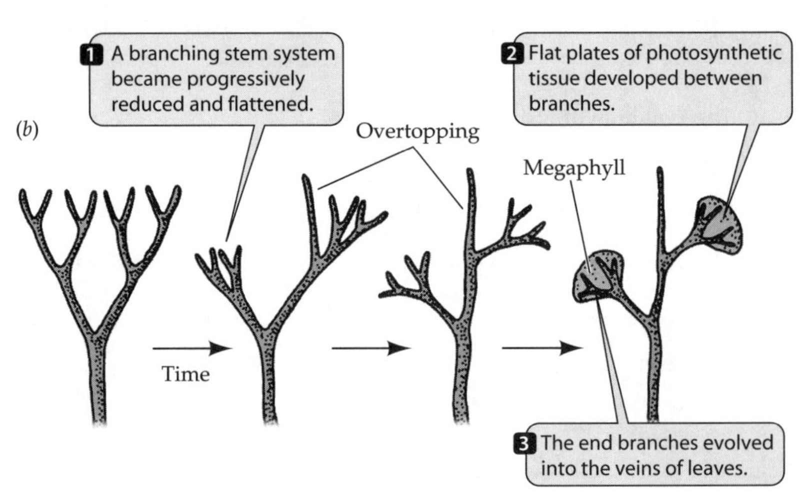
(b)
1 A branching stem system became progressively reduced and flattened.
Overtopping
2 Flat plates of photosynthetic tissue developed between branches.
Megaphyll
Time
3 The end branches evolved into the veins of leaves.

29.13 The Evolution of Leaves *(Page 581)*

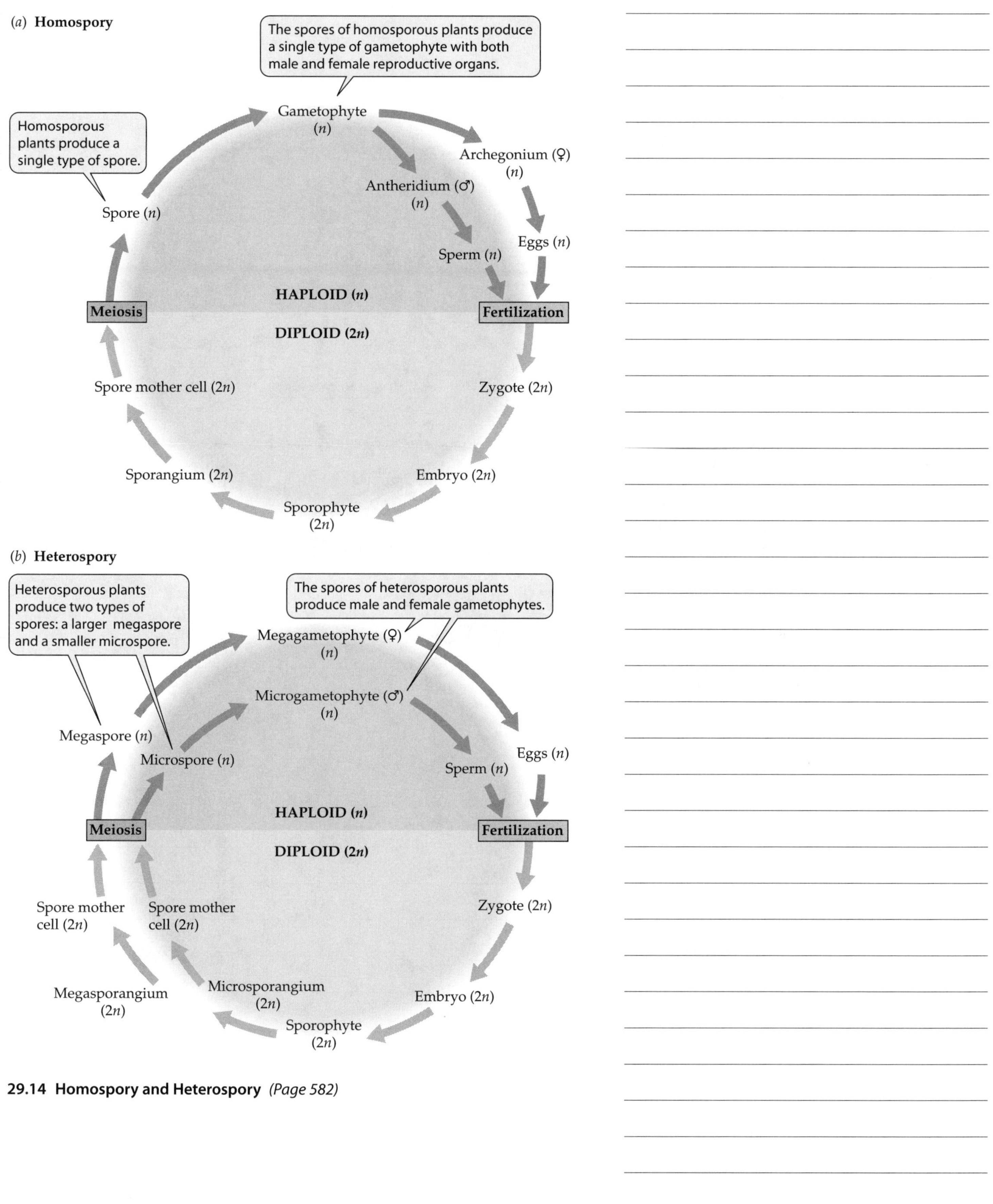

(a) Homospory
The spores of homosporous plants produce a single type of gametophyte with both male and female reproductive organs.
Homosporous plants produce a single type of spore.
Gametophyte (n)
Archegonium (♀) (n)
Antheridium (♂) (n)
Spore (n)
Eggs (n)
Sperm (n)
HAPLOID (n)
Meiosis
Fertilization
DIPLOID (2n)
Spore mother cell (2n)
Zygote (2n)
Sporangium (2n)
Embryo (2n)
Sporophyte (2n)
(b) Heterospory
Heterosporous plants produce two types of spores: a larger megaspore and a smaller microspore.
The spores of heterosporous plants produce male and female gametophytes.
Megagametophyte (♀) (n)
Microgametophyte (♂) (n)
Megaspore (n)
Microspore (n)
Eggs (n)
Sperm (n)
HAPLOID (n)
Meiosis
Fertilization
DIPLOID (2n)
Spore mother cell (2n)
Spore mother cell (2n)
Zygote (2n)
Megasporangium (2n)
Microsporangium (2n)
Embryo (2n)
Sporophyte (2n)

29.14 Homospory and Heterospory *(Page 582)*

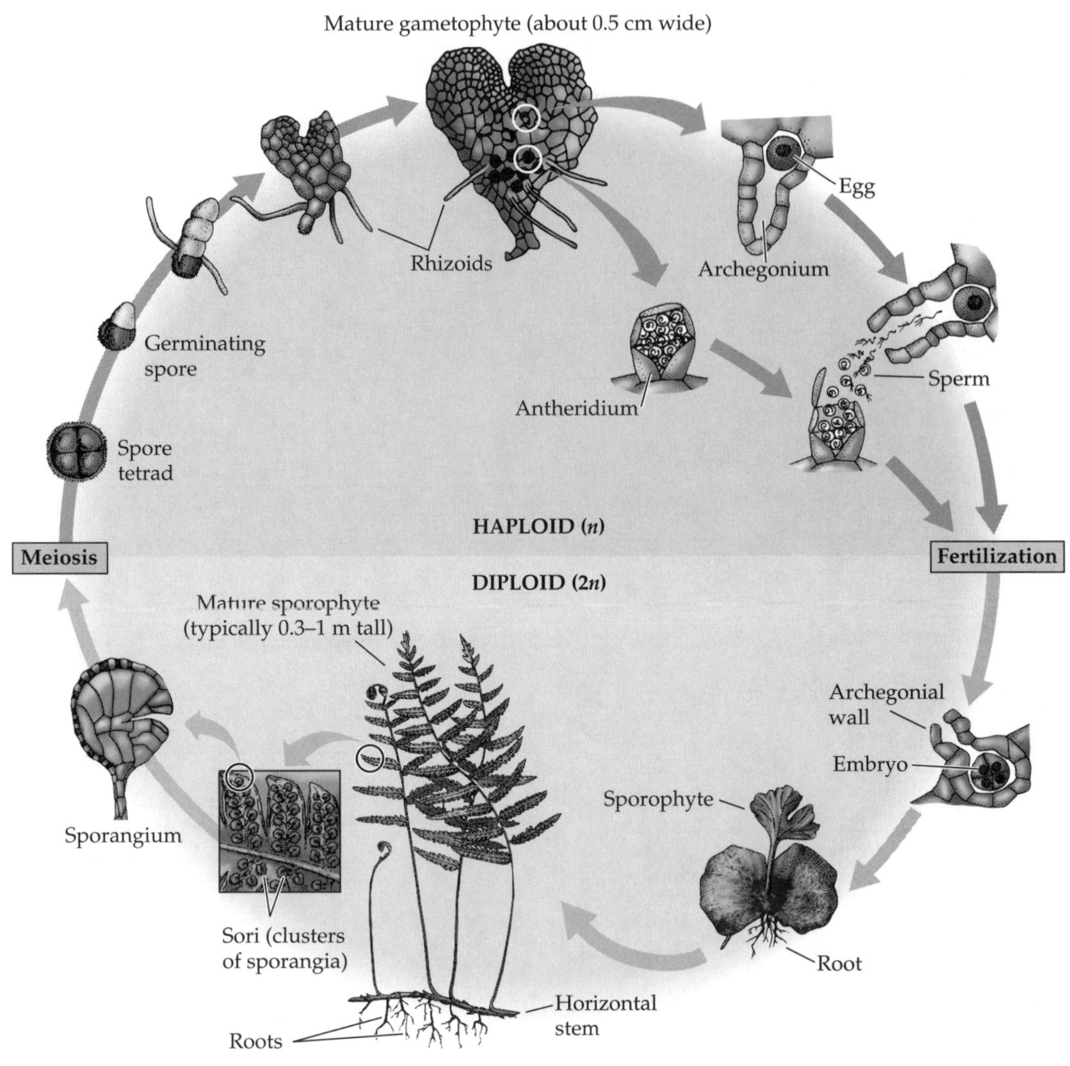
Mature gametophyte (about 0.5 cm wide)
Egg
Archegonium
Rhizoids
Germinating spore
Spore tetrad
Sperm
Antheridium
HAPLOID (n)
Meiosis
Fertilization
DIPLOID (2n)
Mature sporophyte (typically 0.3–1 m tall)
Archegonial wall
Embryo
Sporophyte
Sporangium
Sori (clusters of sporangia)
Root
Horizontal stem
Roots

29.20 The Life Cycle of a Fern *(Page 586)*

30 The Evolution of Seed Plants

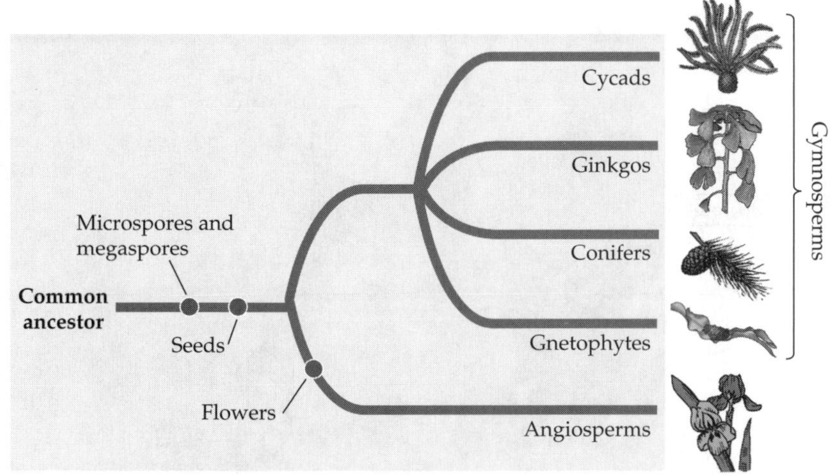

Cycads
Ginkgos
Conifers
Gnetophytes
Angiosperms
Gymnosperms
Microspores and megaspores
Common ancestor
Seeds
Flowers

30.1 The Phyla of Living Seed Plants *(Page 589)*

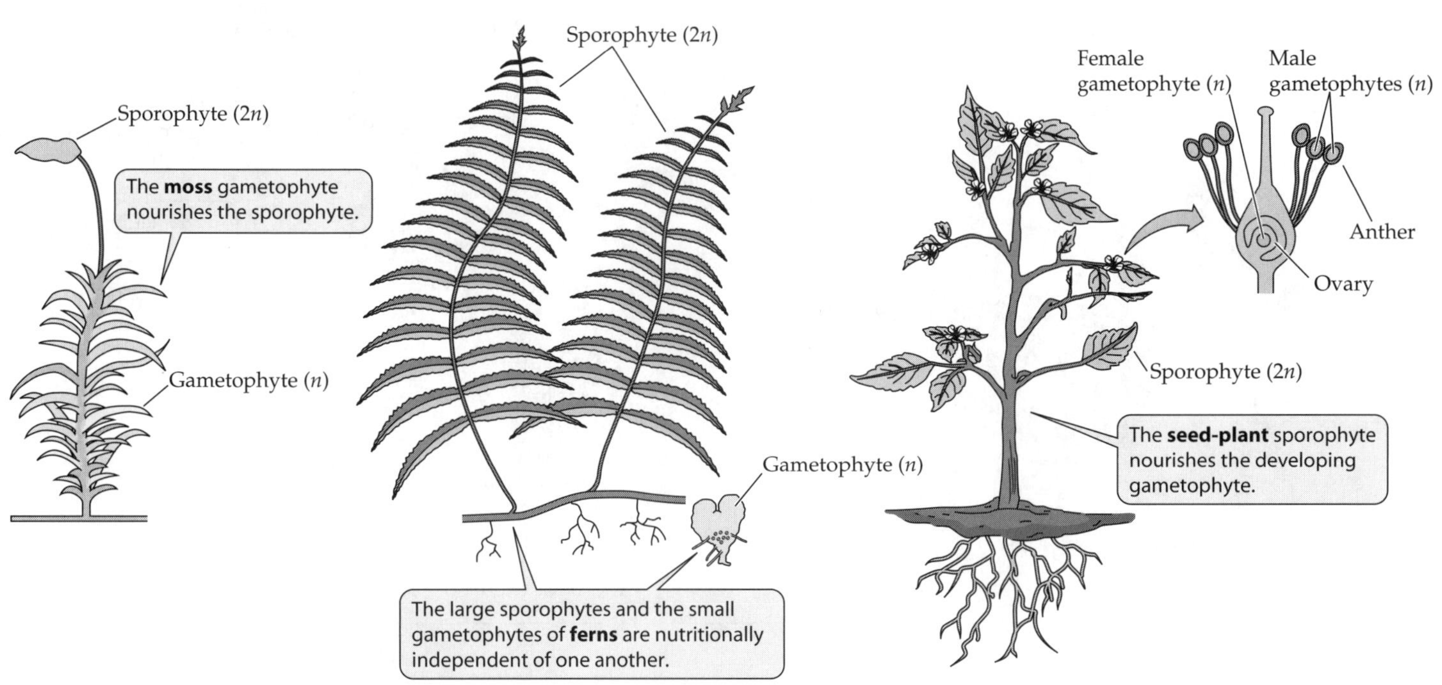

Sporophyte (2n)
The **moss** gametophyte nourishes the sporophyte.
Gametophyte (n)
Sporophyte (2n)
Gametophyte (n)
The large sporophytes and the small gametophytes of **ferns** are nutritionally independent of one another.
Female gametophyte (n)
Male gametophytes (n)
Anther
Ovary
Sporophyte (2n)
The **seed-plant** sporophyte nourishes the developing gametophyte.

30.2 The Relationship between Sporophyte and Gametophyte Has Evolved *(Page 589)*

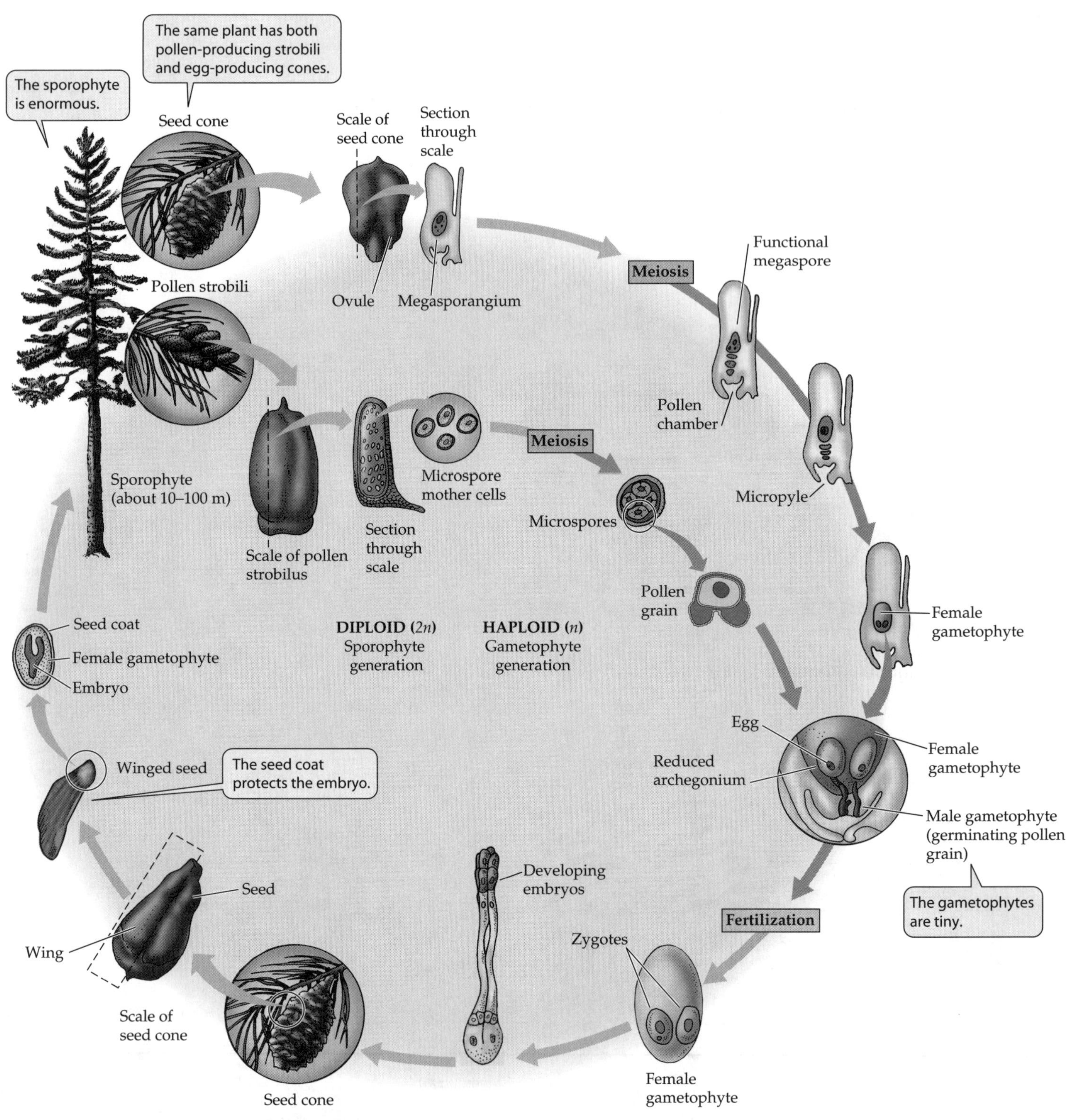
The sporophyte is enormous.
The same plant has both pollen-producing strobili and egg-producing cones.
Seed cone
Scale of seed cone
Section through scale
Meiosis
Functional megaspore
Pollen strobili
Ovule
Megasporangium
Pollen chamber
Meiosis
Microspore mother cells
Sporophyte (about 10–100 m)
Section through scale
Scale of pollen strobilus
Microspores
Micropyle
Pollen grain
Female gametophyte
Seed coat
Female gametophyte
Embryo
DIPLOID (2n) Sporophyte generation
HAPLOID (n) Gametophyte generation
Egg
Reduced archegonium
Female gametophyte
Winged seed
The seed coat protects the embryo.
Male gametophyte (germinating pollen grain)
Developing embryos
Seed
The gametophytes are tiny.
Fertilization
Wing
Zygotes
Scale of seed cone
Seed cone
Female gametophyte

30.6 The Life Cycle of a Pine Tree *(Page 593)*

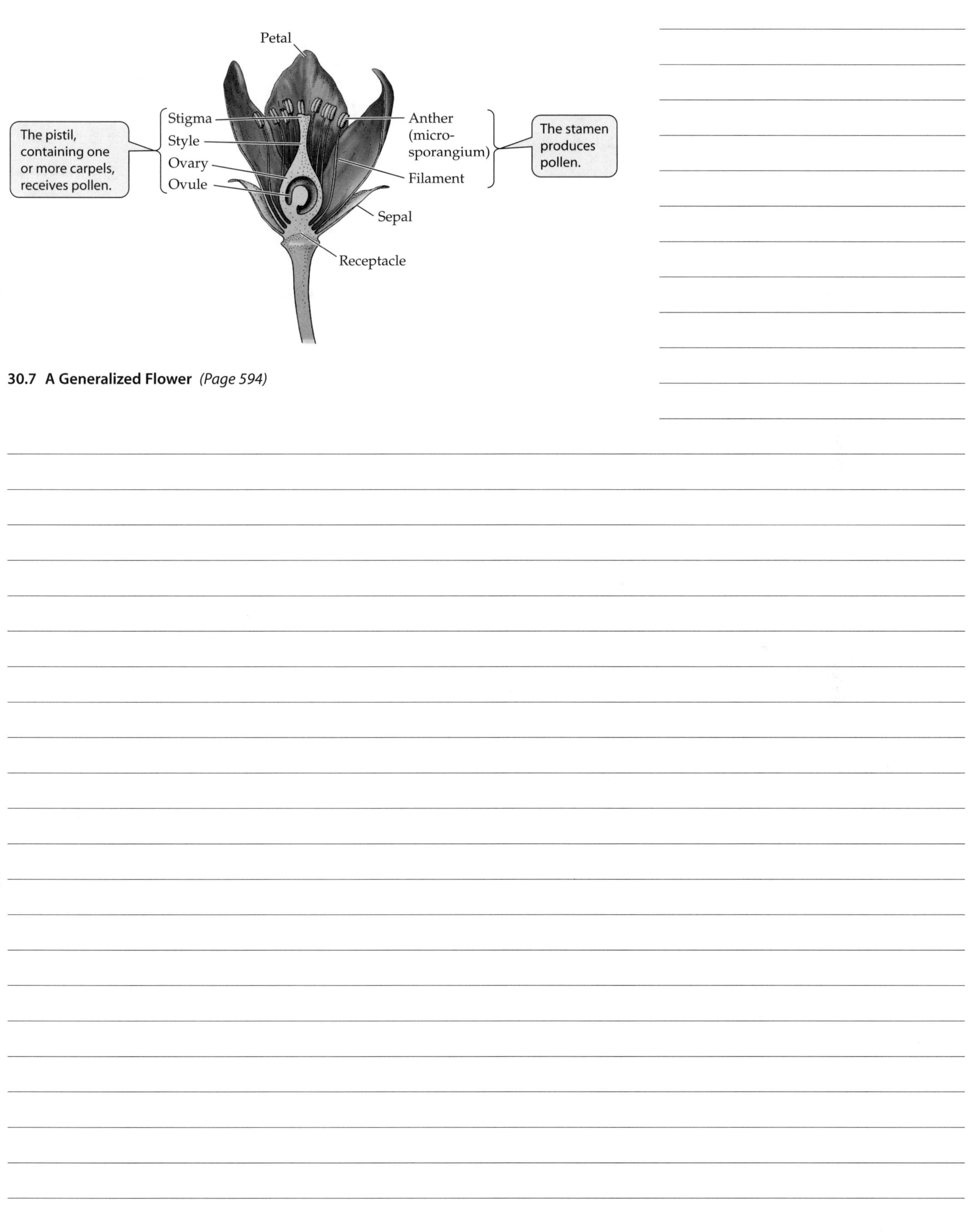
Petal
The pistil, containing one or more carpels, receives pollen.
Stigma
Style
Ovary
Ovule
Anther (micro-sporangium)
Filament
The stamen produces pollen.
Sepal
Receptacle

30.7 A Generalized Flower *(Page 594)*

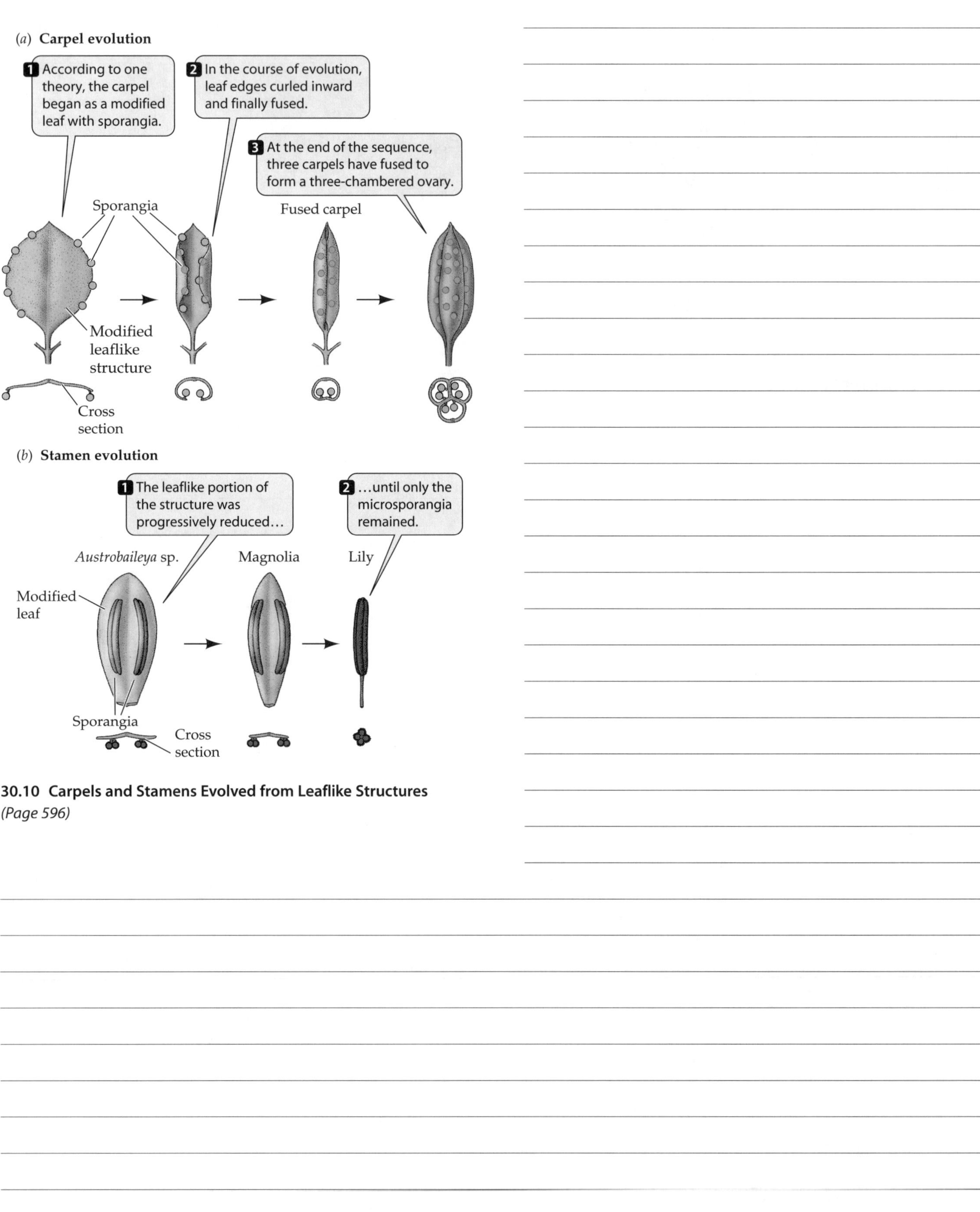
(a) Carpel evolution
1 According to one theory, the carpel began as a modified leaf with sporangia.
2 In the course of evolution, leaf edges curled inward and finally fused.
3 At the end of the sequence, three carpels have fused to form a three-chambered ovary.
Sporangia
Fused carpel
Modified leaflike structure
Cross section
(b) Stamen evolution
1 The leaflike portion of the structure was progressively reduced…
2 …until only the microsporangia remained.
Austrobaileya sp.
Magnolia
Lily
Modified leaf
Sporangia
Cross section

30.10 Carpels and Stamens Evolved from Leaflike Structures
(Page 596)

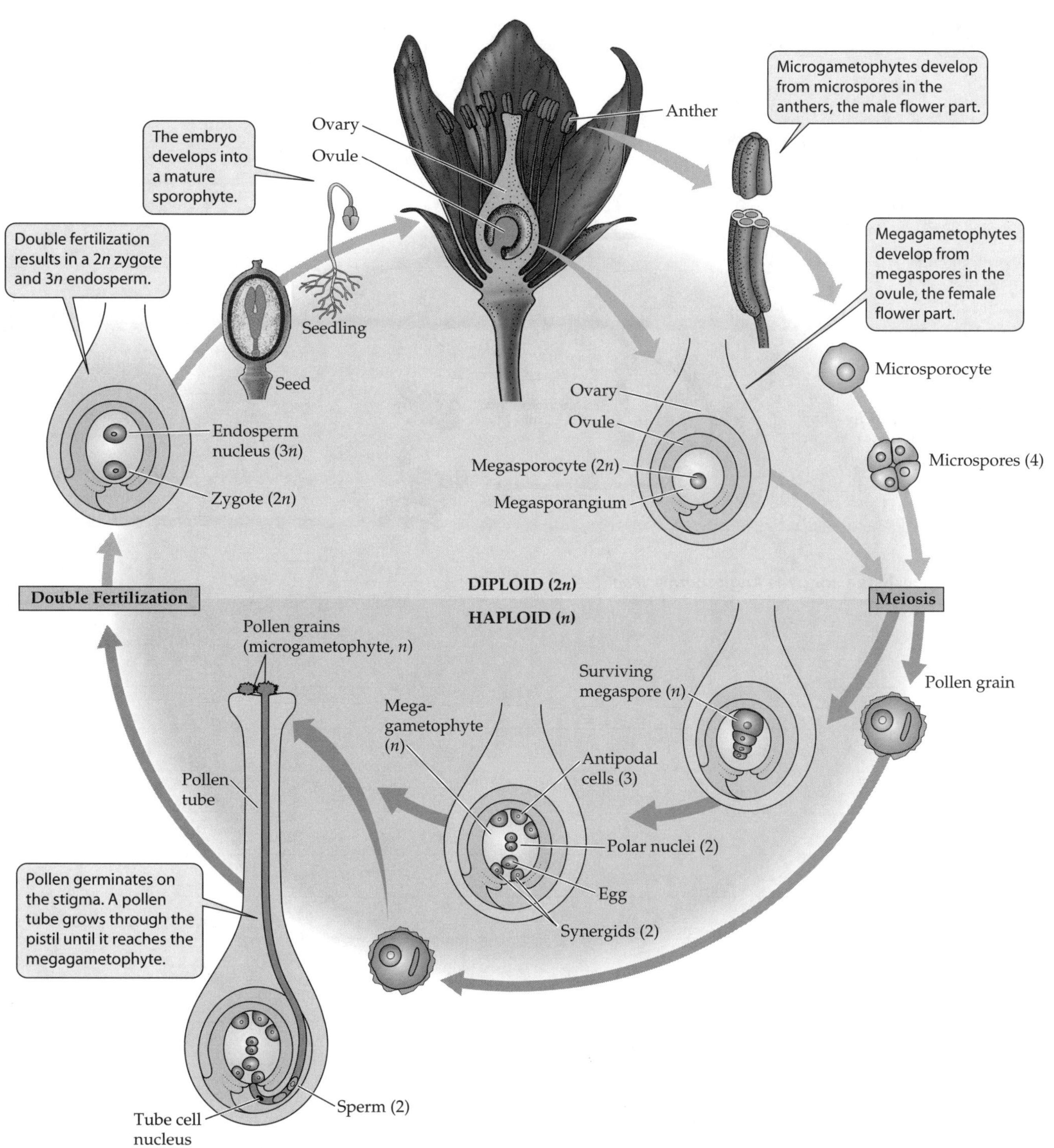

Microgametophytes develop from microspores in the anthers, the male flower part.
Anther
Ovary
Ovule
The embryo develops into a mature sporophyte.
Double fertilization results in a 2n zygote and 3n endosperm.
Megagametophytes develop from megaspores in the ovule, the female flower part.
Seedling
Seed
Microsporocyte
Ovary
Ovule
Endosperm nucleus (3n)
Zygote (2n)
Megasporocyte (2n)
Megasporangium
Microspores (4)
DIPLOID (2n)
Double Fertilization
Meiosis
HAPLOID (n)
Pollen grains (microgametophyte, n)
Surviving megaspore (n)
Pollen grain
Mega-gametophyte (n)
Antipodal cells (3)
Pollen tube
Polar nuclei (2)
Egg
Synergids (2)
Pollen germinates on the stigma. A pollen tube grows through the pistil until it reaches the megagametophyte.
Sperm (2)
Tube cell nucleus

30.11 The Life Cycle of an Angiosperm *(Page 597)*

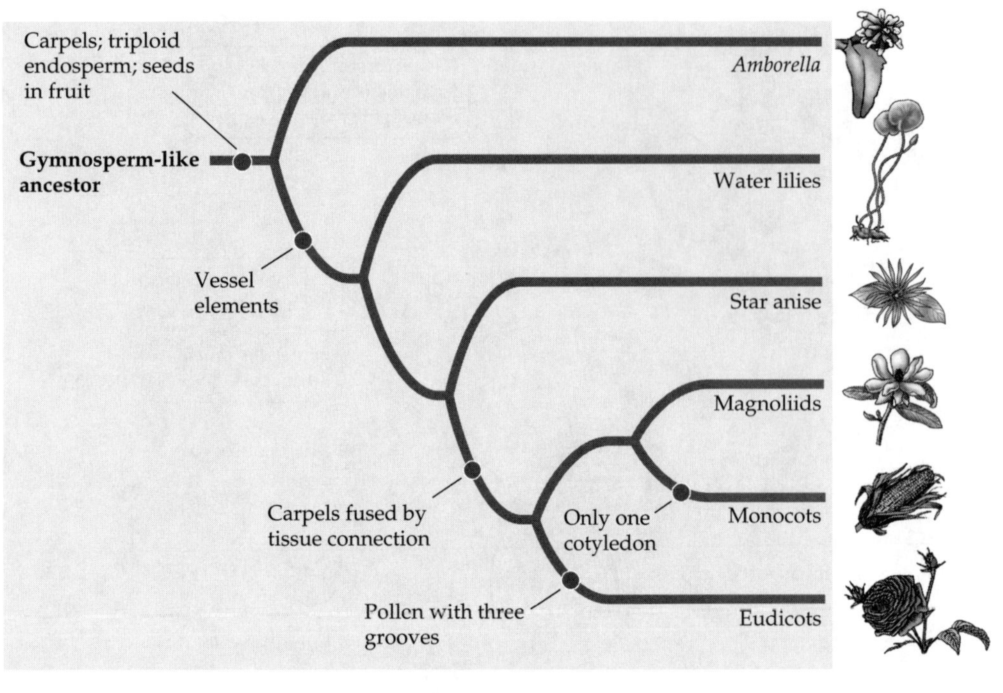
Carpels; triploid endosperm; seeds in fruit
Gymnosperm-like ancestor
Vessel elements
Carpels fused by tissue connection
Only one cotyledon
Pollen with three grooves
Amborella
Water lilies
Star anise
Magnoliids
Monocots
Eudicots

30.13 Evolutionary Relationships among the Angiosperms *(Page 599)*

31 Fungi: Recyclers, Pathogens, Parasites, and Plant Partners

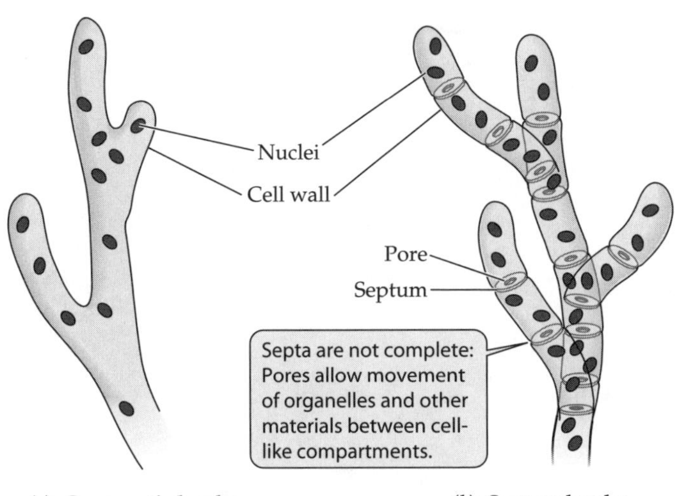
Nuclei
Cell wall
Pore
Septum
Septa are not complete: Pores allow movement of organelles and other materials between cell-like compartments.
(a) Coenocytic hypha
(b) Septate hypha

31.3 Most Hyphae Are Incompletely Divided into Separate Cells *(Page 605)*

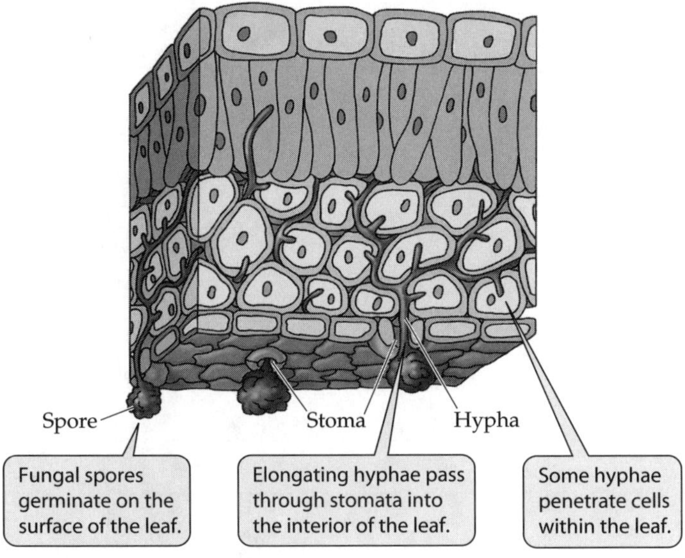
Spore
Stoma
Hypha
Fungal spores germinate on the surface of the leaf.
Elongating hyphae pass through stomata into the interior of the leaf.
Some hyphae penetrate cells within the leaf.

31.4 A Fungus Attacks a Leaf *(Page 605)*

31.1 Classification of Fungi

PHYLUM	COMMON NAME	FEATURES	EXAMPLES
Chytridiomycota	Chytrids	Aquatic; gametes have flagella	*Allomyces*
Zygomycota	Zygote fungi	Zygosporangium; no regularly occurring septa; usually no fleshy fruiting body	*Rhizopus*
Ascomycota	Sac fungi	Ascus; perforated septa	*Neurospora*, baker's yeast
Basidiomycota	Club fungi	Basidium; perforated septa	*Armillariella*, mushrooms

(Page 608)

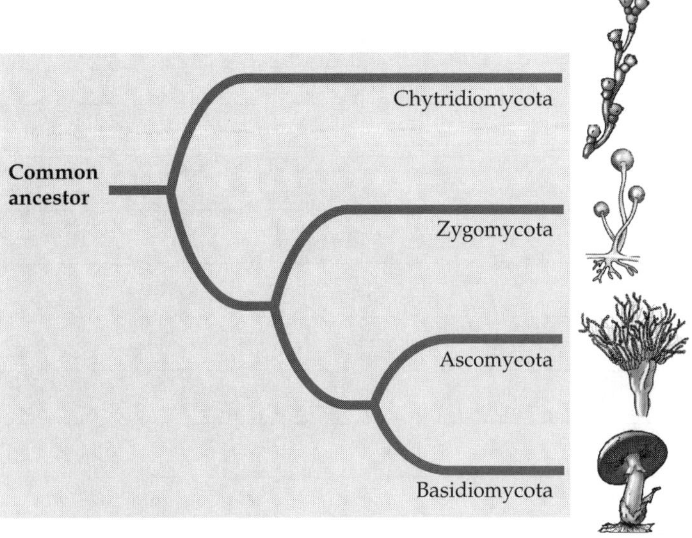

Common ancestor
Chytridiomycota
Zygomycota
Ascomycota
Basidiomycota

31.6 Phylogeny of the Fungi *(Page 608)*

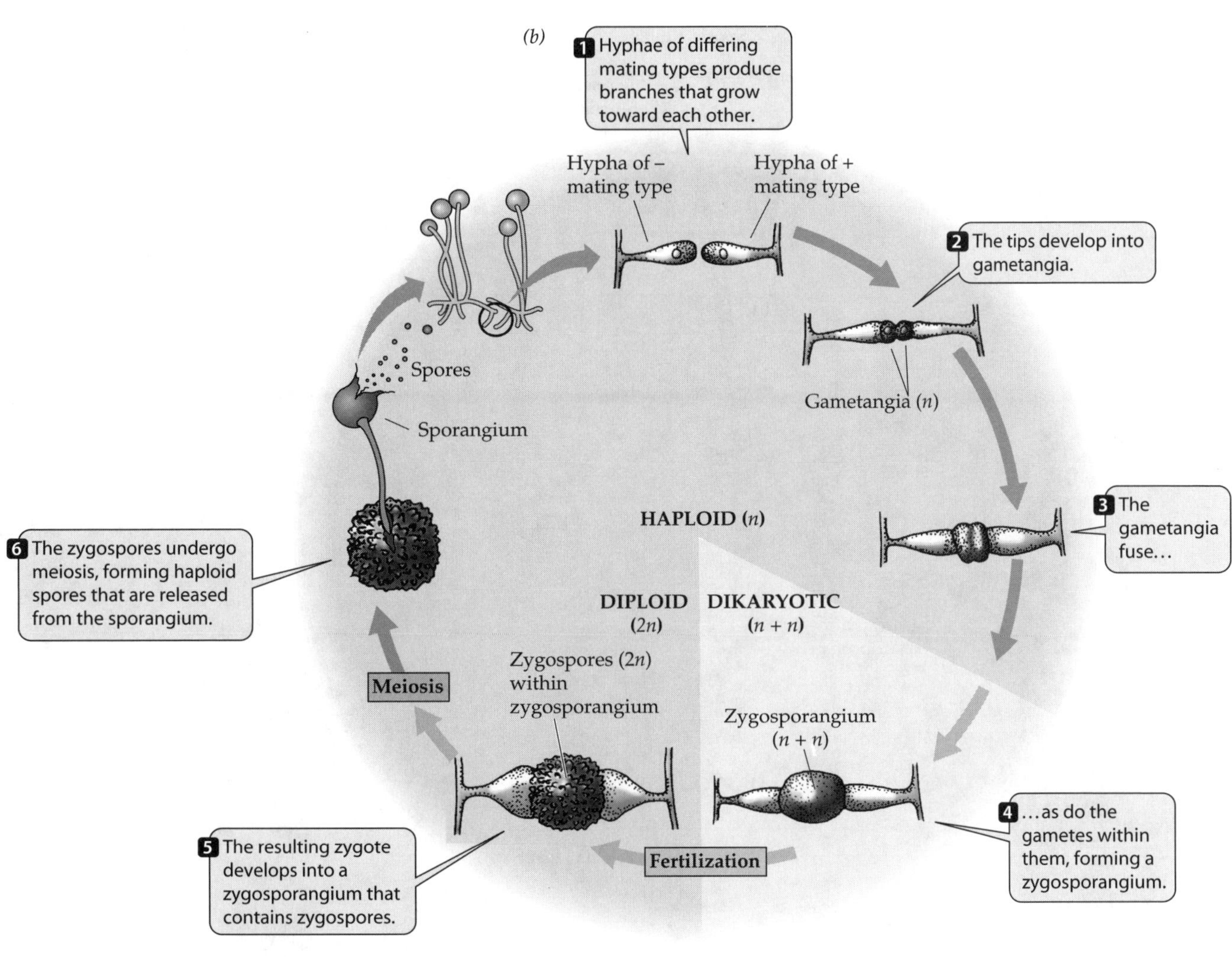
(b)
1 Hyphae of differing mating types produce branches that grow toward each other.
Hypha of – mating type
Hypha of + mating type
2 The tips develop into gametangia.
Gametangia (n)
3 The gametangia fuse...
HAPLOID (n)
DIPLOID (2n)
DIKARYOTIC (n + n)
Zygosporangium (n + n)
4 ...as do the gametes within them, forming a zygosporangium.
Fertilization
Zygospores (2n) within zygosporangium
5 The resulting zygote develops into a zygosporangium that contains zygospores.
Meiosis
6 The zygospores undergo meiosis, forming haploid spores that are released from the sporangium.
Sporangium
Spores

31.9 Sexual Reproduction in a Zygomycete *(Page 610)*

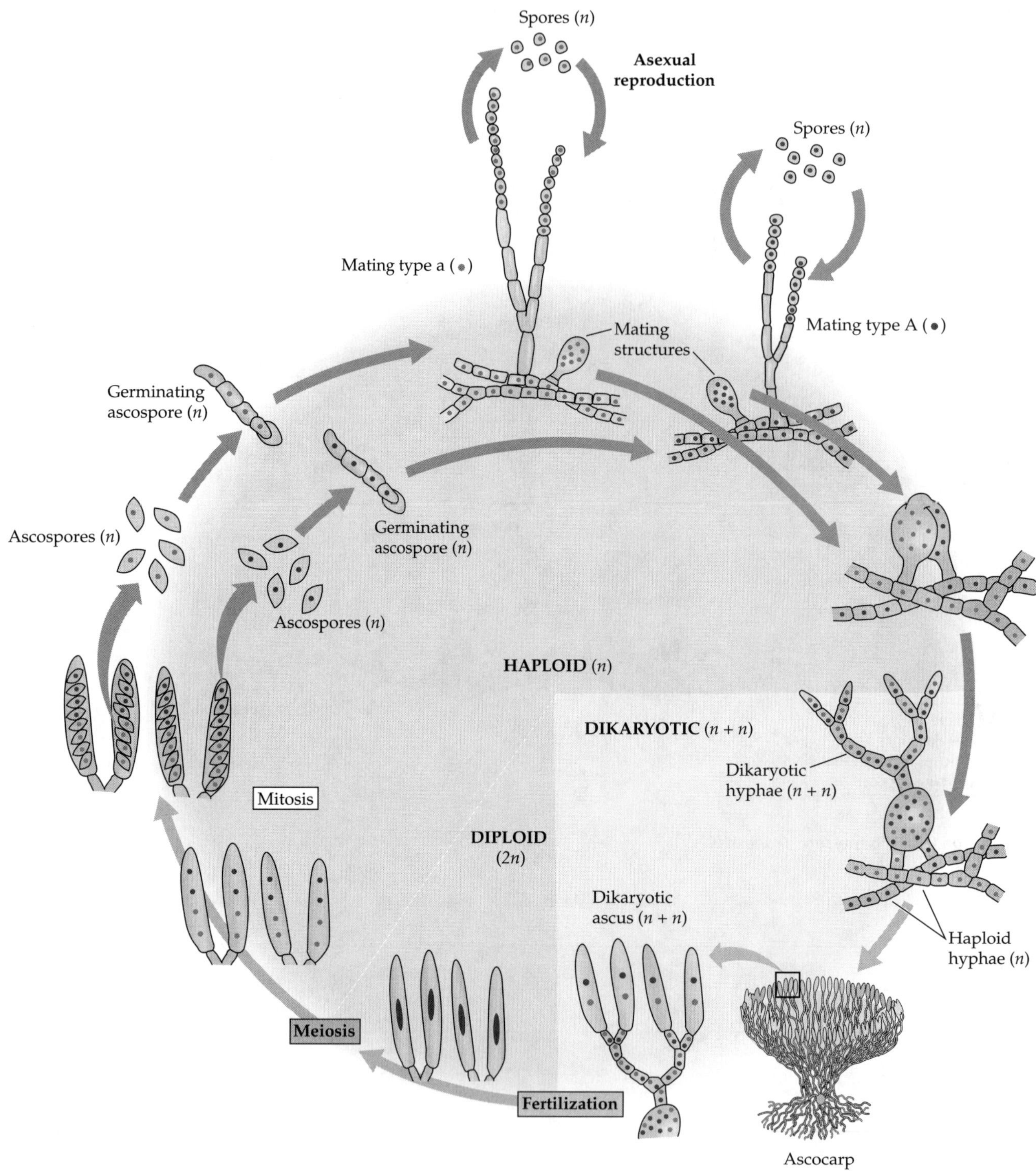
Spores (n)
Asexual reproduction
Spores (n)
Mating type a (•)
Mating type A (•)
Mating structures
Germinating ascospore (n)
Ascospores (n)
Germinating ascospore (n)
Ascospores (n)
HAPLOID (n)
DIKARYOTIC (n + n)
Dikaryotic hyphae (n + n)
Mitosis
DIPLOID (2n)
Dikaryotic ascus (n + n)
Haploid hyphae (n)
Meiosis
Fertilization
Ascocarp

31.13 The Life Cycle of a Euascomycete *(Page 612)*

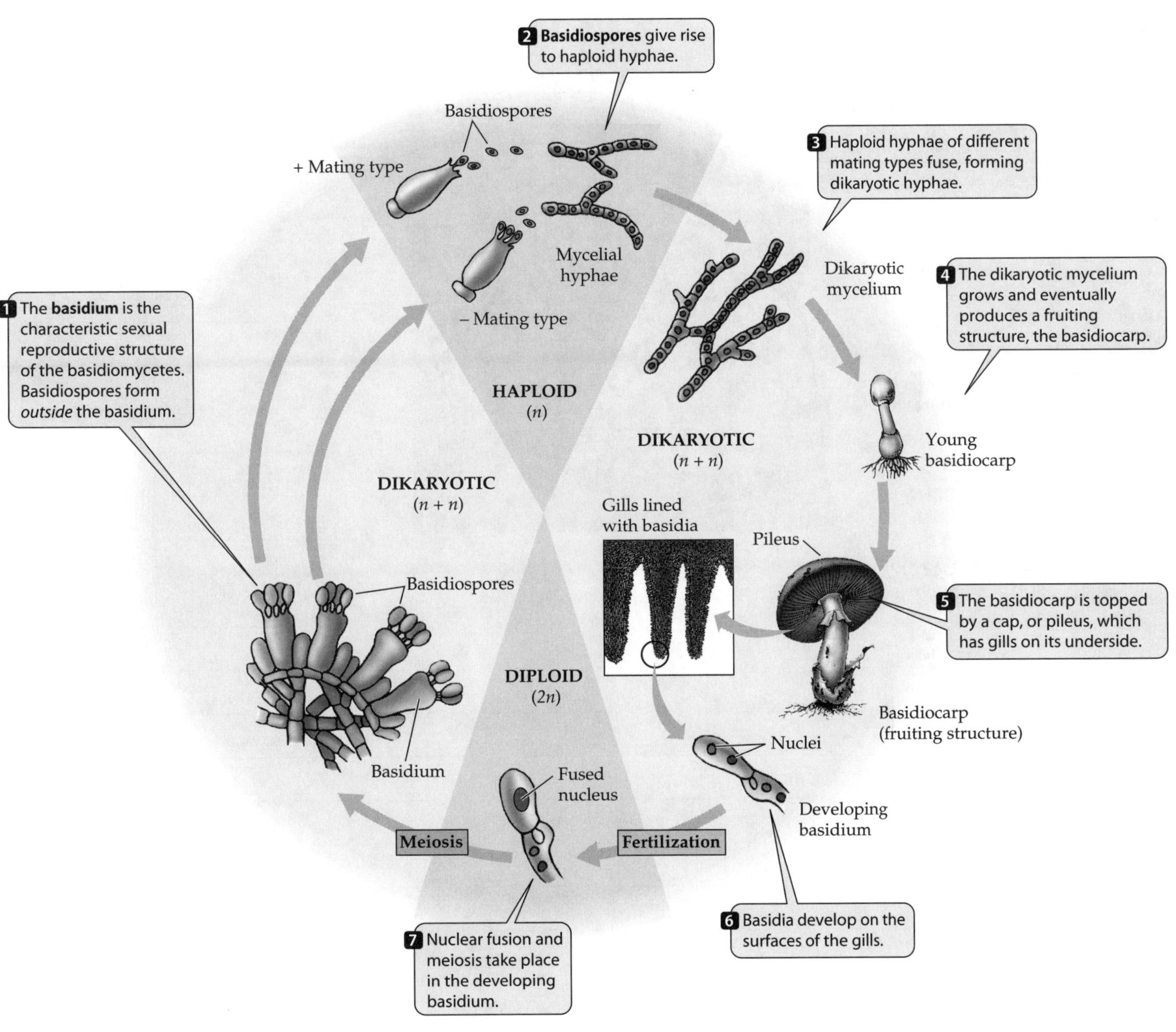
2 Basidiospores give rise to haploid hyphae.
Basidiospores
+ Mating type
Mycelial hyphae
– Mating type
3 Haploid hyphae of different mating types fuse, forming dikaryotic hyphae.
Dikaryotic mycelium
4 The dikaryotic mycelium grows and eventually produces a fruiting structure, the basidiocarp.
1 The basidium is the characteristic sexual reproductive structure of the basidiomycetes. Basidiospores form outside the basidium.
HAPLOID (n)
DIKARYOTIC (n + n)
Young basidiocarp
DIKARYOTIC (n + n)
Gills lined with basidia
Pileus
Basidiospores
5 The basidiocarp is topped by a cap, or pileus, which has gills on its underside.
DIPLOID (2n)
Basidiocarp (fruiting structure)
Nuclei
Basidium
Fused nucleus
Developing basidium
Meiosis
Fertilization
6 Basidia develop on the surfaces of the gills.
7 Nuclear fusion and meiosis take place in the developing basidium.

31.15 The Basidiomycete Life Cycle *(Page 614)*

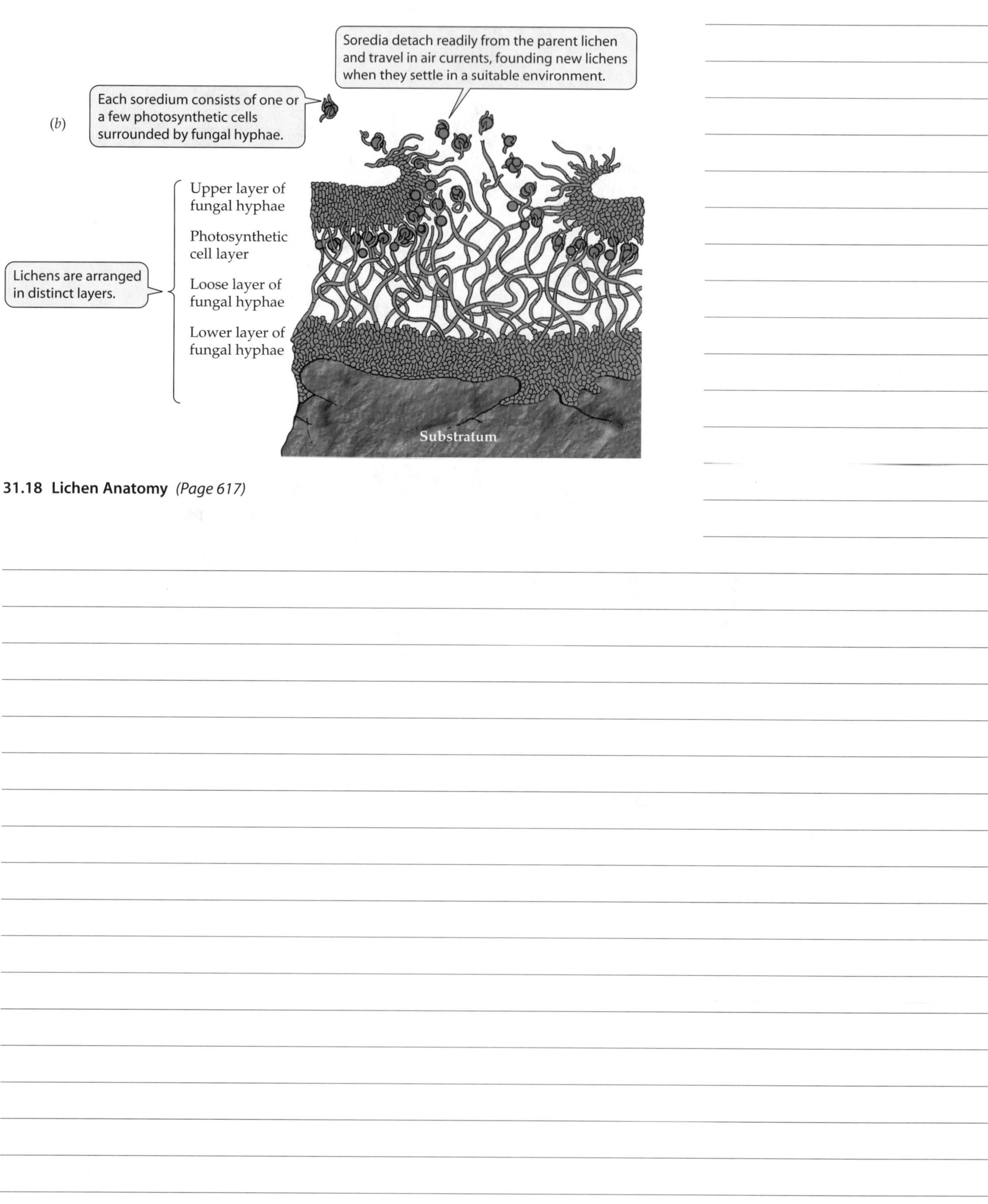
(b)
Each soredium consists of one or a few photosynthetic cells surrounded by fungal hyphae.
Soredia detach readily from the parent lichen and travel in air currents, founding new lichens when they settle in a suitable environment.
Lichens are arranged in distinct layers.
Upper layer of fungal hyphae
Photosynthetic cell layer
Loose layer of fungal hyphae
Lower layer of fungal hyphae
Substratum

31.18 Lichen Anatomy *(Page 617)*

32 Animal Origins and the Evolution of Body Plans

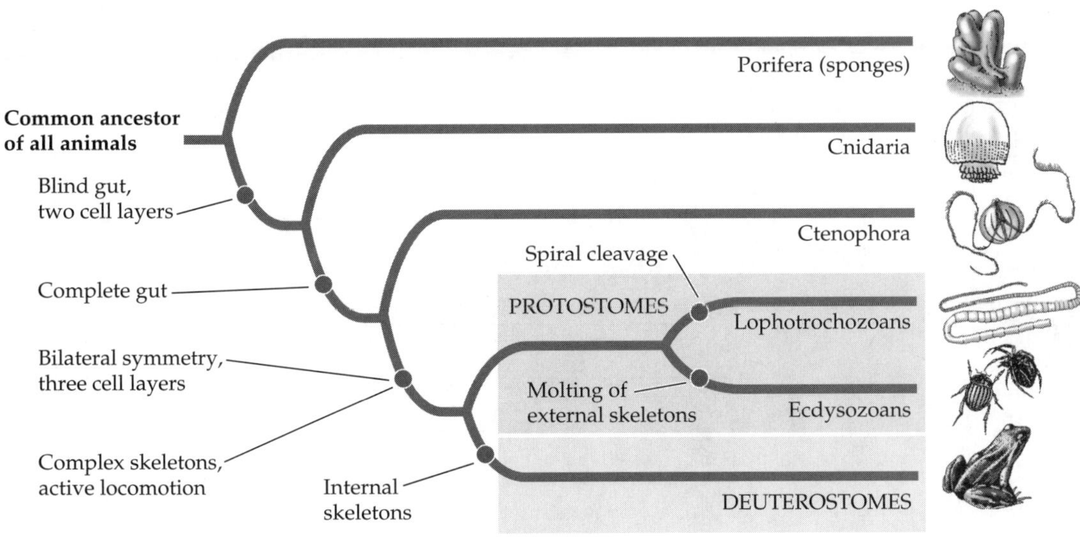
Porifera (sponges)
Common ancestor
of all animals
Cnidaria
Blind gut,
two cell layers
Ctenophora
Spiral cleavage
Complete gut
PROTOSTOMES
Lophotrochozoans
Bilateral symmetry,
three cell layers
Molting of
external skeletons
Ecdysozoans
Complex skeletons,
active locomotion
Internal
skeletons
DEUTEROSTOMES

32.1 A Current Phylogeny of the Animals *(Page 621)*

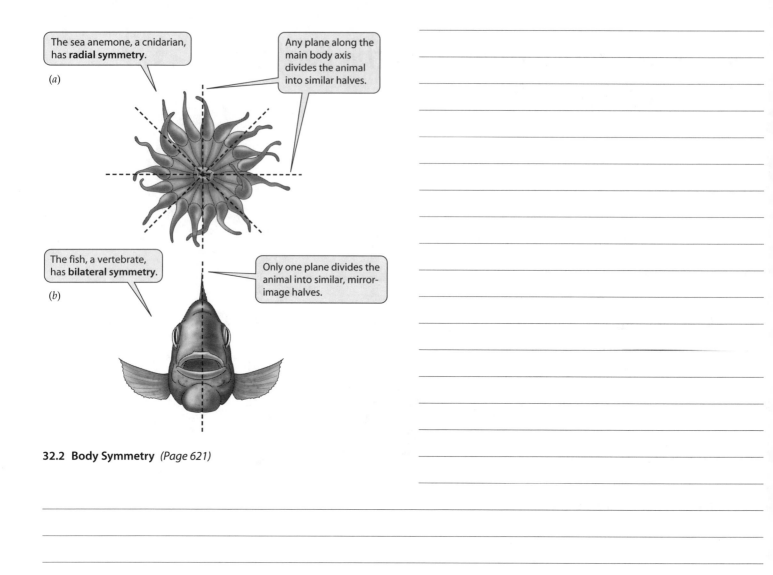

The sea anemone, a cnidarian, has **radial symmetry**.

(*a*)

Any plane along the main body axis divides the animal into similar halves.

The fish, a vertebrate, has **bilateral symmetry**.

(*b*)

Only one plane divides the animal into similar, mirror-image halves.

32.2 Body Symmetry *(Page 621)*

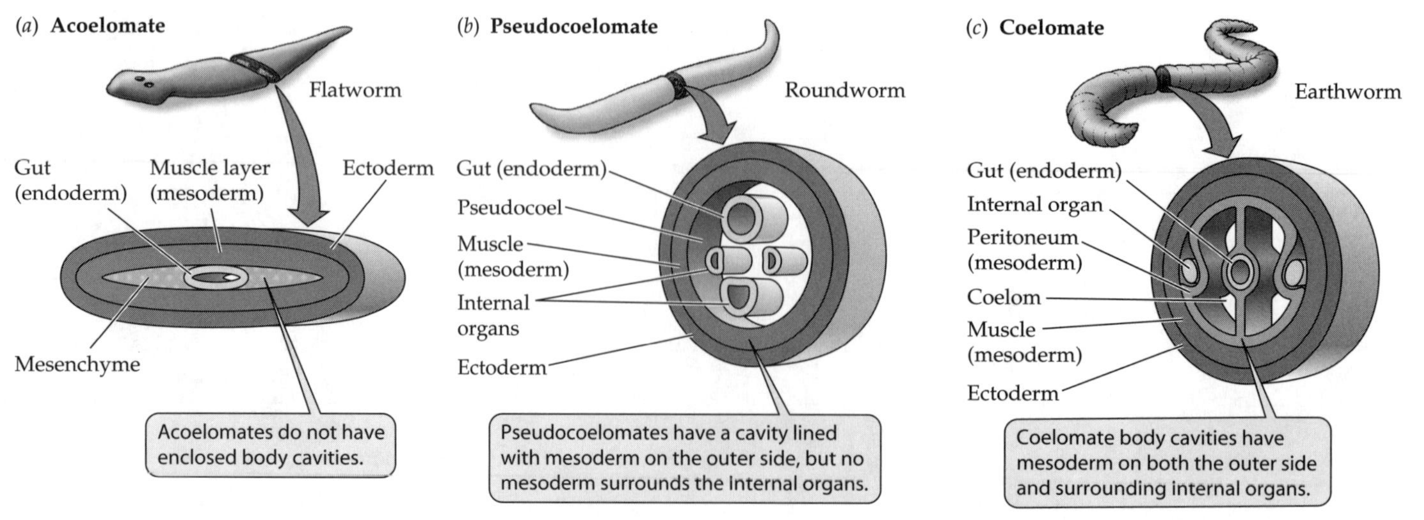
(*a*) **Acoelomate**
Flatworm
Gut (endoderm)
Muscle layer (mesoderm)
Ectoderm
Mesenchyme
Acoelomates do not have enclosed body cavities.
(*b*) **Pseudocoelomate**
Roundworm
Gut (endoderm)
Pseudocoel
Muscle (mesoderm)
Internal organs
Ectoderm
Pseudocoelomates have a cavity lined with mesoderm on the outer side, but no mesoderm surrounds the internal organs.
(*c*) **Coelomate**
Earthworm
Gut (endoderm)
Internal organ
Peritoneum (mesoderm)
Coelom
Muscle (mesoderm)
Ectoderm
Coelomate body cavities have mesoderm on both the outer side and surrounding internal organs.

32.3 Animal Body Cavities *(Page 622)*

Water, carrying food particles, enters through many small pores.
Water flows through the sponge's body and exits through a large opening, the osculum.
Pore
Osculum
Spicules
Choanocyte
Spicules
The inner wall is studded with specialized feeding cells called choanocytes.

32.4 The Body Plan of a Simple Sponge *(Page 623)*

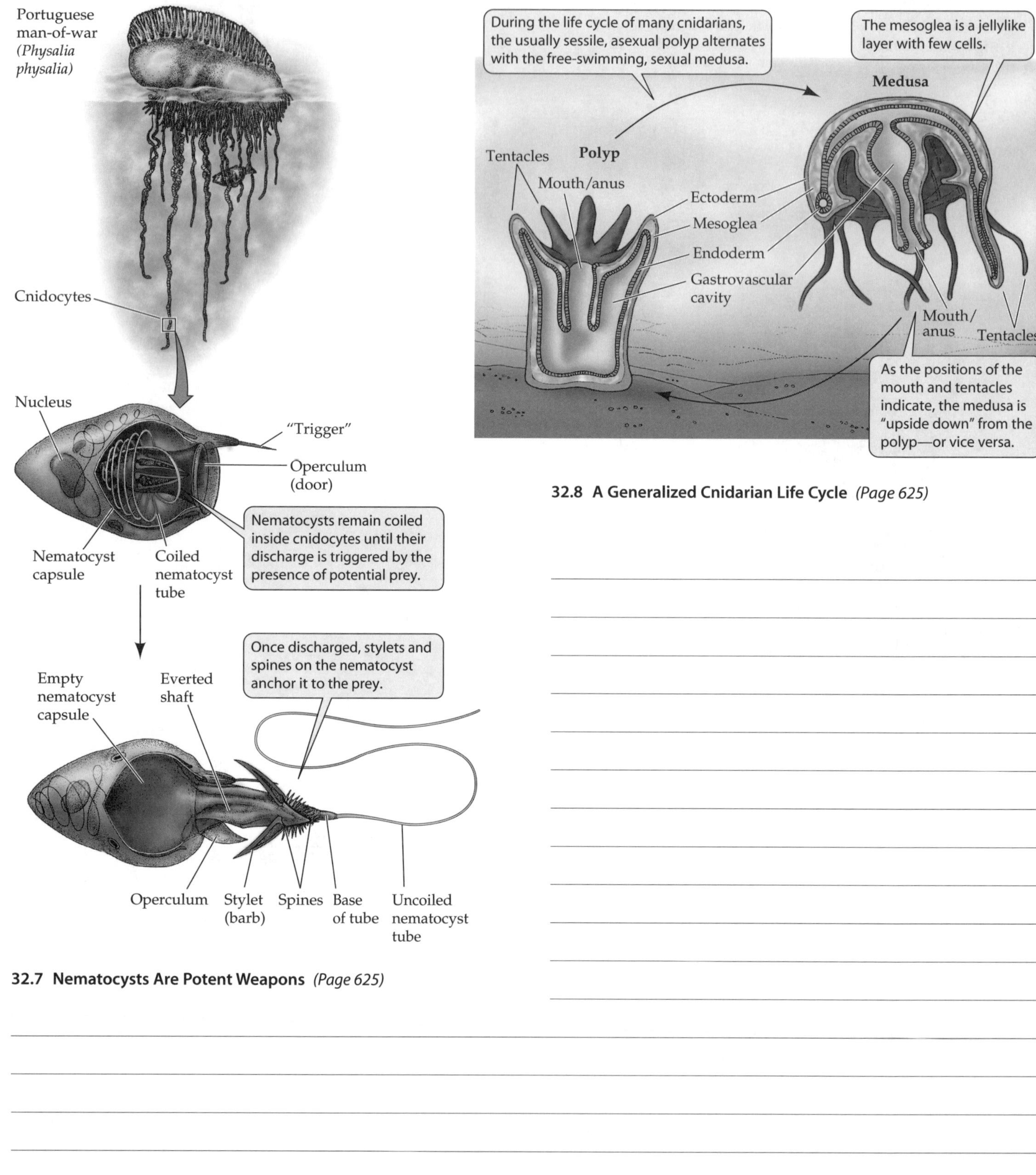
Portuguese man-of-war (*Physalia physalia*)
Cnidocytes
Nucleus
"Trigger"
Operculum (door)
Nematocyst capsule
Coiled nematocyst tube
Nematocysts remain coiled inside cnidocytes until their discharge is triggered by the presence of potential prey.
Once discharged, stylets and spines on the nematocyst anchor it to the prey.
Empty nematocyst capsule
Everted shaft
Operculum
Stylet (barb)
Spines
Base of tube
Uncoiled nematocyst tube
During the life cycle of many cnidarians, the usually sessile, asexual polyp alternates with the free-swimming, sexual medusa.
The mesoglea is a jellylike layer with few cells.
Medusa
Tentacles
Polyp
Mouth/anus
Ectoderm
Mesoglea
Endoderm
Gastrovascular cavity
Mouth/ anus
Tentacles
As the positions of the mouth and tentacles indicate, the medusa is "upside down" from the polyp—or vice versa.

32.8 A Generalized Cnidarian Life Cycle *(Page 625)*

32.7 Nematocysts Are Potent Weapons *(Page 625)*

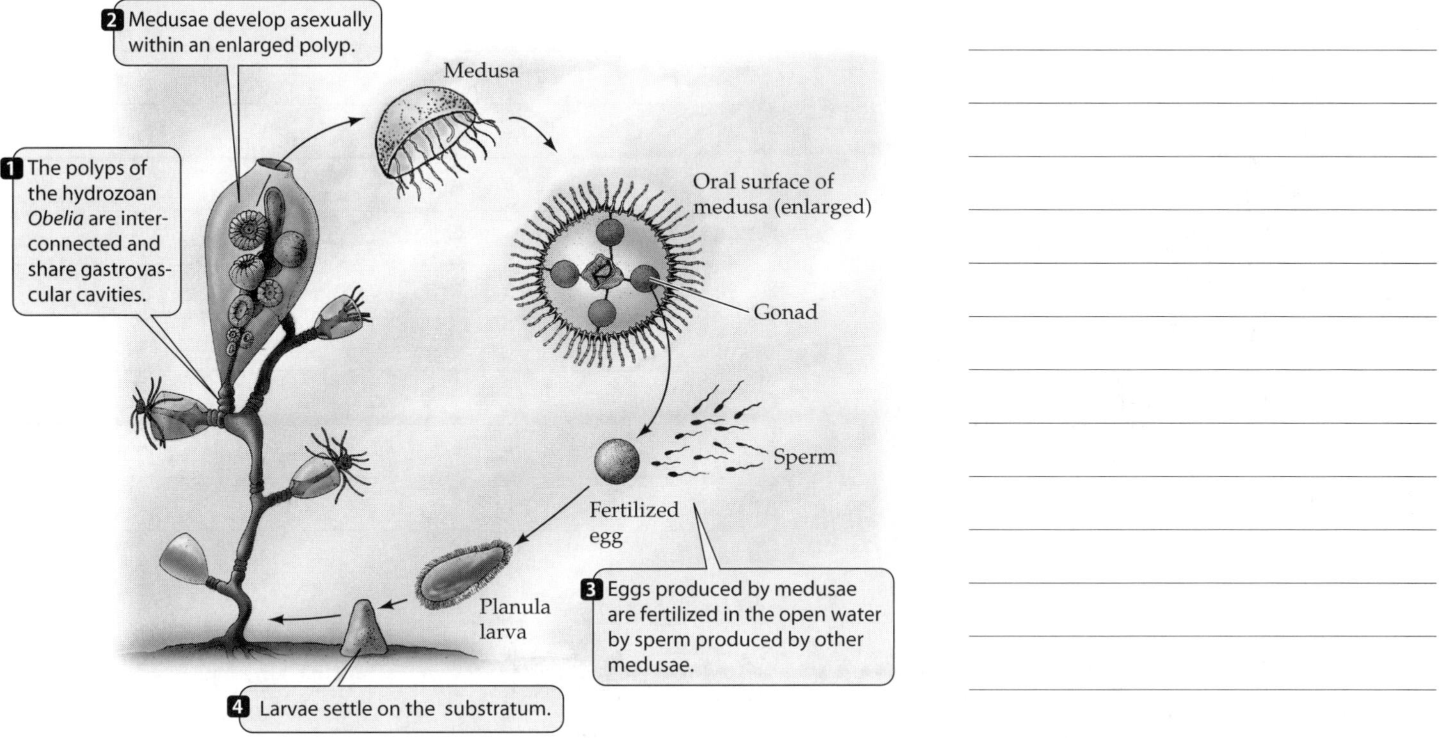
2 Medusae develop asexually within an enlarged polyp.
Medusa
1 The polyps of the hydrozoan *Obelia* are interconnected and share gastrovascular cavities.
Oral surface of medusa (enlarged)
Gonad
Sperm
Fertilized egg
Planula larva
3 Eggs produced by medusae are fertilized in the open water by sperm produced by other medusae.
4 Larvae settle on the substratum.

32.10 **Hydrozoans Often Have Colonial Polyps** *(Page 627)*

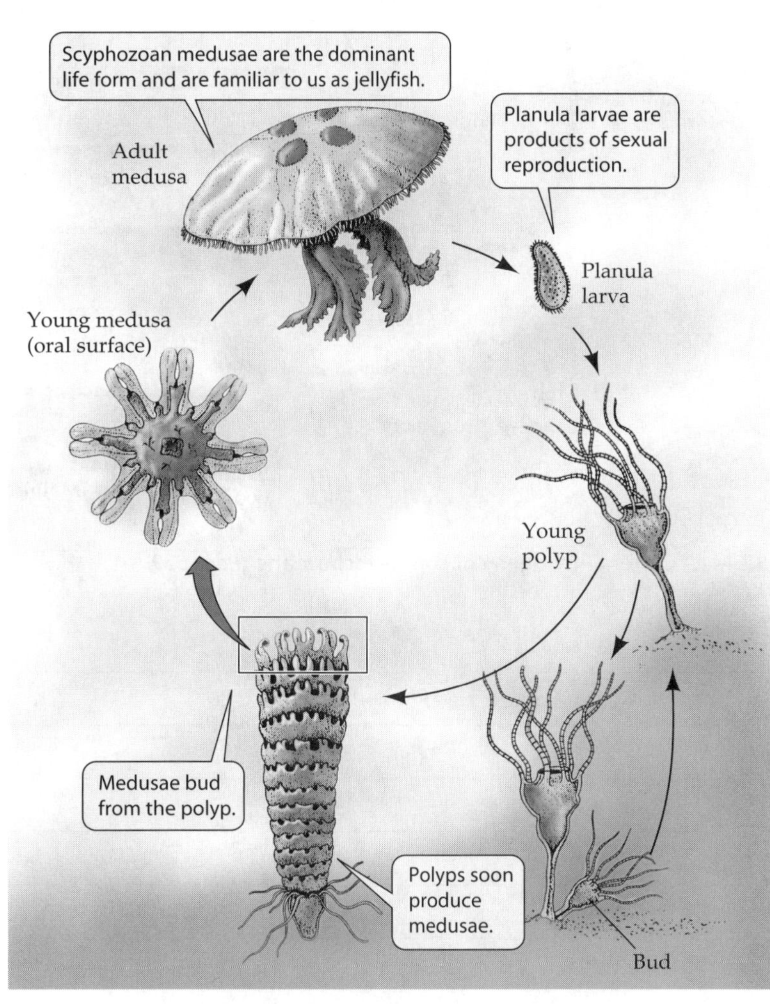
Scyphozoan medusae are the dominant life form and are familiar to us as jellyfish.
Adult medusa
Planula larvae are products of sexual reproduction.
Planula larva
Young medusa (oral surface)
Young polyp
Medusae bud from the polyp.
Polyps soon produce medusae.
Bud

32.11 **Medusae Dominate Scyphozoan Life Cycles** *(Page 627)*

(a)

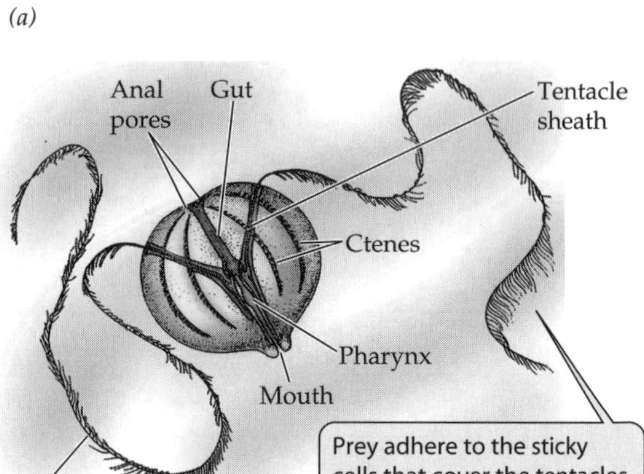
Anal pores
Gut
Tentacle sheath
Ctenes
Pharynx
Mouth
Tentacle
Prey adhere to the sticky cells that cover the tentacles, which lack nematocysts.

32.12 Comb Jellies Feed with Tentacles *(Page 628)*

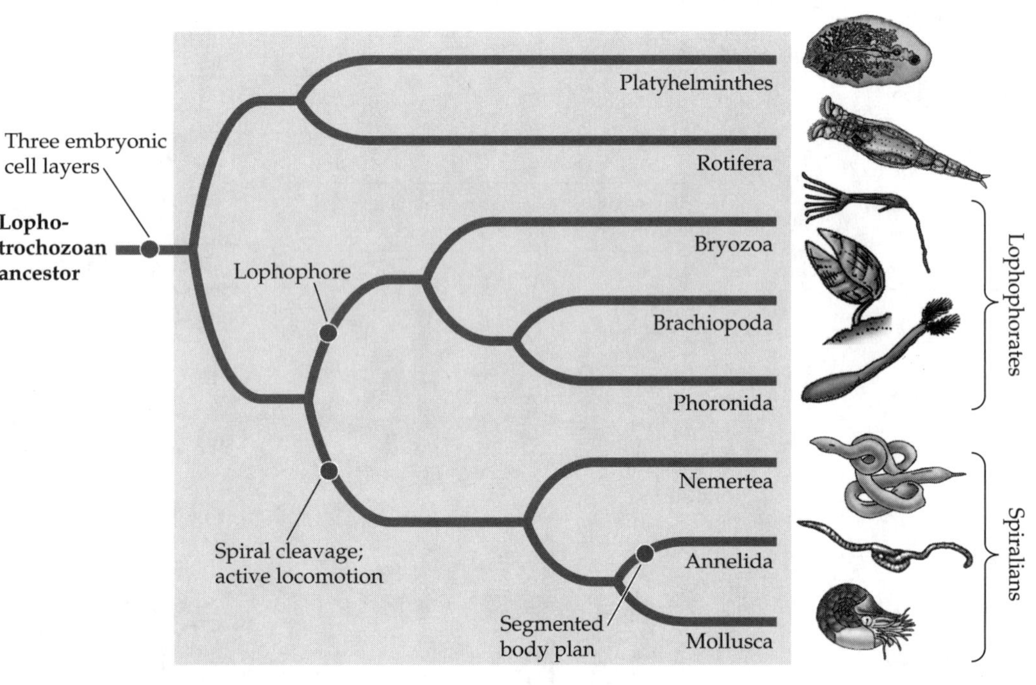
Three embryonic cell layers
Lopho-trochozoan ancestor
Platyhelminthes
Rotifera
Bryozoa
Lophophore
Brachiopoda
Phoronida
Nemertea
Annelida
Mollusca
Spiral cleavage; active locomotion
Segmented body plan
Lophophorates
Spiralians

32.14 A Current Phylogeny of Lophotrochozoans *(Page 629)*

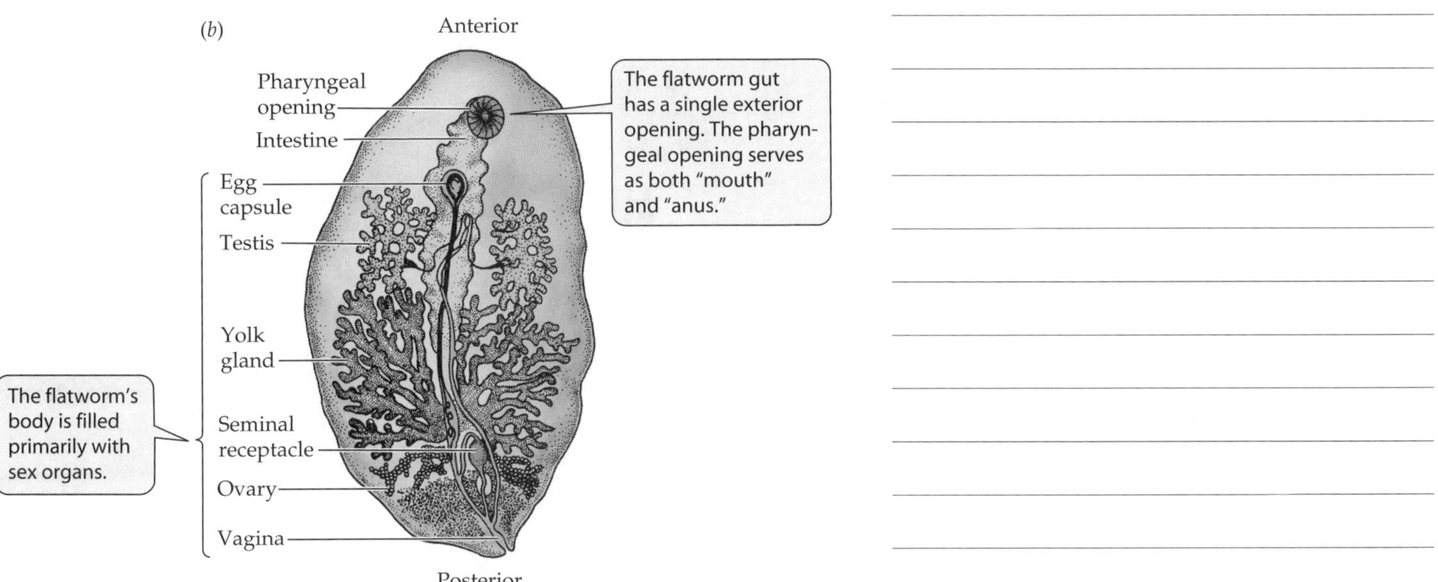
(b)
Anterior
Pharyngeal opening
Intestine
Egg capsule
Testis
Yolk gland
Seminal receptacle
Ovary
Vagina
Posterior
The flatworm gut has a single exterior opening. The pharyngeal opening serves as both "mouth" and "anus."
The flatworm's body is filled primarily with sex organs.

32.15 Flatworms Live Freely and Parasitically *(Page 630)*

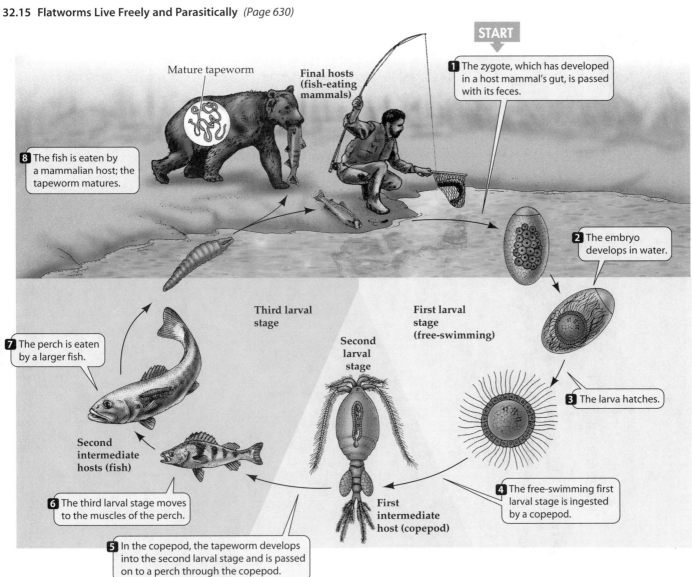

START

1 The zygote, which has developed in a host mammal's gut, is passed with its feces.

Mature tapeworm

Final hosts (fish-eating mammals)

8 The fish is eaten by a mammalian host; the tapeworm matures.

2 The embryo develops in water.

Third larval stage

First larval stage (free-swimming)

Second larval stage

7 The perch is eaten by a larger fish.

3 The larva hatches.

Second intermediate hosts (fish)

4 The free-swimming first larval stage is ingested by a copepod.

First intermediate host (copepod)

6 The third larval stage moves to the muscles of the perch.

5 In the copepod, the tapeworm develops into the second larval stage and is passed on to a perch through the copepod.

32.16 Reaching a Host by a Complex Route *(Page 630)*

(a) *Philadeina roseola*

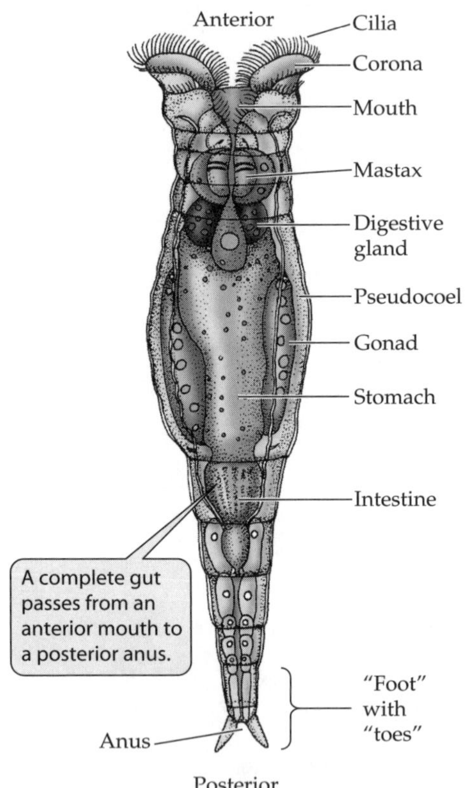
Anterior
Cilia
Corona
Mouth
Mastax
Digestive gland
Pseudocoel
Gonad
Stomach
Intestine
A complete gut passes from an anterior mouth to a posterior anus.
"Foot" with "toes"
Anus
Posterior

(b)

Lophophore spreads
Lophophore oscillates and rotates
Lophophore retracts

32.17 Rotifers *(Page 631)*

32.19 Ectoprocts *(Page 632)*

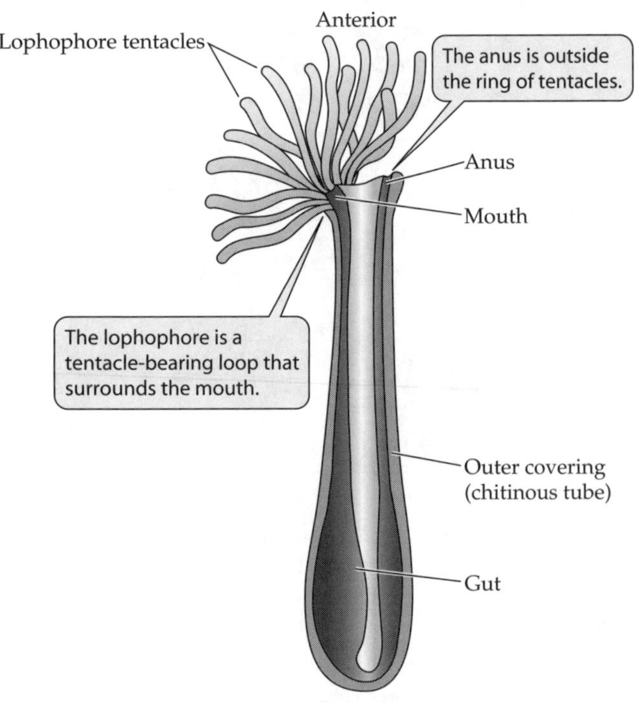
Anterior
Lophophore tentacles
The anus is outside the ring of tentacles.
Anus
Mouth
The lophophore is a tentacle-bearing loop that surrounds the mouth.
Outer covering (chitinous tube)
Gut
Posterior

32.18 Lophophore Artistry *(Page 632)*

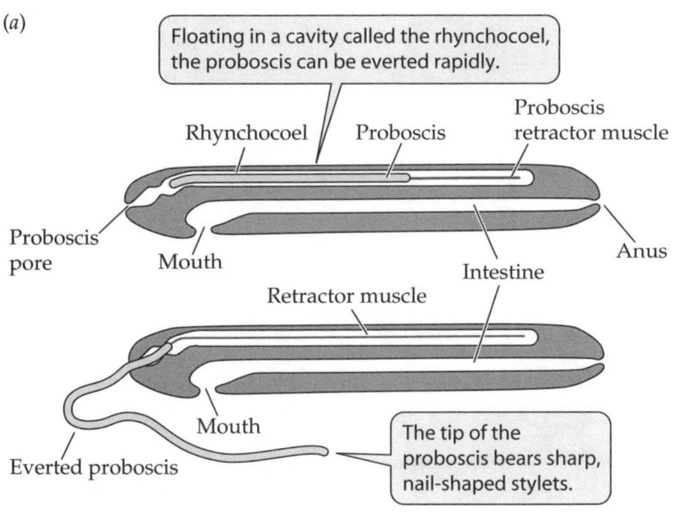
(a)
Floating in a cavity called the rhynchocoel, the proboscis can be everted rapidly.
Rhynchocoel
Proboscis
Proboscis retractor muscle
Proboscis pore
Mouth
Intestine
Anus
Retractor muscle
Mouth
Everted proboscis
The tip of the proboscis bears sharp, nail-shaped stylets.

32.21 Ribbon Worms *(Page 633)*

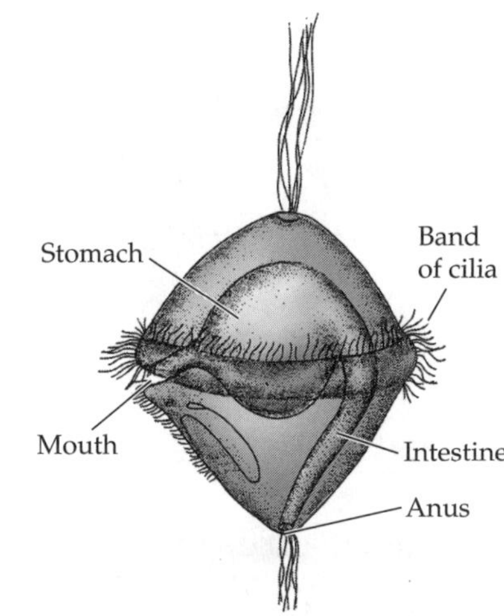
Stomach
Band of cilia
Mouth
Intestine
Anus

32.23 The Trochophore Larva *(Page 634)*

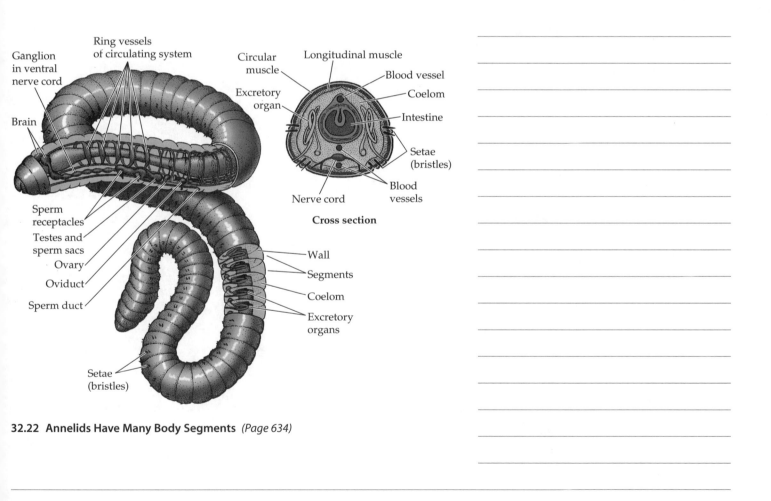

Ring vessels of circulating system
Ganglion in ventral nerve cord
Brain
Sperm receptacles
Testes and sperm sacs
Ovary
Oviduct
Sperm duct
Setae (bristles)
Circular muscle
Longitudinal muscle
Blood vessel
Excretory organ
Coelom
Intestine
Setae (bristles)
Nerve cord
Blood vessels
Cross section
Wall
Segments
Coelom
Excretory organs

32.22 Annelids Have Many Body Segments *(Page 634)*

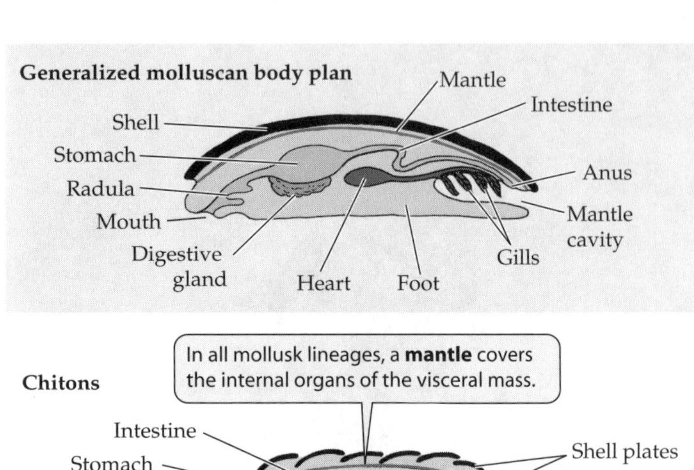
Generalized molluscan body plan
Mantle
Intestine
Shell
Stomach
Radula
Mouth
Anus
Mantle cavity
Digestive gland
Heart
Foot
Gills

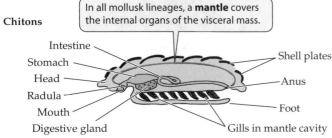

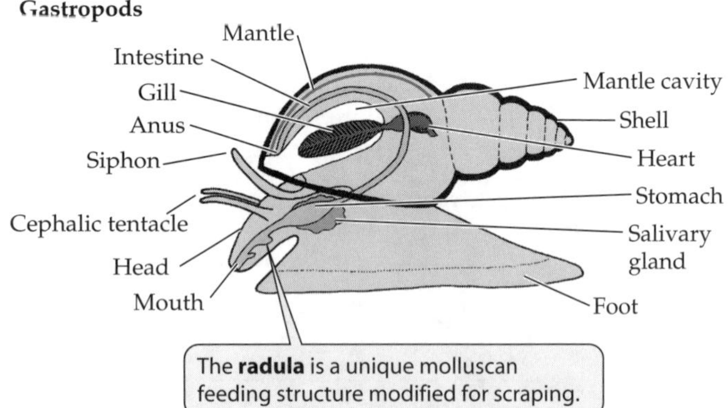
Chitons
In all mollusk lineages, a **mantle** covers the internal organs of the visceral mass.
Intestine
Stomach
Head
Radula
Mouth
Digestive gland
Shell plates
Anus
Foot
Gills in mantle cavity

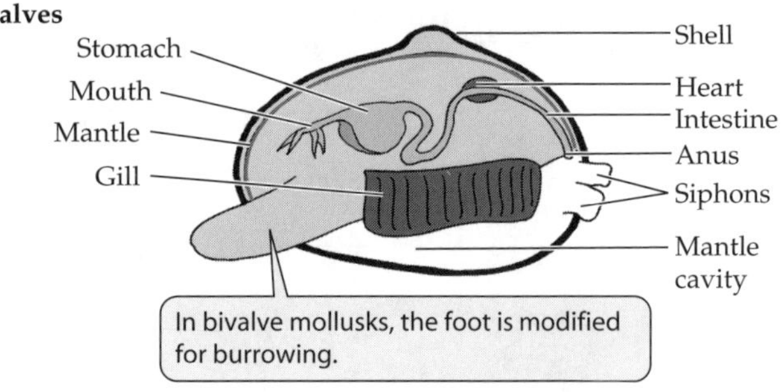
Gastropods
Mantle
Intestine
Gill
Anus
Siphon
Cephalic tentacle
Head
Mouth
Mantle cavity
Shell
Heart
Stomach
Salivary gland
Foot
The **radula** is a unique molluscan feeding structure modified for scraping.

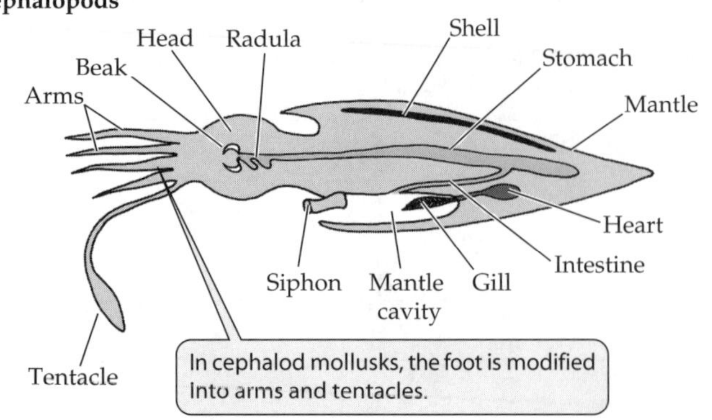
Bivalves
Stomach
Mouth
Mantle
Gill
Shell
Heart
Intestine
Anus
Siphons
Mantle cavity
In bivalve mollusks, the foot is modified for burrowing.
Cephalopods
Head
Radula
Beak
Arms
Shell
Stomach
Mantle
Heart
Intestine
Siphon
Mantle cavity
Gill
Tentacle
In cephalod mollusks, the foot is modified into arms and tentacles.

32.25 Molluscan Body Plans *(Page 636)*

33 Ecdysozoans: The Molting Animals

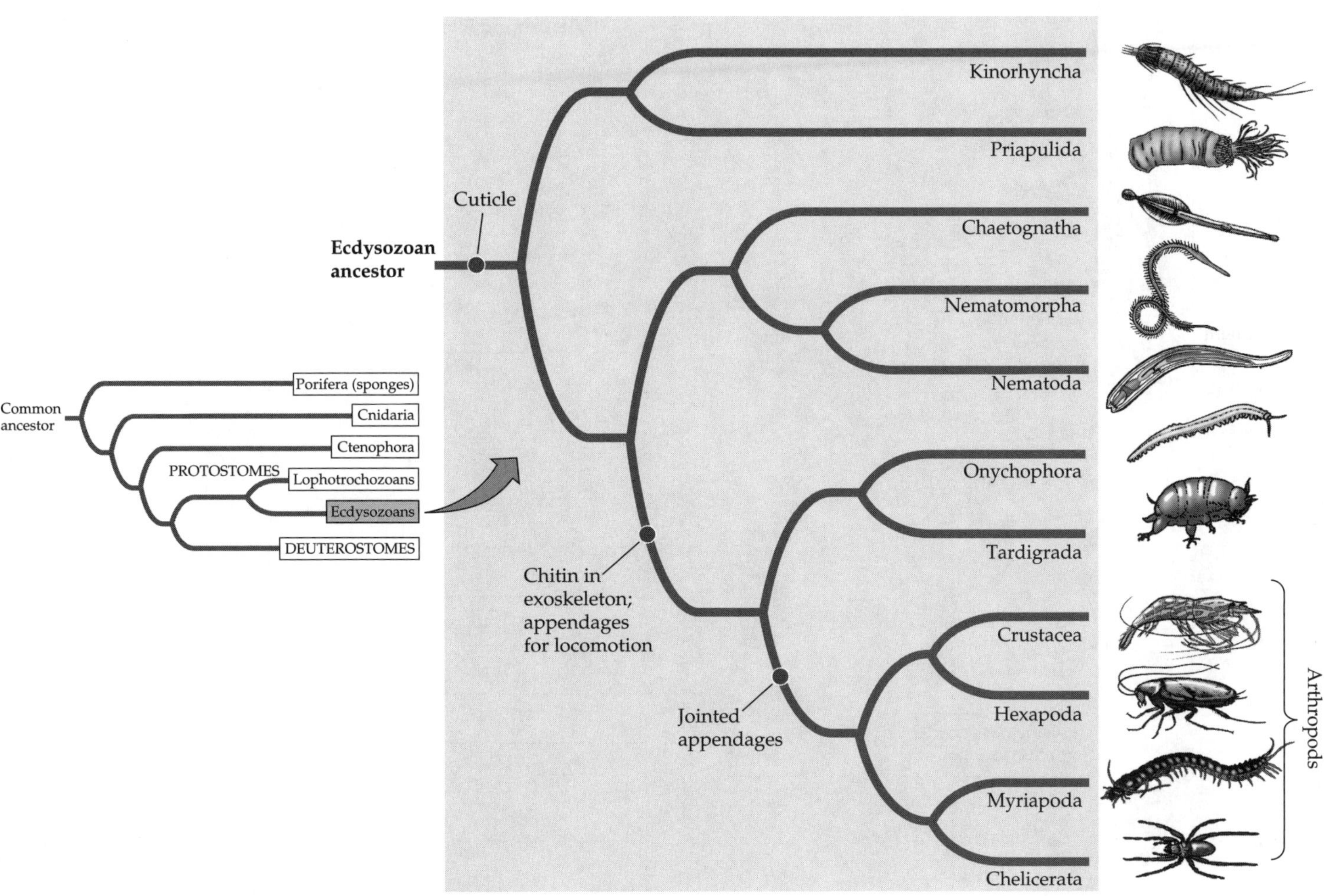

Kinorhyncha
Priapulida
Chaetognatha
Nematomorpha
Nematoda
Onychophora
Tardigrada
Crustacea
Hexapoda
Myriapoda
Chelicerata
Arthropods
Cuticle
Ecdysozoan ancestor
Chitin in exoskeleton; appendages for locomotion
Jointed appendages
Common ancestor
Porifera (sponges)
Cnidaria
Ctenophora
PROTOSTOMES
Lophotrochozoans
Ecdysozoans
DEUTEROSTOMES

33.1 A Current Phylogeny of the Ecdysozoans *(Page 642)*

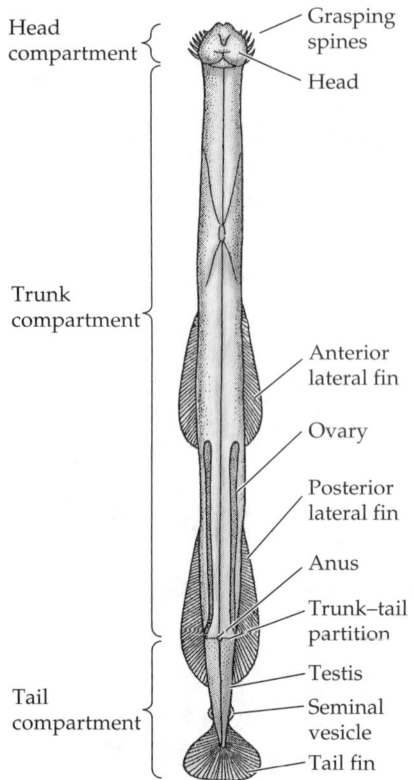
Head compartment
Grasping spines
Head
Trunk compartment
Anterior lateral fin
Ovary
Posterior lateral fin
Anus
Trunk–tail partition
Testis
Tail compartment
Seminal vesicle
Tail fin

33.3 An Arrow Worm *(Page 643)*

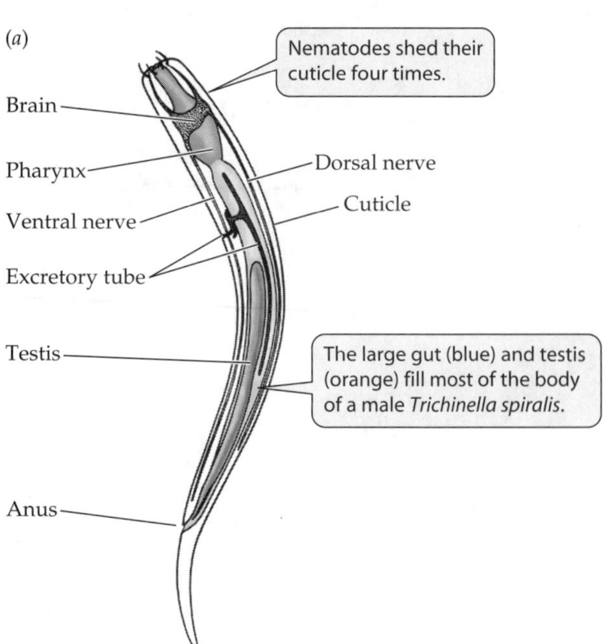
(a)
Nematodes shed their cuticle four times.
Brain
Pharynx
Dorsal nerve
Cuticle
Ventral nerve
Excretory tube
Testis
The large gut (blue) and testis (orange) fill most of the body of a male Trichinella spiralis.
Anus

33.5 Roundworms *(Page 644)*

Hemocoel
Heart
Longitudinal muscle
Dorsoventral muscle
Exoskeleton (cuticle)
Muscles that move appendage
Longitudinal muscle
Ventral nerve cord
Muscles within appendage

33.8 Arthropod Exoskeletons Are Rigid and Jointed *(Page 646)*

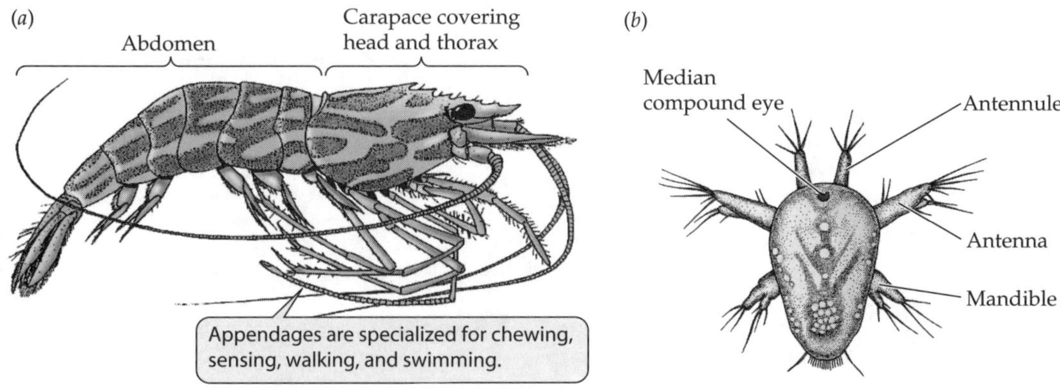

(a)
Abdomen
Carapace covering head and thorax
Appendages are specialized for chewing, sensing, walking, and swimming.
(b)
Median compound eye
Antennule
Antenna
Mandible

33.10 Crustacean Structure *(Page 647)*

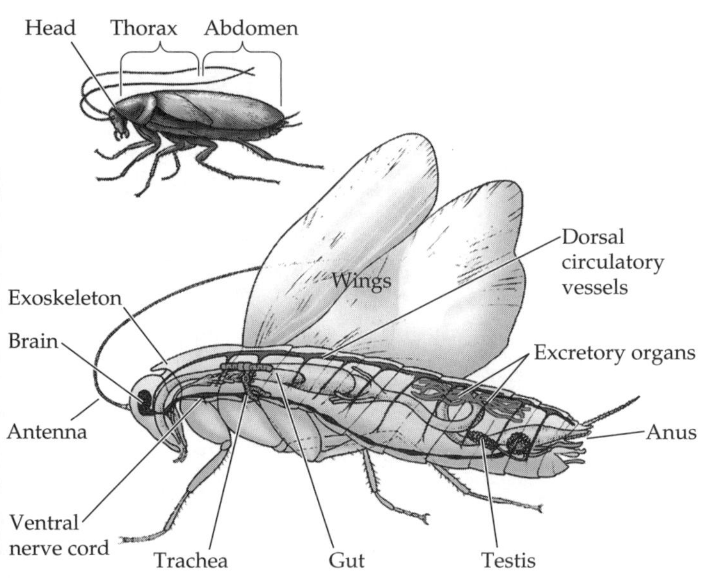

Head
Thorax
Abdomen
Wings
Dorsal circulatory vessels
Exoskeleton
Brain
Excretory organs
Antenna
Anus
Ventral nerve cord
Trachea
Gut
Testis

33.11 Structure of an Insect *(Page 648)*

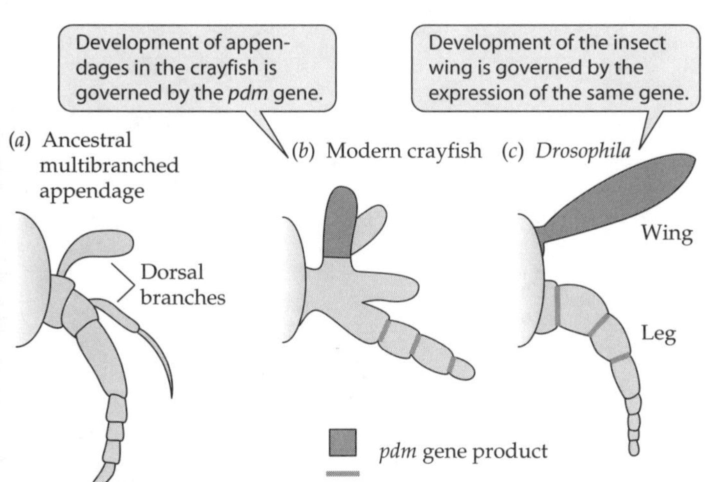

Development of appendages in the crayfish is governed by the pdm gene.
Development of the insect wing is governed by the expression of the same gene.
(a) Ancestral multibranched appendage
(b) Modern crayfish
(c) Drosophila
Dorsal branches
Wing
Leg
pdm gene product

33.14 Origin of Insect Wings *(Page 650)*

33.1 Anatomical Characteristics of the Major Protostomate Phyla

PHYLUM	BODY CAVITY	DIGESTIVE TRACT	CIRCULATORY SYSTEM
Lophotrochozoans			
Platyhelminthes	None	Dead-end sac	None
Rotifera	Pseudocoelom	Complete	None
Bryozoa	Coelom	Complete	None
Brachiopoda	Coelom	Complete in most	Open
Phoronida	Coelom	Complete	Closed
Nemertea	Coelom	Complete	Closed
Annelida	Coelom	Complete	Closed or open
Mollusca	Reduced coelom	Complete	Open except in cephalopods
Ecdysozoans			
Chaetognatha	Coelom	Complete	None
Nematomorpha	Pseudocoelom	Greatly reduced	None
Nematoda	Pseudoceolom	Complete	None
Crustacea	Hemocoel	Complete	Open
Hexapoda	Hemocoel	Complete	Open
Myriapoda	Hemocoel	Complete	Open
Chelicerata	Hemocoel	Complete	Open

Note: All protostomes have bilateral symmetry.

(Page 653)

34 Deuterostomate Animals

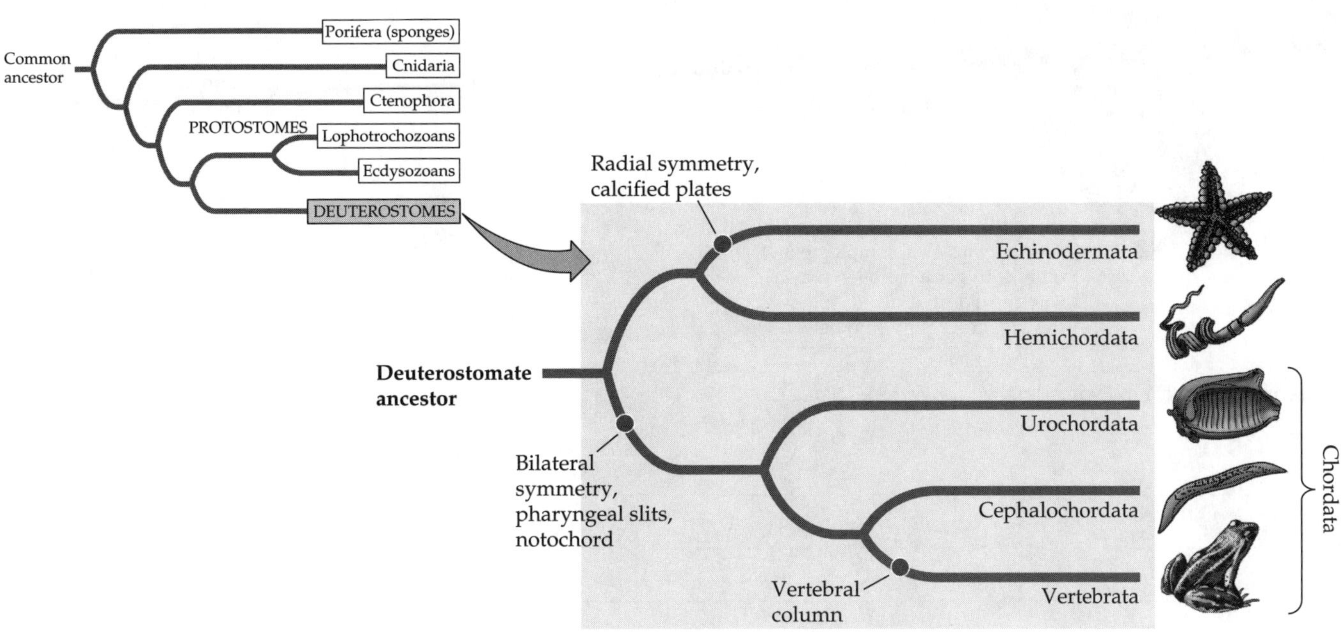

Common ancestor
Porifera (sponges)
Cnidaria
Ctenophora
PROTOSTOMES
Lophotrochozoans
Ecdysozoans
DEUTEROSTOMES
Radial symmetry, calcified plates
Echinodermata
Hemichordata
Deuterostomate ancestor
Urochordata
Bilateral symmetry, pharyngeal slits, notochord
Cephalochordata
Chordata
Vertebral column
Vertebrata

34.2 A Current Phylogeny of the Deuterostomes *(Page 656)*

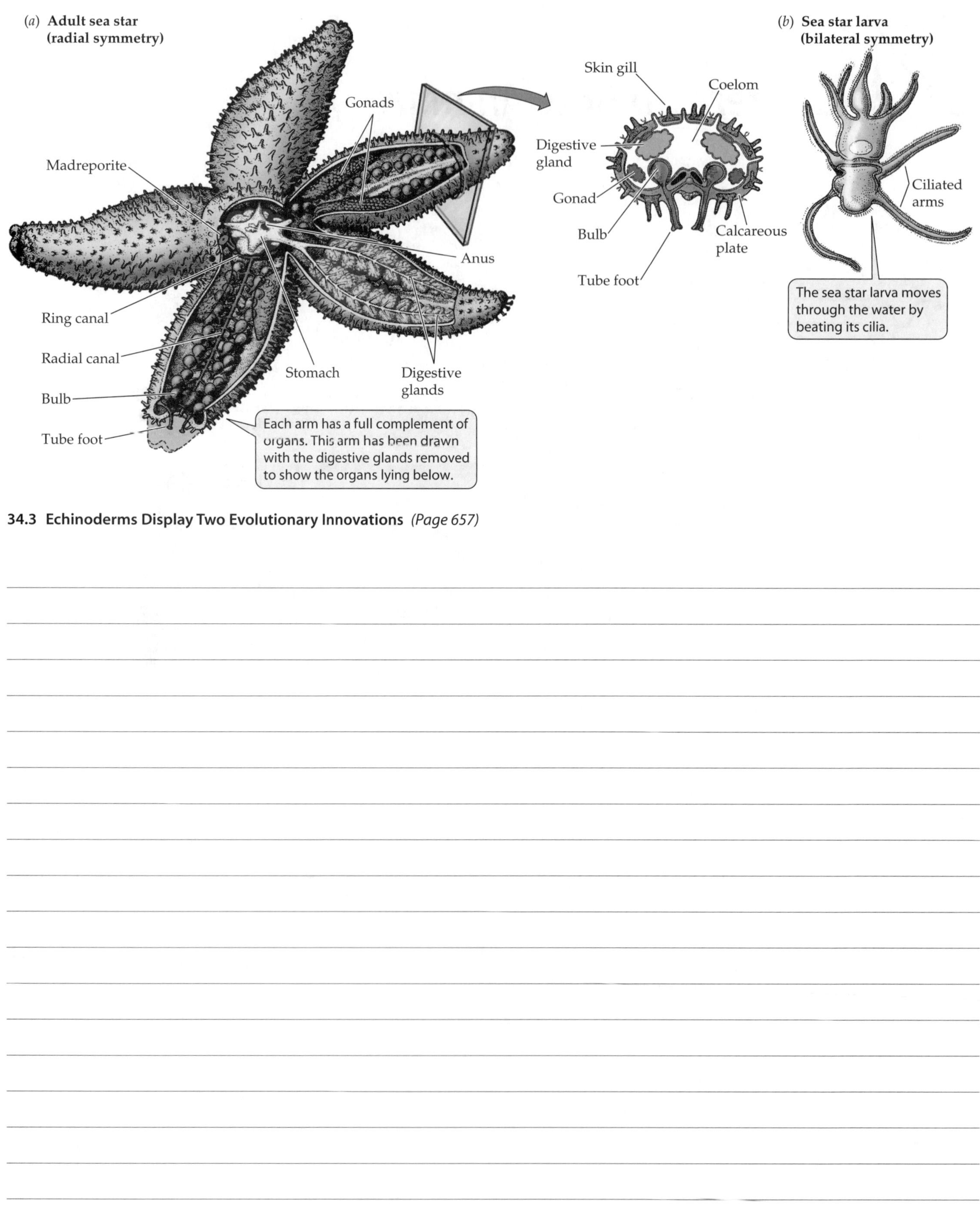
(a) Adult sea star
(radial symmetry)
Gonads
Madreporite
Anus
Ring canal
Radial canal
Stomach
Digestive glands
Bulb
Tube foot
Each arm has a full complement of organs. This arm has been drawn with the digestive glands removed to show the organs lying below.
Skin gill
Coelom
Digestive gland
Gonad
Bulb
Calcareous plate
Tube foot
(b) Sea star larva
(bilateral symmetry)
Ciliated arms
The sea star larva moves through the water by beating its cilia.

34.3 Echinoderms Display Two Evolutionary Innovations *(Page 657)*

34.1 *Summary of Living Members of the Kingdom Animalia*

PHYLUM	NUMBER OF LIVING SPECIES DESCRIBED	MAJOR GROUPS
Porifera: Sponges	10,000	
Cnidaria: Cnidarians	10,000	Hydrozoa: Hydras and hydroids Scyphozoa: Jellyfishes Anthozoa: Corals, sea anemones
Ctenophora: Comb jellies	100	
PROTOSTOMES		
Lophotrochozoans		
Platyhelminthes: Flatworms	20,000	Turbellaria: Free-living flatworms Trematoda: Flukes (all parasitic) Cestoda: Tapeworms (all parasitic) Monogenea (ectoparasites of fishes)
Rotifera: Rotifers	1,800	
Ectoprocta: Bryozoans	4,500	
Brachiopoda: Lamp shells	340	More than 26,000 fossil species described
Phoronida: Phoronids	20	
Nemertea: Ribbon worms	900	
Annelida: Segmented worms	15,000	Polychaeta: Polychaetes (all marine) Oligochaeta: Earthworms, freshwater worms Hirudinea: Leeches
Mollusca: Mollusks	50,000	Monoplacophora: Monoplacophorans Polyplacophora: Chitons Bivalvia: Clams, oysters, mussels Gastropoda: Snails, slugs, limpets Cephalopoda: Squids, octopuses, nautiloids
Ecdysozoans		
Kinorhyncha: Kinorhynchs	150	
Chaetognatha: Arrow worms*	100	
Nematoda: Roundworms	20,000	
Nematomorpha: Horsehair worms	230	
Onychophora: Onychophorans	80	
Tardigrada: Water bears	600	
Chelicerata: Chelicerates	70,000	Merostomata: Horseshoe crabs Arachnida: Scorpions, harvestmen, spiders, mites, ticks
Crustacea	50,000	Crabs, shrimps, lobsters, barnacles, copepods
Hexapoda	1,500,000	Insects
Myriapoda	13,000	Millipedes, centipedes
DEUTEROSTOMES		
Echinodermata: Echinoderms	7,000	Crinoidea: Sea lilies, feather stars Ophiuroidea: Brittle stars Asteroidea: Sea stars Concentricycloidea: Sea daisies Echinoidea: Sea urchins Holothuroidea: Sea cucumbers
Hemichordata: Hemichordates	95	Acorn worms and pterobranchs
Chordata: Chordates	50,000	Urochordata: Sea squirts Cephalochordata: Lancelets Agnatha: Lampreys, hagfishes Chondrichthyes: Cartilaginous fishes Osteichthyes: Bony fishes Amphibia: Amphibians Reptilia: Reptiles Aves: Birds Mammalia: Mammals

* The position of this phylum is uncertain. Many researchers place them in the deuterostomes.

(Page 658)

(b)

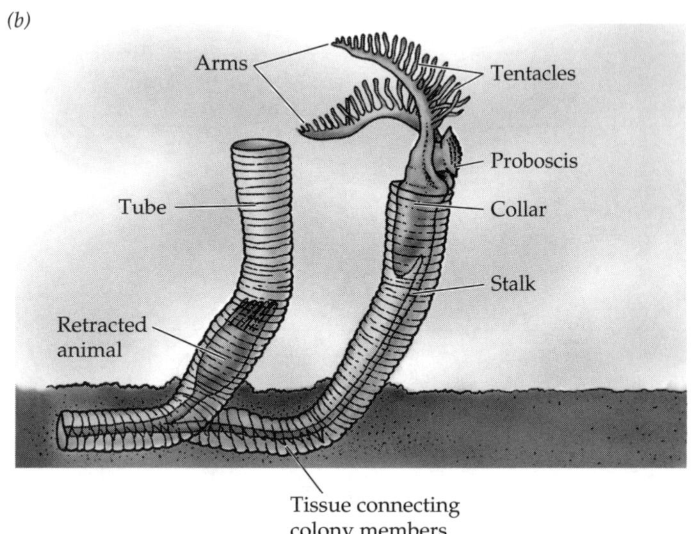
Arms
Tentacles
Proboscis
Tube
Collar
Stalk
Retracted animal
Tissue connecting colony members

34.5 Hemichordates *(Page 660)*

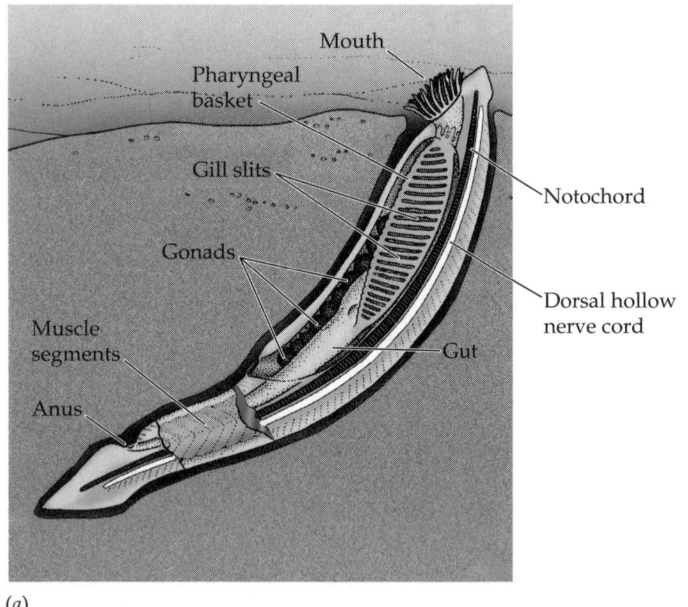
Mouth
Pharyngeal basket
Gill slits
Notochord
Gonads
Dorsal hollow nerve cord
Muscle segments
Gut
Anus

(a)

34.7 Lancelets *(Page 662)*

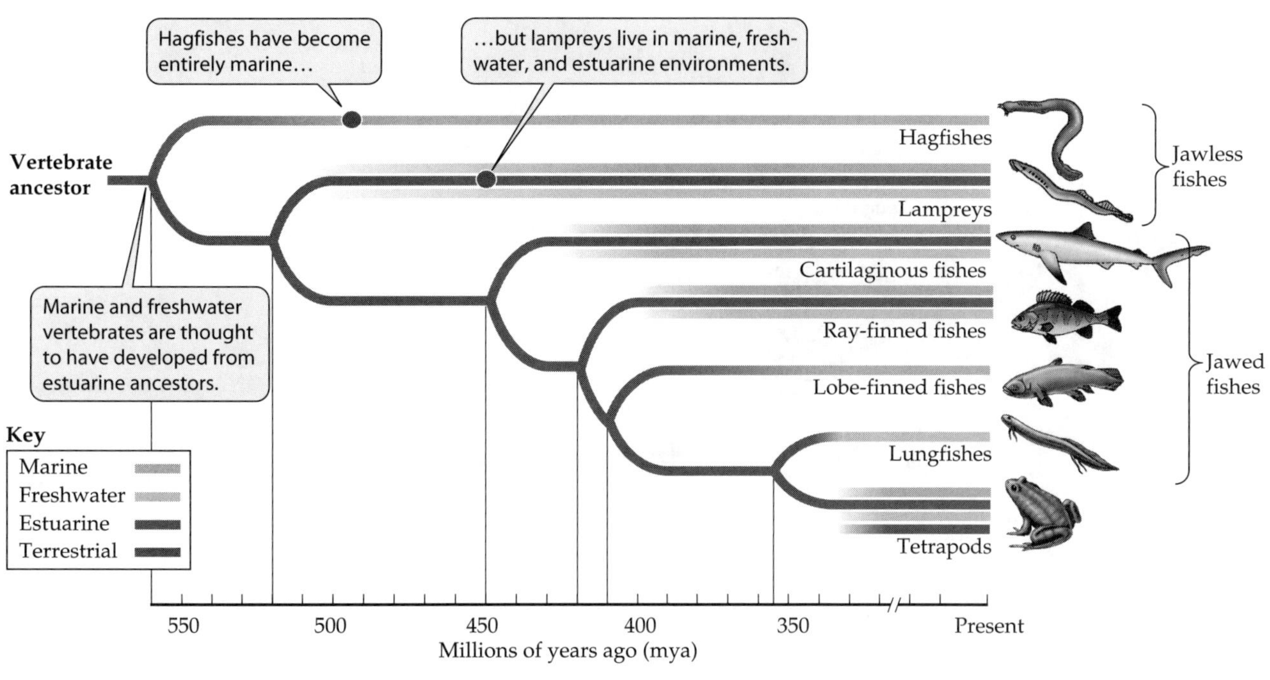
Hagfishes have become entirely marine…
…but lampreys live in marine, freshwater, and estuarine environments.
Vertebrate ancestor
Hagfishes
Lampreys
Cartilaginous fishes
Ray-finned fishes
Lobe-finned fishes
Lungfishes
Tetrapods
Jawless fishes
Jawed fishes
Marine and freshwater vertebrates are thought to have developed from estuarine ancestors.
Key
Marine
Freshwater
Estuarine
Terrestrial
550
500
450
400
350
Present
Millions of years ago (mya)

34.8 A Current Phylogeny of the Vertebrates *(Page 662)*

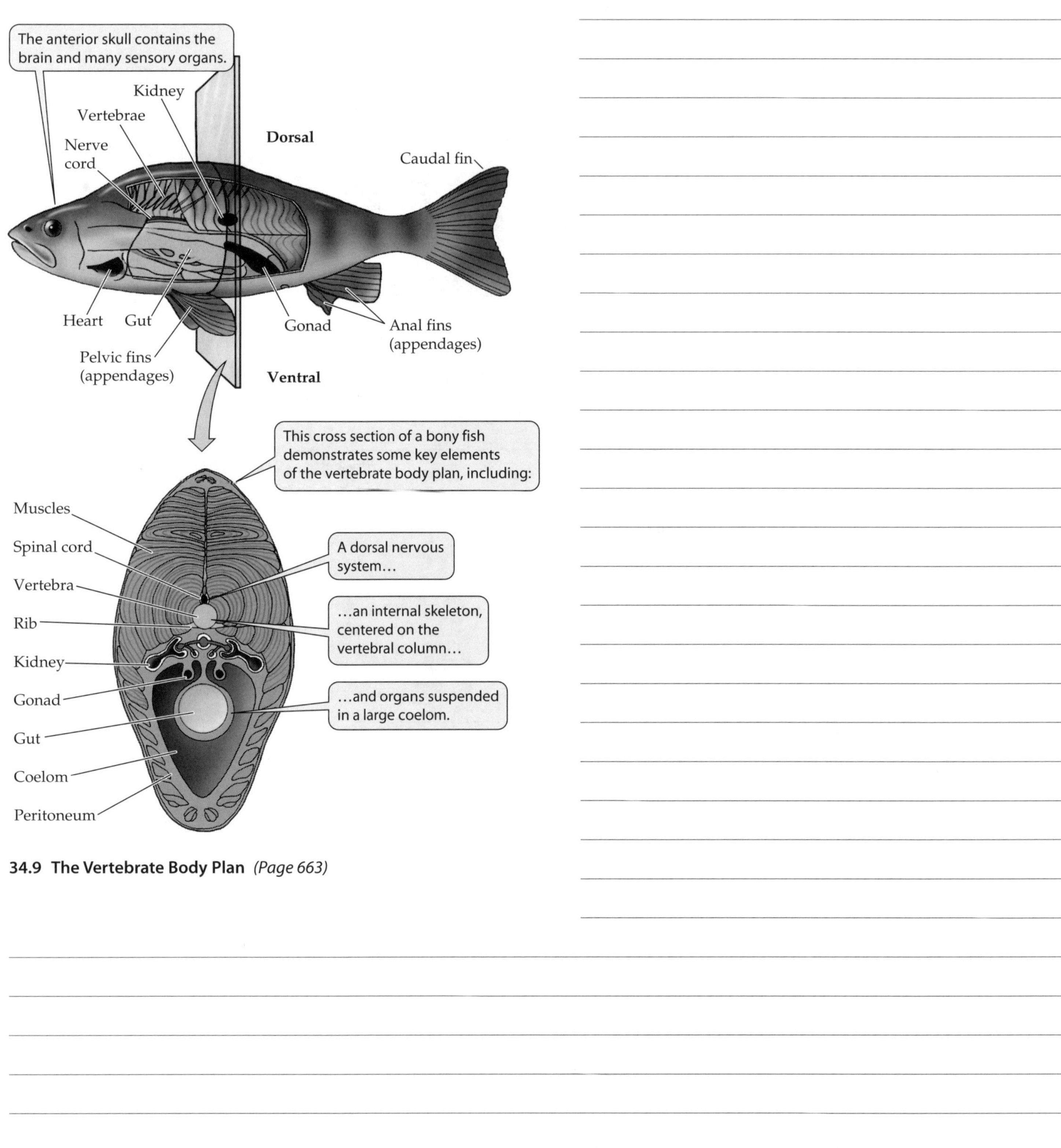
The anterior skull contains the brain and many sensory organs.
Kidney
Vertebrae
Nerve cord
Dorsal
Caudal fin
Heart
Gut
Gonad
Anal fins (appendages)
Pelvic fins (appendages)
Ventral
This cross section of a bony fish demonstrates some key elements of the vertebrate body plan, including:
Muscles
Spinal cord
Vertebra
Rib
Kidney
Gonad
Gut
Coelom
Peritoneum
A dorsal nervous system…
…an internal skeleton, centered on the vertebral column…
…and organs suspended in a large coelom.

34.9 The Vertebrate Body Plan *(Page 663)*

Jawless fishes (agnathans)

Extinct and living forms

Skull (cartilage)

Gill slits

Gill arches made of cartilage supported the gills.

Early jawed fishes (placoderms)

Extinct

Some anterior gill arches became modified to form jaws.

Modern jawed fishes (cartilaginous and bony fishes)

Living forms

Additional gill arches help support heavier, more efficient jaws.

34.11 Jaws from Gill Arches *(Page 664)*

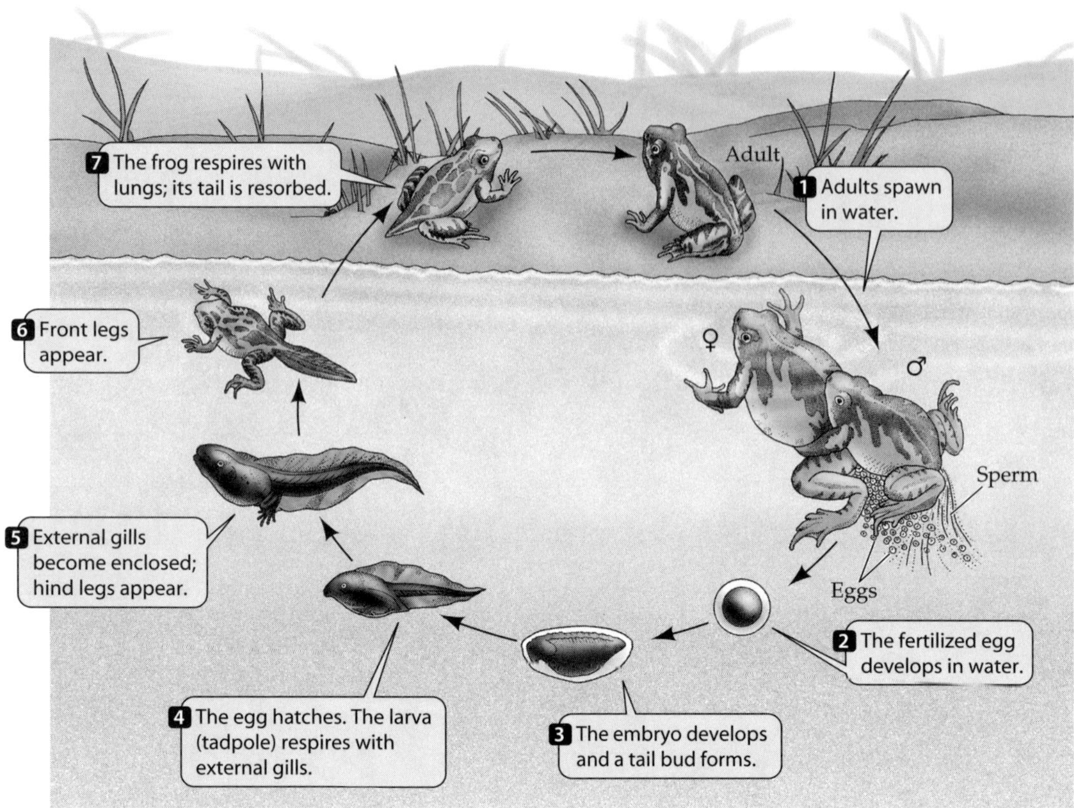

7 The frog respires with lungs; its tail is resorbed.
Adult
1 Adults spawn in water.
6 Front legs appear.
♀
♂
Sperm
5 External gills become enclosed; hind legs appear.
Eggs
2 The fertilized egg develops in water.
4 The egg hatches. The larva (tadpole) respires with external gills.
3 The embryo develops and a tail bud forms.

34.16 In and Out of the Water *(Page 668)*

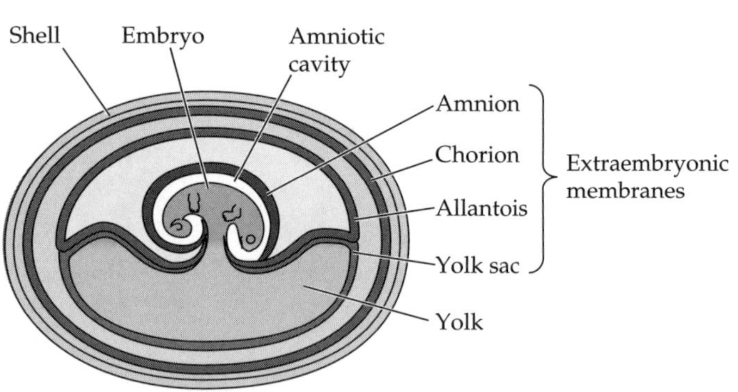

Shell
Embryo
Amniotic cavity
Amnion
Chorion
Allantois
Yolk sac
Yolk
Extraembryonic membranes

34.17 An Egg for Dry Places *(Page 668)*

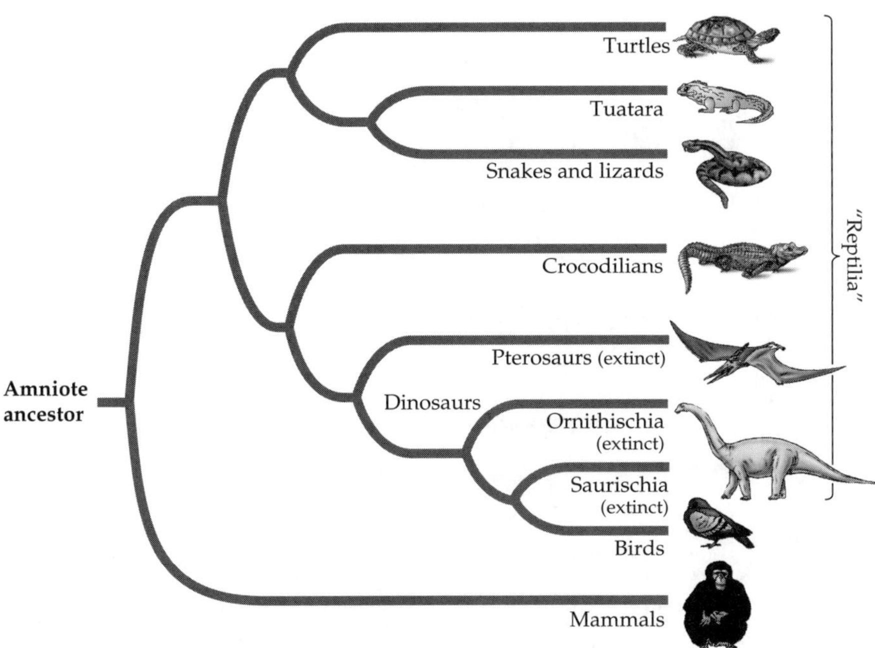
Turtles
Tuatara
Snakes and lizards
Crocodilians
Pterosaurs (extinct)
Dinosaurs
Ornithischia (extinct)
Saurischia (extinct)
Birds
Mammals
Amniote ancestor
"Reptilia"

34.18 The Reptiles Form a Paraphyletic Group *(Page 669)*

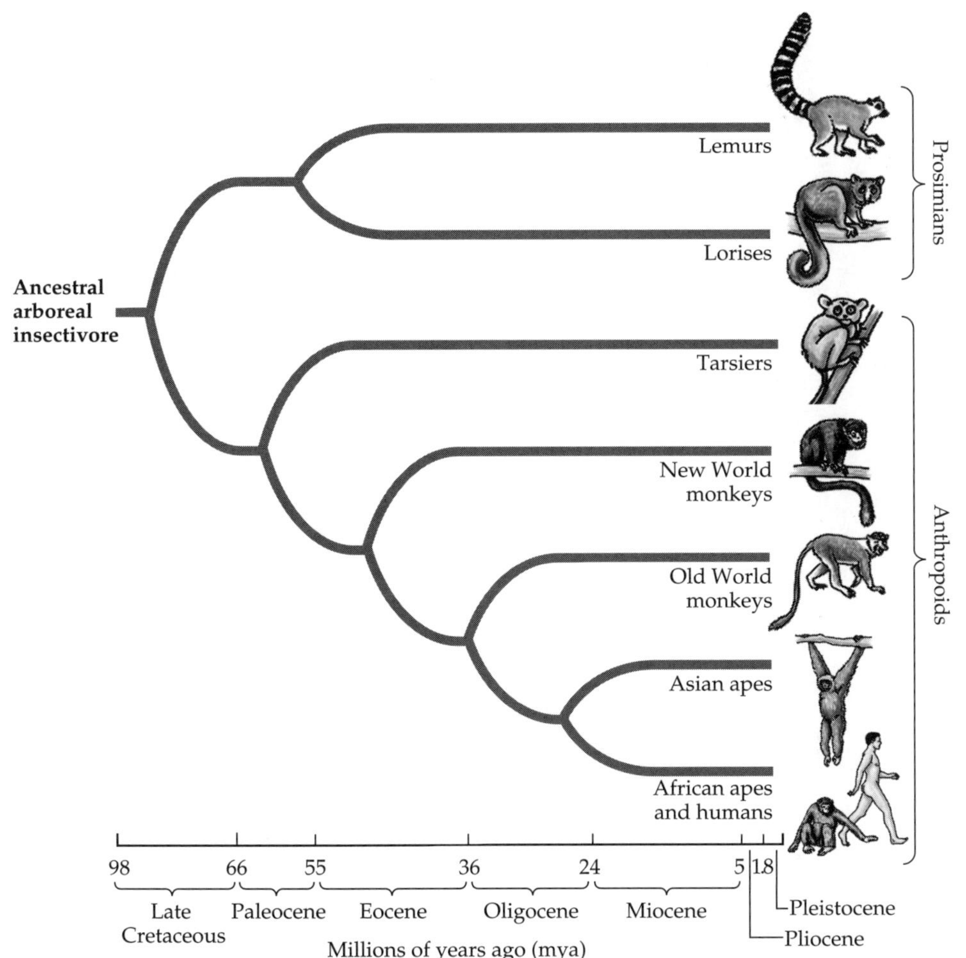
Lemurs
Lorises
Prosimians
Ancestral arboreal insectivore
Tarsiers
New World monkeys
Old World monkeys
Asian apes
African apes and humans
Anthropoids
98
66
55
36
24
5
1.8
Late Cretaceous
Paleocene
Eocene
Oligocene
Miocene
Pleistocene
Pliocene
Millions of years ago (mya)

34.25 A Current Phylogeny of the Primates *(Page 675)*

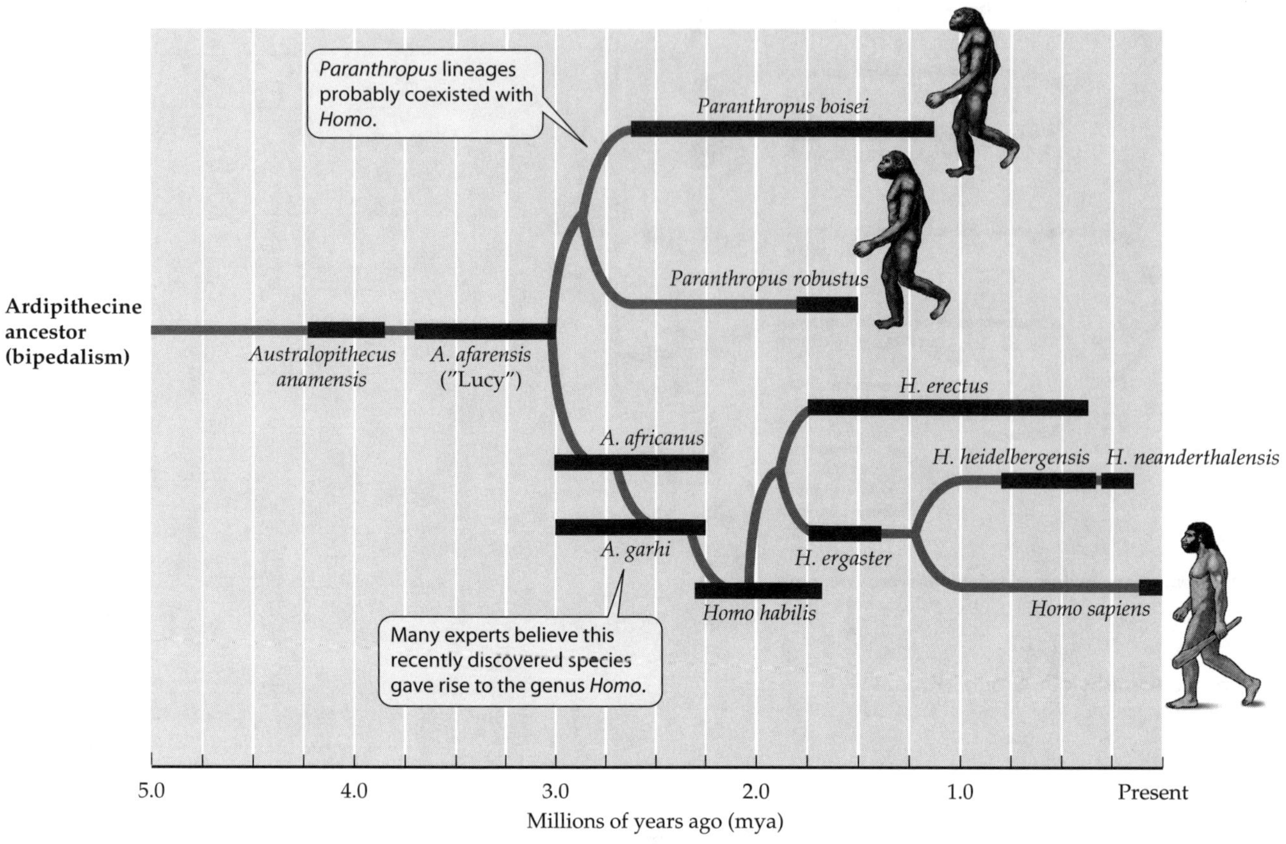
Paranthropus lineages probably coexisted with Homo.
Paranthropus boisei
Paranthropus robustus
Ardipithecine ancestor (bipedalism)
Australopithecus anamensis
A. afarensis ("Lucy")
H. erectus
A. africanus
H. heidelbergensis
H. neanderthalensis
A. garhi
H. ergaster
Homo habilis
Homo sapiens
Many experts believe this recently discovered species gave rise to the genus Homo.
5.0
4.0
3.0
2.0
1.0
Present
Millions of years ago (mya)

34.29 A Current Phylogeny of *Homo sapiens* *(Page 677)*

35 The Plant Body

	Cotyledons	Veins in leaves	Flower parts	Arrangement of primary vascular bundles in stem
Monocots	One	Usually parallel	Usually in multiples of three	Scattered
Eudicots	Two	Usually netlike	Usually in fours or fives	In a ring

35.1 **Monocots versus Eudicots** *(Page 683)*

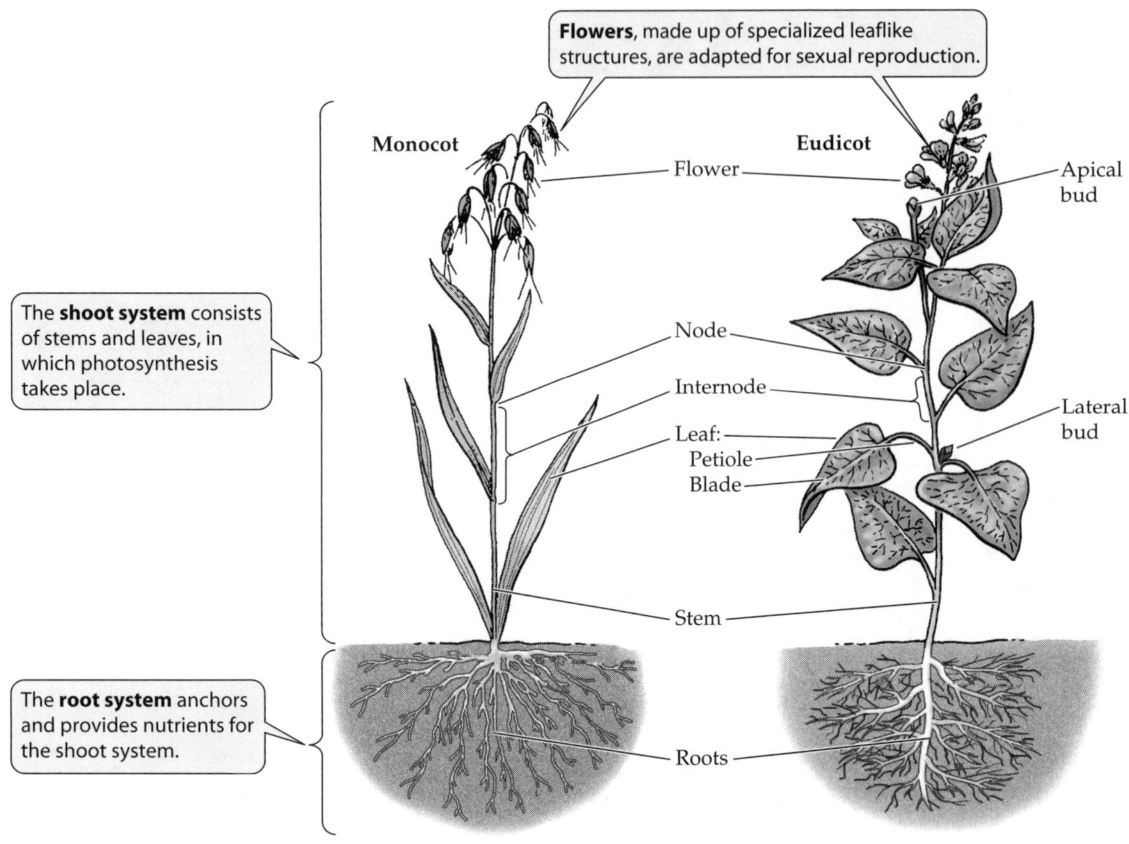
Flowers, made up of specialized leaflike structures, are adapted for sexual reproduction.
Monocot
Eudicot
Flower
Apical bud
The **shoot system** consists of stems and leaves, in which photosynthesis takes place.
Node
Internode
Lateral bud
Leaf:
Petiole
Blade
Stem
The **root system** anchors and provides nutrients for the shoot system.
Roots

35.2 Vegetative Organs and Systems *(Page 684)*

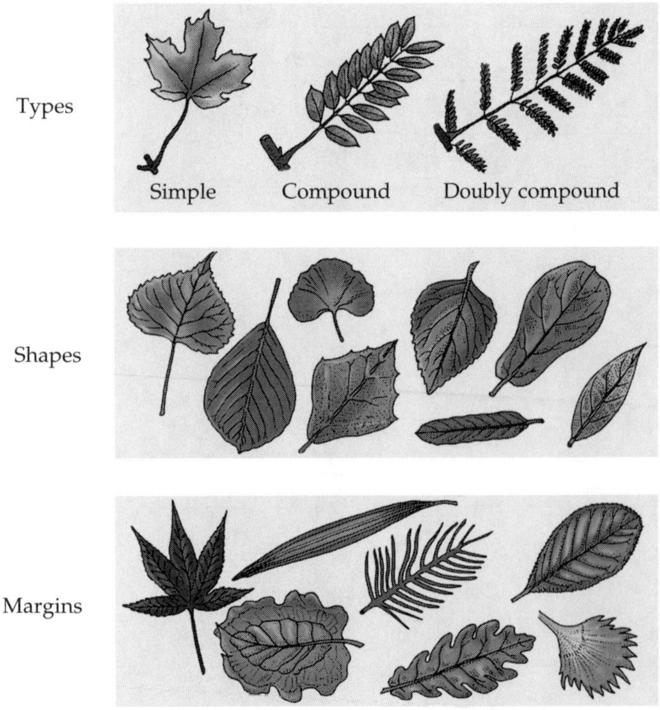

Types

Simple Compound Doubly compound

Shapes

Margins

35.5 The Diversity of Leaf Forms *(Page 686)*

(a)
Primary cell wall
Plasma membrane
The cell plate is the first barrier to form.
(b)
Middle lamella
Each daughter cell deposits a primary wall.
The cells expand.
(c)
Secondary wall
(d)
The primary cell wall thins and fractures.
After the cells stops expanding, they may deposit more layers, forming secondary walls.

35.6 Cell Wall Formation *(Page 686)*

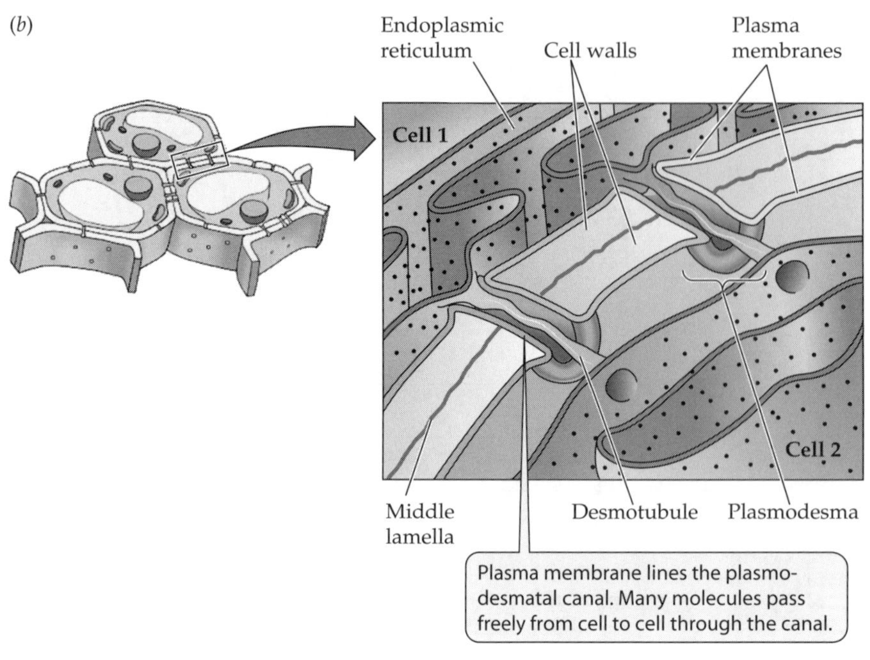
(b)
Endoplasmic reticulum
Cell walls
Plasma membranes
Cell 1
Cell 2
Middle lamella
Desmotubule
Plasmodesma
Plasma membrane lines the plasmo-desmatal canal. Many molecules pass freely from cell to cell through the canal.

35.7 Plasmodesmata *(Page 687)*

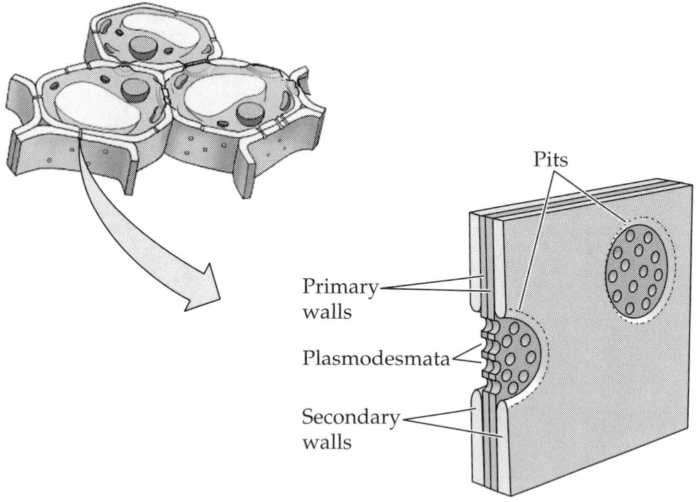
Pits
Primary walls
Plasmodesmata
Secondary walls

35.8 Pits *(Page 687)*

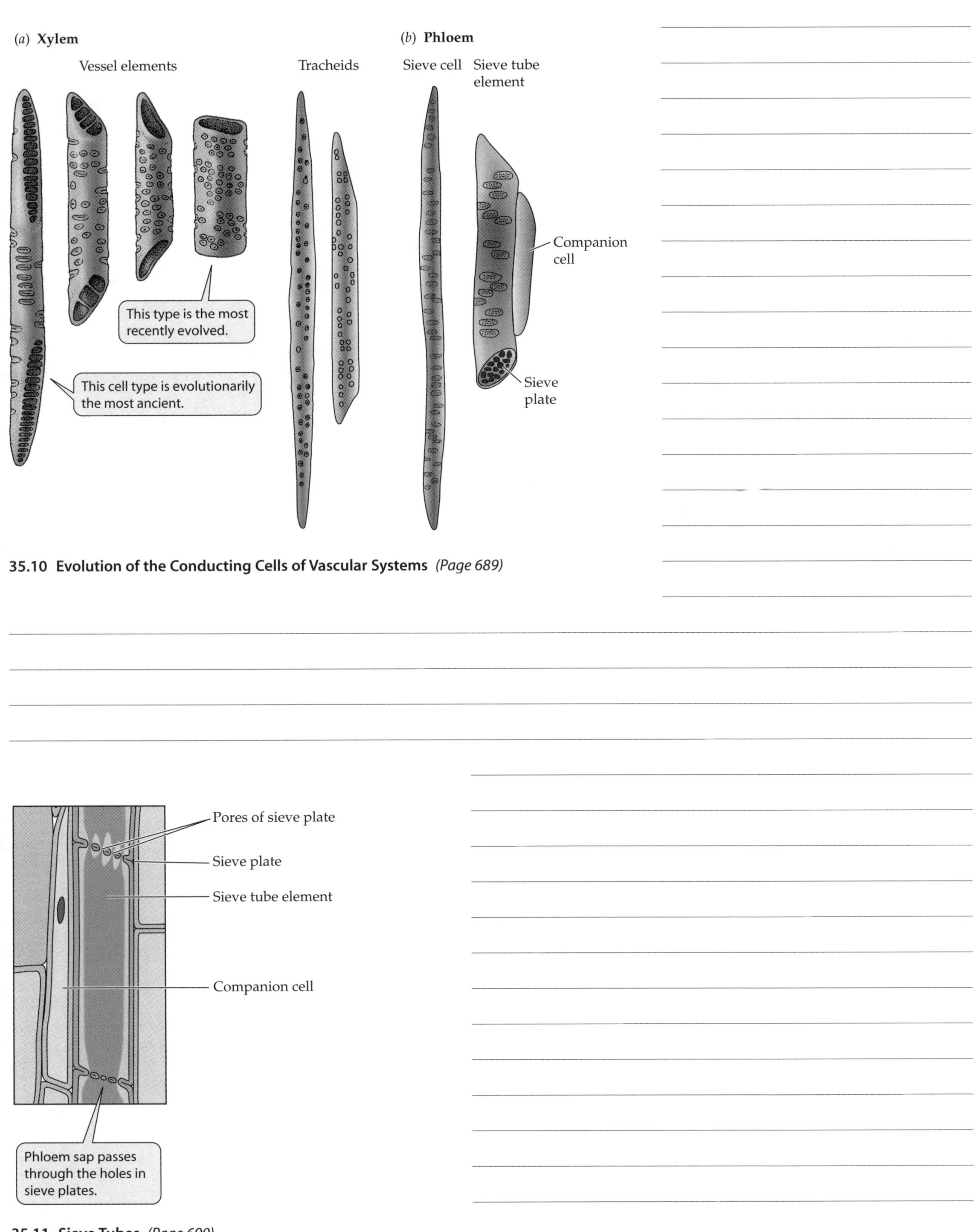
(a) Xylem
Vessel elements
Tracheids
This type is the most recently evolved.
This cell type is evolutionarily the most ancient.
(b) Phloem
Sieve cell
Sieve tube element
Companion cell
Sieve plate

35.10 Evolution of the Conducting Cells of Vascular Systems *(Page 689)*

Pores of sieve plate
Sieve plate
Sieve tube element
Companion cell
Phloem sap passes through the holes in sieve plates.

35.11 Sieve Tubes *(Page 690)*

The **dermal tissue system** is the outer covering of the plant.
The **vascular tissue system** conducts water and solutes throughout the plant.
Leaf
The **ground tissue system** carries out photosynthesis, stores photosynthetic products, and helps support the plant.
Stem
Dermal
Ground
Vascular
Root
Dermal
Ground
Vascular

35.12 Three Tissue Systems Extend throughout the Plant Body
(Page 690)

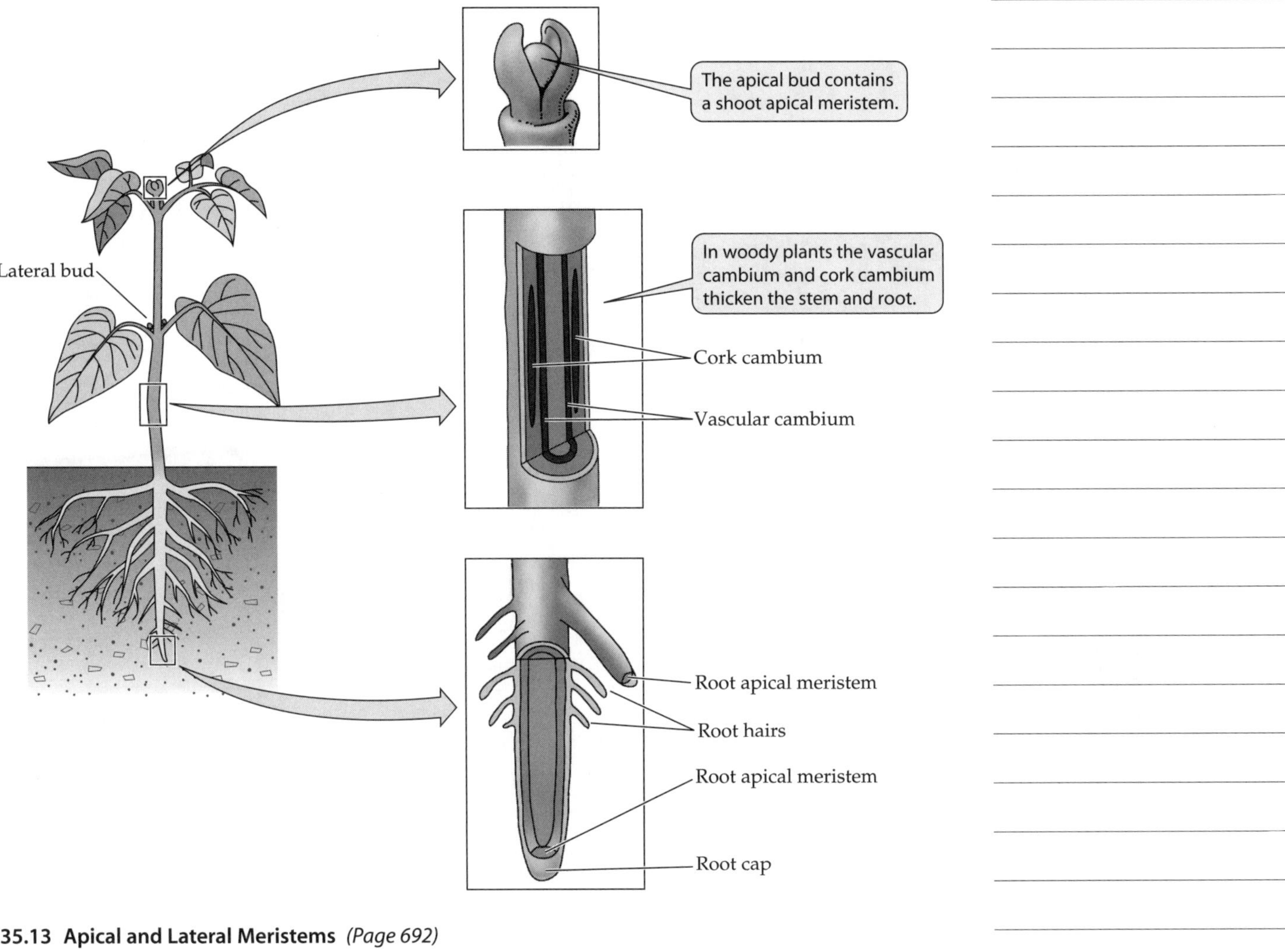
The apical bud contains a shoot apical meristem.
Lateral bud
In woody plants the vascular cambium and cork cambium thicken the stem and root.
Cork cambium
Vascular cambium
Root apical meristem
Root hairs
Root apical meristem
Root cap

35.13 Apical and Lateral Meristems *(Page 692)*

Apical meristems →	Primary meristems →	Tissue systems
Root or shoot apical meristem	Protoderm →	Dermal tissue system
	Ground meristem →	Ground tissue system
	Procambium →	Vascular tissue system

In-Text Art *(Page 692)*

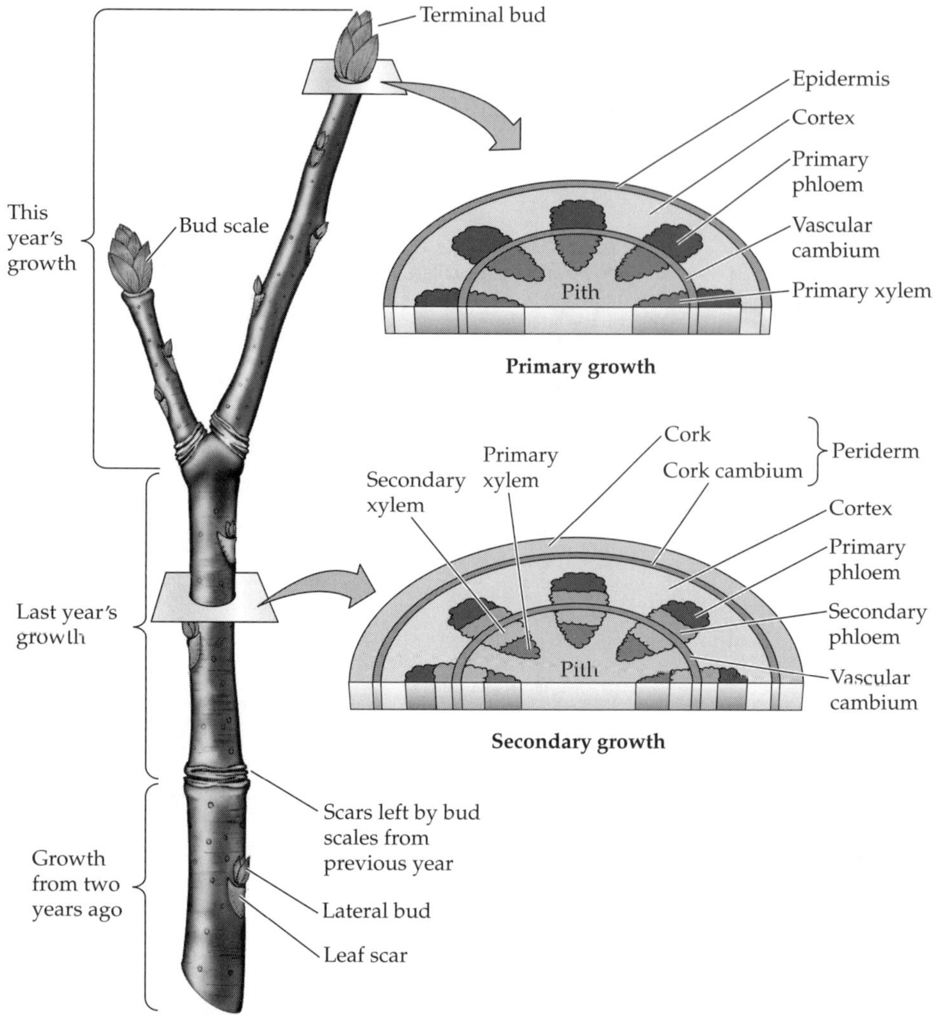

Terminal bud
This year's growth
Bud scale
Epidermis
Cortex
Primary phloem
Vascular cambium
Primary xylem
Pith
Primary growth
Last year's growth
Cork
Cork cambium
Periderm
Primary xylem
Secondary xylem
Cortex
Primary phloem
Secondary phloem
Vascular cambium
Pith
Secondary growth
Growth from two years ago
Scars left by bud scales from previous year
Lateral bud
Leaf scar

35.14 A Woody Twig *(Page 693)*

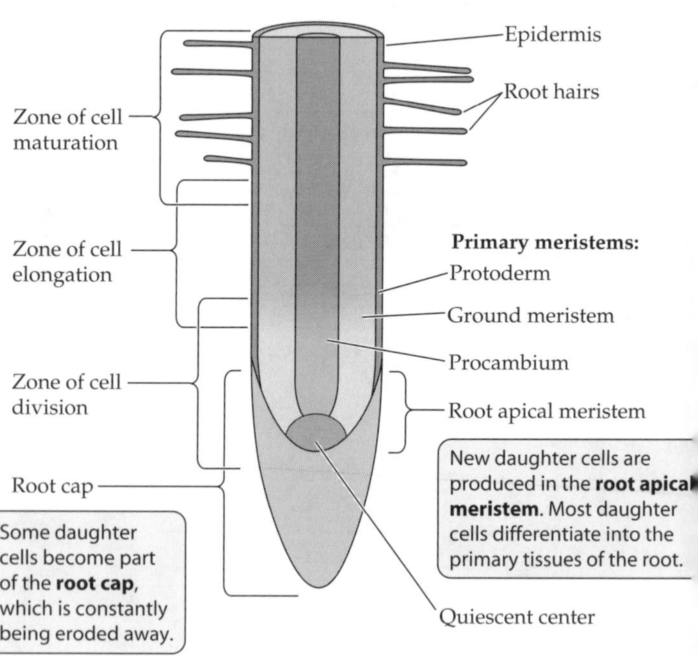
Epidermis
Root hairs
Zone of cell maturation
Zone of cell elongation
Primary meristems:
Protoderm
Ground meristem
Procambium
Zone of cell division
Root apical meristem
New daughter cells are produced in the **root apical meristem**. Most daughter cells differentiate into the primary tissues of the root.
Root cap
Some daughter cells become part of the **root cap**, which is constantly being eroded away.
Quiescent center

35.15 Tissues and Regions of the Root Tip *(Page 693)*

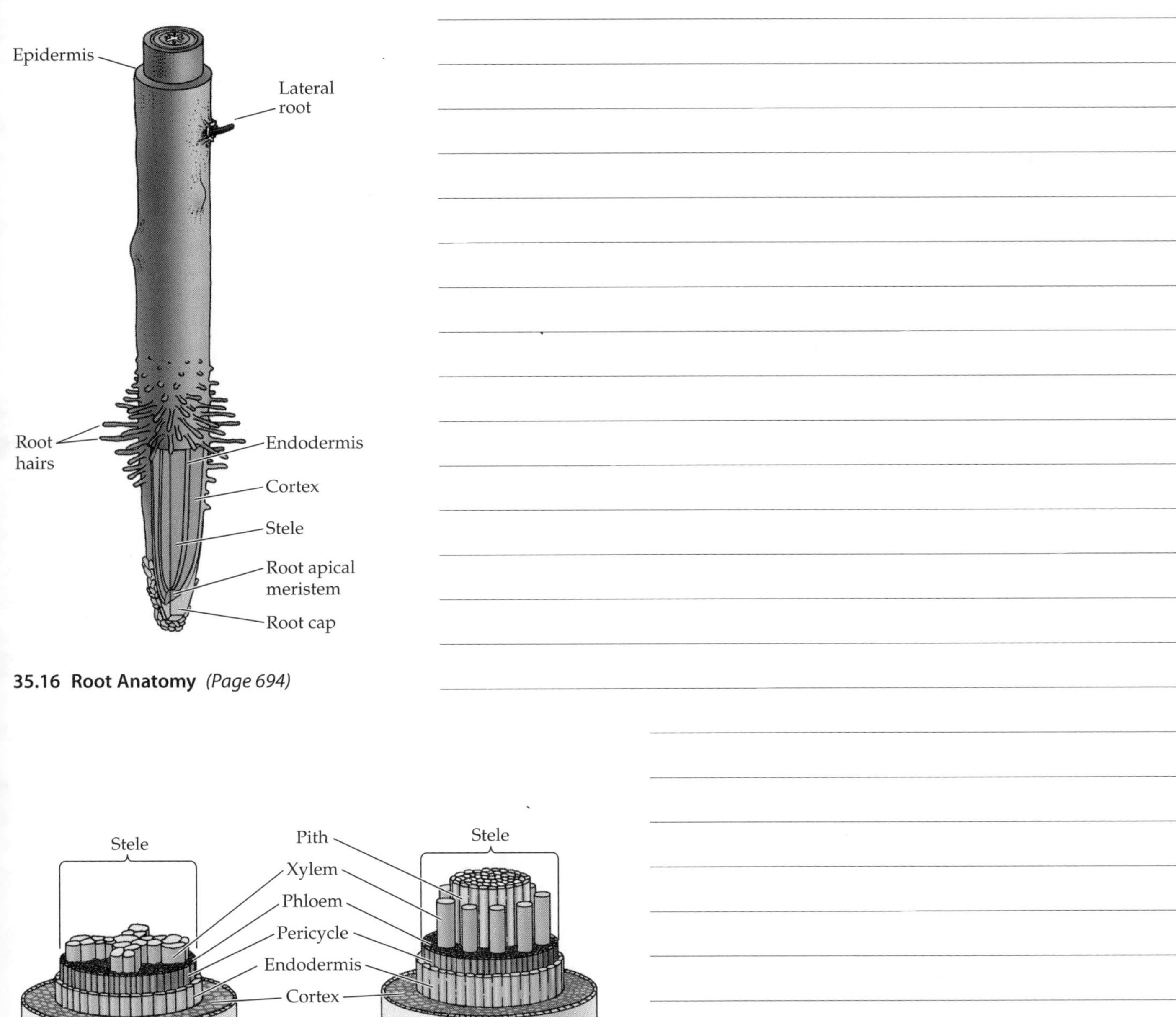
Epidermis
Lateral root
Root hairs
Endodermis
Cortex
Stele
Root apical meristem
Root cap
Stele
Pith
Xylem
Phloem
Pericycle
Endodermis
Cortex
Epidermis
Eudicot root
Stele
Monocot root

35.16 Root Anatomy *(Page 694)*

35.17 The Stele *(Page 695)*

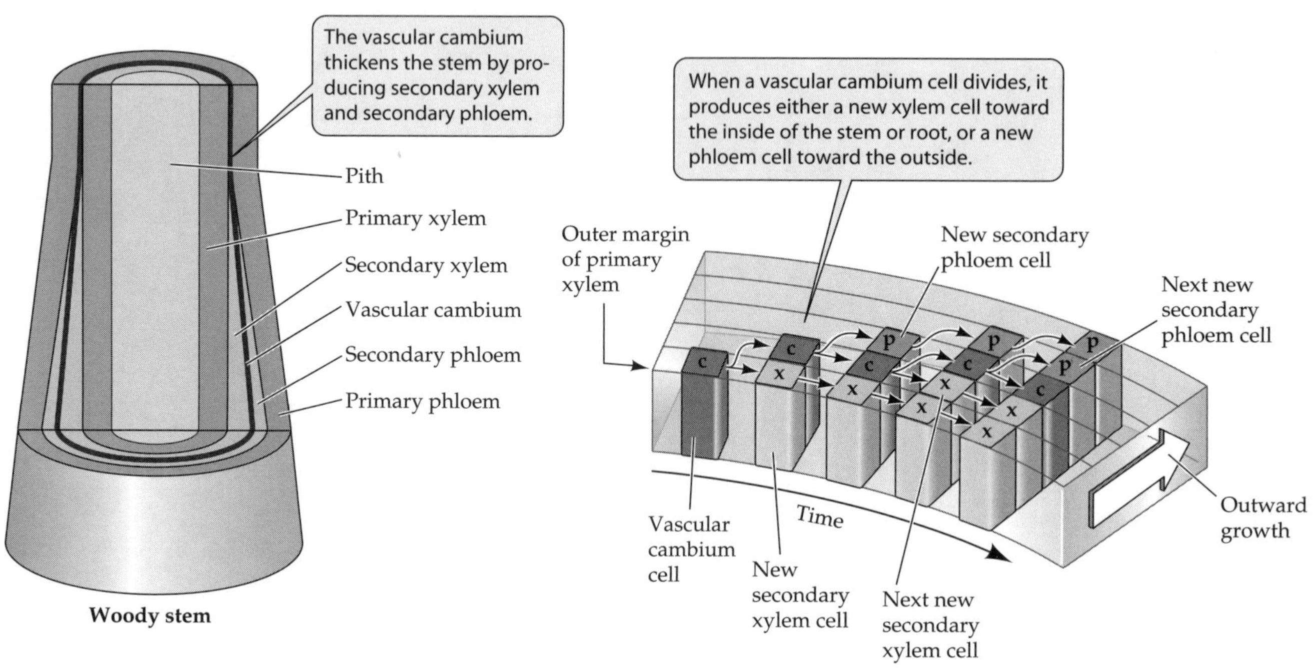
The vascular cambium thickens the stem by producing secondary xylem and secondary phloem.
Pith
Primary xylem
Secondary xylem
Vascular cambium
Secondary phloem
Primary phloem
Woody stem
When a vascular cambium cell divides, it produces either a new xylem cell toward the inside of the stem or root, or a new phloem cell toward the outside.
Outer margin of primary xylem
New secondary phloem cell
Next new secondary phloem cell
c
x
p
Time
Outward growth
Vascular cambium cell
New secondary xylem cell
Next new secondary xylem cell

35.19 Vascular Cambium Thickens Stems and Roots *(Page 696)*

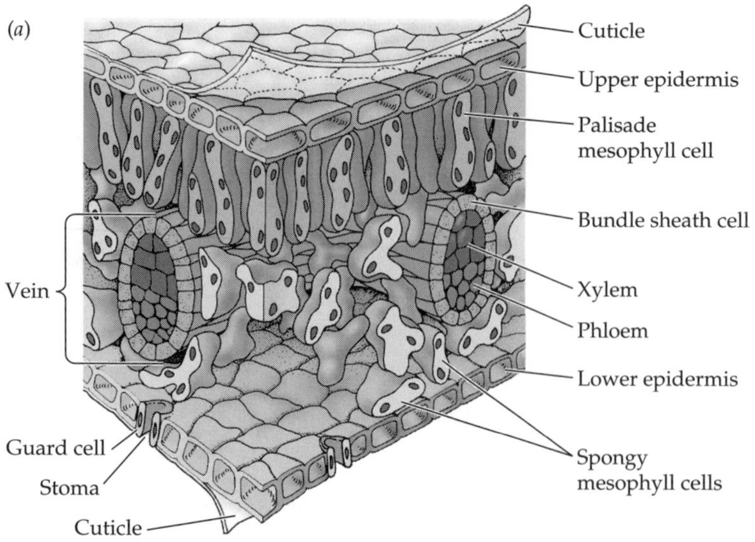

(*a*)

Cuticle

Upper epidermis

Palisade mesophyll cell

Bundle sheath cell

Vein

Xylem

Phloem

Lower epidermis

Guard cell

Stoma

Cuticle

Spongy mesophyll cells

35.23 The Eudicot Leaf *(Page 698)*

36 Transport in Plants

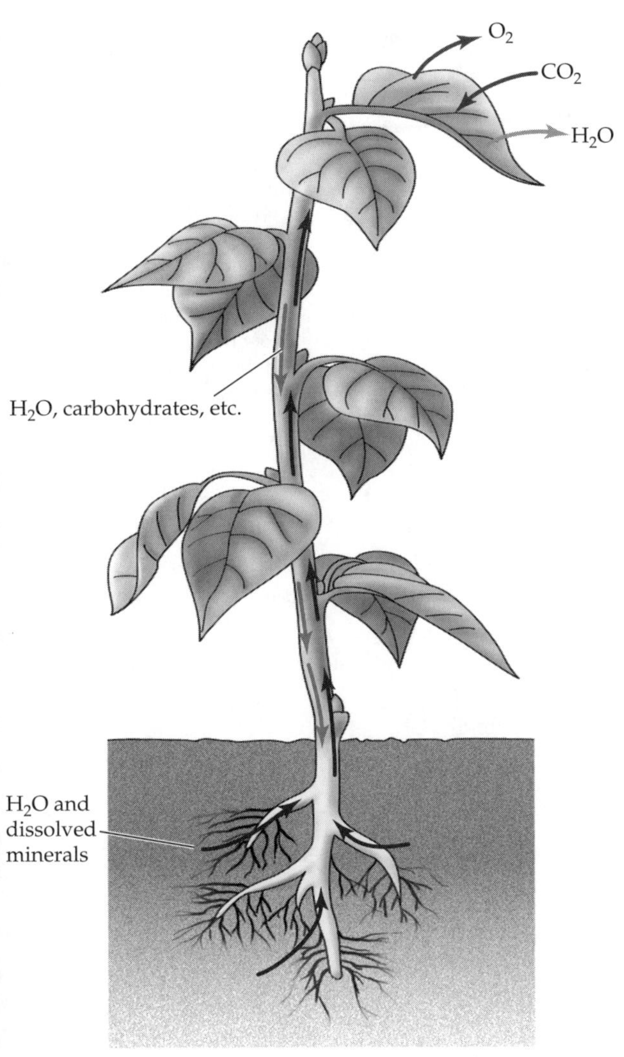
O_2
CO_2
H_2O
H_2O, carbohydrates, etc.
H_2O and dissolved minerals

36.1 The Pathways of Water and Solutes in the Plant
(Page 702)

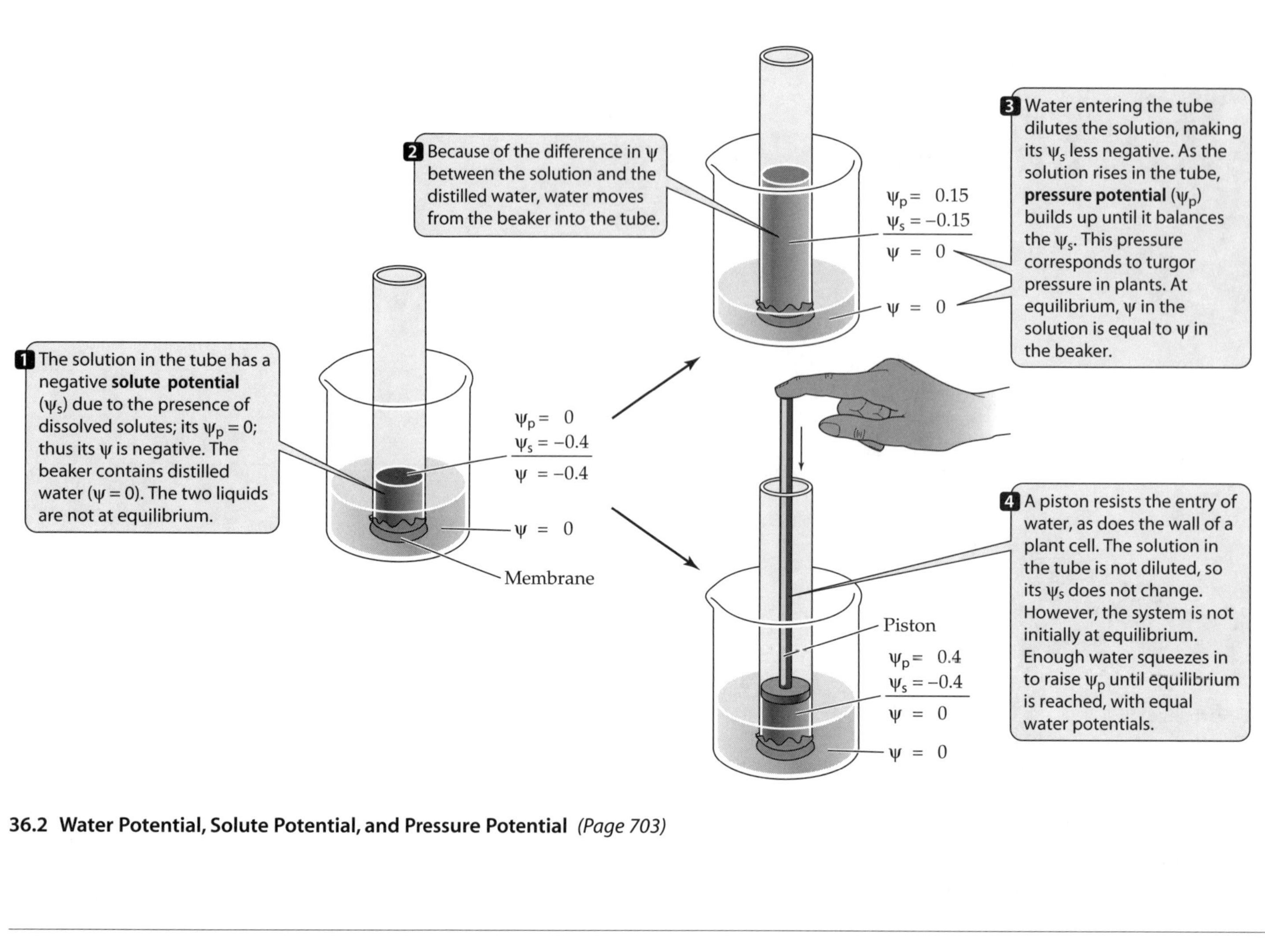

1 The solution in the tube has a negative **solute potential** (ψ_s) due to the presence of dissolved solutes; its $\psi_p = 0$; thus its ψ is negative. The beaker contains distilled water ($\psi = 0$). The two liquids are not at equilibrium.
$\psi_p = 0$
$\psi_s = -0.4$
$\psi = -0.4$
$\psi = 0$
Membrane
2 Because of the difference in ψ between the solution and the distilled water, water moves from the beaker into the tube.
$\psi_p = 0.15$
$\psi_s = -0.15$
$\psi = 0$
$\psi = 0$
3 Water entering the tube dilutes the solution, making its ψ_s less negative. As the solution rises in the tube, **pressure potential** (ψ_p) builds up until it balances the ψ_s. This pressure corresponds to turgor pressure in plants. At equilibrium, ψ in the solution is equal to ψ in the beaker.
4 A piston resists the entry of water, as does the wall of a plant cell. The solution in the tube is not diluted, so its ψ_s does not change. However, the system is not initially at equilibrium. Enough water squeezes in to raise ψ_p until equilibrium is reached, with equal water potentials.
Piston
$\psi_p = 0.4$
$\psi_s = -0.4$
$\psi = 0$
$\psi = 0$

36.2 Water Potential, Solute Potential, and Pressure Potential *(Page 703)*

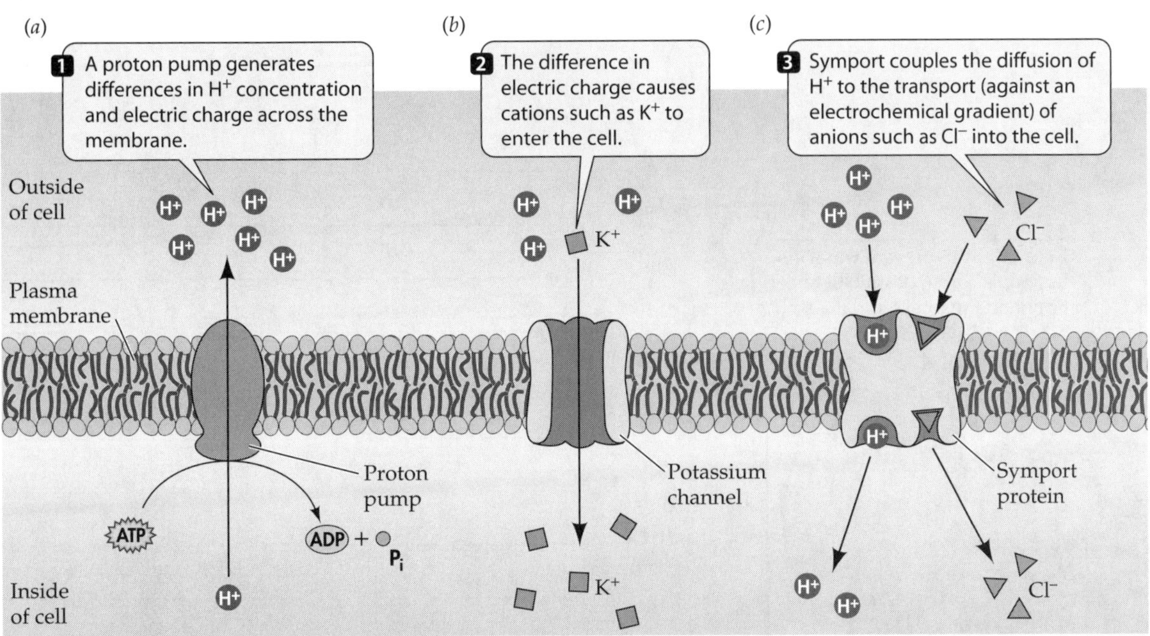
(a)
1 A proton pump generates differences in H+ concentration and electric charge across the membrane.
(b)
2 The difference in electric charge causes cations such as K+ to enter the cell.
(c)
3 Symport couples the diffusion of H+ to the transport (against an electrochemical gradient) of anions such as Cl− into the cell.
Outside of cell
Plasma membrane
H+
K+
Cl−
Proton pump
Potassium channel
Symport protein
ATP
ADP + Pi
Inside of cell

36.3 The Proton Pump in Active Transport of K⁺ and Cl⁻ *(Page 704)*

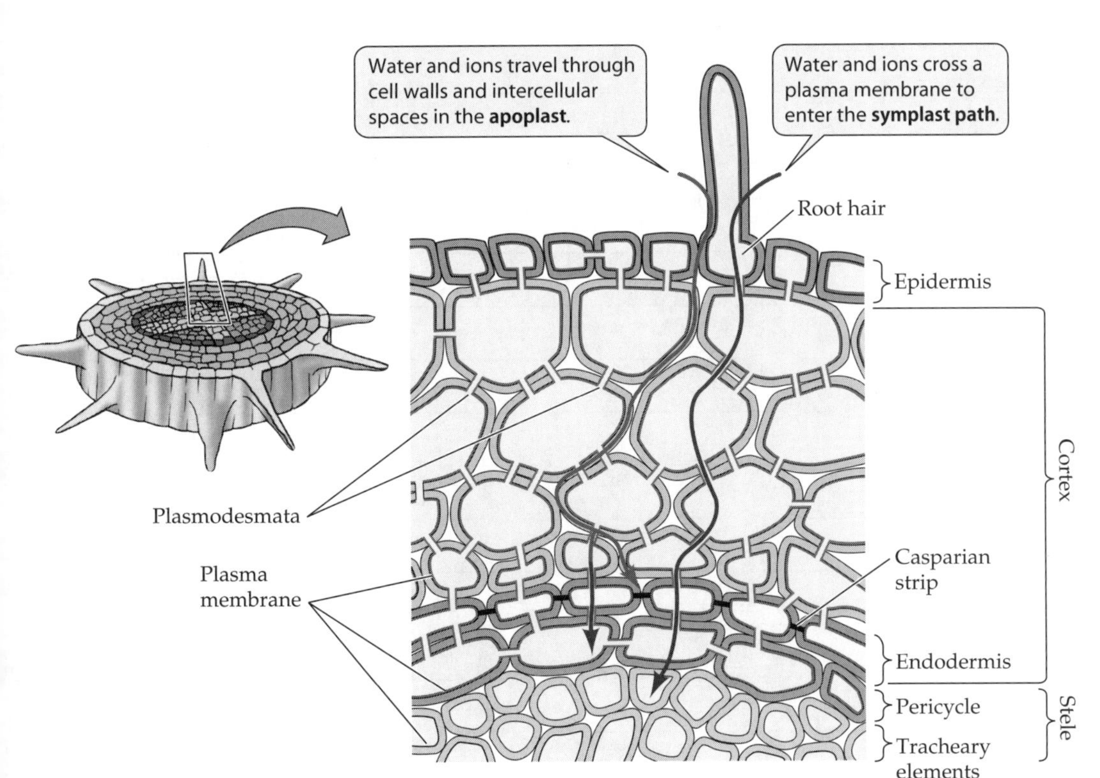
Water and ions travel through cell walls and intercellular spaces in the **apoplast**.
Water and ions cross a plasma membrane to enter the **symplast path**.
Root hair
Epidermis
Cortex
Plasmodesmata
Plasma membrane
Casparian strip
Endodermis
Pericycle
Tracheary elements
Stele

36.4 Apoplast and Symplast *(Page 705)*

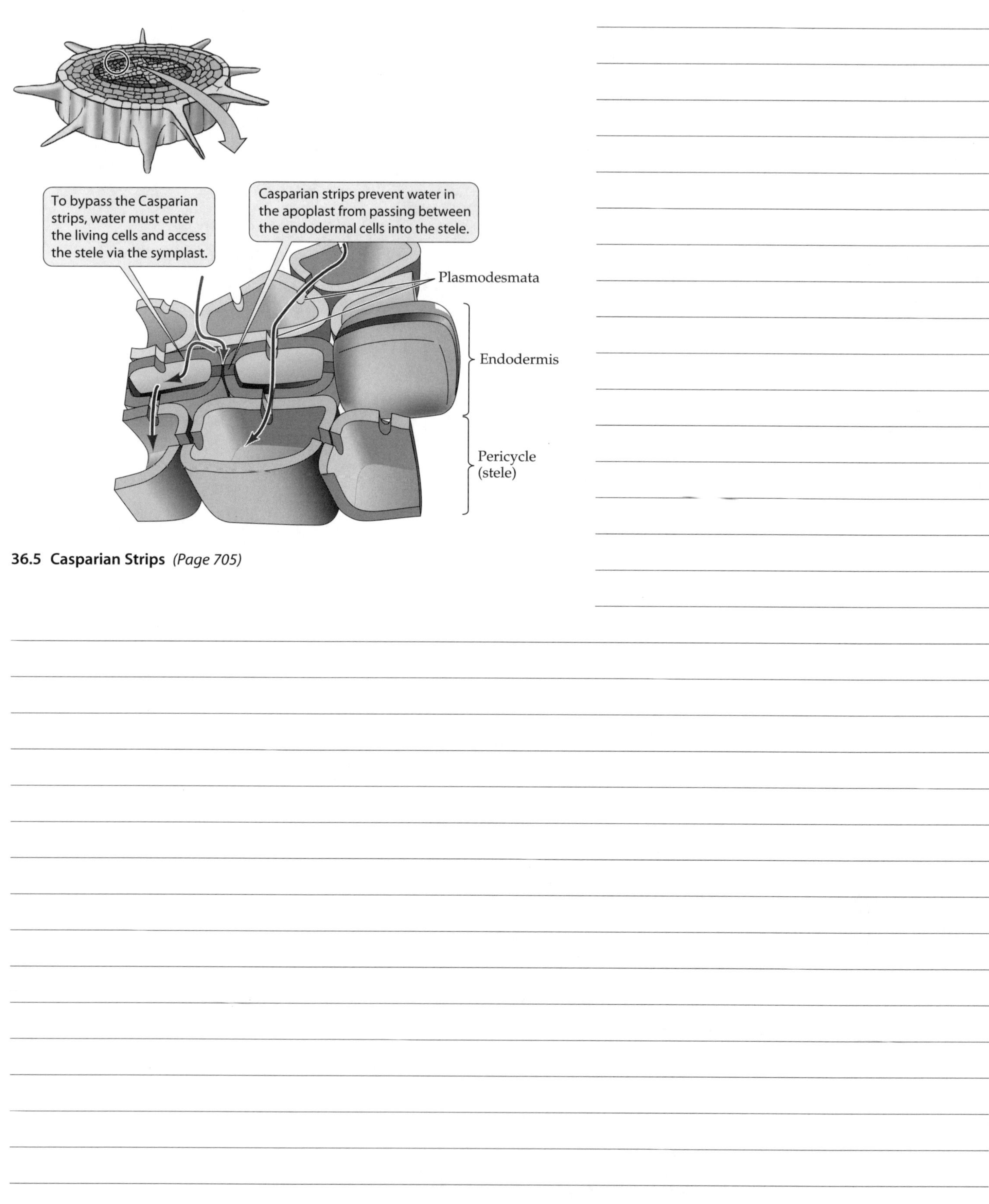
To bypass the Casparian strips, water must enter the living cells and access the stele via the symplast.
Casparian strips prevent water in the apoplast from passing between the endodermal cells into the stele.
Plasmodesmata
Endodermis
Pericycle (stele)

36.5 Casparian Strips *(Page 705)*

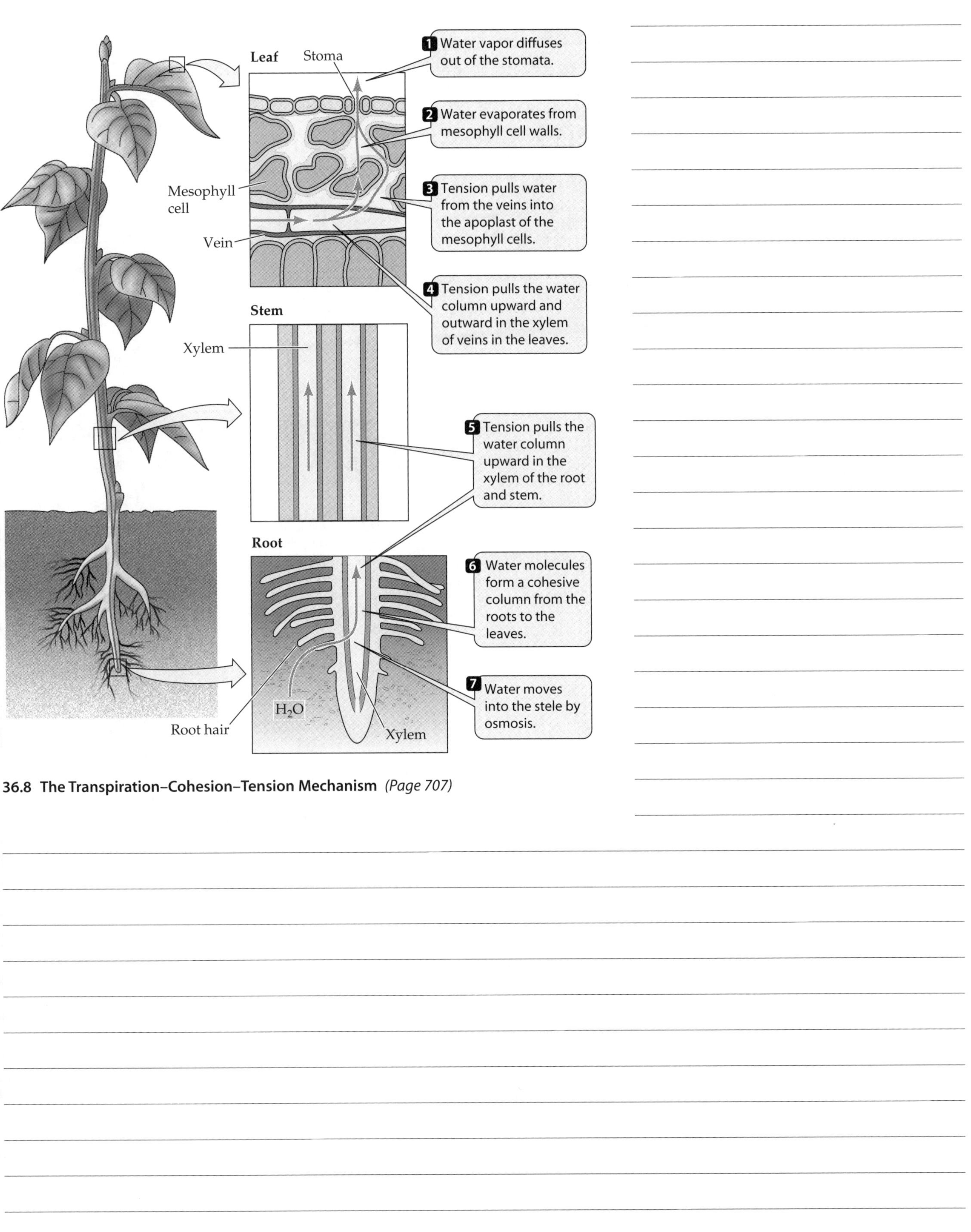
Leaf
Stoma
Mesophyll cell
Vein
1 Water vapor diffuses out of the stomata.
2 Water evaporates from mesophyll cell walls.
3 Tension pulls water from the veins into the apoplast of the mesophyll cells.
4 Tension pulls the water column upward and outward in the xylem of veins in the leaves.
Stem
Xylem
5 Tension pulls the water column upward in the xylem of the root and stem.
Root
6 Water molecules form a cohesive column from the roots to the leaves.
7 Water moves into the stele by osmosis.
H_2O
Xylem
Root hair

36.8 The Transpiration–Cohesion–Tension Mechanism *(Page 707)*

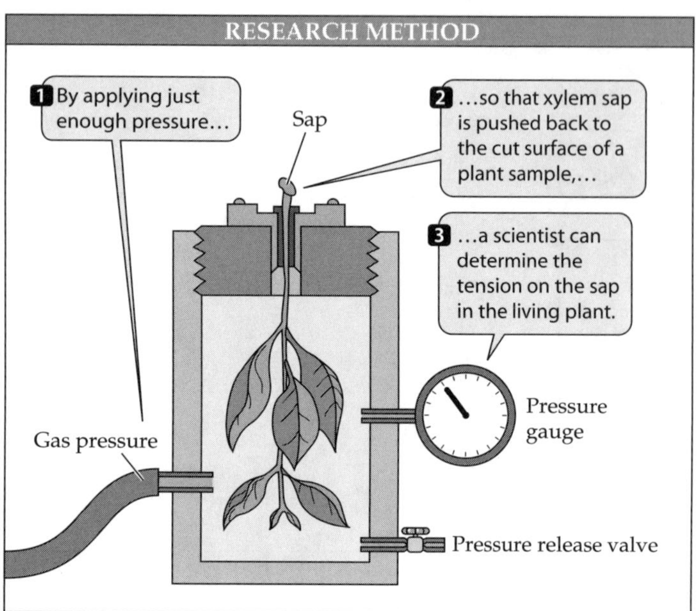
RESEARCH METHOD
1 By applying just enough pressure...
Sap
2 ...so that xylem sap is pushed back to the cut surface of a plant sample,...
3 ...a scientist can determine the tension on the sap in the living plant.
Pressure gauge
Gas pressure
Pressure release valve

36.9 A Pressure Bomb *(Page 708)*

EXPERIMENT

Question: How does K^+ affect xylem flow rate?

METHOD

Two xylem-containing flaps were created on a tobacco plant. One was connected to a source of pure water (the control). The other flap could be connected to either pure water or to a solution containing a known concentration of K^+.

H_2O (control)

K^+ solution *or* H_2O

RESULTS

The addition of the K^+ solution dramatically increased the flow rate. The rate returned to the control level when the K^+ solution was replaced by pure water.

The experimental trace spiked immediately after the injection of K^+ solution.

Return to H_2O alone

K^+ solution injected

H_2O (control)

The control trace (H_2O only) showed little variation in flow rate.

Relative flow rate: 1, 1.5, 2

Time (seconds): 0, 1000, 2000, 3000, 4000

Conclusion: K^+ increases the rate of flow in the xylem.

36.10 Potassium Ions Speed Transport in the Xylem *(Page 709)*

(b)

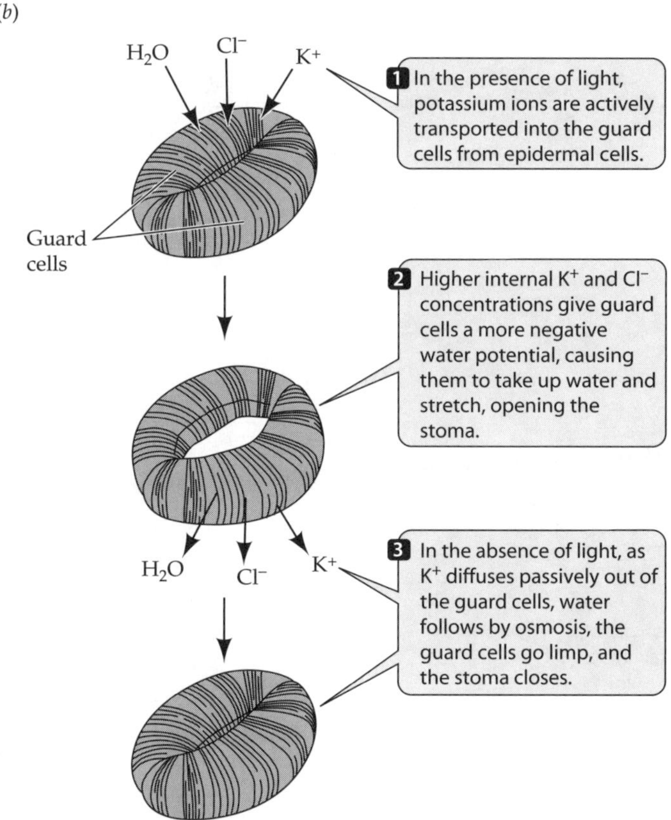
H_2O
Cl^-
K^+
1 In the presence of light, potassium ions are actively transported into the guard cells from epidermal cells.
Guard cells
2 Higher internal K^+ and Cl^- concentrations give guard cells a more negative water potential, causing them to take up water and stretch, opening the stoma.
H_2O
Cl^-
K^+
3 In the absence of light, as K^+ diffuses passively out of the guard cells, water follows by osmosis, the guard cells go limp, and the stoma closes.

36.11 Stomata *(Page 709)*

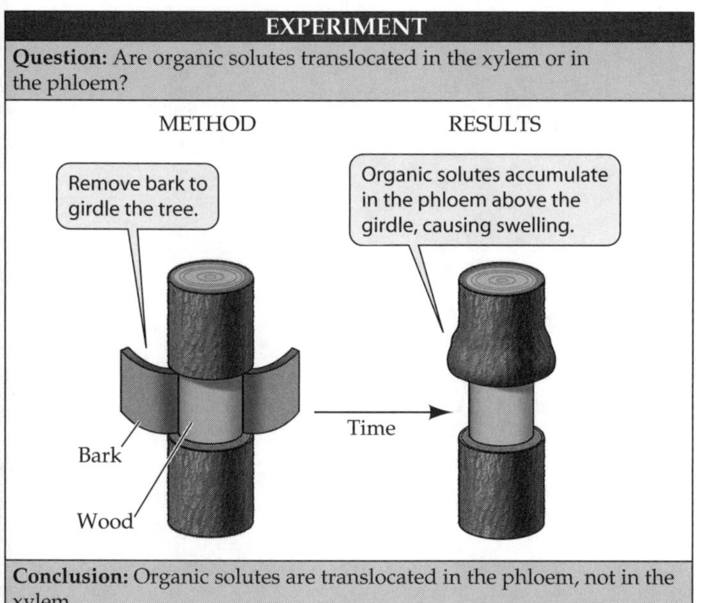
EXPERIMENT
Question: Are organic solutes translocated in the xylem or in the phloem?
METHOD
RESULTS
Remove bark to girdle the tree.
Organic solutes accumulate in the phloem above the girdle, causing swelling.
Time
Bark
Wood
Conclusion: Organic solutes are translocated in the phloem, not in the xylem.

36.12 Girdling Blocks Translocation in the Phloem *(Page 711)*

36.1 Mechanisms of Sap Flow in Plant Vascular Tissues

	XYLEM	PHLOEM
Driving force for bulk flow	Transpiration from leaves	Active transport of sucrose at source
Site of bulk flow	Non-living vessel elements and tracheids (cohesion)	Living sieve tube elements
Pressure potential in sap	Negative (pull from top; tension)	Positive (push from source; pressure)

(Page 712)

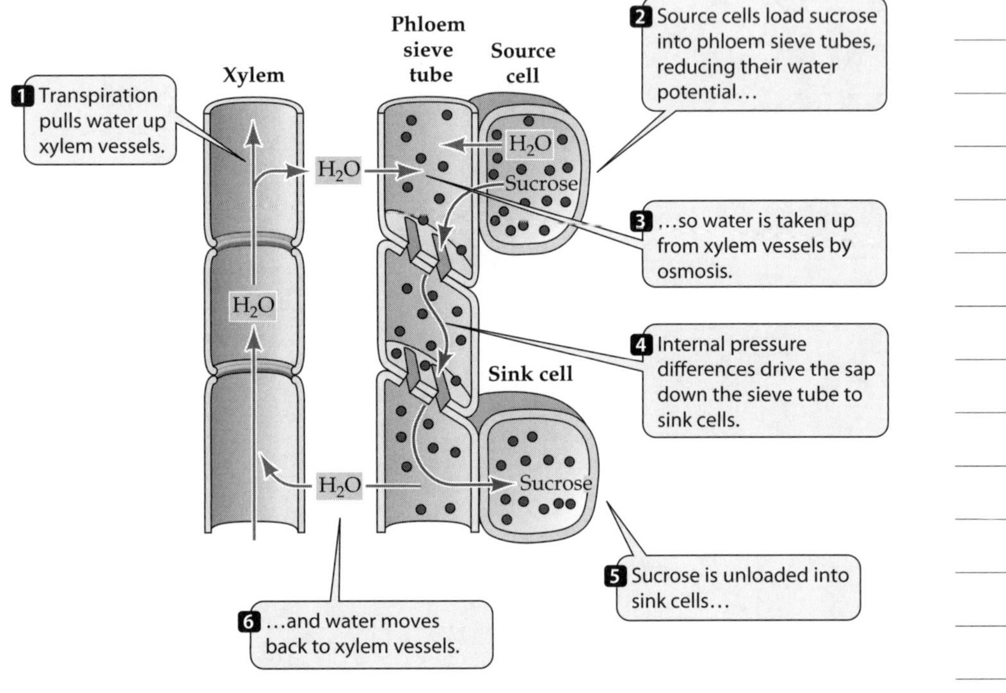

Phloem sieve tube
Source cell
Xylem
1 Transpiration pulls water up xylem vessels.
H_2O
H_2O
Sucrose
2 Source cells load sucrose into phloem sieve tubes, reducing their water potential…
3 …so water is taken up from xylem vessels by osmosis.
H_2O
4 Internal pressure differences drive the sap down the sieve tube to sink cells.
Sink cell
H_2O
Sucrose
5 Sucrose is unloaded into sink cells…
6 …and water moves back to xylem vessels.

36.14 The Pressure Flow Model *(Page 712)*

37 Plant Nutrition

37.1 Mineral Elements Required by Plants

ELEMENT	ABSORBED FORM	MAJOR FUNCTIONS
Macronutrients		
Nitrogen (N)	NO_3^- and NH_4^+	In proteins, nucleic acids, etc.
Phosphorus (P)	$H_2PO_4^-$ and HPO_4^{2-}	In nucleic acids, ATP, phospholipids, etc.
Potassium (K)	K^+	Enzyme activation; water balance; ion balance; stomatal opening
Sulfur (S)	SO_4^{2-}	In proteins and coenzymes
Calcium (Ca)	Ca^{2+}	Affects the cytoskeleton, membranes, and many enzymes; second messenger
Magnesium (Mg)	Mg^{2+}	In chlorophyll; required by many enzymes; stabilizes ribosomes
Micronutrients		
Iron (Fe)	Fe^{2+}	In active site of many redox enzymes and electron carriers; chlorophyll synthesis
Chlorine (Cl)	Cl^-	Photosynthesis; ion balance
Manganese (Mn)	Mn^{2+}	Activation of many enzymes
Boron (B)	$B(OH)_3$	Possibly carbohydrate transport (poorly understood)
Zinc (Zn)	Zn^{2+}	Enzyme activation; auxin synthesis
Copper (Cu)	Cu^{2+}	In active site of many redox enzymes and electron carriers
Nickel (Ni)	Ni^{2+}	Activation of one enzyme
Molybdenum (Mo)	MoO_4^{2-}	Nitrate reduction

(Page 718)

37.2 Some Mineral Deficiencies in Plants

DEFICIENCY	SYMPTOMS
Calcium	Growing points die back; young leaves are yellow and crinkly
Iron	Young leaves are white or yellow with green veins
Magnesium	Older leaves have yellow in stripes between veins
Manganese	Younger leaves are pale with stripes of dead patches
Nitrogen	Oldest leaves turn yellow and die prematurely; plant is stunted
Phosphorus	Plant is dark green with purple veins and is stunted
Potassium	Older leaves have dead edges
Sulfur	Young leaves are yellow to white with yellow veins
Zinc	Young leaves are abnormally small; older leaves have many dead spots

(Page 718)

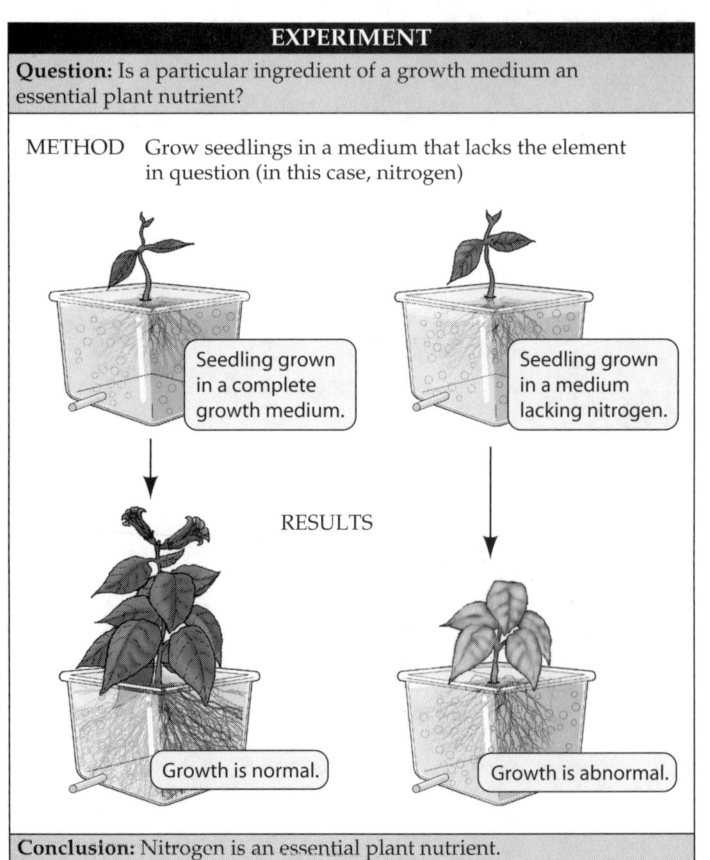
EXPERIMENT
Question: Is a particular ingredient of a growth medium an essential plant nutrient?
METHOD Grow seedlings in a medium that lacks the element in question (in this case, nitrogen)
Seedling grown in a complete growth medium.
Seedling grown in a medium lacking nitrogen.
RESULTS
Growth is normal.
Growth is abnormal.
Conclusion: Nitrogen is an essential plant nutrient.

37.1 Identifying Essential Elements for Plants *(Page 719)*

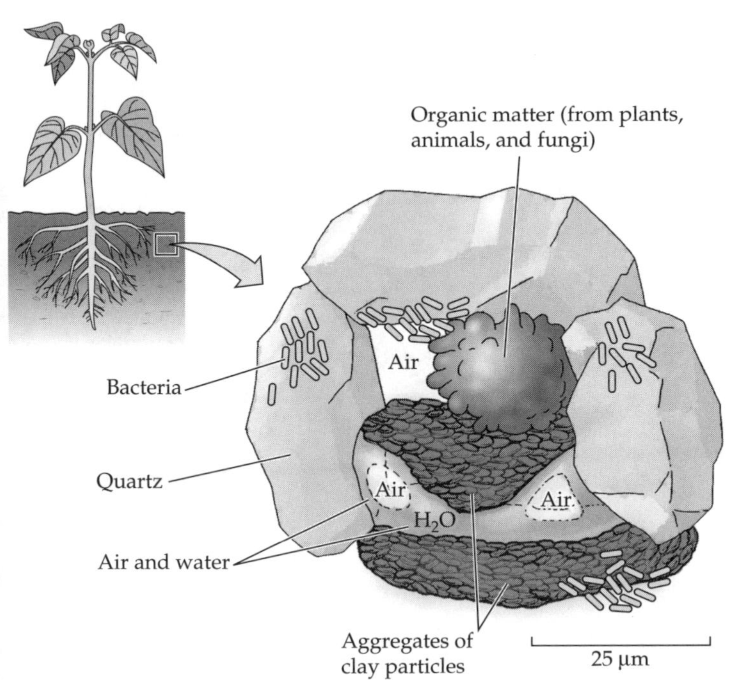
Organic matter (from plants, animals, and fungi)
Air
Bacteria
Quartz
Air
Air
H_2O
Air and water
Aggregates of clay particles
25 µm

37.2 The Complexity of Soil *(Page 720)*

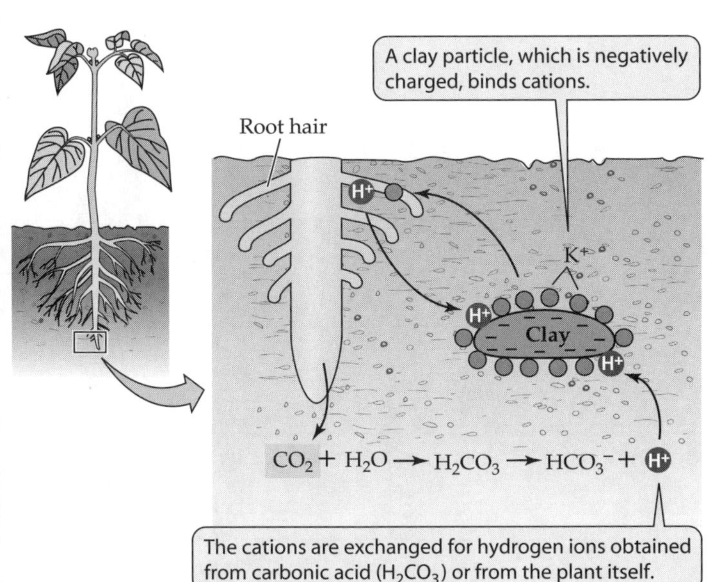
A clay particle, which is negatively charged, binds cations.
Root hair
H^+
K^+
H^+
Clay
H^+
$CO_2 + H_2O \rightarrow H_2CO_3 \rightarrow HCO_3^- + H^+$
The cations are exchanged for hydrogen ions obtained from carbonic acid (H_2CO_3) or from the plant itself.

37.4 Ion Exchange *(Page 721)*

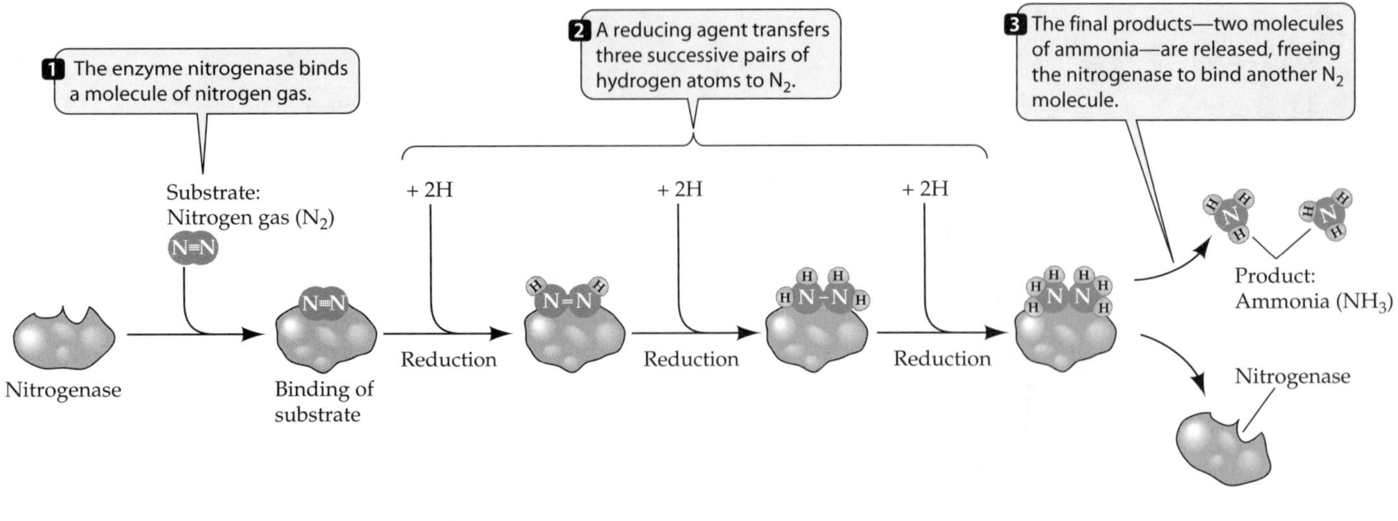

1 The enzyme nitrogenase binds a molecule of nitrogen gas.
2 A reducing agent transfers three successive pairs of hydrogen atoms to N_2.
3 The final products—two molecules of ammonia—are released, freeing the nitrogenase to bind another N_2 molecule.
Substrate: Nitrogen gas (N_2)
Nitrogenase
Binding of substrate
+ 2H
Reduction
+ 2H
Reduction
+ 2H
Reduction
Product: Ammonia (NH_3)
Nitrogenase

37.6 Nitrogenase Fixes Nitrogen *(Page 723)*

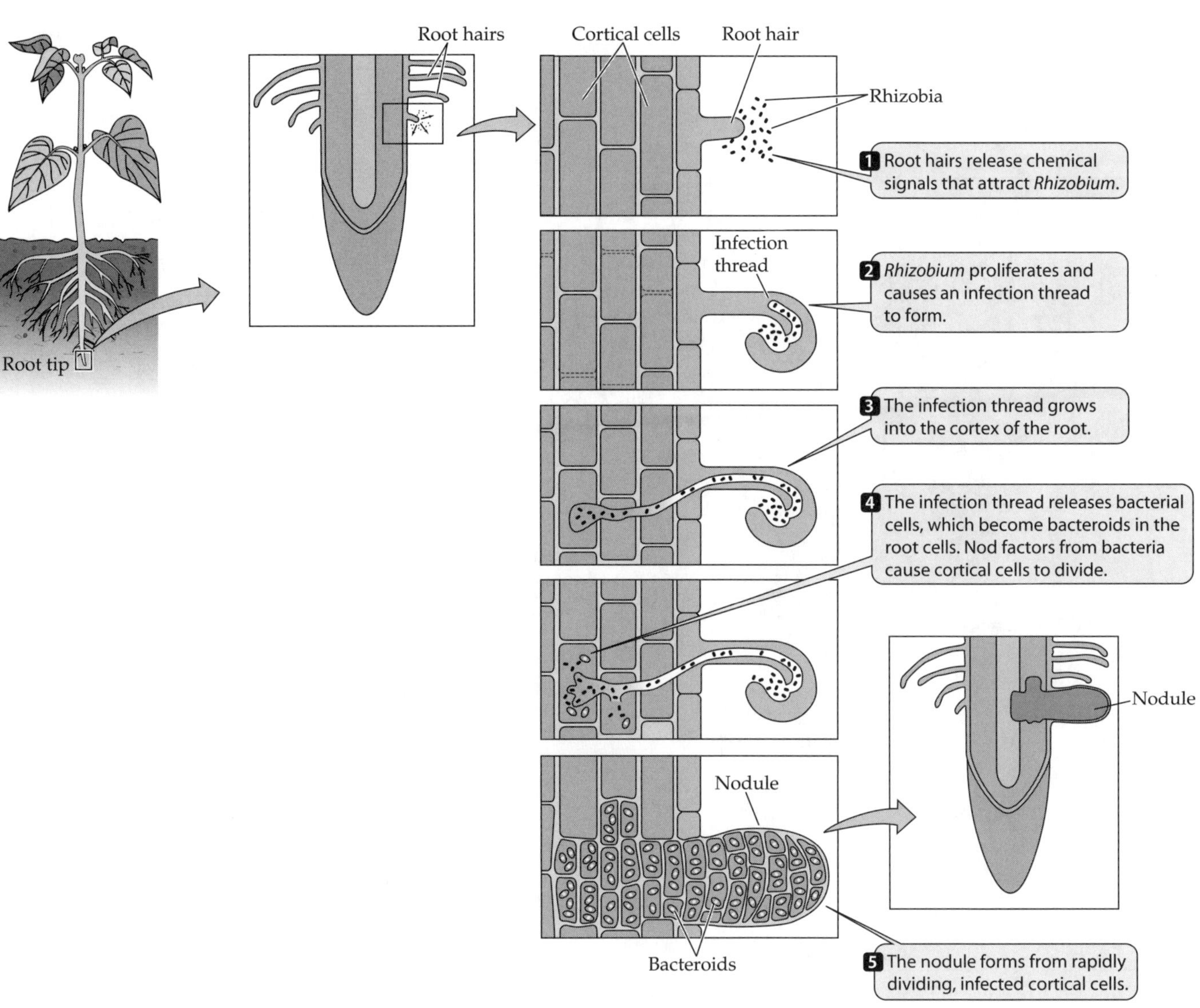
Root hairs
Cortical cells
Root hair
Rhizobia
Root tip
1 Root hairs release chemical signals that attract *Rhizobium*.
Infection thread
2 *Rhizobium* proliferates and causes an infection thread to form.
3 The infection thread grows into the cortex of the root.
4 The infection thread releases bacterial cells, which become bacteroids in the root cells. Nod factors from bacteria cause cortical cells to divide.
Nodule
Nodule
Bacteroids
5 The nodule forms from rapidly dividing, infected cortical cells.

37.7 A Nodule Forms *(Page 724)*

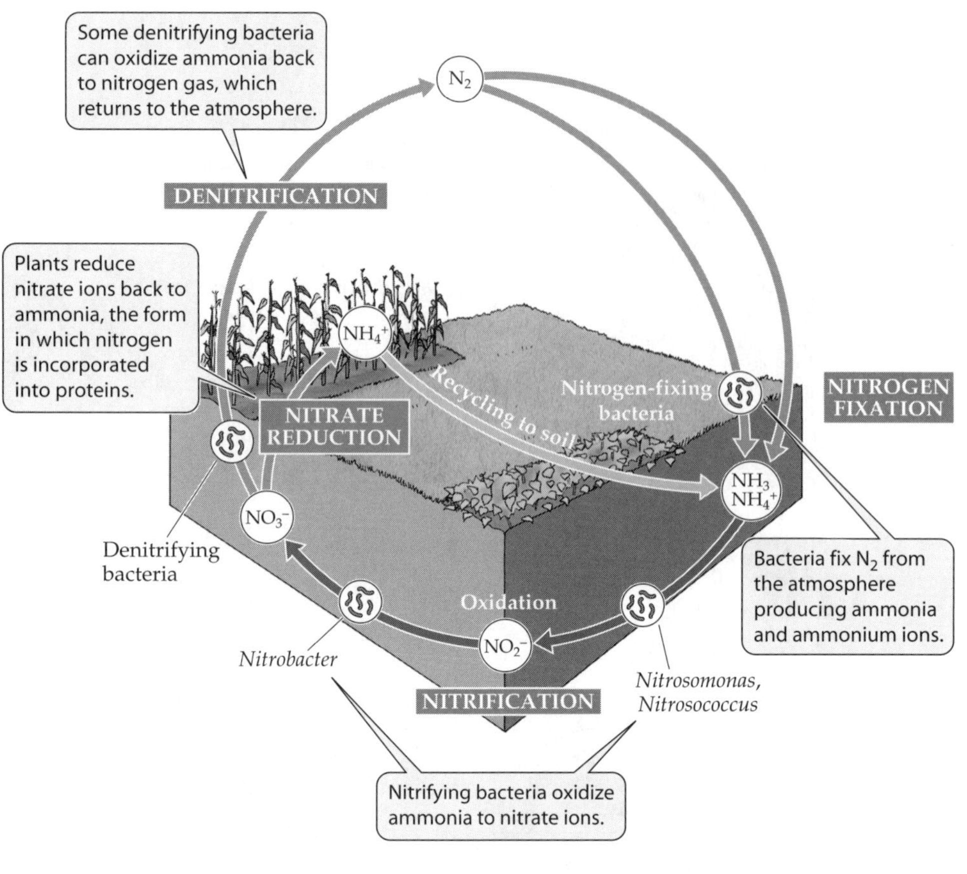
Some denitrifying bacteria can oxidize ammonia back to nitrogen gas, which returns to the atmosphere.
N_2
DENITRIFICATION
Plants reduce nitrate ions back to ammonia, the form in which nitrogen is incorporated into proteins.
NH_4^+
NITRATE REDUCTION
Recycling to soil
Nitrogen-fixing bacteria
NITROGEN FIXATION
NH_3 NH_4^+
NO_3^-
Denitrifying bacteria
Bacteria fix N_2 from the atmosphere producing ammonia and ammonium ions.
Oxidation
NO_2^-
Nitrobacter
NITRIFICATION
Nitrosomonas, Nitrosococcus
Nitrifying bacteria oxidize ammonia to nitrate ions.

37.8 The Nitrogen Cycle *(Page 725)*

38 Regulation of Plant Growth

38.1 Plant Growth Hormones

HORMONE	TYPICAL ACTIVITIES
Abscisic acid	Maintains seed dormancy and winter dormancy; closes stomata
Auxins	Promote stem elongation, adventitious root initiation, and fruit growth; inhibit lateral bud outgrowth and leaf abscission
Brassinosteroids	Promote elongation of stems and pollen tubes; promote vascular tissue differentiation
Cytokinins	Inhibit leaf senescence; promote cell division and lateral bud outgrowth; affect root growth
Ethylene	Promotes fruit ripening and leaf abscission; inhibits stem elongation and gravitropism
Gibberellins	Promote seed germination, stem growth, and fruit development; break winter dormancy; mobilize nutrient reserves in grass seeds

(Page 730)

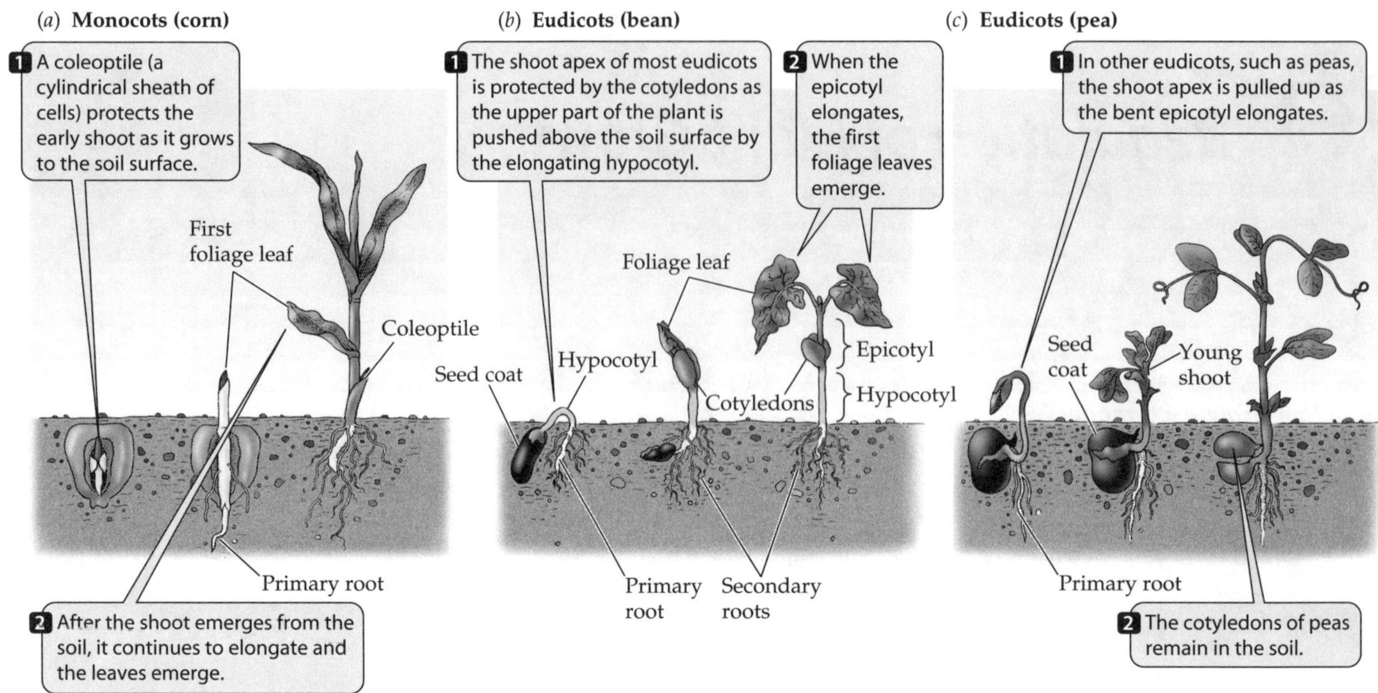

(a) Monocots (corn)
1 A coleoptile (a cylindrical sheath of cells) protects the early shoot as it grows to the soil surface.
First foliage leaf
Coleoptile
Primary root
2 After the shoot emerges from the soil, it continues to elongate and the leaves emerge.
(b) Eudicots (bean)
1 The shoot apex of most eudicots is protected by the cotyledons as the upper part of the plant is pushed above the soil surface by the elongating hypocotyl.
2 When the epicotyl elongates, the first foliage leaves emerge.
Foliage leaf
Hypocotyl
Seed coat
Epicotyl
Hypocotyl
Cotyledons
Primary root
Secondary roots
(c) Eudicots (pea)
1 In other eudicots, such as peas, the shoot apex is pulled up as the bent epicotyl elongates.
Seed coat
Young shoot
Primary root
2 The cotyledons of peas remain in the soil.

38.1 Patterns of Early Shoot Development *(Page 731)*

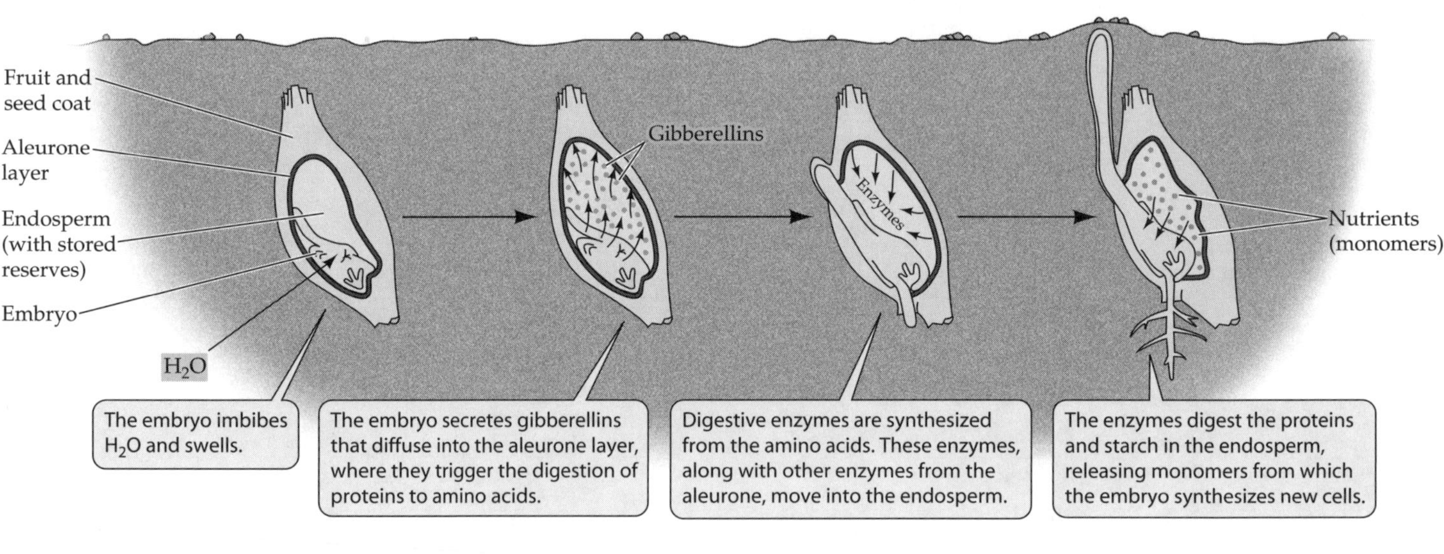

Fruit and seed coat
Aleurone layer
Endosperm (with stored reserves)
Embryo
H_2O
Gibberellins
Enzymes
Nutrients (monomers)
The embryo imbibes H_2O and swells.
The embryo secretes gibberellins that diffuse into the aleurone layer, where they trigger the digestion of proteins to amino acids.
Digestive enzymes are synthesized from the amino acids. These enzymes, along with other enzymes from the aleurone, move into the endosperm.
The enzymes digest the proteins and starch in the endosperm, releasing monomers from which the embryo synthesizes new cells.

38.4 Embryos Mobilize Their Reserves *(Page 734)*

Gibberellin A_1
(important in stem growth)

Gibberellin A_3
(commercially available)

In-Text Art *(Page 734)*

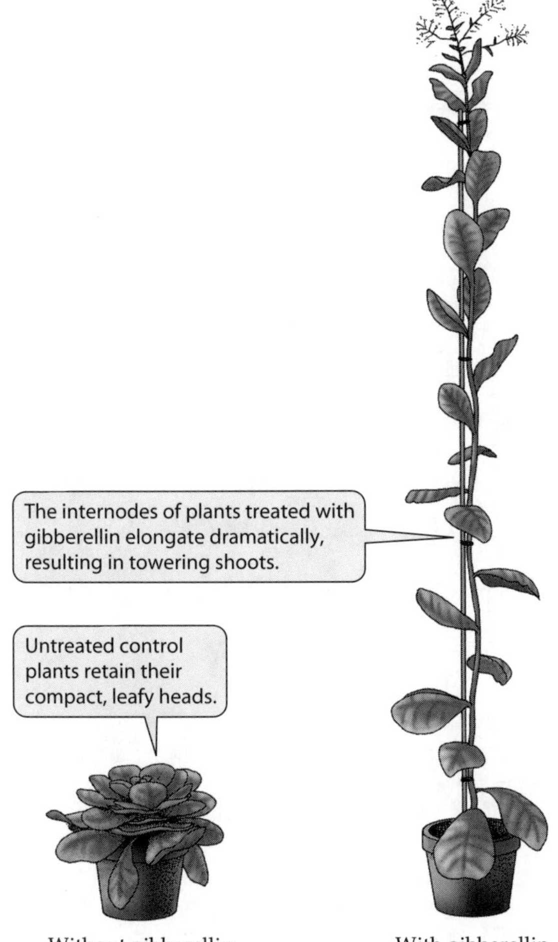
The internodes of plants treated with gibberellin elongate dramatically, resulting in towering shoots.
Untreated control plants retain their compact, leafy heads.
Without gibberellin
With gibberellin

38.6 Bolting *(Page 735)*

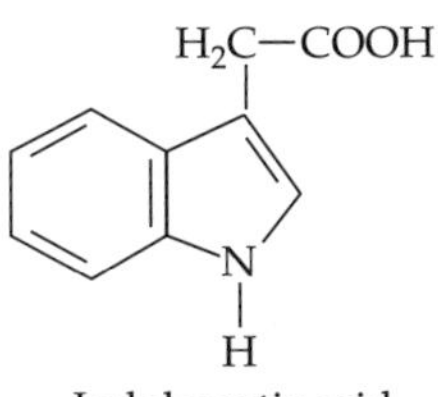
$H_2C-COOH$
N
H

Indoleacetic acid

In-Text Art *(Page 736)*

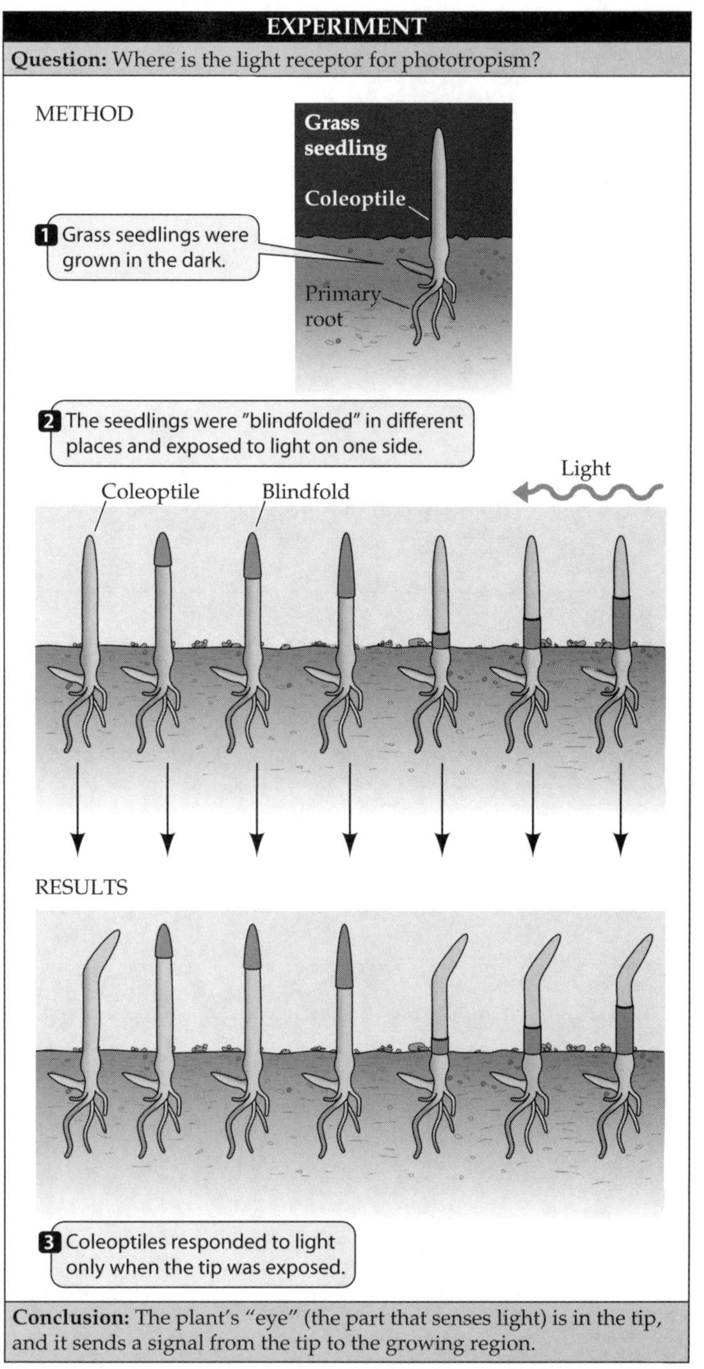
EXPERIMENT
Question: Where is the light receptor for phototropism?
METHOD
Grass seedling
Coleoptile
1 Grass seedlings were grown in the dark.
Primary root
2 The seedlings were "blindfolded" in different places and exposed to light on one side.
Light
Coleoptile
Blindfold
RESULTS
3 Coleoptiles responded to light only when the tip was exposed.
Conclusion: The plant's "eye" (the part that senses light) is in the tip, and it sends a signal from the tip to the growing region.

38.7 The Darwins' Phototropism Experiment *(Page 736)*

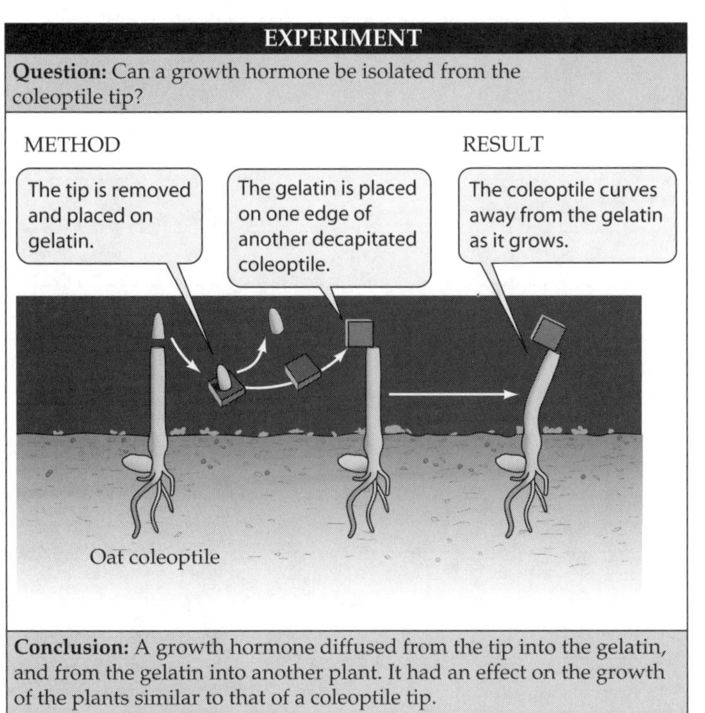
EXPERIMENT
Question: Can a growth hormone be isolated from the coleoptile tip?
METHOD
RESULT
The tip is removed and placed on gelatin.
The gelatin is placed on one edge of another decapitated coleoptile.
The coleoptile curves away from the gelatin as it grows.
Oat coleoptile
Conclusion: A growth hormone diffused from the tip into the gelatin, and from the gelatin into another plant. It had an effect on the growth of the plants similar to that of a coleoptile tip.

38.8 Went's Experiment *(Page 737)*

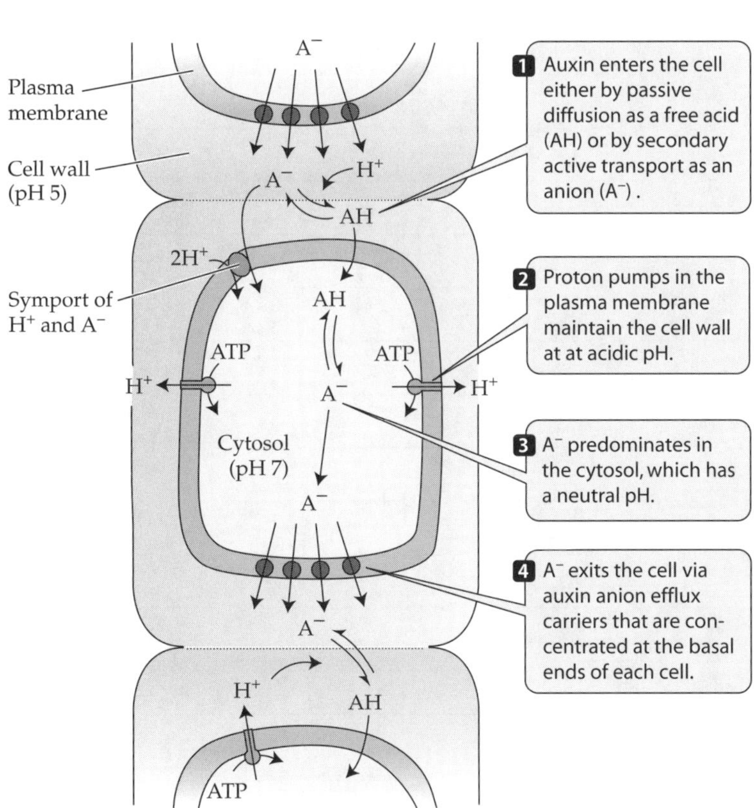
Plasma membrane
Cell wall (pH 5)
Symport of H+ and A−
A−
H+
AH
2H+
ATP
Cytosol (pH 7)
1 Auxin enters the cell either by passive diffusion as a free acid (AH) or by secondary active transport as an anion (A−).
2 Proton pumps in the plasma membrane maintain the cell wall at at acidic pH.
3 A− predominates in the cytosol, which has a neutral pH.
4 A− exits the cell via auxin anion efflux carriers that are concentrated at the basal ends of each cell.

38.9 Polar Transport of Auxin *(Page 737)*

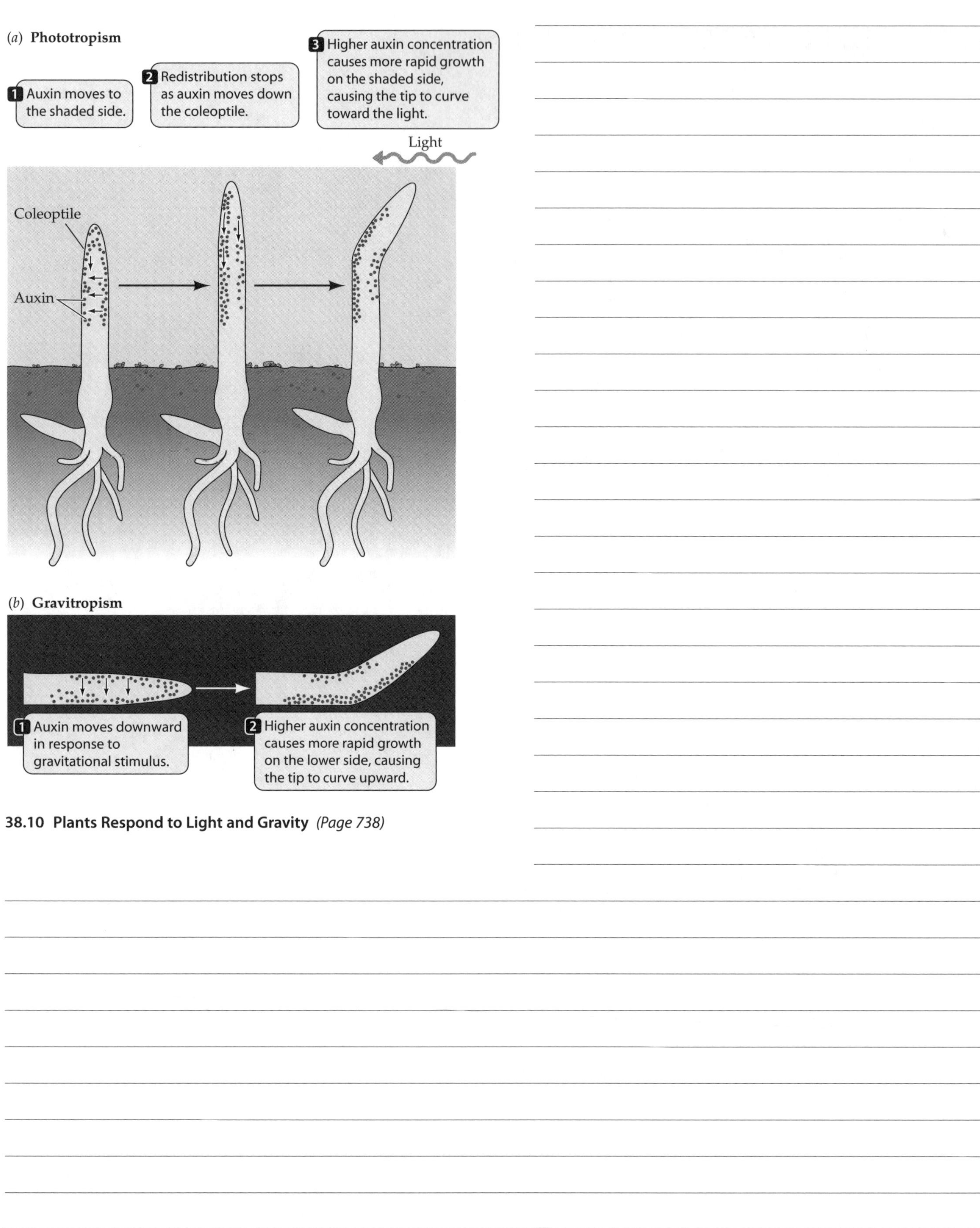

(a) Phototropism
1 Auxin moves to the shaded side.
2 Redistribution stops as auxin moves down the coleoptile.
3 Higher auxin concentration causes more rapid growth on the shaded side, causing the tip to curve toward the light.
Light
Coleoptile
Auxin
(b) Gravitropism
1 Auxin moves downward in response to gravitational stimulus.
2 Higher auxin concentration causes more rapid growth on the lower side, causing the tip to curve upward.

38.10 Plants Respond to Light and Gravity *(Page 738)*

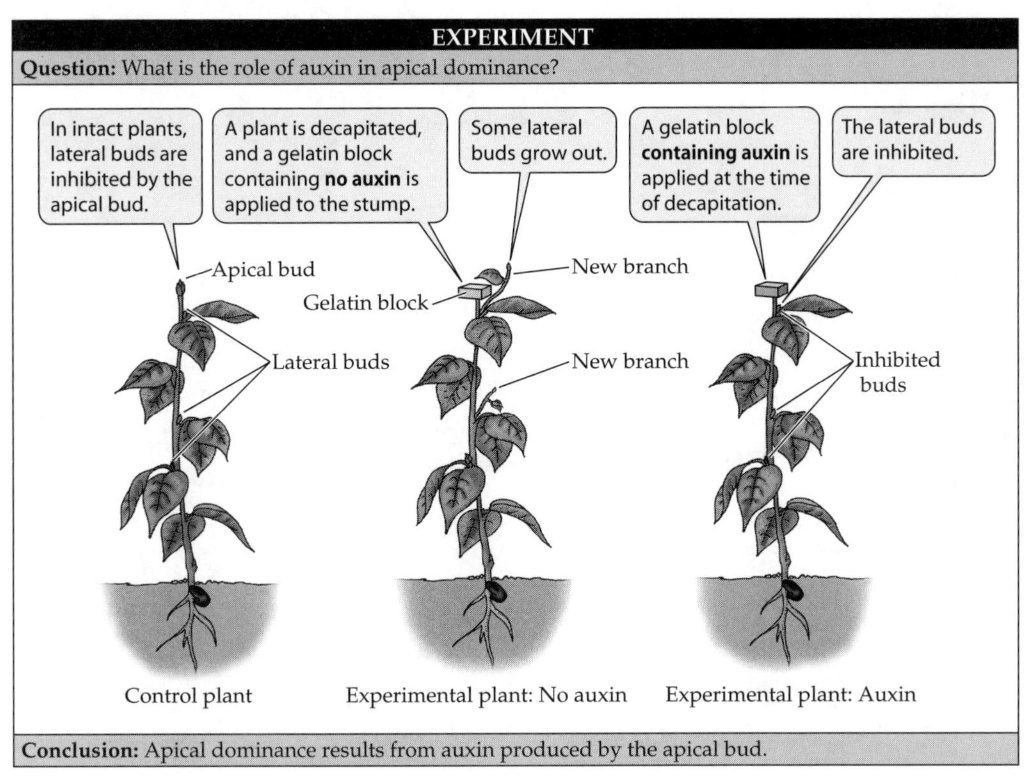
EXPERIMENT
Question: What is the role of auxin in apical dominance?
In intact plants, lateral buds are inhibited by the apical bud.
A plant is decapitated, and a gelatin block containing no auxin is applied to the stump.
Some lateral buds grow out.
A gelatin block containing auxin is applied at the time of decapitation.
The lateral buds are inhibited.
Apical bud
Gelatin block
New branch
Lateral buds
New branch
Inhibited buds
Control plant
Experimental plant: No auxin
Experimental plant: Auxin
Conclusion: Apical dominance results from auxin produced by the apical bud.

38.12 Auxin and Apical Dominance *(Page 739)*

Cellulose microfibrils encircle the cell in a specific orientation and constrain cell expansion.
The cell expands only in the direction of least resistance.

38.14 Plant Cells Expand *(Page 740)*

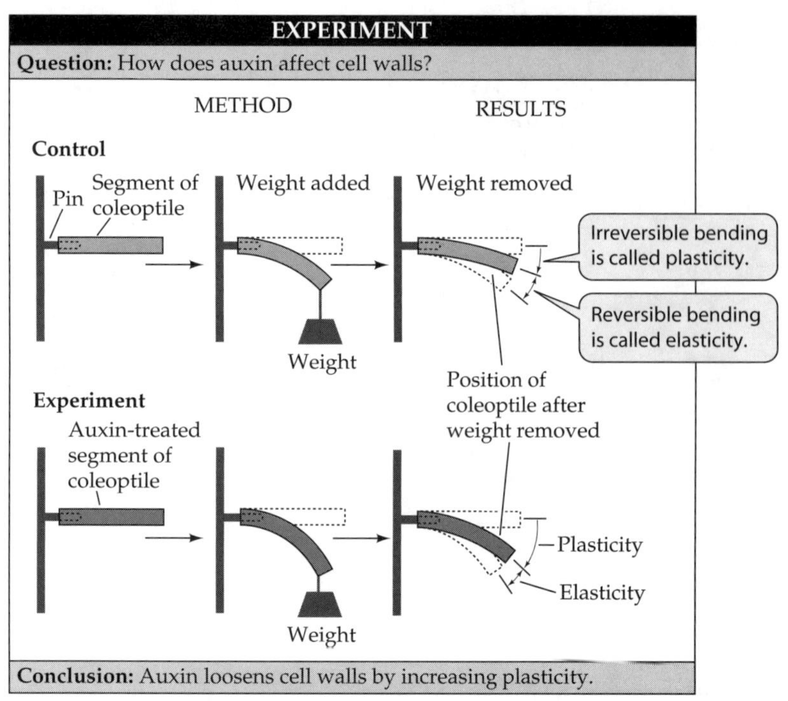
EXPERIMENT
Question: How does auxin affect cell walls?
METHOD
RESULTS
Control
Pin
Segment of coleoptile
Weight added
Weight removed
Irreversible bending is called plasticity.
Reversible bending is called elasticity.
Weight
Position of coleoptile after weight removed
Experiment
Auxin-treated segment of coleoptile
Plasticity
Elasticity
Weight
Conclusion: Auxin loosens cell walls by increasing plasticity.

38.15 Auxin Acts on Cell Walls *(Page 741)*

HN—CH$_2$— O N N N N H

Kinetin

HN—CH$_2$—CH=C CH$_3$ CH$_2$OH N N N N H

Zeatin

In-Text Art *(Page 742)*

Ethylene
(the "senescence hormone")

H H C=C H H

In-Text Art *(Page 742)*

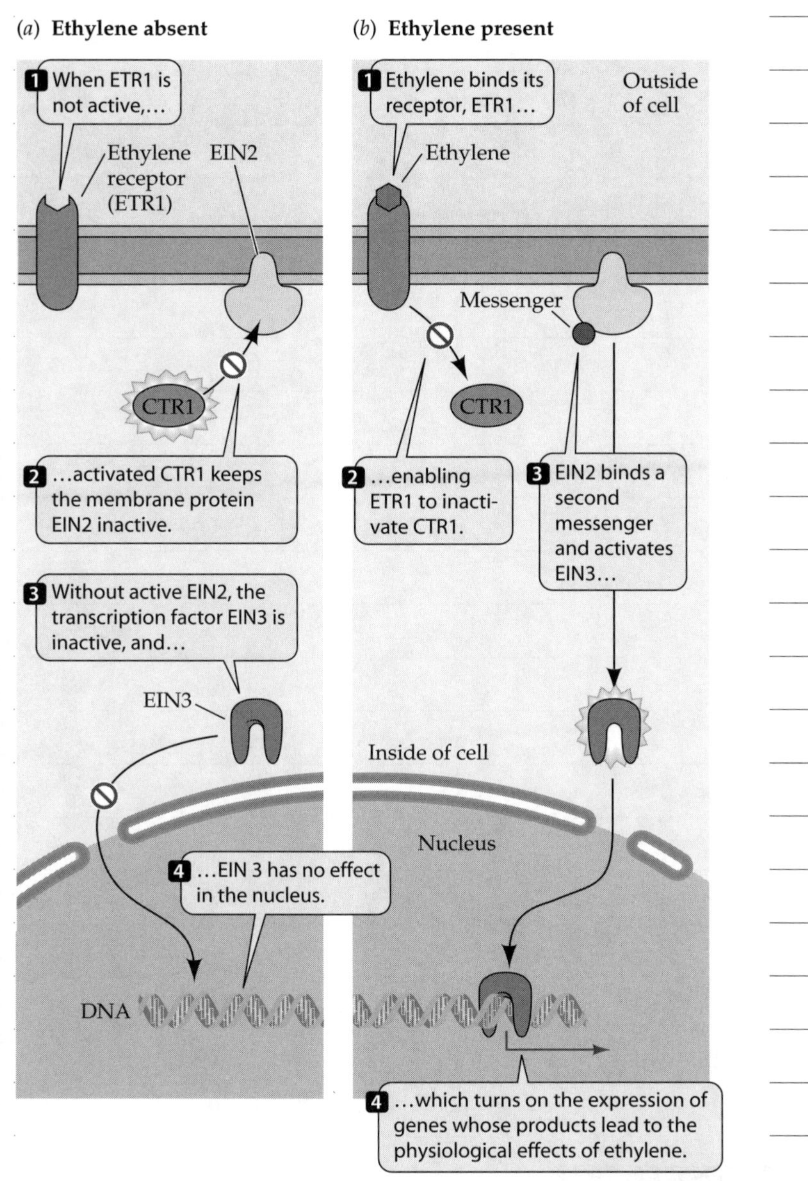
(a) Ethylene absent
1 When ETR1 is not active,...
Ethylene receptor (ETR1)
EIN2
CTR1
2 ...activated CTR1 keeps the membrane protein EIN2 inactive.
3 Without active EIN2, the transcription factor EIN3 is inactive, and...
EIN3
4 ...EIN 3 has no effect in the nucleus.
DNA
(b) Ethylene present
1 Ethylene binds its receptor, ETR1...
Outside of cell
Ethylene
Messenger
CTR1
2 ...enabling ETR1 to inactivate CTR1.
3 EIN2 binds a second messenger and activates EIN3...
Inside of cell
Nucleus
4 ...which turns on the expression of genes whose products lead to the physiological effects of ethylene.

38.17 The Signal Transduction Pathway for Ethylene *(Page 743)*

Abscisic acid
(the "stress hormone")

H_3C CH_3 CH_3 OH CH_3 COOH O

In-Text Art *(Page 744)*

Brassinolide
(a brassinosteroid)

OH CH3 H3C CH3 CH3 OH CH3 CH3 HO HO O O

In-Text Art *(Page 744)*

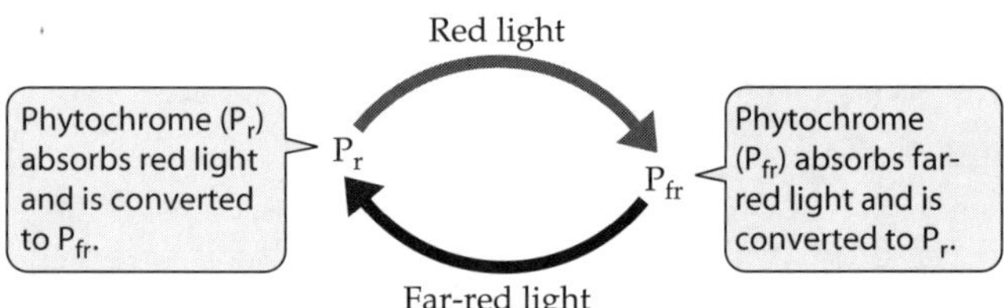
Red light
Phytochrome (P_r) absorbs red light and is converted to P_{fr}.
P_r
P_{fr}
Phytochrome (P_{fr}) absorbs far-red light and is converted to P_r.
Far-red light

In-Text Art *(Page 745)*

EXPERIMENT

Question: How do red and far-red light affect lettuce seed germination?

METHOD

1 Lettuce seeds were exposed to alternate periods of red light R for 1 minute and far-red light FR for 4 minutes.

2 Seeds germinate if the final exposure is to red R …

3 …and remain dormant if the final exposure is to far-red FR.

R

R FR

R FR R FR R FR R

R FR R FR R FR R FR

RESULTS

Most germinate — Few germinate — … — Most germinate — Few germinate

Conclusion: Red light and far-red light reverse each other's effects.

38.18 Sensitivity of Seeds to Red and Far-Red Light *(Page 745)*

39 Reproduction in Flowering Plants

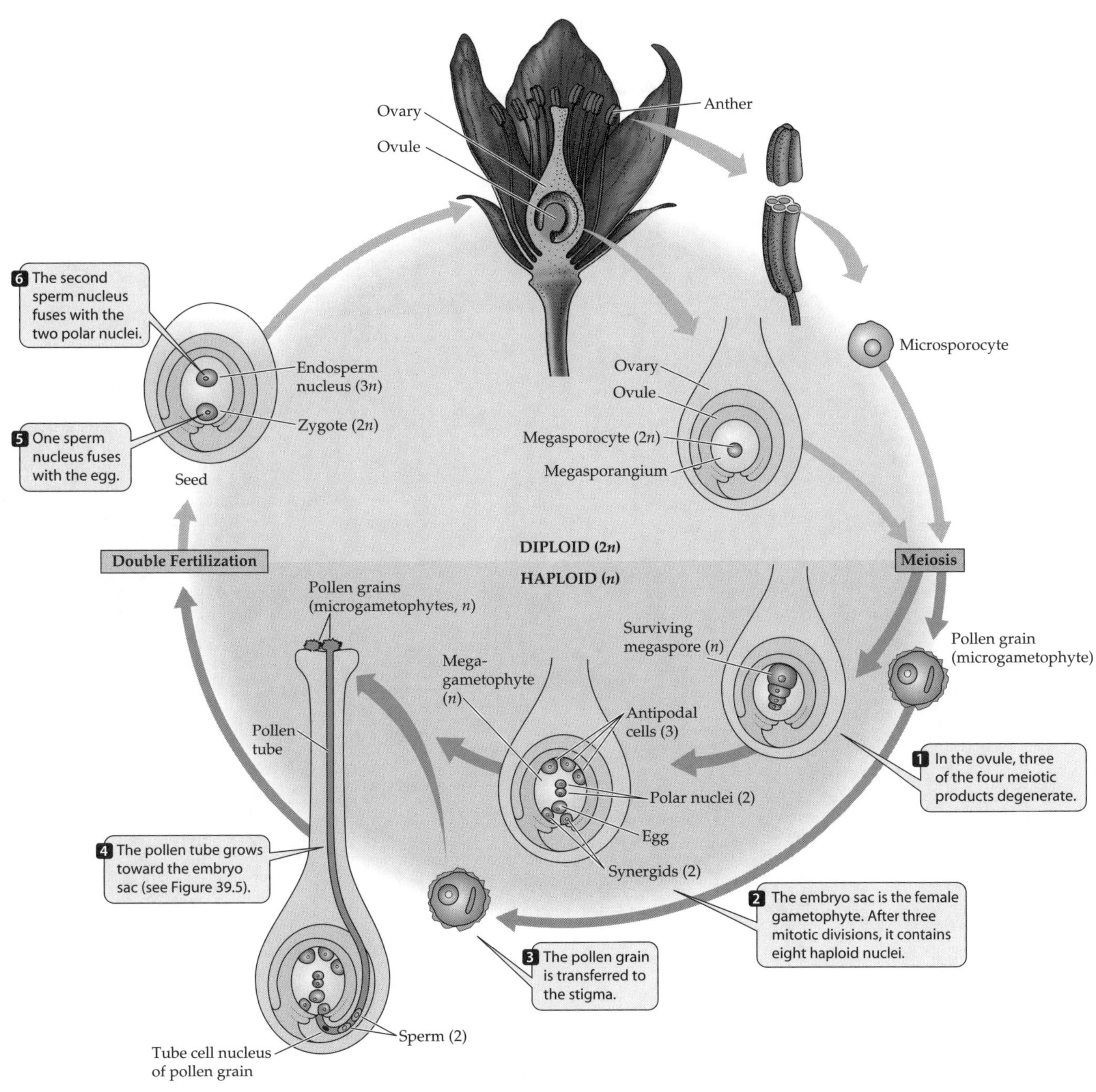

Ovary
Ovule
Anther
The second sperm nucleus fuses with the two polar nuclei.
Endosperm nucleus (3n)
Zygote (2n)
One sperm nucleus fuses with the egg.
Seed
Ovary
Ovule
Megasporocyte (2n)
Megasporangium
Microsporocyte
DIPLOID (2n)
HAPLOID (n)
Double Fertilization
Meiosis
Pollen grains (microgametophytes, n)
Surviving megaspore (n)
Pollen grain (microgametophyte)
Mega-gametophyte (n)
Antipodal cells (3)
Pollen tube
In the ovule, three of the four meiotic products degenerate.
Polar nuclei (2)
Egg
Synergids (2)
The pollen tube grows toward the embryo sac (see Figure 39.5).
The embryo sac is the female gametophyte. After three mitotic divisions, it contains eight haploid nuclei.
The pollen grain is transferred to the stigma.
Sperm (2)
Tube cell nucleus of pollen grain

39.1 Development of Gametophytes and Nuclear Fusion *(Page 751)*

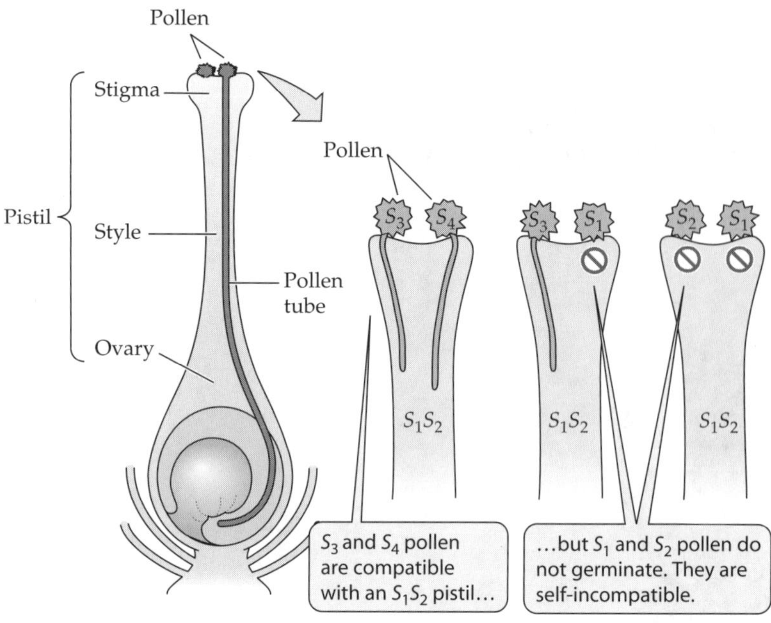
Pollen
Stigma
Pistil
Style
Pollen tube
Ovary
Pollen
S_3 S_4
S_3 S_1
S_2 S_1
S_1S_2
S_1S_2
S_1S_2
S_3 and S_4 pollen are compatible with an S_1S_2 pistil…
…but S_1 and S_2 pollen do not germinate. They are self-incompatible.

39.3 Self-Incompatibility *(Page 752)*

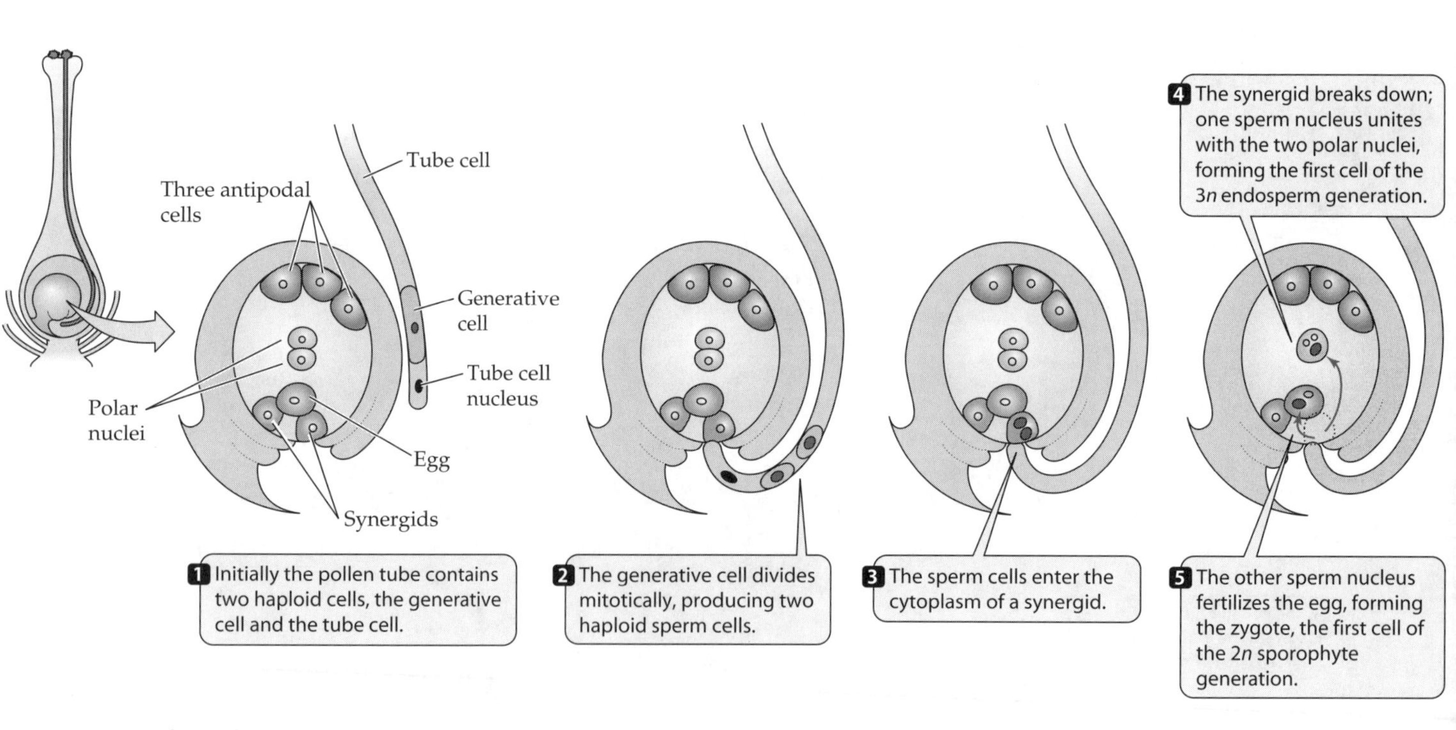
Tube cell
Three antipodal cells
Generative cell
Tube cell nucleus
Polar nuclei
Egg
Synergids
1 Initially the pollen tube contains two haploid cells, the generative cell and the tube cell.
2 The generative cell divides mitotically, producing two haploid sperm cells.
3 The sperm cells enter the cytoplasm of a synergid.
4 The synergid breaks down; one sperm nucleus unites with the two polar nuclei, forming the first cell of the 3n endosperm generation.
5 The other sperm nucleus fertilizes the egg, forming the zygote, the first cell of the 2n sporophyte generation.

39.5 Sperm Nuclei and Double Fertilization *(Page 753)*

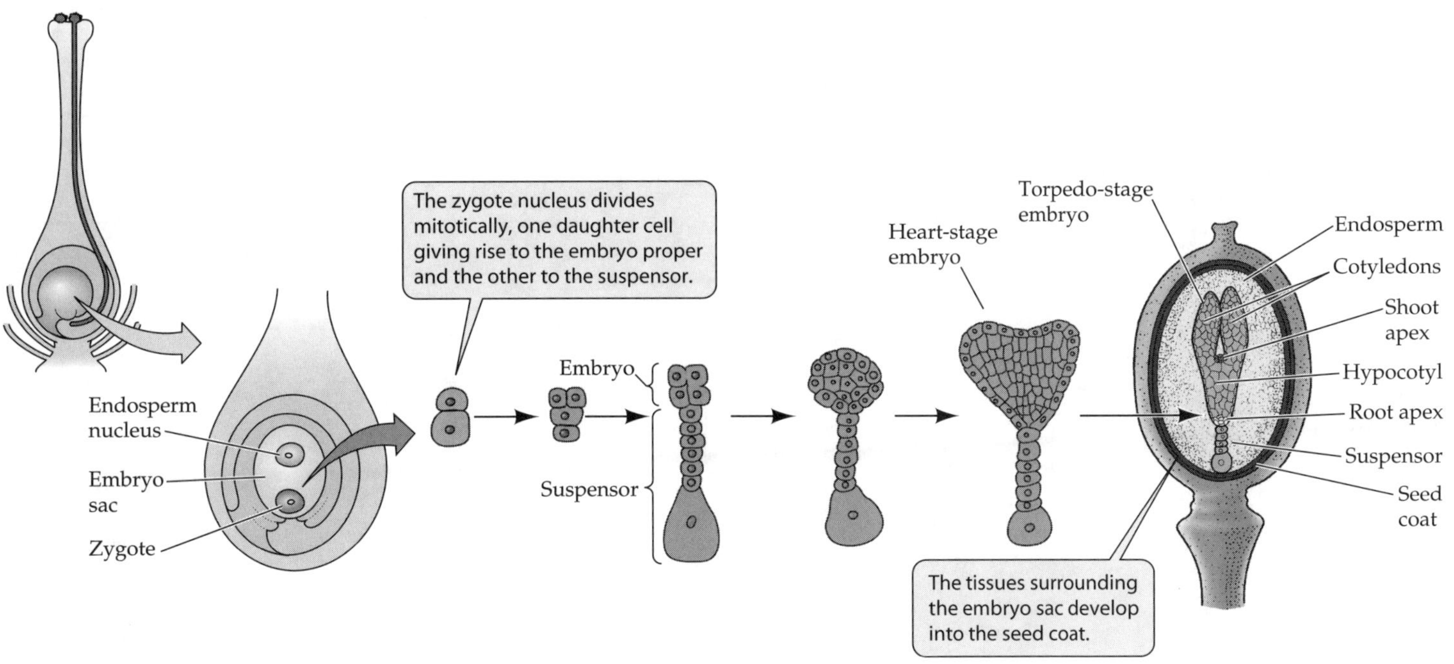
The zygote nucleus divides mitotically, one daughter cell giving rise to the embryo proper and the other to the suspensor.
Torpedo-stage embryo
Heart-stage embryo
Endosperm
Cotyledons
Shoot apex
Hypocotyl
Root apex
Suspensor
Seed coat
Embryo
Suspensor
Endosperm nucleus
Embryo sac
Zygote
The tissues surrounding the embryo sac develop into the seed coat.

39.6 Early Development of a Eudicot *(Page 754)*

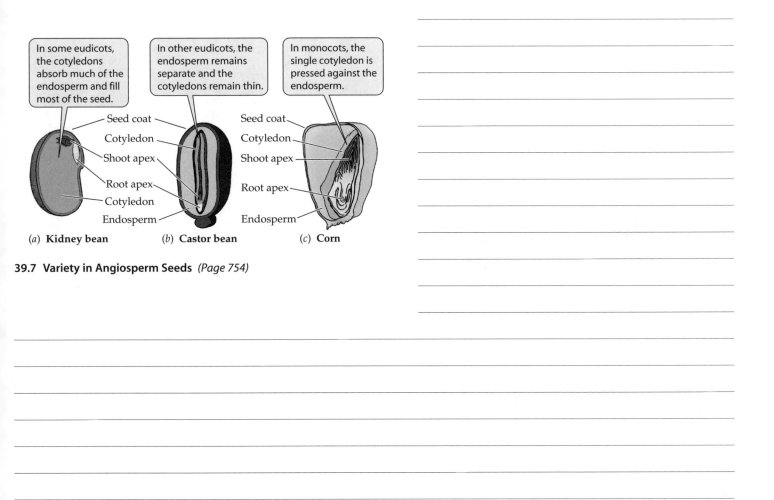

In some eudicots, the cotyledons absorb much of the endosperm and fill most of the seed.

In other eudicots, the endosperm remains separate and the cotyledons remain thin.

In monocots, the single cotyledon is pressed against the endosperm.

Seed coat
Cotyledon
Shoot apex
Root apex
Cotyledon
Endosperm

Seed coat
Cotyledon
Shoot apex
Root apex
Endosperm

(*a*) **Kidney bean** (*b*) **Castor bean** (*c*) **Corn**

39.7 Variety in Angiosperm Seeds *(Page 754)*

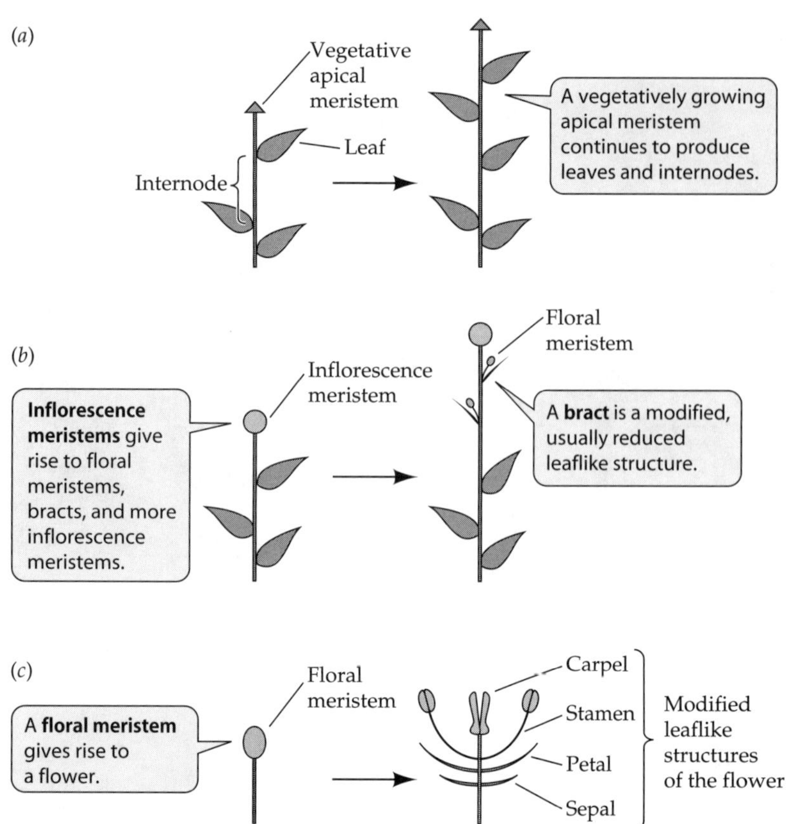

(a)
Vegetative apical meristem
Leaf
Internode
A vegetatively growing apical meristem continues to produce leaves and internodes.
(b)
Inflorescence meristem
Floral meristem
Inflorescence meristems give rise to floral meristems, bracts, and more inflorescence meristems.
A bract is a modified, usually reduced leaflike structure.
(c)
Floral meristem
A floral meristem gives rise to a flower.
Carpel
Stamen
Petal
Sepal
Modified leaflike structures of the flower

39.9 Flowering and the Apical Meristem *(Page 756)*

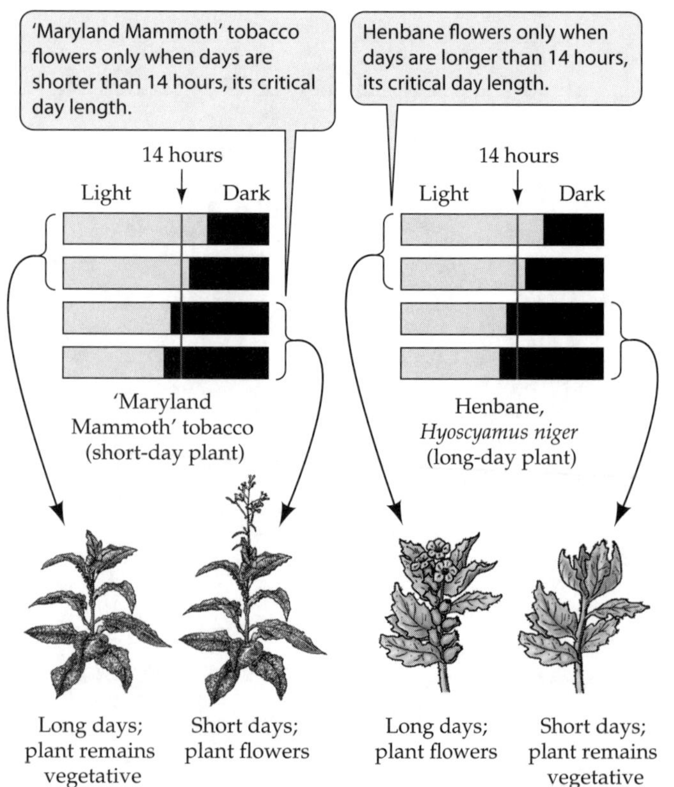

'Maryland Mammoth' tobacco flowers only when days are shorter than 14 hours, its critical day length.
Henbane flowers only when days are longer than 14 hours, its critical day length.
14 hours
Light
Dark
14 hours
Light
Dark
'Maryland Mammoth' tobacco (short-day plant)
Henbane, *Hyoscyamus niger* (long-day plant)
Long days; plant remains vegetative
Short days; plant flowers
Long days; plant flowers
Short days; plant remains vegetative

39.10 Day Length and Flowering *(Page 757)*

EXPERIMENT

Question: Do short-day plants measure day length or night length?

METHOD

Plants were moved between light and dark rooms for specified numbers of hours.

RESULTS

Light constant/Darkness varied

Light	Dark	Result
16	6	No flowering
16	7	No flowering
16	8	No flowering
16	9	Only plants given 9 or more hours of dark flowered.
16	10	Only plants given 9 or more hours of dark flowered.
16	11	Only plants given 9 or more hours of dark flowered.

Light varied/8 or 10 hours of darkness

Light	Dark	Result
8	10	Only plants given 10 hours of dark flowered.
10	10	Only plants given 10 hours of dark flowered.
12	10	Only plants given 10 hours of dark flowered.
8	8	No flowering
10	8	No flowering
12	8	No flowering

Time (hours)

Conclusion: Short-day plants measure the length of the night and could more accurately be called long-night plants.

39.11 Night Length and Flowering *(Page 757)*

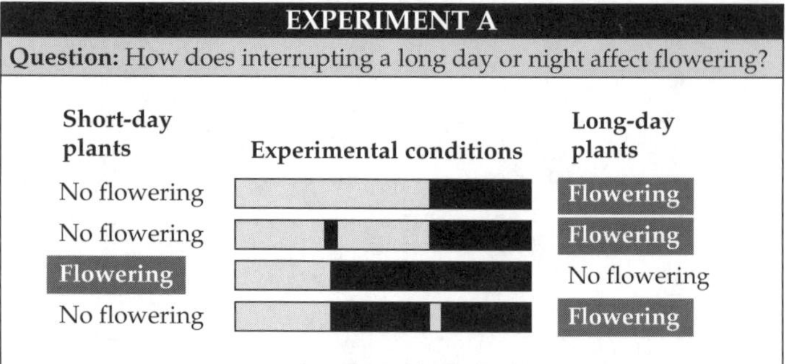
EXPERIMENT A
Question: How does interrupting a long day or night affect flowering?
Short-day plants
Experimental conditions
Long-day plants
No flowering
Flowering
No flowering
Flowering
Flowering
No flowering
No flowering
Flowering
Conclusion: Photoperiodic plants measure the length of the night, not the day. Interrupting a long night with a brief period of light inhibits flowering in short-day plants. Long-day plants flower when the night is short, but interrupting their long day has no effect.

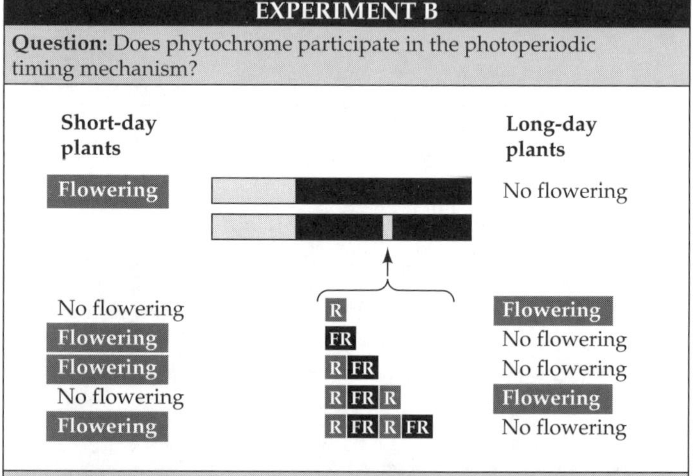
EXPERIMENT B
Question: Does phytochrome participate in the photoperiodic timing mechanism?
Short-day plants
Long-day plants
Flowering
No flowering
No flowering
R
Flowering
Flowering
FR
No flowering
Flowering
R FR
No flowering
No flowering
R FR R
Flowering
Flowering
R FR R FR
No flowering
Conclusion: When plants are exposed to red (R) and far-red (FR) light in alternation, the final treatment determines the effect of the light interruption, suggesting that phytochrome participates in photoperiodic responses.

39.12 The Effect of Interrupted Days and Nights *(Page 758)*

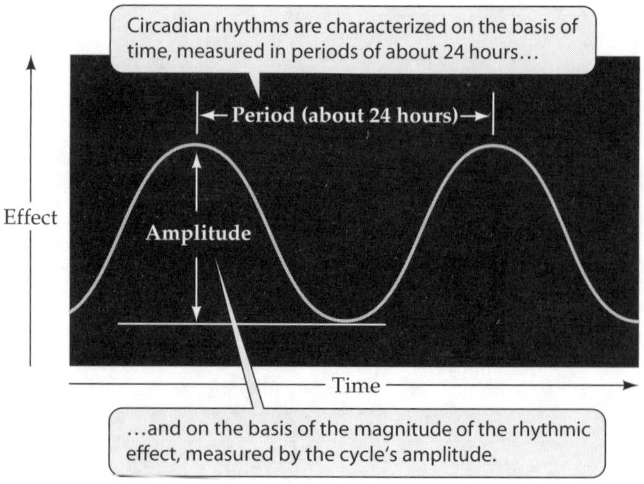
Circadian rhythms are characterized on the basis of time, measured in periods of about 24 hours…
Period (about 24 hours)
Effect
Amplitude
Time
…and on the basis of the magnitude of the rhythmic effect, measured by the cycle's amplitude.

39.13 Features of Circadian Rhythms *(Page 759)*

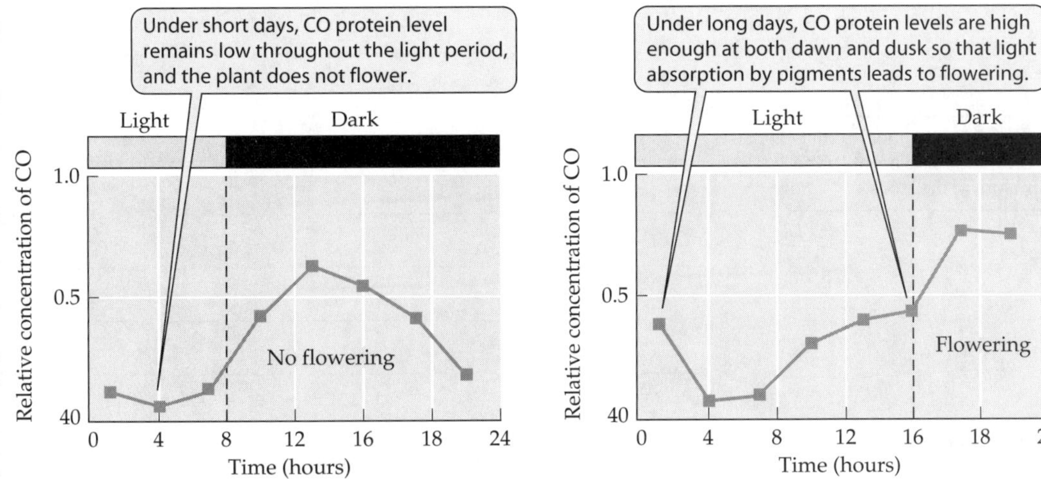
Under short days, CO protein level remains low throughout the light period, and the plant does not flower.
Light
Dark
Relative concentration of CO
1.0
0.5
40
0
4
8
12
16
18
24
Time (hours)
No flowering
Under long days, CO protein levels are high enough at both dawn and dusk so that light absorption by pigments leads to flowering.
Light
Dark
Relative concentration of CO
1.0
0.5
40
0
4
8
12
16
18
24
Time (hours)
Flowering

39.14 Photoreceptors and the Biological Clock Interact in Photoperiodic Plants *(Page 760)*

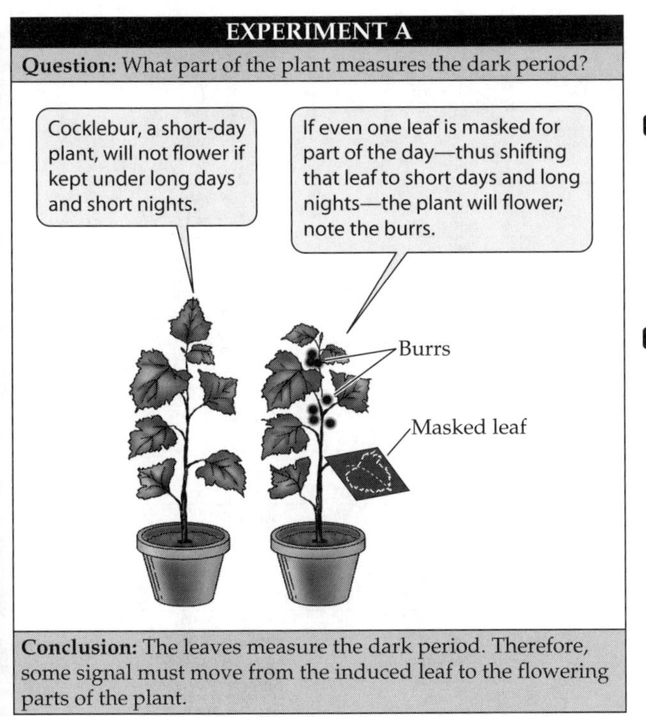
EXPERIMENT A
Question: What part of the plant measures the dark period?
Cocklebur, a short-day plant, will not flower if kept under long days and short nights.
If even one leaf is masked for part of the day—thus shifting that leaf to short days and long nights—the plant will flower; note the burrs.
Burrs
Masked leaf
Conclusion: The leaves measure the dark period. Therefore, some signal must move from the induced leaf to the flowering parts of the plant.

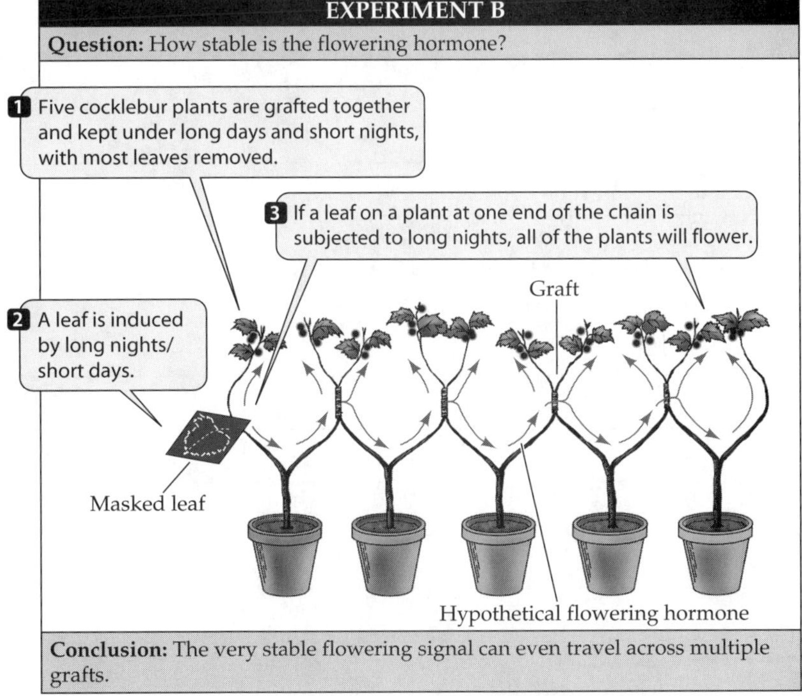
EXPERIMENT B
Question: How stable is the flowering hormone?
1 Five cocklebur plants are grafted together and kept under long days and short nights, with most leaves removed.
2 A leaf is induced by long nights/ short days.
3 If a leaf on a plant at one end of the chain is subjected to long nights, all of the plants will flower.
Graft
Masked leaf
Hypothetical flowering hormone
Conclusion: The very stable flowering signal can even travel across multiple grafts.

39.15 Evidence for a Flowering Hormone *(Page 760)*

Scion

Stock

In grafting, the scion is aligned so that its vascular cambium is adjacent to the vascular cambium in the stock.

39.17 Grafting *(Page 763)*

40 Plant Responses to Environmental Challenges

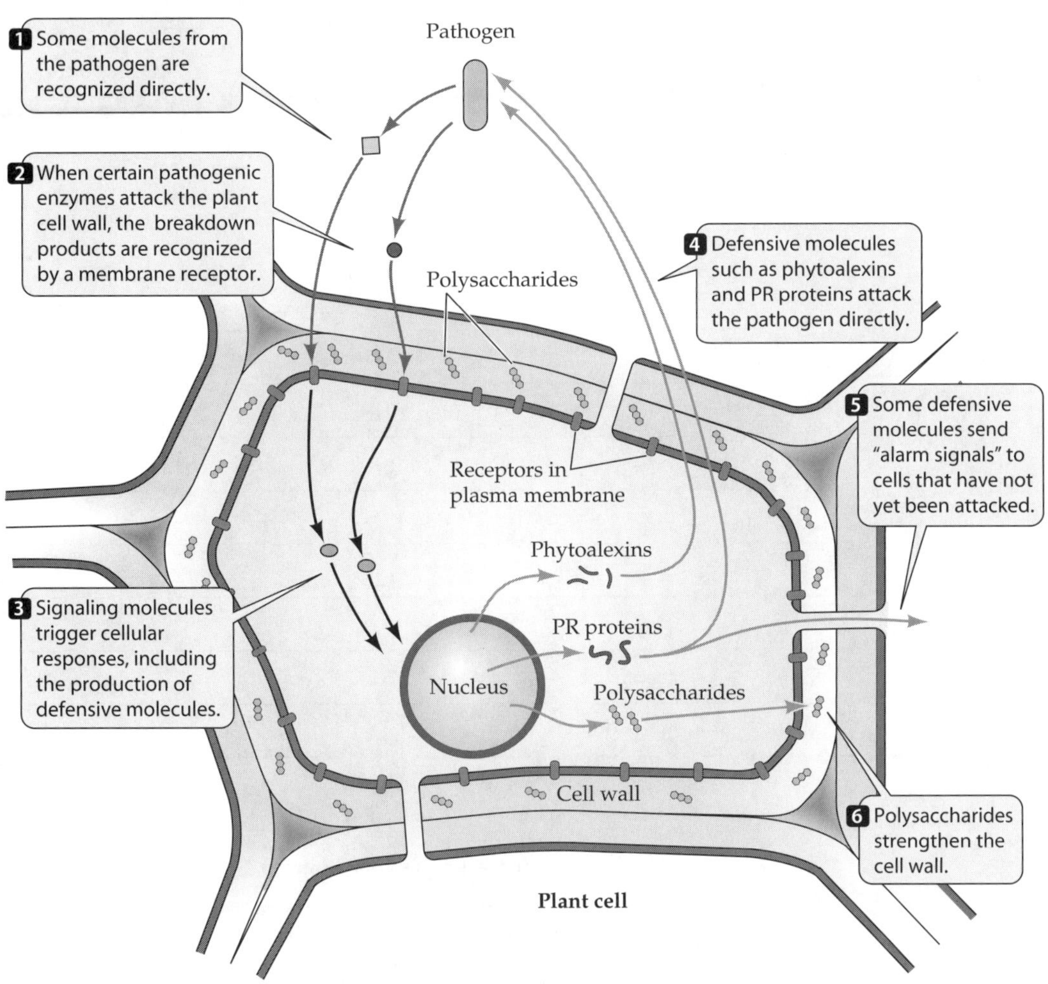
Pathogen
1 Some molecules from the pathogen are recognized directly.
2 When certain pathogenic enzymes attack the plant cell wall, the breakdown products are recognized by a membrane receptor.
Polysaccharides
4 Defensive molecules such as phytoalexins and PR proteins attack the pathogen directly.
5 Some defensive molecules send "alarm signals" to cells that have not yet been attacked.
Receptors in plasma membrane
Phytoalexins
3 Signaling molecules trigger cellular responses, including the production of defensive molecules.
PR proteins
Nucleus
Polysaccharides
Cell wall
6 Polysaccharides strengthen the cell wall.
Plant cell

40.1 Signaling between Plants and Pathogens *(Page 766)*

40.1 Secondary Plant Metabolites Used in Defense

CLASS	TYPE	ROLE	EXAMPLE
Nitrogen-containing	Alkaloids	Affect herbivore nervous system	Nicotine in tobacco
	Glycosides	Release cyanide or sulfur compounds	Dhurrin in sorghum
	Nonprotein amino acids	Disrupt herbivore protein structure	Canavanine in jack bean
Phenolics	Flavonoids	Phytoalexins	Capsidol in peppers
	Quinones	Inhibit competing plants	Juglone in walnut
	Tannins	Herbivore and microbe deterrents	Many woods, such as oak
Terpenes	Monoterpenes	Insecticides	Pyrethroids in chrysanthemums
	Sesquiterpenes	Antiherbivores	Gossypol in cotton
	Steroids	Mimic insect hormones and disrupt insect life cycles	α-Ecdysone in ferns
	Polyterpenes	Feeding deterrent?	Rubber in rubber tree

(Page 767)

Salicylic acid — COOH, OH

In-Text Art *(Page 767)*

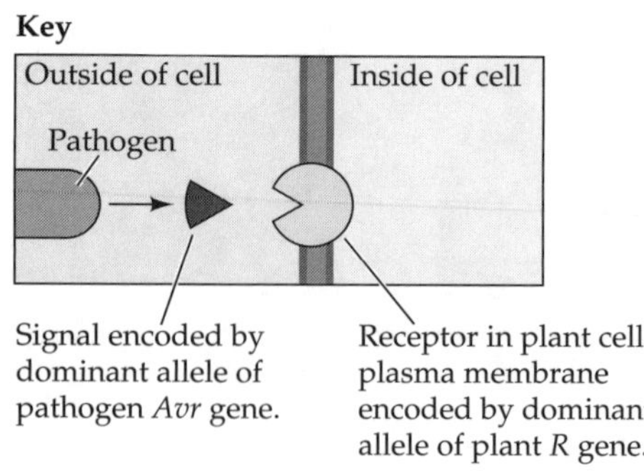
Key
Outside of cell
Inside of cell
Pathogen
Signal encoded by dominant allele of pathogen *Avr* gene.
Receptor in plant cell plasma membrane encoded by dominant allele of plant *R* gene.

Plant resistance gene (*R* gene)

Dominant allele present | Dominant allele absent

Pathogen avirulence gene (*Avr* gene): Dominant allele present — RESISTANT | SUSCEPTIBLE

Dominant allele absent — SUSCEPTIBLE | SUSCEPTIBLE

1 If, in any pair, the host and pathogen both contain a dominant allele, the plant will resist the pathogen…

2 …but if either "partner" lacks the dominant allele, this pair will not cause gene-for-gene resistance.

40.3 Gene-for-Gene Resistance *(Page 768)*

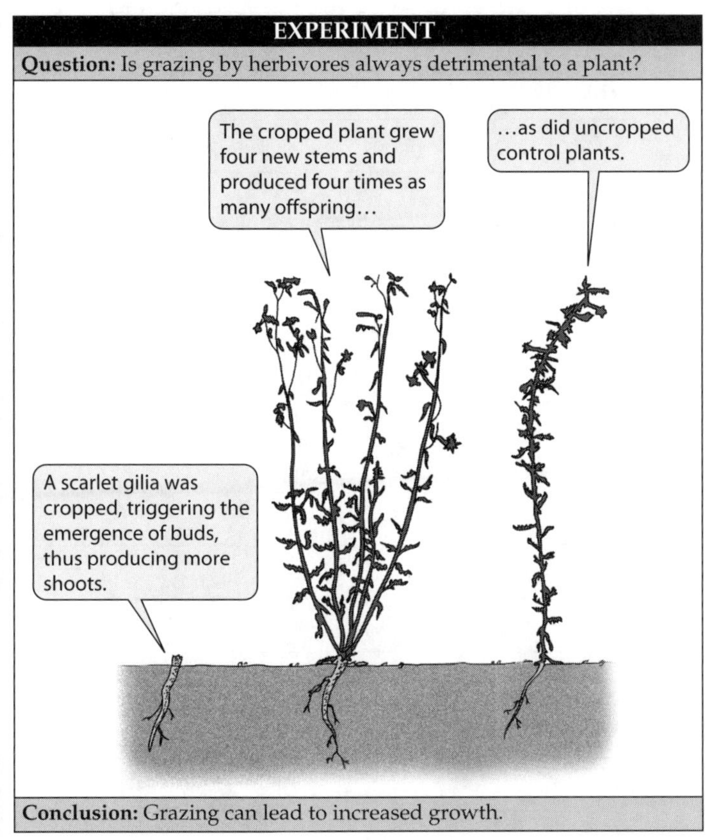
EXPERIMENT
Question: Is grazing by herbivores always detrimental to a plant?
The cropped plant grew four new stems and produced four times as many offspring…
…as did uncropped control plants.
A scarlet gilia was cropped, triggering the emergence of buds, thus producing more shoots.
Conclusion: Grazing can lead to increased growth.

40.4 Overcompensation for Being Eaten *(Page 769)*

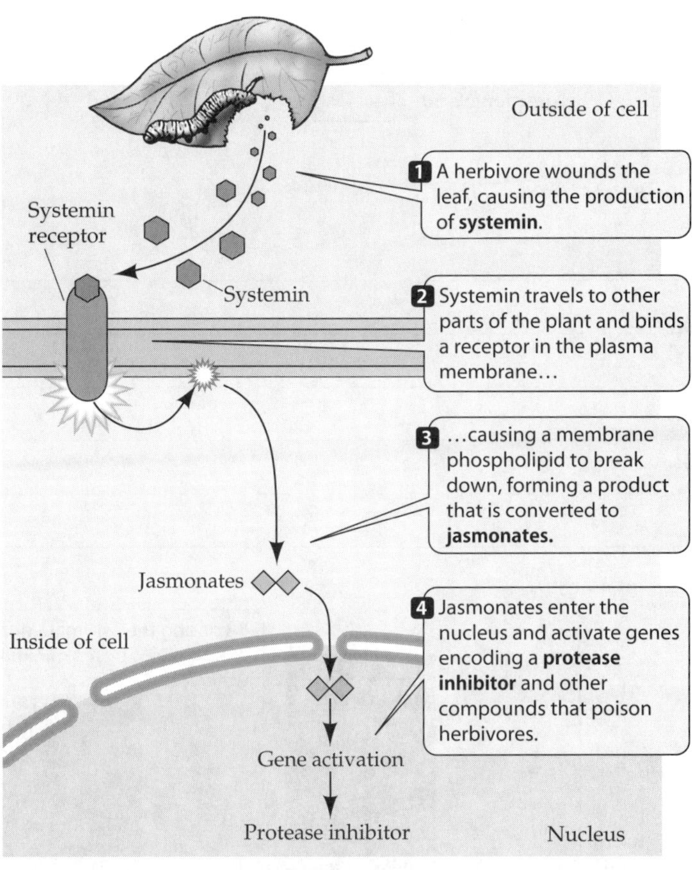
Outside of cell
1 A herbivore wounds the leaf, causing the production of **systemin**.
Systemin receptor
Systemin
2 Systemin travels to other parts of the plant and binds a receptor in the plasma membrane…
3 …causing a membrane phospholipid to break down, forming a product that is converted to **jasmonates.**
Jasmonates
Inside of cell
4 Jasmonates enter the nucleus and activate genes encoding a **protease inhibitor** and other compounds that poison herbivores.
Gene activation
Protease inhibitor
Nucleus

40.5 A Signaling Pathway for Synthesis of a Defensive Secondary Metabolite *(Page 770)*

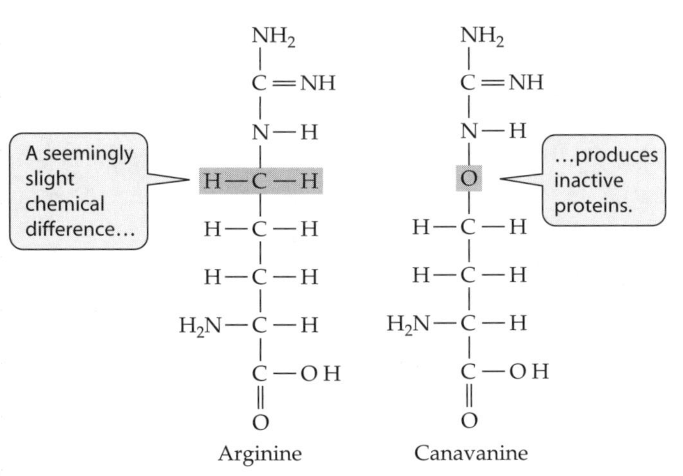
A seemingly slight chemical difference…
…produces inactive proteins.
Arginine
Canavanine

In-Text Art *(Page 770)*

Jasmonic acid

O

COOH

In-Text Art *(Page 771)*

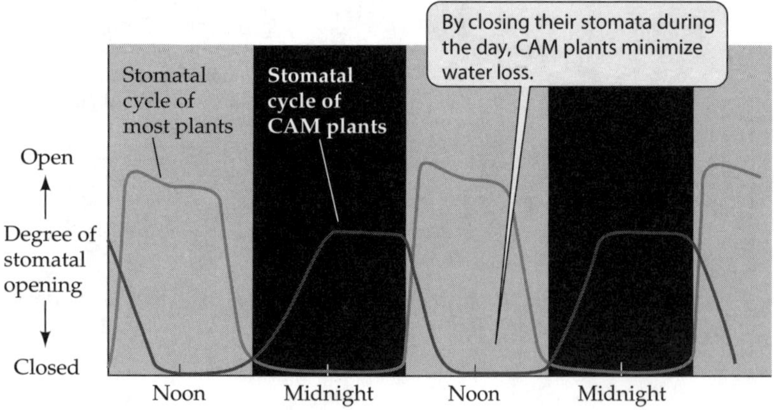
By closing their stomata during the day, CAM plants minimize water loss.
Stomatal cycle of most plants
Stomatal cycle of CAM plants
Open
Degree of stomatal opening
Closed
Noon
Midnight
Noon
Midnight
Light/dark periods

40.14 Stomatal Cycles *(Page 776)*

41 Physiology, Homeostasis, and Temperature Regulation

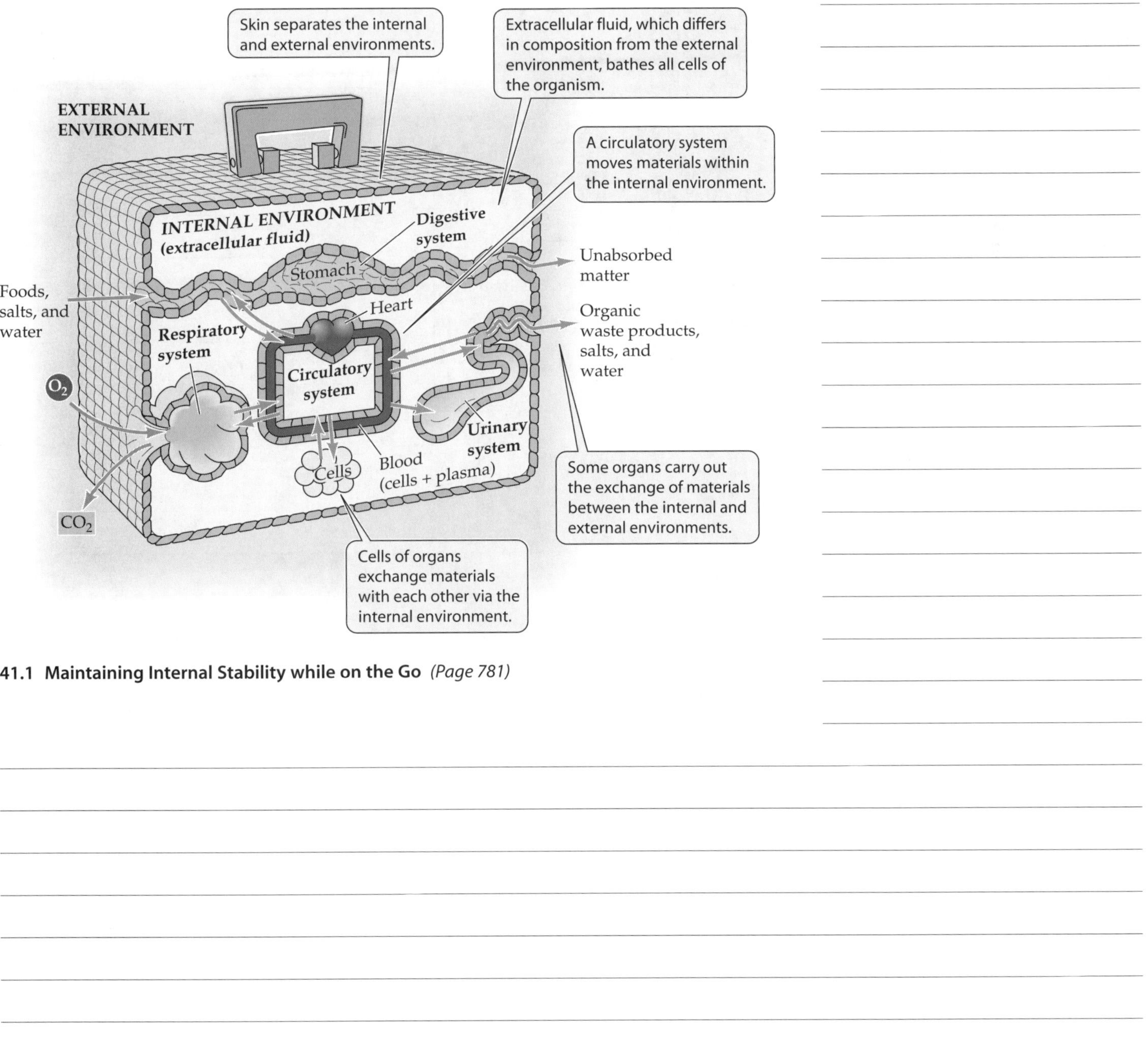
Skin separates the internal and external environments.
Extracellular fluid, which differs in composition from the external environment, bathes all cells of the organism.
EXTERNAL ENVIRONMENT
A circulatory system moves materials within the internal environment.
INTERNAL ENVIRONMENT (extracellular fluid)
Digestive system
Stomach
Unabsorbed matter
Foods, salts, and water
Heart
Organic waste products, salts, and water
Respiratory system
Circulatory system
O_2
Urinary system
Blood (cells + plasma)
Cells
CO_2
Some organs carry out the exchange of materials between the internal and external environments.
Cells of organs exchange materials with each other via the internal environment.

41.1 Maintaining Internal Stability while on the Go *(Page 781)*

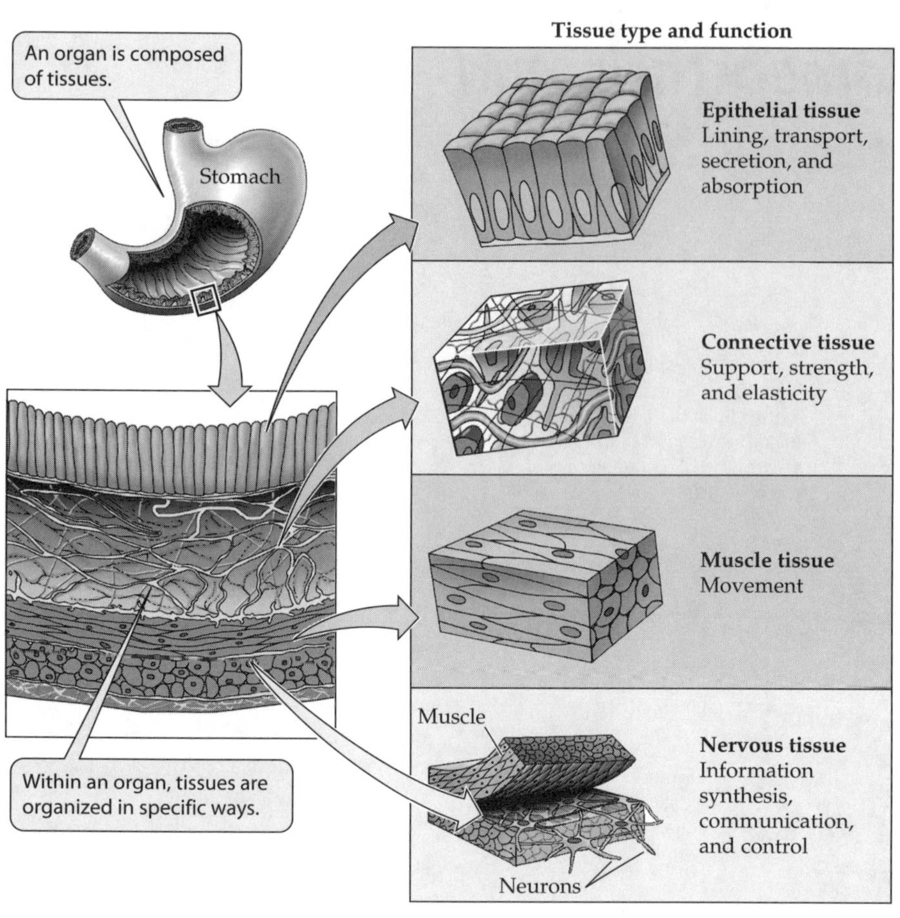
An organ is composed of tissues.
Stomach
Within an organ, tissues are organized in specific ways.
Tissue type and function
Epithelial tissue
Lining, transport, secretion, and absorption
Connective tissue
Support, strength, and elasticity
Muscle tissue
Movement
Muscle
Nervous tissue
Information synthesis, communication, and control
Neurons

41.2 Four Types of Tissue *(Page 782)*

41.1 The Major Organ Systems of Mammals

SYSTEM	TISSUES AND ORGANS	FUNCTIONS
Nervous system	Brain, spinal cord, sensory organs, peripheral nerves	Receives, integrates, stores information and controls muscles and glands (Chapters 44, 45, 46)
Endocrine system	Glands: pituitary, thyroid, parathyroid, pineal, adrenal, testes, ovaries, pancreas	A system of glands releases chemical messages (hormones) that control and regulate other tissues and organs (Chapter 42)
Muscle system	Skeletal muscle, smooth muscle, cardiac muscle	Produces forces and motion (Chapter 47)
Skeletal system	Bones	Provides structural support for the body (Chapter 47)
Reproductive system	Female: ovaries, oviducts, uterus, vagina, mammary glands Male: testes, sperm ducts, accessory glands, penis	Produces sex cells and hormones necessary to procreate and nurture offspring (Chapter 43)
Digestive system	Mouth, esophagus, stomach, intestines, liver, pancreas, rectum, anus	Acquires and digests food, absorbs and stores nutrients, then makes them available to the cells of the body (Chapter 50)
Respiratory system	Airways, lungs, diaphragm	Exchanges respiratory gases with the environment (Chapter 48)
Circulatory system	Heart and blood vessels	Transports respiratory gases, nutrients, hormones, and heat around the body (Chapter 49)
Lymphatic system	Lymph and lymph vessels, lymph nodes, spleen	Brings extracellular fluids back into the circulatory system; helps the immune system fight invading organisms (Chapters 49 and 18)
Immune system	Many types of white blood cells	Fights invading organisms and infections (Chapter 18)
Skin system	Skin, sweat glands, hair	Protects the body from invading organisms and harsh physical conditions, helps regulate body temperature (Chapter 41)
Excretory system	Kidneys, bladder, ureter, urethra	Regulates the composition of the extracellular fluids; excretes waste products (Chapter 51)

(Page 784)

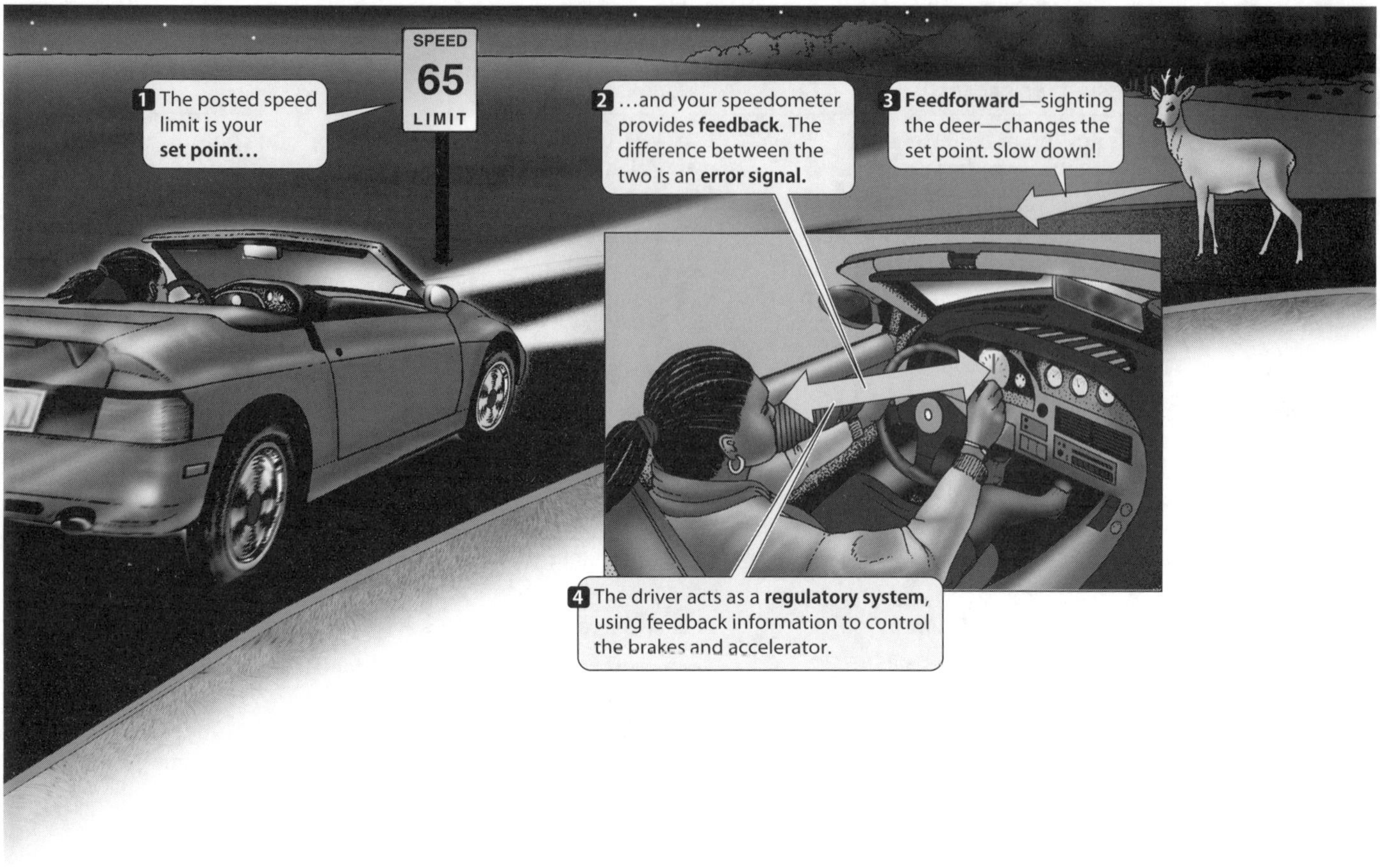

SPEED 65 LIMIT

1 The posted speed limit is your **set point...**

2 ...and your speedometer provides **feedback**. The difference between the two is an **error signal.**

3 **Feedforward**—sighting the deer—changes the set point. Slow down!

4 The driver acts as a **regulatory system**, using feedback information to control the brakes and accelerator.

41.4 Control, Regulation, and Feedback *(Page 785)*

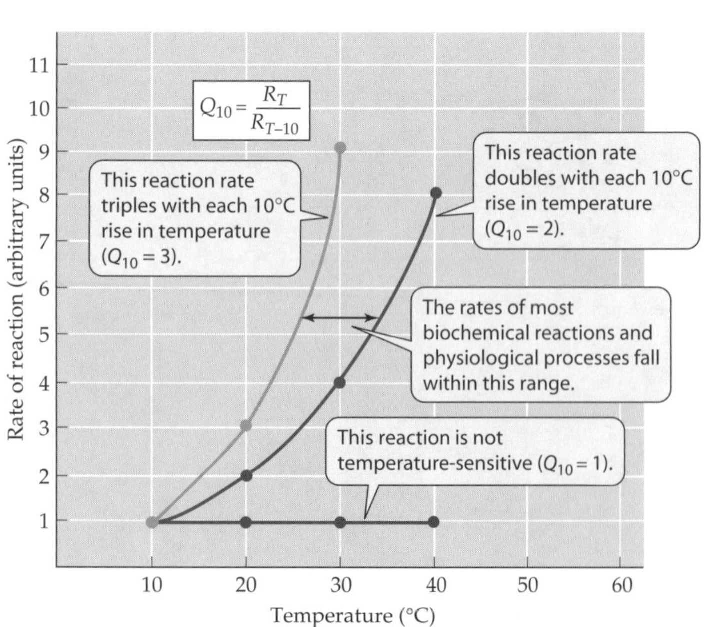
$Q_{10} = \frac{R_T}{R_{T-10}}$
This reaction rate triples with each 10°C rise in temperature ($Q_{10} = 3$).
This reaction rate doubles with each 10°C rise in temperature ($Q_{10} = 2$).
The rates of most biochemical reactions and physiological processes fall within this range.
This reaction is not temperature-sensitive ($Q_{10} = 1$).
Rate of reaction (arbitrary units)
1
2
3
4
5
6
7
8
9
10
11
Temperature (°C)
10
20
30
40
50
60

41.5 Q_{10} and Reaction Rate *(Page 786)*

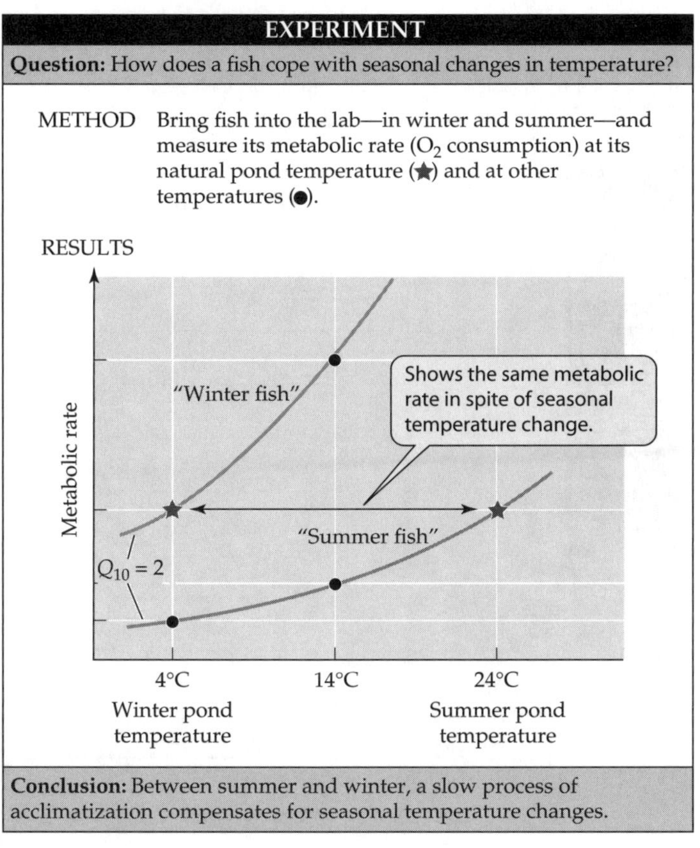
EXPERIMENT
Question: How does a fish cope with seasonal changes in temperature?
METHOD Bring fish into the lab—in winter and summer—and measure its metabolic rate (O_2 consumption) at its natural pond temperature (★) and at other temperatures (●).
RESULTS
Metabolic rate
"Winter fish"
Shows the same metabolic rate in spite of seasonal temperature change.
"Summer fish"
$Q_{10} = 2$
4°C
14°C
24°C
Winter pond temperature
Summer pond temperature
Conclusion: Between summer and winter, a slow process of acclimatization compensates for seasonal temperature changes.

41.6 Metabolic Compensation *(Page 787)*

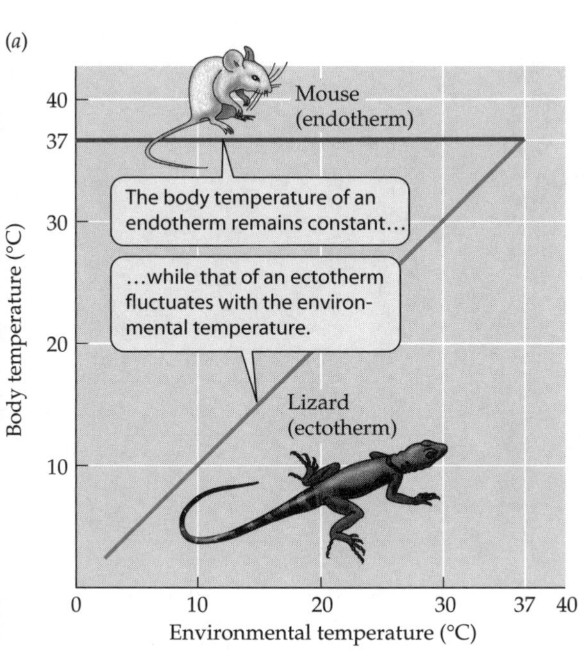
(a)
Mouse (endotherm)
The body temperature of an endotherm remains constant…
…while that of an ectotherm fluctuates with the environmental temperature.
Lizard (ectotherm)
Body temperature (°C)
40
37
30
20
10
0
10
20
30
37
40
Environmental temperature (°C)

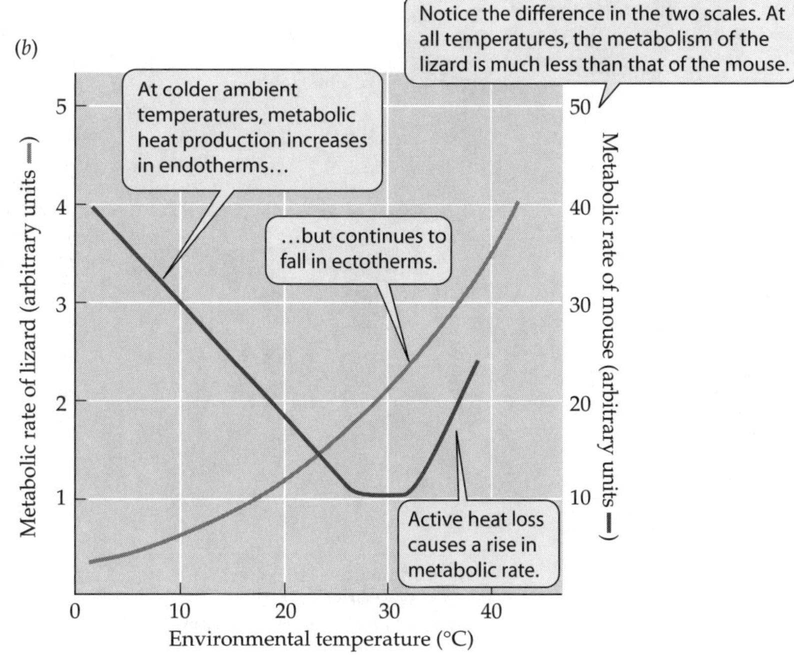
(b)
Notice the difference in the two scales. At all temperatures, the metabolism of the lizard is much less than that of the mouse.
At colder ambient temperatures, metabolic heat production increases in endotherms…
…but continues to fall in ectotherms.
Active heat loss causes a rise in metabolic rate.
Metabolic rate of lizard (arbitrary units —)
Metabolic rate of mouse (arbitrary units —)
5
4
3
2
1
50
40
30
20
10
0
10
20
30
40
Environmental temperature (°C)

41.7 Ectotherms and Endotherms *(Page 788)*

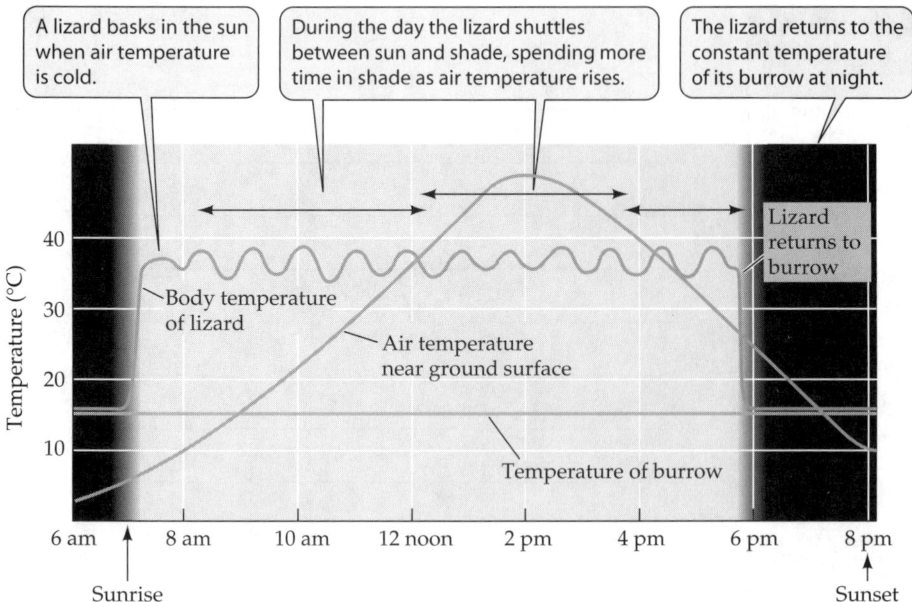
A lizard basks in the sun when air temperature is cold.
During the day the lizard shuttles between sun and shade, spending more time in shade as air temperature rises.
The lizard returns to the constant temperature of its burrow at night.
Lizard returns to burrow
Body temperature of lizard
Air temperature near ground surface
Temperature of burrow
Temperature (°C)
40
30
20
10
6 am
8 am
10 am
12 noon
2 pm
4 pm
6 pm
8 pm
Sunrise
Sunset

41.8 An Ectotherm Uses Behavior to Regulate Its Body Temperature *(Page 788)*

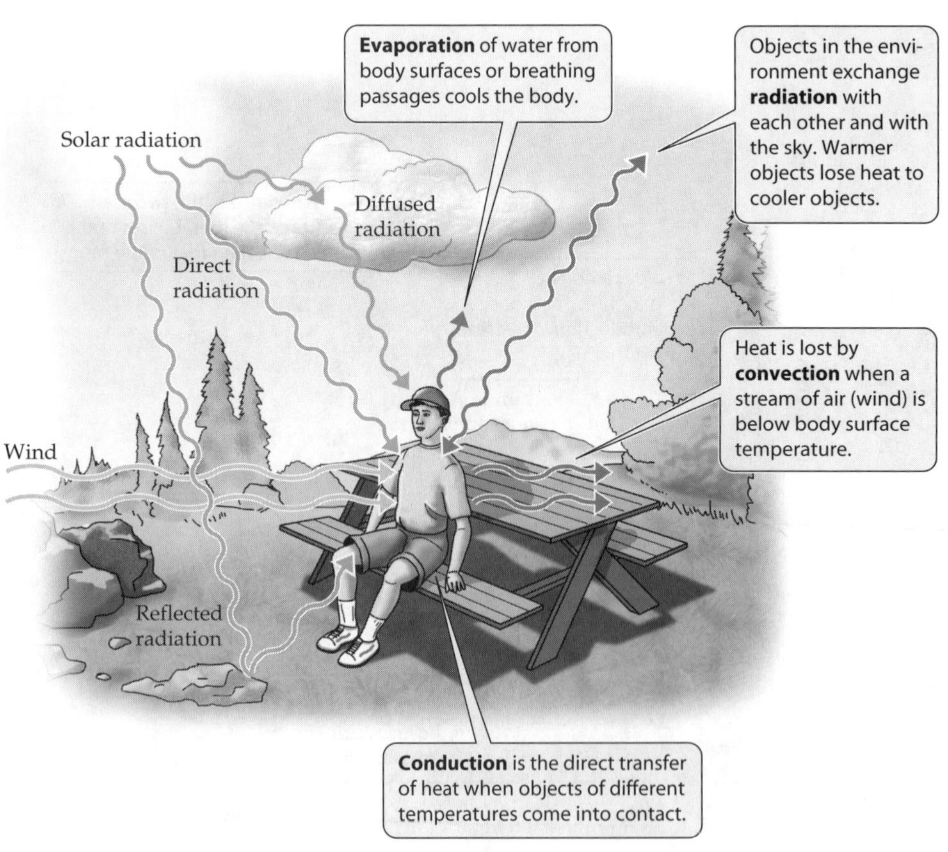
Evaporation of water from body surfaces or breathing passages cools the body.
Objects in the environment exchange **radiation** with each other and with the sky. Warmer objects lose heat to cooler objects.
Solar radiation
Diffused radiation
Direct radiation
Heat is lost by **convection** when a stream of air (wind) is below body surface temperature.
Wind
Reflected radiation
Conduction is the direct transfer of heat when objects of different temperatures come into contact.

41.10 Animals Exchange Heat with the Environment *(Page 789)*

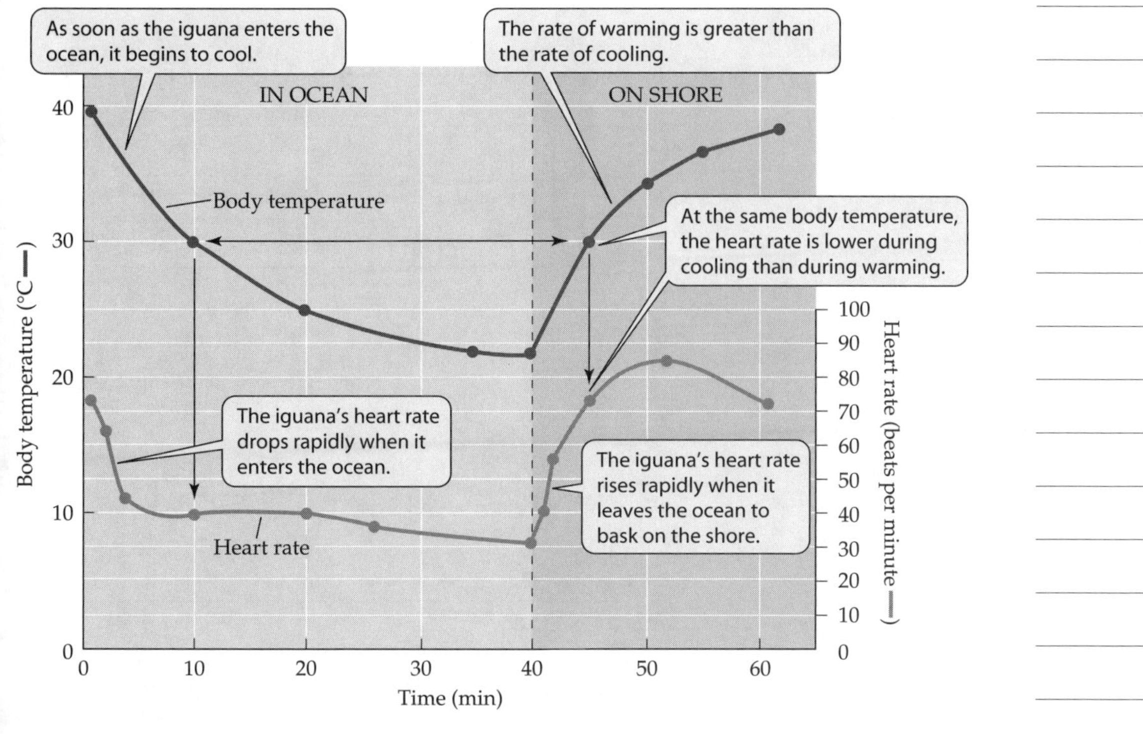

As soon as the iguana enters the ocean, it begins to cool.
The rate of warming is greater than the rate of cooling.
IN OCEAN
ON SHORE
Body temperature
At the same body temperature, the heart rate is lower during cooling than during warming.
The iguana's heart rate drops rapidly when it enters the ocean.
The iguana's heart rate rises rapidly when it leaves the ocean to bask on the shore.
Heart rate
Body temperature (°C —)
Heart rate (beats per minute —)
Time (min)

41.11 Some Ectotherms Regulate Blood Flow to the Skin *(Page 790)*

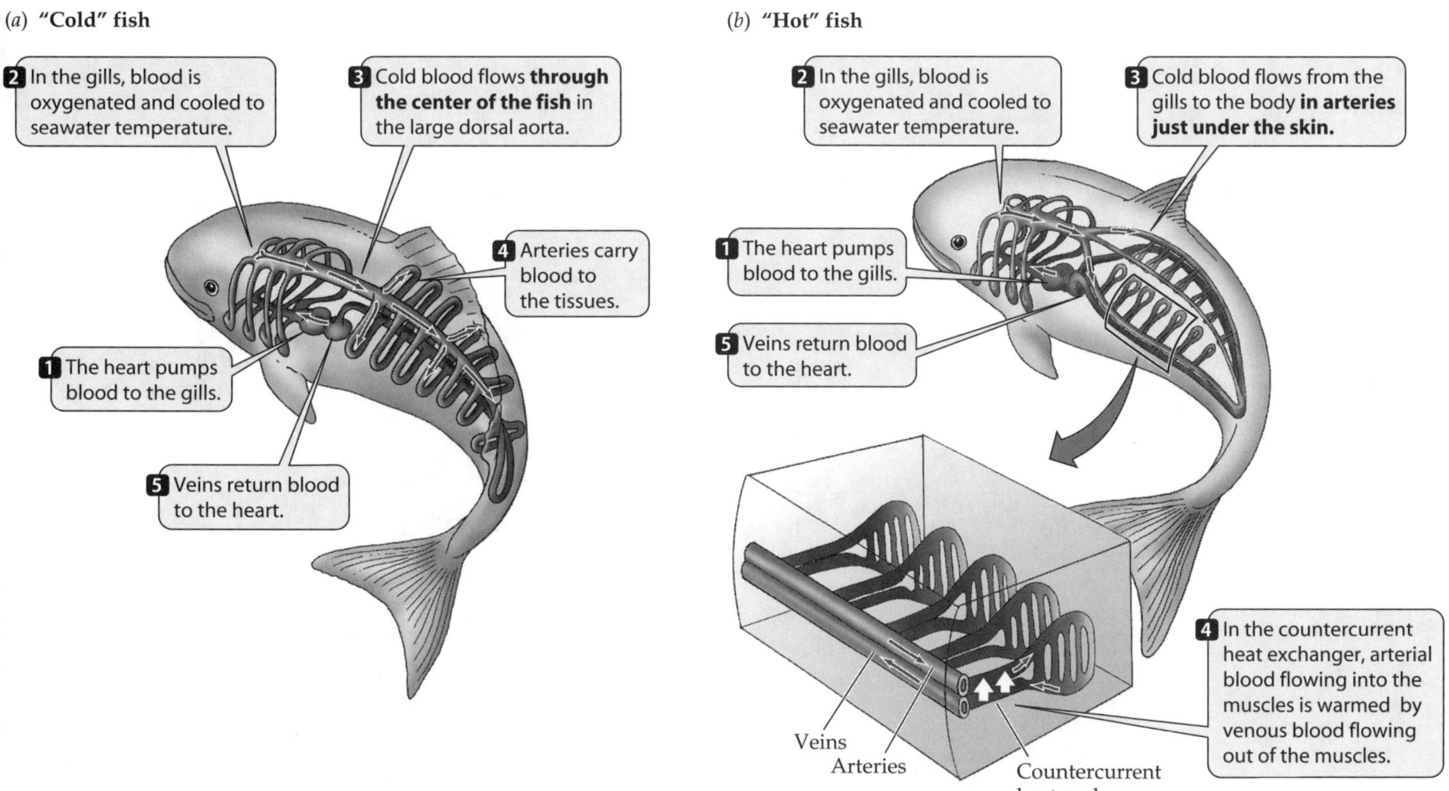

(a) "Cold" fish
2 In the gills, blood is oxygenated and cooled to seawater temperature.
3 Cold blood flows through the center of the fish in the large dorsal aorta.
4 Arteries carry blood to the tissues.
1 The heart pumps blood to the gills.
5 Veins return blood to the heart.
(b) "Hot" fish
2 In the gills, blood is oxygenated and cooled to seawater temperature.
3 Cold blood flows from the gills to the body in arteries just under the skin.
1 The heart pumps blood to the gills.
5 Veins return blood to the heart.
4 In the countercurrent heat exchanger, arterial blood flowing into the muscles is warmed by venous blood flowing out of the muscles.
Veins
Arteries
Countercurrent heat exchanger

41.12 "Cold" and "Hot" Fish *(Page 791)*

ANIMAL	BODY MASS (KG)	TOTAL O_2 CONSUMPTION (LITERS/HR)	O_2 CONSUMPTION (LITERS/HR) PER KG OF BODY MASS ($l\ O_2\ kg^{-1}\ h^{-1}$)
Shrew	0.005	0.036	7.40
Mouse	0.025	0.041	1.65
Rat	0.29	0.250	0.87
Cat	2.5	1.70	0.68
Dog	11.7	3.87	0.33
Sheep	42.7	9.59	0.22
Human	70	14.76	0.21
Horse	650	71.10	0.11
Elephant	3,833	268.00	0.07

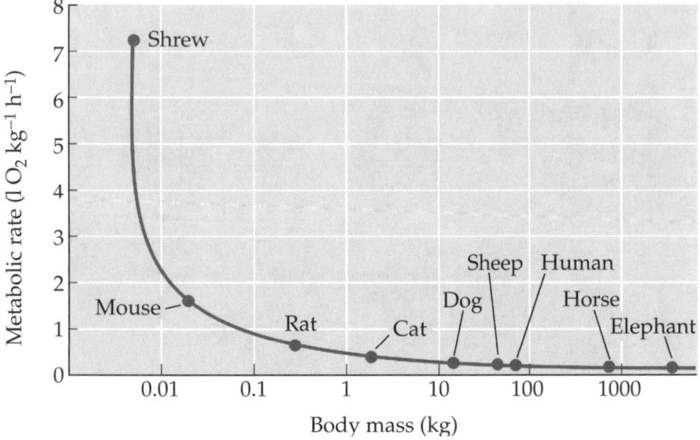

Metabolic rate ($l\ O_2\ kg^{-1}\ h^{-1}$)
Body mass (kg)
Shrew
Mouse
Rat
Cat
Dog
Sheep
Human
Horse
Elephant
0.01
0.1
1
10
100
1000

41.13 The Mouse-to-Elephant Curve *(Page 792)*

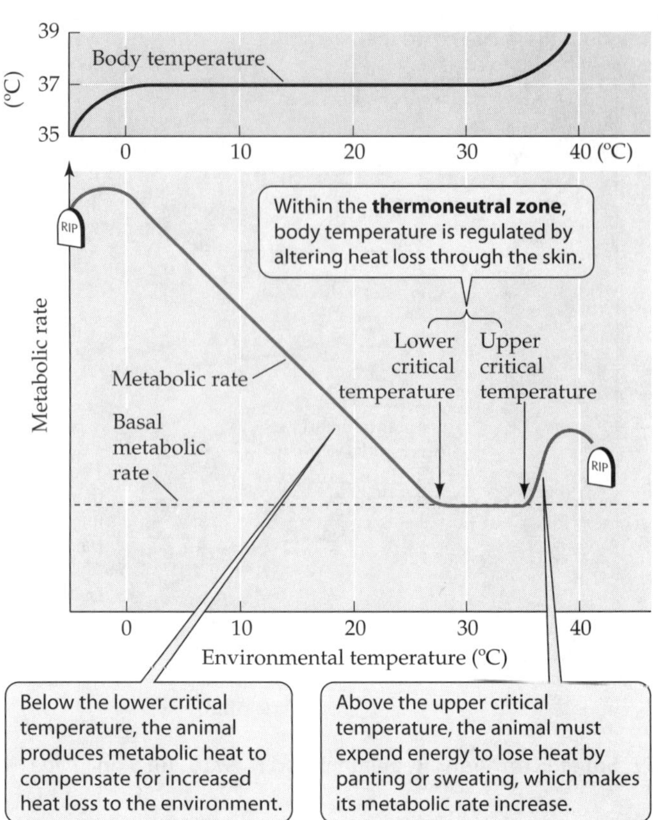

(°C)
Body temperature
Metabolic rate
Within the **thermoneutral zone**, body temperature is regulated by altering heat loss through the skin.
Lower critical temperature
Upper critical temperature
Metabolic rate
Basal metabolic rate
RIP
Environmental temperature (°C)
Below the lower critical temperature, the animal produces metabolic heat to compensate for increased heat loss to the environment.
Above the upper critical temperature, the animal must expend energy to lose heat by panting or sweating, which makes its metabolic rate increase.

41.14 Environmental Temperature and Mammalian Metabolic Rates *(Page 793)*

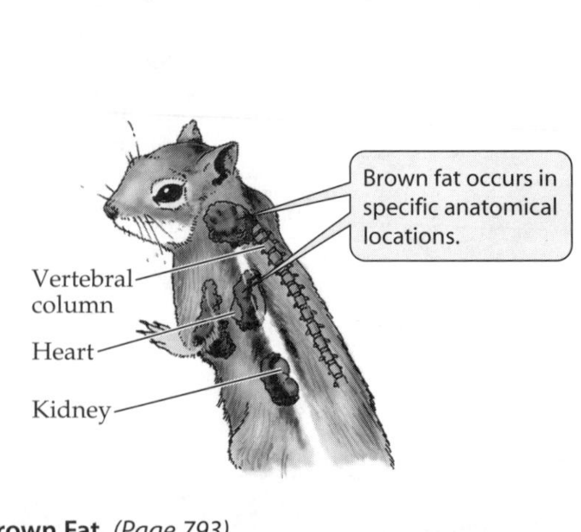

Brown fat occurs in specific anatomical locations.
Vertebral column
Heart
Kidney

41.15 Brown Fat *(Page 793)*

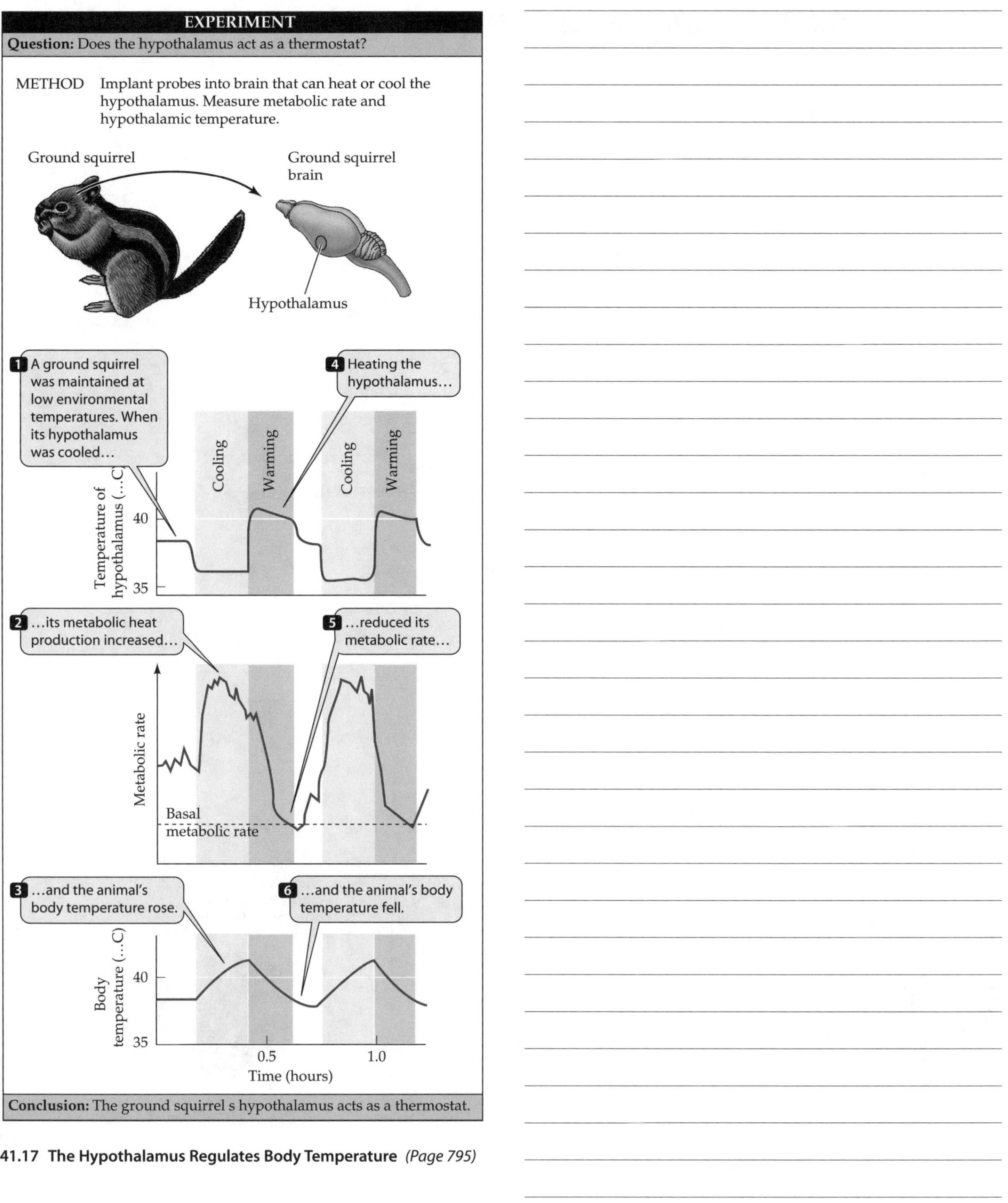
EXPERIMENT
Question: Does the hypothalamus act as a thermostat?
METHOD Implant probes into brain that can heat or cool the hypothalamus. Measure metabolic rate and hypothalamic temperature.
Ground squirrel
Ground squirrel brain
Hypothalamus
1 A ground squirrel was maintained at low environmental temperatures. When its hypothalamus was cooled…
4 Heating the hypothalamus…
Cooling
Warming
Cooling
Warming
Temperature of hypothalamus (…C)
40
35
2 …its metabolic heat production increased…
5 …reduced its metabolic rate…
Metabolic rate
Basal metabolic rate
3 …and the animal's body temperature rose.
6 …and the animal's body temperature fell.
Body temperature (…C)
40
35
0.5
1.0
Time (hours)
Conclusion: The ground squirrel s hypothalamus acts as a thermostat.

41.17 The Hypothalamus Regulates Body Temperature *(Page 795)*

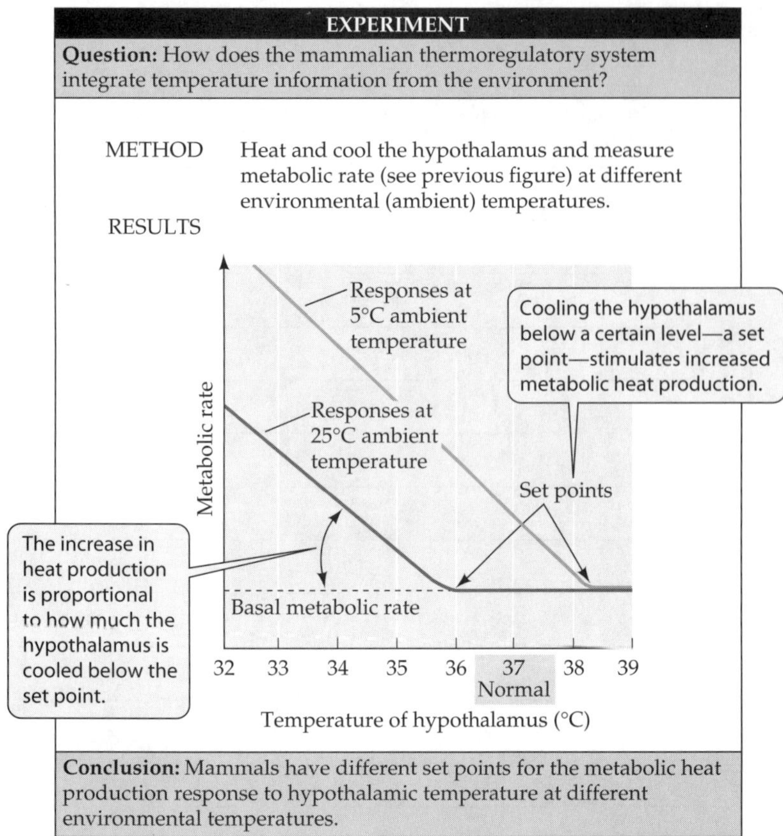
EXPERIMENT
Question: How does the mammalian thermoregulatory system integrate temperature information from the environment?
METHOD Heat and cool the hypothalamus and measure metabolic rate (see previous figure) at different environmental (ambient) temperatures.
RESULTS
Responses at 5°C ambient temperature
Cooling the hypothalamus below a certain level—a set point—stimulates increased metabolic heat production.
Responses at 25°C ambient temperature
Metabolic rate
Set points
The increase in heat production is proportional to how much the hypothalamus is cooled below the set point.
Basal metabolic rate
32 33 34 35 36 37 38 39
Normal
Temperature of hypothalamus (°C)
Conclusion: Mammals have different set points for the metabolic heat production response to hypothalamic temperature at different environmental temperatures.

41.18 Adjustable Set Points *(Page 796)*

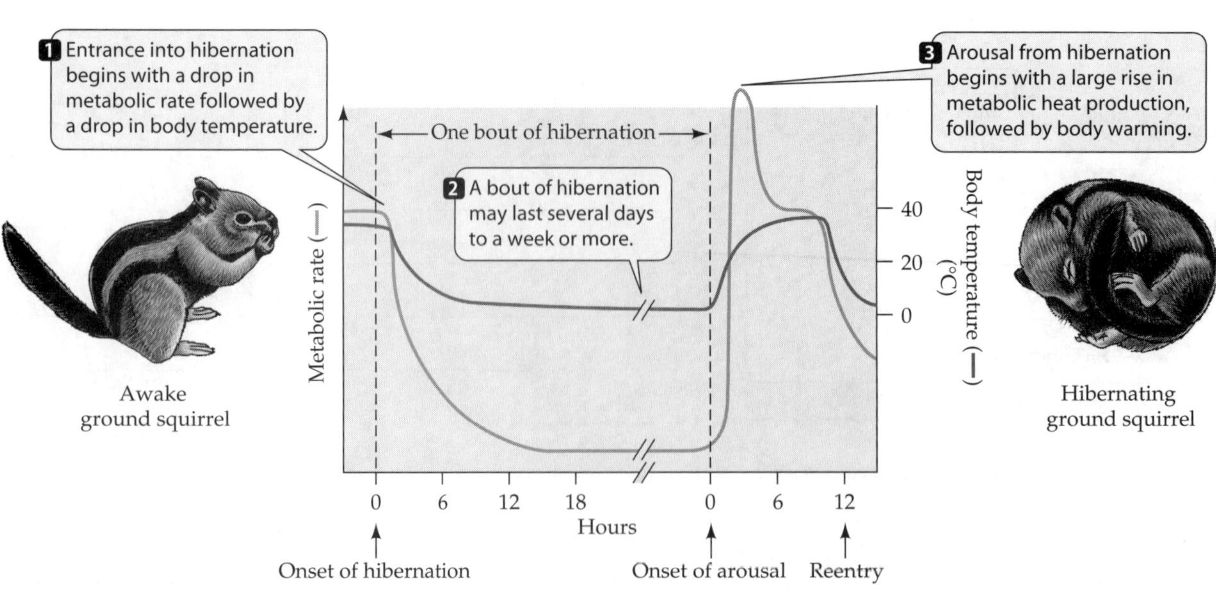
1 Entrance into hibernation begins with a drop in metabolic rate followed by a drop in body temperature.
One bout of hibernation
2 A bout of hibernation may last several days to a week or more.
3 Arousal from hibernation begins with a large rise in metabolic heat production, followed by body warming.
Metabolic rate (—)
Body temperature (—) (°C)
40
20
0
Awake ground squirrel
Hibernating ground squirrel
0 6 12 18 0 6 12
Hours
Onset of hibernation
Onset of arousal
Reentry

41.19 A Ground Squirrel Enters Repeated Bouts of Hibernation during Winter *(Page 797)*

42 Animal Hormones

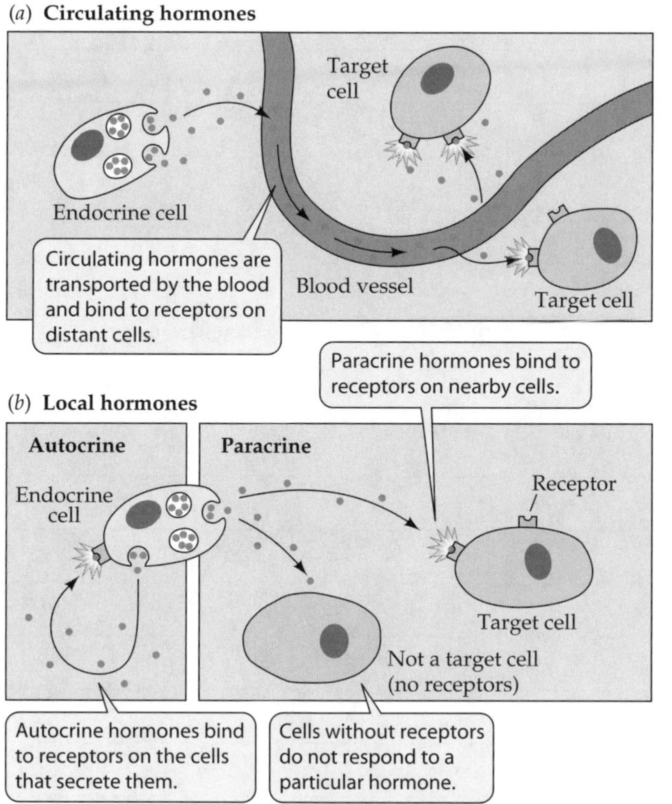

(a) Circulating hormones
Target cell
Endocrine cell
Circulating hormones are transported by the blood and bind to receptors on distant cells.
Blood vessel
Target cell
Paracrine hormones bind to receptors on nearby cells.
(b) Local hormones
Autocrine
Paracrine
Endocrine cell
Receptor
Target cell
Not a target cell (no receptors)
Autocrine hormones bind to receptors on the cells that secrete them.
Cells without receptors do not respond to a particular hormone.

42.1 Chemical Signaling Systems *(Page 800)*

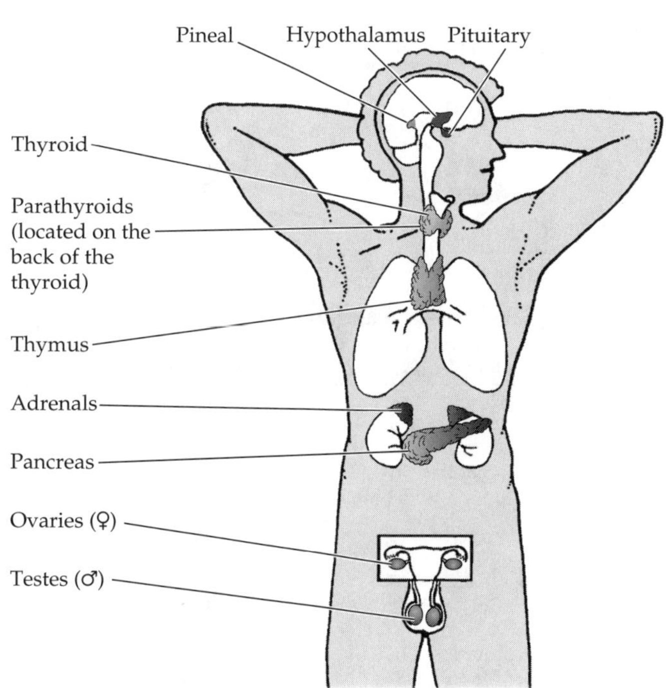

Pineal
Hypothalamus
Pituitary
Thyroid
Parathyroids (located on the back of the thyroid)
Thymus
Adrenals
Pancreas
Ovaries (♀)
Testes (♂)

42.2 The Endocrine System of Humans *(Page 801)*

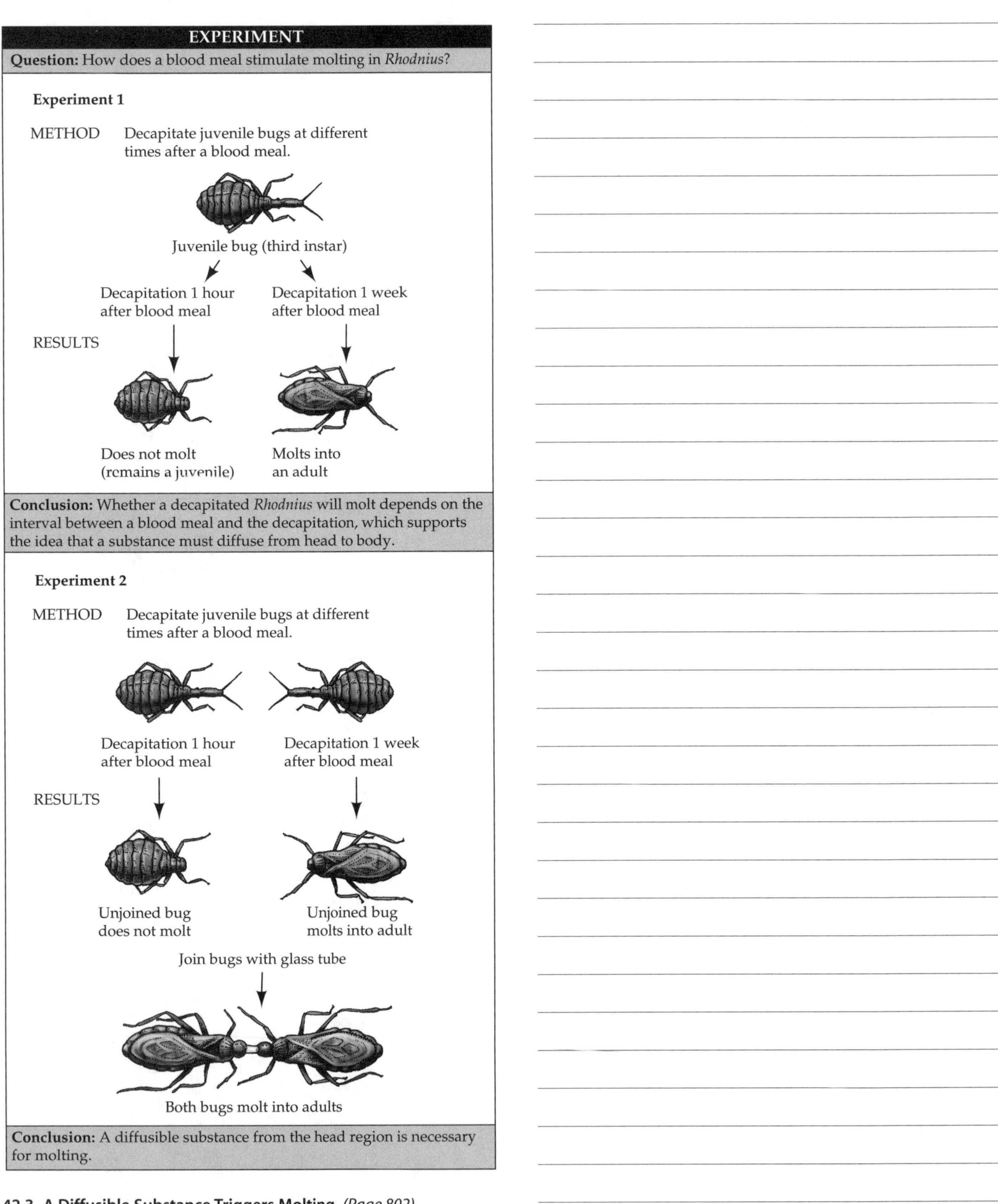
EXPERIMENT
Question: How does a blood meal stimulate molting in Rhodnius?
Experiment 1
METHOD Decapitate juvenile bugs at different times after a blood meal.
Juvenile bug (third instar)
Decapitation 1 hour after blood meal
Decapitation 1 week after blood meal
RESULTS
Does not molt (remains a juvenile)
Molts into an adult
Conclusion: Whether a decapitated Rhodnius will molt depends on the interval between a blood meal and the decapitation, which supports the idea that a substance must diffuse from head to body.
Experiment 2
METHOD Decapitate juvenile bugs at different times after a blood meal.
Decapitation 1 hour after blood meal
Decapitation 1 week after blood meal
RESULTS
Unjoined bug does not molt
Unjoined bug molts into adult
Join bugs with glass tube
Both bugs molt into adults
Conclusion: A diffusible substance from the head region is necessary for molting.

42.3 A Diffusible Substance Triggers Molting *(Page 802)*

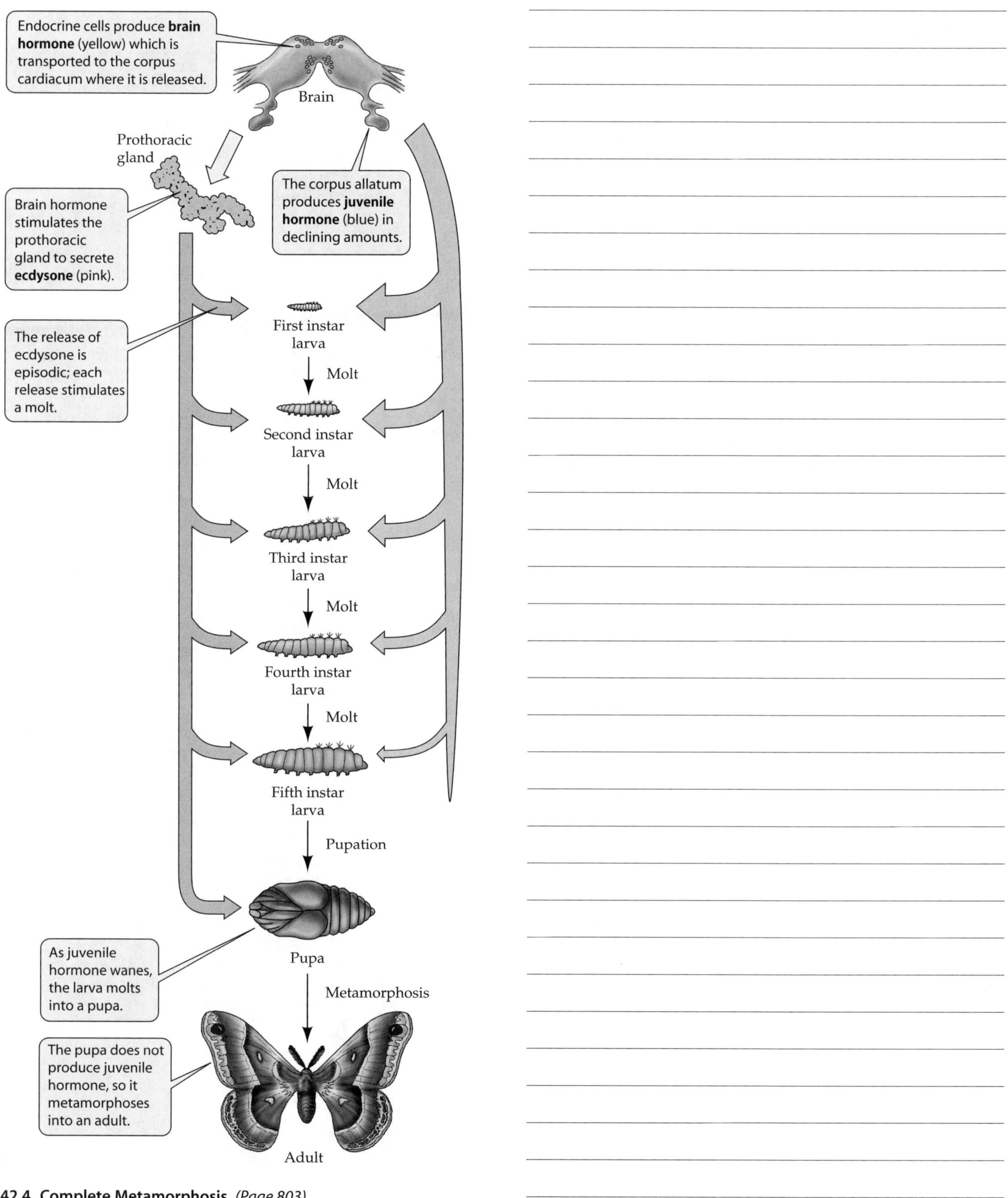

Endocrine cells produce **brain hormone** (yellow) which is transported to the corpus cardiacum where it is released.
Brain
Prothoracic gland
Brain hormone stimulates the prothoracic gland to secrete **ecdysone** (pink).
The corpus allatum produces **juvenile hormone** (blue) in declining amounts.
The release of ecdysone is episodic; each release stimulates a molt.
First instar larva
Molt
Second instar larva
Molt
Third instar larva
Molt
Fourth instar larva
Molt
Fifth instar larva
Pupation
Pupa
As juvenile hormone wanes, the larva molts into a pupa.
Metamorphosis
The pupa does not produce juvenile hormone, so it metamorphoses into an adult.
Adult

42.4 Complete Metamorphosis *(Page 803)*

42.1 Principal Hormones of Humans

SECRETING TISSUE OR GLAND	HORMONE	CHEMICAL NATURE	TARGET(S)	IMPORTANT PROPERTIES OR ACTIONS
Hypothalamus	Releasing and release-inhibiting hormones (see Table 42.2)	Peptides	Anterior pituitary	Control secretion of hormones of anterior pituitary
	Oxytocin, antidiuretic hormone	Peptides	(See Posterior pituitary)	Stored and released by posterior pituitary
Anterior pituitary: Tropic hormones	Thyrotropin	Glycoprotein	Thyroid gland	Stimulates synthesis and secretion of thyroxine
	Adrenocorticotropin (ACTH)	Polypeptide	Adrenal cortex	Stimulates release of hormones from adrenal cortex
	Luteinizing hormone (LH)	Glycoprotein	Gonads	Stimulates secretion of sex hormones from ovaries and testes
	Follicle-stimulating hormone (FSH)	Glycoprotein	Gonads	Stimulates growth and maturation of eggs in females; stimulates sperm production in males
Anterior pituitary: Other hormones	Growth hormone (GH)	Protein	Bones, liver, muscles	Stimulates protein synthesis and growth
	Prolactin	Protein	Mammary glands	Stimulates milk production
	Melanocyte-stimulating hormone	Peptide	Melanocytes	Controls skin pigmentation
	Endorphins and enkephalins	Peptides	Spinal cord neurons	Decreases painful sensations
Posterior pituitary	Oxytocin	Peptide	Uterus, breasts	Induces birth by stimulating labor contractions; causes milk flow
	Antidiuretic hormone (ADH) (vasopressin)	Peptide	Kidneys	Stimulates water reabsorption and raises blood pressure
Thyroid	Thyroxine	Iodinated amino acid derivative	Many tissues	Stimulates and maintains metabolism necessary for normal development and growth
	Calcitonin	Peptide	Bones	Stimulates bone formation; lowers blood calcium
Parathyroids	Parathyroid hormone	Protein	Bones	Resorbs bone; raises blood calcium
Thymus	Thymosins	Peptides	White blood cells	Activate immune responses of T cells in the lymphatic system
Pancreas	Insulin	Protein	Muscles, liver, fat, other tissues	Stimulates uptake and metabolism of glucose; increases conversion of glucose to glycogen and fat
	Glucagon	Protein	Liver	Stimulates breakdown of glycogen and raises blood sugar
	Somatostatin	Peptide	Digestive tract; other cells of the pancreas	Inhibits insulin and glucagon release; decreases secretion, motility, and absorption in the digestive tract

(Pages 804–805)

42.1 Principal Hormones of Humans (continued)

SECRETING TISSUE OR GLAND	HORMONE	CHEMICAL NATURE	TARGET(S)	IMPORTANT PROPERTIES OR ACTIONS
Adrenal medulla	Epinephrine, norepinephrine	Modified amino acids	Heart, blood vessels, liver, fat cells	Stimulate fight-or-flight reactions: increase heart rate, redistribute blood to muscles, raise blood sugar
Adrenal cortex	Glucocorticoids (cortisol)	Steroids	Muscles, immune system, other tissues	Mediate response to stress; reduce metabolism of glucose, increase metabolism of proteins and fats; reduce inflammation and immune responses
	Mineralocorticoids (aldosterone)	Steroids	Kidneys	Stimulate excretion of potassium ions and reabsorption of sodium ions
Stomach lining	Gastrin	Peptide	Stomach	Promotes digestion of food by stimulating release of digestive juices; stimulates stomach movements that mix food and digestive juices
Lining of small intestine	Secretin	Peptide	Pancreas	Stimulate secretion of bicarbonate solution by ducts of pancreas
	Cholecystokinin	Peptide	Pancreas, liver, gallbladder	Stimulates secretion of digestive enzymes by pancreas and other digestive juices from liver; stimulates contractions of gallbladder and ducts
	Enterogastrone	Polypeptide	Stomach	Inhibits digestive activities in the stomach
Pineal	Melatonin	Modified amino acid	Hypothalamus	Involved in biological rhythms
Ovaries	Estrogens	Steroids	Breasts, uterus, other tissues	Stimulate development and maintenance of female characteristics and sexual behavior
	Progesterone	Steroid	Uterus	Sustains pregnancy; helps maintain secondary female sexual characteristics
Testes	Androgens	Steroids	Various tissues	Stimulate development and maintenance of male sexual behavior and secondary male sexual characteristics; stimulate sperm production
Many cell types	Prostaglandins	Modified fatty acids	Various tissues	Have many diverse actions
Heart	Atrial natriuretic hormone	Peptide	Kidneys	Increases sodium ion excretion
Skin	Vitamin D (cholecalciferol)	Sterol	Digestive tract, kidneys, bone	Increases blood calcium levels

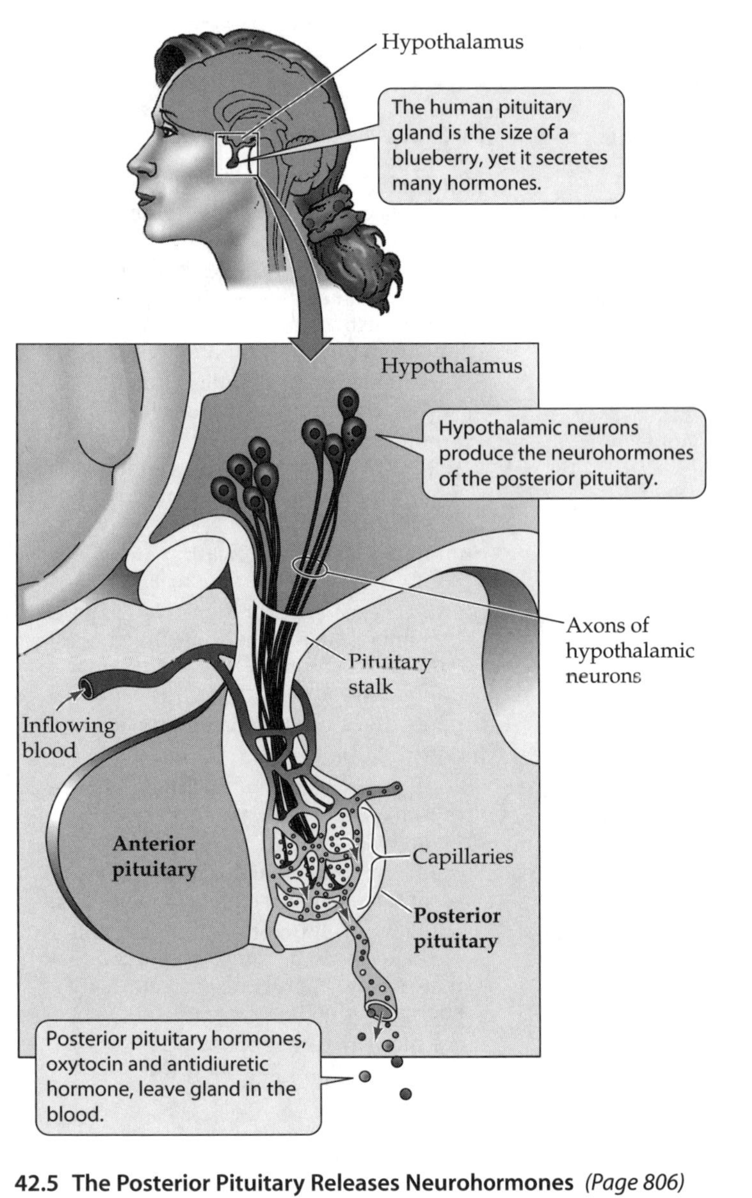
Hypothalamus
The human pituitary gland is the size of a blueberry, yet it secretes many hormones.
Hypothalamus
Hypothalamic neurons produce the neurohormones of the posterior pituitary.
Axons of hypothalamic neurons
Pituitary stalk
Inflowing blood
Anterior pituitary
Capillaries
Posterior pituitary
Posterior pituitary hormones, oxytocin and antidiuretic hormone, leave gland in the blood.

42.5 The Posterior Pituitary Releases Neurohormones *(Page 806)*

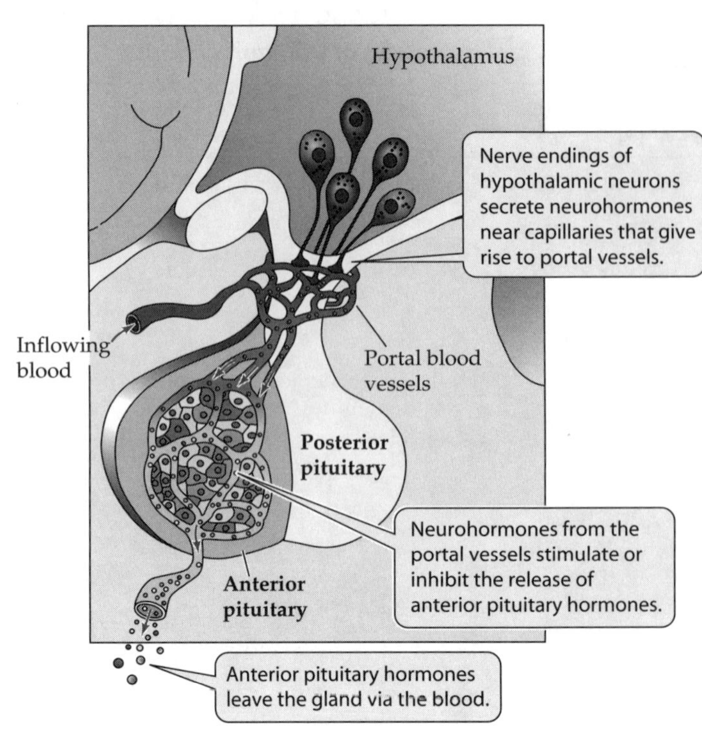
Hypothalamus
Nerve endings of hypothalamic neurons secrete neurohormones near capillaries that give rise to portal vessels.
Inflowing blood
Portal blood vessels
Posterior pituitary
Neurohormones from the portal vessels stimulate or inhibit the release of anterior pituitary hormones.
Anterior pituitary
Anterior pituitary hormones leave the gland via the blood.

42.7 Hormones from the Hypothalamus Control the Anterior Pituitary *(Page 807)*

42.2 *Releasing and Release-Inhibiting Neurohormones of the Hypothalamus*

NEUROHORMONE	ACTION
Thyrotropin-releasing hormone (TRH)	Stimulates thyrotropin release
Gonadotropin-releasing hormone (GnRH)	Stimulates release of follicle-stimulating hormone and luteinizing hormone
Prolactin release-inhibiting hormone	Inhibits prolactin release
Prolactin-releasing hormone	Stimulates prolactin release
Somatostatin (growth hormone release-inhibiting hormone)	Inhibits growth hormone release; interferes with thyrotropin release
Growth hormone-releasing hormone	Stimulates growth hormone release
Adrenocorticotropin-releasing hormone	Stimulates adrenocorticotropin release
Melanocyte-stimulating hormone release-inhibiting hormone	Inhibits release of melanocyte-stimulating hormone

(Page 808)

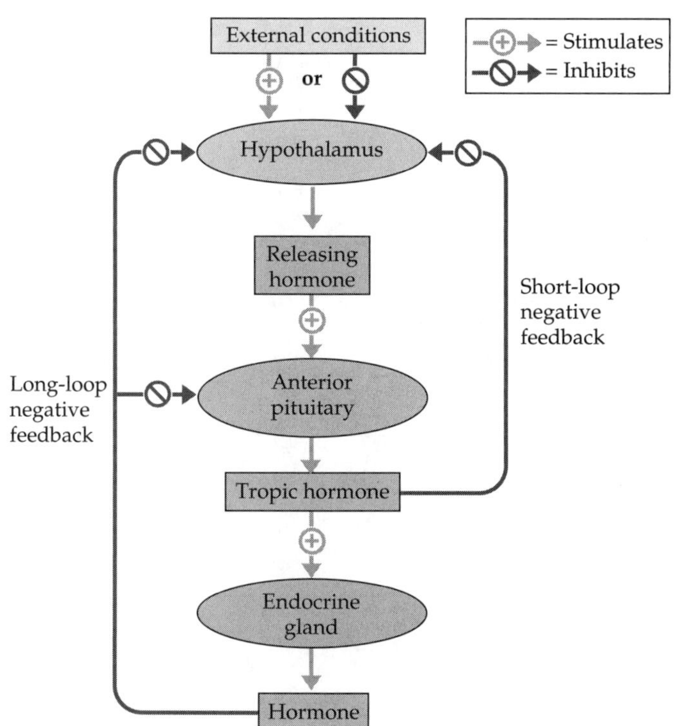
External conditions
or
= Stimulates
= Inhibits
Hypothalamus
Releasing hormone
Short-loop negative feedback
Long-loop negative feedback
Anterior pituitary
Tropic hormone
Endocrine gland
Hormone

42.8 Multiple Feedback Loops Control Hormone Secretion
(Page 808)

Thyroxine (T_4)

In-Text Art *(Page 809)*

Triiodothyronine (T_3)

In-Text Art *(Page 809)*

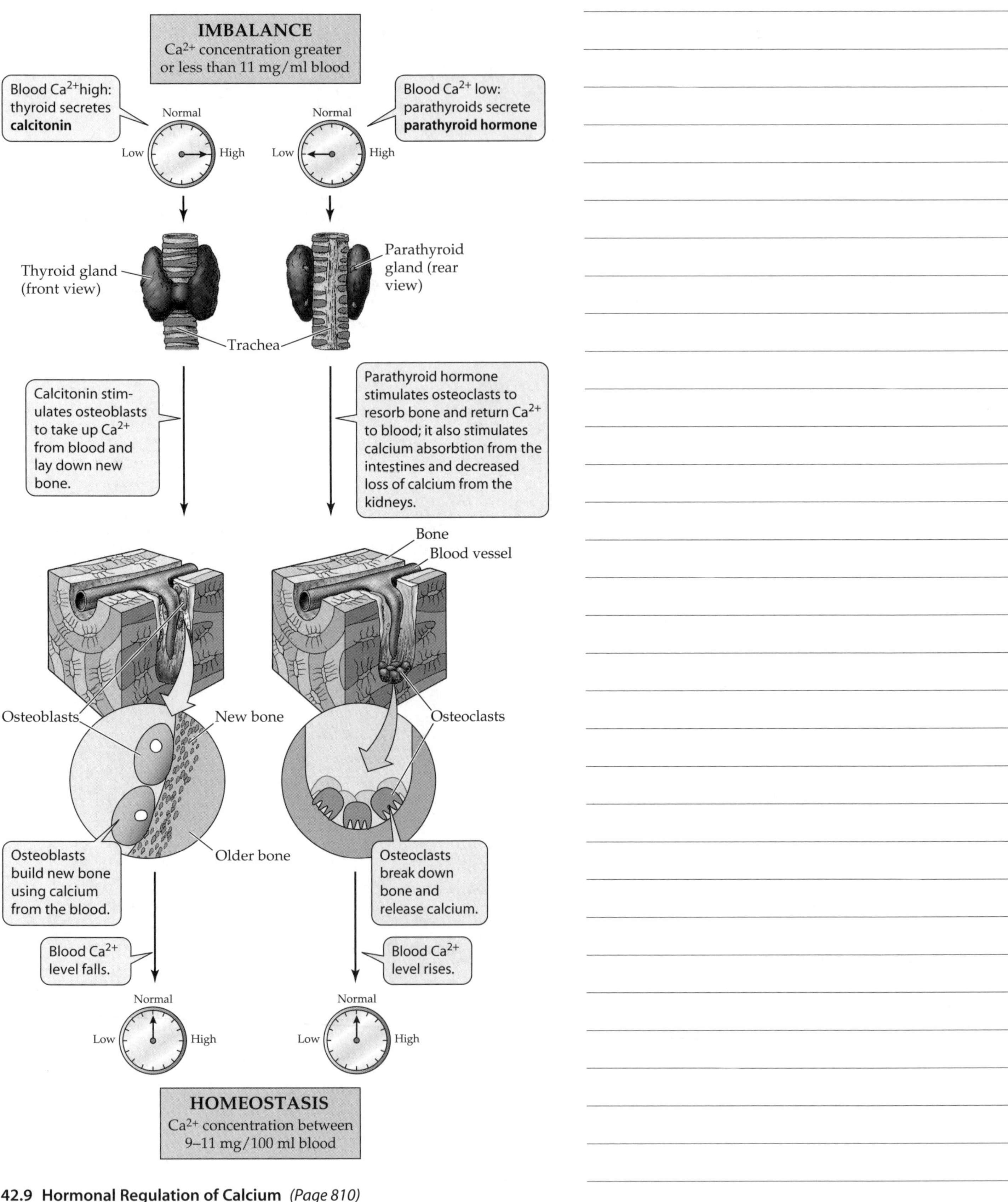

IMBALANCE
Ca^{2+} concentration greater or less than 11 mg/ml blood
Blood Ca^{2+} high: thyroid secretes **calcitonin**
Normal
Low
High
Normal
Low
High
Blood Ca^{2+} low: parathyroids secrete **parathyroid hormone**
Thyroid gland (front view)
Parathyroid gland (rear view)
Trachea
Calcitonin stimulates osteoblasts to take up Ca^{2+} from blood and lay down new bone.
Parathyroid hormone stimulates osteoclasts to resorb bone and return Ca^{2+} to blood; it also stimulates calcium absorbtion from the intestines and decreased loss of calcium from the kidneys.
Bone
Blood vessel
Osteoblasts
New bone
Osteoclasts
Osteoblasts build new bone using calcium from the blood.
Older bone
Osteoclasts break down bone and release calcium.
Blood Ca^{2+} level falls.
Blood Ca^{2+} level rises.
Normal
Low
High
Normal
Low
High
HOMEOSTASIS
Ca^{2+} concentration between 9–11 mg/100 ml blood

42.9 Hormonal Regulation of Calcium *(Page 810)*

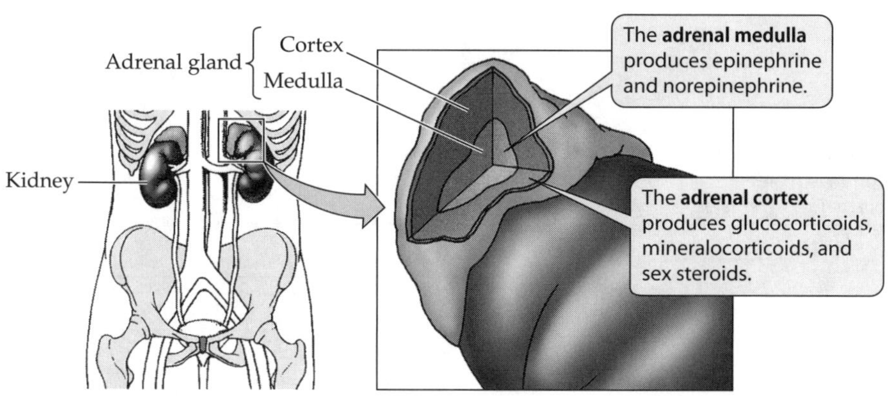
Adrenal gland
Cortex
Medulla
Kidney
The **adrenal medulla** produces epinephrine and norepinephrine.
The **adrenal cortex** produces glucocorticoids, mineralocorticoids, and sex steroids.

42.10 The Adrenal Gland Has an Outer and an Inner Portion *(Page 812)*

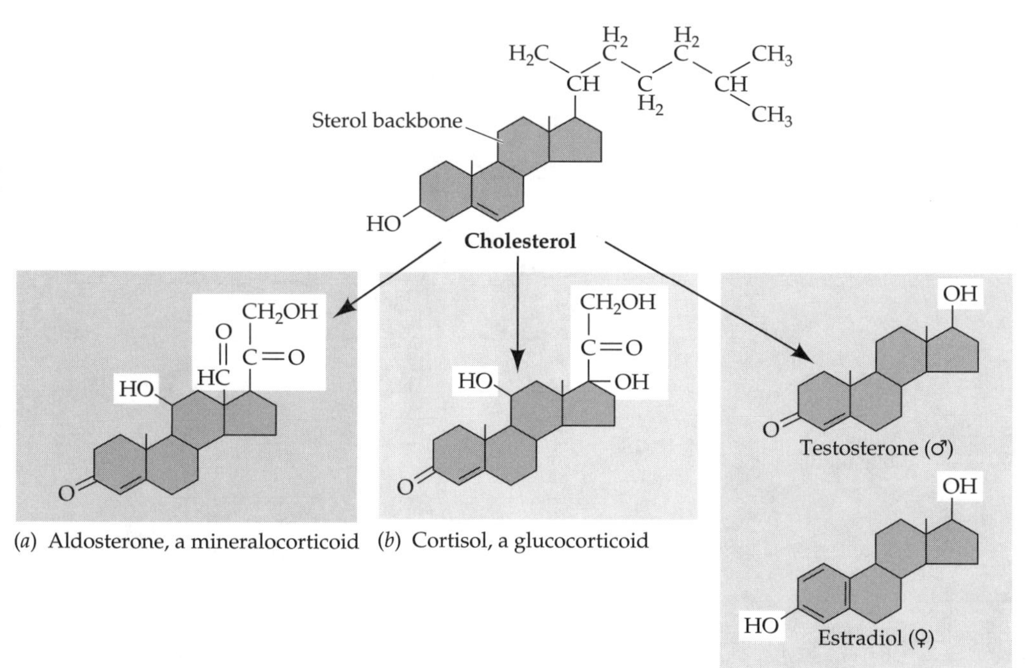
Sterol backbone
Cholesterol
HO
(a) Aldosterone, a mineralocorticoid
(b) Cortisol, a glucocorticoid
Testosterone (♂)
Estradiol (♀)
(c) Sex steroids

42.11 The Corticosteroid Hormones are Built from Cholesterol *(Page 813)*

7 WEEKS

Urethral groove

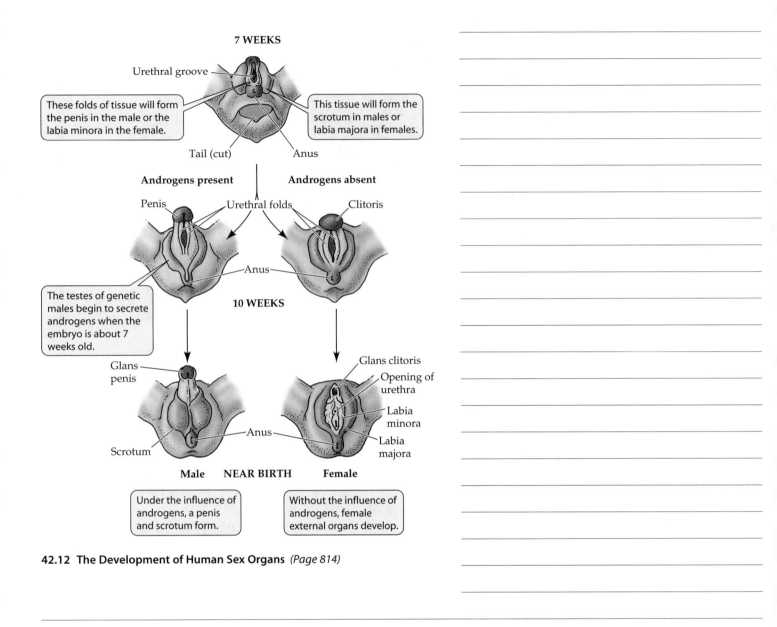

These folds of tissue will form the penis in the male or the labia minora in the female.

This tissue will form the scrotum in males or labia majora in females.

Tail (cut)

Anus

Androgens present

Androgens absent

Penis

Urethral folds

Clitoris

Anus

The testes of genetic males begin to secrete androgens when the embryo is about 7 weeks old.

10 WEEKS

Glans penis

Glans clitoris

Opening of urethra

Labia minora

Anus

Labia majora

Scrotum

Male NEAR BIRTH Female

Under the influence of androgens, a penis and scrotum form.

Without the influence of androgens, female external organs develop.

42.12 The Development of Human Sex Organs *(Page 814)*

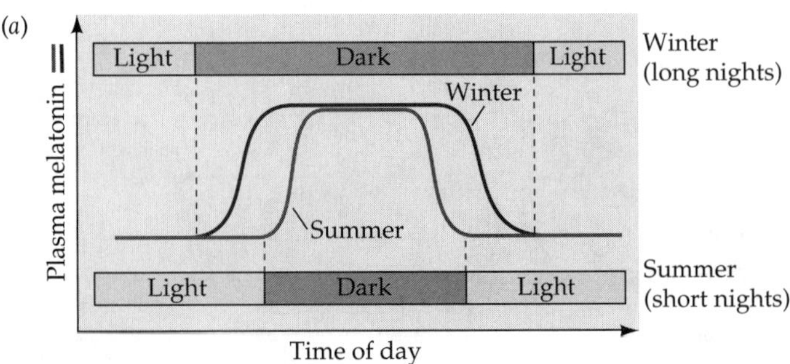
(a)
Light
Dark
Light
Winter
(long nights)
Winter
Summer
Plasma melatonin
Light
Dark
Light
Summer
(short nights)
Time of day

42.13 The Release of Melatonin Regulates Seasonal Changes *(Page 815)*

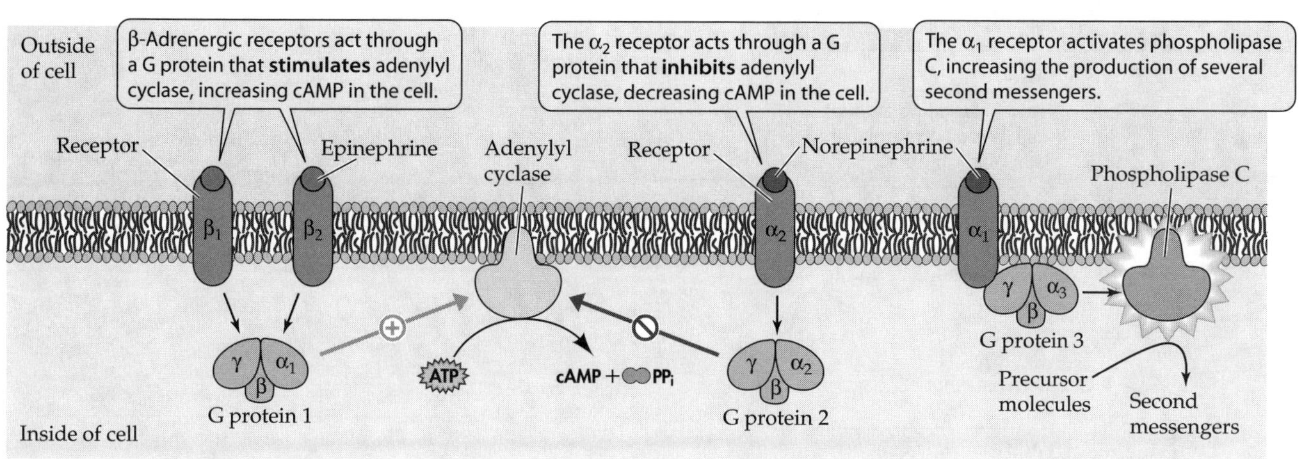
Outside of cell
β-Adrenergic receptors act through a G protein that **stimulates** adenylyl cyclase, increasing cAMP in the cell.
The α_2 receptor acts through a G protein that **inhibits** adenylyl cyclase, decreasing cAMP in the cell.
The α_1 receptor activates phospholipase C, increasing the production of several second messengers.
Receptor
Epinephrine
Adenylyl cyclase
Receptor
Norepinephrine
Phospholipase C
β_1
β_2
α_2
α_1
γ α_3 β
G protein 3
γ α_1 β
ATP
cAMP + PP$_i$
γ α_2 β
Precursor molecules
Second messengers
G protein 1
G protein 2
Inside of cell

42.14 Some Hormones Can Activate a Variety of Signal Transduction Pathways *(Page 816)*

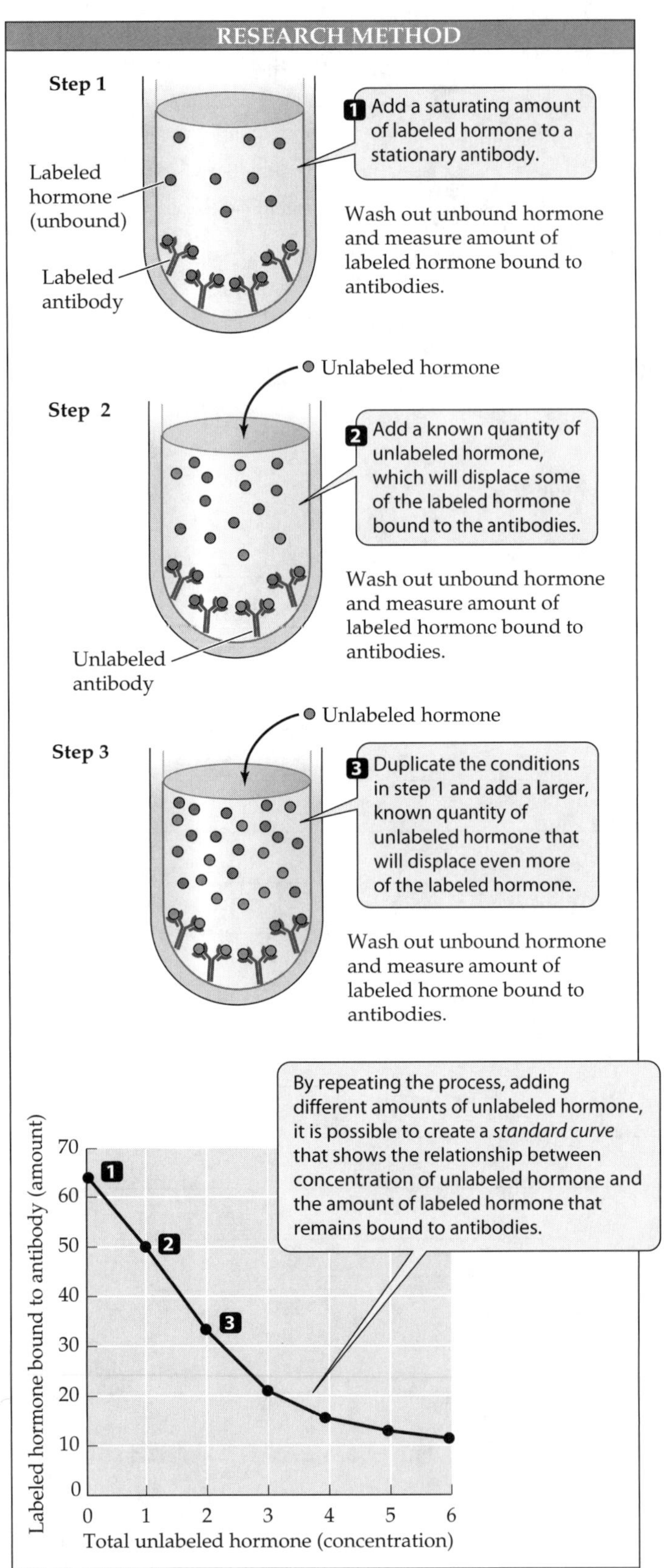
RESEARCH METHOD
Step 1
Labeled hormone (unbound)
Labeled antibody
1 Add a saturating amount of labeled hormone to a stationary antibody.
Wash out unbound hormone and measure amount of labeled hormone bound to antibodies.
Unlabeled hormone
Step 2
2 Add a known quantity of unlabeled hormone, which will displace some of the labeled hormone bound to the antibodies.
Wash out unbound hormone and measure amount of labeled hormone bound to antibodies.
Unlabeled antibody
Unlabeled hormone
Step 3
3 Duplicate the conditions in step 1 and add a larger, known quantity of unlabeled hormone that will displace even more of the labeled hormone.
Wash out unbound hormone and measure amount of labeled hormone bound to antibodies.
By repeating the process, adding different amounts of unlabeled hormone, it is possible to create a standard curve that shows the relationship between concentration of unlabeled hormone and the amount of labeled hormone that remains bound to antibodies.
Labeled hormone bound to antibody (amount)
70
60
50
40
30
20
10
0
0 1 2 3 4 5 6
Total unlabeled hormone (concentration)

42.15 An Immunoassay Measures Hormone Concentration
(Page 817)

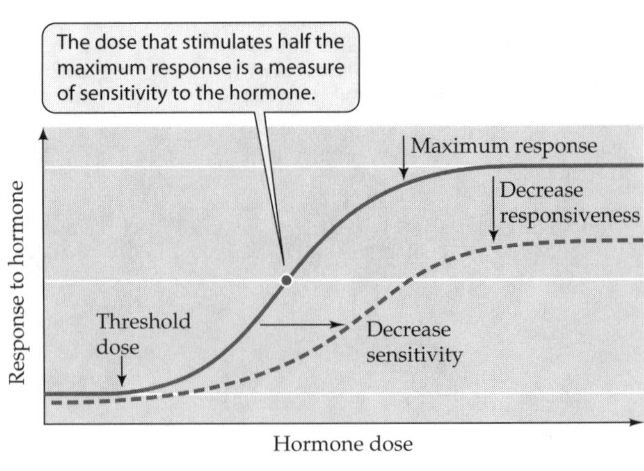
The dose that stimulates half the maximum response is a measure of sensitivity to the hormone.
Maximum response
Decrease responsiveness
Threshold dose
Decrease sensitivity
Response to hormone
Hormone dose

42.16 Dose–Response Curves Quantify Response to a Hormone
(Page 817)

43 Animal Reproduction

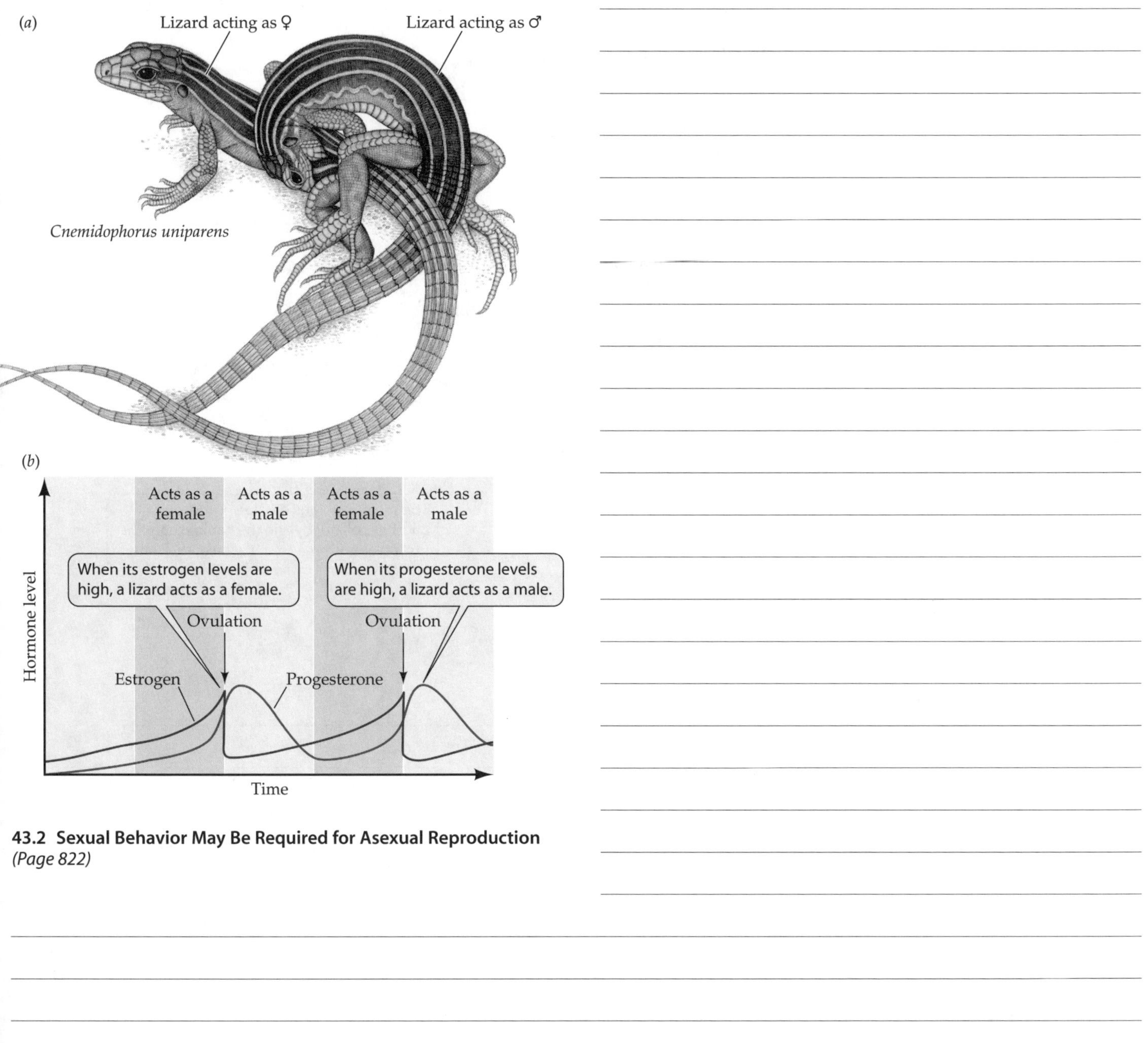

(a)
Lizard acting as ♀
Lizard acting as ♂
Cnemidophorus uniparens
(b)
Acts as a female
Acts as a male
Acts as a female
Acts as a male
When its estrogen levels are high, a lizard acts as a female.
When its progesterone levels are high, a lizard acts as a male.
Ovulation
Ovulation
Hormone level
Estrogen
Progesterone
Time

43.2 Sexual Behavior May Be Required for Asexual Reproduction
(Page 822)

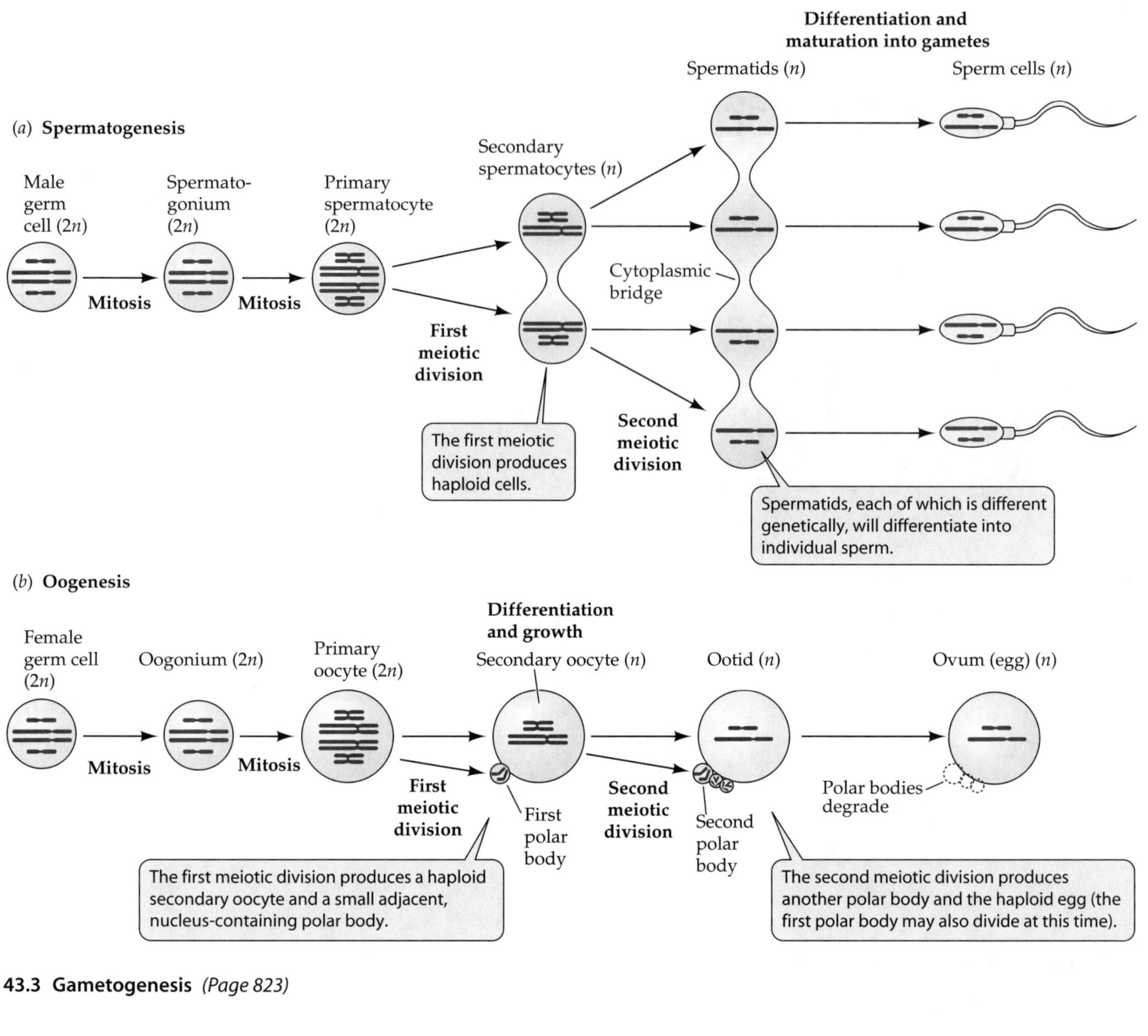

(a) Spermatogenesis
Male germ cell (2n)
Mitosis
Spermatogonium (2n)
Mitosis
Primary spermatocyte (2n)
First meiotic division
Secondary spermatocytes (n)
The first meiotic division produces haploid cells.
Second meiotic division
Cytoplasmic bridge
Differentiation and maturation into gametes
Spermatids (n)
Sperm cells (n)
Spermatids, each of which is different genetically, will differentiate into individual sperm.
(b) Oogenesis
Female germ cell (2n)
Mitosis
Oogonium (2n)
Mitosis
Primary oocyte (2n)
First meiotic division
Differentiation and growth
Secondary oocyte (n)
First polar body
The first meiotic division produces a haploid secondary oocyte and a small adjacent, nucleus-containing polar body.
Second meiotic division
Ootid (n)
Second polar body
Ovum (egg) (n)
Polar bodies degrade
The second meiotic division produces another polar body and the haploid egg (the first polar body may also divide at this time).

43.3 Gametogenesis *(Page 823)*

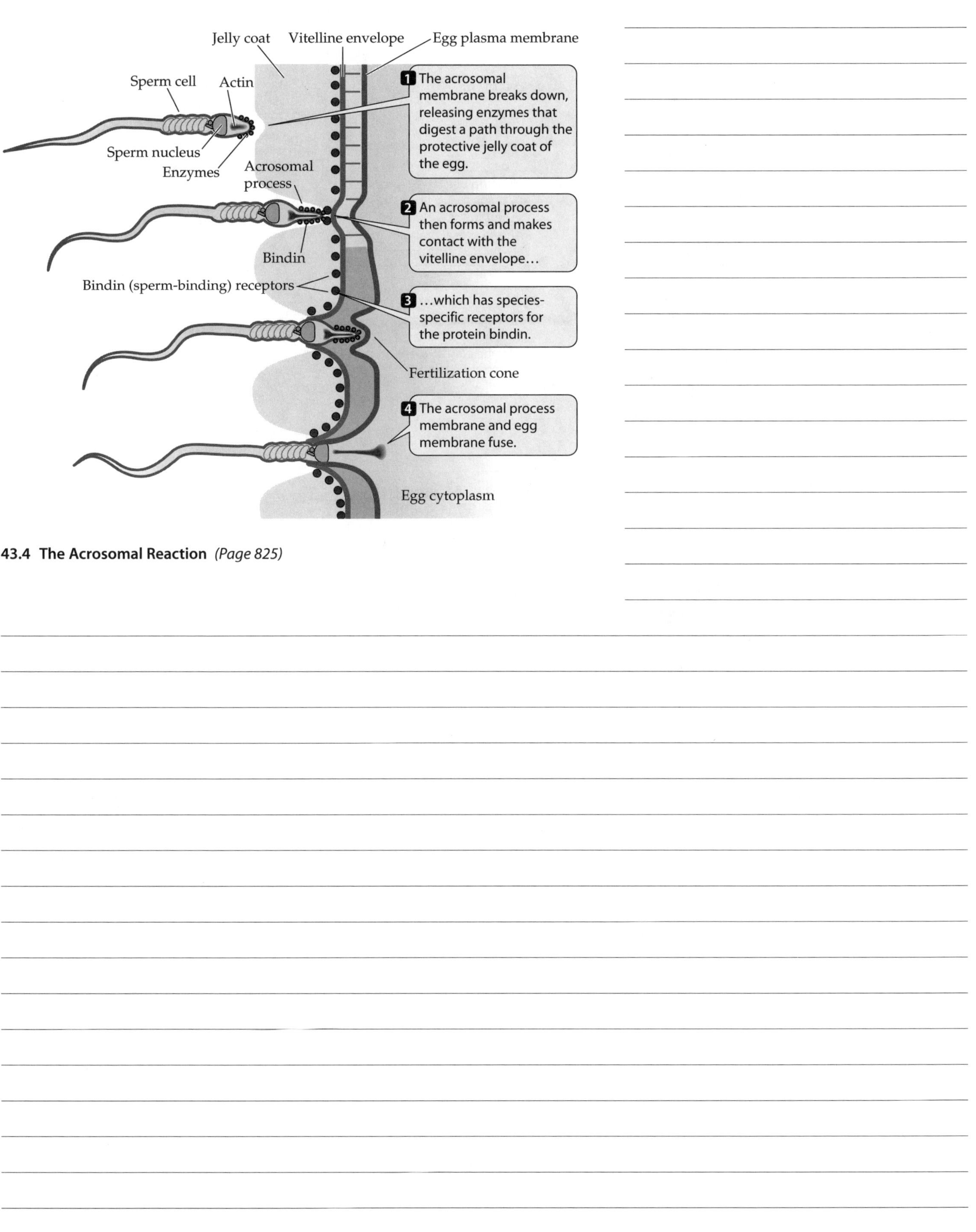

Jelly coat
Vitelline envelope
Egg plasma membrane
Sperm cell
Actin
Sperm nucleus
Enzymes
Acrosomal process
Bindin
Bindin (sperm-binding) receptors
Fertilization cone
Egg cytoplasm
1 The acrosomal membrane breaks down, releasing enzymes that digest a path through the protective jelly coat of the egg.
2 An acrosomal process then forms and makes contact with the vitelline envelope…
3 …which has species-specific receptors for the protein bindin.
4 The acrosomal process membrane and egg membrane fuse.

43.4 The Acrosomal Reaction *(Page 825)*

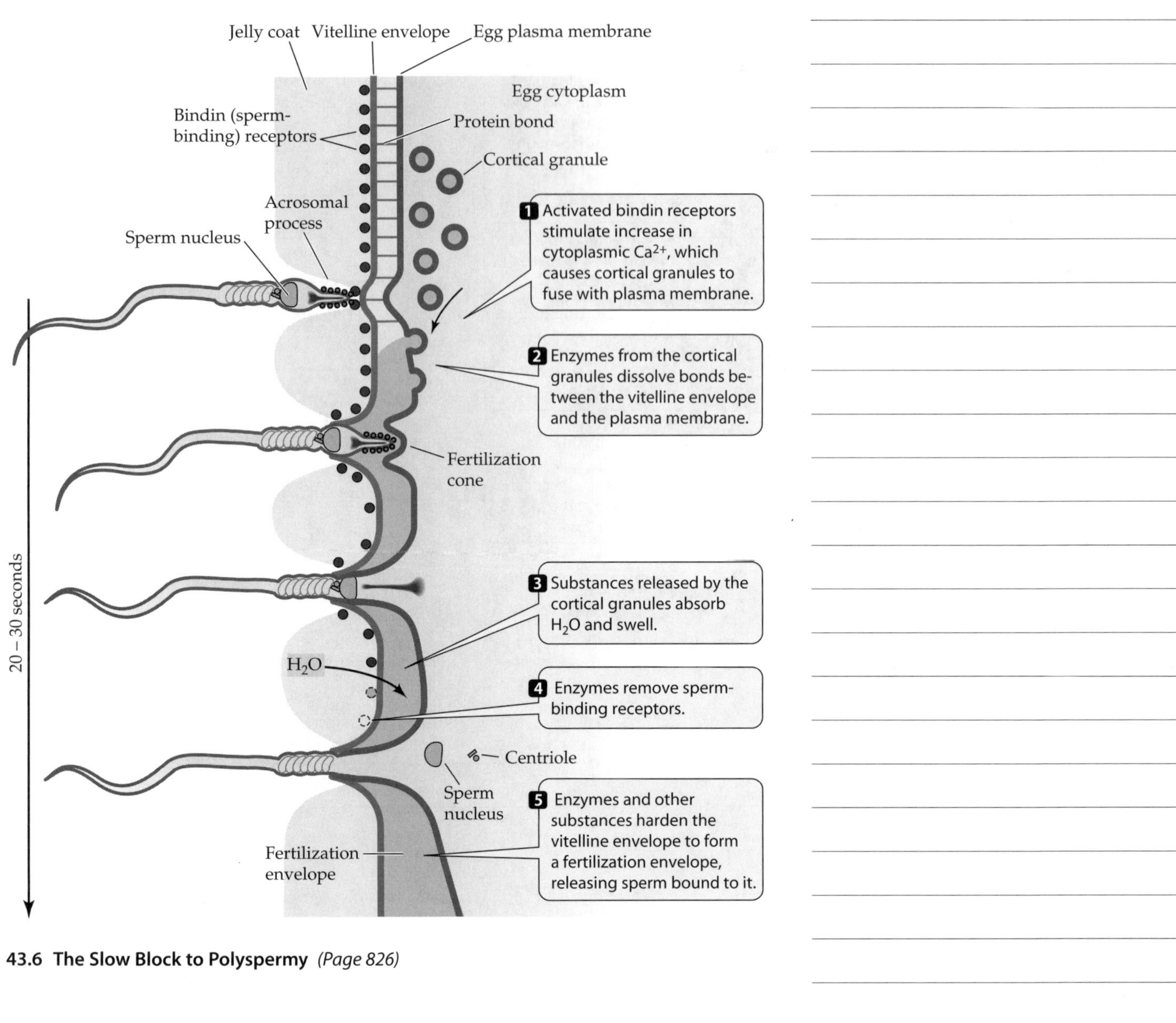

Jelly coat
Vitelline envelope
Egg plasma membrane
Egg cytoplasm
Bindin (sperm-binding) receptors
Protein bond
Cortical granule
Acrosomal process
Sperm nucleus
1 Activated bindin receptors stimulate increase in cytoplasmic Ca^{2+}, which causes cortical granules to fuse with plasma membrane.
2 Enzymes from the cortical granules dissolve bonds between the vitelline envelope and the plasma membrane.
Fertilization cone
20 – 30 seconds
3 Substances released by the cortical granules absorb H_2O and swell.
H_2O
4 Enzymes remove sperm-binding receptors.
Centriole
Sperm nucleus
5 Enzymes and other substances harden the vitelline envelope to form a fertilization envelope, releasing sperm bound to it.
Fertilization envelope

43.6 The Slow Block to Polyspermy *(Page 826)*

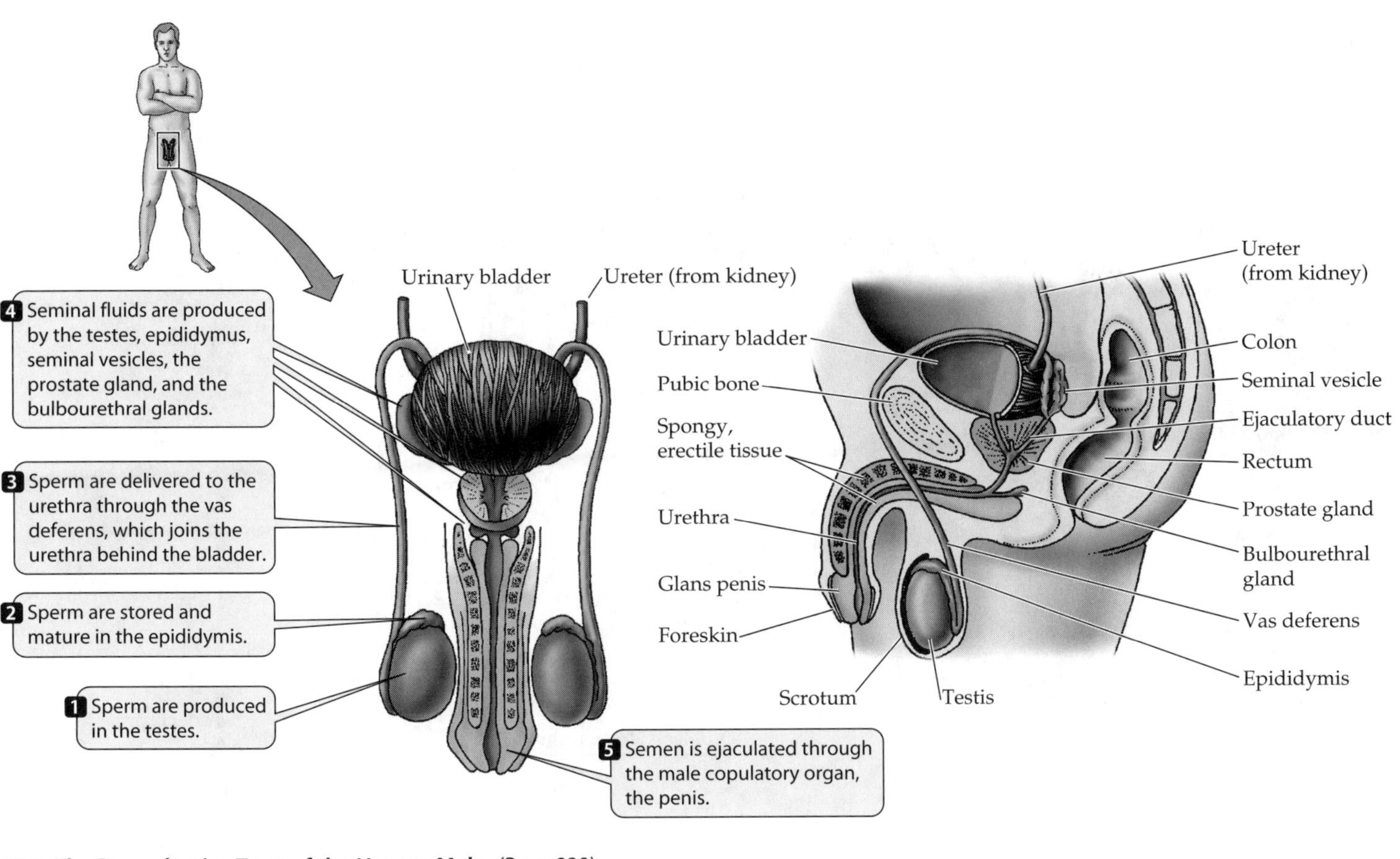

Urinary bladder
Ureter (from kidney)
4 Seminal fluids are produced by the testes, epididymus, seminal vesicles, the prostate gland, and the bulbourethral glands.
3 Sperm are delivered to the urethra through the vas deferens, which joins the urethra behind the bladder.
2 Sperm are stored and mature in the epididymis.
1 Sperm are produced in the testes.
5 Semen is ejaculated through the male copulatory organ, the penis.
Urinary bladder
Pubic bone
Spongy, erectile tissue
Urethra
Glans penis
Foreskin
Scrotum
Testis
Ureter (from kidney)
Colon
Seminal vesicle
Ejaculatory duct
Rectum
Prostate gland
Bulbourethral gland
Vas deferens
Epididymis

43.8 The Reproductive Tract of the Human Male *(Page 829)*

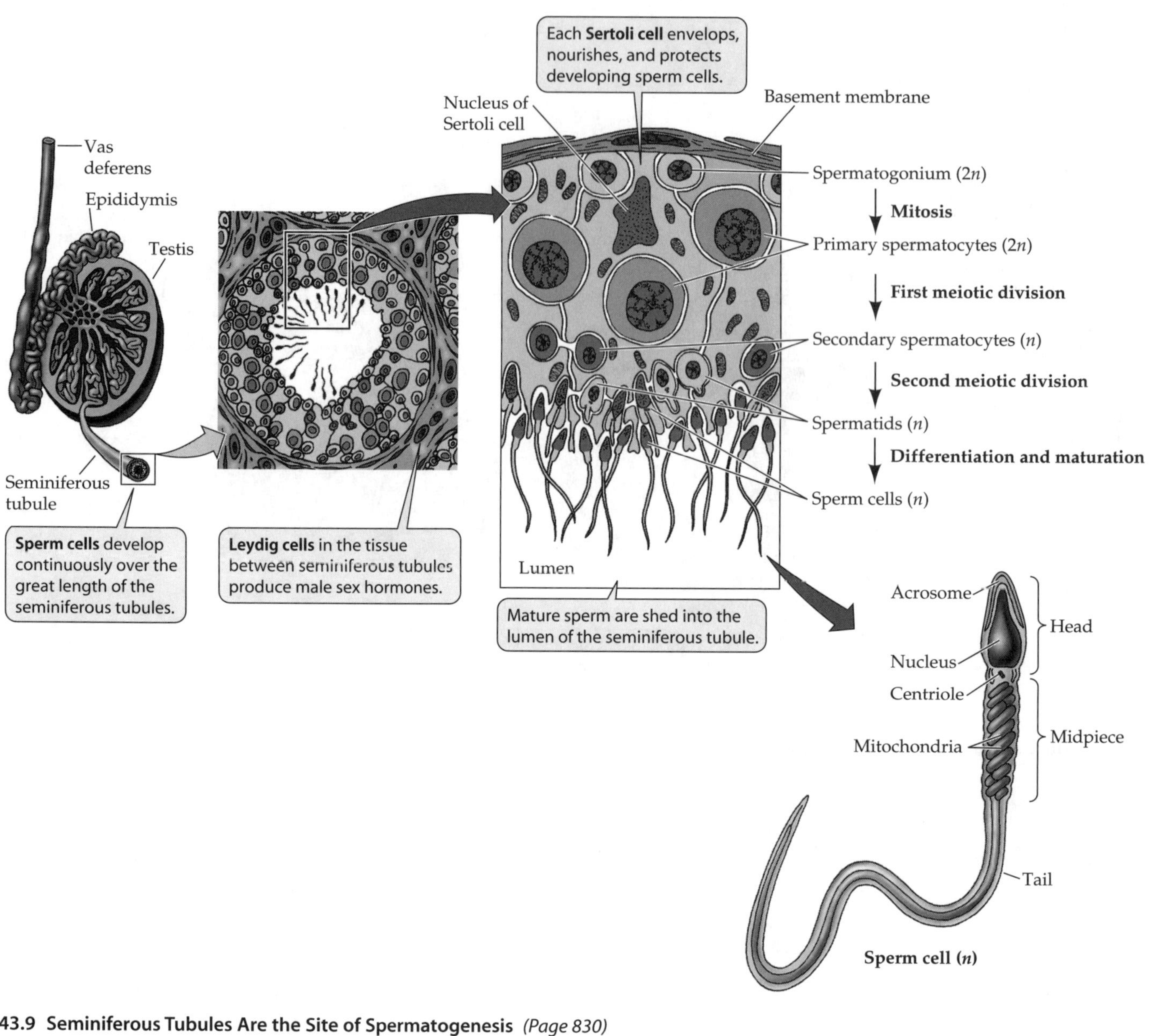
Vas deferens
Epididymis
Testis
Seminiferous tubule
Sperm cells develop continuously over the great length of the seminiferous tubules.
Leydig cells in the tissue between seminiferous tubules produce male sex hormones.
Each **Sertoli cell** envelops, nourishes, and protects developing sperm cells.
Nucleus of Sertoli cell
Basement membrane
Spermatogonium (2*n*)
Mitosis
Primary spermatocytes (2*n*)
First meiotic division
Secondary spermatocytes (*n*)
Second meiotic division
Spermatids (*n*)
Differentiation and maturation
Sperm cells (*n*)
Lumen
Mature sperm are shed into the lumen of the seminiferous tubule.
Acrosome
Nucleus
Head
Centriole
Mitochondria
Midpiece
Tail
Sperm cell (*n*)

43.9 Seminiferous Tubules Are the Site of Spermatogenesis *(Page 830)*

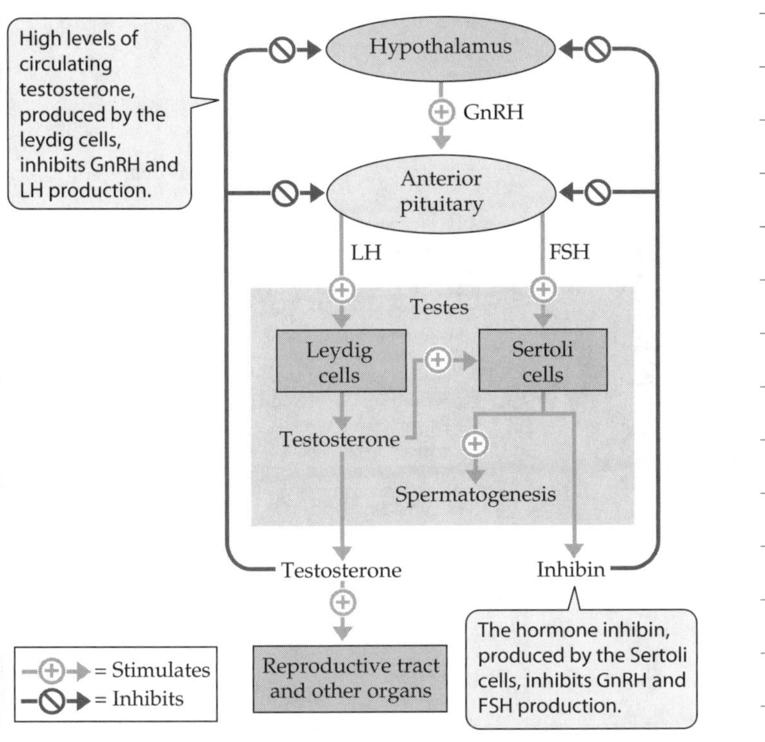
High levels of circulating testosterone, produced by the leydig cells, inhibits GnRH and LH production.
Hypothalamus
GnRH
Anterior pituitary
LH
FSH
Testes
Leydig cells
Sertoli cells
Testosterone
Spermatogenesis
Testosterone
Inhibin
Reproductive tract and other organs
= Stimulates
= Inhibits
The hormone inhibin, produced by the Sertoli cells, inhibits GnRH and FSH production.

43.10 Hormones Control the Male Reproductive System *(Page 831)*

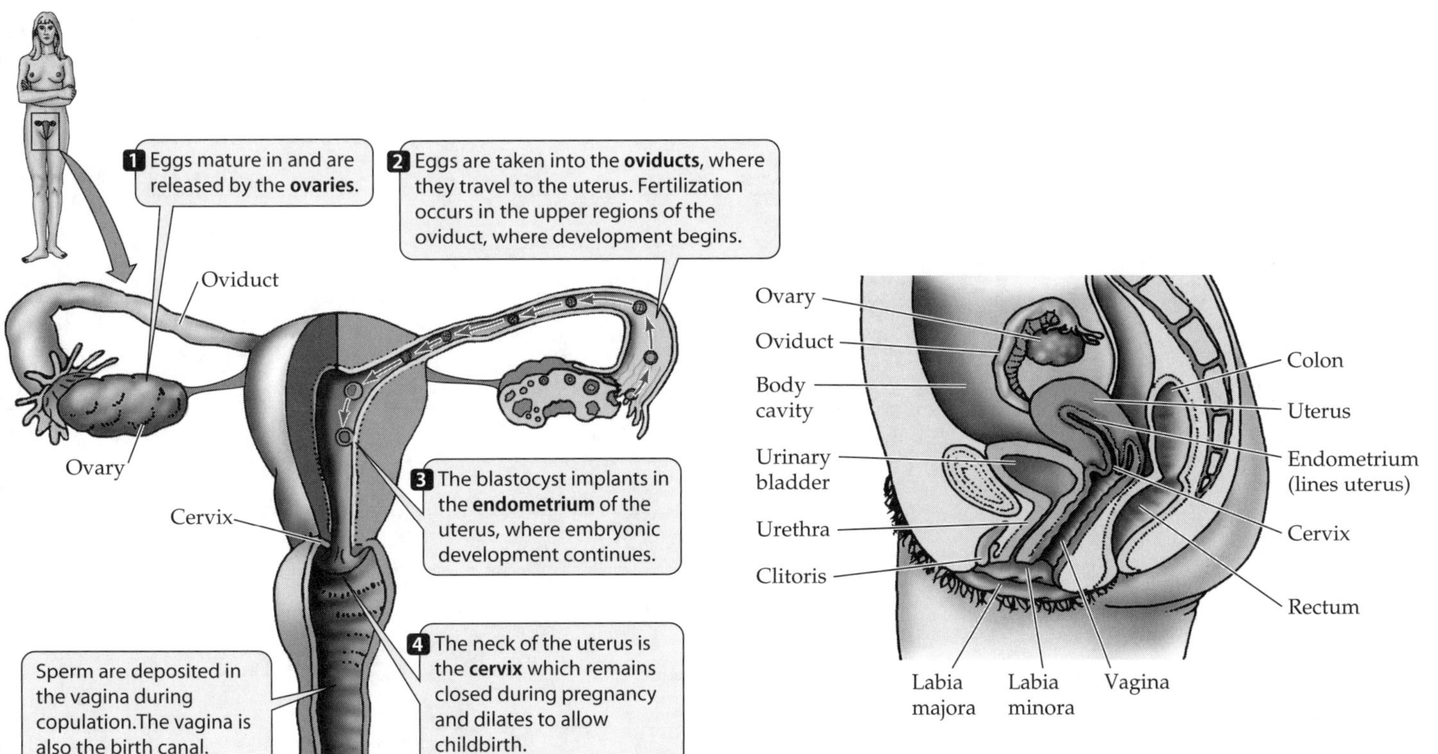
1 Eggs mature in and are released by the **ovaries**.
2 Eggs are taken into the **oviducts**, where they travel to the uterus. Fertilization occurs in the upper regions of the oviduct, where development begins.
Oviduct
Ovary
Cervix
3 The blastocyst implants in the **endometrium** of the uterus, where embryonic development continues.
4 The neck of the uterus is the **cervix** which remains closed during pregnancy and dilates to allow childbirth.
Sperm are deposited in the vagina during copulation.The vagina is also the birth canal.
Ovary
Oviduct
Body cavity
Urinary bladder
Urethra
Clitoris
Labia majora
Labia minora
Vagina
Colon
Uterus
Endometrium (lines uterus)
Cervix
Rectum

43.11 The Reproductive Tract of the Human Female *(Page 832)*

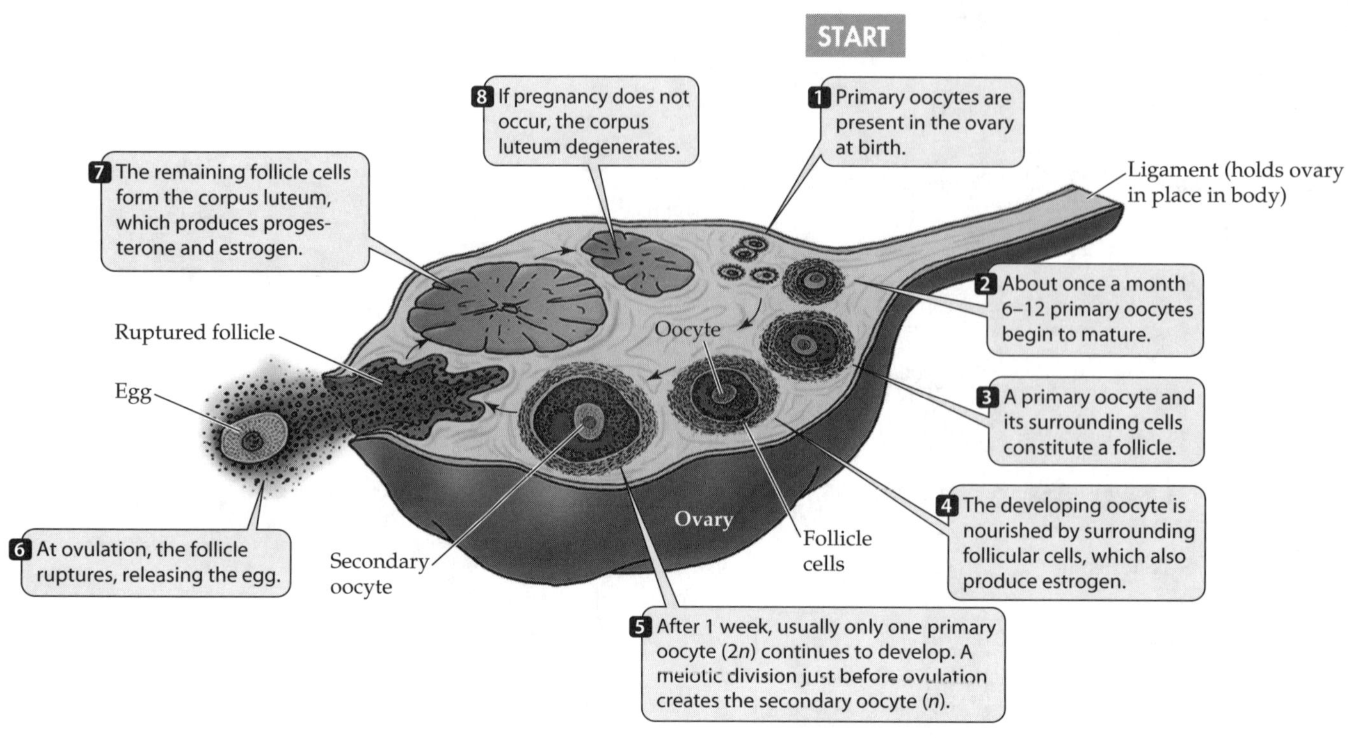
START
1 Primary oocytes are present in the ovary at birth.
Ligament (holds ovary in place in body)
2 About once a month 6–12 primary oocytes begin to mature.
3 A primary oocyte and its surrounding cells constitute a follicle.
4 The developing oocyte is nourished by surrounding follicular cells, which also produce estrogen.
5 After 1 week, usually only one primary oocyte (2n) continues to develop. A meiotic division just before ovulation creates the secondary oocyte (n).
6 At ovulation, the follicle ruptures, releasing the egg.
7 The remaining follicle cells form the corpus luteum, which produces progesterone and estrogen.
8 If pregnancy does not occur, the corpus luteum degenerates.
Ruptured follicle
Egg
Oocyte
Ovary
Follicle cells
Secondary oocyte

43.12 The Ovarian Cycle *(Page 833)*

FSH and LH are under control of GnRH from the hypothalamus and the ovarian hormones estrogen and progesterone.

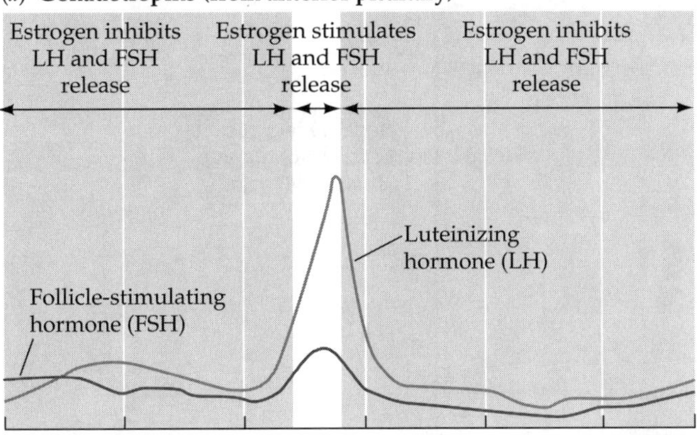
(a) Gonadotropins (from anterior pituitary)
Estrogen inhibits LH and FSH release
Estrogen stimulates LH and FSH release
Estrogen inhibits LH and FSH release
Luteinizing hormone (LH)
Follicle-stimulating hormone (FSH)

FSH stimulates the development of follicles; the LH surge causes ovulation and then the development of the corpus luteum.

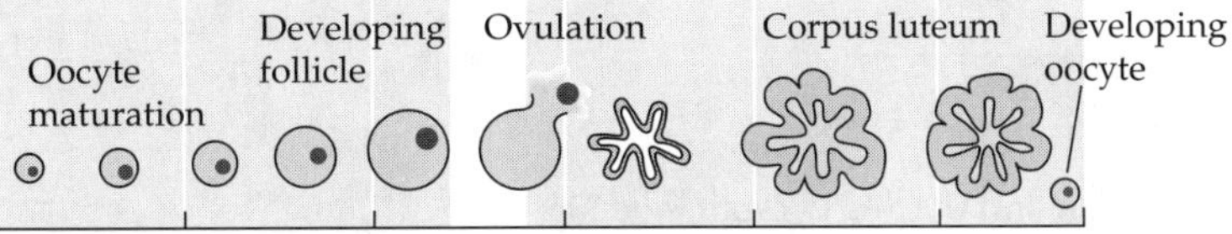
(b) Events in ovary (ovarian cycle)
Oocyte maturation
Developing follicle
Ovulation
Corpus luteum
Developing oocyte

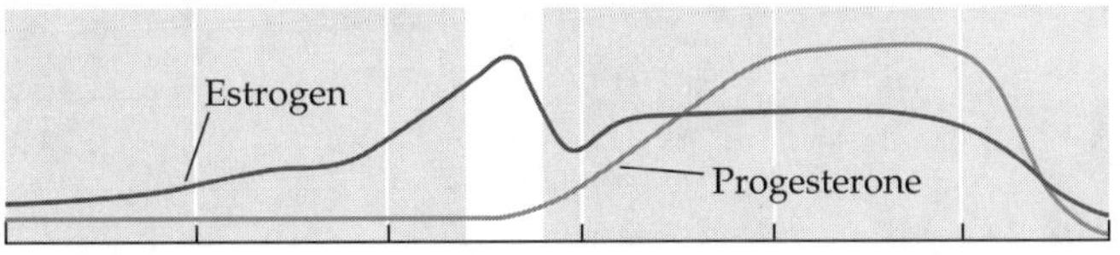
(c) Ovarian hormones and the uterine cycle
Estrogen
Progesterone

Estrogen and progesterone stimulate the development of the endometrium in preparation for pregnancy.

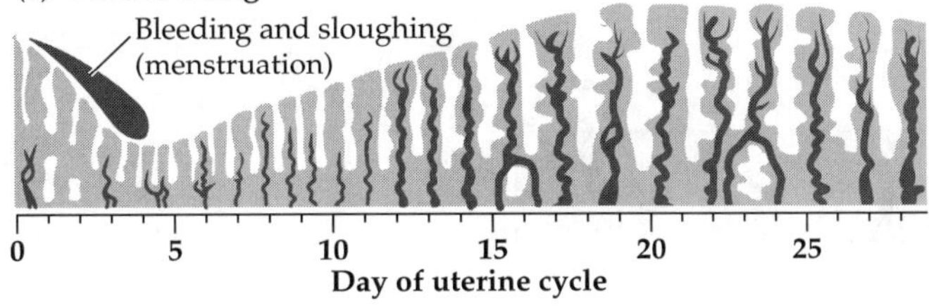
(d) Uterine lining
Bleeding and sloughing (menstruation)
0
5
10
15
20
25
Day of uterine cycle

43.13 The Ovarian and Uterine Cycles *(Page 834)*

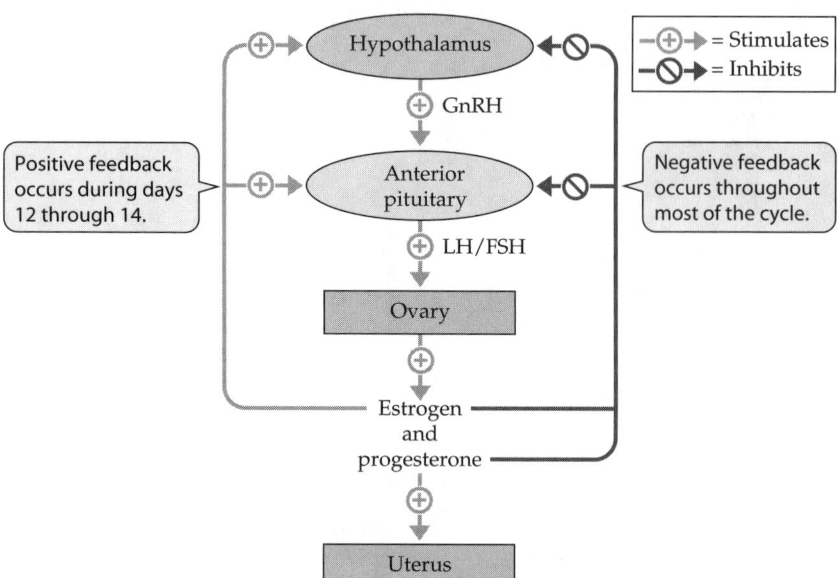
= Stimulates
= Inhibits
Hypothalamus
GnRH
Positive feedback occurs during days 12 through 14.
Anterior pituitary
Negative feedback occurs throughout most of the cycle.
LH/FSH
Ovary
Estrogen and progesterone
Uterus

43.14 Hormones Control the Ovarian and Uterine Cycles *(Page 835)*

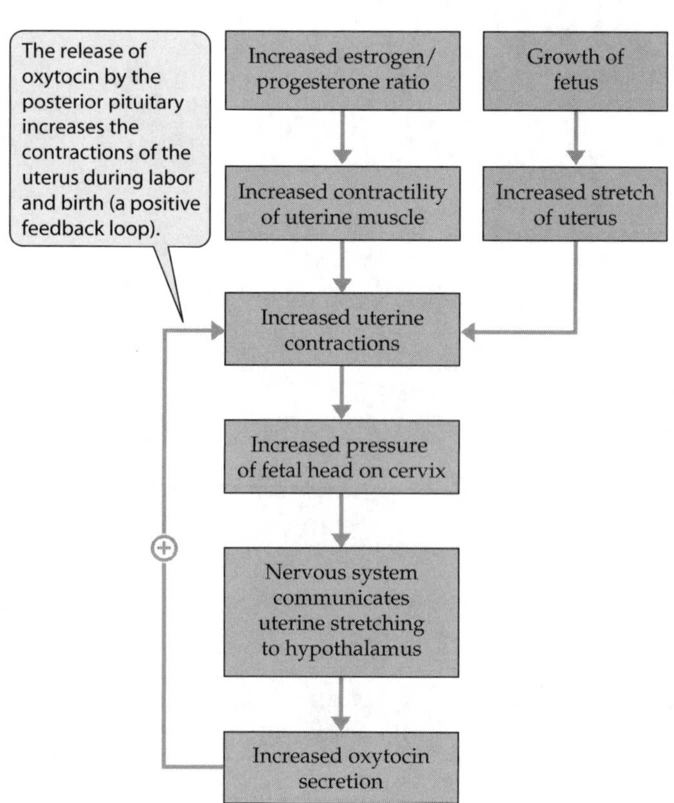
The release of oxytocin by the posterior pituitary increases the contractions of the uterus during labor and birth (a positive feedback loop).
Increased estrogen/ progesterone ratio
Growth of fetus
Increased contractility of uterine muscle
Increased stretch of uterus
Increased uterine contractions
Increased pressure of fetal head on cervix
Nervous system communicates uterine stretching to hypothalamus
Increased oxytocin secretion

43.15 Control of Uterine Contractions and Childbirth *(Page 836)*

43.1 Methods of Contraception

METHOD	MODE OF ACTION	FAILURE RATE[a]
Rhythm method	Abstinence near time of ovulation	15–35
Coitus interruptus	Prevents sperm from reaching egg	10–40
Condom	Prevents sperm from entering vagina	3–20
Diaphragm/jelly	Prevents sperm from entering uterus; kills sperm	3–25
Vaginal jelly or foam	Kills sperm; blocks sperm movement	3–30
Douche	Supposedly flushes sperm from vagina	80
Birth control pills	Prevent ovulation	0–3
Vasectomy	Prevents release of sperm	0.0–0.15
Tubal ligation	Prevents egg from entering uterus	0.0–0.05
Intrauterine device (IUD)	Prevents implantation of fertilized egg	0.5–6
RU-486	Prevents development of fertilized egg	0–15
Unprotected	No form of birth control	85

[a]Number of pregnancies per 100 women per year

(Page 837)

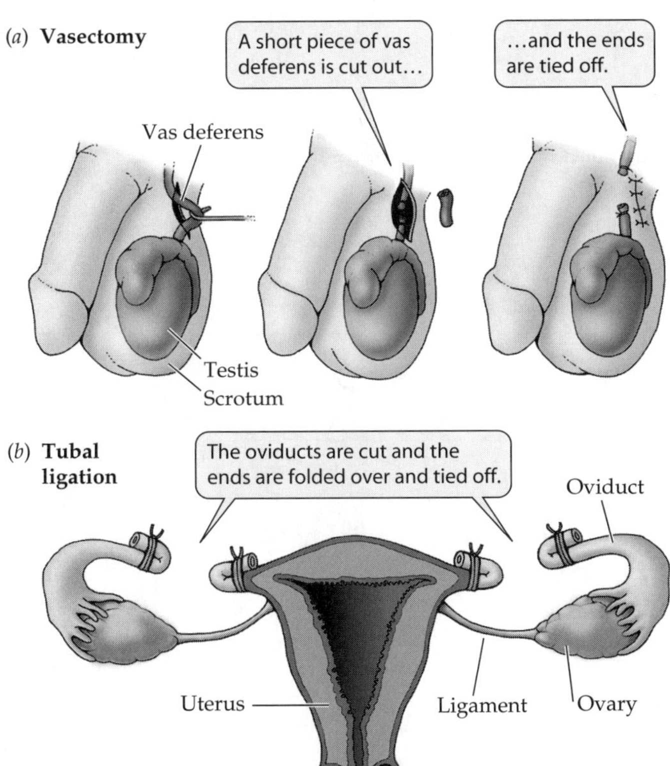
(a) Vasectomy
A short piece of vas deferens is cut out…
…and the ends are tied off.
Vas deferens
Testis
Scrotum
(b) Tubal ligation
The oviducts are cut and the ends are folded over and tied off.
Oviduct
Uterus
Ligament
Ovary

43.16 Sterilization Techniques *(Page 839)*

43.2 Some Sexually Transmitted Diseases

DISEASE	INCIDENCE IN UNITED STATES	SYMPTOMS
Syphilis	80,000 new cases/yr	Primary stage (weeks): skin lesion (chancre) at site of infection Secondary stage (months): skin rash and flu-like symptoms, may be followed by a latent period Tertiary stage (years): deterioration of the cardiovascular and central nervous systems
Gonorrhea	800,000 new cases/yr	Pus-filled discharge from penis or vagina; burning urination. Infection can also start in throat or rectum
Chlamydia	>4,000,000 new cases/yr	Symptoms similar to gonorrhea, although often there are no obvious symptoms. Can result in pelvic inflammatory disease in females (see below)
Genital herpes	500,000 new cases/yr	Small blisters that can cause itching or burning sensations are accompanied by inflammation and by secondary infections
Genital warts	10% of adults infected	Small growths on genital tissues. Increases risk of cervical cancer in women
Hepatitis B	5–20% of population	Fatigue, fever, nausea, loss of appetite, jaundice, abdominal pain, muscle and joint pain. Can lead to destruction of liver or liver cancer
Pelvic inflammatory disease	1,000,000 new cases/yr (females only)	Fever and abdominal pain. Frequently results in sterility
AIDS	Approximately 900,000 cases[a]	Failure of the immune system (see Chapter 18)

[a]AIDS is widespread in other parts of the world, most notably in the southern part of the African continent, where some 9 million people are infected. The virus is also spreading rapidly in India and Southeast Asia.

(Pages 840–841)

CAUSE	MODE OF TRANSMISSION	CURE/TREATMENT
Spirochete bacterium (*Treponema pallidum*) that penetrates mucosal membranes and abraded skin	Intimate sexual contact (including kissing)	Antibiotics
Bacterium (*Neisseria gonorrheoeae*)	Communicated across mucous membranes	Antibiotics (but antibiotic-resistant strains have arisen)
Bacterium (*Chlamydia trachomatis*)	Communicated across mucous membranes	Antibiotics
Herpes simplex virus	Communicated by contact with infected surfaces, which can be mucous membranes or skin	No cure. Symptoms can be alleviated. Antiviral drugs may lessen subsequent outbreaks
Human papillomavirus	Communicated across mucous membranes through sexual contact	No cure for the virus. Warts can be removed surgically or by burning, freezing, or chemical treatment
Virus	Sexual contact or blood transfusions	No cure. Symptoms can be treated. A vaccine is available that can protect only if given before infection occurs
A variety of bacteria that migrate to the uterus and fallopian tubes	Sexual intercourse	Antibiotics
HIV (see Chapter 13)	The virus enters the bloodstream via cuts or abrasions, including minute ones in the genitalia. Spread primarily by intimate sexual contact, but can also be transmitted via contaminated needles	No cure. Treatments with a variety of medications can slow the course of the infection

44 Neurons and Nervous Systems

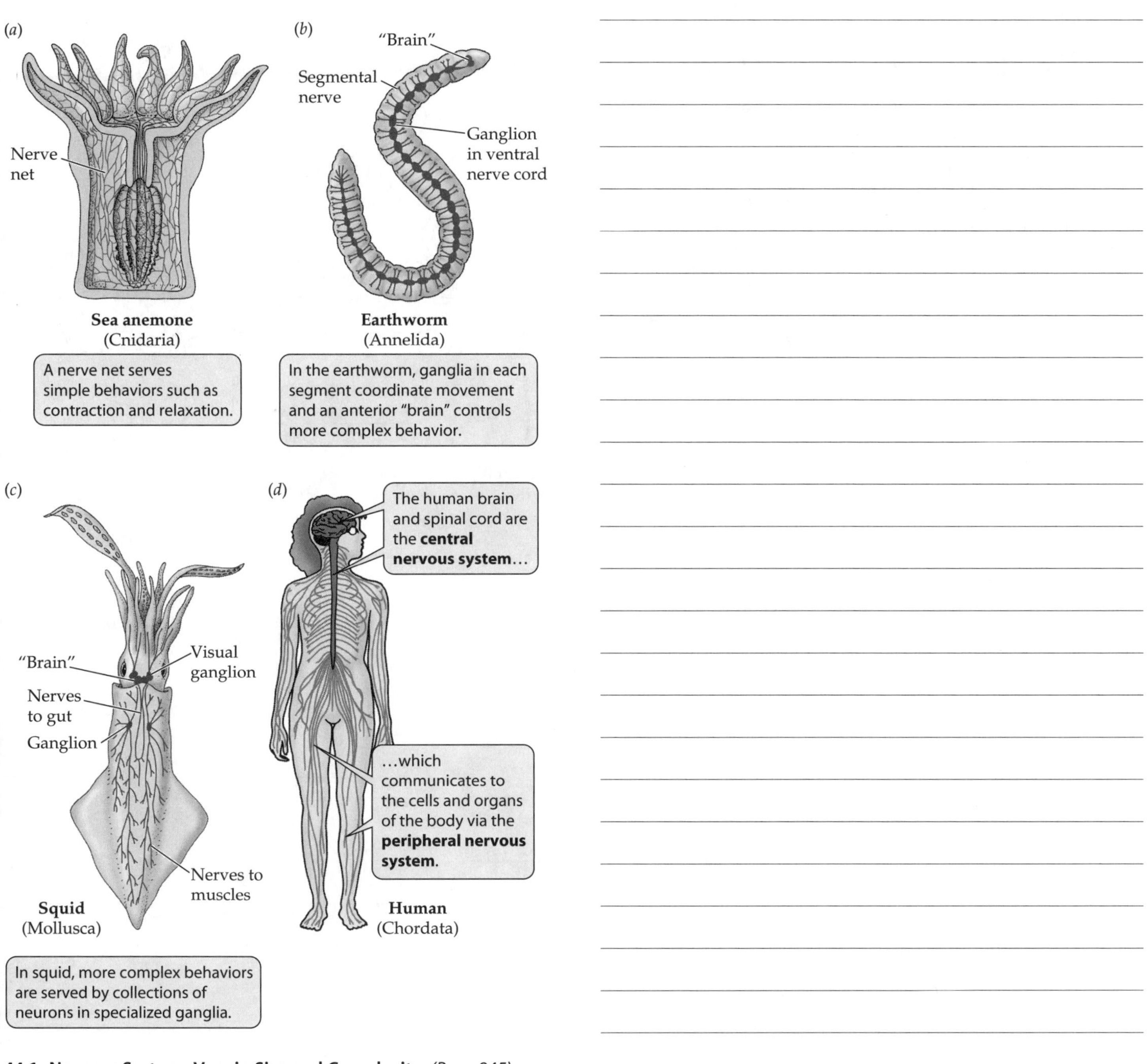

(a)
Nerve net
Sea anemone
(Cnidaria)
A nerve net serves simple behaviors such as contraction and relaxation.
(b)
"Brain"
Segmental nerve
Ganglion in ventral nerve cord
Earthworm
(Annelida)
In the earthworm, ganglia in each segment coordinate movement and an anterior "brain" controls more complex behavior.
(c)
"Brain"
Visual ganglion
Nerves to gut
Ganglion
Nerves to muscles
Squid
(Mollusca)
In squid, more complex behaviors are served by collections of neurons in specialized ganglia.
(d)
The human brain and spinal cord are the **central nervous system**...
...which communicates to the cells and organs of the body via the **peripheral nervous system**.
Human
(Chordata)

44.1 Nervous Systems Vary in Size and Complexity *(Page 845)*

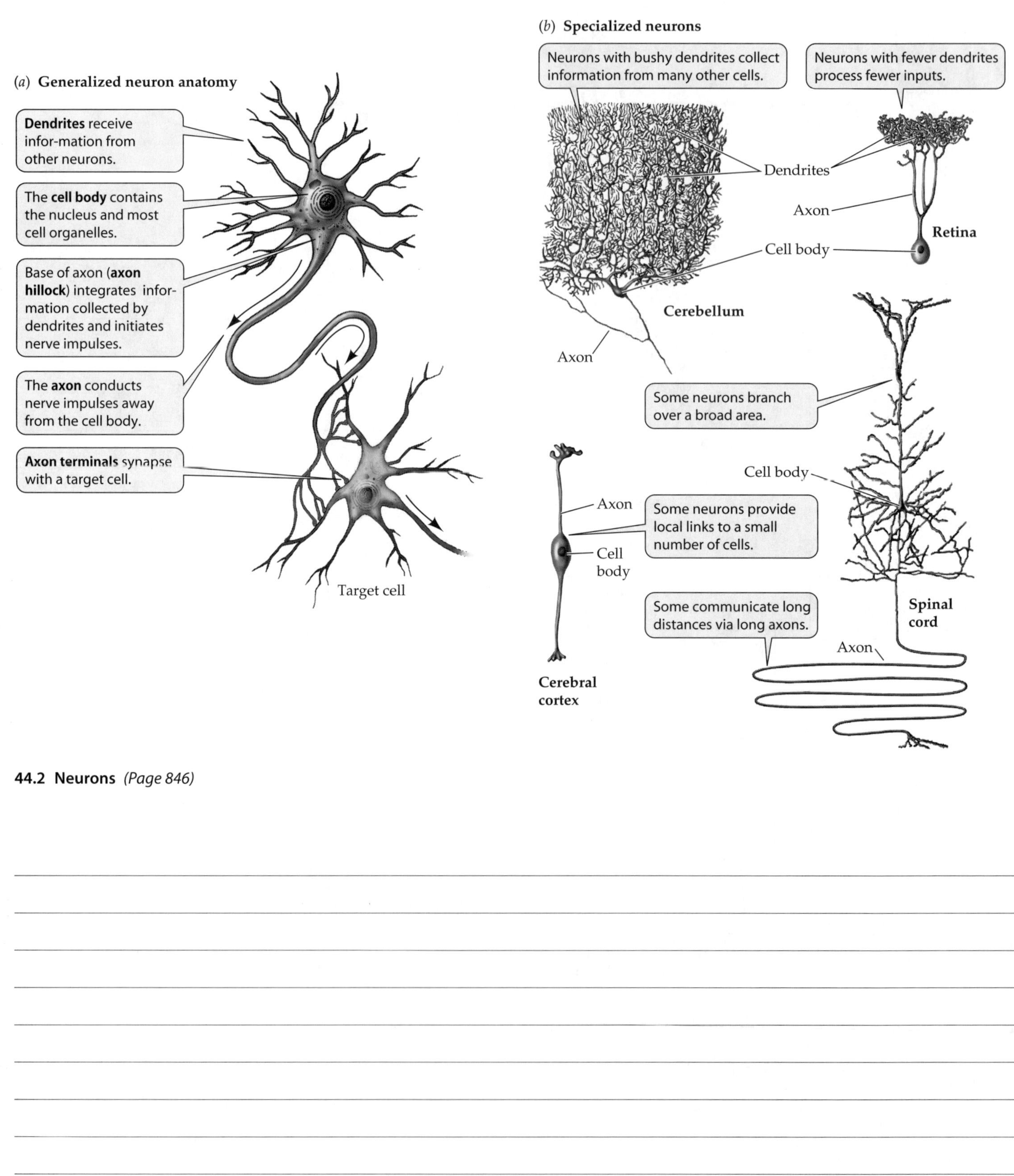
(a) Generalized neuron anatomy
Dendrites receive infor-mation from other neurons.
The cell body contains the nucleus and most cell organelles.
Base of axon (axon hillock) integrates infor-mation collected by dendrites and initiates nerve impulses.
The axon conducts nerve impulses away from the cell body.
Axon terminals synapse with a target cell.
Target cell
(b) Specialized neurons
Neurons with bushy dendrites collect information from many other cells.
Neurons with fewer dendrites process fewer inputs.
Dendrites
Axon
Retina
Cell body
Cerebellum
Axon
Some neurons branch over a broad area.
Cell body
Axon
Some neurons provide local links to a small number of cells.
Cell body
Some communicate long distances via long axons.
Spinal cord
Axon
Cerebral cortex

44.2 Neurons *(Page 846)*

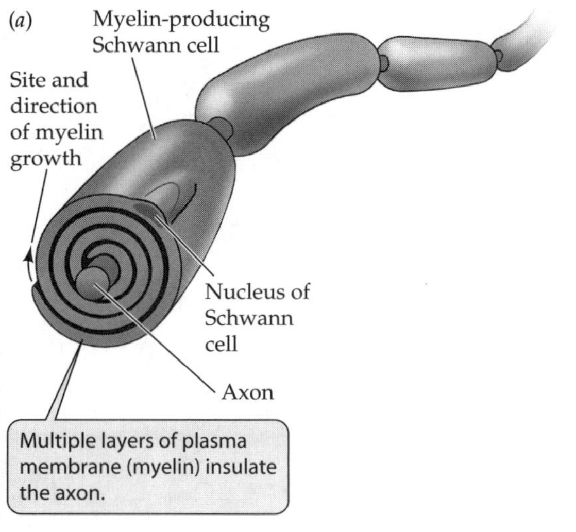
(a)
Myelin-producing Schwann cell
Site and direction of myelin growth
Nucleus of Schwann cell
Axon
Multiple layers of plasma membrane (myelin) insulate the axon.

44.3 Wrapping Up an Axon *(Page 847)*

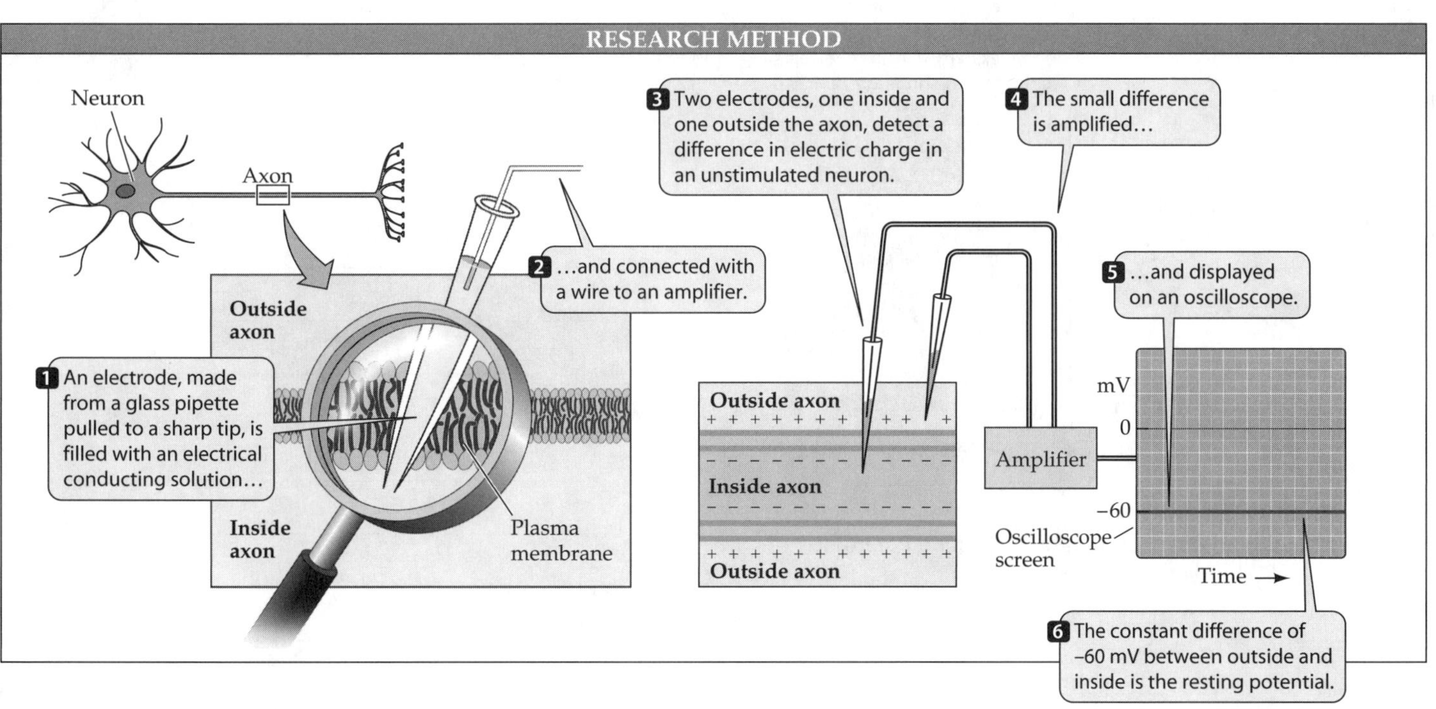
RESEARCH METHOD
Neuron
Axon
Outside axon
Inside axon
Plasma membrane
1 An electrode, made from a glass pipette pulled to a sharp tip, is filled with an electrical conducting solution…
2 …and connected with a wire to an amplifier.
3 Two electrodes, one inside and one outside the axon, detect a difference in electric charge in an unstimulated neuron.
4 The small difference is amplified…
5 …and displayed on an oscilloscope.
Outside axon
Inside axon
Outside axon
Amplifier
mV
0
–60
Oscilloscope screen
Time
6 The constant difference of –60 mV between outside and inside is the resting potential.

44.4 Measuring the Resting Potential *(Page 848)*

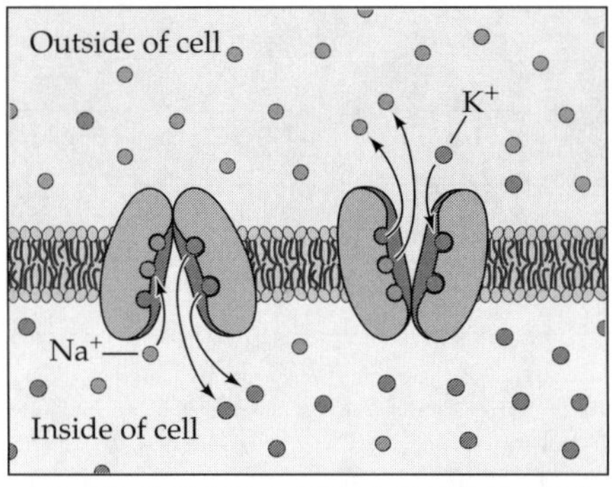
(a) Na+–K+ pump
Outside of cell
K+
Na+
Inside of cell

The Na^+–K^+ pump moves Na^+ and K^+ ions against their concentration gradients.

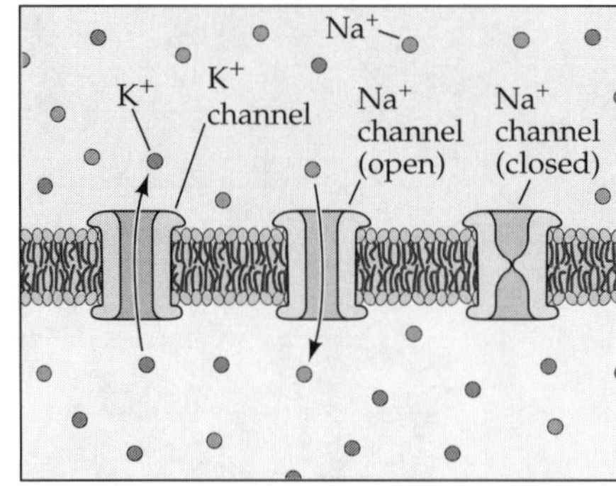
(b) Na+ and K+ channels
Na+
K+
K+ channel
Na+ channel (open)
Na+ channel (closed)

K^+ and Na^+ ions tend to diffuse down their concentration gradients through ion-specific channels.

44.5 Ion Pumps and Channels *(Page 848)*

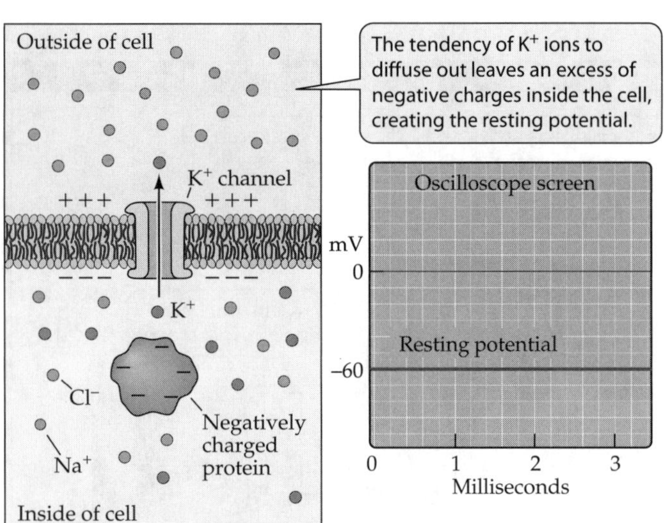
Outside of cell
K+ channel
+++
+++
– – –
– – –
K+
Cl–
Na+
Negatively charged protein
Inside of cell
Oscilloscope screen
mV
0
–60
Resting potential
0
1
2
3
Milliseconds

The tendency of K^+ ions to diffuse out leaves an excess of negative charges inside the cell, creating the resting potential.

44.6 Open Potassium Channels Create the Resting Potential *(Page 849)*

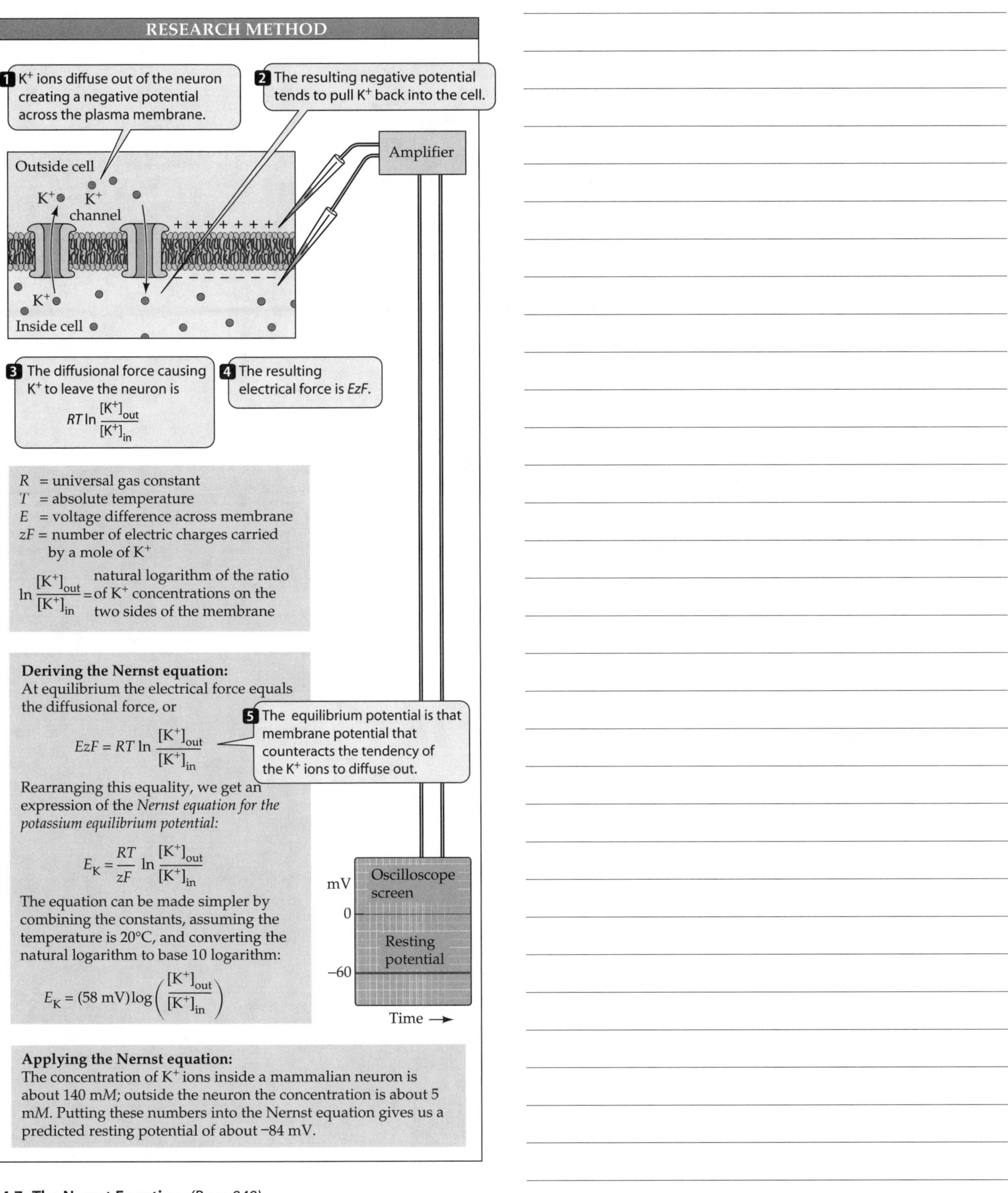
RESEARCH METHOD
1 K^+ ions diffuse out of the neuron creating a negative potential across the plasma membrane.
2 The resulting negative potential tends to pull K^+ back into the cell.
Amplifier
Outside cell
K^+
K^+ channel
+ + + + + + +
– – – – – – –
K^+
Inside cell
3 The diffusional force causing K^+ to leave the neuron is
$RT \ln \frac{[K^+]_{out}}{[K^+]_{in}}$
4 The resulting electrical force is EzF.
R = universal gas constant
T = absolute temperature
E = voltage difference across membrane
zF = number of electric charges carried by a mole of K^+
$\ln \frac{[K^+]_{out}}{[K^+]_{in}}$ = natural logarithm of the ratio of K^+ concentrations on the two sides of the membrane
Deriving the Nernst equation:
At equilibrium the electrical force equals the diffusional force, or
$EzF = RT \ln \frac{[K^+]_{out}}{[K^+]_{in}}$
5 The equilibrium potential is that membrane potential that counteracts the tendency of the K^+ ions to diffuse out.
Rearranging this equality, we get an expression of the Nernst equation for the potassium equilibrium potential:
$E_K = \frac{RT}{zF} \ln \frac{[K^+]_{out}}{[K^+]_{in}}$
The equation can be made simpler by combining the constants, assuming the temperature is 20°C, and converting the natural logarithm to base 10 logarithm:
$E_K = (58\ mV)\log\left(\frac{[K^+]_{out}}{[K^+]_{in}}\right)$
mV
0
–60
Oscilloscope screen
Resting potential
Time →
Applying the Nernst equation:
The concentration of K^+ ions inside a mammalian neuron is about 140 mM; outside the neuron the concentration is about 5 mM. Putting these numbers into the Nernst equation gives us a predicted resting potential of about –84 mV.

44.7 The Nernst Equation *(Page 849)*

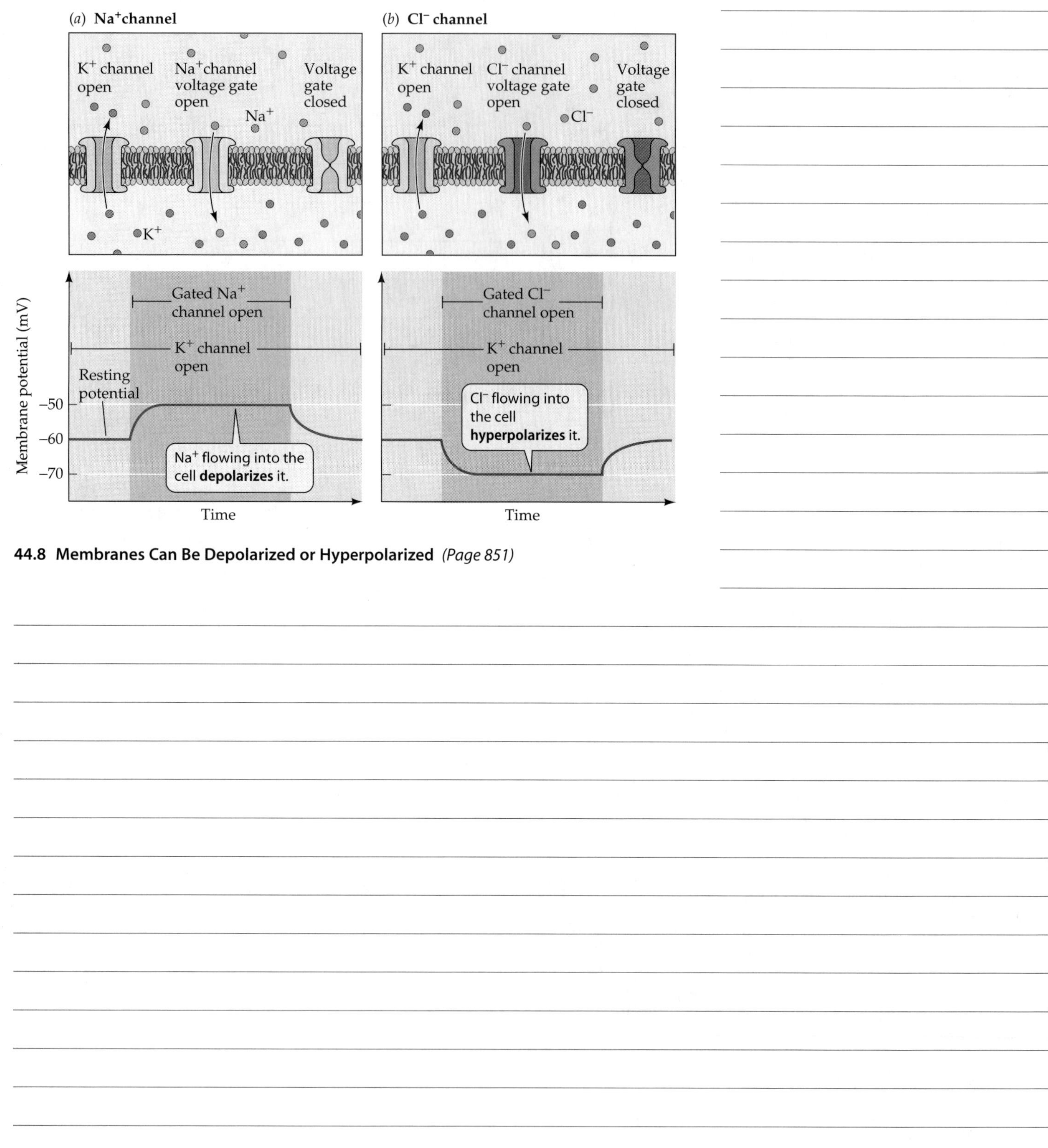
(a) Na⁺channel
K⁺ channel open
Na⁺channel voltage gate open
Voltage gate closed
Na⁺
K⁺
Gated Na⁺ channel open
K⁺ channel open
Resting potential
Membrane potential (mV)
–50
–60
–70
Na⁺ flowing into the cell **depolarizes** it.
Time
(b) Cl⁻ channel
K⁺ channel open
Cl⁻ channel voltage gate open
Voltage gate closed
Cl⁻
Gated Cl⁻ channel open
K⁺ channel open
Cl⁻ flowing into the cell **hyperpolarizes** it.
Time

44.8 Membranes Can Be Depolarized or Hyperpolarized *(Page 851)*

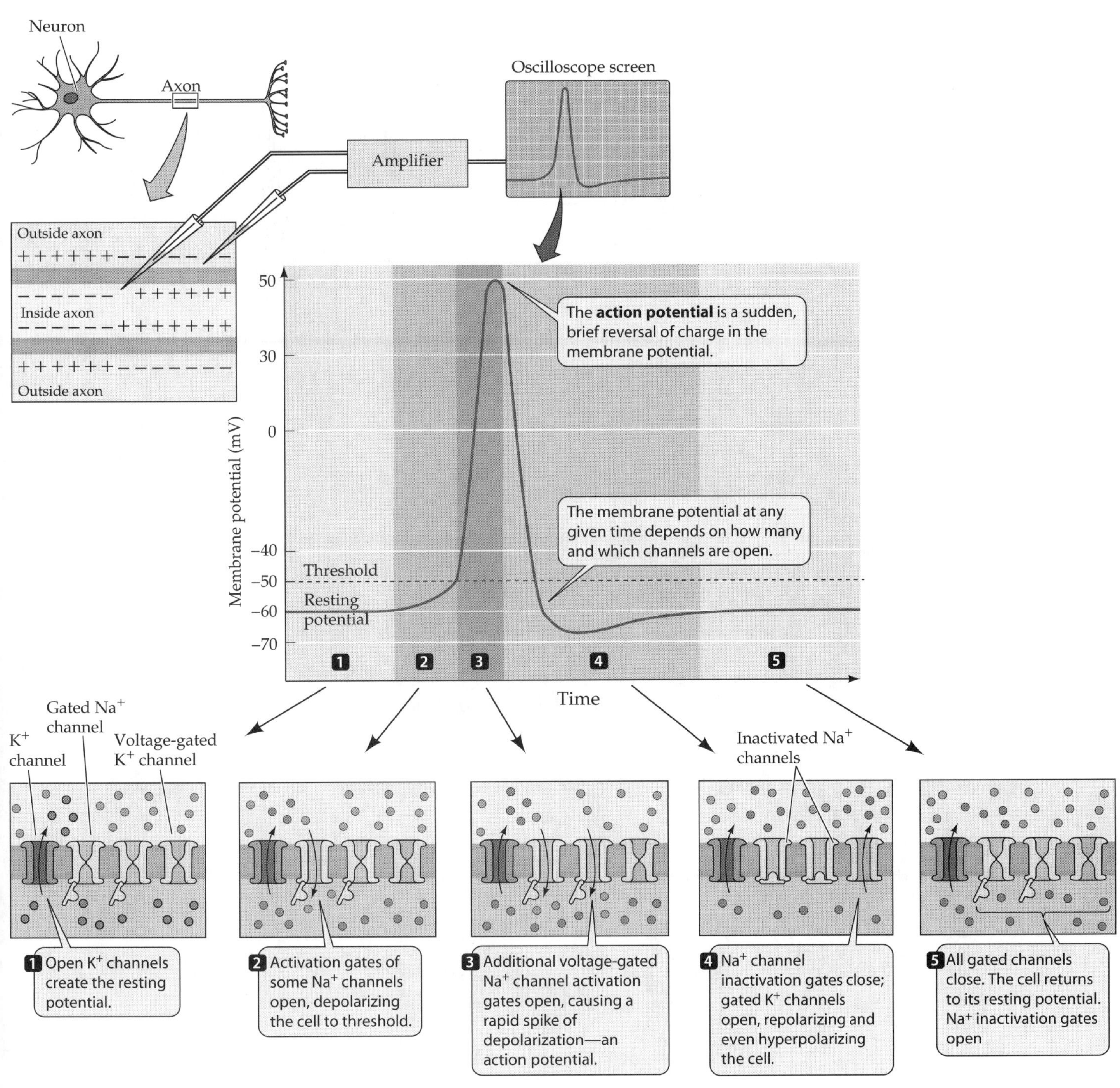
Neuron
Axon
Oscilloscope screen
Amplifier
Outside axon
Inside axon
Outside axon
50
30
0
−40
−50
−60
−70
Membrane potential (mV)
Threshold
Resting potential
The **action potential** is a sudden, brief reversal of charge in the membrane potential.
The membrane potential at any given time depends on how many and which channels are open.
1
2
3
4
5
Time
Gated Na⁺ channel
K⁺ channel
Voltage-gated K⁺ channel
Inactivated Na⁺ channels
1 Open K⁺ channels create the resting potential.
2 Activation gates of some Na⁺ channels open, depolarizing the cell to threshold.
3 Additional voltage-gated Na⁺ channel activation gates open, causing a rapid spike of depolarization—an action potential.
4 Na⁺ channel inactivation gates close; gated K⁺ channels open, repolarizing and even hyperpolarizing the cell.
5 All gated channels close. The cell returns to its resting potential. Na⁺ inactivation gates open

44.9 The Course of an Action Potential *(Page 851)*

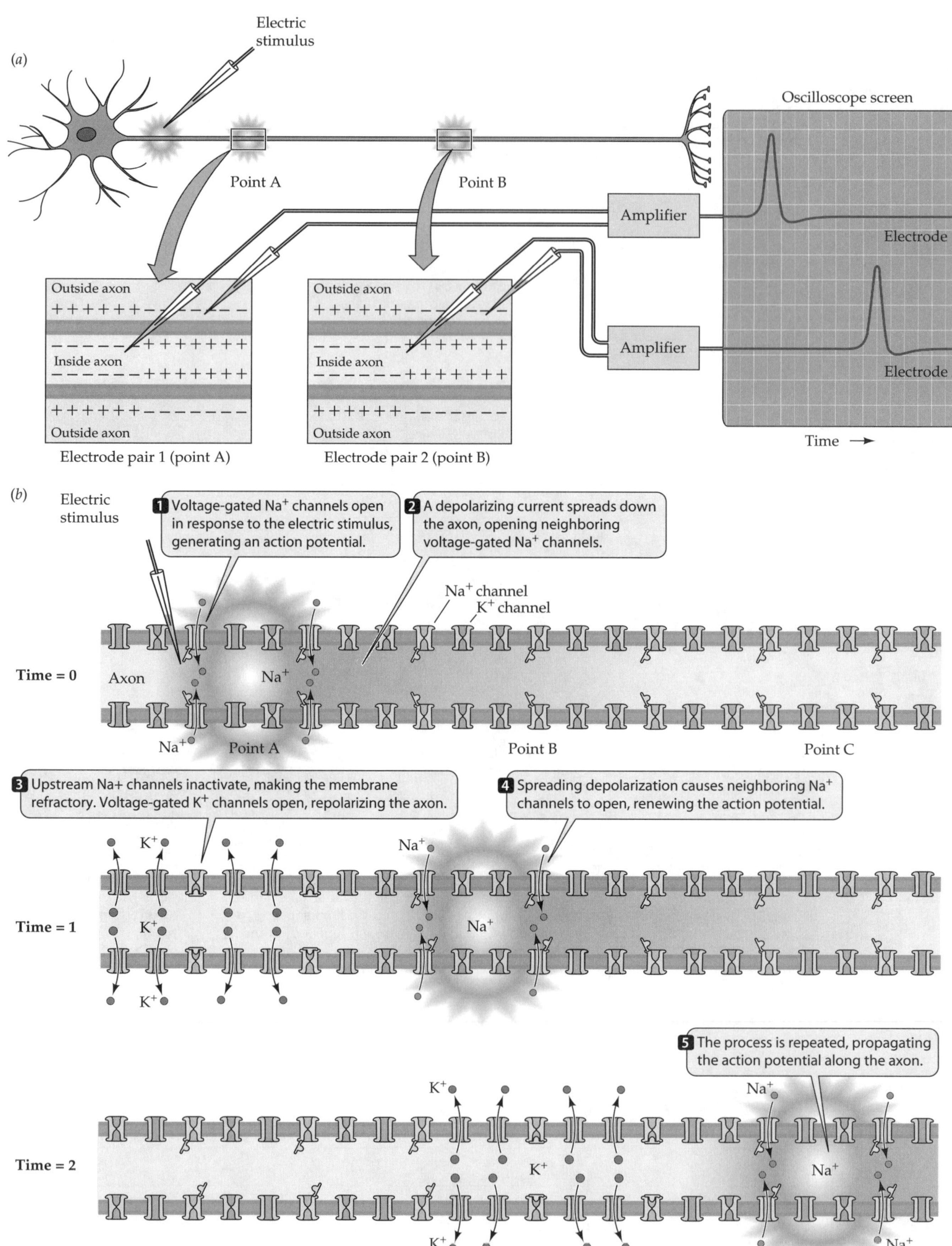
(a)
Electric stimulus
Point A
Point B
Oscilloscope screen
Amplifier
Electrode 1
Amplifier
Electrode 2
Time →
Outside axon
+ + + + + + – – – – – – –
– – – – – – + + + + + + +
Inside axon
– – – – – – + + + + + + +
+ + + + + + – – – – – – –
Outside axon
Electrode pair 1 (point A)
Electrode pair 2 (point B)
(b)
Electric stimulus
1 Voltage-gated Na+ channels open in response to the electric stimulus, generating an action potential.
2 A depolarizing current spreads down the axon, opening neighboring voltage-gated Na+ channels.
Na+ channel
K+ channel
Time = 0
Axon
Na+
Point A
Point B
Point C
3 Upstream Na+ channels inactivate, making the membrane refractory. Voltage-gated K+ channels open, repolarizing the axon.
4 Spreading depolarization causes neighboring Na+ channels to open, renewing the action potential.
K+
Time = 1
5 The process is repeated, propagating the action potential along the axon.
Time = 2

44.10 Action Potentials Travel along Axons *(Page 853)*

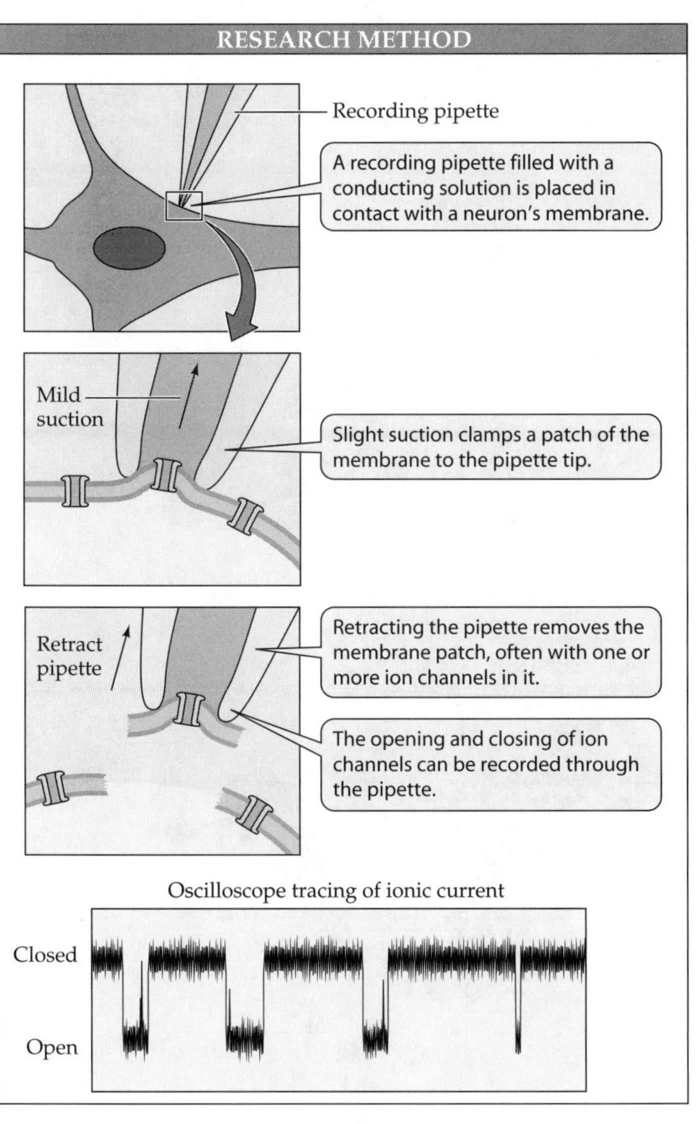
RESEARCH METHOD
Recording pipette
A recording pipette filled with a conducting solution is placed in contact with a neuron's membrane.
Mild suction
Slight suction clamps a patch of the membrane to the pipette tip.
Retract pipette
Retracting the pipette removes the membrane patch, often with one or more ion channels in it.
The opening and closing of ion channels can be recorded through the pipette.
Oscilloscope tracing of ionic current
Closed
Open

44.11 Patch Clamping *(Page 854)*

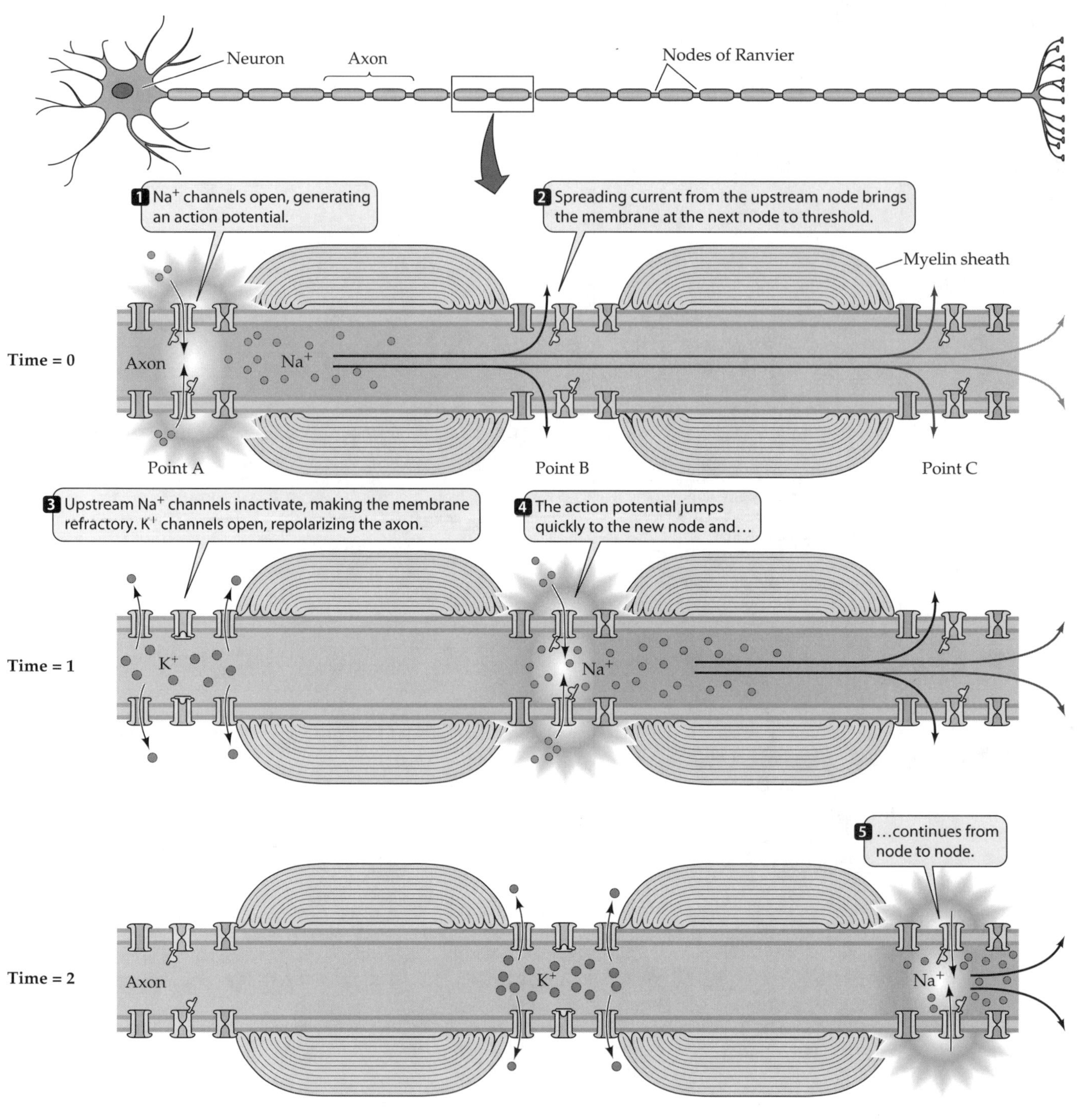
Neuron
Axon
Nodes of Ranvier
1 Na+ channels open, generating an action potential.
2 Spreading current from the upstream node brings the membrane at the next node to threshold.
Myelin sheath
Time = 0
Axon
Na+
Point A
Point B
Point C
3 Upstream Na+ channels inactivate, making the membrane refractory. K+ channels open, repolarizing the axon.
4 The action potential jumps quickly to the new node and...
Time = 1
K+
Na+
5 ...continues from node to node.
Time = 2
Axon
K+
Na+

44.12 Saltatory Action Potentials *(Page 855)*

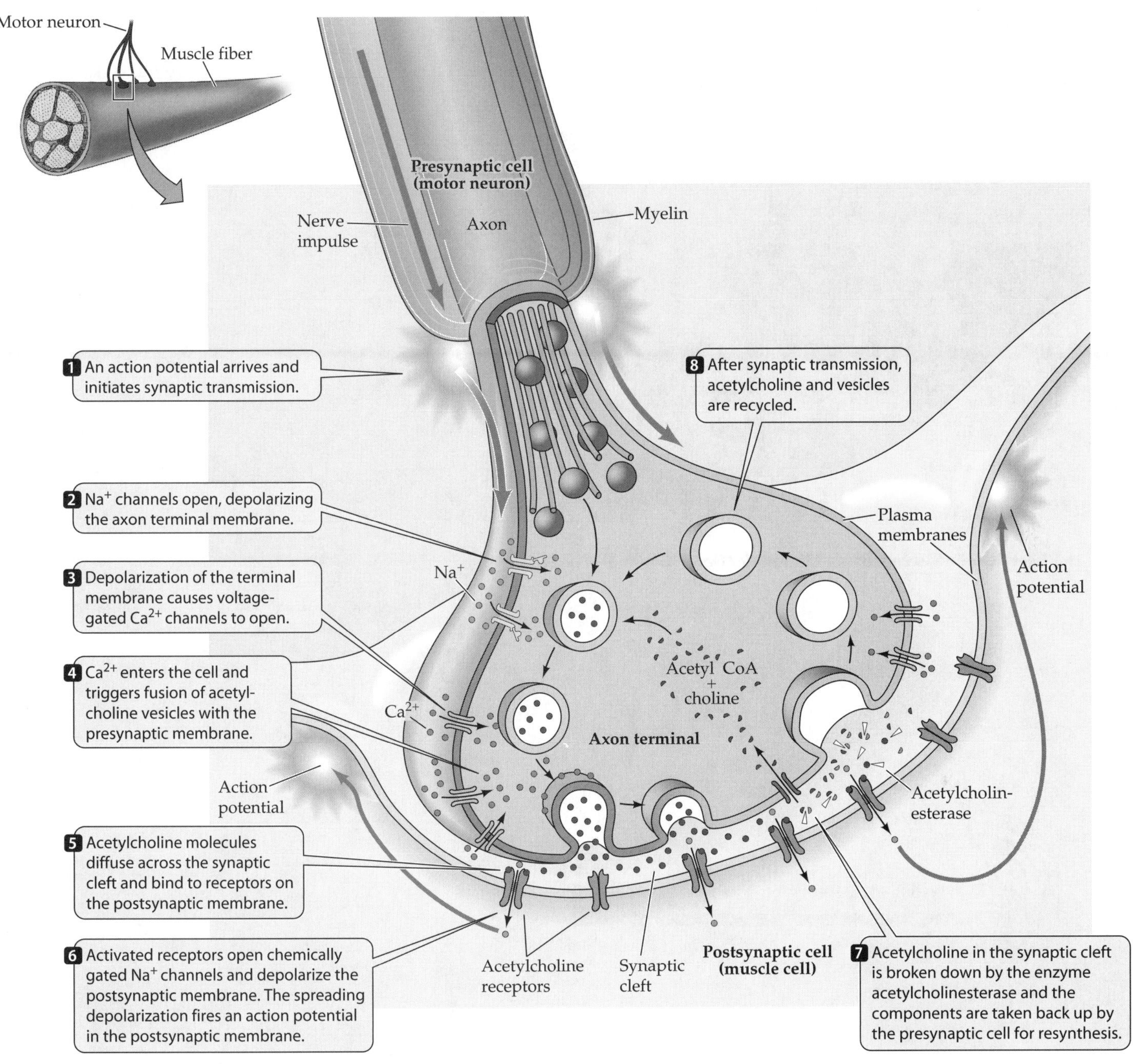

Motor neuron
Muscle fiber
Presynaptic cell (motor neuron)
Nerve impulse
Axon
Myelin
1 An action potential arrives and initiates synaptic transmission.
8 After synaptic transmission, acetylcholine and vesicles are recycled.
2 Na+ channels open, depolarizing the axon terminal membrane.
Plasma membranes
Action potential
Na+
3 Depolarization of the terminal membrane causes voltage-gated Ca2+ channels to open.
Acetyl CoA + choline
4 Ca2+ enters the cell and triggers fusion of acetylcholine vesicles with the presynaptic membrane.
Ca2+
Axon terminal
Action potential
Acetylcholinesterase
5 Acetylcholine molecules diffuse across the synaptic cleft and bind to receptors on the postsynaptic membrane.
6 Activated receptors open chemically gated Na+ channels and depolarize the postsynaptic membrane. The spreading depolarization fires an action potential in the postsynaptic membrane.
Acetylcholine receptors
Synaptic cleft
Postsynaptic cell (muscle cell)
7 Acetylcholine in the synaptic cleft is broken down by the enzyme acetylcholinesterase and the components are taken back up by the presynaptic cell for resynthesis.

44.13 Synaptic Transmission Begins with the Arrival of a Nerve Impulse *(Page 856)*

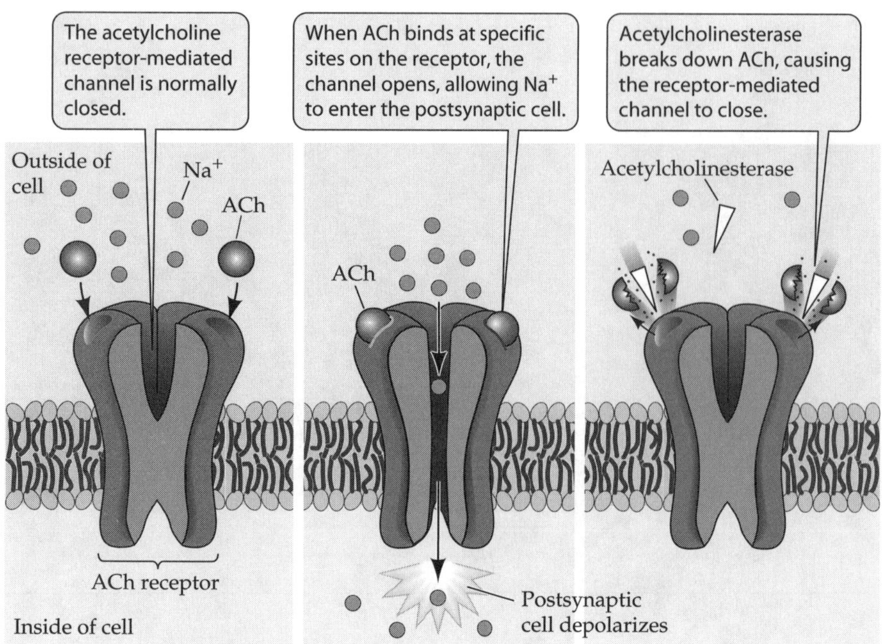

The acetylcholine receptor-mediated channel is normally closed.
When ACh binds at specific sites on the receptor, the channel opens, allowing Na^+ to enter the postsynaptic cell.
Acetylcholinesterase breaks down ACh, causing the receptor-mediated channel to close.
Outside of cell
Na^+
ACh
ACh
Acetylcholinesterase
ACh receptor
Inside of cell
Postsynaptic cell depolarizes

44.14 The Acetylcholine Receptor Is a Chemically Gated Channel *(Page 857)*

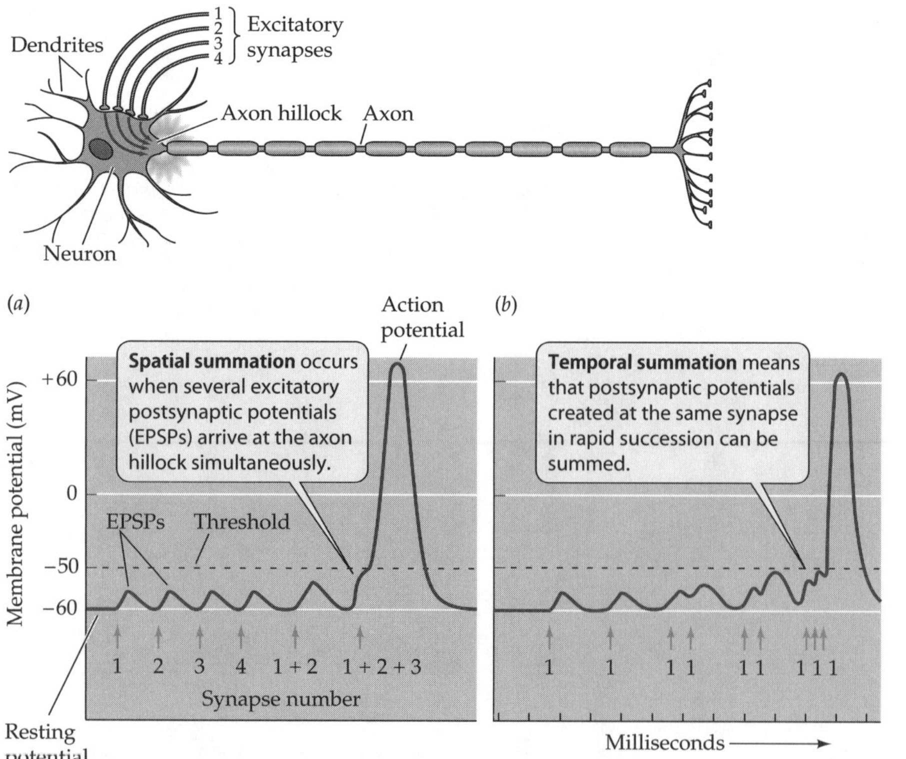

Dendrites
1 2 3 4
Excitatory synapses
Axon hillock
Axon
Neuron
(a)
Action potential
(b)
Spatial summation occurs when several excitatory postsynaptic potentials (EPSPs) arrive at the axon hillock simultaneously.
Temporal summation means that postsynaptic potentials created at the same synapse in rapid succession can be summed.
Membrane potential (mV)
+60
0
–50
–60
EPSPs
Threshold
1 2 3 4 1 + 2 1 + 2 + 3
Synapse number
1 1 1 1 1 1 1 1 1
Resting potential
Milliseconds

44.15 The Postsynaptic Neuron Sums Information *(Page 858)*

Outside of cell
1 Neurotransmitter binds to the receptor.
Neuro-transmitter
Receptor
Ions
Ion channel
β
α
γ
GDP
GTP
G protein
G protein α subunit
2 The receptor activates a G protein.
3 A G protein subunit activates an ion channel directly, or indirectly through a second messenger.
Inside of cell

44.16 Metabotropic Receptors Act through G Proteins *(Page 859)*

44.1 *Some Well-Known Neurotransmitters*

NEUROTRANSMITTER	ACTIONS	COMMENTS
Acetylcholine	The neurotransmitter of vertebrate motor neurons and of some neural pathways in the brain	Broken down in the synapse by acetylcholinesterase; blockers of this enzyme are powerful poisons
Monoamines		
Norepinephrine	Used in certain neural pathways in the brain. Also found in the peripheral nervous system, where it causes gut muscles to relax and the heart to beat faster	Related to epinephrine and acts at some of the same receptors
Dopamine	A neurotransmitter of the central nervous system	Involved in schizophrenia. Loss of dopamine neurons is the cause of Parkinson's disease
Histamine	A minor neurotransmitter in the brain	Thought to be involved in maintaining wakefulness
Serotonin	A neurotransmitter of the central nervous system that is involved in many systems, including pain control, sleep/wake control, and mood	Certain medications that elevate mood and counter anxiety act by inhibitng the reuptake of serotonin
Purines		
ATP	Co-released with many neurotransmitters	Large family of receptors may shape postsynaptic responses to classical neurotransmitters
Adenosine	Transported across cell membranes; not synaptically released	Largely inhibitory effects on postsynaptic cells
Amino acids		
Glutamate	The most common excitatory neurotransmitter in the central nervous system	Some people have reactions to the food additive monosodium glutamate because it can affect the nervous system
Glycine Gamma-aminobutyric acid (GABA)	Common inhibitory neurotransmitters	Drugs called benzodiazepines, used to reduce anxiety and produce sedation, mimic the actions of GABA
Peptides		
Endorphins Enkephalins Substance P	Used by certain sensory nerves, especially in pain pathways	Receptors are activated by narcotic drugs: opium, morphine, heroin, codeine
Gas		
Nitric oxide	Widely distributed in the nervous system	Not a classic neurotransmitter, it diffuses across membranes rather than being released synaptically. A means whereby a postsynaptic cell can influence a presynaptic cell

(Page 860)

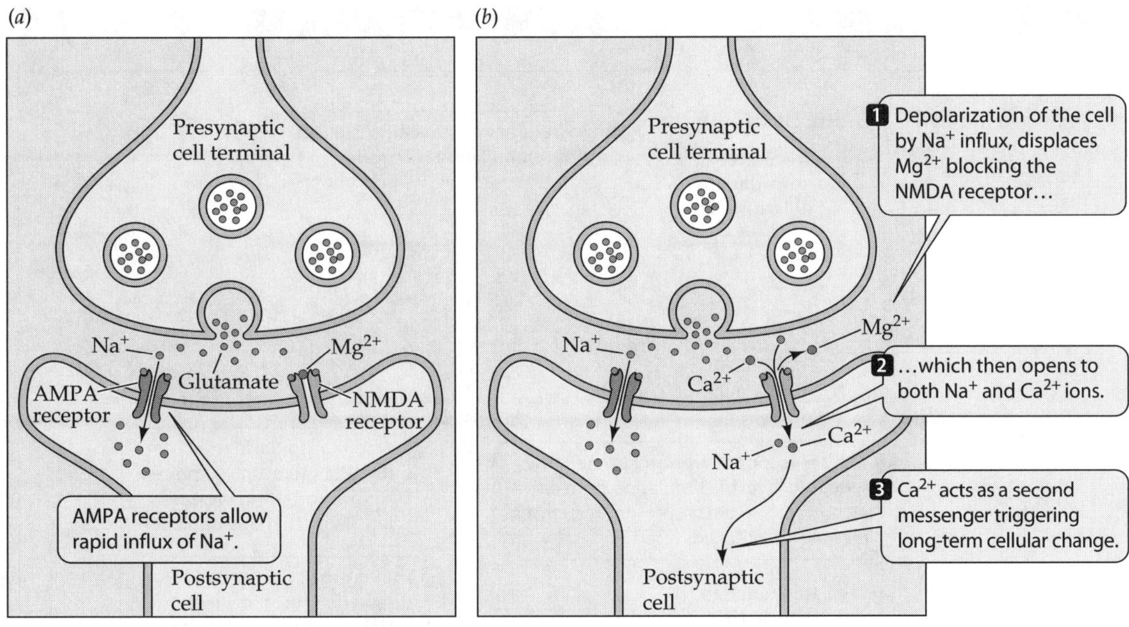
(a)
Presynaptic cell terminal
Na^+
Mg^{2+}
Glutamate
AMPA receptor
NMDA receptor
AMPA receptors allow rapid influx of Na^+.
Postsynaptic cell
(b)
Presynaptic cell terminal
1 Depolarization of the cell by Na^+ influx, displaces Mg^{2+} blocking the NMDA receptor…
Na^+
Ca^{2+}
Mg^{2+}
2 …which then opens to both Na^+ and Ca^{2+} ions.
Na^+
Ca^{2+}
3 Ca^{2+} acts as a second messenger triggering long-term cellular change.
Postsynaptic cell

44.17 Two Ionotropic Glutamate Receptors *(Page 861)*

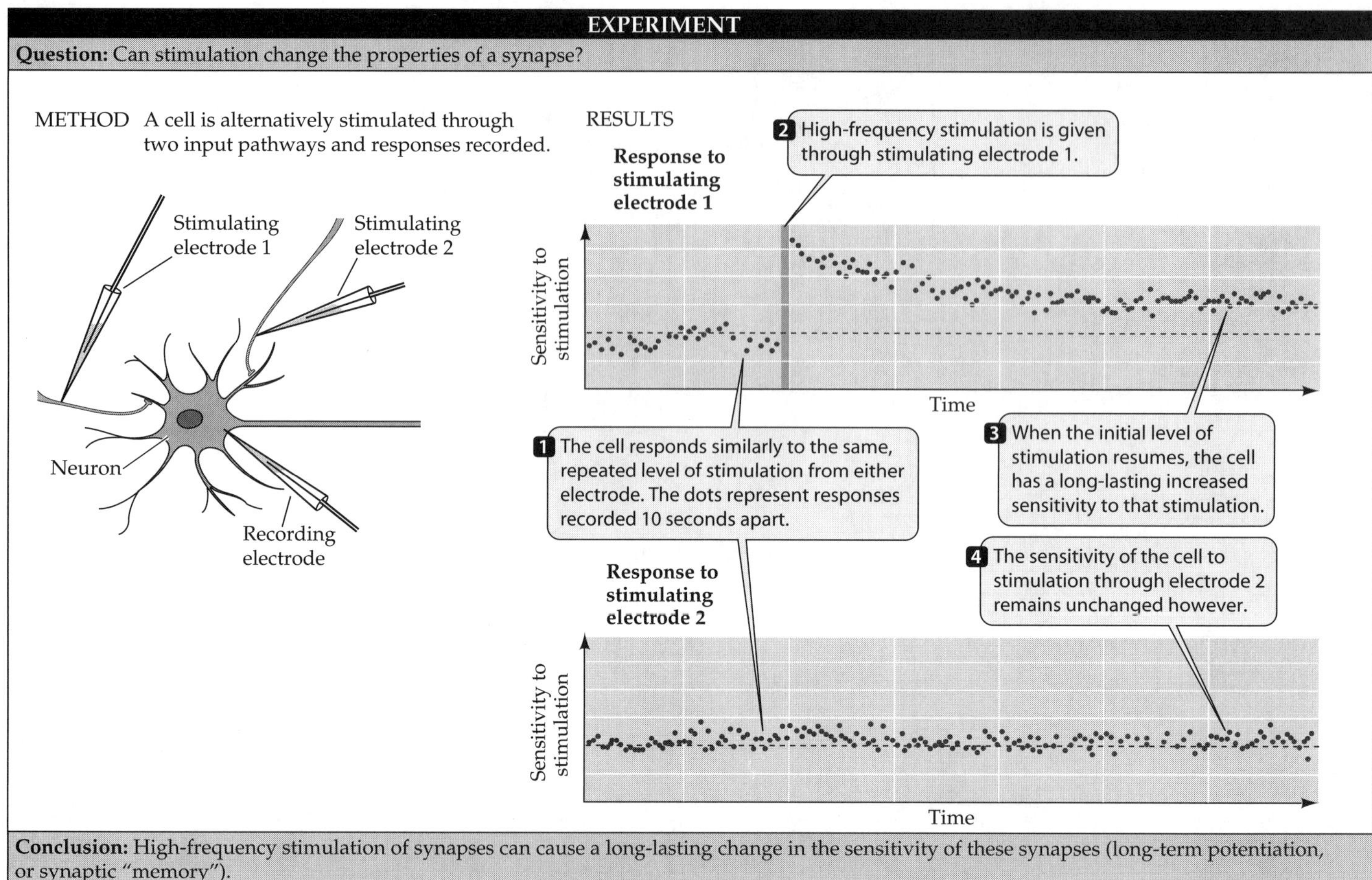
EXPERIMENT
Question: Can stimulation change the properties of a synapse?
METHOD A cell is alternatively stimulated through two input pathways and responses recorded.
Stimulating electrode 1
Stimulating electrode 2
Neuron
Recording electrode
RESULTS
Response to stimulating electrode 1
2 High-frequency stimulation is given through stimulating electrode 1.
Sensitivity to stimulation
Time
1 The cell responds similarly to the same, repeated level of stimulation from either electrode. The dots represent responses recorded 10 seconds apart.
3 When the initial level of stimulation resumes, the cell has a long-lasting increased sensitivity to that stimulation.
4 The sensitivity of the cell to stimulation through electrode 2 remains unchanged however.
Response to stimulating electrode 2
Sensitivity to stimulation
Time
Conclusion: High-frequency stimulation of synapses can cause a long-lasting change in the sensitivity of these synapses (long-term potentiation, or synaptic "memory").

44.18 Repeated Stimulation Can Cause Long-Term Potentiation *(Page 862)*

45 Sensory Systems

Mechanoreceptor
Pressure opens an ion channel.
Outside of cell
Pressure
Plasma membrane
Pressure-sensitive Na⁺ channel
Inside of cell
Chemoreceptor
A molecule binds to a receptor, initiating a signal that controls the ion channel via second messenger cascades.
Taste/smell molecule
Na⁺ or K⁺ channel
Photoreceptor
Light alters a receptor protein, initiating a signal that controls an ion channel.
Light
cGMP-gated Na⁺ channel

45.1 Sensory Cell Membrane Receptor Proteins Respond to Stimuli *(Page 866)*

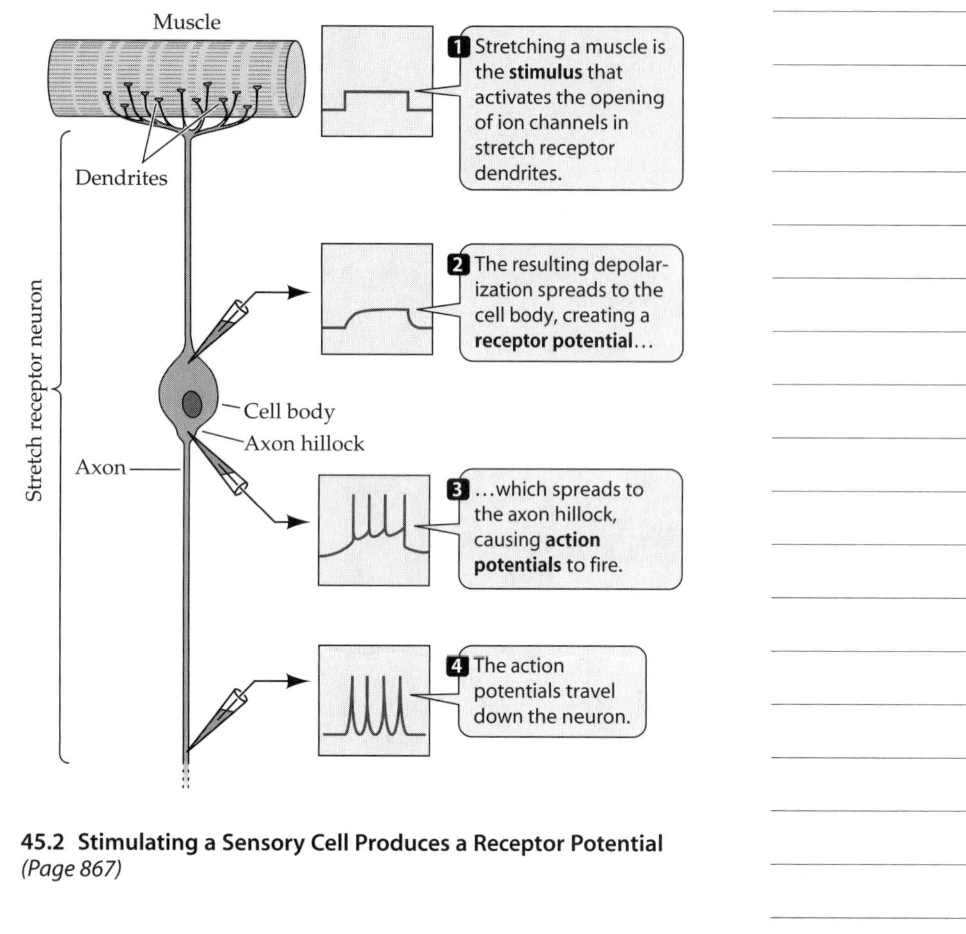
Muscle
Dendrites
Stretch receptor neuron
Cell body
Axon hillock
Axon
1 Stretching a muscle is the **stimulus** that activates the opening of ion channels in stretch receptor dendrites.
2 The resulting depolarization spreads to the cell body, creating a **receptor potential**…
3 …which spreads to the axon hillock, causing **action potentials** to fire.
4 The action potentials travel down the neuron.

45.2 Stimulating a Sensory Cell Produces a Receptor Potential
(Page 867)

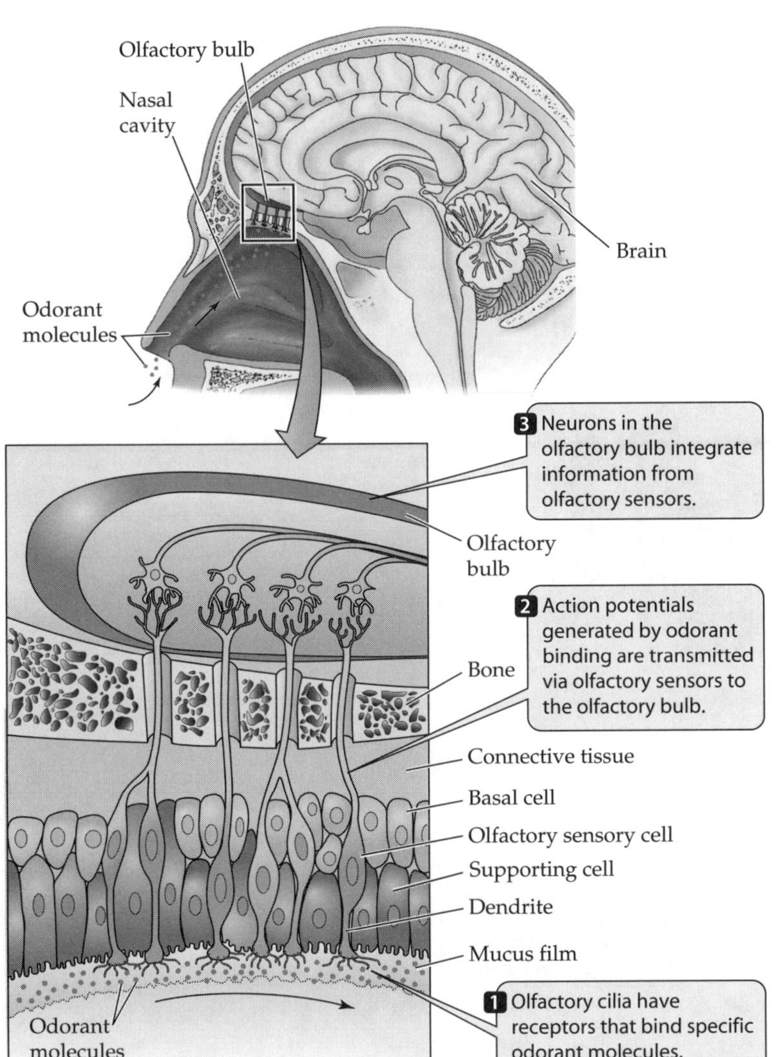
Olfactory bulb
Nasal cavity
Brain
Odorant molecules
3 Neurons in the olfactory bulb integrate information from olfactory sensors.
Olfactory bulb
2 Action potentials generated by odorant binding are transmitted via olfactory sensors to the olfactory bulb.
Bone
Connective tissue
Basal cell
Olfactory sensory cell
Supporting cell
Dendrite
Mucus film
1 Olfactory cilia have receptors that bind specific odorant molecules.
Odorant molecules

45.4 Olfactory Receptors Communicate Directly with the Brain *(Page 869)*

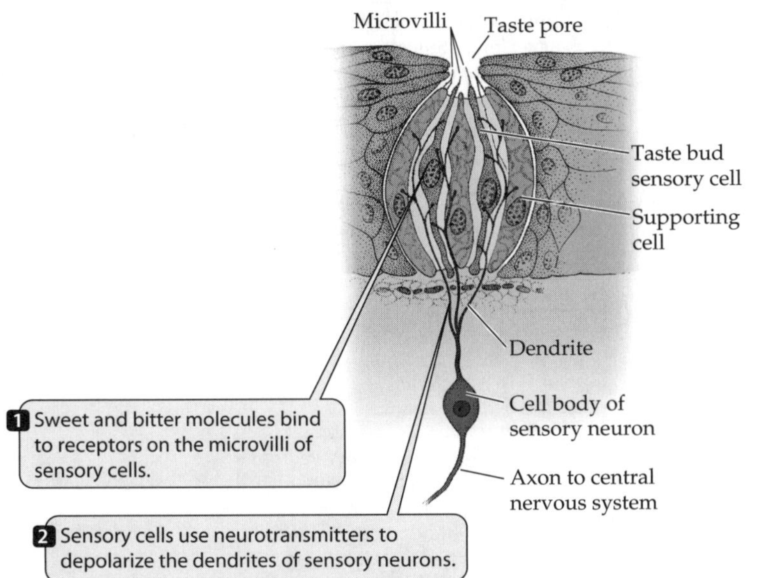
Microvilli
Taste pore
Taste bud sensory cell
Supporting cell
Dendrite
Cell body of sensory neuron
Axon to central nervous system
1 Sweet and bitter molecules bind to receptors on the microvilli of sensory cells.
2 Sensory cells use neurotransmitters to depolarize the dendrites of sensory neurons.

45.5 Taste Buds Are Clusters of Sensory Cells *(Page 871)*

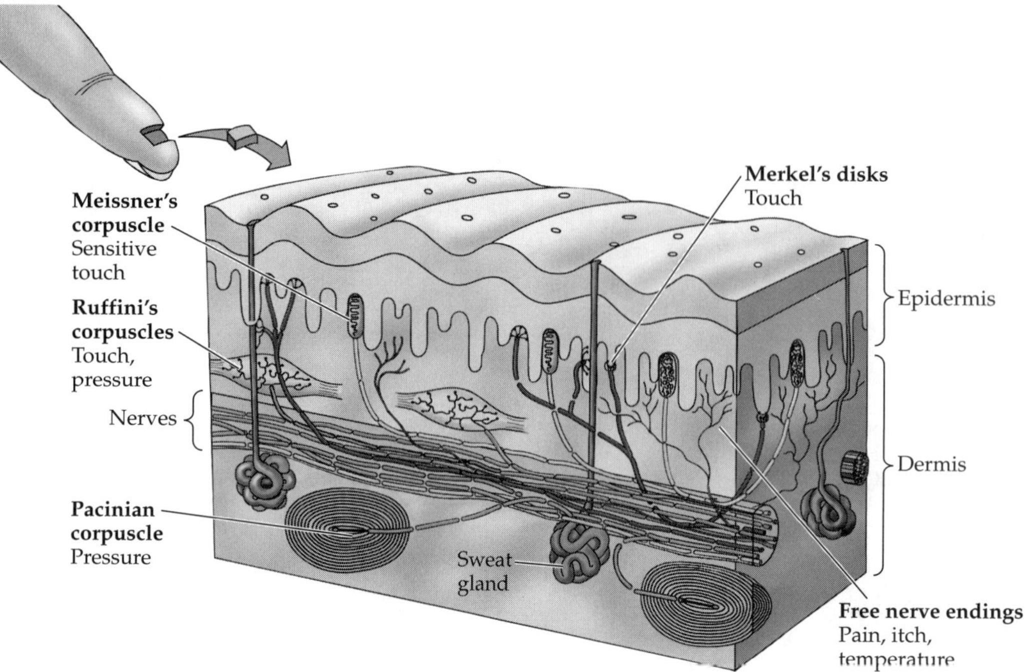

Meissner's corpuscle
Sensitive touch
Ruffini's corpuscles
Touch, pressure
Nerves
Pacinian corpuscle
Pressure
Sweat gland
Merkel's disks
Touch
Epidermis
Dermis
Free nerve endings
Pain, itch, temperature

45.6 The Skin Feels Many Sensations *(Page 871)*

(*a*) **Muscle spindles**

Muscle

Muscle spindle

Sensory neuron

1 Muscle spindles are stretch receptors. When muscle spindles are stretched…

Stretch

Firing of sensory neuron

Time

2 …sensory neurons associated with them transmit action potentials to the CNS. These signals stimulate motor neurons that initiate muscle contraction.

Muscle

Tendon

Load

(*b*) **Golgi tendon organs**

Muscle

Golgi tendon organ

Tendon

Sensory neuron

1 Golgi tendon organs sense load and measure the force of muscle contraction. When contraction becomes too forceful…

Load on muscle

Firing of sensory neuron

Time

2 …the sensory neurons send action potentials to the CNS that inhibit motor neurons, and the muscle relaxes.

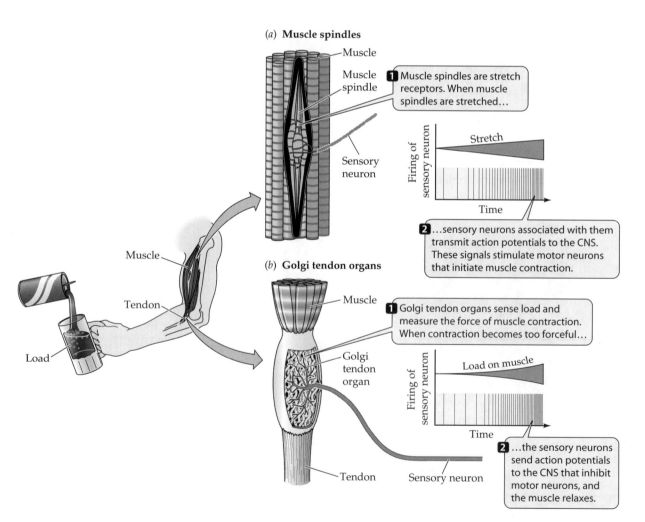

45.7 Stretch Receptors Are Activated when Limbs Are Stretched *(Page 871)*

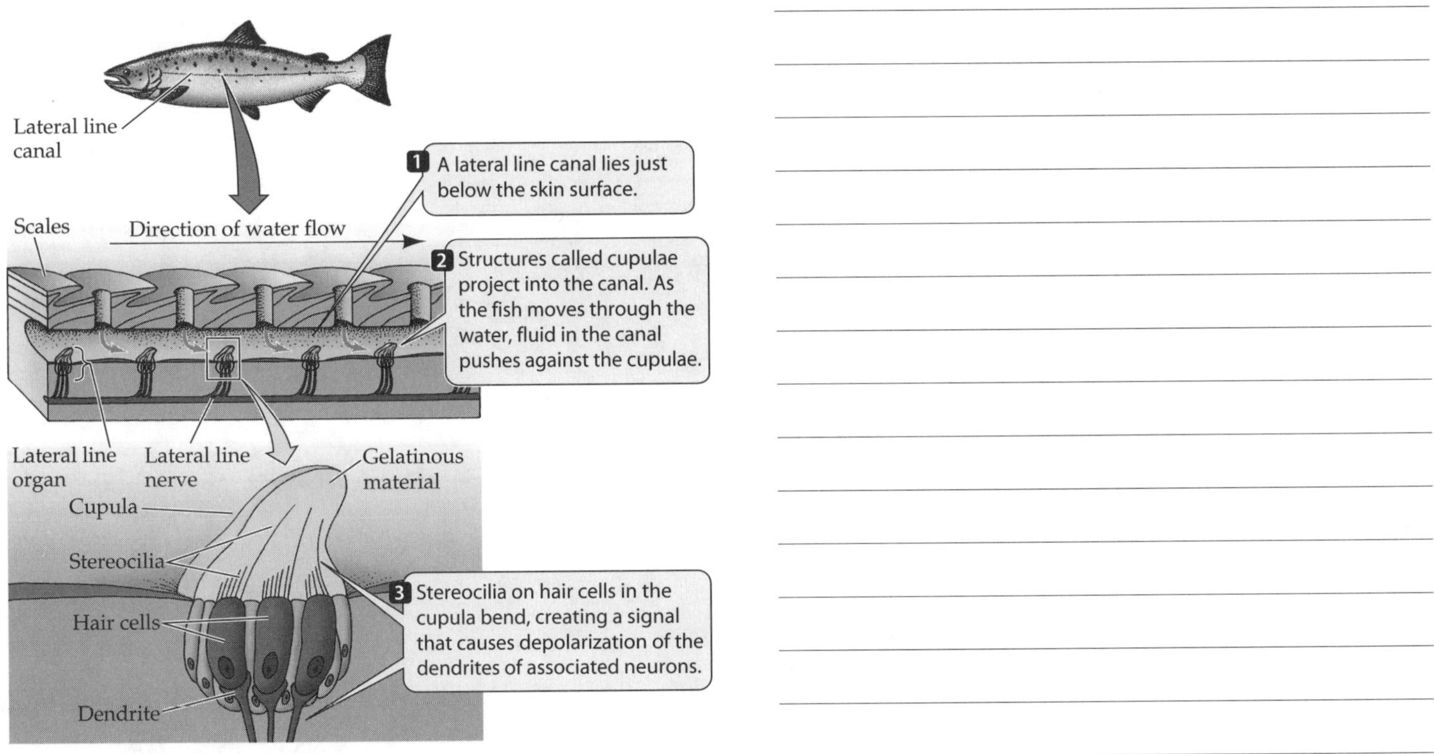
Lateral line canal
1 A lateral line canal lies just below the skin surface.
Scales
Direction of water flow
2 Structures called cupulae project into the canal. As the fish moves through the water, fluid in the canal pushes against the cupulae.
Lateral line organ
Lateral line nerve
Gelatinous material
Cupula
Stereocilia
Hair cells
3 Stereocilia on hair cells in the cupula bend, creating a signal that causes depolarization of the dendrites of associated neurons.
Dendrite

45.8 The Lateral Line System Contains Mechanoreceptors *(Page 872)*

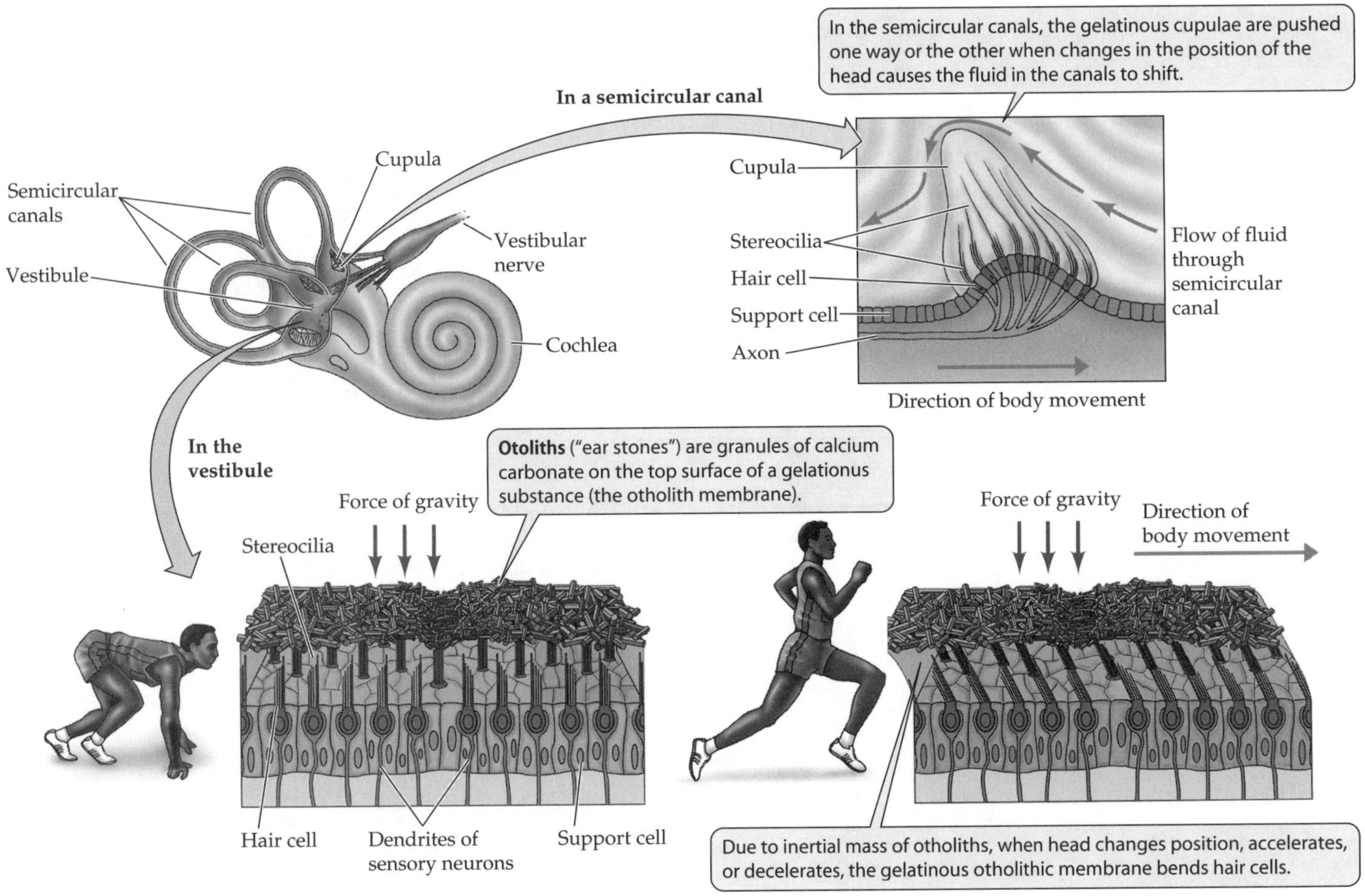

In the semicircular canals, the gelatinous cupulae are pushed one way or the other when changes in the position of the head causes the fluid in the canals to shift.

In a semicircular canal

Cupula

Semicircular canals

Vestibule

Vestibular nerve

Cochlea

Cupula

Stereocilia

Hair cell

Support cell

Axon

Flow of fluid through semicircular canal

Direction of body movement

In the vestibule

Otoliths ("ear stones") are granules of calcium carbonate on the top surface of a gelatinous substance (the otholith membrane).

Force of gravity

Stereocilia

Force of gravity

Direction of body movement

Hair cell

Dendrites of sensory neurons

Support cell

Due to inertial mass of otoliths, when head changes position, accelerates, or decelerates, the gelatinous otholithic membrane bends hair cells.

45.9 Organs in the Inner Ear of Mammals Provide the Sense of Equilibrium *(Page 872)*

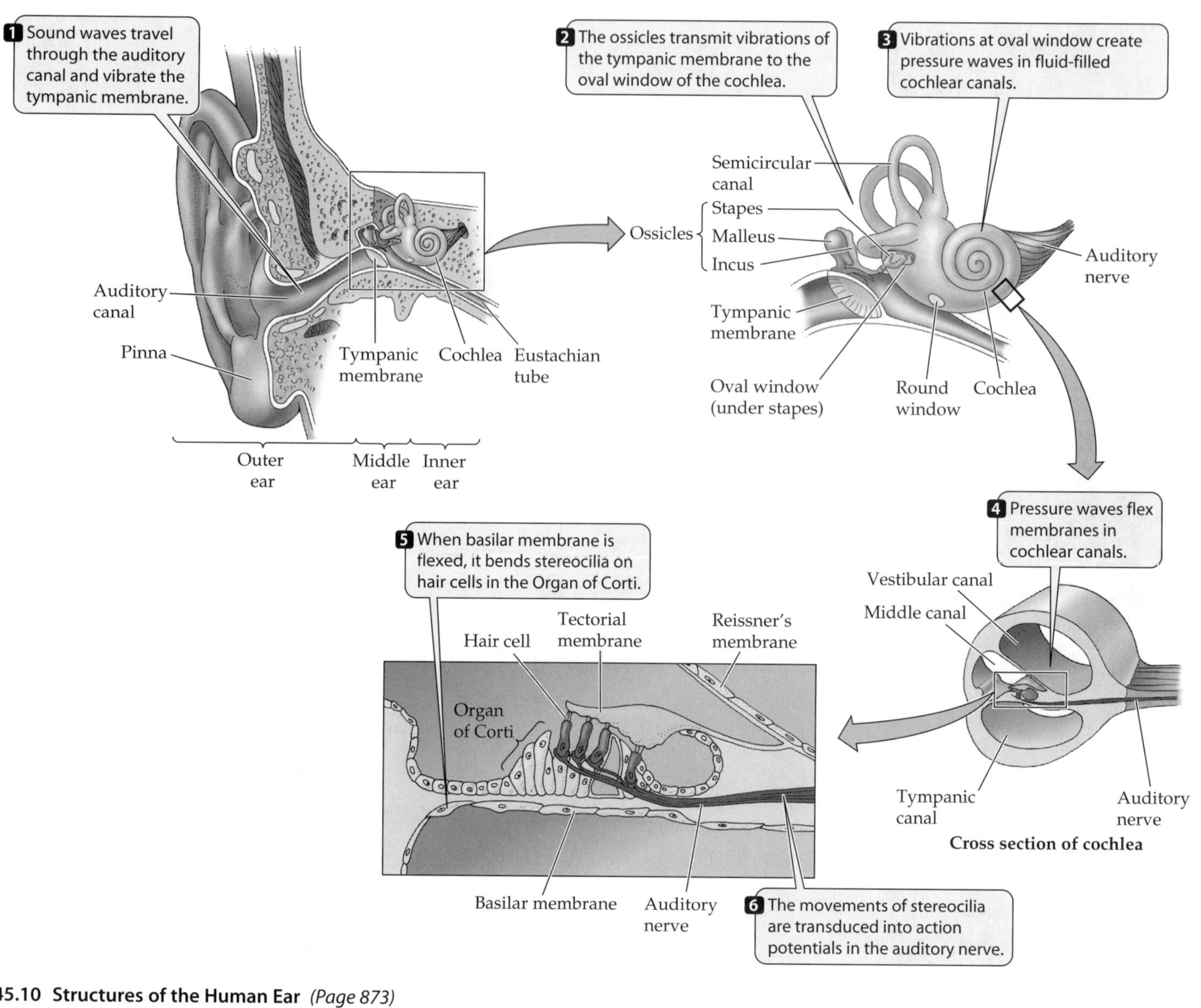

1 Sound waves travel through the auditory canal and vibrate the tympanic membrane.
Auditory canal
Pinna
Tympanic membrane
Cochlea
Eustachian tube
Outer ear
Middle ear
Inner ear
2 The ossicles transmit vibrations of the tympanic membrane to the oval window of the cochlea.
3 Vibrations at oval window create pressure waves in fluid-filled cochlear canals.
Semicircular canal
Ossicles
Stapes
Malleus
Incus
Tympanic membrane
Oval window (under stapes)
Round window
Cochlea
Auditory nerve
4 Pressure waves flex membranes in cochlear canals.
Vestibular canal
Middle canal
Tympanic canal
Auditory nerve
Cross section of cochlea
5 When basilar membrane is flexed, it bends stereocilia on hair cells in the Organ of Corti.
Hair cell
Tectorial membrane
Reissner's membrane
Organ of Corti
Basilar membrane
Auditory nerve
6 The movements of stereocilia are transduced into action potentials in the auditory nerve.

45.10 Structures of the Human Ear *(Page 873)*

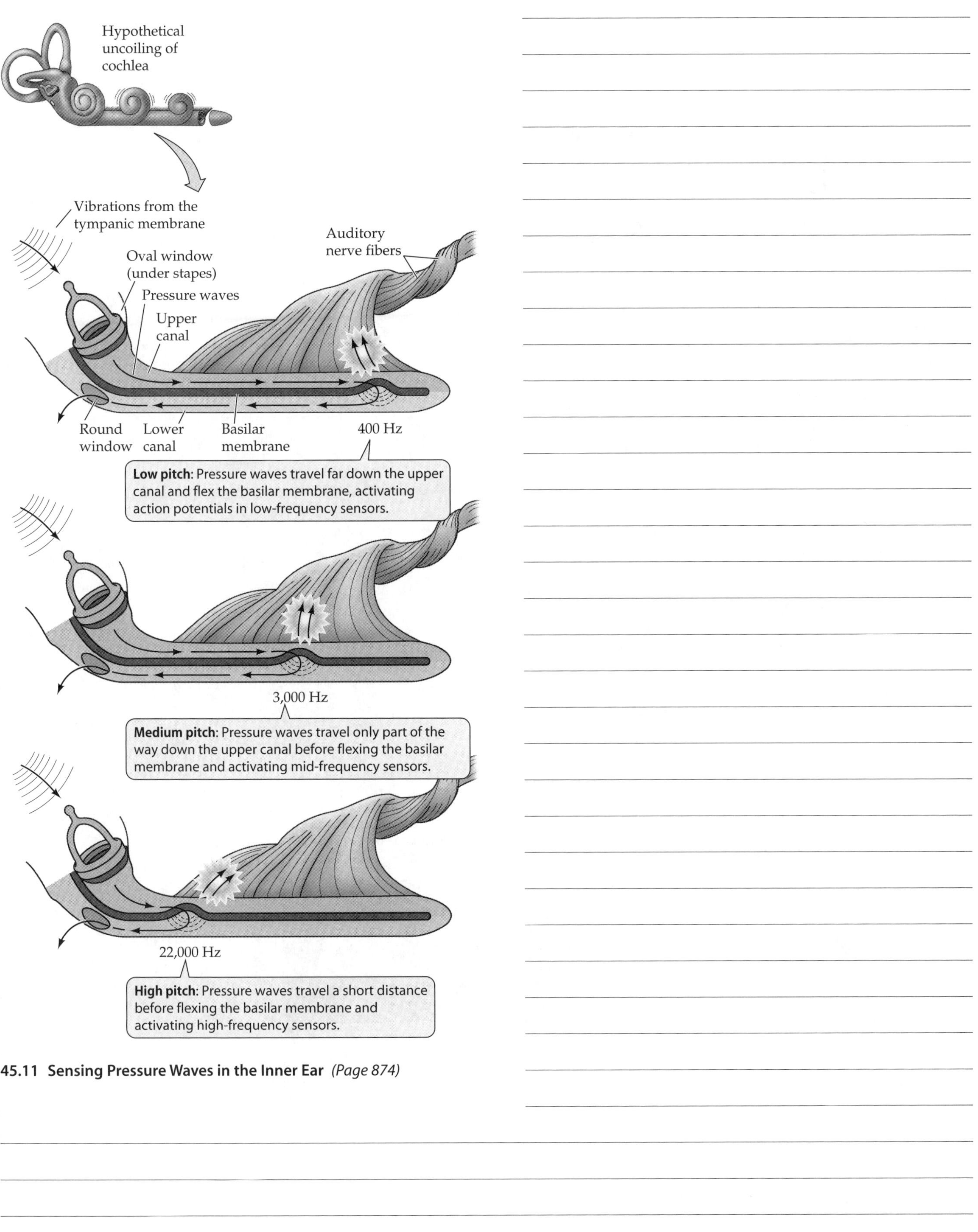

Hypothetical uncoiling of cochlea
Vibrations from the tympanic membrane
Oval window (under stapes)
Pressure waves
Upper canal
Auditory nerve fibers
Round window
Lower canal
Basilar membrane
400 Hz
Low pitch: Pressure waves travel far down the upper canal and flex the basilar membrane, activating action potentials in low-frequency sensors.
3,000 Hz
Medium pitch: Pressure waves travel only part of the way down the upper canal before flexing the basilar membrane and activating mid-frequency sensors.
22,000 Hz
High pitch: Pressure waves travel a short distance before flexing the basilar membrane and activating high-frequency sensors.

45.11 Sensing Pressure Waves in the Inner Ear *(Page 874)*

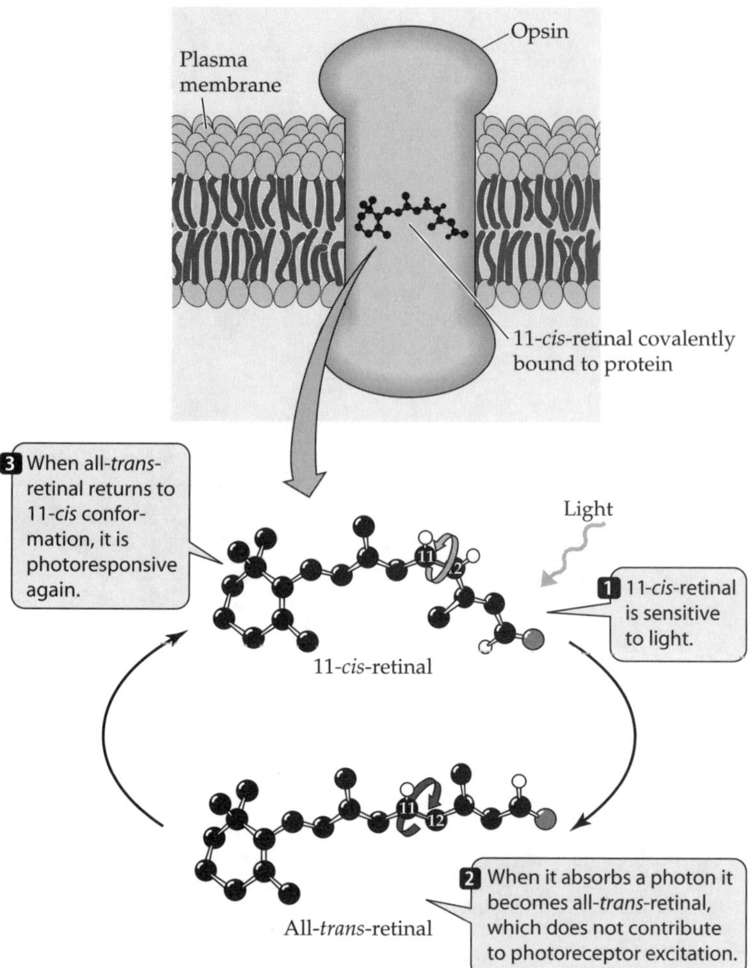
Opsin
Plasma membrane
11-*cis*-retinal covalently bound to protein
3 When all-*trans*-retinal returns to 11-*cis* conformation, it is photoresponsive again.
Light
1 11-*cis*-retinal is sensitive to light.
11-*cis*-retinal
All-*trans*-retinal
2 When it absorbs a photon it becomes all-*trans*-retinal, which does not contribute to photoreceptor excitation.

45.12 Rhodopsin: A Photosensitive Molecule *(Page 875)*

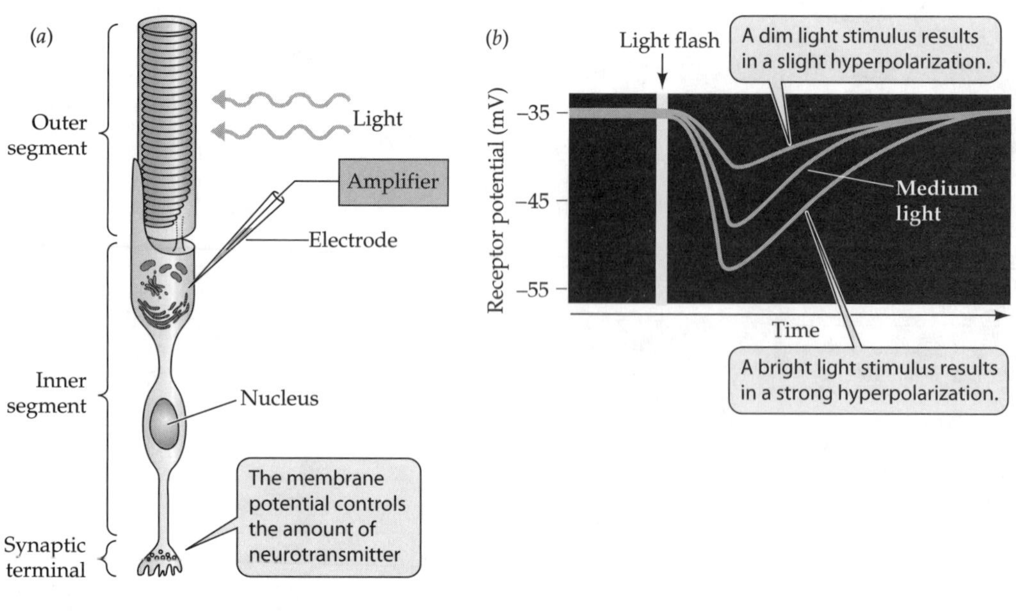
(*a*)
Outer segment
Light
Amplifier
Electrode
Inner segment
Nucleus
Synaptic terminal
The membrane potential controls the amount of neurotransmitter
(*b*)
Light flash
A dim light stimulus results in a slight hyperpolarization.
Receptor potential (mV)
–35
–45
–55
Medium light
Time
A bright light stimulus results in a strong hyperpolarization.

45.13 A Rod Cell Responds to Light *(Page 876)*

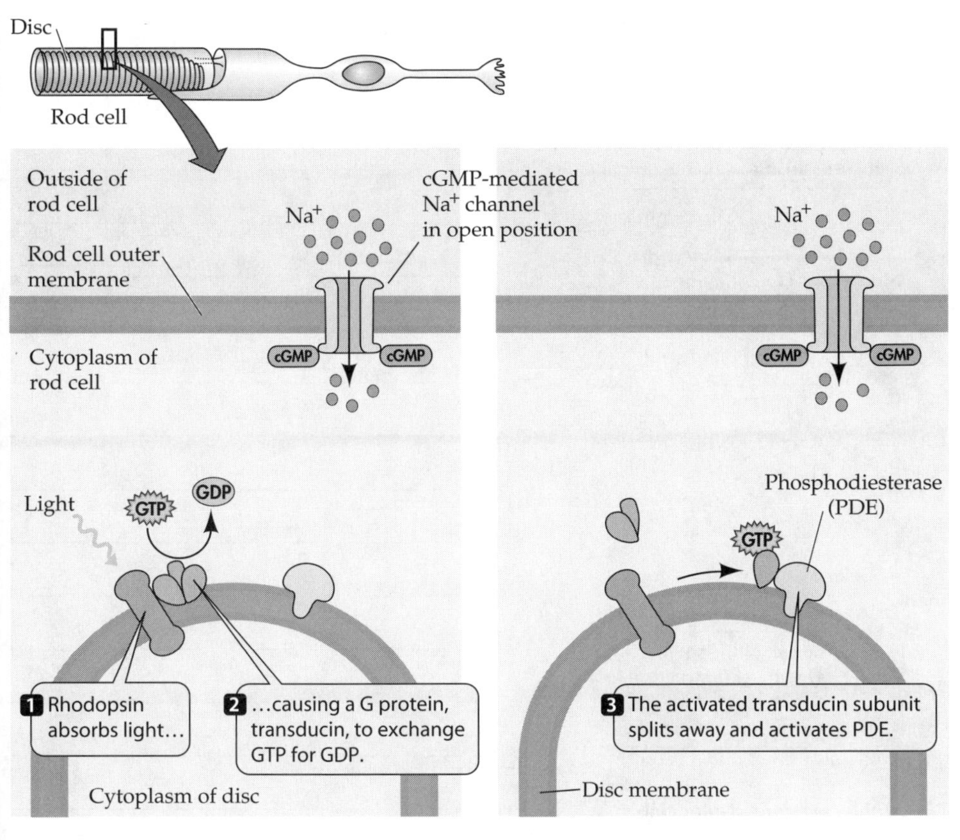

Disc
Rod cell
Outside of rod cell
Rod cell outer membrane
cGMP-mediated Na+ channel in open position
Na+
cGMP
Cytoplasm of rod cell
Light
GTP
GDP
1 Rhodopsin absorbs light…
2 …causing a G protein, transducin, to exchange GTP for GDP.
Cytoplasm of disc
Phosphodiesterase (PDE)
3 The activated transducin subunit splits away and activates PDE.
Disc membrane

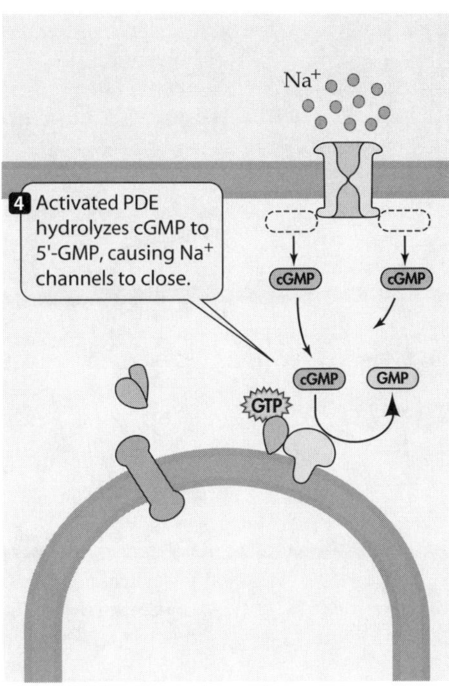

Na+
4 Activated PDE hydrolyzes cGMP to 5'-GMP, causing Na+ channels to close.
cGMP
GTP
GMP

45.14 Light Absorption Closes Sodium Channels *(Page 877)*

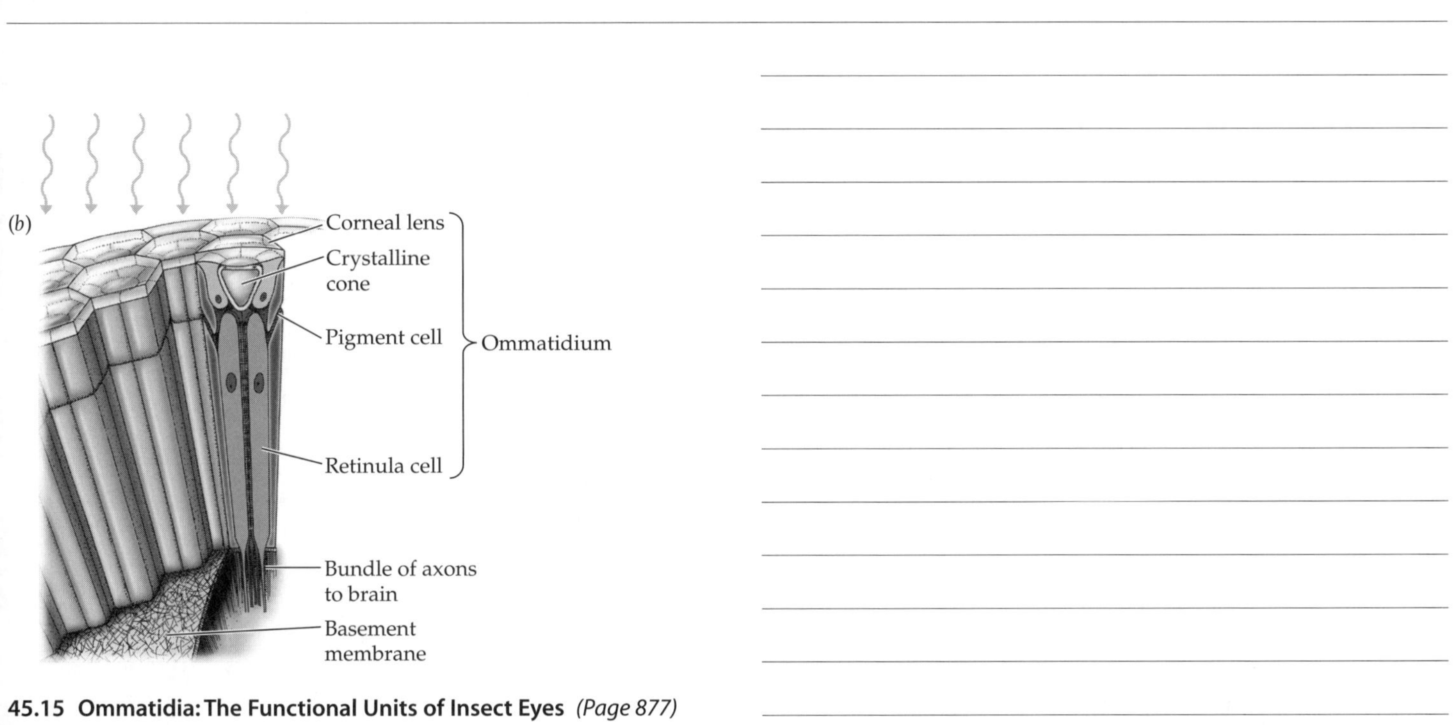

(b)
Corneal lens
Crystalline cone
Pigment cell
Ommatidium
Retinula cell
Bundle of axons to brain
Basement membrane

45.15 Ommatidia: The Functional Units of Insect Eyes *(Page 877)*

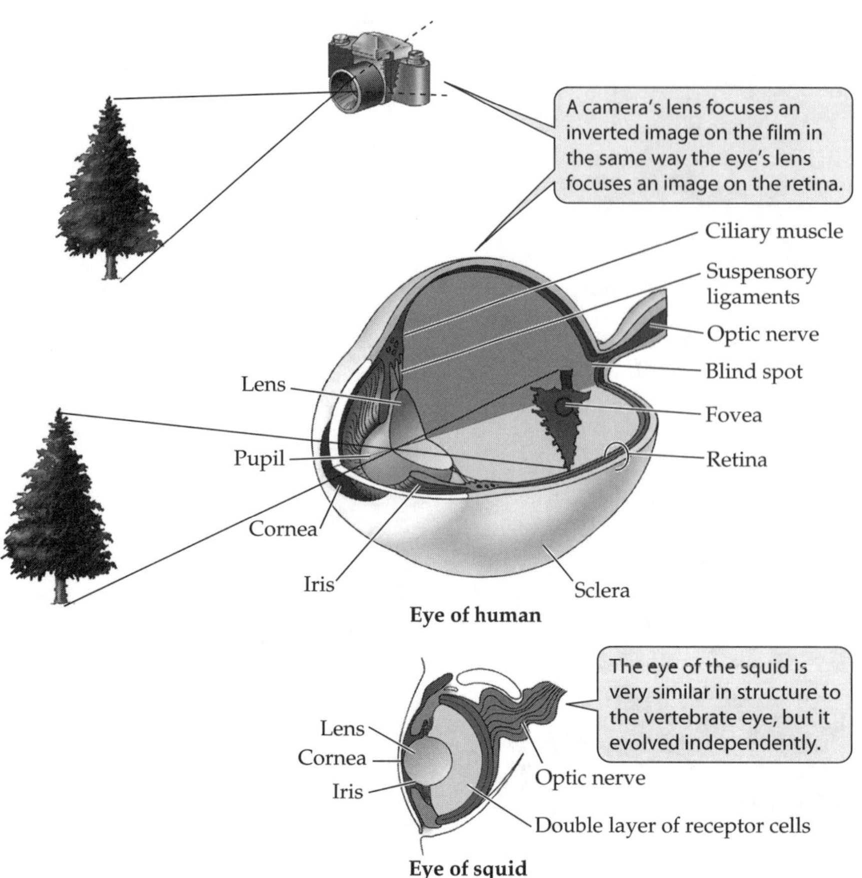
A camera's lens focuses an inverted image on the film in the same way the eye's lens focuses an image on the retina.
Ciliary muscle
Suspensory ligaments
Optic nerve
Blind spot
Fovea
Retina
Lens
Pupil
Cornea
Iris
Sclera
Eye of human
The eye of the squid is very similar in structure to the vertebrate eye, but it evolved independently.
Lens
Cornea
Iris
Optic nerve
Double layer of receptor cells
Eye of squid

45.16 Eyes Like Cameras *(Page 878)*

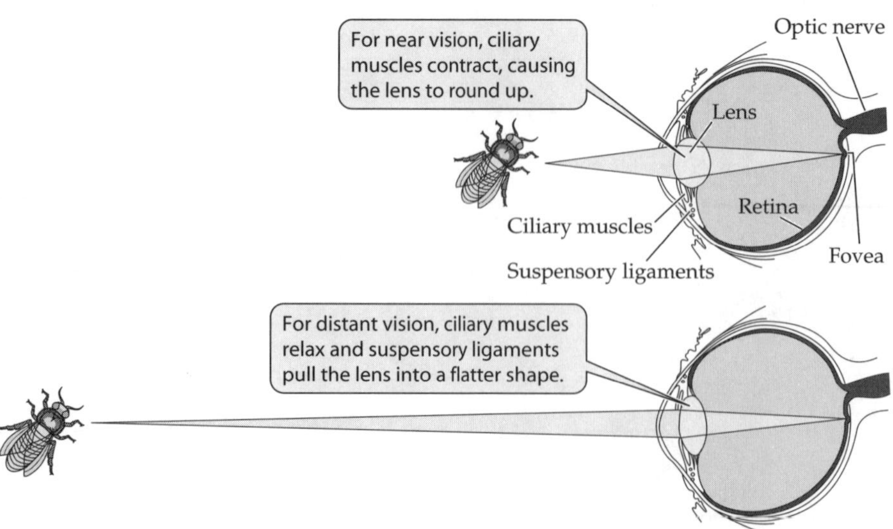
For near vision, ciliary muscles contract, causing the lens to round up.
Optic nerve
Lens
Retina
Ciliary muscles
Suspensory ligaments
Fovea
For distant vision, ciliary muscles relax and suspensory ligaments pull the lens into a flatter shape.

45.17 Staying in Focus *(Page 878)*

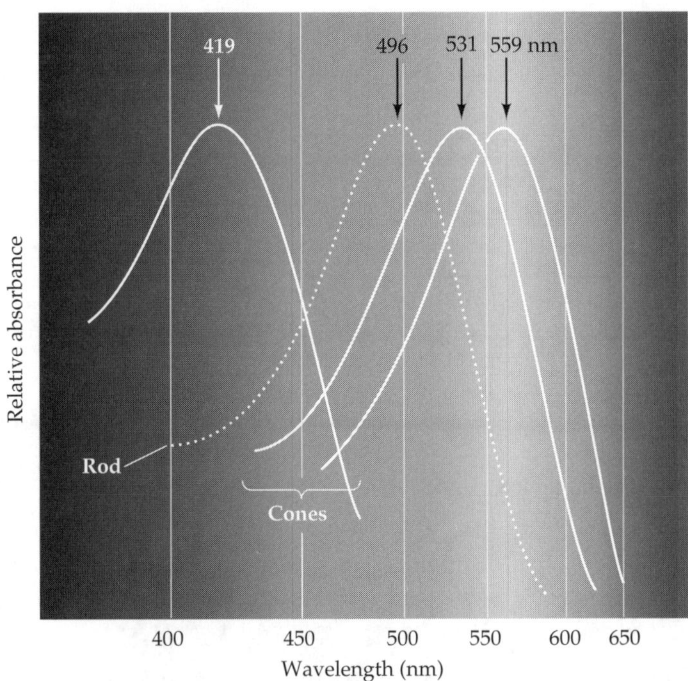
419
496
531
559 nm
Relative absorbance
Rod
Cones
400
450
500
550
600
650
Wavelength (nm)

45.19 Absorption Spectra of Cone Cells *(Page 879)*

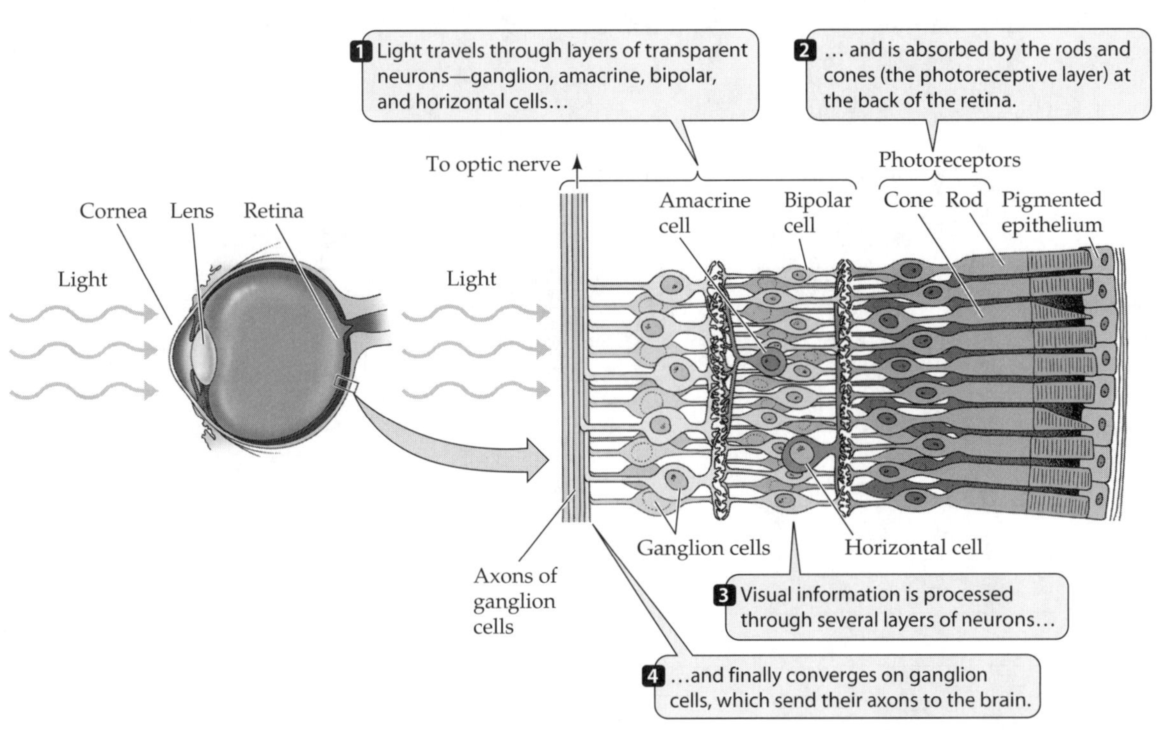
1 Light travels through layers of transparent neurons—ganglion, amacrine, bipolar, and horizontal cells…
2 … and is absorbed by the rods and cones (the photoreceptive layer) at the back of the retina.
To optic nerve
Photoreceptors
Cornea
Lens
Retina
Amacrine cell
Bipolar cell
Cone
Rod
Pigmented epithelium
Light
Light
Ganglion cells
Horizontal cell
Axons of ganglion cells
3 Visual information is processed through several layers of neurons…
4 …and finally converges on ganglion cells, which send their axons to the brain.

45.20 The Retina *(Page 880)*

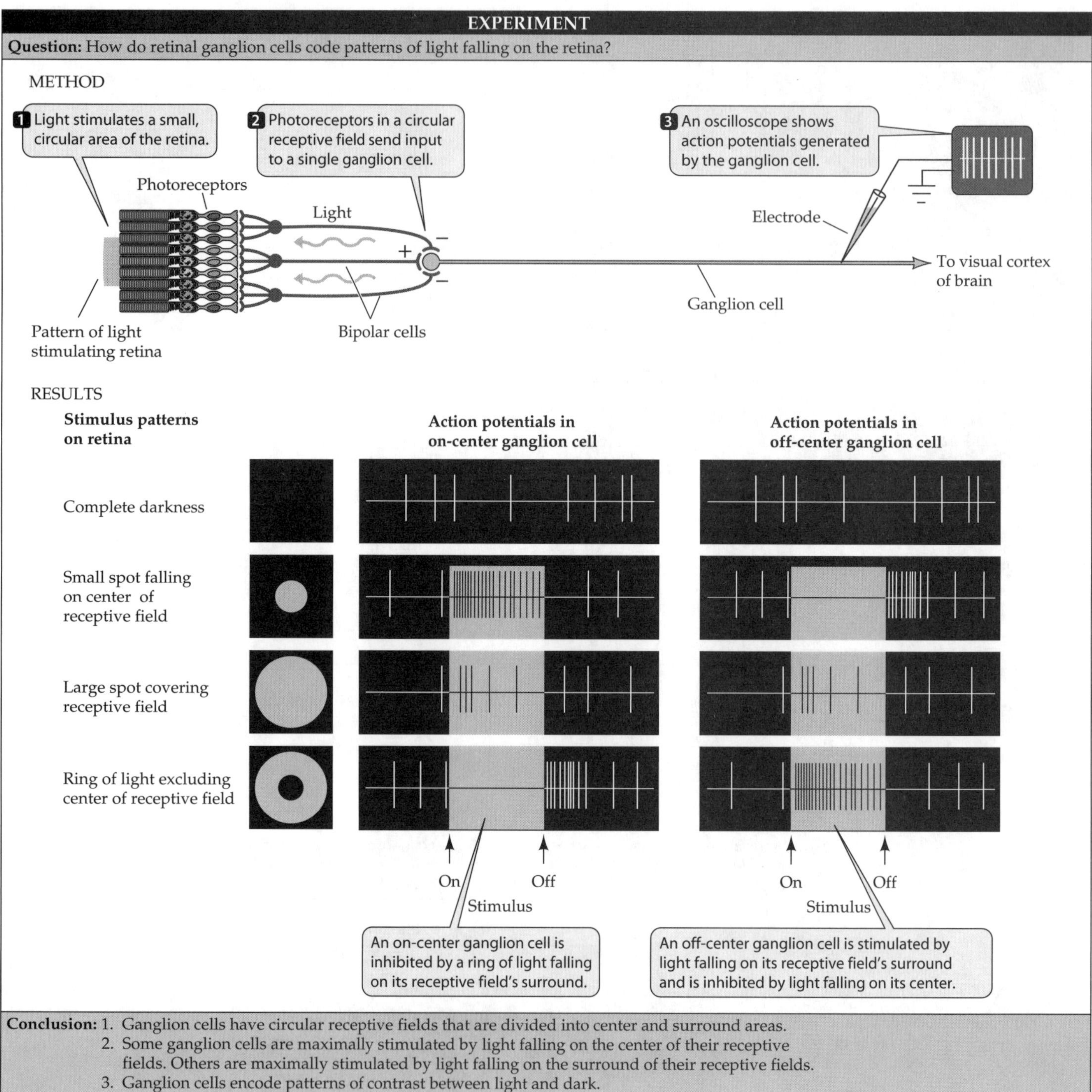
EXPERIMENT
Question: How do retinal ganglion cells code patterns of light falling on the retina?
METHOD
1 Light stimulates a small, circular area of the retina.
2 Photoreceptors in a circular receptive field send input to a single ganglion cell.
3 An oscilloscope shows action potentials generated by the ganglion cell.
Photoreceptors
Light
Electrode
To visual cortex of brain
Ganglion cell
Pattern of light stimulating retina
Bipolar cells
RESULTS
Stimulus patterns on retina
Action potentials in on-center ganglion cell
Action potentials in off-center ganglion cell
Complete darkness
Small spot falling on center of receptive field
Large spot covering receptive field
Ring of light excluding center of receptive field
On
Off
Stimulus
On
Off
Stimulus
An on-center ganglion cell is inhibited by a ring of light falling on its receptive field's surround.
An off-center ganglion cell is stimulated by light falling on its receptive field's surround and is inhibited by light falling on its center.
Conclusion: 1. Ganglion cells have circular receptive fields that are divided into center and surround areas.
2. Some ganglion cells are maximally stimulated by light falling on the center of their receptive fields. Others are maximally stimulated by light falling on the surround of their receptive fields.
3. Ganglion cells encode patterns of contrast between light and dark.

45.21 What Does the Eye Tell the Brain? *(Page 881)*

46 The Mammalian Nervous System: Structure and Higher Functions

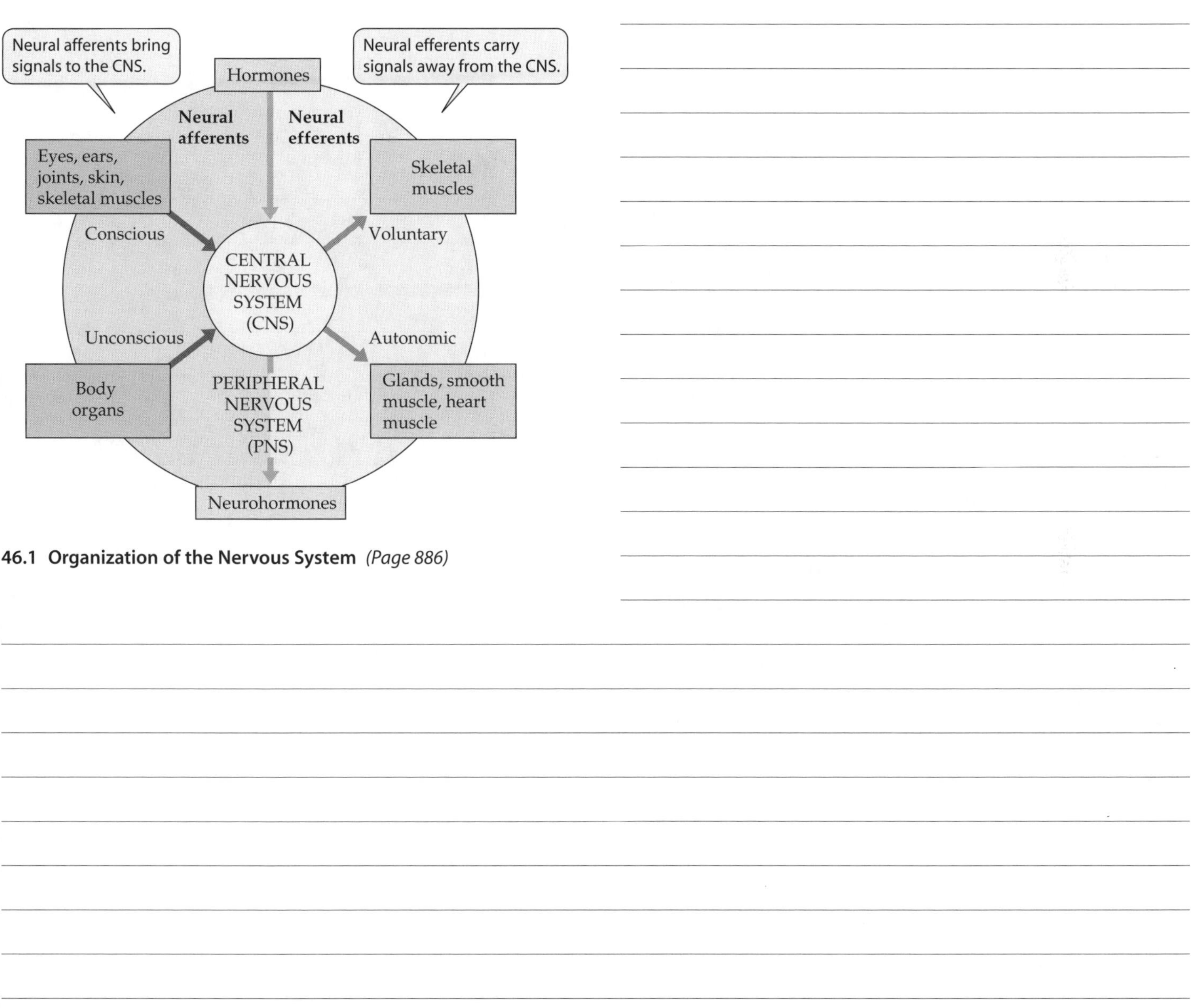

Neural afferents bring signals to the CNS.
Hormones
Neural efferents carry signals away from the CNS.
Neural afferents
Neural efferents
Eyes, ears, joints, skin, skeletal muscles
Skeletal muscles
Conscious
Voluntary
CENTRAL NERVOUS SYSTEM (CNS)
Unconscious
Autonomic
Body organs
PERIPHERAL NERVOUS SYSTEM (PNS)
Glands, smooth muscle, heart muscle
Neurohormones

46.1 Organization of the Nervous System *(Page 886)*

The streched-out neural tube, viewed from above, shows three swellings that will form the adult brain.
Lateral views
Neural tube
25 days
35 days
Forebrain
Midbrain
Hindbrain
Spinal cord
40 days
50 days
Cerebral hemisphere
100 days
Dorsal views
Forebrain
Midbrain
Hindbrain
Spinal cord
25 days
The forebrain develops into two major divisions, the telencephalon and diencephalon.
Telencephalon
Diencephalon
Developing eye
Midbrain
Hindbrain
40 days
Cerebral hemisphere
Thalamus
Hypothalamus
Pituitary
Cerebellum
Pons
Medulla
Adult brain
The hindbrain develops into three major divisions: the cerebellum, pons, and medulla.

46.2 Development of the Human Nervous System *(Page 887)*

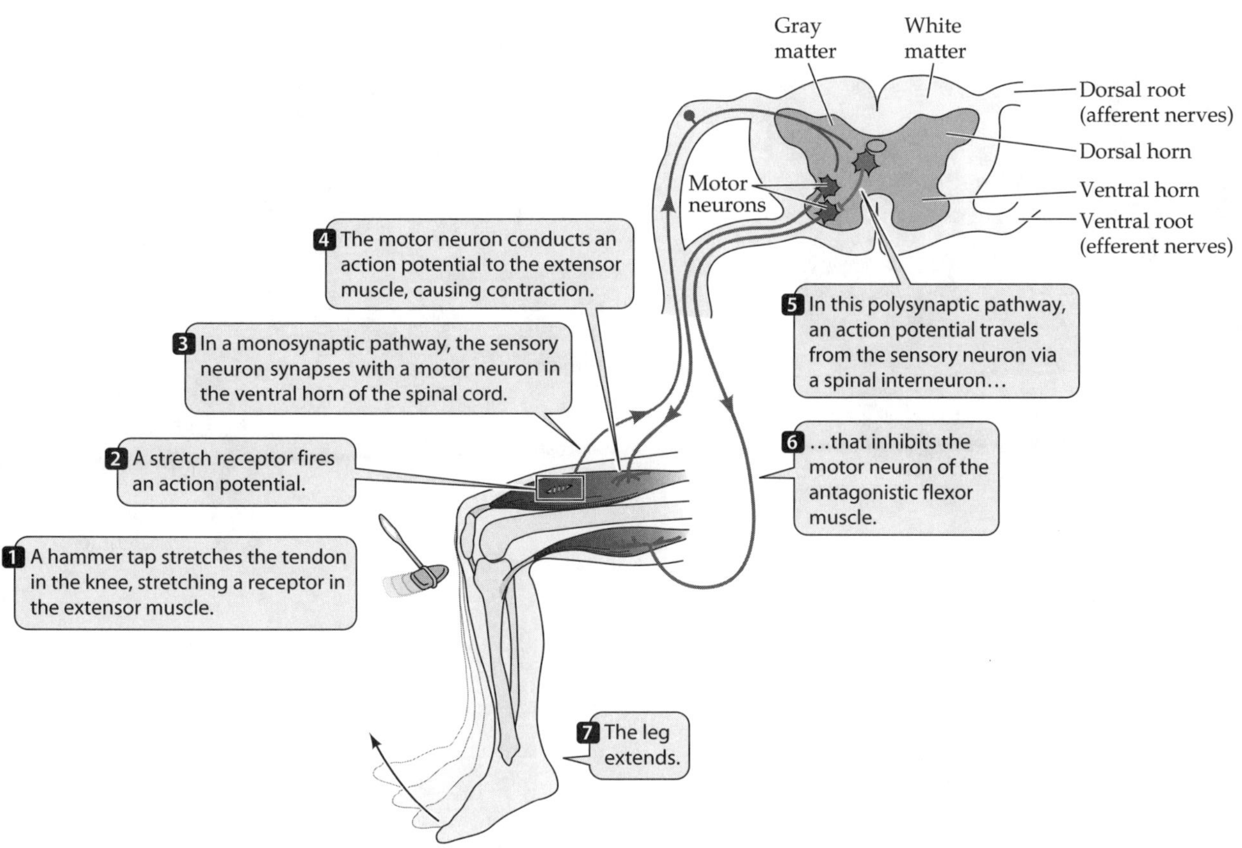
Gray matter
White matter
Dorsal root (afferent nerves)
Dorsal horn
Ventral horn
Ventral root (efferent nerves)
Motor neurons
4 The motor neuron conducts an action potential to the extensor muscle, causing contraction.
3 In a monosynaptic pathway, the sensory neuron synapses with a motor neuron in the ventral horn of the spinal cord.
2 A stretch receptor fires an action potential.
1 A hammer tap stretches the tendon in the knee, stretching a receptor in the extensor muscle.
5 In this polysynaptic pathway, an action potential travels from the sensory neuron via a spinal interneuron…
6 …that inhibits the motor neuron of the antagonistic flexor muscle.
7 The leg extends.

46.3 The Spinal Cord Processes Information *(Page 888)*

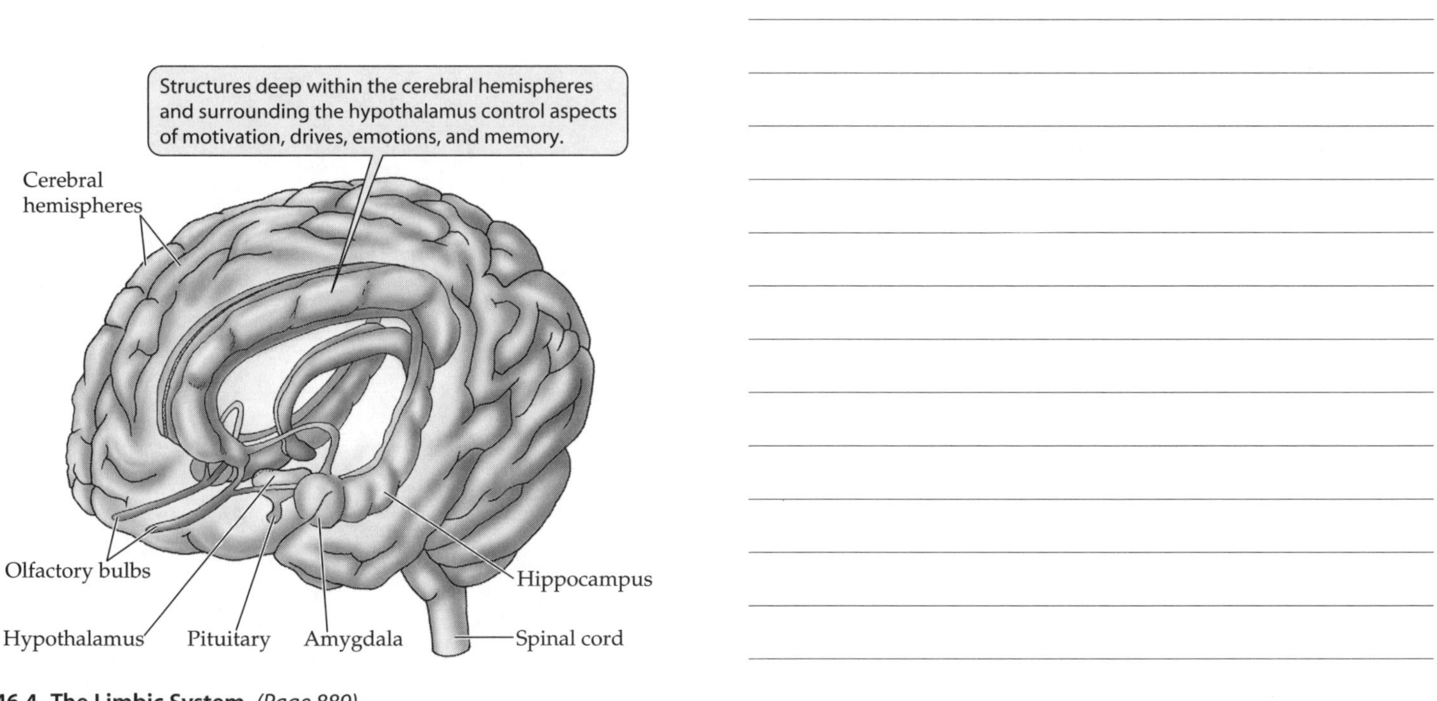
Structures deep within the cerebral hemispheres and surrounding the hypothalamus control aspects of motivation, drives, emotions, and memory.
Cerebral hemispheres
Olfactory bulbs
Hypothalamus
Pituitary
Amygdala
Hippocampus
Spinal cord

46.4 The Limbic System *(Page 889)*

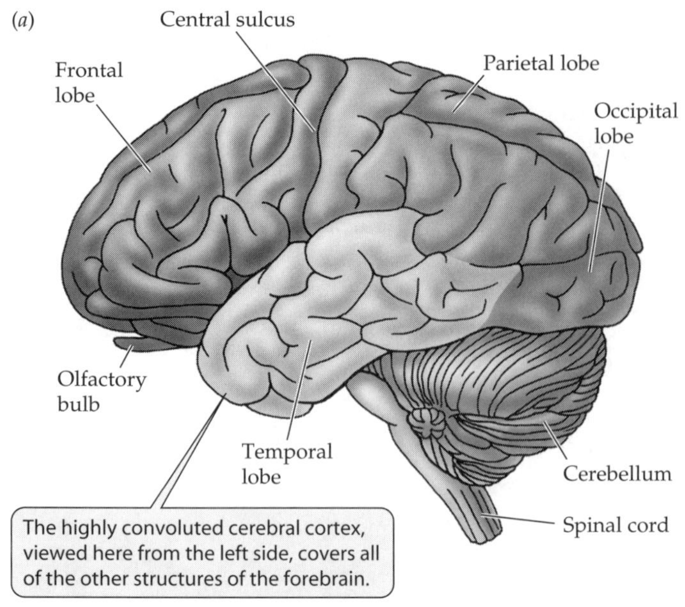
(a)
Central sulcus
Frontal lobe
Parietal lobe
Occipital lobe
Olfactory bulb
Temporal lobe
Cerebellum
Spinal cord
The highly convoluted cerebral cortex, viewed here from the left side, covers all of the other structures of the forebrain.

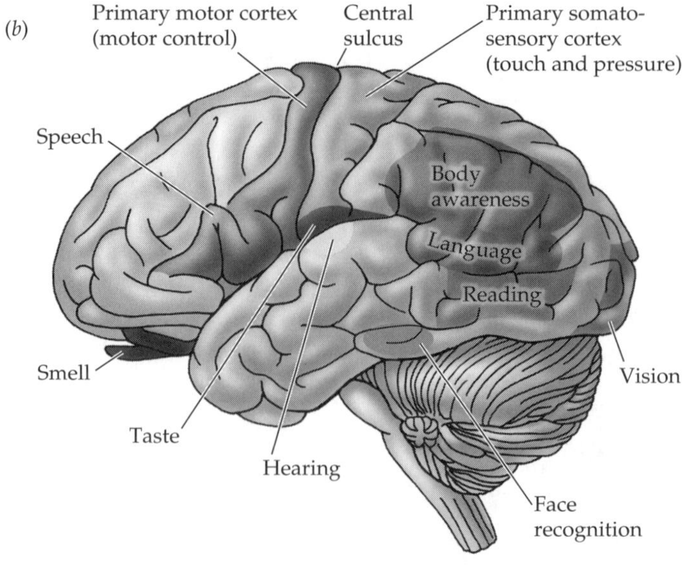
(b)
Primary motor cortex (motor control)
Central sulcus
Primary somatosensory cortex (touch and pressure)
Speech
Body awareness
Language
Reading
Smell
Vision
Taste
Hearing
Face recognition

46.5 The Human Cerebrum *(Page 890)*

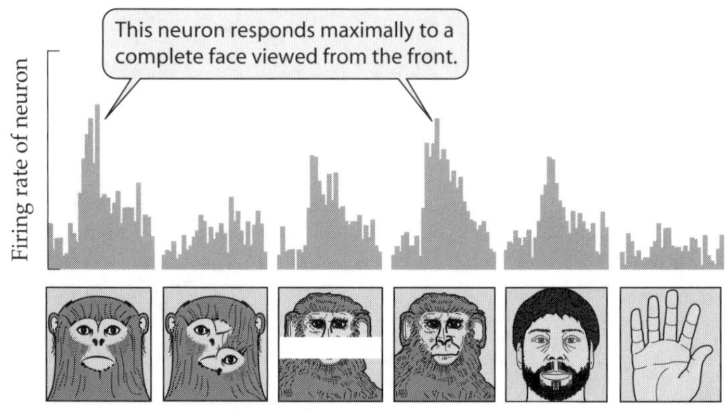
This neuron responds maximally to a complete face viewed from the front.
Firing rate of neuron

46.6 Neurons in One Region of the Temporal Lobe Respond to Faces *(Page 890)*

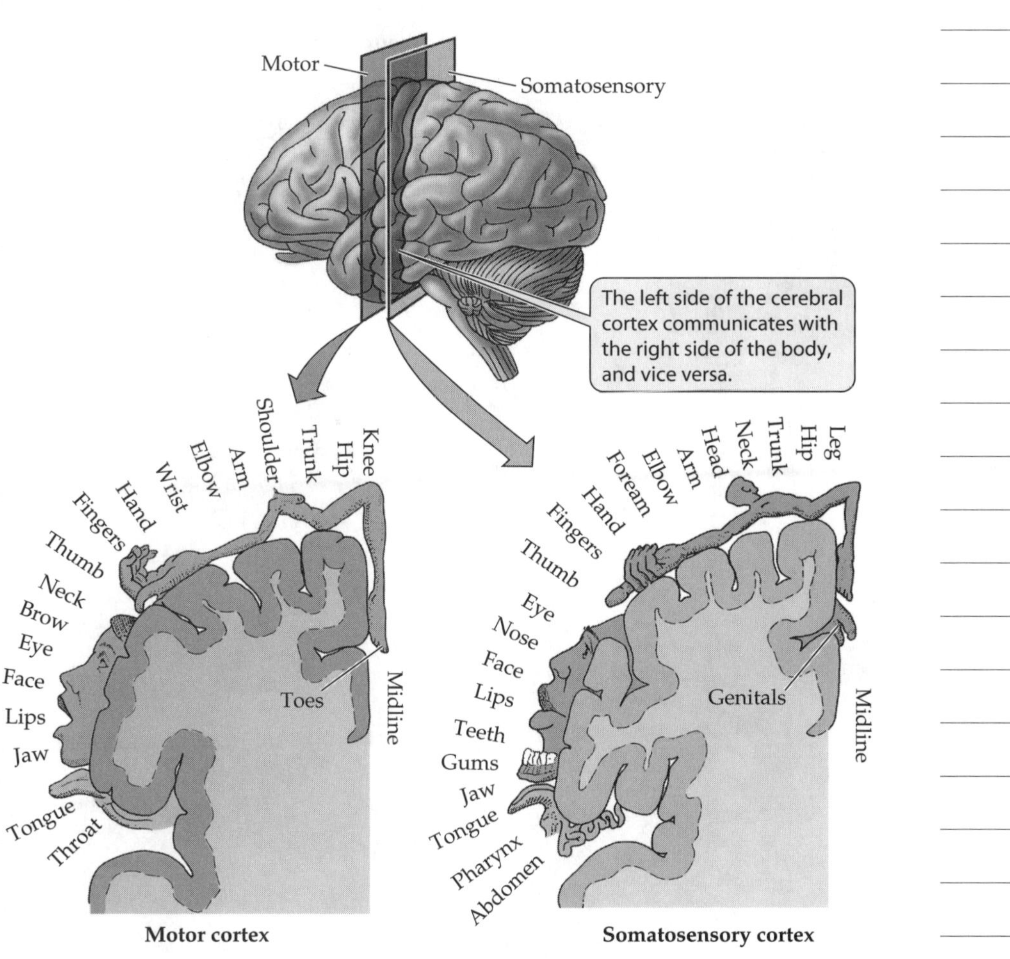

Motor
Somatosensory
The left side of the cerebral cortex communicates with the right side of the body, and vice versa.
Knee
Hip
Trunk
Shoulder
Arm
Elbow
Wrist
Hand
Fingers
Thumb
Neck
Brow
Eye
Face
Lips
Jaw
Tongue
Throat
Toes
Midline
Leg
Hip
Trunk
Neck
Head
Arm
Elbow
Foream
Hand
Fingers
Thumb
Eye
Nose
Face
Lips
Teeth
Gums
Jaw
Tongue
Pharynx
Abdomen
Genitals
Midline
Motor cortex
Somatosensory cortex

46.7 The Body Is Represented in the Primary Motor Cortex and the Primary Somatosensory Cortex *(Page 891)*

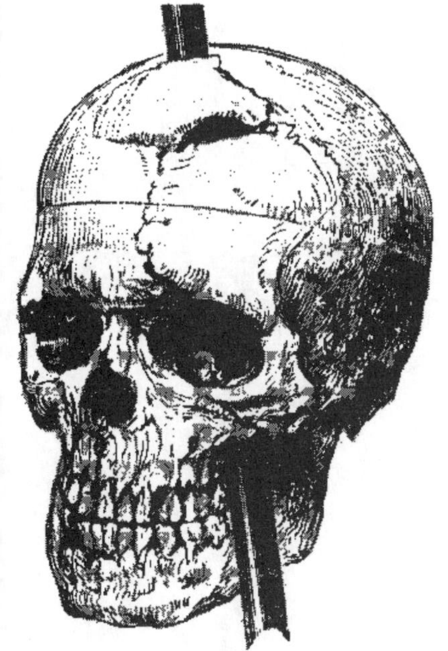

46.8 A Mind-Altering Experience *(Page 891)*

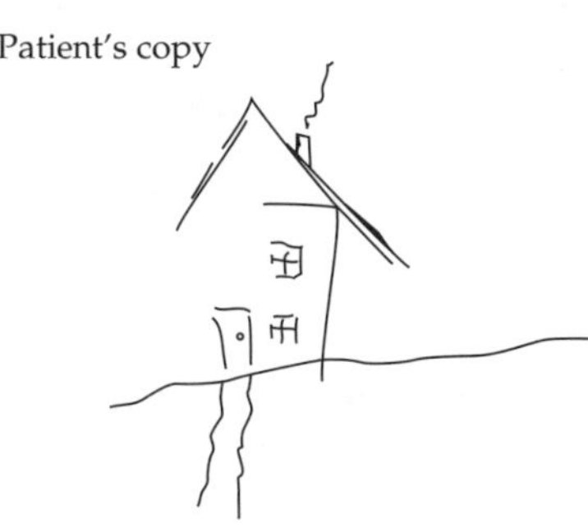

Model
Patient's copy

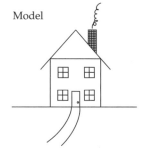

46.9 Contralateral Neglect Syndrome *(Page 892)*

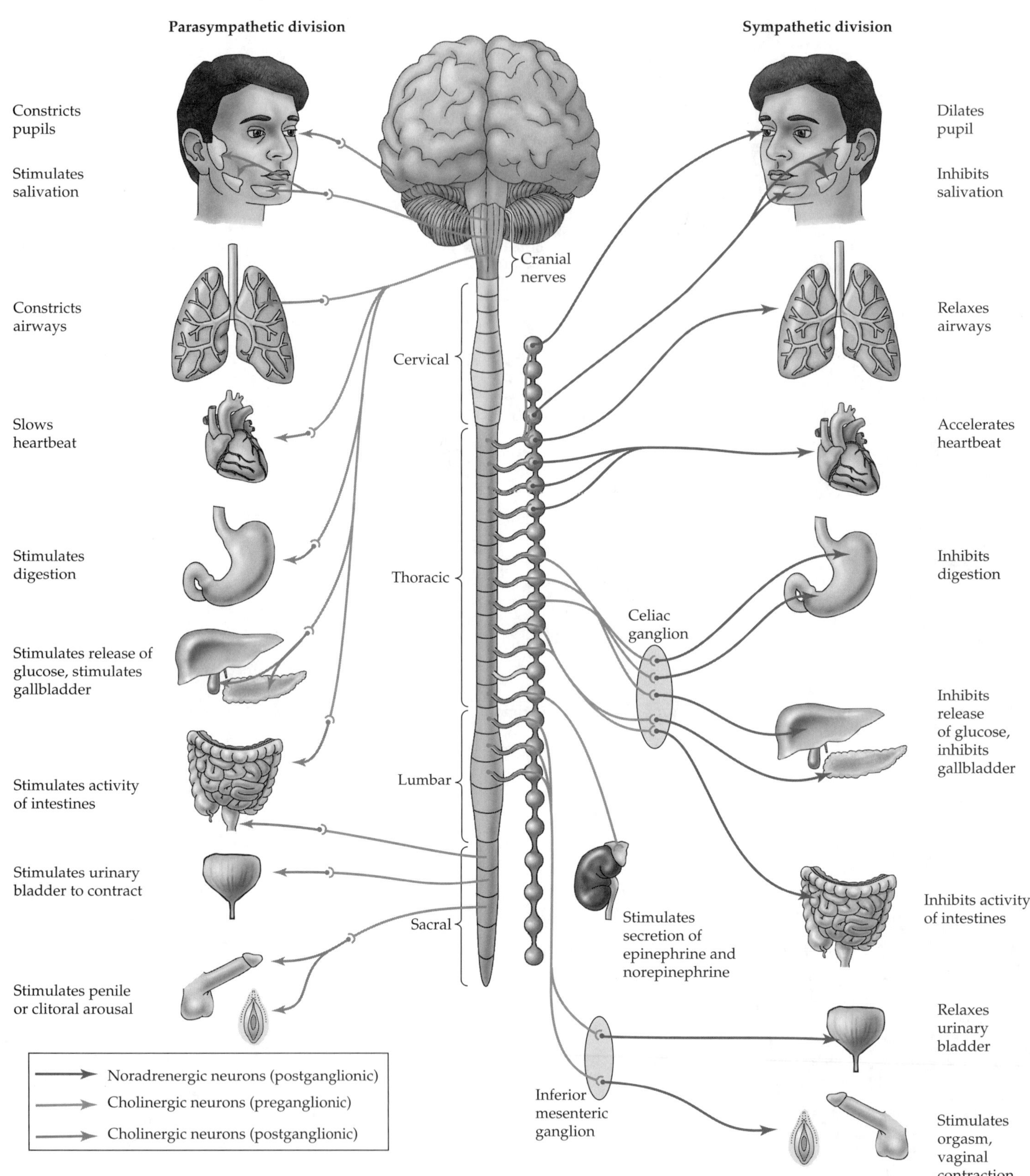

Parasympathetic division
Sympathetic division
Constricts pupils
Stimulates salivation
Constricts airways
Slows heartbeat
Stimulates digestion
Stimulates release of glucose, stimulates gallbladder
Stimulates activity of intestines
Stimulates urinary bladder to contract
Stimulates penile or clitoral arousal
Cranial nerves
Cervical
Thoracic
Lumbar
Sacral
Celiac ganglion
Inferior mesenteric ganglion
Dilates pupil
Inhibits salivation
Relaxes airways
Accelerates heartbeat
Inhibits digestion
Inhibits release of glucose, inhibits gallbladder
Stimulates secretion of epinephrine and norepinephrine
Inhibits activity of intestines
Relaxes urinary bladder
Stimulates orgasm, vaginal contraction
Noradrenergic neurons (postganglionic)
Cholinergic neurons (preganglionic)
Cholinergic neurons (postganglionic)

46.10 Organization of the Autonomic Nervous System *(Page 893)*

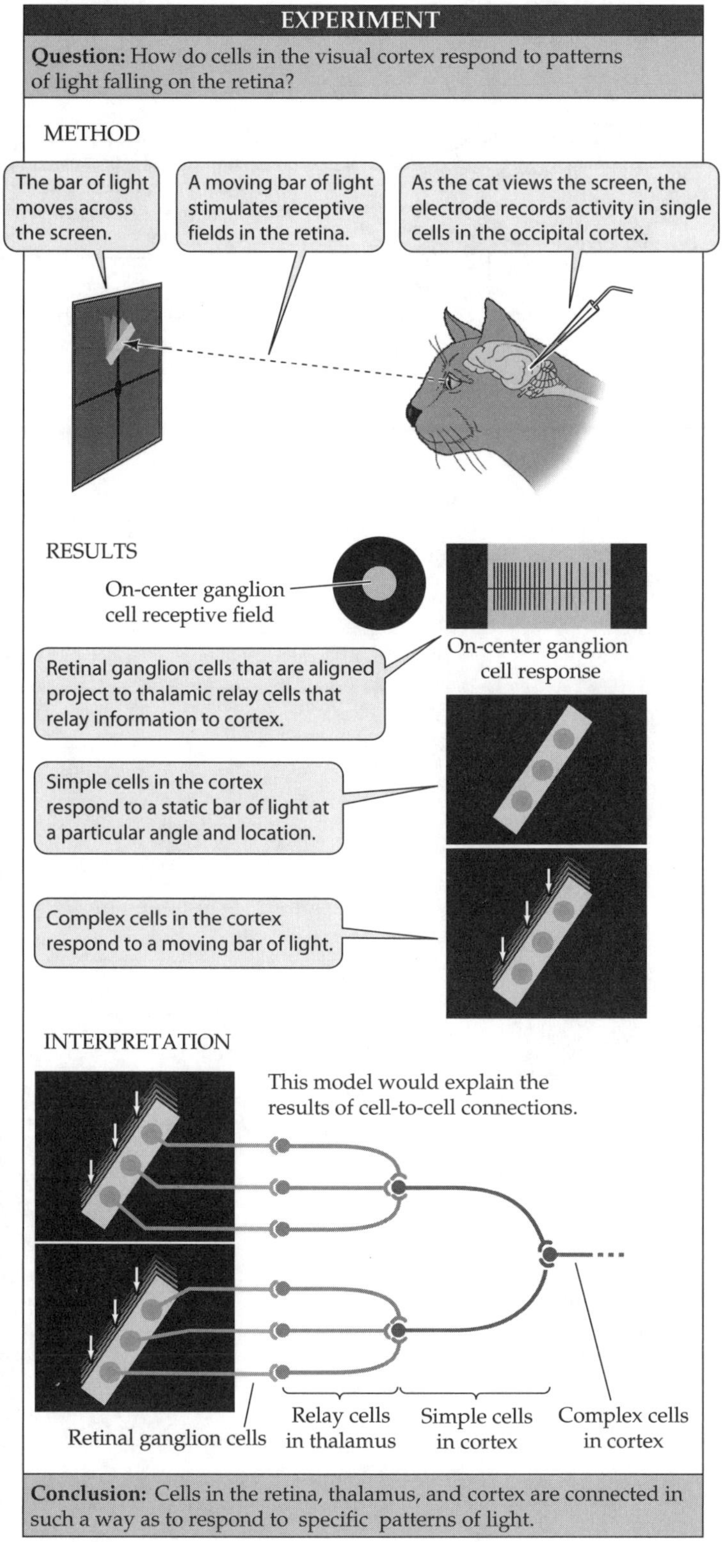

EXPERIMENT
Question: How do cells in the visual cortex respond to patterns of light falling on the retina?
METHOD
The bar of light moves across the screen.
A moving bar of light stimulates receptive fields in the retina.
As the cat views the screen, the electrode records activity in single cells in the occipital cortex.
RESULTS
On-center ganglion cell receptive field
On-center ganglion cell response
Retinal ganglion cells that are aligned project to thalamic relay cells that relay information to cortex.
Simple cells in the cortex respond to a static bar of light at a particular angle and location.
Complex cells in the cortex respond to a moving bar of light.
INTERPRETATION
This model would explain the results of cell-to-cell connections.
Retinal ganglion cells
Relay cells in thalamus
Simple cells in cortex
Complex cells in cortex
Conclusion: Cells in the retina, thalamus, and cortex are connected in such a way as to respond to specific patterns of light.

46.11 Receptive Fields of Cells in the Visual Cortex *(Page 895)*

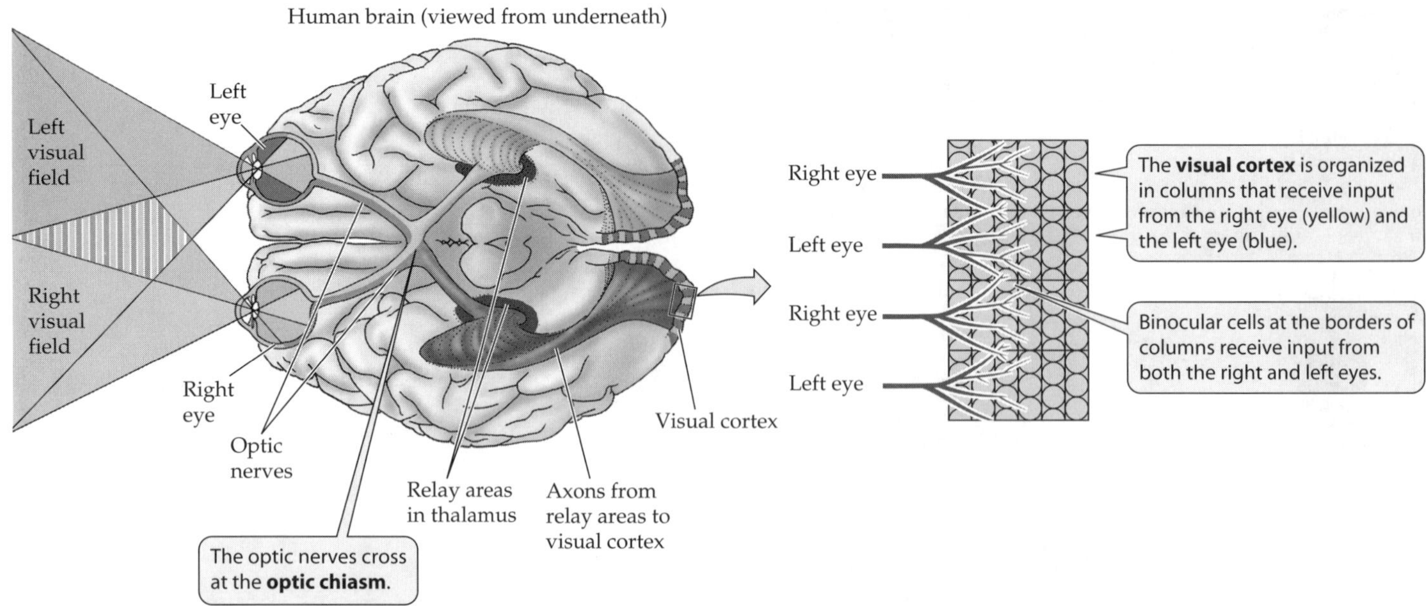
Human brain (viewed from underneath)
Left visual field
Right visual field
Left eye
Right eye
Optic nerves
Relay areas in thalamus
Axons from relay areas to visual cortex
Visual cortex
The optic nerves cross at the **optic chiasm**.
Right eye
Left eye
Right eye
Left eye
The **visual cortex** is organized in columns that receive input from the right eye (yellow) and the left eye (blue).
Binocular cells at the borders of columns receive input from both the right and left eyes.

46.12 The Anatomy of Binocular Vision *(Page 896)*

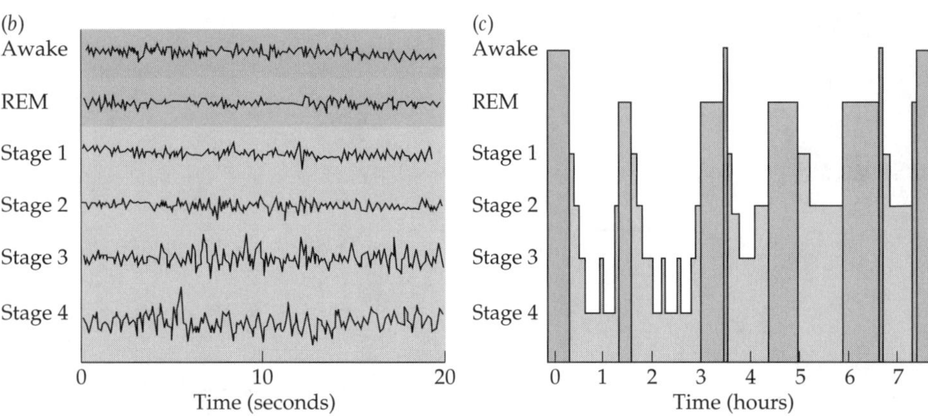
(b)
Awake
REM
Stage 1
Stage 2
Stage 3
Stage 4
0
10
20
Time (seconds)
(c)
Awake
REM
Stage 1
Stage 2
Stage 3
Stage 4
0 1 2 3 4 5 6 7
Time (hours)

46.13 Patterns of Electrical Activity in the Cerebral Cortex Characterize Stages of Sleep *(Page 897)*

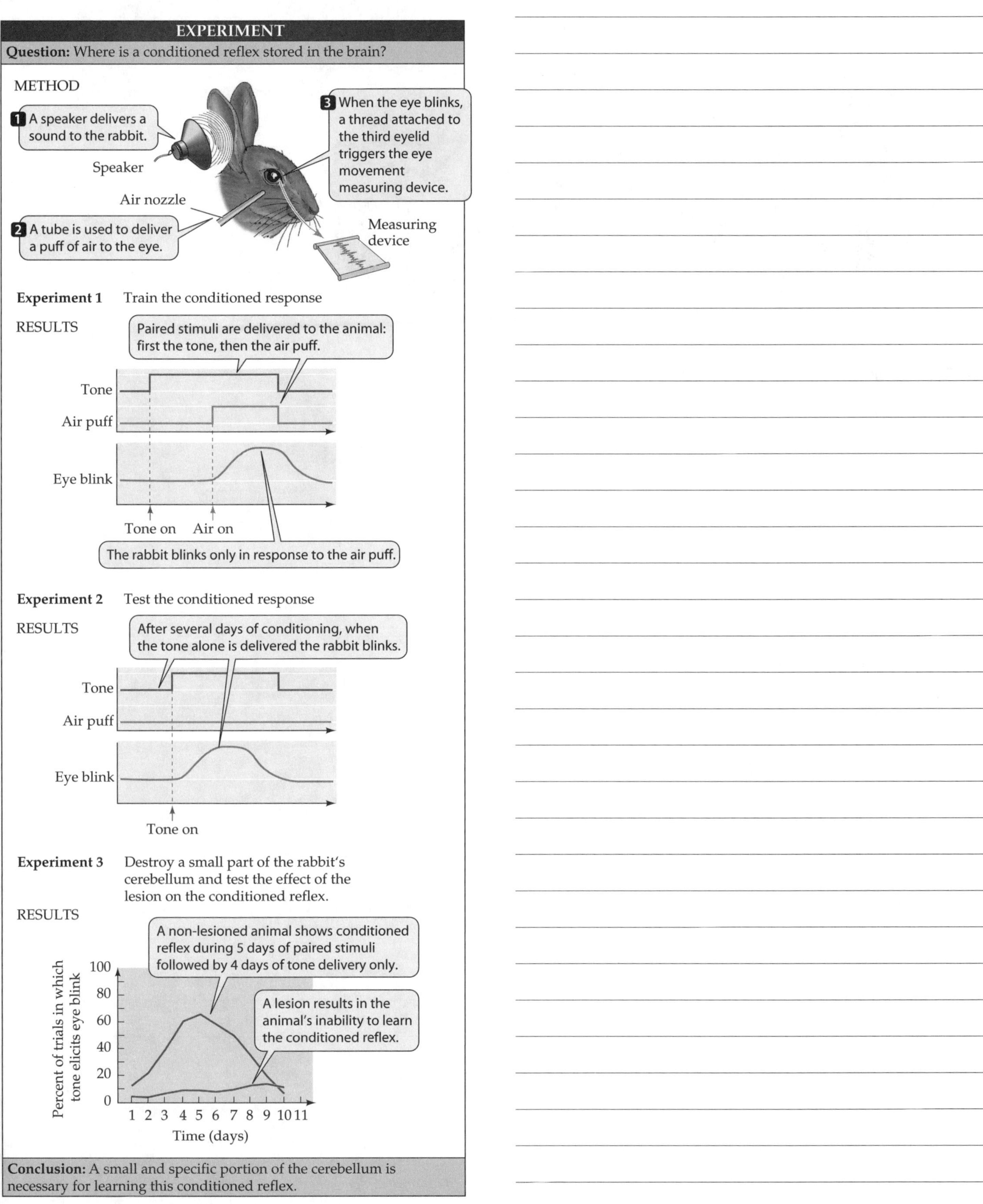
EXPERIMENT
Question: Where is a conditioned reflex stored in the brain?
METHOD
1 A speaker delivers a sound to the rabbit.
Speaker
Air nozzle
2 A tube is used to deliver a puff of air to the eye.
3 When the eye blinks, a thread attached to the third eyelid triggers the eye movement measuring device.
Measuring device
Experiment 1 Train the conditioned response
RESULTS
Paired stimuli are delivered to the animal: first the tone, then the air puff.
Tone
Air puff
Eye blink
Tone on
Air on
The rabbit blinks only in response to the air puff.
Experiment 2 Test the conditioned response
RESULTS
After several days of conditioning, when the tone alone is delivered the rabbit blinks.
Tone
Air puff
Eye blink
Tone on
Experiment 3 Destroy a small part of the rabbit's cerebellum and test the effect of the lesion on the conditioned reflex.
RESULTS
A non-lesioned animal shows conditioned reflex during 5 days of paired stimuli followed by 4 days of tone delivery only.
A lesion results in the animal's inability to learn the conditioned reflex.
Percent of trials in which tone elicits eye blink
100
80
60
40
20
0
1 2 3 4 5 6 7 8 9 10 11
Time (days)
Conclusion: A small and specific portion of the cerebellum is necessary for learning this conditioned reflex.

46.14 The Conditioned Eye-Blink Reflex Depends on a Cerebellar Circuit *(Page 898)*

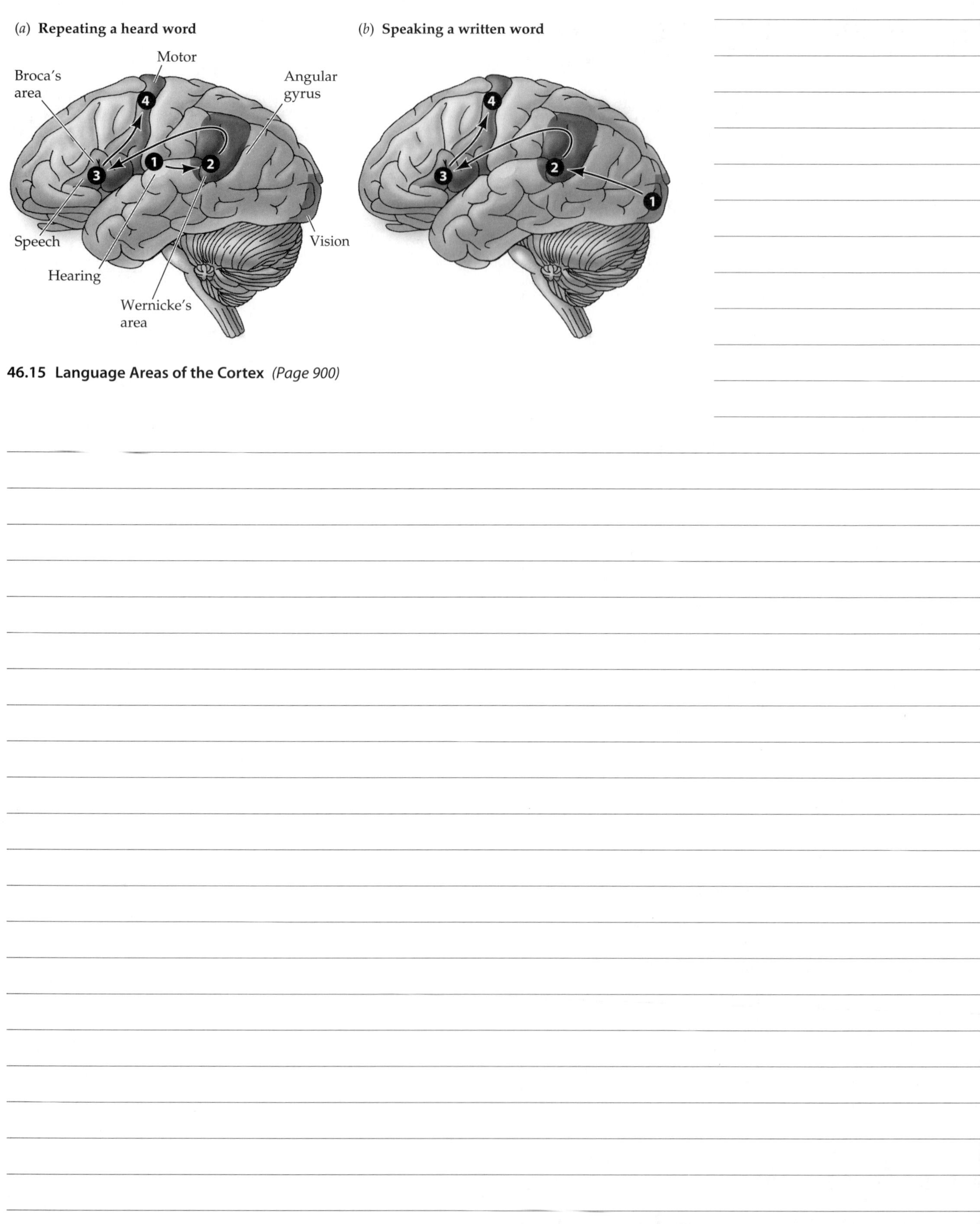
(a) Repeating a heard word
Motor
Broca's area
Angular gyrus
Speech
Hearing
Wernicke's area
Vision
(b) Speaking a written word
1
2
3
4

46.15 Language Areas of the Cortex *(Page 900)*

47 Effectors: Making Animals Move

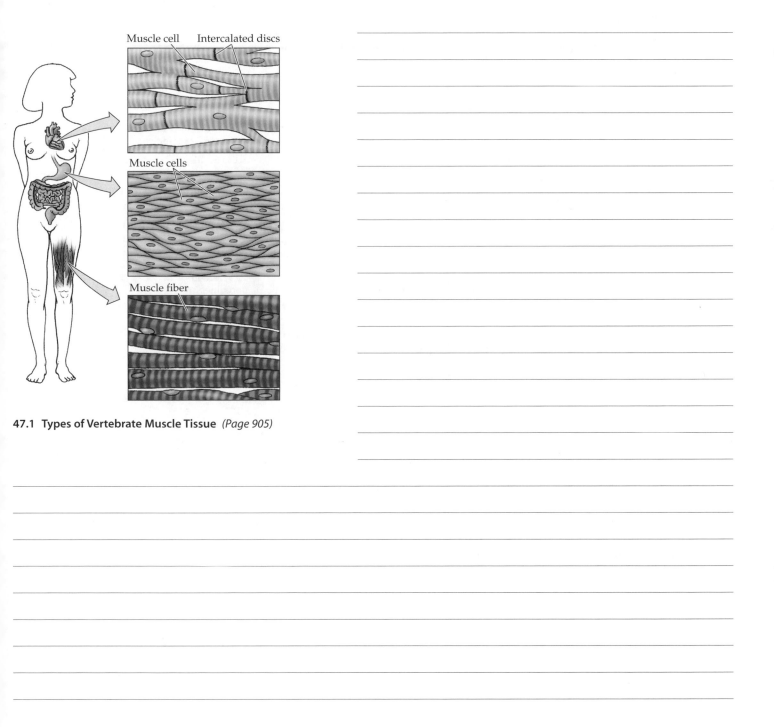

Muscle cell
Intercalated discs
Muscle cells
Muscle fiber

47.1 Types of Vertebrate Muscle Tissue *(Page 905)*

EXPERIMENT

Question: What stimulates contraction of smooth muscle?

METHOD Incubate a strip of smooth (intestinal) muscle in a saline bath. Measure action potentials and force of contraction.

Experiment 1 Stretch intestinal muscle and analyze response.

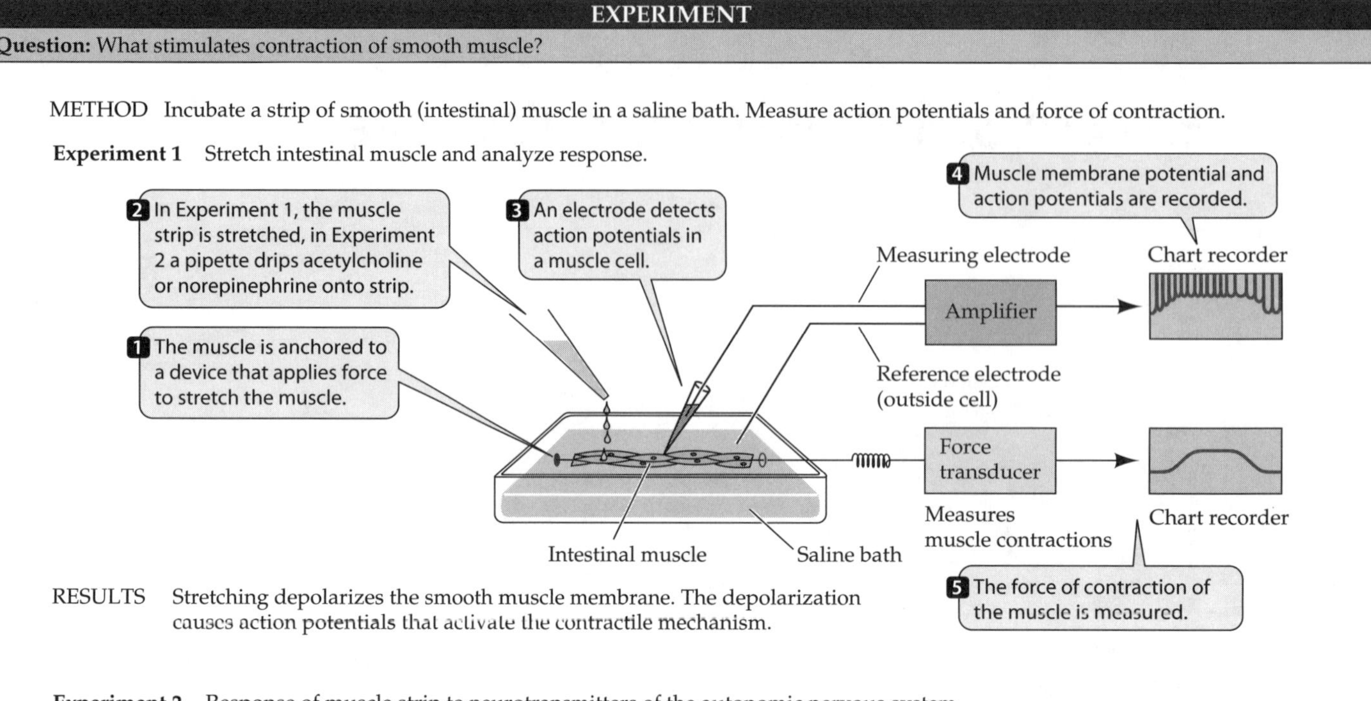
4 Muscle membrane potential and action potentials are recorded.
2 In Experiment 1, the muscle strip is stretched, in Experiment 2 a pipette drips acetylcholine or norepinephrine onto strip.
3 An electrode detects action potentials in a muscle cell.
Measuring electrode
Chart recorder
Amplifier
1 The muscle is anchored to a device that applies force to stretch the muscle.
Reference electrode (outside cell)
Force transducer
Measures muscle contractions
Chart recorder
Intestinal muscle
Saline bath
5 The force of contraction of the muscle is measured.

RESULTS Stretching depolarizes the smooth muscle membrane. The depolarization causes action potentials that activate the contractile mechanism.

Experiment 2 Response of muscle strip to neurotransmitters of the autonomic nervous system.

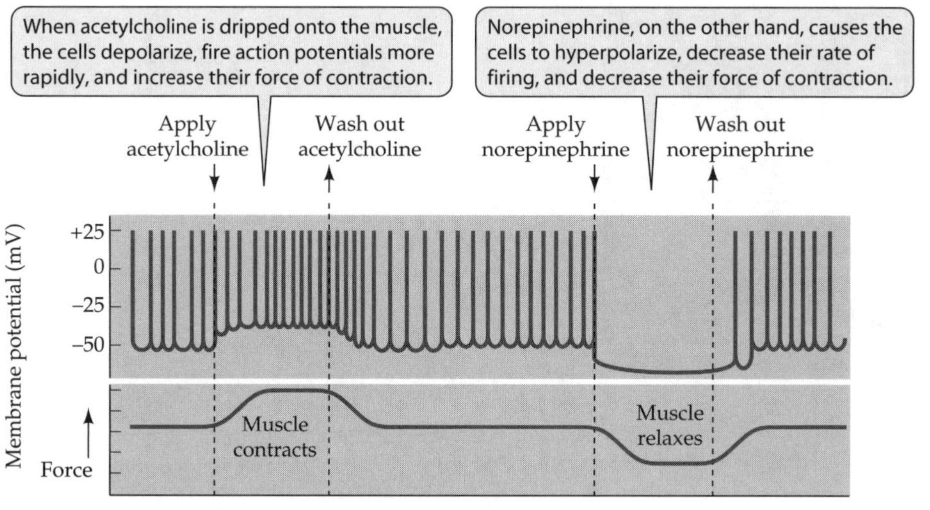
When acetylcholine is dripped onto the muscle, the cells depolarize, fire action potentials more rapidly, and increase their force of contraction.
Norepinephrine, on the other hand, causes the cells to hyperpolarize, decrease their rate of firing, and decrease their force of contraction.
Apply acetylcholine
Wash out acetylcholine
Apply norepinephrine
Wash out norepinephrine
Membrane potential (mV)
+25
0
–25
–50
Force
Muscle contracts
Muscle relaxes

RESULTS Autonomic neurotransmitters alter membrane resting potential and thereby determine the rate that smooth muscle cells fire action potentials.

Conclusion: Smooth muscle contraction is stimulated by stretch and by the parasympathetic neurotransmitter acetylcholine.

47.2 Mechanisms of Smooth Muscle Activation *(Page 906)*

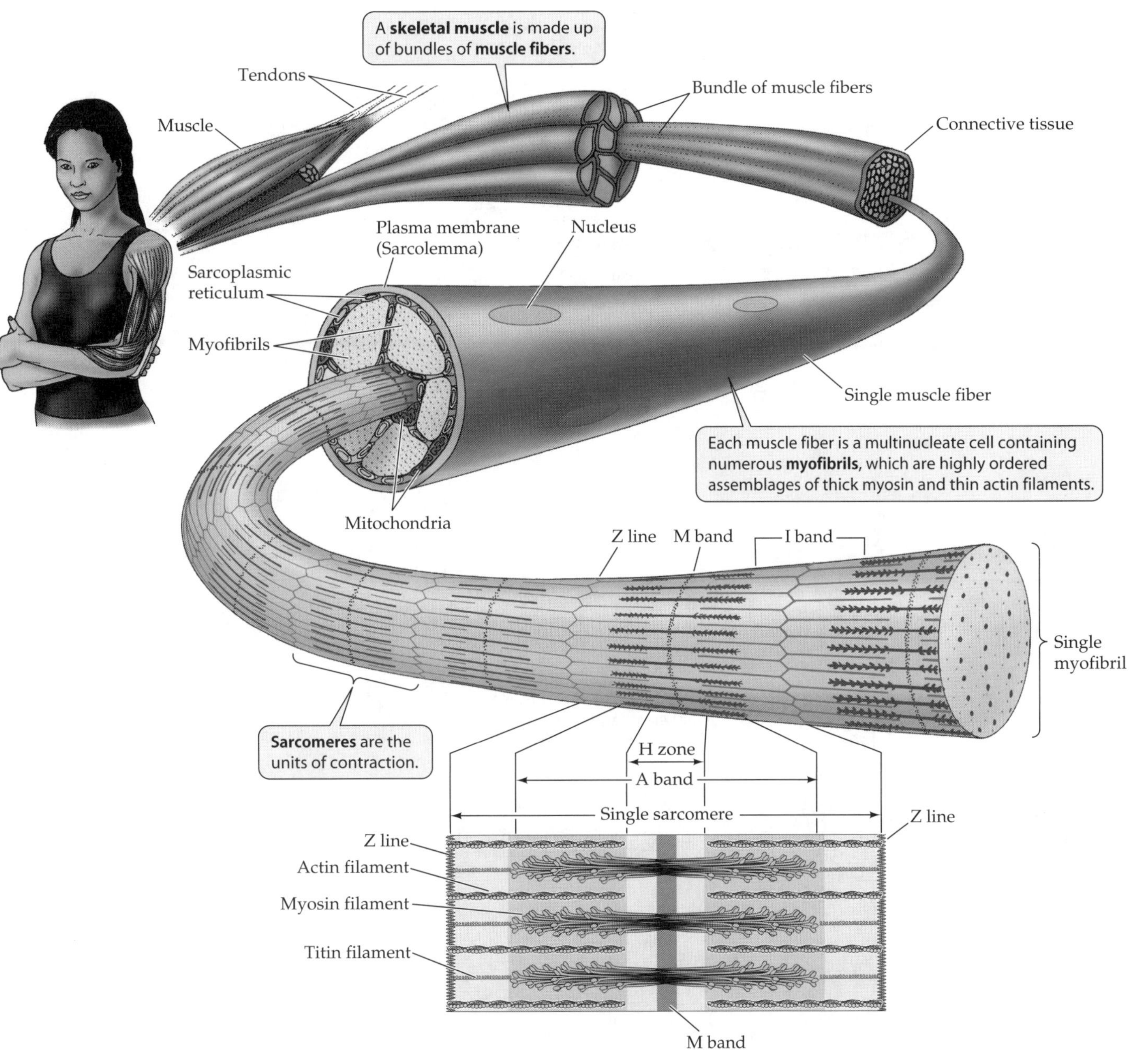
A **skeletal muscle** is made up of bundles of **muscle fibers**.
Tendons
Muscle
Bundle of muscle fibers
Connective tissue
Plasma membrane (Sarcolemma)
Nucleus
Sarcoplasmic reticulum
Myofibrils
Single muscle fiber
Each muscle fiber is a multinucleate cell containing numerous **myofibrils**, which are highly ordered assemblages of thick myosin and thin actin filaments.
Mitochondria
Z line
M band
I band
Single myofibril
Sarcomeres are the units of contraction.
H zone
A band
Single sarcomere
Z line
Z line
Actin filament
Myosin filament
Titin filament
M band

47.3 The Structure of Skeletal Muscle *(Page 907)*

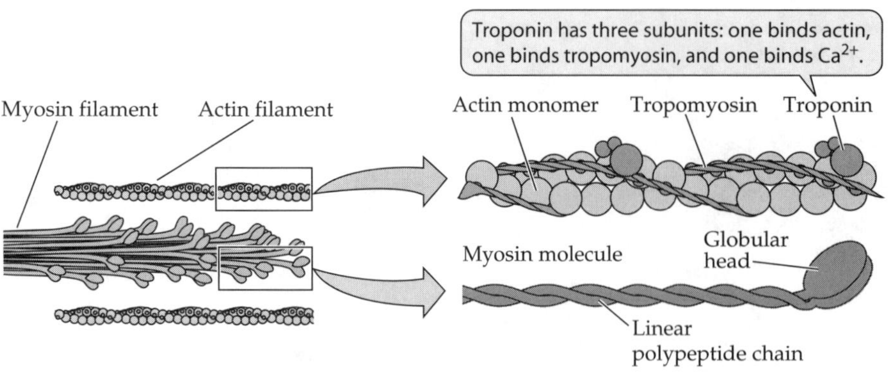
Troponin has three subunits: one binds actin, one binds tropomyosin, and one binds Ca^{2+}.
Myosin filament
Actin filament
Actin monomer
Tropomyosin
Troponin
Myosin molecule
Globular head
Linear polypeptide chain

47.4 Actin and Myosin Filaments Overlap to Form Myofibrils *(Page 908)*

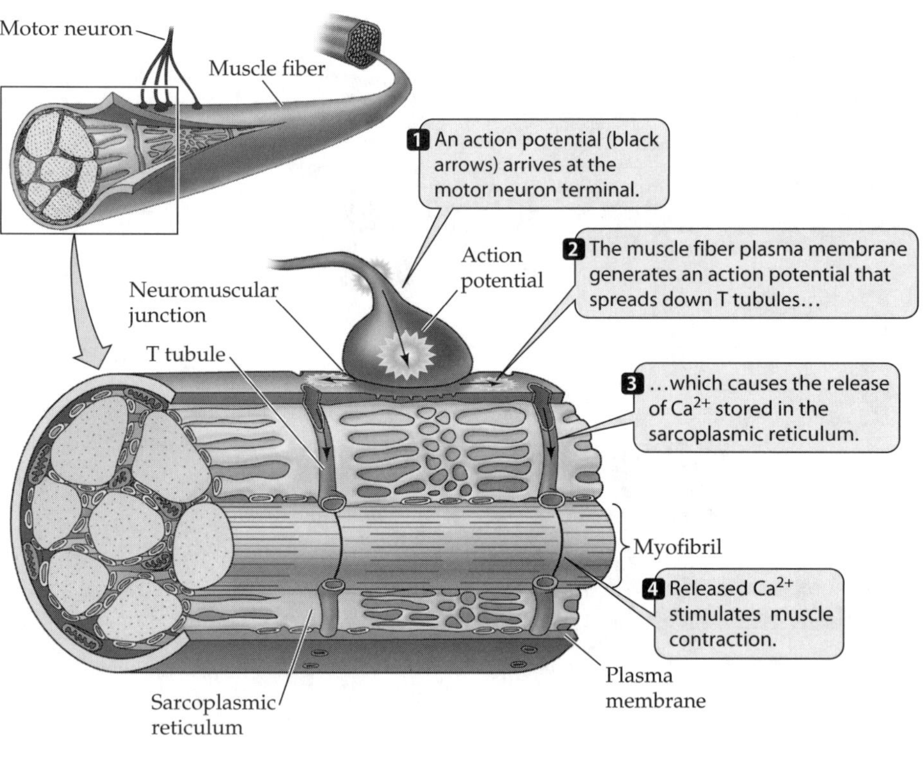
Motor neuron
Muscle fiber
1 An action potential (black arrows) arrives at the motor neuron terminal.
Action potential
Neuromuscular junction
2 The muscle fiber plasma membrane generates an action potential that spreads down T tubules…
T tubule
3 …which causes the release of Ca^{2+} stored in the sarcoplasmic reticulum.
Myofibril
4 Released Ca^{2+} stimulates muscle contraction.
Plasma membrane
Sarcoplasmic reticulum

47.5 T Tubules in Action *(Page 909)*

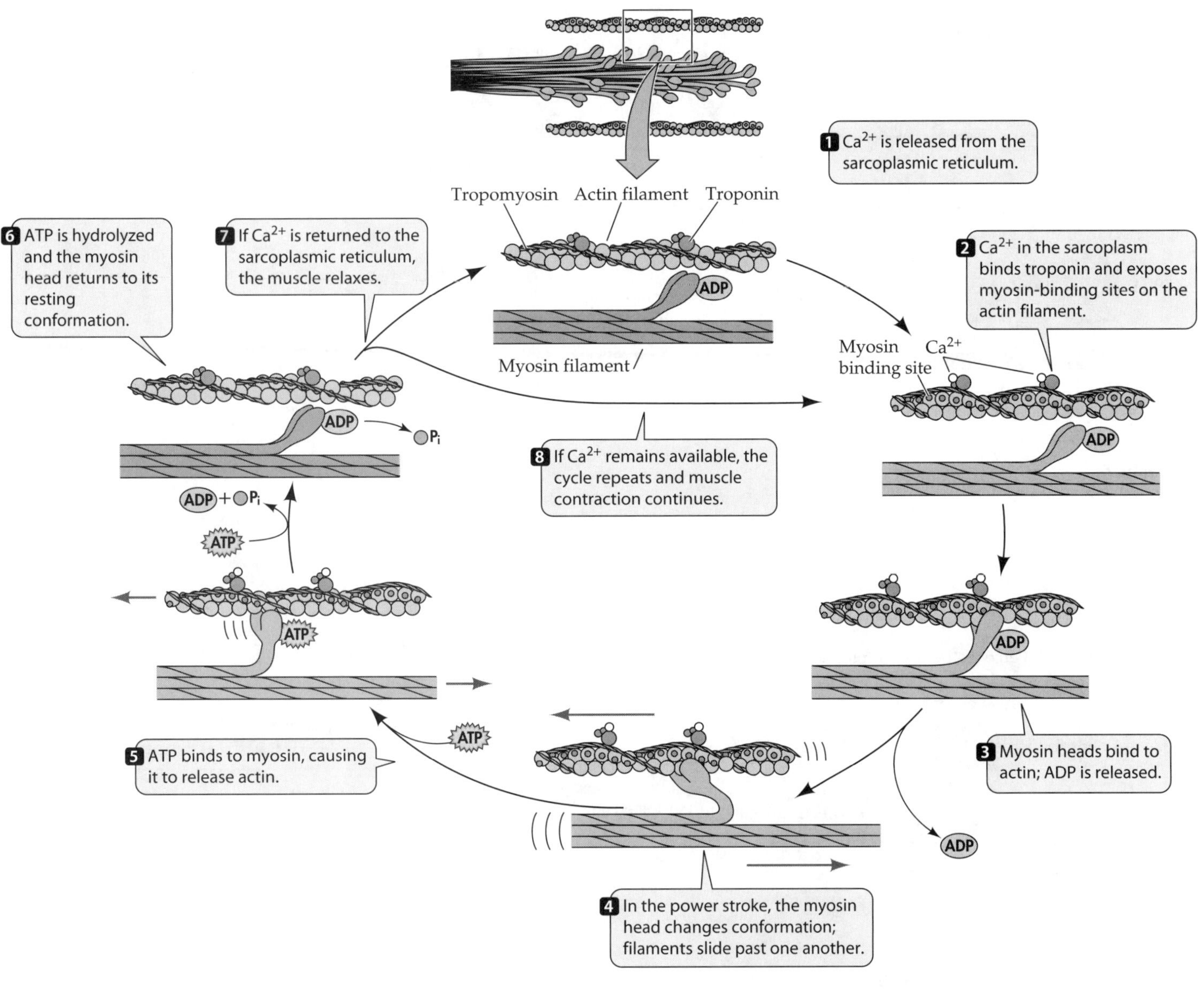
1 Ca^{2+} is released from the sarcoplasmic reticulum.
Tropomyosin
Actin filament
Troponin
ADP
Myosin filament
2 Ca^{2+} in the sarcoplasm binds troponin and exposes myosin-binding sites on the actin filament.
Myosin binding site
Ca^{2+}
ADP
ADP
3 Myosin heads bind to actin; ADP is released.
ADP
4 In the power stroke, the myosin head changes conformation; filaments slide past one another.
ATP
5 ATP binds to myosin, causing it to release actin.
ATP
ATP
ADP + P_i
6 ATP is hydrolyzed and the myosin head returns to its resting conformation.
ADP
P_i
7 If Ca^{2+} is returned to the sarcoplasmic reticulum, the muscle relaxes.
8 If Ca^{2+} remains available, the cycle repeats and muscle contraction continues.

47.6 The Release of Ca^{2+} from the Sarcoplasmic Reticulum Triggers Muscle Contraction *(Page 910)*

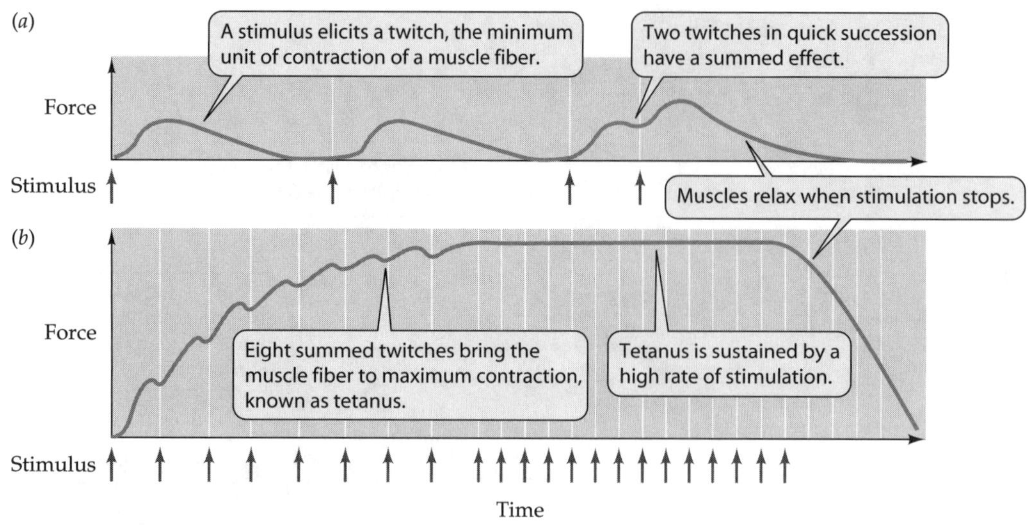
(a)
A stimulus elicits a twitch, the minimum unit of contraction of a muscle fiber.
Two twitches in quick succession have a summed effect.
Force
Stimulus
Muscles relax when stimulation stops.
(b)
Force
Eight summed twitches bring the muscle fiber to maximum contraction, known as tetanus.
Tetanus is sustained by a high rate of stimulation.
Stimulus
Time

47.7 Twitches and Tetanus *(Page 911)*

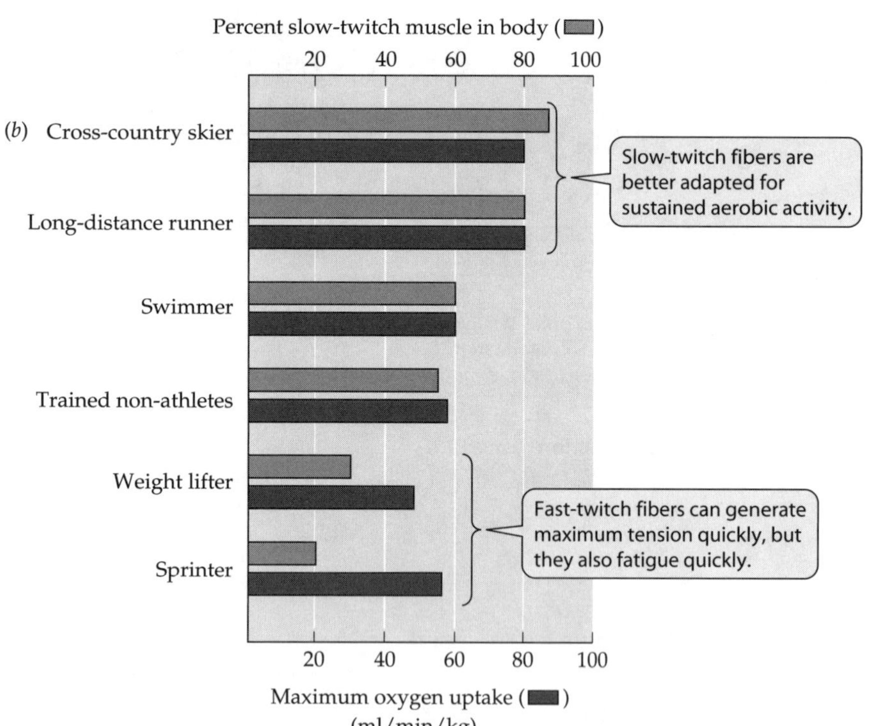
Percent slow-twitch muscle in body
20
40
60
80
100
(b)
Cross-country skier
Long-distance runner
Swimmer
Trained non-athletes
Weight lifter
Sprinter
Slow-twitch fibers are better adapted for sustained aerobic activity.
Fast-twitch fibers can generate maximum tension quickly, but they also fatigue quickly.
20
40
60
80
100
Maximum oxygen uptake
(ml/min/kg)

47.8 Two Types of Muscle Fibers *(Page 912)*

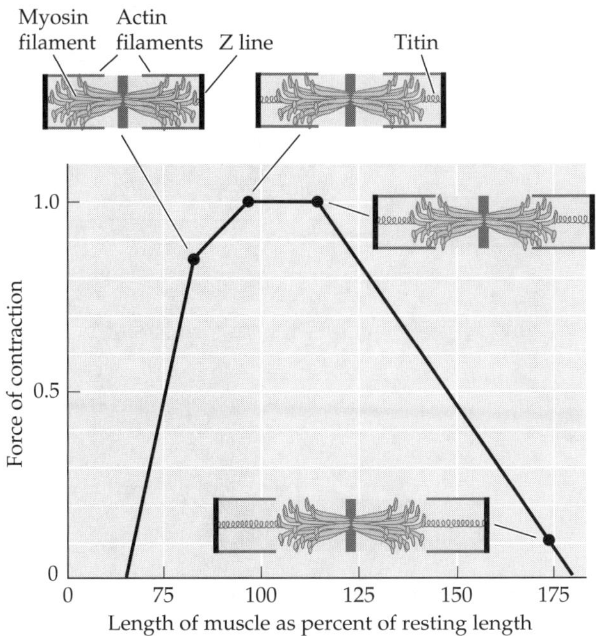
Myosin filament
Actin filaments
Z line
Titin
Force of contraction
1.0
0.5
0
0
75
100
125
150
175
Length of muscle as percent of resting length

47.9 Strength and Length *(Page 913)*

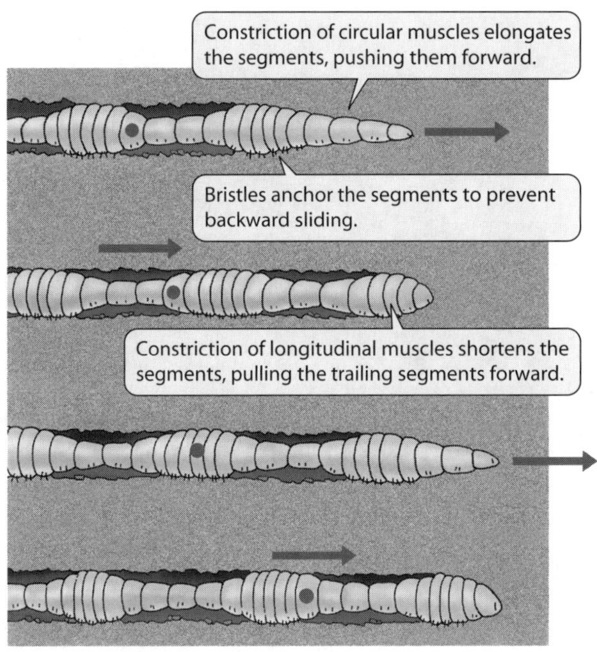
Constriction of circular muscles elongates the segments, pushing them forward.
Bristles anchor the segments to prevent backward sliding.
Constriction of longitudinal muscles shortens the segments, pulling the trailing segments forward.

47.11 A Hydrostatic Skeleton *(Page 914)*

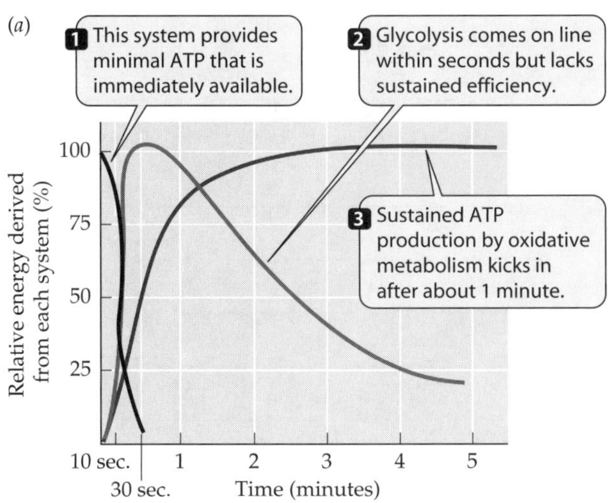
(a)
1 This system provides minimal ATP that is immediately available.
2 Glycolysis comes on line within seconds but lacks sustained efficiency.
3 Sustained ATP production by oxidative metabolism kicks in after about 1 minute.
Relative energy derived from each system (%)
100
75
50
25
10 sec.
30 sec.
1
2
3
4
5
Time (minutes)

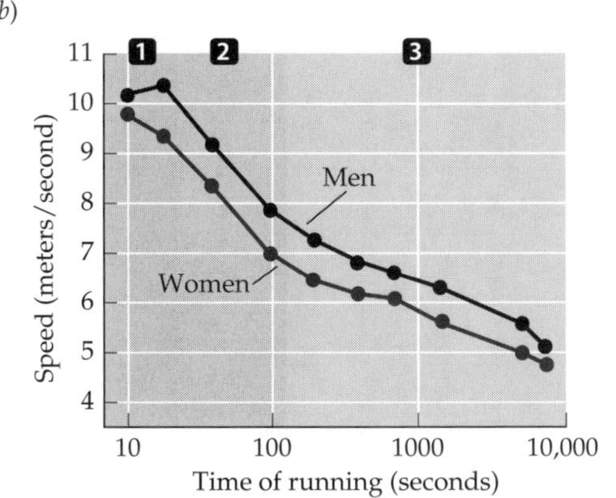
(b)
1
2
3
Speed (meters/second)
11
10
9
8
7
6
5
4
Men
Women
10
100
1000
10,000
Time of running (seconds)

47.10 Supplying Fuel for High Performance *(Page 913)*

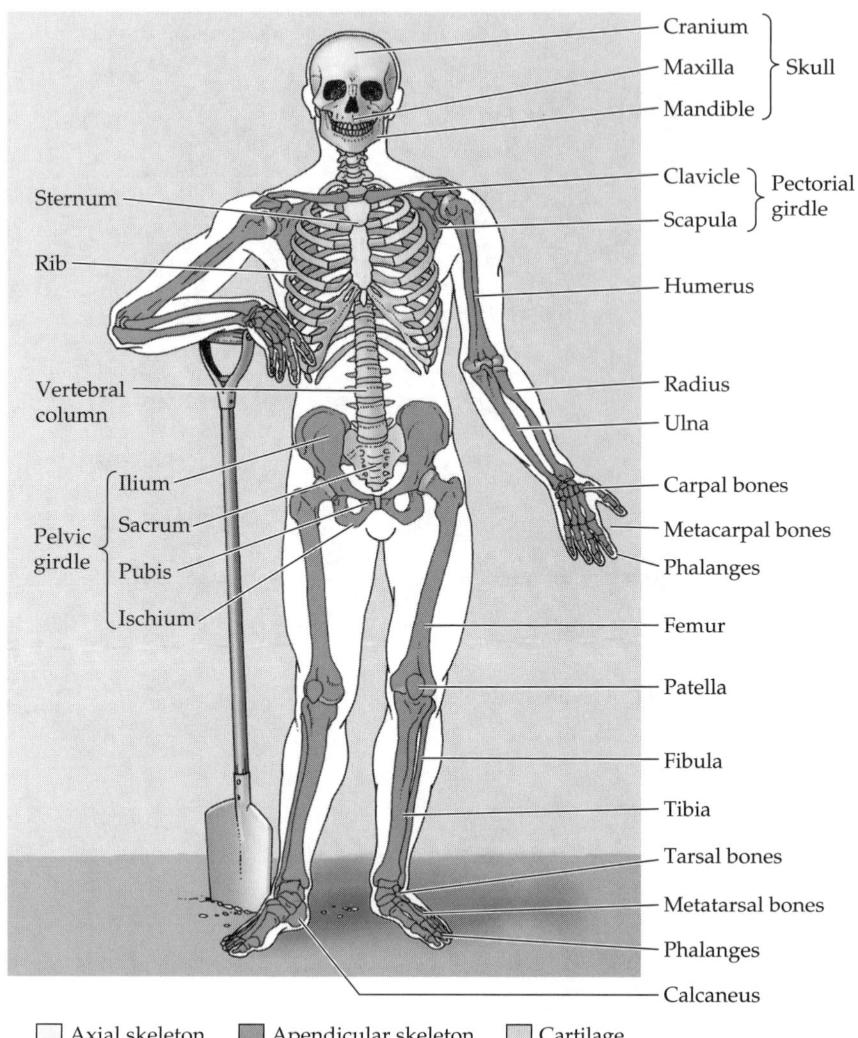
Cranium
Maxilla
Mandible
Skull
Clavicle
Scapula
Pectorial girdle
Sternum
Rib
Humerus
Vertebral column
Radius
Ulna
Ilium
Sacrum
Pubis
Ischium
Pelvic girdle
Carpal bones
Metacarpal bones
Phalanges
Femur
Patella
Fibula
Tibia
Tarsal bones
Metatarsal bones
Phalanges
Calcaneus
Axial skeleton
Apendicular skeleton
Cartilage

47.12 The Human Endoskeleton *(Page 915)*

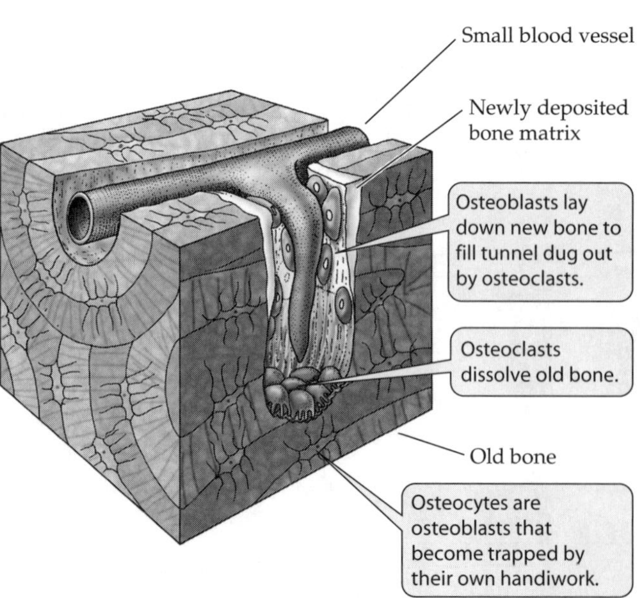
Small blood vessel
Newly deposited bone matrix
Osteoblasts lay down new bone to fill tunnel dug out by osteoclasts.
Osteoclasts dissolve old bone.
Old bone
Osteocytes are osteoblasts that become trapped by their own handiwork.

47.13 Renovating Bone *(Page 916)*

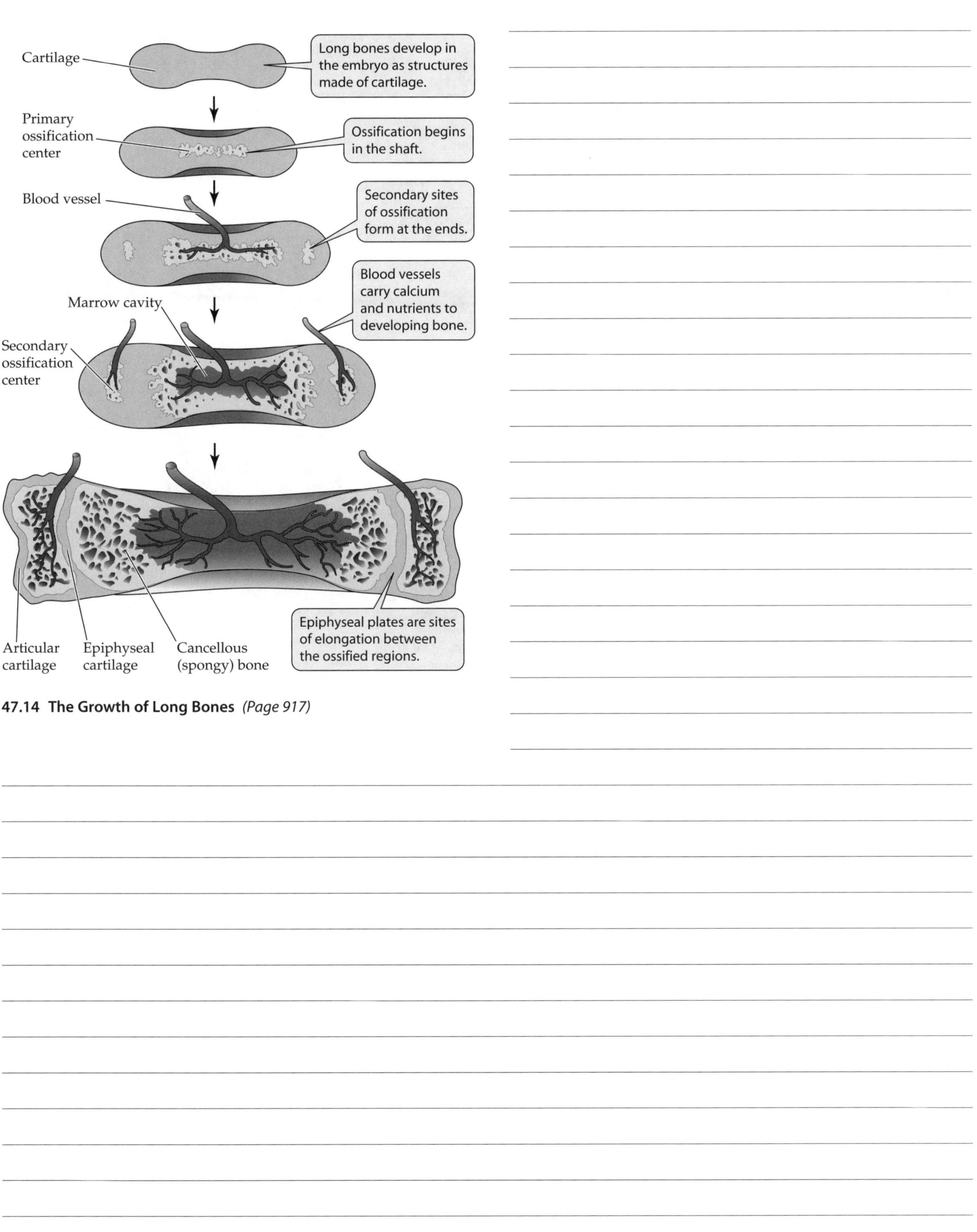
Cartilage
Long bones develop in the embryo as structures made of cartilage.
Primary ossification center
Ossification begins in the shaft.
Blood vessel
Secondary sites of ossification form at the ends.
Marrow cavity
Blood vessels carry calcium and nutrients to developing bone.
Secondary ossification center
Articular cartilage
Epiphyseal cartilage
Cancellous (spongy) bone
Epiphyseal plates are sites of elongation between the ossified regions.

47.14 The Growth of Long Bones *(Page 917)*

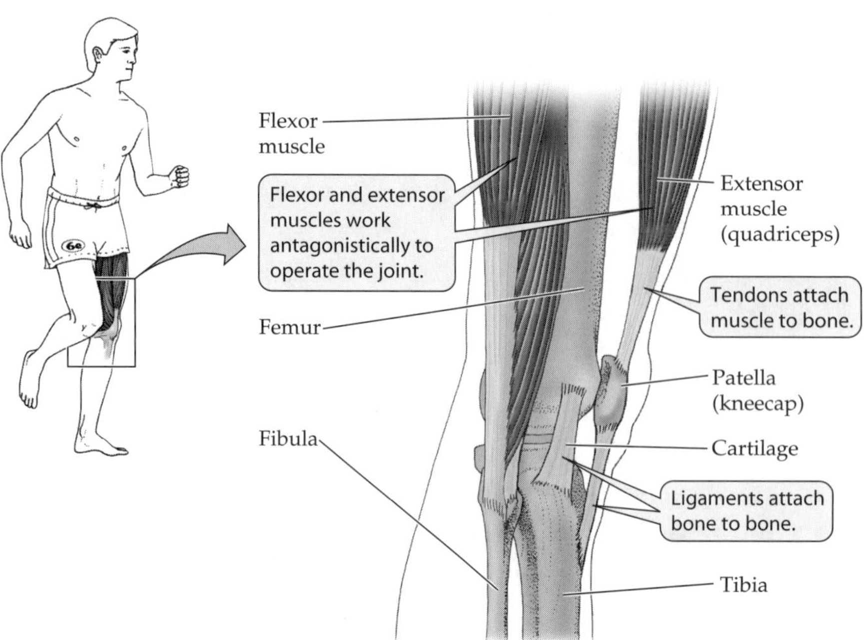
Flexor muscle
Flexor and extensor muscles work antagonistically to operate the joint.
Femur
Fibula
Extensor muscle (quadriceps)
Tendons attach muscle to bone.
Patella (kneecap)
Cartilage
Ligaments attach bone to bone.
Tibia

47.16 Joints, Ligaments, and Tendons *(Page 917)*

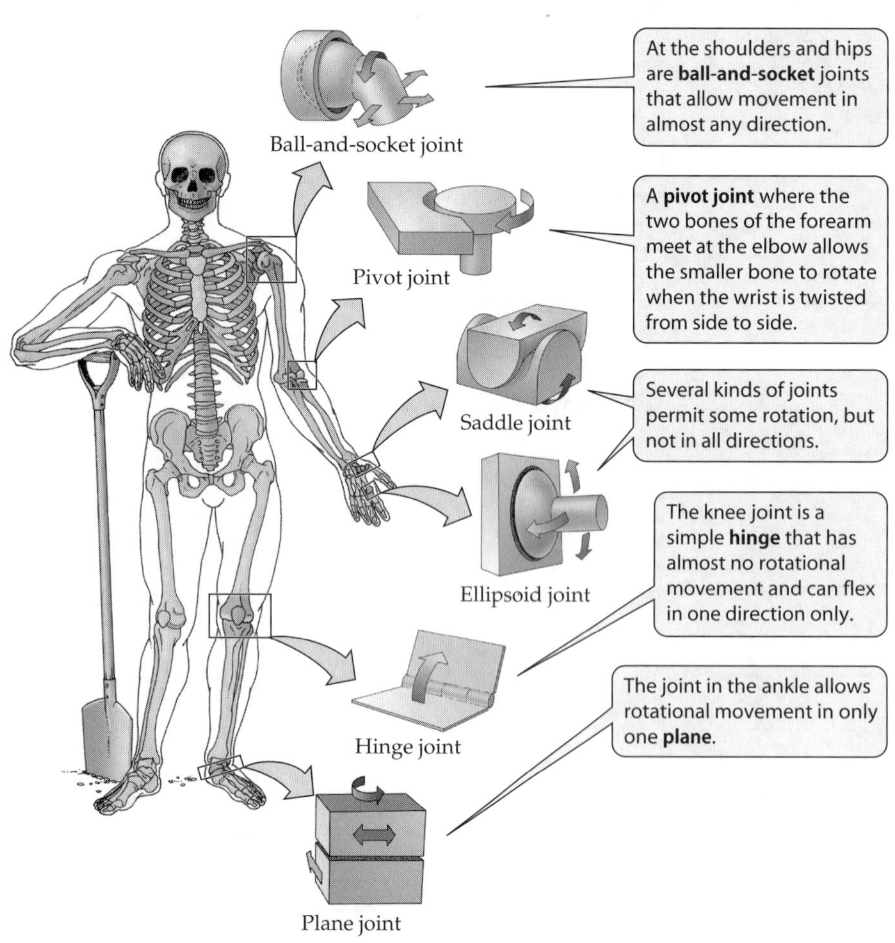
Ball-and-socket joint
At the shoulders and hips are ball-and-socket joints that allow movement in almost any direction.
Pivot joint
A pivot joint where the two bones of the forearm meet at the elbow allows the smaller bone to rotate when the wrist is twisted from side to side.
Saddle joint
Several kinds of joints permit some rotation, but not in all directions.
Ellipsoid joint
The knee joint is a simple hinge that has almost no rotational movement and can flex in one direction only.
Hinge joint
The joint in the ankle allows rotational movement in only one plane.
Plane joint

47.17 Types of Joints *(Page 918)*

Lever system designed for power
Load arm: power arm = 2:1 ratio which generates much force over a small distance.

Lever system designed for speed
Load arm: power arm = 5:1 ratio which moves low weights long distances with speed.

Fulcrum

Power arm = 1

Load arm = 2

Power arm = 1

Load arm = 5

An example of a lever system designed for power is the human jaw. The power arm is long relative to the load arm.

An example of a lever system designed for speed is the human leg. The power arm is short relative to the load arm.

47.18 Bones and Joints Work Like Systems of Levers *(Page 918)*

(*a*)

Chromatophore compacted (animal is pale)
Muscle fibers relaxed
Muscle fibers contracted
Chromatophore spread (animal is dark)
Pigment

47.19 Chromatophores Help Animals Camouflage Themselves or Communicate *(Page 919)*

48 Gas Exchange in Animals

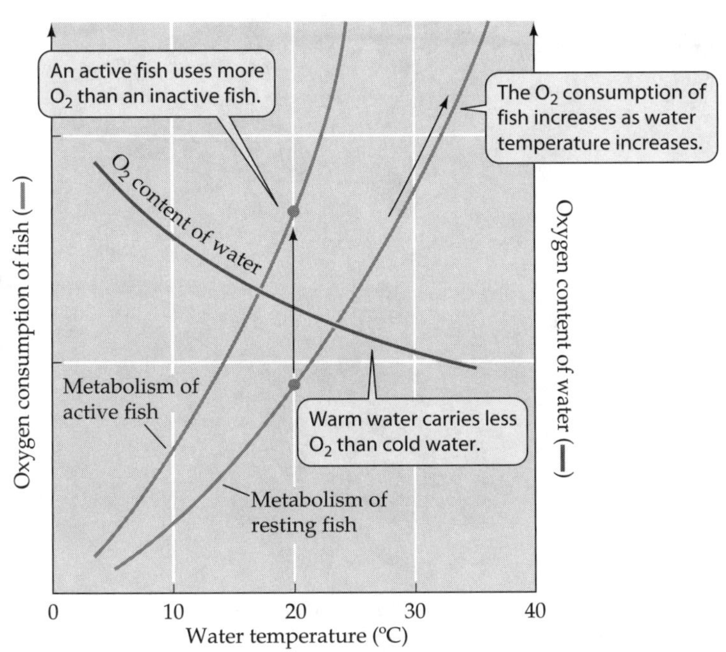

An active fish uses more O_2 than an inactive fish.
The O_2 consumption of fish increases as water temperature increases.
O_2 content of water
Metabolism of active fish
Warm water carries less O_2 than cold water.
Metabolism of resting fish
Oxygen consumption of fish (—)
Oxygen content of water (—)
0
10
20
30
40
Water temperature (°C)

48.2 The Double Bind of Water Breathers *(Page 924)*

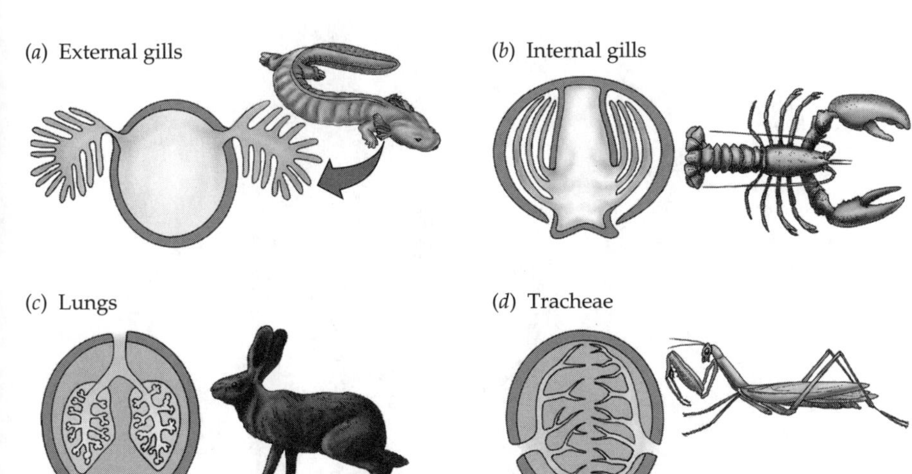

(a) External gills
(b) Internal gills
(c) Lungs
(d) Tracheae

48.3 Gas Exchange Systems *(Page 925)*

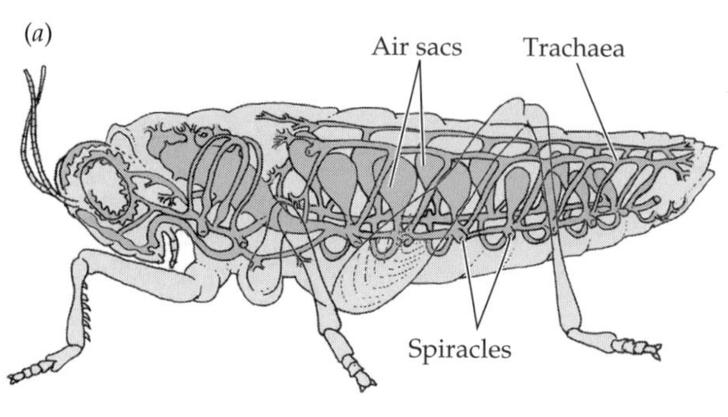
(a)
Air sacs
Trachaea
Spiracles

48.4 The Tracheal Gas Exchange System of Insects *(Page 926)*

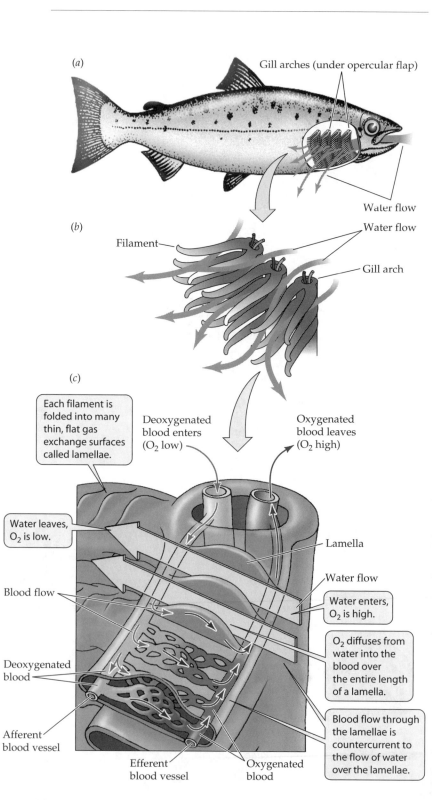

(a)
Gill arches (under opercular flap)
Water flow
(b)
Filament
Water flow
Gill arch
(c)
Each filament is folded into many thin, flat gas exchange surfaces called lamellae.
Deoxygenated blood enters (O_2 low)
Oxygenated blood leaves (O_2 high)
Water leaves, O_2 is low.
Lamella
Water flow
Blood flow
Water enters, O_2 is high.
O_2 diffuses from water into the blood over the entire length of a lamella.
Deoxygenated blood
Blood flow through the lamellae is countercurrent to the flow of water over the lamellae.
Afferent blood vessel
Efferent blood vessel
Oxygenated blood

48.5 Fish Gills *(Page 926)*

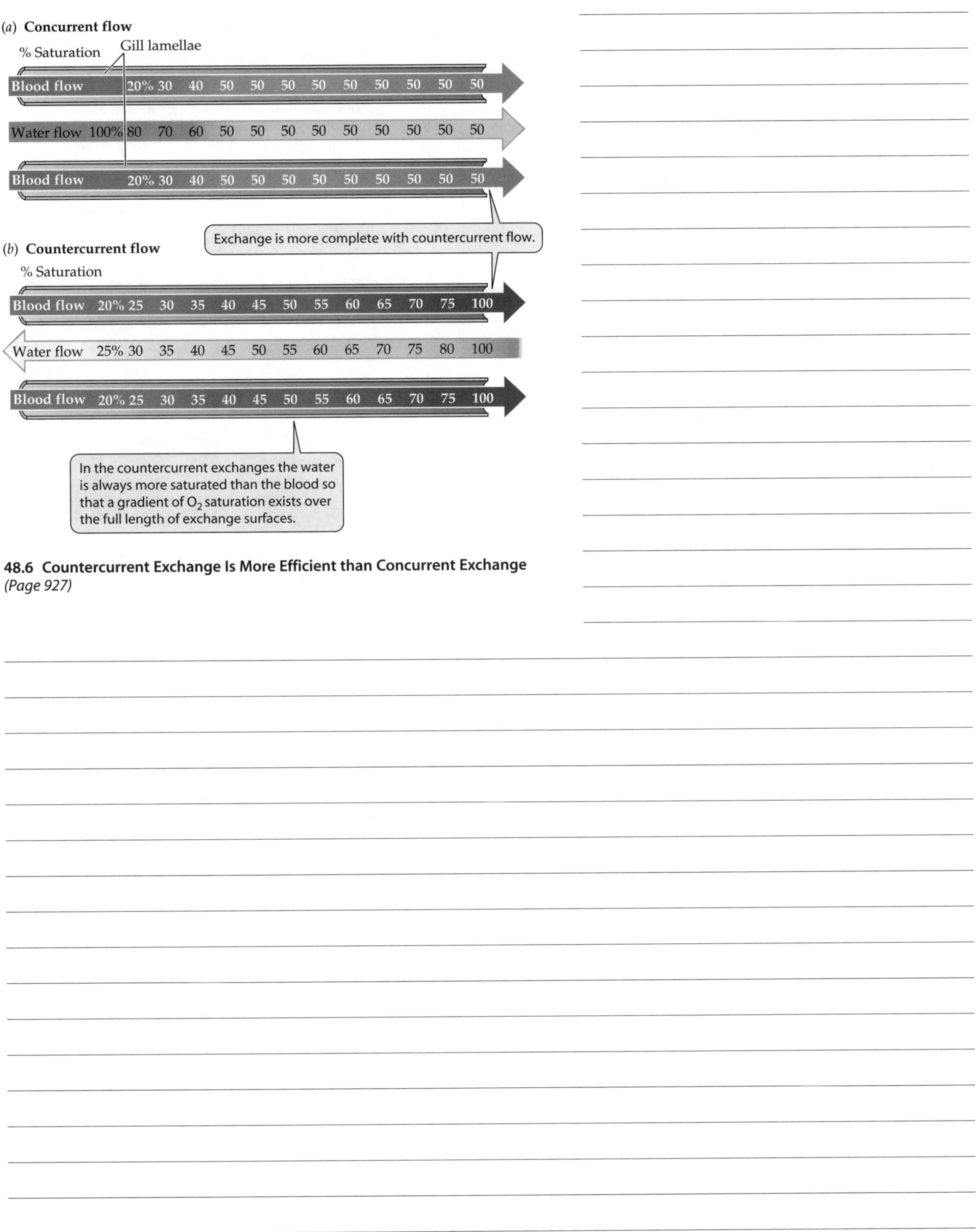
(a) Concurrent flow
% Saturation
Gill lamellae
Blood flow 20% 30 40 50 50 50 50 50 50 50 50 50
Water flow 100% 80 70 60 50 50 50 50 50 50 50 50 50
Blood flow 20% 30 40 50 50 50 50 50 50 50 50 50
Exchange is more complete with countercurrent flow.
(b) Countercurrent flow
% Saturation
Blood flow 20% 25 30 35 40 45 50 55 60 65 70 75 100
Water flow 25% 30 35 40 45 50 55 60 65 70 75 80 100
Blood flow 20% 25 30 35 40 45 50 55 60 65 70 75 100
In the countercurrent exchanges the water is always more saturated than the blood so that a gradient of O2 saturation exists over the full length of exchange surfaces.

48.6 Countercurrent Exchange Is More Efficient than Concurrent Exchange
(Page 927)

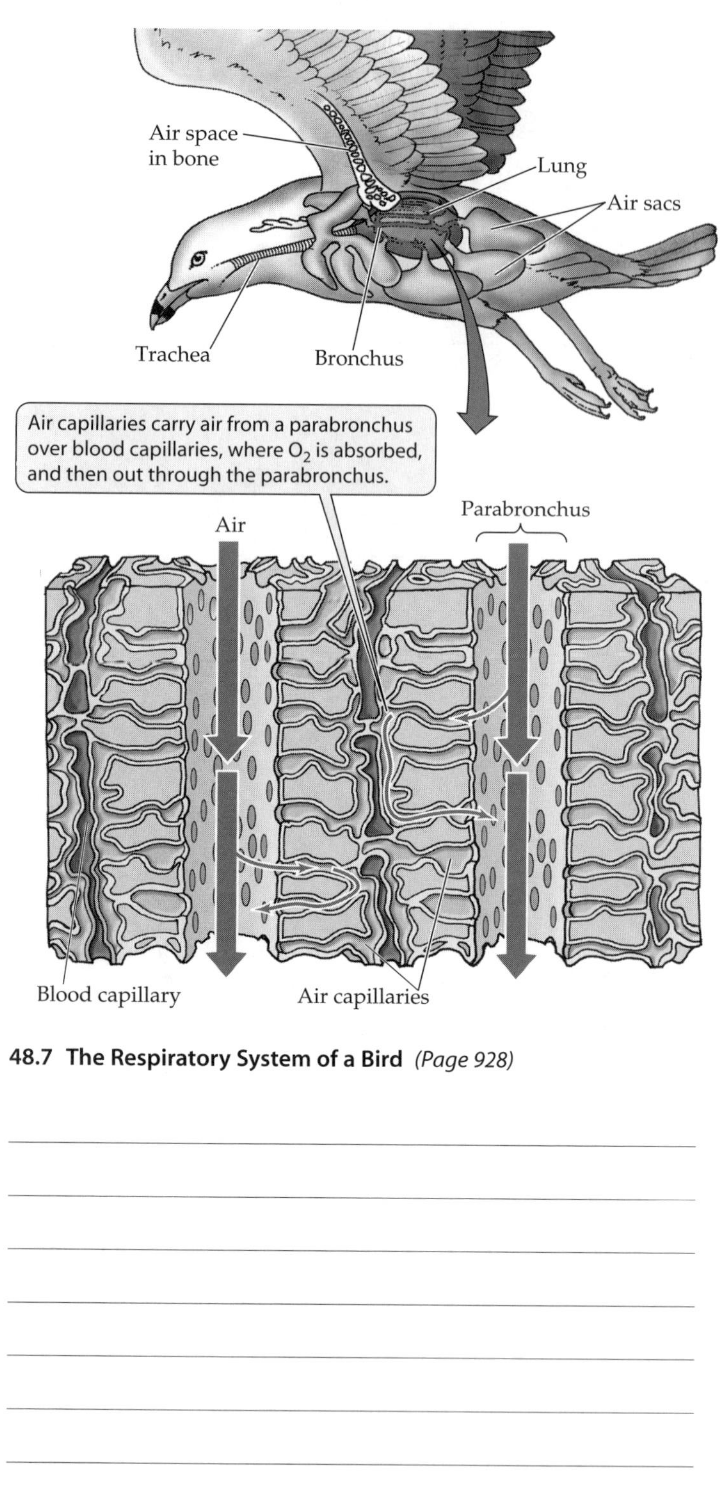
Air space in bone
Lung
Air sacs
Trachea
Bronchus
Air capillaries carry air from a parabronchus over blood capillaries, where O_2 is absorbed, and then out through the parabronchus.
Air
Parabronchus
Blood capillary
Air capillaries

48.7 The Respiratory System of a Bird *(Page 928)*

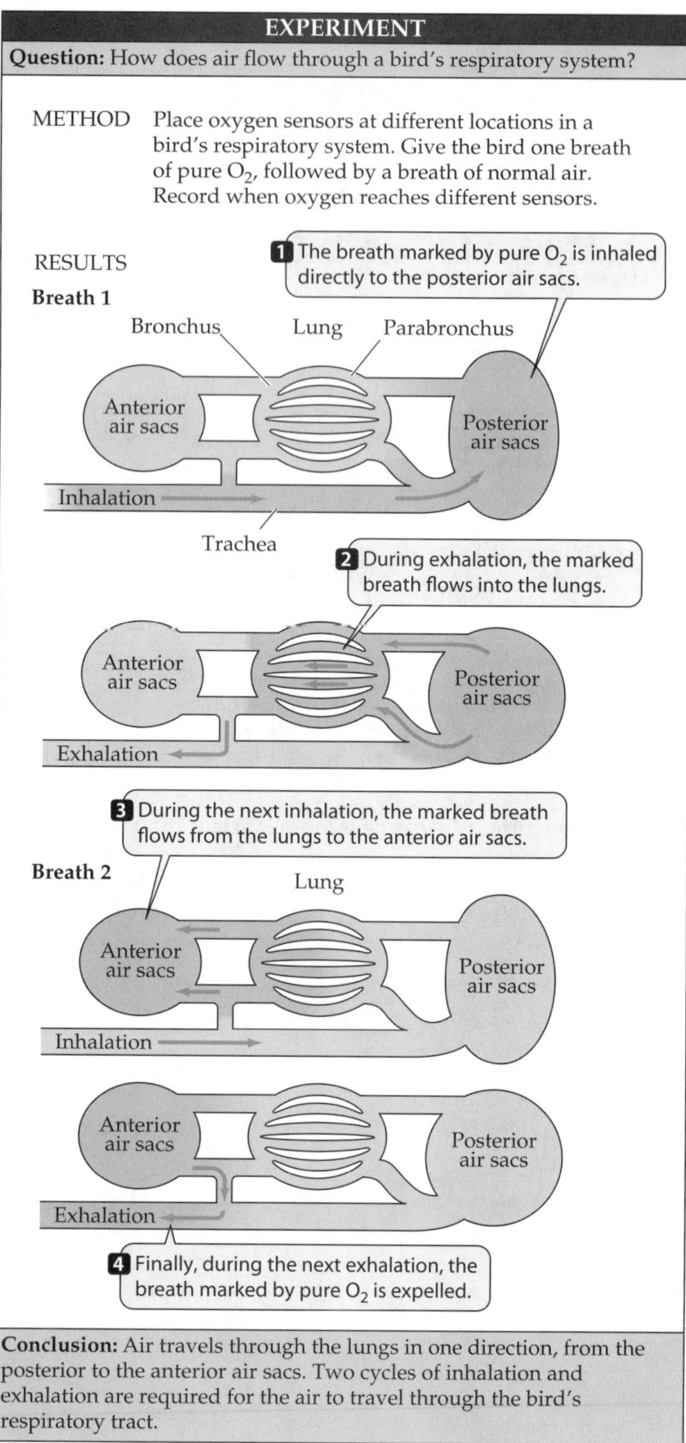
EXPERIMENT
Question: How does air flow through a bird's respiratory system?
METHOD Place oxygen sensors at different locations in a bird's respiratory system. Give the bird one breath of pure O_2, followed by a breath of normal air. Record when oxygen reaches different sensors.
RESULTS
Breath 1
1 The breath marked by pure O_2 is inhaled directly to the posterior air sacs.
Bronchus
Lung
Parabronchus
Anterior air sacs
Posterior air sacs
Inhalation
Trachea
2 During exhalation, the marked breath flows into the lungs.
Anterior air sacs
Posterior air sacs
Exhalation
3 During the next inhalation, the marked breath flows from the lungs to the anterior air sacs.
Breath 2
Lung
Anterior air sacs
Posterior air sacs
Inhalation
Anterior air sacs
Posterior air sacs
Exhalation
4 Finally, during the next exhalation, the breath marked by pure O_2 is expelled.
Conclusion: Air travels through the lungs in one direction, from the posterior to the anterior air sacs. Two cycles of inhalation and exhalation are required for the air to travel through the bird's respiratory tract.

48.8 The Path of Air Flow through Bird Lungs *(Page 928)*

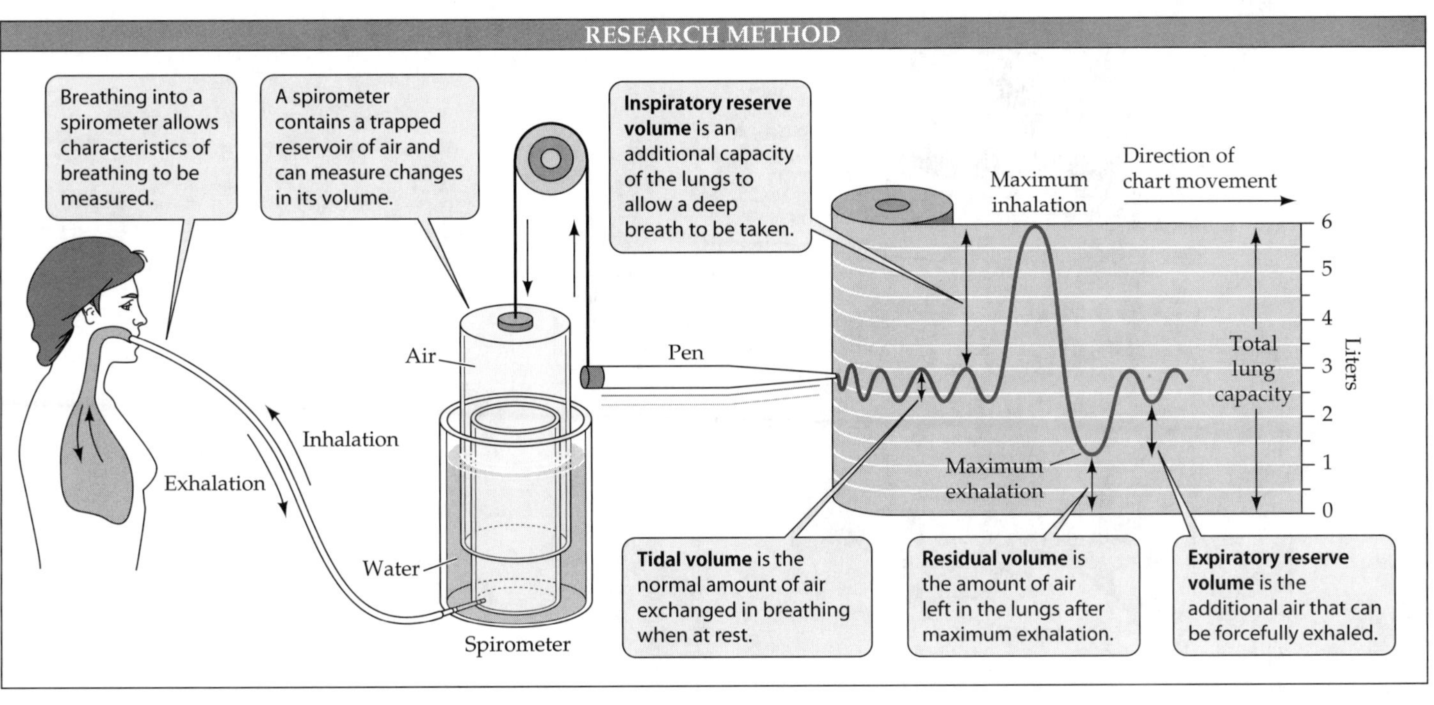
RESEARCH METHOD
Breathing into a spirometer allows characteristics of breathing to be measured.
A spirometer contains a trapped reservoir of air and can measure changes in its volume.
Inspiratory reserve volume is an additional capacity of the lungs to allow a deep breath to be taken.
Maximum inhalation
Direction of chart movement
Air
Pen
Inhalation
Exhalation
Water
Spirometer
Total lung capacity
Liters
6
5
4
3
2
1
0
Maximum exhalation
Tidal volume is the normal amount of air exchanged in breathing when at rest.
Residual volume is the amount of air left in the lungs after maximum exhalation.
Expiratory reserve volume is the additional air that can be forcefully exhaled.

48.9 Measuring Lung Ventilation *(Page 929)*

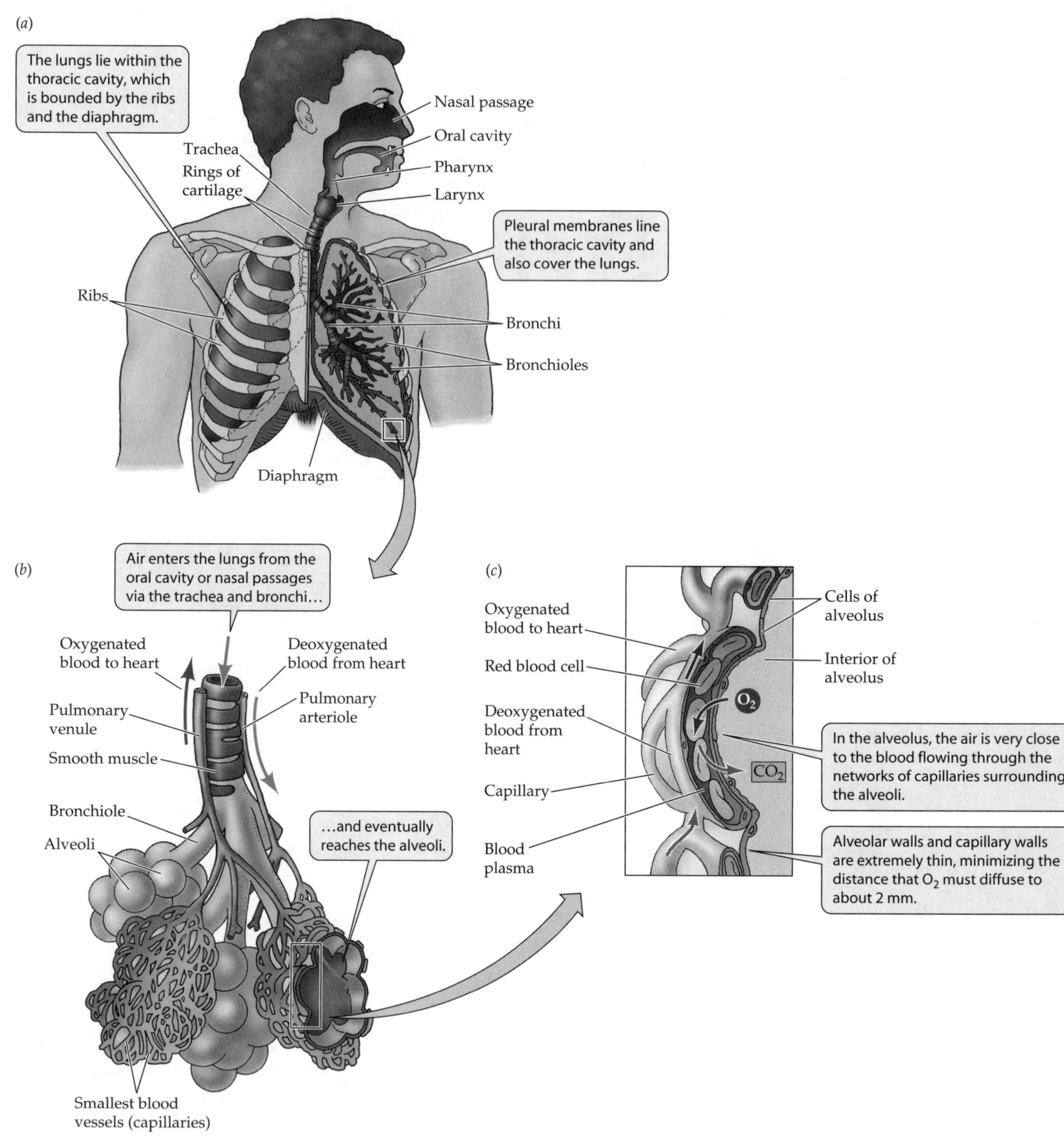
(a)
The lungs lie within the thoracic cavity, which is bounded by the ribs and the diaphragm.
Nasal passage
Oral cavity
Trachea
Rings of cartilage
Pharynx
Larynx
Pleural membranes line the thoracic cavity and also cover the lungs.
Ribs
Bronchi
Bronchioles
Diaphragm
(b)
Air enters the lungs from the oral cavity or nasal passages via the trachea and bronchi…
Oxygenated blood to heart
Deoxygenated blood from heart
Pulmonary venule
Pulmonary arteriole
Smooth muscle
Bronchiole
Alveoli
…and eventually reaches the alveoli.
Smallest blood vessels (capillaries)
(c)
Oxygenated blood to heart
Red blood cell
Deoxygenated blood from heart
Capillary
Blood plasma
Cells of alveolus
Interior of alveolus
O_2
CO_2
In the alveolus, the air is very close to the blood flowing through the networks of capillaries surrounding the alveoli.
Alveolar walls and capillary walls are extremely thin, minimizing the distance that O_2 must diffuse to about 2 mm.

48.10 The Human Respiratory System *(Page 931)*

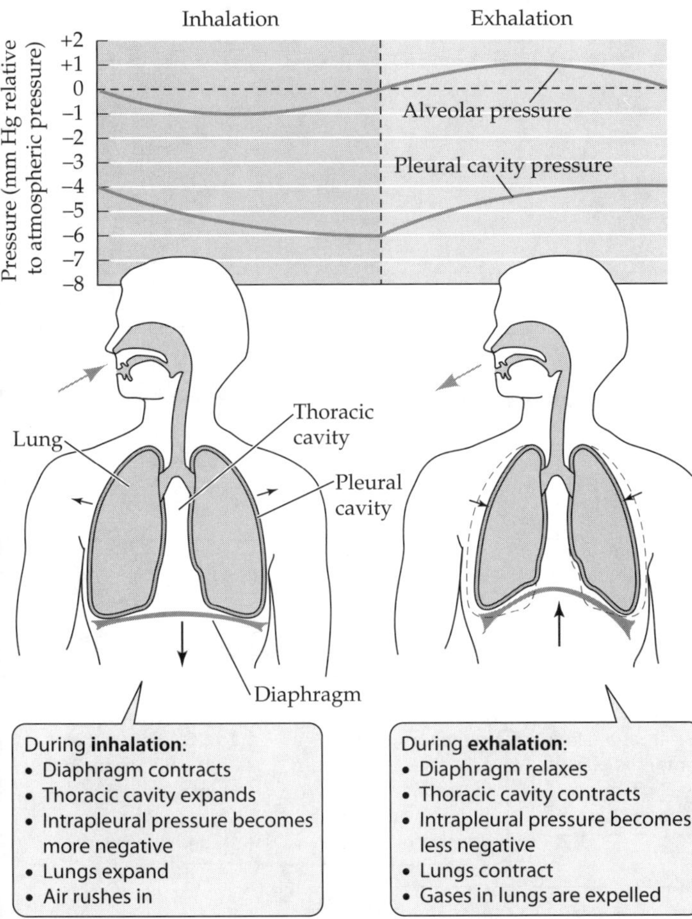
Inhalation
Exhalation
Pressure (mm Hg relative to atmospheric pressure)
+2
+1
0
−1
−2
−3
−4
−5
−6
−7
−8
Alveolar pressure
Pleural cavity pressure
Lung
Thoracic cavity
Pleural cavity
Diaphragm
During **inhalation**:
• Diaphragm contracts
• Thoracic cavity expands
• Intrapleural pressure becomes more negative
• Lungs expand
• Air rushes in
During **exhalation**:
• Diaphragm relaxes
• Thoracic cavity contracts
• Intrapleural pressure becomes less negative
• Lungs contract
• Gases in lungs are expelled

48.11 Into the Lungs and Out Again *(Page 932)*

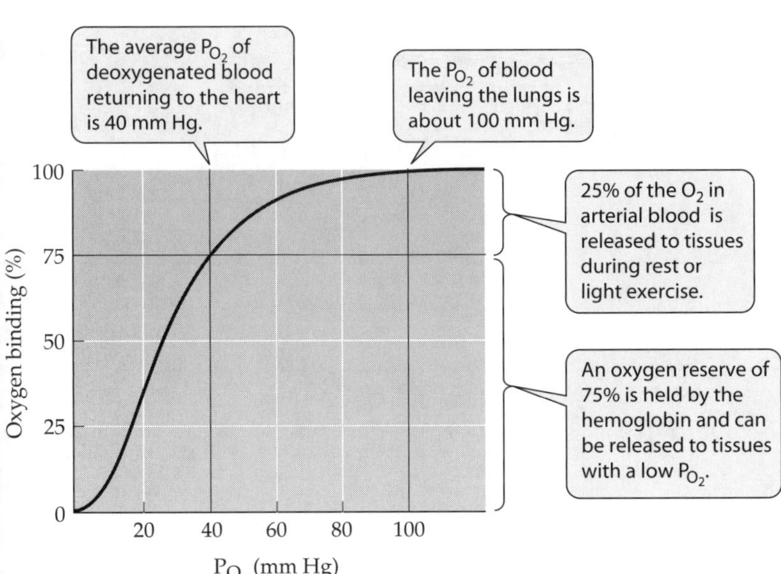
The average P_{O_2} of deoxygenated blood returning to the heart is 40 mm Hg.
The P_{O_2} of blood leaving the lungs is about 100 mm Hg.
Oxygen binding (%)
100
75
50
25
0
20
40
60
80
100
P_{O_2} (mm Hg)
25% of the O_2 in arterial blood is released to tissues during rest or light exercise.
An oxygen reserve of 75% is held by the hemoglobin and can be released to tissues with a low P_{O_2}.

48.12 The Binding of O_2 to Hemoglobin Depends on P_{O_2} *(Page 933)*

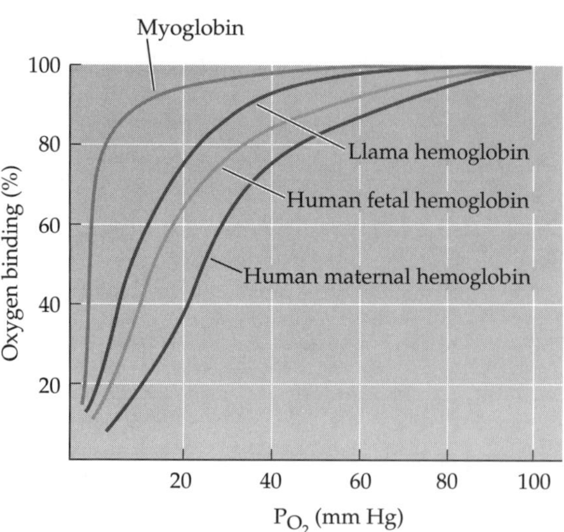
Myoglobin
Llama hemoglobin
Human fetal hemoglobin
Human maternal hemoglobin
Oxygen binding (%)
100
80
60
40
20
20
40
60
80
100
P_{O_2} (mm Hg)

48.13 Oxygen-Binding Adaptations *(Page 934)*

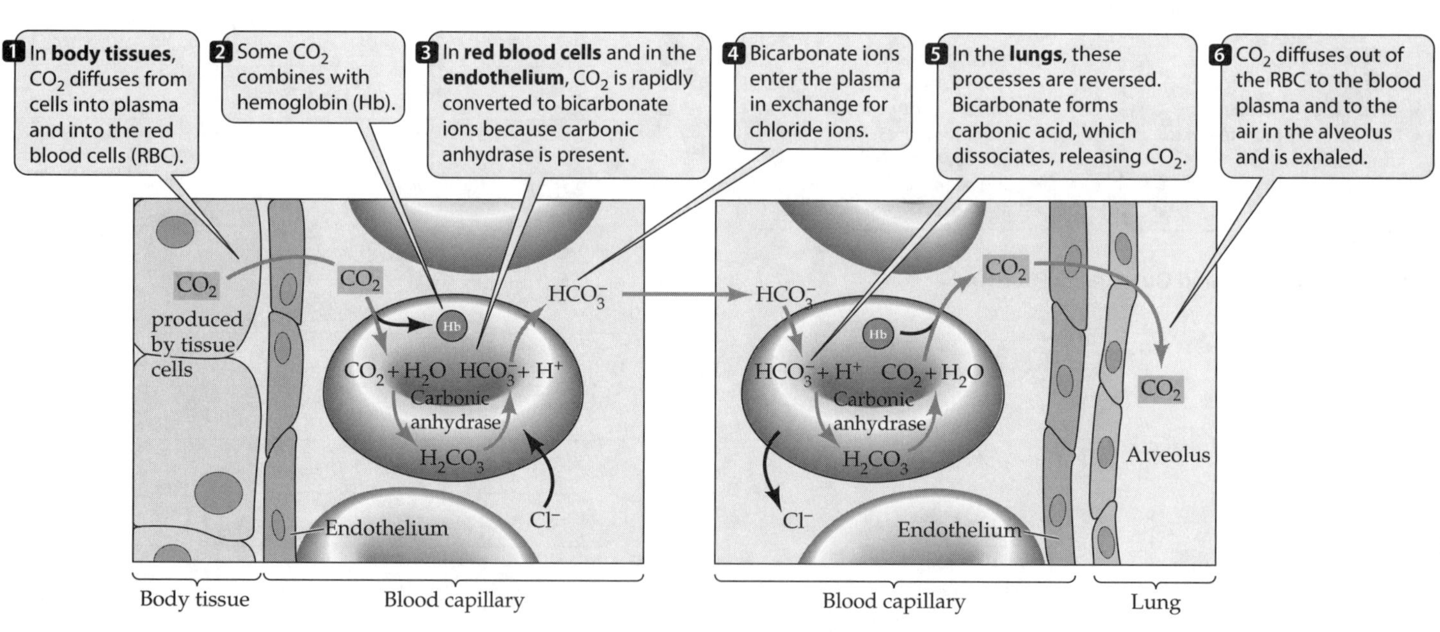
1 In **body tissues**, CO_2 diffuses from cells into plasma and into the red blood cells (RBC).
2 Some CO_2 combines with hemoglobin (Hb).
3 In **red blood cells** and in the **endothelium**, CO_2 is rapidly converted to bicarbonate ions because carbonic anhydrase is present.
4 Bicarbonate ions enter the plasma in exchange for chloride ions.
5 In the **lungs**, these processes are reversed. Bicarbonate forms carbonic acid, which dissociates, releasing CO_2.
6 CO_2 diffuses out of the RBC to the blood plasma and to the air in the alveolus and is exhaled.
CO_2 produced by tissue cells
CO_2
Hb
$CO_2 + H_2O$ $HCO_3^- + H^+$
Carbonic anhydrase
H_2CO_3
HCO_3^-
Cl^-
Endothelium
HCO_3^-
Hb
$HCO_3^- + H^+$ $CO_2 + H_2O$
Carbonic anhydrase
H_2CO_3
Cl^-
Endothelium
CO_2
CO_2
Alveolus
Body tissue
Blood capillary
Blood capillary
Lung

48.14 Carbon Dioxide Is Transported as Bicarbonate Ions *(Page 935)*

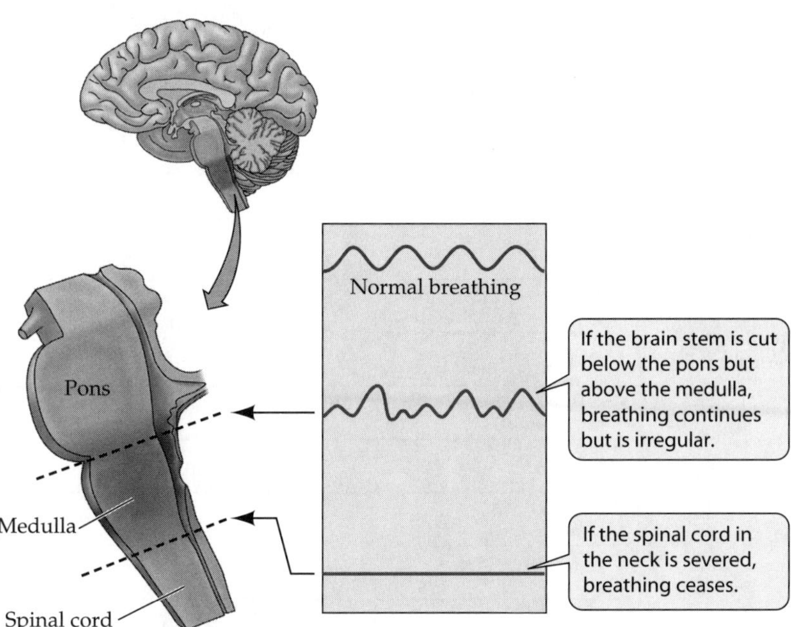
Normal breathing
Pons
Medulla
Spinal cord
If the brain stem is cut below the pons but above the medulla, breathing continues but is irregular.
If the spinal cord in the neck is severed, breathing ceases.

48.15 Breathing is Generated in the Brain Stem *(Page 936)*

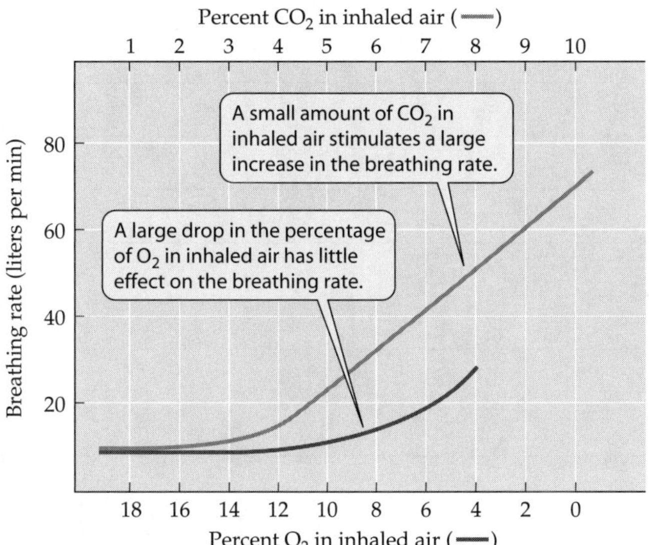
Percent CO_2 in inhaled air (—)
1 2 3 4 5 6 7 8 9 10
A small amount of CO_2 in inhaled air stimulates a large increase in the breathing rate.
A large drop in the percentage of O_2 in inhaled air has little effect on the breathing rate.
Breathing rate (liters per min)
80
60
40
20
18 16 14 12 10 8 6 4 2 0
Percent O_2 in inhaled air (—)

48.16 Carbon Dioxide Affects Breathing Rate *(Page 936)*

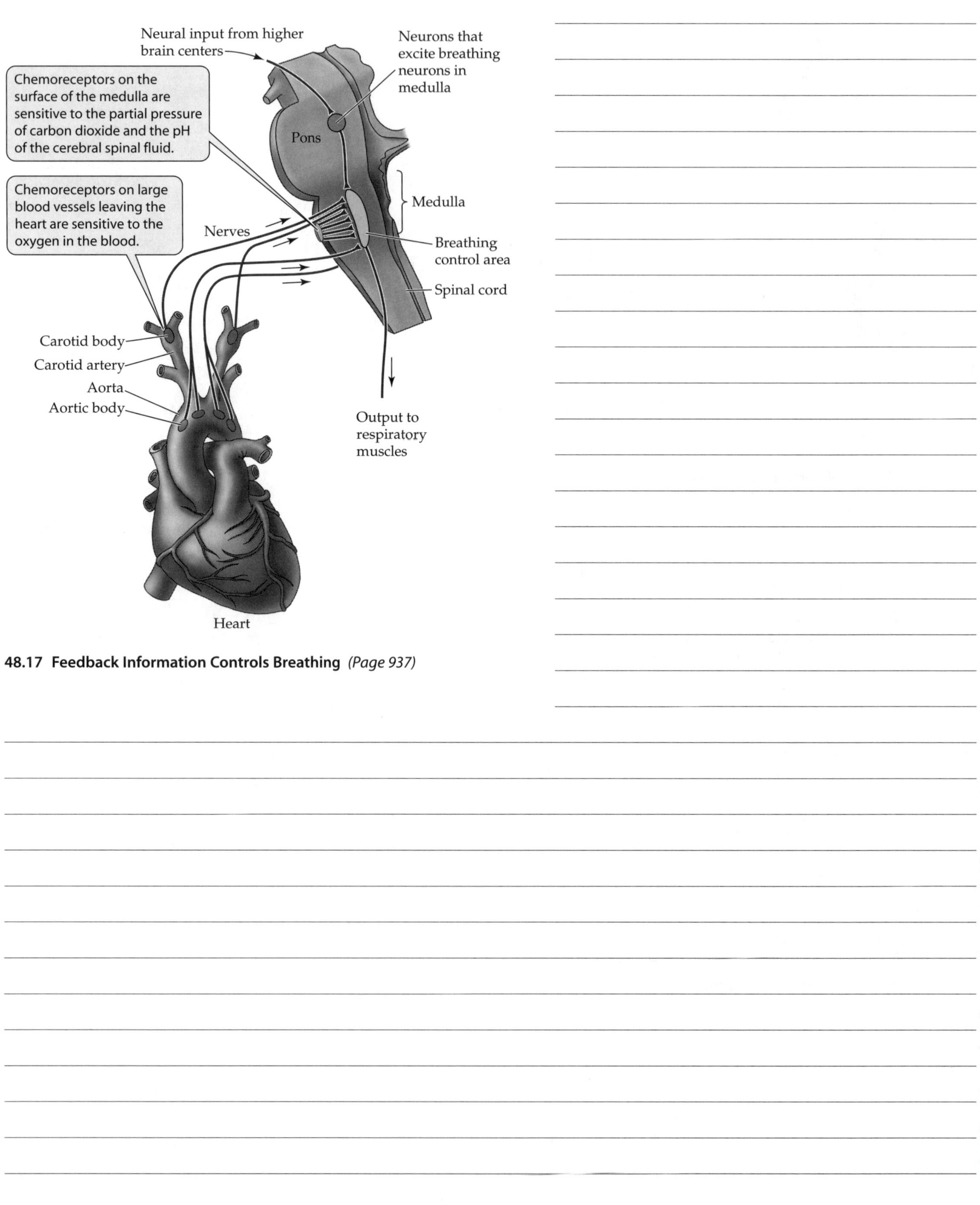
Neural input from higher brain centers
Neurons that excite breathing neurons in medulla
Chemoreceptors on the surface of the medulla are sensitive to the partial pressure of carbon dioxide and the pH of the cerebral spinal fluid.
Pons
Chemoreceptors on large blood vessels leaving the heart are sensitive to the oxygen in the blood.
Medulla
Nerves
Breathing control area
Spinal cord
Carotid body
Carotid artery
Aorta
Aortic body
Output to respiratory muscles
Heart

48.17 Feedback Information Controls Breathing *(Page 937)*

49 Circulatory Systems

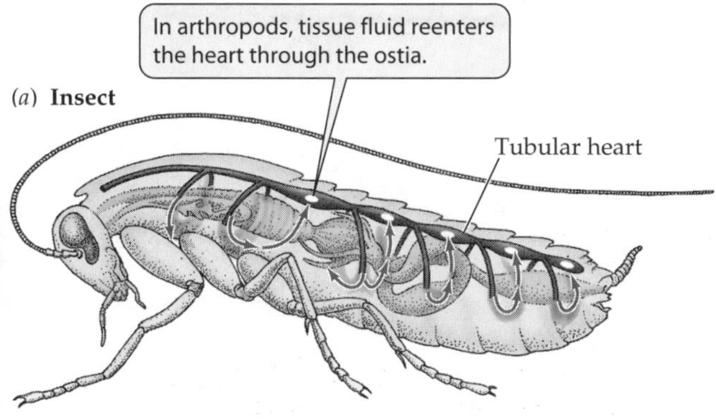

In arthropods, tissue fluid reenters the heart through the ostia.
(a) Insect
Tubular heart

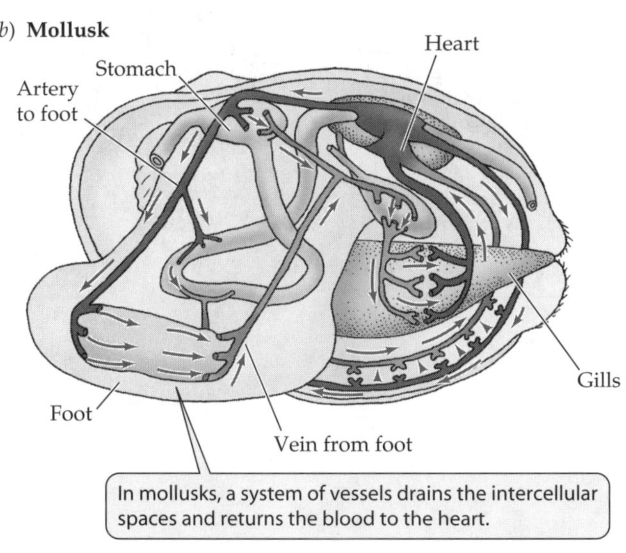

(b) Mollusk
Heart
Stomach
Artery to foot
Gills

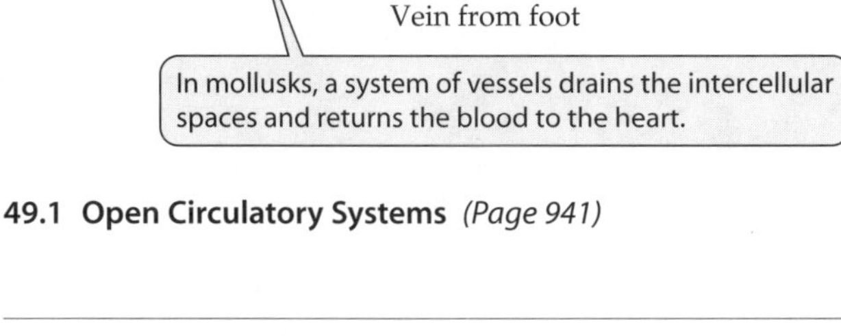

Foot
Vein from foot
In mollusks, a system of vessels drains the intercellular spaces and returns the blood to the heart.

49.1 Open Circulatory Systems *(Page 941)*

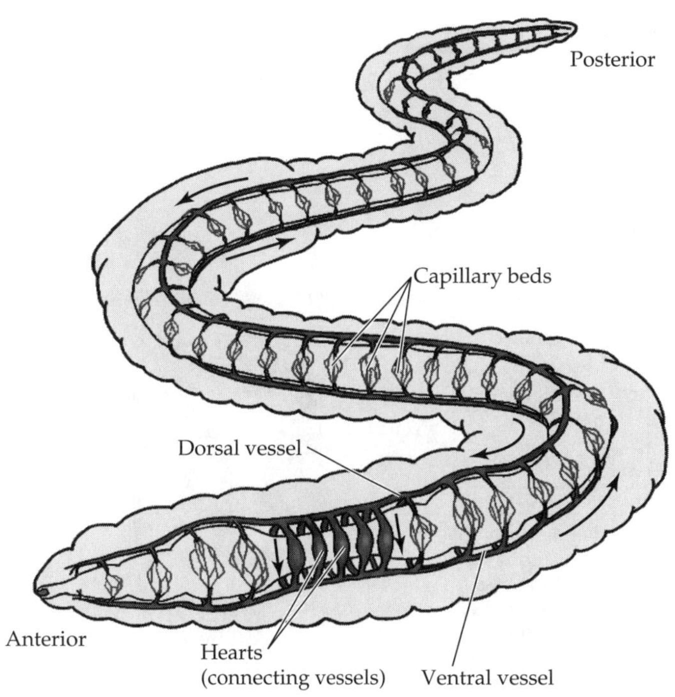

Posterior
Capillary beds
Dorsal vessel
Anterior
Hearts (connecting vessels)
Ventral vessel

49.2 A Closed Circulatory System *(Page 942)*

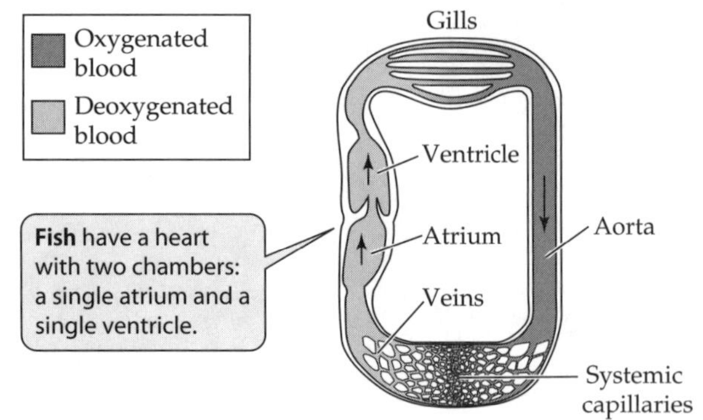
Oxygenated blood
Deoxygenated blood
Gills
Ventricle
Atrium
Aorta
Veins
Systemic capillaries
Fish have a heart with two chambers: a single atrium and a single ventricle.

In-Text Art *(Page 943)*

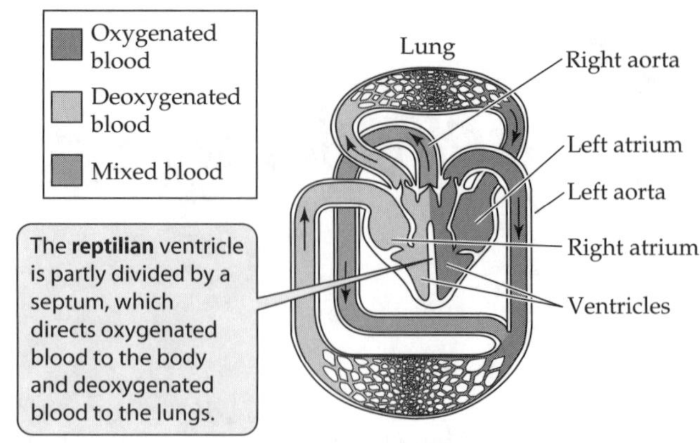
Oxygenated blood
Deoxygenated blood
Mixed blood
Lung
Right aorta
Left atrium
Left aorta
Right atrium
Ventricles
The **reptilian** ventricle is partly divided by a septum, which directs oxygenated blood to the body and deoxygenated blood to the lungs.

In-Text Art *(Page 944)*

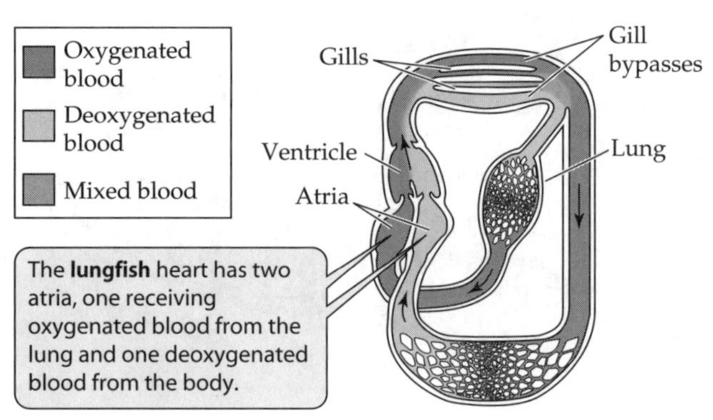
Oxygenated blood
Deoxygenated blood
Mixed blood
Gills
Gill bypasses
Ventricle
Lung
Atria
The **lungfish** heart has two atria, one receiving oxygenated blood from the lung and one deoxygenated blood from the body.

In-Text Art *(Page 943)*

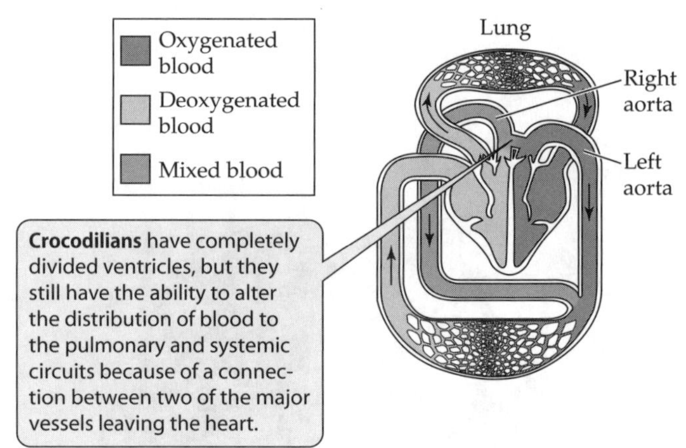
Oxygenated blood
Deoxygenated blood
Mixed blood
Lung
Right aorta
Left aorta
Crocodilians have completely divided ventricles, but they still have the ability to alter the distribution of blood to the pulmonary and systemic circuits because of a connection between two of the major vessels leaving the heart.

In-Text Art *(Page 944)*

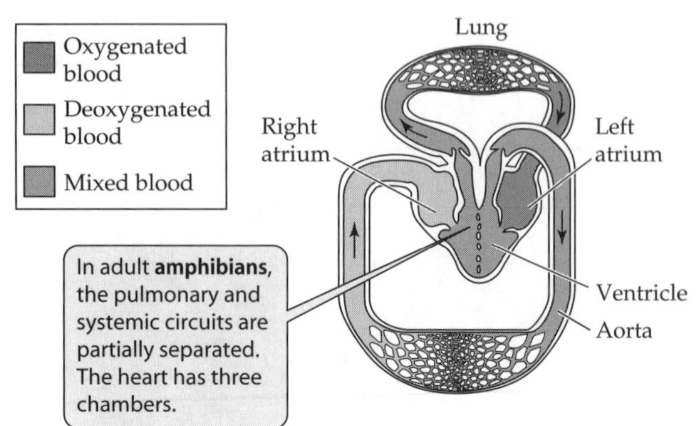
Oxygenated blood
Deoxygenated blood
Mixed blood
Lung
Right atrium
Left atrium
Ventricle
Aorta
In adult **amphibians**, the pulmonary and systemic circuits are partially separated. The heart has three chambers.

In-Text Art *(Page 943)*

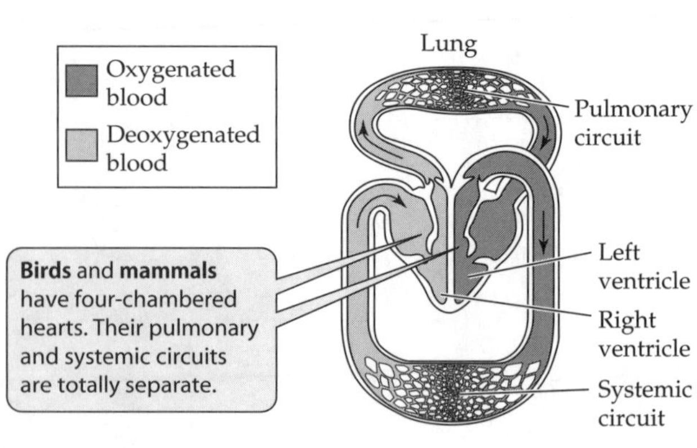
Oxygenated blood
Deoxygenated blood
Lung
Pulmonary circuit
Left ventricle
Right ventricle
Systemic circuit
Birds and **mammals** have four-chambered hearts. Their pulmonary and systemic circuits are totally separate.

In-Text Art *(Page 945)*

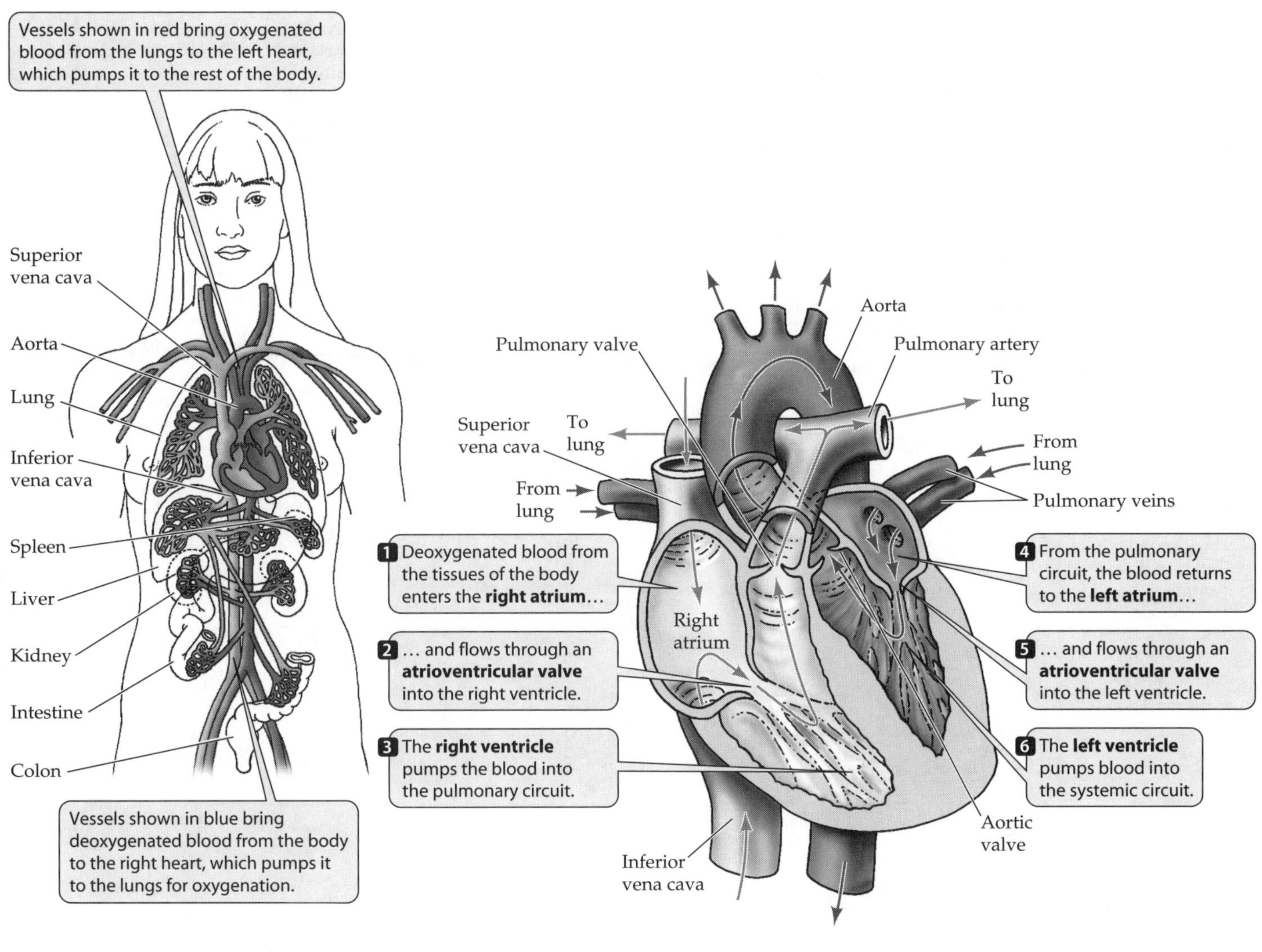

Vessels shown in red bring oxygenated blood from the lungs to the left heart, which pumps it to the rest of the body.
Superior vena cava
Aorta
Lung
Inferior vena cava
Spleen
Liver
Kidney
Intestine
Colon
Vessels shown in blue bring deoxygenated blood from the body to the right heart, which pumps it to the lungs for oxygenation.
Aorta
Pulmonary valve
Pulmonary artery
To lung
Superior vena cava
To lung
From lung
From lung
Pulmonary veins
1 Deoxygenated blood from the tissues of the body enters the **right atrium**…
4 From the pulmonary circuit, the blood returns to the **left atrium**…
Right atrium
2 … and flows through an **atrioventricular valve** into the right ventricle.
5 … and flows through an **atrioventricular valve** into the left ventricle.
3 The **right ventricle** pumps the blood into the pulmonary circuit.
6 The **left ventricle** pumps blood into the systemic circuit.
Aortic valve
Inferior vena cava

49.3 The Human Heart and Circulation *(Page 945)*

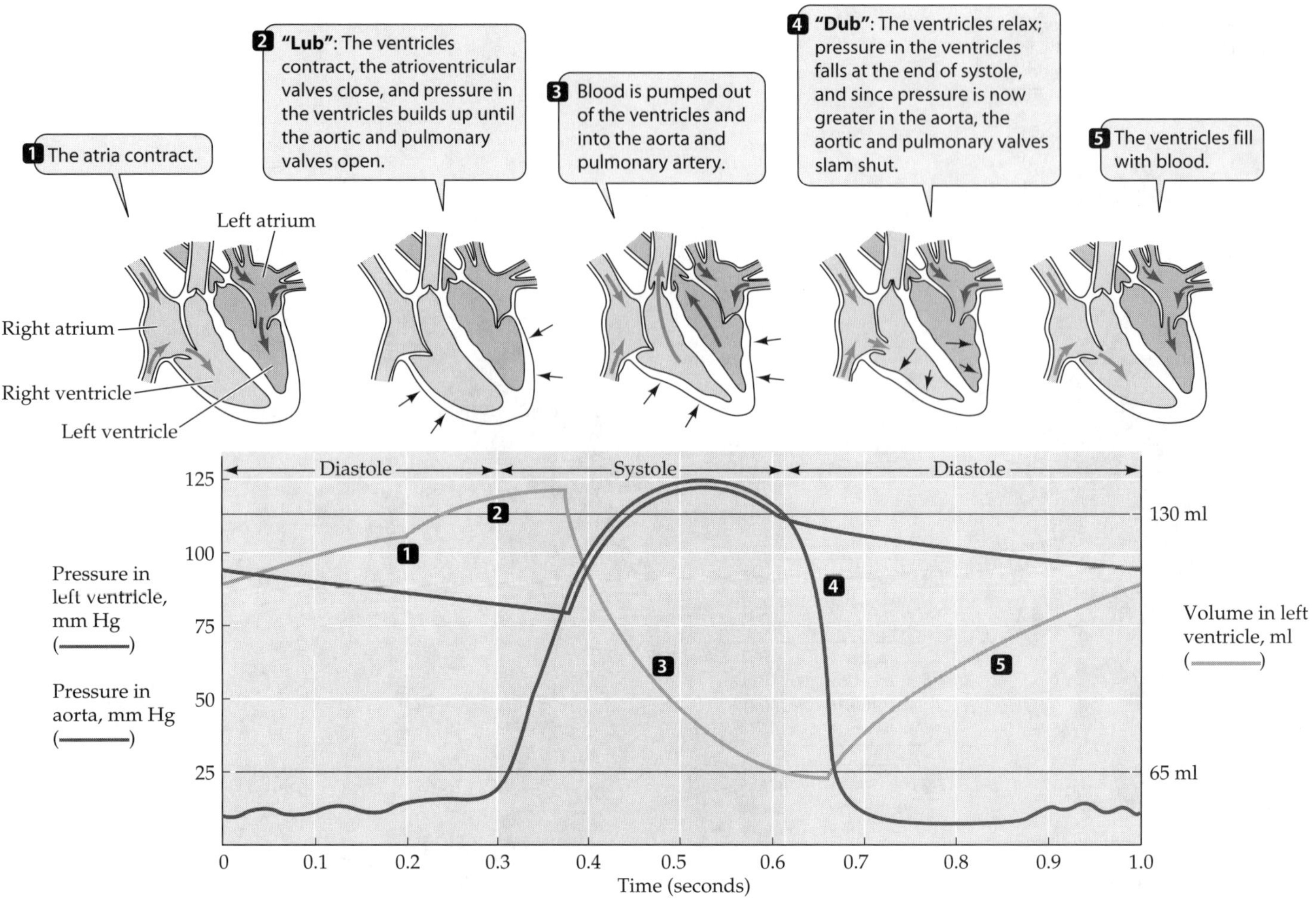
1 The atria contract.
2 "Lub": The ventricles contract, the atrioventricular valves close, and pressure in the ventricles builds up until the aortic and pulmonary valves open.
3 Blood is pumped out of the ventricles and into the aorta and pulmonary artery.
4 "Dub": The ventricles relax; pressure in the ventricles falls at the end of systole, and since pressure is now greater in the aorta, the aortic and pulmonary valves slam shut.
5 The ventricles fill with blood.
Left atrium
Right atrium
Right ventricle
Left ventricle
Diastole
Systole
Diastole
125
100
75
50
25
130 ml
65 ml
Pressure in left ventricle, mm Hg
Pressure in aorta, mm Hg
Volume in left ventricle, ml
0
0.1
0.2
0.3
0.4
0.5
0.6
0.7
0.8
0.9
1.0
Time (seconds)

49.4 The Cardiac Cycle *(Page 946)*

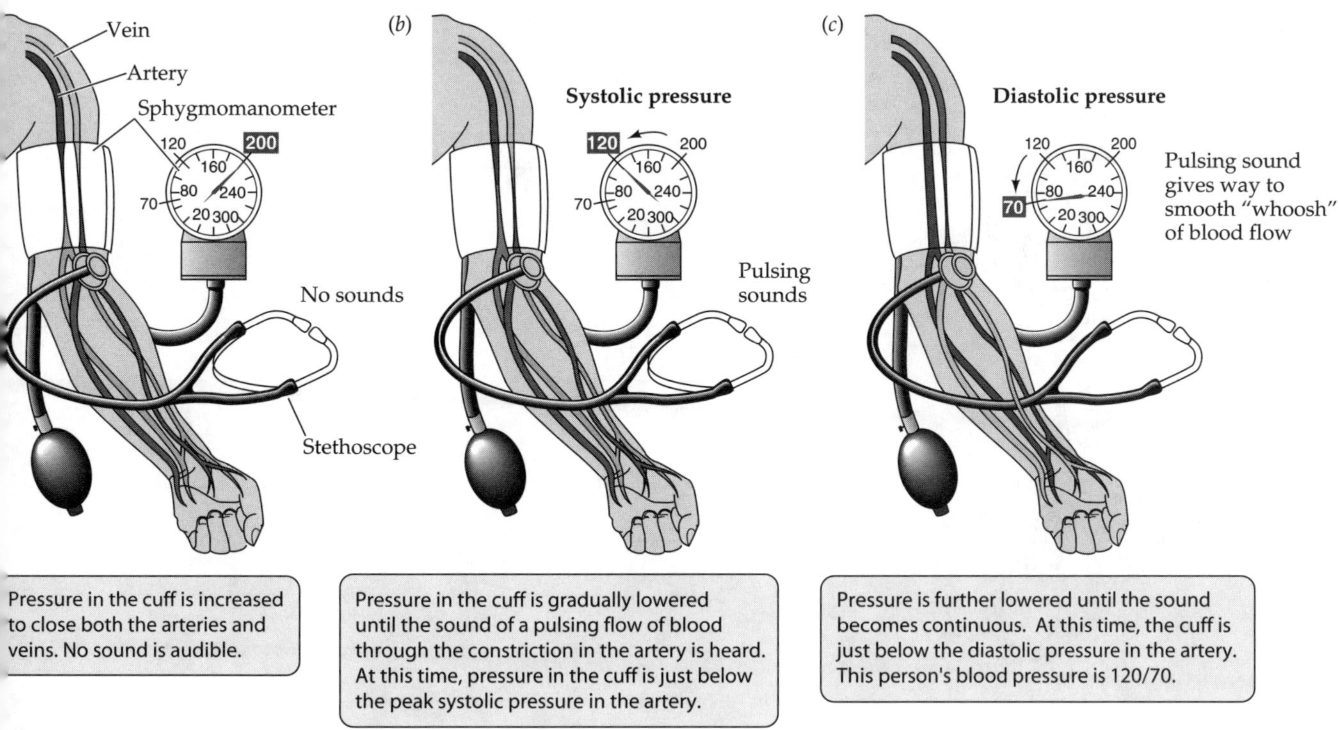

Vein
Artery
Sphygmomanometer
120
200
160
80
240
70
20
300
No sounds
Stethoscope
(b)
Systolic pressure
120
200
160
80
240
70
20
300
Pulsing sounds
(c)
Diastolic pressure
120
200
160
80
240
70
20
300
Pulsing sound gives way to smooth "whoosh" of blood flow
Pressure in the cuff is increased to close both the arteries and veins. No sound is audible.
Pressure in the cuff is gradually lowered until the sound of a pulsing flow of blood through the constriction in the artery is heard. At this time, pressure in the cuff is just below the peak systolic pressure in the artery.
Pressure is further lowered until the sound becomes continuous. At this time, the cuff is just below the diastolic pressure in the artery. This person's blood pressure is 120/70.

49.5 Measuring Blood Pressure *(Page 947)*

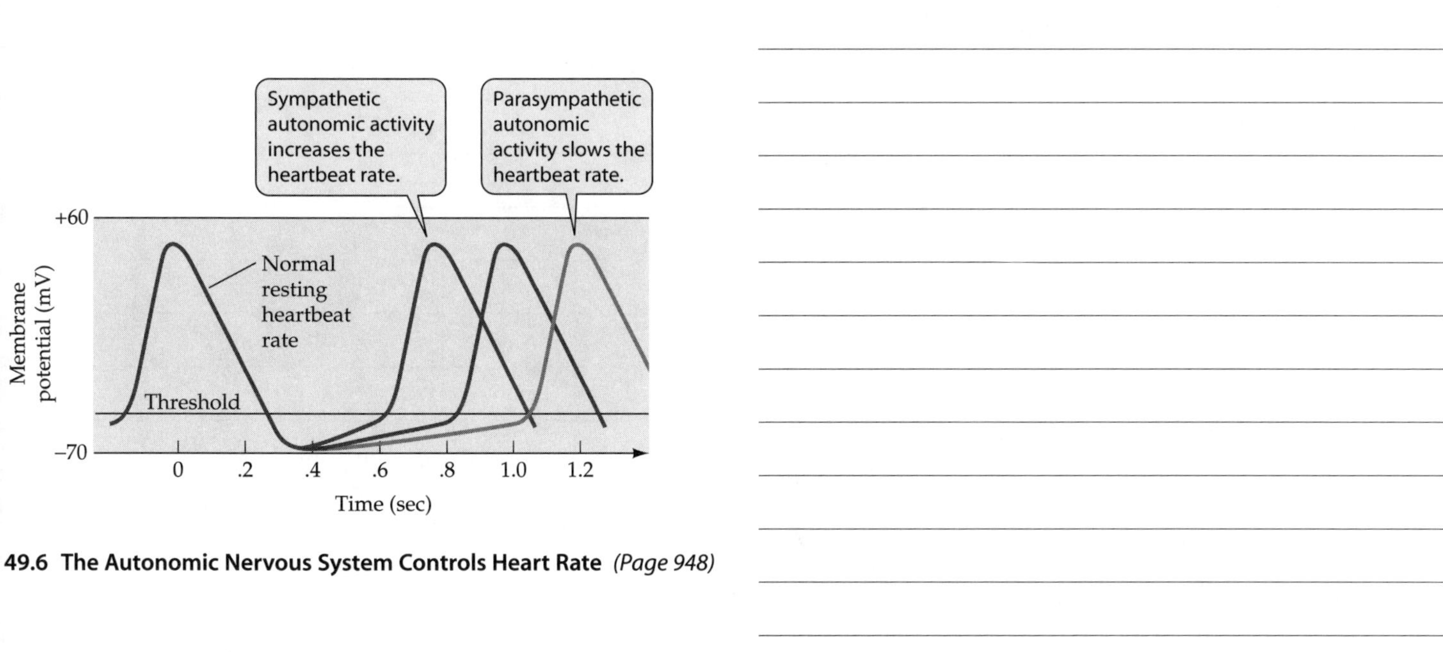

Sympathetic autonomic activity increases the heartbeat rate.
Parasympathetic autonomic activity slows the heartbeat rate.
+60
Normal resting heartbeat rate
Membrane potential (mV)
Threshold
–70
0
.2
.4
.6
.8
1.0
1.2
Time (sec)

49.6 The Autonomic Nervous System Controls Heart Rate *(Page 948)*

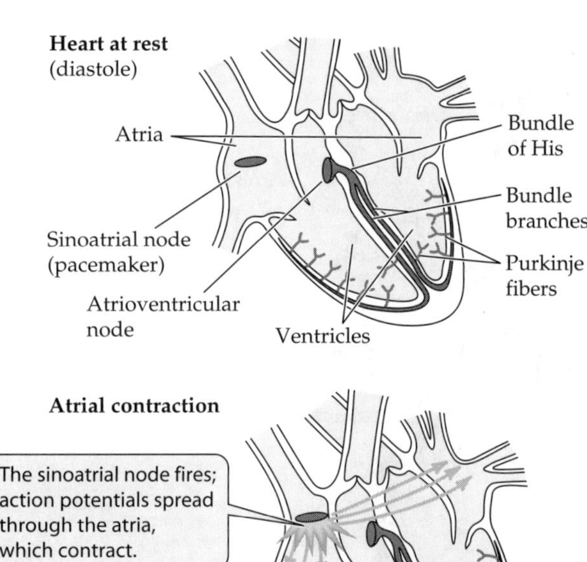

Heart at rest
(diastole)
Atria
Sinoatrial node
(pacemaker)
Atrioventricular
node
Ventricles
Bundle
of His
Bundle
branches
Purkinje
fibers

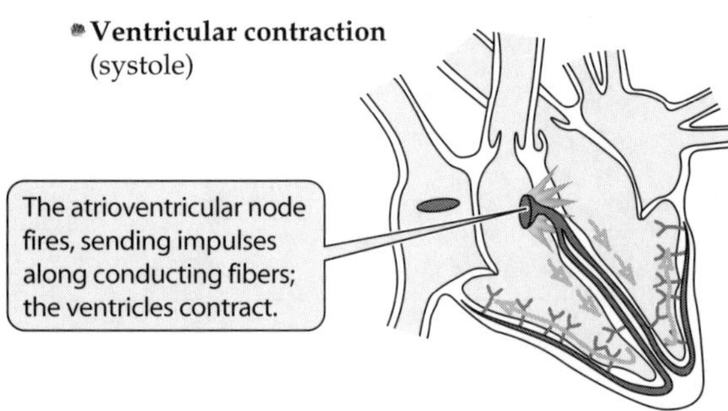

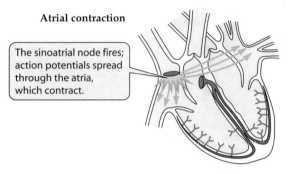

Atrial contraction
The sinoatrial node fires; action potentials spread through the atria, which contract.

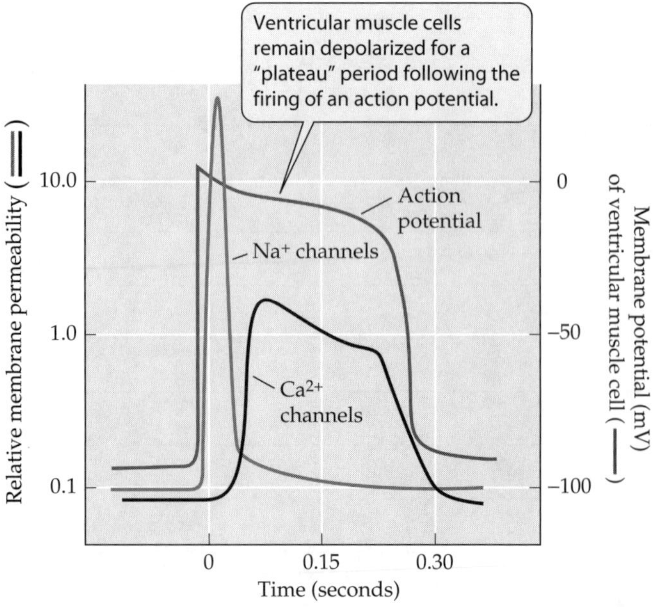

Ventricular contraction
(systole)
The atrioventricular node fires, sending impulses along conducting fibers; the ventricles contract.

49.7 The Heartbeat *(Page 948)*

Ventricular muscle cells remain depolarized for a "plateau" period following the firing of an action potential.
Action
potential
Na+ channels
Ca2+
channels
Relative membrane permeability
Membrane potential (mV)
of ventricular muscle cell
10.0
1.0
0.1
0
−50
−100
0
0.15
0.30
Time (seconds)

49.8 The Action Potential of Ventricular Muscle Fibers *(Page 949)*

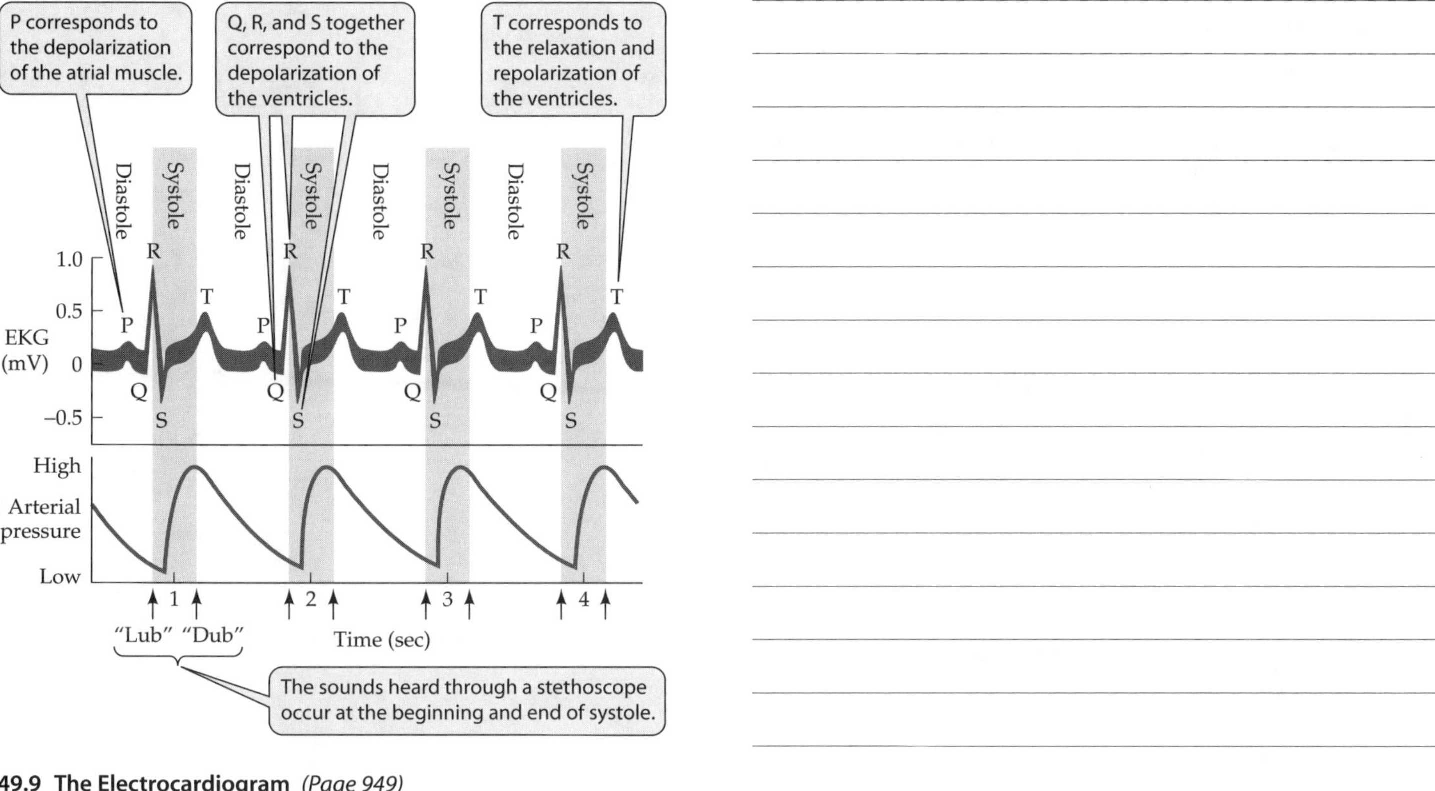

P corresponds to the depolarization of the atrial muscle.
Q, R, and S together correspond to the depolarization of the ventricles.
T corresponds to the relaxation and repolarization of the ventricles.
Diastole
Systole
Diastole
Systole
Diastole
Systole
Diastole
Systole
1.0
0.5
EKG (mV) 0
–0.5
P
Q
R
S
T
High
Arterial pressure
Low
1
2
3
4
"Lub" "Dub"
Time (sec)
The sounds heard through a stethoscope occur at the beginning and end of systole.

49.9 The Electrocardiogram *(Page 949)*

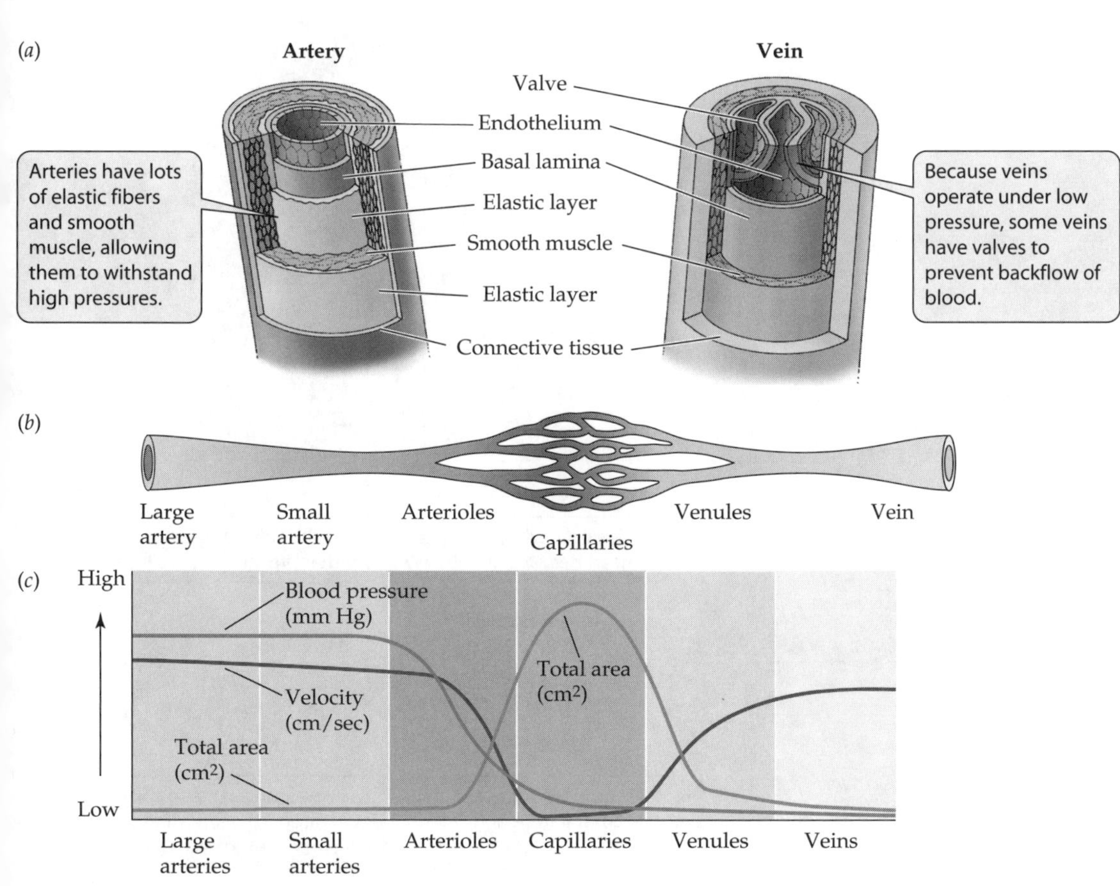

(a)
Artery
Vein
Valve
Endothelium
Basal lamina
Elastic layer
Smooth muscle
Elastic layer
Connective tissue
Arteries have lots of elastic fibers and smooth muscle, allowing them to withstand high pressures.
Because veins operate under low pressure, some veins have valves to prevent backflow of blood.
(b)
Large artery
Small artery
Arterioles
Capillaries
Venules
Vein
(c)
High
Low
Blood pressure (mm Hg)
Velocity (cm/sec)
Total area (cm2)
Total area (cm2)
Large arteries
Small arteries
Arterioles
Capillaries
Venules
Veins

49.10 Anatomy of Blood Vessels *(Page 950)*

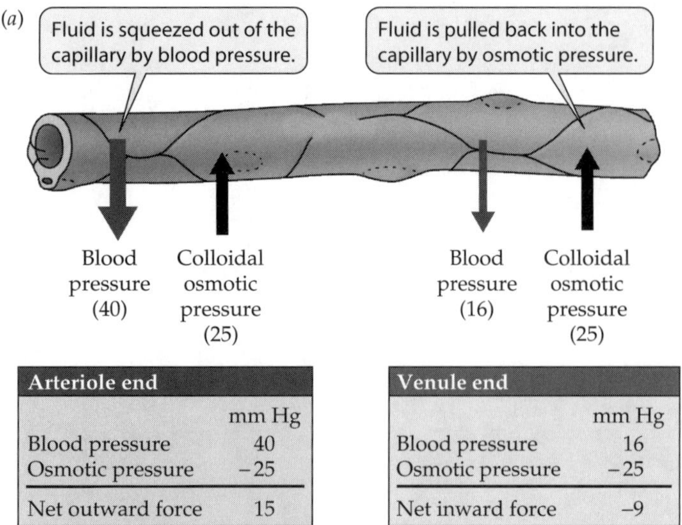
(a)
Fluid is squeezed out of the capillary by blood pressure.
Fluid is pulled back into the capillary by osmotic pressure.
Blood pressure (40)
Colloidal osmotic pressure (25)
Blood pressure (16)
Colloidal osmotic pressure (25)
Arteriole end
mm Hg
Blood pressure 40
Osmotic pressure −25
Net outward force 15
Venule end
mm Hg
Blood pressure 16
Osmotic pressure −25
Net inward force −9

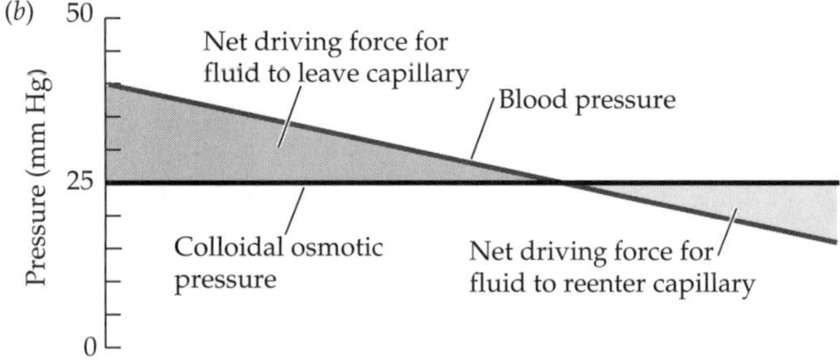
(b)
50
25
0
Pressure (mm Hg)
Net driving force for fluid to leave capillary
Blood pressure
Colloidal osmotic pressure
Net driving force for fluid to reenter capillary

49.12 Starling's Forces *(Page 951)*

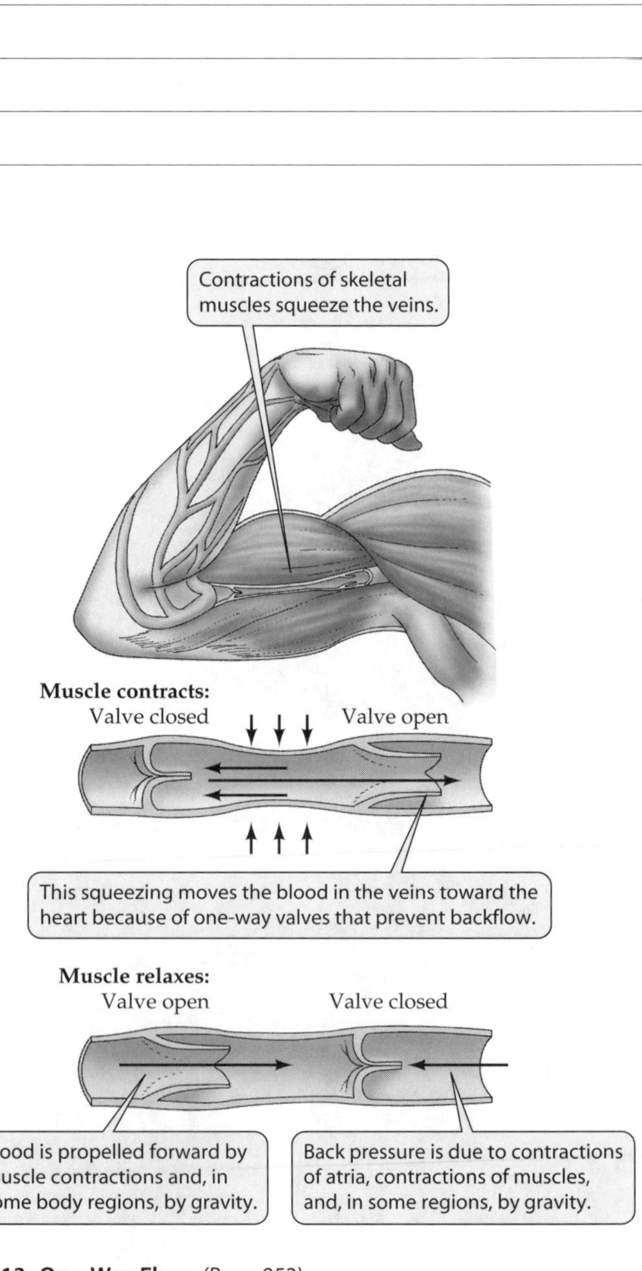
Contractions of skeletal muscles squeeze the veins.
Muscle contracts:
Valve closed
Valve open
This squeezing moves the blood in the veins toward the heart because of one-way valves that prevent backflow.
Muscle relaxes:
Valve open
Valve closed
Blood is propelled forward by muscle contractions and, in some body regions, by gravity.
Back pressure is due to contractions of atria, contractions of muscles, and, in some regions, by gravity.

49.13 One-Way Flow *(Page 952)*

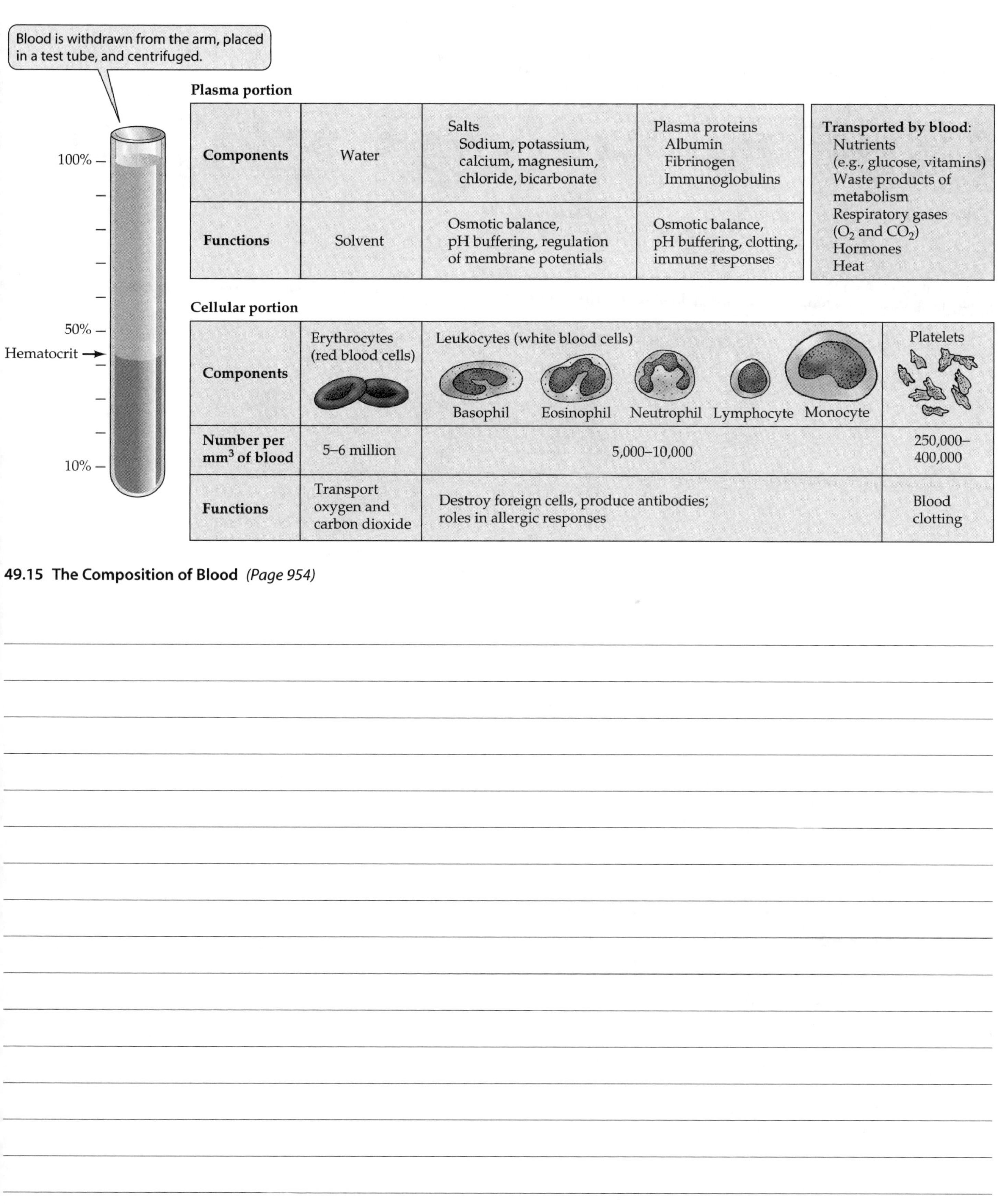
Blood is withdrawn from the arm, placed in a test tube, and centrifuged.
100%
50%
Hematocrit
10%
Plasma portion
Cellular portion

Plasma portion

Components	Water	Salts Sodium, potassium, calcium, magnesium, chloride, bicarbonate	Plasma proteins Albumin Fibrinogen Immunoglobulins	**Transported by blood:** Nutrients (e.g., glucose, vitamins) Waste products of metabolism Respiratory gases (O_2 and CO_2) Hormones Heat
Functions	Solvent	Osmotic balance, pH buffering, regulation of membrane potentials	Osmotic balance, pH buffering, clotting, immune responses	

Cellular portion

Components	Erythrocytes (red blood cells)	Leukocytes (white blood cells) Basophil, Eosinophil, Neutrophil, Lymphocyte, Monocyte	Platelets
Number per mm^3 of blood	5–6 million	5,000–10,000	250,000–400,000
Functions	Transport oxygen and carbon dioxide	Destroy foreign cells, produce antibodies; roles in allergic responses	Blood clotting

49.15 The Composition of Blood *(Page 954)*

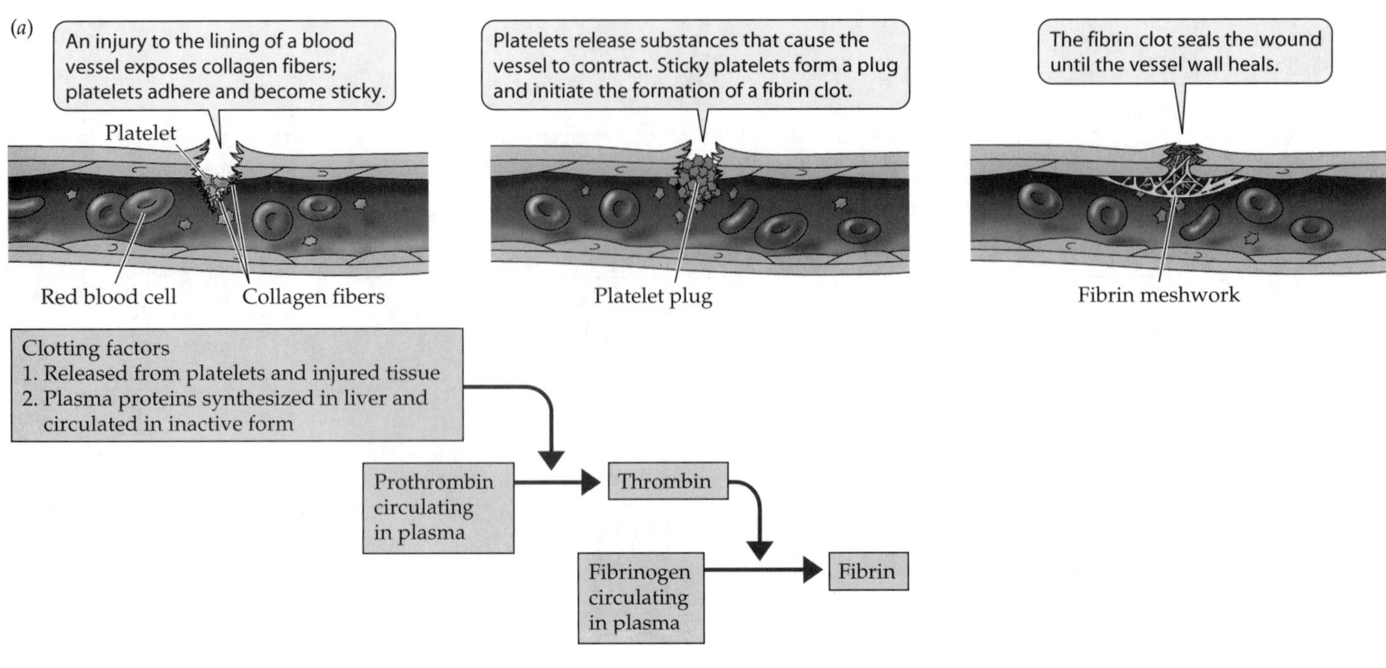
(a)
An injury to the lining of a blood vessel exposes collagen fibers; platelets adhere and become sticky.
Platelets release substances that cause the vessel to contract. Sticky platelets form a plug and initiate the formation of a fibrin clot.
The fibrin clot seals the wound until the vessel wall heals.
Platelet
Red blood cell
Collagen fibers
Platelet plug
Fibrin meshwork
Clotting factors
1. Released from platelets and injured tissue
2. Plasma proteins synthesized in liver and circulated in inactive form
Prothrombin circulating in plasma
Thrombin
Fibrinogen circulating in plasma
Fibrin

49.16 Blood Clotting *(Page 955)*

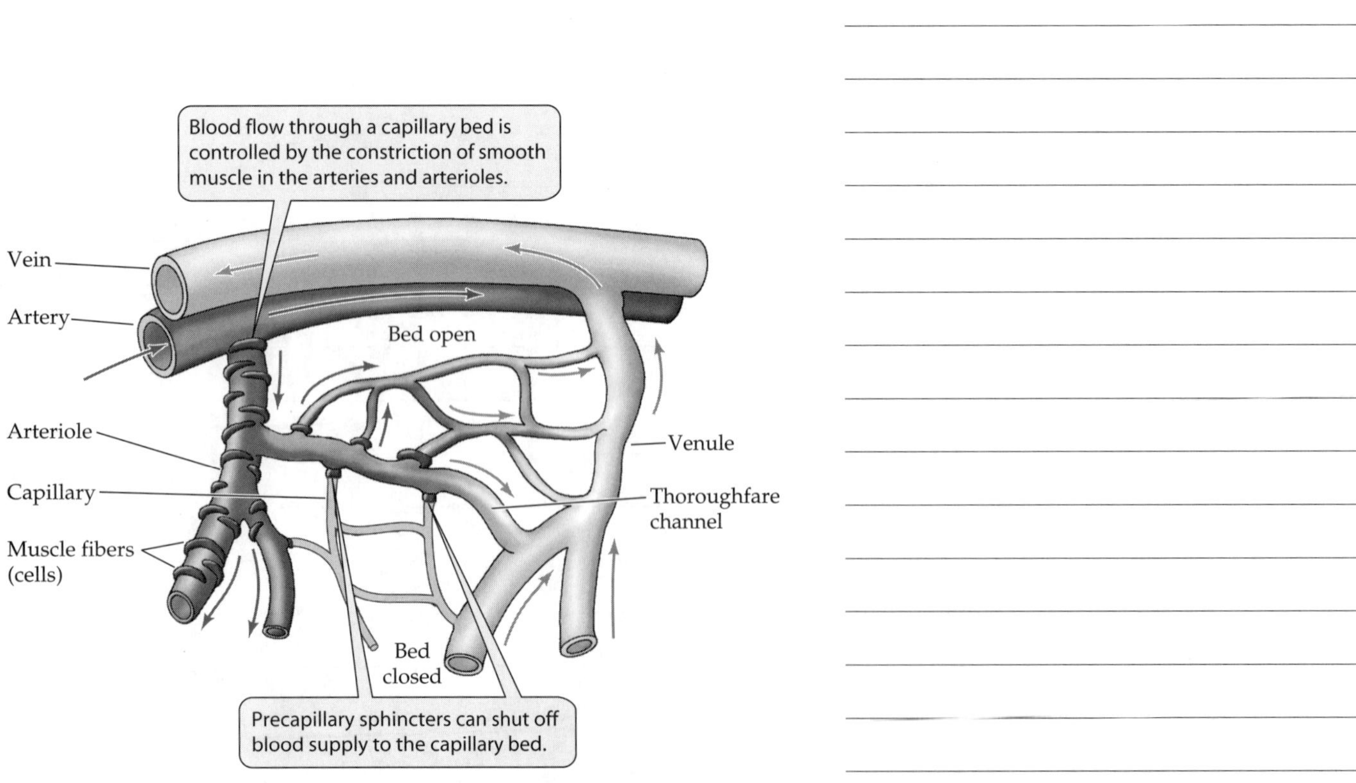
Blood flow through a capillary bed is controlled by the constriction of smooth muscle in the arteries and arterioles.
Vein
Artery
Bed open
Arteriole
Venule
Capillary
Thoroughfare channel
Muscle fibers (cells)
Bed closed
Precapillary sphincters can shut off blood supply to the capillary bed.

49.17 Local Control of Blood Flow *(Page 956)*

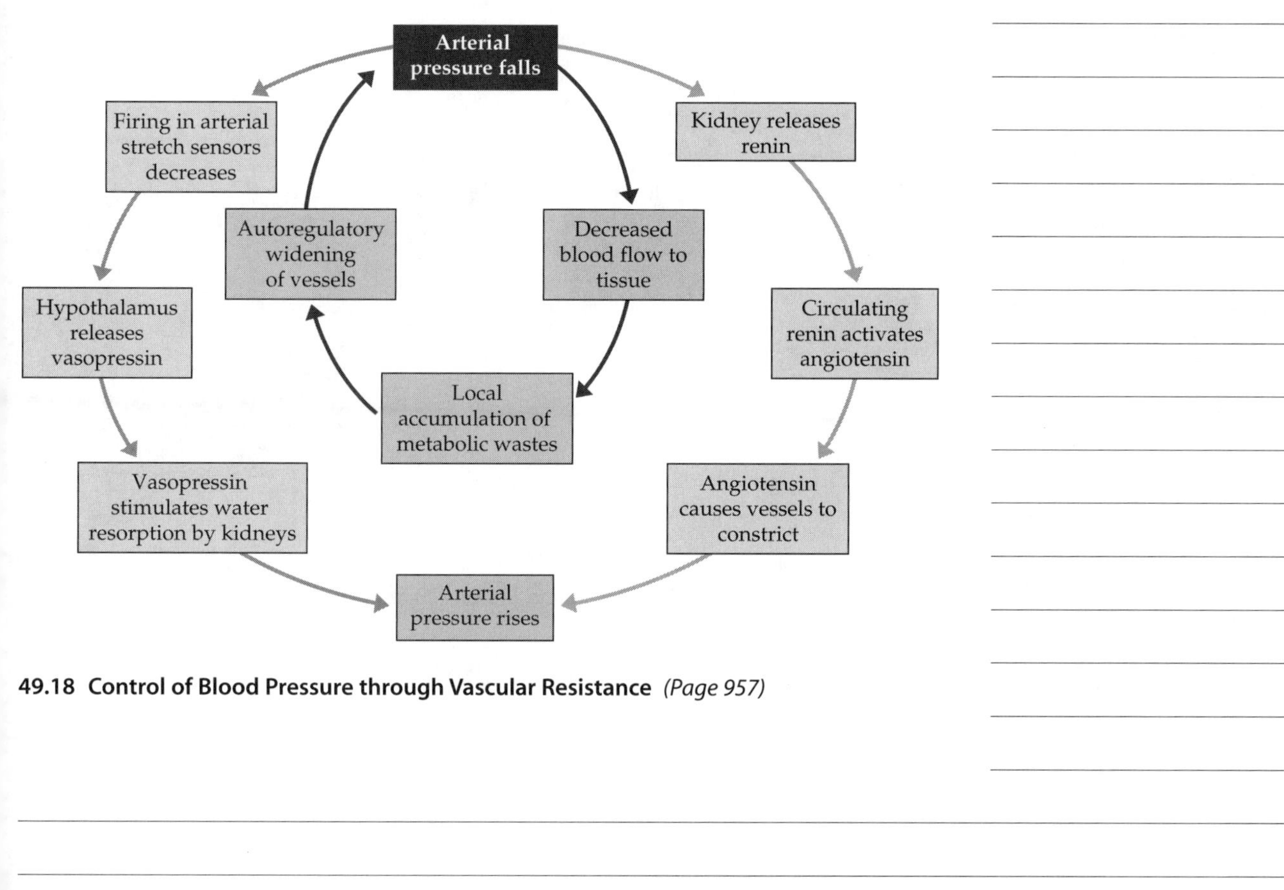
Arterial pressure falls
Firing in arterial stretch sensors decreases
Kidney releases renin
Autoregulatory widening of vessels
Decreased blood flow to tissue
Hypothalamus releases vasopressin
Circulating renin activates angiotensin
Local accumulation of metabolic wastes
Vasopressin stimulates water resorption by kidneys
Angiotensin causes vessels to constrict
Arterial pressure rises

49.18 Control of Blood Pressure through Vascular Resistance *(Page 957)*

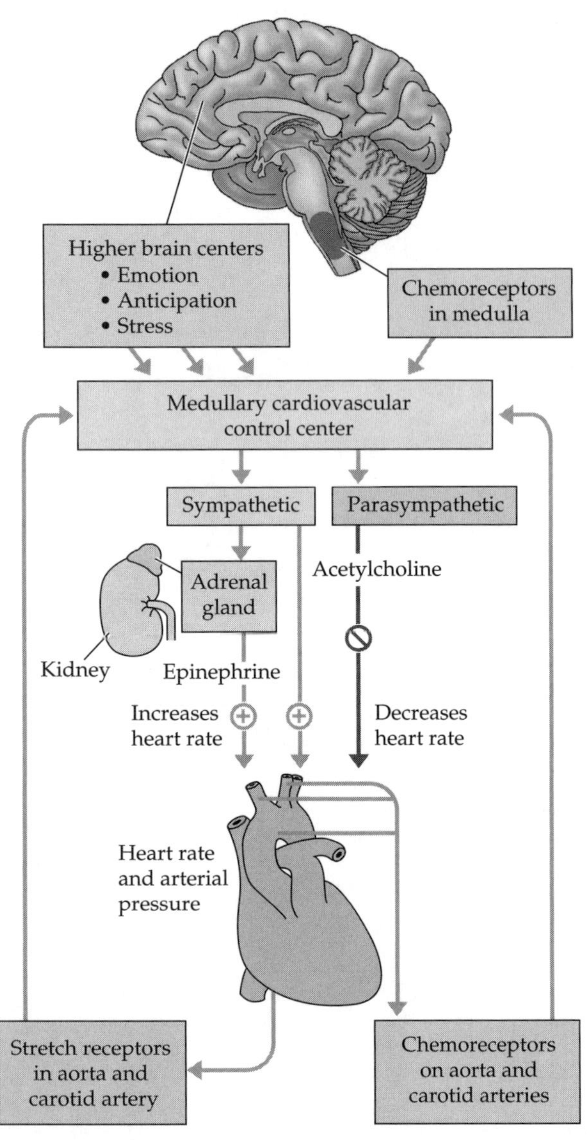

Higher brain centers
• Emotion
• Anticipation
• Stress
Chemoreceptors in medulla
Medullary cardiovascular control center
Sympathetic
Parasympathetic
Adrenal gland
Kidney
Epinephrine
Acetylcholine
Increases heart rate
Decreases heart rate
Heart rate and arterial pressure
Stretch receptors in aorta and carotid artery
Chemoreceptors on aorta and carotid arteries

49.19 Regulating Blood Pressure *(Page 957)*

49.20 The Diving Reflex *(Page 958)*

50 Nutrition, Digestion, and Absorption

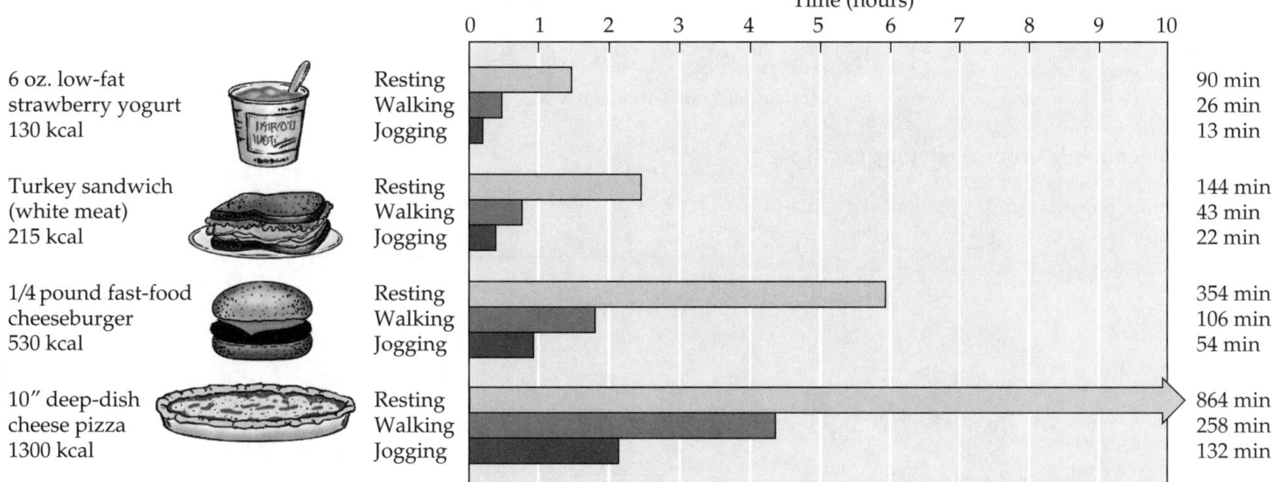

Time (hours)
0 1 2 3 4 5 6 7 8 9 10
6 oz. low-fat strawberry yogurt 130 kcal
Resting
Walking
Jogging
90 min
26 min
13 min
Turkey sandwich (white meat) 215 kcal
Resting
Walking
Jogging
144 min
43 min
22 min
1/4 pound fast-food cheeseburger 530 kcal
Resting
Walking
Jogging
354 min
106 min
54 min
10″ deep-dish cheese pizza 1300 kcal
Resting
Walking
Jogging
864 min
258 min
132 min

50.2 Food Energy and How We Use It *(Page 963)*

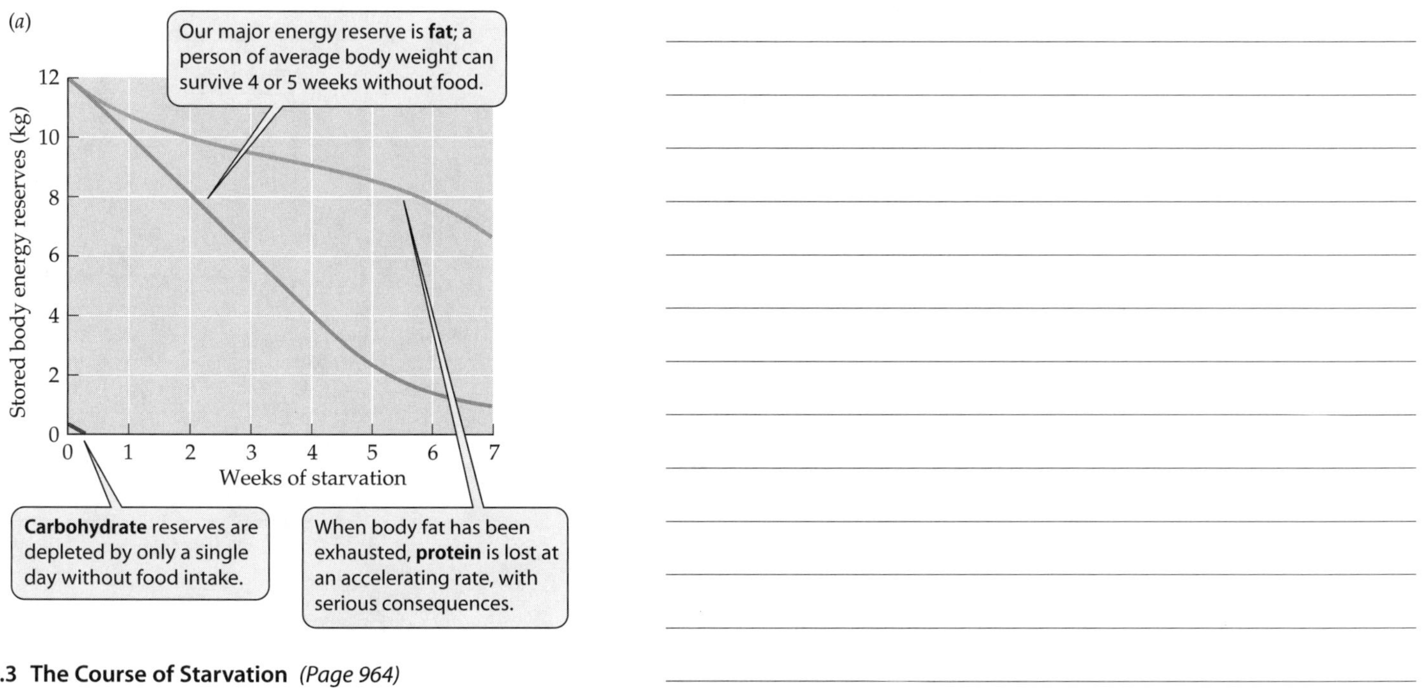

(a)
Our major energy reserve is **fat**; a person of average body weight can survive 4 or 5 weeks without food.
Stored body energy reserves (kg)
0 2 4 6 8 10 12
0 1 2 3 4 5 6 7
Weeks of starvation
Carbohydrate reserves are depleted by only a single day without food intake.
When body fat has been exhausted, **protein** is lost at an accelerating rate, with serious consequences.

50.3 The Course of Starvation *(Page 964)*

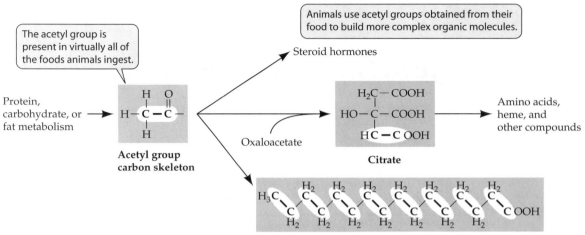

The acetyl group is present in virtually all of the foods animals ingest.

Animals use acetyl groups obtained from their food to build more complex organic molecules.

Protein, carbohydrate, or fat metabolism → Acetyl group carbon skeleton

Steroid hormones

Oxaloacetate

$H_2C-COOH$
$HO-C-COOH$
$HC-COOH$

Citrate → Amino acids, heme, and other compounds

$H_3C-(CH_2)_{14}-COOH$

Acetyl group carbon skeleton

Palmitic acid (and other fatty acids)

50.4 The Acetyl Group Is an Acquired Carbon Skeleton *(Page 964)*

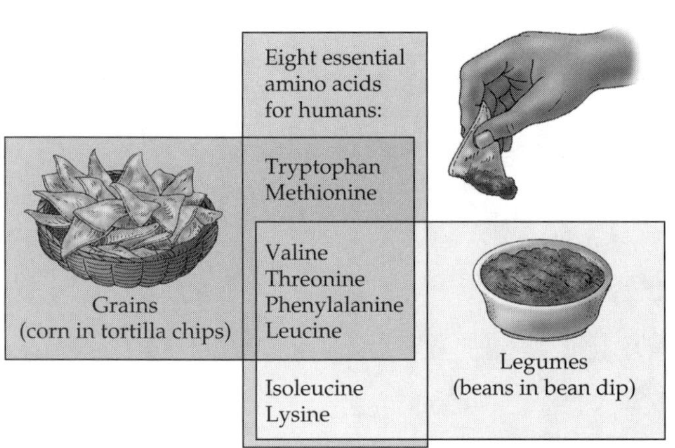
Eight essential amino acids for humans:
Tryptophan
Methionine
Valine
Threonine
Phenylalanine
Leucine
Isoleucine
Lysine
Grains
(corn in tortilla chips)
Legumes
(beans in bean dip)

50.5 A Strategy for Vegetarians *(Page 965)*

50.1 *Mineral Elements Required by Animals*

ELEMENT	SOURCE IN HUMAN DIET	MAJOR FUNCTIONS
MACRONUTRIENTS		
Calcium (Ca)	Dairy foods, eggs, green leafy vegetables, whole grains, legumes, nuts	Found in bones and teeth; blood clotting; nerve and muscle action; enzyme activation
Chlorine (Cl)	Table salt (NaCl), meat, eggs, vegetables, dairy foods	Water balance; digestion (as HCl); principal negative ion in tissue fluid
Magnesium (Mg)	Green vegetables, meat, whole grains, nuts, milk, legumes	Required by many enzymes; found in bones and teeth
Phosphorus (P)	Dairy, eggs, meat, whole grains, legumes, nuts	Found in nucleic acids, ATP, and phospholipids; bone formation; buffers; metabolism of sugars
Potassium (K)	Meat, whole grains, fruits, vegetables	Nerve and muscle action; protein synthesis; principal positive ion in cells
Sodium (Na)	Table salt, dairy foods, meat, eggs, vegetables	Nerve and muscle action; water balance; principal positive ion in tissue fluid
Sulfur (S)	Meat, eggs, dairy foods, nuts, legumes	Found in proteins and coenzymes; detoxification of harmful substances
MICRONUTRIENTS		
Chromium (Cr)	Meat, dairy, whole grains, legumes, yeast	Glucose metabolism
Cobalt (Co)	Meat, tap water	Found in vitamin B_{12}; formation of red blood cells
Copper (Cu)	Liver, meat, fish, shellfish, legumes, whole grains, nuts	Found in active site of many redox enzymes and electron carriers; production of hemoglobin; bone formation
Fluorine (F)	Most water supplies	Found in teeth; helps prevent decay
Iodine (I)	Fish, shellfish, iodized salt	Found in thyroid hormones
Iron (Fe)	Liver, meat, green vegetables, eggs, whole grains, legumes, nuts	Found in active sites of many redox enzymes and electron carriers, hemoglobin, and myoglobin
Manganese (Mn)	Organ meats, whole grains, legumes, nuts, tea, coffee	Activates many enzymes
Molybdenum (Mo)	Organ meats, dairy, whole grains, green vegetables, legumes	Found in some enzymes
Selenium (Se)	Meat, seafood, whole grains, eggs, milk, garlic	Fat metabolism
Zinc (Zn)	Liver, fish, shellfish, and many other foods	Found in some enzymes and some transcription factors; insulin physiology

(Page 966)

50.2 *Vitamins in the Human Diet*

VITAMIN	SOURCE	FUNCTION	DEFICIENCY SYMPTOMS
WATER-SOLUBLE			
B_1, thiamin	Liver, legumes, whole grains,	Coenzyme in cellular respiration	Beriberi, loss of appetite, fatigue
B_2, riboflavin	Dairy, meat, eggs, green leafy vegetables	Coenzyme in FAD	Lesions in corners of mouth, eye irritation, skin disorders
Niacin	Meat, fowl, liver, yeast	Coenzyme in NAD and NADP	Pellagra, skin disorders, diarrhea, mental disorders
B_6, pyridoxine	Liver, whole grains, dairy foods	Coenzyme in amino acid metabolism	Anemia, slow growth, skin problems, convulsions
Pantothenic acid	Liver, eggs, yeast	Found in acetyl CoA	Adrenal problems, reproductive problems
Biotin	Liver, yeast, bacteria in gut	Found in coenzymes	Skin problems, loss of hair
B_{12}, cobalamin	Liver, meat, dairy foods, eggs	Formation of nucleic acids, proteins, and red blood cells	Pernicious anemia
Folic acid	Vegetables, eggs, liver, whole grains	Coenzyme in formation of heme and nucleotides	Anemia
C, ascorbic acid	Citrus fruits, tomatoes, potatoes	Formation of connective tissues; antioxidant	Scurvy, slow healing, poor bone growth
FAT-SOLUBLE			
A, retinol	Fruits, vegetables, liver, dairy	Found in visual pigments	Night blindness
D, calciferol	Fortified milk, fish oils, sunshine	Absorption Ca^{2+} and phosphate	Rickets
E, tocopherol	Meat, dairy foods, whole grains	Muscle maintenance, antioxidant	Anemia
K, menadione	Intestinal bacteria, liver	Blood clotting	Blood-clotting problems

(Page 967)

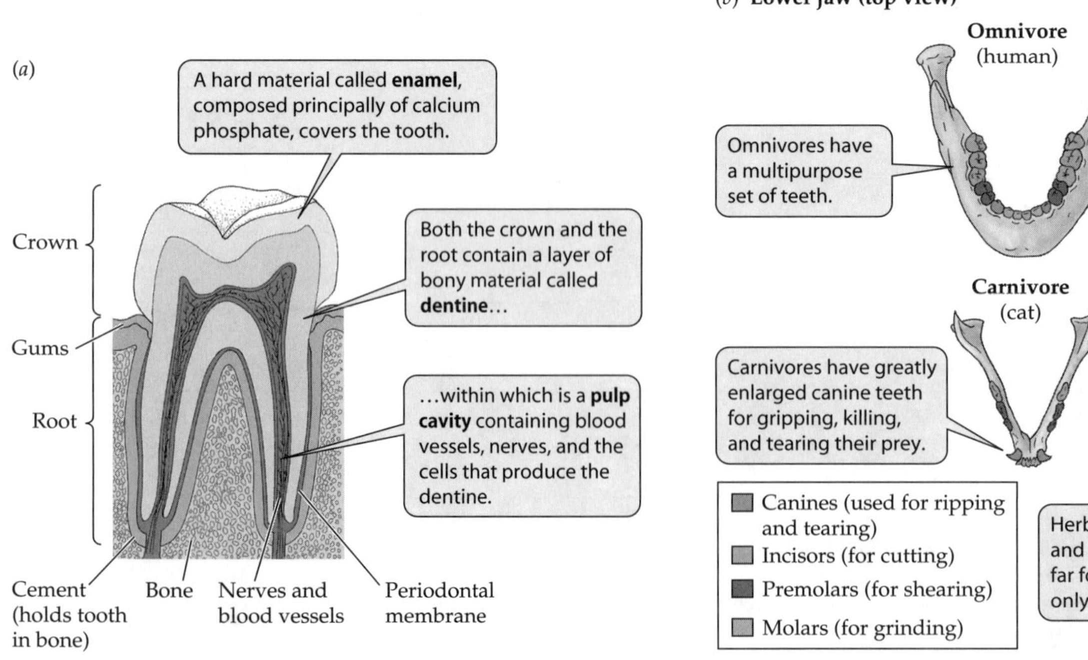
(a)
A hard material called **enamel**, composed principally of calcium phosphate, covers the tooth.
Crown
Both the crown and the root contain a layer of bony material called **dentine**...
Gums
Root
...within which is a **pulp cavity** containing blood vessels, nerves, and the cells that produce the dentine.
Cement (holds tooth in bone)
Bone
Nerves and blood vessels
Periodontal membrane
(b) Lower jaw (top view)
Omnivore (human)
Herbivore (sheep)
Omnivores have a multipurpose set of teeth.
Carnivore (cat)
Carnivores have greatly enlarged canine teeth for gripping, killing, and tearing their prey.
Canines (used for ripping and tearing)
Incisors (for cutting)
Premolars (for shearing)
Molars (for grinding)
Herbivores use their incisors and canines, which are found far forward on the lower jaw only, to tear leaves off of plants.

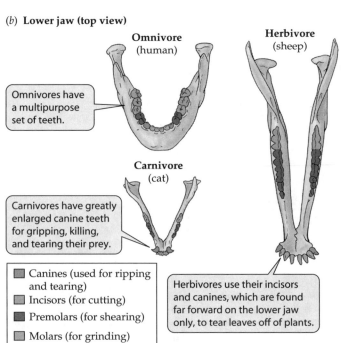

50.7 Mammalian Teeth *(Page 969)*

Earthworm
Mouth
Crop
Intestine
Anus
Pharynx
Esophagus
Gizzard
Cockroach
Esophagus
Crop
Gizzard
Anus
Rectum
Mandibles
Salivary glands
Intestine
Rabbit
Salivary glands
Pancreas
Cecum
Rectum
Teeth
Anus
Small intestine
Esophagus
Liver
Stomach
Large intestine

50.8 Compartments for Digestion and Absorption *(Page 970)*

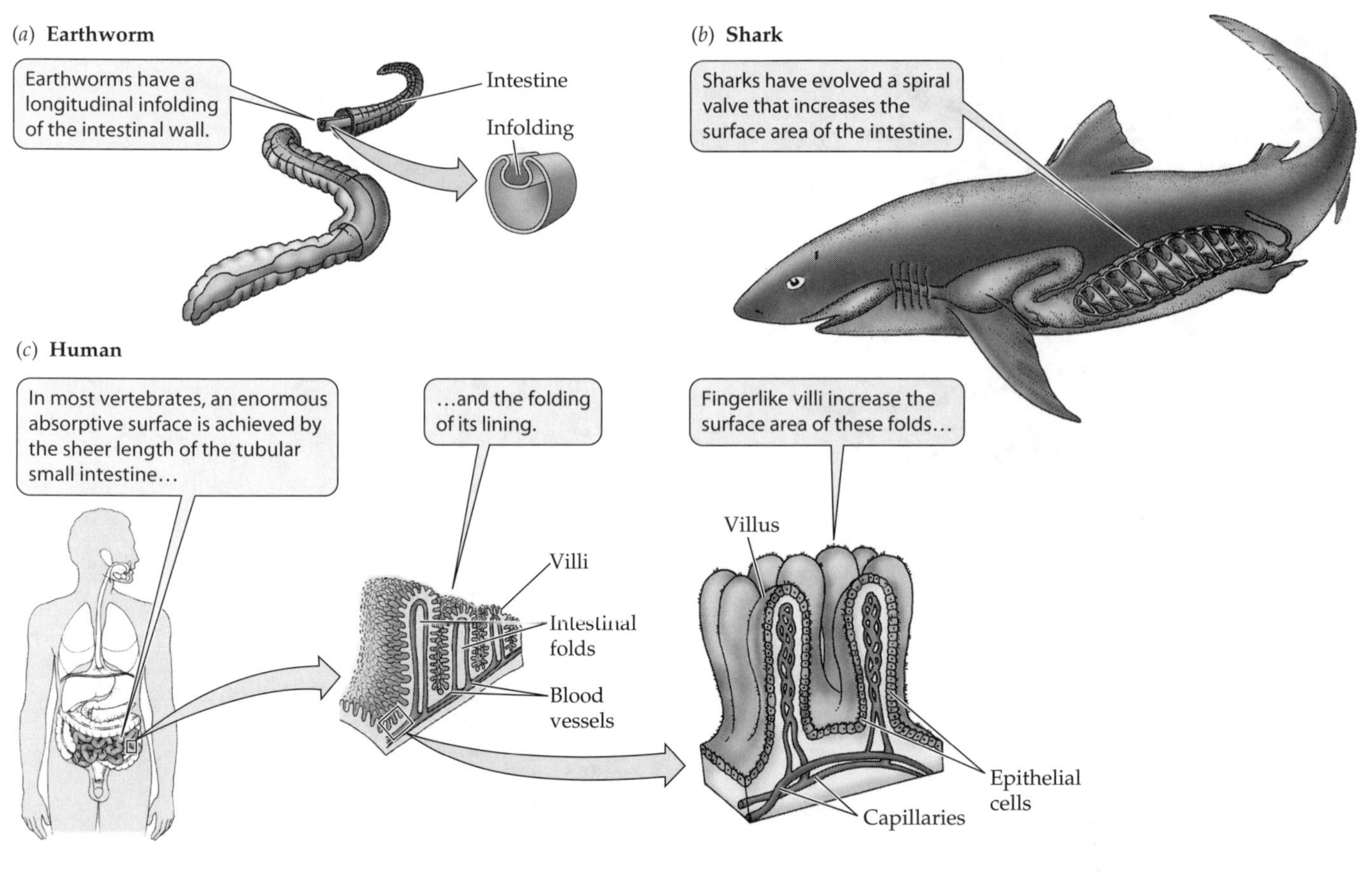
(a) Earthworm
Earthworms have a longitudinal infolding of the intestinal wall.
Intestine
Infolding
(b) Shark
Sharks have evolved a spiral valve that increases the surface area of the intestine.
(c) Human
In most vertebrates, an enormous absorptive surface is achieved by the sheer length of the tubular small intestine…
…and the folding of its lining.
Fingerlike villi increase the surface area of these folds…
Villi
Intestinal folds
Blood vessels
Villus
Epithelial cells
Capillaries

50.9 Greater Intestinal Surface Area Means More Nutrient Absorption *(Page 971)*

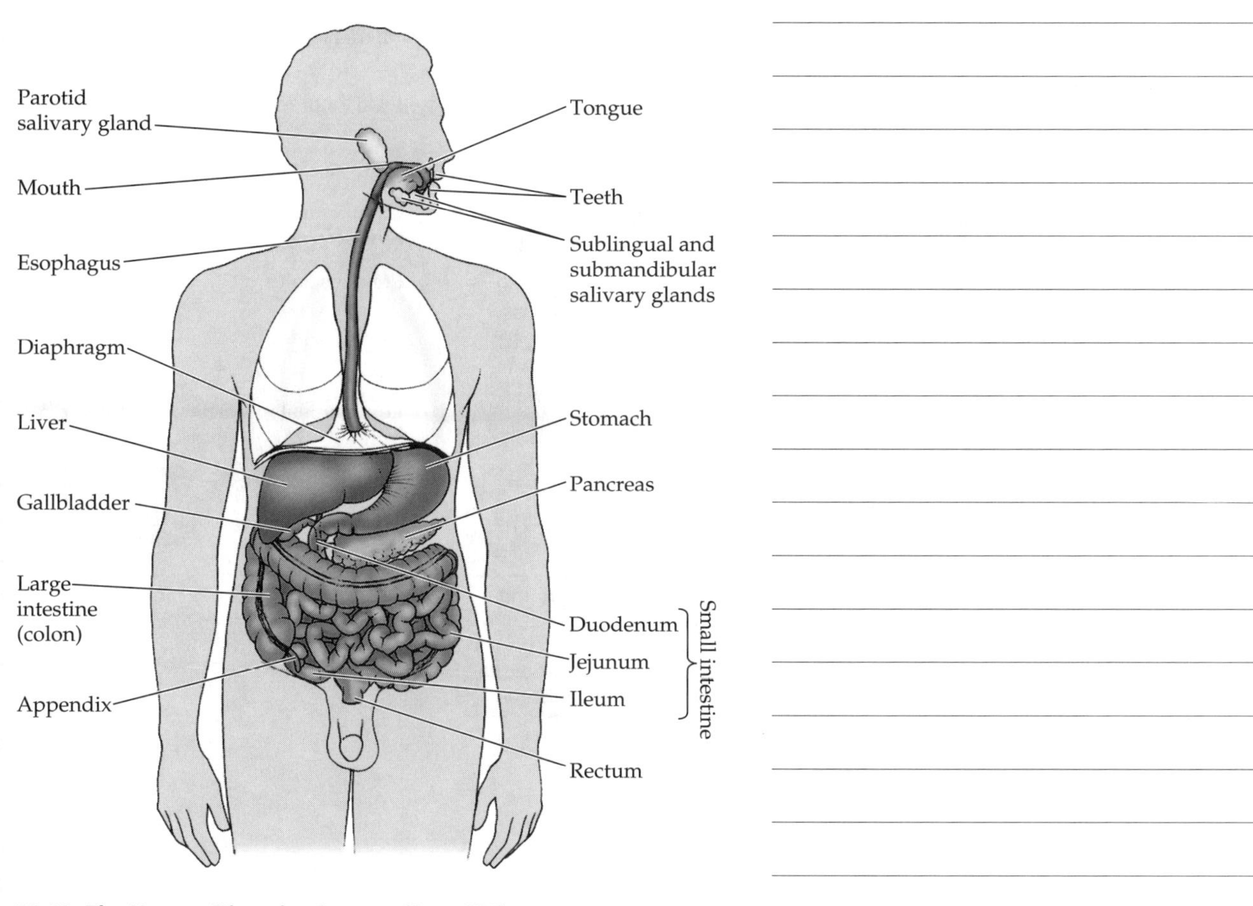
Parotid salivary gland
Mouth
Esophagus
Diaphragm
Liver
Gallbladder
Large intestine (colon)
Appendix
Tongue
Teeth
Sublingual and submandibular salivary glands
Stomach
Pancreas
Duodenum
Jejunum
Ileum
Small intestine
Rectum

50.10 The Human Digestive System *(Page 971)*

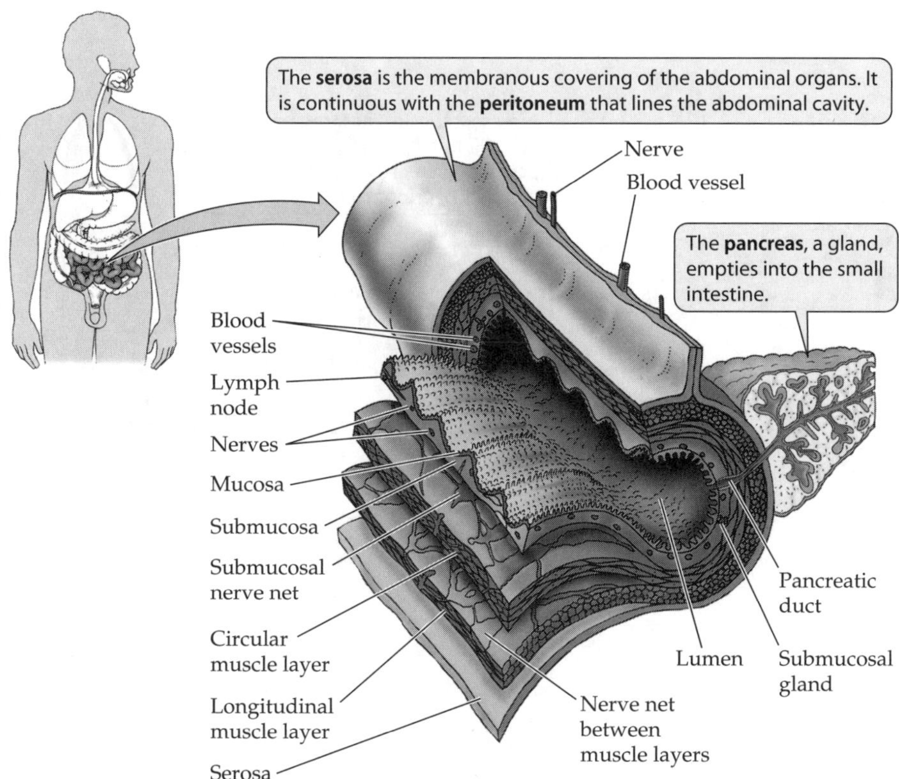
The **serosa** is the membranous covering of the abdominal organs. It is continuous with the **peritoneum** that lines the abdominal cavity.
Nerve
Blood vessel
The **pancreas**, a gland, empties into the small intestine.
Blood vessels
Lymph node
Nerves
Mucosa
Submucosa
Submucosal nerve net
Circular muscle layer
Longitudinal muscle layer
Serosa
Nerve net between muscle layers
Lumen
Pancreatic duct
Submucosal gland

50.11 Tissue Layers of the Vertebrate Gut *(Page 972)*

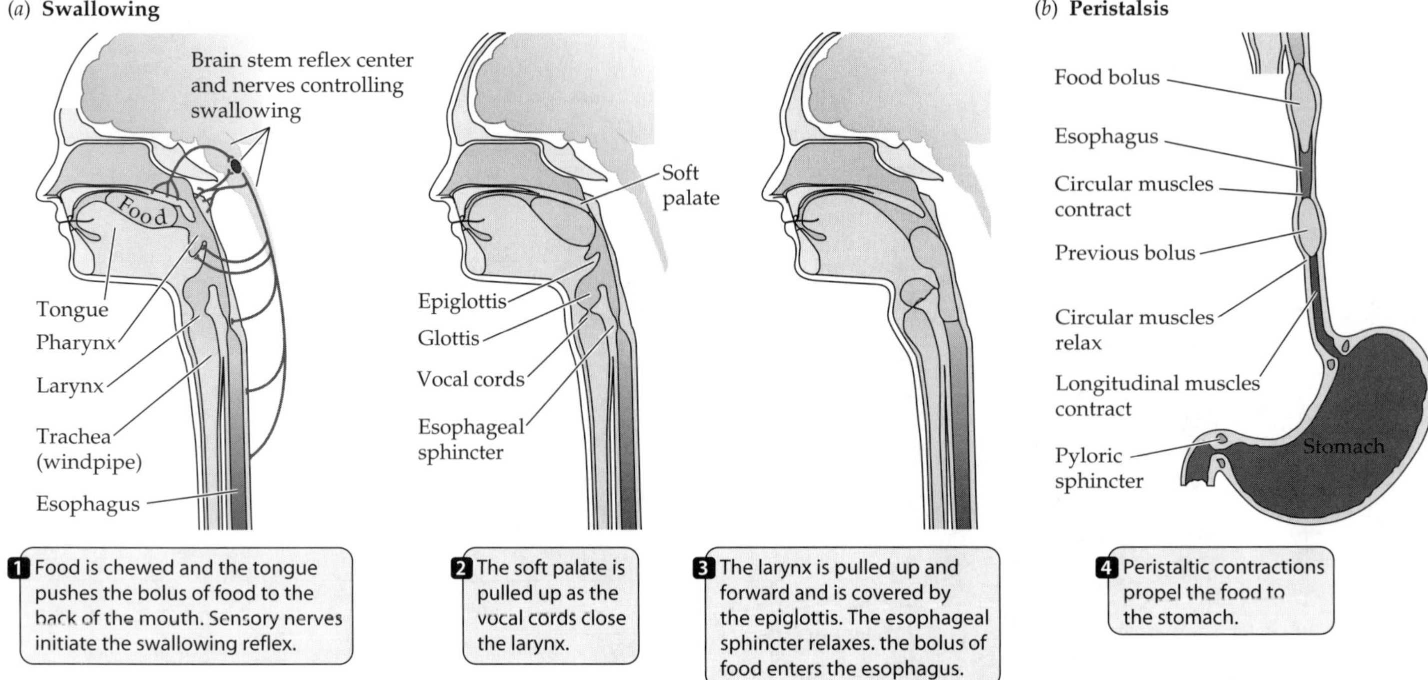
(a) Swallowing
Brain stem reflex center and nerves controlling swallowing
Food
Tongue
Pharynx
Larynx
Trachea (windpipe)
Esophagus
Soft palate
Epiglottis
Glottis
Vocal cords
Esophageal sphincter
(b) Peristalsis
Food bolus
Esophagus
Circular muscles contract
Previous bolus
Circular muscles relax
Longitudinal muscles contract
Pyloric sphincter
Stomach
1 Food is chewed and the tongue pushes the bolus of food to the back of the mouth. Sensory nerves initiate the swallowing reflex.
2 The soft palate is pulled up as the vocal cords close the larynx.
3 The larynx is pulled up and forward and is covered by the epiglottis. The esophageal sphincter relaxes. the bolus of food enters the esophagus.
4 Peristaltic contractions propel the food to the stomach.

50.12 Swallowing and Peristalsis *(Page 973)*

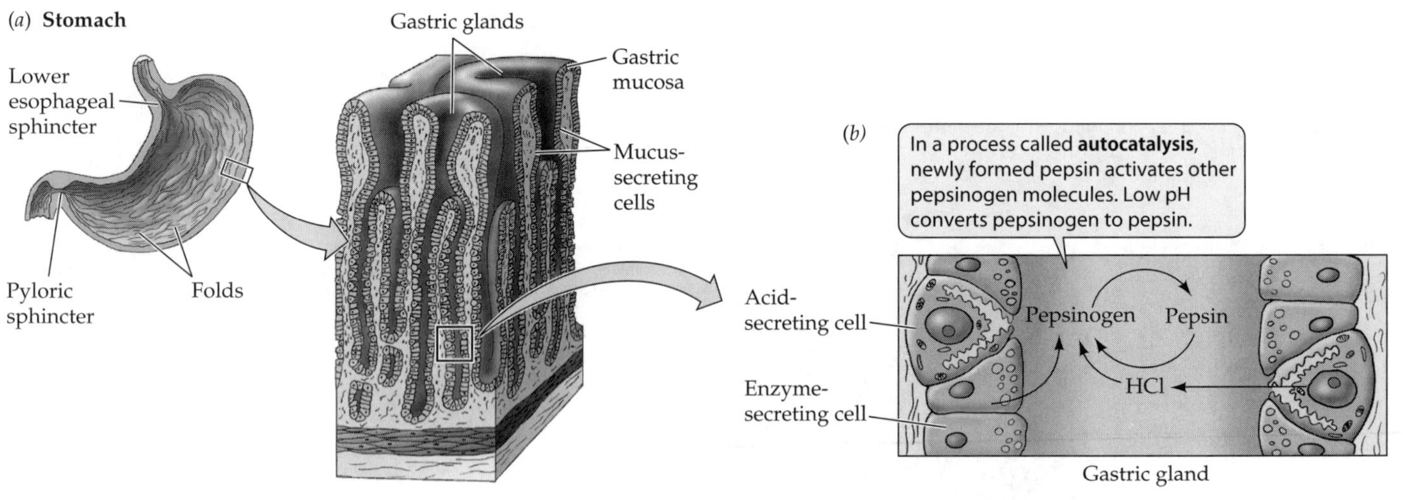
(a) Stomach
Lower esophageal sphincter
Pyloric sphincter
Folds
Gastric glands
Gastric mucosa
Mucus-secreting cells
(b)
In a process called **autocatalysis**, newly formed pepsin activates other pepsinogen molecules. Low pH converts pepsinogen to pepsin.
Acid-secreting cell
Enzyme-secreting cell
Pepsinogen
Pepsin
HCl
Gastric gland

50.13 The Stomach *(Page 973)*

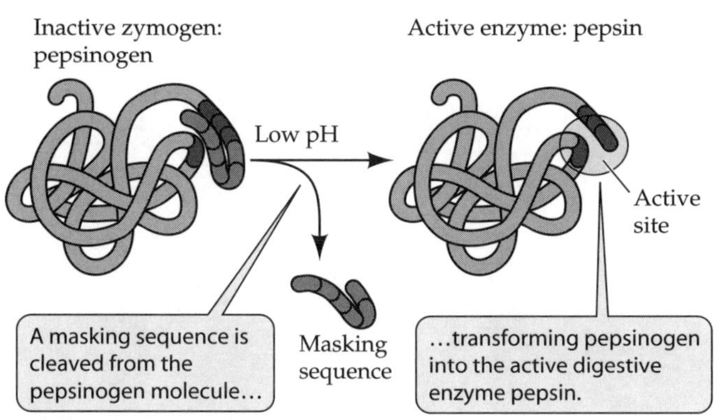
Inactive zymogen: pepsinogen
Active enzyme: pepsin
Low pH
Active site
A masking sequence is cleaved from the pepsinogen molecule…
Masking sequence
…transforming pepsinogen into the active digestive enzyme pepsin.

50.14 Activating a Zymogen *(Page 974)*

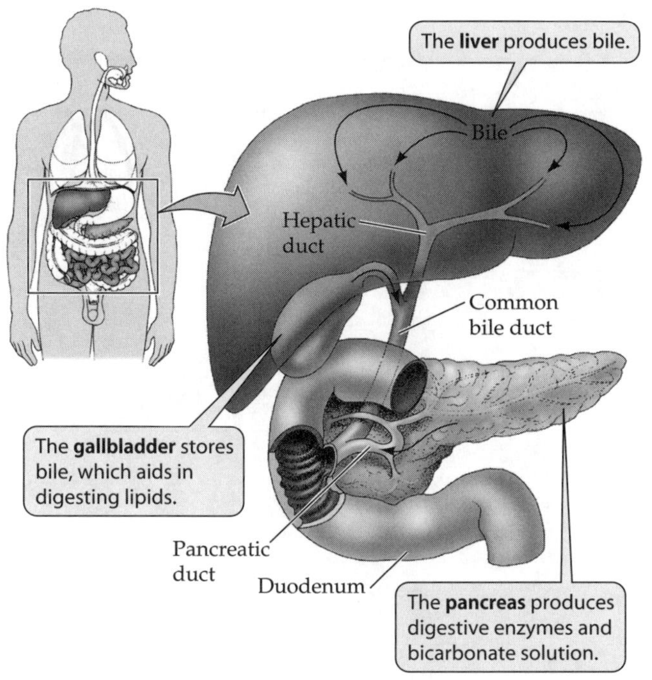
The **liver** produces bile.
Bile
Hepatic duct
Common bile duct
The **gallbladder** stores bile, which aids in digesting lipids.
Pancreatic duct
Duodenum
The **pancreas** produces digestive enzymes and bicarbonate solution.

50.15 The Ducts of the Gallbladder and Pancreas *(Page 974)*

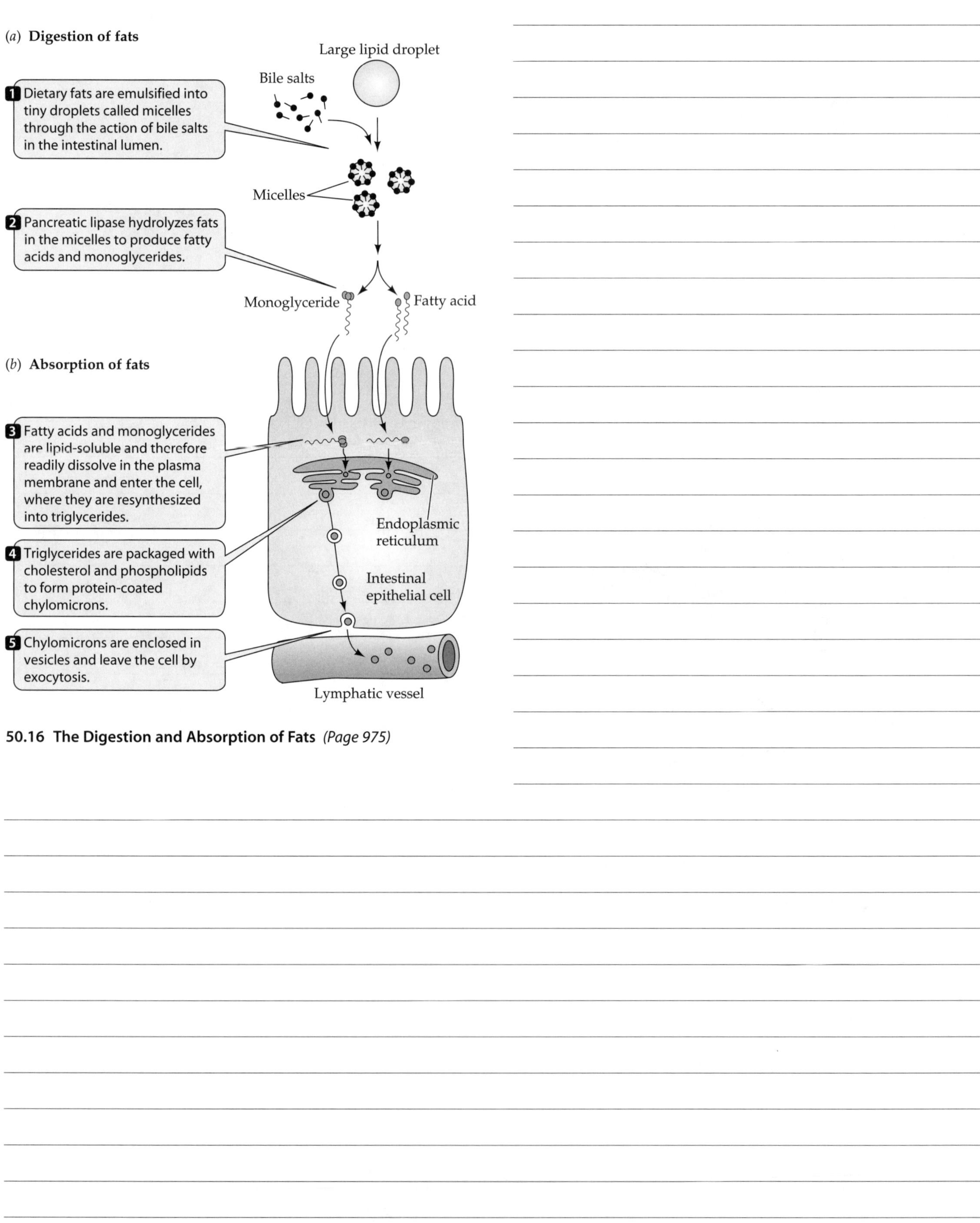
(a) Digestion of fats
Large lipid droplet
Bile salts
1 Dietary fats are emulsified into tiny droplets called micelles through the action of bile salts in the intestinal lumen.
Micelles
2 Pancreatic lipase hydrolyzes fats in the micelles to produce fatty acids and monoglycerides.
Monoglyceride
Fatty acid
(b) Absorption of fats
3 Fatty acids and monoglycerides are lipid-soluble and therefore readily dissolve in the plasma membrane and enter the cell, where they are resynthesized into triglycerides.
Endoplasmic reticulum
4 Triglycerides are packaged with cholesterol and phospholipids to form protein-coated chylomicrons.
Intestinal epithelial cell
5 Chylomicrons are enclosed in vesicles and leave the cell by exocytosis.
Lymphatic vessel

50.16 The Digestion and Absorption of Fats *(Page 975)*

50.3 *Sources and Functions of the Major Digestive Enzymes of Humans*

ENZYME	SOURCE	ACTION
Salivary amylase	Salivary glands	Starch → Maltose
Pepsin	Stomach	Proteins → Peptides; autocatalysis
Pancreatic amylase	Pancreas	Starch → Maltose
Lipase	Pancreas	Fats → Fatty acids and glycerol
Nuclease	Pancreas	Nucleic acids → Nucleotides
Trypsin	Pancreas	Proteins → Peptides; activation of zymogens
Chymotrypsin	Pancreas	Proteins → Peptides
Carboxypeptidase	Pancreas	Peptides → Peptides and amino acids
Aminopeptidase	Small intestine	Peptides → Peptides and amino acids
Dipeptidase	Small intestine	Dipeptides → Amino acids
Enterokinase	Small intestine	Trypsinogen → Trypsin
Nuclease	Small intestine	Nucleic acids → Nucleotides
Maltase	Small intestine	Maltose → Glucose
Lactase	Small intestine	Lactose → Galactose and glucose
Sucrase	Small intestine	Sucrose → Fructose and glucose

(Page 976)

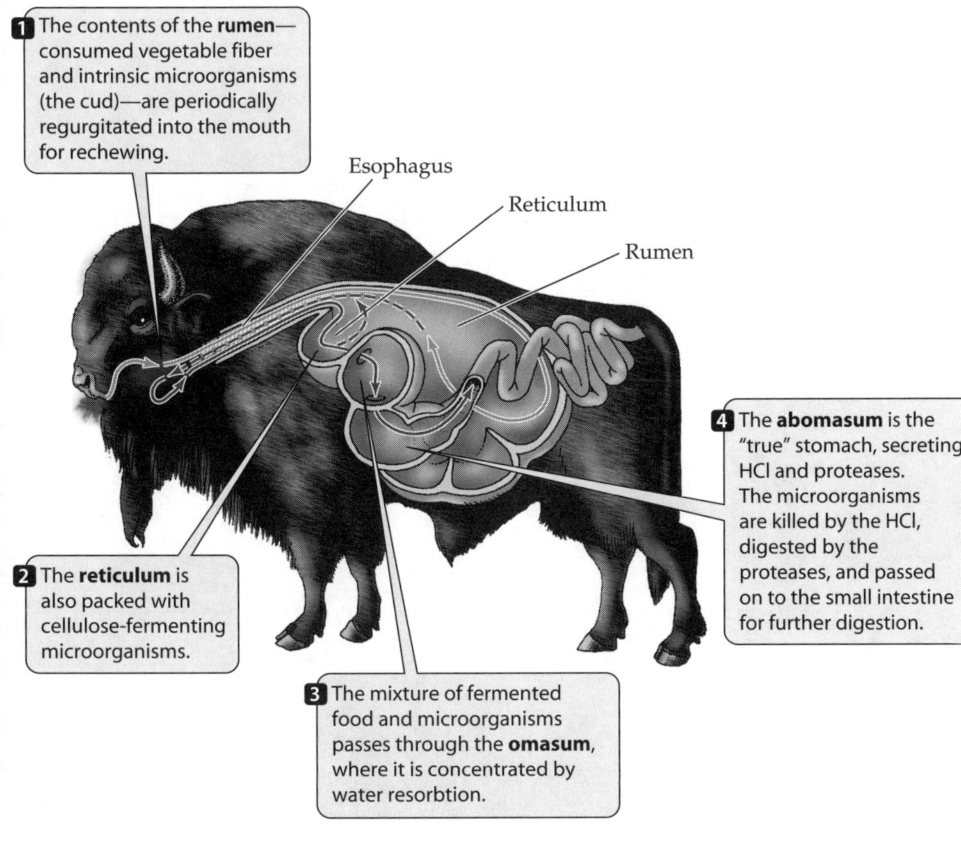
1 The contents of the **rumen**—consumed vegetable fiber and intrinsic microorganisms (the cud)—are periodically regurgitated into the mouth for rechewing.
Esophagus
Reticulum
Rumen
2 The **reticulum** is also packed with cellulose-fermenting microorganisms.
3 The mixture of fermented food and microorganisms passes through the **omasum**, where it is concentrated by water resorbtion.
4 The **abomasum** is the "true" stomach, secreting HCl and proteases. The microorganisms are killed by the HCl, digested by the proteases, and passed on to the small intestine for further digestion.

50.17 A Ruminant's Stomach *(Page 977)*

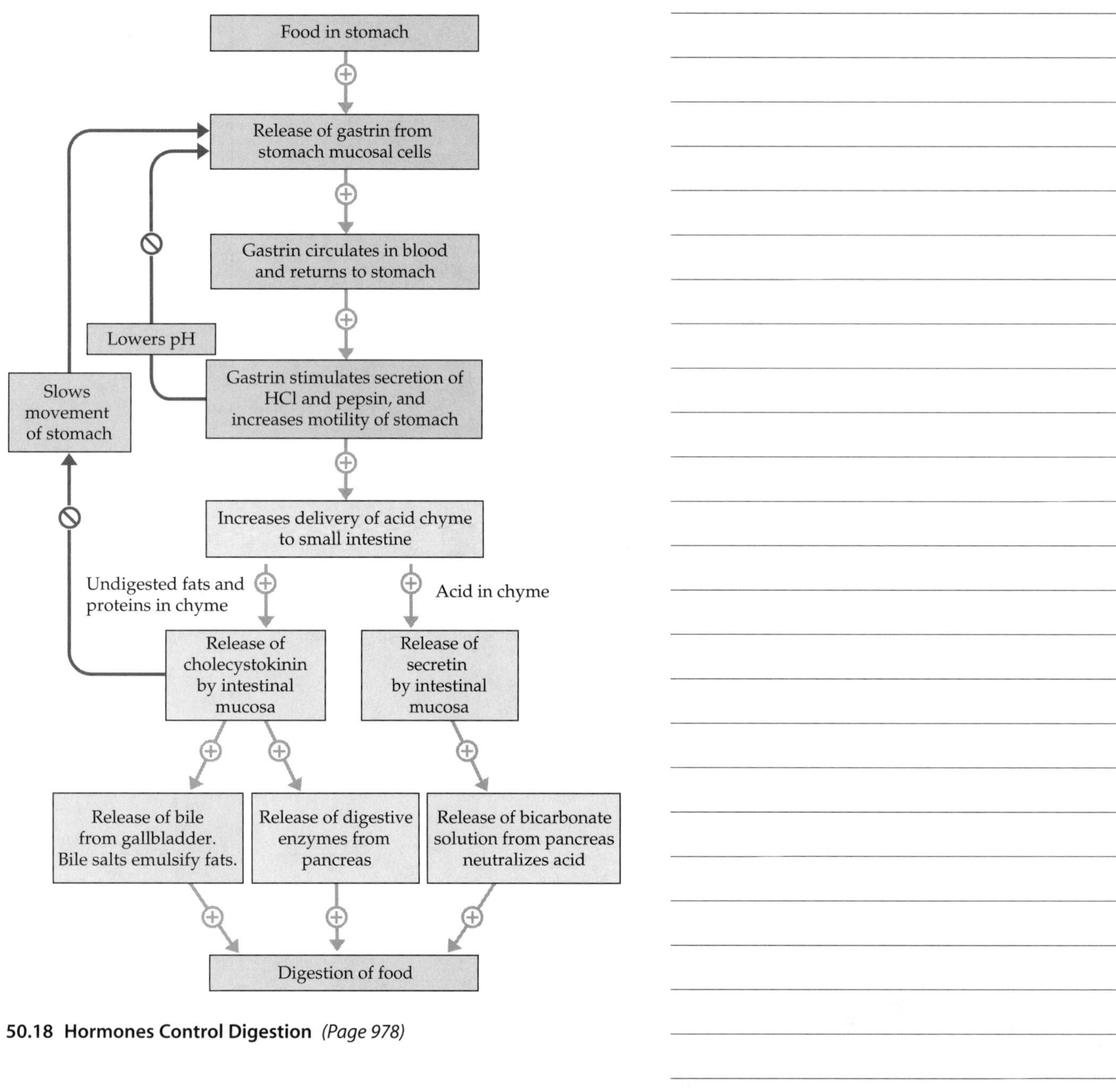
Food in stomach
Release of gastrin from stomach mucosal cells
Gastrin circulates in blood and returns to stomach
Lowers pH
Slows movement of stomach
Gastrin stimulates secretion of HCl and pepsin, and increases motility of stomach
Increases delivery of acid chyme to small intestine
Undigested fats and proteins in chyme
Acid in chyme
Release of cholecystokinin by intestinal mucosa
Release of secretin by intestinal mucosa
Release of bile from gallbladder. Bile salts emulsify fats.
Release of digestive enzymes from pancreas
Release of bicarbonate solution from pancreas neutralizes acid
Digestion of food

50.18 Hormones Control Digestion *(Page 978)*

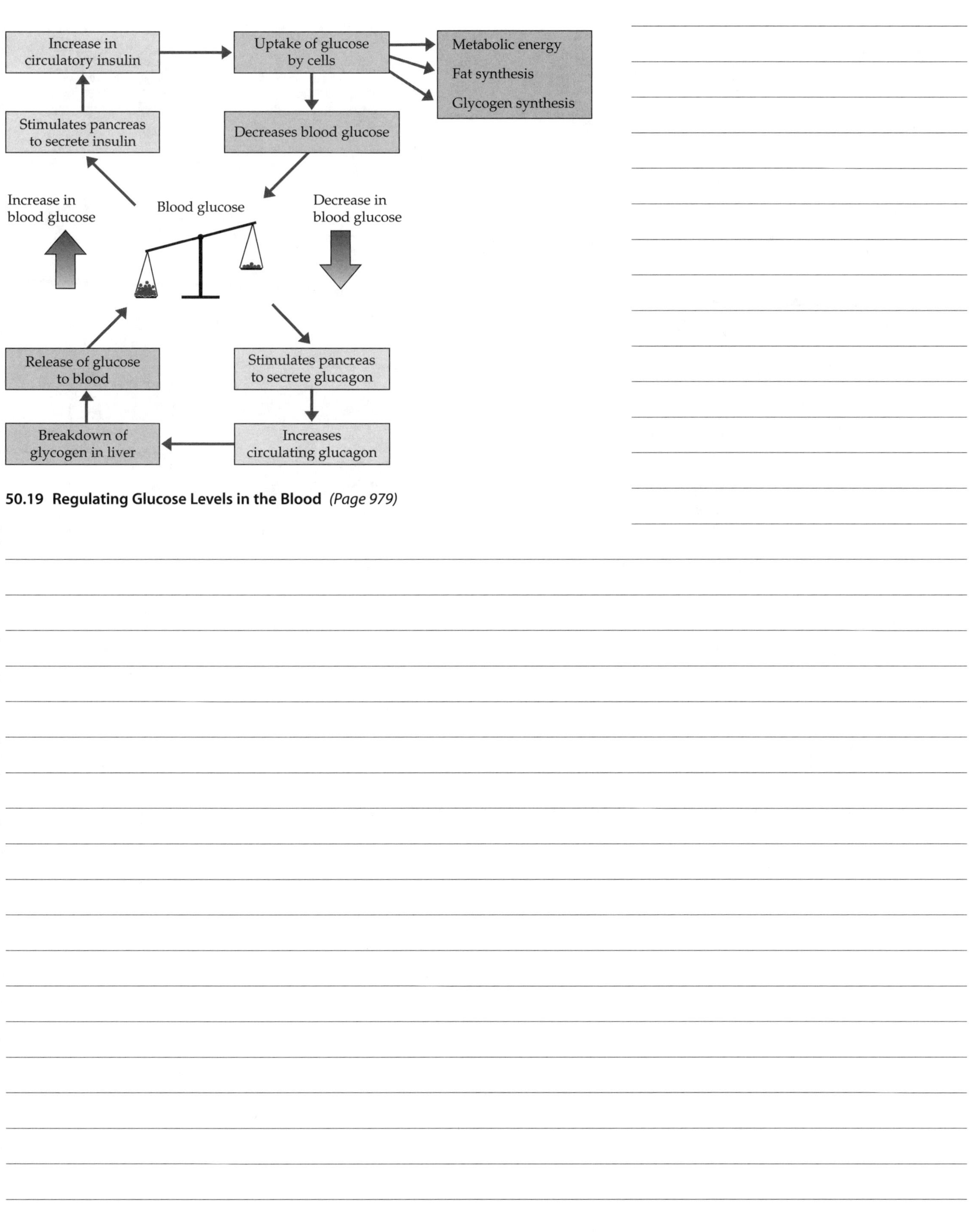
Increase in circulatory insulin
Uptake of glucose by cells
Metabolic energy
Fat synthesis
Glycogen synthesis
Stimulates pancreas to secrete insulin
Decreases blood glucose
Increase in blood glucose
Blood glucose
Decrease in blood glucose
Release of glucose to blood
Stimulates pancreas to secrete glucagon
Breakdown of glycogen in liver
Increases circulating glucagon

50.19 Regulating Glucose Levels in the Blood *(Page 979)*

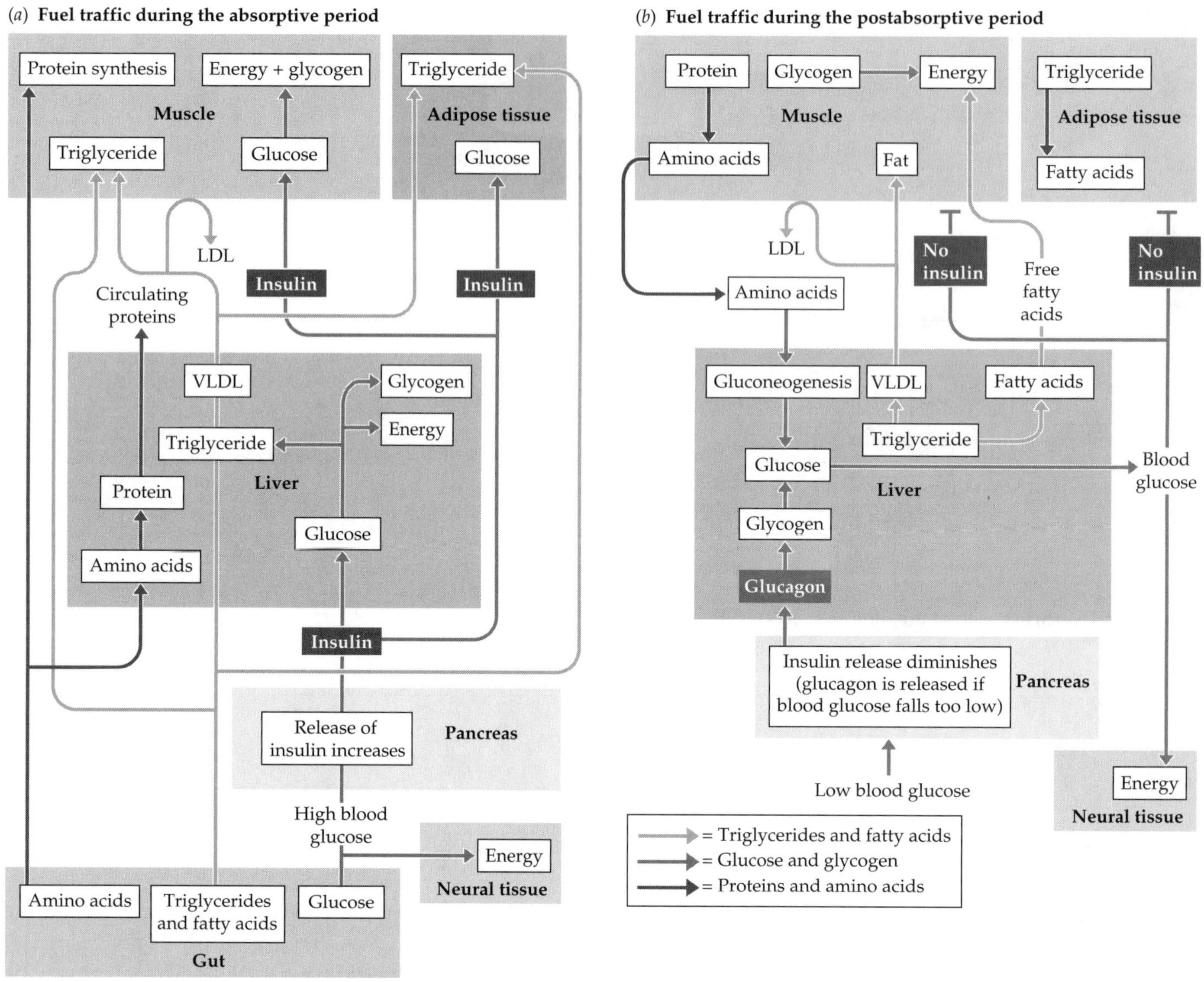
(a) Fuel traffic during the absorptive period
Protein synthesis
Energy + glycogen
Muscle
Triglyceride
Glucose
Triglyceride
Adipose tissue
Glucose
LDL
Insulin
Insulin
Circulating proteins
VLDL
Glycogen
Energy
Triglyceride
Liver
Protein
Glucose
Amino acids
Insulin
Release of insulin increases
Pancreas
High blood glucose
Energy
Neural tissue
Amino acids
Triglycerides and fatty acids
Glucose
Gut
(b) Fuel traffic during the postabsorptive period
Protein
Glycogen
Energy
Muscle
Amino acids
Fat
Triglyceride
Adipose tissue
Fatty acids
LDL
No insulin
Free fatty acids
No insulin
Amino acids
Gluconeogenesis
VLDL
Fatty acids
Triglyceride
Glucose
Blood glucose
Liver
Glycogen
Glucagon
Insulin release diminishes (glucagon is released if blood glucose falls too low)
Pancreas
Low blood glucose
Energy
Neural tissue
= Triglycerides and fatty acids
= Glucose and glycogen
= Proteins and amino acids

50.20 Fuel Molecule Traffic during the Absorptive and Postabsorptive Periods *(Page 980)*

51 Salt and Water Balance and Nitrogen Excretion

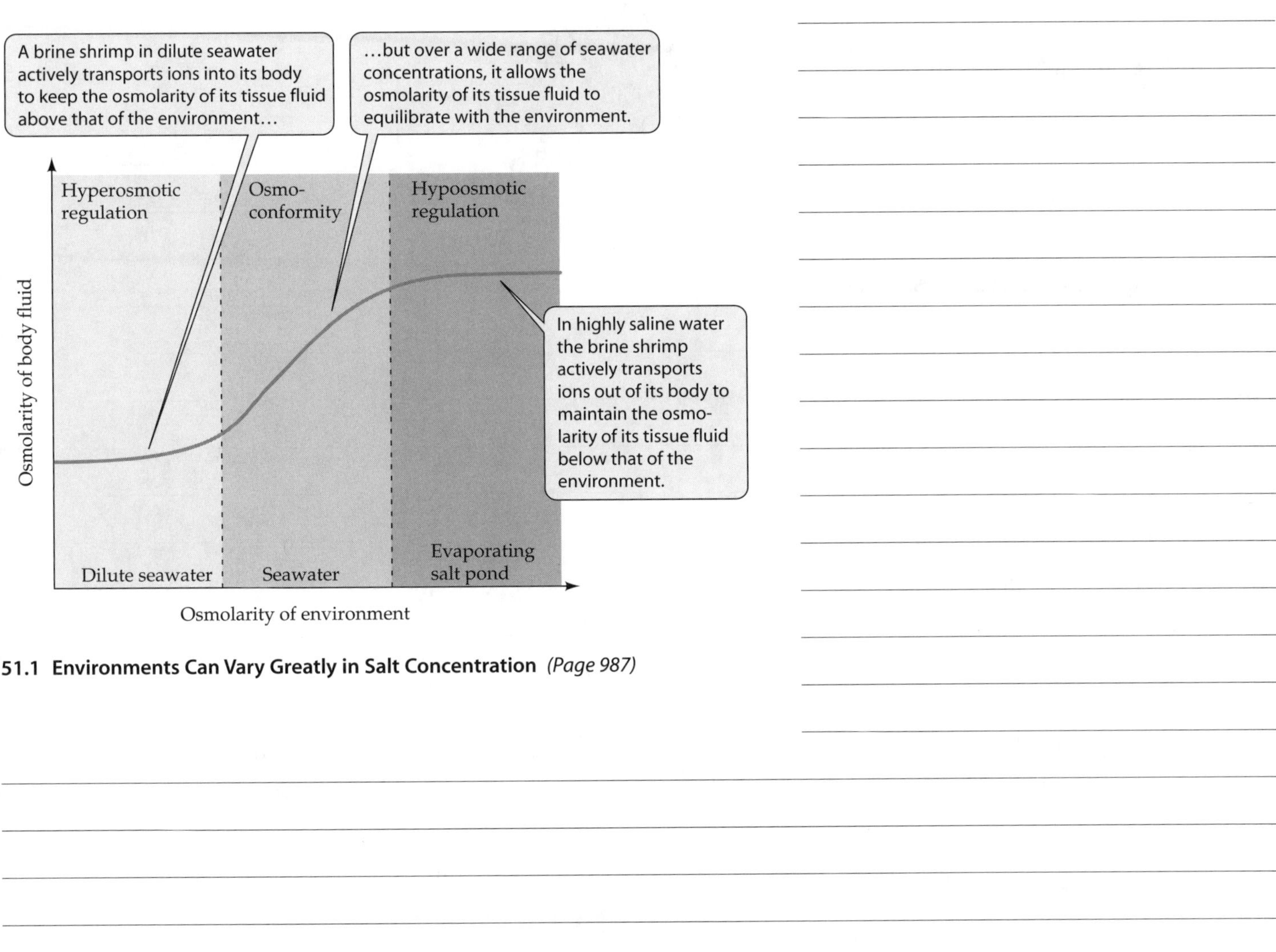
A brine shrimp in dilute seawater actively transports ions into its body to keep the osmolarity of its tissue fluid above that of the environment...
...but over a wide range of seawater concentrations, it allows the osmolarity of its tissue fluid to equilibrate with the environment.
Hyperosmotic regulation
Osmo-conformity
Hypoosmotic regulation
In highly saline water the brine shrimp actively transports ions out of its body to maintain the osmolarity of its tissue fluid below that of the environment.
Osmolarity of body fluid
Dilute seawater
Seawater
Evaporating salt pond
Osmolarity of environment

51.1 Environments Can Vary Greatly in Salt Concentration *(Page 987)*

(a)

Salt glands are located in depressions in the skull above the eyes.

The functional unit of a salt gland consists of secretory tubules that drain into a central canal.

Na^+ ions are carried from the blood to secretory tubules via active transport. Cl^- ions follow. The central canal carries salt solution to the nasal cavity.

Skull

Eye

Nasal cavity

Artery

Vein

Central canal

Secretory tubule

Artery

Vein

Capillaries

NaCl

Central canal

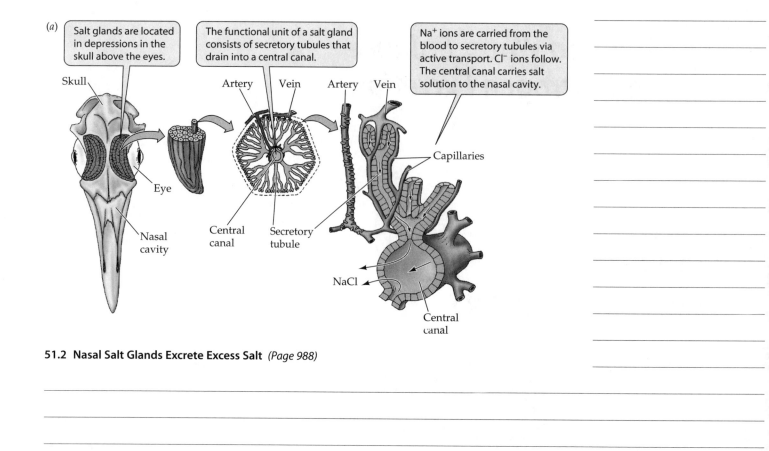

51.2 Nasal Salt Glands Excrete Excess Salt *(Page 988)*

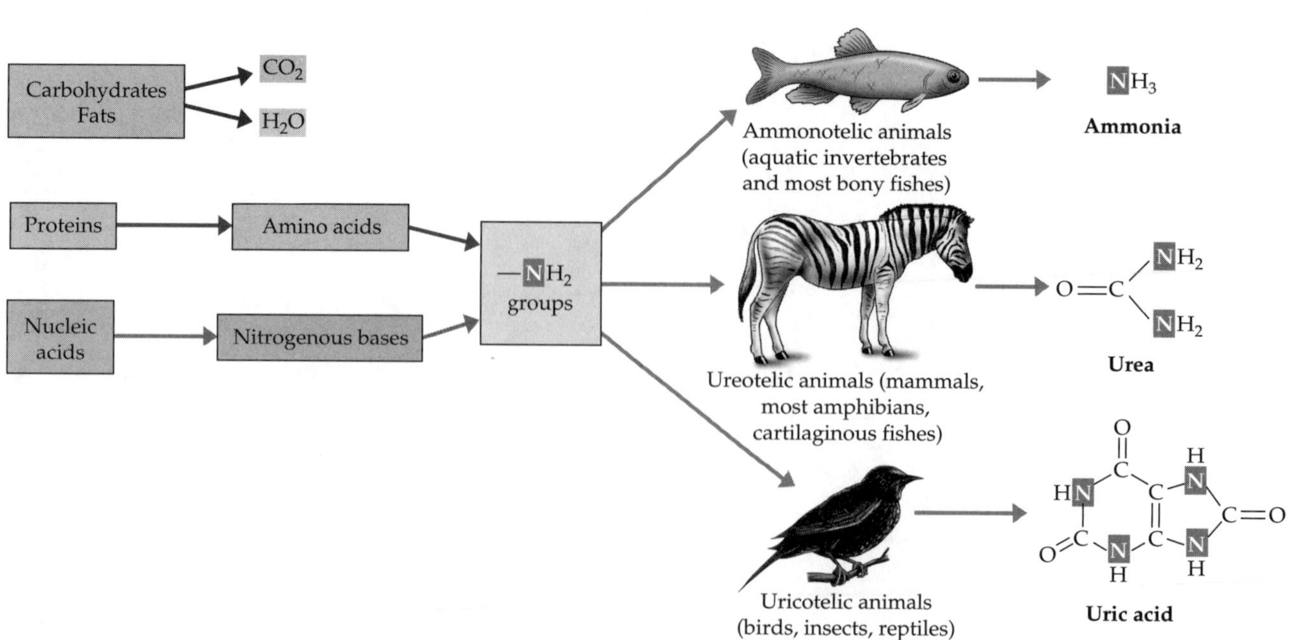
Carbohydrates
Fats
CO_2
H_2O
Proteins
Amino acids
Nucleic acids
Nitrogenous bases
$-NH_2$ groups
Ammonotelic animals (aquatic invertebrates and most bony fishes)
NH_3
Ammonia
Ureotelic animals (mammals, most amphibians, cartilaginous fishes)
Urea
Uricotelic animals (birds, insects, reptiles)
Uric acid

51.3 Waste Products of Metabolism *(Page 988)*

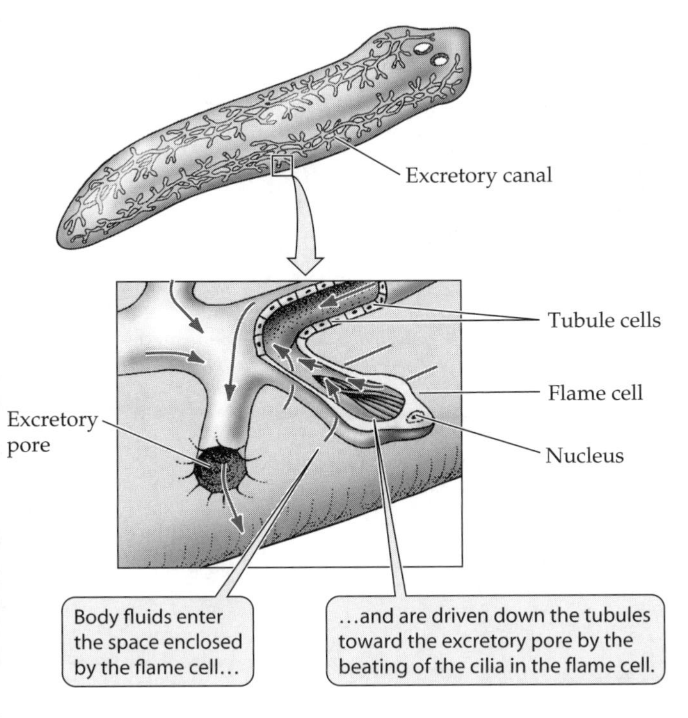
Excretory canal
Tubule cells
Flame cell
Excretory pore
Nucleus
Body fluids enter the space enclosed by the flame cell...
...and are driven down the tubules toward the excretory pore by the beating of the cilia in the flame cell.

51.4 Protonephridia in Flatworms *(Page 989)*

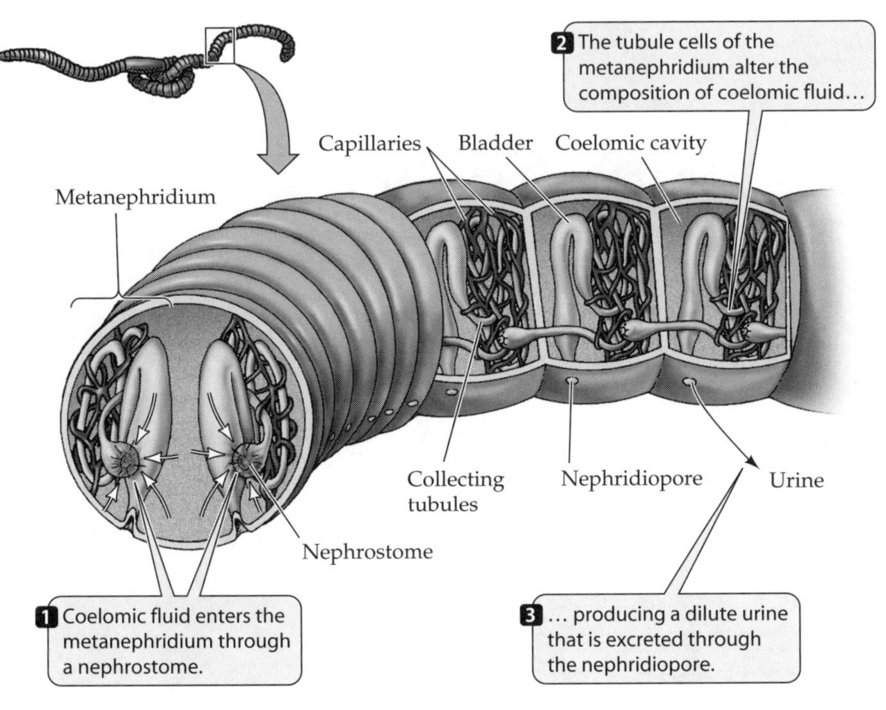
2 The tubule cells of the metanephridium alter the composition of coelomic fluid...
Capillaries
Bladder
Coelomic cavity
Metanephridium
Collecting tubules
Nephridiopore
Urine
Nephrostome
1 Coelomic fluid enters the metanephridium through a nephrostome.
3 ... producing a dilute urine that is excreted through the nephridiopore.

51.5 Metanephridia in Earthworms *(Page 990)*

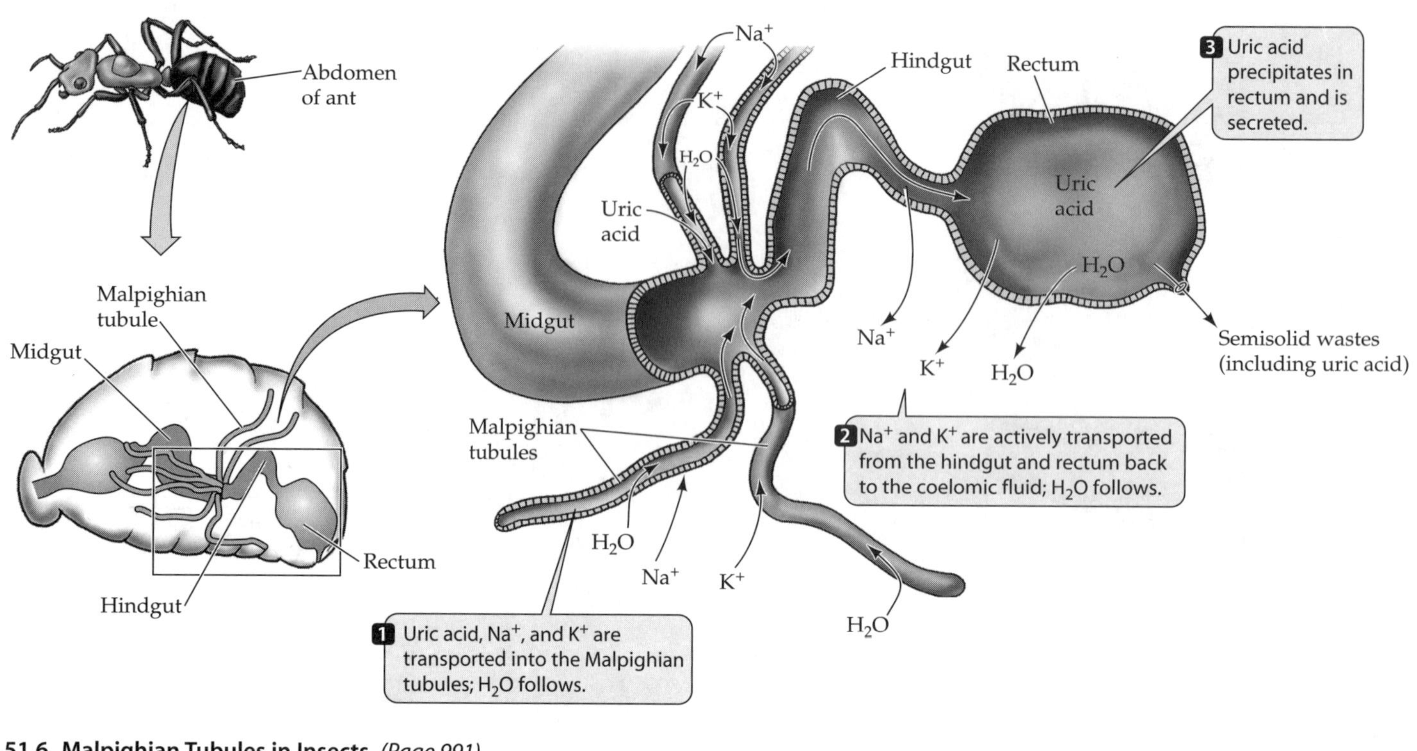
Abdomen of ant
Malpighian tubule
Midgut
Rectum
Hindgut
Na^+
K^+
H_2O
Uric acid
Hindgut
Rectum
3 Uric acid precipitates in rectum and is secreted.
Uric acid
H_2O
Midgut
Na^+
K^+
H_2O
Semisolid wastes (including uric acid)
Malpighian tubules
2 Na^+ and K^+ are actively transported from the hindgut and rectum back to the coelomic fluid; H_2O follows.
H_2O
Na^+
K^+
H_2O
1 Uric acid, Na^+, and K^+ are transported into the Malpighian tubules; H_2O follows.

51.6 Malpighian Tubules in Insects *(Page 991)*

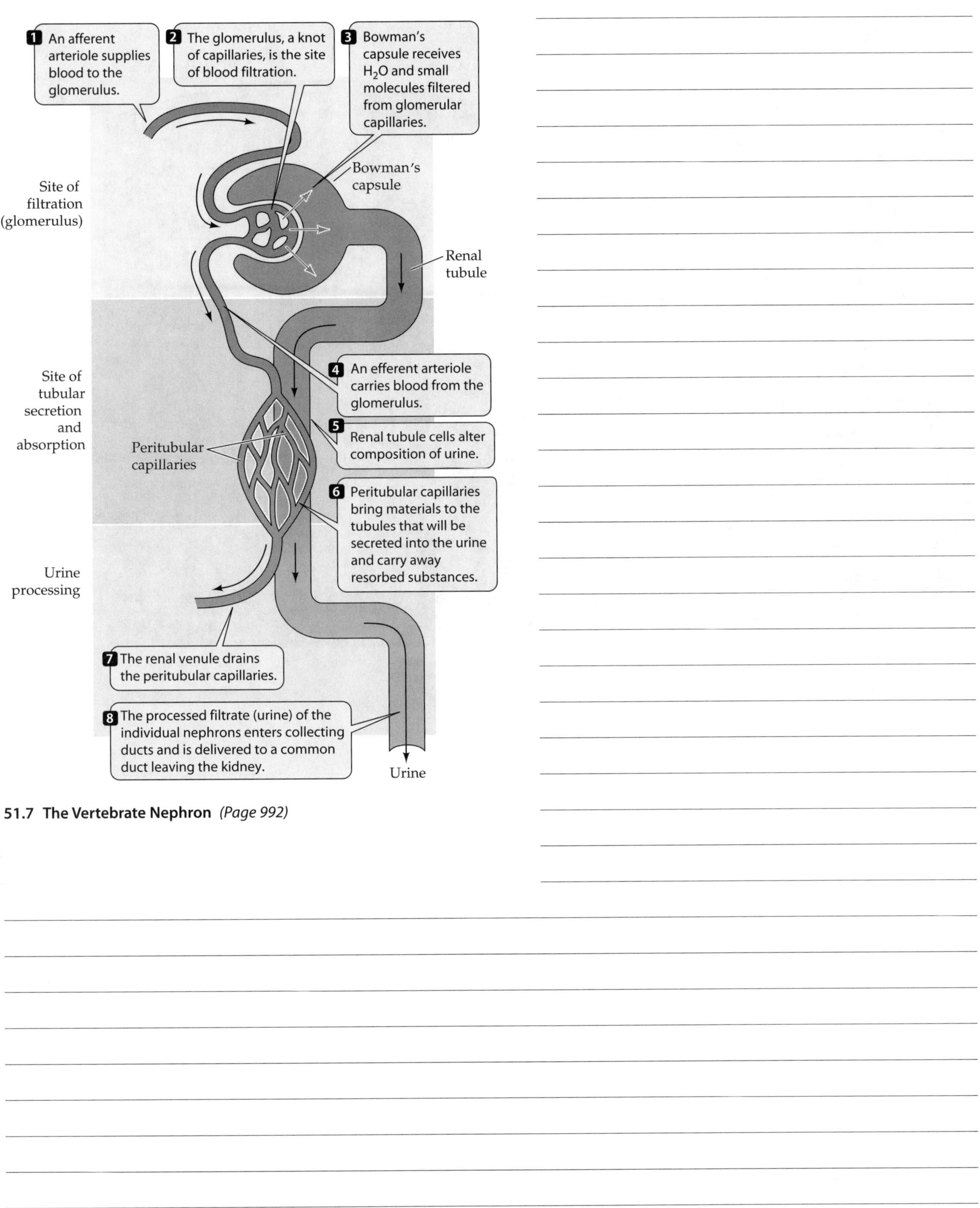
1 An afferent arteriole supplies blood to the glomerulus.
2 The glomerulus, a knot of capillaries, is the site of blood filtration.
3 Bowman's capsule receives H_2O and small molecules filtered from glomerular capillaries.
Site of filtration (glomerulus)
Bowman's capsule
Renal tubule
Site of tubular secretion and absorption
4 An efferent arteriole carries blood from the glomerulus.
5 Renal tubule cells alter composition of urine.
Peritubular capillaries
6 Peritubular capillaries bring materials to the tubules that will be secreted into the urine and carry away resorbed substances.
Urine processing
7 The renal venule drains the peritubular capillaries.
8 The processed filtrate (urine) of the individual nephrons enters collecting ducts and is delivered to a common duct leaving the kidney.
Urine

51.7 The Vertebrate Nephron *(Page 992)*

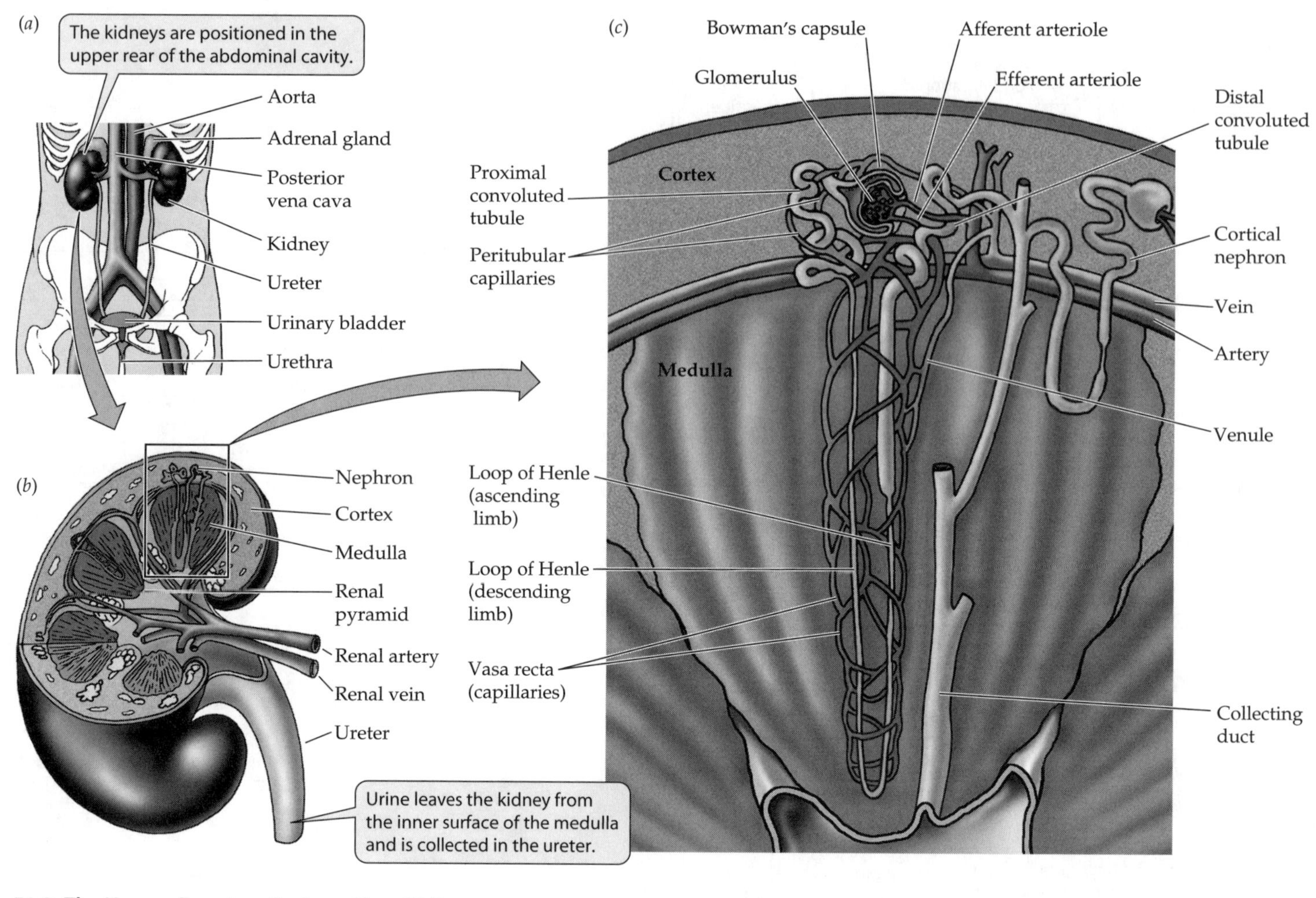
(a)
The kidneys are positioned in the upper rear of the abdominal cavity.
Aorta
Adrenal gland
Posterior vena cava
Kidney
Ureter
Urinary bladder
Urethra
(b)
Nephron
Cortex
Medulla
Renal pyramid
Renal artery
Renal vein
Ureter
Urine leaves the kidney from the inner surface of the medulla and is collected in the ureter.
(c)
Bowman's capsule
Afferent arteriole
Glomerulus
Efferent arteriole
Distal convoluted tubule
Cortex
Proximal convoluted tubule
Peritubular capillaries
Cortical nephron
Vein
Artery
Medulla
Venule
Loop of Henle (ascending limb)
Loop of Henle (descending limb)
Vasa recta (capillaries)
Collecting duct

51.9 The Human Excretory System *(Page 994)*

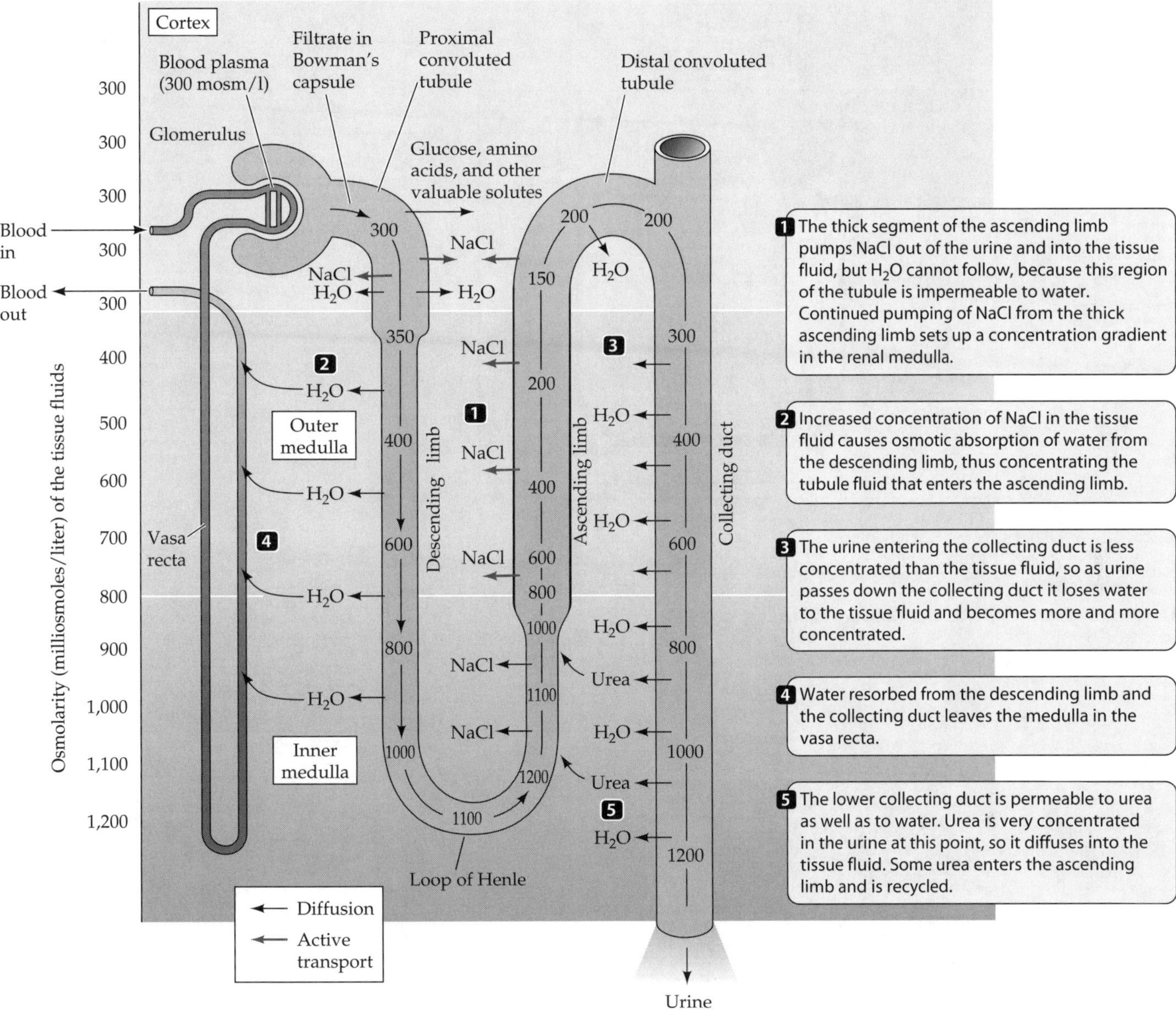

Cortex
Blood plasma (300 mosm/l)
Filtrate in Bowman's capsule
Proximal convoluted tubule
Distal convoluted tubule
Glomerulus
Glucose, amino acids, and other valuable solutes
Blood in
Blood out
Osmolarity (milliosmoles/liter) of the tissue fluids
300
300
300
300
300
400
500
600
700
800
900
1,000
1,100
1,200
NaCl
H_2O
Outer medulla
Inner medulla
Vasa recta
Descending limb
Ascending limb
Collecting duct
Urea
Loop of Henle
Urine
Diffusion
Active transport
1 The thick segment of the ascending limb pumps NaCl out of the urine and into the tissue fluid, but H_2O cannot follow, because this region of the tubule is impermeable to water. Continued pumping of NaCl from the thick ascending limb sets up a concentration gradient in the renal medulla.
2 Increased concentration of NaCl in the tissue fluid causes osmotic absorption of water from the descending limb, thus concentrating the tubule fluid that enters the ascending limb.
3 The urine entering the collecting duct is less concentrated than the tissue fluid, so as urine passes down the collecting duct it loses water to the tissue fluid and becomes more and more concentrated.
4 Water resorbed from the descending limb and the collecting duct leaves the medulla in the vasa recta.
5 The lower collecting duct is permeable to urea as well as to water. Urea is very concentrated in the urine at this point, so it diffuses into the tissue fluid. Some urea enters the ascending limb and is recycled.

51.10 Concentrating the Urine *(Page 996)*

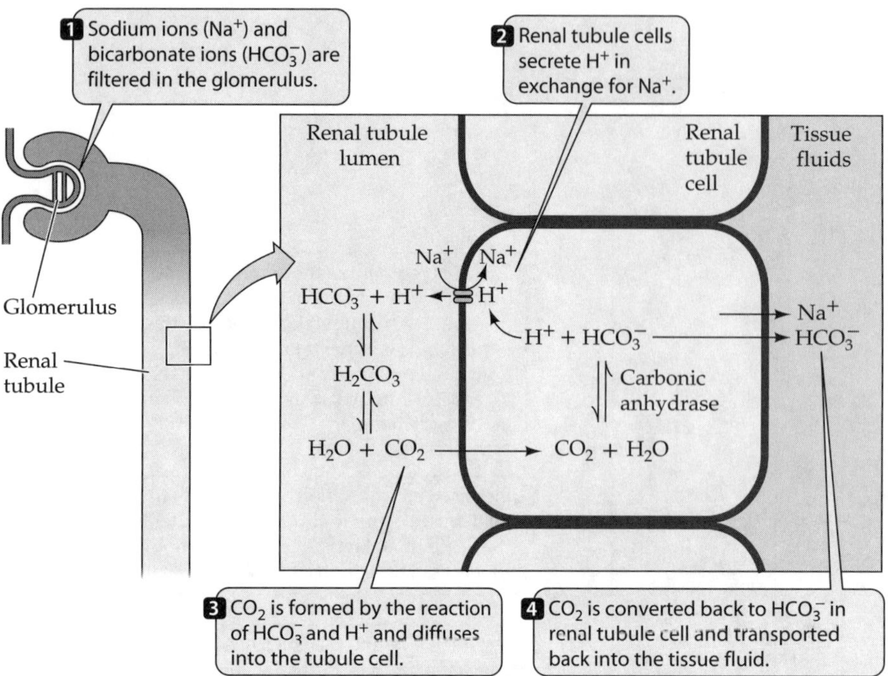
1 Sodium ions (Na^+) and bicarbonate ions (HCO_3^-) are filtered in the glomerulus.
2 Renal tubule cells secrete H^+ in exchange for Na^+.
Glomerulus
Renal tubule
Renal tubule lumen
Renal tubule cell
Tissue fluids
Na^+
Na^+
$HCO_3^- + H^+$
H^+
H_2CO_3
$H_2O + CO_2$
$H^+ + HCO_3^-$
Carbonic anhydrase
$CO_2 + H_2O$
Na^+
HCO_3^-
3 CO_2 is formed by the reaction of HCO_3^- and H^+ and diffuses into the tubule cell.
4 CO_2 is converted back to HCO_3^- in renal tubule cell and transported back into the tissue fluid.

51.12 The Kidney Excretes Acids and Conserves Bases *(Page 998)*

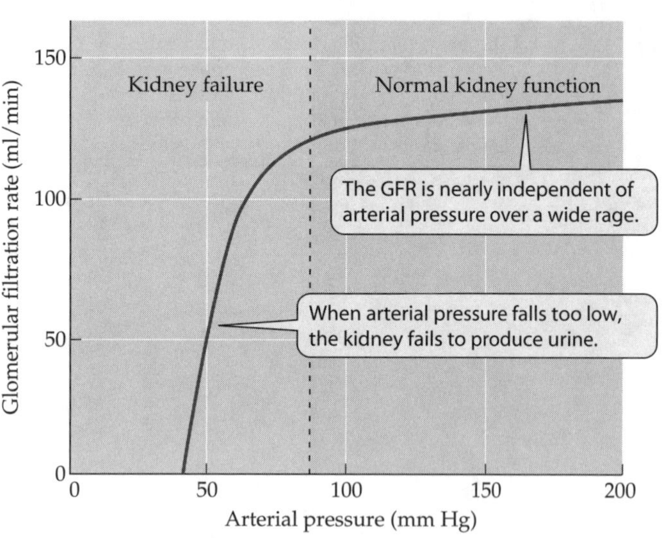
Glomerular filtration rate (ml/min)
150
100
50
0
Kidney failure
Normal kidney function
The GFR is nearly independent of arterial pressure over a wide rage.
When arterial pressure falls too low, the kidney fails to produce urine.
0
50
100
150
200
Arterial pressure (mm Hg)

51.13 Maintaining the Glomerular Filtration Rate *(Page 998)*

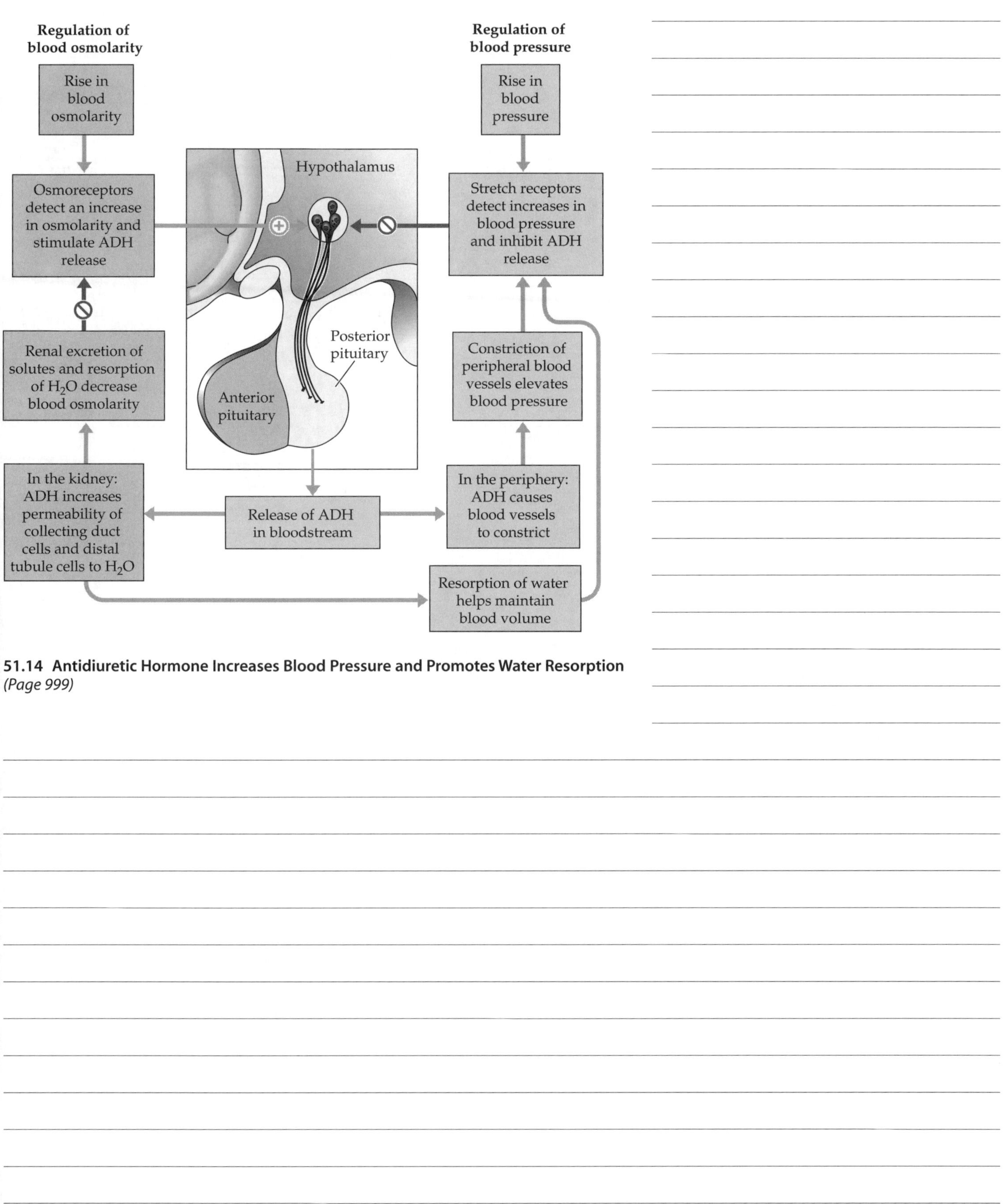
Regulation of blood osmolarity
Rise in blood osmolarity
Osmoreceptors detect an increase in osmolarity and stimulate ADH release
Renal excretion of solutes and resorption of H_2O decrease blood osmolarity
In the kidney: ADH increases permeability of collecting duct cells and distal tubule cells to H_2O
Hypothalamus
Posterior pituitary
Anterior pituitary
Release of ADH in bloodstream
Regulation of blood pressure
Rise in blood pressure
Stretch receptors detect increases in blood pressure and inhibit ADH release
Constriction of peripheral blood vessels elevates blood pressure
In the periphery: ADH causes blood vessels to constrict
Resorption of water helps maintain blood volume

51.14 Antidiuretic Hormone Increases Blood Pressure and Promotes Water Resorption
(Page 999)

52 Animal Behavior

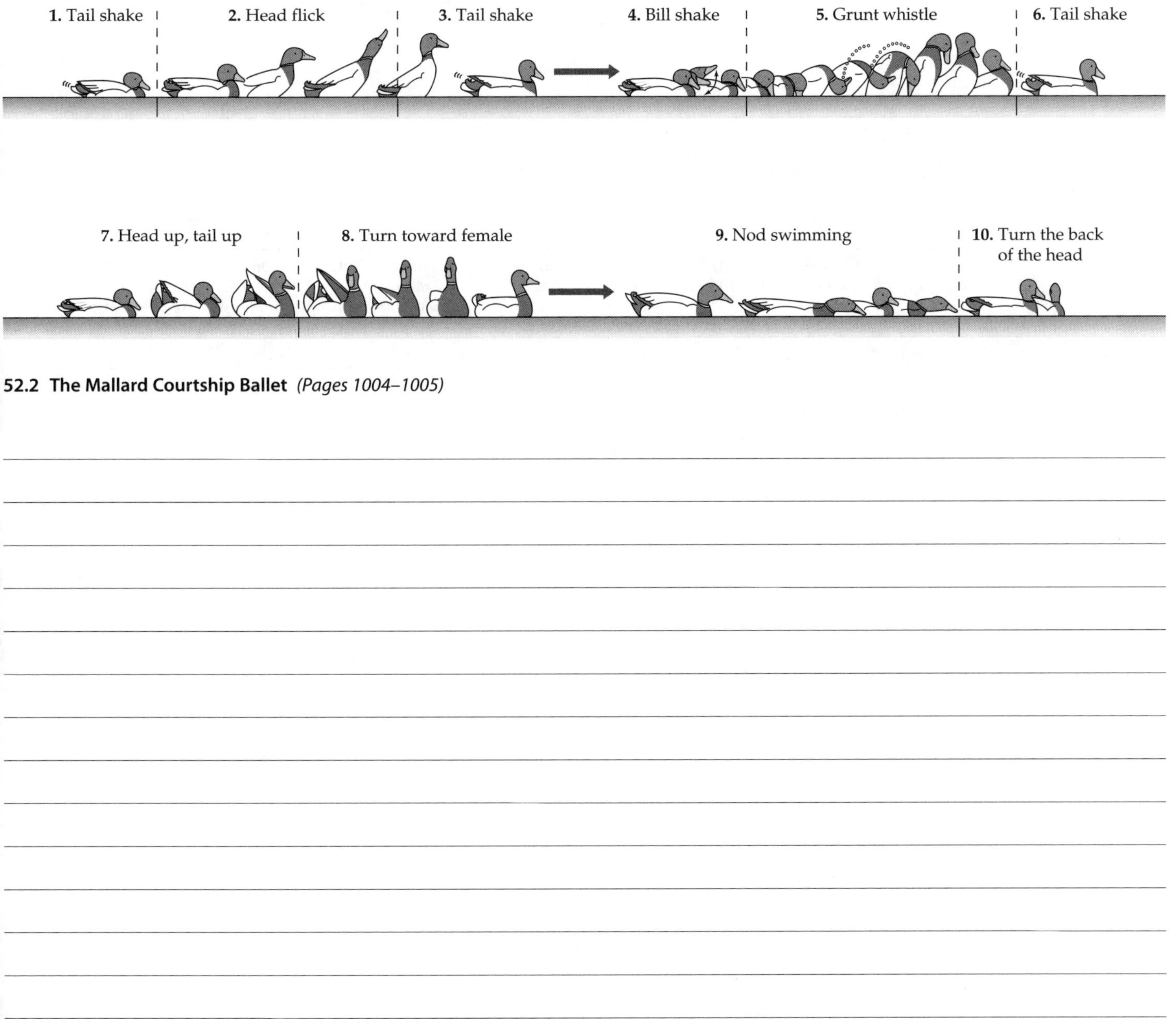

1. Tail shake
2. Head flick
3. Tail shake
4. Bill shake
5. Grunt whistle
6. Tail shake
7. Head up, tail up
8. Turn toward female
9. Nod swimming
10. Turn the back of the head

52.2 The Mallard Courtship Ballet *(Pages 1004–1005)*

EXPERIMENT

Question: What characteristics of the herring gull parent release begging responses from their chicks?

METHOD Paper cut-out models of gull heads with many variations were presented to chicks and their begging pecks in response to each model were counted.

RESULTS

Stimulus:
Presence of red patch on bill
Head shape
Head color
Bill alone
Bill shape
Control:
Pecking responses (% of control)
100
62
35
91
125
98
99
94
64
174
A red dot is more important than a realistic profile.
Head color and shape have little effect on the ability of a red dot to stimulate the begging response.
A red dot elicits a response, even without a head.
Even without a red dot, a long, thin bill elicits the strongest pecking response.

Conclusion: A red dot and/or a long, thin bill release begging responses. Head shape and color have little or no influence.

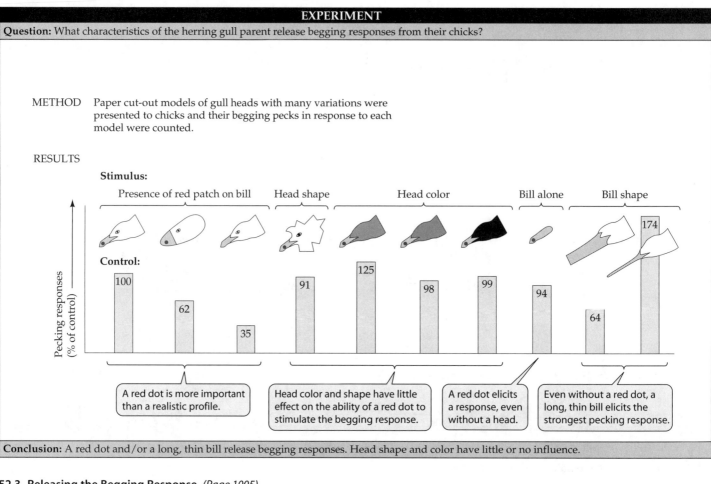

52.3 Releasing the Begging Response *(Page 1005)*

EXPERIMENT

Question: Does a wasp learn to locate her nest by visual cues?

METHOD Surround nest entrance with moveable visual cues, and move them to another location after the wasp leaves the nest and has surveyed her surroundings.

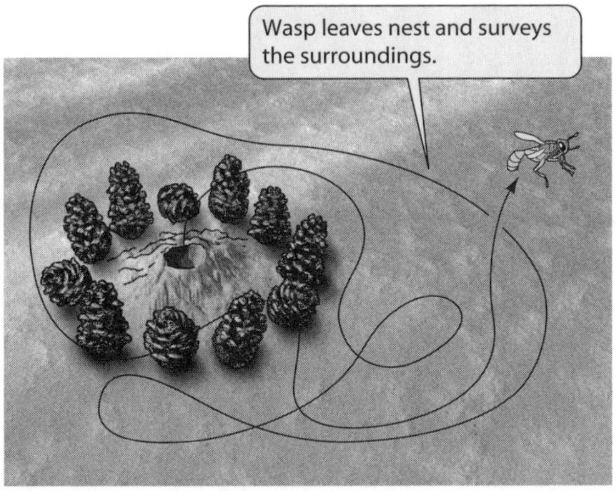

Wasp leaves nest and surveys the surroundings.
Move cues
Wasp looks for nest entrance in relation to visual cues.

RESULTS

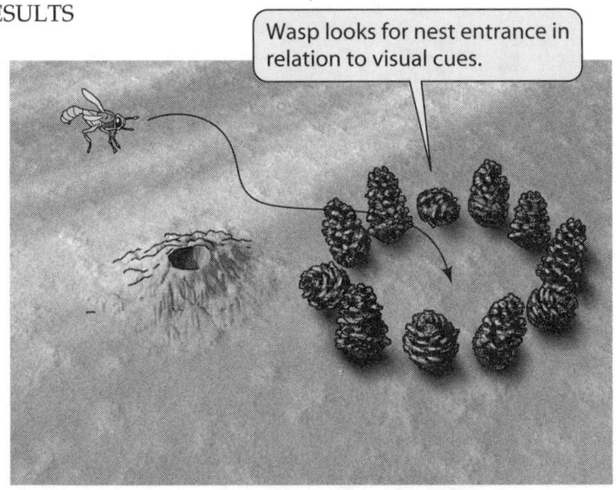

Conclusion: A wasp learns to use objects in its environment to locate her nest.

52.4 Spatial Learning *(Page 1006)*

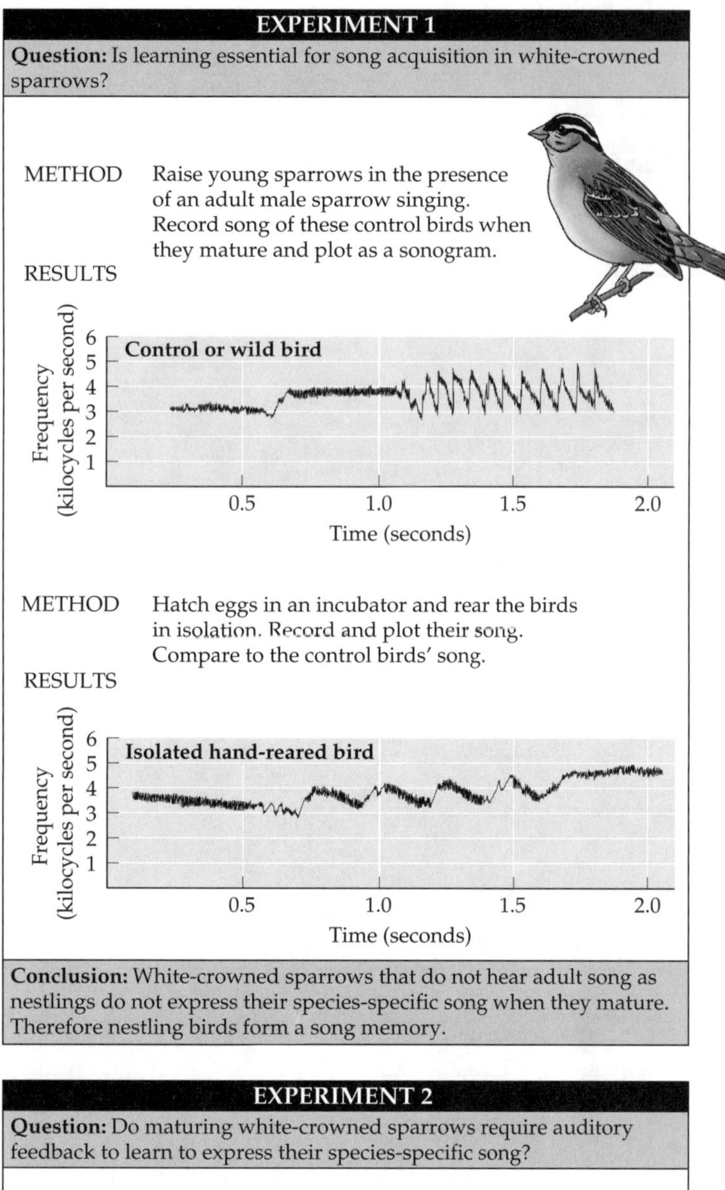

EXPERIMENT 1
Question: Is learning essential for song acquisition in white-crowned sparrows?
METHOD Raise young sparrows in the presence of an adult male sparrow singing. Record song of these control birds when they mature and plot as a sonogram.
RESULTS
Frequency (kilocycles per second)
6
5
4
3
2
1
Control or wild bird
0.5
1.0
1.5
2.0
Time (seconds)
METHOD Hatch eggs in an incubator and rear the birds in isolation. Record and plot their song. Compare to the control birds' song.
RESULTS
Frequency (kilocycles per second)
Isolated hand-reared bird
Time (seconds)
Conclusion: White-crowned sparrows that do not hear adult song as nestlings do not express their species-specific song when they mature. Therefore nestling birds form a song memory.

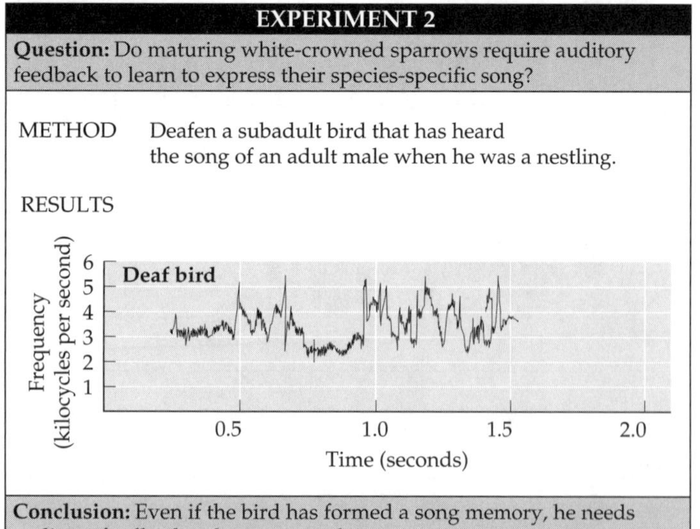

EXPERIMENT 2
Question: Do maturing white-crowned sparrows require auditory feedback to learn to express their species-specific song?
METHOD Deafen a subadult bird that has heard the song of an adult male when he was a nestling.
RESULTS
Frequency (kilocycles per second)
6
5
4
3
2
1
Deaf bird
0.5
1.0
1.5
2.0
Time (seconds)
Conclusion: Even if the bird has formed a song memory, he needs auditory feedback to learn to match it.

52.6 Two Critical Periods for Song Learning *(Page 1007)*

EXPERIMENT

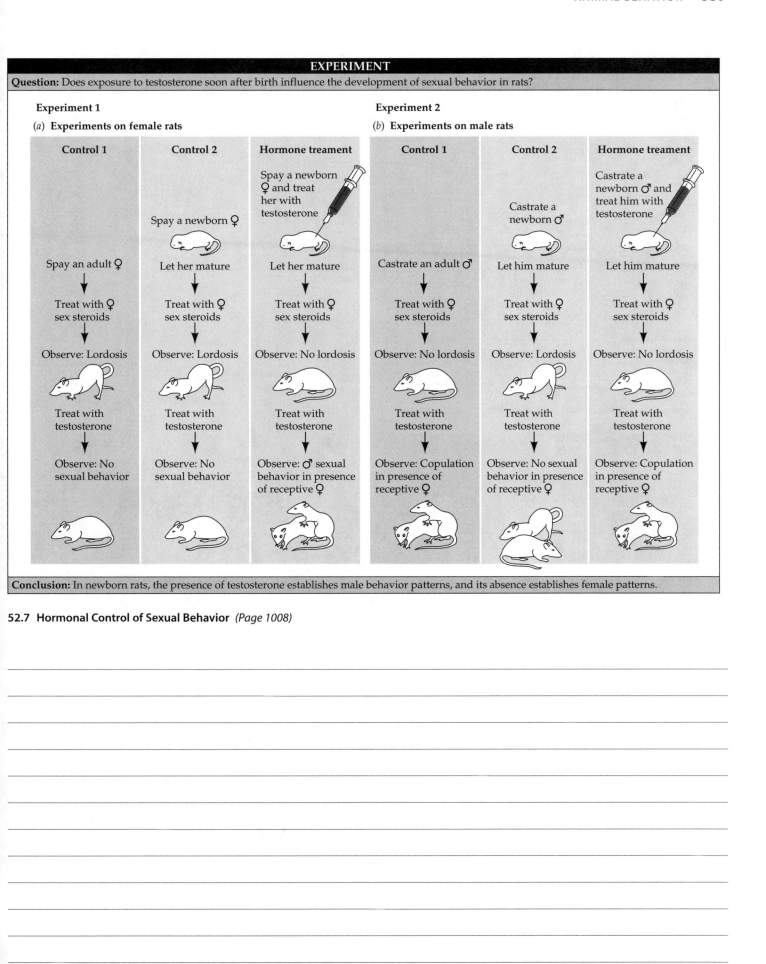

Question: Does exposure to testosterone soon after birth influence the development of sexual behavior in rats?

Experiment 1

(*a*) **Experiments on female rats**

Control 1	Control 2	Hormone treament
	Spay a newborn ♀	Spay a newborn ♀ and treat her with testosterone
Spay an adult ♀	Let her mature	Let her mature
Treat with ♀ sex steroids	Treat with ♀ sex steroids	Treat with ♀ sex steroids
Observe: Lordosis	Observe: Lordosis	Observe: No lordosis
Treat with testosterone	Treat with testosterone	Treat with testosterone
Observe: No sexual behavior	Observe: No sexual behavior	Observe: ♂ sexual behavior in presence of receptive ♀

Experiment 2

(*b*) **Experiments on male rats**

Control 1	Control 2	Hormone treament
	Castrate a newborn ♂	Castrate a newborn ♂ and treat him with testosterone
Castrate an adult ♂	Let him mature	Let him mature
Treat with ♀ sex steroids	Treat with ♀ sex steroids	Treat with ♀ sex steroids
Observe: No lordosis	Observe: Lordosis	Observe: No lordosis
Treat with testosterone	Treat with testosterone	Treat with testosterone
Observe: Copulation in presence of receptive ♀	Observe: No sexual behavior in presence of receptive ♀	Observe: Copulation in presence of receptive ♀

Conclusion: In newborn rats, the presence of testosterone establishes male behavior patterns, and its absence establishes female patterns.

52.7 Hormonal Control of Sexual Behavior *(Page 1008)*

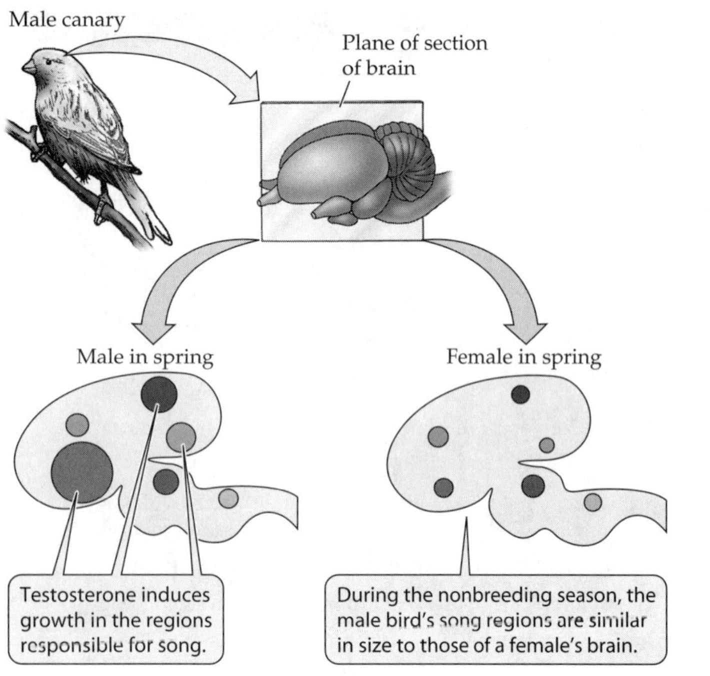
Male canary
Plane of section of brain
Male in spring
Female in spring
Testosterone induces growth in the regions responsible for song.
During the nonbreeding season, the male bird's song regions are similar in size to those of a female's brain.

52.8 Effects of Testosterone on Bird Brains *(Page 1009)*

Nonhygienic bees | Hygienic bees

Hygienic bees uncap cells and remove the dead larvae.

Parental generation ×

Genotype of females: *UURR* | *uurr*

Genotype of males: *UR* | *ur*

Gametes (male or female): *UR* | *ur*

uu is the "uncap" gene. *rr* is the "remove" gene.

F_1 (all nonhygienic)

Genotype of females: *UuRr*

Gametes produced by females: *UR* *Ur* *uR* *ur*

Backcross to males of hygienic strain: *ur*

F_2 generation females

Genotypes	*UuRr*	*Uurr*	*uuRr*	*uurr*
Behaviors	Nonhygienic			Hygienic

These bees are nonhygienic, but will remove dead larvae if cells are uncapped.

These bees are nonhygienic. They will uncap cells of dead larvae, but won't remove them.

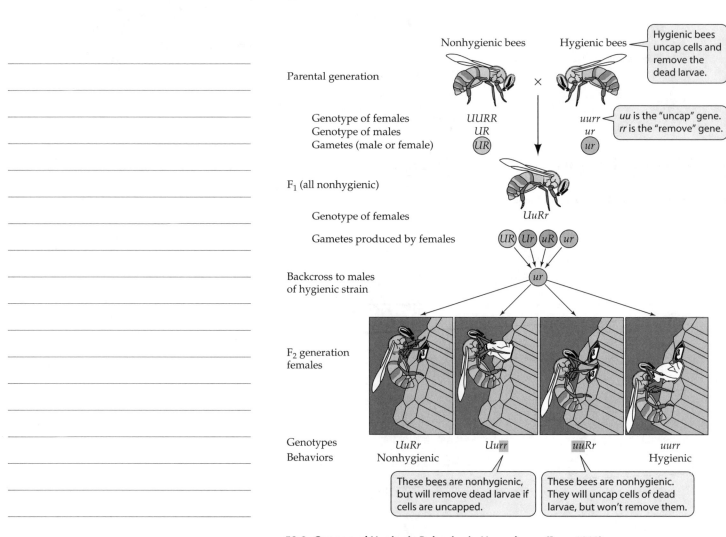

52.9 Genes and Hygienic Behavior in Honeybees *(Page 1010)*

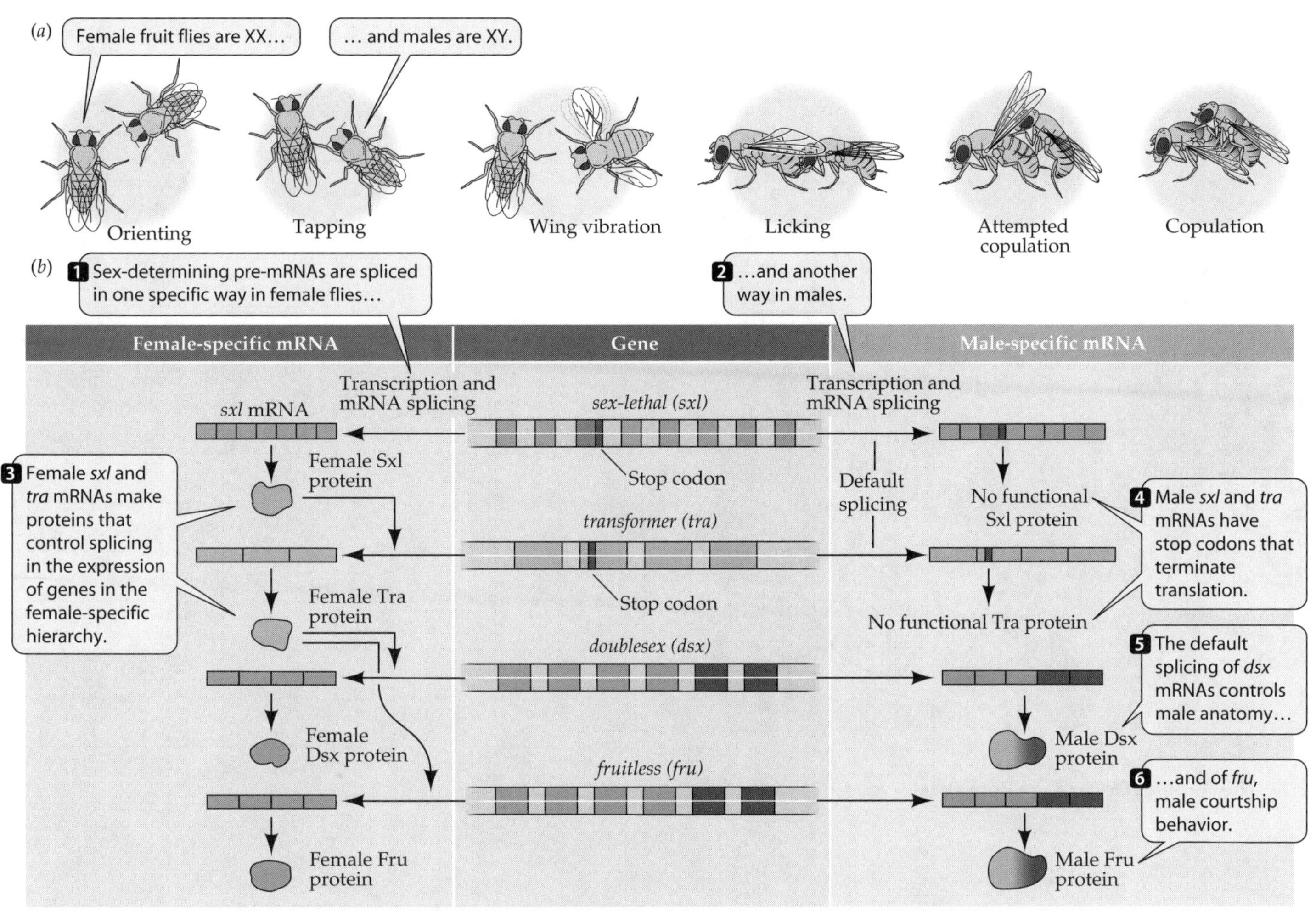

(a)
Female fruit flies are XX…
… and males are XY.
Orienting
Tapping
Wing vibration
Licking
Attempted copulation
Copulation
(b)
1 Sex-determining pre-mRNAs are spliced in one specific way in female flies…
2 …and another way in males.
Female-specific mRNA
Gene
Male-specific mRNA
Transcription and mRNA splicing
Transcription and mRNA splicing
sxl mRNA
sex-lethal (sxl)
Stop codon
Default splicing
Female Sxl protein
No functional Sxl protein
3 Female sxl and tra mRNAs make proteins that control splicing in the expression of genes in the female-specific hierarchy.
transformer (tra)
4 Male sxl and tra mRNAs have stop codons that terminate translation.
Stop codon
Female Tra protein
No functional Tra protein
doublesex (dsx)
5 The default splicing of dsx mRNAs controls male anatomy…
Female Dsx protein
Male Dsx protein
fruitless (fru)
6 …and of fru, male courtship behavior.
Female Fru protein
Male Fru protein

52.10 The *fruitless* Gene Controls Male Courtship Behavior in Fruit Flies *(Page 1011)*

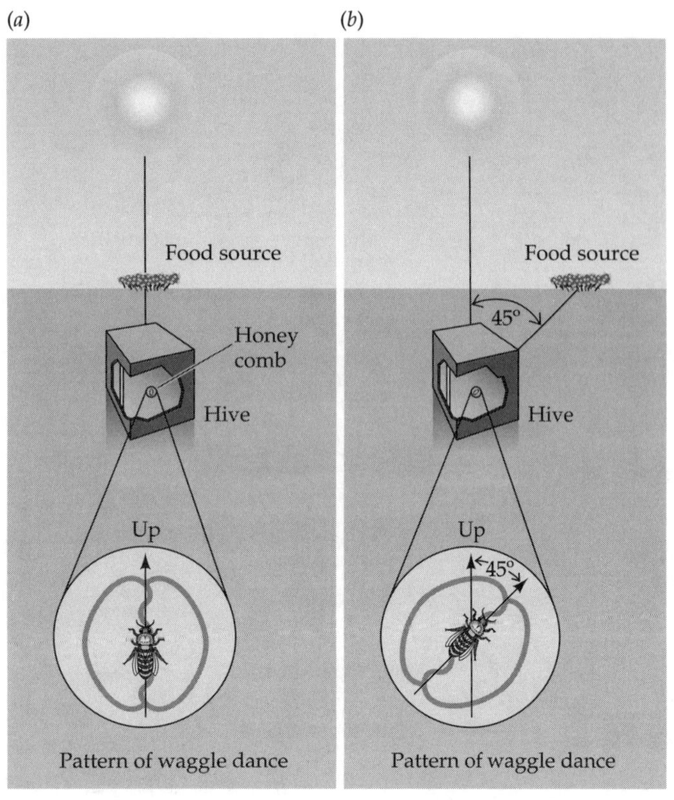
(a)
(b)
Food source
Honey comb
Hive
Up
Pattern of waggle dance
Food source
45°
Hive
Up
45°
Pattern of waggle dance

52.12 The Waggle Dance of the Honeybee *(Page 1013)*

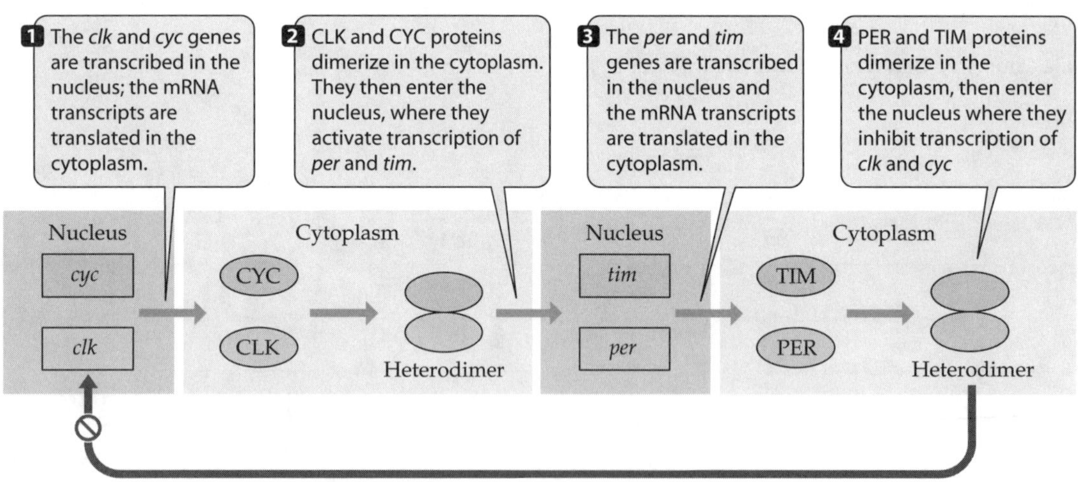
1 The *clk* and *cyc* genes are transcribed in the nucleus; the mRNA transcripts are translated in the cytoplasm.
2 CLK and CYC proteins dimerize in the cytoplasm. They then enter the nucleus, where they activate transcription of *per* and *tim*.
3 The *per* and *tim* genes are transcribed in the nucleus and the mRNA transcripts are translated in the cytoplasm.
4 PER and TIM proteins dimerize in the cytoplasm, then enter the nucleus where they inhibit transcription of *clk* and *cyc*
Nucleus
cyc
clk
Cytoplasm
CYC
CLK
Heterodimer
Nucleus
tim
per
Cytoplasm
TIM
PER
Heterodimer

52.15 Circadian Rhythms Are Generated by a Molecular Clock *(Page 1017)*

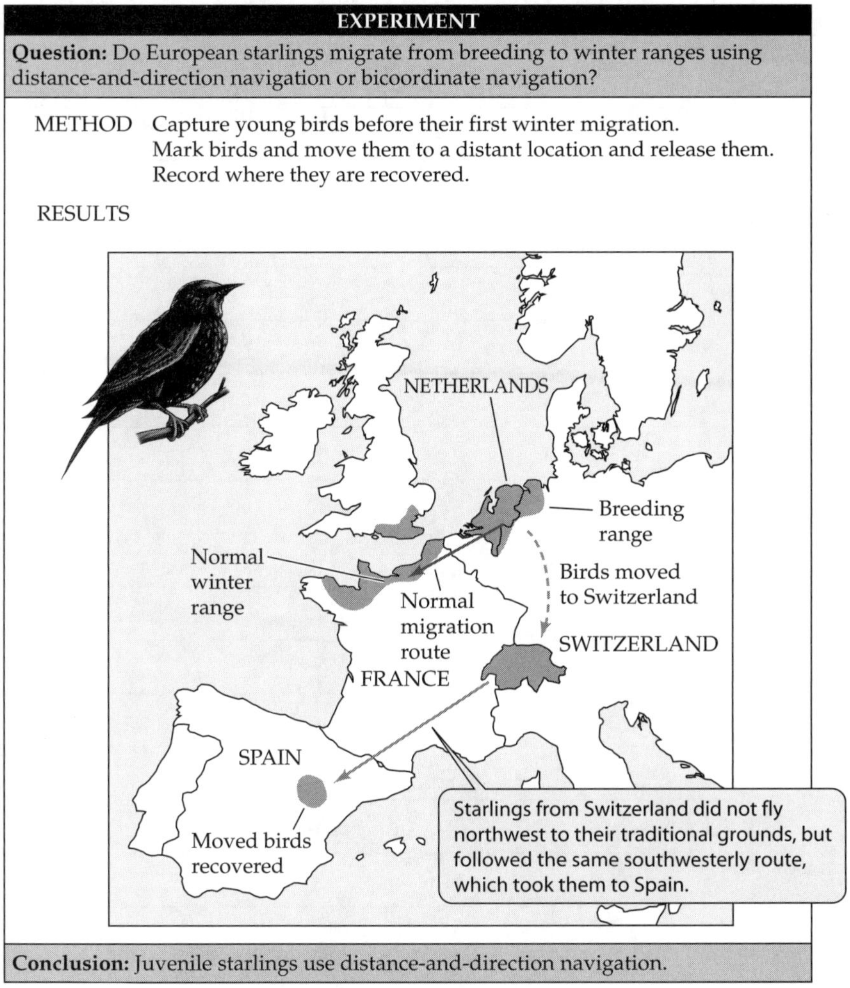

EXPERIMENT

Question: Do European starlings migrate from breeding to winter ranges using distance-and-direction navigation or bicoordinate navigation?

METHOD Capture young birds before their first winter migration.
Mark birds and move them to a distant location and release them.
Record where they are recovered.

RESULTS

NETHERLANDS
Breeding range
Normal winter range
Normal migration route
Birds moved to Switzerland
SWITZERLAND
FRANCE
SPAIN
Moved birds recovered

Starlings from Switzerland did not fly northwest to their traditional grounds, but followed the same southwesterly route, which took them to Spain.

Conclusion: Juvenile starlings use distance-and-direction navigation.

52.17 Distance-and-Direction Navigation *(Page 1019)*

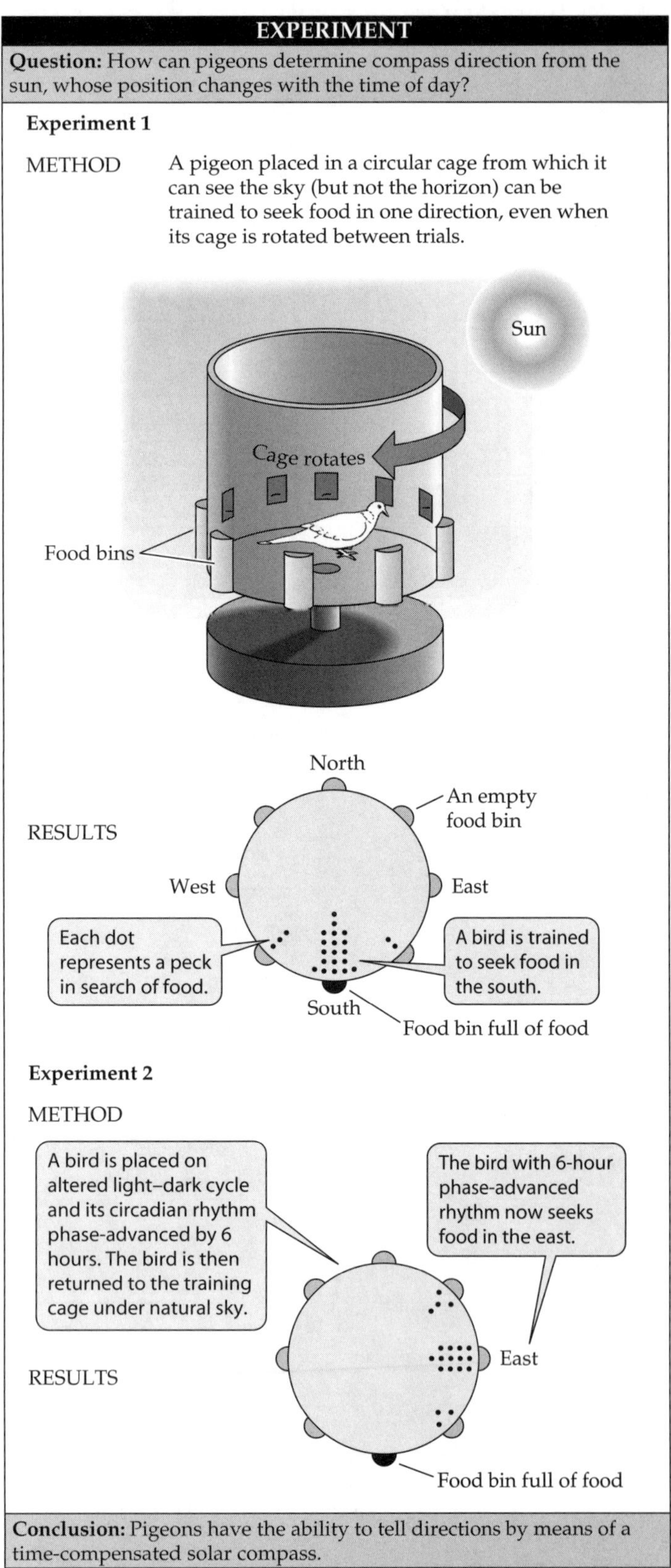
EXPERIMENT
Question: How can pigeons determine compass direction from the sun, whose position changes with the time of day?
Experiment 1
METHOD
A pigeon placed in a circular cage from which it can see the sky (but not the horizon) can be trained to seek food in one direction, even when its cage is rotated between trials.
Sun
Cage rotates
Food bins
RESULTS
North
An empty food bin
West
East
Each dot represents a peck in search of food.
A bird is trained to seek food in the south.
South
Food bin full of food
Experiment 2
METHOD
A bird is placed on altered light–dark cycle and its circadian rhythm phase-advanced by 6 hours. The bird is then returned to the training cage under natural sky.
The bird with 6-hour phase-advanced rhythm now seeks food in the east.
RESULTS
East
Food bin full of food
Conclusion: Pigeons have the ability to tell directions by means of a time-compensated solar compass.

52.18 The Time-Compensated Solar Compass *(Page 1020)*

53 Behavioral Ecology

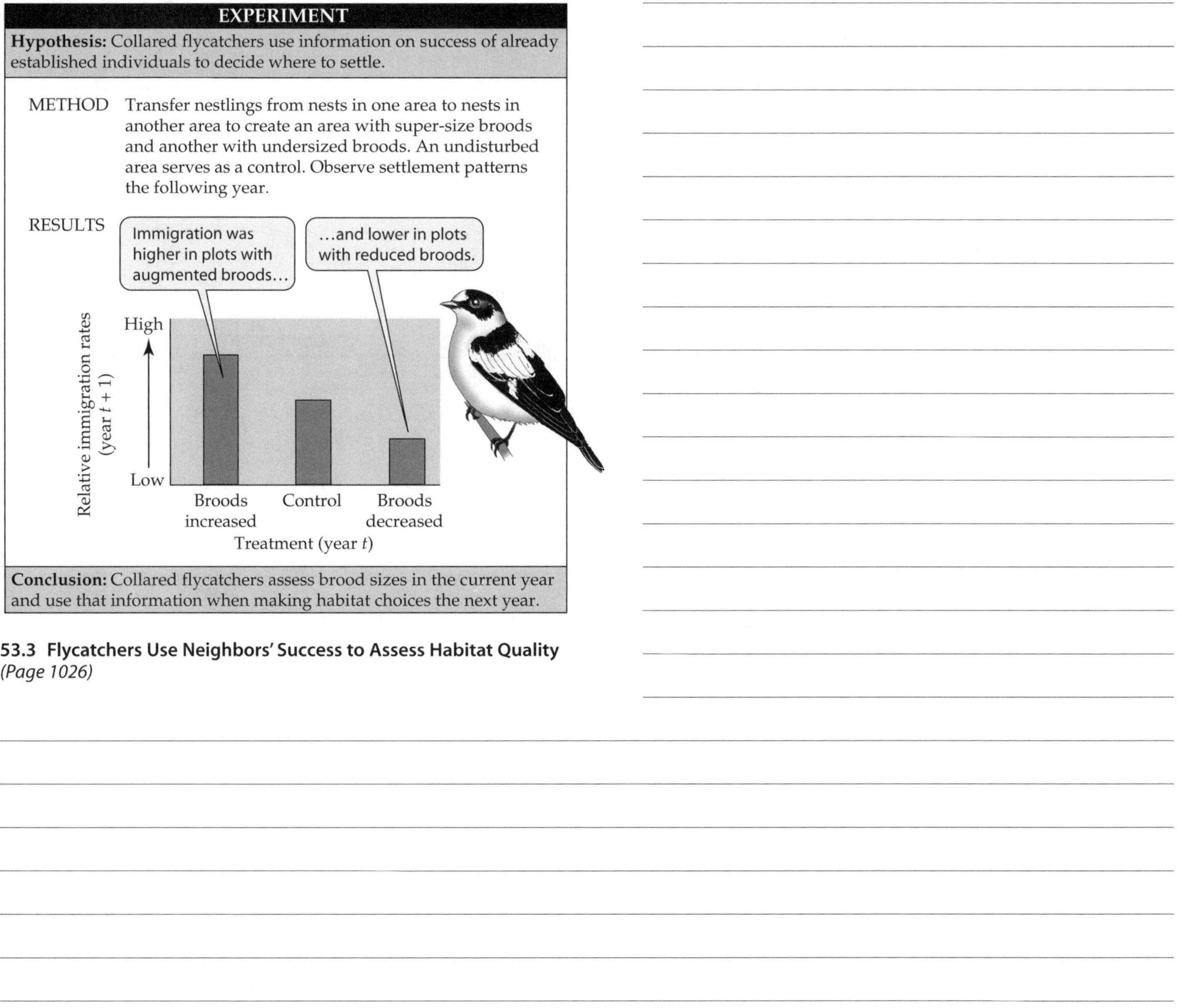
EXPERIMENT
Hypothesis: Collared flycatchers use information on success of already established individuals to decide where to settle.
METHOD Transfer nestlings from nests in one area to nests in another area to create an area with super-size broods and another with undersized broods. An undisturbed area serves as a control. Observe settlement patterns the following year.
RESULTS
Immigration was higher in plots with augmented broods...
...and lower in plots with reduced broods.
Relative immigration rates (year t + 1)
High
Low
Broods increased
Control
Broods decreased
Treatment (year t)
Conclusion: Collared flycatchers assess brood sizes in the current year and use that information when making habitat choices the next year.

53.3 Flycatchers Use Neighbors' Success to Assess Habitat Quality
(Page 1026)

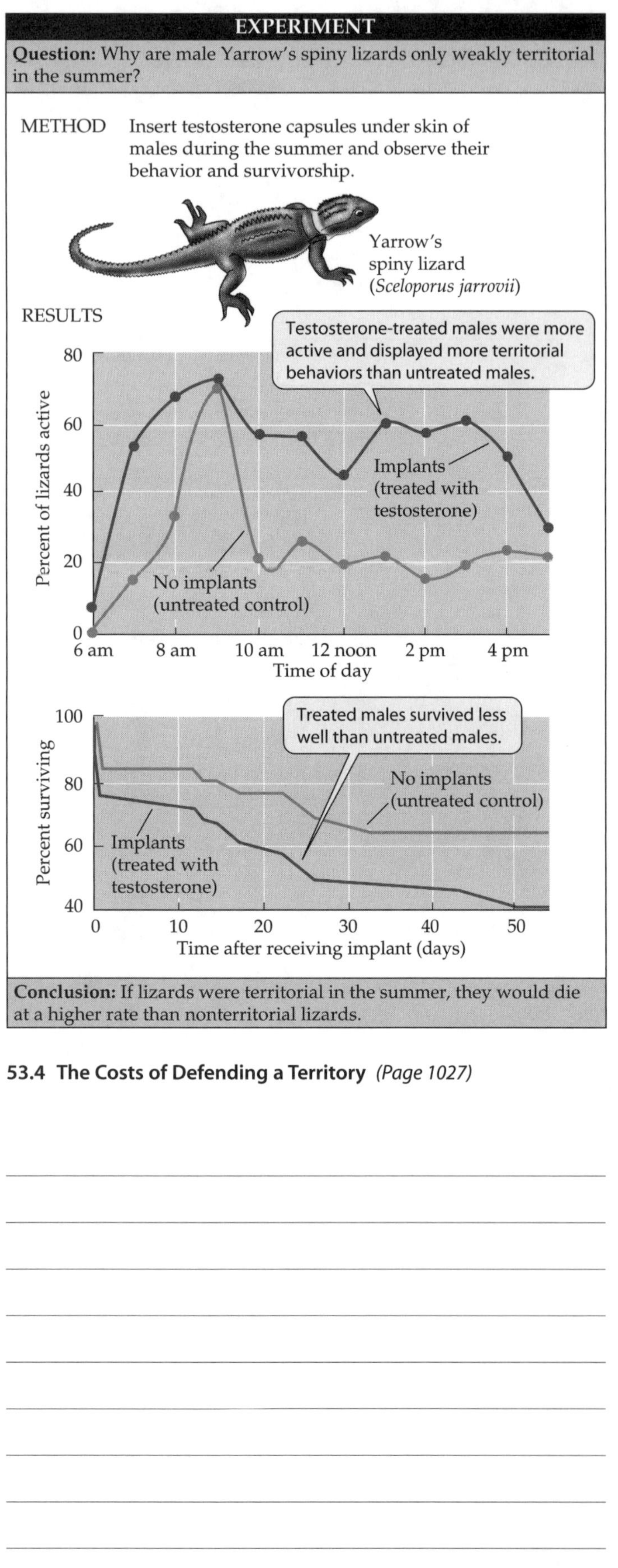
EXPERIMENT
Question: Why are male Yarrow's spiny lizards only weakly territorial in the summer?
METHOD Insert testosterone capsules under skin of males during the summer and observe their behavior and survivorship.
Yarrow's spiny lizard (Sceloporus jarrovii)
RESULTS
Testosterone-treated males were more active and displayed more territorial behaviors than untreated males.
Percent of lizards active
80
60
40
20
0
Implants (treated with testosterone)
No implants (untreated control)
6 am 8 am 10 am 12 noon 2 pm 4 pm
Time of day
Treated males survived less well than untreated males.
Percent surviving
100
80
60
40
No implants (untreated control)
Implants (treated with testosterone)
0 10 20 30 40 50
Time after receiving implant (days)
Conclusion: If lizards were territorial in the summer, they would die at a higher rate than nonterritorial lizards.

53.4 The Costs of Defending a Territory *(Page 1027)*

EXPERIMENT

Question: Do bluegills select prey to maximize their energy intake?

METHOD Provide bluegills with varying proportions of *Daphnia* (water fleas) of different sizes and in differing abundances (density). Compare prey actually eaten with the predictions of an optimality model.

Density of *Daphnia*

Low Medium High

Proportions of *Daphnia* of each size

Large
Medium
Small

RESULTS

Proportions in diet predicted from model

Actual proportions in diet

Similar bar widths show that the diet is very similar to that predicted by the model.

Bluegills were more selective when large *Daphnia* were more abundant.

Daphnia

Bluegill

Conclusion: Bluegills select prey to maximize their rate of energy intake.

53.5 Bluegills Are Energy Maximizers *(Page 1028)*

EXPERIMENT

Hypothesis: Spices commonly used in cooking have antibacterial activity against food-borne bacteria.

METHOD Prepare alcohol extracts of spices and test whether they inhibit growth of food-borne bacteria in culture media.

RESULTS

● Complete inhibition ● Moderate inhibition ● No inhibition

Spice

Mace
Bay leaf
Nutmeg
Garlic
Sage
Cinnamon
Thyme
Paprika
Oregano
Anise
Turmeric
Cardamom
White pepper
Black pepper
Allspice
Rosemary

Staphylococcus aureus *Bacillus stearothermophilus* *Bacillus coagulans* *Vibrio cholerae*

Bacterium tested

Conclusion: Most commonly used spices have strong antibacterial activity against more than one kind of food-borne bacteria.

53.7 Most Spices Have Antimicrobial Activity *(Page 1029)*

EXPERIMENT

Hypothesis: Flocking behavior confers antipredator benefits.

METHOD Release hawks near pigeon flocks of different sizes. Observe whether a hawk captures a pigeon.

Goshawk

Wood pigeon

RESULTS

The more pigeons in the flock, the sooner the hawk is spotted and the pigeons initiate evasive action.

Hawk's distance (meters) when spotted by pigeons ●

Hawk's attack success (%) ●

80
60
40
20

1 2–10 11–50 >50

Number of pigeons in flock

Thus the hawk's attack success is much lower.

Conclusion: Flocking behavior provides protection against predation.

53.10 Groups Provide Protection from Predators *(Page 1031)*

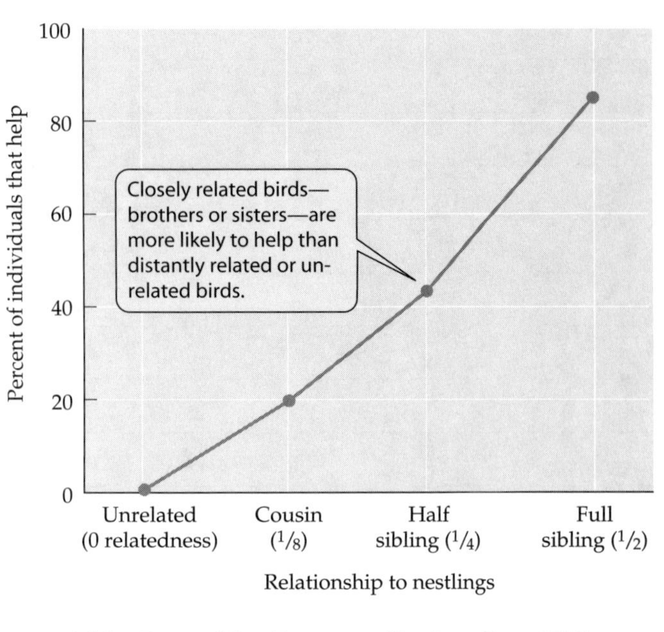
100
80
60
40
20
0
Percent of individuals that help
Closely related birds—brothers or sisters—are more likely to help than distantly related or unrelated birds.
Unrelated (0 relatedness)
Cousin (1/8)
Half sibling (1/4)
Full sibling (1/2)
Relationship to nestlings

53.11 White-Fronted Bee-Eaters are Altruists *(Page 1032)*

54 Population Ecology

54.1 ***Life Table of the 1978 Cohort of the Cactus Finch (Geospiza scandens) on Isla Daphne***

AGE IN YEARS (*X*)	NUMBER ALIVE	SURVIVORSHIP[a]	SURVIVAL RATE[b]	MORTALITY RATE[c]
0	210	1.000	0.434	0.566
1	91	0.434	0.857	0.143
2	78	0.371	0.898	0.102
3	70	0.333	0.928	0.072
4	65	0.309	0.955	0.045
5	62	0.295	0.678	0.322
6	42	0.200	0.548	0.452
7	23	0.109	0.652	0.348
8	15	0.071	0.933	0.067
9	14	0.067	0.786	0.214
10	11	0.052	0.909	0.091
11	10	0.048	0.400	0.600
12	4	0.019	0.750	0.250
13	3	0.014	0.996	

[a]Survivorship = the proportion of newborns who survive to age x.
[b]Survival rate = the proportion of individuals of age x who survive to age $x + 1$.
[c]Mortality rate = the proportion of individuals of age x who die before the age of $x + 1$.

(Page 1039)

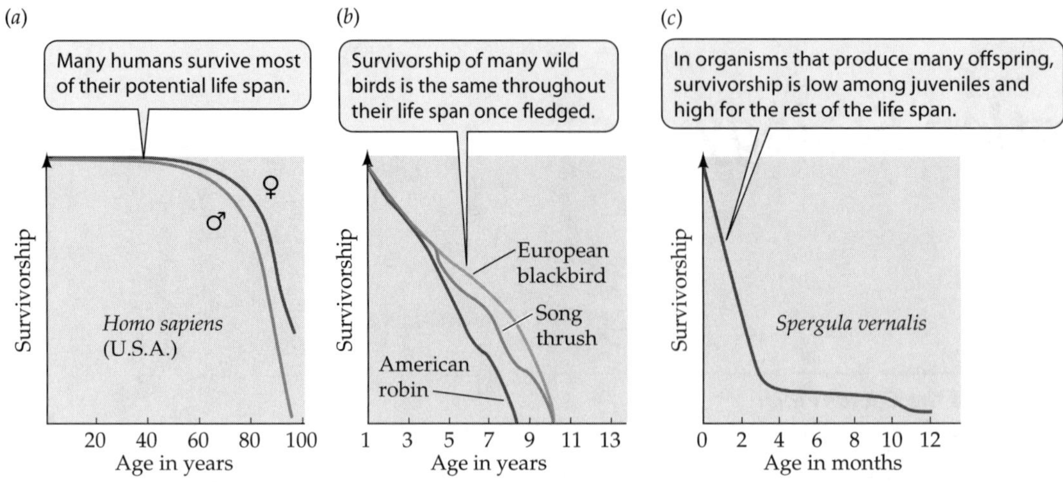
(a)
Many humans survive most of their potential life span.
Survivorship
Homo sapiens (U.S.A.)
20 40 60 80 100
Age in years
(b)
Survivorship of many wild birds is the same throughout their life span once fledged.
Survivorship
European blackbird
Song thrush
American robin
1 3 5 7 9 11 13
Age in years
(c)
In organisms that produce many offspring, survivorship is low among juveniles and high for the rest of the life span.
Survivorship
Spergula vernalis
0 2 4 6 8 10 12
Age in months

54.1 Survivorship Curves *(Page 1039)*

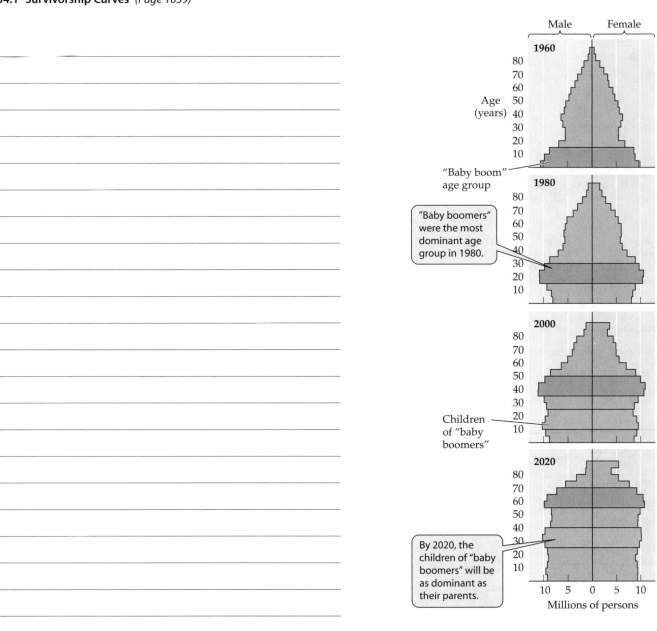

Male
Female
1960
Age (years)
80
70
60
50
40
30
20
10
"Baby boom" age group
1980
"Baby boomers" were the most dominant age group in 1980.
2000
Children of "baby boomers"
2020
By 2020, the children of "baby boomers" will be as dominant as their parents.
10 5 0 5 10
Millions of persons

54.2 Age Distributions Change over Time *(Page 1040)*

54.2 *Types of Ecological Interactions*

		EFFECT ON ORGANISM 2		
		HARM	BENEFIT	NO EFFECT
	HARM	Competition (–/–)	Predation or parasitism (–/+)	Amensalism (–/0)
EFFECT ON ORGANISM 1	BENEFIT	Predation or parasitism (+/–)	Mutualism (+/+)	Commensalism (+/0)
	NO EFFECT	Amensalism (0/–)	Commensalism (0/+)	—

(Page 1040)

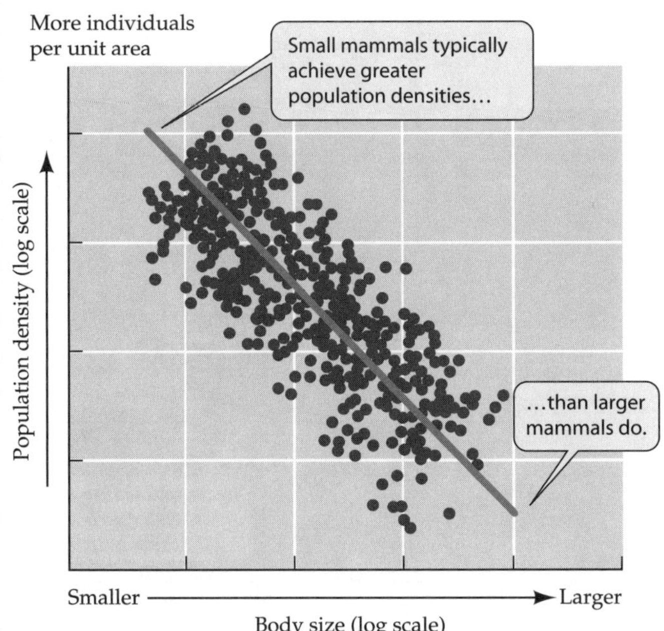
More individuals per unit area
Small mammals typically achieve greater population densities…
…than larger mammals do.
Population density (log scale)
Smaller
Larger
Body size (log scale)

54.4 Population Density Decreases as Body Size Increases
(Page 1041)

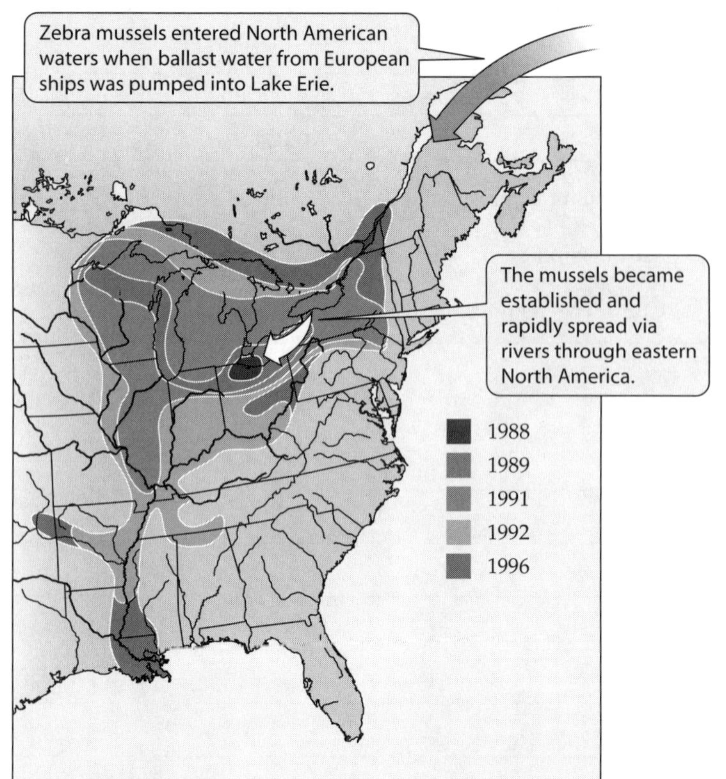
Zebra mussels entered North American waters when ballast water from European ships was pumped into Lake Erie.
The mussels became established and rapidly spread via rivers through eastern North America.
1988
1989
1991
1992
1996

54.5 Introduced Zebra Mussels Have Spread Rapidly *(Page 1042)*

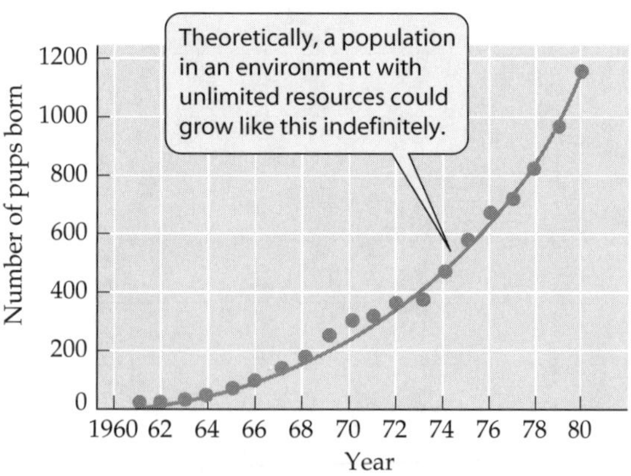
Theoretically, a population in an environment with unlimited resources could grow like this indefinitely.
Number of pups born
1200
1000
800
600
400
200
0
1960 62 64 66 68 70 72 74 76 78 80
Year

54.6 Exponential Population Growth *(Page 1043)*

Carrying capacity of environment (*K*)
Population size
Time
1 The rate accelerates.
2 A maximum growth rate is reached.
3 The rate slows down.

54.7 Logistic Population Growth *(Page 1043)*

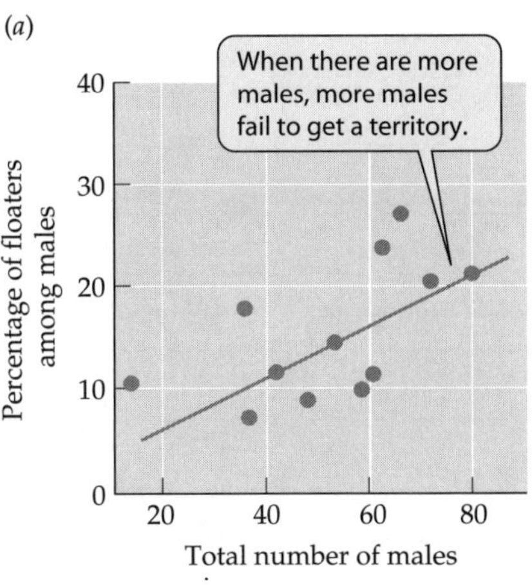
(a)
When there are more males, more males fail to get a territory.
Percentage of floaters among males
Total number of males

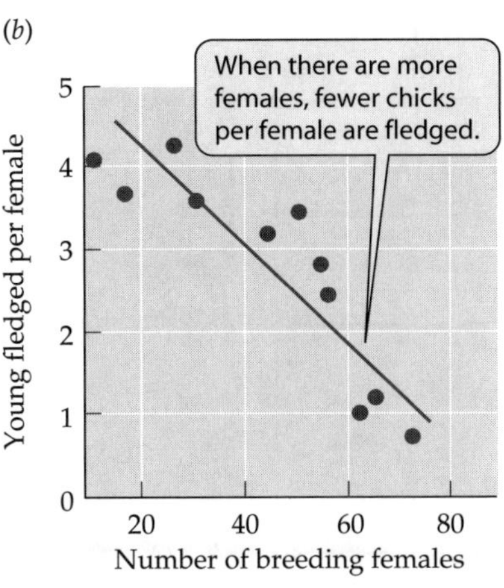
(b)
When there are more females, fewer chicks per female are fledged.
Young fledged per female
Number of breeding females

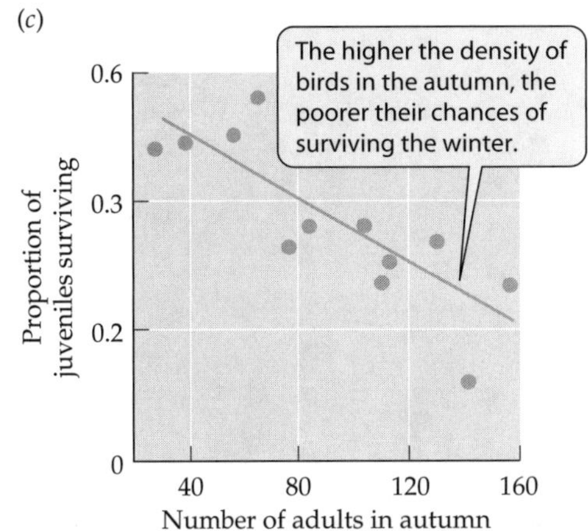
(c)
The higher the density of birds in the autumn, the poorer their chances of surviving the winter.
Proportion of juveniles surviving
Number of adults in autumn

54.8 Regulation of an Island Population of Song Sparrows *(Page 1044)*

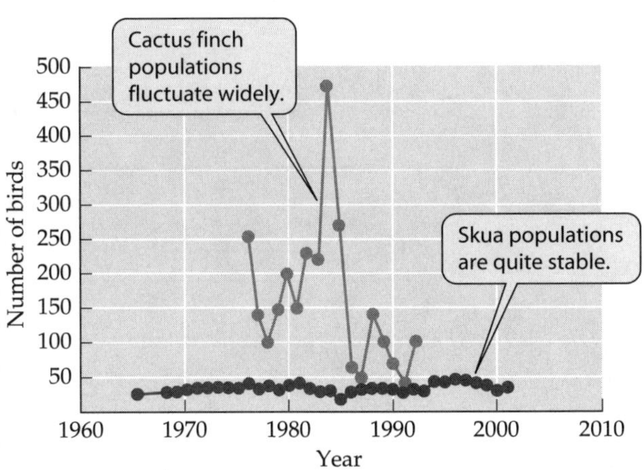
Cactus finch populations fluctuate widely.
Skua populations are quite stable.
Number of birds
Year

54.9 Population Sizes May Be Stable or Highly Variable *(Page 1045)*

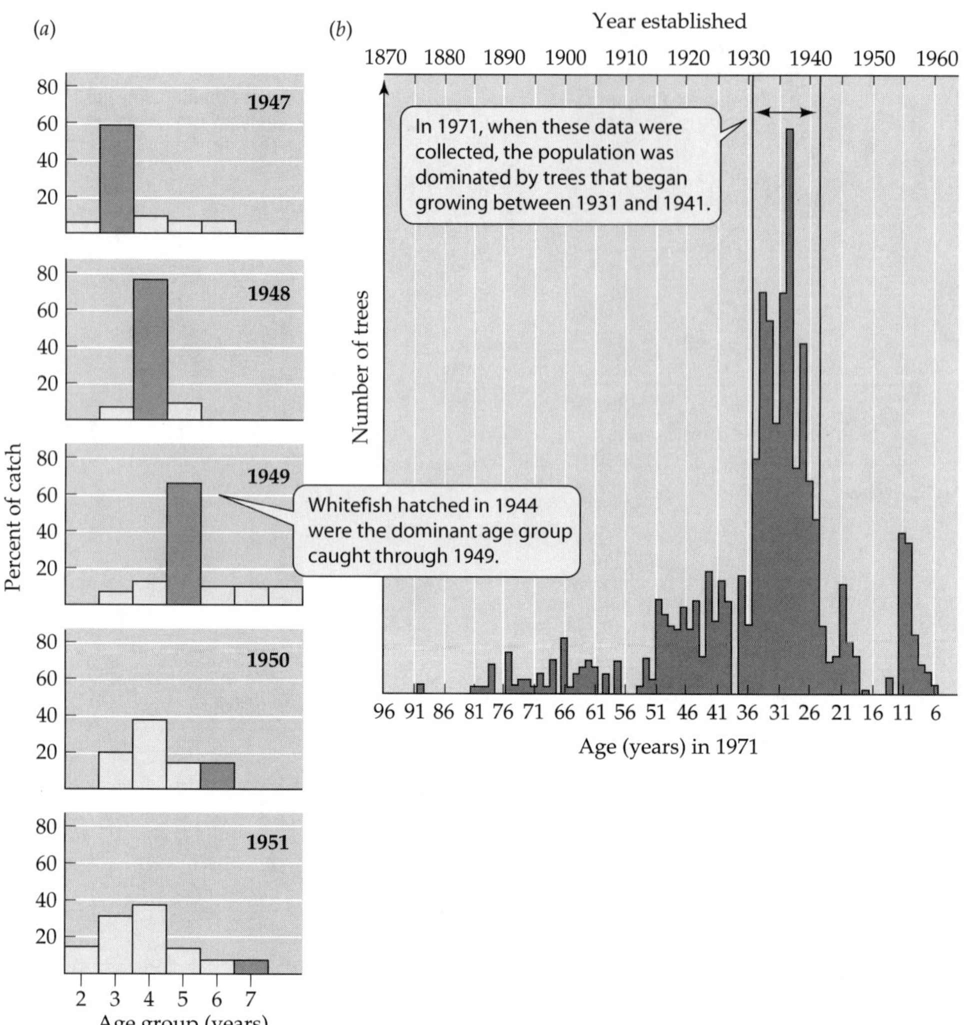
(a)
(b)
Year established
Percent of catch
1947
1948
1949
1950
1951
Age group (years)
Whitefish hatched in 1944 were the dominant age group caught through 1949.
In 1971, when these data were collected, the population was dominated by trees that began growing between 1931 and 1941.
Number of trees
Age (years) in 1971

54.10 Individuals Born During Years of Good Reproduction May Dominate Populations *(Page 1045)*

8

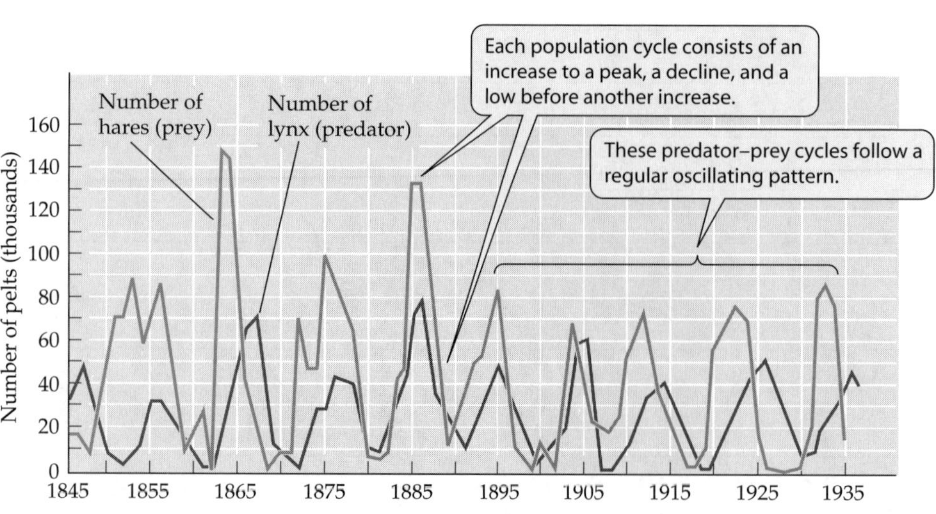
Number of hares (prey)
Number of lynx (predator)
Each population cycle consists of an increase to a peak, a decline, and a low before another increase.
These predator–prey cycles follow a regular oscillating pattern.
Number of pelts (thousands)

54.11 Hare and Lynx Populations Cycle in Nature *(Page 1046)*

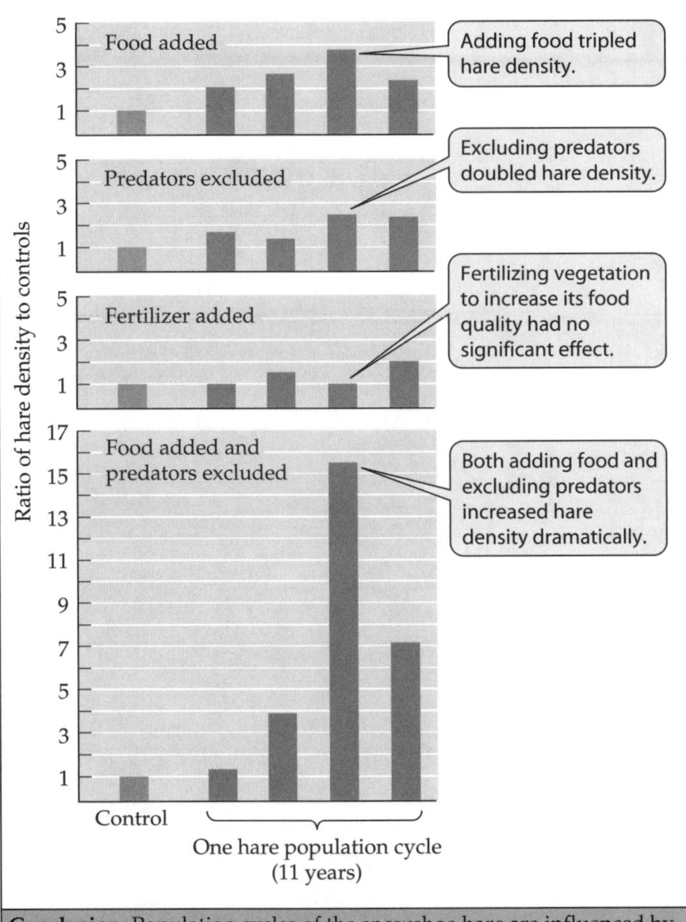
EXPERIMENT
Hypothesis: Population cycles of hares are influenced by both food supply and predators.
METHOD Select 9 1-km² blocks of undisturbed coniferous forest. In two of the blocks give the hares supplemental food year-round. Erect an electric fence around two other blocks, with mesh large enough to allow hares, but not lynxes, to pass. Provide extra food in one of these blocks. In two other blocks add fertilizer to increase food quality. Use three other blocks as unmanipulated controls.
RESULTS
Food added
Adding food tripled hare density.
Predators excluded
Excluding predators doubled hare density.
Fertilizer added
Fertilizing vegetation to increase its food quality had no significant effect.
Food added and predators excluded
Both adding food and excluding predators increased hare density dramatically.
Ratio of hare density to controls
Control
One hare population cycle (11 years)
Conclusion: Population cycles of the snowshoe hare are influenced by their food supply as well as by interactions with their predators.

54.12 Prey Population Cycles May Have Multiple Causes *(Page 1047)*

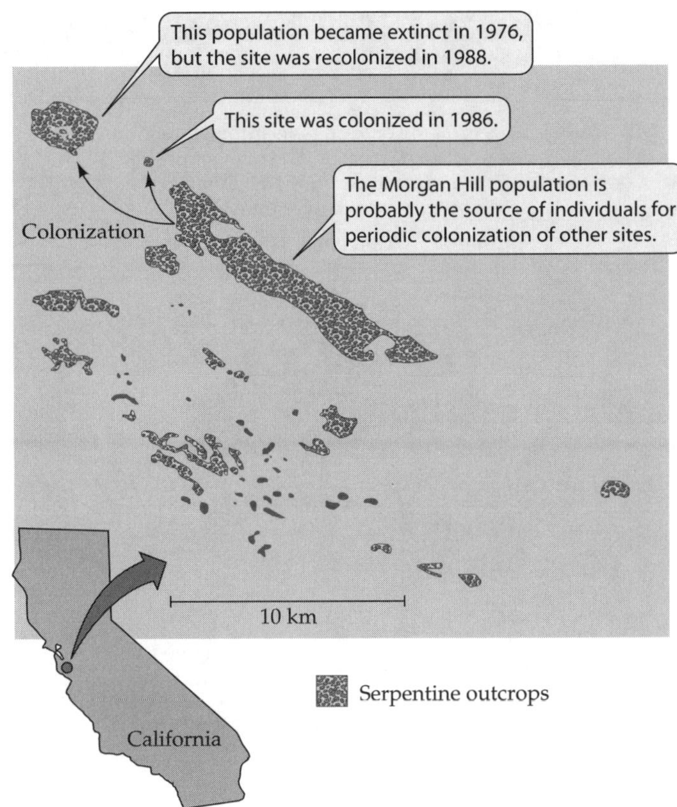
This population became extinct in 1976, but the site was recolonized in 1988.
This site was colonized in 1986.
The Morgan Hill population is probably the source of individuals for periodic colonization of other sites.
Colonization
10 km
California
Serpentine outcrops

54.13 Subpopulation Dynamics *(Page 1047)*

EXPERIMENT

Hypothesis: Even small barriers to dispersal may reduce the number of species in a habitat patch.

METHOD Moss growing on rocks was trimmed to form distinct habitat patches. The number of small organisms (mostly arthropods) living in the patches was observed over time.

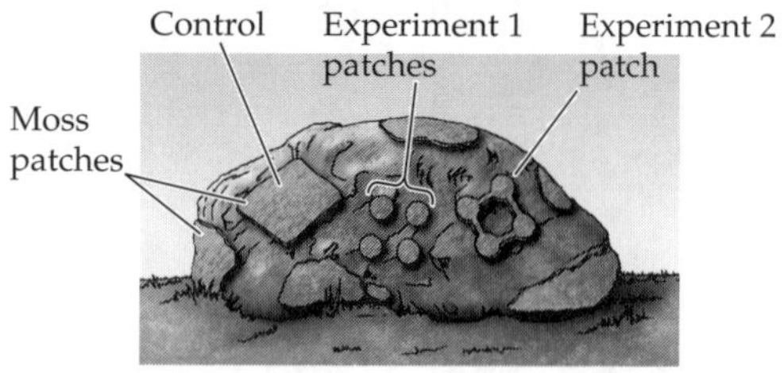

Control
Experiment 1 patches
Experiment 2 patch
Moss patches

Experiment 1

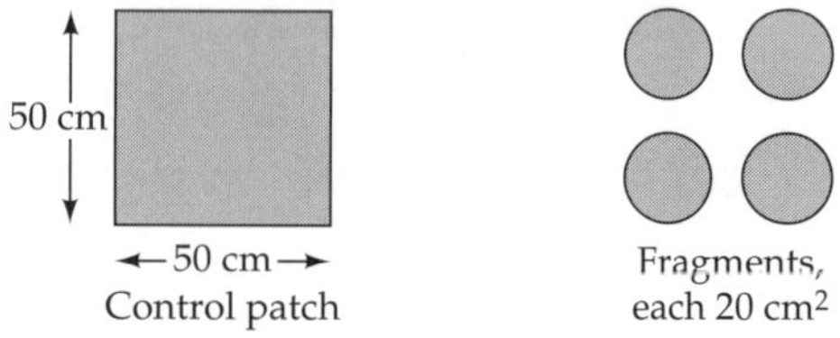

50 cm
50 cm
Control patch
Fragments, each 20 cm²

RESULTS In fragments, 40% of species became extinct after 1 year.

Experiment 2

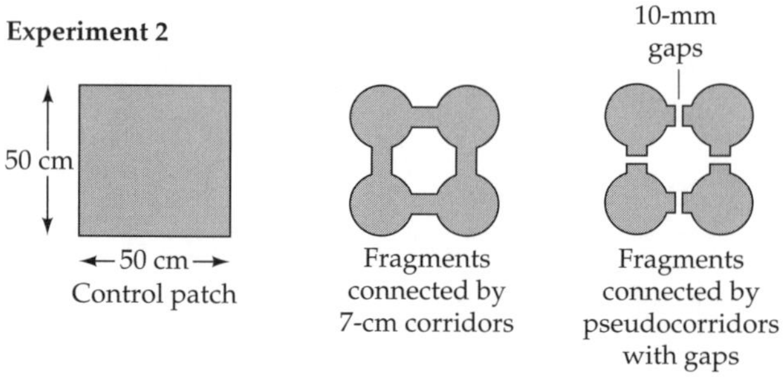

10-mm gaps
50 cm
50 cm
Control patch
Fragments connected by 7-cm corridors
Fragments connected by pseudocorridors with gaps

RESULTS

14% of the species became extinct after 6 months.

41% of the species became extinct after 6 months.

Conclusion: Even small barriers to dispersal raised extinction rates in a fragmented habitat.

54.14 Narrow Barriers Suffice to Separate Arthropod Subpopulations *(Page 1048)*

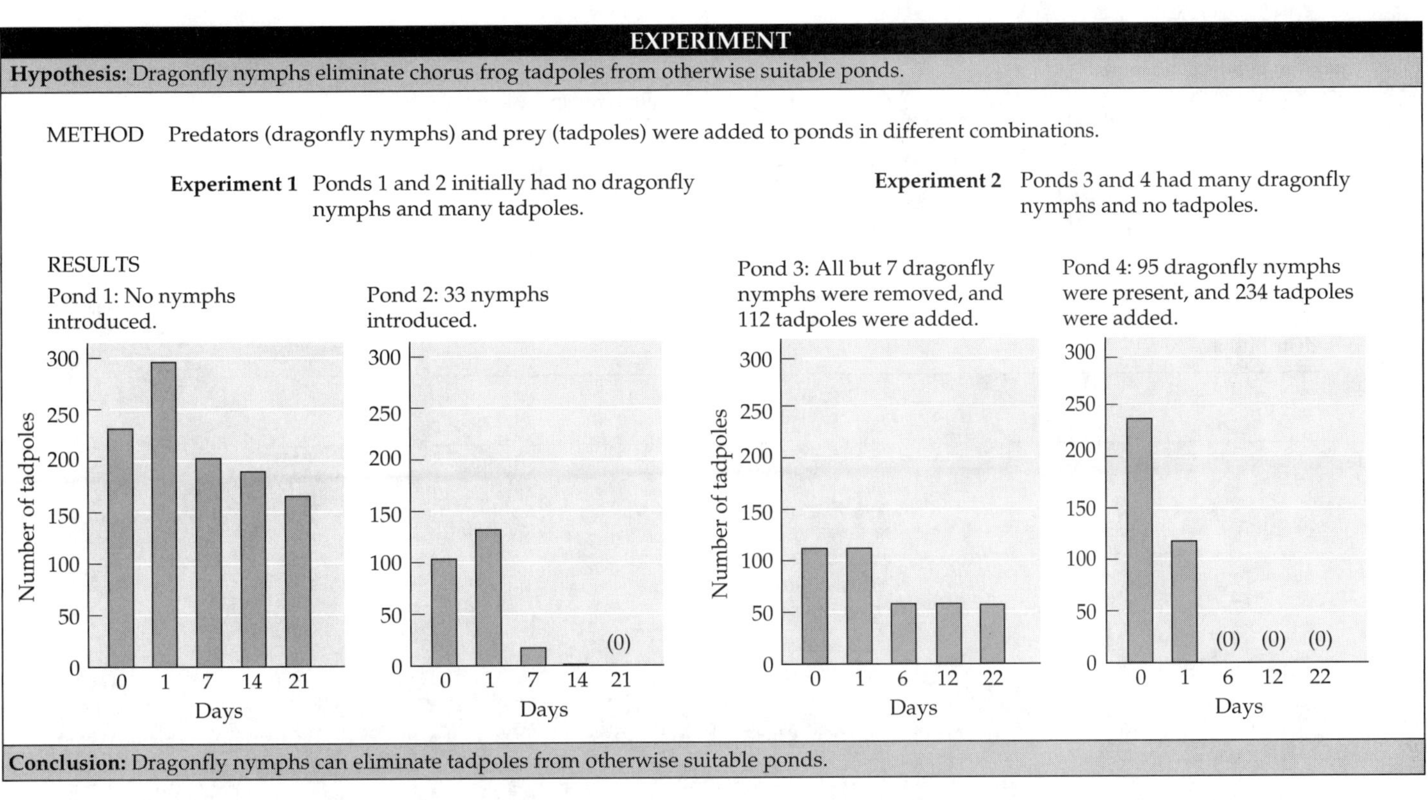
EXPERIMENT
Hypothesis: Dragonfly nymphs eliminate chorus frog tadpoles from otherwise suitable ponds.
METHOD Predators (dragonfly nymphs) and prey (tadpoles) were added to ponds in different combinations.
Experiment 1 Ponds 1 and 2 initially had no dragonfly nymphs and many tadpoles.
Experiment 2 Ponds 3 and 4 had many dragonfly nymphs and no tadpoles.
RESULTS
Pond 1: No nymphs introduced.
Pond 2: 33 nymphs introduced.
Pond 3: All but 7 dragonfly nymphs were removed, and 112 tadpoles were added.
Pond 4: 95 dragonfly nymphs were present, and 234 tadpoles were added.
Number of tadpoles
Days
(0)
(0) (0) (0)
Conclusion: Dragonfly nymphs can eliminate tadpoles from otherwise suitable ponds.

54.16 Predators Exclude Prey from Some Habitats *(Page 1049)*

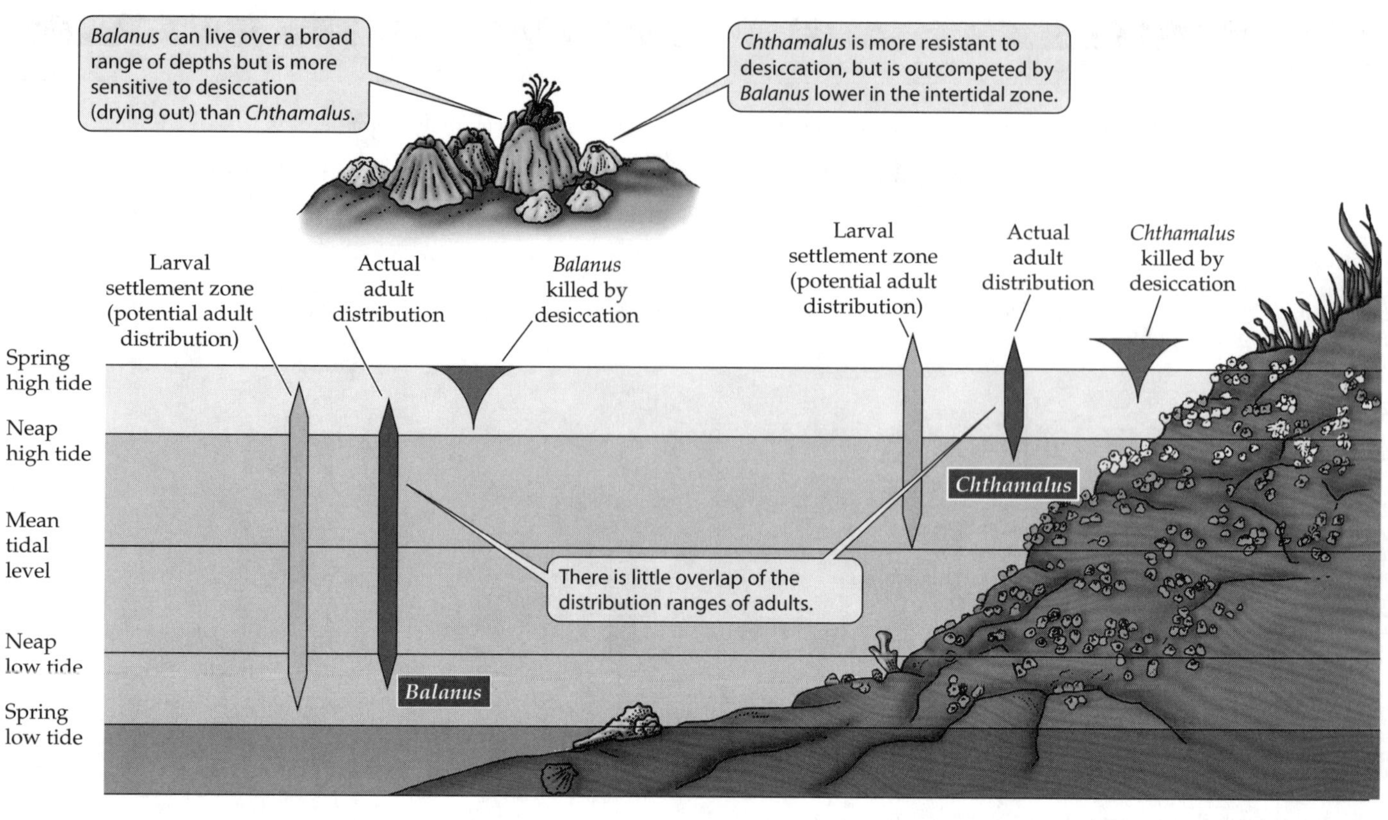

Balanus can live over a broad range of depths but is more sensitive to desiccation (drying out) than *Chthamalus*.
Chthamalus is more resistant to desiccation, but is outcompeted by *Balanus* lower in the intertidal zone.
Larval settlement zone (potential adult distribution)
Actual adult distribution
Balanus killed by desiccation
Larval settlement zone (potential adult distribution)
Actual adult distribution
Chthamalus killed by desiccation
Spring high tide
Neap high tide
Mean tidal level
Neap low tide
Spring low tide
Chthamalus
There is little overlap of the distribution ranges of adults.
Balanus

54.17 Competition Restricts the Intertidal Ranges of Barnacles *(Page 1050)*

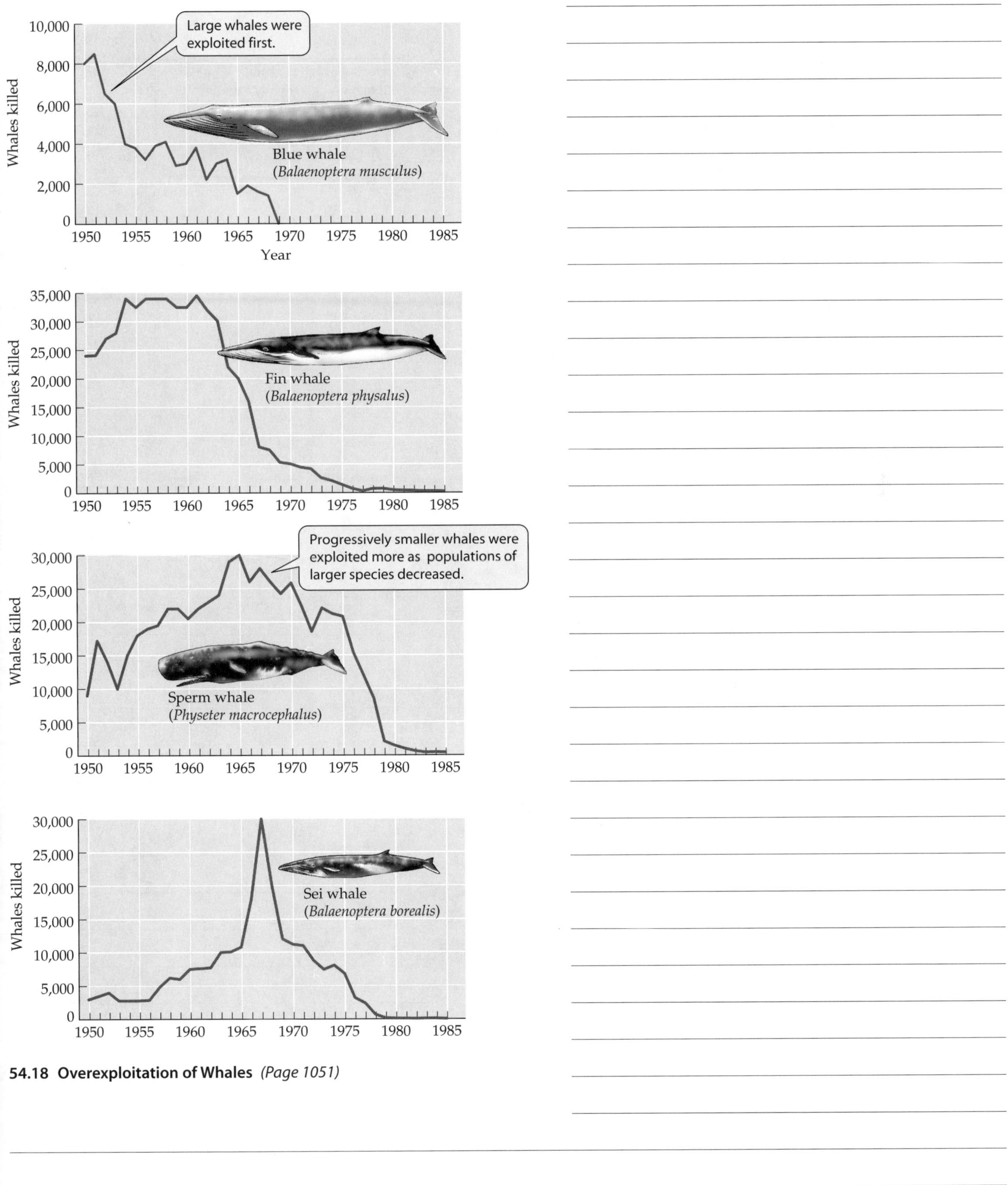

Large whales were exploited first.
Blue whale
(Balaenoptera musculus)
Whales killed
10,000
8,000
6,000
4,000
2,000
0
1950 1955 1960 1965 1970 1975 1980 1985
Year
Fin whale
(Balaenoptera physalus)
35,000
30,000
25,000
20,000
15,000
10,000
5,000
Progressively smaller whales were exploited more as populations of larger species decreased.
Sperm whale
(Physeter macrocephalus)
Sei whale
(Balaenoptera borealis)

54.18 Overexploitation of Whales *(Page 1051)*

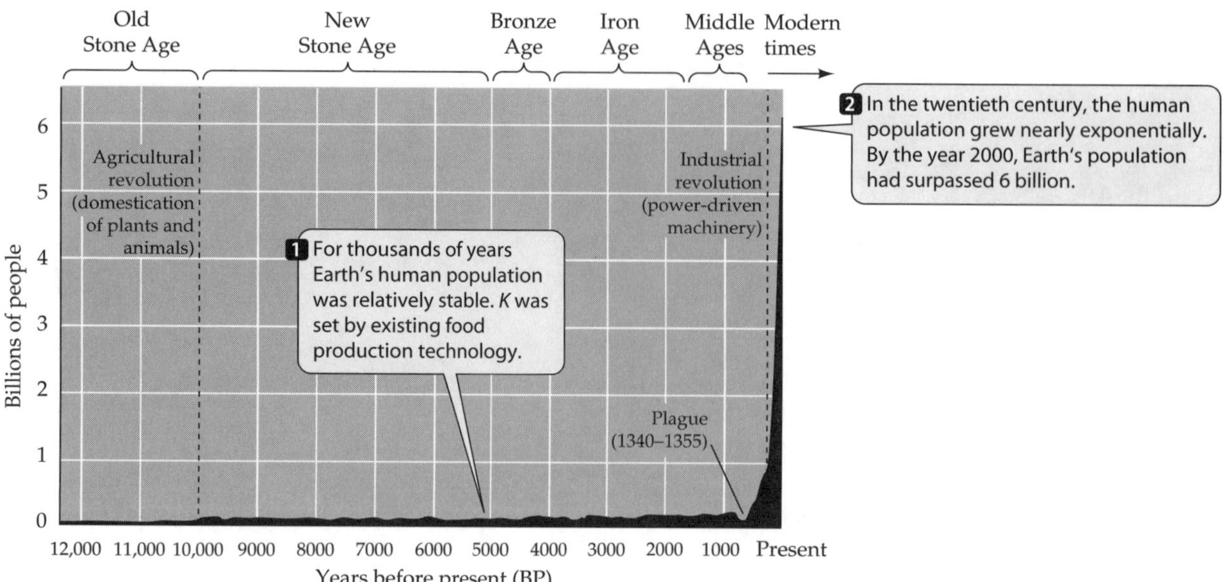

Old Stone Age
New Stone Age
Bronze Age
Iron Age
Middle Ages
Modern times
Agricultural revolution (domestication of plants and animals)
Industrial revolution (power-driven machinery)
1 For thousands of years Earth's human population was relatively stable. K was set by existing food production technology.
2 In the twentieth century, the human population grew nearly exponentially. By the year 2000, Earth's population had surpassed 6 billion.
Plague (1340–1355)
Billions of people
0
1
2
3
4
5
6
12,000 11,000 10,000 9000 8000 7000 6000 5000 4000 3000 2000 1000 Present
Years before present (BP)

54.20 Human Population Growth *(Page 1052)*

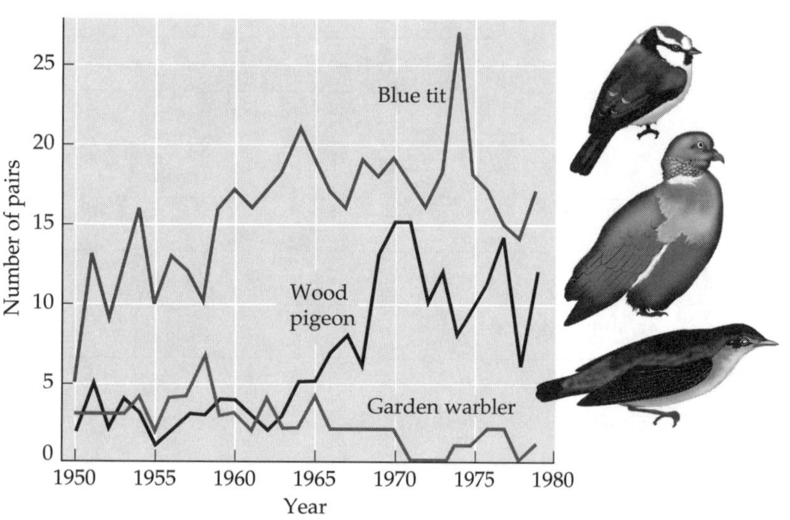

Blue tit
Wood pigeon
Garden warbler
Number of pairs
0
5
10
15
20
25
1950 1955 1960 1965 1970 1975 1980
Year

54.21 Populations May Be Influenced by Remote Events *(Page 1053)*

55 Communities and Ecosystems

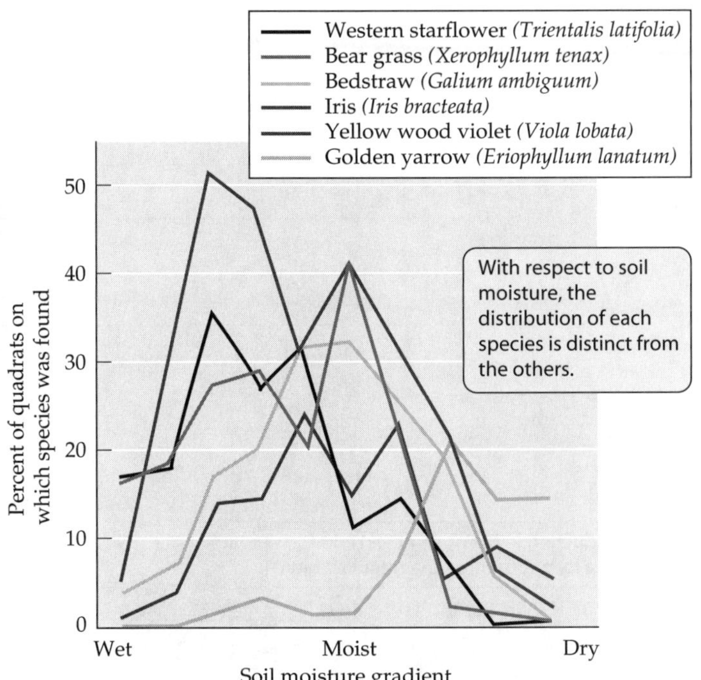

Western starflower (Trientalis latifolia)
Bear grass (Xerophyllum tenax)
Bedstraw (Galium ambiguum)
Iris (Iris bracteata)
Yellow wood violet (Viola lobata)
Golden yarrow (Eriophyllum lanatum)
With respect to soil moisture, the distribution of each species is distinct from the others.
Percent of quadrats on which species was found
0
10
20
30
40
50
Wet
Moist
Dry
Soil moisture gradient

55.1 Plant Distributions along an Environmental Gradient
(Page 1056)

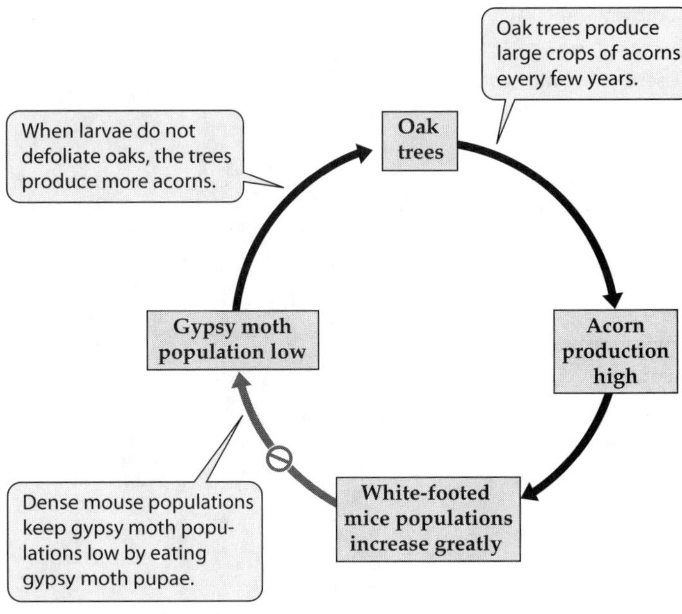

Oak trees produce large crops of acorns every few years.
When larvae do not defoliate oaks, the trees produce more acorns.
Oak trees
Acorn production high
White-footed mice populations increase greatly
Gypsy moth population low
Dense mouse populations keep gypsy moth populations low by eating gypsy moth pupae.

55.2 Interactions within Communities Control Populations
(Page 1057)

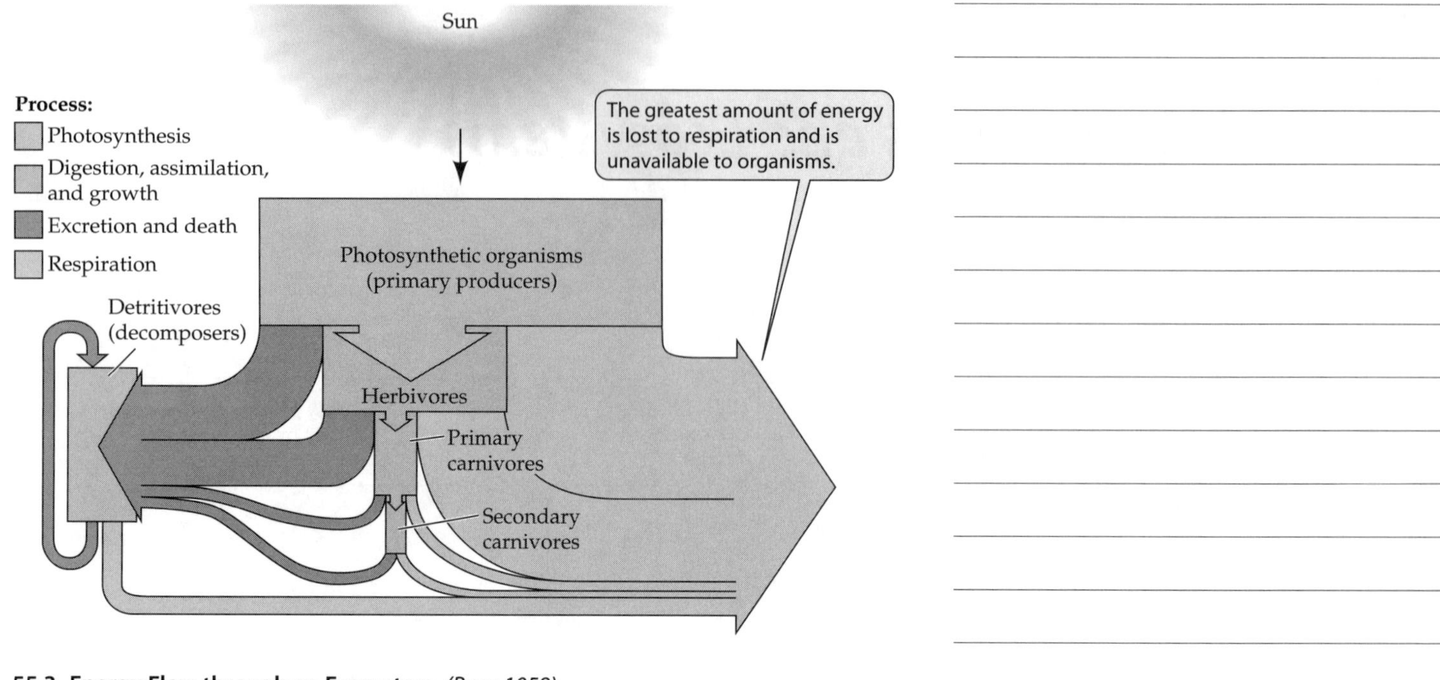
Sun
Process:
Photosynthesis
Digestion, assimilation, and growth
Excretion and death
Respiration
The greatest amount of energy is lost to respiration and is unavailable to organisms.
Photosynthetic organisms (primary producers)
Detritivores (decomposers)
Herbivores
Primary carnivores
Secondary carnivores

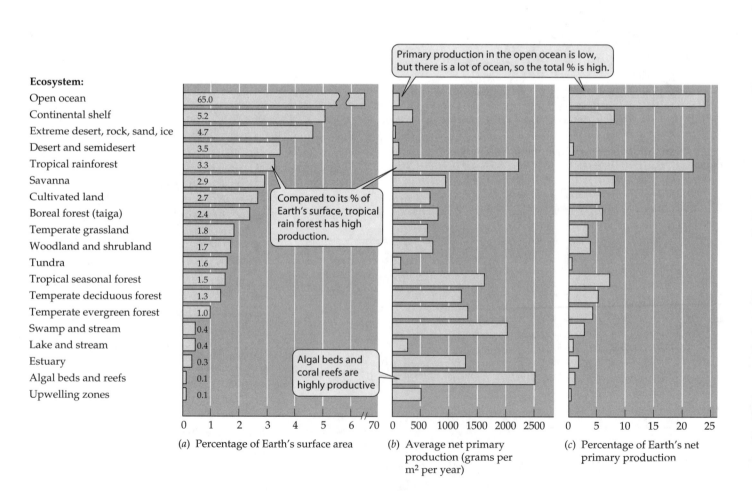

55.3 Energy Flow through an Ecosystem *(Page 1058)*

Primary production in the open ocean is low, but there is a lot of ocean, so the total % is high.

Ecosystem:	(a) Percentage of Earth's surface area
Open ocean	65.0
Continental shelf	5.2
Extreme desert, rock, sand, ice	4.7
Desert and semidesert	3.5
Tropical rainforest	3.3
Savanna	2.9
Cultivated land	2.7
Boreal forest (taiga)	2.4
Temperate grassland	1.8
Woodland and shrubland	1.7
Tundra	1.6
Tropical seasonal forest	1.5
Temperate deciduous forest	1.3
Temperate evergreen forest	1.0
Swamp and stream	0.4
Lake and stream	0.4
Estuary	0.3
Algal beds and reefs	0.1
Upwelling zones	0.1

Compared to its % of Earth's surface, tropical rain forest has high production.

Algal beds and coral reefs are highly productive

0 1 2 3 4 5 6 70

(*a*) Percentage of Earth's surface area

0 500 1000 1500 2000 2500

(*b*) Average net primary production (grams per m^2 per year)

0 5 10 15 20 25

(*c*) Percentage of Earth's net primary production

55.4 Primary Production in Different Ecosystems *(Page 1058)*

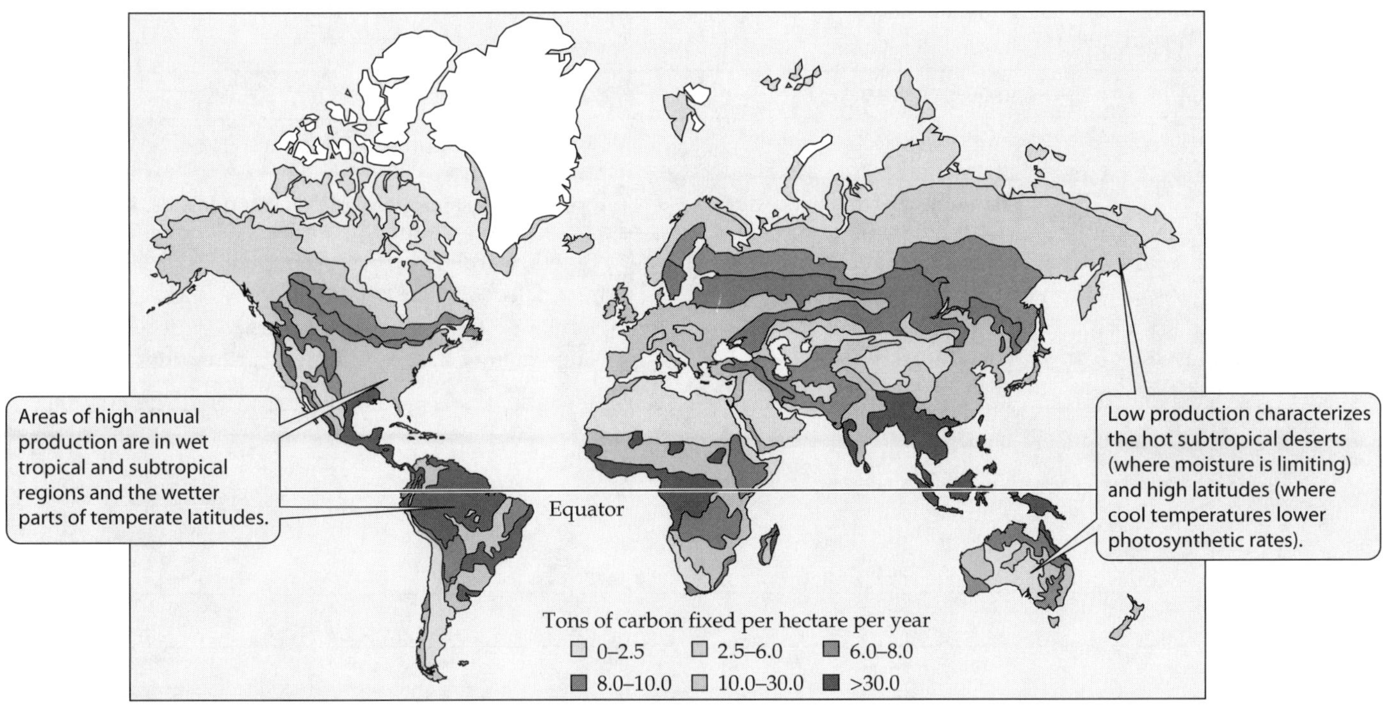

Areas of high annual production are in wet tropical and subtropical regions and the wetter parts of temperate latitudes.
Equator
Low production characterizes the hot subtropical deserts (where moisture is limiting) and high latitudes (where cool temperatures lower photosynthetic rates).
Tons of carbon fixed per hectare per year
0–2.5
2.5–6.0
6.0–8.0
8.0–10.0
10.0–30.0
>30.0

55.5 Net Primary Production of Terrestrial Ecosystems *(Page 1059)*

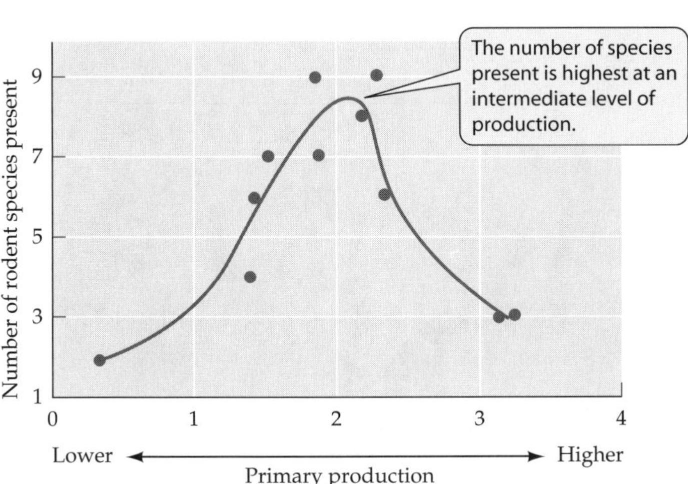

The number of species present is highest at an intermediate level of production.
Number of rodent species present
1
3
5
7
9
0
1
2
3
4
Lower
Higher
Primary production

55.6 Local Species Richness Peaks at Intermediate Productivity *(Page 1059)*

55.1 *The Major Trophic Levels*

TROPHIC LEVEL	SOURCE OF ENERGY	EXAMPLES
Photosynthesizers (primary producers)	Solar energy	Green plants, photosynthetic bacteria and protists
Herbivores	Tissues of primary producers	Termites, grasshoppers, gypsy moth larvae, anchovies, deer, geese, white-footed mice
Primary carnivores	Herbivores	Spiders, warblers, wolves, copepods
Secondary carnivores	Primary carnivores	Tuna, falcons, killer whales
Omnivores	Several trophic levels	Humans, opossums, crabs, robins
Detritivores (decomposers)	Dead bodies and waste products of other organisms	Fungi, many bacteria, vultures, earthworms

(Page 1060)

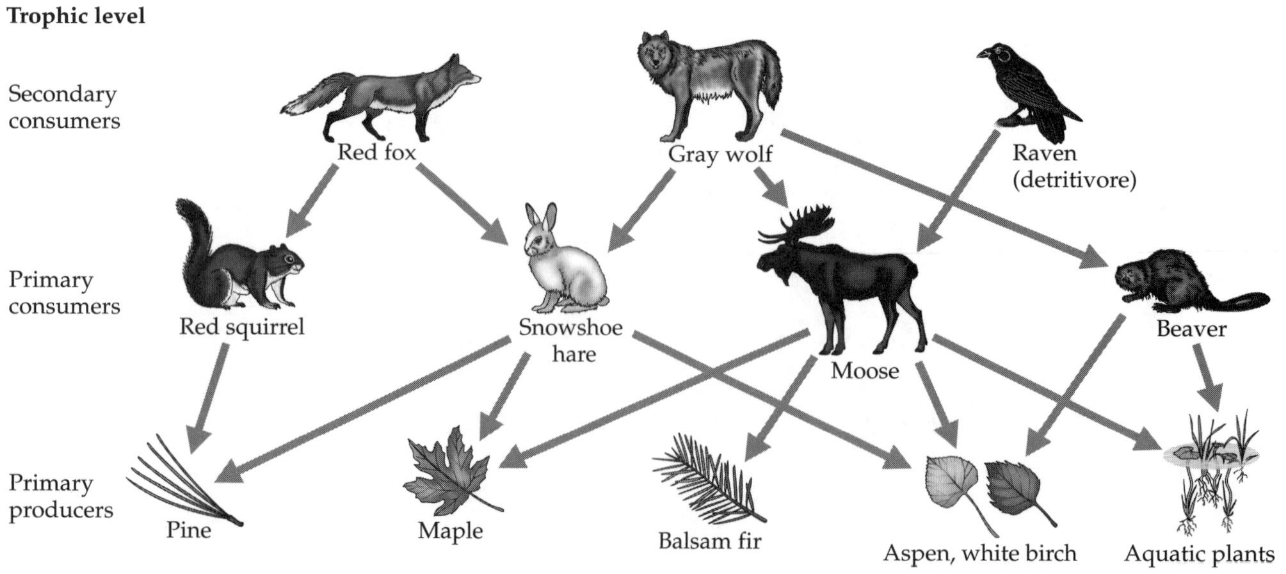
Trophic level
Secondary consumers
Primary consumers
Primary producers
Red fox
Gray wolf
Raven (detritivore)
Red squirrel
Snowshoe hare
Moose
Beaver
Pine
Maple
Balsam fir
Aspen, white birch
Aquatic plants

55.7 Food Web of Isle Royale National Park *(Page 1061)*

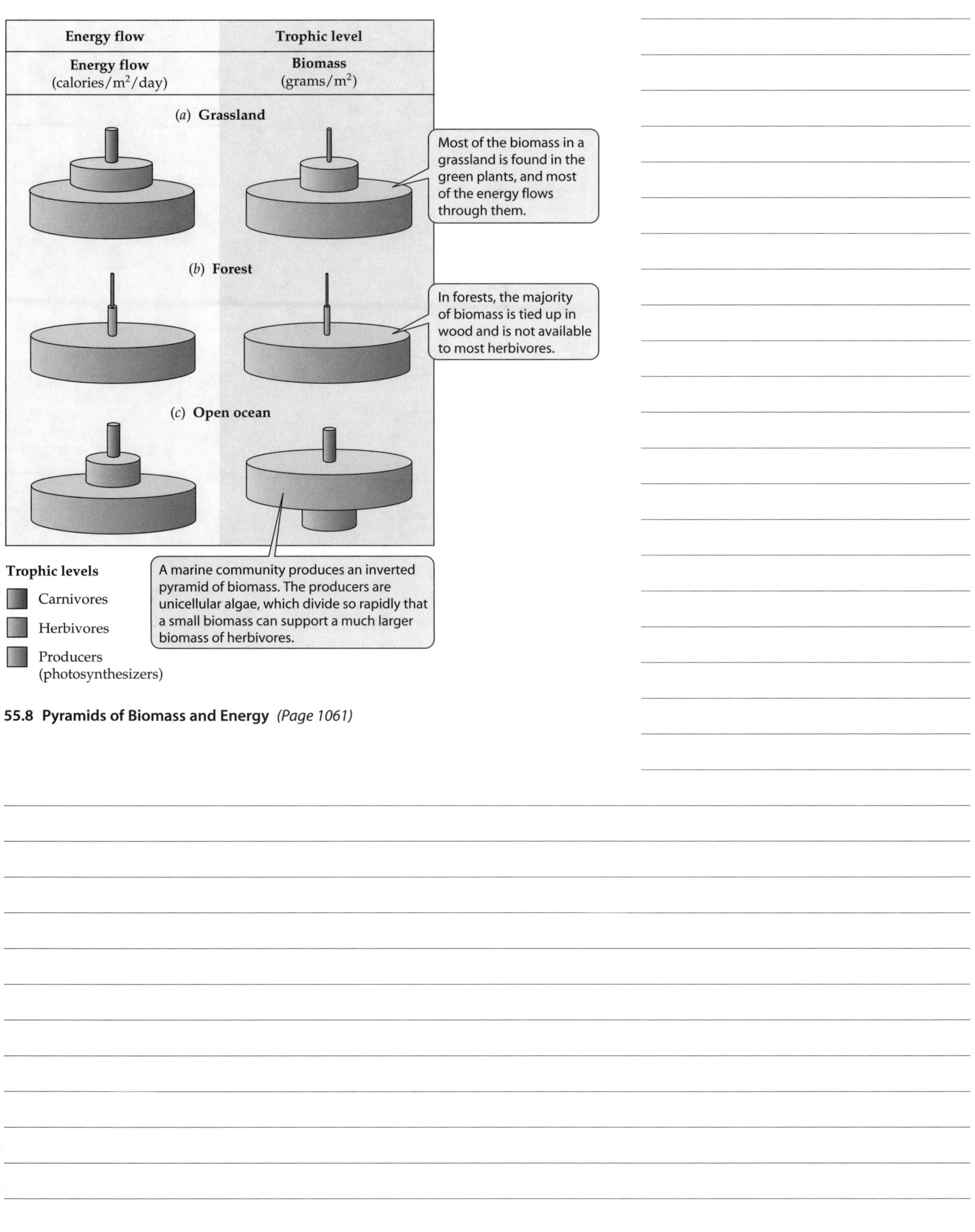
Energy flow
Trophic level
Energy flow (calories/m²/day)
Biomass (grams/m²)
(a) Grassland
Most of the biomass in a grassland is found in the green plants, and most of the energy flows through them.
(b) Forest
In forests, the majority of biomass is tied up in wood and is not available to most herbivores.
(c) Open ocean
A marine community produces an inverted pyramid of biomass. The producers are unicellular algae, which divide so rapidly that a small biomass can support a much larger biomass of herbivores.
Trophic levels
Carnivores
Herbivores
Producers (photosynthesizers)

55.8 Pyramids of Biomass and Energy *(Page 1061)*

EXPERIMENT

Hypothesis: Communities with many species of plants should have higher productivity and stability than communities with few species.

METHOD Clear and plant plots with different numbers and mixtures of grass species. Measure primary production and species composition of the plots over 11 years.

RESULTS

(*a*) Plant cover increases with species richness

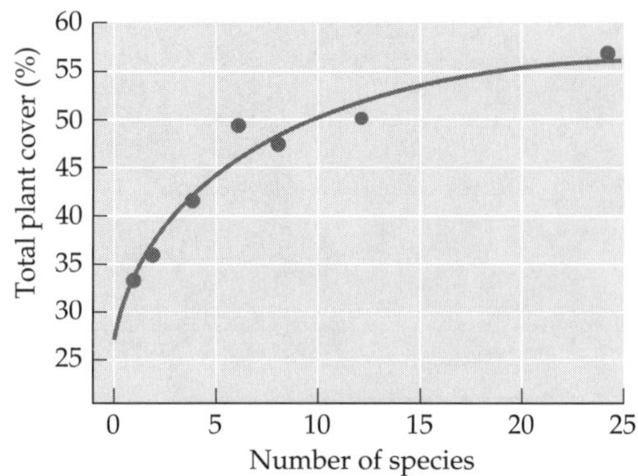
Total plant cover (%)
60
55
50
45
40
35
30
25
0
5
10
15
20
25
Number of species

(*b*) Variation in biomass decreases with species richness

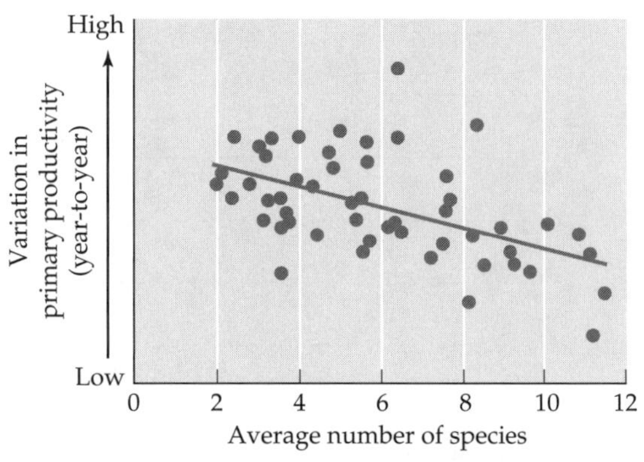
Variation in primary productivity (year-to-year)
High
Low
0
2
4
6
8
10
12
Average number of species

Conclusion: Plots with more species were more productive and varied less in annual net primary production of biomass.

55.9 Species Richness Enhances Community Productivity and Stability *(Page 1062)*

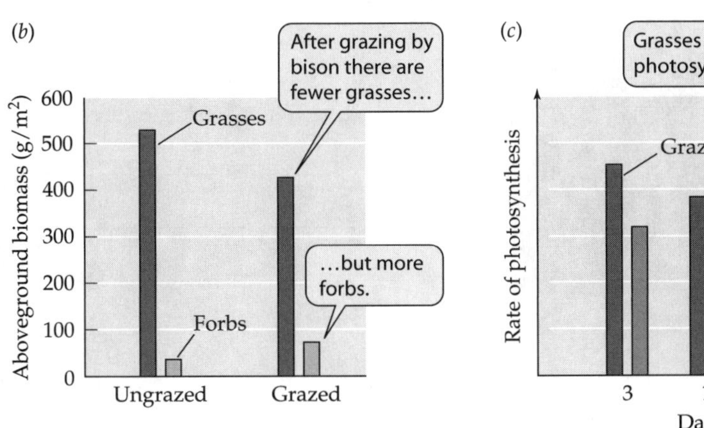
(b)
After grazing by bison there are fewer grasses…
Grasses
…but more forbs.
Forbs
Aboveground biomass (g/m²)
600
500
400
300
200
100
0
Ungrazed
Grazed
(c)
Grasses fertilized by bison urine photosynthesize faster.
Grazed
Ungrazed
Rate of photosynthesis
3
15
22
25
Date in July

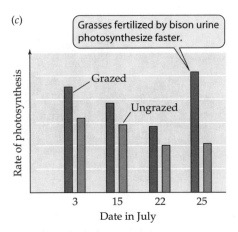

55.10 Grazing Increases Plant Species Richness and Productivity *(Page 1063)*

EXPERIMENT

Hypothesis: Medium-sized boulders have more species growing on them than either smaller or larger boulders because they are subjected to intermediate levels of disturbance.

METHOD Sterilize a number of small boulders. Secure some of them to the natural substratum with glue. Leave other boulders unsecured to serve as controls. Observe accumulation of species on the boulders over time.

RESULTS Secured small boulders accumulate many more species than unsecured small boulders.

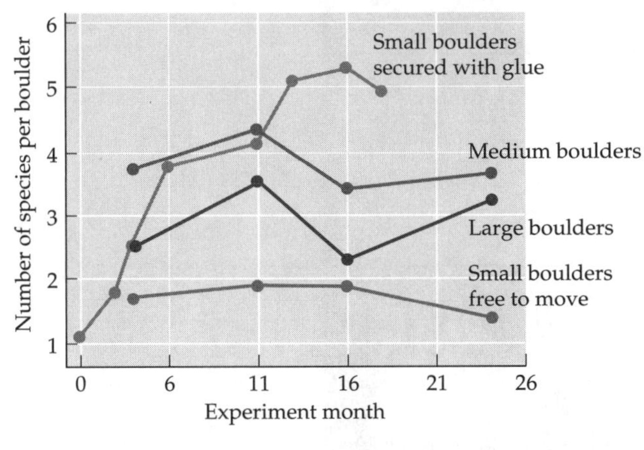
Small boulders secured with glue
Medium boulders
Large boulders
Small boulders free to move
Number of species per boulder
6
5
4
3
2
1
0
6
11
16
21
26
Experiment month

Conclusion: Small boulders have fewer species not because they are unsuitable for local species, but because the high rates at which they are moved by waves prevent many species from surviving on them.

55.13 Species Richness Is Greatest at Intermediate Levels of Disturbance *(Page 1064)*

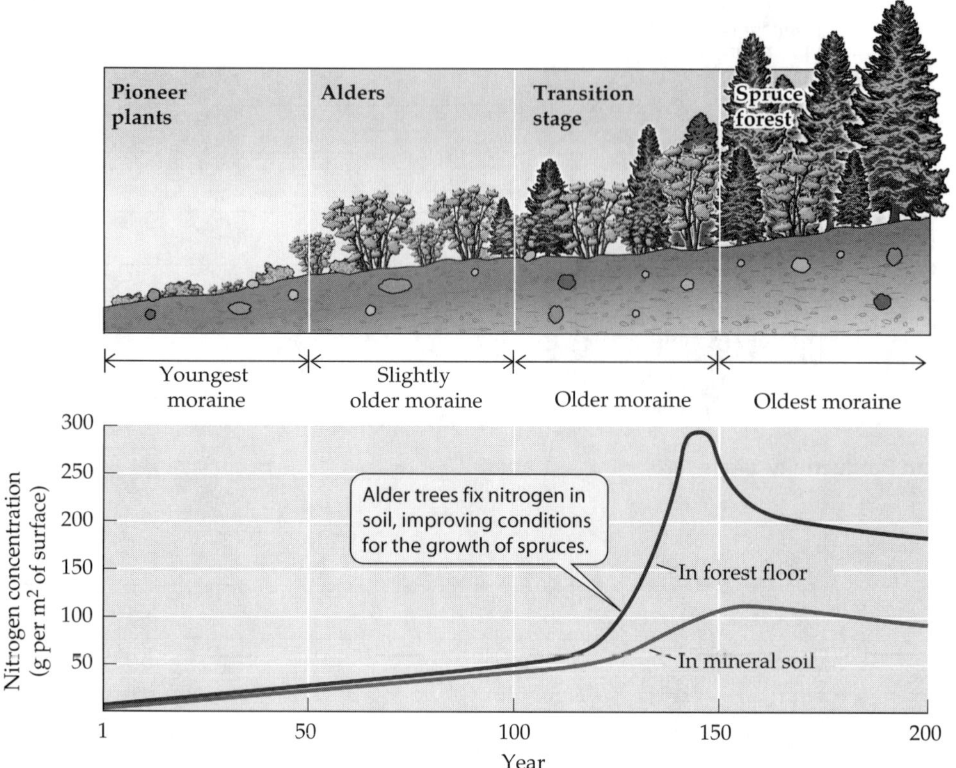
Pioneer plants
Alders
Transition stage
Spruce forest
Youngest moraine
Slightly older moraine
Older moraine
Oldest moraine
Alder trees fix nitrogen in soil, improving conditions for the growth of spruces.
In forest floor
In mineral soil
Nitrogen concentration (g per m² of surface)
300
250
200
150
100
50
1
50
100
150
200
Year

55.14 Primary Succession on a Glacial Moraine *(Page 1065)*

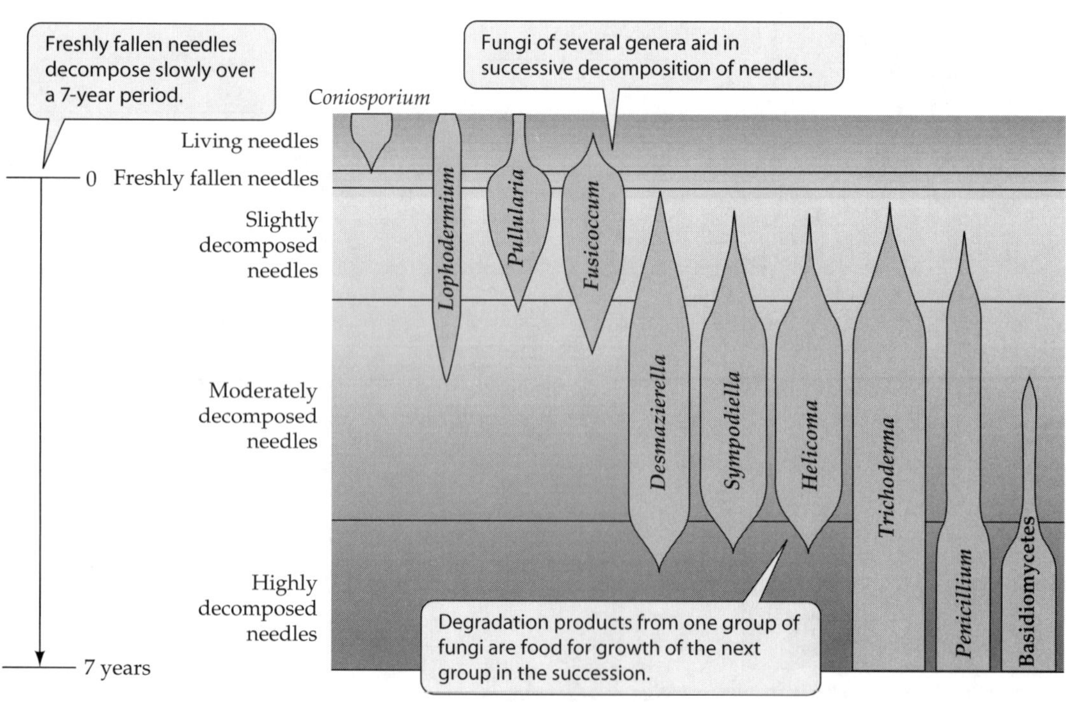
Freshly fallen needles decompose slowly over a 7-year period.
Fungi of several genera aid in successive decomposition of needles.
Coniosporium
Living needles
0 Freshly fallen needles
Slightly decomposed needles
Moderately decomposed needles
Highly decomposed needles
7 years
Lophodermium
Pullularia
Fusicoccum
Desmazierella
Sympodiella
Helicoma
Trichoderma
Penicillium
Basidiomycetes
Degradation products from one group of fungi are food for growth of the next group in the succession.

55.15 Secondary Succession on Pine Needles *(Page 1066)*

56 Biogeography

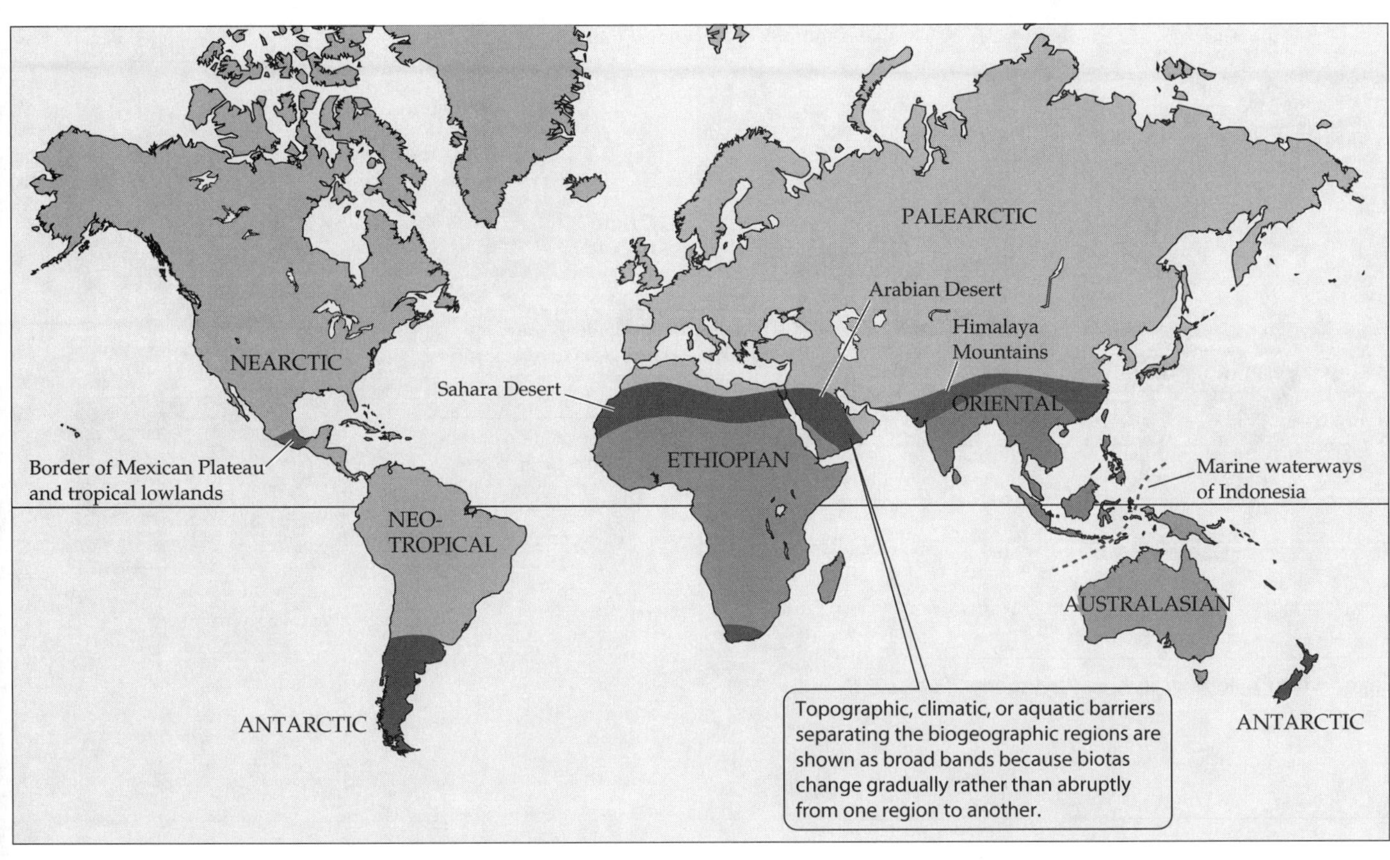

PALEARCTIC
Arabian Desert
Himalaya Mountains
NEARCTIC
Sahara Desert
ORIENTAL
ETHIOPIAN
Border of Mexican Plateau and tropical lowlands
Marine waterways of Indonesia
NEO-TROPICAL
AUSTRALASIAN
ANTARCTIC
Topographic, climatic, or aquatic barriers separating the biogeographic regions are shown as broad bands because biotas change gradually rather than abruptly from one region to another.
ANTARCTIC

56.1 Major Biogeographic Regions *(Page 1070)*

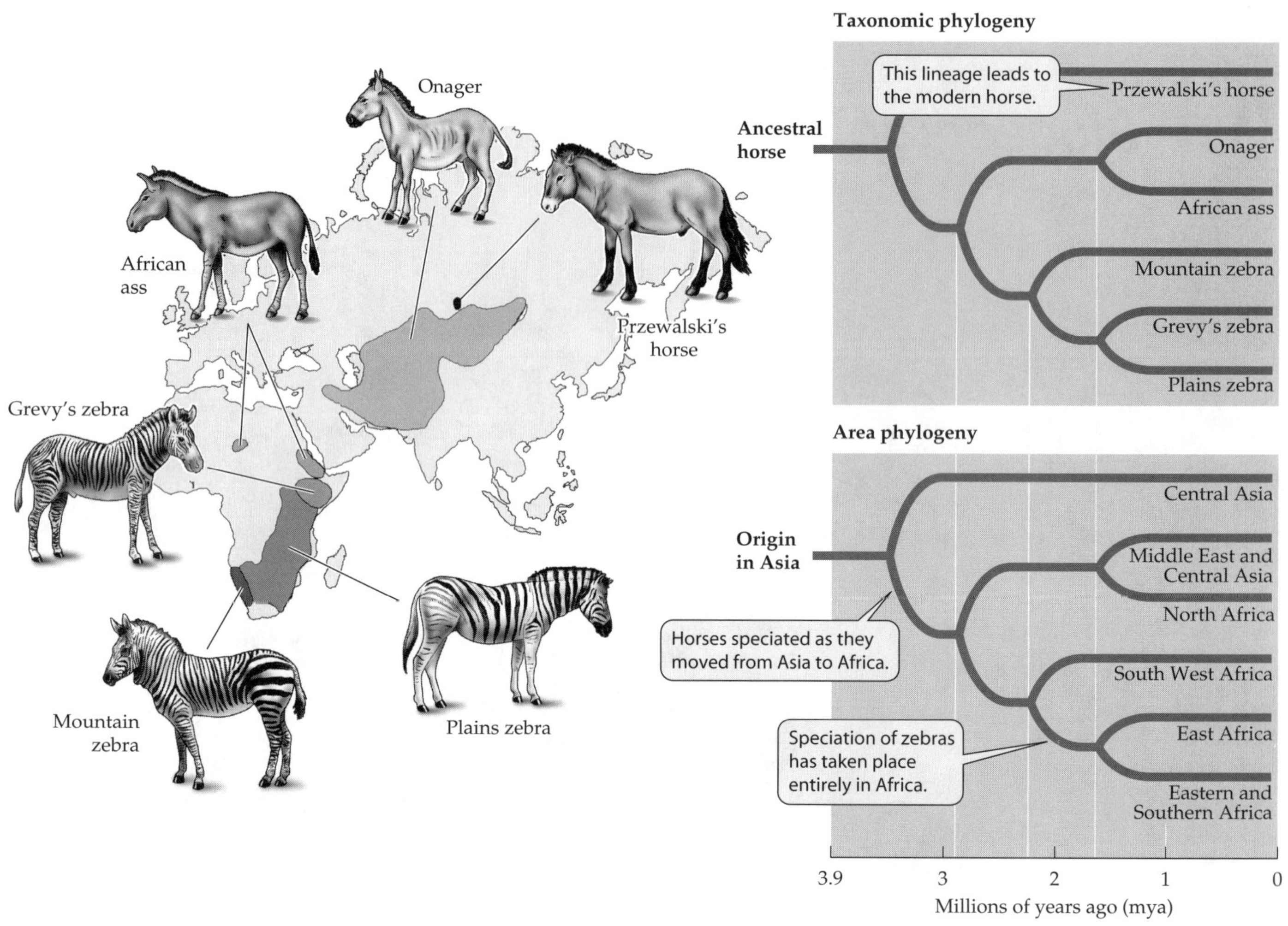
Onager
African ass
Przewalski's horse
Grevy's zebra
Mountain zebra
Plains zebra
Taxonomic phylogeny
This lineage leads to the modern horse.
Ancestral horse
Przewalski's horse
Onager
African ass
Mountain zebra
Grevy's zebra
Plains zebra
Area phylogeny
Origin in Asia
Central Asia
Middle East and Central Asia
North Africa
South West Africa
East Africa
Eastern and Southern Africa
Horses speciated as they moved from Asia to Africa.
Speciation of zebras has taken place entirely in Africa.
3.9
3
2
1
0
Millions of years ago (mya)

56.3 Taxonomic Phylogeny to Area Phylogeny *(Page 1073)*

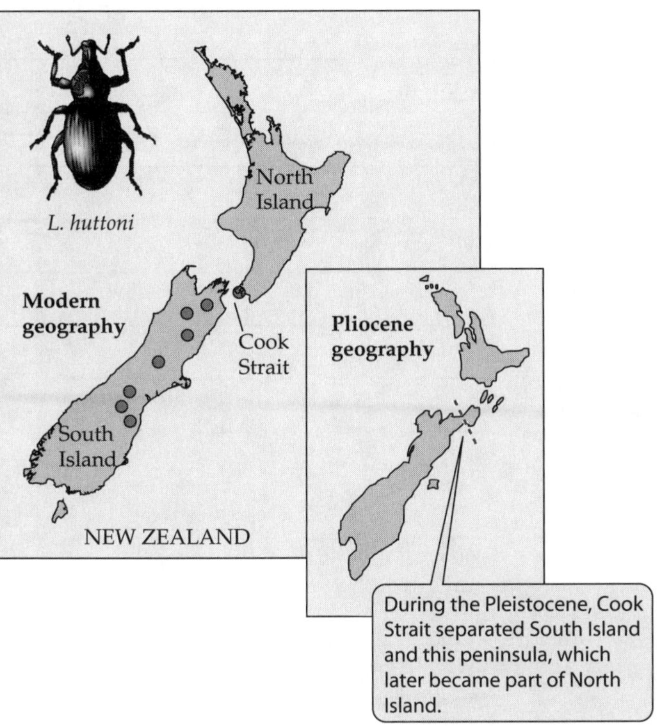
L. huttoni
Modern geography
North Island
Cook Strait
South Island
NEW ZEALAND
Pliocene geography
During the Pleistocene, Cook Strait separated South Island and this peninsula, which later became part of North Island.

56.4 A Vicariant Distribution Explained *(Page 1074)*

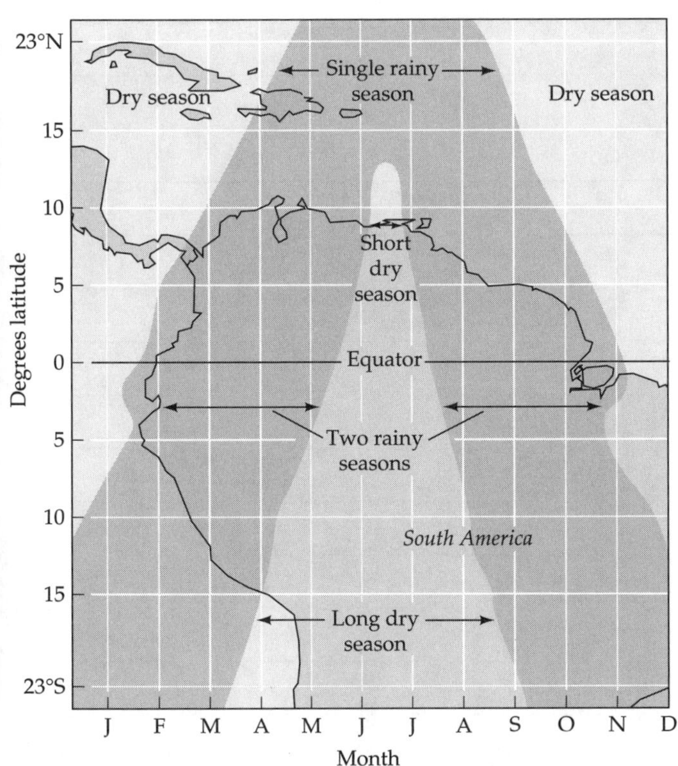
23°N
15
10
5
0
5
10
15
23°S
Degrees latitude
Dry season
Single rainy season
Dry season
Short dry season
Equator
Two rainy seasons
South America
Long dry season
J F M A M J J A S O N D
Month

56.5 Rainy and Dry Seasons Change with Latitude *(Page 1075)*

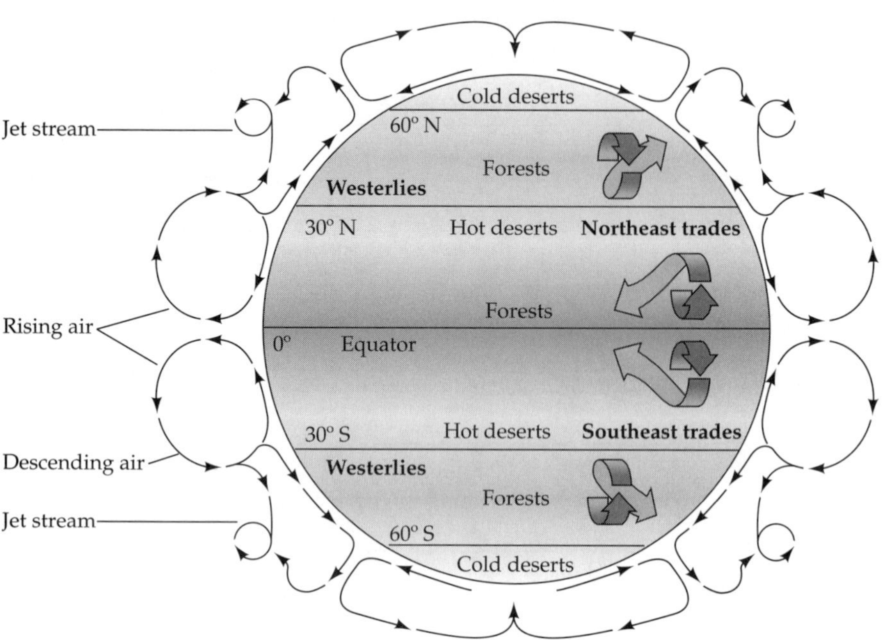
Cold deserts
60° N
Forests
Westerlies
30° N
Hot deserts
Northeast trades
Forests
0°
Equator
30° S
Hot deserts
Southeast trades
Westerlies
Forests
60° S
Cold deserts
Jet stream
Rising air
Descending air
Jet stream

56.6 The Circulation of Earth's Atmosphere *(Page 1075)*

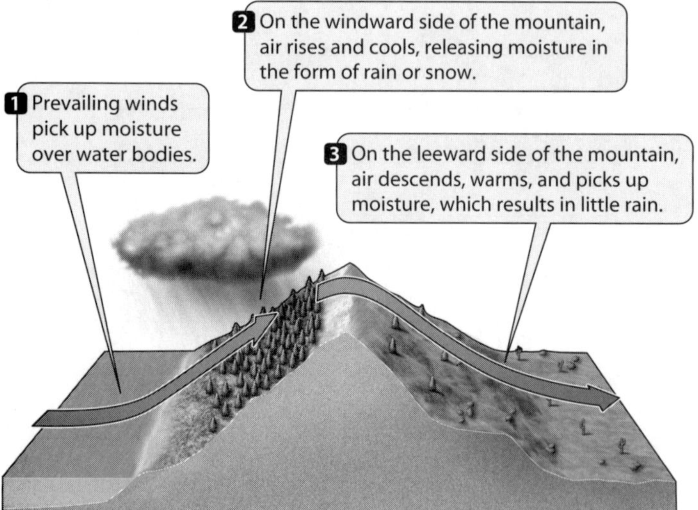
1 Prevailing winds pick up moisture over water bodies.
2 On the windward side of the mountain, air rises and cools, releasing moisture in the form of rain or snow.
3 On the leeward side of the mountain, air descends, warms, and picks up moisture, which results in little rain.

56.7 A Rain Shadow *(Page 1076)*

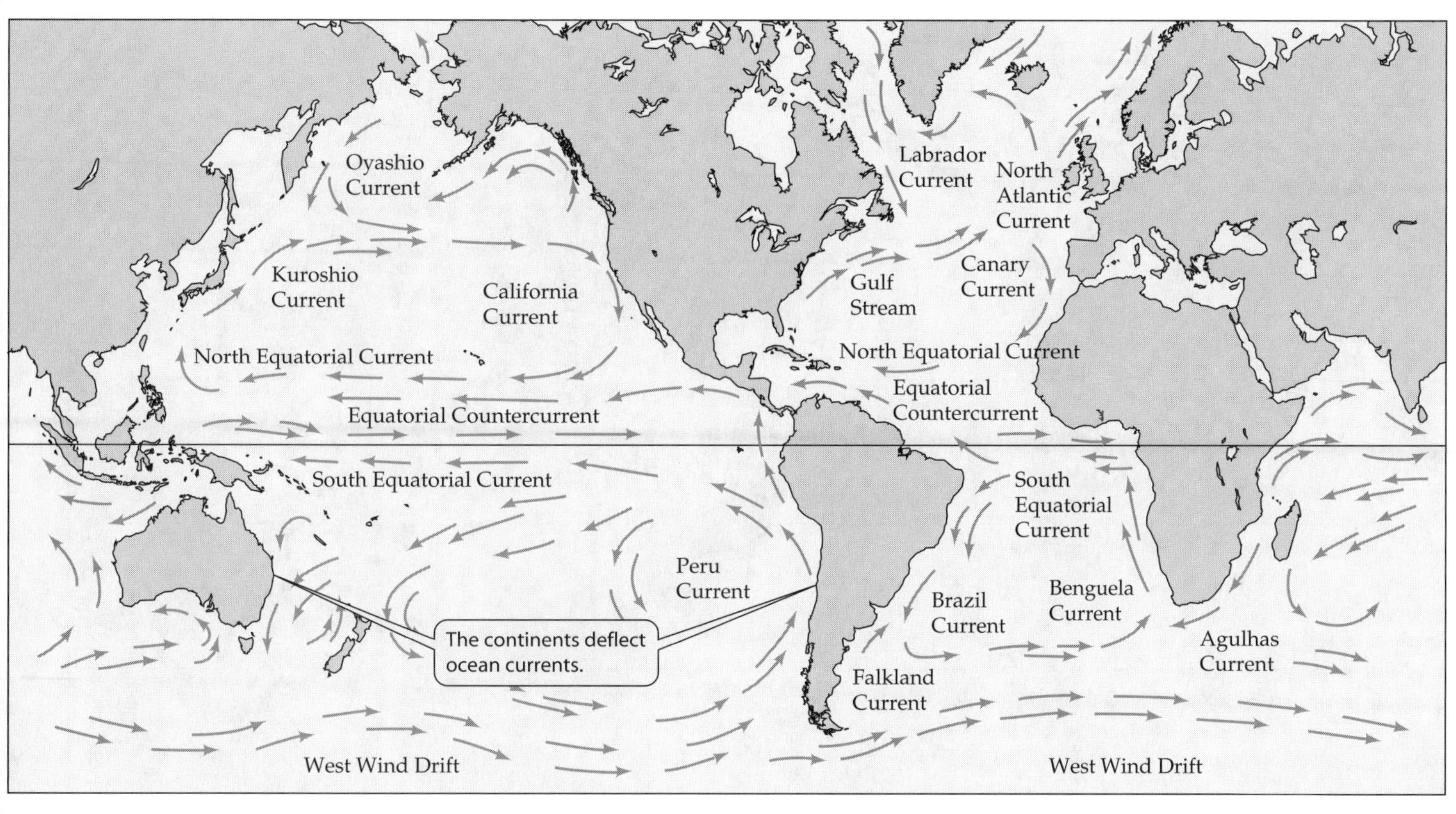
Oyashio Current
Kuroshio Current
California Current
North Equatorial Current
Equatorial Countercurrent
South Equatorial Current
Labrador Current
North Atlantic Current
Canary Current
Gulf Stream
North Equatorial Current
Equatorial Countercurrent
South Equatorial Current
Peru Current
Brazil Current
Benguela Current
Agulhas Current
Falkland Current
The continents deflect ocean currents.
West Wind Drift
West Wind Drift

56.8 Global Oceanic Circulation *(Page 1076)*

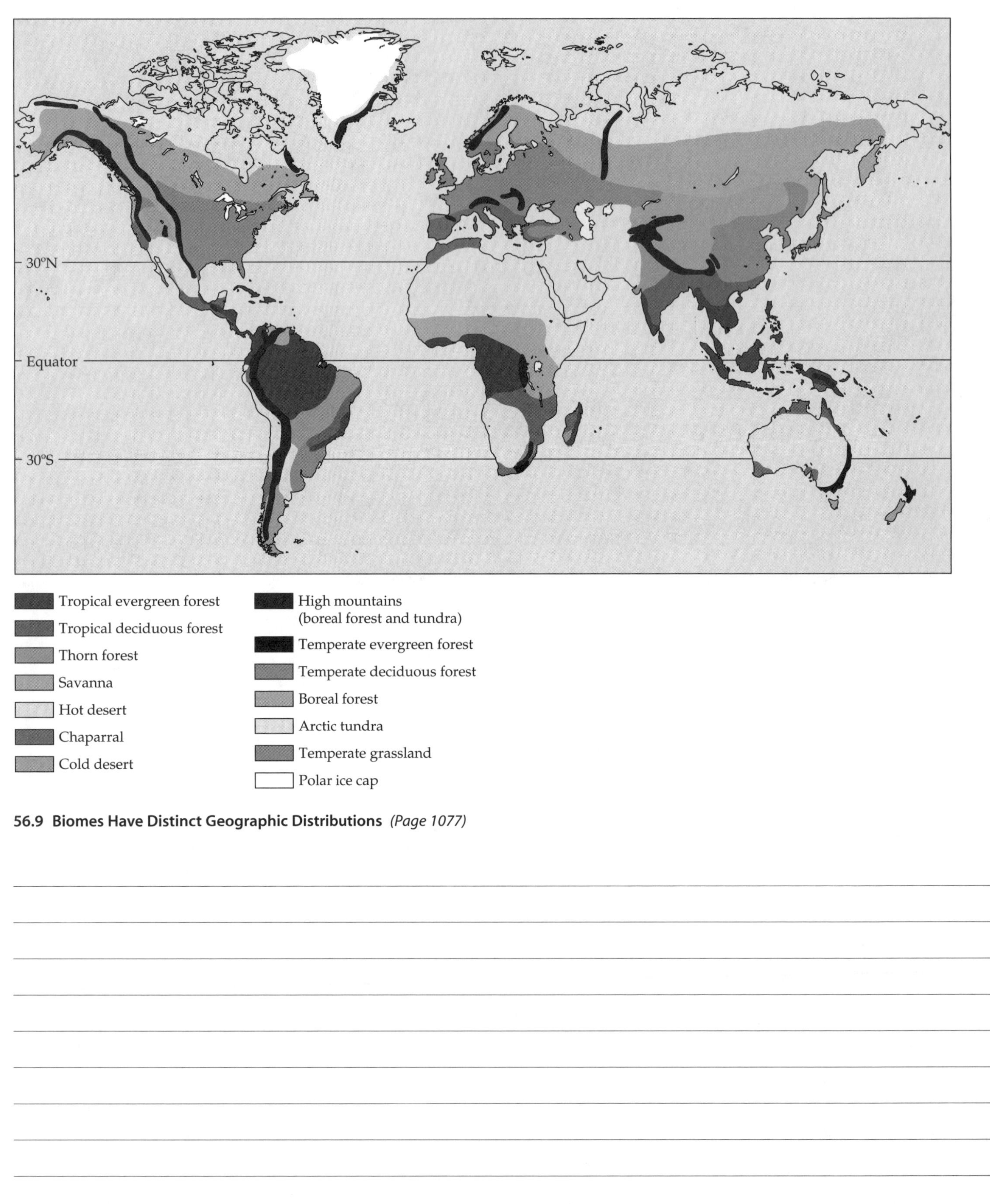
30°N
Equator
30°S
Tropical evergreen forest
Tropical deciduous forest
Thorn forest
Savanna
Hot desert
Chaparral
Cold desert
High mountains
(boreal forest and tundra)
Temperate evergreen forest
Temperate deciduous forest
Boreal forest
Arctic tundra
Temperate grassland
Polar ice cap

56.9 Biomes Have Distinct Geographic Distributions *(Page 1077)*

Equator

20°C is a "comfortable" 68°F.

0°C is the freezing point of water (=32°F).

Temperature

Upernavik, Greenland 73°N

Winter is very cold and long.

Summer is cool and short.

Range 28°C

°C: 20, 15, 10, 5, 0, −5, −10, −15, −20, −25

Jan Jul Dec

5 cm equals just over 2 inches.

Precipitation

cm: 5, 0

Annual total: 23 cm

Jan Jul Dec

Biological Activity

Photosynthesis

Flowering

Fruiting

Mammals

Birds

Insects

Soil Biota

Jan Jul Dec

Community Composition

Dominant Plants

Perennial herbs and small shrubs

Species Richness

Plants: Low; higher in tropical alpine
Animals: Low; many birds migrate in for summer; a few species of insects abundant in summer

Soil Biota

Few species

Tundra *(Page 1078)*

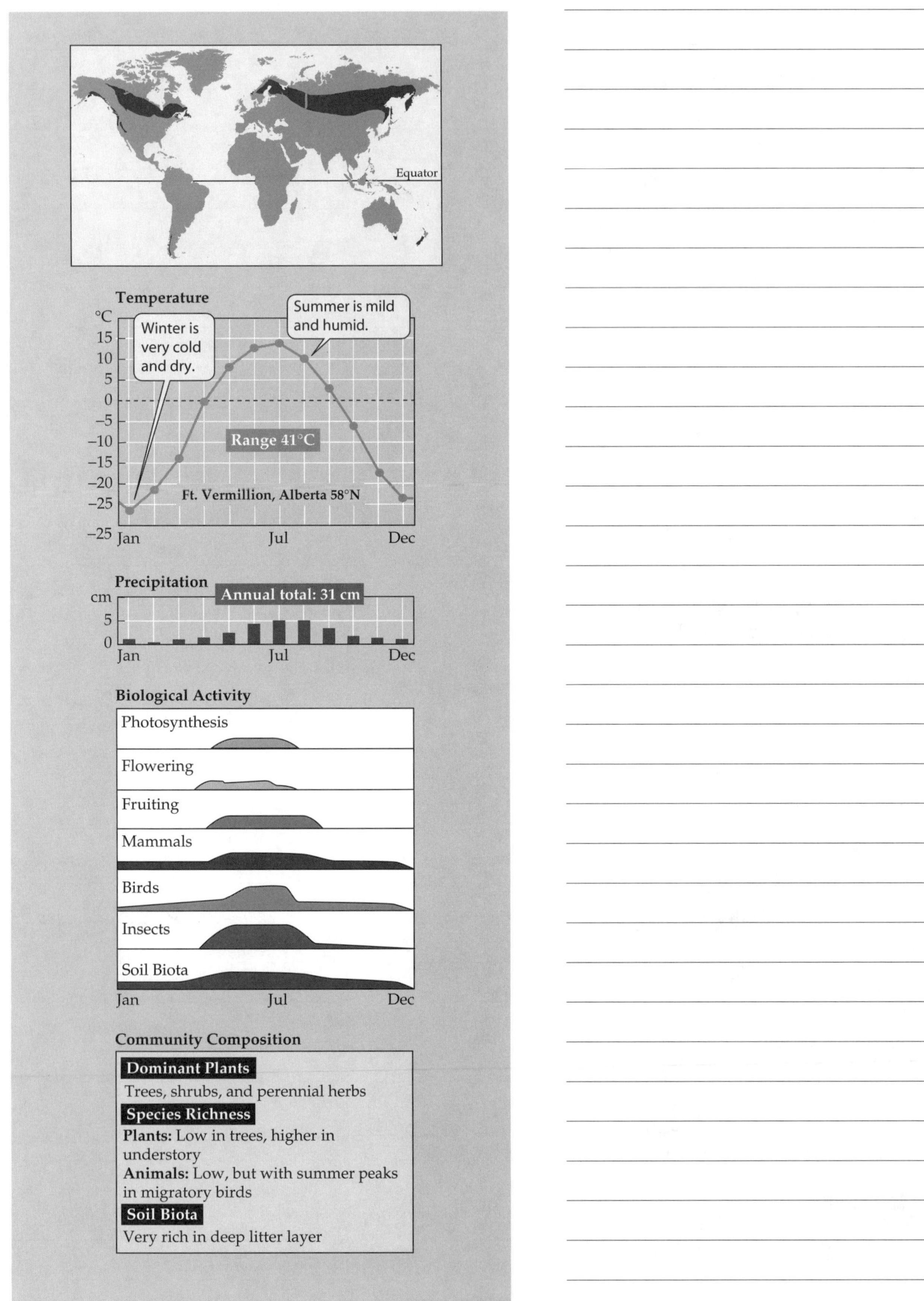
Equator
Temperature
°C
15
10
5
0
−5
−10
−15
−20
−25
−25
Winter is very cold and dry.
Summer is mild and humid.
Range 41°C
Ft. Vermillion, Alberta 58°N
Jan
Jul
Dec
Precipitation
cm
5
0
Annual total: 31 cm
Jan
Jul
Dec
Biological Activity
Photosynthesis
Flowering
Fruiting
Mammals
Birds
Insects
Soil Biota
Jan
Jul
Dec
Community Composition
Dominant Plants
Trees, shrubs, and perennial herbs
Species Richness
Plants: Low in trees, higher in understory
Animals: Low, but with summer peaks in migratory birds
Soil Biota
Very rich in deep litter layer

Boreal Forest *(Page 1079)*

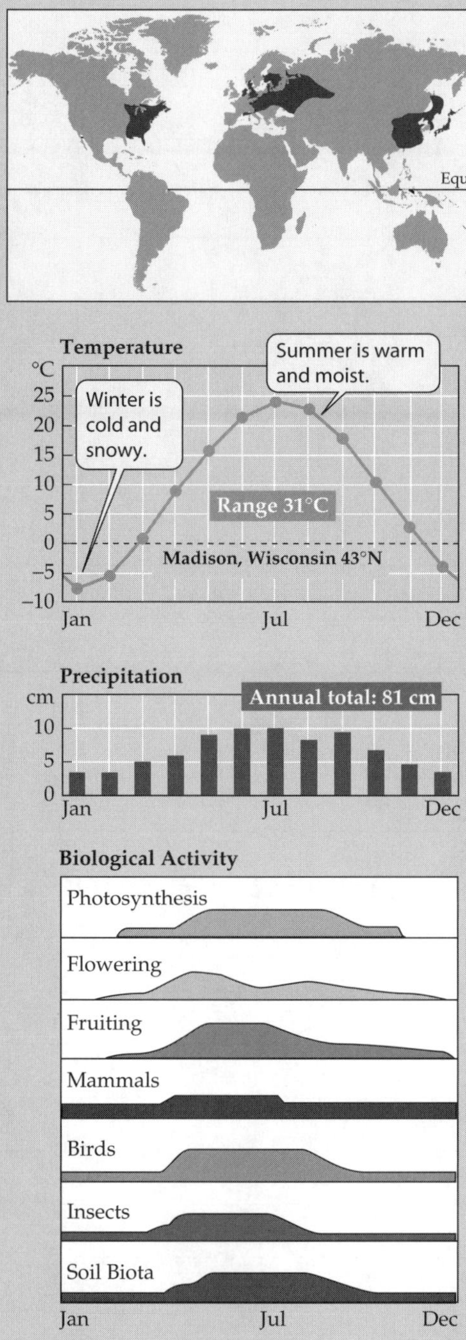

Equator
Temperature
°C
25
20
15
10
5
0
–5
–10
Winter is cold and snowy.
Summer is warm and moist.
Range 31°C
Madison, Wisconsin 43°N
Jan
Jul
Dec
Precipitation
cm
10
5
0
Annual total: 81 cm
Jan
Jul
Dec
Biological Activity
Photosynthesis
Flowering
Fruiting
Mammals
Birds
Insects
Soil Biota
Jan
Jul
Dec
Community Composition
Dominant Plants
Trees and shrubs
Species Richness
Plants: Many tree species in southeastern U.S. and eastern Asia, rich shrub layer
Animals: Rich; many migrant birds, richest amphibian communities on Earth, rich summer insect fauna
Soil Biota
Rich

Temperate Deciduous Forest *(Page 1080)*

Equator
Temperature
°C
30
25
20
15
10
5
0
–5
Summer is warm and wetter.
Winter is cold and dry.
Range 24°C
Pueblo, Colorado 38°N
Jan
Jul
Dec
Precipitation
cm
10
5
0
Annual total: 31 cm
Jan
Jul
Dec
Biological Activity
Photosynthesis
Flowering
Fruiting
Mammals
Birds
Insects
Soil Biota
Jan
Jul
Dec
Community Composition
Dominant Plants
Perennial grasses and forbs
Species Richness
Plants: Fairly high
Animals: Relatively few birds because of simple structure; mammals fairly rich
Soil Biota
Rich

Temperate Grasslands *(Page 1081)*

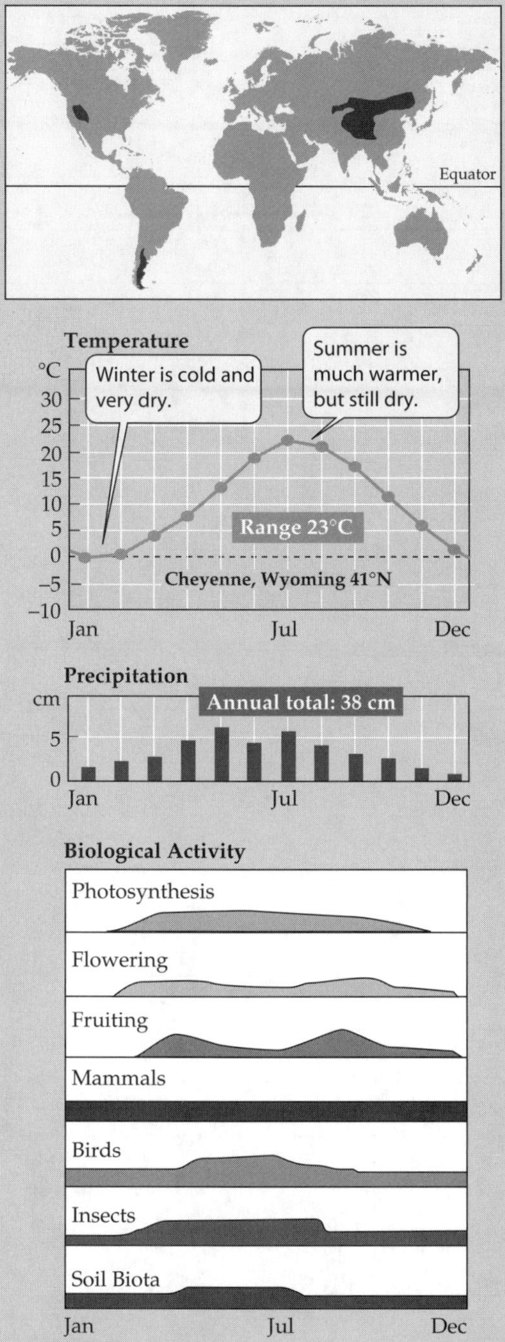

Equator

Temperature

°C

30
25
20
15
10
5
0
–5
–10

Winter is cold and very dry.

Summer is much warmer, but still dry.

Range 23°C

Cheyenne, Wyoming 41°N

Jan Jul Dec

Precipitation

cm

Annual total: 38 cm

5
0

Jan Jul Dec

Biological Activity

Photosynthesis

Flowering

Fruiting

Mammals

Birds

Insects

Soil Biota

Jan Jul Dec

Community Composition

Dominant Plants

Low stature shrubs and herbaceous plants

Species Richness

Plants: Few species

Animals: Rich in seed-eating birds, ants, and rodents; low in all other taxa

Soil Biota

Poor in species

Cold Desert *(Page 1082)*

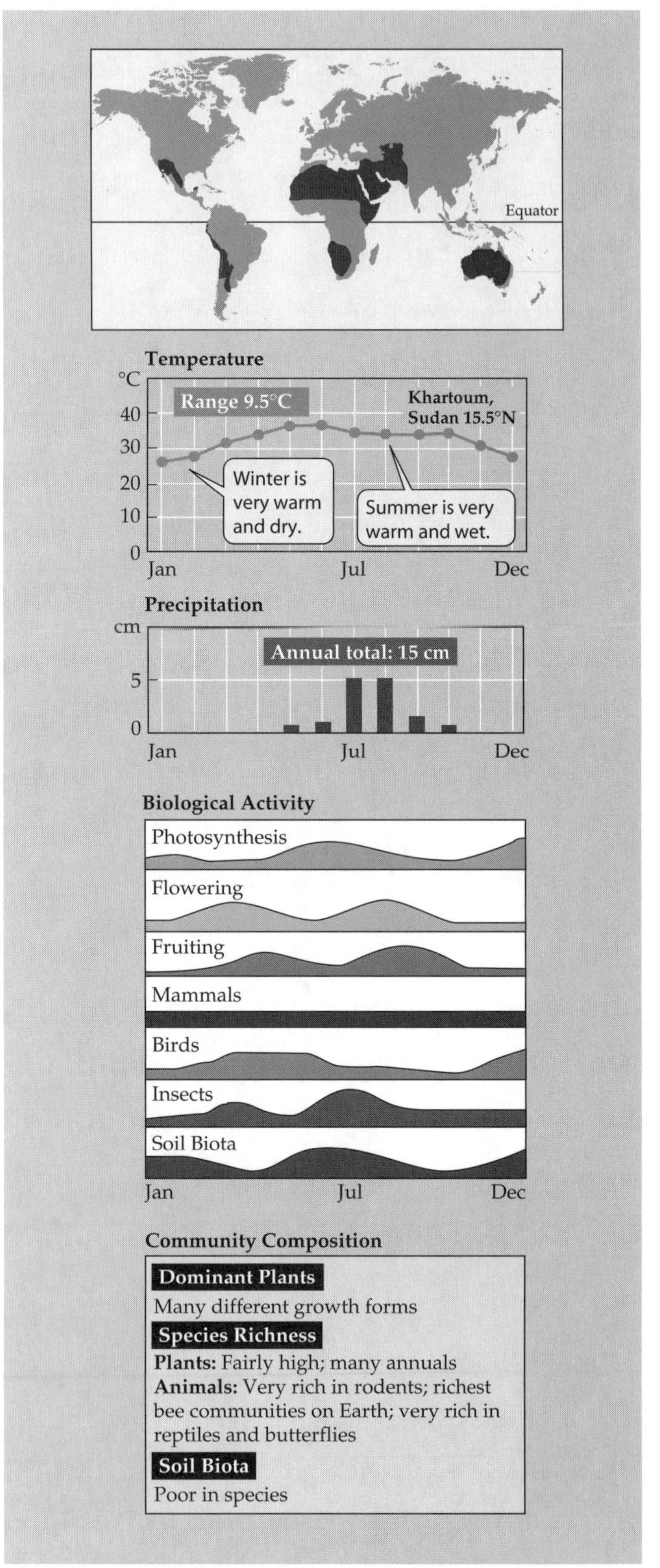
Equator
Temperature
°C
40
30
20
10
0
Range 9.5°C
Khartoum, Sudan 15.5°N
Winter is very warm and dry.
Summer is very warm and wet.
Jan
Jul
Dec
Precipitation
cm
5
0
Annual total: 15 cm
Jan
Jul
Dec
Biological Activity
Photosynthesis
Flowering
Fruiting
Mammals
Birds
Insects
Soil Biota
Jan
Jul
Dec
Community Composition
Dominant Plants
Many different growth forms
Species Richness
Plants: Fairly high; many annuals
Animals: Very rich in rodents; richest bee communities on Earth; very rich in reptiles and butterflies
Soil Biota
Poor in species

Hot Desert *(Page 1083)*

Equator
Temperature
°C
25
20
15
10
5
0
Winter is mild and humid.
Summer is mild and very dry.
Range 7°C
Monterey, California 36°N
Precipitation
cm
10
5
0
Annual total: 42 cm
Jan
Jul
Dec
Biological Activity
Photosynthesis
Flowering
Fruiting
Mammals
Birds
Insects
Soil Biota
Jan
Jul
Dec
Community Composition
Dominant Plants
Low stature shrubs and herbaceous plants
Species Richness
Plants: Extremely high in South Africa and Australia
Animals: Rich in rodents and reptiles; very rich in insects, especially bees
Soil Biota
Moderately rich

Chaparral *(Page 1084)*

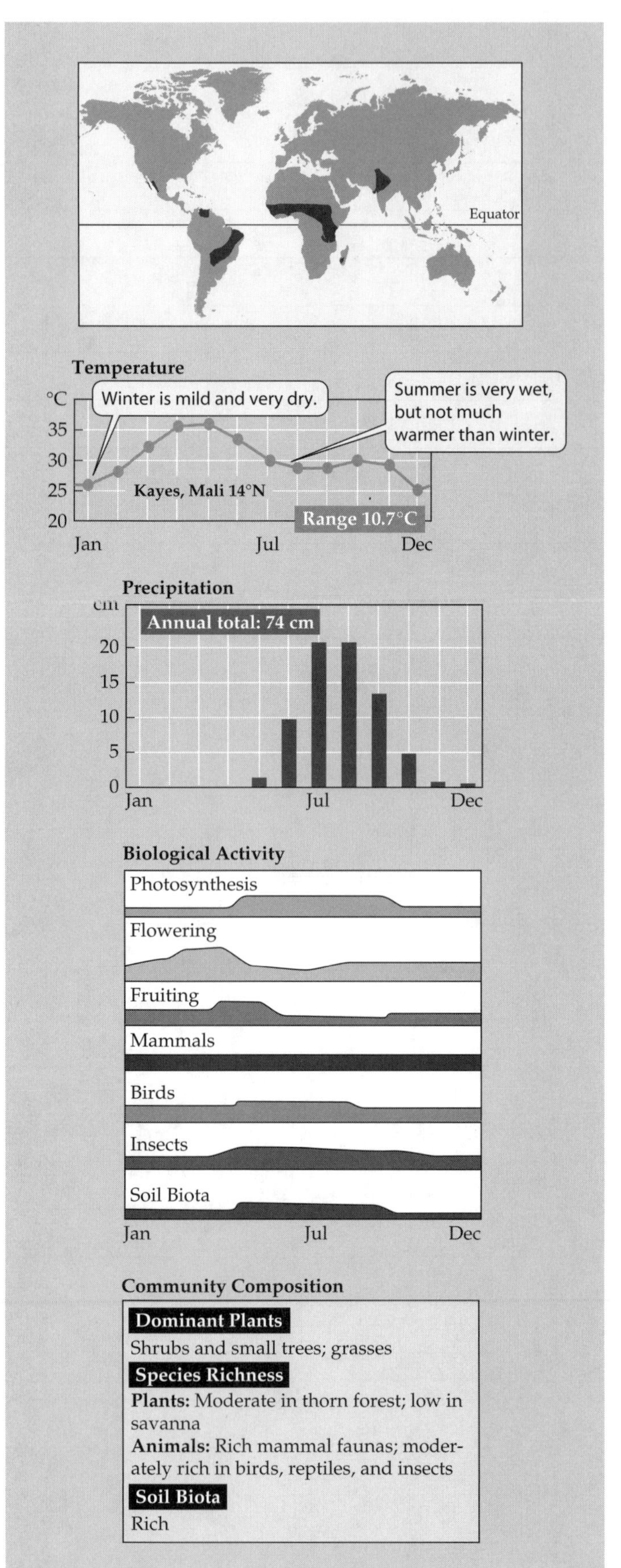
Equator
Temperature
°C
35
30
25
20
Winter is mild and very dry.
Summer is very wet, but not much warmer than winter.
Kayes, Mali 14°N
Range 10.7°C
Jan
Jul
Dec
Precipitation
cm
Annual total: 74 cm
20
15
10
5
0
Jan
Jul
Dec
Biological Activity
Photosynthesis
Flowering
Fruiting
Mammals
Birds
Insects
Soil Biota
Jan
Jul
Dec
Community Composition
Dominant Plants
Shrubs and small trees; grasses
Species Richness
Plants: Moderate in thorn forest; low in savanna
Animals: Rich mammal faunas; moderately rich in birds, reptiles, and insects
Soil Biota
Rich

Thorn Forest and Tropical Savanna *(Page 1085)*

Equator
Temperature
Winter is very warm and dry.
Summer is warm and wet.
°C
30
25
20
Range 5.4°C
Timbo, Guinea 10°N
Jan
Jul
Dec
Precipitation
cm
35
30
25
20
15
10
5
0
Annual total: 163 cm
Jan
Jul
Dec
Biological Activity
Photosynthesis
Flowering
Fruiting
Mammals
Birds
Insects
Soil Biota
Jan
Jul
Dec
Community Composition
Dominant Plants
Deciduous trees
Species Richness
Plants: Moderately rich in tree species
Animals: Rich mammal, bird, reptile, and amphibian communities; rich in insects
Soil Biota
Rich, but poorly known

Tropical Deciduous Forest *(Page 1086)*

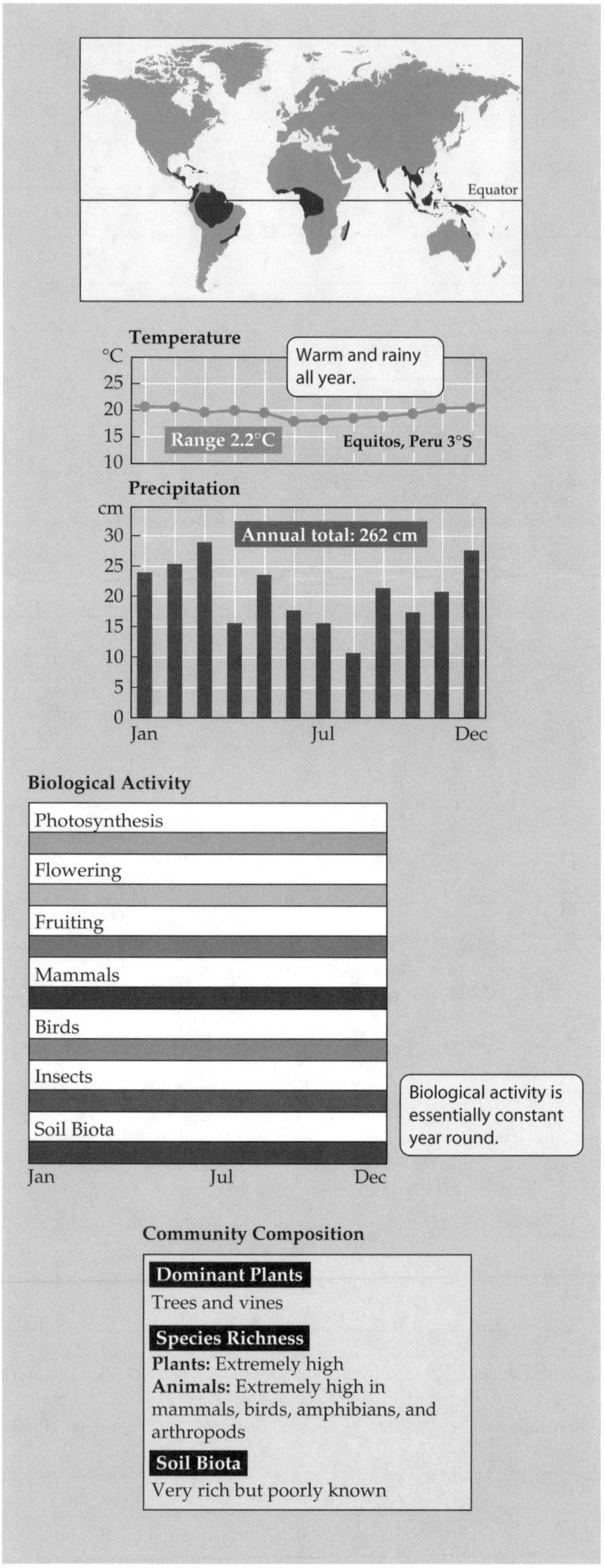

Equator

Temperature

°C

25
20
15
10

Warm and rainy all year.

Range 2.2°C

Equitos, Peru 3°S

Precipitation

cm

30
25
20
15
10
5
0

Annual total: 262 cm

Jan Jul Dec

Biological Activity

Photosynthesis

Flowering

Fruiting

Mammals

Birds

Insects

Soil Biota

Jan Jul Dec

Biological activity is essentially constant year round.

Community Composition

Dominant Plants

Trees and vines

Species Richness

Plants: Extremely high

Animals: Extremely high in mammals, birds, amphibians, and arthropods

Soil Biota

Very rich but poorly known

Tropical Evergreen Forest *(Page 1087)*

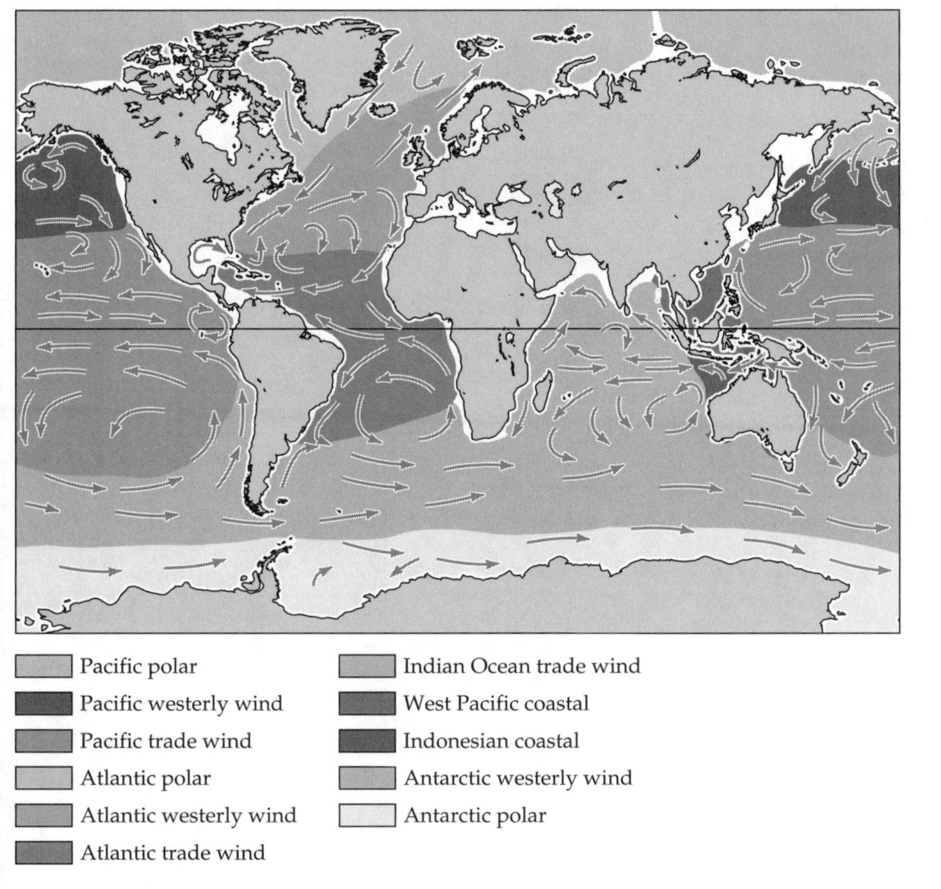
Pacific polar
Pacific westerly wind
Pacific trade wind
Atlantic polar
Atlantic westerly wind
Atlantic trade wind
Indian Ocean trade wind
West Pacific coastal
Indonesian coastal
Antarctic westerly wind
Antarctic polar

56.10 Oceanic Biogeographic Regions are Determined by Ocean Currents *(Page 1088)*

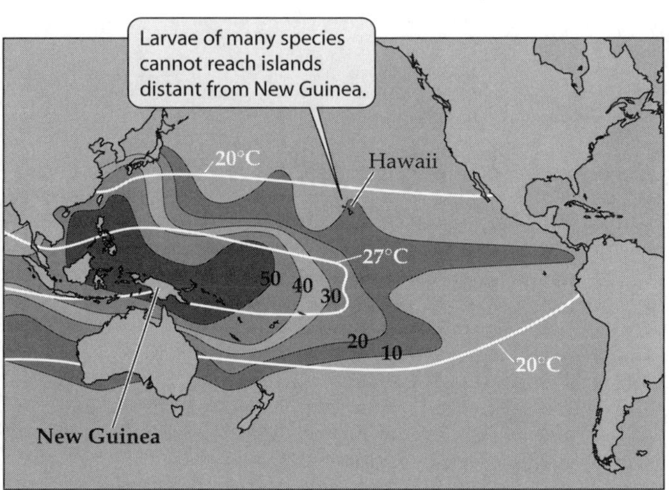
Larvae of many species cannot reach islands distant from New Guinea.
20°C
Hawaii
27°C
50
40
30
20
10
20°C
New Guinea

56.11 Generic Richness of Reef-Building Corals Declines with Distance from Indonesia *(Page 1089)*

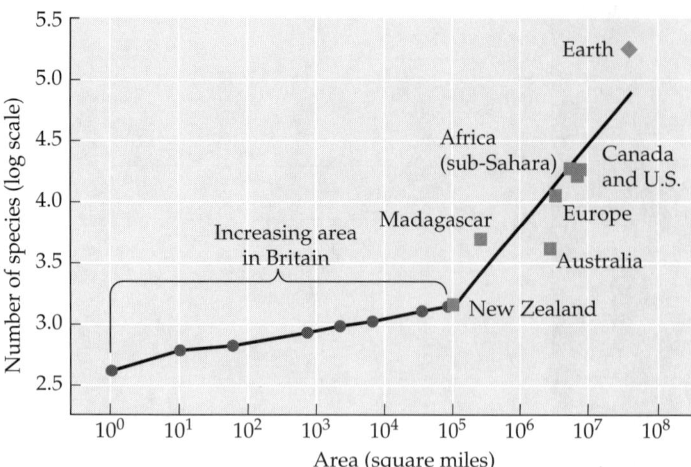

Number of species (log scale)
Area (square miles)
Earth
Africa (sub-Sahara)
Canada and U.S.
Madagascar
Europe
Increasing area in Britain
Australia
New Zealand

56.12 Species Richness Increases with Area Sampled *(Page 1089)*

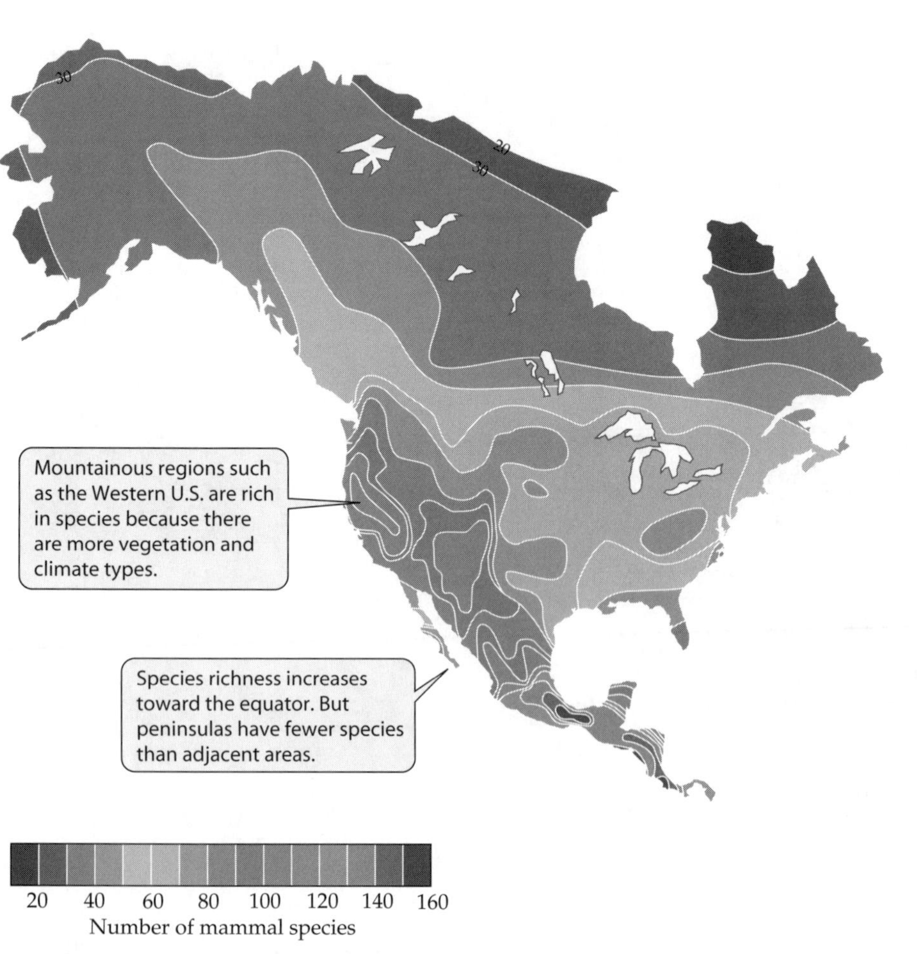

Mountainous regions such as the Western U.S. are rich in species because there are more vegetation and climate types.
Species richness increases toward the equator. But peninsulas have fewer species than adjacent areas.
20 40 60 80 100 120 140 160
Number of mammal species

56.13 The Latitudinal Gradient of Species Richness of North American Mammals *(Page 1089)*

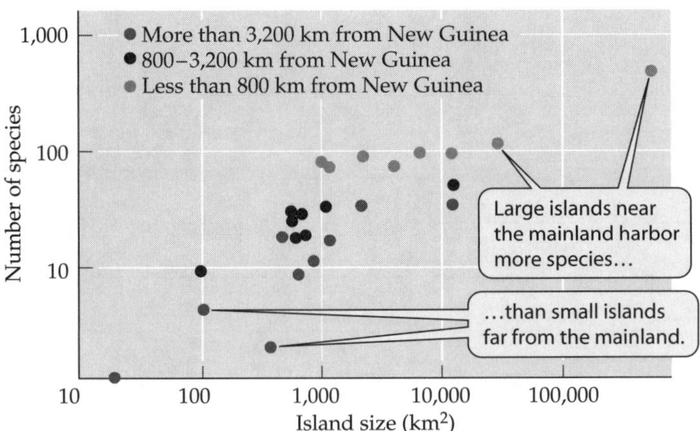
More than 3,200 km from New Guinea
800–3,200 km from New Guinea
Less than 800 km from New Guinea
Large islands near the mainland harbor more species…
…than small islands far from the mainland.
Number of species
1,000
100
10
10
100
1,000
10,000
100,000
Island size (km²)

56.14 Small, Distant Islands Have Fewer Bird Species *(Page 1090)*

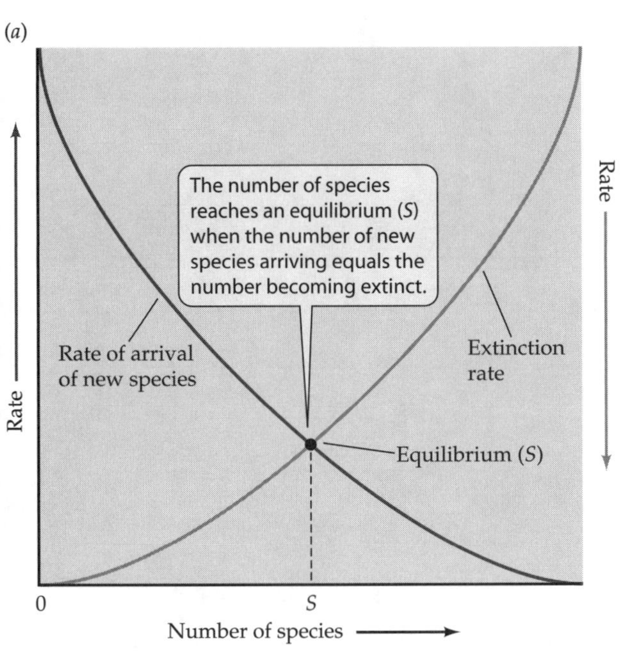
(a)
The number of species reaches an equilibrium (S) when the number of new species arriving equals the number becoming extinct.
Rate of arrival of new species
Extinction rate
Equilibrium (S)
Rate
Rate
0
S
Number of species

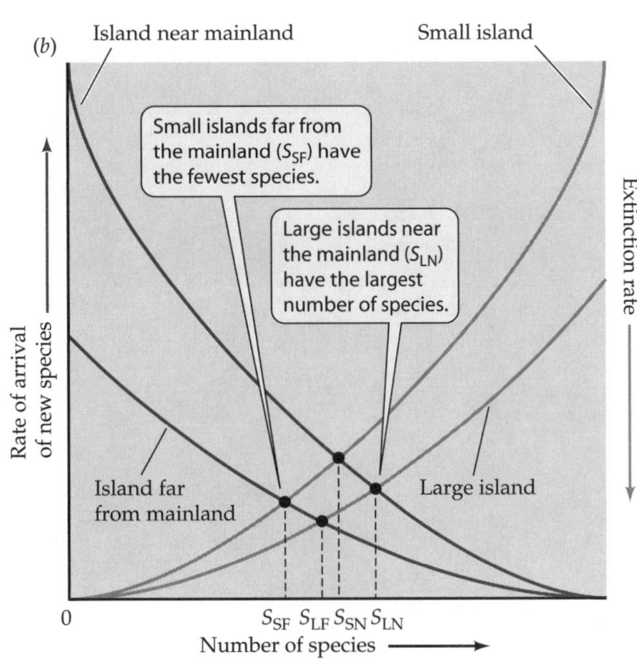
(b)
Island near mainland
Small island
Small islands far from the mainland (S_{SF}) have the fewest species.
Large islands near the mainland (S_{LN}) have the largest number of species.
Island far from mainland
Large island
Rate of arrival of new species
Extinction rate
0
S_{SF} S_{LF} S_{SN} S_{LN}
Number of species

56.15 MacArthur and Wilson's Model of Species Richness on Islands *(Page 1091)*

56.1 Number of Species of Resident Land Birds on Krakatau

PERIOD	NUMBER OF SPECIES	EXTINCTIONS	COLONIZATIONS
1908	13		
1908–1919		2	17
1919–1921	28		
1921–1933		3	4
1933–1934	29		
1934–1951		3	7
1951	33		
1952–1984		4	7
1984–1996	36		

(Page 1091)

57 Conservation Biology

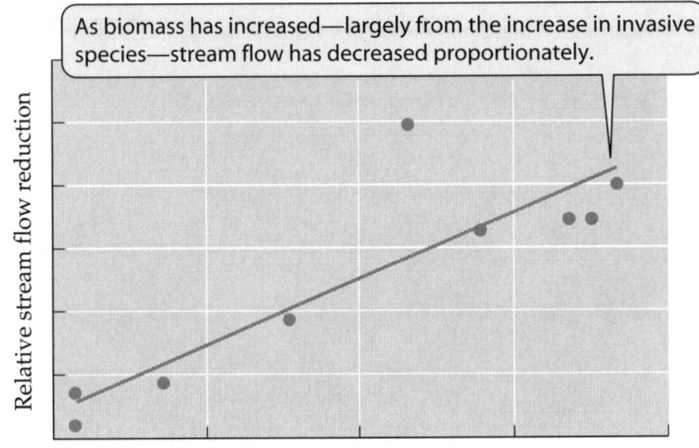

(b) Stream flow from fynbos watersheds
As biomass has increased—largely from the increase in invasive species—stream flow has decreased proportionately.
Relative stream flow reduction
Relative biomass

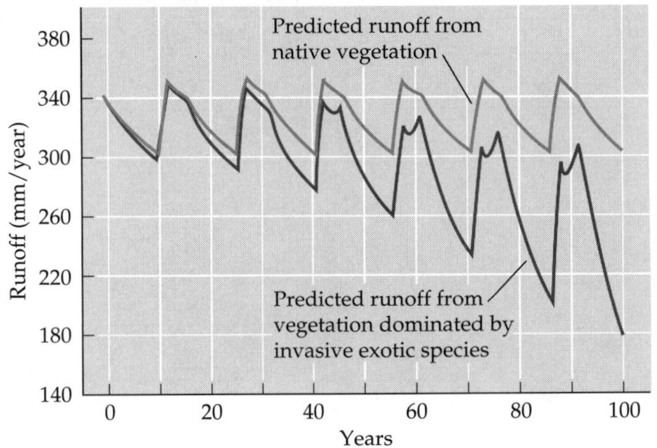

(c) Computer simulation
Predicted runoff from native vegetation
Predicted runoff from vegetation dominated by invasive exotic species
Runoff (mm/year)
380
340
300
260
220
180
140
0
20
40
60
80
100
Years

57.1 Invasive Species Disrupt Ecosystem Function *(Page 1095)*

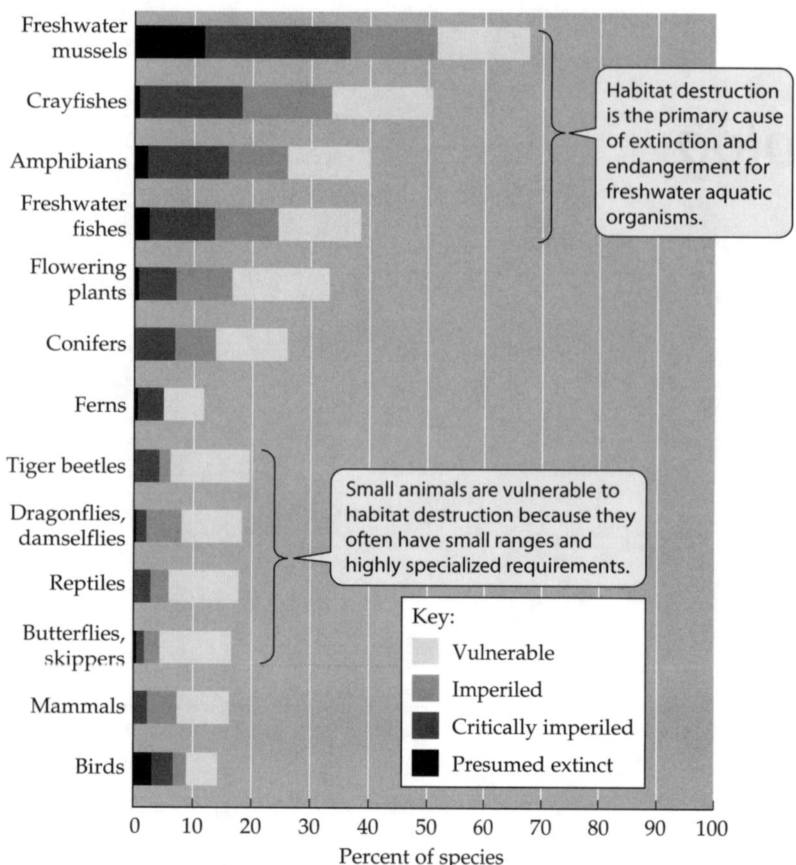
Freshwater mussels
Crayfishes
Amphibians
Freshwater fishes
Flowering plants
Conifers
Ferns
Tiger beetles
Dragonflies, damselflies
Reptiles
Butterflies, skippers
Mammals
Birds
Habitat destruction is the primary cause of extinction and endangerment for freshwater aquatic organisms.
Small animals are vulnerable to habitat destruction because they often have small ranges and highly specialized requirements.
Key:
Vulnerable
Imperiled
Critically imperiled
Presumed extinct
0 10 20 30 40 50 60 70 80 90 100
Percent of species

57.2 Proportions of U.S. Species Extinct or Endangered *(Page 1097)*

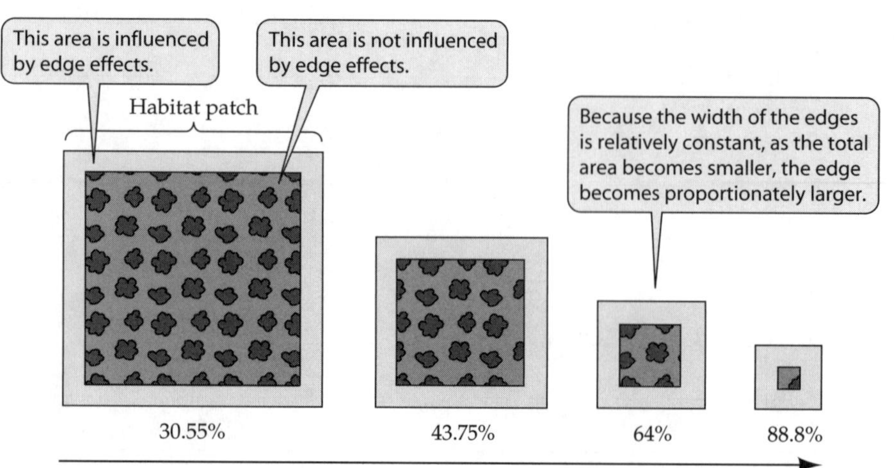
This area is influenced by edge effects.
This area is not influenced by edge effects.
Habitat patch
Because the width of the edges is relatively constant, as the total area becomes smaller, the edge becomes proportionately larger.
30.55%
43.75%
64%
88.8%
Percentage of patch influenced by edge effects

57.3 Edge Effects *(Page 1097)*

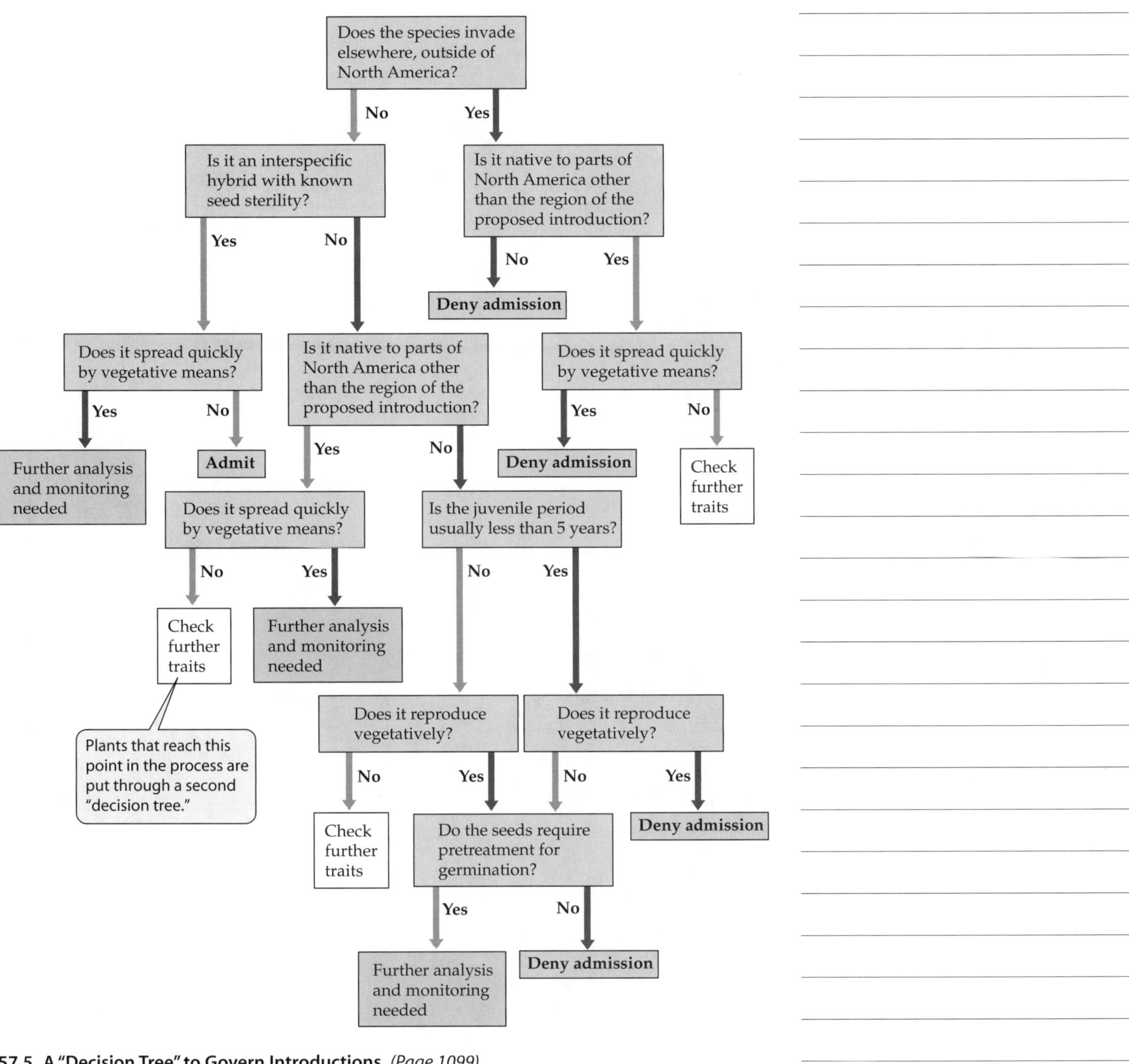
Does the species invade elsewhere, outside of North America?
No
Yes
Is it an interspecific hybrid with known seed sterility?
Yes
No
Is it native to parts of North America other than the region of the proposed introduction?
No
Yes
Deny admission
Does it spread quickly by vegetative means?
Yes
No
Further analysis and monitoring needed
Admit
Is it native to parts of North America other than the region of the proposed introduction?
Yes
No
Does it spread quickly by vegetative means?
Yes
No
Deny admission
Check further traits
Does it spread quickly by vegetative means?
No
Yes
Check further traits
Further analysis and monitoring needed
Plants that reach this point in the process are put through a second "decision tree."
Is the juvenile period usually less than 5 years?
No
Yes
Does it reproduce vegetatively?
No
Yes
Does it reproduce vegetatively?
No
Yes
Check further traits
Do the seeds require pretreatment for germination?
Deny admission
Yes
No
Further analysis and monitoring needed
Deny admission

57.5 A "Decision Tree" to Govern Introductions *(Page 1099)*

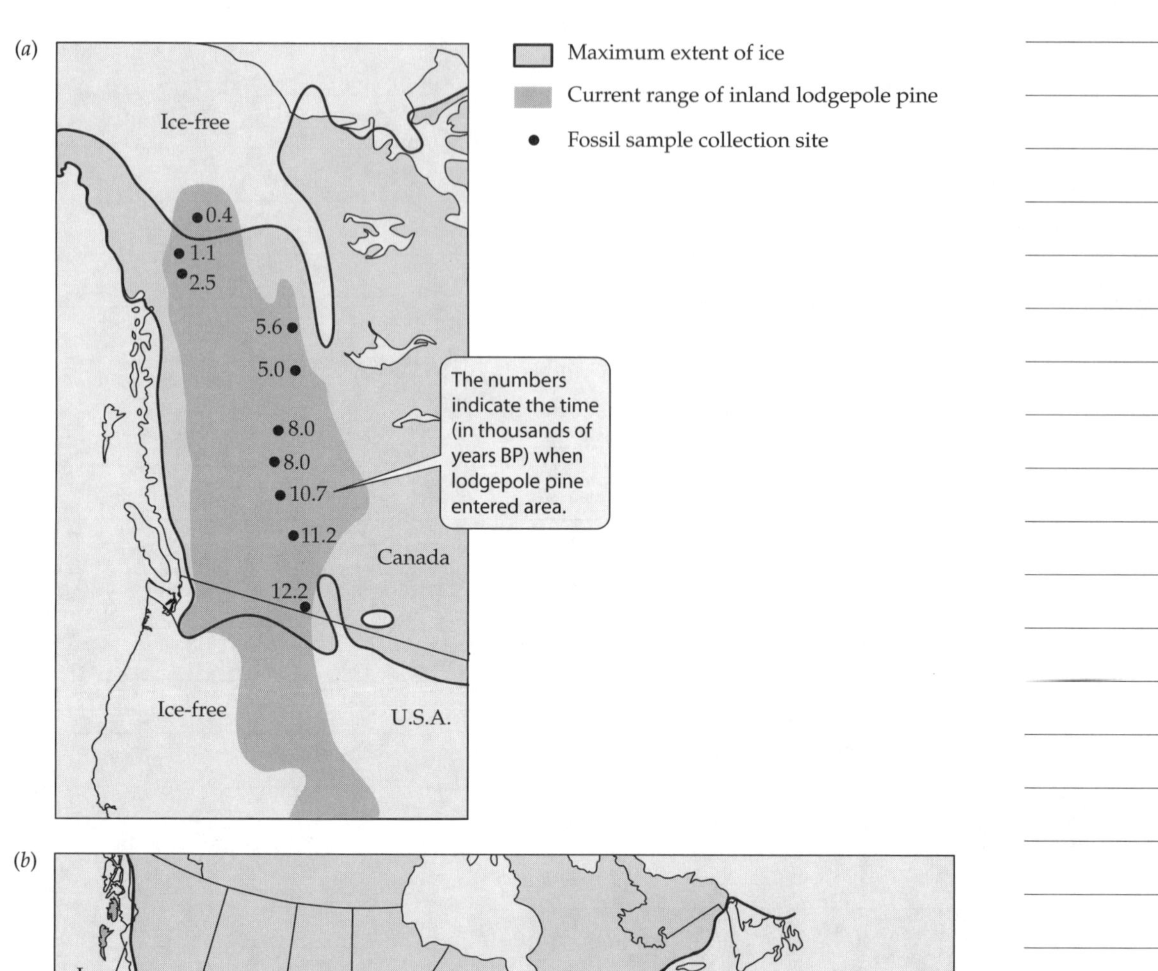
(a)
Maximum extent of ice
Current range of inland lodgepole pine
Fossil sample collection site
Ice-free
0.4
1.1
2.5
5.6
5.0
8.0
8.0
10.7
11.2
12.2
The numbers indicate the time (in thousands of years BP) when lodgepole pine entered area.
Canada
Ice-free
U.S.A.

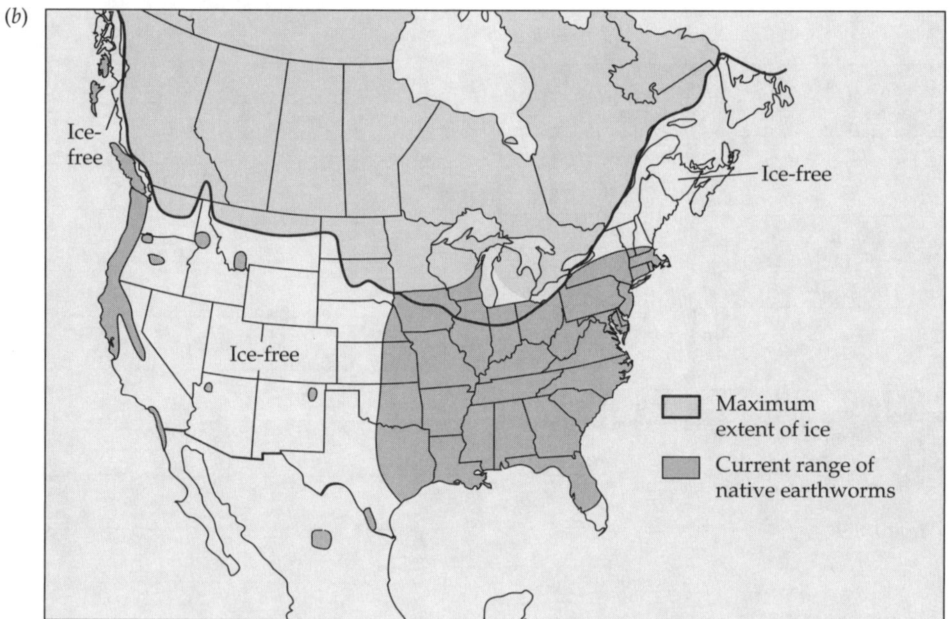
(b)
Ice-free
Ice-free
Ice-free
Maximum extent of ice
Current range of native earthworms

57.7 Some Species Shift Their Ranges in Response to Climate Change *(Page 1101)*

(*a*)

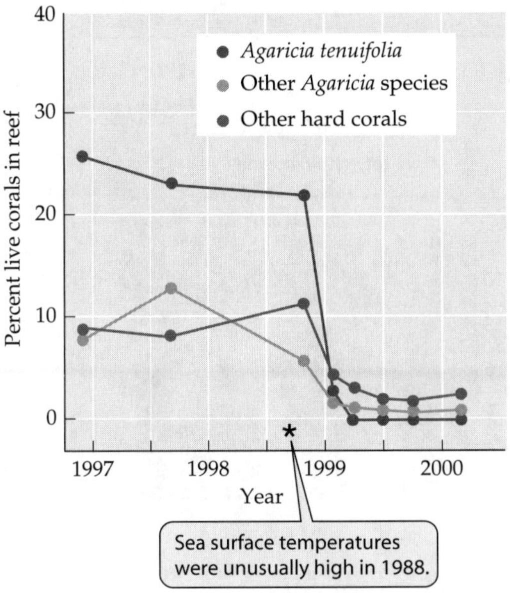

40
30
20
10
0
Percent live corals in reef
Agaricia tenuifolia
Other Agaricia species
Other hard corals
1997
1998
1999
2000
Year
Sea surface temperatures were unusually high in 1988.

57.8 Global Warming Affects Corals *(Page 1102)*

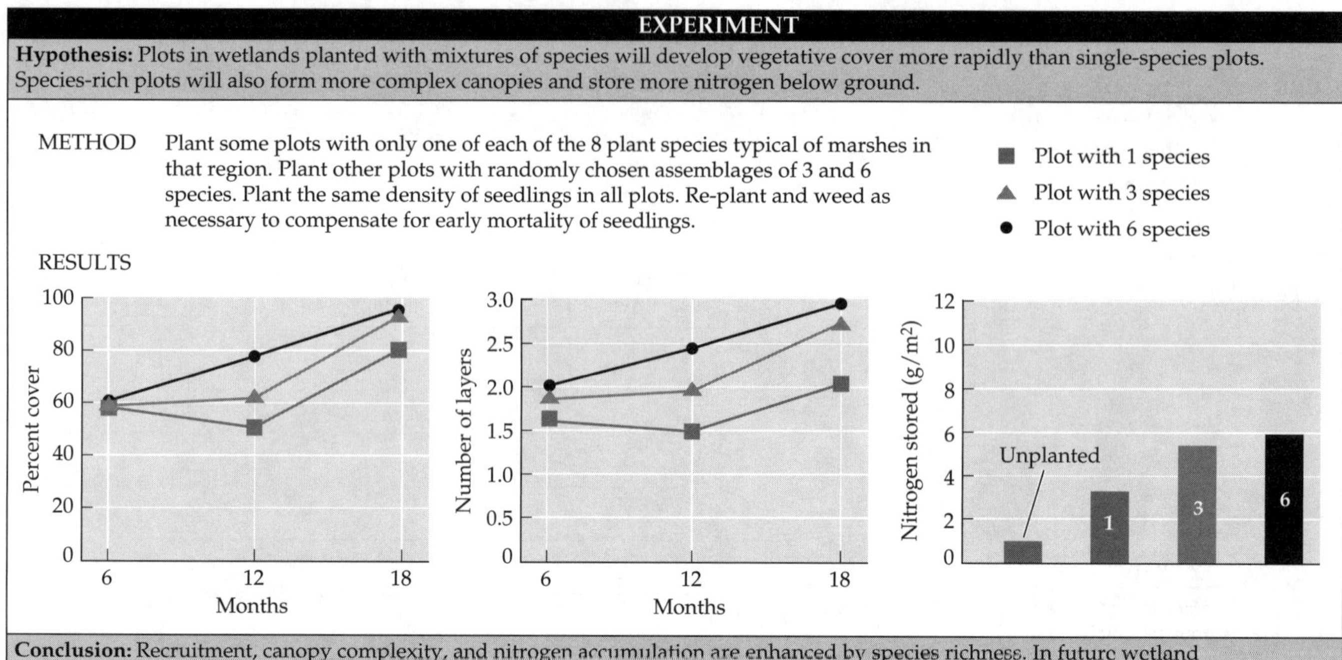
EXPERIMENT
Hypothesis: Plots in wetlands planted with mixtures of species will develop vegetative cover more rapidly than single-species plots. Species-rich plots will also form more complex canopies and store more nitrogen below ground.
METHOD Plant some plots with only one of each of the 8 plant species typical of marshes in that region. Plant other plots with randomly chosen assemblages of 3 and 6 species. Plant the same density of seedlings in all plots. Re-plant and weed as necessary to compensate for early mortality of seedlings.
Plot with 1 species
Plot with 3 species
Plot with 6 species
RESULTS
Percent cover
100
80
60
40
20
0
6
12
18
Months
Number of layers
3.0
2.5
2.0
1.5
1.0
0.5
0
6
12
18
Months
Nitrogen stored (g/m²)
12
10
8
6
4
2
0
Unplanted
1
3
6
Conclusion: Recruitment, canopy complexity, and nitrogen accumulation are enhanced by species richness. In future wetland restoration attempts, a rich mixture of species should be planted.

57.9 Species Richness Enhances Wetland Restoration *(Page 1103)*

58 Earth System Science

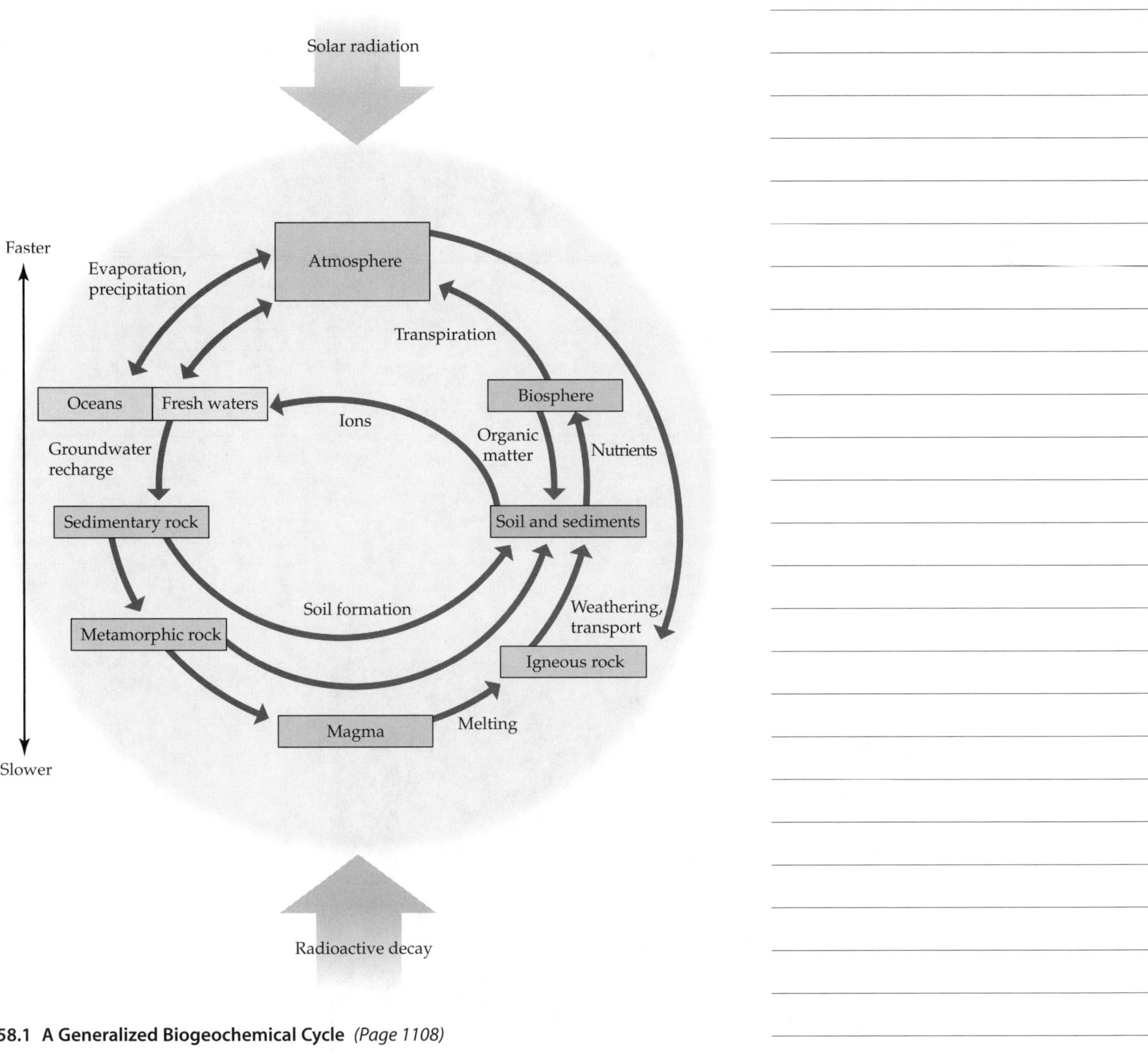
Solar radiation
Faster
Slower
Atmosphere
Evaporation, precipitation
Transpiration
Oceans
Fresh waters
Biosphere
Ions
Organic matter
Nutrients
Groundwater recharge
Sedimentary rock
Soil and sediments
Soil formation
Weathering, transport
Metamorphic rock
Igneous rock
Magma
Melting
Radioactive decay

58.1 A Generalized Biogeochemical Cycle *(Page 1108)*

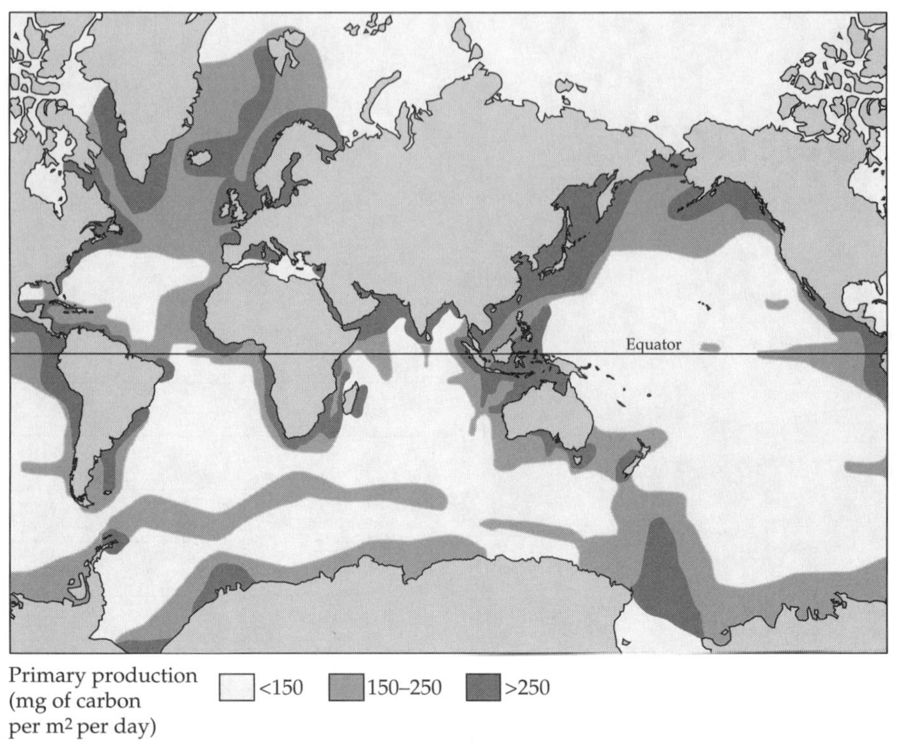
Equator
Primary production (mg of carbon per m2 per day)
<150
150–250
>250

58.2 Primary Production Is High in Zones of Upwelling *(Page 1109)*

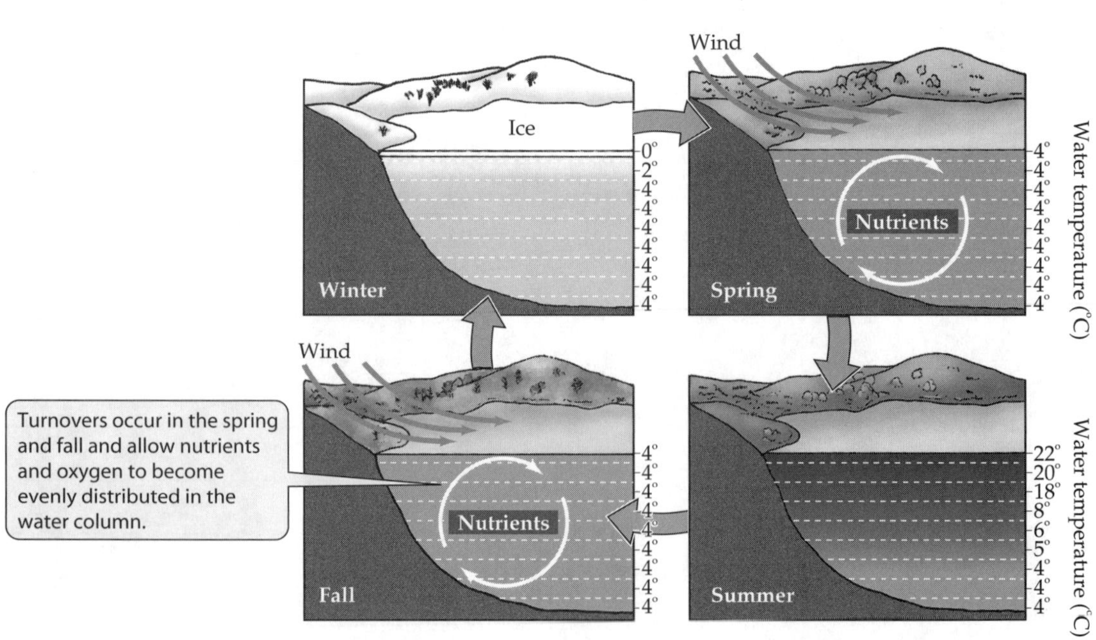
Ice
Wind
0°
2°
4°
Winter
Spring
Nutrients
Water temperature (°C)
Wind
Turnovers occur in the spring and fall and allow nutrients and oxygen to become evenly distributed in the water column.
Fall
Summer
22°
20°
18°
8°
6°
5°
4°

58.3 Annual Temperature and Oxygen Cycles in a Temperate Lake *(Page 1110)*

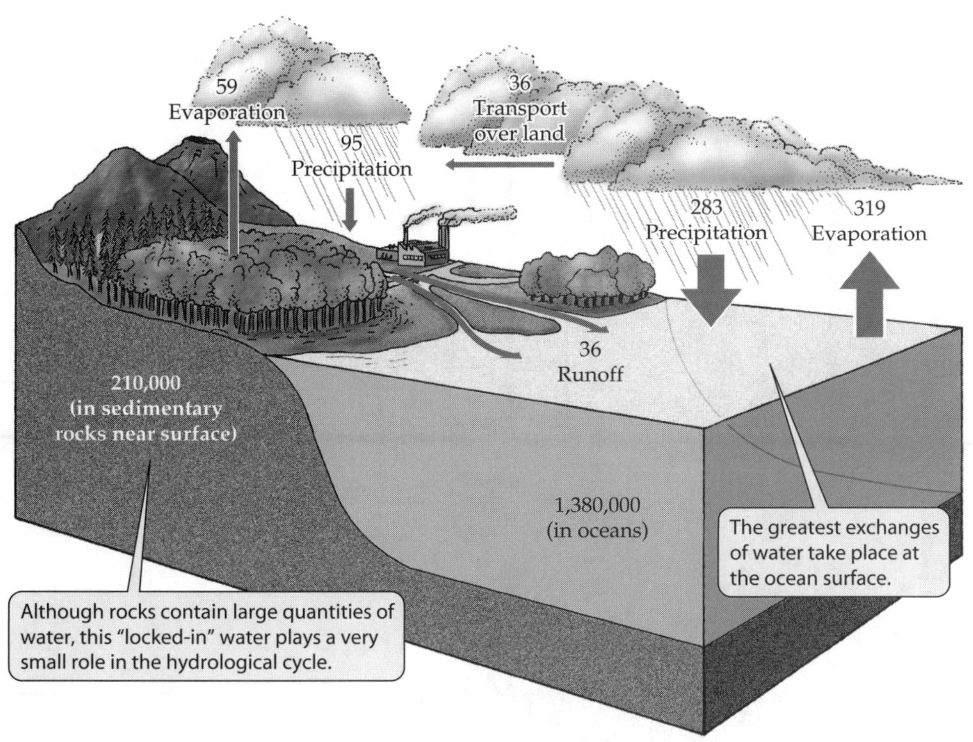
59
Evaporation
95
Precipitation
36
Transport over land
283
Precipitation
319
Evaporation
36
Runoff
210,000 (in sedimentary rocks near surface)
1,380,000 (in oceans)
The greatest exchanges of water take place at the ocean surface.
Although rocks contain large quantities of water, this "locked-in" water plays a very small role in the hydrological cycle.

58.4 The Global Hydrological Cycle *(Page 1111)*

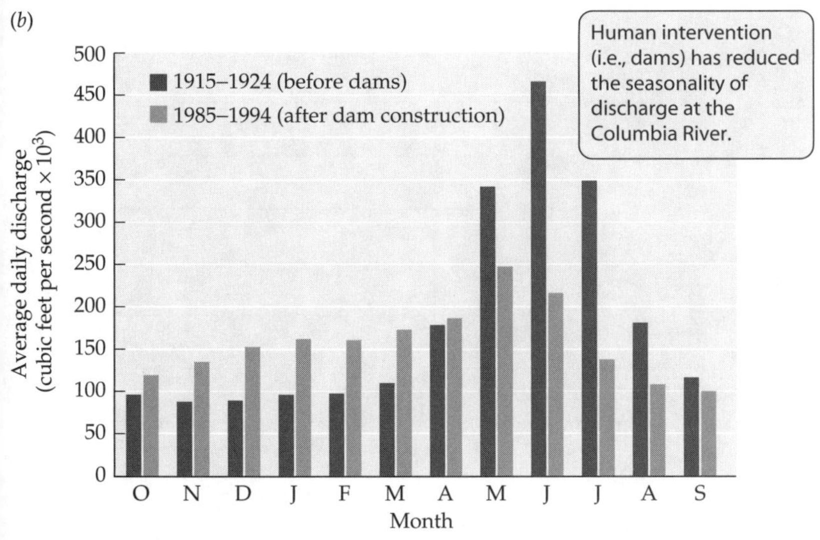
(b)
1915–1924 (before dams)
1985–1994 (after dam construction)
Human intervention (i.e., dams) has reduced the seasonality of discharge at the Columbia River.
Average daily discharge (cubic feet per second × 10³)
500
450
400
350
300
250
200
150
100
50
0
O N D J F M A M J J A S
Month

58.5 Columbia River Flows Have Been Massively Altered *(Page 1112)*

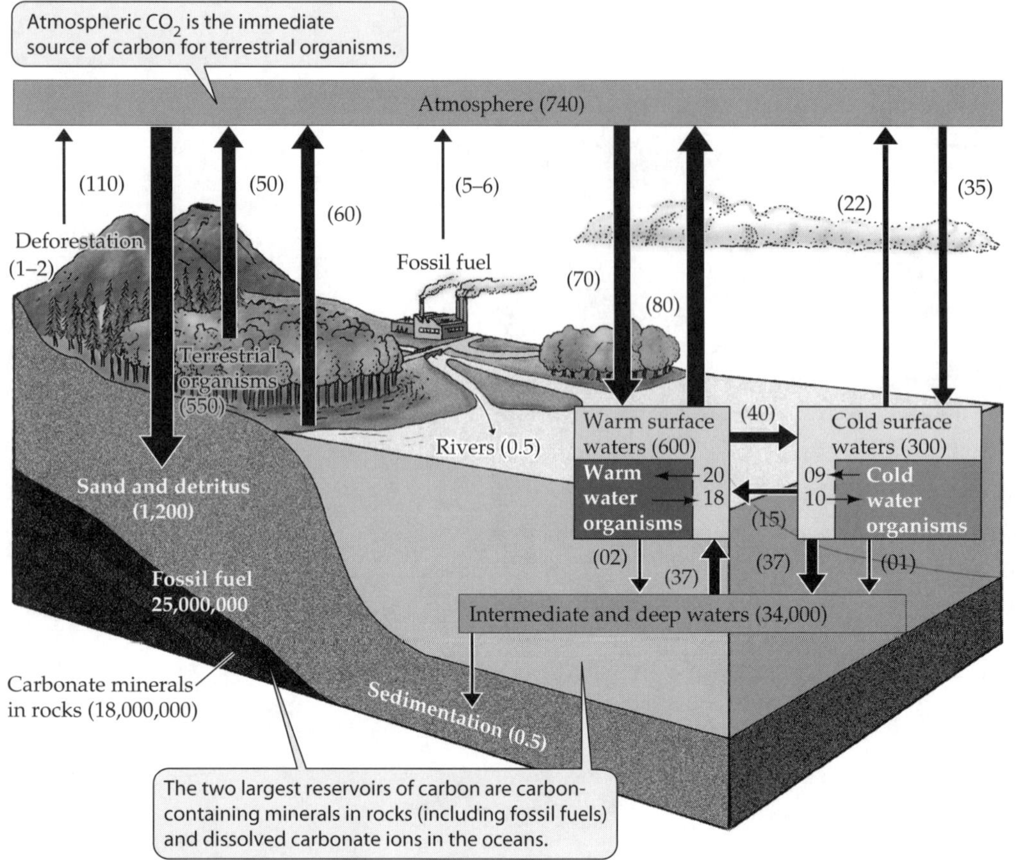
Atmospheric CO_2 is the immediate source of carbon for terrestrial organisms.
Atmosphere (740)
(110)
(50)
(60)
(5–6)
Deforestation (1–2)
Fossil fuel
(70)
(80)
(22)
(35)
Terrestrial organisms (550)
Sand and detritus (1,200)
Fossil fuel 25,000,000
Rivers (0.5)
Warm surface waters (600)
(40)
Cold surface waters (300)
Warm water organisms
20
18
09
10
Cold water organisms
(15)
(02)
(37)
(37)
(01)
Intermediate and deep waters (34,000)
Carbonate minerals in rocks (18,000,000)
Sedimentation (0.5)
The two largest reservoirs of carbon are carbon-containing minerals in rocks (including fossil fuels) and dissolved carbonate ions in the oceans.

58.6 The Global Carbon Cycle *(Page 1113)*

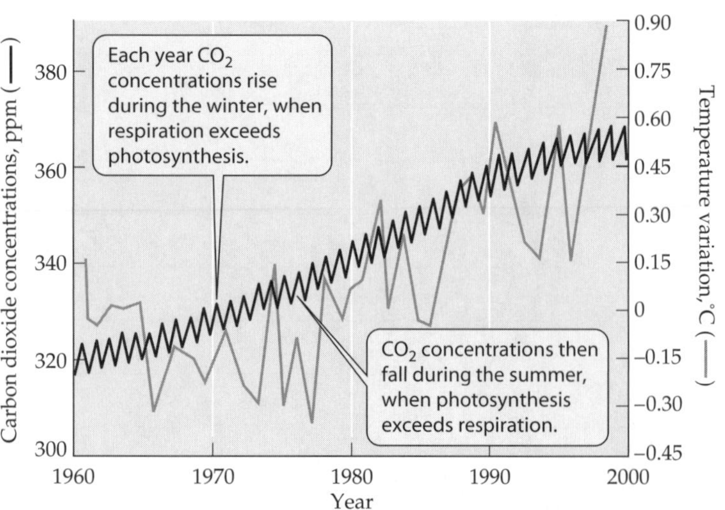
Each year CO_2 concentrations rise during the winter, when respiration exceeds photosynthesis.
CO_2 concentrations then fall during the summer, when photosynthesis exceeds respiration.
Carbon dioxide concentrations, ppm (—)
300
320
340
360
380
Temperature variation, °C (—)
0.90
0.75
0.60
0.45
0.30
0.15
0
–0.15
–0.30
–0.45
1960
1970
1980
1990
2000
Year

58.7 Atmospheric Carbon Dioxide Concentrations Are Increasing *(Page 1113)*

EXPERIMENT

Hypothesis: Higher atmospheric concentrations of CO_2 will result in increased carbon uptake and increased storage of carbon in below-ground carbon pools.

METHOD Establish circular, open-top chambers. Blow CO_2-enriched air into experimental chambers to achieve a concentration of 720 ppm. Blow natural air into other chambers that serve as controls. Measure shoot and surface litter mass in each chamber. Remove soil cores from each chamber and measure carbon content of roots and detritus.

RESULTS

(*a*) Sandstone grassland ecosystem

Above ground
Below ground
Natural CO_2 (controls)
Elevated CO_2 (experimental)

Carbon (grams per cm²)
300
200
100
0
100
200
Shoots
Surface detritus
Roots
Buried detritus
Microbes
Carbon
0
500
1,000
1,500
2,000
2,500
Soil

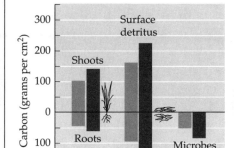

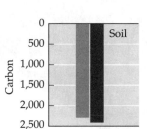

(*b*) Serpentine grassland ecosystem

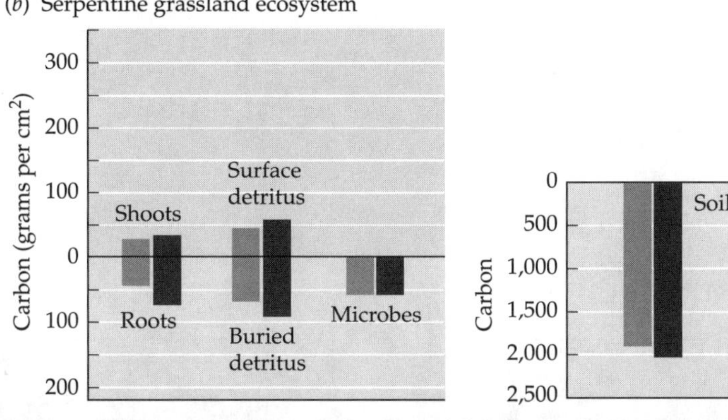
Carbon (grams per cm²)
300
200
100
0
100
200
Shoots
Surface detritus
Roots
Buried detritus
Microbes
Carbon
0
500
1,000
1,500
2,000
2,500
Soil

Conclusion: Experimental doubling of CO_2 did increase carbon uptake, but, contrary to predictions, carbon was partitioned to rapidly cycling carbon pools below ground. As a result, carbon storage did not increase. Thus, increased productivity may not lead to increased long-term storage of carbon.

58.8 Will Increased CO_2 Levels Increase Carbon Storage? *(Page 1114)*

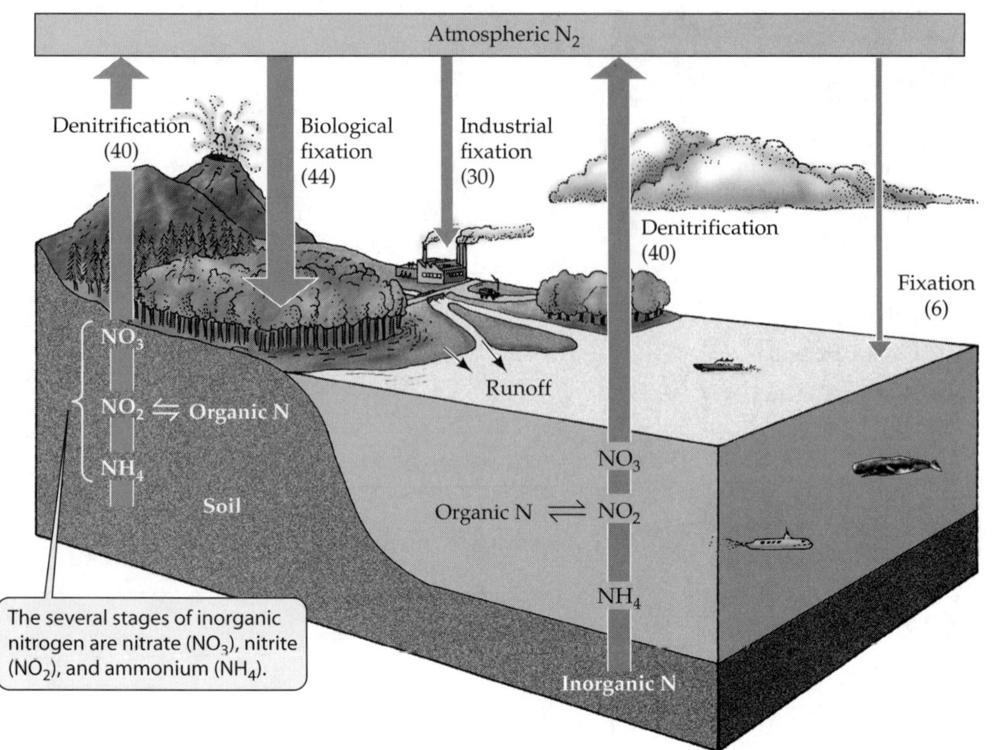

Atmospheric N_2
Denitrification (40)
Biological fixation (44)
Industrial fixation (30)
Denitrification (40)
Fixation (6)
NO_3
NO_2 ⇆ Organic N
NH_4
Soil
Runoff
NO_3
Organic N ⇌ NO_2
NH_4
Inorganic N
The several stages of inorganic nitrogen are nitrate (NO_3), nitrite (NO_2), and ammonium (NH_4).

58.9 The Global Nitrogen Cycle *(Page 1115)*

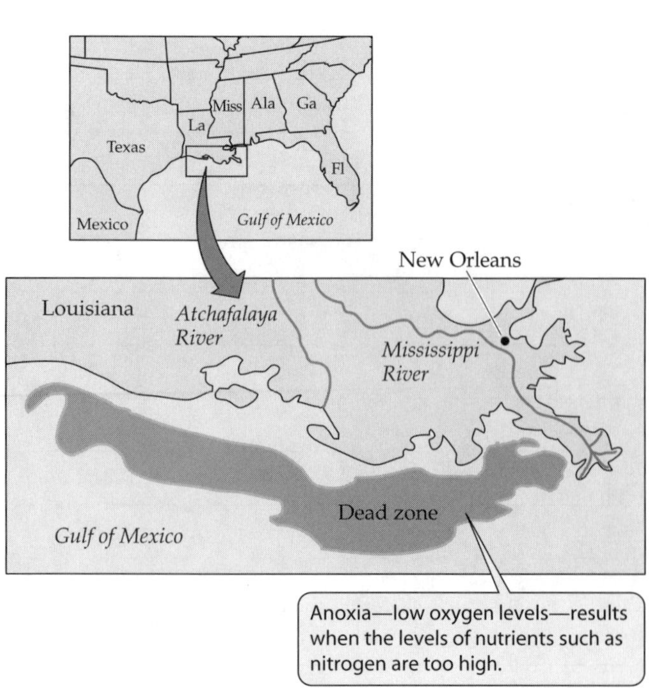

Miss
Ala
Ga
La
Texas
Fl
Mexico
Gulf of Mexico
New Orleans
Louisiana
Atchafalaya River
Mississippi River
Dead zone
Gulf of Mexico
Anoxia—low oxygen levels—results when the levels of nutrients such as nitrogen are too high.

58.10 A "Dead Zone" at the Mouth of the Mississippi River *(Page 1116)*

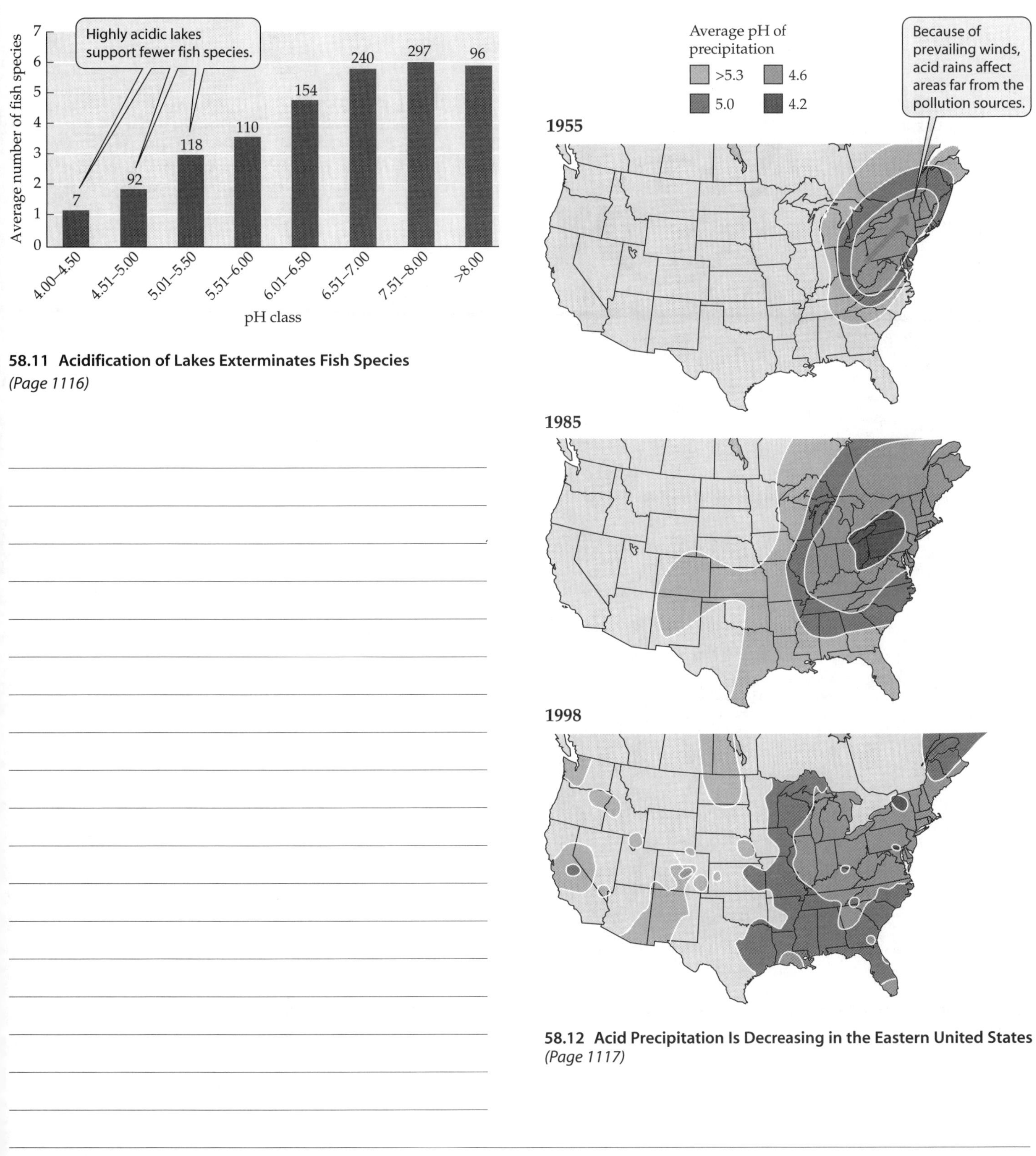
Highly acidic lakes support fewer fish species.
Average number of fish species
7
6
5
4
3
2
1
0
7
92
118
110
154
240
297
96
4.00–4.50
4.51–5.00
5.01–5.50
5.51–6.00
6.01–6.50
6.51–7.00
7.51–8.00
>8.00
pH class
Average pH of precipitation
>5.3
4.6
5.0
4.2
Because of prevailing winds, acid rains affect areas far from the pollution sources.
1955
1985
1998

58.11 Acidification of Lakes Exterminates Fish Species
(Page 1116)

58.12 Acid Precipitation Is Decreasing in the Eastern United States
(Page 1117)

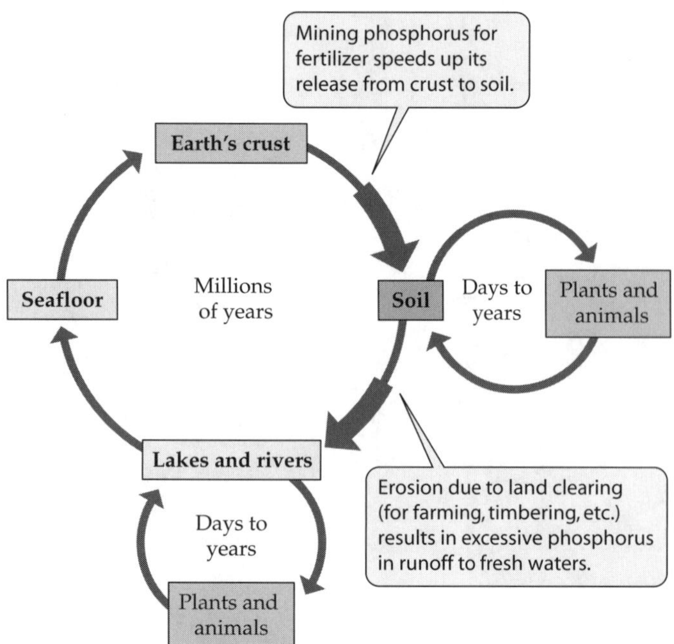
Mining phosphorus for fertilizer speeds up its release from crust to soil.
Earth's crust
Seafloor
Millions of years
Soil
Days to years
Plants and animals
Lakes and rivers
Days to years
Plants and animals
Erosion due to land clearing (for farming, timbering, etc.) results in excessive phosphorus in runoff to fresh waters.

58.13 The Phosphorus Cycle *(Page 1117)*

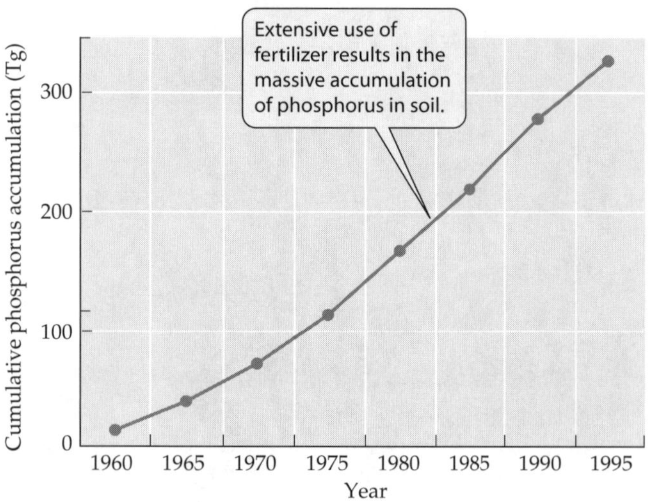
Extensive use of fertilizer results in the massive accumulation of phosphorus in soil.
Cumulative phosphorus accumulation (Tg)
0
100
200
300
1960
1965
1970
1975
1980
1985
1990
1995
Year

58.14 Phosphorus Is Accumulating in Agricultural Soils *(Page 1118)*